FINANCIAL TI

WORLD
DESK
REFERENCE

DORLING KINDERSLEY
LONDON • NEW YORK • MUNICH • MELBOURNE • DELHI

For the very latest information, visit:
www.dk.com and click on the Maps & Atlases icon

A DORLING KINDERSLEY BOOK
www.dk.com

FOR THE FIFTH EDITION

EDITOR-IN-CHIEF	SENIOR CARTOGRAPHIC MANAGER
Andrew Heritage	David Roberts
SENIOR CARTOGRAPHIC EDITOR	SYSTEMS COORDINATOR
Simon Mumford	Philip Rowles

FIFTH EDITION UPDATED AND EDITED BY
Cambridge International Reference on Current Affairs (CIRCA)

PROJECT MANAGER	EDITORIAL SUPERVISION
Catherine Jagger	Roger East
DATABASE	GRAPHICS
Carolyn Postgate, Kate Yedigaroff	Alex Yedigaroff, Elizabeth Postgate

EDITORS
Richard J. Thomas, Philippa Youngman

EDITORIAL AND RESEARCH
Joceline Bury, John Coggins, Ian Gorvin, Helen Hawkins
Lawrence Joffe, Frances Nicholson, Farzana Shaikh

PICTURE RESEARCH
Louise Thomas, ilumi

DORLING KINDERSLEY CARTOGRAPHY

EDITORIAL DIRECTION
Andrew Heritage

MANAGING EDITORS
Ian Castello-Cortes, Wim Jenkins

PROJECT EDITORS
Debra Clapson, Catherine Day,
Jo Edwards, Jane Oliver

EDITORS
Alastair Dougall, Ailsa Heritage,
Nicholas Kynaston, Lisa Thomas,
Susan Turner, Chris Whitwell, Elizabeth Wyse

ADDITIONAL EDITORIAL ASSISTANCE
Sam Atkinson, Louise Keane, Zoë Ellinson,
Caroline Lucas, Sophie Park, Laura Porter,
Jo Russ, Crispian Martin St. Valery,
Sally Wood, Ulrike Fritz-Weltz

READERS
Jane Bruton, Reg Grant, Ann Kramer, Lesley Riley

ART DIRECTION
Chez Picthall, Philip Lord

PROJECT DESIGNERS
Martin Biddulph, Scott David,
Carol Ann Davis, David Douglas,
Yahya El-Droubie, Karen Gregory

DESIGNERS
Tony Cutting, Rhonda Fisher,
Nicola Liddiard, Katy Wall

ADDITIONAL DESIGN ASSISTANCE
Paul Bayliss, Carol Ann Davis,
Adam Dobney, Kenny Laurenson,
Paul Williams

SYSTEMS COORDINATOR
Phil Rowles

PRODUCTION
Michelle Thomas, Wendy Penn

PROJECT CARTOGRAPHERS
Caroline Bowie, Ruth Duxbury,
James Mills-Hicks, John Plumer, Julie Turner

CARTOGRAPHERS
James Anderson, Dale Buckton,
Roger Bullen, Tony Chambers,
Jan Clark, Tom Coulson, Martin Darlison,
Claire Ellam, Julia Lunn, Michael Martin,
Alka Ranger, Peter Winfield, Claudine Zante

PICTURE RESEARCH
Alison McKittrick, Sarah Moule,
Christine Rista, Louise Thomas

DATABASE MANAGER
Simon Lewis

INDEX GAZETTEER
Margaret Hynes, Julia Lynch,
Barbara Nash, Jayne Parsons, Janet Smy

Printed and bound in Italy by Graphicom

First published in Great Britain in 1994
by Dorling Kindersley Limited,
80 Strand, London WC2R 0RL

A Penguin Company

Previously published as the DK World Reference Atlas

Reprinted with revisions 1995.Second Edition 1996. Revised 1998.
Third Edition (revised) 2000. Fourth Edition (revised) 2002, Fifth Edition (revised 2003).

A catalogue record for this book is available from the British Library.
ISBN: 0-7513-4889-9

FOREWORD

THIS DESK REFERENCE is presented to the public in the full knowledge that the world is in a state of continual flux. Political fashions and personalities come and go, while the ebb and flow of peoples and ideas across the face of the planet creates constant shifts in the cultural landscape. All the material assembled for this book has been researched from the most up-to-date and authoritative sources; our team of consultants and contributors, designers, editors, and cartographers have endeavored not only to explain the meaning of this material, to place it in a useful and clear context, but also to present it in a way that has a lasting value and relevance, regardless of the turmoil of daily events. This new edition, bearing the imprimatur of the *Financial Times*, has been completely revised and updated, to reflect the global changes of the past few years. It includes the latest statistical data, and over 60 new photographs.

The publishers would like to thank the many consultants and contributors whose diligence, perseverance, and attention to detail made this book possible.

GENERAL CONSULTANTS
Anthony Goldstone, Senior Editor Asia-Pacific, *The Economist* Intelligence Unit, London
Professor Jack Spence, Director of Studies, The Royal Institute of International Affairs, London

REGIONAL CONSULTANTS

ASIA
Anthony Goldstone, London

USA
Michael Elliot, Diplomatic Editor, *Newsweek*, Washington DC

AFRICA
James Hammill, Lecturer in African Politics, University of Leicester
Kaye Whiteman, Editor-in-Chief, *West Africa Magazine*, London

EUROPE
John Ardagh, London
Rory Clarke, Senior Editor Europe, *The Economist* Intelligence Unit, London
Charles Powell, Centre for European Studies, St Antony's College, Oxford

RUSSIA AND CIS
Martin McCauley, Senior Lecturer, School of Slavonic and East European Studies, University of London

MIDDLE EAST
John Whelan, Ex Editor-in-Chief, *Middle East Economic Digest*

CENTRAL AND SOUTH AMERICA
Nick Caistor, Producer, Latin American Section, BBC World Service

PACIFIC
Jim Boutilier, Professor in History, Royal Roads Military College, Victoria, Canada

CARIBBEAN
Canute James, *Financial Times*, Kingston, Jamaica

CONTRIBUTORS

Janice Bell, School of Slavonic and East European Studies, University of London
Gerry Bourke, Asia Correspondent, *The Guardian*, Islamabad
Vincent Cable, Director, International Economics Programme
P K Clark, MA, Former Chief Map Research Officer, Ministry of Defence
Ken Davies, Senior Editor, *The Economist* Intelligence Unit, London
Roger Dunn, Analyst, Control Risks Group, London
Aidan Foster-Carter, Senior Lecturer in Sociology, University of Leeds
Professor Murray Forsyth, Centre for Federal Studies, University of Leicester
Natasha Franklin, School of Slavonic and East European Studies, London
Adam Hannestad, *Blomberg Business News*, Copenhagen
Peter Holden, *The Economist* Research Department, London
Tim Jones, Knight Ritter, Brussels
Angella Johnstone, Home Affairs Correspondent, *The Guardian*, London
Oliver Keserü, International Chamber of Commerce, Paris
Robert Macdonald, *The Economist* Intelligence Unit
William Mader, Former Europe Bureau Chief, *Time Magazine*, Washington DC
Professor Brian Matthews, Institute of Commonwealth Studies, London
Nick Middleton, Oriel College, Oxford
Professor Mya Maung, Department of Finance, Boston College, Massachusetts
Judith Nordby, Leeds University
Simon Orme, London

Professor Richard Overy, Department of History, King's College, London
Steve Percy, East Asia Service, BBC World Service
Douglas Rimmer, Honorary Senior Research Fellow, Centre for West African Studies, University of Birmingham
Donna Rispoli, Linacre College, Oxford
Ian Rodger, *The Financial Times*, Zürich
The Royal Institute of International Affairs, London
Struan Simpson, St. James Research, London
Julie Smith, Brasenose College, Oxford
Elizabeth Spencer, London
Michiel Van Kuyen, Erasmus University, Rotterdam
Steven Whitefield, Pembroke College, Oxford
Georgina Wilde, Regional Director, Asia-Pacific, *The Economist* Intelligence Unit, London
H P Willmott, Visiting Professor, Dept. of Military Strategy & Operations, The National War College, Washington DC
Andrew Wilson, Sydney Sussex College, Cambridge
Tom Wingfield, *Reuters*, Bangkok
The World Conservation Monitoring Centre, Cambridge
Cambridge International Reference on Current Affairs (CIRCA)

CONTENTS

1
WORLD FACTFILE

2
THE NATIONS
OF THE WORLD

OVERSEAS TERRITORIES & DEPENDENCIES

3
INDEX ~ GAZETTEER

—— COVER FLAPS ——
KEY TO SYMBOLS, ICONS, AND
ABBREVIATIONS USED IN THE ATLAS

KEY TO CHARTS AND ICONS

ICONS AND TREND INDICATORS vary. Not all variations are shown in the key below, but where they do occur the symbols have been "stacked."

COUNTRY PROFILES

 Date of country's independence, or formation.

CLIMATE

 Indication of the climatic types and zones found in each country.

 Statistics are given for the national capital. They represent maximum summer and minimum winter averages.

TRANSPORTATION

 Indicates on which side of the road vehicles are driven in each country.

 The country's principal international airport with annual passenger numbers.

 Total size of national merchant or cargo fleet.

THE TRANSPORTATION NETWORK
National communications infrastructure given in kilometers and miles.

 Extent of national paved road network

 Extent of expressways, freeways, or major highways

 Extent of commercial rail network

 Extent of inland waterways navigable by commercial craft

TOURISM

 The ratio of foreign visitors to population.

Number of visitors per year, including business travelers.

Indicators showing trend in recent visitor numbers (up/level/down).

 The state of each nation's tourism is explained, with reasons given when there is no significant tourist industry. The chart shows the percentage of total visitors by country of origin.

 Date when the country's current borders were established.

 National Day

 Vehicle country identifying code

PEOPLE

An easy indication of the population density in each country (high/medium/low).

 Main languages spoken, including official languages.

 Population density. This is an average over the whole country.

 The pie chart proportions show the religious affiliations of those who profess a belief.

 This pie chart illustrates the ethnic origin of the country's population.

89%   **11%**

This graph represents the proportion of the population living in urban areas (gray) and rural areas (green).

 This chart shows the breakdown of the population by age groupings, providing an interesting insight into the country's demography.

POLITICS

Indicates the type of elections held within each country.

 Dates of last and next legislative elections for Lower (L.) and Upper (U.) Houses.

 Name of head of state. In many cases this is a nominal position and does not indicate that this is the country's most powerful person.

A graphic representation of the political makeup of the country's government, based on each party's showing at the last election. Where there are two houses, the more important elected body is shown first.

WORLD AFFAIRS

 Indication of membership of the UN (United Nations), and date of entry.

 Abbreviations indicate membership of international organizations.

 Nonmembership of additional international organizations.

AID

 Indication as to which countries are aid givers (donors) or aid recipients.

 The amount of net international aid given or received is shown in US$. Undisclosed military aid is not included.

 Symbols indicate whether aid payments or receipts are rising, level, or declining.

DEFENSE

 An indication of the status of conscription and mandatory military service.

 The defense budget, the country's annual expenditure (in US$) on arms and military personnel.

 Symbols indicate if the trend in defense spending is rising, level, or declining.

THE ARMED FORCES
Icons represent the main branches of the national armed forces.

 Army: equipment and personnel

 Navy: equipment and personnel

 Air force: equipment and personnel

 Nuclear capability: armaments

ECONOMICS

An indication of the average rate of inflation per annum, over the past decade.

 Gross National Product (GNP) – the total value of goods and services produced by a country.

 Exchange rates against the US$ at the start and end of the last year.

❏ World GNP Ranking.....................24th	
❏ GNP per Capita............................$24,388	
❏ Balance of Payments.....................$2.2bn	
❏ Inflation..5.6%	
❏ Unemployment...............................10.6%	

The score cards are intended to give a broad picture of the country's economy. Gross National Product (GNP), unlike GDP, includes income from investments and businesses held abroad. Balance of payments is the difference between a country's payments to and receipts from abroad.

This graph shows year-on-year variations in GDP and consumer prices.

 This pie chart gives a broad picture of the country's principal import trading partners.

 This pie chart gives a broad picture of the country's principal export trading partners.

RESOURCES

▷ Indicates the capacity of the combined national electricity generating sources (in kilowatts).

 Oil produced in barrels per day (b/d). Refining capacity and oil reserves are given where applicable.

 Estimated livestock resources.

 Main mineral reserves are listed in descending order of economic importance.

 Fish catch per year (where fishing is a major industry).

% of total generation by type

Percentages of the different energy sources used for the generation of electricity are represented graphically ("Combustion" indicates the burning of fossil fuels, wood etc.). An account of the country's resource base is given in the text.

ENVIRONMENT

 The 2002 Index of progress toward environmental sustainability, based on 22 core indicators. Compiled by the World Economic Forum taskforce.

 Percentage of land which is protected or conserved by law. Protection is often only theoretical.

Trend in total CO_2 emissions since 1990 (up/level/down) and current emissions per capita.

ENVIRONMENTAL TREATIES
National parties to international environmental treaties.

 Ramsar: (wetlands)

 Montreal Protocol: (CFC emissions)

 CITES: (endangered species)

 CBD: (biological diversity)

 Basel: (hazardous wastes)

 Kyoto: (greenhouse gases)

MEDIA

▷ Indicates the average rates of television ownership across the country.

 Media free to express critical views.

 Partial controls or constraints on media freedom.

 Severe restrictions on media freedom.

PUBLISHING AND BROADCAST MEDIA
National broadcast and print media, by size and ownership.

 Main national newspapers

 Television stations: state-owned/independent

 Radio: state-owned/ independent

CRIME

▷ An indication of the status of capital punishment and the death penalty.

 Prison population statistics

 Symbols show general trend in crime figures.

This section records official crime figures only. Reported statistics are normally lower than the actual figures.

CHRONOLOGY

Beginning at a significant date in the recent history of the country, the outline chronology continues through to the present day, and highlights key dates and turning points.

EDUCATION

 Displays the age until which children are legally required to attend school.

 Literacy rate. UNESCO defines as literate anyone who can read and write a short statement.

The number of students in all forms of tertiary education within that country.

This graph shows, for each level of education, the total enrollment, regardless of age, as a percentage of the population of the age group that officially corresponds to that level.

HEALTH

▷ An indication of the existence of health benefits provided by the state.

 Ratio of the number of people per doctor is given as a national average.

 Major causes of death are listed.

SPENDING

▷ Indicates the trend in GDP per capita over the previous decade.

 Levels of car ownership (per 1000 head of population)

 Rates of telephone connectivity (per 1000 head of population)

Defense, Health, Education spending as % of GDP

Percentage of the country's GDP that is spent on defense, education, and health.

WORLD RANKING

Schooling, educational attainment, and human development rankings are based on the UN Human Development Index (which covers 172 countries and Hong Kong).

SOURCES OF STATISTICAL DATA
USED IN THIS BOOK

Airports Council International

Amnesty International

Automobile Association (AA)

British Petroleum:
World Energy Data

Cambridge International Reference on Current Affairs (CIRCA)

Commonwealth Secretariat:
Small States Economic Review and Basic Statistics

Dorling Kindersley

Europa World Yearbook

European Bank for Reconstruction and Development (EBRD)

Financial Times

Fischer Weltalmanach

Food and Agriculture Organization (FAO)

International Atomic Energy Agency (IAEA)

International Institute for Strategic Studies (IISS):
The Military Balance

International Labor Organization (ILO):
World Labor Report

International Monetary Fund (IMF):
Balance of Payments Statistics Yearbook,
Direction of Trade Statistics Yearbook,
Government Financial Statistics Yearbook,
International Financial Statistics,
World Economic Outlook

International Road Federation

International Union for Conservation of Nature (IUCN)

International Union of Railways

INTERPOL International Crime Statistics

Lloyd's Register of Shipping

Organization for Economic Cooperation Development (OECD):
Economic surveys

OECD Development Assistance Committee (DAC):
Development Cooperation Report

Organization of Petroleum Exporting Countries (OPEC)

Ramsar Convention Bureau

Royal Automobile Club (RAC)

United Nations Demographic Yearbook

United Nations Energy Statistics Yearbook

United Nations Industrial Commodity Statistics Yearbook

United Nations International Trade Statistics Yearbook

United Nations Statistical Yearbook

United Nations Statistical Yearbook of Asia and the Pacific

United Nations Children's Fund (UNICEF)

United Nations Development Program (UNDP):
Human Development Report

United Nations Educational, Scientific, and Cultural Organization
(UNESCO): Statistical Yearbook

United Nations Environment Program (UNEP):
Ozone Secretariat
Secretariat of the Basel Convention
Secretariat of the Convention on Biological Diversity (CBD)
Secretariat of the Convention on International Trade in Endangered
Species (CITES)

United Nations Framework Convention on Climate Change

United Nations Population Fund (UNFPA):
The State of World Population

United States Central Intelligence Agency (CIA)

World Bank: World Development Indicators,
World Development Report, World Bank Atlas

World Conservation Monitoring Center (WCMC):
Biodiversity Data Sourcebook

World Economic Forum

World Health Organization (WHO)

World Prison Population List (Research, Development, and Statistics
Directorate, UK Home Office)

World Tourist Organization (WTO)

Worldwide Fund for Nature (WWF)

1

WORLD
FACTFILE

THE PHYSICAL WORLD

THE EARTH'S SURFACE IS constantly being transformed: it is uplifted, folded, and faulted by tectonic forces; weathered and eroded by wind, water, and ice. Sometimes change is dramatic, the spectacular results of earthquakes or floods. More often it is a slow process lasting millions of years. A physical map of the world represents a snapshot of the ever-evolving architecture of the Earth. This terrain map shows the whole surface of the Earth, both above and below the sea. The size of the Earth can be measured in different ways. When taken from the Equator, the diameter of the Earth measures 12,756 km (7927 miles); when taken from pole to pole, the diameter measures 12,714 km (7900 miles). Two-thirds of the Earth's surface is covered by oceans. The landscape of the ocean floor, like the surface of the land, has been shaped by movements of the Earth's crust over millions of years to form volcanic mountain ranges, deep trenches, basins, and plateaus. Ocean currents constantly redistribute warm and cold water around the world. The largest ocean in the world is the Pacific, which covers an area of over 181 million sq. km (70 million sq. miles).

Continental plate

Plate boundary, most tectonic activity takes place here

Oceanic plate

Shield area, center of plate, little tectonic activity occurs here

NORTHERN HEMISPHERE

MOST OF THE LAND on Earth is concentrated in the northern hemisphere, although Europe and North America are the only continents that lie entirely in the north.

THE DYNAMIC EARTH

THE EARTH'S CRUST is made up of eight major (and several minor) rigid continental and oceanic tectonic plates, which constantly move relative to one another. It is this movement which causes volcanic eruptions, earthquakes, and sometimes tsunamis along the plate boundaries. The largest volcanoes formed by this process are Aconcagua in Argentina at 6959 m (22,831 ft) and Kilimanjaro in Tanzania at 5895 m (19,340 ft), both of which are now extinct. Plate tectonics are responsible for the formation of the Himalayas – which were created by two colliding plates – and the Hawaiian Islands, created by the Pacific plate's movement over a "hot spot" of magma.

GEOGRAPHICAL REGIONS

- ice
- tundra
- needleleaf forest
- broadleaf forest
- cultivated land
- hot desert
- cold desert
- tropical grassland
- tropical rainforest
- mountain
- submarine regions

PHYSICAL WORLD FACTFILE

HIGHEST MOUNTAINS

1 Everest	8848 m	(29,028 ft)
2 K2	8611 m	(28,251 ft)
3 Kangchenjunga I	8590 m	(28,169 ft)
4 Makalu I	8463 m	(27,766 ft)
5 Cho Oyu	8201 m	(26,906 ft)

LONGEST RIVERS

1 Nile	6695 km	(4160 mi.)
2 Amazon	6516 km	(4048 mi.)
3 Yangtze	6380 km	(3964 mi.)
4 Mississippi /Missouri	6019 km	(3740 mi.)
5 Ob'-Irtysh	5570 km	(3461 mi.)

LARGEST DESERTS

1 Sahara	9,065,000 km²	(3,263,400 mi²)
2 Australian	3,750,000 km²	(1,350,000 mi²)
3 Gobi	1,295,000 km²	(466,200 mi²)
4 Arabian	750,000 km²	(270,000 mi²)
5 Sonoran	311,000 km²	(111,960 mi²)

SOUTHERN HEMISPHERE

OCEANS DOMINATE the southern hemisphere. Australia and Antarctica are the only continental landmasses that lie entirely in the south.

THE POLITICAL WORLD

IN 2002, EAST TIMOR joined the international community, becoming the world's 193rd recognized independent state. In 1950 there were only 82. With the exception of Antarctica, where territorial claims have been deferred by international treaty, every land area of the Earth's surface either belongs to, or is claimed by, one country or another. Some 60 overseas dependent territories remain, administered variously by Australia, Denmark, France, the Netherlands, New Zealand, Norway, the UK, and the US. Over the last half-century, national self-determination has been a driving force for many states with a history of colonialism or oppression. While some new states on gaining independence moved peacefully to establish a democracy, many others have been torn by religious or ethnic conflicts or became submerged in power struggles resulting in dictatorship by a military regime or an individual despot.

OLDEST COUNTRIES

Denmark
950 CE

China
960 CE

Portugal
1139 CE

France
987 CE

Thailand
1238 CE

ARCTIC OCEAN

Alaska (part of US)
Arctic Circle
Bering Sea
Aleutian Is (part of US)

CANADA
Hudson Bay
Lake Winnipeg

Baffin Bay

Greenland (to Denmark)

Ja (to

ICELAND
Faeroe Isla (to Denma
U
KI

REPUBLIC OF IRELAND
Isle of Ma (to UK)
Channel Isla (to UK)

PACIFIC OCEAN

UNITED STATES OF AMERICA

Midway Islands (to US)
Tropic of Cancer
Hawaii (part of US)

Johnston Atoll (to US)

Guadalupe (part of Mexico)

MEXICO

Gulf of Mexico

BAHAMAS

Revillagigedo Islands (part of Mexico)

Bermuda (to UK)

ATLANTIC OCEAN

Azores (part of Portugal)

PORTUGA

Gibraltar (to UK)
Ceuta (part of Spain) –
Melilla (part of Spain)
Madeira (part of Portugal)

Canary Islands (part of Spain)

WESTERN SAHARA (disputed)

MAURITANI

CUBA
Turks & Caicos Is (to UK)
Puerto Rico (to US)
Virgin Is (to US)
Cayman Is (to UK)
JAMAICA
BELIZE
Navassa I (to US)
HAITI
DOM. REP.
British Virgin Is (to UK)
Anguilla (to UK)
ANTIGUA & BARBUDA
ST KITTS & NEVIS
Montserrat (to UK)
Guadeloupe (to France)
DOMINICA
Martinique (to France)
ST LUCIA
BARBADOS
ST VINCENT & THE GRENADINES
GRENADA
TRINIDAD & TOBAGO

CAPE VERDE

SENEGAL
GAMBIA
GUINEA-BISSAU
GUINEA

SIERRA LEONE
LIBERIA
D'I
EQUAT
G

GUATEMALA
HONDURAS
EL SALVADOR
NICARAGUA
COSTA RICA
PANAMA

Netherlands Antilles (to Neth.)
Aruba (to Neth.)
Caribbean Sea

VENEZUELA
COLOMBIA
GUYANA
SURINAME
French Guiana (to France)

Kingman Reef (to US)
Palmyra Atoll (to US)
Baker & Howland Is (to US)
Equator
Jarvis (to US)

Galapagos Is (part of Ecuador)

ECUADOR

PERU

BRAZIL

Fernando de Noronha (part of Brazil)

ATLANTIC OCEAN

Ascension (to St Heler

KIRIBATI

Tokelau (to NZ)
SAMOA
Wallis & Futuna (to France)
American Samoa (to US)
Cook Islands (to NZ)
TONGA
Niue (to NZ)

French Polynesia (to France)

PACIFIC OCEAN

Lake Titicaca
BOLIVIA

PARAGUAY

Trindade (part of Brazil)

Tropic of Capricorn

Pitcairn Islands (to UK)

Easter Island (to Chile)

Sala y Gomez (part of Chile)

San Felix Island (part of Chile)
San Ambrosio Island (part of Chile)

ARGENTINA

URUGUAY

Tristan da Cunh (to St Helena)

Kermadec Islands (part of NZ)

Juan Fernandez Islands (part of Chile)

Gough Isla (part of Tristan d

Chatham Islands (part of NZ)

Falkland Islands (to UK)

South Georgia & South Sandwich Islands (to UK)

SO
T
H
E

South Shetland Islands
South Orkney Islands

Antarctic Circle

Peter I Island (to Norway)

KEY

——————	Full borders
- - - - - - - -	Disputed borders
···········	Undefined borders
- · - · - ·	Extent of dependent island territories
– – – – –	Extent of country boundaries for island territories
Tristan da Cunha (to St Helena)	Dependent territory with self-government
Gough Island (part of Tristan da Cunha)	Territory without self-government (the state it belongs to is given in brackets)

INTERNATIONAL BORDERS

BOUNDARIES BETWEEN states fall into three categories. Full borders are internationally recognized territorial boundaries. Undefined borders exist where no fixed boundary has been demarcated. A disputed border is where a de facto boundary exists which is not agreed upon or is subject to arbitration. Disputed borders exist throughout the world, such as the land borders between India and China and between Ethiopia and Eritrea, and the maritime border between Samoa and American Samoa.

COUNTRIES WITH THE MOST LAND BORDERS

1 **China:** *14* (Afghanistan, Bhutan, Burma, India, Kazakhstan, North Korea, Kyrgyzstan, Laos, Mongolia, Nepal, Pakistan, Russian Federation, Tajikistan, Vietnam)

Russian Federation: *14* (Azerbaijan, Belarus, China, Estonia, Finland, Georgia, Kazakhstan, North Korea, Latvia, Lithuania, Mongolia, Norway, Poland, Ukraine)

2 **Brazil:** *10* (Argentina, Bolivia, Colombia, French Guiana, Guyana, Paraguay, Peru, Suriname, Uruguay, Venezuela)

YOUNGEST COUNTRIES

East Timor 2002

Slovakia 1993

Palau 1994

Czech Rep. 1993

Eritrea 1993

MARS

- ⊖ **Diameter:** 6786 km
- ● **Mass:** 642 billion billion tons
- ○ **Temperature:** –137 to 37°C
- ◑ **Distance from Sun:** 228 million km
- ◐ **Length of year:** 1.88 years
- ⊖ **Surface gravity:** 1 kg = 0.38 kg

EARTH

- ⊖ **Diameter:** 12,756 km
- ● **Mass:** 5976 billion billion tons
- ○ **Temperature:** –70 to 55°C
- ◑ **Distance from Sun:** 150 million km
- ◐ **Length of year:** 365.25 days
- ⊖ **Surface gravity:** 1kg = 1 kg

THE EARTH

GASES SUCH AS CARBON dioxide are known as "greenhouse gases" because they prevent shortwave solar radiation from entering the Earth's atmosphere, but help to stop longwave radiation from escaping. This traps heat, raising the Earth's temperature. An excess of these gases traps more heat and can lead to global warming.

Incoming shortwave solar radiation

Greenhouse gases prevent the escape of longwave radiation

Longwave radiation deflected by the Earth heats the atmosphere

VENUS

- ⊖ **Diameter:** 12,102 km
- ● **Mass:** 4870 billion billion tons
- ○ **Temperature:** 457°C
- ◑ **Distance from Sun:** 108 million km
- ◐ **Length of year:** 224.7 days
- ⊖ **Surface gravity:** 1 kg = 0.88 kg

MERCURY

- ⊖ **Diameter:** 4878 km
- ● **Mass:** 330 billion billion tons
- ○ **Temperature:** –173 to 427°C
- ◑ **Distance from Sun:** 58 million km
- ◐ **Length of year:** 87.97 days
- ⊖ **Surface gravity:** 1 kg = 0.38 kg

THE SOLAR SYSTEM

THE SOLAR SYSTEM CONSISTS of the nine major planets, their moons, the asteroids, and the comets that orbit around the Sun. The Sun itself is composed of 70% hydrogen and 30% helium, and at its core nuclear fusion reactions turning hydrogen into helium produce the heat and light which make life possible on Earth. Of the planets, the inner four (Mercury, Venus, Earth, and Mars) are termed terrestrial, while the next four (Jupiter, Saturn, Uranus, and Neptune) are termed gas giants. Pluto, at the edge of the solar system, is much smaller, and made of rock. The largest natural satellite in the Solar System is Ganymede (5262 km – 3270 miles – in diameter), which orbits around Jupiter, the largest planet. Halley's comet is the brightest comet when seen from Earth, and orbits the Sun once every 76 years. The largest asteroid is named Ceres (940 km – 584 miles – in diameter), which is found in the asteroid belt between Mars and Jupiter. The planet Earth is unique within the solar system (and possibly the universe), being the only planet capable of sustaining life.

JUPITER

- ⊖ **Diameter:** 142,984 km
- ● **Mass:** 1,900,000,000 billion billion tons
- ○ **Temperature:** –153°C
- ◑ **Distance from Sun:** 778 million km
- ◐ **Length of year:** 11.86 years
- ⊖ **Surface gravity:** 1 kg = 2.53 kg

SATURN

- **Diameter:** 120,660 km
- **Mass:** 570,000 billion billion tons
- **Temperature:** –185°C
- **Distance from Sun:** 1427 million km
- **Length of year:** 29.46 years
- **Surface gravity:** 1 kg = 1.07 kg

URANUS

- **Diameter:** 51,118 km
- **Mass:** 102,000 billion billion tons
- **Temperature:** –214°C
- **Distance from Sun:** 2870 million km
- **Length of year:** 84.01 years
- **Surface gravity:** 1 kg = 0.92 kg

NEPTUNE

- **Diameter:** 49,528 km
- **Mass:** 13 billion billion tons
- **Temperature:** –225°C
- **Distance from Sun:** 4497 million km
- **Length of year:** 164.79 years
- **Surface gravity:** 1 kg = 1.18 kg

PLUTO

- **Diameter:** 2300 km
- **Mass:** 13 billion billion tons
- **Temperature:** –236°C
- **Distance from Sun:** 5900 million km
- **Length of year:** 248.54 years
- **Surface gravity:** 1 kg = 0.30 kg

MOON AND TIDES

TIDES ARE CREATED by the pull of the Sun and the Moon's gravity on the surface of the oceans. Waves are formed by wind blowing over the surface of the oceans. The highest tides occur when the Earth, the Moon, and the Sun are aligned (*below left*). The lowest tides are experienced when the Sun and Moon align at right angles to one another (*below right*).

NEAR SIDE OF THE MOON

FAR SIDE OF THE MOON

HIGHEST HIGH TIDES

LOWEST HIGH TIDES

Earth

Moon

Sun

Tidal bulge created by gravitational pull

Timeline of Space Exploration

1957: USSR launches *Sputnik I* - first artificial satellite

Apr 12, 1961: Yuri Gagarin (USSR) first person in space

Feb 13, 1966: *Luna 9* first probe to land on Moon

1976: Missions of *Viking 1* and *2* analyze surface of Mars

Feb 20, 1986: Launch of space station *Mir*

Apr 24, 1990: Launch of Hubble Space Telescope

2000: *Mir* brought to earth. Dennis Tito is first space tourist. 100th shuttle mission completed

2001: *Near* probe lands on Eros asteroid

2005: Provisional date for the launch of the first Chinese manned space flight

1955 1960 1970 1980 1990 2000 2010

Oct 10, 1959: *Luna 3* sends back first pictures of dark side of the Moon

Jul 10, 1962: Launch of *Telstar I*, first commercial communications satellite

Jul 21, 1969: Neil Armstrong and Buzz Aldrin first people to land on Moon

Jan 28, 1986: *Challenger* shuttle explodes; all seven crew members killed

Aug 25, 1989: *Voyager 2* probe passes Neptune on way out of Solar System

Nov 20, 1998: Launch of first part of International Space Station

2004: Provisional date for a probe landing on Titan (Saturn's largest moon)

THE CLIMATE

T HE EARTH'S CLIMATIC REGIONS consist of stable patterns
of weather conditions averaged out over a long period of time.
Different climates are categorized according to particular
combinations of temperature and humidity. By contrast,
weather consists of short-term fluctuations in wind,
temperature, and humidity conditions. Different
climates are determined by latitude, altitude,
the prevailing wind, and circulation of ocean
currents. Longer-term changes in climate, such
as global warming or the onset of ice ages, are
punctuated by shorter-term events which
comprise the day-to-day weather of a
region, such as frontal depressions,
hurricanes, and blizzards.

CLIMATE ZONES

Ice cap	Mediterranean
Tundra	Semi-arid
Subarctic	Arid
Cool continental	Tropical
Warm humid	Humid equatorial

OCEAN CURRENTS	PREVAILING WINDS	LOCAL WINDS
Warm	Warm	→ Warm
Cold	Cold	→ Cold
		Seasonal*

* (seasonal winds which
can either be warm or cold)

TEMPERATURE

THE WORLD CAN BE DIVIDED into three major
climatic zones, stretching like large belts across
the latitudes: the tropics which are warm, the cold
polar regions, and the temperate zones which lie
between them. Temperature is also controlled by
altitude: mountainous regions are typically colder
than those at sea level.

below - 30°C (-22°F)	-10 to 0°C (14 to 32°F)	20 to 30°C (68 to 86°F)
-30 to - 20°C (-22 to -4°F)	0 to 10°C (32 to 50°F)	above 30°C (86°F)
-20 to - 10°C (-4 to 14°F)	10 to 20°C (50 to 68°F)	

AVERAGE JULY TEMPERATURE

AVERAGE JANUARY TEMPERATURE

AVERAGE JULY RAINFALL

0–25 mm (0–1 in) 100–200 mm (4–8 in) 400–500 mm (16–20 in)
25–50 mm (1–2 in) 200–300 mm (8–12 in) above 500 mm (20 in)
50–100 mm (2–4 in) 300–400 mm (12–16 in)

AVERAGE JANUARY RAINFALL

RAINFALL

WHEN AIR IS HEATED, it expands, rises, and cools, and the water vapor it carries condenses to form clouds. Heavy, regular rainfall is characteristic of the equatorial region, while the poles, being cold, receive only slight snowfall. Tropical regions have marked dry and rainy seasons, while in temperate regions rainfall is unpredictable.

CLIMATE FACTFILE

DRIEST INHABITED PLACES
1 **Aswân** (EGYPT)
  0.5 mm .... (0.02 in)
2 **Luxor** (EGYPT)
  0.7 mm .... (0.03 in)
3 **Arica** (CHILE)
  1.1 mm .... (0.04 in)
4 **Ica** (PERU)
  2.3 mm .... (0.1 in)
5 **Antofagasta** (CHILE)
  4.9 mm ....(0.2 in)

WETTEST INHABITED PLACES
1 **Buenaventura** (COLOMBIA)
  6734 mm ..... (266 in)
2 **Monrovia** (LIBERIA)
  5131 mm ..... (202 in)
3 **Pago Pago** (AMERICAN SAMOA)
  4990 mm ..... (197 in)
4 **Moulein** (BURMA)
  4852 mm ..... (191 in)
5 **Lae** (PAPUA NEW GUINEA)
  4645 mm ..... (183 in)

THE ENVIRONMENT

THE EARTH CAN BE DIVIDED into a series of biogeographic regions, or biomes – ecological communities where certain species of plant and animal coexist within particular climatic conditions. Within these broad classifications, other factors affect the local distribution of species in each biome, such as soil richness, altitude, and human activities such as urbanization, intensive agriculture, and deforestation. Apart from the polar ice caps, there are few areas which have not been colonized by animals or plants over the course of the Earth's history. Because of all animals' reliance on plants for survival, plants are known as primary producers. The availability of nutrients and the temperature of an area define its primary productivity, which affects the number and type of animals which are able to live there; the level of humidity or aridity is also a determining factor.

BIODIVERSITY

THE NUMBER OF PLANT AND ANIMAL SPECIES, and the range of genetic diversity within the populations of each species, make up the Earth's biodiversity. The plants and animals which are endemic to a region – that is, those which are found nowhere else in the world – are also important in determining levels of biodiversity. Human settlement and intervention have encroached on many areas of the world once rich in endemic plant and animal species. Increasing international efforts are being made to monitor and conserve the biodiversity of the Earth's remaining wild places.

ANIMALS

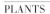

Number of animal species per country

- More than 2000
- 1000–1999
- 700–999
- 400–699
- 200–399
- 100–199
- 0–99
- Data not available

ANIMAL ADAPTATION

THE DEGREE OF AN ANIMAL'S ADAPTABILITY to different climates and conditions is extremely important in ensuring its success as a species. Many animals, particularly the largest mammals, are becoming restricted to ever-smaller regions as human development and modern agricultural practices reduce their natural habitats. In contrast, humans have been responsible – both deliberately and accidentally – for the spread of some of the world's most successful species. Many of these introduced species are now more numerous than the indigenous animal populations.

PLANTS

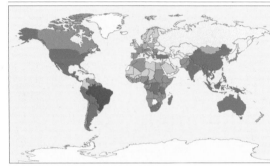

Number of plant species per country

- More than 50,000
- 7000–49,999
- 3000–6999
- 2000–2999
- 1000–1999
- 600–999
- 0–599
- Data not available

PLANT ADAPTATION

ENVIRONMENTAL CONDITIONS, such as climate, soil type, and competition with other organisms, influence the development of plants into distinctive forms. Similar conditions in different parts of the world create similar adaptations in the plants, which may then be modified by other, local, factors specific to the region.

BIOME TYPES
- Mountains
- Polar regions
- Tundra
- Tropical rainforests
- Dry woodlands
- Savanna
- Temperate grasslands
- Mediterranean
- Coniferous forests
- Temperate rainforests
- Broadleaf forests
- Cold deserts
- Hot deserts
- Wetlands

ENVIRONMENT FACTFILE

LARGEST PROTECTED AREAS

Seychelles	95%
Ecuador	44%
Kiribati	39%
Venezuela	37%
Denmark	33%

HIGHEST ANNUAL DEFORESTATION

Brazil	23,090 km²	(8936 mi²)
China	18,060 km²	(6989 mi²)
Indonesia	13,120 km²	(5077 mi²)
Sudan	9590 km²	(3711 mi²)
Zambia	8510 km²	(3293 mi²)

IDENTIFIED SPECIES

Microorganisms	5800
Invertebrates	1,021,000
Plants	322,500
Fish	19,100
Reptiles & amphibians	12,000
Mammals	4000

ENDANGERED SPECIES

Mammals	484
Birds	403
Reptiles	100
Amphibians	49
Fish	291
Invertebrates	763

POPULATION
■ over 5 000 000
▣ over 1 000 000
◉ over 500 000
● over 100 000
• over 50 000
○ under 50 000

LAND HEIGHT
4000m/13 124ft
3000m/9845ft
2000m/6562ft
1000m/3281ft
200m/656ft
Sea Level

Line of cross-section

NORTH AMERICA

North America's climate is as varied as its topography: much of Canada is snowbound or clothed in forest, its sparse population congregating along the US border. Along the continent's western flank are the spectacular Rocky Mountains. To the east lie the older, wooded Appalachians. Between these are the Great Plains – grazed by herds of livestock or sown with cereals. These plains were once home to tribes of native Americans, supplanted by incoming white settlers. The population and industry of the US are concentrated in the temperate northeast, while the drier south and west are rural and thinly populated. North America is rich in minerals and oil. Mexico has the world's largest Spanish-speaking population. Central America and the Caribbean contain some 20 countries and numerous small islands. The climate is tropical and prone to storms, the landscape mountainous and volcanic.

CONTINENTAL FACTS

PHYSICAL FEATURES

LARGEST LAKE: Lake Superior, Canada/ US 85,270 km² (32,140 mi²)

LONGEST RIVER: Mississippi-Missouri, US 6019 km (3740 miles)

HIGHEST POINT: Mt. McKinley (Denali), Alaska, US 6194 m (20,322 ft)

LOWEST POINT: Death Valley, California, US 86 m (282 ft) below sea level

POLITICAL FEATURES

TOTAL POPULATION: 482.9 million

LARGEST CITY AND ITS POPULATION: Mexico City, Mexico 18.1 million

COUNTRY WITH HIGHEST POPULATION DENSITY: Barbados 625 people/km² (1614 people/mi²)

LARGEST COUNTRY: Canada 9,976,140 km² (5,851,788 mi²)

SMALLEST COUNTRY: Grenada 340 km² (151 mi²)

Scale : 1:32 500 000

CROSS-SECTION THROUGH NORTH AMERICA: 43° 30'N, 126°W–65°W

SOUTH AMERICA

THE WORLD'S fourth-largest continent includes one of its most important resources – the Amazonian rainforest. It is a major source of oxygen and includes half of all known living species, while the Amazon – the world's second-longest river – contains one-fifth of the world's fresh water. The Andes mountain chain reaches down South America's western flank, sheltering the prairies of the Gran Chaco, the Pampas, and the wastes of the far south. Most South Americans are *mestizo* – of mixed European and Amerindian descent – and live in the coastal regions. Spanish is the most widely spoken language, and over 90% of South Americans are Roman Catholic. South America has massive mineral resources, many exploited by US and European multinationals.

LAND HEIGHT

6000m/19 686ft
4000m/13 124ft
3000m/9843ft
2000m/6562ft
1000m/3281ft
200m/656ft
Sea Level

Line of cross-section

POPULATION

over 5 000 000
over 1 000 000
over 500 000
over 100 000
over 50 000
under 50 000

CONTINENTAL FACTS

PHYSICAL FEATURES

LARGEST LAKE:
Lake Titicaca,
Bolivia/Peru 8340 km²
(3220 mi²)

LONGEST RIVER: Amazon,
Brazil 6516 km
(4050 miles)

HIGHEST POINT: Cerro
Aconcagua, Argentina
6959 m (22,835 ft)

LOWEST POINT: Península
Valdés, Argentina
40 m (131 ft)
below sea level

POLITICAL FEATURES

TOTAL POPULATION:
350.6 million

LARGEST CITY AND
ITS POPULATION: São Paulo,
Brazil 17.8 million

COUNTRY WITH HIGHEST
POPULATION DENSITY:
Ecuador 47 people/km²
(121 people/mi²)

LARGEST COUNTRY:
Brazil 8,511,965 km²
(3,286,469 mi²)

SMALLEST COUNTRY:
Suriname 161,470 km²
(62,344 mi²)

Scale : 1:26 500 000

0 250 500 km
0 250 500 miles

FALKLAND ISLANDS
(to UK)

CROSS-SECTION THROUGH SOUTH AMERICA: 18°S, 75°W–36°W

8000m
4000m
Sea Level
-4000m
-8000m

PACIFIC OCEAN
Peru-Chile Trench
Hot Desert
Atacama Desert
Cordillera Occidental
Mountain Vegetation
Altiplano
Andes
Cordillera Central
Tropical Dry Forest
Cordillera Oriental
Tropical Grassland
Paraguay
Pantanal
Temperate Rainforest
Brazilian Highlands
São Francisco
ATLANTIC OCEAN

GREENLAND
(Danish province)

JAN MAYEN
(to Norway)

Denmark Strait

REYKJAVIK ICELAND Akureyri

Norwegian Sea

FAEROE IS
(to Denmark)

Shetland Is

Orkney Is

Outer
Hebrides

EUROPE

THE SMALLEST CONTINENT AFTER AUSTRALIA, Europe has a wide variety of climates and landscapes. The tundra of the far north gives way to a cool, wet, heavily forested region. The North European Plain is well-drained, fertile, and rich in oil, coal, and natural gas. The shores of the Mediterranean are generally warm, dry, and hilly, ideal for cultivating olives, citrus fruit, and grapes. A great curve of mountain ranges, including the Pyrenees, Alps, and Carpathians, divides north from south. To the east, the rolling plains of European Russia and the Ukraine, clad in coniferous forests or cultivated for wheat, run up to the Ural Mountains. Europeans are mainly Christian – Catholic, Orthodox, or Protestant – and speak a variety of languages, most of which spring from Latin (Romance), Germanic, or Slavic roots.

Edinburgh

Belfast

REPUBLIC
OF
IRELAND ISLE OF MAN
(to UK) UNITED
KINGDOM

DUBLIN Manchester NETHER

Cork Birmingham AMSTERDA

Cardiff LONDON Rotterdam

English Channel BRUSSELS
BEL

GUERNSEY (to UK)
JERSEY (to UK) Seine

PARIS

Nantes Loire

FRANCE

Ger

Bay of
Biscay Bordeaux Massif
Central

A Coruña Rhône

Toulouse

ANDORRA Marsei

Porto Pyrenees

Zaragoza Barcelona

SPAIN Ebro

■ MADRID

Valencia Mallorca Menor
Palma

LISBON Eivissa Balearic

Sevilla M e

Málaga

GIBRALTAR
(to UK) Ceuta (to Spain)

Melilla (to Spain)

ALGERI

MOROCCO

POPULATION

- ■ over 5 000 000
- ▣ over 1 000 000
- ◉ over 500 000
- ● over 100 000
- ◦ over 50 000
- ○ under 50 000

LAND HEIGHT

- 3000m/9843ft
- 2000m/6562ft
- 1000m/3281ft
- 200m/656ft
- Sea Level

Azores
(to Portugal)

A T L A N T I C   O C E A N

Line of cross-section

Scale : 1:22 500 000

0 — 250 — 500 km

0 — 250 — 500 miles

CROSS-SECTION THROUGH EUROPE: 46°N, 5°W–48°E

| | Broadleaf Forest | Mountain Vegetation | Broadleaf Forest | Mountain Vegetation | Broadleaf Forest | Temperate Grassland | Cold Desert |

8000m

4000m

Sea Level

-4000m

-8000m

ATLANTIC
OCEAN Alps Alföld Carpathian
Mountains Black
Sea Crimea Sea of
Azov CASPIAN
SEA

Rhône Danube Carpatii Occidentali Volga Delta

0° 11°E 22°E 33°E 44°E

CONTINENTAL FACTS

PHYSICAL FEATURES

LARGEST LAKE: Ladoga, European Russia 18,390 km² (7100 mi²)

LONGEST RIVER: Volga, European Russia 3688 km (2290 miles)

HIGHEST POINT: El' brus, Caucasus Mountains, European Russia 5642 m (18,510 ft)

LOWEST POINT: Volga Delta, Caspian Sea, European Russia 28 m (92 ft) below sea level

POLITICAL FEATURES

TOTAL POPULATION: 701.9 million

COUNTRY WITH HIGHEST POPULATION DENSITY: Monaco 16,253 people/km² (42,095 people/mi²)

LARGEST CITY AND ITS POPULATION: Moscow, European Russia 9.3 million

LARGEST COUNTRY: European Russia 3,955,818 km² (1,527,341 mi²)

SMALLEST COUNTRY: Vatican City, Italy 0.44 km² (0.17 mi²)

AFRICA

A FRICA IS THE SECOND-LARGEST CONTINENT after Asia. It is dominated by the Sahara in the north and the Great Rift Valley in the east. The Mediterranean climate of the extreme north and south enables cultivation of grapes and other fruit. A belt of tropical rainforest lies along the Equator, while Africa's great tropical grasslands provide grazing for herds of wild animals and domestic livestock. A narrow strip of Egypt is watered by the world's longest river, the Nile, which has sustained communities from prehistoric times. The center and south of the continent are rich in minerals. Just over one-tenth of the world's population lives in Africa – a wide variety of peoples with their own distinctive languages and cultures. Although Islam and Christianity are widespread, many Africans adhere to their own local customs and religious beliefs.

Scale : 1:56 000 000

LAND HEIGHT

4000m/15 124ft
3000m/9845ft
2000m/6562ft
1000m/5281ft
200m/656ft
Sea Level

POPULATION

over 5 000 000
over 1 000 000
over 500 000
over 100 000
over 50 000
under 50 000

Line of cross-section

CONTINENTAL FACTS

PHYSICAL FEATURES

LARGEST LAKE: Lake Victoria, Kenya/Tanzania/Uganda 68,880 km² (26,560 mi²)

LONGEST RIVER: Nile, Uganda/Sudan/Egypt 6695 km (4160 miles)

HIGHEST POINT: Kilimanjaro, Tanzania 5895 m (19,341 ft)

LOWEST POINT: Lac' Assal, Djibouti 156 m (512 ft) below sea level

POLITICAL FEATURES

TOTAL POPULATION: 811.6 million

LARGEST CITY AND POPULATION: Lagos, Nigeria, 15.4 million

COUNTRY WITH HIGHEST POPULATION DENSITY: Mauritius 645 people/km² (1671 people/mi²)

LARGEST COUNTRY: Algeria 2,505,810 km² (967,493 mi²)

SMALLEST COUNTRY: Seychelles 455 km² (175 mi²)

CROSS-SECTION THROUGH AFRICA 7°N, 15°W–55°E

LAND HEIGHT

- 6000m/19 686ft
- 4000m/13 124ft
- 3000m/9843ft
- 2000m/6562ft
- 1000m/3281ft
- 200m/656ft
- Sea Level

POPULATION

- ■ over 5 000 000
- ◨ over 1 000 000
- ◉ over 500 000
- ● over 100 000
- ◦ over 50 000
- ○ under 50 000

Line of cross-section

WEST ASIA

WEST ASIA, or the Middle East, lies at the meeting point of Europe, Asia, and Africa. The world's first towns and cities were built in the Fertile Crescent, which extended from the Mediterranean to the land between the Tigris and Euphrates rivers. The cradle of Judaism, Christianity, and Islam, the region has long been a center of religious and ethnic strife – a situation exacerbated by the presence of some of the world's richest oil fields, the main source of income for several Middle Eastern countries. Much of the region is desert, or rugged mountains and plateaus, rising in the east to the Hindu Kush. Cultivation today depends on sophisticated irrigation techniques.

CROSS-SECTION THROUGH WEST ASIA: 20°N, 35°E–62°E

K A Z A K H S T A N

Aral Sea

Kyzyl Kum

BISHKEK

Karakol
Ozero Issyk-Kul'

Kirghiz Range

KYRGYZSTAN

TASHKENT

Namangan

Osh

Tien Shan

C H I N A

U Z B E K I S T A N

Dashkhovuz

Urganch

Ozero Aydarkul'

Khūjand

Pamirs

Samarqand

K a r a k u m y

Qarshi

TAJIKISTAN

DUSHANBE

Gäncä

Caspian Sea

AZERBAIJAN

BAKU

Krasnovodsk

Nebîtdag

Chardzhev

Amu

Kūlob

Khorugh

K2
8611m

Karakoram Range

Indus

T U R K M E N I S T A N

Länkäran

Khrebet Kopetdag

ASHGABAT

Mary

Kurgan-Tyube

Mazâr-e-Sharîf

Tabrîz

Rasht

Gorgän

Mashhad

Baghlän

Hindu Kush

Jaläläbäd

Peshäwar

ISLAMABAD

ycheh-ye miyeh

Reshteh-ye Kuhhä-ye Alborz

TEHRÄN

Hamadän

Qom

Dasht-e-Kavir

Herät

KÄBUL

Räwalpindi

Gujränwäla

Lahore

Chenab

Faisaläbäd

Bäkhtarän

I R A N

Eşfahän

Plateau of Iran

A F G H A N I S T A N

Hämün-e Şäberi

Kandahär

Helmand

Multän

Ahväz

Zagros Mountains

Başrah

Âbâdân

Shîräz

Kermän

Zähedän

Quetta

Indus

Thar Desert

KUWAIT

KUWAIT CITY

Sukkur

P A K I S T A N

I N D I A

The Gulf

Bandar-e 'Abbäs

Strait of Hormuz

BAHRAIN

MANAMA

Dubai

Sharjah

Hyderäbäd

RIYADH

Al Hufüf

QATAR

DOHA

ABU DHABI

Şuḩär

Gulf of Oman

Karächi

Tropic of Cancer

Ḩaraḍ

UNITED ARAB EMIRATES

Ar Rustäq

MUSCAT

Nazwä

A r a b i a n

S e a

R A B I A

ub' al Khäli

O M A N

Khalij Maşirah

Şür

Scale : 1:17 000 000

0   200   400 km

0   200   400 miles

E M E N

Şalälah

Hadhramaut

I N D I A N   O C E A N

Al Mukallä

Aden

Socotra (to Yemen)

CONTINENTAL FACTS

PHYSICAL FEATURES

LARGEST LAKE: Caspian Sea 371,000 km² (143,205 mi²)

LONGEST RIVER: Euphrates, Turkey/ Syria/Iraq 2815 km (1750 miles)

HIGHEST POINT: K2, Kashmir, India/Pakistan 8611m (28,252 ft)

LOWEST POINT: Dead Sea, Israel/Jordan 392 m (1286 ft) below sea level

POLITICAL FEATURES

TOTAL POPULATION: 463.3 million

LARGEST CITY AND ITS POPULATION: Karachi, Pakistan 11.8 million

COUNTRY WITH HIGHEST POPULATION DENSITY: Bahrain 924 people/km² (2388 people/mi²)

LARGEST COUNTRY: Saudi Arabia 1,960,582 km² (756,980 mi²)

SMALLEST COUNTRY: Bahrain 620 km² (239 mi²)

CONTINENTAL FACTS

PHYSICAL FEATURES

- **LARGEST LAKE:** Aral Sea, Kazakhstan/Uzbekistan 66,500 km² (25,700 mi²)
- **LONGEST RIVER:** Chang Jiang (Yangtze), China 6380 km (3965 miles)
- **HIGHEST POINT:** Xixabangma Feng, China 8012 m (26,286 ft)
- **LOWEST POINT:** Turpan Hami (Turfan Basin), China 154 m (505 ft) below sea level

POLITICAL FEATURES

- **TOTAL POPULATION:** 1552 million
- **LARGEST CITY AND ITS POPULATION:** Tokyo, Japan 26.4 million
- **COUNTRY WITH HIGHEST POPULATION DENSITY:** Taiwan 688 people/km² (1782 people/mi²)
- **LARGEST COUNTRY:** Asiatic Russia 13,119,582 km² (5,065,471 mi²)
- **SMALLEST COUNTRY:** Taiwan 52,260 km² (12,456 mi²)

over 5 000 000
over 1 000 000
over 500 000
over 100 000
over 50 000
under 50 000

LAND HEIGHT

6000m/19 686ft
4000m/13 124ft
3000m/9843ft
2000m/6562ft
1000m/3281ft
200m/656ft
Sea Level

Line of cross-section

Scale : 1:31 000 000

0   250   500 km

0   250   500 miles

NORTH ASIA

THE URAL MOUNTAINS form a huge natural boundary between European and Asian Russia. East of the Urals, vast areas of windswept grassland – the steppes – stretch as far as the Pacific Ocean. Much of Siberia is a frozen wilderness, covered by pine forests, but rich in oil and precious stones. To the south, the deserts of Kazakhstan merge with the Tien Shan range. The central Asian republics lie along the ancient Silk Road between Asia and Europe, where traditional nomadic lifestyles are being replaced by agriculture and mineral prospecting. Tibet lies on a remote plateau, the world's highest region. In Mongolia, the world's largest land-locked country, people still depend on herding. One-fifth of the world's population lives in China, the majority along the coast and in the warm, fertile river valleys. China remains a predominantly rural country, but has extensive reserves of coal and iron ore. Japan's four main islands are mountainous, heavily forested, and have few raw materials, most of the large population living along the coastal fringes.

CROSS-SECTION THROUGH NORTH ASIA: 38°N, 52°E–145°E

SOUTH ASIA

Dominated in the north by the Himalayas, the highest mountain range in the world, India is isolated from the rest of Asia, forming a densely populated subcontinent. Its climate and topography range from the mountains of Kashmir in the north to coral beaches in the south. It is the birthplace of Hinduism, Buddhism, and Sikhism. Much of mainland southeast Asia is mountainous and forested, the people living in the river valleys and fertile coastal plains. Tropical rainforests, rich in species, cover much of the region. Indonesia forms a huge arc of over 13,000 volcanic islands. The Philippines, the region's only Christian country until 2002, comprises over 7000 mountainous islands.

Cross-section through South Asia: 28°N, 60°E–124°E

CONTINENTAL FACTS

PHYSICAL FEATURES

LARGEST LAKE: Tônlé Sap, Cambodia 2850 km² (1000 mi²)

LONGEST RIVER: Mekong, China/Burma/Laos/Thailand/Cambodia/Vietnam 4425 km (2750 miles)

HIGHEST POINT: Mount Everest, Nepal 8848 m (29,030 ft)

LOWEST POINT: About half of Bangladesh lies between sea level and 8 m (25 ft)

POLITICAL FEATURES

TOTAL POPULATION: 1740 million

LARGEST CITY AND ITS POPULATION: Mumbai, India 18.1 million

COUNTRY WITH HIGHEST POPULATION DENSITY: Singapore 6721 people/km² (17,408 people/mi²)

LARGEST COUNTRY: India 3,287,590 km² (1,269,338 mi²)

SMALLEST COUNTRY: Maldives 300 km² (116 mi²)

POPULATION

over 5 000 000
over 1 000 000
over 500 000
over 100 000
over 50 000
under 50 000

LAND HEIGHT

6000m/19 686ft
4000m/13 124ft
3000m/9843ft
2000m/6562ft
1000m/3281ft
200m/656ft
Sea Level

Line of cross-section

Scale : 1:25 000 000

0   250   500 km

0   250   500 miles

33

Tropic of Cancer

TAIWAN

South
China Sea

Philippine
Sea

PHILIPPINES

NORTHERN
MARIANA ISLANDS
(to US)

P A C I F I C

WAKE I. (to US)

MARSHALL ISLANDS

Saipan

GUAM (to US)

Ratak Chain

Ralik Chain

MALAYSIA
BRUNEI

Sulu
Sea

Celebes
Sea

KOROR
Babelthuap

Yap

PALAU

MICRONESIA

Chuuk Is

Pohnpei
KOLONIA

Kosrae

Majuro

BAIRIKI

Tarawa

NAURU

TUVALU

Nanumea

Nukufeta
FONGAFALE
Nukul

Tungar

Equator

I N D O N E S I A

Bismarck
Archipelago

Bismarck
Sea

Mt Wilhelm
4509m

PAPUA

New
Guinea

NEW GUINEA

PORT MORESBY

New
Britain

Rabaul

Bougainville

Solomon Sea

SOLOMON ISLANDS
Santa Isabel

Malaita

Guadalcanal HONIARA

Rennell

Santa Cruz Is

San Cristobal

Banda Sea

EAST TIMOR

Arafura
Sea

Torres Strait

Timor
Sea

ASHMORE & CARTIER ISLANDS
(to Australia)

Darwin

Gulf
of
Carpentaria

Coral Sea

CORAL SEA
ISLANDS
(to Australia)

VANUATU

Espiritu Santo

Malekula

PORT-VILA Éfaté

NEW CALEDONIA
(to France)

Îles Loyauté

New
Caledonia

NOUMÉA

Vanua
Viti Levu
SU

FIJ

INDIAN
OCEAN

Broome

NORTHERN
TERRITORY

Great
Sandy
Desert

Cairns

Townsville

MacKay

Great Barrier Reef

Rockhampton

NORFOLK ISLAND
(to Australia)

A U S T R A L I A

Alice Springs

QUEENSLAND

Brisbane
Gold Coast
Toowoomba

Lord Howe I. (to Australia)
Ball's Pyramid (to Australia)

WESTERN
AUSTRALIA

Gibson Desert Musgrave Ranges

Simpson
Desert

Lake Eyre

Great
Victoria
Desert

SOUTH
AUSTRALIA

Lake
Torrens

Darling

Great
Dividing Range

NEW SOUTH
WALES

Newcastle
Sydney
Wollongong

CANBERRA

Tasman

Sea

NEW
ZEALAND

Auckland
Hamilton

WELLINGTON

Geraldton

Kalgoorlie

Great Australian Bight

Port Lincoln

Adelaide

Murray

AUSTRALIAN
CAPITAL
TERRITORY

Perth

Esperance

VICTORIA
Bendigo
Geelong Melbourne

Christchurch

Bunbury

Albany

Bass Strait

Dunedin

Bounty Islands
(to NZ)

TASMANIA

Launceston

Hobart

Antipodes Islands
(to NZ)

Auckland Islands
(to NZ)

LAND HEIGHT		POPULATION	
	3000m/9843ft	■	over 5 000 000
	2000m/6562ft	▣	over 1 000 000
	1000m/3281ft	◉	over 500 000
	200m/656ft	●	over 100 000
	Sea Level	◎	over 50 000
		○	under 50 000

Scale : 1:40 000 000

0   500   1000 km

0   500   1000 miles

AUSTRALASIA & OCEANIA

OCEANIA, A CONTINENT OF ISLANDS stretching across a vast area of the Pacific Ocean, is home to only 0.5 per cent of the world's population. Dominated by Australia, it includes few other countries with significant land mass apart from New Zealand, Papua New Guinea, and Fiji, but a myriad volcanic and coral islands in three main groups, Micronesia, Melanesia, and Polynesia. Australia, flat and dry, is sparsely populated, most people living along the coastal lowlands, especially in the southeast. Its first inhabitants, the Aboriginal peoples, retain some of their original lands in the interior, but the European and Asian settlers of recent centuries form most of the population. Australia is rich in minerals, such as gold, uranium, and iron ore, which are the basis of its prosperity. Mountainous Papua New Guinea is covered in tropical rainforest, while New Zealand is temperate, rugged, and volcanic in the north. Owing to their isolation, these countries' flora and fauna have evolved many unique species. The peoples of Oceania colonized the Pacific by 1100 CE, and the many insular farming and fishing communities have developed distinctive cultures, the Maoris of New Zealand being among the most notable.

CONTINENTAL FACTS

PHYSICAL FEATURES

- **LARGEST LAKE:** Lake Eyre, Australia 9583 km² (3700 mi²)
- **LONGEST RIVER:** Murray-Darling, Australia 3750 km (2330 miles)
- **HIGHEST POINT:** Mt. Wilhelm, Papua New Guinea 4509 m (14,794 ft)
- **LOWEST POINT:** Lake Eyre, Australia 16 m (52 ft) below sea level

POLITICAL FEATURES

- **TOTAL POPULATION:** 30.4 million
- **LARGEST CITY AND ITS POPULATION:** Sydney, Australia 3.6 million
- **COUNTRY WITH HIGHEST POPULATION DENSITY:** Nauru 564 people/km² (1461 people/mi²)
- **LARGEST COUNTRY:** Australia 7,686,850 km² (2,967,892 mi²)
- **SMALLEST COUNTRY:** Nauru 21 km² (8 mi²)

CROSS-SECTION THROUGH AUSTRALIA: 27°S, 112°E–160°W

CHRONOLOGY OF WORLD HISTORY

THIS TABLE PRESENTS A SUMMARY of the world's crucial historical events, from the first evidence of settlement and agriculture until 2000 CE. Each of the six columns is shaded a different color, with each color representing a particular continent. Reading across the columns, one can follow the development of cultures across the major landmasses of the world. By reading downwards, each continent's particular cultural history can be seen, from its first steps toward civilization, through periods of migration, empire, and revolution, to its involvement in the global wars and political diplomacy of the late 20th century.

NORTH AMERICA

- ❏ **15,000** BCE Evidence of human settlement in North America
- ❏ **4000** BCE Earliest cultivation of corn in Central America
- ❏ **100** CE Teotihuacan becomes capital of largest state in Mesoamerica
- ❏ **c.300** CE Start of classic Mayan civilization in Yucatan
- ❏ **900** Toltecs rise to power after Teotihuacan and Mayan states collapse
- ❏ **c.1000** Vikings colonize Greenland and discover America (Vinland)
- ❏ **1200** Aztecs enter Valley of Mexico
- ❏ **1325** Tenochtitlan founded by Aztecs
- ❏ **1492** Columbus reaches Caribbean
- ❏ **c.1500** Inuit peoples found throughout Arctic region
- ❏ **1502** Introduction of African slaves to Caribbean
- ❏ **1519** Cortes begins conquest of Aztec empire
- ❏ **1565** First African slaves arrive on mainland North America
- ❏ **1607** First permanent English settlement in North America (Jamestown, Virginia)
- ❏ **1608** French colonists found Québec
- ❏ **1620** Puritans on *Mayflower* land in New England
- ❏ **1759** British capture Québec
- ❏ **1776** American Declaration of Independence
- ❏ **1789** George Washington becomes first president of US
- ❏ **1791** Revolution in Haiti
- ❏ **1803** Louisiana Purchase nearly doubles size of US
- ❏ **1810** Revolution in Mexico
- ❏ **1819** US buys Florida from Spain
- ❏ **1821** Mexico gains independence
- ❏ **1828** Federalist–Centralist wars in Mexico (to 1859)
- ❏ **1845** Texas annexed by US
- ❏ **1846** US–Mexican War (to 1848)
- ❏ **1848** Californian Gold Rush

SOUTH AMERICA

- ❏ **c.20,000** BCE First settlers arrive
- ❏ **c.11,000** BCE Evidence of settlement at Monte Verde in present-day Chile
- ❏ **c.4500** BCE Evidence of agriculture and herding in central Andes
- ❏ **c.2500** BCE Masonry building and temple architecture on Pacific coast
- ❏ **c.1800** BCE Ceremonial center of La Florida built in Peru
- ❏ **c.450** CE Nazca culture flourishing; lines and giant figures drawn in desert
- ❏ **c.900** CE Wari and Tiwanaku flourish as capitals of first competing empires
- ❏ **1100** Emergence of Chimu state on north coast of Peru
- ❏ **1380** Beginning of Inca empire in central highlands
- ❏ **1475** Chimu conquered by Incas
- ❏ **1494** Treaty of Tordesillas divides western hemisphere between Spain and Portugal
- ❏ **1500** Cabral sights Brazilian coast
- ❏ **1502** First expedition sent from Portugal to exploit coast of Brazil
- ❏ **c.1510** First African slaves to South America
- ❏ **1525** Civil war in Inca empire
- ❏ **1532** Pizarro begins defeat of Incas (to 1540)
- ❏ **1562** War and disease kill much of Amerindian population of Brazil (to 1563)
- ❏ **1630** Dutch establish New Holland, covering much of northern Brazil
- ❏ **1654** Portuguese regain control of Brazil
- ❏ **1663** Brazil becomes viceroyalty
- ❏ **1695** Gold discovered in Brazil
- ❏ **1739** Viceroyalty of New Granada established to defend Spanish interests on Caribbean coast
- ❏ **1750** Treaty of Madrid defines boundary between Spanish colonies and Brazil

EUROPE

- ❏ **c.6500** BCE Farming spreads rapidly into central Europe
- ❏ **c.2800** BCE Building of Stonehenge begins
- ❏ **c.1600** BCE Minoan Palace civilization on Crete
- ❏ **c.750** BCE Beginnings of Greek city-states
- ❏ **510** BCE Roman Republic founded
- ❏ **451** BCE Outbreak of Peloponnesian War between Sparta and Athens
- ❏ **218** BCE Carthaginians invade Italy under command of Hannibal
- ❏ **49** BCE Julius Caesar conquers Gaul
- ❏ **43** CE Roman invasion of Britain
- ❏ **238** CE Goths begin to invade borders of Roman Empire
- ❏ **330** Constantinople becomes new capital of Roman Empire
- ❏ **410** Invasion and pillage of Rome by Visigoths
- ❏ **711** Spain invaded by Muslims
- ❏ **793** Viking raids across Europe
- ❏ **800** Charlemagne becomes first Holy Roman Emperor
- ❏ **950** Harold Bluetooth consolidates the unification of Denmark
- ❏ **987** Feudal lords elect first Capetian king of France
- ❏ **1066** Norman conquest of England
- ❏ **1236** Russia invaded by Mongols
- ❏ **1337** Onset of Hundred Years War
- ❏ **1453** Byzantine Empire collapses as Ottoman Turks capture Constantinople
- ❏ **1478** Ivan III first czar of Russia
- ❏ **c.1500** Italian Renaissance
- ❏ **1521** Beginning of Protestant Reformation
- ❏ **1534** Henry VIII of England breaks with Rome
- ❏ **1588** Spanish Armada defeated by English
- ❏ **1618** Onset of Thirty Years War
- ❏ **1642** English Civil War (to 1649)
- ❏ **1756** Onset of Seven Years War
- ❏ **1789** French Revolution

AFRICA

- **c.400,000** BCE First evidence of *Homo sapiens* (modern humans) in Rift Valley
- **3100** BCE King Narmer unifies Upper and Lower Egypt and becomes first pharaoh
- **c.2650** BCE Start of great pyramid building in Egypt
- **2040** BCE Beginning of Middle Kingdom in Egypt
- **1352** BCE Pharaoh Akhenaten promotes Aten (sun) worship in Egypt
- **814** BCE Foundation of Phoenician colony of Carthage
- **146** BCE Rome conquers Carthage
- **31** BCE Cleopatra's death marks end of Ptolemaic dynasty in Egypt
- **c.600** CE Kingdom of Ghana founded
- **641** CE Muslims conquer Egypt
- **c.900** CE Emergence of Great Zimbabwean state
- **1067** Almoravids destroy kingdom of Ghana
- **c.1300** Emergence of empire of Benin (Nigeria)
- **1390** Formation of kingdom of Kongo
- **1443** Portuguese begin mass export of slaves from Atlantic coast to Europe
- **1498** Vasco da Gama rounds Cape of Good Hope
- **1502** First slaves taken to New World
- **1570** Establishment of Portuguese colony of Angola
- **1652** Dutch establish colony at Cape of Good Hope
- **1787** British establish Sierra Leone for freed slaves
- **1795** British capture Cape of Good Hope from the Dutch
- **1798** Occupation of Egypt by Napoléon Bonaparte
- **1816** Shaka leads expansion of Zulu
- **1822** Freed black slaves found colony of Liberia
- **1830** French invasion of Algeria
- **1836** Start of Boer Great Trek
- **1848** Boers found Orange Free State
- **1853** Livingstone finds Victoria Falls (Musi-o-Tunya)
- **1869** Opening of Suez Canal
- **1875** Stanley establishes source of Nile

ASIA AND THE MIDDLE EAST

- **c.12,000** BCE Beginnings of farming in Palestine
- **c.4800** BCE 'Ubaid culture builds towns in Mesopotamia
- **c.3500** BCE Fortified towns built throughout northern China
- **c.2800** BCE Emergence of city states in Indus Valley
- **Wheel used in Mesopotamia**
- **c.1750** BCE Foundation of Old Babylonian Empire under Hammurabi
- **c.1200** BCE Traditional date for exodus of the Jews from Egypt
- **c.1100** BCE Phoenician civilization spreads throughout Mediterranean
- **c.660** BCE Japanese empire founded
- **550** BCE Persian Empire founded
- **c.480** BCE Death – or *parinibbana* – of Gautama Buddha
- **334** BCE Alexander the Great invades Asia Minor
- **332** BCE Foundation of Mauryan empire in India
- **202** BCE Han dynasty begins in China
- **c.112** BCE "Silk Road" links China to West
- **c.30** CE Crucifixion of Jesus of Nazareth, founder of Christianity
- **c.350** CE Huns invade Persia and India
- **622** Mohammed, founder of Islam, flees Mecca; start of Muslim calendar
- **960** China united under Sung dynasty
- **1044** Foundation of Burma
- **1099** Jerusalem sacked in First Crusade
- **1185** Minamoto shoguns rule Japan
- **1206** Mongols begin to conquer Asia under Genghis Khan
- **1238** Foundation of first Thai kingdom
- **1258** Baghdad sacked by Mongols
- **1264** Yuan dynasty founded in China by Kublai Khan
- **1275** Marco Polo arrives in China
- **1333** Civil war in Japan
- **1368** Ming dynasty begins in China
- **1392** Korea proclaims independence
- **1498** Vasco da Gama completes first European voyage to India
- **1526** Foundation of Mughal empire in India
- **1600** Charter granted to East India Company
- **1609** Beginning of Tokugawa shogunate in Japan
- **1619** Dutch found Batavia (Jakarta)

AUSTRALASIA AND OCEANIA

- **c.60,000** BCE First people arrive in Australia
- **c.30,000** BCE Aboriginal rock art begins to appear
- **c.8000-6000** BCE Rising sea level covers New Guinea land bridge
- **c.6000** BCE Migrations from southeast Asia give rise to Austronesian culture
- **c.4000** BCE Austronesians reach southwestern Pacific islands
- **c.1000** BCE Emergence of archaic Polynesian society in Fiji, Tonga, and Samoa
- **c.500** CE Easter Island settled
- **c.600** CE Polynesians arrive in Hawaii
- **1520** Magellan enters Pacific
- **1526** Jorge de Meneses first European to sight New Guinea
- **1606** Torres sails through strait that now bears his name; proves New Guinea is an island
- **1642** Tasman, searching for a southern continent, finds Tasmania and New Zealand
- **1688** Dampier first Englishman to visit Australia
- **1768** Cook's first voyage
- **1773** Cook crosses Antarctic Circle and explores Southern Ocean (to 1775)
- **1779** Cook killed in Hawaii on third voyage
- **1788** First penal settlement established at Port Jackson (Sydney)
- **1802** Flinders circumnavigates Australia (to 1803)
- **1818** Start of Maori "Musket Wars" in New Zealand
- **1819** Bellingshausen's expedition sights Antarctica
- **1829** Britain annexes western and final third of Australian continent
- **1830** A mere 200 foreigners, mostly British, permanently resident in New Zealand
- **1840** Treaty of Waitangi grants sovereignty over New Zealand to British
- **1841** New Zealand becomes a separate Crown colony
- **1845** Northern War in New Zealand (to 1846)

NORTH AMERICA (CONTINUED)

- **1861** US Civil War (to 1865)
- **1863** Emancipation Proclamation
- **1865** Lee surrenders to Grant at Appomattox
- Assassination of President Lincoln
- Slavery abolished in US
- **1867** US buys Alaska from Russia
- Dominion of Canada established
- **1869** 15th Amendment gives vote to freed slaves in US
- **1871** Start of Apache Wars
- **1876** Battle of Little Big Horn: Sioux warriors kill 250 US soldiers
- **1890** Massacre of Sioux warriors at Wounded Knee ends Amerindian wars
- **1896** Klondike Gold Rush, Alaska
- **1898** Spanish–American War
- **1899** Spain cedes Cuba and Puerto Rico to US
- **1910** Mexican Revolution begins
- **1921** US restricts immigration
- **1929** Wall Street Crash
- **1933** President Roosevelt introduces New Deal
- **1940s** Race riots in Harlem, Los Angeles, Detroit, and Chicago
- **1941** US enters war against Germany and Japan
- **1945** End of World War II
- **1949** Formation of NATO
- Cold War begins
- **1950** US supports south in Korean War (to 1953)
- **1959** Cuban Revolution
- **1962** Cuban missile crisis
- **1963** Martin Luther King leads march on Washington D.C.
- Assassination of President Kennedy
- **1964** US Congress approves sending first troops to Vietnam
- **1968** Assassination of Martin Luther King sparks riots in 124 US cities
- **1969** Neil Armstrong becomes first person on moon
- **1973** US withdraws from Vietnam
- **1974** President Nixon resigns over Watergate scandal
- **1979** Civil war in Nicaragua (to 1990)
- Civil war in El Salvador (to 1992)
- **1991** US leads alliance in Gulf War to end Iraq's invasion of Kuwait
- **1994** North American Free Trade Agreement (NAFTA) established
- **1999** President Clinton survives impeachment
- **2000** George W. Bush wins presidential elections but does not receive a majority of votes

SOUTH AMERICA (CONTINUED)

- **1811** Bolívar starts fight to liberate Venezuela
- Paraguay independent
- **1817** San Martin wins decisive victory over Spanish and liberates Chile
- **1821** Peru independent
- **1822** Brazil independent
- **1823** Slavery abolished in Chile
- **1825** Bolivia independent
- **1828** Uruguay independent
- **1830** Ecuador, Colombia, and Venezuela (formerly Gran Colombia) become separate states
- **1851** Slavery abolished in Colombia
- **1853** Slavery abolished in Ecuador, Argentina, and Uruguay
- **1854** Slavery abolished in Bolivia and Venezuela
- **1864** Paraguayan War: Brazil, Argentina, and Uruguay defeat Paraguay
- **1870** Slavery abolished in Paraguay
- **1888** Slavery abolished in Brazil
- **1900** Major Italian migration to Argentina
- **1914** Panama Canal opens
- **1930** Military revolution in Brazil
- **1932** Chaco War between Bolivia and Paraguay (to 1935); Paraguay defeats Bolivia
- **1937** "New State" in Brazil launched by Vargas
- **1946** Peron comes to power in Argentina
- **1955** Peron ousted by military coup; returns to power in 1973
- **1968** Tupamaros urban guerrilla group founded in Uruguay
- Military junta takes over Peru
- **1970** Allende elected president of Chile
- **1973** US backs Pinochet coup against elected government in Chile; Allende assassinated
- **1976** "Dirty War" of right-wing death squads in Argentina
- **1982** Falklands War between Argentina and UK
- **1983** Democracy restored in Argentina
- **1985** Democracy restored in Brazil and Uruguay
- **1989** Democracy restored in Chile
- **1999** Panama takes control of Panama Canal
- **2000** President Fujimori ousted amid massive corruption scandal in Peru

EUROPE (CONTINUED)

- **1804** Napoléon becomes emperor of France
- **1815** Napoléon defeated
- Treaty of Vienna
- **1845** Beginning of Irish potato famine
- **1854** Crimean War (to 1856)
- **1861** Italy unified
- Emancipation of serfs in Russia
- **1870** Franco-Prussian War
- **1871** Germany unified
- **1914** World War I (to 1918)
- **1917** Russian Revolution
- **1922** Mussolini comes to power in Italy after Fascist "March on Rome"
- **1933** Nazis take power in Germany; Hitler is elected chancellor
- **1936** Spanish Civil War (to 1939)
- **1939** Germany invades Poland precipitating World War II
- **1941** German forces invade Russia
- **1944** British and US troops land in Normandy; Russians advance into eastern Europe
- **1945** Defeat of Germany
- **1949** Formation of NATO
- Cold War begins
- **1957** Treaties of Rome establish European Economic Community
- **1961** Building of the Berlin Wall
- Yuri Gagarin first person in space
- **1968** Troubles in Northern Ireland
- **1973** UK and Ireland join European Communities
- **1975** End of dictatorship in Spain with death of Gen. Franco
- **1986** Explosion at Chernobyl nuclear power reactor
- Soviet launch of *Mir* space station
- **1989** Democratic revolutions in eastern Europe
- Berlin Wall demolished
- **1990** Reunification of Germany
- **1991** The Soviet Union splits into its component countries
- Slovenia and Croatia claim their independence
- **1992** Civil war in Bosnia & Herzegovina (to 1995)
- **1993** Velvet Divorce: separation of Czech Republic and Slovakia
- **1994** Outbreak of war in Chechnya
- **1995** European Union expands to 15 members as Austria, Finland, and Sweden join
- **1999** "Ethnic cleansing" of Albanians in Kosovo leads to NATO airstrikes against Yugoslavia
- **2000** President Milošević is ousted in Yugoslavia (Serbia & Montenegro) in a popular revolution

AFRICA (CONTINUED)

- ❑ 1879 British defeat Zulus
- ❑ 1881 French occupy Tunisia
- ❑ 1882 Britain occupies Egypt
- ❑ 1883 France begins conquest of Madagascar
- ❑ 1889 Colonization of "Rhodesia"
- ❑ 1890 Land connection between Angola and Mozambique ended
- ❑ 1894 Britain occupies Uganda
- ❑ 1896 Ethiopian emperor Menelik II defeats Italians at Adawa
- ❑ 1899 Boer War (to 1902)
- ❑ 1910 Formation of Union of South Africa
- ❑ 1911 Italian conquest of Libya
- ❑ 1935 Second Italian invasion of Ethiopia
- ❑ 1942 British halt German advance at El Alamein
- ❑ 1948 Pro-apartheid National Party wins power in South Africa
- ❑ 1956 UK fails to block Egypt's nationalization of Suez Canal
- ❑ 1960 Outbreak of civil war in Belgian Congo
- ❑ Fifteen countries gain independence
- ❑ 1962 Algeria gains independence
- ❑ 1963 Zambia and Malawi granted independence
- ❑ Organization of African Unity (OAU) founded
- ❑ 1964 Nelson Mandela sentenced to life imprisonment
- ❑ 1974 Emperor Haile Selassie of Ethiopia deposed
- ❑ 1975 Angola and Mozambique gain independence; civil wars ensue
- ❑ 1980 Black majority rule established in Zimbabwe
- ❑ 1981 President Sadat of Egypt assassinated
- ❑ 1984 Worst recent famine in Ethiopia
- ❑ 1990 Mandela released: apartheid begins to be dismantled
- ❑ Namibia becomes independent
- ❑ 1991 Civil war in Sierra Leone (to 2001)
- ❑ 1994 South Africa holds first multiracial election; Mandela wins presidency
- ❑ Attempted genocide of Tutsis by Hutu in Rwanda
- ❑ 1997 Overthrow of Mobutu in Zaire (Democratic Republic of the Congo – DRC)
- ❑ 1998 Start of civil war in DRC
- ❑ 1999 Nigeria returns to democracy
- ❑ 2000 Ethiopia–Eritrea conflict

ASIA AND THE MIDDLE EAST (CONTINUED)

- ❑ 1757 East India Company defeats Nawab of Bengal's forces at Plassey
- ❑ 1842 Opium Wars (to 1854), Britain compels China to open Treaty Ports and annexes Hong Kong
- ❑ 1851 Taiping rebellion in China, 20 million killed (to 1864)
- ❑ 1868 Meiji Restoration in Japan
- ❑ 1877 Queen Victoria proclaimed empress of India
- ❑ 1911 Manchu dynasty overthrown in China, republic declared
- ❑ 1922 The last Ottoman sultan is deposed; Turkey proclaimed a republic
- ❑ 1932 Kingdom of Saudi Arabia founded
- ❑ 1937 Japanese forces invade China
- ❑ 1941 Pearl Harbor attacked by Japan
- ❑ 1945 Atom bombs dropped on Hiroshima and Nagasaki; c.210,000 killed, Japan surrenders
- ❑ 1947 Partition of India: Pakistan and India independent
- ❑ 1948 Burma and Ceylon (Sri Lanka) proclaim their independence
- ❑ Establishment of Israel
- ❑ 1949 People's Republic of China proclaimed
- ❑ Indonesia independent
- ❑ 1950 Korean War (until 1953)
- ❑ 1954 Laos, Cambodia, and Vietnam proclaim their independence
- ❑ 1959 China occupies Tibet
- ❑ 1965 US combat troops in Vietnam
- ❑ 1966 Cultural Revolution in China
- ❑ 1971 East Pakistan (Bangladesh) claims independence
- ❑ 1975 Fall of Saigon ends Vietnam War
- ❑ Civil war in Lebanon (to 1989)
- ❑ 1979 Overthrow of shah in Iran, Islamic Republic founded
- ❑ Vietnam pushes Khmer Rouge from Cambodia
- ❑ 1980 Iran–Iraq War (to 1988)
- ❑ 1982 Israeli invasion of Lebanon
- ❑ 1986 Marcos deposed in Philippines
- ❑ 1989 Massacre in Tiananmen Square
- ❑ 1990 Invasion of Kuwait by Iraq
- ❑ 1991 Gulf War
- ❑ 1996 *Taliban* take over in Afghanistan
- ❑ 1997 Hong Kong is returned to China
- ❑ 1997 Asian financial crisis
- ❑ 1998 Suharto regime collapses in Indonesia
- ❑ India and Pakistan test nuclear weapons
- ❑ 2000 Palestinians begin new *intifada*

AUSTRALASIA AND OCEANIA (CONTINUED)

- ❑ c.1850 Migrant workers from China, Japan, and Philippines start arriving in Hawaii
- ❑ 1851 Gold discovered in New South Wales
- ❑ 1858 King Movement demands Maori state and opposes further land sales
- ❑ 1860 European settlers outnumber Maoris in New Zealand
- ❑ 1862 Second Maori War
- ❑ 1864 First French convict settlers in New Caledonia
- ❑ 1865 1000 Chinese brought to Tahiti to work cotton plantation (to 1866)
- ❑ 1869 Last convict ship arrives in Australia
- ❑ 1870 Maori resistance crushed
- ❑ Germans start to buy up large tracts of Western Samoa
- ❑ 1874 Indian sugarcane workers arrive in Fiji
- ❑ 1888 Chile starts colonization of Easter Island
- ❑ 1890 Gold discovered in Western Australia
- ❑ 1898 US annexes Hawaii and seizes Guam from Spain
- ❑ 1901 Australia is self-governing federation within British Empire
- ❑ 1912 Amundsen's expedition reaches South Pole
- ❑ 1914 Over 60,000 Australian troops and more than 15,000 New Zealanders lose their lives in World War I (to 1918)
- ❑ 1930s Australia hit hard by global Depression
- ❑ 1942 Australia under threat of invasion as Japanese bomb Darwin
- ❑ 1946 US begins nuclear tests at Eniwetok and Bikini atolls in Micronesia
- ❑ 1952 UK begins nuclear tests (to 1991) on mainland Australia
- ❑ 1966 France begins nuclear tests in Tuamotu Islands
- ❑ 1975 Restrictions imposed on immigrants to Australia
- ❑ 1985 South Pacific Forum declares nuclear-free Pacific
- ❑ 1988 Bicentennial celebrations in Australia occasion Aboriginal protests
- ❑ 1996 France halts nuclear testing in the Pacific
- ❑ 1998 Ethnic conflict in Solomon Islands (to 2000)
- ❑ 1999 Australian referendum rejects proposal on becoming republic
- ❑ 2000 Coup in Fiji ousts first ethnic Indian government

THE FORMATION OF THE MODERN WORLD

THE WORLD AS WE KNOW IT today, like all of the species that inhabit it, is the product of many thousands of years of evolution. The political and cultural map of the globe bears the hallmark of many varied courses of human development the world over. Nevertheless, much of the modern human geography of the planet can be traced to developments in the relatively recent past. The following pages chart the rise and fall of the various states and empires of the early modern and modern ages. Beginning with the first great achievement of European exploration, the "discovery" of the Americas in 1492, the maps show the way in which various European and Asian powers expanded their cultural and political influence and control down to the present day. This process left indelible cultural imprints in the form of language, religion, education, and systems of government on every part of the planet.

MAJOR MIGRATIONS SINCE 1500

SCANDINAVIA

RUSSIA
1 million Russians to Siberia

GREAT BRITAIN

IRELAND   GERMANY

CANADA

CENTRA

from Asia

UNITED STATES OF AMERICA

PORTUGAL   SPAIN   FRANCE   ITALY

WEST INDIES

ATLANTIC OCEAN

ARABIA

CENTRAL AMERICA

11 million African slaves to the Americas

WEST AFRICA

slaves

EAST AFRICA

PACIFIC OCEAN

BRAZIL

SOUTH AMERICA

slaves

SOUTHERN AFRICA

3 million Europeans to Australia & New Zealand

40 million Europeans to North America

KEY

- Europeans
- Russians
- Africans
- East Asians
- South Asians

LANGUAGES OF THE WORLD

KEY

- Arabic
- Chinese
- English
- French
- Portuguese
- Russian
- Spanish
- Hindi
- Others

LANGUAGES OF THE WORLD

THERE ARE OVER 3000 LANGUAGES or "speech communities" in the world today; some are spoken by many millions, some by only dozens. Many people speak more than one language. The diffusion of the major languages throughout the world during the modern era has seen the emergence of a few dominant languages (shown on the map). In many areas, the language of a colonial power has been maintained either as an official language or has become the *lingua franca* of the region. The largest single language, encompassing many dialects, is Chinese, with over one billion speakers; Hindi (400 million) and Arabic (200 million), are the next largest first languages. The most successful colonial languages are English (estimated at up to 1500 million, including those using English as a second language), French (200 million), and Spanish (270 million). While the last is now estimated to be the world's fastest-growing language, owing to Latin America's burgeoning population growth, both English and French are spoken in a wide variety of patois, pidgins, and creoles, thus achieving unique levels of cultural penetration.

MAJOR MIGRATIONS SINCE 1500

THE LAST FIVE HUNDRED YEARS have witnessed a dramatic redistribution of the world's population, which occurred in a series of waves. The first of these involved, in the 16th–18th centuries, the mass transshipment of captive peoples from sub-Saharan Africa to supply the slave markets of West Asia and to work newly founded European plantations in the Americas. The rapidly growing populations of Europe and Asia encouraged a heavy flow of migration. The Cantonese from southern China spread throughout southeast Asia, while from the 16th century millions of Europeans emigrated to the "New Worlds" of the Americas and, later, Australasia. This European diaspora reached a peak at the end of the 19th century. Then, as the colonial empires coalesced in the early years of the 20th century, there was a final wave of global movement within them, when south and east Asians migrated to fill labor markets and exploit opportunities in Africa and the Americas. While homogeneous societies have developed in North America and Australia, many diverse ethnic communities remain scattered across the world.

THE WORLD IN 1492

WHEN CHRISTOPHER COLUMBUS sailed west from Europe, seeking a quicker route to Asia, he launched a process of discovery that was eventually to bring the disparate regions of the world into closer contact, to form the global map we know today. The largest political entity in the world at that time was the Chinese Ming empire. Culturally, the Islamic faith had forged a bond of religious unity which extended in a broad swathe from southeast Asia to the Atlantic coast of north Africa. Europe was a mêlée of rival monarchies; sub-Saharan Africa a patchwork of trading kingdoms; the Americas, a separate world of rich tribal cultures, with empires established only in Central America and the central Andes.

GLOBAL STATES AND TERRITORIES

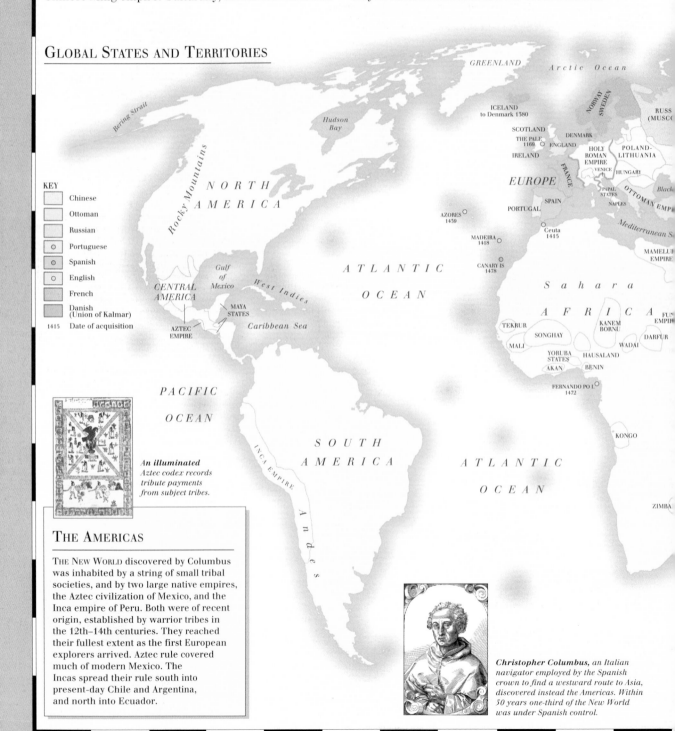

KEY

- Chinese
- Ottoman
- Russian
- ⊙ Portuguese
- ⊙ Spanish
- ○ English
- French
- Danish (Union of Kalmar)

1415   Date of acquisition

An illuminated Aztec codex records tribute payments from subject tribes.

THE AMERICAS

THE NEW WORLD discovered by Columbus was inhabited by a string of small tribal societies, and by two large native empires, the Aztec civilization of Mexico, and the Inca empire of Peru. Both were of recent origin, established by warrior tribes in the 12th–14th centuries. They reached their fullest extent as the first European explorers arrived. Aztec rule covered much of modern Mexico. The Incas spread their rule south into present-day Chile and Argentina, and north into Ecuador.

Christopher Columbus, an Italian navigator employed by the Spanish crown to find a westward route to Asia, discovered instead the Americas. Within 50 years one-third of the New World was under Spanish control.

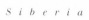

EUROPE

THOUGH CHRISTIAN EUROPE later transformed the exploration and settlement of the world, the Europe from which Columbus sailed was an unstable, violent continent, threatened by invaders from Asia to the east, and from the Ottoman Empire to the south. Civil wars and dynastic conflict resulted in shifting frontiers and small, militarily weak states. Only France, united by the late 15th century, Spain, a single monarchy from the 1490s, Portugal, and England were close to their modern forms.

The Portuguese caravel, buoyant, sturdy, and lateen-rigged, was an ideal ocean-going vessel.

EAST ASIA

THE MOST POWERFUL STATE in the world in 1492 was Ming China. Set up in 1386 after the collapse of Mongol power, the Ming dynasty ruled an area from Manchuria in the north to the borders of Vietnam in the south. Based on a traditional structure of bureaucratic control, the Ming emperors controlled their vast empire from Peking (Beijing), from where they launched punitive wars against the Mongols and Japanese pirates along the coast. Chinese culture and trade spread throughout east and southeast Asia, and Chinese navigators reached the Red Sea and the east African coast.

Chinese junks plied the China seas, and traded as far as the East Indies, Ceylon (Sri Lanka), and east Africa.

SOUTH ASIA AND OCEANIA

THE ETHNIC, POLITICAL, and religious map of southeast Asia was largely in place by the late 15th century. However, the largest state was the vast Srivijayan Hindu–Buddhist empire, which spanned the East Indies archipelago. Muslim traders were in the process of incorporating this rich region into an Indian Ocean trading empire. Further east, the scattered island groups of the Pacific were being successively colonized by waves of Melanesians.

The outrigger canoe was the vehicle of Pacific colonization.

Arab dhows built a trading network around the Indian Ocean.

MIDDLE EAST AND AFRICA

AFTER CENTURIES OF INVASION from the Christian West and Asian nomadic empires, the Middle Eastern world stabilized around a revival of the Ottoman Empire. Vassal states extended across north Africa to Morocco, which linked the trading kingdoms of sub-Saharan Africa with the markets of Asia. The great cities of the Middle East surpassed those of Europe in wealth and learning.

The magnetic compass, in use since the 13th century, was a primary navigational tool for the first ocean-going explorers, although early compasses were not always reliable, and ships often went astray. Accurate navigation only came later with the invention of the chronometer.

Map labels

Bering Strait
Siberia
A S I A
Gobi
ANATE CRIMEA
Aral Sea
Uzbekh Khanate
ian Sea
KOYUNLU
TIMURID PERSIA
The Gulf
Himalayas
NEPAL
SULTANATE OF DELHI
TIBET
Sea of Japan
JAPAN
KOREA
MING EMPIRE
YEMEN
Arabian Sea
Bay of Bengal
AVA
PEGU
LAOS
ANNAM
CAMBODIA
SIAM
South China Sea
PACIFIC OCEAN
HIOPIA
VIJAYANAGAR
Ceylon
Micronesia
INDIAN OCEAN
SRIVIJAYAN EMPIRE
East Indies
Melanesia
Madagascar
AUSTRALIA
NEW ZEALAND

THE AGE OF DISCOVERY: 1492–1648

THE FIRST STATE to take advantage of the new age of exploration was Spain. By the middle of the 16th century, under the Emperor Charles V, Spain was established as the foremost European colonial power, and one of the richest and most powerful kingdoms in Europe. Spanish rule was extended over the whole of Central America, much of South America, Florida, and the Caribbean; in Asia, Spanish rule was established in the Philippines. Spain led the way in establishing European settler colonies overseas. By the middle of the 17th century, British, Dutch, and French colonists began to challenge Spanish dominance in the Americas and east Asia, while pirates around the world plundered Spain's wealthy merchant convoys.

GLOBAL STATES AND TERRITORIES

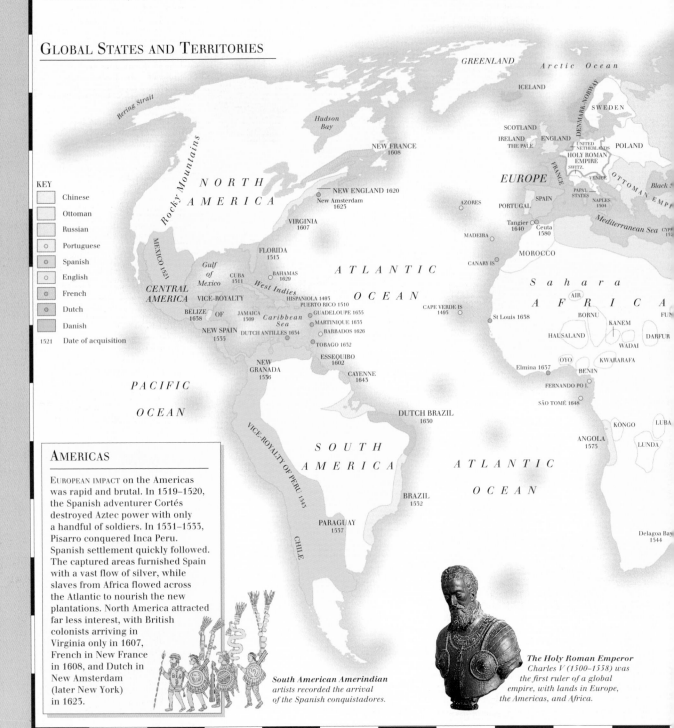

KEY

	Chinese
	Ottoman
	Russian
○	Portuguese
◉	Spanish
○	English
◉	French
◉	Dutch
	Danish
1521	Date of acquisition

GREENLAND · Arctic Ocean · ICELAND · DENMARK-NORWAY · SWEDEN · Hudson Bay · SCOTLAND · IRELAND · ENGLAND · THE PALE · UNITED NETHERLANDS · POLAND · HOLY ROMAN EMPIRE · SWITZ. · NEW FRANCE 1608 · EUROPE · FRANCE · VENICE · OTTOMAN EMPIRE · Bering Strait · NORTH AMERICA · Rocky Mountains · NEW ENGLAND 1620 · New Amsterdam 1623 · AZORES · SPAIN · PORTUGAL · PAPAL STATES · Black S · NAPLES 1504 · VIRGINIA 1607 · Tangier 1640 · Ceuta 1580 · Mediterranean Sea · CYP · MADEIRA · MOROCCO · FLORIDA 1513 · MEXICO 1521 · Gulf of Mexico · CUBA 1511 · BAHAMAS 1629 · ATLANTIC OCEAN · CANARY IS · Sahara · AIR · AFRICA · CENTRAL AMERICA · VICE-ROYALTY · West Indies · HISPANIOLA 1495 · PUERTO RICO 1510 · CAPE VERDE IS 1495 · St Louis 1638 · BORNU · FUN · BELIZE 1638 · OF · JAMAICA 1509 · Caribbean Sea · GUADELOUPE 1635 · MARTINIQUE 1635 · KANEM · NEW SPAIN 1535 · DUTCH ANTILLES 1634 · BARBADOS 1626 · HAUSALAND · DARFUR · WADAI · TOBAGO 1652 · Elmina 1657 · OYO · KWARARAFA · ESSEQUIBO 1602 · BENIN · NEW GRANADA 1556 · CAYENNE 1645 · FERNANDO PO I. · PACIFIC OCEAN · SÃO TOMÉ 1648 · KONGO · LUBA · DUTCH BRAZIL 1630 · ANGOLA 1575 · LUNDA · VICE-ROYALTY OF PERU 1545 · SOUTH AMERICA · ATLANTIC OCEAN · BRAZIL 1532 · CHILE · PARAGUAY 1537 · Delagoa Bay 1544

AMERICAS

EUROPEAN IMPACT on the Americas was rapid and brutal. In 1519–1520, the Spanish adventurer Cortés destroyed Aztec power with only a handful of soldiers. In 1531–1533, Pisarro conquered Inca Peru. Spanish settlement quickly followed. The captured areas furnished Spain with a vast flow of silver, while slaves from Africa flowed across the Atlantic to nourish the new plantations. North America attracted far less interest, with British colonists arriving in Virginia only in 1607, French in New France in 1608, and Dutch in New Amsterdam (later New York) in 1623.

South American Amerindian artists recorded the arrival of the Spanish conquistadores.

The Holy Roman Emperor *Charles V (1500–1558) was the first ruler of a global empire, with lands in Europe, the Americas, and Africa.*

EUROPE

FOR MORE THAN A CENTURY after Martin Luther inspired the Protestant Reformation in the 1520s, Europe was torn by religious wars. Scandinavia, England, and Scotland adopted the new beliefs, but elsewhere bitter civil conflicts led to the prolonged warfare and persecution known as the Thirty Years War. This ended in 1648; it destroyed wide areas of central Europe and decimated the German population, but resulted in a religious settlement which continues to the present. The Dutch Republic and northern Germany became Protestant while southern Germany, Poland, and southwest Europe remained Roman Catholic.

Printing, using movable type, was a key development in the dissemination of ideas, knowledge, and commerce in early modern Europe.

ASIA

IN 1480, THE SMALL PRINCIPALITY OF MUSCOVY (Moscow) threw off Mongol control, and proceeded to expand Muscovite power over the whole of the area from the Arctic Ocean to the Caspian Sea. In the 1550s, the conquest of Kazan brought Russian power to the Urals, and over the next century it spread across Siberia, reaching the Pacific coast by 1649. Much of the area remained uninhabited, but to the south this new empire jostled uneasily with a string of central Asian Muslim khanates, and with the newly established Manchurian Ch'ing dynasty, which wrested control of China from the Ming in 1644.

European navigators and surveyors produced accurate maps and charts of their voyages.

The Indian Mughal ruler Shahjahan (1592–1648), builder of the Taj Mahal.

SOUTH ASIA AND OCEANIA

THE PORTUGUESE and the Spanish were the first European powers to open trade with the powerful Asian states of Mughal India and Ch'ing China. The Spanish opened trans-Pacific routes between Central America, the Philippines, and China. But the establishment of the Dutch and British East India companies in the early 17th century announced the advent of two new maritime powers.

Map labels

USSIAN EMPIRE
Siberia
Bering Strait
KAZAKHSTAN
A S I A
Aral Sea
KHWARIZM
KHOKAND KHANATE
KASHGAR KHANATE
UZBEKISTAN
Sea of Japan
JAPAN
KOREA
MANCHU (CH'ING) EMPIRE
SAFAVID PERSIA
The Gulf
Himalayas
TIBET
NEPAL
Deshima 1641
MUGHAL EMPIRE
Hooghly 1640
BURMA
PACIFIC OCEAN
OMAN 1508
Diu 1555
Surat 1608
Daman 1559
Bombay 1534
ARAKAN
Bay of Bengal
FORMOSA 1624
Macao 1557
Arabian Sea
Goa 1510
LAOS
ANNAN
PHILIPPINES from 1565
Masulipatam 1611
Madras 1639
SIAM
South China Sea
CEYLON 1505
THIOPIA
Galle 1640
Micronesia
Malacca 1641
Melanesia
MOLUCCAS from 1605
INDIAN OCEAN
RTUGUESE ST AFRICA rom 1505
Makassar 1607
Batavia 1619
East Indies
1610
TIMOR 1618
Madagascar
AUSTRALIA
NEW ZEALAND

West African trading kingdoms produced artifacts such as this bronze Portuguese soldier from Benin.

AFRICA AND THE MIDDLE EAST

WHILE EUROPE WAS DIVIDED by the Reformation, Islam experienced a remarkable resurgence in the 16th century. The revival of the Ottoman Empire brought Islamic rule over much of southeast Europe. Islam spread along trade routes to sub-Saharan Africa. In east Africa, it spread south along the coast. Further east, Muslim rulers established new imperial states in Persia (Iran) and India.

The sextant allowed navigators to take accurate measurements of heavenly bodies in relation to the horizon, thus allowing latitude to be calculated correctly. Early sextants had to be hand-held and were often used on shore rather than on board ship.

THE AGE OF EXPANSION: 1648–1789

THE YEARS FROM the middle of the 17th century to the end of the 18th century saw a massive consolidation of European discovery and exploration, which took the form of colonial settlement and political expansion. This period also witnessed the beginning of a sharp rise in European population and in its economic strength, accompanied by rapid developments in the arts and sciences. All these factors powered European expansion – a process that would bring European culture to every part of the globe, gradually filling in the world map, and bringing it into often fatal contact with less robust indigenous cultures. By the last quarter of the 18th century, with Europe poised on the brink of political turmoil, only Africa and Australasia remained largely unmolested by European attentions.

GLOBAL STATES AND TERRITORIES

KEY

	Chinese
	Ottoman
	Russian
○	Portuguese
○	Spanish
○	British
○	French
○	Dutch
	Danish
	United States

1776 Date of acquisition

The plantations of the New World grew valuable crops such as cotton and sugar.

THE AMERICAS

NORTH AMERICA WAS DIVIDED between the great European empires, Britain, France, Spain, and Russia, but most of it remained unexplored or unsettled. Fish and furs were the main interest. By the 18th century, European rivalry and conflicts with the native American population led to a series of wars. French influence was gradually reduced and excluded by British conquest. In 1776, colonists rebelled against British rule, launching a war of independence which ended with the creation of a republican United States of America. Canada remained under British rule. Modern North America was gradually taking shape.

Peter the Great (1672–1725) oversaw the consolidation of the Russian Empire, and sought to emulate the prestige, learning, and sophistication of the western European monarchies.

EUROPE

AFTER THE CRISIS of the Thirty Years War, Europe began to develop a more settled state system as successful dynastic houses imposed more centralized rule. The Habsburgs acquired control over Hungary and much of central Europe. Russia's frontiers pushed into Poland and the Ukraine. The French Bourbon monarchy became the most powerful in Europe. Its material wealth and culture made it a rival to the older empires of Asia. French became the common language of educated Europeans, and French philosophy led to the intellectual "Enlightenment."

Isaac Newton (1642–1727), the leading scientist of Europe's Age of Reason.

ASIA

THE CH'ING DYNASTY forged the shape of modern China. By 1658 the whole of southern China was under Manchu control. Formosa (Taiwan) was occupied in 1683, outer Mongolia in 1697. A protectorate was established over Tibet in 1751. Over the course of this expansion, the population of China tripled and the economy boomed through trade in tea, porcelain, and silk with Russia and the West. Manchu China was powerful enough to resist incursions by the European empires, avoiding the fate of the crumbling Mughal empire in India, where Britain and France competed for trade and territory.

Dutch and British East Indiamen carried the vast European trade with Asia.

Maori New Zealand was one of the few indigenous cultures to remain untouched by European contact until the 19th century.

OCEANIA

SOUTHEAST ASIA AND OCEANIA were areas of small, warring kingdoms, increasingly prey to the ambitions of European traders, first Spanish and Portuguese, then Dutch and British. Yet, by the late 18th century, there was still little formal colonization. Officially discovered by Europeans in the early 17th century, most of Australasia was still unexplored and unsettled, except for a number of small penal colonies set up by the British in New South Wales (1788) and Tasmania (1804).

African slavers marched their human cargo from the interior to the coast for transshipment.

AFRICA

DURING THE 17TH AND 18TH CENTURIES Africa was regarded by the rest of the world as a source of two things: gold and slaves. Some 13.5 million slaves were shipped in the 1700s, from the west coast and from Portuguese Angola. African dealers sold to European middlemen, who in turn sold on the surviving slaves. In northern and northeastern Africa, Arab slavers traded with the Ottoman Empire. But the rest of Africa remained isolated from the outside world.

Harrison's chronometer, invented in 1762, allowed navigators to measure time accurately, and thus calculate longitude correctly. This greatly reduced the risk of shipwreck and heralded the beginning of accurate mapping of the world.

Map labels

RUSSIAN EMPIRE

Bering Strait

KAZAKHSTAN
Aral Sea
A S I A
MONGOLIA 1697
KHOKAND
SINKIANG 1760
TURKESTAN
MANCHU (CH'ING) EMPIRE
Sea of Japan
JAPAN
KOREA
Caspian Sea
PERSIA
AFGHANISTAN
Himalayas
TIBET 1751
Deshima
PACIFIC OCEAN
The Gulf
BALUCHISTAN
NEPAL
BENGAL 1757
BURMA 1688
MARATHA CONFEDERACY
Chandernagore
FORMOSA 1683
Surat
Diu
Daman
Bombay 1661
Bay of Bengal
SIAM
Macao
PHILIPPINES
MARIANAS 1668
Arabian Sea
Goa
NORTHERN CIRCARS 1756
ANNAM
South China Sea
ETHIOPIA
Mahé 1725
MADRAS
ANDAMAN IS 1789
CAROLINE IS 1686
Karikal 1738
Pondicherry 1674
Galle
CEYLON 1658
Penang 1786
MALAYA
Micronesia
INDIAN OCEAN
MOLUCCAS
Melanesia
CHAGOS IS 1784
DUTCH EAST INDIES
PORTUGUESE AFRICA
TIMOR
Madagascar
RÉUNION 1662
Fort Dauphin 1766
AUSTRALIA
LORD HOWE I. 1788
NEW SOUTH WALES 1788
NEW ZEALAND

THE AGE OF REVOLUTION: 1789–1830

IN 1789 ROYAL POWER was shattered by the French Revolution. The collapse of the most powerful monarchy in Europe reverberated worldwide. The revolutions in France and America ushered in the idea of the modern nation state, and of popular representative government. Revolutionary outbreaks occurred elsewhere in Europe, and overseas colonies in Latin America won their independence. At the same time, an industrial revolution was taking place in Europe, transforming the old trading economy into a manufacturing base which would require a global supply of raw materials and a global market to fuel it. The revolutionary years thus marked the beginning of the modern political and economic world order.

GLOBAL STATES AND TERRITORIES

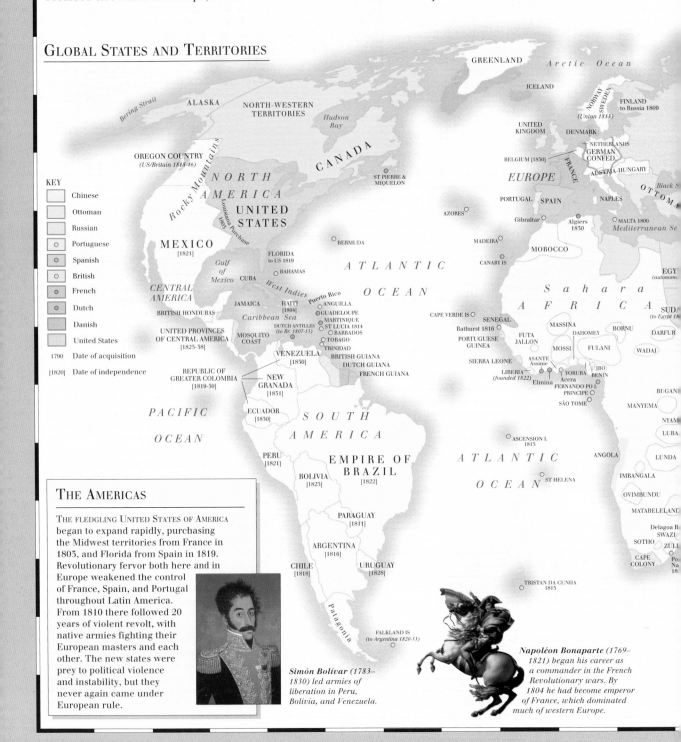

KEY

- Chinese
- Ottoman
- Russian
- Portuguese
- Spanish
- British
- French
- Dutch
- Danish
- United States

1790   Date of acquisition

[1820]   Date of independence

GREENLAND   *Arctic Ocean*

ICELAND

ALASKA   NORTH-WESTERN TERRITORIES   *Hudson Bay*

NORWAY SWEDEN (Union 1815)   FINLAND to Russia 1809

UNITED KINGDOM   DENMARK

NETHERLANDS   GERMAN CONFED.

BELGIUM [1830]   FRANCE   AUSTRIA-HUNGARY

OREGON COUNTRY (US/Britain 1818-46)   *Bering Strait*

Rocky Mountains   NORTH AMERICA   CANADA

ST PIERRE & MIQUELON

EUROPE   OTTOMA   *Black S*

PORTUGAL   SPAIN   NAPLES

Gibraltar   Algiers 1830   MALTA 1800   *Mediterranean Se*

Louisiana Purchase 1803   UNITED STATES

MEXICO [1821]   *Gulf of Mexico*

AZORES

MADEIRA   MOROCCO

BERMUDA

FLORIDA to US 1819   CANARY IS

CENTRAL AMERICA   CUBA   BAHAMAS   *West Indies*

ATLANTIC OCEAN

EGY (autonomo

Sahara

AFRICA   SUDA (to Egypt 18

BRITISH HONDURAS   JAMAICA   HAITI [1804]   Puerto Rico   ANGUILLA   GUADELOUPE   MARTINIQUE   *Caribbean Sea*

CAPE VERDE IS   SENEGAL   MASSINA   BORNU   DAHOMEY   DARFUR

UNITED PROVINCES OF CENTRAL AMERICA [1825-38]   MOSQUITO COAST   DUTCH ANTILLES to Br. 1807-15   ST LUCIA 1814   BARBADOS   TOBAGO   TRINIDAD

Bathurst 1816   PORTUGUESE GUINEA   FUTA JALLON   MOSSI   FULANI   WADAI

VENEZUELA [1830]   BRITISH GUIANA   DUTCH GUIANA   FRENCH GUIANA

SIERRA LEONE   ASANTE Assinie   IBO   BENIN

REPUBLIC OF GREATER COLOMBIA [1819-30]   NEW GRANADA [1831]

LIBERIA (founded 1822)   Elmina   YORUBA Accra   FERNANDO PO I.   PRINCIPE   SÃO TOME

BUGAN

ECUADOR [1830]

PACIFIC OCEAN

SOUTH AMERICA

MANYEMA   NYAM   LUBA

ASCENSION I. 1815

PERU [1821]   EMPIRE OF BRAZIL [1822]   BOLIVIA [1825]

ANGOLA   LUNDA   IMBANGALA

ATLANTIC OCEAN   ST HELENA

OVIMBUNDU   MATABELELAND

PARAGUAY [1811]

ARGENTINA [1816]

Delagoa B SWAZI SOTHO ZULU CAPE COLONY Po Na 18

CHILE [1818]   URUGUAY [1828]

TRISTAN DA CUNHA 1815

Patagonia

FALKLAND IS (to Argentina 1820-33)

THE AMERICAS

THE FLEDGLING UNITED STATES OF AMERICA began to expand rapidly, purchasing the Midwest territories from France in 1803, and Florida from Spain in 1819. Revolutionary fervor both here and in Europe weakened the control of France, Spain, and Portugal throughout Latin America. From 1810 there followed 20 years of violent revolt, with native armies fighting their European masters and each other. The new states were prey to political violence and instability, but they never again came under European rule.

Simón Bolívar (1783–1830) led armies of liberation in Peru, Bolivia, and Venezuela.

Napoléon Bonaparte (1769–1821) began his career as a commander in the French Revolutionary wars. By 1804 he had become emperor of France, which dominated much of western Europe.

EUROPE

UNDER NAPOLÉON BONAPARTE, France subordinated a large part of Europe and destroyed the old feudal order. Napoléon helped to shape the new nation states that emerged in 19th-century Europe – Belgium, Italy, and Germany. He gave much of Europe its modern legal code and systems of measurement, education, and local government.

Steam-powered engines transformed the European industrial economy.

ASIA

THE PRINCIPAL COLONIAL POWER in Asia was Russia, whose consolidation of its empire in northern and central Asia continued throughout the 19th century. But now the Dutch began to extend their control of the East Indies, while a bitter struggle between the British and the French was conducted in and around the Indian Ocean. France was gradually forced to concede many of its footholds in India, where the British East India Company rapidly extended its interests by a mixture of diplomacy and military force. But the elusive key to Asia's largest markets remained the slumbering giant of Ch'ing China, whose Manchu rulers, like the shoguns of Japan, remained unimpressed by European overtures.

The spices of the East Indies, such as pepper, were among the most highly valued trade commodities from Asia.

James Cook (1728–1779) charted much of the Pacific.

OCEANIA

THOUGH PORTUGUESE and Dutch explorers had confirmed the existence of Australasia in the 17th century, it was not until the voyages of Captain Cook in the 1770s that the geography of the Pacific was established, and the fertile eastern coast of Australia was explored and charted. Over the next 30 years, small settlements were established around the coast; by 1829, Britain had brought the whole of Australia under the British flag.

RUSSIAN EMPIRE

A S I A

Aral Sea

Caspian Sea

MONGOLIA

Bering Strait

MANCHU (CH'ING) CHINA

TIBET
(Chinese protectorate from 1750)

Himalayas

NEPAL

BHUTAN

Sea of Japan

JAPAN

KOREA

PERSIA

The Gulf

ARABIA

AFGHAN-ISTAN

BURMA

ANNAM

SIAM

INDIA

Macao

FORMOSA

PACIFIC

OCEAN

MARIANAS

OMAN

Diu
Daman

Arabian Sea

Goa

Bay of Bengal

South China Sea

PHILIPPINES

Mahé

Pondicherry

TENASSERIM 1826

Karikal

ANDAMAN IS

CAROLINE IS

LACCADIVE IS 1791

Micronesia

ETHIOPIA

Ceylon

MALAYA

Malacca 1824

Melanesia

MALDIVE IS 1887

SINGAPORE 1819

New Guinea

ZANZIBAR (to Oman)

SEYCHELLES 1794

CHAGOS IS

DUTCH EAST INDIES

PORTUGUESE EAST AFRICA

INDIAN

Timor

Madagascar

HOVA KINGDOM

MAURITIUS 1810

RÉUNION

OCEAN

WESTERN AUSTRALIA 1829

NEW SOUTH WALES

AUSTRALIA

LORD HOWE I.

The first European migrants to Africa settled in Cape Colony.

AFRICA

THE NORTHERN REGIONS OF AFRICA were part of the vast Islamic Ottoman Empire; from here Islam spread south to west Africa and the Horn of Africa. Holy wars (or *jihads*) in the late 18th and early 19th centuries completed the conversion to Islam of much of Saharan and sub-Saharan Africa. Large tribal kingdoms flourished in the Congo basin and southern Africa.

NEW ZEALAND

CHATHAM IS 1791

TASMANIA (Van Diemen's Land)

AUCKLAND IS 1806

MACQUARIE IS 1811

The development during the European industrial revolution of mechanized manufacturing plant and machinery, such as power looms, gave Europe effective control of a booming global trade in raw materials and mass-manufactured commodities.

THE AGE OF EMPIRE: 1830–1914

THE 19TH CENTURY was dominated by the spread of modern industry and transportation, and the expansion of European trade and influence worldwide. Industry made Europe rich and powerful; its capital cities were monuments to the self-confidence of the new European age. Railroads and steamships revolutionized communications, bringing a stream of industrial goods, technical know-how, and European settlers across America, Africa, and Asia. Modern industry and weapons brought Europe to the summit of global influence. In these developments lay the origins of the division of the world into rich and poor regions; a developed, prosperous north and an underdeveloped, dependent south.

GLOBAL STATES AND TERRITORIES

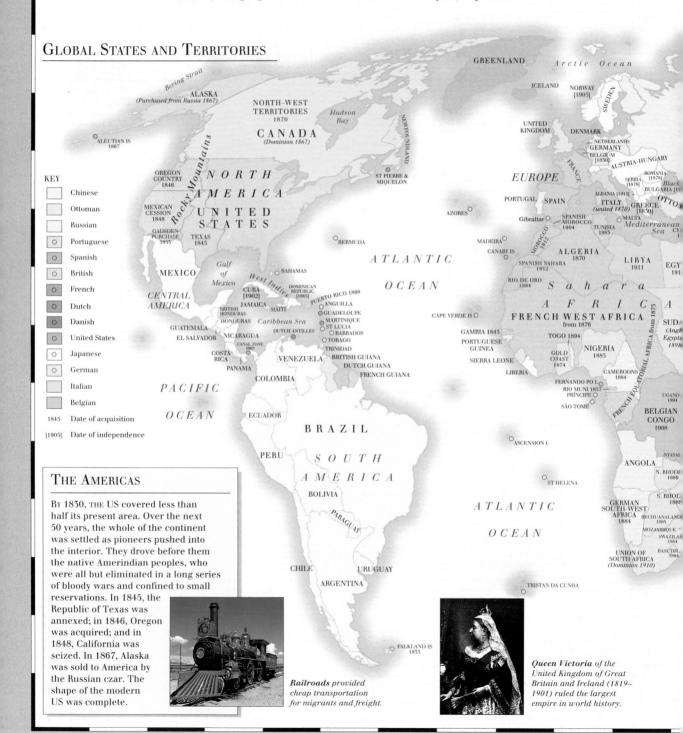

KEY

☐	Chinese
☐	Ottoman
☐	Russian
⊙	Portuguese
⊙	Spanish
⊙	British
⊙	French
⊙	Dutch
⊙	Danish
⊙	United States
⊙	Japanese
⊙	German
☐	Italian
☐	Belgian
1845	Date of acquisition
[1905]	Date of independence

GREENLAND
ICELAND · NORWAY [1905] · SWEDEN
UNITED KINGDOM · DENMARK · NETHERLANDS · GERMANY · BELGIUM [1830] · AUSTRIA-HUNGARY
EUROPE · FRANCE · ROMANIA [1878] · SERBIA [1878] · Black · BULGARIA [18
PORTUGAL · SPAIN · ALBANIA [1913] · ITALY (united 1870) · GREECE [1830] · OTTO
Gibraltar · SPANISH MOROCCO 1904 · MALTA · TUNISIA 1885 · Mediterranean Sea · CY
MADEIRA · ALGERIA 1870 · LIBYA 1911 · EGY 191
CANARY IS · MOROCCO 1912
SPANISH SAHARA 1912 · Sahara
RIO DE ORO 1884 · AFRICA
CAPE VERDE IS · FRENCH WEST AFRICA from 1876 · SUD (Angl Egypt 189
GAMBIA 1845 · TOGO 1894 · NIGERIA 1885
PORTUGUESE GUINEA · GOLD COAST 1874 · CAMEROONS 1884
SIERRA LEONE · LIBERIA · FERNANDO PO 1 · RIO MUNI 1857 · PRÍNCIPE · UGAND 1894
SÃO TOMÉ · BELGIAN CONGO 1908
ASCENSION I. · FRENCH EQUATORIAL AFRICA from 1875
NYASAI
ANGOLA · N. RHODE 1889
ST HELENA · S. RHOL 1888
GERMAN SOUTH-WEST AFRICA 1884 · BECHUANALAND 1885
ATLANTIC OCEAN · MOZAMBIQUE · SWAZILA
TRISTAN DA CUNHA · BASUTOL 1884
UNION OF SOUTH AFRICA (Dominion 1910)

Bering Strait
ALASKA (Purchased from Russia 1867)
NORTH-WEST TERRITORIES 1870
Hudson Bay
NEWFOUNDLAND
CANADA (Dominion 1867)
ALEUTIAN IS 1867
OREGON COUNTRY 1846
ST PIERRE & MIQUELON
Rocky Mountains
NORTH AMERICA
UNITED STATES
MEXICAN CESSION 1848
GADSDEN PURCHASE 1855 · TEXAS 1845
BERMUDA
AZORES
MEXICO
Gulf of Mexico
West Indies
BAHAMAS
CUBA [1902] · JAMAICA · DOMINICAN REPUBLIC [1865] · PUERTO RICO 1898 · ANGUILLA
BRITISH HONDURAS · HAITI · GUADELOUPE · MARTINIQUE · ST LUCIA · BARBADOS
CENTRAL AMERICA
GUATEMALA · EL SALVADOR · HONDURAS · Caribbean Sea · DUTCH ANTILLES · TOBAGO
NICARAGUA · CANAL ZONE 1905 · TRINIDAD
COSTA RICA · VENEZUELA · BRITISH GUIANA · DUTCH GUIANA · FRENCH GUIANA
PANAMA · COLOMBIA
PACIFIC OCEAN
ECUADOR
BRAZIL
PERU
SOUTH AMERICA
BOLIVIA
PARAGUAY
ATLANTIC OCEAN
CHILE · URUGUAY
ARGENTINA
FALKLAND IS 1833
Arctic Ocean

THE AMERICAS

BY 1830, THE US covered less than half its present area. Over the next 50 years, the whole of the continent was settled as pioneers pushed into the interior. They drove before them the native Amerindian peoples, who were all but eliminated in a long series of bloody wars and confined to small reservations. In 1845, the Republic of Texas was annexed; in 1846, Oregon was acquired; and in 1848, California was seized. In 1867, Alaska was sold to America by the Russian czar. The shape of the modern US was complete.

Railroads provided cheap transportation for migrants and freight.

Queen Victoria of the United Kingdom of Great Britain and Ireland (1819–1901) ruled the largest empire in world history.

EUROPE

IN THE 19TH CENTURY, Europe was transformed into an industrial economy. In the new industrial cities, pressure developed for liberal reforms and parliamentary politics. Nationalists created new states in Germany, Italy, Greece, Serbia, and Belgium. While the modern map of Europe gradually began to take shape, European imperialists brought still more areas of the world under their control.

Sailing ships carried most oceanic trade until 1900.

ASIA

BUILDING ON COLONIAL INTERESTS that stretched back into the 18th century, Britain and France transformed the political world of south Asia. Britain extended its rule in India and, in 1885, Burma was brought under British control. The Vietnamese and Chinese Empires were pressured by Europeans anxious to trade and to spread Christianity: the Ch'ing empire conceded areas of influence; the Vietnamese empire resisted and was brought under French domination by force. By the 1890s the whole of southern Asia except for Siam was dominated by Europe, which created the modern state structure of the region.

The Japanese emperor Meiji (1852–1912) opened Japan to Western trade and influence.

The colonization of Australia and New Zealand was based on sheep farming.

OCEANIA

DURING THE 19TH century, Australia and New Zealand remained closely tied to the British homeland. British settlers came to farm and later to prospect for gold and other valuable minerals. In 1840, New Zealand came under British rule and the native Maoris were forced off the land. Not until 1872 was the continent of Australia traversed, and not until 1901 was a single state, the Commonwealth of Australia, proclaimed.

Map labels

RUSSIAN EMPIRE

Bering Strait

KAZAKHSTAN 1854
Aral Sea
BUKHARA 1868
TURKESTAN 1895
TURKMENISTAN 1885
Caspian Sea
A S I A
AFGHANISTAN
PERSIA
The Gulf
EMPIRE
BAHRAIN 1861
A R A B I A
OMAN

MONGOLIA (autonomous 1912)
MANCHURIA
AMUR 1858
USSURI 1860
Port Arthur 1905
Weihaiwei 1898
Tsingtao 1898
TIBET [1912]
Himalayas
NEPAL BHUTAN
CHINA
BURMA

SAKHALIN 1855 1905
KURILE IS 1875
Sea of Japan
JAPAN
KOREA 1905
RYUKYU IS 1874

PACIFIC OCEAN

Chandernagore
INDIA
Diu Daman
Arabian Sea
Goa
Mahé Pondicherry
Karikal
LACCADIVE IS
CEYLON
NICOBAR IS 1869
ANDAMAN IS
MALDIVE IS

Macao
Hong Kong 1841
FRENCH INDO-CHINA
SIAM
South China Sea

FORMOSA 1895
PHILIPPINES 1898

MARIANAS 1899
GUAM 1898
CAROLINE IS 1899
Micronesia

ERITREA 1889
HADHRAMAUT 1888
Aden 1839
SOCOTRA 1886
BRITISH SOMALILAND 1884
FRENCH SOMALILAND 1884
ETHIOPIA
ITALIAN SOMALILAND
BRITISH EAST AFRICA 1888
ZANZIBAR 1890
GERMAN EAST AFRICA 1885
COMORO IS 1886
SEYCHELLES
CHAGOS IS
MADAGASCAR 1882

MALAYA
BRITISH NORTH BORNEO 1881
SARAWAK 1888
DUTCH EAST INDIES

NEW GUINEA
BISMARCK ARCHIPELAGO 1884
PAPUA 1906
SOLOMON IS 1893
NAURU 1888
Melanesia

INDIAN OCEAN
COCOS IS 1857
CHRISTMAS I. 1888
TIMOR
MAURITIUS
RÉUNION

NEW CALEDONIA 1853

A U S T R A L I A (Commonwealth 1901)
NORFOLK ISLAND

NEW ZEALAND 1840 (Dominion 1907)
CHATHAM IS.
TASMANIA
AUCKLAND IS
MACQUARIE IS

Quinine – the cure for malaria.

New medicines made the colonization of Africa possible.

AFRICA

THE POLITICAL STRUCTURE of independent Africa was torn up by encroaching European empires. As native societies reacted violently to European intrusion, so European military and political power was increased to secure European interests. In 1884, in Berlin, the European powers divided Africa between them. The "Partition of Africa" established the modern frontiers of many states.

The Gatling gun, the most successful of the hand-crank-operated machine guns of the 19th century.

The European imperial powers maintained control of their often far-flung colonies by military superiority. Native forces were rarely a match for the large, highly trained armies, powerful navies, and technically advanced weaponry which the Europeans had at their disposal.

THE AGE OF GLOBAL WAR: 1914–1945

IN 1914, IMPERIAL AND MILITARY rivalry in Europe provoked the first of two world wars, the largest and most destructive wars in human history. At the end of the first war, in 1918, the old international order was dead. The Russian Empire collapsed in revolution and was transformed by a communist minority into the Soviet Union. The German, Habsburg, and Ottoman empires were dismembered. A fragile peace ensued but the old equilibrium was gone. The rise of strident nationalism in Germany, Japan, and Italy destroyed the peace once again in 1939. The second war cost the lives of 50 million people and ravaged Europe and Asia. At its end, in 1945, the US and the Soviet Union had emerged as the new superpowers.

GLOBAL STATES AND TERRITORIES

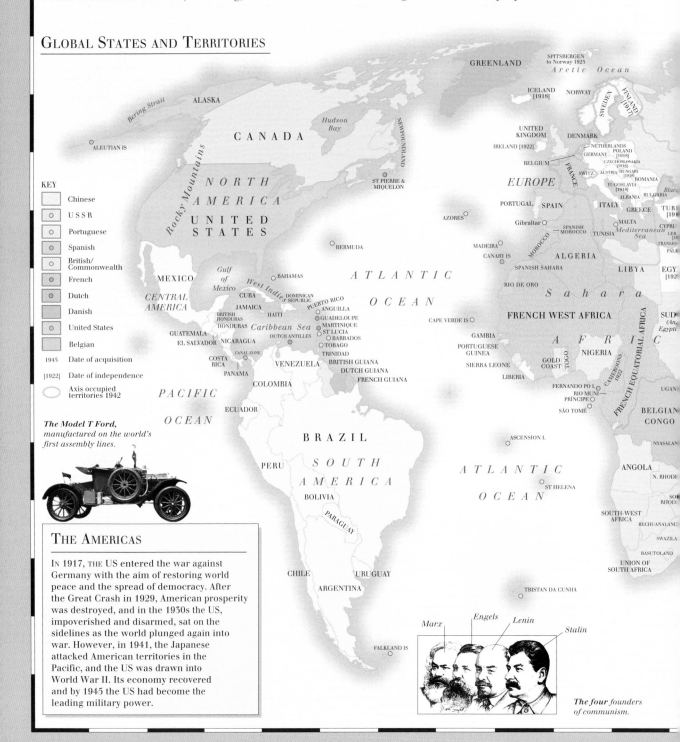

KEY

	Chinese
○	USSR
○	Portuguese
◉	Spanish
○	British/Commonwealth
◉	French
◉	Dutch
	Danish
◉	United States
	Belgian
1945	Date of acquisition
[1922]	Date of independence
⬭	Axis occupied territories 1942

The Model T Ford, manufactured on the world's first assembly lines.

The four founders of communism.

Marx Engels Lenin Stalin

THE AMERICAS

IN 1917, THE US entered the war against Germany with the aim of restoring world peace and the spread of democracy. After the Great Crash in 1929, American prosperity was destroyed, and in the 1930s the US, impoverished and disarmed, sat on the sidelines as the world plunged again into war. However, in 1941, the Japanese attacked American territories in the Pacific, and the US was drawn into World War II. Its economy recovered and by 1945 the US had become the leading military power.

EUROPE

BOTH WORLD WARS had their origins in Europe. In 1914, Germany invaded Belgium; Britain, France, and Russia combined to defeat it, with US help. In 1918 new nation states were established in eastern Europe. But, by 1939, revived German nationalism started a second world war; much of western Europe came under a German "New Order" until the Soviet Union, Britain, and the US developed sufficient military strength to reconquer Europe and defeat Germany.

World War II was decided by mechanical and industrial superiority.

ASIA

THE COLLAPSE OF THE CHINESE EMPIRE in 1911, followed in 1917 by the disappearance of the Russian Empire, produced instability across Asia. Full-scale war broke out between Japan and China in 1937, with Japan trying to conquer China. The Soviet Union was the victim of German aggression from 1941. Both Japan and Germany were held at bay by communist forces which eventually succeeded in imposing stable politics on Asia. By 1945, the Soviet Union had reconquered its lost territories and dominated eastern Europe. In China, communist armies filled the vacuum left by the Japanese defeat.

Mahatma Gandhi (1868–1948) led India to independence through peaceful noncooperation and protest.

OCEANIA

FOR THE ONLY TIME in its history, Australia was faced with the very real prospect of invasion. In World War II, Japanese armies reached the island of New Guinea, and bombed towns in northern Australia. Japanese submarines attacked Sydney harbor. The Battle of the Coral Sea, in May 1942, saved Australia, but it took almost three years to clear Japanese forces from the South Pacific, where they hung on grimly to the rich oil and mineral resources they had captured.

Haile Selassie (1892–1975), ruler of Ethiopia, the only independent empire in Africa.

Japan promoted itself as the liberator of Asia from the chains of European colonialism.

MIDDLE EAST

IN 1918, THE OTTOMAN EMPIRE disappeared after being in existence for 400 years. The modern map of north Africa and the Middle East was carved out of its ruins by the victors of World War I. The genocide of Europe's Jews by Nazi Germany during World War II accelerated the foundation of a new state of Israel in 1948, leading inexorably to conflict between displaced native Arabs and Jewish migrants.

The conquest of the air was the most important technological achievement of the period. It added a devastating dimension to warfare, in the form of aerial bombing, while transforming civil transportation.

A German Zeppelin airship of the 1930s.

THE MODERN AGE: 1945–PRESENT DAY

TWO NUCLEAR EXPLOSIONS in Japan finally ended World War II and ushered in a new era of global rivalry. Total destruction by nuclear weapons was a real possibility as the two superpowers, the Soviet Union and the US, became locked with their allies in a fearful Cold War. Various side conflicts were staged in the ruins of the old empires where new nation states, with old colonial borders, were emerging.

By 1991, communist power in the Soviet Union and Eastern Europe had crumbled, while the capitalism of the West had gained in strength.

The US is now the sole global superpower and has used its status to push Western goods and ideals to every corner of the globe, provoking often violent reactions among proud, conservative cultures. The age of globalization has begun.

GLOBAL STATES AND TERRITORIES

KEY

○	Portuguese
○	Spanish
○	British
◉	French
◉	Dutch
	Danish
○	US
[1972]	Date of independence

US President John F. Kennedy (1917–1963) personified American postwar optimism.

THE AMERICAS

NORTH AND SOUTH AMERICA have taken very different routes since 1945. The US and Canada enjoyed an economic boom unrivalled in the world with standards of living soaring. "American" culture took the world by storm.

In contrast, the countries of South and Central America struggled through more than 40 years of political uncertainty. Military dictatorships and socialist governments vied for control. Whole generations were decimated as dissidents were "disappeared." Since the 1970s, South America has begun to embrace democracy. Development has been hamstrung by an unstable economic climate while large areas are dominated by the powerful narcotics trade.

In 1985, the Soviet leader Mikhail Gorbachev launched a program of economic and political reform which brought Soviet communism to an end.

The Berlin Wall, symbol of the Cold War division of Europe, was demolished in 1989.

EUROPE

IN 1945, EUROPE LAY IN RUINS. The iron curtain descended to divide capitalist West from communist East. The West underwent an economic boom which restored widespread prosperity and political stability. It progressed toward economic and political unity under the EU. The East, meanwhile, labored under planned economies and quickly sought to catch up with the West after 1989, with most countries now aspiring to join the EU.

ASIA

IN SOUTHERN ASIA, popular nationalist movements came to power in India, Burma, Malaya, and Indonesia; in China and Indo-China, power passed to native communist movements whose roots went back to the 1920s. After 1949, China under Mao Zedong became, with its vast population and large military forces, a second communist superpower. Japan, meanwhile, was Asia's capitalist "miracle." Its economy and cities laid waste by bombing in 1945, it rebuilt with US aid so successfully that by the 1980s, it was the world's second-largest economy. China's potential economic growth, however, could put Japan's past achievements in the shade.

Chinese communism, based on the mobilization of peasants and workers, has nevertheless recognized the need for economic reforms.

A treaty banning the testing of nuclear bombs in the Pacific was signed in 1986.

OCEANIA

THE POSTWAR economies of Japan, the US, and Australia had by the 1990s created a new industrial and trading network around the Pacific Rim. Cheap labor and low overheads drew younger states – South Korea, Taiwan, Singapore, Malaysia, Indonesia – into the system and much of the world's manufacturing is now concentrated there, creating a consequent shift in the balance of the global economy.

RUSSIAN FEDERATION

KAZAKHSTAN [1991]

A S I A

MONGOLIA

UZBEKISTAN [1991]
Aral Sea
KYRGYZSTAN [1991]
TURKMENISTAN [1991]
TAJIKISTAN [1991]
Caspian Sea
AZERBAIJAN [1991]
ARMENIA [1991]
IRAN
AFGHANISTAN

CHINA

Sea of Japan

JAPAN

S. KOREA [1948]

Himalayas

IRAQ
QATAR [1971]
SAUDI ARABIA
PAKISTAN [1947]
NEPAL
BHUTAN
The Gulf
U. A. E. [1961]
BANGLADESH [1971] (formerly E. Pakistan)

ERITREA [1993]
OMAN
Arabian Sea
INDIA [1947]
Bay of Bengal
(Vietnam united 1976)
South China Sea

(Yemen united 1990)
THAILAND

MARIANAS

PACIFIC OCEAN

GUAM

DJIBOUTI [1977]
ETHIOPIA

BRUNEI [1984]
PALAU [1995]
MICRONESIA [1991]

SEYCHELLES [1976]
CHAGOS IS

I N D O N E S I A [1949]

Melanesia
SOLOMON IS [1978]

I N D I A N
O C E A N

E. TIMOR

MADAGASCAR [1960]

RÉUNION

Gamal Abd al-Nasser (1918–1970) of Egypt, galvanized the Arab states to resist the West.

NEW CALEDONIA

A U S T R A L I A

AFRICA AND THE MIDDLE EAST

THE COLONIAL POWERS, weakened by war, faced an irresistible wave of demands for self-determination. Between 1958 and 1975, 41 African countries gained independence. In north Africa and throughout the Middle East a new form of anti-imperialism emerged in the 1970s in the form of Islamic fundamentalism. In South Africa, white rule and the apartheid system ended in 1994.

NEW ZEALAND

AUCKLAND IS (to N Z)
MACQUARIE IS (to Australia)

From the 1950s to the 1970s, superpower rivalry focused on space exploration. The Soviets put the first man in space in 1961, and the Americans landed on the moon in 1969. Since then, both manned and unmanned missions have become almost everyday events.

POPULATION

THE WORLD'S POPULATION – 6.2 billion in 2002 – is likely to reach nearly 10 billion by 2050. Better nutrition, health care, and sanitation mean fewer infant deaths and longer life expectancy, although around 800 million people in the developing world are malnourished and over one billion live in extreme poverty. In much of Africa in particular, the AIDS epidemic is so severe that the population is set to fall significantly. Elsewhere it is lower birthrates, already familiar in most industrialized countries, that have slowed the rate of growth. The result is a rapidly aging population: it is thought that by 2050 there will be around two billion people over the age of 60. The distribution of population is very uneven, dependent on climate, terrain, natural resources, and economic factors. The great majority of people live in coastal zones and along river valleys. Urbanization is on the increase, and by 2002 just under half of the world's population lived in cities – most of them in Asia. The mass migration of people from rural areas in search of work has resulted in the growth of huge sprawling squatter camps on the edge of many Third World cities.

POPULATION

- City over 5 million inhabitants

POPULATION DENSITY
(People/mi^2)

	Below 3
	3–13
	13–29
	30–51
	52–130
	131–260
	261–520
	Above 520

INFANT MORTALITY

WORLD INFANT
MORTALITY RATES
(deaths per
1000 live births)

- Above 100
- 61–100
- 31–60
- 15–30
- Below 15
- Data not available

INFANT MORTALITY

INFANT MORTALITY RATES are highest in Africa, South America, and south Asia, where poverty and disease are rife, and where average standards of health care are not as high as in North America or Europe. The country with the highest infant mortality rate is Afghanistan, where years of conflict have devastated communities.

LIFE EXPECTANCY

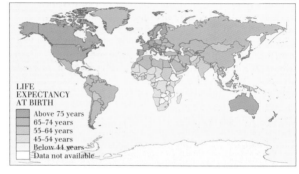

LIFE
EXPECTANCY
AT BIRTH

- Above 75 years
- 65–74 years
- 55–64 years
- 45–54 years
- Below 44 years
- Data not available

LIFE EXPECTANCY

LIFE EXPECTANCY IS poorest in Africa, for reasons similar to those noted above. In western Europe and North America, life expectancy is increasing at such a rate that each successive generation may expect to live longer than the last. In the developed world, people can now expect to live twice as long as they did a century ago.

POPULATION FACTFILE	
LARGEST POPULATIONS	
1 **China**	1285 million people
2 **India**	1025 million people
3 **USA**	281 million people
LARGEST COUNTRIES BY AREA	
1 **Russian Federation**	16,995,800 km² (6,562,100 mi²)
2 **China**	9,326,410 km² (3,600,927 mi²)
3 **Canada**	9,220,970 km² (3,560,217 mi²)
MOST DENSELY POPULATED COUNTRIES	
1 **Monaco**	16,253 people/km² (42,095 people/mi²)
2 **Singapore**	6721 people/km² (17,408 people/mi²)
3 **Malta**	1225 people/km² (3173 people/mi²)
MOST SPARSELY POPULATED COUNTRIES	
1 **Mongolia**	2 people/km² (4 people/mi²)
2 **Namibia**	2 people/km² (6 people/mi²)
3 **Australia**	3 people/km² (7 people/mi²)

Map labels (lower left globe):

Arctic Circle
Tehran
Lahore
Tianjin
Beijing
Seoul
Tokyo
Osaka
Delhi
Wuhan
Shanghai
Chongqing
Karachi
Dhaka
Calcutta (Kolkata)
Hong Kong
Mumbai (Bombay)
Hyderabad
Bangkok
Manilla
Bangalore
Chennai (Madras)
Tropic of Cancer
Equator
Jakarta
Tropic of Capricorn
Antarctic Circle

WORLD ECONOMY

THE WEALTHY COUNTRIES of the developed world, with their aggressive, market-led economies and their access to productive new technologies and international markets, dominate the world economic system. At the other extreme, many of the countries of the developing world are locked in a cycle of unrepayable debt, rising populations, and unemployment. The state-managed economies of the former communist bloc began to be dismantled during the 1990s, and China is emerging as a major economic power following decades of isolation. Since the late 1980s, technological advances have enabled transactions between financial centers to occur at even greater speed, and new markets have sprung up throughout the world.

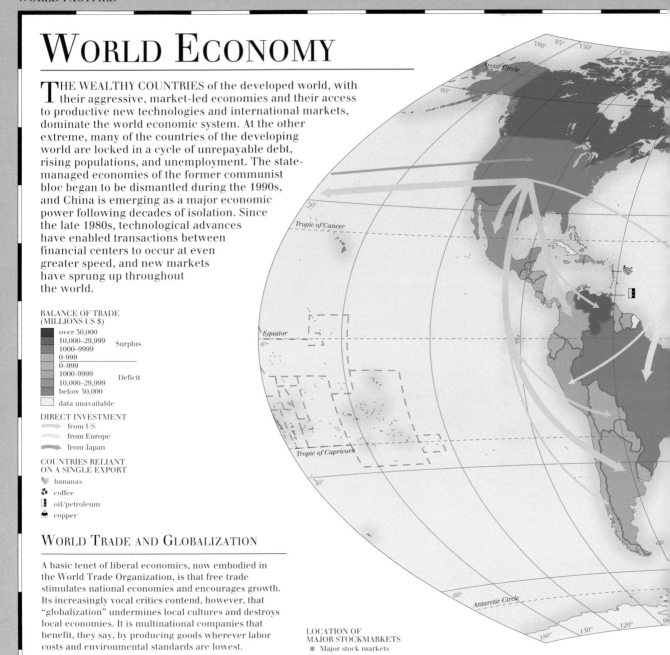

BALANCE OF TRADE
(MILLIONS US $)

over 30,000	
10,000–29,999	Surplus
1000–9999	
0–999	
0–999	
1000–9999	Deficit
10,000–29,999	
below 30,000	
data unavailable	

DIRECT INVESTMENT
from US
from Europe
from Japan

COUNTRIES RELIANT
ON A SINGLE EXPORT
bananas
coffee
oil/petroleum
copper

WORLD TRADE AND GLOBALIZATION

A basic tenet of liberal economics, now embodied in the World Trade Organization, is that free trade stimulates national economies and encourages growth. Its increasingly vocal critics contend, however, that "globalization" undermines local cultures and destroys local economies. It is multinational companies that benefit, they say, by producing goods wherever labor costs and environmental standards are lowest.

LOCATION OF
MAJOR STOCKMARKETS
● Major stock markets

INTERNATIONAL TRADE

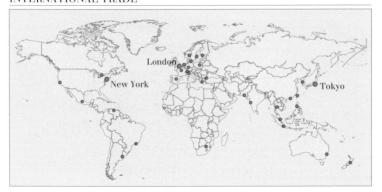

WORLD ECONOMIES

HIGHEST GNP PER CAPITA

1 Liechtenstein $50,000
2 Luxembourg $42,060
3 Switzerland $38,140
4 Japan $33,470
5 Norway $32,050

LOWEST GNP PER CAPITA

1= Somalia $100
1= Ethiopia $100
3= Burundi $110
3= Congo, Dem. Rep. $110
5   Sierra Leone $150

TRADE BLOCS

INTERNATIONAL TRADE BLOCS are formed when groups of countries, often already enjoying close military and political ties, join together to offer mutually preferential terms of trade for both imports and exports. Increasingly, global trade is dominated by three main blocs: the EU, NAFTA, and ASEAN. They are supplanting older trade blocs such as the Commonwealth, a legacy of colonialism.

TRADE BLOCS

EU	NAFTA	MERCOSUR
ASEAN	SADC	ECOWAS

TRADE BLOCS

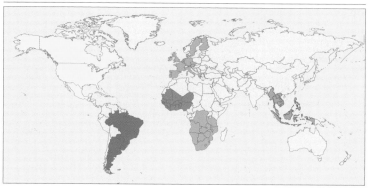

WORLD ECONOMY

THE SIZE OF A COUNTRY'S economy does not relate directly to its population or even its resources. Japan, for example, has a much "bigger" economy than China, India, Russia, or Latin America as a whole. Such imbalances usually occur because countries differ enormously in their living standards, the education and skills of their workforces, the productivity of their agriculture, and the value of their markets. A country's economic performance can be evaluated by calculating its gross national product (GNP). This is the total value of both the goods and the services (including so-called "invisible exports" – financial services, tourism, and so on) that it produces. Most trade (62% of the global total by value) is in manufactured goods, but during the last three decades the most rapidly growing sector has been services – banking, insurance, tourism, consultancy, accountancy, films, music and other cultural services, airlines, and shipping. Accounting for 20% of the total, services now exceed the value of trade in food and raw materials.

COMPARATIVE WORLD WEALTH

A global assessment of GNP by country reveals great disparities. The developed world, with only a quarter of the world's population, has 80% of the world's manufacturing income. This imbalance is maintained as war and political instability undermine poor countries' prospects.

Mass-market tourism is now an all-important source of revenue in many countries.

AVERAGE GDP
PER CAPITA (IN $US)
- Above 10,000
- 2000–10,000
- 500–1999
- Below 500
- Data unavailable

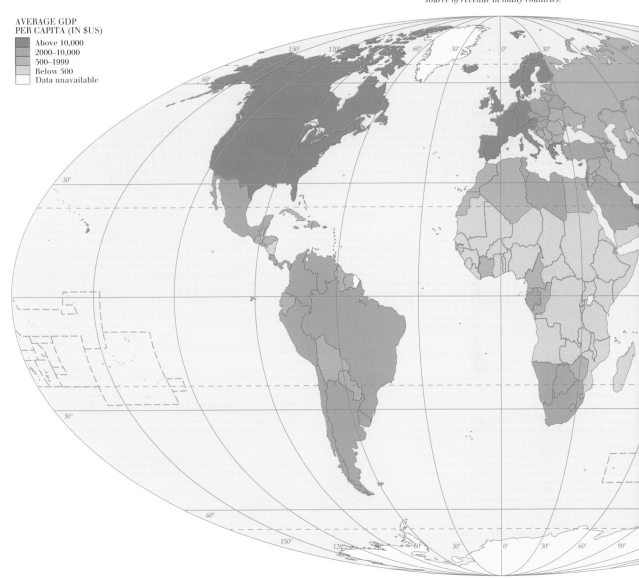

DEBT OF POOR AND MIDDLE-INCOME COUNTRIES

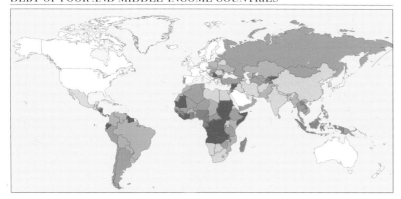

INTERNATIONAL DEBT (AS PERCENTAGE OF GNP)

- over 100%
- 70–100%
- 50–69%
- 30–49%
- below 30
- negligible
- not applicable
- data unavailable

INTERNATIONAL DEBT

In response to unsustainable levels of debt in the developing world, the IMF and World Bank have introduced a program to help heavily indebted poor countries (HIPCs) manage their repayments. The Jubilee 2000 campaign, advocating debt cancellation, has won some qualified support among creditor countries.

WORLD'S 20 LARGEST CORPORATIONS

2001, $ millions

	Corporation	
1	Wal-Mart Stores (US)	.$219,812
2	Exxon Mobil (US)	$191,581
3	General Motors (US)	$177,260
4	BP (UK)	$174,218
5	Ford Motor Company (US)	$162,412
6	Enron (US)	$138,718
7	DaimlerChrysler (US/Germany)	$136,897
8	Royal Dutch/Shell Group (UK/Neth.)	$135,211
9	General Electric (US)	$125,913
10	Toyota Motor Company (Japan)	$120,814
11	Citigroup (US)	$112,022
12	Mitsubishi (Japan)	$105,814
13	Mitsui (Japan)	$101,206
14	ChevronTexaco (US)	$99,699
15	Total Fina Elf (France)	$94,312
16	Nippon Telegraph and Telephone (Japan)	$93,425
17	Itochu (Japan)	$91,177
18	Allianz (Germany)	$85,929
19	Intl. Business Machines (US)	$85,866
20	ING Group (Neth.)	$82,999

NEWLY INDUSTRIALIZED COUNTRIES

In the 1990s, the fast-growing export-oriented "Asian tiger" economies, such as Singapore, South Korea, and Taiwan, offered exciting prospects for foreign investors. The 1997–1998 Asian financial crisis came as a severe shock. Some countries quickly returned to growth, but investors became more cautious about their exposure in NICs around the world.

NEWLY INDUSTRIALIZED AND INDUSTRIALIZING COUNTRIES

GLOBAL TOURISM

TOURISM IS THE WORLD'S biggest industry. In 2000 there were a record 697 million tourists worldwide, a number expected to rise to over one billion by 2010, despite the setbacks to the industry in 2001. France is the world's most popular destination country, with 77 million visitors annually, but with cheaper flights, improved transportation, and increased leisure time, many of the countries of the developing world are rapidly becoming tourist meccas. Since the 1960s, mass tourism has become increasingly specialized, encompassing sporting and adventure holidays as well as ecological tours. Tourism employs 200 million people worldwide – 8% of the working population. However, the benefits of tourism are not always felt at a local level, where jobs are often low-paid and menial.

TOURIST ARRIVALS

- Data not available
- Less than 50,000
- 50,000 - 99,999
- 100,000 - 399,999
- 400,000 - 1 million
- 1 million - 2.5 million
- 2.5 million - 10 million
- More than 10 million

CARIBBEAN
- BARBADOS
- GRENADA
- ST LUCIA
- TRINIDAD & TOBAGO
- DOMINICA
- ANTIGUA & BARBUDA
- GUADELOUPE (Fr.)
- ST KITTS & NEVIS
- BERMUDA (UK)
- MARTINIQUE (Fr.)
- ST VINCENT & THE GRENADINES
- BAHAMAS

Even in the remote Himalayas, rubbish discarded by trekkers and mountaineers pollutes the landscape.

"ECO-TOURISM"

Spectacular and ecologically fascinating natural features such as tropical rainforests and coral reefs are becoming increasingly popular as tourist attractions. Holidays in such destinations can put serious pressure on the very ecological systems that bring in the visitors in the first place. Strenuous efforts are being made to limit the environmental impact of this so-called "eco-tourism."

EUROPE
- MONACO
- SAN MARINO
- LIECHTENSTEIN
- GIBRALTAR (to UK)
- LUXEMBOURG
- MALTA
- ANDORRA

AFRICA
- CAPE VERDE
- SAO TOME & PRINCIPE

The beautiful island of Phuket, Thailand, has been taken over by tourist developments.

A TOURIST PARADISE?

THE MOST REMOTE CORNERS of the world are now being penetrated by tourists in their quest for the exotic. In many parts of the developing world, tourism can be described as a form of "neocolonialism;" hotels and beaches are owned by multinational companies, and most of the profits are taken outside the country. Tourism frequently alienates local people from their own land, and has a negative impact on the local culture and environment.

Eco-tourists travel to the distant Antarctic, where they observe its rich wildlife.

PACIFIC OCEAN
- FIJI
- MICRONESIA
- NAURU
- SOLOMON ISLANDS
- VANUATU
- SAMOA
- TONGA
- KIRIBATI

MIDDLE EAST
- BAHRAIN

INDIAN OCEAN
- COMOROS
- MALDIVES
- MAURITIUS
- SEYCHELLES

ASIA
- SINGAPORE

GLOBAL SECURITY

THE ENDING OF THE COLD WAR in 1989 greatly reduced the risk of another global (and possibly nuclear) war, but did little to end localized tensions and conflicts. Since then territorial disputes, and particularly ethnic tensions, have undermined peace and security around the world. Conflict is now more frequent within states than between them. The first post-Cold War international war, following the invasion of Kuwait by Iraq in 1990, set the tone for joint action by more powerful states to punish apparent injustices elsewhere. Economic sanctions are usually employed as the first stage of punitive measures. The UN, with its laborious decision-making process, has tried to fill the role of global peacemaker wherever possible, without becoming involved in military action. On the other hand, regional organizations such as NATO (with the US and the UK to the fore), the Russian-dominated Commonwealth of Independent States, or ECOWAS in west Africa have taken direct action against the forces of the supposed aggressor. Attacks on the US in 2001 underlined that ideological fanaticism is the greatest threat to global security.

USA

The September 11, 2001 terrorist atrocities, targeting the Pentagon and World Trade Center, prompted the US to launch a "war on terrorism," initially against Islamic militants in Afghanistan.

COLOMBIA

For more than 40 years left- and right-wing guerrillas have fought a vicious civil war against government troops and one another. Civilians have all too often been caught in the crossfire. Idealistic notions of revolution have been superseded by a simple desire to defend the massive profits accumulated through the narcotics trade. The US has taken an active role in aiding the Colombian government's efforts to regain control and restore peace.

LIBERIA

Vicious civil wars fought by diamond-hungry militias against autocratic governments continue to disrupt hundreds of thousands of people in west Africa. The latest conflict began in Liberia in 2001, just as a brutal ten-year war ended in neighboring Sierra Leone.

CENTRAL AFRICA

Surrounding countries have become involved in the fighting in what is now the Democratic Republic of the Congo, where 2.5 million people have died since 1996. Ethnic conflict between Hutus and Tutsis in Rwanda and Burundi continues to threaten stability.

CONFLICTS AND INTERNATIONAL DISPUTES SINCE 2000

- Countries involved in active external conflict
- Active territorial or border disputes
- Countries involved in internal conflict
- Active territorial or border disputes and internal conflict

Map labels: ICELAND, Rockall, Northern Ireland, Gibraltar, Ceuta, MOROCCO, WESTERN SAHARA, SENEGAL, SIERRA LEONE, LIBERIA, UNITED STATES OF AMERICA, CUBA, Guantánamo Bay, HAITI, BELIZE, GUATEMALA, EL SALVADOR, NICARAGUA, VENEZUELA, GUYANA, SURINAME, French Guiana, COLOMBIA, Falkland Islands

Conflict in the Modern Era

1899–1902: Boer War in South Africa

1910: Start of Mexican Revolution

Jun 1914: Assassination of Archduke Franz Ferdinand, heir to the Austro-Hungarian throne, at Sarajevo, precipitates start of World War I

1932–1935: Paraguay defeats Bolivia in Gran Chaco War

1940: Germany invades Norway, Denmark, Belgium, the Netherlands, and France

1941: Japan attacks US at Pearl Harbor; Germany attempts to invade Russia

1949: Mao Zed Chinese commu win civil wi

1904–1905: Russo-Japanese War; series of Russian defeats

1911: Start of Chinese revolution

1917: Russian revolution

1918: End of World War I

1936–1939: Spanish Civil War

1939: Germany invades Poland; start of World War II

1945: German surrender e war in Europe; atomic bom Japan end war in Pacific

Timeline: 1900, 1910, 1920, 1930, 1940, 1950

NEPAL

Maoist rebels launched a republican insurgency in 1999, gathering support in rural areas. Attempts to engage the rebels in the political process collapsed in 2001 and fighting intensified. In response the government has launched a determined effort to defeat the rebels, leaving heavy casualties on both sides and ruining the country's all-important tourist industry.

CHECHNYA

Russian troops first entered the breakaway republic of Chechnya in 1994, launching a bloody war to crush Islamic separatists there. Russia claimed to have definitively taken control of the republic in 2002, but Muslim guerrillas continue their struggle.

IRAQ

The regime of President Saddam Hussein has been besieged by the US and its allies since they failed to topple the dictator in their liberation of Kuwait in 1991. A strict economic embargo has been in place since the so-called Gulf War, and US and UK warplanes have carried out punitive airstrikes within strictly policed "no-fly" zones across the north and south of the country since 1998.

Svalbard

RUSSIAN FEDERATION

UKRAINE
MOLD.
Chechnya
CEDONIA
GEORG.
AZERB.
UZBEKISTAN
KYRGYZSTAN
GREECE
TURKEY
CYPRUS
SYRIA
AFGHANISTAN
ISRAEL
IRAQ
IRAN
Golan
Heights
EGYPT
SAUDI
ARABIA
UAE
CHAD
ERITREA
Hamish
Islands
ERIA
SUDAN
SOMALIA
ETHIOPIA
CAMEROON
UGANDA
DEM. REP.
CONGO
KENYA
CONGO
RWANDA
BURUNDI
ANGOLA
COMOROS
MADAGASCAR
NAMIBIA
ZIMBABWE
MAURITIUS
BOTSWANA

Kurile
Islands
Liancourt Rocks
NORTH KOREA
JAPAN
CHINA
SOUTH
KOREA
Aksai Chin
PAKISTAN
Jammu and Kashmir
Sênkaku Islands
NEPAL
TAIWAN
BANGLADESH
INDIA
BURMA
(MYANMAR)
VIETNAM
Wake Island
Paracel
Islands
PHILIPPINES
LAOS
THAI.
Spratly
Islands
SRI LANKA
Sipidan and Ligitan
MALAYSIA
SINGAPORE
INDONESIA
SOLOMON
ISLANDS

SRI LANKA

From 1983, until a tentative peace deal in 2002, ethnic Tamil militants known as the Tamil Tigers (LTTE) fought against government forces in an attempt to secure their own state in the north and east of the country. More than 60,000 people died. A change of government in 2001 prompted a determined push for peace, and talks began the following year.

1950–1953: Korean War

1960: Outbreak of civil war in Belgian Congo

1964: US Congress approves war with Vietnam

1968: Troubles begin in Northern Ireland

1975: US withdraws from Vietnam

1979: Civil war in El Salvador (–1992)

1980: Iran–Iraq War (–1988)

1990–1991: Iraqi invasion of Kuwait and Gulf War

1994: Massacre of Tutsis in Rwanda

1996: Start of conflict in Dem. Rep. Congo (former Zaire)

2002: US declares 'War on Terrorism'

1960
1970
1980
1990
2000

1954: Algerian war of independence begins

1965: India-Pakistan War over sovereignty of Kashmir

1975: Angola and Mozambique gain independence; civil wars ensue

1979: Civil war in Nicaragua (–1990)

1982: Falklands War between UK and Argentina

1992: Civil war in Bosnia-Herzegovina

1999: 'Ethnic cleansing' of Albanians in Kosovo by Serbs. Russian offensive in Chechnya

2001: Terror attacks on US

TIME ZONES

Map column headings: −2 −1 0 +1 +2 +3 +4 +5 +6 +7 +8 +9

SVALBARD (to Norway) +1

JAN MAYEN (to Norway)

Franz Josef Land

Severnaya Zemlya +7

New Siberian Island

Novaya Zemlya +5

+10

ICELAND 0

FAEROE ISLANDS (to Denmark) 0

NORWAY

SWEDEN

FINLAND +2

RUSSIAN FEDERATION

+10

European Russia +3

+4

+5

Asian Russia +7

+9

UNITED KINGDOM 0

DENMARK

EST. RUS. FED. LAT. LITH.

+4

+6

+8

REPUBLIC OF IRELAND 0

NETH.

POLAND

BELA.

CHANNEL ISLANDS (to UK) 0

BELG. LUX. GERMANY

CZ.REP. SLVK.

UKRAINE +2

+4

KAZAKHSTAN

MONGOLIA

+10

FRANCE SWITZ. LIECH.

AUT. HUNG.

MOLD.

+3

+6

Azores (to Portugal)

ANDORRA MONACO

SLVN. CRO.

S.& MON. (YUGO.)

ROM.

UZBEK. +5

KYRG. +6

+8

+10

PORTUGAL

S.M. VAT. CITY

B.&H.

BULG.

GEORGIA +4

AZERB.

TURKMEN. +5

TAJ.

N. KOREA

JAPAN

SPAIN

ITALY

ALB. MACED.

ARMENIA +4

AZERB.

+6

CHINA +8

S. KOREA +9

−1

GIBRALTAR (to UK)

GREECE +2

TURKEY +2

SYRIA

CYPRUS

LEB.

IRAN +3½

AFGH. +4½

+5

Madeira (to Portugal)

TUNISIA

MALTA

ISRAEL

JORDAN

IRAQ +3

NEPAL +5¾

BHUTAN +6

Canary Islands (to Spain) 0

MOROCCO

ALGERIA +1

LIBYA +2

EGYPT

KUWAIT

BAHRAIN QATAR

SAUDI ARABIA

OMAN

PAKISTAN +5

BANGLADESH +6

BURMA (MYANMAR) +6½

LAOS

TAIWAN

WESTERN SAHARA 0

SUDAN

ERITREA

YEMEN

INDIA +5½

THAI. +7

VIETNAM +7

Paracel Islands (disputed)

NORTH MARI. ISLA.

GUAM (to US) +10

CAPE VERDE −1

MAURITANIA 0

MALI 0

NIGER +1

CHAD

Laccadive Islands (to India)

Andaman Islands (to India)

CAMB. +8

MICRON. +10

SENEGAL −1

BURKINA

NIGERIA

DJIBOUTI

SRI LANKA

MALAYSIA +8

BRUNEI +8

PALAU +9

GAMBIA −1

GUINEA-BISSAU

GUINEA

BENIN

C.A.R.

ETHIOPIA +3

SINGAPORE +7½

SIERRA LEONE

LIBERIA

EQ. GUINEA

CAMEROON

UGANDA

SOMALIA

MALDIVES

INDONESIA +8

PAPUA NEW GUIN.

IVORY COAST (CÔTE D'IVOIRE) 0

GHANA 0

TOGO

GABON

CONGO

RWANDA

BURUNDI

KENYA

ASCENSION ISLAND (to St Helena) 0

SAO TOME & PRINCIPE

DEM. REP. CONGO

TANZANIA

SEYCHELLES +4

BRITISH INDIAN OCEAN TERRITORY (to UK)

CHRISTMAS ISLAND (to Australia) +7

EAST TIMOR

ST HELENA (to UK) 0

ANGOLA

ZAMBIA

MALAWI

COMOROS

MAYOTTE (to France)

COCOS (KEELING) ISLANDS (to Australia) +6½

ASHMORE & CARTIER ISLANDS (to Australia)

ATLANTIC OCEAN

NAMIBIA +1

ZIMB.

BOTS.

MOZAMBIQUE

MADAGASCAR +3

MAURITIUS

REUNION (to France)

+9½

SOUTH AFRICA +2

SWAZILAND

LESOTHO

INDIAN OCEAN

AUSTRALIA +8

TRISTAN DA CUNHA (to St Helena) 0

+9½

Gough Island (to Tristan da Cunha) 0

+11

FRENCH SOUTHERN & ANTARCTIC TERRITORIES (to France) +5

Greenwich Meridian

Prince Edward Islands (to South Africa)

+5

+5

Tasm.

HEARD & McDONALD ISLANDS (to Australia)

Clock labels: 11:00 12:00 13:00 14:00 15:00 16:00 17:00 18:00 19:00 20:00 21:00

ANTARCTICA
(except for Graham Land, Antarctica is not subject to any time zone)

CHRONOLOGY 2001–2002

THE 20th CENTURY came to a definitive end on September 11, 2001. The momentum of the terrorist attacks in the US revolutionized relations between old enemies, helped to boost right-wing politics around the Western world – including the new US administration – and opened the 21st century with a global conflict. However, in some places peace prospects did improve, with agreements ending a 19-year war in Sri Lanka and the 27-year civil conflict in Angola. Into this altered landscape came the world's newest country, East Timor, and out of it crept "Yugoslavia," as the name was abandoned for good.

NORTH AMERICA

❑ **February 12, 2001** The human genome is mapped by a team of international scientists.

❑ **April 28, 2001** Dennis Tito from the US becomes the first space tourist, aboard the International Space Station.

❑ **May 1, 2001** US president George W. Bush declares the 1972 antiballistic missile (ABM) treaty a thing of the past, opening the way for development of the controversial missile defense shield.

❑ **July 23, 2001** The US is the only industrialized country not to sign up to a new global environmental treaty on greenhouse gas emissions.

❑ **September 11, 2001** Around 3000 people die in a concerted attack against US targets by Islamic militants: hijacked aircraft are flown into New York's celebrated World Trade Center (WTC) and the Pentagon in Washington D.C.

❑ **October 31. 2001** Figures show that the longest economic boom in the US has come to an end.

❑ **December 19, 2001** Fires in the ruins of the WTC are extinguished.

❑ **February 1, 2002** President Bush labels Iran, Iraq, and North Korea an "axis of evil."

❑ **June 17, 2002** Asteroid 2002MN whistles past the earth but is only detected three days later.

❑ **July 21, 2002** Just seven months after the collapse of the Enron corporation, telecommunications firm WorldCom files the largest corporate bankruptcy in US history amid revelations about false accounting.

CENTRAL AND SOUTH AMERICA

❑ **January 13, 2001** The first in a month of massive earthquakes hits El Salvador, killing over 1000.

❑ **July 9, 2001** Charges against former Chilean dictator Gen. Augusto Pinochet are dropped due to his deteriorating mental condition.

❑ **November 1, 2001** The first direct flights for 40 years between Cuba and the US are inaugurated.

❑ **November 11, 2001** Hurricane Michelle devastates Cuba, prompting trade with the US for essential supplies.

❑ **January 1, 2002** Eduard Duhalde is sworn in as Argentina's fifth president in 12 days. The country continues to suffer economic chaos.

❑ **February 20, 2002** Peace talks in Colombia are abandoned: it returns to full-scale civil war.

❑ **April 14, 2002** President Hugo Chávez is returned to power in Venezuela after a 48-hour coup.

❑ **June 30, 2002** Brazil wins the soccer World Cup for a record fifth time.

EUROPE

❑ **February 21, 2001** The first registered case of foot-and-mouth precipitates a one-year epidemic which devastates livestock across the UK.

❑ **April 10, 2001** The Netherlands becomes the first country in the world to legalize euthanasia.

❑ **May 13, 2001** Right-wing businessman Silvio Berlusconi returns as prime minister of Italy, bringing "postfascists" and northern separatists back into government.

❑ **July 12, 2001** A movement led by ex-king Simeon II wins elections in Bulgaria. He becomes prime minister, taking the surname Saxecoburggotski.

❑ **September 6, 2001** A peace deal in Macedonia ends a seven-month insurrection by ethnic Albanian rebels.

❑ **September 23, 2001** The historic Solidarity party is wiped out of the Polish Sejm (parliament) in elections.

❑ **November 10, 2001** A right-wing coalition wins elections in Norway. Ten days later a conservative coalition triumphs in Denmark.

❑ **January 1, 2002** The euro is fully introduced at midnight in 12 EU countries. Six months later the franc, deutsche mark, drachma, and other old currencies are consigned to the melting pot of history.

❑ **February 12, 2002** The trial begins in The Hague of former Serbian dictator Slobodan Milošević.

❑ **March 3, 2002** Switzerland votes to apply to become the 190th member of the UN, but pledges to maintain its neutrality.

❑ **March 14, 2002** The constituent governments of Yugoslavia agree to dissolve the union and replace it with "Serbia and Montenegro."

❑ **April 18, 2002** Russian president Vladimir Putin declares the Chechen war over. Guerrilla fighting continues.

❑ **April 21, 2002** Far-right leader Jean-Marie Le Pen shocks the world by coming second in the first round of the French presidential elections.

❑ **May 6, 2002** Controversial Dutch anti-immigrant politician Pim Fortuyn is shot dead. His party goes on to form part of the new government.

❑ **May 28, 2002** A Russia–NATO council is established, giving Russia more input into NATO policy.

❑ **August 14, 2002** Floodwaters recede from Prague. Rising river levels cause havoc across central Europe.

AFRICA

❑ **January 16, 2001** President Laurent Kabila of the Democratic Republic of Congo (DRC) is assassinated by a bodyguard. He is succeeded by his son Joseph, raising hopes of peace.

❑ **January 31, 2001** Abdelbaset al-Megrahi of Libya is convicted of the 1988 bombing of a PanAm airplane over Lockerbie, Scotland.

❑ **April 2, 2001** Ten years of brutal civil war are brought to an end in Sierra Leone.

❑ **April** 18, **2001** Violence breaks out in the ethnic Berber-dominated region of Kabylia in northeastern Algeria.

❑ **May** 28, **2001** A coup attempt in the Central African Republic turns into a pitched battle; thousands flee.

❑ **November** 20, **2001** A previously unknown rebel group launches an insurrection in Liberia.

❑ **January** 28, **2002** Over 1000 people die trying to escape a fire at a munitions dump in Lagos, Nigeria.

❑ **February** 22, **2002** Marc Ravalomanana declares himself the elected president of Madagascar, prompting a tense and increasingly violent four-month standoff with the incumbent, Didier Ratsiraka.

❑ **April** 4, **2002** The death of UNITA leader Joseph Savimbi prompts a cease-fire, ending 27 years of civil war in Angola.

❑ **July** 9, **2002** The African Union is created as the OAU's successor in Durban, South Africa.

❑ **July** 22, **2002** Moroccan soldiers leave the disputed Spanish islet of Perejil after a ten-day standoff.

WEST ASIA/MIDDLE EAST

❑ **February** 6, **2001** Right-wing Likud leader Ariel Sharon is elected prime minister of Israel in the last direct elections for the post. Under his leadership, confrontation with the Palestinians intensifies.

❑ **May** 6, **2001** Pope John Paul II becomes the first pope to pray in a mosque, during his visit to Syria.

❑ **December** 6, **2001** Kandahar, the spiritual capital of the *taliban* in Afghanistan, is "liberated" by the Northern Alliance, signaling the collapse of the Islamic regime.

❑ **February** 14, **2002** Bahrain is declared a constitutional monarchy.

❑ **March** 19, **2002** US forces complete Operation Anaconda in Afghanistan, leaving more than 500 al-Qaida and *taliban* fighters dead.

❑ **May** 10, **2002** The five-week siege in Bethlehem of the Church of the Nativity is finally lifted; 39 Palestinian militants involved go into exile.

❑ **June** 19, **2002** Israeli forces begin the construction of a wall between Israel proper and the Palestinian West Bank.

NORTH AND EAST ASIA

❑ **March** 21, **2001** Amnesty International declares that China executes more people than the rest of the world put together. Up to 40 people a week are killed as part of the anticorruption "Strike Hard" campaign.

❑ **April** 1, **2001** A US reconnaissance plane collides with a Chinese fighter near Hainan, causing a major diplomatic crisis.

❑ **July** 16, **2001** China and Russia sign their first friendship treaty since the Sino-Soviet split of the early 1960s. They both oppose the proposed US missile defense system.

❑ **August** 31, **2001** New populist Japanese prime minister Junichiro Koizumi inflames regional tensions by visiting the controversial Yasakuni war shrine.

❑ **September** 26, **2001** The Communist Party of China (CCP) adopts the "Three Represents" doctrine encouraging the inclusion of capitalists in the Maoist party.

❑ **November** 10, **2001** The World Trade Organization (WTO) ratifies China's membership. The next day Taiwan's accession is also ratified .

❑ **April** 11, **2002** The Chinese government admits that there are around 850,000 people in China who are HIV-positive. The UN estimates that the figure is already likely to be more than one million.

❑ **June** 7, **2002** Figures show that a six-month recession has come to an end in Japan.

❑ **August** 23, **2002** Reclusive North Korean leader Kim Jong Il meets Russian president Vladimir Putin in Vladivostok in only his third foreign visit in eight years.

SOUTH ASIA

❑ **January** 26, **2001** A powerful earthquake in Gujarat, India, kills an estimated 30,000 people and obliterates infrastructure.

❑ **June** 1,**2001** King Birendra of Nepal, and nine other members of the royal family, are murdered by Crown Prince Dipendra in a drunken rage. Dipendra also kills himself. His uncle Gyanendra becomes king.

❑ **November** 26, **2001** A state of emergency is declared in Nepal as the government launches a counteroffensive against Maoist rebels.

❑ **February** 28, **2002** The murder of 58 Hindus in Gujarat sparks sectarian violence across the state, leaving over 500 people – mostly Muslims – dead.

❑ **May** 6, **2002** Aung San Suu Kyi, the leader of the democratic opposition in Burma, is released from house arrest.

❑ **May** 14, **2002** Kashmiri separatists kill 34 people in a terrorist attack, raising tensions between India and Pakistan and bringing them to the brink of possible nuclear war.

SOUTHEAST ASIA

❑ **July** 23, **2001** President Wahid of Indonesia is ousted by parliament and replaced by his deputy, Megawati Sukarnoputri.

❑ **January** 1, **2002** The Indonesian provinces of Aceh and Papua (formerly Irian Jaya) are granted limited autonomy.

❑ **February** 21, **2002** A cease-fire in Sri Lanka promise peace after 19 years of civil war.

❑ **May** 20, **2002** East Timor becomes an independent country.

❑ **June** 22, **2002** Prime Minister Mahathir Mohamed of Malaysia announces he will resign in 2003 after 22 years in power.

AUSTRALIA AND OCEANIA

❑ **March** 23, **2001** The 15-year-old *Mir* space station, the last remnant of the Cold-War era of space exploration, is brought down into the Pacific.

❑ **August** 29, **2001** Australian troops board the *Tampa*, a Norwegian ship carrying over 400 mainly Afghan refugees, whose forcible relocation to camps in other Pacific countries heralds the start of the controversial "Pacific Solution."

❑ **August** 31, **2001** A final peace agreement is reached in Bougainville, Papua New Guinea, ending a 13-year independence struggle.

❑ **November** 10, **2001** The right-wing government of Prime Minister John Howard is unexpectedly reelected in Australia.

❑ **May** 14, **2002** AusAid predicts that 40% of Papua New Guinea's adults could die of AIDS by 2020.

❑ **July** 2, **2002** US adventurer Steve Fosset lands in Australia after completing the first solo circumnavigation of the Earth in a hot air balloon.

see pages 36–39 for CHRONOLOGY OF WORLD HISTORY

INTERNATIONAL ORGANIZATIONS

THIS LISTING GIVES the full names of all international organizations referred to, often by acronym, in the World Desk Reference (political parties are to be found under the Politics heading within each country entry). The full names are followed by the date of the establishment or foundation, an indication of membership, where appropriate, and a summary of the organization's aims and functions.

ACC
Arab Cooperation Council
established 1989
members – Egypt, Iraq, Jordan, Yemen
Promotes Arab economic cooperation

ACP
African, Caribbean, and Pacific Countries
established 1976
members – 78 developing countries and territories
Preferential economic and aid relationship with the EU under the Lomé Convention

ACS
Association of Caribbean States
established 1994
members – 25 Caribbean countries
Promotes economic, scientific, and cultural cooperation in the region

ADB
Asian Development Bank
established 1966
members – 43 Asia–Pacific countries and territories, 17 nonregional countries
Encourages regional development

AfDB
African Development Bank
established 1964
members – 53 African countries, 24 non-African countries
Encourages African economic and social development

AFESD
Arab Fund for Economic and Social Development
established 1968
members – 21 Arab countries (including Palestine)
Promotes social and economic development in Arab states

AL
League of Arab States (Arab League)
established 1945
members – 22 Arab countries (including Palestine)
Forum to promote Arab cooperation on social, political, and military issues

ALADI
Latin American Integration Association
established 1960
members – 12 Central and South American countries
Promotes trade and regional integration

AMCC
Amazonian Cooperation Council
established 1978
members – Bolivia, Brazil, Colombia, Ecuador, Guyana, Peru, Suriname, Venezuela
Promotes the harmonious development of the Amazon region

AMF
Arab Monetary Fund
established 1977
members – 21 Arab countries (including Palestine)
Promotes monetary and economic cooperation

AMU
Arab Maghreb Union
established 1989
members – Algeria, Libya, Mauritania, Morocco, Tunisia
Promotes integration and economic cooperation among north African Arab states

ANZUS
Australia–New Zealand–United States Security Treaty
established 1951
members – Australia, New Zealand, US
Trilateral security agreement. Security relations between the US and New Zealand were suspended in 1984 over the issue of US nuclear-powered or potentially nuclear-armed naval vessels visiting New Zealand ports. High-level contacts between the US and New Zealand were resumed in 1994

AP
Andean Pact
(Acuerdo de Cartegena), also known as Andean Community
established 1969
members – Bolivia, Colombia, Ecuador, Peru, Venezuela
Promotes development through integration

APEC
Asia–Pacific Economic Cooperation
established 1989
members – 20 Pacific Rim countries
Promotes regional economic cooperation

ASEAN
Association of Southeast Asian Nations
established 1967
members – Brunei, Burma, Cambodia, Indonesia, Laos, Malaysia, Philippines, Singapore, Thailand, Vietnam
Promotes economic, social, and cultural cooperation

AU
African Union
established 2002
members – 52 African countries and Western Sahara
Promotes unity and cooperation in Africa. (The African Union was, until 2002, the Organization of African Unity, which had been founded in 1963)

BADEA
Arab Bank for Economic Development in Africa
established 1973
members – 18 Arab countries (including Palestine)
Established as an agency of the Arab League to promote economic development in Africa

BDEAC
Central African States Development Bank
established 1975
members – Cameroon, Central African Republic, Chad, Congo, Equatorial Guinea, France, Gabon, Germany, Kuwait
Furthers economic development

Benelux
Benelux Economic Union
established 1960
members – Belgium, Luxembourg, Netherlands
Develops economic ties between member countries

BOAD
West African Development Bank
established 1973
members – Benin, Burkina, Guinea-Bissau, Ivory Coast, Mali, Niger, Senegal, Togo
Promotes economic development and integration in West Africa

BSEC
Organization of the Black Sea Economic Cooperation
established 1992
members – Albania, Armenia, Azerbaijan, Bulgaria, Georgia, Greece, Moldova, Romania, Russia, Turkey, Ukraine
Furthers regional stability through economic cooperation

CAEU
Council of Arab Economic Unity
established 1957
members – 12 Arab countries (including Palestine)
Encourages economic integration

Caricom
Caribbean Community and Common Market
established 1973
members – 13 Caribbean countries and Montserrat
Fosters economic ties in the Caribbean

CBSS
Council of the Baltic Sea States
established 1992
members – Denmark, Estonia, Finland, Germany, Iceland, Latvia, Lithuania, Norway, Poland, Russia, Sweden
Promotes cooperation among Baltic Sea states

CDB
Caribbean Development Bank
established 1969
members – 17 Caribbean countries/dependencies, 8 non-Caribbean countries
Promotes regional development

CE
Council of Europe
established 1949
members – 44 European countries
Promotes unity and quality of life in Europe

CEFTA
Central European
Free Trade Agreement
established 1992
members – Bulgaria,
Czech Republic, Hungary,
Poland, Romania, Slovakia,
Slovenia
Promotes trade and
cooperation

CEI
Central European Initiative
established 1989
members – 17 eastern and
central European countries:
Albania, Austria, Belarus,
Bosnia and Herzegovina,
Bulgaria, Croatia, Czech
Republic, Hungary, Italy,
Macedonia, Moldova,
Poland, Romania, Serbia
and Montenegro, Slovakia,
Slovenia, Ukraine
Evolved from the Hexagonal
Group; promotes economic
and political cooperation,
within the OSCE

CEMAC
Central African Economic
and Monetary Community
established 1994
members – Cameroon,
Central African Republic,
Chad, Congo, Equatorial
Guinea, Gabon
Aims to promote
subregional integration,
by economic and monetary
union
(replaced UDEAC)

CEPGL
Economic Community of
the Great Lakes Countries
established 1976
members – Burundi,
Democratic Republic
of the Congo, Rwanda
Promotes regional
economic cooperation

CERN
European Organization
for Nuclear Research
established 1954
members – 20 European
countries
Provides for collaboration
in nuclear research for
peaceful purposes

CILSS
Permanent Interstate
Committee for Drought
Control in the Sahel
established 1973
members – 9 African
countries in the Sahel
region
Promotes prevention
of drought and crop
failure in the region

CIS
Commonwealth of
Independent States
established 1991
members – Armenia,
Azerbaijan, Belarus,
Georgia, Kazakhstan,
Kyrgyzstan, Moldova,
Russia, Tajikistan,
Turkmenistan,
Ukraine, Uzbekistan
Promotes interstate
relationships among
former republics of
the Soviet Union

CMCA
Central American
Monetary Council
established 1960
members – Costa Rica,
El Salvador, Guatemala,
Honduras, Nicaragua
Now a subsystem of SICA.
Furthers economic ties
between members; one
of its institutions is the
BCIE – Central American
Bank for Economic
Integration

COI
Indian Ocean Commission
established 1982
members – Comoros, France
(representing Réunion),
Madagascar, Mauritius,
Seychelles
Promotes regional
cooperation

COMESA
Common Market for Eastern
and Southern Africa
established 1993
members – 20 African
countries
Promotes economic
development and
cooperation (replaced PTA)

Comm
Commonwealth
established 1931
members – 54 countries
(although Pakistan and
Zimbabwe are currently
suspended). Members are
chiefly former members
of the British Empire
Develops relationships
and contacts between
members

CP
Colombo Plan
established 1950
members – Four donor
countries: Australia, Japan,
New Zealand, US; and
20 Asia–Pacific countries
Encourages economic
and social development
in Asia–Pacific region

CPLP
Community of Portuguese-
speaking Countries
established 1996
members – Portugal, Brazil,
and five Portuguese-speaking
African countries – Angola,
Cape Verde, Guinea-Bissau,
Mozambique, São Tomé and
Príncipe
To promote political and
diplomatic links between
member states, and
cooperation on economic,
social, cultural, judicial,
and scientific development
among Portuguese-speaking
countries

Damasc
Damascus Declaration
established 1991
members – Bahrain,
Egypt, Kuwait, Oman,
Qatar, Saudi Arabia, Syria,
United Arab Emirates
A loose association, formed
after the Gulf War, which
aims to secure the stability
of the region

EAC
East African Community
established 2001
members – Kenya,
Tanzania, Uganda
Promotes economic
cooperation

EAPC
Euro-Atlantic Partnership
Council
established 1991
members – The 19 members
of NATO plus 27 eastern
European countries
Forum for cooperation on
political and security issues
(successor to the NACC,
North Atlantic Cooperation
Council)

EBRD
European Bank for
Reconstruction and
Development
established 1991
members – 60 countries
Helps transition of former
communist European states
to market economies

ECO
Economic Cooperation
Organization
established 1985
members – Afghanistan,
Azerbaijan, Iran, Kazakhstan,
Kyrgyzstan, Pakistan,
Tajikistan, Turkey,
Turkmenistan, Uzbekistan
Aims at cooperation
in economic, social,
and cultural affairs

ECOWAS
Economic Community
of West African States
established 1975
members – 15 west
African countries
Promotes regional
economic cooperation

EEA
European Economic Area
established 1994
members – The 15 members
of the EU, and Iceland,
Liechtenstein, and Norway
Aims to include EFTA members
in the EU single market

EEC
Eurasian Economic Community
established 2001
members – Belarus, Kazakhstan,
Kyrgyzstan, Russia, Tajikistan
Coordinates regional trade

EFTA
European Free Trade
Association
established 1960
members – Iceland,
Liechtenstein, Norway,
Switzerland
Promotes economic cooperation

ESA
European Space Agency
established 1973
members – 15 European
countries
Promotes cooperation in space
research for peaceful purposes

EU
European Union
established 1992
members – 15 European
countries
Aims to integrate the economies
of member states and promote
cooperation and coordination
of policies. Previously the
European Communities (EC),
established in 1957 by the
Treaties of Rome

FZ
Franc zone
established Not applicable
members – France (including
overseas departments and
territories), Monaco, and
15 African countries
Aims to form monetary
union among countries
whose currencies are
linked to that of France

G3
Group of 3
established 1987
members – Colombia,
Mexico, Venezuela
Aims to ease trade restrictions

G7
Group of 7
established 1975
members – The seven major industrialized countries: Canada, France, Germany, Italy, Japan, UK, US
Summit meetings of the seven major industrialized countries, originally for economic purposes. For political purposes summit meetings are now held as the G8, including Russia

G8
Group of 8
established 1994
members – Canada, France, Germany, Italy, Japan, Russia, UK, US
Global forum of world's major powers, which holds regular summit meetings

G10
Group of 10
established 1962
members – 11 members: G7, plus Belgium, the Netherlands, Sweden, and Switzerland
Ministers meet to discuss monetary issues

G15
Group of 15
established 1989
members – 19 developing countries
Meets annually to further cooperation among developing countries

G24
Group of 24
established Not applicable
members – 24 developing countries within the IMF
Promotes the interests of developing countries on monetary and development issues.

GCC
Gulf Cooperation Council
established 1981
members – Bahrain, Kuwait, Oman, Qatar, Saudi Arabia, UAE
Promotes cooperation in social, economic, and political affairs

GEPLACEA
Latin American and Caribbean Sugar Exporting Countries
established 1974
members – 23 countries
A forum for consultation on the production and sale of sugar

GGC
Gulf of Guinea Commission
established 2001
members – Angola, Cameroon, Congo, Democratic Republic of the Congo, Equatorial Guinea, Gabon, Nigeria, São Tomé and Príncipe
Promotes regional cooperation

IAEA
International Atomic Energy Agency
established 1957
members – 134 countries
Promotes and monitors peaceful use of atomic energy

IBRD
International Bank for Reconstruction and Development
(also known as the World Bank)
established 1945
members – 184 countries
UN agency providing economic development loans

ICRC
International Committee of the Red Cross
established 1863
members – Up to 25 Swiss nationals form the international committee. Red Cross or Red Crescent societies exist in 175 countries
Coordinates all international humanitarian activities of the International Red Cross and Red Crescent Movement, giving legal and practical assistance to the victims of wars and disasters. It works through national committees of Red Cross or Red Crescent societies

IDB
Inter-American Development Bank
established 1959
members – 28 American countries and 18 nonregional countries
Promotes development in Latin America and the Caribbean through the financing of economic and social development projects and the provision of technical assistance

IGAD
Intergovernmental Authority on Development
established 1996
members – Djibouti, Eritrea, Ethiopia, Kenya, Somalia, Sudan, Uganda
Promotes cooperation on food security, infrastructure, and other development issues (supersedes IGADD, founded 1986, to promote cooperation on drought-related matters)

IMF
International Monetary Fund
established 1945
members – 184 countries. The voting rights of the Democratic Republic of the Congo are currently suspended.
Promotes international monetary cooperation, the balanced growth of trade, and exchange rate stability; provides credit resources to members experiencing balance-of-trade difficulties.

ISDB
Islamic Development Bank
established 1975
members – 53 countries (including Palestine)
Promotes economic development on Islamic principles among Muslim communities (agency of the OIC)

IWC
International Whaling Commission
established 1946
members – 48 countries
Reviews conduct of whaling throughout world; coordinates and funds whale research

LCBC
Lake Chad Basin Commission
established 1964
members – Cameroon, Central African Republic, Chad, Niger, Nigeria
Encourages economic and environmental development in Lake Chad region

Mekong River
Mekong River Commission
established 1995
members – Cambodia, Laos, Thailand, Vietnam
Accord on the sustainable development of Mekong River basin (replacing the 1958 interim Mekong Secretariat)

Mercosur
Southern Common Market
established 1991
members – Argentina, Brazil, Paraguay, Uruguay
Promotes economic integration, free trade, and common external tariffs

MRU
Mano River Union
established 1973
members – Guinea, Liberia, Sierra Leone
Aims to create customs and economic union in order to promote development

NAFTA
North American Free Trade Agreement
established 1994
members – Canada, Mexico, US
Free trade zone

NAM
Non-Aligned Movement
established 1961
members – 115 countries (including Palestine). Serbia and Montenegro has been suspended since 1992
Fosters political and military cooperation away from traditional Eastern or Western blocs

NATO
North Atlantic Treaty Organization
established 1949
members – 19 countries
Promotes mutual defense cooperation. Since January 1994, NATO's Partnerships for Peace program has provided a loose framework for cooperation with former members of the Warsaw Pact and the ex-Soviet republics. A historic Founding Act signed between Russia and NATO in May 1997 allowed for the organization's eastward expansion, under which the Czech Republic, Hungary, and Poland were the first three countries to join

NC
Nordic Council
established 1952
members – Denmark, Finland, Iceland, Norway, Sweden
Promotes cultural and environmental cooperation in Scandinavia

OAPEC
Organization of Arab Petroleum Exporting Countries
established 1968
members – 10 Arab countries: Algeria, Bahrain, Egypt, Iraq, Kuwait, Libya, Qatar, Saudi Arabia, Syria, UAE
Aims to promote the interests of member countries and increase cooperation in the petroleum industry

OAS
Organization of American States
established 1948
members – 35 American countries (although Cuba has been suspended since 1962)
Promotes security, economic, and social development in the Americas

OAU
Organization of African Unity
Predecessor of the AU.

OECD
Organization for Economic
Cooperation and Development
established 1961
members – 30 industrialized
democracies
Forum for coordinating
economic policies among
industrialized countries

OECS
Organization of Eastern
Caribbean States
established 1981
members – 7 Caribbean
countries/dependencies:
Antigua and Barbuda, Dominica,
Grenada, Montserrat, St. Kitts
and Nevis, St. Lucia, St. Vincent
and the Grenadines
Promotes political, economic,
and defense cooperation

OIC
Organization of the
Islamic Conference
established 1971
members – 57 countries
(including Palestine)
Furthers Islamic
solidarity and cooperation

OIF
International Organization
of Francophony
established 1970
members – 48 countries and the
governments of Québec, New
Brunswick, and the French
Community of Belgium were
represented at the 1999 summit.
The Intergovernmental
Agency of Francophony
has 49 members.
To promote cooperation and
cultural and technical links
among French-speaking
countries and communities

OMVG
Gambia River Development
Organization
established 1978
members – Gambia, Guinea,
Guinea-Bissau, Senegal
Promotes integrated develop-
ment of the Gambia River basin

Opanal
Agency for the Prohibition
of Nuclear Weapons in Latin
America and the Caribbean
established 1969
members – 32 countries (plus
Cuba, who has signed but not
ratified the Treaty of Tlatelolco)
Aims to ensure compliance
with the Treaty of Tlatelolco
(banning nuclear weapons
from the region)

OPEC
Organization of the Petroleum
Exporting Countries
established 1960
members – 11 oil producers:
Algeria, Indonesia, Iran, Iraq,
Kuwait, Libya, Nigeria, Qatar,
Saudi Arabia, United Arab
Emirates, Venezuela
Aims to coordinate oil
policies to ensure fair
and stable prices

OSCE
Organization for Security
and Cooperation in Europe
established 1972
members – 55 countries
Aims to strengthen democracy
and human rights, and settle
disputes peacefully
(formerly CSCE; renamed 1994)

Partnerships for Peace (PfP)
see NATO
established Not applicable
members – 27 members:
eastern European and
former Soviet countries,
Sweden, Finland, Malta,
Austria, and Switzerland

PC
Pacific Community
(formerly South Pacific
Commission)
established 1948
members – 27 countries
and territories
A forum for dialogue
between Pacific countries
and powers administering
Pacific territories

PIF
Pacific Islands Forum
(formerly the South
Pacific Forum)
established 1971
members – 16 countries and
self-governing territories
Develops regional political
cooperation

RG
Rio Group
established 1987
members – 19 Latin American
and Caribbean countries
Forum for Latin American
and Caribbean issues
(evolved from Contadora
Group, established 1948)

SAARC
South Asian Association
for Regional Cooperation
established 1985
members – Bangladesh,
Bhutan, India, Maldives,
Nepal, Pakistan, Sri Lanka
Encourages economic,
social, and cultural
cooperation

SACU
Southern African
Customs Union
established 1969
members – 5 southern African
countries: Botswana, Lesotho,
Namibia, South Africa,
Swaziland
Promotes cooperation in trade
and customs matters among
southern African states

SADC
Southern African
Development Community
established 1992
members – 14 southern
African countries
Promotes economic
integration

San José
San José Group
established 1988
members – Costa Rica,
El Salvador, Guatemala,
Honduras, Nicaragua,
Panama
A 'complementary, voluntary,
and gradual' economic union

SCO
Shanghai Cooperation
Organization
(formerly Shanghai Five)
established 1996
members – China,
Kazakhstan, Kyrgyzstan,
Russia, Tajikistan,
Uzbekistan
Promotes regional
security and cooperation

SELA
Latin American Economic
System
established 1975
members – 28 countries
Promotes economic and
social development through
regional cooperation

SICA
Central American
Integration System
established 1991
members – 6 countries:
Costa Rica, El Salvador,
Guatemala, Honduras,
Nicaragua, Panama
Coordinates the political,
economic, social, and
environmental integration
of the region

UEMOA
West African Economic
and Monetary Union
established 1994
members – 8 West
African countries
Aims for convergence
of monetary policies
and economic union

UN
United Nations
established 1945
members – 191 countries;
permanent members of the
Security Council – China,
France, Russia, UK, US
Aims to maintain international
peace and security and to
promote cooperation over
economic, social, cultural,
and humanitarian problems

Agencies include the regional
commissions of the UN's
Economic and Social Council:
ECA (Economic Commission
for Africa – established 1958);
ECE (Economic Commission
for Europe – established 1947);
ECLAC (Economic Commission
for Latin America and the
Caribbean – established 1948);
ESCAP (Economic and Social
Commission for Asia and the
Pacific – established 1947);
ESCWA (Economic and Social
Commission for Western Asia
– established 1973).

Other bodies of the UN, in
which most members
participate, include
UNICEF (UN Children's Fund);
UNCTAD (UN Conference on
Trade and Development);
UNDP (UN Development
Program);
UNHCR (UN High
Commissioner for Refugees);
UNFPA (UN Population Fund);
IDA (International
Development Association)

Taiwan and the Vatican
City do not belong to the UN.

WEU
Western European Union
established 1955
members – 10 countries
A forum for European
military cooperation

WTO
World Trade Organization
established 1995
members – 143 countries
(including Hong Kong
and Macao) and the EU
Aims to liberalize trade
through multilateral
trade agreements (as
the successor to GATT
(the General Agreement
on Tariffs and Trade))

2

THE NATIONS OF THE WORLD

THE NATIONS OF THE WORLD
• AFGHANISTAN ~ ZIMBABWE
OVERSEAS TERRITORIES & DEPENDENCIES

A

AFGHANISTAN

OFFICIAL NAME: Islamic State of Afghanistan **CAPITAL:** Kabul
POPULATION: 22.5 million **CURRENCY:** Afghani **OFFICIAL LANGUAGES:** Pashtu and Dari

AFGHANISTAN LIES LANDLOCKED in central Asia, three-quarters of its territory inaccessible terrain. Its political system, economy, and infrastructure have been devastated by decades of armed conflict. In the 1980s Islamic *mujahideen* factions defeated the Soviet-backed communist regime, but rivalries undermined their fragile power-sharing agreement and the hard-line *taliban* militia swept to power in 1996. Islamic dress codes and behavior were vigorously enforced and women left with few rights or opportunities. The *taliban* regime crumbled in the face of the US-led "war on terrorism" launched in late 2001.

The Band-i-Amir River, *in the Hindu Kush. Afghanistan is mountainous and arid. Many Afghans are nomadic sheep farmers.*

CLIMATE ▷ Mountain/cold desert

WEATHER CHART FOR KABUL

Afghanistan has the world's widest temperature range, with lows of –50°C (–58°F) and highs of 53°C (127°F). Severe drought, a frequent problem, affected half the population in 2000.

TRANSPORTATION ▷ Drive on right

 Kabul International Has no fleet

THE TRANSPORTATION NETWORK

2793 km (1735 miles)	None
25 km (16 miles)	1200 km (746 miles)

The repair and reconstruction of the roads, severely damaged by war, and the modernization of the air traffic control system are the most urgent priorities. The rebuilding of roads is usually carried out by local communities. However, neighboring Pakistan has undertaken to rebuild a number of key routes, including the Kabul–Peshawar link, which will benefit its own trade with central Asia.

Securing key supply routes was a crucial factor in intra-*mujahideen* feuding and in the *taliban*'s efforts to gain control over the whole country. Anti-*taliban* forces relied heavily on supplies from the north. Much of Afghanistan's outlying territory is sown with land mines.

TOURISM ▷ Visitors : Population 1:5625

4000 visitors | No change in 1995–1998

MAIN TOURIST ARRIVALS

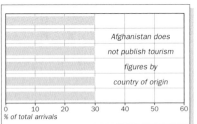

Afghanistan does not publish tourism figures by country of origin

% of total arrivals

Afghanistan is a war zone. Restrictions on the activities of UN and aid agency personnel were lifted in 2001 following the overthrow of the *taliban* regime. Few hotels are open in Kabul and travel is extremely dangerous. Air Ariana, the Afghan national airline, lost six of its eight aircraft in the 2001 US bombing.

The lack of a formal economy means that Afghanistan gets few visits from businessmen, and any expatriates who were previously in Kabul left during the *taliban* period.

PEOPLE ▷ Pop. density low

Pashtu, Tajik, Dari, Farsi, Uzbek, Turkmen | 35/km² (89/mi²)

THE URBAN/RURAL POPULATION SPLIT

22% | 78%

RELIGIOUS PERSUASION

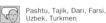

- Other 1%
- Shi'a Muslim 15%
- Sunni Muslim 84%

ETHNIC MAKEUP

- Other 3%
- Uzbek and Turkmen 15%
- Pashtun 38%
- Hazara 19%
- Tajik 25%

Ethnic divisions largely determined intra-*mujahideen* feuding after 1992. Pashtuns have traditionally been the rulers of Afghanistan, and dominated the *taliban*. The fall of that regime in 2001 provided the opportunity for the Tajik–Uzbek alliance to enforce a power-sharing agreement. Pashtuns have since faced reprisal attacks, particularly in the north, where they form a minority. Religious differences between Sunnis and Shi'as became particularly acute under the Sunni *taliban* regime.

Some two million of the country's population were killed in the ten-year conflict which followed the invasion by Soviet Union forces in 1979 and in the post-1992 civil war. As many people again were maimed. A further six million people were forced to flee to neighboring Pakistan and Iran; many returned, but the fighting in 2001 created a fresh wave of refugees and left hundreds of thousands more people internally displaced.

Women had few rights under the rigid Islamic regime of the *taliban*. The interim government has lifted the ban on women in employment and allowed girls to resume schooling.

POPULATION AGE BREAKDOWN

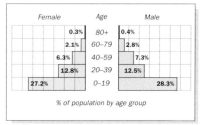

Female	Age	Male
0.3%	80+	0.4%
2.1%	60–79	2.8%
6.3%	40–59	7.3%
12.8%	20–39	12.5%
27.2%	0–19	28.3%

% of population by age group

POLITICS ▷ In transition

 1988/2004

 President Hamid Karzai

AT THE LAST ELECTION

House of Representatives (dissolved)

Following the downfall of Najibullah's regime in April 1992, both houses were dissolved and an interim *mujahideen* legislature formed.

Senate (dissolved)

The political system had virtually collapsed in Afghanistan prior to the *taliban* takeover in 1996. Rival *mujahideen* factions had been in control since April 1992, when the communist President Najibullah, whose regime had been backed until 1990 by the Soviet Union, was forced to step down.

According to the 1993 Islamabad peace accord, elections were to be held by the end of the year. These were indefinitely postponed owing to factional fighting, which led in early 1995 to the emergence of a potent new force, the extremist Islamic *taliban*. Initially the product of fundamentalist Islamic schools established in the refugee camps on the Pakistani border,

the *taliban* were mostly ethnic Pashtuns, and their support was strongest around Kandahar in the south. Kabul fell to them in 1996, and an opposition Northern Alliance headed by the former president, Burhanuddin Rabbani of the Jamiat-i-Islami, failed to reverse the situation. On assuming power the *taliban* declared Afghanistan a "complete" Islamic state, and imposed a strict Islamic code.

The US-led intervention in late 2001 helped replace the *taliban* regime with groups allied to the Northern Alliance, which succeeded in gaining control of most major cities, including Kabul. Later that year UN-sponsored talks led to the formation of an interim government headed by Pashtun leader Hamid Karzai, who is committed to implementing the decisions of a Loya Jirga (grand council), pending elections in 2004. Ex-king Zahir Shah returned in April 2002 to Kabul, where he endorsed Karzai's election as head of state by the Loya Jirga in June. The fragility of the political situation was underlined by the assassination in July of Vice President Haji Abdul Qadir.

Hamid Karzai was confirmed as head of state in 2002 by the traditional Loya Jirga council.

WORLD AFFAIRS ▷ Joined UN in 1946

 CP   ECO   IBRD   NAM OIC

Under the Islamic extremist *taliban* government Afghanistan was a pariah state, with only three Islamic countries maintaining diplomatic ties. The regime became the first target in the US-led "war on terrorism" in October 2001, leading to its destruction. The new interim government is consequently heavily dependent on the countries which effectively installed it, making good relations with them essential for future aid. Pakistan was the last country to break off links with the *taliban* and remains Afghanistan's most delicately placed partner, with a history of religious as well as ethnic sympathies with the ousted, Pashtun-dominated, Islamic regime.

AFGHANISTAN

Total Land Area : 647 500 sq. km
(250 000 sq. miles)

LAND HEIGHT		POPULATION	
	3000m/9843ft	over 1 000 000	▣
	2000m/6562ft	over 100 000	◎
	1000m/3281ft	over 50 000	○
	500m/1640ft	over 10 000	●
	200m/656ft	under 10 000	●

0 ___ 100 km
0 ___ 100 miles

CHRONOLOGY

The foundations of an Afghan state of Pashtun peoples were laid in the mid-18th century, when Durrani Ahmad Shah became paramount chief of the Abdali Pashtun peoples.

❏ **1838–1842** First Anglo-Afghan war.
❏ **1878** Second British invasion of Afghan territory.
❏ **1879** Under Treaty of Gandmak signed with Amir Yaqub Ali Khan, various Afghan areas annexed by Britain. Yaqub Ali Khan later exiled. New treaty signed with Amir Abdul Rahman, establishing the Durand line, a contentious boundary between Afghanistan and Pakistan.
❏ **1919** Independence declared.
❏ **1933** Mohammed Zahir Shah ascends throne.
❏ **1953–1963** Mohammed Daud Khan, prime minister, resigns after king rejects proposals for democratic reforms.

⇨

CHRONOLOGY *continued*

- ❑ **1965** Elections held, but monarchy retains power. Marxist Party of Afghanistan (PDPA) formed and banned. PDPA splits into the Parcham and Khalq factions.
- ❑ **1973** Daud mounts a coup, abolishes monarchy, and declares republic. *Mujahideen* rebellion begins. Refugees flee to Pakistan.
- ❑ **1978** Opposition to Daud from PDPA culminates in Saur revolution. Revolutionary Council under Mohammad Taraki takes power. Daud assassinated.
- ❑ **1979** Taraki ousted. Hafizullah Amin takes power. Amin killed in December coup backed by USSR. 80,000 Soviet Army troops invade Afghanistan. *Mujahideen* rebellion stepped up into full-scale guerrilla war, with US backing.
- ❑ **1980** Babrak Karmal, leader of Parcham PDPA, installed as head of Marxist regime.
- ❑ **1986** Najibullah replaces Karmal as head of government.
- ❑ **1989** Soviet Army withdraws. Najibullah remains in office.
- ❑ **1992** Najibullah hands over power to *mujahideen* factions.
- ❑ **1993** *Mujahideen* agree on formation of government.
- ❑ **1994** Power struggle between Rabbani and Hekmatyar.
- ❑ **1996** *Taliban* take power and impose strict Islamic regime.
- ❑ **1998** Earthquake in northern regions kills thousands.
- ❑ **1999** Power-sharing agreement between *taliban* and Northern Alliance breaks down.
- ❑ **2000** Worst drought in 30 years. UN imposes sanctions in response to *taliban* support for Osama bin Laden.
- ❑ **2001** *Taliban* government falls after intense US-led air strikes from October – first campaign in "war on terrorism." Interim government formed under Hamid Karzai; peacekeepers deployed in Kabul.
- ❑ **2002** Earthquakes kill thousands. Ex-king Zahir Shah returns from exile. Loya Jirga convenes, elects Karzai head of state.

AID

 Recipient

💲 $141m (receipts)   ⬇ Down 1% in 2000

The fall of the *taliban* regime opened the floodgates for massive inflows of aid to rebuild the country, notably from the EU, Japan, Saudi Arabia, and the US. However, relations between foreign aid agencies and the authorities can be fraught, and working conditions for aid personnel are extremely hazardous.

DEFENSE

 Compulsory military service

💲 $245m   ⬇ Down 2% in 2000

AFGHAN ARMED FORCES

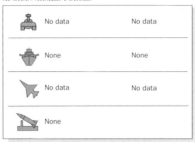

🛡	No data	No data
🚢	None	None
✈	No data	No data
🚀	None	

The new regime is establishing a national army; private armies have been banned. An international peacekeeping force maintains security in Kabul. Heavy ground offensives against the remaining al-Qaida and *taliban* forces were launched by the US in March 2002; hundreds were killed, including non-combatants.

The US–Russian agreement in 1991 to suspend military supplies to rival Afghan groups marked an end to active superpower involvement for a decade. The Kabul communists had been almost totally dependent on Moscow for arms, even after the Soviet military withdrawal in 1989. The covert arms trade expanded with the activity of Islamic militants from abroad. The bulk of these arms originate in eastern Europe and the former Soviet Union. The movement of Islamist militants and weapons between Tajikistan and Afghanistan is tackled by CIS troops.

Afghanistan holds hundreds of *Stinger* missiles given to the *mujahideen* by the US in the 1980s. The US, worried that they might be used against civilian airliners, has offered to buy them back, but none have so far been returned.

ECONOMICS

 Not available

📊 $5.87bn   💲 4750 afghanis

SCORE CARD

❑ World GNP Ranking	104th
❑ GNP per Capita	$270
❑ Balance of Payments	–$143m
❑ Inflation	56.7%
❑ Unemployment	8%

EXPORTS

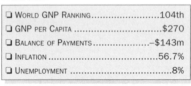

- Finland 7%
- India 8%
- UK 9%
- Belgium 13%
- Pakistan 27%
- Other 36%

IMPORTS

- Kenya 7%
- Japan 9%
- Kazakhstan 10%
- South Korea 15%
- Pakistan 21%
- Other 38%

STRENGTHS
Very few, apart from illicit opium trade. Agriculture still the largest sector. Overseas assets unfrozen from 2002.

WEAKNESSES
Decades of fighting; agriculture, industry, and infrastructure in ruins. Communication links damaged by earthquakes and devastated by bombing in 2001.

PROFILE
The protracted fighting has left Afghanistan one of the poorest and least developed countries in the world. Estimates suggest that $4 billion is needed to rebuild the country and that over 80% of infrastructure has been destroyed. Agricultural activity has fallen back from pre-1979 levels; the Soviets' "scorched earth" policy laid waste large areas, and much of the rural population fled to the cities. Many farmers turned back to growing poppies for opium production, but saw little profit from the trade, despite Afghanistan being one of the world's largest opium sources. Poppy cultivation was eventually banned by the *taliban*, and remains prohibited under the new regime.

ECONOMIC PERFORMANCE INDICATOR

Consumer Price Index — GDP

Consumer price index 1990=100

GDP unavailable

1988 1989 1990 1991 1992

Much of Afghanistan's *infrastructure, already scarred by decades of civil war, was reduced to ruins by heavy US bombardment in 2001.*

RESOURCES

 Electric power 494,000 kw

 1200 tonnes

 Not an oil producer and has no refineries

14.3m sheep, 2.2m goats, 1.5m cattle, 7.2m chickens

Natural gas, salt, coal, copper, lapis lazuli, barytes, talc

ELECTRICITY GENERATION

Hydro 65% (315m kwh)	
Combustion 35% (170m kwh)	
Nuclear 0%	
Other 0%	

% of total generation by type

Natural gas and coal are the most important strategic resources. In 2002 Afghanistan signed a gas agreement with Pakistan and Turkmenistan, which could generate revenue estimated at $300 million. Restoring the power

generation system is a government priority. The construction of dams on the Kunar and Laghman rivers is being considered. Coal production has fallen from prewar levels and mines are in urgent need of rehabilitation.

AFGHANISTAN : LAND USE

Cropland
Forest
Pasture
Desert
High mountain regions
Sheep
Cereals

ENVIRONMENT

 Not available

0.3% (0.2% partially protected)

0.05 tonnes per capita

ENVIRONMENTAL TREATIES

No   Yes   No

No   No   No

Environmental priorities are low, given Afghanistan's relatively anarchic conditions. However, the country's lack of industry, even in Kabul, means that industrial pollution is minimal. The biggest problem facing Afghanistan is land mines: over ten million have been laid, and the UN estimates that it will take 100 years to make the country safe for civilians.

MEDIA

 TV ownership low

Daily newspaper circulation 6 per 1000 people

PUBLISHING AND BROADCAST MEDIA

	Dailies such as *Anis*, *Hewaad* and the English-language *Kābul Times* now seek to build an independent post-*taliban* role
	1 state-owned service
	1 state-owned service, Radio Afghanistan, and several independent stations

Rural Afghan factions run newspapers and radio stations. Television and the Internet were banned under the *taliban*, as were video cassette recorders and satellite dishes, but regular television broadcasts restarted in November 2001 in Kabul. The BBC, which broadcasts in Pashtu and Dari, is popular, especially for its soap operas, which convey information on welfare issues.

CRIME

 Death penalty in use

Afghanistan does not publish prison figures

Levels of all crimes remain very high

CRIME RATES

No statistics for murders, rapes, or thefts are published due to the war situation

Gun law operates widely. Anyone traveling in rural areas is vulnerable to armed robbery. Cities vary according to which faction is dominant and the level of disruption which has been caused by war.

EDUCATION

 School leaving age: 13

37%

12,800 students

THE EDUCATION SYSTEM

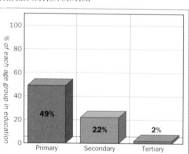

% of each age group in education

Primary 49%
Secondary 22%
Tertiary 2%

Under the *taliban* regime, education for women was strictly limited and segregation rigidly enforced; the literacy rate for women is now the lowest in the world. The Northern Alliance announced the lifting of restrictions on female education in November 2001.

Kabul University, which was closed in 1992, has been partially reopened.

HEALTH

 No welfare state health benefits

1 per 10,000 people

Infectious, parasitic, respiratory, and digestive diseases

The health service has collapsed completely and almost all medical professionals have left the country. Infant and maternal mortality rates are among the highest in the world, and life expectancy is very low.

Parasitic diseases and infections are a particular problem. The UN organized a program for the chlorination of well water, following an outbreak of cholera in Kabul, and launched a mass measles vaccination program in January 2002.

Under the *taliban* regime most women in Afghanistan had very little access to health care; their admission to hospital was strongly discouraged, as was the employment of female medical staff.

SPENDING

GDP/cap. decrease

CONSUMPTION AND SPENDING

0.1 per 1000 population

1 per 1000 population

Defense 13%	
Education 2%	
Health 1.6%	

Defense, Health, Education spending as % of GDP

The vast majority of Afghans live in conditions of extreme poverty. The country does not have the resources to feed its people at present – a situation exacerbated by the severe drought of 2000, the 2001 fighting, and the 2002 earthquakes. The return of refugees from neighboring Pakistan and Iran would make Afghanistan even more dependent on outside assistance for its rehabilitation.

A number of *mujahideen* leaders accumulated personal fortunes during the civil war. These derive in part from the substantial foreign aid that was once available and, in some cases, from the trafficking of opium.

WORLD RANKING

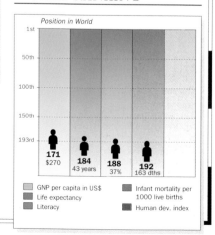

Position in World

1st
50th
100th
150th
193rd

171 $270
184 43 years
188 37%
192 163 dths

GNP per capita in US$
Life expectancy
Literacy
Infant mortality per 1000 live births
Human dev. index

ALBANIA

OFFICIAL NAME: Republic of Albania **CAPITAL:** Tirana
POPULATION: 3.1 million **CURRENCY:** Lek **OFFICIAL LANGUAGE:** Albanian

L YING AT THE SOUTHEASTERN end of the Adriatic Sea, opposite the heel of Italy, Albania is a poverty-stricken, mountainous country which became a one-party communist state in 1944. The "land of the eagles," as it is known locally, first held multiparty elections in 1991. Economic collapse provoked uprisings and virtual anarchy in 1997, and the country was only stabilized by the intervention of OSCE troops.

CLIMATE
▷ Mediterranean/continental

WEATHER CHART FOR TIRANA

The coastal climate is Mediterranean, but rather wet in winter. Heavy rain or snow falls in winter in the mountains.

TRANSPORTATION
▷ Drive on right

Tiranë (Rinas) • 34 ships 23,928 grt

THE TRANSPORTATION NETWORK

5400 km (3355 miles)	None
440 km (273 miles)	43 km (27 miles)

The transportation infrastructure is poor. The rail network is limited and roads are in disrepair. Private cars were first allowed in 1991. Buses and private vans are the main means of transportation.

TOURISM
▷ Visitors : Population 1:79

39,000 visitors • Up 39% in 1999

MAIN TOURIST ARRIVALS

Italy 12%
Egypt 8%
UK 7%
Other 73%

0 10 20 30 40 50 60 70 80 90 100
% of total arrivals

Instability and then the war in Kosovo stalled plans to exploit Albania's scenic beauty. Facilities remain very limited, especially outside Tirana.

PEOPLE
▷ Pop. density medium

Albanian, Greek • 113/km² (293/mi²)

THE URBAN/RURAL POPULATION SPLIT

42%   58%

RELIGIOUS PERSUASION

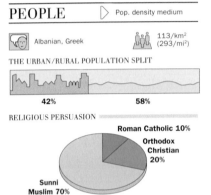

Roman Catholic 10%
Orthodox Christian 20%
Sunni Muslim 70%

The existence of ethnic minorities in Albania was only officially acknowledged in 1989. The Greek minority strongly contests statistics which state that 98% of the population are Albanian. Located mainly in the south and identifying with Athens rather than Tirana, the Greeks claim to make up 10% of the population. They suffer considerable discrimination.

Under communism, Albania was the only officially atheist state in the world. Many Albanians maintained their beliefs in private. Religious worship is now permitted and mosques have reopened. Society is traditional and male-dominated. The extended family remains a strong influence.

In 1999 the country temporarily sheltered nearly half a million ethnic Albanian refugees fleeing "ethnic cleansing" in neighboring Kosovo.

City of a thousand windows. Berat was preserved as a museum while a new town was built further down the valley.

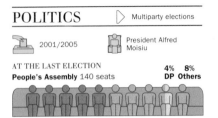

POLITICS
▷ Multiparty elections

2001/2005 • President Alfred Moisiu

AT THE LAST ELECTION
People's Assembly 140 seats

52% PSS   33% BF   3% PDS   4% DP   8% Others

PSS = Socialist Party of Albania **BF** = Union for Victory (led by the Democratic Party – **PD**) **DP** = Democrat Party (splinter from PD) **PDS** = Social Democratic Party

Albania was dominated for more than 40 years by communist ruler Enver Hoxha, who died in 1985. An exodus of Albanians in 1991 finally persuaded the regime to call multiparty elections. The resulting center-right coalition failed, however, to create a Western-style liberal state.

Many people were ruined in 1997 by the collapse of the "pyramid" savings schemes, prompting a rebellion in the south and forcing the government to resign. The new coalition led by the socialist PSS won elections later that year, and a further term in 2001, as Prime Minister Ilir Meta claimed credit for restoring a measure of hope and security. The poll, though disputed by the PD, was praised internationally as the fairest yet. In 2002 the PSS was wracked by internal disputes. By July, party leader Fatos Nano had won through to become prime minister.

WORLD AFFAIRS
▷ Joined UN in 1955

WTO CE OSCE OIC PfP

Foreign policy in the late 1990s was dominated by the fate of Kosovo, the predominately ethnic Albanian province in neighboring Serbia. Ethnic Albanian separatism also erupted in Macedonia in 2001, although Albania was not directly involved. Membership of NATO and the EU is a long-term goal. In 2001 an interim EU "stabilization and association agreement" was offered.

AID
▷ Recipient

$319m (receipts) • Down 35% in 2000

Since 1991 the West has provided aid. Food aid was stepped up in 1997, when anarchy swept the country, and again in 1999 to help cope with the hundreds of thousands of refugees arriving from Kosovo, in neighboring Serbia. EU aid now focuses on helping structural reforms.

A

DEFENSE

 Compulsory military service

 $111m

Down 21% in 2000

Officer ranks were reestablished in 1991, and a ten-year reconstruction program was launched in 2000. Military service of 18 months is mandatory. The Kosovo crisis prompted Albania in 1999 to make its airspace available to NATO.

ECONOMICS

 Inflation 39% p.a. (1990–2000)

 $3.83bn

142.75–136.80 lekë

SCORE CARD

- ❑ WORLD GNP RANKING 122nd
- ❑ GNP PER CAPITA $1120
- ❑ BALANCE OF PAYMENTS....................–$156m
- ❑ INFLATION 0.1%
- ❑ UNEMPLOYMENT................................16%

STRENGTHS

Oil and gas reserves. Significant economic growth achieved since 1997 collapse. Progress with privatization.

WEAKNESSES

Rudimentary public services and energy, transportation, and water networks. Pyramid schemes wiped out savings.

EXPORTS

IMPORTS

RESOURCES

 Electric power 1.9m kw

 3055 tonnes

7378 b/d (reserves 183m barrels)

1.94m sheep, 1.12m goats, 4m chickens

Chromium, oil, coal, natural gas, copper, nickel

Albania needs huge capital investment to develop its minerals and to create a modern electricity supply system.

ENVIRONMENT

 Sustainability rank: 24th

 3% (0.8% partially protected)

0.5 tonnes per capita

Toxic waste pollution from communist-era heavy industry is among the worst in Europe. Years of shortages mean that most materials are recycled.

MEDIA

 TV ownership medium

Daily newspaper circulation 37 per 1000 people

PUBLISHING AND BROADCAST MEDIA

 There are 4 daily newspapers, including *Rilindja Demokratike*, *Zëri i Popullit*, and *Koha Jonë*, the best-selling newspaper

 1 state-run service, 75 private stations

1 state-run service, 30 private stations

Media freedom is improving. A 1998 law banned political or religious control of TV stations. Newspaper sales are falling; some dailies are still party-run.

CRIME

 Death penalty not used in practice

 2922 prisoners

Down 6% in 1999

Lawlessness is widespread; guns are easily available after the anarchy of 1997. Cannabis is widely grown.

ALBANIA

Total Land Area : 28 748 sq. km (11 100 sq. miles)

POPULATION
- ◎ over 100 000
- ◉ over 50 000
- ● over 10 000
- • under 10 000

LAND HEIGHT
- 2000m/6562ft
- 1000m/3281ft
- 500m/1640ft
- 200m/656ft
- Sea Level

CHRONOLOGY

Albania gained independence in 1912 for the first time in its history.

- ❑ **1924–1939** Ahmet Zogu in power; crowned King Zog in 1928.
- ❑ **1939–1943** Occupied by Italy.
- ❑ **1944** Communist state; led by Enver Hoxha until 1985.
- ❑ **1991** First multiparty elections.
- ❑ **1997** Economic chaos as failure of pyramid schemes causes revolt.
- ❑ **1999** Refugee influx from Kosovo.
- ❑ **2001** PSS wins second term.

EDUCATION

 School leaving age: 14

85%

34,257 students

The system is derived from the Soviet, Chinese, and Italian models. Albania has eight universities.

HEALTH

Welfare state health benefits

1 per 769 people

Heart, respiratory, and digestive diseases, cancers

The health service is rudimentary, and dependent on Western aid for most drugs and medical supplies.

SPENDING

GDP/cap. increase

CONSUMPTION AND SPENDING

29 per 1000 population

39 per 1000 population

Defense 3%

Education 3.1%

Health 2%

Defense, Health, Education spending as % of GDP

Wealth is limited to a few private-sector entrepreneurs. Poverty is worst in northern rural areas but also acute in slum settlements around Tirana and other cities.

WORLD RANKING

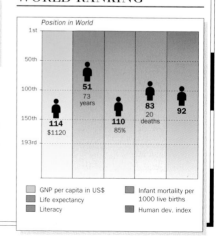

- GNP per capita in US$
- Life expectancy
- Literacy
- Infant mortality per 1000 live births
- Human dev. index

A

ALGERIA

OFFICIAL NAME: People's Democratic Republic of Algeria   **CAPITAL:** Algiers
POPULATION: 30.8 million   **CURRENCY:** Algerian dinar   **OFFICIAL LANGUAGES:** Arabic and Tamazight

1962 · 1962 · Nov 1 · DZ · +1 · +213 · .dz

AFRICA'S SECOND-LARGEST country, which extends from a densely populated Mediterranean coast to the empty northern Sahara, Algeria won independence from France in 1962. The military blocked radical Islamists from taking power after winning elections in 1991, setting up a new civilian regime and fighting a bloody terrorist conflict ever since. Algeria has one of the youngest populations, and highest birthrates, in the region.

CLIMATE

▷ Hot desert/ Mediterranean

WEATHER CHART FOR ALGIERS

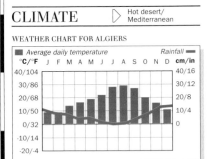

Coastal areas have a warm, temperate climate. The whole area to the south of the Atlas Mountains is hot desert.

TRANSPORTATION

▷ Drive on right

Houari Boumedienne, Algiers
2.78m passengers

142 ships
960,824 grt

THE TRANSPORTATION NETWORK

71,656 km (44,525 miles)	640 km (398 miles)
3973 km (2469 miles)	None

There are five international airports. Rail is the quickest way to travel between the main urban centers.

TOURISM

▷ Visitors : Population 1:36

866,000 visitors

Up 16% in 2000

MAIN TOURIST ARRIVALS

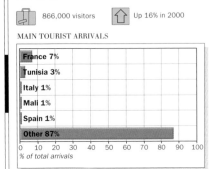

France 7%
Tunisia 3%
Italy 1%
Mali 1%
Spain 1%
Other 87%

% of total arrivals

The once-popular desert safaris are now rare. Tourists are a target for militant Islamist groups.

PEOPLE

▷ Pop. density low

Arabic, Tamazight (Kabyle, Shawia, Tamashek), French

13/km² (33/mi²)

THE URBAN/RURAL POPULATION SPLIT

60%   40%

RELIGIOUS PERSUASION

Christian and Jewish 1%
Sunni Muslim 99%

ETHNIC MAKEUP

European 1%
Berber 24%
Arab 75%

Algeria's population is predominantly Arab, under 30 years of age, and urban; around a quarter are Berber. More than 85% speak Arabic and 99% are Sunni Muslim. Mosques also provide social and medical services. Of the million or so French who settled in Algeria before independence, only about 6000 remain. Most Berbers consider the mountainous Kabylia region their homeland. Demonstrations there have met with violent police crackdowns, particularly in the Berber Spring of 1980, and since its anniversary in 2001. The Berber language, Tamazight, was recognized as an official language in 2002.

POPULATION AGE BREAKDOWN

Female	Age	Male
0.7%	80+	0.6%
2.4%	60–79	2.2%
6.3%	40–59	6.3%
15.3%	20–39	15.7%
24.7%	0–19	25.8%

% of population by age group

POLITICS

▷ Multiparty elections

L. House 2002/2007
U. House 2000/2003

President Abdelaziz Bouteflika

AT THE LAST ELECTION

National People's Assembly 389 seats

8% Ind   3% Others

51% FLN   12% RND   11% MRN   10% MSP   5% PT

FLN = National Liberation Front   **RND** = National Democratic Rally   **MRN** = Movement for National Reform   **MSP** = Movement for a Peaceful Society   **Ind** = Independents   **PT** = Workers' Party   **App** = Appointed   **FFS** = Front of Socialist Forces

Council of the Nation 144 seats

3% FFS

53% RND   33% App   9% FLN   2% MSP

96 seats are elected, 48 are appointed by the president

Algeria has a multiparty parliamentary system.

PROFILE

Until 1988, Algeria was a single-party socialist-style regime. The aging ruling elite then adopted privatization policies strongly opposed by radical Islamists. The Islamic Salvation Front (FIS) won elections in 1991, but the results were canceled. Since 1992 the country has been scarred by appalling violence and tens of thousands of people have been killed in a terrorist campaign and by ruthless state counterterrorism. Hopes for peace rose when Abdelaziz Bouteflika became president in 1999, although the election was marred by the withdrawal of all other candidates. Parliamentary elections in 2002, won by the FLN, also suffered from a low turnout and a partial opposition boycott. Violence continues, with an annual death toll which is still in the thousands.

MAIN POLITICAL ISSUES
Islamic fundamentalism

Radical Islamists want Algeria to be a theocracy. The imposition of military rule to prevent them taking office in 1992 unleashed violence spearheaded by the extremist Armed Islamic Group (GIA).

The market economy

The FIS's 1991 election victory was in part due to popular reaction against economic reforms. After a brief suspension following the army takeover in 1992, the liberalization program was revived under pressure from the IMF and the World Bank.

WORLD AFFAIRS ▷ Joined UN in 1962

Algeria's struggle for independence from France lasted from 1954 until 1962. Throughout the 1960s and 1970s, Algeria's success in rejecting a colonial power made it a champion for the developing world. It had a leading voice within the UN, the Arab League, and the Organization of African Unity. However, relations with the West remained essentially stable. Algeria was increasingly seen by the diplomatic community as a useful bridge between the West and Iran. In 1981, Algerian diplomats helped to secure the release of US hostages being held in Tehran. Algeria also attempted to act in a mediating role during the 1980–1988 Iran–Iraq War.

Algeria's influence overseas has diminished as the country has become increasingly unstable politically. A victory for the fundamentalist FIS in Algeria would greatly encourage Islamist militants in neighboring Morocco and Tunisia, and further undermine Egypt's embattled government.

France fears the spill-over of terrorism and has been shocked by the killings, especially those of seven French priests and of the French Roman Catholic bishop of Oran.

European governments are anxious to help stabilize the regime to avoid refugees seeking entry into France, Spain, and Italy.

AID ▷ Recipient

💲 $162m (receipts) ⬆ Up 82% in 2000

As a major oil producer, Algeria receives relatively small quantities of aid. During the 1980s, its economy became dependent on eastern European manufactures, which were swapped for oil. The collapse of this trade in the 1990s led Algeria to turn to the West for loans. The growing weight of Western economic involvement in turn fortified the regime against criticism of its hard-line methods against Islamic opponents. The IMF has provided loans to help Algeria meet payments on its debt, on condition that it move toward a market-oriented economy.

ALGERIA

Total Land Area :
2 381 740 sq. km
(919 590 sq. miles)

POPULATION

over 500 000	◉
over 100 000	◎
over 50 000	○
over 10 000	●
under 10 000	•

LAND HEIGHT

2000m/6562ft
1000m/3281ft
500m/1640ft
200m/656ft
Sea Level

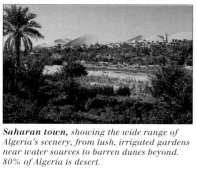
Saharan town, *showing the wide range of Algeria's scenery, from lush, irrigated gardens near water sources to barren dunes beyond. 80% of Algeria is desert.*

Abdelaziz Bouteflika, *who was elected president in 1999.*

Abassi Madani, *leader of the Islamic Salvation Front (FIS).*

CHRONOLOGY

The conquest of Algeria by France began in 1830. By 1900, French settlers occupied most of the best land. In 1954, war was declared on the colonial administration by the National Liberation Front (FLN).

❏ **1962** Cease-fire agreed, followed by independence of Algerian republic.

❏ **1965** Military junta topples government of Ahmed Ben Bella. Revolutionary council set up.

❏ **1966** Judiciary "Algerianized." Tribunals try "economic crimes."

❏ **1971** Oil industry nationalized. President Boumedienne continues with land reform, a national health service, and "socialist" management.

❏ **1976** National Charter establishes a socialist state.

❏ **1980** Ben Bella released after 15 years' detention. Agreement with France whereby latter gives incentives for return home of 800,000 Algerian immigrants.

❏ **1981** Algeria helps to negotiate release of hostages from US embassy in Tehran, Iran.

❏ **1985** Two most popular Kabyle (Berber) singers given three-year jail sentences for opposing regime.

❏ **1987** Limited economic liberalization. Cooperation agreement with Soviet Union.

❏ **1988** Anti-FLN violence; state of emergency. Algeria negotiates release of Kuwaiti hostages from aircraft; Shi'a hijackers escape.

❏ **1989** Constitutional reforms diminish power of FLN. New political parties founded, including Islamic Salvation Front (FIS). AMU established.

❏ **1990** Political exiles permitted to return. FIS is victorious in municipal elections.

❏ **1991** FIS leaders Abassi Madani and Ali Belhadj arrested. FIS wins most seats in National People's Assembly.

❏ **1992** President Chadli overthrown by military. President Boudiaf assassinated. Madani and Belhadj given 12 years in jail.

❏ **1994** Political violence led by GIA.

❏ **1995** Democratic presidential elections won by Liamine Zéroual.

❏ **1996** Murders continue, notably of Catholic clergy and GIA leader.

❏ **1997** Madani released from jail but debarred from active politics.

❏ **1999** Abdelaziz Bouteflika elected president in poll boycotted by opposition candidates.

❏ **2001** Fresh investment in oil and gas benefits economy. Resurgence of Berber protests.

❏ **2002** Berber language Tamazight recognized as national language. FLN election victory.

DEFENSE

 ▷ Compulsory military service

 $2.93bn ⬇ Down 5% in 2000

The National Liberation Army (NLA), equipped with Russian weapons, is the dominant power in politics. There have been fears that parts of the army would forge an alliance with Muslim militants; the extreme rebel Armed Islamic Group, which has split from the FIS, is led by former army officers. However, the military are also suspected of taking part in reprisal killings of large numbers of Islamists.

ALGERIAN ARMED FORCES

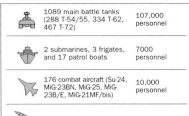

1089 main battle tanks (288 T-54/55, 334 T-62, 467 T-72)	107,000 personnel
2 submarines, 3 frigates, and 17 patrol boats	7000 personnel
176 combat aircraft (Su-24, MiG-23BN, MiG-25, MiG-23B/E, MiG-21MF/bis)	10,000 personnel
None	

ECONOMICS

 ▷ Inflation 18% p.a. (1990–2000)

 $47.9bn 73.42–76.86 Algerian dinars

SCORE CARD

❏ WORLD GNP RANKING..........................50th
❏ GNP PER CAPITA $1580
❏ BALANCE OF PAYMENTS.....................$8.9bn
❏ INFLATION 2.6%
❏ UNEMPLOYMENT..................................30%

EXPORTS

Brazil 7%
Other 35%
France 11%
Spain 12%
USA 12%
Italy 23%

IMPORTS

Spain 5%
Germany 6%
Italy 9%
Other 39%
USA 10%
France 31%

ECONOMIC PERFORMANCE INDICATOR

Consumer Price Index — GDP

Consumer price index 1995=100 / GDP 1996=100

1996 1997 1998 1999 2000

STRENGTHS

Oil and gas. Recent collaboration with Western oil companies should see improvements in productivity. Natural gas is supplied to Europe, with plans in hand for the construction of a third undersea pipeline.

WEAKNESSES

Political turmoil threatens many new projects and has led to an exodus of European and other expatriate workers important to the economy. Lack of skilled labor coupled with high unemployment. Limited agriculture. Shortages of basic foodstuffs. Thriving black market.

PROFILE

Centralized socialist planning dominated the Algerian economy until the late 1980s, when the economic collapse of the Soviet Union led to a change in policy, and Algeria began moving toward a market economy. Reforms were frozen, however, following the military takeover in 1992, although many have since been resumed under pressure from the IMF and the World Bank. The majority of the economy's most productive sectors remain under state control, although private investment is encouraged in the oil industry and, since early 2001, in telecommunications. A number of Western oil companies have signed exploration contracts with Algeria since it has accepted more competitive production-sharing agreements. However, Western investment levels are likely to remain small as long as the political situation is unstable.

ALGERIA : MAJOR BUSINESSES

Algiers
Constantine
Oran
Annaba
Hassi R'Mel
Hassi Messaoud

♠ Oil/gas
❀ Textiles
⬡ Iron & steel
✿ Heavy engineering
◎ Light engineering
♠ Chemicals/petrochemicals
▯ Food processing

0   200 km
0   200 miles

RESOURCES

 Electric power 6m kw

 105,943 tonnes

 1.56m b/d (reserves 9.2bn barrels)

19.3m sheep, 3.5m goats, 1.7m cattle, 110m chickens

Oil, natural gas, iron, phosphates, lead, zinc, silver, copper, gold

ELECTRICITY GENERATION

- Hydro 0%
- Combustion 100% (24bn kwh)
- Nuclear 0%
- Other 0%

% of total generation by type

Crude oil and natural gas, Algeria's main resources, have been produced since the 1950s. Algeria also has diverse minerals, including iron ore, zinc, silver, copper ore, lead, gold, and phosphates. In the 1960s and 1970s, Algeria sought to become a major manufacturer, with investments in building materials, refined products, and steel; none of these sectors is competitive on world markets. Agriculture employs one-quarter of Algeria's workforce, but its importance to the economy is diminishing. State forests cover some 2% of Algeria's land. Most are brushwood, but some areas include cork oak trees, Aleppo pine, evergreen oak, and cedar. Algeria has a large fishing fleet. Sardines, anchovies, tuna, and shellfish are the major species caught commercially.

ENVIRONMENT

 Sustainability rank: 70th

 3% (0.1% partially protected)

3.6 tonnes per capita

ENVIRONMENTAL TREATIES

Yes	Yes	Yes
Yes	Yes	No

Since most of Algeria is desert or semidesert, over 90% of the population is forced to live on what remains – some 20% of the land. The desert is moving northward. Vegetation has been stripped for use as firewood and animal fodder, leaving fragile soils exposed which then require expensive specialist care to conserve them. Techniques for water purification are substandard, and rivers are being increasingly contaminated by untreated sewage, industrial effluent, and wastes from petroleum refining.

MEDIA

 TV ownership medium

Daily newspaper circulation 38 per 1000 people

PUBLISHING AND BROADCAST MEDIA

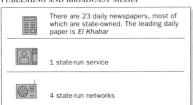

There are 23 daily newspapers, most of which are state-owned. The leading daily paper is *El Khabar*

1 state-run service

4 state-run networks

Newspapers, TV, and radio are state-controlled and permit no criticism of government actions. TV is broadcast in Arabic, French, and Tamazight, received by about two million sets. The five main daily newspapers have a combined circulation of 1.3 million. However, distribution is limited outside the major cities.

CRIME

 Death penalty in use

 35,737 prisoners

Down 70% 1993–1999

CRIME RATES

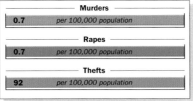

Murders
0.7 per 100,000 population

Rapes
0.7 per 100,000 population

Thefts
92 per 100,000 population

Thousands of people have been killed by radical Islamists since 1992, while human rights groups have accused pro-government death squads of brutal reprisal killings and of persecuting suspected Islamist militants.

EDUCATION

 School leaving age: 15

 68%

347,410 students

THE EDUCATION SYSTEM

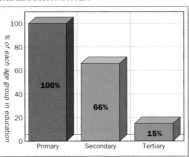

% of each age group in education

Primary	Secondary	Tertiary
100%	66%	15%

Over three-quarters of the school-age population receive a formal education, and the literacy rate is rising.

Since 1973, the curriculum has been Arabicized and the teaching of French has been restricted. New legislation in 1996 enforced the use of Arabic in public life, including for teaching.

Ten universities, seven polytechnics, and several technical colleges provide higher education.

ALGERIA : LAND USE

S A H A R A

- Forest
- Pasture
- Cropland
- Desert
- Sheep
- Potatoes
- Dates

0   200 km
0   200 miles

HEALTH

Welfare state health benefits

1 per 1000 people

Respiratory, heart, and cerebrovascular diseases, malaria

Since 1974 all Algerians have had the right to free health care. Primary health care is rudimentary outside main cities. Because the formal health care system is overburdened, many people turn to alternative forms of medicine. The infant mortality rate is well below the average for north Africa, and life expectancy is just above the regional average.

SPENDING

GDP/cap. increase

CONSUMPTION AND SPENDING

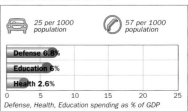

25 per 1000 population

57 per 1000 population

- Defense 6.8%
- Education 6%
- Health 2.6%

Defense, Health, Education spending as % of GDP

There is great disparity in wealth between the political elite and the rest of the population. Those with connections in the military are the wealthiest group. Most Algerians have had to contend with soaring prices for basic necessities.

WORLD RANKING

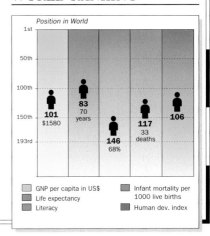

Position in World

- 101 $1580
- 83 70 years
- 146 68%
- 117 33 deaths
- 106

- GNP per capita in US$
- Life expectancy
- Literacy
- Infant mortality per 1000 live births
- Human dev. index

ANDORRA

OFFICIAL NAME: Principality of Andorra   **CAPITAL:** Andorra la Vella
POPULATION: 66,800   **CURRENCY:** Euro (French franc & Spanish peseta until 2002)   **OFFICIAL LANGUAGE:** Catalan

A TINY, LANDLOCKED principality between France and Spain, Andorra lies high in the eastern Pyrenees. From the 13th century, French and Spanish co-princes (today the President of France and the Bishop of Urgell) have ruled Andorra. In December 1993, the principality held its first full elections. Andorra's spectacular scenery, alpine climate, and duty-free shopping have made tourism, especially skiing, its main source of income.

Andorra's outstanding mountain scenery attracts skiers in winter, walkers in summer.

CLIMATE   ▷ Mountain

WEATHER CHART FOR ANDORRA LA VELLA

Springs are cool and wettest in May; summers are relatively dry and warm. Snowfalls in December and January provide snow for good skiing up to March. Andorra's climate supports an abundance of wild flowers.

TRANSPORTATION   ▷ Drive on right

✈ None   🚢 Has no fleet

THE TRANSPORTATION NETWORK

198 km (123 miles)   None

None   None

The road from France to Spain climbs to 2704 m (8875 ft) through one of the most dramatic mountain passes in Europe. Traffic congestion is a major problem in Andorra la Vella, especially in the summer. In 2001 plans for an overhead rail system were announced.

TOURISM   ▷ Visitors : Population 44:1

🧳 2.95m visitors   ⬆ Up 26% in 2000

MAIN TOURIST ARRIVALS

Spain 68%
France 26%
Other 6%

0 10 20 30 40 50 60 70 80 90 100
% of total arrivals

Most tourists visit Andorra to ski or shop. There is a traditional trade in day trippers from France and Spain, drawn by the many tax-free designer-label boutiques.

Five resorts offer Alpine skiing facilities, and specialize in Nordic skiing. In summer they cater instead for mountain hikers; Andorra's wide variety of wild flowers attract many visitors, but there is also much for the birdwatcher to see.

Although not strongly promoted, hunting of wild boar is popular, and the goat-like chamois can be hunted under special license.

PEOPLE   ▷ Pop. density medium

Spanish, Catalan, French, Portuguese   144/km² (372/mi²)

THE URBAN/RURAL POPULATION SPLIT

63%   37%

RELIGIOUS PERSUASION

Other 6%
Roman Catholic 94%

Immigration is restricted to French and Spanish nationals intending to work in Andorra. Divorce is illegal and the marriage rate is low.

POLITICS   ▷ Multiparty elections

2001/2005   Co-Princes Jacques Chirac and Joan Martí Alanis

AT THE LAST ELECTION

General Council of the Valleys 28 seats

54% PLA   21% PSD   18% PD   7% UL

PLA = Liberal Party of Andorra   **PSD** = Social Democratic Party   **PD** = Democratic Party   **UL** = Unió Laurediana

14 members are elected on a national list and 14 are elected in 7 dual-member parishes.

Andorra was a semifeudal state until 1993, when a referendum approved measures which legalized political parties and the right to strike, and altered relations with the co-princes. The ruling PLA, led by Marc Forné, has twice been returned to power.

ANDORRA

Total Land Area : 468 sq. km (181 sq. miles)

N

POPULATION
● over 10 000
• under 10 000

LAND HEIGHT

2000m/6562ft
1500m/4921ft
1000m/3281ft
above 800m

0   5 km
0   5 miles

WORLD AFFAIRS

 Joined UN in 1993

 CE OSCE

In 1991 Andorra became a member of the EU customs union and adopted the euro as its official currency in 2002. It joined the UN in 1993. Andorra's status as a tax haven has prompted criticism from the OECD.

AID

 Not applicable

 Andorra has no aid receipts or donations

Not applicable

The principality of Andorra neither receives nor provides aid, and has no plans to do so.

DEFENSE

No compulsory military service

 Andorra has no defense budget

Not applicable

Andorra has no defense budget; France and Spain provide protection. The last military action was intervention by French *gendarmes* to restore order after a royalist coup in 1933.

ECONOMICS

Not available

$1.02bn   1.0651–1.1231 euros

SCORE CARD

❏ WORLD GNP RANKING	154th
❏ GNP PER CAPITA	$15,600
❏ BALANCE OF PAYMENTS	Included in Spanish total
❏ INFLATION	1.62%
❏ UNEMPLOYMENT	Low unemployment

STRENGTHS

Tourism is the basis of the economy. Strict banking secrecy laws make Andorra an important tax haven. Healthy luxury retail sector. Farming: cereals, potatoes, and tobacco are the major products.

WEAKNESSES

France and Spain effectively decide economic policy. Dependence on imported food and raw materials.

EXPORTS

Greece 2%
Germany 2%
Hong Kong 2%
Other 7%
France 26%
Spain 61%

IMPORTS

Japan 3%
Germany 4%
Italy 4%
Other 13%
Spain 49%
France 27%

RESOURCES

 Not available

 None

Not an oil producer and has no refineries

1586 sheep, 1181 cattle, 1037 horses, 854 goats

None

Water is a major resource, hydropower providing most energy needs. However, Andorra has to import twice as much electricity as it produces, and plans to develop wind power. A third of the country is designated forest.

ENVIRONMENT

 Not available

 None

Not available

The impact of millions of visitors each year on Andorra's alpine environment is of great concern. The development of hotels, ski resorts, and transportation links threaten to despoil the country's picturesque mountain landscape. It is also endangering the remarkable flora, and creates pressure to clear forested areas.

Hunting is no longer promoted but remains a popular attraction; the wild boar and the Pyrenean chamois are particularly targeted. Some restrictions have been introduced to preserve rarer animal species.

MEDIA

 TV ownership high

 Daily newspaper circulation 60 per 1000 people

PUBLISHING AND BROADCAST MEDIA

There are 2 daily newspapers, *Diari d'Andorra* and *El Periódic d'Andorra*

1 independent commercial channel

6 independent commercial stations

Andorra has one full domestic station, Radio i Televisio d'Andorra, and receives TV broadcasts from France and Spain. A Spanish TV company broadcasts one hour a day of programs for Andorra.

CRIME

 No death penalty

 20 prisoners

Little change in 1999

Tourists are natural targets for thieves, most of whom are not Andorran. Thefts of expensive cars for resale in France and Spain are on the increase.

Andorra's two criminal courts are known as the *Tribunals de Corts*.

EDUCATION

 School leaving age: 14

 99%   3186 students

There are around 30 schools in Andorra, with instruction in Catalan, French, and Spanish. The University of Andorra specializes in distance learning using the Internet.

CHRONOLOGY

Since 1278, Andorra has been autonomous, ruled by French and Spanish co-princes.

- ❏ **1970** Women get the vote.
- ❏ **1982** First constitution enshrines popular sovereignty.
- ❏ **1983** General Council votes in favor of income tax.
- ❏ **1984** Government resigns over attempt to introduce indirect taxes.
- ❏ **1991** EU customs union comes into effect.
- ❏ **1992** Political demonstrations demanding constitutional reform. Government resigns.
- ❏ **1993** Referendum approves new constitution.
- ❏ **1994** Government falls; replaced by center-right Liberal cabinet, which is reelected in 1997 and 2001.

HEALTH

Welfare state health benefits

1 per 455 people

Heart and cerebrovascular diseases

Andorra has one public hospital. Health spas are popular, and the hot springs at les Escaldes attract rheumatism sufferers.

SPENDING

GDP/cap. increase

CONSUMPTION AND SPENDING

529 per 1000 population

515 per 1000 population

Defense None
Education No data
Health 10.6%

0   5   10   15   20   25
Defense, Health, Education spending as % of GDP

Hotel owners are the wealthiest group of citizens in Andorran society; many choose to live across the border in neighboring Spain.

WORLD RANKING

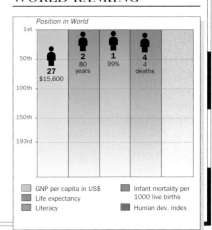

Position in World

1st
50th
27
$15,600
2
80 years
1
99%
4
4 deaths
100th
150th
193rd

GNP per capita in US$
Life expectancy
Literacy
Infant mortality per 1000 live births
Human dev. index

ANGOLA

SOUTHERN AFRICA   Africa

OFFICIAL NAME: Republic of Angola   **CAPITAL:** Luanda
POPULATION: 13.5 million   **CURRENCY:** Readjusted kwanza   **OFFICIAL LANGUAGE:** Portuguese

AN OIL- AND DIAMOND-RICH country in southwest Africa, Angola has suffered almost continuous civil war since independence from Portugal in 1975. During the Cold War the West supported UNITA rebels against the Soviet-backed MPLA government. After many failed peace initiatives, the latest, in 2002, has raised hopes yet again of a more permanent end to the violence.

Angola's capital, Luanda. Founded in 1575 by the Portuguese, it became a transshipment point for slaves en route to Brazil.

CLIMATE   ▷ Tropical/steppe

WEATHER CHART FOR LUANDA

The climate varies from temperate to tropical. Rainfall decreases from north to south. The Benguela Current makes the coast unusually cool and dry.

TRANSPORTATION   ▷ Drive on right

Luanda International
766,077 passengers

125 ships
66,335 grt

THE TRANSPORTATION NETWORK

19,156 km (11,903 miles)	Much of this infrastructure has been destroyed by civil war.
2952 km (1834 miles)	1295 km (805 miles)

War has destroyed infrastructure, restricted movement of people and goods, and devastated port traffic. UN peacekeepers have tried to clear mines and repair roads, bridges, and railroads.

TOURISM   ▷ Visitors : Population 1:265

51,000 visitors   Up 13% in 2000

MAIN TOURIST ARRIVALS

Portugal 34%	
France 8%	
USA 6%	
Other 52%	

0   10   20   30   40   50   60
% of total arrivals

Since independence Angola has been a war zone. Until the late 1990s the few visitors were journalists or employees of the oil multinationals in Cabinda.

PEOPLE   ▷ Pop. density low

Portuguese, Umbundu, Kimbundu, Kikongo

11/km² (28/mi²)

THE URBAN/RURAL POPULATION SPLIT

34%   66%

ETHNIC MAKEUP

Ovimbundu 37%
Bakongo 13%
Other 25%
Kimbundu 25%

UNITA gave itself the role of sole representative of the predominantly rural-dwelling Ovimbundu, in order to attack the mainly urban-based, largely Kimbundu or mixed race (Portuguese–African) MPLA. Religion has revived since the MPLA abandoned its Marxist tenets in the 1980s. Some 20% of the population are internal refugees.

POLITICS   ▷ In transition

1992/1998 (postponed)

President José Eduardo dos Santos

AT THE LAST ELECTION
National Assembly 223 seats

3% 1% 3%
PRS PLD Others

58% MPLA–PT   32% UNITA   2% FNLA   1% Vacant

MPLA–PT = Popular Movement for the Liberation of Angola–Workers' Party   **UNITA** = National Union for the Total Independence of Angola   **PRS** = Social Renovated Party   **FNLA** = Angolan National Liberation Front   **PLD** = Liberal Democratic Party
The seats allotted to members from abroad remained vacant

In power since 1975, the MPLA abandoned one-party rule in 1991 and won Angola's first multiparty elections in 1992. Jonas Savimbi's defeated UNITA responded by reopening the civil war. The rivals signed a peace protocol in Lusaka, Zambia, in 1994, which stipulated the formation of a government of national unity (GURN). UNITA ministers eventually joined the GURN in 1997, but were expelled in 1998 as fighting escalated. Savimbi's death in February 2002 led to a renewed peace initiative, and a "definitive cease-fire" in April.

ANGOLA

Total Land Area : 1 246 700 sq. km (481 351 sq. miles)

POPULATION
over 1 000 000
over 100 000
over 50 000
over 10 000
under 10 000

LAND HEIGHT
2000m/6562ft
1000m/3281ft
500m/1640ft
200m/656ft
Sea Level

0   200 km
0   200 miles

A

WORLD AFFAIRS

▷ Joined UN in 1976

Angola was a key Cold War frontier in the 1980s, but after 1992 UNITA lost international support. Post-apartheid South Africa joined the US, Russia, and Portugal as guarantors of Angola's

peace process after 1994. Relations with Zambia are tense, as fighting has on occasion spilled across the border. Controls on the diamond trade were tightened in early 2000 in support of UN trade sanctions on UNITA-held regions. UNITA moved to secure territory and influence in the DRC in 2001.

AID

▷ Recipient

 $307m (receipts)   Down 21% in 2000

Donors from the World Bank, the EU, the USA, Japan, and Norway pledged aid in the mid-1990s to rebuild the

shattered economy. However, resumed fighting strained their resources. Attacks on aid convoys prompted many NGOs to leave the country altogether. UN humanitarian flights ceased in 1999.

DEFENSE

▷ No compulsory military service

 $1.25bn   Up 24% in 2000

By 2002 government forces had asserted control over more than 90% of the country. Under the cease-fire, 5000 UNITA troops were integrated into the regular army. The remaining 80,000 were demobilized. The government has agreed to ratify the 1997 Ottawa treaty on the elimination of antipersonnel mines.

ECONOMICS

▷ Inflation 741% p.a. (1990–2000)

$3.85bn   16.70–31.86 readjusted kwanza

SCORE CARD

❑ World GNP Ranking	121st
❑ GNP per Capita	$290
❑ Balance of Payments	–$4m
❑ Inflation	325%
❑ Unemployment	50%

STRENGTHS
Oil an important foreign exchange earner; hopes for advanced offshore drilling. Some of the richest mineral deposits in Africa.

WEAKNESSES
Severe drought. Civil war: destruction of infrastructure. Ten million land mines laid nationwide.

EXPORTS

IMPORTS

RESOURCES

▷ Electric power 462,000 kw

 177,497 tonnes   731,000 b/d (reserves 5.4bn barrels)

4.04m cattle, 2.15m goats, 800,000 pigs, 6.8m chickens   Oil, diamonds, iron, copper, lead, zinc, gold, manganese

Deepwater oil fields have been found. The rich alluvial diamond deposits were controlled by UNITA during the civil war.

ENVIRONMENT

▷ Sustainability rank: 111th

 7% (1% partially protected)   0.5 tonnes per capita

War has damaged the water supply. UNITA has been accused of widescale ivory poaching.

MEDIA

▷ TV ownership low

 Daily newspaper circulation 11 per 1000 people.

PUBLISHING AND BROADCAST MEDIA

There are 2 daily newspapers, *O Jornal de Angola* and *Diário da República*

1 state-controlled service   3 services: 1 state-owned, 2 independent

Many foreign journalists have been arrested. UN officials accuse the MPLA of spreading "hostile war propaganda."

CRIME

▷ No death penalty

 4884 prisoners   Up 92% 1997–1999

Murder, theft, corruption, and diamond smuggling are commonplace. Rural areas are effectively controlled by gangs. Both the MPLA and UNITA have poor human rights records. Global companies assisted UNITA in sanctions-busting.

EDUCATION

▷ School leaving age: 10

40%   8337 students

A government-backed initiative, Adra, is making progress in reviving schooling in cities.

CHRONOLOGY

The Portuguese first established coastal forts in 1482.

- ❑ **1975** Independence. Civil war between MPLA and UNITA.
- ❑ **1979** José Eduardo dos Santos (MPLA) becomes president.
- ❑ **1991** UN-brokered peace.
- ❑ **1992** MPLA election victory provokes UNITA to resume fighting.
- ❑ **1994** Lusaka peace agreement.
- ❑ **1998** Civil war reerupts.
- ❑ **2000** Fighting spreads as UNITA increases guerrilla activity.
- ❑ **2002** UNITA leader Jonas Savimbi killed. April, cease-fire signed.

HEALTH

▷ No welfare state health benefits

 1 per 10,000 people   Malaria, diarrheal and respiratory diseases, severe malnutrition

Angola's health system is barely able to cope with casualties of war and the threat of epidemics. Angola has the greatest number of amputees (caused by land mines) in the world.

Immunization of three million children against polio began in 2002.

SPENDING

▷ GDP/cap. increase

CONSUMPTION AND SPENDING

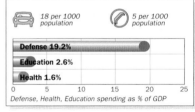

18 per 1000 population   5 per 1000 population

Defense 19.2%
Education 2.6%
Health 1.6%

Defense, Health, Education spending as % of GDP

State officials enjoy various luxuries, such as access to cars and other consumer goods, while the majority struggles to survive. The MPLA accuses its own generals of illicit diamond mining.

WORLD RANKING

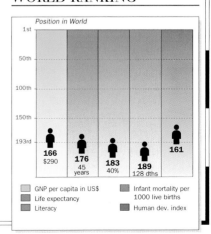

Position in World

166 $290	
176 45 years	
183 40%	
189 128 dths	
161	

- ☐ GNP per capita in US$
- ☐ Life expectancy
- ☐ Literacy
- ☐ Infant mortality per 1000 live births
- ☐ Human dev. index

A

ANTARCTICA

OFFICIAL NAME: Antarctica CAPITAL: None
POPULATION: None CURRENCY: None OFFICIAL LANGUAGE: None

T HE WORLD'S fifth-largest continent, Antarctica is almost entirely covered by ice over 2000 m (6560 ft) thick. The area sustains a varied wildlife, including seals, whales, and penguins. The Antarctic Treaty, signed in 1959 and in force in 1961, provides for international governance of Antarctica. To gain Consultative Status, countries have to set up a program of scientific research into the continent. Following a 1994 international agreement, a whale sanctuary was established around Antarctica.

CLIMATE ▷ Freezing

WEATHER CHART

Antarctica is the windiest as well as the coldest continent. Powerful winds create a narrow storm belt around the continent, which brings cloud, fog, and severe blizzards. Icebergs barricade more than 90% of the coastline, and climate change has seen an increase in their number and size in recent years. Antarctica contains over 80% of the world's fresh water.

TRANSPORTATION ▷ Not applicable

 Airstrips to some stations

 Has no fleet

Ships are the main mode of transportation to Antarctica. They are also used for marine research projects. Air traffic from Chile is growing, and France and the UK are building new airstrips. Most planes have to be equipped with skis.

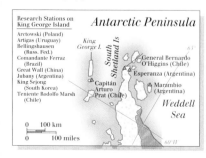

Research Stations on King George Island
Arctowski (Poland)
Artigas (Uruguay)
Bellingshausen (Russ. Fed.)
Comandante Ferraz (Brazil)
Great Wall (China)
Jubany (Argentina)
King Sejong (South Korea)
Teniente Rodolfo Marsh (Chile)

Antarctic Peninsula

TOURISM ▷ Not applicable

🧳 9400 visitors

⇕ No change in 1999

Tourism is mainly by cruise ship to the Antarctic Peninsula, the Ross Sea, and the sub-Antarctic islands. In 1983, the Chileans began flights to King George Island, where an 80-bed hotel has been built. Main attractions are the wildlife, skiing, and visits to scientific stations and historic huts. The growth of tourism has disrupted scientific programs and official regulation of tourism is now essential.

PEOPLE ▷ Not applicable

English, Spanish, French, Norwegian, Chinese, Polish, Russian, German, Japanese

Not applicable

ETHNIC MAKEUP

Antarctica has a transient population of Americans, English, French, Norwegians, Argentinians, Chileans, Chinese, Russians, Poles, and Japanese. Most are involved in research. Few stay more than two years.

Antarctica has no indigenous population. Around 80 Chilean settlers live at any one time in the continent's only permanent community on King George Island. The rest of the population are scientists and logistical staff working at the 40 permanent, and as many as 100 temporary, research stations. Most stations are too far apart for direct contact between different nationalities.

ANTARCTICA

Total Land Area : 14 000 000 sq. km (5 405 000 sq. miles)

△ Research station

▨ Permanent Sea Ice

☐ Ice Cap

TERRITORIAL CLAIMS

△ Australian claim
△ Argentinian claim
△ British claim
△ Chilean claim
△ French claim
△ New Zealand claim
△ Norwegian claim
△ Brazilian zone of interest

A

Neumayer Channel, Antarctica. *Many states are pressing for the whole of Antarctica to be protected as an international park.*

POLITICS Not applicable

 Not applicable Consultative Parties to Antarctic Treaty

NO LEGISLATIVE OR ADVISORY BODIES

The Antarctic Treaty of 1959 was signed by 12 countries. Consultative meetings are held most years to discuss scientific, environmental, and political matters.

There are 27 parties to the Antarctic Treaty and 17 nations with observer status. There are territorial claims by Australia, France, New Zealand, and Norway, and overlapping claims in the Antarctic Peninsula by Argentina, Chile, and the UK. Other states do not recognize these claims.

Of main concern is the adoption of a wide range of environmental protection measures. Proposals include the monitoring of all scientific activities and also the prosecution of any country if it were demonstrated that its research would lead to detrimental global change.

WORLD AFFAIRS Not a UN member

Rivalries exist between nations wishing to preserve Antarctica as a world park and those pursuing territorial claims.

AID Recipient

 Research is funded by governments Subject to individual government budgets

Scientific programs in the Antarctic are almost entirely funded by government agencies in the home countries. Funding is occasionally provided by scientific institutions and universities.

DEFENSE Not applicable

 No defense force Not applicable

Under the Antarctic Treaty, Antarctica can be used only for peaceful purposes. Any military personnel present perform purely scientific or logistic roles.

ECONOMICS Not applicable

 Not applicable Antarctica has no currency

Research is government-funded and is therefore subject to reductions. The exploitation of marine stocks provides no income for Antarctica.

RESOURCES Not applicable

 Included in national fish catch totals 　 Not an oil producer and has no refineries

None 　 Mineral extraction is banned

Antarctica's main resources are its marine stocks, including fin fish, seals, and whales. A campaign by environmental groups, supported by Australia and France, to ban mining and declare Antarctica a world park was rewarded with an agreement in 1991 to impose a 50-year ban on mining, and in 1994 by the approval of a whale sanctuary. Prospects for energy sources alternative to fossil fuels, such as solar power and wind generators, are being explored.

ENVIRONMENT Not applicable

Most of Antarctica is protected 　 Not available

Antarctica is one of the Earth's last great wildernesses. Its layer of ice, 4000 m (13,120 ft) thick in places, has formed over thousands of years. Its ecosystem is such that a "footprint" will leave its mark for many years. Several species are unique to the continent, including king penguins. The blood of polar fish contains antifreeze agents. Ecological concerns include overfishing, particularly of krill, cod, and squid; the disintegration of ice shelves; the depletion of the ozone layer over Antarctica; and the various knock-on effects of global warming. In 1994 the IWC agreed to a French proposal to create an Antarctic whale sanctuary which, together with the Indian Ocean sanctuary, protects the feeding grounds of 90% of the world's whales.

MEDIA 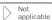 Not applicable

There are no daily newspapers produced in Antarctica

A few bases publish newssheets for local consumption. Local radio stations are found at some of the larger bases.

CRIME Not applicable

 There are no prisons in Antarctica 　 Crime is negligible

Each person in Antarctica is subject to their national laws. Occasional petty theft from stations is linked to visits from tourists.

CHRONOLOGY

The Russian explorer, Thaddeus von Bellingshausen, was the first to sight Antarctica, in 1820. The South Pole was first reached by the Norwegian, Roald Amundsen, in December 1911.

❑ **1957–1958** International Geophysical Year launches scientific exploration of Antarctica.
❑ **1959** Antarctic Treaty signed by 12 countries. Territorial claims frozen.
❑ **1978** Convention limiting seal hunting comes into force.
❑ **1985** Ozone depletion disclosed.
❑ **1994** Establishment of Antarctic whale sanctuary.
❑ **1998** Agreement on 50-year ban on mineral extraction comes into force.

EDUCATION Not applicable

Not applicable 　 None

Schoolhouses exist on the Chilean base, Villa Las Estrellas, and the Argentinian base, Esperanza. Teaching is geared to the relevant national system. Some researchers' studies contribute to higher degrees.

Antarctic-based research has resulted in a number of scientific breakthroughs, including the discovery of the depletion of the ozone layer.

HEALTH 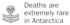 Not applicable

1 medical officer per station 　 Deaths are extremely rare in Antarctica

Each station has its own medical officer. The problems usually associated with polar conditions, such as frostbite and snow blindness, are very rare. All personnel are medically screened before arrival. If serious illnesses develop, patients have to be evacuated by air, including in recent years an acute case of gallstones and a doctor who self-diagnosed breast cancer.

SPENDING 　 Not applicable

Any evidence of wealth disparities reflects the different levels of base funding. US bases are the best-funded, while the budgets of other bases are subject to domestic politics. Most stations have TVs and video recorders. Telephone systems operate only within stations. Computers are supplied for scientific research.

WORLD RANKING

The UN Human Development Index conditions are not applicable to Antarctica.

A

ANTIGUA & BARBUDA

CARIBBEAN

OFFICIAL NAME: Antigua and Barbuda **CAPITAL:** St. John's
POPULATION: 66,400 **CURRENCY:** Eastern Caribbean dollar **OFFICIAL LANGUAGE:** English

 1981   1981   Nov 1   AG -4.5 +1268   .ag

PART OF THE Leeward Islands chain, Antigua was in turn a Spanish, French, and British colony. British influence is still strong and most clearly revealed in the Antiguans' passion for cricket. Antigua has two dependencies: Barbuda, 50 km (30 miles) to the north, sporting a magnificent beach; and Redonda, 40 km (25 miles) west, an uninhabited rock with its own king.

CLIMATE
▷ Tropical oceanic

WEATHER CHART FOR ST. JOHN'S

Antigua is less humid than other Caribbean islands. Year-round trade winds moderate the heat.

TRANSPORTATION
▷ Drive on left

V. C. Bird International, St. John's
727,292 passengers

759 ships
4.22m grt

THE TRANSPORTATION NETWORK

| 250 km (155 miles) | None |
| 77 km (48 miles) | None |

Recent multimillion EC$ projects have expanded the international airport and provided 140 km (90 miles) of roads with all-weather surfaces.

TOURISM
▷ Visitors : Population 3.1:1

207,000 visitors

Down 11% in 2000

MAIN TOURIST ARRIVALS

| USA 31% |
| UK 27% |
| Canada 7% |
| Other 35% |

% of total arrivals

Antigua is especially popular with US cruise ship tourists. Among other draws are the annual international tennis championship, the islands' three golf courses, and the attraction of large duty-free shopping malls.

PEOPLE
▷ Pop. density medium

English, English patois

151/km² (391/mi²)

THE URBAN/RURAL POPULATION SPLIT

37%   63%

RELIGIOUS PERSUASION

Rastafarian 1%   Other 2%
Roman Catholic 10%
Anglican 45%
Other Protestant 42%

Most of Antigua's population is descended from Africans, brought over between the 16th and 19th centuries. There are, in addition, a few Europeans and South Asians. Racial tensions are few. Life is based around the extended family. Since the 1960s the status of women has risen as a result of their greater access to education, and many are now entering the legal, financial, and medical professions. By Caribbean standards, wealth disparities are small.

ANTIGUA & BARBUDA

Total Land Area : 442 sq. km (170 sq. miles)

POLITICS
▷ Multiparty elections

L. House 1999/2004
U. House 1999/2004

H.M. Queen Elizabeth II

AT THE LAST ELECTION

House of Representatives 17 seats

71% ALP   23% UPP   6% BPM

ALP = Antigua Labour Party **UPP** = United Progressive Party
BPM = Barbuda People's Movement

Senate 17 seats

The members of the Senate are appointed. 11 are chosen by the prime minister, 4 by the leader of the opposition, 1 by the governor-general, and 1 by the Barbuda Council

Antigua's multiparty democracy has been dominated for four decades by the Bird family. Vere Bird Sr., the veteran prime minister and ALP leader, retired in 1994, and a battle between his two sons to succeed him was won by Lester Bird; his elder brother Vere Jr. had in 1990 been removed from public office, accused of involvement in gun-running. Lester Bird led the ALP into its sixth consecutive term in 1999.

POPULATION
over 10 000 ●
under 10 000 ·

LAND HEIGHT
200m/656ft
Sea Level

WORLD AFFAIRS ▷ Joined UN in 1981

 ACS 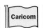 Caricom | Comm | OECS | OAS

Antigua has tended to back US policy in the region, supporting both the US invasion of Grenada in 1983 and economic sanctions against Cuba.

AID ▷ Recipient

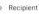 US$10m (receipts) | Down 9% in 2000

Donors include the US, the UK, France, Kuwait, Japan, and the EU. In 1998 the opposition accused the Bird regime of manipulating the distribution of hurricane relief for electoral gain.

DEFENSE ▷ No compulsory military service

US$4m | No change in 2000

There is a 170-strong defense force. Reports that the government had privately imported weaponry from the US in 1998 were denied. Two military bases are leased to the US.

ECONOMICS ▷ Inflation 2.5% p.a. (1990–2000)

US$642m | 2.7 Eastern Caribbean dollars

SCORE CARD

❏ WORLD GNP RANKING	165th
❏ GNP PER CAPITA	US$9440
❏ BALANCE OF PAYMENTS	–US$79m
❏ INFLATION	1.6%
❏ UNEMPLOYMENT	7%

STRENGTHS

Tourism and construction of tourist hotels and infrastructure. Financial and communications services linked to offshore financial sector.

WEAKNESSES

Very little diversification makes Antigua vulnerable to downturns in the world tourism market. Evidence of offshore money laundering harms the country's reputation.

EXPORTS

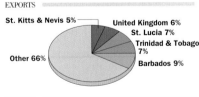

St. Kitts & Nevis 5%
United Kingdom 6%
St. Lucia 7%
Trinidad & Tobago 7%
Barbados 9%
Other 66%

IMPORTS

Trinidad & Tobago 3%
Canada 3%
UK 10%
Barbados 2%
USA 27%
Other 55%

Nelson's Dockyard. *Luxury yachts fitted with state-of-the-art gadgetry contrast with the 18th-century St. John's harbor.*

RESOURCES ▷ Electric power 26,000 kw

 3185 tonnes | Not an oil producer

 15,700 cattle, 12,200 sheep, 90,000 chickens | None

Antigua has no strategic or commodity resources and has to import almost all its energy requirements.

ENVIRONMENT ▷ Not available

 9% | 5 tonnes per capita

Sewage from hotels causes major problems. Untreated effluent pollutes the sea, while uncontrolled disposal has killed valuable inshore fish stocks in the mangrove swamps, whose whole ecosystem is under threat from poorly planned hotel development.

MEDIA ▷ TV ownership hgh

 Daily newspaper circulation 91 per 1000 people

PUBLISHING AND BROADCAST MEDIA

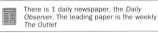

There is 1 daily newspaper, the *Daily Observer*. The leading paper is the weekly *The Outlet*

2 services: 1 state-owned, 1 independent | 5 services: 1 state-owned, 4 independent

Arson was alleged when the printing equipment of opposition weekly *The Outlet* was destroyed in 1998. Of the privately owned broadcasting services, one is affiliated to the opposition.

CRIME ▷ Death penalty in use

 186 prisoners | Down 2% 1992–1996

Murder is rare. Rape, armed robbery, and burglary are main concerns, as is offshore money laundering.

EDUCATION ▷ School leaving age: 16

 95% | 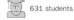 631 students

Education is based on the former British selective system. Students go on to the University of the West Indies, or to study in the UK or the US.

CHRONOLOGY

In 1667, Antigua became a British colony. Barbuda, formerly owned privately by the Codrington family, was annexed in 1860.

- ❏ **1951** Universal adult suffrage introduced.
- ❏ **1981** Independence from Britain; opposed by Barbudan secessionist movement.
- ❏ **1983** Supports US invasion of Grenada.
- ❏ **1994** Lester Bird elected prime minister.
- ❏ **1995** New taxes provoke protests.
- ❏ **1999** ALP wins sixth consecutive elections; Lester Bird remains prime minister.

HEALTH ▷ Welfare state health benefits

 1 per 877 people | Heart and respiratory diseases, cancers

By Caribbean standards, the health system is efficient, with easy access to the state-run clinics and hospitals. A new hospital has been built in St. John's.

SPENDING ▷ GDP/cap. increase

CONSUMPTION AND SPENDING

229 per 1000 population | 499 per 1000 population

Defense 0.6%
Education 2.7%
Health 0.4%

0   5   10   15   20   25
Defense, Health, Education spending as % of GDP

Wealthy Antiguans are active in running the thriving tourist industry; some are allegedly also involved in money laundering. Unemployment is relatively low and the average per capita income is among the highest in the Caribbean.

WORLD RANKING

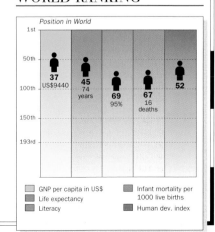

Position in World

1st
50th
100th
150th
193rd

37 US$9440 | 45 74 years | 69 95% | 67 16 deaths | 52

GNP per capita in US$ | Infant mortality per 1000 live births
Life expectancy |
Literacy | Human dev. index

SOUTH AMERICA

ARGENTINA

OFFICIAL NAME: Republic of Argentina **CAPITAL:** Buenos Aires
POPULATION: 37.5 million **CURRENCY:** Argentine peso **OFFICIAL LANGUAGE:** Spanish

OCCUPYING MOST OF THE southern half of South America, Argentina extends 3460 km (2150 miles) from Bolivia to Cape Horn. The Andes mountains in the west run north–south, forming a natural border with Chile. To the east they slope down to the fertile central pampas, the region known as Entre Ríos. Agriculture, especially beef, wheat, and fruit, and energy resources are Argentina's main sources of wealth. Politics in Argentina was characterized in the past by periods of military rule, but in 1983, Argentina returned to a system of multiparty democracy.

Herding cattle in the northeast, near Corrientes. Beef, Argentina's initial source of wealth, remains a major export.

CLIMATE ▷ Mountain/steppe/ subtropical

WEATHER CHART FOR BUENOS AIRES

■ Average daily temperature	Rainfall ■
°C/°F J F M A M J J A S O N D	cm/in
40/104	40/16
30/86	30/12
20/68	20/8
10/50	10/4
0/32	0
-10/14	
-20/-4	

The northeast is near-tropical. The Andes are semiarid in the north and snowy in the south. The western lowlands are desert, while the pampas have a mild climate with heavy summer rains.

TRANSPORTATION ▷ Drive on right

Jorge Newberry, Buenos Aires
7.04m passengers

493 ships
464,300 grt

THE TRANSPORTATION NETWORK

63,553 km (39,490 miles)	734 km (456 miles)
33,000 km (20,506 miles)	11,000 km (6835 miles)

Air travel is expensive, and inadequate connections between provinces frustrate business and tourism. The national airline, Aerolineas Argentinas, was privatized in 1990, but in 2000 was the object of a rescue plan; the 37 airports are privately operated. The privatized railroad, one of the largest in the world, is primarily used for freight, but Buenos Aires' subway and commuter lines have attracted strong investment and heavy use. Some 9500 km (6000 miles) of roads are privatized, and tolls are among the highest worldwide. The six main terminals in the port of Buenos Aires are privately run. A $20 billion national infrastructure program announced in 2001 has been stalled by the economic crisis.

TOURISM ▷ Visitors : Population 1:13

2.95m visitors   Up 2% in 2000

MAIN TOURIST ARRIVALS

| Chile 19% |
| Paraguay 18% |
| Uruguay 18% |
| Brazil 16% |
| Europe 12% |
| Other 17% |

0   10   20   30   40
% of total arrivals

Tourism has been undersold, and the government, working with business, has launched a massive international marketing campaign. Visitors, still mostly from neighboring countries, are attracted by Buenos Aires' rich city life, the fashionable Atlantic coastal resort of Mar del Plata, ski stations such as Bariloche and Las Leñas in the Andes, and wineries around Mendoza. Other major attractions are the Iguazú National Park, Antarctic cruises, and the renowned whale-watching hot spot off Peninsula Valdés.

PEOPLE ▷ Pop. density low

Spanish, Italian, Amerindian languages

14/km² (35/mi²)

THE URBAN/RURAL POPULATION SPLIT

90%   10%

RELIGIOUS PERSUASION

Protestant 2%   Other 6%
Jewish 2%
Roman Catholic 90%

ETHNIC MAKEUP

Amerindian 1%   Mestizo 14%
Indo-European 85%

Most Argentinians of European descent are from recent 20th-century migrations; over one-third are of Italian origin. Indigenous peoples now form a tiny minority, living mainly in Andean regions in the Gran Chaco. Argentina also has communities of Welsh, Lebanese, Syrians, Armenians, Japanese, and Koreans.

POPULATION AGE BREAKDOWN

Female	Age	Male
1%	80+	0.5%
6.5%	60–79	5.2%
10.3%	40–59	9.9%
14.1%	20–39	14.1%
18.9%	0–19	19.5%

% of population by age group

The vast majority of Argentinians are urban dwellers, with some 40% of the population living in Buenos Aires, one of the largest cities in Central or South America.

Catholicism and the extended family remain strong in Argentina. In addition, the family forms the basis of many successful businesses.

Women have a higher profile than in most Latin American states, and were enfranchised in 1947. Today, many enter the professions and rise to positions of influence in service businesses such as the media. The exception is party politics. Eva Perón, who inspired the musical *Evita*, did help to push women into a more active political role in the 1940s and 1950s, but this trend was reversed under military rule.

POLITICS

 Multiparty elections

 L. House 1999/2003
U. House 2001/2003

President Eduardo
Duhalde

Argentina is a multiparty democracy; the president is head of government.

PROFILE

The Peronists dominated politics from the 1940s. The party, founded on mass working-class and left-wing intellectual support, was inimical to the military; coups were staged in 1955, 1966, and 1976. President Carlos Menem won elections in 1989 on a populist platform but steered the Peronists toward a right-wing free-market agenda. The UCR tended to stay in opposition, only winning when the electorate registered a protest vote, as in 1983. In 1999 it captured the presidency as part of a center-left alliance with FREPASO. However, President Fernando de la Rúa was brought down by the 2001 economic crisis, and the Peronists, the largest party in the legislature since the mid-term elections, got their candidate Eduardo Duhalde installed as interim president.

MAIN POLITICAL ISSUES

The legitimacy of the political system

A shaky coalition of the main parties charged the Duhalde caretaker administration in January 2002 with rescuing Argentina from the worst economic and social crisis in a generation before handing power to an elected successor in 2003. From the outset his stabilization project was threatened by the unprecedented rejection of the political system, manifested in the waves of protest that had deposed the de la Rúa government and resulted in three stopgap presidents prior to Duhalde. Public opinion pointed to a major shake-up to restore public confidence in politicians, the judiciary, and business.

The economic crisis

Keeping the peso pegged to the dollar was blamed for prolonging economic recession and was ditched by Duhalde. The severe effects of industrial collapse and fiscal austerity, coupled with hated restrictions on bank withdrawals to defend the peso and head off inflation, hit domestic consumption and reduced a near cashless population to bartering goods. The IMF, however, made the release of funds dependent on fiscal, debt, and exchange rate performance.

AT THE LAST ELECTION

Chamber of Deputies 257 seats

| 46% PJ | 34% Alliance | 7% ARI | 11% Others | 2% AR |

PJ = Justicialist Party (Peronists) **Alliance** = Alliance for Work, Justice, and Education (Radical Civic Union–**UCR** and National Solidarity Front Alliance–**FREPASO**) **ARI** = Alternative for a Republic of Equals **AR** = Action for the Republic

Senate 72 seats

| 55% PJ | 35% Alliance | 10% Others |

ARGENTINA

Total Land Area : 2 766 890 sq. km
(1 068 296 sq. miles)

POPULATION		LAND HEIGHT	
over 1 000 000	▣	4000m/13124ft	
over 500 000	◉	2000m/6562ft	
over 100 000	◎	1000m/3281ft	
over 50 000	○	200m/656ft	
over 10 000	●	Sea Level	

N

| 0 | 200 km |
| 0 | 200 miles |

Carlos Menem: authoritarian rule overshadowed free-market reforms.

Eduardo Duhalde, elected in 2002, became the fifth president in 12 days.

WORLD AFFAIRS

▷ Joined UN in 1945

Argentina takes a pro-Western stance and has deployed its armed forces in a series of UN actions. Solid relations with potential aid donors and trade partners were made all the more essential following the collapse of the Argentine economy in 2001. Ties to the US are strong, although US pressure for an open-skies agreement and demands for royalty payments on patented drugs create some tensions.

Friction with Brazil over trade rules complicates Argentina's membership of Mercosur. It wants Mercosur to be strengthened, and widened to include Chile, currently an associate member.

The normalizing of relations with the UK in 1998 sidelined Argentina's claim to sovereignty over the nearby Falklands Islands (known locally as Las Islas Malvinas), the focus of the 1982 war between the two countries.

In 2002, Argentina requested the extradition of former Bolivian president Hugo Banzer (who later died) in relation to Plan Cóndor, which saw thousands kidnapped and murdered under military regimes in Argentina, Chile, and Bolivia in the 1970s.

AID

 Recipient

$76m (receipts)   Down 24% in 2000

Although one massive "financial shield" was agreed in 2001, the IMF's refusal of a further rescue package precipitated the crash in December.

CHRONOLOGY

The Spanish first established settlements in the Andean foothills in 1543. The indigenous Amerindians, who had stopped any Inca advance into their territory, also prevented the Spaniards from settling in the east until the 1590s.

❑ **1816** United Provinces of Río de la Plata declare independence; 70 years of civil war follow.

❑ **1835–1852** Dictatorship of Juan Manuel Rosas.

❑ **1853** Federal system set up.

❑ **1857** Europeans start settling the pampas; six million by 1930.

❑ **1877** First refrigerated ship starts frozen beef trade to Europe.

❑ **1878–1883** War against pampas Amerindians (almost exterminated).

❑ **1916** Hipólito Yrigoyen wins first democratic presidential elections.

❑ **1930** Military coup.

❑ **1943** New military coup. Gen. Juan Perón organizes trade unions.

❑ **1946** Perón elected president, with military and labor backing.

❑ **1952** Eva Perón, charismatic wife of Juan Perón, dies of leukemia.

❑ **1955** Military coup ousts Perón. Inflation, strikes, unemployment.

❑ **1973** Perón reelected president.

❑ **1974** Perón dies; succeeded by his third wife "Isabelita," who is unable to exercise control.

❑ **1976** Military junta seizes power. Political parties are banned. Brutal repression during "dirty war" sees "disappearance" of over 15,000 "left-wing suspects."

❑ **1981** Gen. Galtieri president.

❑ **1982** Galtieri orders invasion of Falkland Islands. UK retakes them.

❑ **1983** Pro-human rights candidate Raúl Alfonsín (UCR) elected president. Hyperinflation.

❑ **1989** Carlos Menem (Peronist) president.

❑ **1992** Inflation down to 25%.

❑ **1995** Economy enters recession.

❑ **1998–1999** Argentina weathers financial crisis in Brazil.

❑ **1999** Fernando de la Rúa of UCR–FREPASO alliance elected president.

❑ **2001** Weakened by corruption scandals, government is brought down by economic crisis.

❑ **2002** January, Eduardo Duhalde appointed president by Peronist-dominated Congress.

DEFENSE

 No compulsory military service

$4.69bn   Down 14% in 2000

ARGENTINIAN ARMED FORCES

200 main battle tanks (TAM)	41,400 personnel	
3 submarines, 5 destroyers, 8 frigates, 15 patrol boats	16,200 personnel	
130 combat aircraft (Mirage V & III/EA, Dagger Nesher, Fightinghawk)	12,500 personnel	
None		

The end of dictatorship led to trials and prison for the top brass, but subsequent immunity laws were meant to placate the military and close the chapter on the "dirty war" (1976–1983), during which some 15,000 to 30,000 were killed or "disappeared." The military made public admissions of guilt in 1995. A 2001 ruling in a kidnap and murder case, however, said that such immunity was unconstitutional and cleared the way for the trial of many more military personnel for human rights crimes. Former junta leader Gen. Leopoldo Galtieri was arrested in 2002. The armed forces now see themselves as modernized, participating in UN peacekeeping and cooperating on defense within the framework of the Mercosur trade bloc (Brazil, Argentina, Uruguay, and Paraguay).

ECONOMICS

 Inflation 5.2% p.a. (1990–2000)

$276bn   0.9984–1 Argentine pesos

SCORE CARD

❑ WORLD GNP RANKING............................17th
❑ GNP PER CAPITA $7460
❑ BALANCE OF PAYMENTS....................–$8.9bn
❑ INFLATION ....–0.9%
❑ UNEMPLOYMENT..................................16%

EXPORTS

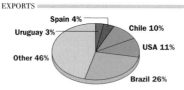

Spain 4%
Uruguay 3%
Other 46%
Chile 10%
USA 11%
Brazil 26%

IMPORTS

Germany 5%
Italy 4%
Other 39%
France 5%
USA 20%
Brazil 27%

STRENGTHS

Rich and varied agricultural base. Powerful agribusiness (mainly beef, soya, wheat, fruit, and wine) and wealth of energy resources. Net exporter of oil.

WEAKNESSES

End of currency stability based on pegging of peso to dollar. Vulnerability to external shocks and downturns in Brazil (largest single export market). Investors scared by changes in assessment of country risk in emerging markets. Heavy debts, public and private, to refinance. Weak banking sector. Global fluctuations in prices of vital non-oil commodities. High unemployment and risk of unrest. Endemic tax evasion. Subsidies and trade barriers bar agricultural produce from US and EU.

ECONOMIC PERFORMANCE INDICATOR

Consumer Price Index   GDP
Consumer price index 1995=100
GDP 1996=100
1996 1997 1998 1999 2000

PROFILE

The "miracle" recovery of the 1990s was based on stabilizing the peso (by pegging it to the US dollar) and on a combination of neoliberal reforms accompanied by privatization. Argentina rode out the Mexican crisis of 1995, but was hit by damage to foreign investor confidence and a shrinking Brazilian market in 1998–1999. Regional recession in 2001 brought economic crisis, and the world's largest default on international debt.

ARGENTINA : MAJOR BUSINESSES

Salta
San Salvador de Jujuy
Corrientes
Córdoba
Santa Fé
Mendoza
Buenos Aires
Viedma

Wine
Textiles
Agribusiness
Metals
Oranges
Tobacco
Vehicle assembly
Light engineering
Cattle/Meat packing
Heavy engineering

0   400 km
0   400 miles   * significant multinational ownership

RESOURCES Electric power 23m kw

 1.03m tonnes

822,000 b/d (reserves 3bn barrels)

50.2m cattle, 13.5m sheep, 4.2m pigs, 110m chickens

Oil, natural gas, coal, iron, zinc, lead, uranium, tin, silver, copper, gold

ELECTRICITY GENERATION

Hydro 36% (27bn kwh)

Combustion 54% (40bn kwh)

Nuclear 10% (7.5bn kwh)

Other 0%

% of total generation by type

Known oil and gas reserves are still underexploited, and copper and gold mining are just taking off; only one-

ARGENTINA : LAND USE

Cropland
Pasture
Forest
Wetlands
High mountain regions
Wheat - cash crop
Cattle

0   400 km
0   400 miles

third of the country has been properly surveyed. Wheat, maize, and oilseeds drive exports rather than trade in beef, which has stagnated, while its image has been weakened by foot-and-mouth disease.

HEALTH Welfare state health benefits

1 per 370 people

Heart diseases, cancers, accidents

Buenos Aires has 33 hospitals, but free state provision suffers from underfunding, poorly paid staff, and long queues. Government-sponsored vaccination, mother-and-child schemes, feeding programs, and rural health projects barely tackle such problems in the poorest provinces as malnutrition, lack of decent water and sewers, and threadbare medical cover. A health care deregulation bill was decreed in 2001 to dismantle trade unions' monopoly of health insurance schemes.

SPENDING GDP/cap. increase

CONSUMPTION AND SPENDING

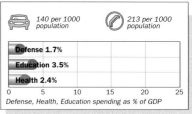

140 per 1000 population

213 per 1000 population

Defense 1.7%

Education 3.5%

Health 2.4%

Defense, Health, Education spending as % of GDP

Members of the wealthy elite, who travel in private jets to *estancias* (country estates), vacation in Europe and the US, and hold dollar accounts offshore to avoid tax, escaped the worst of the economic collapse in 2001–2002. Middle-income groups squeezed after years of free-market reforms lost out in the crisis which forced some 15 million below the poverty line, with four million in extreme poverty. Emergency government aid offered in 2002 to one million unemployed people with children under 18 was at a level below half the legal minimum salary. The precrisis figure of some 40% of workers in the low-wage black economy was set to balloon. Millions of cashless and poor people resorted to exchanging goods at barter clubs.

ENVIRONMENT Sustainability rank: 15th

2% (2% partially protected)

3.8 tonnes per capita

ENVIRONMENTAL TREATIES

Yes   Yes   Yes

Yes   Yes   Yes

Environmental protection has low governmental priority. Legislation is weak and largely ignored by states which retain a good deal of autonomy. Political parties typically shy away from the level of public spending needed to tackle major environmental problems, and a corrupt judiciary had meant poor enforcement of existing laws. Key problems are hazardous waste, poor urban water and air quality and inadequate sewers, pesticide contamination due to agribusiness, deforestation, and illegal hunting.

CRIME Death penalty not used in practice

38,604 prisoners   Up 4% in 1999

CRIME RATES

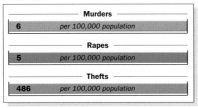

Murders
6   *per 100,000 population*

Rapes
5   *per 100,000 population*

Thefts
486   *per 100,000 population*

Half of Buenos Aires households own a gun for fear of violent crime. Law enforcement in the interior is weak. The judiciary and the police command little respect. Overcrowded prisons lead to frequent riots and criminal cases can take over a year to reach court.

MEDIA TV ownership high

Daily newspaper circulation 56 per 1000 people

PUBLISHING AND BROADCAST MEDIA

There are 181 daily newspapers. *Clarín* and *Crónica* are market leaders

15 stations owned by provincial or national authorities, 29 independent channels

122 stations: 37 state-controlled, 4 provincial, 3 university-run, 3 municipal, and 75 independent

Many journalists were murdered by the military in the 1970s, but the press was liberated under the UCR (1983–1989) and harries governments, especially on corruption. Investigative journalists, however, still face intimidation. The Internet had reached over 10% of the population by 2001.

EDUCATION School leaving age: 14

97%   1.34m students

THE EDUCATION SYSTEM

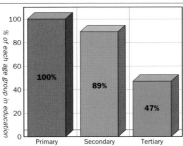

% of age group in education

Primary 100%
Secondary 89%
Tertiary 47%

Public schooling is free and compulsory to the age of 14. Huge numbers of poor students drop out of the education system near or after this age, and 60–70% of middle-class students drop out by the close of five-year courses at cash-strapped free state universities. A fifth of government spending on education finds its way to private institutions.

WORLD RANKING

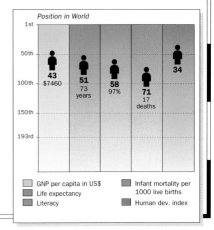

Position in World

1st
50th
100th
150th
193rd

43   $7460
51   73 years
58   97%
71   17 deaths
34

GNP per capita in US$
Life expectancy
Literacy
Infant mortality per 1000 live births
Human dev. index

A

ARMENIA

OFFICIAL NAME: Republic of Armenia **CAPITAL:** Yerevan
POPULATION: 3.8 million **CURRENCY:** Dram **OFFICIAL LANGUAGE:** Armenian

1991	1991	Sept 21	ARM	+4	+374	.am

EUROPE

LANDLOCKED IN THE Lesser Caucasus Mountains, Armenia is the smallest of the former Soviet Union's republics. It was the first to adopt Christianity as its state religion. It is bordered by Muslim states to the south, east, and west. Keen to strengthen links with the rest of the CIS, Armenia has kept to a path of radical economic reform, including privatization. The confrontation with Azerbaijan over the enclave of Nagorno Karabakh has dominated national life since 1988.

Landscape near Yerevan. Armenia's very dry climate results in expanses of semidesert. Its famous vineyards flourish in sheltered areas.

CLIMATE ▷ Mountain

WEATHER CHART FOR YEREVAN

Armenia has a continental climate, with little rainfall in the lowlands. Winters can be very cold.

TRANSPORTATION ▷ Drive on right

Yerevan International

Has no fleet

THE TRANSPORTATION NETWORK

15,998 km (9941 miles)	7567 km (4702 miles)
796 km (495 miles)	None

Public transportation was badly hit by a war-induced fuel crisis. Road and rail links with Georgia, connecting with the main east–west corridor, need upgrading.

TOURISM ▷ Visitors : Population 1:127

30,000 visitors   Down 27% in 2000

MAIN TOURIST ARRIVALS

CIS 40%
USA 13%
Iran 11%
Other 36%

% of total arrivals

The 1700th anniversary of Armenian Christianity in 2001 boosted the war-damaged tourist industry. Most visitors are diaspora Armenians.

PEOPLE ▷ Pop. density medium

Armenian, Russian   128/km² (330/mi²)

THE URBAN/RURAL POPULATION SPLIT

70%   30%

ETHNIC MAKEUP

Other 2%   — Azeri 3%
Russian 2%
Armenian 93%

Minority nationalities are well integrated in Armenia. There are strong contacts with the many Armenian emigrants, numbering nine million in the US, France, and Syria.

Conflict with Azerbaijan forced 350,000 Armenians in Azerbaijan to return to Armenia and 190,000 Azeris in Armenia to return to Azerbaijan.

POLITICS ▷ Multiparty elections

1999/2003   President Robert Kocharian

AT THE LAST ELECTION

National Assembly 131 seats

49% UB	21% Ind	8% CP	5% LUB	1% Vacant	5% ARF–D	11% Others

UB = Unity Bloc (alliance of the Republican Party of Armenia – **RPA** and the People's Party of Armenia – **PPA**)
Ind = Independents
CP = Communist Party **LUB** = Law and Unity Bloc
ARF–D = Armenian Revolutionary Federation–Dashnaktsutyun

Armenia became an independent multiparty democracy in 1991 and held its first parliamentary elections in July 1995. A new constitution approved by referendum set up a presidential republic, and in 1996 President Levon Ter-Petrossian was reelected for a five-year term. The war with Azerbaijan, over the issue of whether the Armenian enclave of Nagorno Karabakh inside Azerbaijan should become part of Armenia, has simmered since a 1994 cease-fire. Ter-Petrossian resigned in 1998, after parliament opposed his softer line in search of peace. He was succeeded by Robert Kocharian, a former premier and ex-governor of Nagorno Karabakh. In 1999 the prime minister, RPA leader Vazgen Sarkissian, was shot dead in a dramatic attack on parliament. He was succeeded first by his brother Aram, then in 2000 by Andranik Markarian.

ARMENIA

Total Land Area : 29 800 sq. km (11 506 sq. miles)

POPULATION

▣	over 1 000 000
◉	over 100 000
○	over 50 000
●	over 10 000
•	under 10 000

LAND HEIGHT

- 3000m/9843ft
- 2000m/6562ft
- 1000m/3281ft
- 500m/1640ft

0   50 km
0   50 m

A

WORLD AFFAIRS

▷ Joined UN in 1992

 BSEC CIS CE EAPC OSCE

Armenia joined the Council of Europe in 2001. The continuing conflict with Azerbaijan is diplomatically damaging.

AID

▷ Recipient

 $216m (receipts)   Up 3% in 2000

The EBRD and the World Bank back infrastructure projects. Expatriate Armenians such as US billionaire Kirk Kerkorian are a major source of funds.

DEFENSE

▷ Compulsory military service

$149m   Down 6% in 2000

Successes in the fighting over Nagorno Karabakh have increased the profile and autonomy of the army, which includes conscripts on 24-month national service. A cease-fire has broadly held since 1994, but peace talks, revived in 1999–2000, remain inconclusive.

ECONOMICS

▷ Inflation 213% p.a. (1990–2000)

 $1.99bn   551.4–562.2 drams

SCORE CARD

❏ WORLD GNP RANKING	137th
❏ GNP PER CAPITA	$520
❏ BALANCE OF PAYMENTS	–$278m
❏ INFLATION	–0.8%
❏ UNEMPLOYMENT	12%

STRENGTHS
Strong ties with Armenian emigrants. Deposits of rare metals, currently unexploited. Machine building and manufacturing – includes textiles and bottling of mineral water.

WEAKNESSES
Dependent on imported energy, raw materials, and semi-finished goods. High unemployment and economic stagnation. Widespread corruption.

EXPORTS

Georgia 5%
Iran 10%
Other 32%
USA 13%
Russia 15%
Belgium 25%

IMPORTS

UK 7%
Belgium 9%
Other 48%
Iran 9%
USA 12%
Russia 15%

A Turkish trade embargo has been in place since 1988. Russia signed a treaty of friendship in 1997 but has been less pro-Armenian since the Azeris rejoined the CIS. Work began on a gas pipeline between Iran and Armenia in 1998.

RESOURCES

▷ Electric power 3m kw

 877 tonnes   Minimal oil production

497,155 sheep, 485,000 cattle, 4.3m chickens   Coal, oil, natural gas, rare metals

Energy resources are negligible, and mismanagement of the energy industry was deemed to have cost the country $200 million in the 1990s. Vegetables and fruit are grown in fertile lowlands, and grains in the hills; agriculture accounts for as much as a third of GDP.

ENVIRONMENT

▷ Sustainability rank: 38th

 8%   0.9 tonnes per capita

Environmental groups, backed by the EU, demand the closure of the Medzamor nuclear power station, declared unsafe after the 1988 earthquake, but reopened in 1995 owing to the energy crisis. HEP generation near Lake Sevan has seriously lowered its water level.

MEDIA

▷ TV ownership medium

 Daily newspaper circulation 24 per 1000 people

PUBLISHING AND BROADCAST MEDIA

There are 11 daily newspapers, including *Azg*, *Haiastan*, and *Ankakhutiun*	
1 state-controlled service, several independent stations	1 state-controlled service, several independent stations

Numerous TV and broadcasting stations assist media freedom. Independent journals and newspapers depend on the government-controlled paper industry.

CRIME

▷ Death penalty in use

 7428 prisoners   Down 7% in 1999

Reforms to the legal system introduced in 1999 included the replacement of the Supreme Court by an appeals court. Assassinations of political figures are common.

EDUCATION

▷ School leaving age: 14

98%   39,770 students

The education system, previously conforming to that of the USSR, now emphasizes Armenian history and culture; 12% of the population have received higher education.

Armenia lost its autonomy in the 14th century. In 1639, Turkey took the west and Persia the east; Persia ceded its part to Russia in 1828.

- ❏ **1877–1878** Massacre of Armenians during Russo-Turkish war.
- ❏ **1915** Ottomans exile 1.75 million Turkish Armenians; most die.
- ❏ **1920** Independence.
- ❏ **1922** Becomes a Soviet republic.
- ❏ **1988** Earthquake kills 25,000. Conflict with Azerbaijan over Nagorno Karabakh begins.
- ❏ **1991** Independence from USSR.
- ❏ **1994** Cease-fire with Azerbaijan.
- ❏ **1995** First parliamentary elections.
- ❏ **1998** Kocharian elected president.
- ❏ **1999** Shooting of prime minister in attack on parliament.

HEALTH

▷ No welfare state health benefits

 1 per 313 people   Circulatory diseases, cancers, accidents, violence

Hospitals suffer from the erratic electricity supply. Poor sewerage and other services have led to a rise in hepatitis, tuberculosis, and cholera.

SPENDING

▷ GDP/cap. decrease

CONSUMPTION AND SPENDING

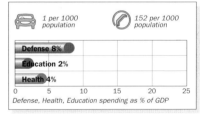

1 per 1000 population   152 per 1000 population

Defense 8%
Education 2%
Health 4%

Defense, Health, Education spending as % of GDP

The richest Armenian people are those living away from Armenia itself, particularly in the US and France. The many refugees from Baku, Azerbaijan, are the poorest.

WORLD RANKING

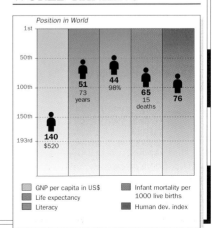

Position in World

51 — 73 years
44 — 98%
65 — 15 deaths
76
140 — $520

GNP per capita in US$
Life expectancy
Literacy
Infant mortality per 1000 live births
Human dev. index

AUSTRALIA

OFFICIAL NAME: Commonwealth of Australia   **CAPITAL:** Canberra
POPULATION: 19.3 million   **CURRENCY:** Australian dollar   **OFFICIAL LANGUAGE:** English

1901 | 1901 | Jan 26 | AUS | +8 to +11 | +61 | .au

T HE WORLD'S SIXTH-LARGEST country, Australia is an island continent located between the Indian and Pacific Oceans. Its varied landscapes include tropical rainforests, the deserts of the arid "red center," snowcapped mountains, rolling tracts of pastoral land, and magnificent beaches. Famous natural features include Uluru (Ayers Rock) and the Great Barrier Reef. Most Australians live on the coast, and all the state capitals, including Sydney, host of the 2000 Olympics, are coastal cities. Only Canberra, the national capital, lies inland. The vast interior is dotted with large reserves, sparsely inhabited by communities from the small Aboriginal population.

Uluru (Ayers Rock), Northern Territory.
The renaming of Ayers Rock reflects growing Aboriginal influence in Australia.

CLIMATE ▷ Hot desert/steppe/ tropical/Mediterranean

WEATHER CHART FOR CANBERRA

The interior, west, and south are arid or semiarid and very hot in summer; central desert temperatures can reach 50°C (122°F). The north, around Darwin and Cape York Peninsula, is hot all year and humid during the summer monsoon. Only the east and southeast, within 400 km (250 miles) of the coast, and the southwest, around Perth, are temperate: most Australians live in these areas.

TRANSPORTATION ▷ Drive on left

 Kingsford Smith, Sydney
25.8m passengers

 631 ships
1.91m grt

THE TRANSPORTATION NETWORK

353,331 km (219,549 miles)	13,630 km (8469 miles)
36,026 km (22,387 miles)	8368 km (5200 miles)

Air transportation is well developed and vital to Australia's sparsely populated center and west. Sydney suffers from air congestion, but proposals for a new West Sydney airport remain controversial. A proposed high-speed train network on the east coast has been shelved due to cost. Most long-distance freight in Australia travels in massive trucks known as "road trains." Improvements in urban transportation are a priority and gained impetus in Sydney from the 2000 Olympic Games.

TOURISM ▷ Visitors : Population 1:3.9

4.95m visitors   Up 11% in 2000

MAIN TOURIST ARRIVALS

New Zealand 16%	
Japan 16%	
UK 12%	
USA 9%	
Singapore 6%	
Other 41%	

% of total arrivals

Tourism is now Australia's largest single foreign exchange earner. Faster, cheaper air travel and highly successful government marketing campaigns, both on a national and a state level, attract tourists in increasing numbers. The focus during the 1990s on drawing tourists from nearby Asian countries left the Australian tourist industry vulnerable after the Asian financial crisis of 1997. In recent years New Zealanders have equaled the Japanese as the largest single category of visitors to Australia. However, tourists from southeast Asia continue to arrive in significant numbers.

The country's many attractions include wildlife, swimming and surfing off Pacific and Indian Ocean beaches, skin diving along the Great Barrier Reef, and skiing in the Australian Alps. Uluru, Aboriginal culture, and the town of Alice Springs are among the outback's attractions. The far north has tropical resorts, the northwest, pearl fishing. The vineyards of the south and southeast attract many visitors, as do the cultural life of Melbourne and Sydney and the arts festivals held in state capitals. Sydney's famous landmarks and cosmopolitan feel, as well as the world-renowned Bondi Beach, make it a favorite.

The mid-1980s saw a phenomenal boom; tourist arrivals almost tripled in five years to reach two million in 1990. Even though growth slowed during the early 1990s, by 2000 the number of visitors had reached almost five million, boosted greatly by the celebrated Sydney Olympic Games.

AUSTRALIA

Total Land Area : 7 686 850 sq. km (2 967 893 sq. miles)

POPULATION		LAND HEIGHT
◉	over 1 000 000	1000m/3281ft
◎	over 100 000	500m/1640ft
○	over 50 000	200m/656ft
●	over 10 000	Sea Level
•	under 10 000	-200m/-656ft

PEOPLE

▷ Pop. density low

English, Italian, Cantonese, Greek, Arabic, Vietnamese, Aboriginal languages

3/km² (7/mi²)

THE URBAN/RURAL POPULATION SPLIT

85% **15%**

The first settlers arrived in Australia at least 100,000 years ago. Their modern descendants, the Aborigines, today make up less than 3% of the population. European settlement began in 1788 and was dominated by British and Irish immigrants – some of whom were convicts – until the gold rushes of the 1850s. Immigrants of other nationalities – including many

Chinese – arrived to prospect for gold, then settled in the cities, especially Melbourne and Sydney. When the new federal government was installed in 1901, one of its first acts was to prevent further Chinese immigration. The act set out the "White Australia" policy, which conditioned attitudes to immigration for almost 70 years.

A massive immigration drive after World War II brought many more British settlers to Australia in the 1950s. Further government initiatives to "populate or perish" saw the arrival of large numbers of Italians and Greeks.

From the late 1960s, the "White Australia" policy was progressively wound down. It was officially ended during the 1972–1975 Whitlam administration. Ever since, up to 50% of immigrants each year have come from Asia, transforming Australia from an almost exclusively European enclave into a

RELIGIOUS PERSUASION

Roman Catholic 26%, Anglican 24%, Other Protestant 6%, United Church 8%, Nonreligious 13%, Other 23%

ETHNIC MAKEUP

Aboriginal and Other 3%, Asian 5%, European 92%

multicultural society in which immigrant groups are encouraged to maintain connections with their own cultures and languages.

Aborigines, the exception in an otherwise integrated society, number around 410,000. Economically and socially marginalized, they face considerable discrimination. Until the mid-1960s, they were denied the vote and full social benefits. Their land had been occupied as *terra nullius* – belonging to no one. Since the 1970s, Aborigines have made a more organized stand on land and civil rights. Native title to land was recognized in 1993, although controversies continue over the extent of its application. Civil rights campaigns have moved on from the initial phase of antiracist protests to demand greater equality in areas such as health, housing, and education. Average life expectancy is still 20 years lower than the rest of the population. Alcoholism is a pervasive problem both in towns and rural areas. Aborigines in urban areas may be relatively better housed but face particular problems in asserting their cultural identity.

During the 1950s and 1960s, Catholic–Protestant differences were sufficient to cause a rift in the ALP. However, a subsequent policy encouraging mixed denomination schooling, coupled with a decline in religious observance, has largely neutralized the issue.

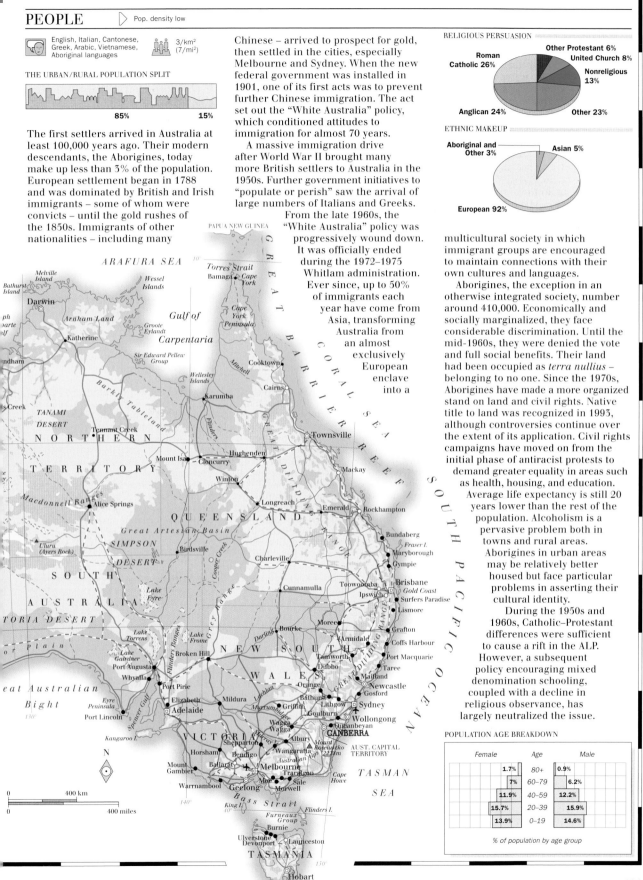

POPULATION AGE BREAKDOWN

Female	Age	Male
1.7%	80+	0.9%
7%	60–79	6.2%
11.9%	40–59	12.2%
15.7%	20–39	15.9%
13.9%	0–19	14.6%

% of population by age group

A

CHRONOLOGY

Dutch, Portuguese, French, and – decisively – British incursions throughout the 17th and 18th centuries signaled the end of a millennia of Aboriginal isolation. Governor Arthur Philip raised the British Union Flag at Sydney Cove on 26 January 1788.

❏ **1901** Inauguration of Commonwealth of Australia.

❏ **1915** Australian troops suffer heavy casualties at Gallipoli.

❏ **1939** Prime Minister Menzies announces Australia will follow Britain into war with Germany.

❏ **1942** Fall of Singapore to Japanese army. Japanese invasion of Australia seems imminent. Government turns to US for help.

❏ **1950** Australian troops committed to UN/US Korean War against North Korean communists.

❏ **1962** Menzies government commits Australian aid to war in Vietnam.

❏ **1966** Adopts decimal currency.

❏ **1972** Whitlam government elected. Aid to South Vietnam ceases.

❏ **1975** Whitlam government dismissed by Governor-General Sir John Kerr. Malcolm Fraser forms coalition government.

❏ **1983** Bob Hawke becomes prime minister at the head of an ALP administration.

❏ **1985** Corporate boom followed by deepening recession.

❏ **1992** Paul Keating defeats Hawke in leadership vote, becomes prime minister; announces "Turning toward Asia" policy. High Court's "Mabo Judgment" recognizes Aboriginal land rights.

❏ **1993** Against most predictions, Keating's ALP government reelected. Native Title Act provides compensation for Aboriginal rights extinguished by existing land title.

❏ **1996** Defeat of Keating government. Liberal John Howard becomes prime minister. Shooting of 35 people by gunman in Port Arthur, Tasmania, prompts tightening of gun control laws. First death under Northern Territory's controversial euthanasia legislation; legislation later overruled at federal level.

❏ **1998** Elections: Howard's Liberal and National coalition retains power with reduced majority; fears of right-wing One Nation party breakthrough prove unfounded.

❏ **1999** Referendum rejects proposals to replace Queen as head of state by indirectly elected president.

❏ **2000** Olympic Games in Sydney.

❏ **2001** Surprise reelection of Liberal–National coalition.

POLITICS

 Multiparty elections

 L. House 2001/2004
U. House 2001/2004   H.M. Queen Elizabeth II

AT THE LAST ELECTION

House of Representatives 150 seats

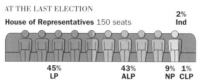

| 45% LP | 43% ALP | 9% NP | 1% CLP | 2% Ind |

LP = Liberal Party   **ALP** = Australian Labor Party   **NP** = National Party   **Ind** = Independents   **CLP** = County–Liberal Party   **AD** = Australian Democrats   **G** = Greens

Senate 76 seats

| 45% LP–NP | 37% ALP | 10% AD | 3% G | 5% Others |

12 seats in the Senate are apportioned to each of the country's constituent states and two each to the Northern Territory and the Australian Capital Territory

Australia is a parliamentary democracy on the British model. There are six state governments, all but one (Queensland) bicameral. The Northern Territory became self-governing in 1978.

PROFILE

The ALP and the Liberal and National parties have dominated Australian politics since 1945. The latter two, politically to the right, work together in coalition and broadly represent big business and agricultural interests. The ALP gained some of this support in the 1980s, adopting free-market policies and blurring the differences between parties, but 13 years of ALP rule ended in 1996, when a Liberal–National coalition took office under John Howard. It retained power with a much reduced majority in the 1998 poll, and again in 2001, due largely to popularity for its hard-line stance on immigration and terrorism. However, by March 2002 it faced ALP regional governments in all states and territories, highlighting the fragility of its hold on power.

MAIN POLITICAL ISSUES
Immigration

Public concern at images of boatloads of would-be immigrants entering Australian waters has boosted right-wing policies. Dramatic attempts by asylum seekers to protest over the conditions in which they are kept have only served further to alienate public opinion.

The Howard government has courted public mistrust of refugees, focusing on "unacceptable" behavior. Its allegations that immigrants had thrown children overboard in a desperate attempt to secure asylum were proved groundless in 2002, and its notorious "Pacific solution" – encouraging tiny Pacific states to house asylum applicants temporarily in return for financial aid – has been roundly criticized by the international community.

Vineyards in South Australia. Wine-making has been one of Australia's greatest agricultural success stories in recent years.

The far right

Although consistently unsuccessful at the federal level (it failed to win a single seat in the 2001 general election), the local success of Pauline Hanson's One Nation party has done much to legitimize many right-wing policies. In response conservative parties have adopted a notably anti-immigrant position while the ALP has conspicuously toned down its support for immigration and embraced market economics.

Republicanism

Despite international press coverage, the republican issue is not of major importance to most Australians. The former ALP government played down the debate in order to avoid inflaming monarchist groups. In his 1996 oath of office, Prime Minister Howard swore allegiance to Queen Elizabeth II, but not her successors. His government's Constitutional Convention of February 1998 discussed various republican models, including the popular election of a head of state. The alternative to monarchy which was eventually chosen to put to referendum in 1999 – where the president would be chosen by parliament, requiring a two-thirds majority – was rejected by the voters.

Paul Keating, resigned as leader of the ALP after his 1996 election defeat.

John Howard, leader of the LP, was elected prime minister in 1996.

Aden Ridgeway, Aboriginal senator and deputy leader of the AD.

WORLD AFFAIRS

▷ Joined UN in 1945

Australia's international focus has shifted from Europe and the US toward Asia. Geopolitically it is in an ambiguous position. It lost its place as a major UK trading partner after the UK joined the EU, but was still regarded as a Western outsider by the Asian nations with which it wanted closer links. Australia took practical steps to redefine its role, backing the 1989 Asia–Pacific Economic Cooperation forum (APEC) to create a multilateral regional trading bloc similar to the EU and NAFTA. The US was a strong supporter, seeing APEC as a means of promoting free-market economics in Asia. Australia's ambition is for APEC to become the leading association in the region. However, the move toward market liberalization slowed, following the 1997 Asian financial crisis.

Relations with the US are tense on questions of trade. Australia objects to subsidized US wheat undercutting its own exports, particularly in the key Chinese market. It sees the EU and the US as its main competitors in selling to southeast Asia.

On security issues Australia still supports the West. It sent troops to the 1991 Gulf War and to assist in the "war on terrorism" in Afghanistan in 2001. Its commitment to the Pacific region also remains strong. In the post-Cold War era this is expressed mainly through development aid rather than defense arrangements, but Australian troops led the UN force in East Timor in 1999.

Australia objects to longline fishing, which kills large numbers of dolphins, and employs submarine patrols in an effort to regulate the industry. There have been a number of minor skirmishes with Indonesian and Japanese boats.

Australia has been strongly criticized by the international community and the UN for its treatment of the Aboriginal population and asylum seekers. The "Pacific Solution" to the immigration problem prompted condemnation on its adoption in late 2001.

AID

▷ Donor

💲 US$987m (donations)   ⬆ Up 1% in 2000

Australia spends only 0.25% of its GNP on aid programs. Most is spent in the Asia–Pacific region. Particular areas of focus are those of HIV/AIDS programs and nongovernmental organizations. The recipient of by far the greatest amount is Papua New Guinea, where Australian companies such as Broken Hill Proprietary have major mining operations.

DEFENSE

▷ No compulsory military service

💲 US$6.95bn   ⬇ Down 11% in 2000

Strategic ties with the US remain important. Australia has defense arrangements with the Philippines, Brunei, and Thailand among others. It dramatically increased expenditure in 2001–2002 to promote its position as a regional power. Updating military equipment and exploiting information technologies are priorities. 1550 soldiers were despatched to assist in the "war on terrorism" in late 2001.

AUSTRALIAN ARMED FORCES

🛡	71 main battle tanks (*Leopard* 1A3)	25,150 personnel
🚢	5 submarines, 1 destroyer, 9 frigates, 15 patrol boats	12,500 personnel
✈	141 combat aircraft (35 F-111, 71 F/A-18)	14,050 personnel
⚓	None	

ECONOMICS

▷ Inflation 1.5% p.a. (1990–2000)

📊 US$388bn   💲 1.7997–1.9535 Australian dollars

SCORE CARD

❑ WORLD GNP RANKING	15th
❑ GNP PER CAPITA	US$20,240
❑ BALANCE OF PAYMENTS	–US$15.5bn
❑ INFLATION	4.5%
❑ UNEMPLOYMENT	7%

ECONOMIC PERFORMANCE INDICATOR

EXPORTS

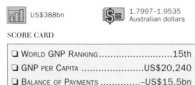

China 6%
New Zealand 6%
South Korea 7%
USA 10%
Japan 20%
Other 51%

IMPORTS

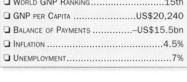

Germany 5%
UK 6%
China 8%
Japan 13%
USA 20%
Other 48%

STRENGTHS

Efficient agricultural and mining industries. Vast mineral deposits. Highly profitable tourist industry with record of dramatic growth. Good history regarding both economic growth and inflation.

WEAKNESSES

May suffer from EU and NAFTA protectionist policies. Political and financial instability in export markets in southeast Asia. Competition from Asian economies with lower wage rates and poorer working conditions. Unemployment likely to remain high. Balance-of-payments deficit.

PROFILE

Companies concentrated during the 1990s on the Asian market, which grew to represent 60% of Australia's trade. They were hit hard when the 1997 Asian financial crisis tipped the region into recession. In order to compete in Asia, Australia's economy has been undergoing massive structural adjustment. The Howard government, like its ALP predecessor, has been dismantling the tariffs that had made Australia one of the most protected OECD economies. Higher unemployment and the collapse of many businesses accompanied this change. However, the economy appeared to have weathered the global economic slowdown in 2001, posting positive growth and lower unemployment by the end of that year.

AUSTRALIA : MAJOR BUSINESSES

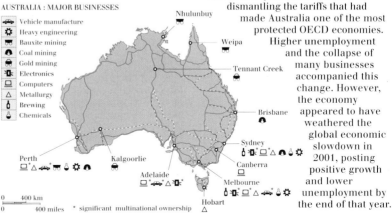

🚚 Vehicle manufacture
⚙ Heavy engineering
🚛 Bauxite mining
⛏ Coal mining
⛏ Gold mining
🔌 Electronics
🖥 Computers
△ Metallurgy
🍺 Brewing
⚗ Chemicals

Nhulunbuy
Weipa
Tennant Creek
Brisbane
Sydney
Canberra
Melbourne
Hobart
Perth
Kalgoorlie
Adelaide

0 400 km
0 400 miles   * significant multinational ownership

A

RESOURCES

 Electric power 39.7m kw

250,075 tonnes

733,000 b/d (reserves 3.5bn barrels)

120m sheep, 27.6m cattle, 96m chickens

Coal, iron, bauxite, zinc, lead, copper, nickel, opals, gold, uranium

ELECTRICITY GENERATION

Hydro 8% (16bn kwh)	
Combustion 92% (179bn kwh)	
Nuclear 0%	
Other 0%	

0   20   40   60   80   100

% of total generation by type

Australia has one of the world's most important mining industries. It is a world leader in exports of coal, iron ore, gold, bauxite, and copper. Minerals account for a tenth of Australia's GDP and over half of its merchandise export earnings. Since the first discoveries of coal in 1798, mineral production in Australia has risen every year; in the decade to 1992 it doubled. Growth slowed but continued during the 1990s: many new projects are planned. The share of minerals in the total economy is expected to continue growing, but, having benefited from Australia's location close to the markets of southeast Asia, it was left vulnerable following the regional crisis of 1997.

While minerals underpin much of Australia's wealth, there is growing concern at the environmental cost of extraction. There is also ongoing uncertainty over the possibility of Aboriginal claims to land holding valuable minerals. The 1992 "Mabo Judgment" recognized Aboriginal land rights predating European settlement. The 1993 Native Title Act confirmed these rights, and in 1996 the High Court's historic "Wik decision" enabled claims to be made over land which was subject to a "pastoral" lease. But legislation passed in 1998 cut back Aborigines' rights to make such claims.

AUSTRALIA : LAND USE

GIBSON DESERT

GREAT VICTORIA DESERT

Darling

Murray

GREAT DIVIDING RANGE

Cropland
Pasture
Forest
Wetlands
Desert
Wheat - cash crop
Sheep
Cattle

0   500 km
0   500 miles

Green Island in the far north of Queensland. It is part of the Great Barrier Reef which stretches 1995 km (1240 miles) down the coast.

ENVIRONMENT

 Sustainability rank: 16th

7% (2% partially protected)

17.7 tonnes per capita

ENVIRONMENTAL TREATIES

Yes	Yes	Yes
Yes	Yes	No

Australians are among the most environmentally conscious in the world, but the government has strongly resisted any commitment to cutting greenhouse gas emissions to help limit global warming. It has not even endorsed the 1997 Kyoto protocol, which would in fact allow Australia to increase emissions by up to 8% until 2010, whereas most industrialized countries had to commit to cuts.

Green issues are dominated by Greenpeace and the less radical Australian Conservation Foundation (ACF), which has been highly critical of Australia's record of destroying natural habitats.

MEDIA

 TV ownership high

Daily newspaper circulation 296 per 1000 people

PUBLISHING AND BROADCAST MEDIA

There are 65 daily newspapers, mostly state circulation. The leading papers are the *Sydney Morning Herald* and *The Australian*

1 state-owned service, 44 independent commercial stations

6 state-owned networks, 166 independent commercial stations

The Australian press is firmly in the grip of press barons such as Rupert Murdoch and Kerry Packer, although cross-media rules prevent the ownership of newspapers and TV channels in the same city. Public-sector broadcasting remains dominated by the politically neutral Australian Broadcasting Corporation (ABC), which receives complaints about its coverage from all main parties.

CRIME

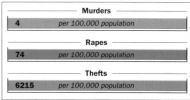 No death penalty

21,126 prisoners

Little change in 1999

CRIME RATES

Murders	
4	per 100,000 population

Rapes	
74	per 100,000 population

Thefts	
6215	per 100,000 population

Each state has its own police force and court system. The High Court and Family Court both have national jurisdiction. Since the 1970s, the legal system has been placing greater emphasis on individual rights. The disproportionate number of Aboriginal deaths in custody is of concern, as are rising narcotics-related offenses. Australia is active in narcotics control throughout southeast Asia. Gun control laws were strengthened following the 1996 Port Arthur shooting, when a lone gunman killed 35 people. In 1997 the Wood inquiry uncovered widespread police corruption in New South Wales and led to major reforms.

EDUCATION

 School leaving age: 15

99%

1.04m students

THE EDUCATION SYSTEM

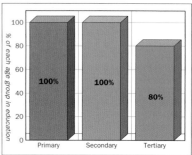

% of each age group in education

100
80
60
40
20
0

Primary 100%
Secondary 100%
Tertiary 80%

Education in Australia is a state responsibility, except in Canberra (where it is funded by the federal government). State education departments run the government schools and set the policies for educational practice and standards for all schools. Nongovernment schools, run by religious and other groups, exist in all states. Special provision is made for inaccessible outback areas, with recent moves to bring new technologies to the bush.

Schooling is compulsory from age 5–6 to age 15–16 in all states. After their final year at school, students sit for the Higher School Certificate. Universities are independent of state control and are funded by the federal government.

ABORIGINAL RIGHTS

CENTRAL TO ABORIGINAL culture and beliefs is the relationship with tribal lands. The Aborigines' land rights campaign, like their civil rights campaign, began gathering momentum in the 1960s, when they were first included in the census and won the right to vote. Now a key political issue, it has far-reaching implications for national identity, for the country's most disadvantaged people, and for powerful mining and farming interests.

BACKGROUND

In the millennia before the arrival of Europeans, Aboriginal peoples ranged widely over their land. Their number was then greatly reduced by exposure to disease and by conflict. Settlers acquired land without reference to any preexisting rights of tribal peoples, based on the legal doctrine that before 1788 the land was *terra nullius* – belonging to no one. Those Aborigines still living a traditional lifestyle were largely confined to reserves.

Land rights campaigns won an initial breakthrough when land councils were set up, first in the Northern Territory in 1976, then in other states. These provided a structure for holding freehold title to land in trust for its tribal inhabitants. However, in the absence of a uniform national land rights policy conservative states such as Queensland – with the largest Aboriginal population – resisted moves to consolidate Aborigine ownership. Particular flashpoints arose when mining companies were granted concessions to exploit sacred sites.

NATIVE TITLE ACTS IN THE 1990s

The first nationwide Native Title Act was introduced in 1993. It recognized the new situation created by a crucial 1992 court ruling. The so-called Mabo Judgment established that rights to

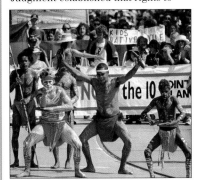

Native dancers from Kuranda, northern Queensland, perform a traditional dance in front of protesters marching for increased Aboriginal land rights.

An Aboriginal elder surveys his native land in Western Australia.

land ownership did indeed exist in common law based on native title, effectively reversing the concept of *terra nullius* and recognizing the pre-1788 Aboriginal occupancy.

The 1993 law specified that native title existed for all Crown land held by federal or by state government, unless it had specifically been extinguished. Native title met powerful resistance from mining companies, especially in Western Australia. In 1996, a court case brought by the Wik people of the Cape York Peninsula in Queensland took the matter further. The Wik ruling said that native title still coexisted with the rights of farmers who had long leases from the Crown on huge tracts of grazing land, granted earlier in the century. Changes of use on such land (to allow mining, cash crops, and tourist developments) would therefore require consultation with Aborigines.

The Liberal–National government responded with a plan to protect leaseholders. Its own Native Title Amendment Bill, tightly restricting Aborigine claims, split the country but was eventually passed in 1998.

As Australia's indigenous population (410,000 and rising) develops a growing sense of pride and assertiveness, court battles continue over mining developments on sensitive tribal sites. Meanwhile, Liberal prime minister John Howard has resisted pressure to go beyond expressions of "regret" and make a formal government apology for the way in which white settlers mistreated Aboriginal people. Campaigners for such an apology see it as central to a national reconciliation agreement, especially after the recent controversy over the "stolen generation" – Aboriginal children forcibly removed from their families and brought up by whites under an assimilation policy which continued until the late 1960s.

HEALTH

 Welfare state health benefits

1 per 400 people

Heart, cerebrovascular, and respiratory diseases, cancers

Australia's extensive public health service has among the highest standards in the world. Hospital waiting lists are short. Outback areas are serviced by the efficient Royal Flying Doctor Service. While vigilance continues in the areas of hygiene, nutrition, and general living standards, health authorities have targeted heart disease, injury prevention, personal fitness, Aboriginal health, and the prevention of cancers – particularly lung, cervical, breast, and skin cancers – as current priorities. Incentives to encourage private health insurance, introduced during the 1990s, sparked fears over public health funding and quality.

SPENDING

 GDP/cap. increase

CONSUMPTION AND SPENDING

485 per 1000 population

525 per 1000 population

Defense	1.9%
Education	4.8%
Health	6%

Defense, Health, Education spending as % of GDP

Australians traditionally enjoyed reasonable equality of wealth distribution. A large proportion of families own two cars and have relatively high disposable incomes, and a benign climate helps most people to live comfortably. However, high unemployment during the 1990s recession widened the gap between rich and poor, and Australia slipped down the world standard of living list for a few years. The incidence of homelessness, critical poverty, and child neglect due to poverty have increased in recent years.

WORLD RANKING

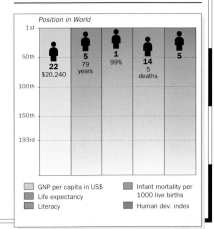

Position in World

22 $20,240	5 79 years	1 99%	14 5 deaths	5

GNP per capita in US$
Life expectancy
Literacy

Infant mortality per 1000 live births
Human dev. index

AUSTRIA

OFFICIAL NAME: Republic of Austria   **CAPITAL:** Vienna   **POPULATION:** 8.1 million
CURRENCY: Euro (Schilling until 2002)   **OFFICIAL LANGUAGE:** German

L YING IN THE HEART OF EUROPE, Austria is dominated by the Alps in the west, while fertile plains make up the east and north. A separate republic after the collapse of the Austro-Hungarian Empire, Austria was absorbed into Hitler's Germany in 1938. It regained independence in 1955 after the departure of the last Soviet troops from the Allied Occupation Force. Its economy encompasses successful high-tech sectors, a tourist industry which attracts wealthier visitors, and a strong agricultural base. Joining the EU in 1995, in 2002 it was one of 12 EU states to adopt the euro.

CLIMATE

> Mountain/continental

WEATHER CHART FOR VIENNA

Austria has a temperate continental climate. Alpine areas experience colder temperatures and higher precipitation.

TRANSPORTATION

> Drive on right

Schwechat, Vienna
11.9m passengers

24 ships
89,600 grt

THE TRANSPORTATION NETWORK

200,000 km (124,274 miles)	1634 km (1015 miles)
5740 km (3567 miles)	358 km (222 miles)

Austria's central location in Europe has encouraged the development of a sophisticated communications and transportation network.

The Tirol is situated in the heart of Austria's Alps. It is the most mountainous region of all and attracts both winter and summer visitors.

TOURISM

> Visitors : Population 2.2:1

 18.2m visitors   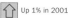 Up 1% in 2001

MAIN TOURIST ARRIVALS

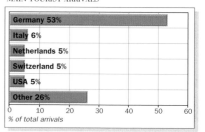

Germany 53%
Italy 6%
Netherlands 5%
Switzerland 5%
USA 5%
Other 26%

% of total arrivals

Well-developed Alpine skiing and winter sports resorts account for almost one-third of the country's total tourist earnings. Many resorts, such as St. Anton and Kitzbühel, cater for the top end of the market. In the summer season, which peaks in July and August, tourists visit the scenic Tirol and the lakes around Bad Ischl. Year-round major attractions are Vienna, with its coffee houses and the Prater park (whose Ferris wheel was immortalized in *The Third Man*), and Salzburg, Austria's second city. The latter is internationally famous for its summer music festival and as the birthplace of Mozart.

AUSTRIA

Total Land Area : 83 858 sq. km (32 378 sq. miles)

LAND HEIGHT

3000m/9843ft
2000m/6562ft
1000m/3281ft
500m/1640ft
200m/656ft
Sea Level

POPULATION

over 1 000 000
over 500 000
over 100 000
over 50 000
over 10 000

PEOPLE　▷ Pop. density medium

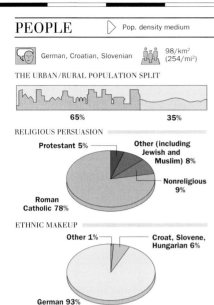

German, Croatian, Slovenian　　98/km² (254/mi²)

THE URBAN/RURAL POPULATION SPLIT

65%　　　35%

RELIGIOUS PERSUASION

Protestant 5%　Other (including Jewish and Muslim) 8%
Nonreligious 9%
Roman Catholic 78%

ETHNIC MAKEUP

Other 1%　Croat, Slovene, Hungarian 6%
German 93%

Austrian society is homogeneous. Almost all Austrians are German speakers, although Austrians like to stress their distinctive identity in relation to Germany. Minorities are few; there are some ethnic Slovenes, Croats, and Hungarians in the south and east, as well as some Roma communities. These minorities have been augmented by large numbers of immigrants from eastern Europe and one-time refugees from the former Yugoslavia. The result has been a perceptible increase in ethnic tension, particularly as the far right claims that migrants are taking jobs from the local population.

The nuclear family is the norm in Austria. It is common for both parents to work. While gender equality is enshrined in the constitution, in practice society is still strongly patriarchal. Compared with the rest of Europe, few women enter politics.

Young Austrians tend to live in their parental home until they marry. This practice reflects the long time taken to complete university degrees, for which students do not receive maintenance grants. Austrians marry at a younger age than the European average. Nominally a Catholic country, Austria is socially less conservative than some Catholic German *Länder*.

POPULATION AGE BREAKDOWN

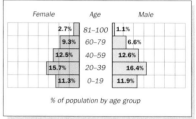

Female	Age	Male
2.7%	81–100	1.1%
9.3%	60–79	6.6%
12.5%	40–59	12.6%
15.7%	20–39	16.4%
11.3%	0–19	11.9%

% of population by age group

POLITICS　▷ Multiparty elections

L. House 1999/2003
U. House varies by province

President Thomas Klestil

AT THE LAST ELECTION

National Council 183 seats

36% SPÖ　28% FPÖ　28% ÖVP　8% GA

SPÖ = Social Democratic Party of Austria **FPÖ** = Freedom Party of Austria **ÖVP** = Austrian People's Party **GA** = Green Alternative

Federal Council 64 seats

42% ÖVP　34% SPÖ　24% FPÖ

Austria is a federal, multiparty democracy. The chancellor (premier) holds real executive power.

PROFILE

The long-established "grand coalition" between the socialist SPÖ and the conservative ÖVP governed Austria almost without interruption from the 1950s until 1999. Their hold on power, with no real alternative, reached into all areas of public life. The far-right FPÖ achieved a breakthrough in elections in October 1999, when it came equal second with the ÖVP. A new right-wing coalition of the ÖVP and the FPÖ was met with regional and international criticism. FPÖ leader Jörg Haider drew strong condemnation, having openly expressed admiration for some of Hitler's policies. He remained a driving force behind the party despite resigning formal leadership in May 2000. Later in that year, local elections showed dwindling support for the FPÖ,

prompting talk of Haider's return to the forefront of Austrian politics. By 2002 the ruling coalition was under strain. The FPÖ adopted a far more skeptical approach to EU enlargement, calling for a veto of the Czech Republic's membership bid unless the Czech Temelín nuclear energy plant was closed.

The nine provincial assemblies and governments have considerable powers. Vienna, with provincial status, has long been dominated by the SPÖ.

MAIN POLITICAL ISSUE
Relations with EU partners

Austria's population is divided over the merits of EU membership. While the farming lobby remains apprehensive of EU agricultural policies, the country as a whole has benefited from lower food prices and greater consumer choice. Fears that membership is eroding national identity, as well as encouraging an influx of cheap east European labor, provided an electoral basis for the nationalist FPÖ. In turn, EU states responded to the FPÖ's membership of the ruling coalition in 2000 by imposing diplomatic sanctions for seven months.

Dr. Thomas Klestil, *the ÖVP candidate, was elected president in 1992.*

Wolfgang Schüssel, *ÖVP chancellor since 2000, in coalition with the far right.*

WORLD AFFAIRS　▷ Joined UN in 1955

EU　CE　PfP　OECD　OSCE

Despite the importance of relations with Germany, Austria's powerful northern neighbor and main trading partner, there has been a conscious policy of stressing Austria's independence and creating some diplomatic distance. Austria is keen to maintain its relationship with the US, which is reinforced by Austria's role as supplier of small arms to the US Army. The inclusion of the far right in the Austrian government in 2000 provoked the imposition of diplomatic sanctions by EU states for seven months.

Austria's status as a neutral state has begun to be questioned since it joined the EU and NATO's Partnerships for Peace program. Austria is part of the Schengen Convention, ending border controls with other participating EU members. Its geopolitical position gives it considerable influence in eastern Europe, and exports to the region trebled in the 1990s. The ÖVP has strongly supported the eastward enlargement of the EU, but the FPÖ is more hostile.

AID　▷ Donor

$423m (donations)　Down 20% in 2000

New projects are now assessed for their impact on the environment and on gender issues. Poland is the largest recipient of official aid and Bosnia and Herzegovina the second. A major exporter to the former Yugoslavia before the wars there in the 1990s, Austria has a key role in regional reconstruction.

CHRONOLOGY

Austria came under the control of the Habsburgs in 1273. In 1867, the Dual Monarchy of Austria-Hungary was formed under Habsburg rule. Defeat in World War I in 1918 led to the breakup of the Habsburg empire and the formation of the Republic of Austria.

❑ **1934** Chancellor Dollfuss dismisses parliament and starts imprisoning social democrats, communists, and National Socialist (Nazi) Party members. Nazis attempt coup.

❑ **1938** The Anschluss – Austria incorporated into Germany by Hitler.

❑ **1945** Austria occupied by Soviet, British, US, and French forces. Elections result in ÖVP–SPÖ coalition.

❑ **1950** Attempted coup by Communist Party fails. Marshall Aid helps economic recovery.

❑ **1955** Occupying troops withdrawn. Austria recognized as a neutral sovereign state.

❑ **1971** SPÖ government formed under Chancellor Bruno Kreisky who dominates Austrian politics for 12 years.

❑ **1983** Socialists and the FPÖ form a coalition government under Fred Sinowatz.

❑ **1986** Kurt Waldheim, former UN secretary-general, elected president, despite war crimes allegations. Franz Vranitzky replaces Sinowatz as federal chancellor. Nationalist Jörg Haider becomes FPÖ leader, prompting the SPÖ to pull out of government. Elections produce stalemate. Return to "grand coalition" of SPÖ–ÖVP.

❑ **1990** ÖVP loses support in elections.

❑ **1992** Thomas Klestil (ÖVP) elected president. Elections confirm some traditional ÖVP supporters defecting to FPÖ.

❑ **1995** Austria joins EU. Elections after coalition disagreement over budget; SPÖ and ÖVP increase representation; "grand coalition" re-forms in early 1996.

❑ **1997** Vranitzky resigns; replaced by Viktor Klima.

❑ **1998** Klestil reelected president.

❑ **1999** Haider's FPÖ wins 40% of votes in Carinthia regional poll, is equal second with ÖVP in general election in October; SPÖ remains as largest party.

❑ **2000** ÖVP accepts FPÖ into coalition, with Wolfgang Schüssel as chancellor; political crisis. EU imposes diplomatic sanctions, lifted after seven months.

❑ **2002** Euro fully adopted.

DEFENSE

 Compulsory military service

💲 $1.61bn   ⬇ Down 5% in 2000

AUSTRIAN ARMED FORCES

🚙	277 main battle tanks (163 M-60A3, 114 *Leopard* 2A4)	34,600 personnel
🚢	None	
✈	52 combat aircraft (23 SAAB J-350e)	6500 personnel
🚀	None	

Under the 1955 State Treaty, which granted Austria its full independence, Austria is neutral, although it has participated in NATO's Partnerships for Peace program since 1995.

Despite the small size of its own forces, the Austrian arms industry is strong. It not only meets most of the hardware needs of its own army, but also exports arms to the US and other countries.

ECONOMICS

▷ Inflation 2% p.a. (1990–2000)

📊 $205bn   💱 1.0651–1.1231 euros

SCORE CARD

❑ WORLD GNP RANKING	22nd
❑ GNP PER CAPITA	$25,220
❑ BALANCE OF PAYMENTS	–$5.21bn
❑ INFLATION	2.4%
❑ UNEMPLOYMENT	6%

ECONOMIC PERFORMANCE INDICATOR

EXPORTS

IMPORTS

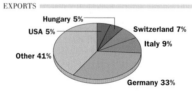

STRENGTHS

Large manufacturing base. Strong chemical and petrochemical industries. Electrical engineering sector, textiles, and wood processing industries. Highly skilled labor force. Tourism is an important foreign currency earner.

WEAKNESSES

Lacks natural resources. Reliant on imported raw materials, particularly oil and gas. Process of introducing greater competition and deregulation has been slow.

PROFILE

Austria's industrial and high-tech sector is highly developed and contributes over a quarter of GDP. Some services, notably tourism, are highly sophisticated and profitable, although tourism receipts have been down in recent years.

A recession in the early 1990s was reversed by a rapid increase in exports to eastern Europe and Germany and by increased domestic demand.

There have been benefits from EU membership since 1995. Prices for many products, particularly food and books, have fallen. The Austrian labor market has also seen an influx of migrant labor more willing to accept flexible working arrangements and lower wages. Foreign investment has increased, as more multinationals locate their headquarters for east European operations in Austria. A far-reaching fiscal stabilization program instituted after accession enabled Austria to meet the economic convergence criteria necessary for it to join the final stage of economic and monetary union, fully adopting the euro in January 2002.

AUSTRIA : MAJOR BUSINESSES

❀	Textiles
◉	Chemicals
△	Metallurgy
🔌	Electronics
📇	Iron & steel
✒	Pharmaceuticals
◉	Light engineering
✿	Heavy engineering

RESOURCES

 Electric power 18.3m kw

 3502 tonnes

19,228 b/d (reserves 77m barrels)

3.43m pigs, 2.16m cattle, 11.1m chickens

Iron, coal, magnesite, zinc, lead

ELECTRICITY GENERATION

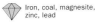

Hydro 67% (39bn kwh)

Combustion 33% (19bn kwh)

Nuclear 0%

Other 0%

0   20   40   60   80   100

% of total generation by type

Austria has few resources. It lacks significant oil, coal, and gas deposits and has to import a large amount of its energy. Russia is a key energy supplier, and gas is provided via pipelines running through the Czech Republic and Slovakia. Oil is imported

AUSTRIA : LAND USE

Cropland
Pasture
Forest
High mountain regions
Pigs
Cattle
Wheat

0   100 km
0   100 miles

up the Danube. Russia and Germany are the major suppliers of iron ore and raw steel for Austria's industry.

ENVIRONMENT

 Sustainability rank: 7th

 30% (24% partially protected)

7.9 tonnes per capita

ENVIRONMENTAL TREATIES

Yes   Yes   Yes

Yes   Yes   Yes

Environmental awareness is high and the government invests nearly 3% of GDP in environmental protection. Roughly half of all domestic waste is separated for recycling, with heavy fines for failing to observe regulations. The safety of nuclear reactors in the neighboring Czech Republic, Slovakia, and Slovenia is a major concern.

MEDIA

 TV ownership high

Daily newspaper circulation 296 per 1000 people

PUBLISHING AND BROADCAST MEDIA

There are 17 daily newspapers, including the *Neue Kronen-Zeitung*. The *Wiener Zeitung* is the world's oldest daily paper

2 state-owned channels

1 state-owned service, broadcasting 3 national and 9 regional stations

TV and radio are controlled by the Austrian Broadcasting Company (ÖRF), which has a politically appointed general director. Private radio stations are restricted to regional broadcasting. Cable TV is licensed by ÖRF to prevent it taking viewers from existing stations. A German-language satellite channel is run jointly with German and Swiss TV.

CRIME

 No death penalty

 6896 prisoners

Up 3% in 1999

CRIME RATES

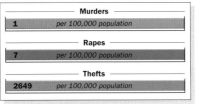

Murders

1   per 100,000 population

Rapes

7   per 100,000 population

Thefts

2649   per 100,000 population

Austria's crime rate is below Europe's average. However, the number of burglaries is rising. The arrival of the Russian mafia in Vienna has led to an increase in money laundering.

EDUCATION

 School leaving age: 15

 99%

230,030 students

THE EDUCATION SYSTEM

100

80

60

40

20

0

% of each age group in education

100%   96%   50%

Primary   Secondary   Tertiary

Of total government expenditure some 10% is spent on education. Children are streamed into two types of secondary school according to their academic ability. Those in a *Gymnasium* (11–18) take the *Reifeprüfung* or *Matura* which determines entry to university, but children in a *Hauptschule* (11–15) do not. The universities are oversubscribed; students take six years or more to finish their first degrees.

HEALTH

 Welfare state health benefits

1 per 333 people

Heart and cerebrovascular diseases, cancers

Austria has relatively high levels of spending on health, accounting for 13% of total government expenditure. However, private spending on health is increasing and accounts for nearly one-third of the total, the highest proportion of any EU state, as patients increasingly choose to use the private health sector to avoid waiting lists for operations.

SPENDING

 GDP/cap. increase

CONSUMPTION AND SPENDING

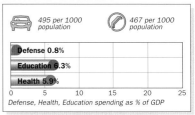

495 per 1000 population

467 per 1000 population

Defense 0.8%

Education 6.3%

Health 5.9%

0   5   10   15   20   25

Defense, Health, Education spending as % of GDP

Despite four decades of mainly left-of-center government, Austria has retained many of its traditional social divisions. Inherited wealth is still respected above earned wealth, and social mobility is lower than in neighboring Germany. Austrians have the highest savings rate of any OECD country. Relatively few Austrians own stocks and shares, and most companies are dominated by a single shareholder. Limited amounts are invested in property. Legislation in 2000 banned anonymous savings accounts, a system unique in the EU to Austria which, it had been argued, encouraged money laundering and insider dealing. Government bonds offer low rates of interest and the property market is weak; many people, particularly in Vienna, tend to rent rather than buy their apartments. Refugees from conflicts in the former Yugoslavia form the poorest group in Austrian society.

WORLD RANKING

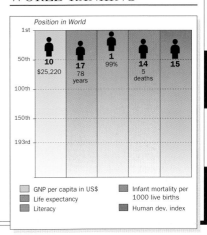

Position in World

1st

50th

100th

150th

193rd

10   $25,220

17   78 years

1   99%

14   5 deaths

15

GNP per capita in US$
Life expectancy
Literacy

Infant mortality per 1000 live births
Human dev. index

A

AZERBAIJAN

OFFICIAL NAME: Republic of Azerbaijan   **CAPITAL:** Baku
POPULATION: 8.1 million   **CURRENCY:** Manat   **OFFICIAL LANGUAGE:** Azerbaijani

1991 1991 May 28 AZ +4 +994 .az

SITUATED ON THE WEST COAST of the Caspian Sea, Azerbaijan was the first Soviet republic to declare independence. The issue of the disputed enclave of Nagorno Karabakh, whose Armenian population seeks secession, led to full-scale war (1988–1994) and is still a dominant concern. Over 200,000 refugees, and more than twice as many internally displaced, added to the problems of the troubled economy. Azerbaijan's oil wealth, however, gives it long-term potential.

Landscape typical of the Lesser Caucasus mountains near Qazax in the extreme northwest of Azerbaijan.

CLIMATE   ▷ Mountain/steppe

WEATHER CHART FOR BAKU

Average daily temperature														Rainfall
°C/°F	J F M A M J J A S O N D													cm/in
40/104														40/16
30/86														30/12
20/68														20/8
10/50														10/4
0/32														0
-10/14														
-20/4														

Coastal areas are subtropical, but bitter winters inland have become a life-or-death issue for thousands of refugees.

TRANSPORTATION   ▷ Drive on right

✈ Baku   🚢 284 ships
                    647,022 grt

THE TRANSPORTATION NETWORK

23,057 km (14,327 miles)	None
2116 km (1315 miles)	None

Buses provide the most efficient public transportation. Access to the Naxçivan enclave is by air or via Iran.

TOURISM   ▷ Visitors : Population 1:12

🧳 681,000 visitors   ⬆ Up 13% in 2000

MAIN TOURIST ARRIVALS

CIS 62%	
Iran 29%	
Turkey 4%	
Other 5%	

0 10 20 30 40 50 60 70 80
% of total arrivals

Because of the Nagorno Karabakh conflict, and strong anti-Western feelings (Azerbaijan perceives the West as taking the Armenian side), there are few visitors, most of them on business.

PEOPLE   ▷ Pop. density medium

Azerbaijani, Russian   👥 94/km² (242/mi²)

THE URBAN/RURAL POPULATION SPLIT

57%   43%

ETHNIC MAKEUP

Armenian 2%   Russian 3%
Other 2%   Dagestani 3%

Azeri 90%

Before independence Armenians and Russians accounted for 11% of the population. The violence of the early 1990s and accompanying Azeri nationalism caused thousands to flee. Racial hostility against those who remain is increasing.

Women, once prominent in the ruling party, have lost their political status, and their general status is declining.

The once effective social security system is now under great strain.

POLITICS   ▷ Multiparty elections

🗳 2000/2005   President Heydar Aliyev

AT THE LAST ELECTION

National Assembly 125 seats   2% VBP   1% Vacant

61% YAP   21% Ind   5% AKC   10% Others

YAP = New Azerbaijan Party   **Ind** = Independents
AKC = Azerbaijan Popular Front   **VBP** = Civic Solidarity Party
Vacant = Seat reserved for member from Nagorno Karabakh

The determination of the Nagorno Karabakh enclave to secede from Azerbaijan led to war from 1988 to 1994, with Armenian forces gaining control over 20% of Azeri territory. Peace talks have yet to reach an accord.

The YAP replaced the communists in 1995 and again dominated elections in 2000, which observers criticized as seriously flawed. It supports Heydar Aliyev, first elected president in 1993 (at the age of 70) and reelected in 1998.

AZERBAIJAN

Total Land Area : 86 600 sq. km (33 436 sq. miles)

POPULATION
▫ over 1 000 000
◎ over 100 000
○ over 50 000
● over 10 000
• under 10 000

LAND HEIGHT
4000m/13 124ft
3000m/9843ft
2000m/6562ft
1000m/3281ft
500m/1640ft
200m/656ft
Sea Level

A

WORLD AFFAIRS

 Joined UN in 1992

 CIS CE EAPC OIC OSCE

Relations with Armenia remain the central foreign policy issue. Turkey, with its common history and culture, is a natural ally. The West, neighboring Iran (with a large Azeri population), and Russia are interested in Azeri oil fields in the Caspian Sea. Azerbaijan joined the Council of Europe in 2001.

AID

▷ Recipient

 $139m (receipts)   ⬇ Down 18% in 2000

World Bank aid has grown steadily. From 1992 to 2002 a pro-Armenian US Congress allowed limited humanitarian aid only.

DEFENSE

▷ Compulsory military service

$213m   ⬆ Up 5% in 2000

Azerbaijan has been a member of NATO's Partnership for Peace program since 1994. Its naval forces operate under CIS control.

ECONOMICS

▷ Inflation 199% p.a. (1990–2000)

$4.85bn   4456–4475 manats

SCORE CARD

❑ WORLD GNP RANKING	113th
❑ GNP PER CAPITA	$600
❑ BALANCE OF PAYMENTS	–$168m
❑ INFLATION	1.8%
❑ UNEMPLOYMENT	1%

STRENGTHS
Extensive oil and natural gas reserves starting to come on stream. Iron, copper, lead, and salt deposits. Cotton and silk.

WEAKNESSES
Antiquated Soviet-era industry. Poor infrastructure and corruption threaten development. Oil pipeline to Turkey not yet built. Fallout from war in Nagorno Karabakh still drains state resources.

EXPORTS

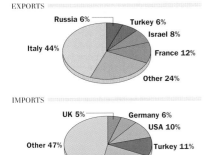

Russia 6%
Turkey 6%
Israel 8%
Italy 44%
France 12%
Other 24%

IMPORTS

UK 5%
Germany 6%
USA 10%
Other 47%
Turkey 11%
Russia 21%

RESOURCES

▷ Electric power 5.2m kw

 4935 tonnes   300,000 b/d (reserves 7bn barrels)

5.56m sheep, 2m cattle, 14.5m chickens   Iron, bauxite, copper, lead, zinc, limestone, salt, oil, gas

Relatively neglected in the Soviet period, Azerbaijan's Caspian Sea oil fields have attracted international interest. The shallowwater Guneshli field alone has over four million barrels of reserves. Offshore natural gas is also plentiful.

ENVIRONMENT

▷ Sustainability rank: 114th

 6%   ⬇ 4.9 tonnes per capita

Under the Soviet regime oil pollution devastated the Caspian Sea, and pesticides were massively overused in agriculture. Major rivers suffer heavy pollution from Georgia and Armenia. Lack of funds restricts action.

MEDIA

▷ TV ownership high

✕ Daily newspaper circulation 10 per 1000 people

PUBLISHING AND BROADCAST MEDIA

There are 6 daily newspapers, including *Bakinskii Rabochii*, *Khalg Gazeti*, and *Respublika*

1 state-controlled service, 1 independent station   1 state-controlled service

A 1998 decree abolished censorship, but freedom is limited by newsprint controls, license restrictions, and intimidation.

CRIME

▷ No death penalty

 24,881 prisoners   ⬇ Down 6% in 1999

The judicial system returned to political control in 1993. Criminality is a particular problem in camps for those displaced in the Nagorno Karabakh conflict. Elsewhere, there is a low rate of violent crime, but assaults in the street have become less rare.

EDUCATION

▷ School leaving age: 17

 96%   156,800 students

When it came to power in the mid-1990s, the YAP began reversing communist control over education policy, which had been particularly noticeable in the teaching of history. Baku State, the largest of eight universities, is investing in modernizing its facilities.

HEALTH

▷ Welfare state health benefits

 1 per 278 people   Heart, cerebrovascular and respiratory diseases, cancers

The already poor health care system effectively collapsed as a result of war and the transition to a market economy.

CHRONOLOGY

Under consecutive Persian, Ottoman, and Russian influences, Azerbaijan, one of the world's major oil producers in 1900, attained independence in 1918.

❑ **1920** Red Army invades. Soviet republic established.

❑ **1922** Incorporated in Transcaucasian Soviet Federative Socialist Republic (TSFSR).

❑ **1930** Forced collectivization of agriculture.

❑ **1936** TSFSR disbanded; Azerbaijan a full union republic (ASSR).

❑ **1945** Attempted annexation of Azeri region of Iran.

❑ **1985** Gorbachev tackles corruption in Communist Party of Azerbaijan.

❑ **1988** Nagorno Karabakh seeks unification with Armenia.

❑ **1990** Nagorno Karabakh attempts secession. Soviet troops move in.

❑ **1991** Independence.

❑ **1993** Heydar Aliyev president; reelected in 1998.

❑ **1994** Cease-fire in war with Armenia over Nagorno Karabakh.

❑ **1995, 1999** General elections. Noncommunist YAP in power.

SPENDING

▷ GDP/cap. decrease

CONSUMPTION AND SPENDING

38 per 1000 population   104 per 1000 population

Defense 4.5%
Education 3.4%
Health 1%

0   5   10   15   20   25
Defense, Health, Education spending as % of GDP

New oil revenues are threatening to create a nouveau riche elite without reaching the 60% of Azerbaijan's population currently living in poverty.

WORLD RANKING

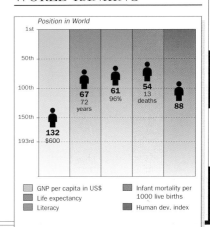

Position in World
1st
50th
100th
150th
193rd

67 / 72 years
61 / 96%
54 / 13 deaths
88
132 / $600

❑ GNP per capita in US$
❑ Life expectancy
❑ Literacy
❑ Infant mortality per 1000 live births
❑ Human dev. index

BAHAMAS

OFFICIAL NAME: Commonwealth of the Bahamas **CAPITAL:** Nassau
POPULATION: 308,000 **CURRENCY:** Bahamian dollar **OFFICIAL LANGUAGE:** English

B

CARIBBEAN
North America
USA · BAHAMAS · Cuba

| 1973 | 1973 | July 10 | BS | -5 | +1242 | .bs |

THE BAHAMAS, ALTHOUGH OFTEN bracketed with Caribbean countries, actually lies northeast of Cuba in the western Atlantic. The archipelago has 700 islands and 2400 cays: just 30 are inhabited. Long established as a tourist resort, the Bahamas today is also a major offshore financial center. It has one of the world's largest open-registry fleets; only a tiny fraction is owned by Bahamian nationals.

CLIMATE ▷ Tropical oceanic

WEATHER CHART FOR NASSAU

■ Average daily temperature   Rainfall ■

The whole of the Bahamas chain has a typically subtropical climate with consistently mild winters. Hurricanes may occur from July to December.

TRANSPORTATION ▷ Drive on left

✈ Freeport International
1.23m passengers

🚢 31.4m grt

THE TRANSPORTATION NETWORK

🛣 1546 km
(961 miles)

None

🚆 None

None

While traveling around and between the major islands is relatively easy, transportation links for the many "Out Islands" are greatly restricted.

TOURISM ▷ Visitors : Population 5.1:1

1.58m visitors   ⬆ Up 1% in 2000

MAIN TOURIST ARRIVALS

| USA 83% |
| Europe 9% |
| Canada 5% |
| Other 3% |

0 10 20 30 40 50 60 70 80 90 100
% of total arrivals

The tourist industry, built around beaches, casinos, and cruise ships, employs over 40% of the population. Larger hotel complexes on the main islands compete with small, family-run guesthouses in the outlying destinations.

BAHAMAS
Total Land Area: 13 940 sq. km (5382 sq. miles)

POPULATION
◎ over 100 000
● over 10 000
• under 10 000

LAND HEIGHT
■ 200 m/656ft
Sea level

0   100 km
0   100 miles

PEOPLE ▷ Pop. density low

English, English Creole, French Creole

31/km²
(80/mi²)

THE URBAN/RURAL POPULATION SPLIT

89%   11%

RELIGIOUS PERSUASION

Methodist 6%
Church of God 6%
Baptist 32%
Other 17%
Anglican 20%
Roman Catholic 19%

Africans first arrived as slaves in the 16th century; their descendants constitute most of the population, alongside a rich white minority. Small families are the norm. Absentee fathers are fairly common, especially in outlying fishing communities. More women are now entering the professions.

POLITICS ▷ Multiparty elections

L. House 2002/2007
U. House 2002/2007

H.M. Queen Elizabeth II

AT THE LAST ELECTION

House of Assembly 40 seats

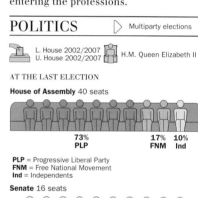

73%
PLP

17%
FNM

10%
Ind

PLP = Progressive Liberal Party
FNM = Free National Movement
Ind = Independents

Senate 16 seats

The members of the Senate are appointed by the governor-general on the recommendation of the prime minister and the leader of the opposition

Twenty-five years of unbroken rule by the PLP under Prime Minister Lynden Pindling were brought to an end by the 1992 elections. His legacy of steering the Bahamas to independence and ending white political domination was undermined by allegations of narcotics-related corruption. There followed a decade of FNM government under Hubert Ingraham, who emphasized tightening up ministerial accountability, introduced legislation to counter money laundering, and achieved relative economic success. However, his privatization drive proved deeply unpopular and the FNM was roundly defeated in 2002 by a resurgent PLP now led by Perry Christie.

B

WORLD AFFAIRS ▷ Joined UN in 1973

The Bahamas is seen as a center for transshipping narcotics and a money-laundering risk, for which G7 blacklisted the country for a year in 2000. The return of unauthorized immigrants to Haiti and Cuba dominates regional relations.

AID ▷ Recipient

 US$6m (receipts)   Down 50% in 2000

Aid is modest. The IDB and the US provide soft development loans. China in 1998 loaned Nassau US$17 million for a convention and theater complex.

DEFENSE ▷ No compulsory military service

US$25m   Down 4% in 2000

The UK is the main trainer of and supplier for the small naval defense force. The interception of narcotics and illegal immigrants is the force's main activity. There is no land army.

ECONOMICS ▷ Inflation 2.7% p.a. (1990–2000)

US$4.53bn   1 Bahamian dollar

SCORE CARD

❏ World GNP Ranking	115th
❏ GNP per Capita	US$14,960
❏ Balance of Payments	–US$438m
❏ Inflation	1.6%
❏ Unemployment	9%

STRENGTHS

Major international financial services sector, including banking, insurance, and business trade center. Major tourism and cruise ship destination. Growing container port. International ship registration.

WEAKNESSES

Growing competition in financial services and tourism from the countries of the Caribbean.

EXPORTS

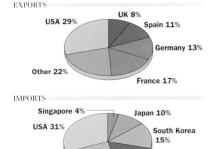

USA 29% · UK 8% · Spain 11% · Germany 13% · France 17% · Other 22%

IMPORTS

Singapore 4% · Japan 10% · USA 31% · South Korea 15% · Italy 17% · Other 23%

Archetypal island paradise. Its natural beauty attracts more than five tourists per inhabitant to the Bahamas every year.

RESOURCES ▷ Electric power 401,000 kw

 10,474 tonnes   Not an oil producer

13,852 goats, 6418 sheep, 4873 pigs, 4.5m chickens   Salt, aragonite

The Bahamas has no strategic resources. A 13.5 MW electricity generating plant was opened in 1998.

ENVIRONMENT ▷ Not available

9%   6.1 tonnes per capita

As on many Caribbean islands, hotel overdevelopment is a major cause for concern. Environmental groups have also pointed out the potential for accidents posed by the Bahamas' enormous oil storage depots.

MEDIA ▷ TV ownership medium

Daily newspaper circulation 99 per 1000 people

PUBLISHING AND BROADCAST MEDIA

There are 4 daily newspapers, the *Nassau Guardian*, the *Tribune*, the *Bahama Journal*, and the *Freeport News*

1 state-owned service   5 services: 1 state-owned, 4 independent

The state-owned TV channel faces very stiff competition from Florida-based US broadcasters.

CRIME ▷ Death penalty in use

1401 prisoners   Down 29% in 1999

The death penalty remains in force. Violent crime, ranging from narcotics-related murders to serious vandalism, is on the increase. Tourists can be targets for petty thefts. Illegal weapons are readily available.

EDUCATION ▷ School leaving age: 15

96%   5305 students

Schooling follows the former British selective system. Tertiary students attend the University of the West Indies or colleges in the US.

CHRONOLOGY

Once an English pirate base, the Bahamas, which gained its first parliament in 1729, formally became a British colony in 1783.

❏ **1920–1933** US prohibition laws turn Bahamas into prosperous bootlegging center.
❏ **1959–1962** Introduction of male suffrage; women gain the vote.
❏ **1973** Independence.
❏ **1983** Narcotics-smuggling scandals involving the government.
❏ **1992** FNM wins elections, ending 25 years of PLP rule.
❏ **2002** PLP returned to power.

HEALTH ▷ Welfare state health benefits

1 per 658 people   Obstetric causes, heart diseases, cancers, murders, accidents

The Bahamian health service combines state and private systems. In the outlying islands access to care relies on the Flying Doctor Service and around 50 local health centers. There are two private clinics on New Providence.

SPENDING ▷ GDP/cap. increase

CONSUMPTION AND SPENDING

46 per 1000 population   376 per 1000 population

Defense 0.6%
Education 3.6%
Health 2.5%

Defense, Health, Education spending as % of GDP

There are marked wealth disparities: urban professionals who work in the financial sector are at one end of the scale, and the poor fishermen from the outlying islands are near the other. Cuban and Haitian refugees, who have no legal status, are the poorest group of all.

WORLD RANKING

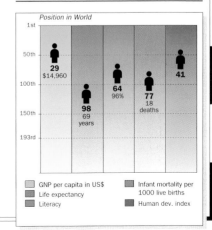

Position in World

29 $14,960 · 98 69 years · 64 96% · 77 18 deaths · 41

GNP per capita in US$ · Life expectancy · Literacy · Infant mortality per 1000 live births · Human dev. index

BAHRAIN

OFFICIAL NAME: State of Bahrain **CAPITAL:** Manama
POPULATION: 652,000 **CURRENCY:** Bahraini dinar **OFFICIAL LANGUAGE:** Arabic

  1971 1971 Dec 16 BRN +3 +973 .bh

BAHRAIN IS AN ARCHIPELAGO state situated between the Qatar peninsula and the Saudi Arabian mainland. Only three of its islands are inhabited. Bahrain Island is connected to Saudi Arabia's Eastern Province by a causeway opened in 1986. Bahrain was the first Gulf emirate to export oil; its reserves are now almost depleted. Services such as offshore banking, insurance, and tourism are major employment sectors for skilled Bahrainis.

CLIMATE 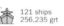 Hot desert

WEATHER CHART FOR MANAMA

Temperatures soar to 40°C (104°F) in June–September. In December–March the weather is pleasantly warm.

TRANSPORTATION Drive on right

 Bahrain International, Manama
3.93m passengers

121 ships
256,235 grt

THE TRANSPORTATION NETWORK

2433 km (1512 miles)	None
None	None

The King Fahd Causeway connects Bahrain to Saudi Arabia, and ferries travel regularly to Iran. Buses are the main form of public transportation.

TOURISM Visitors : Population 3.1:1

1.99m visitors

Up 14% in 1999

MAIN TOURIST ARRIVALS

Saudi Arabia 68%	
India 6%	
UK 4%	
Other 22%	

% of total arrivals

Bahrain's relatively liberal lifestyle has made it something of a magnet for visitors from neighboring Gulf states. Bahrain has a modern airport and is a center for business conventions.

PEOPLE Pop. density high

Arabic

924/km²
(2388/mi²)

THE URBAN/RURAL POPULATION SPLIT

92% **8%**

ETHNIC MAKEUP

European 2%
Other Arab 4%
Iranian, Indian, Pakistani 24%
Bahraini 70%

Bahrain is the smallest and most densely populated Arab state. The key division is between Sunni and Shi'a Muslims, about 30% and 70% of the population respectively. Sunnis hold the best jobs in business and government. Shi'a Muslims tend to do menial work and have a lower standard of living. The most impoverished Shi'a Muslims tend to be of Iranian descent.

Bahrain has a smaller expatriate population than many other Arab countries. The ruling al-Khalifa family has responded to declining oil reserves by diversifying the economy to provide service industry jobs for Bahrainis.

Bahrain is the most liberal of the Gulf states. Alcohol is freely available. Women have access to education and are not obliged to wear the veil. Since 2000 they have been entitled to participate in the Consultative Council.

The Grand Mosque, Manama. It is the largest building in Bahrain and can accommodate 7000 people.

POLITICS In transition

 Not applicable/2002

H.M. Shaikh Hamad bin Isa al-Khalifa

LEGISLATIVE OR ADVISORY BODIES
Chamber of Deputies 40 seats (to be elected)

A bicameral legislative body is to be formed. The lower Chamber of Deputies is to be elected in October 2002, and the Consultative Council, first appointed in 1993 and elected in 2001, is to form the appointed upper house.

Consultative Council 40 seats

The al-Khalifa family has dominated politics since 1783 by means of an effectively autocratic system, although the amir has been advised since 1993 by a Consultative Council. A charter of political reforms, approved by the council in 2000, was backed by a popular referendum in 2001. Under the charter, Bahrain was declared a constitutional monarchy in February 2002. H.M. Shaikh Hamad, amir since 1999, supports the economic liberalization begun by his father. The repeal of the State Security Law in 2001 promised an end to the detention of political dissidents. Many were Shi'a opponents of the regime, who are traditionally backed by Iran.

WORLD AFFAIRS Joined UN in 1971

 AL Damasc GCC OIC OAPEC

Bahrain has good relations with the UK and the US, but is keen to restore relations with Iraq. There is residual tension with Iran. A dispute with Qatar over the Hawar islands was resolved in Bahrain's favor in 2001.

AID Recipient

$49m (receipts)

 Up sharply in 2000

Bahrain receives low levels of aid, but takes the lion's share from the offshore oil field shared with Saudi Arabia, effectively a subsidy from the latter.

DEFENSE No compulsory military service

$435m

Down 1% in 2000

The strong defense force includes a small but well-equipped air force. US air bases on Bahrain have been used in recent conflicts, including the "war on terrorism" in Afghanistan in 2001, and the US 5th Fleet is also based there.

ECONOMICS
▷ Inflation –0.1% p.a. (1990–1999)

 $4.9bn

 0.3770–0.3771 Bahraini dinars

SCORE CARD

❏ WORLD GNP RANKING	110th
❏ GNP PER CAPITA	$7640
❏ BALANCE OF PAYMENTS	$113m
❏ INFLATION	–0.4%
❏ UNEMPLOYMENT	15%

STRENGTHS
Oil and gas. Arab world's major offshore banking sector since 1980s. Inward investment. Tourism. Aluminum production.

WEAKNESSES
Depleted oil reserves and insufficient diversification. High unemployment. High levels of government borrowing.

BAHRAIN

Total Land Area : 620 sq. km (239 sq. miles)

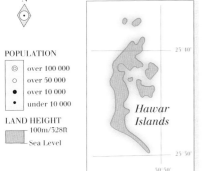

POPULATION

◎ over 100 000
○ over 50 000
● over 10 000
• under 10 000

LAND HEIGHT

▨ 100m/328ft
— Sea Level

EXPORTS

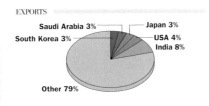

Saudi Arabia 3%
South Korea 3%
Japan 3%
USA 4%
India 8%
Other 79%

IMPORTS

Germany 4%
UK 6%
France 6%
USA 13%
Other 42%
Saudi Arabia 29%

RESOURCES
▷ Electric power 1.1m kw

 10,295 tonnes

 37,473 b/d (reserves 219m barrels)

 17,500 sheep, 16,300 goats, 465,000 chickens

Oil, natural gas

Bahrain remains dependent on its oil and gas industry. Production of crude oil has declined sharply since the 1970s, and there are fears that reserves may run out by 2010. As oil has declined, so gas has assumed greater importance. Most is used to supply local industries, particularly the aluminum plant, which was established in 1972.

ENVIRONMENT
▷ Not available

 None

 29.1 tonnes per capita

Local marine life, particularly the dugong, is vulnerable to upstream oil pollution from the Gulf. Bahrain and Abu Dhabi (UAE) signed an agreement in 2000 on environmental concerns.

MEDIA
▷ TV ownership high

 Daily newspaper circulation 117 per 1000 people

PUBLISHING AND BROADCAST MEDIA

 There are 5 daily newspapers, *Akhbar al-Khaleej, Gulf Daily News, Khaleej Times, Bahrain Tribune,* and *Al-Ayam*

1 state-owned service

 2 services: 1 state-owned, 1 independent

Bahrain has a less authoritarian media regime than most of the Gulf, although government critics have been prosecuted. CNN and BBC satellite TV are freely available.

CRIME
▷ Death penalty in use

 911 prisoners

Down 62% 1996–1998

Crime is minimal, and theft and muggings are rare. Suspected political dissidents are monitored by the police.

CHRONOLOGY
Bahrain has been ruled since 1783 by the al-Khalifa family.

- ❏ **1971** Independence from Britain.
- ❏ **1981** Founder member of GCC.
- ❏ **1990–1991** Bahrain supports US-led action expelling Iraq from Kuwait.
- ❏ **1994–1996** Shi'a unrest.
- ❏ **1999** Accession to throne of Shaikh Hamad bin Isa al-Khalifa.
- ❏ **2001** Referendum approves transition to democracy.

EDUCATION
▷ School leaving age: 15

 88%

 7676 students

Female literacy rates are above the Gulf average. Lack of funding has held up plans for a university.

HEALTH
▷ Welfare state health benefits

 1 per 1000 people

Circulatory diseases, perinatal deaths, injury, poisonings

The high-quality health service is free to Bahraini nationals. Some go abroad for advanced care. The Muharraq Health Center was upgraded in 2001.

SPENDING
▷ GDP/cap. increase

CONSUMPTION AND SPENDING

255 per 1000 population

250 per 1000 population

Defense 6.4%
Education 4.4%
Health 2.6%

0 | 5 | 10 | 15 | 20 | 25
Defense, Health, Education spending as % of GDP

Beneficiaries of the amir's extensive patronage form the wealthiest group in society. Bahrain's largest religious community, the Shi'a Muslims, is also the poorest.

WORLD RANKING

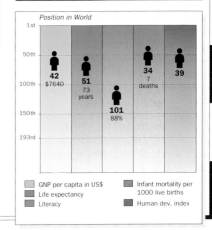

Position in World

1st
50th
100th
150th
193rd

42
$7640

51
73 years

101
88%

34
7 deaths

39

■ GNP per capita in US$
■ Life expectancy
■ Literacy
■ Infant mortality per 1000 live births
■ Human dev. index

BANGLADESH

OFFICIAL NAME: People's Republic of Bangladesh **CAPITAL:** Dhaka
POPULATION: 140.4 million **CURRENCY:** Taka **OFFICIAL LANGUAGE:** Bengali

SOUTH ASIA

LOCATED AROUND the confluence of the mighty Ganges and Jamuna Rivers, Bangladesh is the eastern half of historic Bengal. Most of the country is composed of fertile alluvial plains; the north and northeast are mountainous, as is the Chittagong region in the southeast. After seceding from Pakistan in 1971, Bangladesh had a troubled history of political instability, with periods of emergency rule. Effective democracy was restored in 1991. Bangladesh's major economic sectors are jute production, textiles, and agriculture. Its climate can wreak havoc – in 1991 a massive cyclone killed more than 140,000 people.

CLIMATE

▷ Tropical/subtropical

WEATHER CHART

During the monsoon, the water level generally rises 6 m (20 ft) above normal, flooding up to two-thirds of the country. The floods are made much worse when the Ganges, Jamuna, and Meghna rivers, which converge in a huge delta in Bangladesh, are swollen by the melting of the Himalayan snows and heavy rain in India. Cyclones build up regularly in the Bay of Bengal, with sometimes devastating effects on the flat coastal region.

TRANSPORTATION

▷ Drive on left

 Zia International, Dhaka
2.52m passengers

 310 ships
370,100 grt

THE TRANSPORTATION NETWORK

 19,112 km
(11,876 miles)

None

 2705 km
(1681 miles)

 8433 km
(5240 miles)

Most transportation in Bangladesh is by water, although government policy is now concentrating on developing road and rail links, including the reopening in mid-2000 of a passenger rail service into India. The Bangabandhu bridge across the Jamuna River, which bisects Bangladesh from north to south, was inaugurated in June 1998, after numerous delays. Bangladesh's two major ports, Mungla and Chittagong, are being upgraded to take advanced container ships.

Begum Khaleda Zia, *reelected as prime minister in 2001.*

Sheikh Hasina Wajed, *AL leader and former prime minister.*

TOURISM

▷ Visitors : Population
1:702

 200,000 visitors

 Up 16% in 2000

MAIN TOURIST ARRIVALS

India 36%	
UK 13%	
USA 6%	
Pakistan 5%	
Japan 4%	
Other 36%	

% of total arrivals

The Mughal architecture in Dhaka and the Pala dynasty (7th–10th centuries) city of Sonargaon, just to the southeast, are major attractions, but tourists may be deterred by social and political unrest. Most visitors are Indian businessmen or Bangladeshis living overseas who return to visit relatives.

Traders on the Meghna River. *Life is governed by the vast network of rivers. The floodplains are among the most fertile in the world.*

PEOPLE

▷ Pop. density high

 Bengali, Urdu, Chakma, Marma (Magh), Garo, Khasi, Santhali, Tripuri, Mro

1048/km²
(2716/mi²)

THE URBAN/RURAL POPULATION SPLIT

25%   75%

RELIGIOUS PERSUASION

Other 1%
Hindu 12%
Muslim (mainly Sunni) 87%

ETHNIC MAKEUP

Other 2%
Bengali 98%

Bangladesh is one of the most densely populated countries in the world, despite the fact that three-quarters of the population is rural. As in India, there is considerable Muslim–Hindu tension; in 2001 thousands of Hindus and other religious minorities claimed persecution from the new nationalist government.

Although more than 50% of Bangladeshis, rural and urban, still live below the poverty line, there has been an improvement in living standards over the past decade.

By providing them with an independent income, the textile trade has been a factor in the growing emancipation of Bangladeshi women. They are now included in official employment statistics and are the main customers of the most successful rural bank. Women have led both the government and the opposition. However, Bangladesh was criticized by Amnesty International in 2000 for insufficiently protecting women's rights, and a UN report later that year revealed that almost 50% of Bangladeshi women are victims of domestic violence.

POPULATION AGE BREAKDOWN

Female	Age	Male
0%	80+	0%
2.5%	60–79	3%
8.2%	40–59	9.6%
16.1%	20–39	15.4%
21.8%	0–19	23.4%

% of population by age group

POLITICS ▷ Multiparty elections

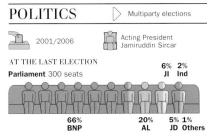

2001/2006

Acting President
Jamiruddin Sircar

AT THE LAST ELECTION

Parliament 300 seats

| | 6% JI | 2% Ind |

66% BNP **20% AL** **5% JD** **1% Others**

BNP = Bangladesh Nationalist Party and allies
AL = Awami League **JI** = Jamaat-e-Islami
JD = Jatiya Dal (Ershad) **Ind** = Independents

Bangladesh returned to multiparty
democracy in 1991, following a
period of military rule.

PROFILE

Between 1975 and 1990 the military
was in power in Bangladesh. The
overthrow of President Ershad in 1990
saw a return to multiparty politics; the
army remains poised, however, to
intervene in the event of a breakdown
in the rule of law. Bangladesh's first
woman prime minister, Begum
Khaleda Zia, head of the BNP,

was elected in 1991. A change from a
presidential to a prime-ministerial
system of government followed.

The AL, which had steered
Bangladesh to independence in 1971,
mounted a sustained campaign against
Khaleda Zia's regime, forcing a rerun
of elections in 1996. It won the largest
number of seats and its leader, Sheikh
Hasina Wajed, became the first prime
minister to complete a full term. The
revived BNP returned to power in the
October 2001 poll, amid much electoral
violence, and formed a coalition with
the hard-line Islamic JI.

MAIN POLITICAL ISSUES
The state sector
Bangladesh is coming under mounting
pressure from multilateral lending
institutions, which account for the
vast majority of the country's capital
inflows, to cut costs in the state sector.
Simultaneously, state-sector workers
are demanding wage increases in
line with inflation.

Autonomy for Chittagong Hill Tracts
Buddhist Mongol groups – the Chakma
– continue to voice demands for greater
autonomy, although the low-level
guerrilla war they have waged
since 1974 has been contained. Many
Chakmas fear persecution by Bengali
Muslim settlers, despite a peace treaty
signed in 1997 which provides for local
autonomy, amnesty, and the return of
refugees from India.

BANGLADESH

Total Land Area : 144 000 sq. km (55 598 sq. miles)

POPULATION

▣ over 1 000 000
◉ over 500 000
◎ over 100 000
○ over 50 000
● over 10 000

LAND HEIGHT

500m/1640ft
200m/656ft
Sea Level

0   100 km
0   100 miles

WORLD AFFAIRS ▷ Joined UN in 1974

| Comm | NAM | OIC | SAARC | WTO |

Good relations with the West, the main
source of essential aid, are a priority.
Relations with Pakistan have slowly
improved since Pakistan's agreement in
1991 to accept the 250,000 pro-Pakistan
Bihari Muslims in Bangladeshi refugee
camps since 1971. Relations with India
are improving. The damaging effects of
the construction of the Farakka Dam on
the Ganges, which deprived Bangladesh
of irrigation water, have been alleviated
by a 30-year agreement signed in 1996
guaranteeing the right of both parties
to share the Ganges water. Bilateral
relations eased further with the 1997
Chittagong Hill Tracts treaty; India had
been accused of fomenting unrest.

AID ▷ Recipient

$1.17bn (receipts)   Down 4% in 2000

Aid disbursements to Bangladesh each
year are substantially greater than the
annual value of foreign investment
in the country. Aid also finances the
bulk of state capital spending. The
Bangladesh Development Aid
Consortium meets annually to discuss
aid spending under the auspices of the
World Bank. One result of the level of
aid is that Bangladesh has fallen into
one of the traps of an aid-dependent
economy: the large middle class has
a vested interest in perpetuating a
system which provides its members
with lucrative contracts and access to
external resources.

CHRONOLOGY

Bengal was the first part of the
Indian subcontinent to come under
British rule when the East India
Company was made the *Diwani*
(tax collector) by the Mughal
emperor in 1765.

❏ **1905** Muslims persuade British
rulers to partition state of Bengal,
to create a Muslim-dominated
East Bengal.
❏ **1906** Muslim League established
in Dhaka.
❏ **1912** Partition of 1905 reversed.
❏ **1947** British withdrawal from India.
Partition plans establish a largely
Muslim state of East (present-
day Bangladesh) and West
Pakistan, separated by 1600 km
(1000 miles) of Indian, and largely
Hindu, territory.
❏ **1949** AL founded to campaign for
autonomy from West Pakistan.
❏ **1968** Gen. Yahya Khan heads
government in Islamabad. ▷

B

B

CHRONOLOGY *continued*

- ❑ **1970** Elections give AL, under Sheikh Mujibur Rahman, clear majority. Rioting and guerrilla warfare following Yahya Khan's refusal to convene assembly. Year ends with worst recorded storms in Bangladesh's history – between 200,000 and 500,000 dead.
- ❑ **1971** Civil war, as Sheikh Mujib and AL declare unilateral independence. Ten million Bangladeshis flee to India. Pakistani troops defeated in 12 days by Mukhti Bahini – the Bengal Liberation Army.
- ❑ **1972** Sheikh Mujib elected prime minister. Nationalization of key industries, including jute and textile. Bangladesh achieves international recognition and joins Commonwealth. Pakistan withdraws in protest.
- ❑ **1974** Severe floods damage rice crop.
- ❑ **1975** Sheikh Mujib assassinated. Military coups end with General Zia ur-Rahman taking power. Institution of single-party state.
- ❑ **1976** Banning of trade union federations.
- ❑ **1977** Gen. Zia assumes presidency. Islam adopted as first principle of the constitution.
- ❑ **1981** Gen. Zia assassinated.
- ❑ **1982** Gen. Ershad takes over.
- ❑ **1983** Democratic elections restored by Ershad. Ershad assumes presidency.
- ❑ **1986** Elections. AL and BNP fail to unseat Ershad.
- ❑ **1987** Ershad announces state of emergency.
- ❑ **1988** Islam becomes constitutional state religion.
- ❑ **1990** Ershad resigns following demonstrations.
- ❑ **1991** Elections won by BNP. Khaleda Zia becomes prime minister. Ershad imprisoned. Role of the president reduced to ceremonial functions. Floods kill 140,000 people.
- ❑ **1994** Author Taslima Nasreen, who is accused of blasphemy, escapes to Sweden.
- ❑ **1996** Election returns BNP to power. Results are rejected by opposition parties, which force fresh elections. Sheikh Hasina Wajed of AL then takes power.
- ❑ **2001** Supreme Court declares issuing of religious decrees (*fatwas*) to be a criminal offense. BNP returned to power following violence-marred elections.

DEFENSE

 ▷ No compulsory military service

 $670m   ⬦ No change in 2000

The military, which dominated politics between 1975 and 1990, still wields considerable influence, despite the restoration of civilian government. Although there is greater emphasis on poverty alleviation programs, spending on defense is disproportionately high. Plans to improve security by constructing a 4000-km (2500-mile) road along the border with India were announced in mid-2000.

BANGLADESHI ARMED FORCES

🛡	200 main battle tanks (100 PRC Type–59/69, 100 T–54/55)	120,000 personnel
🚢	4 frigates and 33 patrol boats	10,500 personnel
✈	83 combat aircraft (18 A–5C *Fantan*, 16 F–6, 23 F–7M/FT–7B)	6500 personnel
🚀	None	

ECONOMICS

 ▷ Inflation 4% p.a. (1990–2000)

 $47.9bn   54.1–56.95 taka

SCORE CARD

- ❑ WORLD GNP RANKING..........................51st
- ❑ GNP PER CAPITA $370
- ❑ BALANCE OF PAYMENTS....................$–306m
- ❑ INFLATION 2.3%
- ❑ UNEMPLOYMENT.....................................3%

ECONOMIC PERFORMANCE INDICATOR

EXPORTS

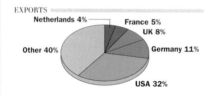

Netherlands 4% France 5% UK 8% Other 40% Germany 11% USA 32%

IMPORTS

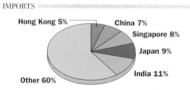

Hong Kong 5% China 7% Singapore 8% Japan 9% India 11% Other 60%

STRENGTHS

80% of the world's jute fiber exports come from Bangladesh. Low wages ensure a competitive and expanding textile industry, which provides over three-quarters of manufacturing export earnings.

WEAKNESSES

The agricultural sector, employing the majority of Bangladeshis, is vulnerable to the violent and unpredictable climate.

PROFILE

Government ministers like to portray Bangladesh as an emerging NIC, but its economy is still overwhelmingly dependent on agriculture and large aid inflows. Agriculture, which provides jute and tobacco, is productive; Bangladesh's soils, fed by the Ganges, Jamuna, and Meghna rivers, are highly fertile. However, the effects of the weather can be devastating, frequently destroying a whole year's crop. Agricultural wages are among the lowest in the world.

The state sector, which owns large, inefficient and massively loss-making companies, is in difficulty. The World Bank, the source of most aid, wishes to see loss-making concerns cut their workforces or close down. A privatization program was announced in 2002.

Textiles and garments are currently the healthiest sectors. Economic zones (export processing zones) with special concessions have attracted foreign investment, as well as helped to promote a small indigenous electronics industry. Bangladesh receives generous textile import quotas from the EU and NAFTA.

BANGLADESH : MAJOR BUSINESSES

- 🍶 Brewing
- 🚬 Tobacco
- Cotton milling
- Jute processing
- ⚙ Light engineering

0 — 100 km
0 — 100 miles

RESOURCES Electric power 3.5m kw

 1.54m tonnes

20 b/d (reserves 70,000 barrels)

34.1m goats, 23.9m cattle, 140m chickens

Salt, oil, natural gas, limestone

Bangladesh is the world's major jute producer, accounting for 80% of world jute fiber exports and about 50% of world jute manufactures exports.

Bangladesh holds world-class gas reserves, estimated to last as much as 200 years at the present extraction rate. Natural gas from the Bay of Bengal,

ELECTRICITY GENERATION

Hydro 6% (0.9bn kwh)
Combustion 94% (13bn kwh)
Nuclear 0%
Other 0%

0   20   40   60   80   100

% of total generation by type

exploited by the state-owned Bangladesh Oil, Gas, and Minerals Corporation, came on stream in 1988.

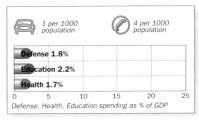

BANGLADESH : LAND USE

Cropland
Wetlands
Forest
Rice
Jute - cash crop

0   100 km
0   100 miles

ENVIRONMENT Sustainability rank: 86th

 1% (0.7% partially protected)

0.2 tonnes per capita

Bangladesh's climate gives rise to devastating floods and cyclones, with consequent huge death tolls and substantial damage to crops. The country is too poor to finance environmental initiatives.

ENVIRONMENTAL TREATIES

Yes		Yes		Yes
Yes		Yes		Yes

HEALTH Welfare state health benefits

 1 per 5000 people

Parasitic, diarrheal, and communicable diseases

More resources are needed to boost health care in rural areas. Although primary health care in these areas improved in the 1990s, Bangladesh's health problems remain severe and are exacerbated by a shortage of medical staff and facilities. Priority for birth control programs has helped reduce the population growth rate by more than 20% over the last 15 years. Half the population is exposed to high levels of arsenic in drinking water.

SPENDING GDP/cap. increase

CONSUMPTION AND SPENDING

1 per 1000 population

4 per 1000 population

Defense 1.8%
Education 2.2%
Health 1.7%

0   5   10   15   20   25

Defense, Health, Education spending as % of GDP

MEDIA TV ownership low

 Daily newspaper circulation 9 per 1000 people

PUBLISHING AND BROADCAST MEDIA

There are 37 daily newspapers. *Dainik Ittefaq, Dainik Inquilab,* and *Dainik Janakantha* have the highest circulations

1 state-controlled service, 1 independant service

1 state-controlled service

Press freedom, which emerged after the fall of President Ershad in 1990, has subsequently been gradually eroded under successive civilian governments. Of the daily newspapers, the ten English-language titles appeal mainly to the urban elite. Among political weeklies, the most prominent is *Holiday.* The vast majority – over 70% – of TV programs are produced locally by the state-run services. Foreign satellite channels are increasingly available, but are subject to government bans.

Average incomes in Bangladesh remain very low, but wealth disparities are not quite as marked as in India or Pakistan. State officials tend to be among the better-off members of society.

WORLD RANKING

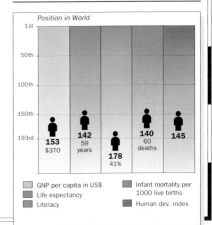

Position in World

1st
50th
100th
150th
193rd

153 $370
142 59 years
178 41%
140 60 deaths
145

GNP per capita in US$
Life expectancy
Literacy

Infant mortality per 1000 live births
Human dev. index

CRIME Death penalty used

 70,000 prisoners

Up 17% 1996–1998

CRIME RATES

Murders
3   per 100,000 population

Rapes
2   per 100,000 population

Thefts
13   per 100,000 population

Rising levels of political and religious violence have led to the enforcement of antiterrorism legislation, containing provisions for summary justice and heavy penalties, including death. There has been a recent sharp rise in crimes against women, including murder, rape, abduction, and acid attacks. Deaths in Bangladeshi prisons are common, and in addition the human rights record of the security forces, especially the paramilitary Bangladesh Rifles, has been attacked by Amnesty International.

EDUCATION School leaving age: 10

41%   434,309 students

THE EDUCATION SYSTEM

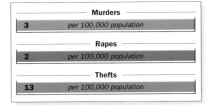

100
80
60
40
20
0

100%   47%   5%

Primary   Secondary   Tertiary

% of each age group in education

Education issues in Bangladeshi society have been poorly addressed, although successive governments have promised to improve literacy levels by increasing spending. A dramatic reduction in the 1990s in the number of child workers meant an accompanying rise in school attendance. Exam cheating is a serious problem. The seven universities are frequently beset by political violence.

BARBADOS

B

OFFICIAL NAME: Barbados **CAPITAL:** Bridgetown
POPULATION: 268,000 **CURRENCY:** Barbados dollar **OFFICIAL LANGUAGE:** English

CARIBBEAN

 1966   1966   Nov 30   BDS -4 +1246   .bb

SITUATED TO THE NORTHEAST of Trinidad, Barbados is the most easterly of the West Indian Windward Islands. In the 16th century, the Portuguese were the first Europeans to reach the island, then inhabited by Arawak Indians. However, Barbados was not colonized until the 1620s, when British settlers arrived. Popularly referred to by its neighbors as "little England," Barbados now seeks to forge a new national identity for itself.

CLIMATE
▷ Tropical oceanic

WEATHER CHART FOR BRIDGETOWN

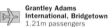

Barbados has a moderate tropical climate and is sunnier and drier than its more mountainous Caribbean neighbors. Hurricanes may occur in the rainy season.

TRANSPORTATION
▷ Drive on left

 Grantley Adams International, Bridgetown 1.21m passengers   79 ships 733,319 grt

THE TRANSPORTATION NETWORK

 1578 km (981 miles) None

None None

A multimillion dollar expansion program has upgraded facilities at the international airport. Piers at Bridgetown's port have been improved with foreign aid, as have the island's paved roads. There are bus routes over most of the island.

House of Assembly, Trafalgar Square, Bridgetown. Barbados's parliament, the third oldest in the Commonwealth, dates from 1639.

TOURISM
▷ Visitors : Population 2.1:1

556,000 visitors ⬆ Up 8% in 2000

MAIN TOURIST ARRIVALS

UK 39%	
USA 20%	
Canada 11%	
Other 30%	

% of total arrivals

Tourists, essential to the Barbadian economy, are attracted by the white sandy beaches and sporting activities, as well as the post-colonial ambience and the generally laid-back culture.

PEOPLE
▷ Pop. density high

 Bajan (Barbadian English), English 623/km² (1614/mi²)

THE URBAN/RURAL POPULATION SPLIT

50%   50%

RELIGIOUS PERSUASION

- Roman Catholic 4%
- Anglican 40%
- Methodist 7%
- Pentecostal 8%
- Nonreligious 17%
- Other 24%

Most Bajans are the descendants of Africans brought to the island between the 16th and 19th centuries; there are also small groups of south Asians and Europeans, mainly expatriates from the UK, many of whom take up residence on retirement. There is some latent tension between the white community, which controls most of the economy, and the majority black population, although this rarely spills over into violence. Increasing social mobility has allowed many black Bajans to move into the professions and the civil service. Barbados enjoys a higher standard of living than most Caribbean countries.

POLITICS
▷ Multiparty elections

L. House 1999/2004
U. House 1999/2004   H.M. Queen Elizabeth II

AT THE LAST ELECTION
House of Assembly 28 seats

93% **BLP**   7% **DLP**

BLP = Barbados Labour Party
DLP = Democratic Labour Party

Senate 21 seats

The members of the Senate are appointed. Twelve are chosen by the prime minister, 2 by the leader of the opposition, and 7 independents by the governor-general.

Barbados is a multiparty democracy. A primarily European, affluent elite finances the parties and has an indirect influence on government policy. The BLP swept to power in 1994 and won a further landslide in 1999. Owen Arthur, BLP leader and prime minister, prioritizes economic growth and international competitiveness. He has pledged to transform Barbados into a republic, while remaining a member of the Commonwealth.

WORLD AFFAIRS
▷ Joined UN in 1966

 ACS   Comm   Caricom   NAM OAS

Considered to be an international tax haven, Barbados has been under pressure to implement reforms.

AID
▷ Recipient

No net receipts   Down sharply in 1999–2000

Most aid comes from the EU, the UN and the UK, mainly in the form of development project loans and balance-of-payments support.

DEFENSE
▷ No compulsory military service

US$13m   Up 8% in 2000

The small Barbadian army and the constabulary benefit from financial support and training from the US and UK governments, which also supply equipment. Barbados is the headquarters of the Regional Security System, established in 1982 by the Windward and Leeward Islands, a body which acts as a multinational security force for its members.

ECONOMICS

 Inflation 3.1% p.a. (1990–2000)

 US$2.47bn

1.99 Barbados dollars

SCORE CARD

❏ WORLD GNP RANKING	133rd
❏ GNP PER CAPITA	US$9250
❏ BALANCE OF PAYMENTS	–US$126m
❏ INFLATION	2.4%
❏ UNEMPLOYMENT	10%

STRENGTHS

Well-developed tourism based on climate and accessibility. Sugar industry. Information processing and financial services are important new growth sectors.

WEAKNESSES

Narrow economic base, vulnerable to downturns in tourism, failures of sugar harvest, and the sector's dependency on loans and secure markets. Relatively high manufacturing costs.

EXPORTS

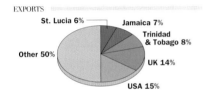

St. Lucia 6%
Jamaica 7%
Trinidad & Tobago 8%
Other 50%
UK 14%
USA 15%

IMPORTS

Canada 4%
Japan 5%
UK 9%
USA 41%
Trinidad & Tobago 15%
Other 26%

BARBADOS

Total Land Area : 430 sq. km (166 sq. miles)

POPULATION
over 10 000 ●
under 10 000 ·

LAND HEIGHT
200m/656ft
Sea Level

N

0   5 km
0   5 miles

RESOURCES

 Electric power 156,000 kw

3206 tonnes

1604 b/d (reserves 7.6m barrels)

41,000 sheep, 33,000 pigs, 3.6m chickens

Oil, natural gas

Barbados has few strategic resources. The domestic petroleum industry provides about one-third of the country's energy requirements.

ENVIRONMENT

 Not available

1%

3.4 tonnes per capita

Oil slicks created by waste dumped from passing ships are polluting the encircling reef and adversely affecting the life cycle of the flying fish, Barbados's main fish stock.

MEDIA

 TV ownership high

 Daily newspaper circulation 199 per 1000 people

PUBLISHING AND BROADCAST MEDIA

There are 2 daily newspapers, the *Barbados Advocate* and the *Nation*

1 state-owned service with subscription option

3 services: 1 state-owned, 2 independent

There is no political interference in the media. The two daily newspapers are privately owned. Multichannel TV is available on subscription.

CRIME

 Death penalty in use

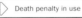 772 prisoners

Up 1% in 1999

A firearms amnesty announced in 1998 proposed heavy fines and prison sentences for illegal possession of guns; the measure was in response to growing drugs-related and other violent crime. Money laundering is a problem.

EDUCATION

 School leaving age: 16

98%

6317 students

A seven-year plan to restructure education includes the rehabilitation of schools and a computerization program. Barbados hosts a campus of the University of the West Indies.

CHRONOLOGY

Colonized by the British in 1627, Barbados grew rich in the 18th century from sugar produced using slave labor.

- ❏ **1951** Universal adult suffrage introduced.
- ❏ **1961** Full internal self-government.
- ❏ **1966** Independence from the UK.
- ❏ **1983** Barbados supports and provides a base for the US invasion of Grenada.
- ❏ **1994, 1999** The BLP wins two successive general elections.

HEALTH

 Welfare state health benefits

1 per 800 people

Heart and cerebrovascular diseases, cancers

The health system is based on subsidized government-run clinics and hospitals, supplemented by more expensive private clinics and private doctors. Facilities are within easy reach of all Bajans.

SPENDING

 GDP/cap. increase

CONSUMPTION AND SPENDING

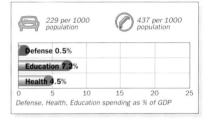

229 per 1000 population

437 per 1000 population

Defense 0.5%
Education 7.2%
Health 4.5%

0   5   10   15   20   25
Defense, Health, Education spending as % of GDP

A significant disparity exists between most Bajans and a small affluent group, its members usually of European origin, which owns and controls business and industry, and parades status symbols such as yachts. Prime Minister Arthur stated in 1998 that "abject poverty" existed in the country.

WORLD RANKING

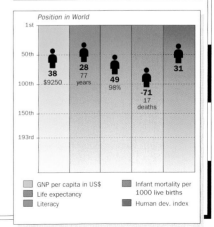

Position in World

1st
50th
100th
150th
193rd

38 $9250
28 77 years
49 98%
-71 17 deaths
31

GNP per capita in US$
Life expectancy
Literacy
Infant mortality per 1000 live births
Human dev. index

BELARUS

OFFICIAL NAME: Republic of Belarus **CAPITAL:** Minsk **POPULATION:** 10.1 million
CURRENCY: Belarussian rouble **OFFICIAL LANGUAGES:** Belarussian and Russian

EUROPE

| 1991 | 1991 | July 3 | BY | +2 | +375 | .by |

BELARUS LITERALLY MEANS "white Russia," a color associated in Slavic culture with freedom, and a reference to the fact that the country was never conquered by the Mongol Golden Horde. Devastated in World War II, and with few resources other than agriculture, Belarus only reluctantly became independent of Moscow in 1991, and President Aleksandr Lukashenka has maintained close links with Russia. The Chernobyl nuclear disaster in Ukraine in 1986 has had lasting effects on the environment and the health of Belarussians.

CLIMATE ▷ Continental

WEATHER CHART FOR MINSK

Average daily temperature Rainfall
°C/°F J F M A M J J A S O N D cm/in
40/104 — — 40/16
30/86 — — 30/12
20/68 — — 20/8
10/50 — — 10/4
0/32 — — 0
-10/14
-20/4

Belarus has a continental climate somewhat moderated by the influence of the nearby Baltic Sea. Temperatures in winter drop well below freezing, however, while summers can be hot and humid. Summer is also the main season for rainfall.

Much of southern Belarus is marshy and sparsely populated. It includes the vast Pripet Marshes and the Dnieper lowlands.

TRANSPORTATION ▷ Drive on right

Minsk International
401,239 passengers

Has no fleet

THE TRANSPORTATION NETWORK

| 60,567 km (37,635 miles) | None |
| 5523 km (3432 miles) | Extensive canal and river systems |

Belarus has no direct access to the sea, but is close to the Baltic ports. Railroad communications are good.

TOURISM ▷ Visitors : Population 1:28

355,000 visitors Up 42% in 1998

MAIN OVERSEAS ARRIVALS

| Russia 59% |
| Ukraine 8% |
| Germany 4% |
| Poland 4% |
| UK 3% |
| Other 22% |

0 10 20 30 40 50 60
% of total arrivals

Belarus has fewer tourists than its neighbors. Many of its historic buildings were destroyed during World War II. Minsk was totally flattened, and is now characterized by Stalinist and other high-rise buildings. There is little of mass appeal on which to build a tourist industry.

BELARUS

Total Land Area : 207 600 sq. km
(80 154 sq. miles)

POPULATION

over 1 000 000
over 500 000
over 100 000
over 50 000
over 10 000
under 10 000

LAND HEIGHT

200m/656ft
100m/328ft

B

PEOPLE ▷ Pop. density low

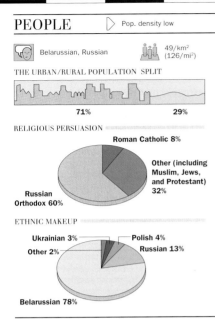

Belarussian, Russian

49/km² (126/mi²)

THE URBAN/RURAL POPULATION SPLIT

71% 29%

RELIGIOUS PERSUASION

Roman Catholic 8%

Other (including Muslim, Jews, and Protestant) 32%

Russian Orthodox 60%

ETHNIC MAKEUP

Ukrainian 3% Polish 4%

Other 2% Russian 13%

Belarussian 78%

Only 2% of the population is non-Slav and there is little ethnic tension. Under a law passed in 1992, the entire population has an automatic right to Belarussian citizenship. Most people speak Russian, and only 11% of the population are fluent in Belarussian, which is used mainly in rural areas. A 1995 referendum and 1998 legislation declared that both languages have equal status.

POPULATION AGE BREAKDOWN

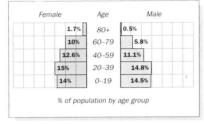

Female		Age	Male	
	1.7%	80+	0.5%	
	10%	60–79	5.8%	
	12.6%	40–59	11.1%	
	15%	20–39	14.8%	
	14%	0–19	14.5%	

% of population by age group

POLITICS ▷ Multiparty elections

L. House 2000/2004
U. House 2000/2004

President Aleksandr Lukashenka

AT THE LAST ELECTION

House of Representatives 110 seats

97% PKB 2% Opp 1% Ind

PKB = Party of Communists of Belarus and government supporters   **Opp** = Minor opposition parties
Ind = Independents

Council of the Republic 64 seats

The Council of the Republic is indirectly elected

Under the 1994 constitution, amended in 1996, Belarus has a directly elected president and a bicameral parliament.

PROFILE

Belarus, by far the slowest of the former Soviet states to implement political reform, has struggled to find an identity since 1991. A post-Soviet constitution was not adopted until 1994, and only in 1995 was the first fully fledged post-Soviet parliament elected, dominated by the PKB and its Agrarian Party ally. There is no strong pluralist culture to check authoritarian moves by the president.

Aleksandr Lukashenka was unexpectedly elected as Belarus's first president in 1994. In 1996 a new constitution significantly strengthened his powers. Pro-Lukashenka deputies then voted to replace the Supreme Council with a new House of Representatives. Lukashenka has also extended his own term of office. The opposition contests the legitimacy

of this, as does the EU. A clampdown on political opponents effectively invalidated parliamentary elections in late 2000/early 2001, and Lukashenka's reelection in late 2001 was immediately condemned by observers.

MAIN POLITICAL ISSUES
Relationship with Russia

In 1994 an accord (reinforced in 1999 and 2000) was signed on future monetary union with Russia. Lukashenka has sought ever closer relations leading toward a joint presidency. A union treaty was signed in late 1999, but is more symbolic than practical. With Russian reticence prevailing over Belarus's enthusiasm for the union, the treaty has produced no significant developments toward joint state institutions or economic programs.

The environment

The 1986 Chernobyl nuclear disaster continues to cast a shadow. The cleanup operation, slow and laborious, will take decades. It is a major drain on state finances, accounting for 25% of spending. A "Chernobyl tax" is levied on businesses.

President Aleksandr Lukashenka *seeks closer ties with the Russian Federation.*

Vladimir Goncharik, *Lukashenka's main rival in presidential elections in 2001.*

WORLD AFFAIRS ▷ Joined UN in 1945

 EAPC   CIS   IAEA CEI OSCE

B

Relations with Russia are paramount. Numerous bilateral agreements were signed after independence in 1991. Ties have been strengthened further by the pro-Russian president Aleksandr Lukashenka, although many in Russia fear that closer links will drain Moscow's resources for little strategic gain.

In mid-1998 diplomats were evicted from many official residences. The US, the EU, and others temporarily withdrew their ambassadors in protest. The March 2001 clampdown on political opponents ended moves to relax EU sanctions, introduced four years previously over the authoritarian changes in the constitution.

AID ▷ Recipient

$40m (receipts) Up 3% in 2000

Although both the World Bank and the IMF provided loans for Belarus in the early 1990s, the lack of structural reforms since Lukashenka's administration came to power in 1994 has meant that further aid has been stalled. Some US bilateral aid continued, but the EU in particular has made it clear that support will depend on human rights improvements and the reversal of authoritarian threats to democracy.

Both the US and the EU extended credits to Belarus to assist in the conversion of the defense industry to nonmilitary production. Belarus also still requires aid to combat the effects of radiation pollution in the wake of the Chernobyl nuclear accident of 1986.

CHRONOLOGY

After forming part of medieval Kievan Rus, Belarus experienced rule by three of its neighbors – Lithuania, Poland, and Russia – before incorporation into the USSR.

- ❑ **1918** Belarussian Bolsheviks stage coup. Independence as Belorussian Soviet Socialist Republic (BSSR).
- ❑ **1919** Invaded by Poland.
- ❑ **1920** Minsk retaken by Red Army. Eastern Belorussia reestablished as Soviet Socialist Republic.
- ❑ **1921** Treaty of Riga – Western Belorussia incorporated into Poland.
- ❑ **1922** BSSR merges with Soviet Russia and Ukraine to form USSR.
- ❑ **1929** Stalin implements collectivization of agriculture.
- ❑ **1939** Western Belorussia reincorporated into USSR when Soviet Red Army invades Poland. ▷

B

CHRONOLOGY *continued*

- ❑ **1941–1944** Occupied by Germany during World War II.
- ❑ **1945** Founding member of UN.
- ❑ **1965** K. T. Mazurau, Communist Party of Belorussia (PKB) leader, becomes first deputy chair of Soviet government.
- ❑ **1986** Radioactive fallout after Chernobyl accident affects 70% of country.
- ❑ **1988** Evidence revealed of mass executions (over 300,000) by Soviet military between 1937 and 1941 near Minsk. Popular outrage fuels formation of nationalist Belorussian Popular Front (BPF), with Zyanon Paznyak as president. PKB authorities crush demonstration.
- ❑ **1989** Belarussian adopted as republic's official language.
- ❑ **1990** PKB prevents BPF from participating in elections to Supreme Soviet. BPF members join other opposition groups in Belorussian Democratic Bloc (BDB). BDB wins 25% of seats. PKB bows to opposition pressure and issues Declaration of the State Sovereignty of BSSR.
- ❑ **1991** March, 83% vote in referendum to preserve union with USSR. April, strikes against PKB and its economic policies. August, independence declared. Republic of Belarus adopted as official name. Stanislau Shushkevich elected chair of Supreme Soviet. December, Belarus, Russia, and Ukraine establish CIS.
- ❑ **1992** Supreme Soviet announces that Soviet nuclear weapons must be cleared from Belarus by 1999. Help promised from US.
- ❑ **1993** Belarussian parliament ratifies START-I and nuclear nonproliferation treaties.
- ❑ **1994** New presidential constitution approved; Aleksandr Lukashenka defeats conservative prime minister Vyacheslav Kebich in elections. Monetary union (reentry into rouble zone) agreed with Russia.
- ❑ **1995** First fully fledged post-Soviet parliament elected.
- ❑ **1996** Referendum approves changes to constitution strengthening Lukashenka's powers.
- ❑ **1997** Belarus and Russia ratify union treaty and Charter.
- ❑ **1998** Eviction from embassies sparks withdrawal of Western ambassadors.
- ❑ **1999** Union treaty with Russia.
- ❑ **2000–2001** Disputed parliamentary elections; clampdown on PKB's political opponents.
- ❑ **2001** Lukashenka reelected. Observers label election seriously flawed.

DEFENSE

 Compulsory military service

 $366m | Down 21% in 2000

BELARUSSIAN ARMED FORCES

🚜	1683 main battle tanks (1539 T-72, 49 T-55, 95 T-80)	43,600 personnel
🚢	None	
✈	177 combat aircraft (29 Su-24, 80 Su-25, 45 MiG-29, 23 Su-27)	12,000 personnel
🚀	None	

After the breakup of the Soviet Union in 1991, Belarus briefly adopted a policy of neutrality. It also committed itself to disposing of its inherited nuclear capability. Tactical nuclear weapons were removed by 1993 and strategic nuclear weapons by 1996.

Despite joining the CIS collective security agreement in 1993, Belarus joined NATO's Partnerships for Peace program in 1995. Lukashenka has not developed NATO ties further, preferring to establish stronger military links with Moscow, which include a 1997 Treaty on Military Cooperation and a 1999 agreement on joint procedures for arms exports. Belarus bears some of the costs of Russian troops stationed on its territory.

ECONOMICS

Inflation 355% p.a. (1990–2000)

📊 $28.7bn | 1218–1603 Belarussian roubles

SCORE CARD

❑ WORLD GNP RANKING	60th
❑ GNP PER CAPITA	$2870
❑ BALANCE OF PAYMENTS	–$162m
❑ INFLATION	168.6%
❑ UNEMPLOYMENT	2%

EXPORTS

- Lithuania 5%
- Latvia 6%
- Germany 3%
- Ukraine 8%
- Russia 51%
- Other 27%

IMPORTS

- Poland 3%
- Ukraine 4%
- Italy 2%
- Germany 7%
- Other 19%
- Russia 65%

STRENGTHS

Low unemployment combined with relative social stability. Potential of forestry and agriculture.

WEAKNESSES

Lack of economic restructuring; support for outmoded businesses. Few natural resources. Dependence on Russia for energy and raw materials. Cleanup costs of Chernobyl.

PROFILE

After 1991, Belarus adopted a slower pace of economic reform than did other former Soviet states, since attempts to move more quickly to a market economy were thwarted by the largely conservative parliament. Upon election in 1994, Lukashenka suspended privatization moves, resuming them only halfheartedly in 1995, under a policy of "market socialism." Traditional industries continued to receive big subsidies as the government printed money to increase production. A currency crisis in 1998, along with rampant inflation, combined with two successive bad harvests in 1998–1999 to create a major threat to the economy. Inflation receded in 2000–2001, but the rate of economic growth also slowed.

ECONOMIC PERFORMANCE INDICATOR

Consumer Price Index — GDP
4757

BELARUS : MAJOR BUSINESSES

- Minsk
- Barysaw
- Vitsyebsk
- Orsha
- Mahilyow
- Hrodna
- Babruysk
- Brest
- Homyel'
- Pinsk
- Mazyr
- Rechytsa

Micro electronics		Textiles	
Shipbuilding parts		Chemicals	
Heavy engineering		Computers	
Food processsing		Saw milling	
Vehicle assembly			
Manufacturing		0 — 100 km	
Consumer goods		0 — 100 miles	

* significant multinational ownership

B

RESOURCES

  Electric power 7.5m kw

5809 tonnes

36,692 b/d (reserves 201m barrels)

4.22m cattle, 3.43m pigs, 32m chickens

Oil, natural gas, coal, rock salt

ELECTRICITY GENERATION

Hydro 0%	
Combustion 100% (23bn kwh)	
Nuclear 0%	
Other 0%	

0 — 20 — 40 — 60 — 80 — 100

% of total generation by type

Belarus has no significant strategic resources and is heavily dependent on the Russian Federation for fuel and energy supplies. Small quantities of oil and natural gas exist close to the Polish border.

BELARUS : LAND USE

Cropland
Forest
Pasture
Wetlands
Flax - cash crop
Cereals
Cattle

0 — 100 km
0 — 100 miles

ENVIRONMENT

Sustainability rank: 49th

6%

6 tonnes per capita

ENVIRONMENTAL TREATIES

Yes		Yes		Yes	
Yes		No		No	

The massive leak from Ukraine's Chernobyl nuclear reactor in 1986 released a huge cloud of radiation. Some 70% of the fallout fell on Belarus, including Minsk; 2.3 million people were immediately affected, and cases of leukemia and cancer continue to emerge. Farmland, forests, and water were all contaminated, including underwater streams feeding rivers in eastern Poland. Some areas in the fallout zone are still being farmed. The cleanup program swallows 25% of government finances, despite substantial Western aid.

The Belovezha primeval forest, on the border with Poland, is Europe's largest nature reserve. It is a habitat of the European bison or wisent.

MEDIA

  TV ownership high

  Daily newspaper circulation 173 per 1000 people

PUBLISHING AND BROADCAST MEDIA

There are 20 daily newspapers, mostly published in Russian, while weekly papers tend to be published in Belarussian

1 state-controlled service; there are a number of small independent stations

1 state-controlled service; some independent stations

There are some independent media outlets, but government critics face harassment. Press freedom is curbed; state-backed publications predominate.

CRIME

  Death penalty in use

 56,000 prisoners

Up 6% in 1999

CRIME RATES

Murders	
12	per 100,000 population

Rapes	
5	per 100,000 population

Thefts	
566	per 100,000 population

As elsewhere in the former Soviet Union, economic hardship and a general breakdown in law and order have resulted in a significant rise in crime. The prison population exceeds the intended capacity of 40,000. Belarus has become a transshipment point for illegal narcotics destined for western Europe, while locally produced opium supplies the internal market.

EDUCATION

  School leaving age: 15

99%

 301,753 students

THE EDUCATION SYSTEM

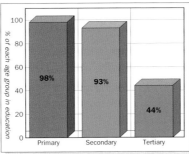

% of each age group in education

Primary	Secondary	Tertiary
98%	93%	44%

Education is officially compulsory for nine years, and teaching is mainly in Russian. Activists complain that because of political bias there is inadequate provision for the teaching of Belarussian. University education – taught in Russian – is of a fairly high standard.

HEALTH

  Welfare state health benefits

1 per 227 people

  Heart attacks, cancers, accidents, violence

The Chernobyl nuclear disaster placed Belarus's health service, which had hitherto been adequate, under enormous strain. A Chernobyl tax funds assistance for victims of the accident. The number of cancer and leukemia cases has soared, and extra wards and specialist units have had to be built. Many Belarussian doctors are being trained in the latest bone-marrow techniques in Europe and the US.

HIV/AIDS is a growing problem; as at end-2001, 0.3% of the adult population were estimated to be affected.

SPENDING

GDP/cap. decrease

CONSUMPTION AND SPENDING

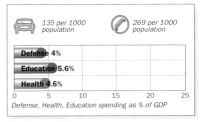

135 per 1000 population

269 per 1000 population

Defense 4%	
Education 5.6%	
Health 4.6%	

0 — 5 — 10 — 15 — 20 — 25

Defense, Health, Education spending as % of GDP

The deteriorating economic situation has resulted in an overall drop in living standards. Wealth is concentrated among a small, communist elite which is opposed to market mechanisms. Now that this elite has the upper hand, its members have strengthened their grip on the state's resources. Thus far Belarus has not seen the expansion of entrepreneurial activity found in other former Soviet bloc countries such as Poland or Russia.

Wage increases in 2001 – an election year – brought salaries to levels which were unaffordable, so that enterprise profitability and investment were severely affected.

WORLD RANKING

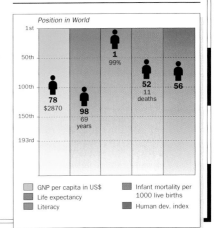

Position in World

1st
50th
100th
150th
193rd

78 $2870	
98 69 years	
1 99%	
52 11 deaths	
56	

- GNP per capita in US$
- Life expectancy
- Literacy
- Infant mortality per 1000 live births
- Human dev. index

BELGIUM

OFFICIAL NAME: Kingdom of Belgium **CAPITAL:** Brussels **POPULATION:** 10.3 million
CURRENCY: Euro (Belgian franc until 2002) **OFFICIAL LANGUAGES:** Dutch, French, and German

 1830   1919   July 21 B +1 +32 .be

EUROPE

LOCATED BETWEEN GERMANY, France, and the Netherlands, Belgium has a short coastline on the North Sea. The south includes the forested Ardennes region, while the north is crisscrossed by canals.
Belgium has been fought over many times in its history; it was occupied by Germany in both world wars. Tensions have existed between the Dutch-speaking Flemings and French-speaking Walloons since the 1830s. These have been somewhat defused by Belgium's move to a federal political structure and the national consensus on the benefits of EU membership.

CLIMATE ▷ Maritime

WEATHER CHART FOR BRUSSELS

Belgium has a typical maritime climate and is influenced by the Gulf Stream. Temperatures are mild, accompanied by heavy cloud cover and much rain. Widely fluctuating weather conditions, caused by cyclonic disturbances, can disrupt the climate on the coast. Summers tend to be short.

TRANSPORTATION ▷ Drive on right

 Brussels International
21.6m passengers

 182 ships
143,900 grt

THE TRANSPORTATION NETWORK

 117,701 km
(73,136 miles)

1702 km
(1058 miles)

 3472 km
(2158 miles)

2043 km
(1269 miles)

Belgium can be crossed within four hours by car or train. The expressway network is extensive, and although the railroad system has been reduced since 1970, it still constitutes one of the world's densest networks. Using high-speed TGV lines, Paris is just 80 minutes from Brussels, and London via the Channel Tunnel is 2 hours 40 minutes.

Antwerp is Europe's second-largest port. National carrier Sabena collapsed dramatically in November 2001, amid a slump in the aviation industry. A section of Sabena subsequently formed the basis for a new private airline, SN Brussels Airlines.

TOURISM ▷ Visitors : Population 1:1.6

6.46m visitors   Up 1% in 2000

MAIN TOURIST ARRIVALS

UK	20%
Netherlands	16%
France	15%
Germany	12%
USA	6%
Other	31%

% of total arrivals

Belgium's main attractions are its historic cities and the museums of Flemish art. Bruges, the capital of West Flanders, is often referred to as the "Venice of the North." With Gothic and Renaissance architecture and a complex canal system, it has become a favored destination for British weekend trippers and Japanese honeymooners. In Brussels, the famous "Grande Place," a cluster of Gothic, Renaissance, and Baroque buildings in a cobbled square, survived bombing during World War II. Much of the rest of the old city center, however, was destroyed. Belgium has 15 resorts on its 62-km (38-mile) coastline, with a single tramline running its entire length. Forests in the Ardennes to the south attract hikers.

The Ardennes, in the southeast, are famous for their forests, cuisine, and lakes. Rivers, such as the Meuse and Semois, dissect the region.

PEOPLE ▷ Pop. density high

Dutch, French, German

314/km²
(813/mi²)

THE URBAN/RURAL POPULATION SPLIT

97%   3%

RELIGIOUS PERSUASION

Muslim 2%   Other 10%
Roman Catholic 88%

ETHNIC MAKEUP

Italian 2%   Other 6%
Moroccan 1%   Walloon 33%
Fleming 58%

Belgium has been marked by the divisions between its Flemish and Walloon communities. The majority Dutch-speaking Flemings are concentrated in Flanders. Wallonia is French-speaking and Brussels is 85% francophone. French-speakers were in the ascendancy for many years, their greater economic wealth reinforced by a constitution giving them political control; tensions between Walloons and Flemings occasionally erupted into violence. In the past three decades, however, the situation has been reversed: Wallonia's industries have declined and Flanders is now the wealthier region. To defuse tensions, Belgium began in 1980 to change from being the most centralist to the most federal state in Europe; each community now controls most of its affairs and has its own government. A small German-speaking community in the east has extensive autonomy in educational and cultural matters.

Belgium has a sizable immigrant population. Women account for 40% of the workforce and 19% of administrators and managers.

POPULATION AGE BREAKDOWN

Female	Age	Male
2.7%	80+	1.1%
9.7%	60–79	7.9%
12.4%	40–59	12.5%
14.6%	20–39	15.1%
11.7%	0–19	12.3%

% of population by age group

B

BELGIUM

Total Land Area : 30 510 sq. km
(11 780 sq. miles)

POPULATION

- ▣ over 1 000 000
- ◎ over 100 000
- ○ over 50 000
- ● over 10 000

LAND HEIGHT

- 500m/1640ft
- 200m/656ft
- Sea Level

B

POLITICS ▷ Multiparty elections

L. House 1999/2003
U. House 1999/2003   H.M. King Albert II

AT THE LAST ELECTION

Chamber of Representatives 150 seats

| 28% VLD/PRL | 22% SP/PS | 21% CVP/PSC | 10% VB | 7% Ecolo | 6% Others | 6% Agalev |

VLD/PRL = Flemish Liberals Party/Liberal Reform Party
(Walloon) **SP/PS** = Socialist Party (Flemish)/Socialist Party
(Walloon) **CVP/PSC** = Christian People's Party (Flemish)/
Christian Social Party (Walloon) **VB** = Vlaams Blok
Ecolo = French Greens **Agalev** = Flemish Ecologists
Co-op = Co-opted members

Senate 71 seats

| 44% Co-op | 15% VLD/PRL | 11% SP/PS | 4% Ecolo | 3% Others | 13% CVP/PSC | 6% VB | 4% Agalev |

The Senate has 40 directly elected members and 31
co-opted members.

Until 1970, Belgium was a unitary state.
Tensions between language groups led
to four waves of federalist reforms from
1980, which culminated in the St. Michel
Accords of 1993, confirming the state as
a federal monarchy.

PROFILE
Belgian politics are defined along lines
of language. Apart from this, a high
degree of consensus exists over the
benefits of membership of the EU and
monetary union. In recent years, support
has increased for the racist VB, which
objects to Belgium's Turkish, Moroccan,
and other African minorities. VB won
28% of the vote in local elections in
Antwerp in 1994 and 33% in 2000.

The ruling centrist coalition of the
Socialist and Christian Democrat
parties of the Flemish and Walloon
communities had a parliamentary
majority, but had difficulty in
securing the necessary majority

for the
constitutional
reforms enacted in
the St. Michel Accords.
These gave the regional governments
– Flanders, Wallonia, and Brussels –
significant powers under a federal
government. The government of Jean-
Luc Dehaene was defeated in the 1999
poll, and a new coalition composed of
the Liberals, Socialists, and Greens was
formed by the VLD's Guy Verhofstadt.

MAIN POLITICAL ISSUES
Language
Tensions between the two language
groups are receding. However, the
divisions remain strong. Each
community has a right–liberal party
(the VLD in Flanders, the PRL in
Wallonia), a socialist party (SP/PS), a
Christian democratic party (CVP/PSC),
and a green party (Agalev/Ecolo).

Police handling of pedophile case
Apparent police incompetence, cover-
ups, and corruption in combating the
activities of pedophile rings has provoked
public anger and protest, focusing in
particular on failure to save children
from pedophile killer Marc Dutroux.
Arrested in 1996, he escaped briefly in
1998, causing two ministers to resign.

King Albert II,
succeeded his brother
King Baudouin who
died in 1993.

Guy Verhofstadt,
youthful leader of the
VLD/PRL and prime
minister since 1999.

WORLD AFFAIRS ▷ Joined UN in 1945

| Benelux | CE | EU | OECD | NATO |

Belgium's key concern is its role in the
EU. It is a keen supporter of economic
and monetary union. As a frequent
victim of wars between France and
Germany, Belgium sees the EU as a
guarantor of western European peace.
It is also perceived as an important
foundation for Belgium's own federalist
structure, without which many fear
that Belgium could split into two.

Belgium has little in the way of
an independent foreign policy, but
frequently contributes troops to the
UN's operations. Belgian soldiers have
served in the Democratic Republic of
the Congo and the Middle East in
recent years, and a number were
killed in Rwanda in 1994.

AID ▷ Donor

$820m (donations)   Up 8% in 2000

Some 0.33% of GNP goes in overseas
development aid. Belgian aid focuses
on education and agricultural projects
in Africa. The major beneficiaries
are the former Belgian colonies of
Burundi, Rwanda, and the DRC.

B

CHRONOLOGY

Formerly ruled by the French dukes of Burgundy, Belgium became a Habsburg possession in 1477. It passed to the Austrian Habsburgs in 1713. Belgium was incorporated into France in 1797.

- ❏ **1814–1815** Congress of Vienna; European powers decide to merge Belgium with the Netherlands under King William I of Orange.
- ❏ **1830** Revolt against Dutch; declaration of independence.
- ❏ **1831** European powers place Leopold Saxe Coburg as king.
- ❏ **1865** Leopold II crowned king.
- ❏ **1885** Berlin Conference gives Congo basin to Leopold as colony.
- ❏ **1914** German armies invade. Belgium occupied until 1918.
- ❏ **1921** Belgo-Luxembourg Economic Union formed. Belgian and Luxembourg currencies locked.
- ❏ **1932** Dutch language accorded equal official status with French.
- ❏ **1936** Belgium declares neutrality.
- ❏ **1940** Leopold III capitulates to Hitler. Belgium occupied till 1944.
- ❏ **1948** Customs union with Netherlands and Luxembourg (Benelux) formed.
- ❏ **1950** King abdicates in favor of son, Baudouin.
- ❏ **1957** Becomes one of six original signatories of Treaty of Rome, the principal foundation of what develops into the EU.
- ❏ **1992** Christian Democrat and Socialist government led by Jean-Luc Dehaene takes over federal government.
- ❏ **1993** Culmination of reforms creating federal state. Greater powers for regions and city governments. Death of Baudouin. Succeeded by Albert II.
- ❏ **1995** Allegations of corruption and murder involving French-speaking PS force resignations of Walloon premier, federal deputy premier, and Willy Claes, NATO secretary-general.
- ❏ **1996** Murder and disappearance of young girls arouse fears of international kidnapping and pedophile ring.
- ❏ **1999** February, Claes found guilty of bribery in connection with defense contract to buy helicopters. June, VLD/PRL wins general election. New coalition formed, including Greens for first time.
- ❏ **2001** November, collapse of national airline Sabena.
- ❏ **2002** January, euro fully adopted, with introduction of coins and notes, and withdrawal of Belgian franc from circulation.
  May, legalization of euthanasia.

DEFENSE

▷ No compulsory military service

 $3.34bn

 Down 3% in 2000

BELGIAN ARMED FORCES

	132 main battle tanks (*Leopard* 1A5)	26,400 personnel
	3 frigates	2560 personnel
	135 combat aircraft (F-16A, F-16B)	8600 personnel
	None	

Belgium spends less on defense than the NATO average of 2.2% of GDP.

In 1994, as part of Belgium's program to reduce government debt, all three military services were targeted for cuts. The government abolished conscription and undertook to cut troop levels. The defense budget was frozen for five years.

However, spending on paratroopers and transport planes has increased. The aim is to allow Belgian forces to fulfill their role in NATO's new rapid reaction forces. It will also make Belgian forces more useful to the UN's worldwide operations. In 1996, the Belgian and Netherlands' navies were brought together under a joint operational command based at Den Helder, the Netherlands.

ECONOMICS

▷ Inflation 1.9% p.a. (1990–2000)

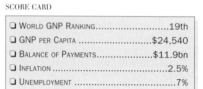 $252bn

1.0651–1.1231 euros

SCORE CARD

- ❏ WORLD GNP RANKING.........................19th
- ❏ GNP PER CAPITA $24,540
- ❏ BALANCE OF PAYMENTS...................$11.9bn
- ❏ INFLATION 2.5%
- ❏ UNEMPLOYMENT 7%

EXPORTS

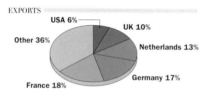

- USA 6%
- UK 10%
- Other 36%
- Netherlands 13%
- Germany 17%
- France 18%

IMPORTS

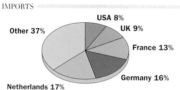

- USA 8%
- UK 9%
- Other 37%
- France 13%
- Germany 16%
- Netherlands 17%

ECONOMIC PERFORMANCE INDICATOR

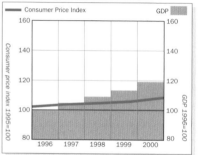

Consumer Price Index — GDP

PROFILE

Recession and rising unemployment in the early 1990s prompted the introduction of work-sharing schemes and benefit reforms. Unemployment and the massive public debt are declining gradually. Against a background of a downturn in the world economy, the government is committed to both greater fiscal stringency, in pursuit of a budget surplus, and reduction in taxation over 2002–2005.

STRENGTHS

One of world's most efficient producers of metal products and textiles. Flanders is a world leader in new high-tech industries. Successful chemicals industry. Highly educated and motivated multilingual workforce: estimates suggest productivity is 20% above that of Germany. Location attractive for US multinationals. Good sea outlets and access to Rhine inland waterway from Antwerp and Ghent.

WEAKNESSES

Public debt of around 100% of GDP, well over EU target of 60%. High long-term and low-skill joblessness with sharp local variations. Early retirement of large numbers of workers results in high state pension bill. Bureaucracy larger than European average.

BELGIUM : MAJOR BUSINESSES

- Gent
- Antwerpen
- Liège
- Kortrijk
- Brussels
- Charleroi

🔲 Electronics		🚗 Vehicle manufacture
💉 Pharmaceuticals		🌸 Petrochemicals
✈ Aerospace industry		❋ Textiles
⚙ Heavy engineering		
☎ Telecommunications		

0   50 km
0   50 miles

RESURCES

 Electric power 15m kw

 30,722 tonnes

Not an oil producer; refines 607,000 b/d

7.35m pigs, 3.25m cattle, 38m chickens

Coal, natural gas, shale, marble, sandstone, dolomite

ELECTRICITY GENERATION

Hydro 2% (1.5bn kwh)
Combustion 44% (38bn kwh)
Nuclear 54% (46bn kwh)
Other 0%

% of total generation by type

Belgium has few natural resources and depends largely on the export of goods and services. The once-rich coal mines of Wallonia are almost depleted. There is some deciduous and conifer forestry in the Ardennes region.

BELGIUM : LAND USE

Cropland
Pasture
Forest
Pig
Wheat

0   50 km
0   50 miles

ENVIRONMENT

 Sustainability rank: 125th

 3%

 9.9 tonnes per capita

ENVIRONMENTAL TREATIES

Yes   Yes   Yes
Yes   Yes   Yes

Flanders is concerned about the pollution of groundwater supplies through acid rain, heavy metals, fertilizers, and pesticides. Its government operates an environmental management plan to raise standards. Wallonia has strict laws against illegal tipping of waste, and regulations on air quality and emissions. Awareness of environmental issues is reflected in the rise of the two green parties, which entered government for the first time in the coalition formed in 1999.

MEDIA

 TV ownership high

Daily newspaper circulation 161 per 1000 people

PUBLISHING AND BROADCAST MEDIA

There are 30 daily newspapers, published in Dutch, French, and German, including *Het Nieuwsblad, Le Soir,* and *De Standaard*

3 state-owned services, broadcasting in Dutch, French, and German, and 5 independent commercial services

3 state-owned services, broadcasting in Dutch, French, and German, and numerous private stations

Newspapers tend to be regional and divided by language. Circulation is low: the most widely read paper has a circulation of only 370,000. Control of broadcasting is divided along linguistic lines between two major corporations. Over 95% of Belgians have cable TV, with access to channels from all over Europe. Commercial TV only began in 1989, with the Flemish station VTM.

CRIME

 No death penalty

 8671 prisoners

Up 19% 1996–1998

CRIME RATES

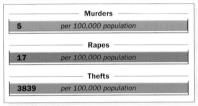

Murders
5   *per 100,000 population*

Rapes
17   *per 100,000 population*

Thefts
3839   *per 100,000 population*

Brussels has one of the lowest murder rates for any capital city. In urban areas pickpocketing is rising and car theft is a problem. Penalties for illegal use of narcotics are strict.

EDUCATION

 School leaving age: 18

 99%   358,214 students

THE EDUCATION SYSTEM

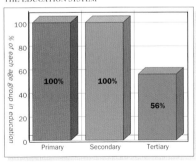

% of each age group in education

Primary 100%
Secondary 100%
Tertiary 56%

In Belgium, parents can choose between schooling provided by the two main language communities, by public authorities, or by private interests. Roman Catholic schools constitute the greatest number of "free" (privately organized) establishments. Since 1989 the system has been administered by the governments of the two main language groups. All universities are split by language.

HEALTH

 Welfare state health benefits

 1 per 263 people

Heart and respiratory diseases, cancers, car accidents

The quality of health care is among the best in the world, and government spending is high. Belgium is a world leader in fertility treatment and heart and lung transplants. Treatment is not free, but Belgians are able to claim up to 75% of their costs.

Only heart disease and cancer rank before car accidents as a cause of death. There were an estimated 8500 people living with HIV/AIDS at end-2001. In May 2002 Belgium became the second country (after the Netherlands) to legalize euthanasia.

SPENDING

GDP/cap. increase

CONSUMPTION AND SPENDING

448 per 1000 population
498 per 1000 population

Defense 1.4%
Education 5.6%
Health 6.3%

Defense, Health, Education spending as % of GDP

Despite high levels of state debt and failing traditional industries, Belgium is one of Europe's richest countries. GDP per capita is lower than for Germany but higher than for Italy or the UK.

This statistic masks considerable regional differences. In Flanders, with its many high-tech businesses, the level of unemployment is only half that in Wallonia.

The presence of highly paid EU officials and international company employees and bankers has made Brussels a distinctly wealthy, and expensive, city. The recession of the early 1990s prompted Belgians to save a higher proportion of their income, but the level of savings has fallen since then as consumer confidence has recovered.

WORLD RANKING

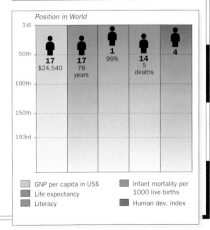

Position in World

1st
50th
100th
150th
193rd

17 $24,540
17 78 years
1 99%
14 5 deaths
4

GNP per capita in US$
Life expectancy
Literacy
Infant mortality per 1000 live births
Human dev. index

BELIZE

B

OFFICIAL NAME: Belize **CAPITAL:** Belmopan
POPULATION: 200,000 **CURRENCY:** Belizean dollar **OFFICIAL LANGUAGE:** English

FORMERLY BRITISH HONDURAS, Belize was the last Central American country to gain its independence, in 1981. It lies on the southeastern shore of the Yucatan peninsula and shares a border with Mexico along the River Hondo. Belize is Central America's least populous country, and almost half of its land area is still forested. Its swampy coastal plains are protected from flooding by the world's second-largest barrier reef.

Small fishing village near Belize City. About 500 tonnes of Caribbean spiny lobster, the main inshore species, are caught every year.

CLIMATE
▷ Tropical equatorial

WEATHER CHART FOR BELMOPAN

Conditions are hot and humid. Coastal regions are affected by hurricanes, notably Hurricane Iris in late 2001.

TRANSPORTATION
▷ Drive on right

Phillip S. W. Goldson, Belize City
272,000 passengers

1660 ships
2.25m grt

THE TRANSPORTATION NETWORK

488 km (303 miles)	None
None	825 km (513 miles)

A US$16 million IDB loan in 1998 helped improve the country's road network and its feeder roads. A terminal and runway extension have been completed at the international airport near Belize City.

TOURISM
▷ Visitors : Population 1:1.1

181,000 visitors

Up 2% in 1999

MAIN TOURIST ARRIVALS

USA 40%	
Guatemala 24%	
Mexico 10%	
Other 26%	

0   10   20   30   40
% of total arrivals

The barrier reef, good beaches, and Mayan ruins draw visitors. "Eco" attractions need conservation.

PEOPLE
▷ Pop. density low

English Creole, Spanish, English, Mayan, Garifuna (Carib)

9/km² (23/mi²)

THE URBAN/RURAL POPULATION SPLIT

54%   46%

ETHNIC MAKEUP

Other 4%
Asian Indian 4%
Garifuna 7%
Maya 11%
Creole 30%
Mestizo 44%

Over 80% of Belizeans are of mixed descent involving African, Amerindian, and European strands. Along with the *mestizo* and Creole populations there are the Afro-Carib *garifuna*. Christianity is dominant, and the Roman Catholic, Anglican, and Methodist Churches run most of the schools.

POLITICS
▷ Multiparty elections

L. House 1998/2003
U. House 1998/2003

H.M. Queen Elizabeth II

AT THE LAST ELECTION

House of Representatives 29 seats

90%
PUP

10%
UDP

PUP = People's United Party
UDP = United Democratic Party

Senate 8 seats

The members of the Senate are appointed by the governor-general

The desire for independence dominated politics until the 1980s. The PUP, under George Price, negotiated this with the British in 1981. During the 1984–1989 UDP administration, the main concerns were to maintain a pro-US line and the fear of communism in the region. In the absence of any major ideological or policy distinctions, the UDP lost power to the PUP in 1989, winning it back in 1993; the pendulum swung back to the PUP again in 1998. Growth, job creation, "economic citizenship" for foreigners, and border tension with Guatemala are key issues.

BELIZE

Total Land Area : 22 966 sq. km (8867 sq. miles)

POPULATION
● over 10 000
• under 10 000

LAND HEIGHT
1000m/3281ft
500m/1640ft
200m/656ft
Sea Level

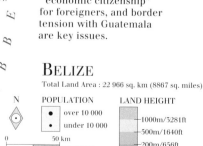

0   50 km
0   50 miles

WORLD AFFAIRS

 Joined UN in 1981

 ACS   Comm   Caricom   NAM   OAS

The major concern is Guatemala's historic claim to over half of Belize. Suspended in 1986, the claim was revived in early 2000 as border tension rose.

AID

 Recipient

 US$15m (receipts)   Down 67% in 2000

In 1999 the IDB, the Commonwealth Development Corporation, the European Investment Bank, and the CDB invested in citrus farms. Belize is one of the highest per capita recipients of US aid.

DEFENSE

 No compulsory military service

 US$17m   No change in 2000

The small Belize Defense Force took over full responsibility from the UK in 1994 for the country's defense. The UK withdrew its garrison in the same year, but continues to maintain a jungle training school.

ECONOMICS

 Inflation 2.7% p.a. (1990–2000)

US$746m   1.97 Belizean dollars

SCORE CARD

- ❏ WORLD GNP RANKING.........................160th
- ❏ GNP PER CAPITA ....US$3110
- ❏ BALANCE OF PAYMENTS.................–US$139m
- ❏ INFLATION 0.6%
- ❏ UNEMPLOYMENT.....................................13%

STRENGTHS

Sugar, textile manufacture, citrus fruits, bananas, shellfish, forestry, and considerable tourist potential. Sustainable public debt; fair access to concessionary foreign finance.

WEAKNESSES

Narrow export base dependent on preferential market access; reliance on imports of processed foods. Poor fiscal management in late 1990s.

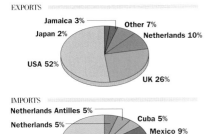

EXPORTS

Jamaica 3% — Other 7%
Japan 2% — Netherlands 10%
USA 52%
UK 26%

IMPORTS

Netherlands Antilles 5% — Cuba 5%
Netherlands 5% — Mexico 9%
USA 50% — Other 26%

RESOURCES

 Electric power 43,000 kw

 43,103 tonnes   Not an oil producer

45,000 cattle, 24,000 pigs, 1.35m chickens   None

Hopes of finding significant oil and gas deposits in the north of the country have so far proved fruitless.

ENVIRONMENT

 Not available

 5% partially protected   1.8 tonnes per capita

Tourist developments and logging have depleted the dense tropical forests. Mahogany is endangered so all exports and transshipments now require a certificate of origin. Global warming poses a major threat to the corals of the barrier reef.

MEDIA

 TV ownership medium

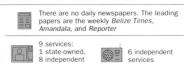 There are no daily newspapers

PUBLISHING AND BROADCAST MEDIA

There are no daily newspapers. The leading papers are the weekly *Belize Times*, *Amandala*, and *Reporter*

9 services: 1 state-owned, 8 independent   6 independent services

Belize has not suffered the degree of press interference experienced in neighboring states, but successive governments have remained sensitive to even minor criticisms. The two radio stations of the public Broadcasting Corporation of Belize were sold in 1998 to two local stations, but the government has retained ownership of the transmitters. Two official newspapers compete with party-political and independent publications.

CRIME

 Death penalty in use

 1097 prisoners   Increase in gun-related crime

Belize is a major transit point to the US for cocaine, despite being decertified in 1997 for its antinarcotics efforts. Drugs-related crime is high. Armed robberies by criminal gangs based in neighboring Guatemala are also a major concern. A government ombudsman was appointed in 2000 to investigate police brutality and corruption.

EDUCATION

 School leaving age: 14

93%   2853 students

Although most schools are run by the different Churches, a handful are funded by the government, particularly those catering for special needs. The University College of Belize provides for higher education.

CHRONOLOGY

The Mayan heartland included what is now Belize. Between 1798 and 1981 it was effectively a British colony.

- ❏ **1919** Demands for more political rights by black Belizeans returning from World War I.
- ❏ **1936** New constitution.
- ❏ **1950** PUP formed. Voting age limit for women reduced from 30 to 21.
- ❏ **1954** Full adult suffrage.
- ❏ **1972** Guatemala threatens invasion. Britain sends troops.
- ❏ **1981** Full independence.
- ❏ **1998** PUP wins crushing general election victory.
- ❏ **2000** Guatemala revives claim to half of Belize.

HEALTH

 Welfare state health benefits

 1 per 1818 people   Respiratory, heart, and cerebrovascular diseases

The health service provided by the government includes seven hospitals, more than 30 regional health centers, and numerous mobile clinics. Water supplies and sanitation have been improved; most homes in Belmopan now have both.

SPENDING

GDP/cap. increase

CONSUMPTION AND SPENDING

44 per 1000 population   149 per 1000 population

Defense 2.4%
Education 5%
Health 2.3%

0   5   10   15   20   25
Defense, Health, Education spending as % of GDP

The European Development Fund in 1999 granted 3.5 million Belizean dollars toward the reduction of rural poverty. Narcotics trading remains a source of wealth.

WORLD RANKING

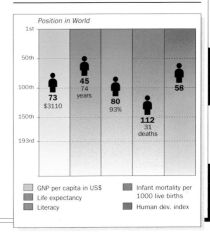

Position in World

1st
50th
100th
150th
193rd

73 $3110
45 74 years
80 93%
112 31 deaths
58

GNP per capita in US$ — Infant mortality per 1000 live births
Life expectancy — Human dev. index
Literacy

BENIN

OFFICIAL NAME: Republic of Benin **CAPITAL:** Porto-Novo
POPULATION: 6.4 million **CURRENCY:** CFA franc **OFFICIAL LANGUAGE:** French

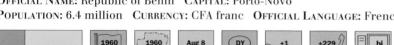

1960 | 1960 | Aug 8 | DY | +1 | +229 | .bj

BENIN STRETCHES NORTH from the west African coast, with a 100-km (60-mile) shoreline on the Bight of Benin. Formerly the kingdom of Dahomey, Benin was under French colonial rule, becoming part of French West Africa, until independence in 1960. In 1990 Benin was a pioneer of multipartyism in Africa, ending 17 years of one-party Marxist–Leninist rule. Benin's economy is based on well-diversified agriculture.

CLIMATE ▷ Tropical wet & dry

WEATHER CHART FOR PORTO-NOVO

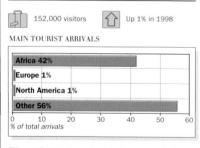

There are two rainy seasons. The hot, dusty *harmattan* wind characterizes the December to February dry season.

TRANSPORTATION ▷ Drive on right

Cotonou
310,459 passengers | 7 ships
1100 grt

THE TRANSPORTATION NETWORK

| 1357 km (843 miles) | 10 km (6 miles) |
| 458 km (285 miles) | None |

The joint Benin–Niger railroad runs only as far as Parakou. The Cotonou–Porto-Novo line reopened in 1999.

TOURISM ▷ Visitors : Population 1:42

152,000 visitors | Up 1% in 1998

MAIN TOURIST ARRIVALS

Africa 42%	
Europe 1%	
North America 1%	
Other 56%	
0 10 20 30 40 50 60	
% of total arrivals	

The industry is not well developed; there are plans to increase package tourism. There is safari tourism in the north, in the Atakora Mountains in particular. Benin is popular for weekend breaks for tourists vacationing in Nigeria.

PEOPLE ▷ Pop. density medium

Fon, Bariba, Yoruba, Adja, Houeda, Somba, French | 58/km² (150/mi²)

THE URBAN/RURAL POPULATION SPLIT

42% | 58%

RELIGIOUS PERSUASION

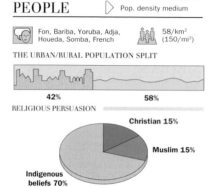

Christian 15%
Muslim 15%
Indigenous beliefs 70%

Benin is politically dominated by the southern Fon people. There is some north–south tension, partly because the south is more developed, and partly reflecting a Muslim–Christian divide. Women tend to wield power and influence in the retail trade.

BENIN

Total Land Area :
112 620 sq. km
(43 483 sq. miles)

POPULATION
◎ over 100 000
○ over 50 000
● over 10 000
• under 10 000

LAND HEIGHT
500m/1640ft
200m/656ft
Sea Level

0   100 km
0   100 miles

WEST AFRICA
Burkina
Togo
Nigeria
BENIN
Africa

POLITICS ▷ Multiparty elections

 1999/2003 | President Mathieu Kérékou

AT THE LAST ELECTION

National Assembly 83 seats

7% MADEP

33% PRB | 13% PRD | 12% FARD | 11% PSD | 5% IPD | 19% Others

PRB = Benin Renaissance Party **PRD** = Party of Democratic Renewal **FARD** = Action Front for Renewal and Development **PSD** = Social Democrat Party **MADEP** = African Movement for Democracy and Progress **IPD** = Impetus for Progress and Democracy

Benin's image as a leader in African democratization was tarnished by allegations of fraud over the 2001 presidential election. Democratization had begun at the National Conference of 1990, when Mathieu Kérékou agreed to hold multiparty elections after years of military one-party rule. The main political parties in Benin tend to be regionally based and depend on the leadership of individuals influential in local communities. Alliances change constantly. Kérékou became the first of the African one-party leaders to hand over power peacefully, to Nicéphore Soglo, a former World Bank official, after elections in 1991. Soglo did not have an automatic majority in the National Assembly, and was forced to include members of the opposition parties in his government. The main political issue became his World Bank-style deregulation of the economy. He was defeated in a controversial election in 1996 which brought Kérékou back to power as president. Kérékou dismissed claims of vote rigging in the presidential election in 2001, saying that democracy was "alive and kicking." He easily won reelection following Soglo's withdrawal from the race.

WORLD AFFAIRS ▷ Joined UN in 1960

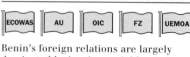

ECOWAS | AU | OIC | FZ | UEMOA

Benin's foreign relations are largely dominated by its giant neighbor, Nigeria, which is by far the most powerful state in the region. The continuation of good relations with France, which is currently the main source of financial aid, is considered to be critical.

B

AID

 Recipient

 $239m (receipts)   ⬆ Up 13% in 2000

Benin's poverty is such that the maintenance of aid is at the top of the political agenda. France, the main protector of Benin since independence in 1960, is the major aid donor. Other donors include the World Bank, the US, Germany, the EU, Denmark, Japan, and Italy. Almost all development finance comes from aid, and some has been used to finance debt servicing. There is the usual problem of finding suitable projects, although Benin has a large, well-educated (if top-heavy) civil service, making implementation easier than in many parts of Africa.

DEFENSE

 Compulsory military service

 $36m   ⬆ Up 6% in 2000

The 4500-strong army is actively involved in the attempt to curb smuggling on the Nigerian border. In 1989 the army was employed internally against rioters.

ECONOMICS

 Inflation 8.3% p.a. (1990–2000)

 $2.35bn   698.7–736.7 CFA francs

SCORE CARD

- ❏ WORLD GNP RANKING.......................135th
- ❏ GNP PER CAPITA $370
- ❏ BALANCE OF PAYMENTS...................–$168m
- ❏ INFLATION 4.2%
- ❏ UNEMPLOYMENT.................................2%

STRENGTHS

Agriculture-based economy, with good product diversification. Long-overdue devaluation of the CFA franc in January 1994 made exports more competitive.

WEAKNESSES

Large-scale smuggling. Power failures caused by drought brought major economic problems in 1998, and resultant slowdown in GDP growth. Top-heavy civil service.

EXPORTS

Thailand 5%
Indonesia 6%
Brazil 14%
Other 39%
India 18%
Italy 18%

IMPORTS

Thailand 4%
Togo 4%
UK 5%
France 16%
Other 43%
China 28%

Flat landscape near Cotonou, characteristic of Benin's coastal region. Numerous lagoons lie behind its short coastline.

RESOURCES

 Electric power 15,000 kw

38,542 tonnes   1243 b/d (reserves 30m barrels)

1.5m cattle, 1.18m goats, 23m chickens   Oil, limestone, marble, gold

Since 1988 most electricity – which previously had to be imported from Ghana – has been generated by the Nangbeto Dam on the River Mono.

ENVIRONMENT

 Sustainability rank: 95th

 7%   ⬆ 0.1 tonnes per capita

Desertification in the north is the major problem. Benin has been used in the past as a dumping ground for toxic waste.

MEDIA

 TV ownership low

Daily newspaper circulation 2 per 1000 people

PUBLISHING AND BROADCAST MEDIA

There are 18 daily newspapers, including *Le Matinal* and *La Nation*

5 services: 1 state-owned, 4 independent   19 services: 1 state-owned, 18 independent

Benin publishes over 50 newspapers and periodicals. The press has considerable freedom, and a media code of practice was introduced in 1999.

CRIME

 Death penalty in use

4961 prisoners   ⬆ Up 219% 1996–1998

Armed crime has risen sharply since 1995, despite the reintroduction of the death penalty. Smuggling, including child trafficking, is a major problem.

EDUCATION

 School leaving age: 12

40%   16,284 students

More is spent on education than on defense, and this is reinforced by Benin's active intellectual community, the "Latin Quarter of Africa." The university at Abomey-Calavi is rated highly in medicine and law.

In 1625 the Fon, indigenous slave traders, founded the kingdom of Dahomey. Dahomey in turn conquered the neighboring kingdoms of Dan, Allada, and the coast around Porto-Novo.

- ❏ **1857** French establish trading post at Grand-Popo.
- ❏ **1889** French defeat King Behanzin.
- ❏ **1892** French protectorate.
- ❏ **1904** Part of French West Africa.
- ❏ **1960** Full independence.
- ❏ **1975** Renamed Benin.
- ❏ **1989** Marxism–Leninism abandoned as official ideology.
- ❏ **1996** Former ruler Kérékou defeats Soglo in controversial election.
- ❏ **2001** Kérékou reelected to presidency amid claims of electoral fraud.

HEALTH

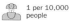 No welfare state health benefits

1 per 10,000 people   Communicable and diarrheal diseases, malaria

Outside the major towns, health services and doctors are scarce. It is forecast that by 2030, one million people will have died from AIDS.

SPENDING

 GDP/cap. increase

CONSUMPTION AND SPENDING

7 per 1000 population   8 per 1000 population

Defense 1.4%
Education 2.6%
Health 1.6%

0   5   10   15   20   25
Defense, Health, Education spending as % of GDP

Substantial differences in wealth reflect the strongly hierarchical nature of society, especially in the south. French cars are considered to be status symbols.

WORLD RANKING

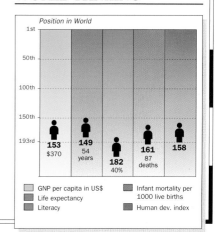

Position in World

1st
50th
100th
150th
193rd

153 $370
149 54 years
182 40%
161 87 deaths
158

- ☐ GNP per capita in US$
- ☐ Life expectancy
- ☐ Literacy
- ☐ Infant mortality per 1000 live births
- ☐ Human dev. index

BHUTAN

OFFICIAL NAME: Kingdom of Bhutan   **CAPITAL:** Thimphu
POPULATION: 2.1 million   **CURRENCY:** Ngultrum   **OFFICIAL LANGUAGE:** Dzongkha

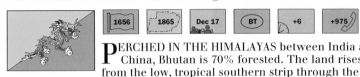

| 1656 | 1865 | Dec 17 | BT | +6 | +975 | .bt |

PERCHED IN THE HIMALAYAS between India and China, Bhutan is 70% forested. The land rises from the low, tropical southern strip through the fertile central valleys to the high Himalayas, inhabited by seminomadic yak herders. Formally a Buddhist state where power is shared by the king and the government, Bhutan began modernizing in the 1960s, but has chosen to do so gradually, and remains largely closed to the outside world.

PEOPLE

▷ Pop. density low

Dzongkha, Nepali, Assamese

45/km²
(116/mi²)

THE URBAN/RURAL POPULATION SPLIT

7% | 93%

RELIGIOUS PERSUASION

Other 6%
Hindu 24%
Mahayana Buddhism 70%

The majority of the population, the Drukpa peoples, originated from Tibet and are devoutly Buddhist. The Hindu minority is made up of Nepalese who settled in the south from 1910 to 1950. Bhutan has 20 languages. Dzongkha, the language of western Bhutan, native to just 16% of people, was made the official language in 1988. The Nepalese community has reacted strongly, regarding this act as "cultural imperialism."

CLIMATE

▷ Mountain

WEATHER CHART FOR THIMPHU

■ Average daily temperature   Rainfall ■
°C/°F   J F M A M J J A S O N D   cm/in
40/104 — 40/16
30/86 — 30/12
20/68 — 20/8
10/50 — 10/4
0/32 — 0
-10/14
-20/-4

The south is tropical, the north alpine, cold, and harsh. The central valleys are warmer in the east than in the west. The summer monsoon affects all parts.

TRANSPORTATION

▷ Drive on left

✈ **Paro International**
19,939 passengers

🚢 Has no fleet

THE TRANSPORTATION NETWORK

| 1994 km (1239 miles) | None |
| None | None |

The main surfaced road runs east–west across central Bhutan. Two others run south into India. Only the national airline, Druk Air, flies into Bhutan.

TOURISM

▷ Visitors : Population 1:339

6200 visitors

⬆ Up 4% in 1999

MAIN TOURIST ARRIVALS

USA 24%
Japan 17%
UK 11%
Other 48%

0   10   20   30   40   50   60
% of total arrivals

Tourism is restricted to protect Bhutan's culture and natural environment; entry has been easier since the industry was privatized in 1991. Most monasteries are closed to tourists. In 1998, fire damaged the famous Taktsang monastery.

Less than 10% of Bhutan is arable, but its fertility allows almost any crop to grow. The diversity of wild plant species inspired its old name: Southern Valleys of the Medicinal Herbs.

POLITICS

▷ No multiparty elections

Not applicable

H.M. Druk Gyalpo (Dragon King) Jigme Singye Wangchuk

LEGISLATIVE OR ADVISORY BODIES

National Assembly 150 seats

There are no legal political parties; members are elected individually to the National Assembly, to advise the king, who rules as an absolute monarch

The modernization of Bhutan's absolute monarchy began in 1961. Under further changes proposed in 1998, the king relinquished his right to appoint the government in favor of a cabinet elected by the National Assembly. The National Assembly was also empowered to pass a vote of no confidence against the king. These proposals came as a response to a prodemocracy movement fueled by ethnic Nepalese opposed to the Drukpa-dominated political system.

BHUTAN

Total Land Area :
47 000 sq. km (18 147 sq. miles)

LAND HEIGHT

6000m/19686ft
4000m/13124ft
2000m/6562ft
1000m/3281ft
500m/1640ft
200m/656ft
160m/252ft

POPULATION

● over 10 000
• under 10 000

N

0   50 km
0   50 miles

B

WORLD AFFAIRS

 Joined UN in 1971

 CP   IBRD   NAM   SAARC   ADB

Bhutan's closest links are with India. Relations with China are cordial, and negotiations to settle the China–Bhutan border have progressed smoothly since 1984. There is tension with Nepal over Bhutan's treatment of its ethnic Nepalese minority and the influx of Bhutanese refugees into Nepal.

AID

 Recipient

 $53m (receipts)   Down 21% in 2000

Bhutan relies on foreign aid for about half of its annual budget. The largest single donor is Japan.

DEFENSE

 No compulsory military service

$20m   No change in 2000

The army is under the command of the king, and is trained by Indian military instructors. India provides de facto military protection and is obliged to defend Bhutan against attack.

ECONOMICS

 Inflation 9.5% p.a. (1990–2000)

 $479m   46.68–48.22 ngultrum

SCORE CARD

- ❑ World GNP Ranking.........................171st
- ❑ GNP per Capita $590
- ❑ Balance of Payments......................–$87m
- ❑ Inflation 6.8%
- ❑ Unemployment ....Low rate

Strengths

New development of cash crops for Asian markets (cardamoms, apples, oranges, apricots). Hardwoods in south, especially teak, but exploitation currently tightly controlled. Large hydroelectric potential.

Weaknesses

Dependence on Indian workers for many public-sector jobs from road building to teaching. The majority of the population are dependent on agriculture. Cultivated land is extremely restricted because of steep mountain slopes. Very little industry. Few mineral resources.

EXPORTS

Other 5%

India 95%

IMPORTS

Other 25%

India 75%

RESOURCES

 Electric power 356,000 kw

 330 tonnes

Not an oil producer and has no refineries

 435,000 cattle, 74,900 pigs, 310,000 chickens

Talc, gypsum, coal, limestone, slate, dolomite

Bhutan's forests remain largely intact, and logging is very strictly controlled. Hydroelectric potential is considerable, but few dams have been built. Power is sold to India from the Chhukha Dam, bringing in substantial foreign earnings.

ENVIRONMENT

 Sustainability rank: 30th

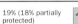 19% (18% partially protected)

0.5 tonnes per capita

Bhutan's forests stabilize the steep mountainsides and supply the bulk of its fuel needs. Road building, begun in the 1960s, is the biggest cause of deforestation, which has led to topsoil erosion. The high northern pastures are at risk from overgrazing by yaks. Traditional Buddhist values instilling respect for nature and forbidding the killing of animals are still observed.

MEDIA

 TV ownership low

 There are no daily newspapers

PUBLISHING AND BROADCAST MEDIA

 There are no daily newspapers. *Kuensel* is published weekly by the government in Dzongkha, English, and Nepali

 1 state-owned service

 1 state-owned service

Until 1999 TV was banned, in order to protect cultural values. An Internet café opened in Thimphu in 2000.

CRIME

 Death penalty not used in practice

 Bhutan does not publish prison figures

Little variation from year to year

Levels of violent crime and theft are low. In 1991, *Driglam namzha*, an ancient code of conduct including the requirement to wear traditional dress, was revived, with imprisonment or fines for those not in compliance.

EDUCATION

 Schooling is not compulsory

 47%   2683 students

Education is free. A small minority of children attend secondary school. Teaching is in English and Dzongkha. There are no universities.

CHRONOLOGY

The Drukpa, originally from Tibet, united Bhutan in 1656. In 1865 the Drukpa lost the Duars Strip to British India.

- ❑ **1907** Monarchy established.
- ❑ **1949** Independence.
- ❑ **1953** National Assembly inaugurated.
- ❑ **1968** King forms first cabinet.
- ❑ **1971** Bhutan joins UN.
- ❑ **1990** Ethnic Nepalese launch campaign for minority rights.
- ❑ **1998** King proposes to reform government.
- ❑ **1999** First TV service inaugurated.

HEALTH

 Welfare state health benefits

 1 per 6510 people

Diarrheal, respiratory diseases, tuberculosis, malaria, infant deaths

Free clinics, along with Thimphu's hospital, provide basic health care. Progress is being made in child immunization, and monks have been persuaded to teach hygiene. Infant mortality is high. Bhutanese, Tibetan, and Chinese traditional medicines are widely practiced.

SPENDING

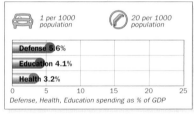 GDP/cap. increase

CONSUMPTION AND SPENDING

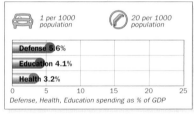

1 per 1000 population   20 per 1000 population

Defense 5.6%
Education 4.1%
Health 3.2%

0   5   10   15   20   25
Defense, Health, Education spending as % of GDP

Most of Bhutan's people are chronically poor, although starvation is virtually unknown. There is a small middle class, consisting of public employees and storekeepers.

WORLD RANKING

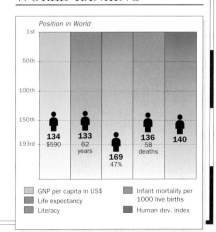

Position in World

1st
50th
100th
150th
193rd

134 $590
133 62 years
169 47%
136 58 deaths
140

- GNP per capita in US$
- Life expectancy
- Literacy
- Infant mortality per 1000 live births
- Human dev. index

BOLIVIA

B

OFFICIAL NAME: Republic of Bolivia **CAPITALS:** La Paz (administrative); Sucre (judicial)
POPULATION: 8.5 million **CURRENCY:** Boliviano **OFFICIAL LANGUAGES:** Spanish, Quechua, and Aymara

 1825 1938 Aug 6 BOL -4 +591 .bo

BOLIVIA LIES LANDLOCKED high in central South America, and is one of the continent's poorest nations. Over half of the population lives on the *altiplano*, the windswept plateau between two ranges of the Andes, 3500 m (11,500 ft) above sea level. La Paz, the highest capital in the world, has spawned a neighboring large twin, El Alto. Bolivia has the world's highest golf course, ski run, and soccer stadium. The eastern lowland regions are tropical and underdeveloped but are rapidly being colonized.

CLIMATE

▷ Tropical/mountain

WEATHER CHART FOR LA PAZ

The Andean *altiplano* has an extreme tropical highland climate with winter night frosts. Annual rainfall in the west is only 25 cm (10 in). The hot eastern lowlands receive most rain in summer.

TRANSPORTATION

▷ Drive on right

✈ **El Alto, La Paz**   🚢 1 ship 15,800 dwt

THE TRANSPORTATION NETWORK

2872 km (1785 miles)	27 km (17 miles)
3698 km (2297 miles)	14,000 km (8699 miles)

Obtaining port access to the Pacific coast for landlocked Bolivia is important. Only 4% of roads are paved. The national railroad was privatized in 1996. Domestic airlines are generally reliable.

Potato harvest on the altiplano. *Migration to the more fertile lands in the east has been encouraged.*

TOURISM

▷ Visitors : Population 1:25

🧳 342,000 visitors   ⬇ Down 21% in 1999

MAIN TOURIST ARRIVALS

Peru 14%
Argentina 12%
USA 12%
Brazil 7%
Chile 6%
Other 49%

% of total arrivals

Foreign tourists are drawn by the traditional festivals, especially carnivals in February or March, the variety of Bolivia's scenery, and its Spanish colonial architecture. Major attractions include the Silver Mountain at Potosí, and Lake Titicaca, the highest navigable lake in the world, covering an area of 8970 sq. km (3463 sq. miles). Recent political stability encouraged some growth in tourism in the 1990s, but potential is limited by Bolivia's isolation, the rugged, inaccessible terrain, and the limited infrastructure.

Copacabana on the shores of Lake Titicaca. It lies on a large headland owned by Bolivia on the Peruvian side of the lake.

BOLIVIA

Total Land Area : 1 098 580 sq. km (424 162 sq. miles)

POPULATION

over 500 000	⊙
over 100 000	◎
over 50 000	○
over 10 000	●
under 10 000	•

LAND HEIGHT

4000m/13124ft	
2000m/6562ft	
1000m/5281ft	
500m/1640ft	
200m/656ft	
Sea Level	

PEOPLE   ▷ Pop. density low

Aymara, Quechua, Spanish   8/km² (20/mi²)

THE URBAN/RURAL POPULATION SPLIT

63%   37%

RELIGIOUS PERSUASION

- Other 7%
- Roman Catholic 93%

ETHNIC MAKEUP

- Other 8%
- European 10%
- Mixed 13%
- Quechua 37%
- Aymara 32%

Two-thirds of Bolivians are Quechua and Aymara Amerindians who historically have been marginalized. In recent years, however, they have played a more active role in politics by supporting new populist parties.

Wealthy city elites, dating back to Spanish colonial rule, retain great influence, but new entrepreneurs with political ambitions have appeared. Most Bolivians are subsistence farmers, miners, small traders, or artisans earning low incomes. Government schemes, spontaneous colonization, and the collapse of tin mining have led to large-scale migration from the Andes to lowland eastern regions in the last few decades.

Family life tends to be close-knit; Amerindians practice Roman Catholicism mixed with their own traditions and culture. Women have low status. There are some 130,000 lowland Amerindians in western regions.

POPULATION AGE BREAKDOWN

Female	Age	Male
1.2%	80+	1%
2.1%	60–79	1.8%
7.4%	40–59	6.9%
14.6%	20–39	14.3%
24.9%	0–19	25.8%

% of population by age group

POLITICS   ▷ Multiparty elections

L. House 2002/2007
U. House 2002/2007

President Gonzalo Sanchez de Lozada

AT THE LAST ELECTION

Chamber of Deputies 130 seats

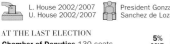

| 27% MNR | 21% MAS | 20% MIR | 19% NFR | 5% MIP | 4% UCS | 3% ADN | 1% PS |

MNR = Nationalist Revolutionary Movement
MAS = Movement for Socialism   **MIR** = Movement of the Revolutionary Left   **NFR** = New Republican Force
MIP = Pachakuti Indigenous Movement   **UCS** = Union for Civic Solidarity   **ADN** = Nationalist Democratic Action
PS = Socialist Party

Chamber of Senators 27 seats

| 41% MNR | 30% MAS | 18% MIR | 7% NFR | 4% ADN |

Bolivia is a multiparty democracy.

PROFILE

From 1825 to the early 1980s, Bolivia experienced, on average, more than one armed coup a year, punctuated by a national revolution in 1952 which delivered important reforms. The fragmented and drug-tainted military finally stepped down in 1982, but full elections were delayed until 1985.

New populist parties have emerged to challenge traditional politics, and even the drift to free-market economics. Coalitions are unstable, nepotism remains rife, and the narcotics trade, the profits of which underpin the economy, is frequently implicated in political corruption scandals. The main trade union federation COB was the traditional focus of opposition, but coca growers and other popular groups have now assumed this role. Peasant leader Evo Morales even gathered enough support to come a close second in presidential elections in 2002.

The austerity policies of the MNR government were continued after its defeat in 1997 by former dictator Hugo Banzer. Ill health forced his resignation in 2001, and the MNR was returned to power in 2002 with Gonzalo Sanchez resuming the presidency.

MAIN POLITICAL ISSUES
Disillusionment with political system
Public opinion in the run-up to the 2002 presidential elections pointed to a lack of faith in promised electoral reform. The country's veteran politicians are widely viewed as a self-serving, self-perpetuating class tainted by corruption.

Coca growers
Poor farmers oppose the government's forced eradication of coca crops in an anticocaine drive to ensure more US aid. Serious clashes continued in 2001–2002 as troops operating in eastern coca-growing areas removed farmers' road blocks. The *cocaleros* complain that modest government subsidies to switch crops take no account of low prices for bananas, pineapples, and palm hearts.

WORLD AFFAIRS   ▷ Joined UN in 1945

AP   AmCC   NAM   OAS   RG

B

Gaining improved access to the Pacific via Peru and Chile is Bolivia's historic foreign policy concern. The prospect of a Bolivia–Peru gas pipeline is a long-term project. Current gas export projects, coupled with the opening of the telecoms sector, and the extension of the Apta preferential trade act with the US should strengthen economic growth. Current US aid is conditional on the Bolivian government taking measures to destroy the cocaine-producing and -trafficking industry, involving military and police attacks on impoverished coca growers.

A pointer to the future is that Bolivia is an associate member, along with Chile, of Mercosur. This southern grouping, under the leadership of Brazil, will act as a strong negotiating bloc in upcoming talks for a Free Trade Area of the Americas (FTAA), due in 2005.

Hugo Banzer Suárez, former dictator, who was elected president (1997–2001).

Gonzalo Sanchez de Lozada, president for 1993–1997, was reelected in 2002.

CHRONOLOGY

The Aymara civilization was conquered by the Incas in the late 1400s. Fifty years later, the Incas were defeated by the *conquistadores* and Upper Peru, as it became, was governed by Spain from Lima.

- ❑ **1545** Cerro Rico, the Silver Mountain, discovered at Potosí. Provides Spain with vast wealth.
- ❑ **1776** Upper Peru becomes part of Viceroyalty of Río de la Plata centered on Buenos Aires.
- ❑ **1809** Simón Bolívar inspires first revolutionary uprisings in Latin America at Chuquisaca (Sucre), La Paz, and Cochabamba, but they fail.
- ❑ **1824** Spaniards suffer final defeat by Bolívar's general, José de Sucre.
- ❑ **1825** Independence.
- ❑ **1836–1839** Union with Peru fails. Internal disorder.
- ❑ **1864–1871** Ruthless rule of Mariano Melgarejo. Three Amerindian revolts over seizure of ancestral lands. ⇨

CHRONOLOGY *continued*

- ❏ **1879–1883** War of the Pacific, won by Chile. Bolivia left landlocked.
- ❏ **1880–1930** Period of stable governments. Exports from revived mining industry bring prosperity.
- ❏ **1903** Acre province ceded to Brazil.
- ❏ **1914** Republican Party founded.
- ❏ **1920** Amerindian rebellion.
- ❏ **1923** Miners bloodily suppressed.
- ❏ **1932–1935** Chaco War with Paraguay. Bolivia loses three-quarters of Chaco. Rise of radicalism and labor movement.
- ❏ **1951** Víctor Paz Estenssoro of MNR elected president. Military coup.
- ❏ **1952** Revolution. Paz Estenssoro and MNR brought back. Land reforms improve Amerindians' status. Education reforms, universal suffrage, tin mines nationalized.
- ❏ **1964** Military takes over in coup.
- ❏ **1967** Che Guevara killed while trying to mobilize Bolivian workers.
- ❏ **1969–1979** Military regimes rule with increasing severity. 1979 coup fails. Interim civilian rule.
- ❏ **1980** Military takes over again.
- ❏ **1982** President-elect Siles Zuazo finally heads leftist civilian MIR government. Inflation 24,000%.
- ❏ **1985** MNR wins elections. Austerity measures bring down inflation.
- ❏ **1986** Tin market collapses. 21,000 miners sacked.
- ❏ **1989** MIR takes power after close-run elections. President Paz Zamora makes pact with 1970s dictator Gen. Hugo Banzer, leader of ADN.
- ❏ **1990** 1.6 million hectares (4 million acres) of rainforest recognized as Amerindian territory.
- ❏ **1993** MNR back in power. Gonzalo Sanchez de Lozada president.
- ❏ **1997** Banzer wins presidency.
- ❏ **1999** Opposition demands inquiry into Banzer's role in regional military repression in 1970s.
- ❏ **2001** Banzer resigns due to ill health.
- ❏ **2001–2002** Government program of coca eradication and bans on coca trading and marketing provoke clashes with peasant farmers.
- ❏ **2002** MNR wins elections. Sanchez returns as president.

AID ▷ Recipient

 $477m (receipts)   Down 16% in 2000

Most aid comes from the US and depends on progress in coca crop eradication. Smaller amounts come from western European countries. Poor rural areas get project aid from Western religious organizations, NGOs, and charities. The IDB provided emergency aid after flash floods destroyed infrastructure and badly damaged the historic center of La Paz in 2002.

DEFENSE ▷ Compulsory military service

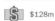 $128m   Down 14% in 2000

BOLIVIAN ARMED FORCES

36 light tanks (36 SK-105 *Kuerassier*)	25,000 personnel
60 riverine craft	3500 personnel
37 combat aircraft (18 AT-33AN)	3000 personnel
None	

Although the military has not actively interfered in politics for more than two decades, it is frequently used to quell internal dissent. The army is the main focus of defense spending, with weaponry bought almost entirely from the US. The Bolivian navy consists mainly of gunboats on Lake Titicaca, which borders Peru, and on the Pilcomayo River. The army has worked with US forces against the cocaine business, although its integrity is questioned due to its past associations with narcotics trafficking. The main ambition of the military, apart from protecting its own interests and privileges, is the unrealizable aim of recapturing territory that would allow Bolivia access to the Pacific.

Military service lasts for one year.

ECONOMICS ▷ Inflation 8.5% p.a. (1990–2000)

$8.21bn   6.36–6.82 bolivianos

SCORE CARD

- ❏ World GNP Ranking...........................93rd
- ❏ GNP per Capita $990
- ❏ Balance of Payments....................–$464m
- ❏ Inflation 4.6%
- ❏ Unemployment 4%

EXPORTS

- Ecuador 8%
- Brazil 13%
- Other 33%
- UK 13%
- Colombia 17%
- USA 16%

IMPORTS

- Peru 6%
- Other 25%
- Chile 11%
- USA 17%
- Brazil 24%
- Argentina 17%

STRENGTHS

Mineral riches: gold, silver, zinc, lead, tin. Newly discovered oil and natural gas deposits attracting foreign investment.

WEAKNESSES

Raw materials vulnerable to fluctuating world prices. Lack of processed or manufactured exports with higher added value. Lack of integration between economic sectors and regions. Poor infrastructure.

PROFILE

Traditionally, the state used earnings from the publicly owned state mining sector to control the economy. Years of deep recession in the 1980s, accompanied by accelerating inflation and a collapsing currency, saw the introduction of severe, IMF-approved, austerity policies. These, along with the introduction of a new currency and tax reform, succeeded in curbing inflation, reducing public spending, and restoring international loans, but at the price of great social unrest. Growth was restored in the 1990s and a controversial "capitalization" program was launched which allowed for the 50/50 sell-off of shares in all six state companies on attractive terms to investors and employees. Narcotics revenues remain important for the economy.

ECONOMIC PERFORMANCE INDICATOR

Consumer Price Index — GDP

BOLIVIA : MAJOR BUSINESSES

- Riberalta
- Cochabamba
- Santa Cruz
- La Paz
- Oruro
- Potosí

Textiles
Saw milling
Food processing
Pharmaceuticals
Tin mining
Narcotics
Agribusiness

0   200 km
0   200 miles

B

RESOURCES

 Electric power 1m kw

 6450 tonnes

34,846 b/d (reserves 114m barrels)

8.75m sheep, 6.73m cattle, 2.8m pigs, 74m chickens

Tin, natural gas, oil, zinc, tungsten, gold, antimony, silver, lead

ELECTRICITY GENERATION

Hydro 44% (1.6bn kwh)

Combustion 56% (2.1bn kwh)

Nuclear 0%

Other 0%

| 0 | 20 | 40 | 60 | 80 | 100 |

% of total generation by type

Bolivia is the world's fifth-largest tin producer. The government is allowing foreign companies to prospect for more oil, and to increase sales of natural gas to Brazil and Argentina.

BOLIVIA : LAND USE

Cropland
Pasture
Forest
Wetlands
High mountain regions
Coca - cash crop
Sugarcane
Cattle

0   200 km
0   200 miles

ENVIRONMENT

 Sustainability rank: 21st

 14% (5% partially protected)

1.5 tonnes per capita

ENVIRONMENTAL TREATIES

Yes   Yes   Yes

Yes   Yes   Yes

Deforestation is Bolivia's major ecological problem, as throughout the Amazon region. Land clearances are running at 164,000 hectares (398,000 acres) a year. Much of the cleared land is turned over to cattle ranching or the growing of coca. Pesticide and fertilizer overuse in the coca business is a concern. The industry is effectively uncontrolled and rivers in Amazonia have high pollution levels.

Pollution problems are compounded by waste chemicals used in minerals industries. Mercury, used in the extraction of silver, has been found in dangerous quantities in river systems.

MEDIA

 TV ownership medium

Daily newspaper circulation 55 per 1000 people

PUBLISHING AND BROADCAST MEDIA

There are 18 daily newspapers, including *Presencia*, *El Diario*, and *La Razón*

1 state-owned service with 9 stations, 36 independent stations

1 state-owned service, 145 independent stations

Bolivia has strict defamation laws and considerable self-censorship. One of the TV stations is university-run, broadcasting mainly educational programs.

CRIME

 Death penalty not used in practice

 8315 prisoners

Crime is rising in narcotics-trafficking centers

CRIME RATES

Murders
29   *per 100,000 population*

Rapes
21   *per 100,000 population*

Thefts
107   *per 100,000 population*

Violent crime is centered on narcotics-trafficking towns in the eastern lowlands, particularly Santa Cruz. Main cities are much safer for tourists, and have lower crime rates than cities in neighboring Peru. The police and army have a history of mistreating poor farmers and miners.

EDUCATION

 School leaving age: 14

 86%   220,000 students

THE EDUCATION SYSTEM

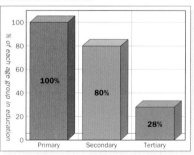

% of each age group in education

100%   Primary
80%   Secondary
28%   Tertiary

IMF targets for increased school attendance are being met, but education, based on a combination of the French and US systems, is seriously underfunded.

Although the majority of people speak indigenous languages, most teaching is in Spanish. Bolivia has one of the lowest literacy rates in South America. Reform and multilateral aid have led to some improvements.

HEALTH

 No welfare state health benefits

 1 per 769 people

Influenza, tuberculosis, other communicable diseases, malaria

Bolivia has one of the highest child mortality rates in the western hemisphere. Nearly 40% of children up to three years of age suffer from chronic malnutrition, and fewer than half the children under one year are immunized; diseases preventable by vaccination are a major cause of death.

Formal health services are costly, so that over half the population does not use them. Some 60% of births take place at home with no trained assistant present. High maternal mortality was targeted in a UNICEF-backed initiative in the 1990s.

Official figures showed that as at the end of 2001, 4600 people were living with HIV/AIDS.

SPENDING

 GDP/cap. increase

CONSUMPTION AND SPENDING

29 per 1000 population

61 per 1000 population

Defense 1.4%

Education 4.9%

Health 4.1%

| 0 | 5 | 10 | 15 | 20 | 25 |

Defense, Health, Education spending as % of GDP

Havoc created by economic reforms has widened the already huge gap between rich and poor. Generally, the indigenous population who form the rural poor are the worst off. The Andean highlands suffer from grinding poverty that has hardly changed in generations. Migrants to more prosperous eastern regions have faired better, but skewed land ownership remains a big problem. Poor housing, and lack of utilities and regular income are common to urban poverty. Only some 5% of people have bank accounts.

WORLD RANKING

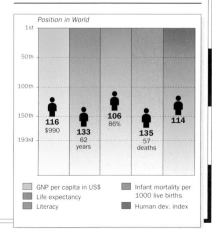

Position in World

1st
50th
100th
150th
193rd

116    $990
133   62 years
106   86%
135   57 deaths
114

GNP per capita in US$
Life expectancy
Literacy

Infant mortality per 1000 live births
Human dev. index

BOSNIA & HERZEGOVINA

EUROPE

OFFICIAL NAME: Bosnia and Herzegovina **CAPITAL:** Sarajevo
POPULATION: 4.1 million **CURRENCY:** Marka **OFFICIAL LANGUAGE:** Serbo-Croat

  1992 1992 March 1 BIH +1 +387 .ba

BOSNIA AND HERZEGOVINA is a mountainous country in southeastern Europe. A narrow corridor south of Mostar provides access to the Adriatic Sea. The post-1945 Yugoslav regime promoted coexistence between Bosniak Muslims, Croats, and Serbs, but, with Yugoslavia collapsing after 1990, ethnic rivalries led to violent conflict. Around 250,000 people died and more than two million were displaced before the 1995 Dayton peace accord ended three years of war.

CLIMATE
▷ Continental

WEATHER CHART FOR SARAJEVO

Bosnia has a continental climate with warm summers and bitterly cold winters, often with snow.

TRANSPORTATION
▷ Drive on right

Sarajevo International
354,996 passengers

Has no fleet

THE TRANSPORTATION NETWORK

11,425 km (7099 miles)	None
1031 km (641 miles)	None

War severely damaged the transportation network, resulting in wrecked bridges, roads, and railroads. De-mining and reconstruction have reopened most main routes. Sarajevo remains the hub of the communications network.

TOURISM
▷ Visitors : Population 1:37

110,000 visitors

Up 24% in 2000

MAIN TOURIST ARRIVALS

Bosnia & Herzegovina does not publish tourism figures by country of origin

% of total arrivals

Even before the war, Bosnia did not have much tourism infrastructure; only the adventurous have begun to visit.

PEOPLE
▷ Pop. density medium

Serbo-Croat

80/km² (208/mi²)

THE URBAN/RURAL POPULATION SPLIT

43%   57%

RELIGIOUS PERSUASION

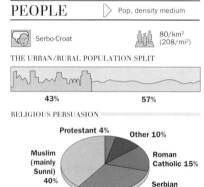

Protestant 4%
Other 10%
Muslim (mainly Sunni) 40%
Roman Catholic 15%
Serbian Orthodox 31%

In 1990, the population was 44% ethnic Bosniak (mostly Muslim), 31% Serb, and 17% Croat, while 8% described themselves as Yugoslav or of other ethnicity. Intermarriage was common and ethnic violence rare. Society was largely secular and materialistic. However, civil war and "ethnic cleansing" displaced some 60% of the population. Bosniak, Serb, and Croat areas became more distinct and separate.

BOSNIA & HERZEGOVINA

Total Land Area : 51 129 sq. km (19 741 sq. miles)

POPULATION
◎ over 100 000
○ over 50 000
● over 10 000
· under 10 000

LAND HEIGHT
2000m/6562ft
1000m/3281ft
500m/1640ft
200m/656ft
Sea Level

N

0   50 km
0   50 miles

POLITICS
▷ Multiparty elections

L. House 2000/2002
U. House 2000/2002

Chair of the Presidency
Beriz Belkic

AT THE LAST ELECTION

House of Representatives 42 seats

22% SDP	19% SDA	14% SDS	12% SBiH	12% HDZ	21% Others

SDP = Social Democratic Party **SDA** = Party of Democratic Action **SDS** = Serb Democratic Party **SBiH** = Party for Bosnia & Herzegovina **HDZ** = Croatian Democratic Union

The House of Representatives comprises the legislatures of the Federation of Bosnia-Herzegovina and Republika Srpska

House of Peoples 15 seats

10 members are appointed from the Federation of Bosnia-Herzegovina and 5 from the Republika Srpska

Bosnia is subdivided into the Muslim–Croat Federation, the Serbian Republika Srpska (RS), and the multiethnic district of Brčko. As such, there are two levels of government: the overall Republic – with a rotating three-member collective presidency – and the constituent republics, with their own presidents, governments, and parliaments. A UN High Representative retains ultimate control. Nationalist parties have substantial support, although moderate policies are favored in government. Bosniaks have led the way in supporting non-ethnic parties. Croats in the Federation have agitated for greater autonomy.

B

WORLD AFFAIRS
 Joined UN in 1992

An international High Representative administers Bosnia under the 1995 Dayton agreement. The Stabilization Force (SFOR) is NATO-led; its 35 participant countries include Russia.

AID
 Recipient

 $737m (receipts)   Down 31% in 2000

Aid for reconstruction and refugee return, totaling several billion dollars, has come mainly from the EU and the US. Major objectives have included developing a market economy and job creation, housing, and infrastructure.

DEFENSE
 Compulsory military service

 $183m   Down 50% in 2000

The international Stabilization Force (SFOR), in place since 1996, has been progressively downsized. The merger of Bosnian and Croat forces under the Dayton accord was completed in 1997.

ECONOMICS
Inflation 2.3% p.a. (1992–2000)

 $4.9bn   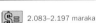 2.083–2.197 maraka

SCORE CARD

- ❏ WORLD GNP RANKING.........................111th
- ❏ GNP PER CAPITA $1230
- ❏ BALANCE OF PAYMENTS....................–$565m
- ❏ INFLATION 8%
- ❏ UNEMPLOYMENT......................................40%

STRENGTHS
Potential to revive prewar industries and become a thriving market economy, with solid manufacturing base. Growth in 2000 of 10%. End of cross-border tariffs with Slovenia from 2002.

WEAKNESSES
War damage of $20–$40 billion. Lowest foreign investment in entire region. Corruption. High unemployment.

EXPORTS

IMPORTS

RESOURCES
 Electric power 2.7m kw

 2500 tonnes   Not an oil producer

 640,000 sheep, 440,000 cattle, 4.74m chickens   Coal, lignite, iron, bauxite, cement

Bosnia's land is not well suited to agriculture, but has mineral deposits, forests, and hydroelectric potential.

ENVIRONMENT
Sustainability rank: 57th

 1%   1.2 tonnes per capita

Apart from war damage, Bosnia faces the effects of industrial pollution incurred during the communist regime.

MEDIA
 TV ownership medium

 Daily newspaper circulation 152 per 1000 people

PUBLISHING AND BROADCAST MEDIA

There are 6 daily newspapers. *Oslobodjenje* (*Liberation*) was published daily throughout the war

8 independent services   10 independent services

Violence against journalists is common. Hundreds of print and broadcast outlets, mainly controlled by government or parties, are highly partisan on ethnic lines.

CRIME
 Death penalty not used in practice

 1852 prisoners   Crime is rising

All sides in the war, but especially the Serbs, have been accused of war crimes. The International Criminal Tribunal for the former Yugoslavia (ICTY) in The Hague has indicted over 60 suspects relating to crimes committed in Bosnia. Serb leader Radovan Karadjic and over 20 others have evaded arrest. The RS finally passed laws on cooperating with the tribunal in October 2001.

EDUCATION
 School leaving age: 15

 93%   40,000 students

Ethnic bias and educational segregation are being combated by the development of a new coordinated curriculum.

The Muslim town of Mostar. Its 16th-century bridge at a strategic river crossing, and much of the old town, were destroyed in the war.

CHRONOLOGY
In 1945, Bosnia and Herzegovina became one of Yugoslavia's six constituent republics.

- ❏ **1990** Nationalists defeat communists in multiparty elections.
- ❏ **1991** Parliament announces republican sovereignty.
- ❏ **1992** EU and USA recognize Bosnia. Serbs declare "Serbian Republic." Civil war begins.
- ❏ **1995** NATO air strikes on Serbs; US-brokered Dayton peace accord.
- ❏ **1996** First international war crimes trial since 1945 opens in The Hague. Elections held under Dayton accord.
- ❏ **1998–2000** Elections: dwindling support for nationalist parties.
- ❏ **2001** Ethnic Croats briefly establish autonomy in Herzegovina (in south).

HEALTH
 No welfare state health benefits

 1 per 714 people   Cholera and diphtheria epidemics, violence, deaths from war-stress

War strained health services severely and many died for lack of basic care. Reconstruction includes reform of the primary health care system.

SPENDING
GDP/cap. increase

CONSUMPTION AND SPENDING

27 per 1000 population   103 per 1000 population

Defense 3.7%
Education No data
Health 8%

Defense, Health, Education spending as % of GDP

With some 60% of the prewar population displaced, housing, jobs, and reintegration are key challenges, while a lack of significant investment means that postwar poverty remains.

WORLD RANKING

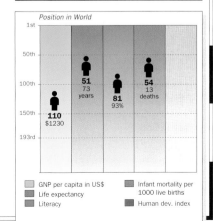

BOTSWANA

B

OFFICIAL NAME: Republic of Botswana **CAPITAL:** Gaborone
POPULATION: 1.6 million **CURRENCY:** Pula **OFFICIAL LANGUAGE:** English

SOUTHERN AFRICA

| 1966 | 1966 | Sept 30 | RB | +2 | +267 | .bw |

ARID AND LANDLOCKED, Botswana's central plateau separates the populous eastern grasslands from the Kalahari Desert and the swamps of the Okavango Delta in the west. Diamonds provide Botswana with a prosperous economy, but rain is an even more precious resource, honored in the name of the currency, the pula. Botswana has the world's highest rate of HIV infection among adults: a staggering 38.8% at end-2001.

The Okavango Delta, "jewel of the Kalahari," is the largest inland river delta in the world and home to a rich variety of wildlife.

CLIMATE
▷ Steppe/hot desert

WEATHER CHART FOR GABORONE

■ *Average daily temperature* *Rainfall* ■

The subtropical climate is dry and prone to drought. Rainfall declines from 64 cm (25 in) in the north to under 10 cm (4 in) in the Kalahari Desert in the west.

TRANSPORTATION
▷ Drive on left

Sir Seretse Khama International, Gaborone
240,526 passengers

Has no fleet

THE TRANSPORTATION NETWORK

| 4343 km (2699 miles) | None |
| 888 km (552 miles) | None |

The opening of the trans-Kalahari road to Namibia in 1998 has reduced Botswana's dependence on South African ports. Upgrading existing rail and road networks is a priority.

TOURISM
▷ Visitors : Population 1:2.1

750,000 visitors   Up 3% in 1998

MAIN TOURIST ARRIVALS

| South Africa 48% |
| Zimbabwe 26% |
| UK 2% |
| Other 24% |

% of total arrivals

Tourism is aimed at wealthy wildlife enthusiasts and focuses on safaris, especially to the Okavango Delta.

PEOPLE
▷ Pop. density low

Tswana, English, Shona, San, Khoikhoi, Ndebele

3/km² (7/mi)²

THE URBAN/RURAL POPULATION SPLIT

50%   50%

ETHNIC MAKEUP

Other 2%
Tswana 98%

Botswana's stability reflects its ethnic homogeneity and the power of traditional authorities, notably the village *kgotla*, or parliament. Almost the whole population is Tswana, with the Bamangwato forming the largest Tswana group. The indigenous San (or Bushmen) of the Kalahari were ordered to abandon their nomadic lifestyle in 2002. Whites continue to dominate the professions.

POLITICS
▷ Multiparty elections

1999/2004   President Festus Mogae

AT THE LAST ELECTION

National Assembly 46 seats

13% App

72% BDP   13% BNF   2% BCP

BDP = Botswana Democratic Party **BNF** = Botswana National Front **App** = Appointed **BCP** = Botswana Congress Party

In addition to 40 elected members, 4 are co-opted and the president and the attorney general are ex-officio members

Although Botswana is formally a multiparty democracy, it has been ruled by the BDP since independence. In 1994, however, economic problems, corruption scandals, and increasing urbanization led to the mainly town-based BNF gaining seats at the expense of the BDP, which nevertheless retained its absolute parliamentary majority. Power transferred smoothly from President Masire to Festus Mogae in 1998. The opposition BNF split in two, and the 1999 elections confirmed the BDP's hold on power.

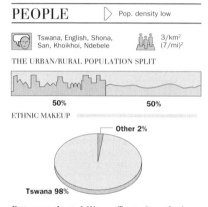

BOTSWANA

Total Land Area : 600 370 sq. km (231 803 sq. miles)

POPULATION
- over 500 000 ⊙
- over 50 000 ○
- over 10 000 ●
- under 10 000 ·

LAND HEIGHT
- 1000m/3281ft
- 500m/1640ft

0   200 km
0   200 miles

WORLD AFFAIRS ▷ Joined UN in 1966

Comm NAM AU SADC WTO

Botswana has strongly backed a politically and economically stable postapartheid South Africa, and in 1994 appointed its first ambassador to Pretoria since 1966. Potential South African domination of the SADC is a concern. Traditionally pro-Western in orientation, Botswana cherishes its relations with the UK and the US.

AID ▷ Recipient

 $31m (receipts)   Down 49% in 2000

Botswana's political and economic record has made it a favored aid recipient, notably from the EU, Japan, Germany, and the UK. Some 90% of EU aid goes to projects which try to balance wildlife needs with rural development. Aid also targets transportation projects.

DEFENSE ▷ No compulsory military service

 $245m   Down 5% in 2000

Reforms of the armed forces in mid-2000 aimed at improving morale included the raising of the compulsory retirement age and the enlistment of women.

ECONOMICS ▷ Inflation 9.7% p.a. (1990–2000)

 $5.28bn   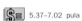 5.37–7.02 pula

SCORE CARD

❑ WORLD GNP RANKING	109th
❑ GNP PER CAPITA	$3300
❑ BALANCE OF PAYMENTS	$517m
❑ INFLATION	8.6%
❑ UNEMPLOYMENT	40%

STRENGTHS

Diamonds: one of the world's top producers. Economic growth among the highest in the world, averaging 8.5% between 1980 and 1998. Prudent management, large financial reserves, and exchange control liberalization. Lucrative exports of copper, nickel, assembly-produced vehicles, beef.

WEAKNESSES

Spread of AIDS. Overdependence on diamonds. Agriculture and industry weak. Small population, water shortages, and drought. Impact of beef industry on environment. High transportation costs to coast. Widespread unemployment.

EXPORTS

Zimbabwe 2% Other 3%
Other Africa 1% SACU 10%
Other Europe 18%
UK 66%

IMPORTS

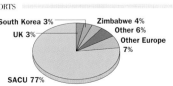

South Korea 3% Zimbabwe 4%
UK 3% Other 6%
Other Europe 7%
SACU 77%

RESOURCES ▷ Electric power: Included in South African total

 2000 tonnes   Not an oil producer and has no refineries

 2.4m cattle, 2.25m goats, 370,000 sheep, 4m chickens   Diamonds, copper, coal, nickel, soda ash, gold

The Orapa 2000 diamond mining project began in May 2000. Large coal deposits are the basis of power grid expansion. Water is Botswana's scarcest resource.

ENVIRONMENT ▷ Sustainability rank: 13th

 19% (2% partially protected)   2.4 tonnes per capita

Botswana is trying to help communities to earn a living from wildlife protection. A campaign has been launched to curb the use of agrochemicals.

MEDIA ▷ TV ownership low

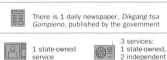 Daily newspaper circulation 27 per 1000 people

PUBLISHING AND BROADCAST MEDIA

There is 1 daily newspaper, *Dikgang tsa Gompieno*, published by the government

1 state-owned service   3 services: 1 state-owned, 2 independent

A government-funded TV service began broadcasting in 2000. The government bias of radio and the one daily paper is offset in the many journals.

CRIME ▷ Death penalty in use

 6339 prisoners   Down 11% 1992–1996

President Mogae warned of a "crime wave" in 1999. Official corruption and diamond smuggling stay major concerns. Human rights are generally respected.

EDUCATION ▷ School leaving age: 15

 77%   9595 students

Revenues from the diamond industry have helped to fund educational programs which have improved the country's literacy rate.

CHRONOLOGY

From 1600, Tswana migrations slowly displaced San people. In 1895, at local request, the UK set up the Bechuanaland Protectorate to preempt annexation by South Africa.

- ❑ **1965** BDP, led by Sir Seretse Khama, wins first general election and all subsequent general elections.
- ❑ **1966** Independence declared.
- ❑ **1980** Vice President Quett (later Ketumile) Masire succeeds the late Sir Seretse as president.
- ❑ **1985–1986** South African raids.
- ❑ **1992–1993** Strikes and corruption scandals prompt resignations of senior BDP figures.
- ❑ **1994** BDP support eroded in general election.
- ❑ **1998** Vice President Festus Mogae succeeds Masire as president.

HEALTH ▷ Welfare state health benefits

 1 per 5000 people   AIDS, tuberculosis, heart diseases, pneumonia

With the world's highest rate of HIV-positive adults (38.8%), Botswana in 2001 became the first sub-Saharan country to provide antiretroviral drugs through its public health service.

SPENDING ▷ GDP/cap. increase

CONSUMPTION AND SPENDING

30 per 1000 population   93 per 1000 population

Defense 5.5%
Education 9.1%
Health 2.5%

0   5   10   15   20   25
Defense, Health, Education spending as % of GDP

GNP per capita is among Africa's highest, but about half the population live below the poverty line. Economic growth has exacerbated wealth inequalities.

WORLD RANKING

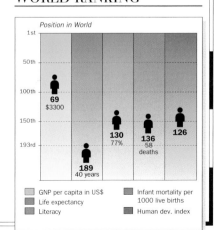

Position in World
1st
50th
100th
150th
193rd

69 $3300
189 40 years
130 77%
136 58 deaths
126

GNP per capita in US$ | Infant mortality per 1000 live births
Life expectancy
Literacy | Human dev. index

BRAZIL

SOUTH AMERICA

OFFICIAL NAME: Federative Republic of Brazil   **CAPITAL:** Brasília
POPULATION: 172.6 million   **CURRENCY:** Real   **OFFICIAL LANGUAGE:** Portuguese

THE LARGEST COUNTRY in South America, Brazil became independent of Portugal in 1822. Today, it is renowned as the site of the world's largest tropical rainforest, the threat to which led to the UN's first international environment conference, held in Rio de Janeiro in 1992. Covering one-third of Brazil's total land area, the rainforest grows around the massive Amazon River and its delta. Apart from the basin of the River Plate in the south, the rest of the country consists of highlands. The mountainous northeast is part forested and part desert. Brazil is the world's leading coffee producer and also has rich reserves of gold, diamonds, oil, and iron ore. Cattle ranching is an expanding industry. The city of São Paulo is the world's fourth-biggest conurbation, with some 18 million inhabitants.

BRAZIL

Total Land Area : 8 511 965 sq. km
(3 286 470 sq. miles)

POPULATION	
▣	over 5 000 000
▢	over 1 000 000
◉	over 500 000
◎	over 100 000
●	over 50 000

LAND HEIGHT

- 2000m/6562ft
- 1000m/3281ft
- 500m/1640ft
- 200m/656ft
- Sea Level

CLIMATE   ▷ Tropical equatorial/hot & dry/subtropical/steppe

WEATHER CHART FOR BRASÍLIA

Brazil's share of the Amazon basin, occupying half of the country, has a model equatorial climate. Its 150–200 cm (59–79 in) of rain are spread throughout the year. Temperatures are high, with almost no seasonal variation, but scarcely ever rise above 38°C (100°F).

The Brazilian plateau, occupying most of the rest of the country, has far greater temperature ranges. Rain falls mainly between October and April. The northeast, the least productive region of Brazil, is very dry and in recent years has been prone to severe drought. However, torrential rains hit Pernambuco and Alagoas in mid-2000, causing the worst flooding for 25 years. The southern states have hot summers and cool winters, when frost may occur.

TRANSPORTATION   ▷ Drive on right

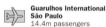

Guarulhos International, São Paulo
14.4m passengers

505 ships
3.81m grt

THE TRANSPORTATION NETWORK

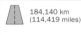

184,140 km (114,419 miles)	5000 km (3107 miles)
20,500 km (12,739 miles)	50,000 km (31,069 miles)

Air travel is expensive, while roads are clogged in cities and poor outside.

Proposed cross-border construction projects include a railroad, an intercontinental highway, and a 3442-km (2140-mile) waterway.

Parati, in Rio state, was one of Brazil's major gold-exporting ports in the 17th century. Its colonial architecture is well preserved.

TOURISM   ▷ Visitors : Population 1:32

 5.31m visitors   ⬆ Up 4% in 2000

MAIN TOURIST ARRIVALS

Argentina 30%	
USA 11%	
Paraguay 10%	
Uruguay 8%	
Germany 6%	
Other 35%	

% of total arrivals (0 10 20 30 40)

Brazil is underperforming in tourism, with revenues equivalent to 2.5% of GDP in 1996, compared with a world average of around 10%. The situation is slowly improving.

Such attractions as Atlantic beaches stretching 2000 km (1250 miles), the Amazon River basin, the Pantanal – the vast wetland region in the west – and Brazil's world-famous carnivals are offset by the limited availability of medium- to low-cost travel and budget hotels, a factor which deters both domestic and foreign travelers.

In the virtual absence of low-cost charter flights domestic air travel is expensive. This is blamed on high airport charges and inertia in Brazil's aviation department, which is controlled by the air force.

Average overnight hotel rates are higher than in Europe and the US, and the quality of service is generally poor. Basic infrastructure, such as sanitation and water supply, is also deficient.

B

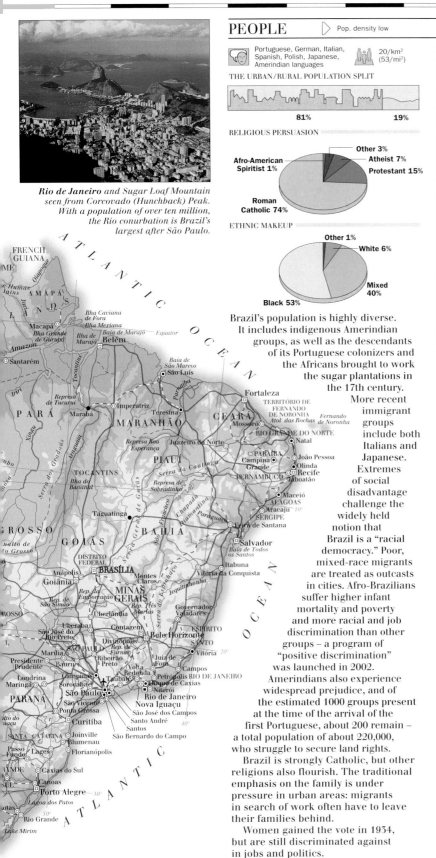

Rio de Janeiro and Sugar Loaf Mountain seen from Corcovado (Hunchback) Peak. With a population of over ten million, the Rio conurbation is Brazil's largest after São Paulo.

PEEPLE

▷ Pop. density low

Portuguese, German, Italian, Spanish, Polish, Japanese, Amerindian languages

20/km²
(53/mi²)

THE URBAN/RURAL POPULATION SPLIT

81%   **19%**

RELIGIOUS PERSUASION

- Other 3%
- Atheist 7%
- Protestant 15%
- Afro-American Spiritist 1%
- Roman Catholic 74%

ETHNIC MAKEUP

- Other 1%
- White 6%
- Mixed 40%
- Black 53%

POPULATION AGE BREAKDOWN

Female	Age	Male
0.6%	80+	0.4%
4.1%	60–79	3.4%
9.7%	40–59	9%
16.5%	20–39	16%
19.9%	0–19	20.4%

% of population by age group

Brazil's population is highly diverse. It includes indigenous Amerindian groups, as well as the descendants of its Portuguese colonizers and the Africans brought to work the sugar plantations in the 17th century. More recent immigrant groups include both Italians and Japanese. Extremes of social disadvantage challenge the widely held notion that Brazil is a "racial democracy." Poor, mixed-race migrants are treated as outcasts in cities. Afro-Brazilians suffer higher infant mortality and poverty and more racial and job discrimination than other groups – a program of "positive discrimination" was launched in 2002. Amerindians also experience widespread prejudice, and of the estimated 1000 groups present at the time of the arrival of the first Portuguese, about 200 remain – a total population of about 220,000, who struggle to secure land rights.

Brazil is strongly Catholic, but other religions also flourish. The traditional emphasis on the family is under pressure in urban areas: migrants in search of work often have to leave their families behind.

Women gained the vote in 1934, but are still discriminated against in jobs and politics.

CHRONOLOGY

The first Portuguese, Pedro Alvares Cabral, arrived in Brazil in 1500. By the time Portugal took control of the region, in 1580, it was a thriving colony drawing its wealth from sugar plantations in the northeast, worked by imported Africans, or Amerindians captured from farther inland.

❏ **1637–1654** Dutch control sugar-growing areas.

❏ **1763** Rio becomes capital.

❏ **1788** Inconfidência rebellion, led by Tiradentes, fails.

❏ **1807** French invade Portugal. King João VI flees to Brazil with British naval escort. In return, Brazil's ports opened to foreign trade.

❏ **1821** King returns to Portugal; his son Pedro made regent of Brazil.

❏ **1822** Pedro I declares independence and is made emperor of Brazil.

❏ **1828** Brazil loses Uruguay.

❏ **1831** Military revolt after war with Argentina (1825–1828). Emperor abdicates; his five-year-old son succeeds him as Pedro II.

❏ **1835–1845** Rio Grande secedes.

❏ **1865–1870** Brazil wins war of Triple Alliance with Argentina and Uruguay against Paraguay.

❏ **1888** Pedro II abolishes slavery; landowners and military turn against him.

❏ **1889** First Republic established. Emperor goes into exile in Paris. Increasing prosperity as result of international demand for coffee.

❏ **1891** Federal constitution established.

❏ **1914–1918** World War I causes coffee exports to slump.

❏ **1920s** Working-class and intellectual movements call for end to oligarchic rule.

❏ **1930** Coffee prices collapse. Revolt led by Dr. Getúlio Vargas, the "Father of the Poor," who becomes president. Fast industrial growth.

❏ **1937** Vargas's position as benevolent dictator formalized in "New State," based on fascist model.

❏ **1942** Declares war on Germany. ▷

Map labels

FRENCH GUIANA
ATLANTIC OCEAN
AMAPÁ
Macapá
Ilha Caviana de Fora
Ilha Mexiana
Ilha Grande de Gurupá
Ilha de Marajó
Baía de Marajó
Belém
Equator
Santarém
Amazon
Baía de São Marcos
São Luís
Fortaleza
TERRITÓRIO DE FERNANDO DE NORONHA
Atol das Rochas
Fernando de Noronha
Imperatriz
Teresina
PARÁ
Marabá
MARANHÃO
CEARÁ
Mossoró
RIO GRANDE DO NORTE
Natal
Represa Boa Esperança
Juazeiro do Norte
PARAÍBA
João Pessoa
Campina Grande
Olinda
Recife
PIAUÍ
Serra da Caatinga
PERNAMBUCO
Jaboatão
Represa de Sobradinho
Maceió
ALAGOAS
Taguatinga
Aracaju
SERGIPE
Feira de Santana
BAHIA
GOIÁS
Salvador
Baía de Todos os Santos
Itabuna
DISTRITO FEDERAL
BRASÍLIA
Anápolis
Goiânia
Montes Claros
Vitória da Conquista
MINAS GERAIS
Uberlândia
Governador Valadares
São José do Rio Preto
Contagem
ESPÍRITO SANTO
Belo Horizonte
Divinópolis
Vitória
Marília
Bauru
Ribeirão Preto
Juiz de Fora
Volta Redonda
Campos
Presidente Prudente
Campinas
Petrópolis
RIO DE JANEIRO
Londrina
Maringá
Sorocaba
Taubaté
Duque de Caxias
São Paulo
Niterói
PARANÁ
São Vicente
Rio de Janeiro
Ponta Grossa
Nova Iguaçu
Curitiba
São José dos Campos
Santo André
Santos
São Bernardo do Campo
SANTA CATARINA
Joinville
Blumenau
Lages
Florianópolis
Passo Fundo
Caxias do Sul
Canoas
Porto Alegre
Rio Grande
Lagoa dos Patos
Lake Mirim
ATLANTIC OCEAN

B

POLITICS

 Multiparty elections

| | L. House 1998/2002 | President Fernando |
| | U. House 1998/2002 | Henrique Cardoso |

AT THE LAST ELECTION

Chamber of Deputies 513 seats

| 21% PFL | 20% PT | 19% PSDB | 16% PMDB | 6% PTB | 12% PPB | 6% Others |

PFL = Liberal Front Party **PT** = PT–leftist coalition (includes the Workers' Party – **PT**, the Democratic Labor Party – **PDT**, and the Brazilian Socialist Party – **PSB**)
PSDB = Brazilian Social Democratic Party **PMDB** = Brazilian Democratic Movement Party **PPB** = Brazilian Progressive Party **PTB** = Brazilian Labor Party
Others include the Popular Socialist Party (**PPS**)

Federal Senate 81 seats

| 33% PMDB | 24% PFL | 20% PSDB | 16% PT | 1% PTB | 6% PPB |

President Fernando Cardoso, who took office in January 1995.

Luís Ignacio da Silva, "Lula," former leader of the left-wing Workers' Party.

Getúlio Vargas, the socialist president (1930–1945 and 1950–1954) who was known as the "Father of the Poor."

Brazil is a democratic federal republic with 27 regional parliaments and a national Congress. In 1993, Brazilians voted to retain the direct election of their president.

PROFILE

Military rule between 1964 and 1985 led to gross human rights abuses, against Amazon Amerindians in particular, and to economic mismanagement, which left Brazil with a legacy of huge debts and inefficient state industries.

Brazil's young democracy is characterized by a weak party system, centered around personalities. Parties do not have set ideological programs, but form shaky coalitions and engage in horsetrading to get legislation through the Congress. A preponderance of small parties and corruption adds to the problems. Former President Collor de Mello was impeached in 1992 on fraud charges.

Dissatisfaction with the center-right provided a boost for the left, led by the influential Luís da Silva, who came second in the 1989 presidential elections. However, his failure to beat Fernando Henrique Cardoso in 1994 or 1998 revealed a lack of fresh ideas and direction in his Workers' Party. Cardoso, the father of the successful anti-inflation plan for the real, held his shaky coalition together until 2002 and emergency fiscal adjustments saved Brazil from a return to persistent economic crisis. Hard-won stability permitted Brazil to avoid the worst effects of the economic crisis in Argentina in 2001–2002.

MAIN POLITICAL ISSUES
Political stability

Cardoso, who was constitutionally barred from seeking a third term in the 2002 presidential elections, has left his mark on politics, since few parties are prepared to deviate far from his largely successful policy mix which has provided stability and boosted consumer and investor confidence. A ruling by the Supreme Electoral Tribunal, instructing parties forming electoral alliances to retain them from the federal level down to the state, promises cleaner campaigns.

Economic management

Rapid recovery from the 1999 currency crisis relied on a tight monetary and fiscal regime backed by a sound macroeconomic policy. This permitted modest growth in 2001 and 2002 despite the global downturn and energy rationing at home. Plaudits from the IMF separated Brazil in investors' minds from the economic turmoil in Argentina. Tax reform and improved energy resources are priorities for the incoming government.

Diplomatic role

Brazil is a leader in South America in the promotion of greater geopolitical unity. Talks with the US on the proposed Free Trade Area of the Americas (FTAA), and a similar trade area with the EU and Mercosur, are among priorities.

Coffee plantation, São Paulo state. Coffee was introduced into Brazil in the early 18th century. It is declining in importance and now accounts for less than 4% of export revenues.

WORLD AFFAIRS

▷ Joined UN in 1945

AmCC | G24 | Mercosr | OAS | RG

Brazil's regional influence recently helped to abort coups in Paraguay and end one in Ecuador, and settle a border dispute between Peru and Ecuador. Brazil has ambitions to act as main broker in future talks with the US on the FTAA, which is due to be signed in 2005. To bolster its position, it has revived negotiations on a bilateral trade agreement with Mexico, its chief rival for the crown of regional leader. Free trade talks between the EU and Mercosur have raised the prospect of movement by Europe on agriculture, long the main bone of contention between the two regions. Brazil has also strengthened ties with the Andean Community and in 2002 hosted a conference attended by 34 countries invited to join the FTAA.

AID

▷ Recipient

💲 $322m (receipts) ⬆ Up 72% in 2000

Aid, mainly from Japan and EU states, funds environmental, basic sanitation, road-building, and antipoverty projects. As a consequence of the Argentine economic crisis, the IMF loaned Brazil $10 billion in 2002, and promised $30 billion more in the following years.

DEFENSE

▷ Compulsory military service

💲 $17.5bn ⬆ Up 10% in 2000

BRAZILIAN ARMED FORCES

🛡	178 main battle tanks (87 *Leopard* 1, 91 M-60A3)	189,000 personnel
🚢	4 submarines, 1 carrier, 14 frigates, 4 corvettes, 50 patrol boats	48,600 personnel
✈	281 combat aircraft (53 AT-26, 47 F-5E/B/F, 50 AMX, 18 *Mirage* F-103E/D)	50,000 personnel
🚀	None	

Brazil's dictatorship ended in 1985, but the military still has an important internal security role, for instance controlling vast regions of the north. The arms industry is large, but Brazil states that it has no intention of using its nuclear energy for military purposes: the Comprehensive Test Ban and Nuclear Non-Proliferation treaties were signed in 1998.

Membership of the Mercosur trade bloc has led to increased military cooperation with Argentina, Paraguay, and Uruguay. Brazilian troops have participated in UN peacekeeping, most recently in East Timor.

ECONOMICS

▷ Inflation 208% p.a. (1990–2000)

📊 $610bn   💲 1.950–2.311 reals

SCORE CARD

❑ WORLD GNP RANKING	9th
❑ GNP PER CAPITA	$3580
❑ BALANCE OF PAYMENTS	–$24.6bn
❑ INFLATION	7%
❑ UNEMPLOYMENT	7%

EXPORTS

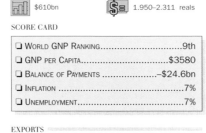

Germany 4% | Japan 4% | Netherlands 5% | Argentina 11% | Other 52% | USA 24%

IMPORTS

Italy 4% | Japan 5% | Germany 8% | Argentina 12% | Other 48% | USA 23%

STRENGTHS

Dominant economy in region. Strong foreign direct investment flows. Huge growth potential: immense natural resources; major producer of coffee, soy, sugar, oranges; large deposits of gold, silver, and iron; major steel producer; developing oil industry. Development aided by cross-border infrastructure projects and modernization of telecoms.

WEAKNESSES

Expensive domestic borrowing. Weak local capital markets. Vulnerability to external shocks and commodity price fluctuations. Modest productivity. Heavy debt burden. Great social inequalities threaten social unrest. Electricity blackouts caused by badly maintained transmission network or failure of hydropower supply due to drought.

PROFILE

Brazil is the world's ninth-largest economy. Average growth from the start of the 20th century to the early 1970s was over 5%, second only to Japan over a comparable period. Diversification and industrialization transformed Brazil into a producer of cars, computers, and aircraft, but profligate spending produced heavy debts in the 1980s. International lenders demanded belt tightening in return for rescheduling and a steep recession followed in 1990–1992.

The launching of the new currency, the real, in 1994, was the fifth attempt at monetary stabilization since 1986; it contributed to a dramatic fall in inflation. Economic growth in 1994

ECONOMIC PERFORMANCE INDICATOR

— Consumer Price Index   GDP ▨

Consumer price index 1995=100

GDP 1996=100

1996 1997 1998 1999 2000

boosted regional confidence and facilitated the launch of Mercosur. In 1995 a fractious Congress blocked reforms of the tax and social security systems, but finally agreed to end state monopolies in such sectors as telecommunications and oil, thus reviving the privatization program.

The economy grew strongly through 1996 and 1997, but was seriously threatened in 1998 by an international financial crisis. A $41.5 billion rescue package was arranged by the IMF, but foreign currency reserves were heavily depleted in a bid to support the real, which was devalued in 1999 due to speculative pressures. A deep recession was avoided, however, by the successful application over 18 months of tight fiscal and monetary policies, restoring domestic and international confidence.

Modest growth was posted in 2001, but the economic meltdown in Argentina and the effects of global downturn brought the value of the real down in 2002. An end to energy rationing promised to revive industrial production and activity, but deregulation of the labor market, reforms of the social security and tax systems, and high debt-servicing costs remained concerns.

BRAZIL : MAJOR BUSINESSES

🗄	Banking
💻	Computers
📖	Publishing
🔌	Electronics
🪚	Saw milling
↓	Sugarcane refining
🚗	Vehicle assembly
☕	Coffee processing

Manaus · Belém · Amazon · Recife · Salvador · Brasília · Belo Horizonte · Rio de Janeiro · São Paulo · Porto Alegre

0   1000 km
0   1000 miles

* significant multinational ownership

B

RESOURCES

 Electric power 65.2m kw

 774,750 tonnes

 1.34m b/d (reserves 8.5bn barrels)

 172m cattle, 29.4m pigs, 15m sheep, 1006m chickens

Iron, manganese, coal, bauxite, nickel, oil, tin, silver, diamonds, gold

ELECTRICITY GENERATION

Hydro 91% (291bn kwh)

Combustion 8% (27bn kwh)

Nuclear 1% (3.3bn kwh)

Other 0%

% of total generation by type

Brazil pumps in gas from Argentina and Bolivia, and has similar plans with Venezuela and Uruguay. Nuclear power has been dogged by controversy and high costs. Hydropower, which already accounts for over 90% of electricity generation, is being expanded further, although output is vulnerable to drought. Producing ethanol from sugar represents an attempt to reduce gasoline imports, and the welcoming of foreign companies in areas of exploration and production is set to increase known oil and natural gas reserves. Exploration of the Amazon's biodiversity was brought under government control in mid-2000, with all new ventures involving any living thing requiring official approval.

BRAZIL : LAND USE

- Cropland
- Forest
- Pasture
- Cattle
- Coffee – cash crop
- Oranges

0   1000 km

0   1000 miles

Equatorial vegetation near Manaus in the center of Amazonas state. The brown waters of the Rio Solimões and the black waters of the Rio Negro meet near Manaus.

ENVIRONMENT

 Sustainability rank: 20th

4% (1% partially protected)

1.8 tonnes per capita

ENVIRONMENTAL TREATIES

Yes	Yes	Yes
Yes	Yes	No

Federal agencies charged with protecting the Amazon rainforest are underfunded, understaffed, and accused of corruption.

The forest contains an estimated 90% of all the world's plant and animal species.

However, 38 million animals a year are smuggled out of the region, and the forest's destruction, usually for cattle pasture and logging, proceeds at a rate of up to 23,000 sq. km (9000 sq. miles) a year; vital genetic diversity is being lost. The government's latest and biggest campaign to stop illegal logging and burning was announced in 2000.

Opencast bauxite mines pollute rivers and threaten indigenous Amerindians, while in 2000 the worst oil spill in 25 years devastated the Iguaçu River. Urban industrial pollution and untreated sewage are major problems.

MEDIA

 TV ownership high

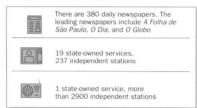 Daily newspaper circulation 46 per 1000 people

PUBLISHING AND BROADCAST MEDIA

There are 380 daily newspapers. The leading newspapers include *A Folha de São Paulo*, *O Dia*, and *O Globo*

19 state-owned services, 237 independent stations

1 state-owned service, more than 2900 independent stations

TV and radio licenses are notoriously awarded as political favors. The huge Globo group dominates the home market, with radio, press, and online interests: its TV network is the fifth-largest in the world. However, it is being challenged by the Internet and the growth of new media companies, involving foreign multinationals. A constitutional amendment before the Congress allows foreign groups to take 30% stakes in TV, radio, and the press.

CRIME

 Death penalty not used in practice

 194,074 prisoners

Rising in 2000–2002

CRIME RATES

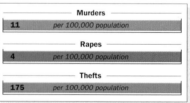

Murders

11 per 100,000 population

Rapes

4 per 100,000 population

Thefts

175 per 100,000 population

Crime levels are among the world's highest, especially in cities, with armed robbery and narcotics-related crime uppermost. Badly paid police are frequently accused of extortion, violence and the murder of citizens. Death squads, thought to be linked to the police, have targeted street children in major cities. A combination of atrocious conditions and overcrowding mean that violent disturbances in prisons are common. An explosion in muggings and kidnappings was a key issue in the 2002 elections.

In the countryside, landless squatters and indigenous peoples have been wounded and murdered in the process of being driven off land by gunmen funded by large landowners. In Roraima state, the discovery of large gold deposits has led to the homelands of Brazil's largest tribe, the Yanomami, being invaded by thousands of gun-toting prospectors, *garimpeiros*.

EDUCATION

 School leaving age: 14

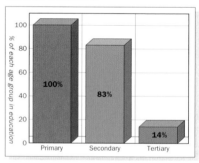

85%

1.95m students

THE EDUCATION SYSTEM

% of each age group in education

Primary 100%

Secondary 83%

Tertiary 14%

The average time spent at school is less than that in other Latin American countries. The portion of GDP spent on education is comparable to that of European countries, but it is misapplied, so that basic primary education remains weak, while many children of wealthy families receive excellent tuition at free public universities. Of Brazil's 95 universities, 55 are administered by the state. Nearly three million children hardly receive any schooling, especially those living in the northeast and in Amazonia, and the urban poor, including street children. However, the number of seven- to 14-year-olds in school has increased during the last decade. Despite an anti-illiteracy campaign, begun in 1971, the adult illiteracy rate is 15%.

EXPLOITATION VERSUS PROTECTION IN AMAZONIA

THE VAST AMAZON RIVER basin, extending across northern Brazil, contains one of the world's last great remaining virgin rainforests. Major road-building projects mean that it no longer enjoys the protection of inaccessibility. Since 1970, when it was still 99% intact, 600,000 sq. km (over 230,000 sq. miles) of Amazonia, more than 15% of the total, have been deforested. The rate of destruction did stop accelerating and actually slow down slightly in 2001, but even that year the official record showed a loss of nearly 16,000 sq. km (over 6000 sq. miles). US and Brazilian researchers warned in January 2001 that the government's massive "Advance

Brazil" road- and dam-building project would by 2020 leave at best 28% of the forest untouched. Some of the damage is done by mining, particularly for gold, and the associated pollution of soil and water. Land clearance for ranching is a major cause of forest loss, occurring wherever the building of access roads takes place.

Logging, too, poses a continuing threat. Amazonian tropical hardwoods are highly attractive to the timber industry. A score of foreign-owned multinational companies dominate the timber export business, and some 2500 logging companies and sawmills operate in the Brazilian Amazon. Brazilian mahogany is now protected by a moratorium on exploitation, but illegal logging is widespread; the government estimates that it accounts for 80% of all timber being extracted from the Amazon today.

ENVIRONMENTAL CONSEQUENCES
The rainforest typically has poor and shallow soil. Its ecology depends on the recycling of minerals in the leaf litter on the forest floor. In a damp environment, the leaf litter is broken down rapidly by soil organisms. Once exposed by the loss of tree cover, however, it dries out rapidly and can easily be washed away by the next rain. The forest cannot regenerate growth where substantial areas have been cleared. The deforestation has far-reaching implications. Forests act as "carbon sinks," fixing carbon dioxide produced by the burning of fuel, and thus helping to counteract the buildup of the "greenhouse gases" linked with global warming.

Amazonia is also immensely rich in native plant and animal species. Its resource value, taking account of the potential significance of its genetic pool, for example in medicine or agriculture, was estimated in 2001 by a government research project as $2000 billion. This biodiversity, including much as yet undiscovered by the international scientific community, is dependent on proper protection of rainforest habitat.

INDIGENOUS PEOPLES
Colonization has brought violence and devastating epidemics to its indigenous peoples, along with the loss of their lands, forced removal, confinement to reservations, and the destruction of their lifestyle and culture. Many of them suffered extinction in the last half century. There remain about 200 distinct known forest-dwelling indigenous peoples. One-fifth of the Amazon is Amerindian land and under the 1988 constitution, forest peoples have the right to inhabit their ancestral lands, but they do not have legal title to them. Out of 580 identified Amerindian territories, over a third remain as yet undemarcated. The National Indian Foundation (FUNAI), a government agency responsible for demarcation, has been heavily criticized internationally for failing to provide adequate protection. When a territory is demarcated, its inhabitants have legal rights to protect it from logging, mining, and other industrial activities.

HEALTH
 Welfare state health benefits

 1 per 769 people   Heart diseases, cancers, accidents, violence

Federal health is underfunded. Fewer than 20% of hospitals are state-run, and they need modernization. Private care is beyond the means of the majority. On average only 15% of the health budget goes to child health, immunization, and other preventive programs. However, infant mortality, at 95 per 1000 children in 1970, had dropped to one-third of that level by 2000; access to potable water increased from 74% of the population in 1992 to 87% in 2000. In 2001 international drug companies and the US dropped patent infringement claims against Brazil for distributing anti-AIDS drugs free to more than 100,000 HIV patients.

SPENDING
▷ GDP/cap. increase

CONSUMPTION AND SPENDING

79 per 1000 population   182 per 1000 population

Defense 2.8%
Education 4.6%
Health 2.9%

Defense, Health, Education spending as % of GDP

Brazil's income distribution is among the most skewed in the world. IDB figures in 1999 showed that the richest 10% of the population take 50% of the income and the poorest 50% only 10%. Governments have failed to tackle the problem of homelessness and street children in large cities. An estimated one to five million families remain landless, while nearly 80% of farmland is owned by 10% of landowners. In 2000 the government announced that vast tracts of land of unproven ownership would be redistributed, but widespread invasions by landless workers have continued.

WORLD RANKING

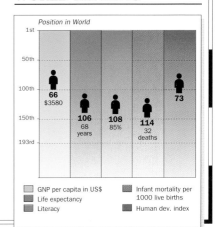

Position in World

- 66 $3580
- 106 / 68 years
- 108 / 85%
- 114 / 32 deaths
- 73

GNP per capita in US$
Life expectancy
Literacy
Infant mortality per 1000 live births
Human dev. index

BRUNEI

B

OFFICIAL NAME: Sultanate of Brunei **CAPITAL:** Bandar Seri Begawan
POPULATION: 335,000 **CURRENCY:** Brunei dollar **OFFICIAL LANGUAGE:** Malay

| 1984 | 1984 | Feb 23 | BRU | +8 | +673 | .bn |

L YING ON THE NORTHWESTERN coast of the island
of Borneo, Brunei is divided in two by a strip of the
surrounding Malaysian state of Sarawak. The interior is mostly rainforest.
Independent from the UK since 1984, Brunei is ruled by decree of the
sultan. It is undergoing increasing Islamicization. Oil and gas reserves
have brought one of the world's highest standards of living.

CLIMATE ▷ Tropical equatorial

WEATHER CHART FOR BANDAR SERI BEGAWAN

Just 480 km (300 miles) north of
the equator, Brunei has a long rainy
season with extremely high humidity.

TRANSPORTATION ▷ Drive on left

**Brunei International,
Bandar Seri Begawan**
982,041 passengers

97 ships
361,700 grt

THE TRANSPORTATION NETWORK

| 399 km (248 miles) | None |
| 19 km (12 miles) | 209 km (130 miles) |

Interest-free loans for civil servants,
subsidized gasoline, and limited
public transportation account
for the high rates of
car ownership.

TOURISM ▷ Visitors : Population 2.9:1

964,000 visitors

Up 13% in 1998

MAIN TOURIST ARRIVALS

| Malaysia 80% |
| Philippines 3% |
| UK 3% |
| Other 14% |

% of total arrivals

Although the goverment is keen to
protect Bruneians from Western
influence, it wants to develop quality
tourism as part of its diversification
program. Promoted as the "Gateway
to Borneo," Brunei's rainforests
could be developed for tourism. A
former attraction was the Churchill
Museum, founded by the late sultan.
This has now been superseded by the
Museum of Royal Regalia.

PEOPLE ▷ Pop. density medium

Malay, English, Chinese

64/km²
(165/mi²)

THE URBAN/RURAL POPULATION SPLIT

72%   28%

ETHNIC MAKEUP

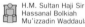

Indigenous 6%
Other 11%
Chinese 16%
Malay 67%

Malays are the beneficiaries of positive
discrimination; many in the Chinese
community are either stateless or hold
British protected person passports.
Among indigenous groups, the Murut
and Dusuns are favored over the Ibans.
Women, less restricted than in some
Muslim states, are obliged to wear
headscarves but not the veil. Many hold
influential posts in the civil service.

POLITICS ▷ No legislative elections

Not applicable

H.M. Sultan Haji Sir
Hassanal Bolkiah
Mu'izzadin Waddaulah

LEGISLATIVE OR ADVISORY BODIES

Brunei is an absolute monarchy; the sultan consults four
advisory councils: Religious Council, Privy Council, Council
of Cabinet Ministers, and Council of Succession, which he
appoints. Political parties have been banned since 1988.

Since a failed rebellion in 1962, a state
of emergency has been in force and the
sultan has ruled by decree. Hopes for
democracy were dashed when political
parties were banned in 1988. In 1990,
"Malay Muslim Monarchy" was
introduced, promoting Islamic
values as the state ideology. This
further alienated the large Chinese
and expatriate communities. Power is
closely tied to the royal family. One of
the sultan's brothers holds the foreign
affairs portfolio and the sultan himself
looks after defense and finance.

WORLD AFFAIRS ▷ Joined UN in 1984

| APEC | ASEAN | Comm | OIC | WTO |

Brunei leads calls for a regional free
trade area. Political exiles opposed to
the government and based in Malaysia
are a main concern. Relations with the
UK, the ex-colonial power, are good.

BRUNEI

Total Land Area :
5770 sq. km
(2228 sq. miles)

LAND HEIGHT

POPULATION

○ over 50 000
● over 10 000
• under 10 000

1500m/4921ft
1000m/3281ft
500m/1640ft
200m/656ft
Sea Level

0 20 km
0 20 miles

The magnificent Omar Ali Saifuddin Mosque is surrounded by an artificial lagoon.

AID
 Donor

 Ad hoc handouts from the sultan   Not applicable

Aid spending is largely ad hoc. It has included donations to the Contras in Nicaragua, the Bosnian Muslims, and the homeless of New York.

DEFENSE
 No compulsory military service

 US$348m   Down 13% in 2000

As well as being head of the 5900-strong armed forces, the sultan has a personal bodyguard of 2000 UK-trained Gurkhas. The UK and Singapore are close defense allies.

ECONOMICS
 Inflation 1.1% p.a. (1990–1999)

 US$7.8bn   1.734–1.847 Brunei dollars

SCORE CARD

❏ WORLD GNP RANKING	95th
❏ GNP PER CAPITA	US$24,762
❏ BALANCE OF PAYMENTS	US$2.09bn
❏ INFLATION	1%
❏ UNEMPLOYMENT	5%

STRENGTHS
Nearly 20 years of known oil reserves; over 30 years of gas. Earnings from massive overseas investments, mainly in the US and Europe, now exceed oil and gas revenues.

WEAKNESSES
Single-product economy. Failure of diversification programs could lead to problems in the future.

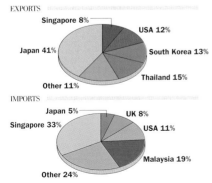

EXPORTS

Singapore 8%
USA 12%
Japan 41%
South Korea 13%
Thailand 15%
Other 11%

IMPORTS

Japan 5%
UK 8%
Singapore 33%
USA 11%
Malaysia 19%
Other 24%

RESOURCES
 Electric power 479,000 kw

 3308 tonnes

195,000 b/d (reserves 1.4bn barrels)

40,000 ducks, 8000 cattle, 6000 pigs, 4.7m chickens

 Oil, natural gas

Oil and gas are the major resources. Energy policy is now focused on regulating output in order to conserve stocks, since reserves are of limited duration. Almost all food is imported.

ENVIRONMENT
 Not available

14% (5% partially protected)

17.1 tonnes per capita

The Forestry Strategic Plan aims to protect Brunei's forests (which take up 80% of its land area). It has allocated 64% of their area for protection and recreation and the prevention of soil erosion. However, Brunei's mangrove swamps, the largest on the island of Borneo, remain unprotected.

MEDIA
 TV ownership medium

 Daily newspaper circulation 69 per 1000 people

PUBLISHING AND BROADCAST MEDIA

There are 3 daily newspapers, the *Borneo Bulletin*, the *Daily News Digest*, and the *News Express*

1 state-owned service   1 state-owned service

The state effectively controls all media. Brunei's television service now has a heavy religious content.

CRIME
 Death penalty not used in practice

333 prisoners   Up 10% in 1999

Crime levels are low. Most crime involves petty theft or is linked to alcohol and narcotics (both banned). A stolen car often makes TV news headlines. The state of emergency enables the government to detain without charge or trial for indefinitely renewable two-year periods.

EDUCATION
 School leaving age: 12

92%   1878 students

Free schooling is available to the entire population, with the exception of the stateless Chinese, who do not qualify. The University of Brunei Darussalam was opened in 1985.

HEALTH
 Welfare state health benefits

1 per 939 people   Heart diseases, cancers

The health service is free, although if major surgery is required Bruneians tend to travel to Singapore.

CHRONOLOGY
Under British control since 1841, Brunei became a formal British Protectorate in 1888.

❏ **1929** Oil extraction begins.
❏ **1959** First constitution enshrines Islam as state religion. Internal self-government.
❏ **1962** Prodemocracy rebellion. State of emergency; sultan rules by decree.
❏ **1984** Independence from Britain. Brunei joins ASEAN.
❏ **1990** Ideology of "Malay Muslim Monarchy" introduced.
❏ **1991** Imports of alcohol banned.
❏ **1992** Joins Non-Aligned Movement.
❏ **1998** Sultan's son, Prince Al-Muhtadee Billah, made crown prince.

SPENDING
 GDP/cap. increase

CONSUMPTION AND SPENDING

529 per 1000 population   245 per 1000 population

Defense 5.8%
Education 3.1%
Health 5.7%

0 5 10 15 20 25
Defense, Health, Education spending as % of GDP

The wealthiest people in Brunei are those close to the sultan, one of the world's richest men. A generally high standard of living, along with a degree of social mobility among Malays, keeps discontent to a minimum. Bruneians are major consumers of high-tech hi-fi and video equipment, designer-label watches, and Western designer clothes. The sultan's younger brother, Prince Jefri, in 2001 auctioned his possessions, ranging from fire engines to marble baths, after the failure of his business left him with debts of US$3 billion.

WORLD RANKING

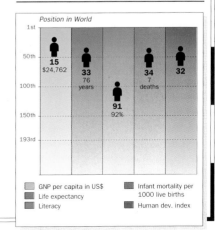

Position in World

1st
50th — 15 $24,762 — 33 76 years — 34 — 32
100th — 91 92%
150th
193rd

❏ GNP per capita in US$ ❏ Infant mortality per 1000 live births
❏ Life expectancy   ❏ Human dev. index
❏ Literacy

B

BULGARIA

B

OFFICIAL NAME: Republic of Bulgaria **CAPITAL:** Sofia
POPULATION: 7.9 million **CURRENCY:** Lev **OFFICIAL LANGUAGE:** Bulgarian

| 1908 | 1947 | March 3 | BG | +2 | +359 | .bg |

LOCATED IN SOUTHEASTERN EUROPE, Bulgaria is a mainly mountainous country. The River Danube forms the northern border, while the popular resorts of the Black Sea lie to the east. The most populated areas are around Sofia in the west, Plovdiv in the south, and along the Danube plain. Bulgaria was directly ruled by the Turks from 1396 until 1878. In 1908, it became an independent kingdom and was under communist rule from 1947, with Todor Zhivkov in power from 1954 to 1989. The 1990s brought political instability as Bulgaria adjusted to democracy and economic reconstruction.

***Rila Monastery** in the Rila Mountains. It is famous for its 1200 National Revival Period frescoes dating from the mid-19th century.*

CLIMATE
▷ Mediterranean/continental

WEATHER CHART FOR SOFIA

The central valley and the lowlands have warm summers and cold, snowy winters, but hot or cold winds from Russia can bring spells of more extreme weather. The hotter summers on the Black Sea coast have encouraged the growth of tourist resorts. Snow may lie on the high mountain peaks until June.

TRANSPORTATION
▷ Drive on right

Sofia International
1.13m passengers

164 ships
989,600 grt

THE TRANSPORTATION NETWORK

33,818 km (21,013 miles)		319 km (198 miles)	
4290 km (2666 miles)		470 km (292 miles)	

At the crossroads between Europe and Asia, Bulgarian railroads and expressways were underfunded under Zhivkov (when north–south routes were left undeveloped) and in the economically uncertain 1990s. Funding for modernizing key routes is now in place. Ferries are used for most cross-Danube traffic. In 2000 agreement was reached with Romania on building a second bridge across the river, scheduled to open in 2005.

TOURISM
▷ Visitors : Population 1:2.8

2.8m visitors

Up 13% in 2000

MAIN TOURIST ARRIVALS

Romania 25%	
Macedonia 19%	
Turkey 10%	
Serbia & Montenegro 8%	
Greece 8%	
Other 30%	

% of total arrivals

The tourist industry formerly catered for the east European mass-market. Western tourists are attracted by low prices for skiing and beach vacations. Bulgaria is now privatizing the industry and seeks to move it upmarket by stressing its heritage. Since the mid-1990s a slump in earnings has been reversed: Russians are returning in larger numbers and there is a growth in tours from western Europe, especially from Germany.

BULGARIA

Total Land Area : 110 910 sq. km
(42 822 sq. miles)

POPULATION

over 1 000 000	▣
over 100 000	◎
over 50 000	○
over 10 000	●

LAND HEIGHT

2000m/6562ft
1000m/3281ft
500m/1640ft
200m/656ft
Sea Level

PEOPLE   ▷ Pop. density medium

Bulgarian, Turkish, Macedonian, Romany   71/km² (185/mi²)

THE URBAN/RURAL POPULATION SPLIT

70%   30%

RELIGIOUS PERSUASION

- Jewish 1%
- Roman Catholic 1%
- Other 1%
- Muslim 13%
- Bulgarian Orthodox 84%

ETHNIC MAKEUP

- Roma 3%
- Macedonian 3%
- Turkish 9%
- Bulgarian 85%

The communist era was marked by assimilationist policies suppressing minority cultural identities. In the 1970s, Bulgarian Muslims, or Pomaks, were forced to change Muslim names to Bulgarian ones. Bulgarian Turks were particularly targeted in the 1980s.

Linguistic and religious freedom was granted in 1989, but 300,000 ethnic Turks, or 40%, still left for Turkey. Their farming skills have traditionally been important to agriculture. Recent privatizations have left many Turks landless and have provoked new emigration. The MRF, supported mainly by ethnic Turks, is a minority party whose influence can be strong, a fact borne out by its inclusion in the governing coalition formed in 2001.

Macedonians and Roma each account for 3% of the population. The latter minority suffers discrimination at all levels. Women have equal rights in theory, but society remains patriarchal, especially among Turks.

POPULATION AGE BREAKDOWN

Female	Age	Male
1.4%	80+	0.9%
10.5%	60–79	8.6%
13.5%	40–59	12.9%
13.7%	20–39	14%
11.9%	0–19	12.6%

% of population by age group

POLITICS   ▷ Multiparty elections

2001/2005   President Georgi Purvanov

AT THE LAST ELECTION

National Assembly 240 seats

50% NMS II	21% UDF	20% BSP	9% MRF

NMS II = National Movement Simeon II
UDF = Union of Democratic Forces
BSP = Bulgarian Socialist Party
MRF = Movement for Rights and Freedoms

Bulgaria is a multiparty democracy.

PROFILE

Having moved falteringly to a pluralist democratic system after the fall of the communist Zhivkov regime in 1989, Bulgaria suffered during the 1990s from successive weak governments, each brought down by no-confidence votes. The UDF, a broad anticommunist alliance, fell from office in 1992, and by the time of the 1994 general election the former communist BSP appeared to be firmly in the ascendant, winning an overall majority. The BSP government resisted political and economic change. The result was one of the slowest privatization programs in eastern Europe, with the old communist web of patronage still intact. A new UDF government in 1997 launched free-market reforms backed up by the IMF. Its considerable success, and reorientation of policy toward the goals of EU and NATO membership, allowed the UDF to approach the June 2001 elections with some confidence, despite a surge in support for a monarchist party launched by ex-king Simeon II (who had left Bulgaria as a small child in 1946). The poll, however, left the UDF with less than a quarter of the National Assembly seats, exactly half of which went to the NMS II. Uncertainty over whether the ex-king himself would take on the role of prime minister ended when he formed a coalition with the MRF (representing mainly the ethnic Turkish minority) and was sworn in in July. He took the oath to the republic (although he had not formally given up his claim to the throne) as Simeon Saxecoburggotski, promising "spiritual and economic revival," clean government, tax cuts, and more privatization, with integration with the EU and NATO his top priorities.

***Ex-king Simeon II** (Saxecoburggotski) returned as prime minister in 2001.*

***Georgi Purvanov**, of the BSP, elected president against expectations in 2001.*

WORLD AFFAIRS   ▷ Joined UN in 1955

BSEC   CE   EAPC   CEFTA   OSCE

B

Bulgaria is among six "second wave" candidates for EU membership which began negotiations in 2000. It hopes to join NATO.

Bulgaria conscientiously adhered to UN sanctions against Yugoslavia, despite the costs of lost trade. Relations with Russia are no longer close, but are maintained carefully because of heavy Bulgarian dependence on Russia for oil and gas. Relations with Turkey have greatly improved since the tensions of the final years of communist rule.

AID   ▷ Recipient

$311m (receipts)   Up 15% in 2000

IMF, World Bank, and EBRD loans are mainly intended for infrastructure improvements. Large-scale EU assistance toward reforms, in preparation for Bulgaria's eventual EU membership, is estimated at 2% of GDP. Humanitarian aid focuses mainly on medical provision and children's homes.

CHRONOLOGY

Bulgaria was part of the Ottoman Empire for five centuries until its independence in 1908. Under King Ferdinand, it sided with Germany during World War I, and subsequently lost valuable territory to Greece and Serbia. Under King Boris, Bulgaria once again sided with Germany in World War II.

❑ **1943** Child king Simeon II accedes.
❑ **1944** Allies firebomb Sofia. Soviet army invades. Antifascist Fatherland Front coalition, including Agrarian Party and Bulgarian Communist Party (BCP), takes power in bloodless coup. Kimon Georgiev prime minister.
❑ **1946** September, referendum abolishes monarchy. Republic proclaimed. October, general election results in BCP majority.
❑ **1947** Prime Minister Georgi Dmitrov discredits Agrarian Party leader Nikola Petkov. Petkov arrested and sentenced to death. International recognition of Dmitrov government. Soviet-style constitution adopted; one-party state established. Country renamed People's Republic of Bulgaria. Nationalization of economy begins.
❑ **1949** Dmitrov dies, succeeded as prime minister by Vasil Kolarov.
❑ **1950** Kolarov dies. "Little Stalin" Vulko Chervenkov replaces him ⇨

B

CHRONOLOGY *continued*

and begins BCP purge and collectivization.

❑ **1953** Stalin dies; Chervenkov's power begins to wane.

❑ **1954** Chervenkov yields power to Todor Zhivkov. Zhivkov sets out to make Bulgaria an inseparable part of the Soviet system.

❑ **1955–1960** Zhivkov exonerates victims of Chervenkov's purges.

❑ **1965** Plot to overthrow Zhivkov discovered by Soviet agents.

❑ **1968** Bulgarian troops aid Soviet army in invasion of Czechoslovakia.

❑ **1971** New constitution. Zhivkov becomes president of State Council and resigns as premier.

❑ **1978** Purge of BCP: 30,000 members expelled.

❑ **1984** Turkish minority forced to take Slavic names.

❑ **1989** June–August, exodus of 300,000 Bulgarian Turks. November, Zhivkov ousted as BCP leader and head of state. Replaced by Petur Mladenov. Mass protest in Sofia for democratic reform. December, Union of Democratic Forces (UDF) formed.

❑ **1990** Economic collapse. Zhivkov arrested. BCP loses constitutional role as leading political party, changes name to Bulgarian Socialist Party (BSP). Elections: BSP victory. Parliament chooses Zhelyu Zhelev, UDF leader, as president. Country renamed Republic of Bulgaria; communist symbols removed from national flag.

❑ **1991** February, price controls abolished; steep price rises. July, new constitution adopted. October, UDF wins elections.

❑ **1992** Continued political and social unrest. October, UDF resigns after losing vote of confidence. December, Movement for Rights and Freedoms (MRF) forms government. Zhivkov convicted of corruption and human rights abuses.

❑ **1993** Ambitious privatization program begins.

❑ **1994** General elections return BSP to power.

❑ **1995** BSP leader, Zhan Videnov, heads coalition government.

❑ **1996** Financial crisis and collapse of lev. Presidential elections won by opposition UDF candidate, Peter Stoyanov.

❑ **1997** General election won by UDF, whose leader Ivan Kostov becomes prime minister.

❑ **2001** Despite economic upturn, voters turn to new party headed by ex-king, who, as Simeon Saxecoburggotski, becomes prime minister. November, BSP leader Georgi Purvanov elected president.

DEFENSE

 Compulsory military service

 $347m

⬇ Down 11% in 2000

BULGARIAN ARMED FORCES

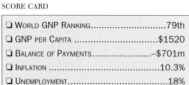

🛡	1475 main battle tanks (1042 T–55, 433 T–72)	42,400 personnel
🚢	1 submarine, 1 frigate, and 23 patrol boats	5260 personnel
✈	181 combat aircraft (Su–25, MiG–21/23/29)	18,300 personnel
	None	

Defense spending fell from 14% of GDP in 1985 to 2.8% in 2000. Plans to join NATO, which were announced in 1997, involved a major reorientation in defense thinking.

In late 1999 the government adopted "Plan 2004," which embodies a radical acceleration of its previous plans to restructure the armed forces. The new plan envisages downsizing to only 45,000 personnel by 2004, and moving to an emphasis on rapid reaction capabilities. It was considered that this smaller but combat-ready force would be less costly to maintain in the long run than a larger force.

In 1999 the crisis in Kosovo prompted Bulgaria to make its airspace available to NATO.

ECONOMICS

▷ Inflation 103% p.a. (1990–2000)

📊 $12.4bn

💲 2.084–2.187 leva

SCORE CARD

❑ WORLD GNP RANKING............................79th
❑ GNP PER CAPITA $1520
❑ BALANCE OF PAYMENTS.....................–$701m
❑ INFLATION 10.3%
❑ UNEMPLOYMENT....................................18%

EXPORTS

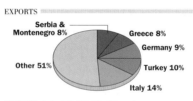

- Serbia & Montenegro 8%
- Greece 8%
- Germany 9%
- Turkey 10%
- Italy 14%
- Other 51%

IMPORTS

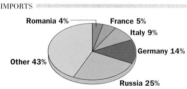

- Romania 4%
- France 5%
- Italy 9%
- Germany 14%
- Russia 25%
- Other 43%

ECONOMIC PERFORMANCE INDICATOR

own companies outright. Trade has shifted toward the EU, while that with the former Soviet Union has fallen sharply. The UDF government, returned in 1997, followed IMF advice, with free-market reforms, backed by foreign loans, successfully bringing inflation under control. After a setback in 1999, attributed to the regional impact of the crisis over Kosovo, forecasts of economic growth suggested Bulgaria's continuing recovery.

STRENGTHS

Coal and natural gas. Good agricultural production, especially grapes for well-developed wine industry, and tobacco. Strong expertise in computer software.

WEAKNESSES

Outdated infrastructure and equipment, and outstanding debt throughout industry. Slow privatization and structural reform until 1998.

PROFILE

Restructuring the economy is linked to privatization – a process delayed for political and technical reasons until the late 1990s. A financial crisis in 1996 triggered the collapse of the national currency, the lev. Foreign investment is still low, despite laws that since 1992 have allowed foreign firms to

BULGARIA : MAJOR BUSINESSES

Symbol	Industry	Symbol	Industry
🍷	Wine	🚬	Tobacco
	Steel	💻	Computers
✳	Textiles		Oil refining
	Shipbuilding		
◤	Leather tanning		
	Food processing		
△	Metal processing		
🚗	Vehicle assembly		
✿	Heavy engineering		

0   200 km
0   200 miles

B

RESOURCES

 Electric power 12.1m kw

 18,336 tonnes

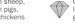 662 b/d (reserves 15m barrels)

2.29m sheep, 1.14m pigs, 15m chickens

Coal, iron, copper, lead, zinc, natural gas, oil, manganese

ELECTRICITY GENERATION

Hydro 8% (3.3bn kwh)

Combustion 52% (21bn kwh)

Nuclear 40% (17bn kwh)

Other 0%

% of total generation by type

Bulgaria has modest oil reserves and rather larger ones of coal and natural gas, but still has to import about 70% of its primary energy needs, much of it from the CIS. Unreliable supplies in the past led to frequent winter power cuts. These have largely disappeared since the mid-1990s, as reduced production in heavy industry and improved domestic supply have lowered import demand. Bulgaria is partly reliant on nuclear power. The EU is providing aid to upgrade two reactors at Kozloduy in return for the promised closure of other reactors there which pose particular safety risks. The government also plans to extend the controversial Chaira Dam, which began generating hydropower in 1993. Bulgaria has the northern hemisphere's largest manganese mine.

ENVIRONMENT

 Sustainability rank: 71st

 5%

5.7 tonnes per capita

ENVIRONMENTAL TREATIES

Yes   Yes   Yes

Yes   Yes   No

Environmental degradation led to the foundation in 1989 of the *Ecoglasnost* party. It circulated information on pollution and nuclear waste dump locations, and brought polluters to court. The Kozloduy nuclear complex, east of Lom, restarted in 1995 despite safety concerns, is under pressure to close its oldest reactors. Air pollution has diminished, but problems remain. NATO bombing of Serbian chemical and oil refineries on the Danube in 1999 led to downriver pollution in Bulgaria.

MEDIA

 TV ownership high

Daily newspaper circulation 254 per 1000 people

PUBLISHING AND BROADCAST MEDIA

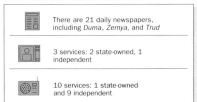

There are 21 daily newspapers, including *Duma*, *Zemya*, and *Trud*

3 services: 2 state-owned, 1 independent

10 services: 1 state-owned and 9 independent

Although it was liberalized in 1989, media freedom remains relative, and under new electronic media laws in 1998 state-run broadcasters retain an effective news monopoly. One group dominates the newspaper market, and of the others most are party-owned. Internet providers are regulated. Journalists also complain of the tough libel laws.

CRIME

 No death penalty

9424 prisoners   Down 11% in 1999

CRIME RATES

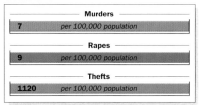

Murders

7   *per 100,000 population*

Rapes

9   *per 100,000 population*

Thefts

1120   *per 100,000 population*

In the 1990s Bulgaria became a key narcotics trafficking route to western Europe. Former security agents, party officials, and prestigious ex-athletes have moved into protection rackets, counterfeiting, and similar activities. Violations of minority rights are a sensitive political issue.

EDUCATION

 School leaving age: 16

 98%   261,321 students

THE EDUCATION SYSTEM

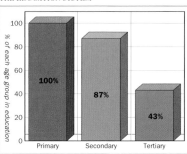

% of each age group in education

Primary 100%
Secondary 87%
Tertiary 43%

Education is free and compulsory between the ages of seven and 16. The system has been changed from a Soviet-inspired to a west European-style model. Government spending on education fell sharply until the 1999 budget. Teaching standards continue to be lowest in the rural and Turkish communities.

BULGARIA : LAND USE

Cropland
Pasture
Forest
Sheep
Cereals
Tobacco

0   200 km
0   200 miles

HEALTH

 Welfare state health benefits

 1 per 286 people

Heart and cerebrovascular diseases, cancers

Hospital facilities have kept pace with population growth, but the 1997 economic crisis brought the health service to the brink of collapse. A new health policy was formulated in 1999. The plan of action emphasizes primary care. The Bulgarian Red Cross assists in health administration.

SPENDING

GDP/cap. decrease

CONSUMPTION AND SPENDING

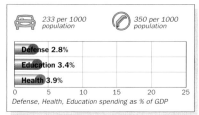

233 per 1000 population

350 per 1000 population

Defense 2.8%

Education 3.4%

Health 3.9%

Defense, Health, Education spending as % of GDP

Conquering the hyperinflation of 1995–1997 has remedied the most acute crisis, and associated hardship has been reduced. Turks and Roma remain the poorest people.

WORLD RANKING

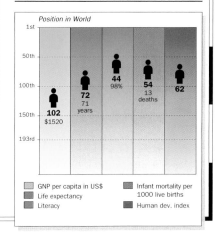

Position in World

102 $1520

72 71 years

44 98%

54 13 deaths

62

GNP per capita in US$
Life expectancy
Literacy

Infant mortality per 1000 live births
Human dev. index

BURKINA

B

OFFICIAL NAME: Burkina Faso **CAPITAL:** Ouagadougou
POPULATION: 11.9 million **CURRENCY:** CFA franc **OFFICIAL LANGUAGE:** French

  1960 1960 Dec 11 BF 0 +226 .bf

LANDLOCKED IN WEST AFRICA, Burkina (formerly Upper Volta) gained independence from France in 1960. The majority of Burkina lies in the arid fringe of the Sahara known as the Sahel. Ruled by military dictators for much of its postindependence history, Burkina became a multiparty state in 1991. However, much power still rests with President Blaise Compaoré. Burkina's economy remains largely based on agriculture.

CLIMATE ▷ Tropical/steppe

WEATHER CHART FOR OUAGADOUGOU

The tropical climate comprises two seasons – unreliable rains from June to October, and a long dry season.

TRANSPORTATION ▷ Drive on right

 Ouagadougou International 232,312 passengers Has no fleet

THE TRANSPORTATION NETWORK

| 2001 km (1243 miles) | None |
| 622 km (387 miles) | None |

The railroad to the port of Abidjan in Ivory Coast provides the main commercial route to the sea. Roads through Benin, Togo, and Ghana provide alternative access.

TOURISM ▷ Visitors : Population 1:55

 218,000 visitors Up 36% in 1999

MAIN TOURIST ARRIVALS

| France 28% |
| Ivory Coast 8% |
| Mali 4% |
| Other 60% |

% of total arrivals

Some potential exists for safari tourism, and the cities offer an attractive mix of colonial and African architecture. Big game hunting is allowed in some areas.

PEOPLE ▷ Pop. density low

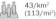 Mossi, Fulani, French, Tuareg, Dyula, Songhai 43/km² (113/mi²)

THE URBAN/RURAL POPULATION SPLIT

19%   81%

RELIGIOUS PERSUASION

- Other Christian 1%
- Roman Catholic 9%
- Muslim 35%
- Traditional beliefs 55%

No ethnic group is dominant in Burkina, although the Mossi people who live in the area of their old empire around Ouagadougou are the most numerous and have always played an important role in government. Burkina's first president, Maurice Yameogo, and Blaise Compaoré, leader since 1987, are both Mossi. The people from the west are much more ethnically mixed.

The extended family is important, both in rural areas and in the towns and cities. Extreme poverty has led to a strong sense of egalitarianism within society.

The absence of women in public life belies their real power and influence, particularly within the traditional framework of the extended family. However, most women are still denied access to education and senior professional positions.

Camel plowing. Burkina's poor soils and frequent droughts lead many young men to emigrate seasonally in search of work.

POLITICS ▷ Multiparty elections

L. House 2002/2007 President Blaise
U. House varies Compaoré

AT THE LAST ELECTION

National Assembly 111 seats

- 51% CDP
- 15% ADF–RDA
- 9% PDP
- 4% PAREN
- 21% Others

CDP = Congress for Democracy and Progress
ADF–RDA = Alliance for Democracy and Federation–African Democratic Rally **PDP** = Party for Democracy and Progress
PAREN = Party for National Renewal

House of Representatives 178 seats

Members of the House of Representatives are appointed or indirectly elected on a nonparty basis by provincial councils and various communities

A multiparty democracy in theory, Burkina is still dominated in practice by former military dictator Blaise Compaoré, and the army remains influential behind the scenes. Compaoré has been in power since the assassination in 1987 of Capt. Thomas Sankara, his former superior. Several of Compaoré's close military colleagues have been murdered. His grip on power in Burkina appears to be solid, and he was reelected president in 1998 with almost 90% of the vote. Most opposition leaders are still living in exile, but opposition parties made small gains in the 2002 polls.

The CDP and the government came under unexpected pressure in 1998 and 1999 after the assassination of a popular newspaper editor, Norbert Zongo, in which leading establishment figures were implicated.

WORLD AFFAIRS ▷ Joined UN in 1960

 CILSS ECOWAS AU OIC FZ

Burkina's landlocked position means that good relations with countries to the south are a major concern. The US used the threat of sanctions in 2000 to stem the illegal trade in Burkina of diamonds from Sierra Leone.

AID ▷ Recipient

 $336m (receipts) Down 16% in 2000

External aid, mostly from the World Bank and France, is important to the economy. The large number of NGOs has caused organizational problems; there is often difficulty in finding suitable projects for all the prospective donors.

BURKINA

Total Land Area : 274 200 sq. km
(105 869 sq. miles)

POPULATION

- ◎ over 100 000
- ○ over 50 000
- ● over 10 000
- ∙ under 10 000

LAND HEIGHT

500m/1640ft
200m/656ft
Sea Level

B

CHRONOLOGY

Ruled by Mossi kings from the 16th century, Upper Volta, a province of French West Africa in the late 19th century, gained independence in 1960.

- ❏ **1980** Ousting of military ruler; Col. Saye Zerbo becomes president.
- ❏ **1982** Capt. Thomas Sankara takes power. People's Salvation Council (PSC) begins radical reforms.
- ❏ **1984** Renamed Burkina.
- ❏ **1987** Sankara assassinated, Capt. Blaise Compaoré takes power.
- ❏ **1991** New constitution. Compaoré elected president.
- ❏ **1997** CDP landslide election victory.
- ❏ **1999** Biggest gold mine closed. General strike.
- ❏ **2001** Meningitis epidemic. HIV infection rate second highest in West Africa.

DEFENSE

▷ No compulsory military service

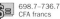 $68m

⬇ Down 9% in 2000

The main role of the 5600-strong army has been maintaining internal security. Burkina is reliant on France for most equipment and training.

ECONOMICS

▷ Inflation 3.8% p.a. (1990–2000)

$2.42bn

698.7–736.7 CFA francs

SCORE CARD

❏ WORLD GNP RANKING	134th
❏ GNP PER CAPITA	$210
❏ BALANCE OF PAYMENTS	–$65m
❏ INFLATION	–0.3%
❏ UNEMPLOYMENT	1%

STRENGTHS

Remittances from plantation workers in Ghana and the Ivory Coast. Strongly improved economic management. Low debt burden. Ability to attract foreign aid. Cotton.

WEAKNESSES

Landlocked. Natural resources not in the main economically viable. Donors' fears over political instability. Food crop fluctuations. Prone to drought. Migrants' remittances have halved to about $80 million since 1988.

EXPORTS

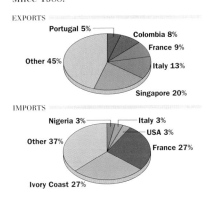

Portugal 5%
Colombia 8%
France 9%
Other 45%
Italy 13%
Singapore 20%

IMPORTS

Nigeria 3%
Italy 3%
USA 3%
Other 37%
France 27%
Ivory Coast 27%

RESOURCES

▷ Electric power 78,000 kw

 7660 tonnes

Not an oil producer and has no refineries

8.65m goats, 6.78m sheep, 22.4m chickens

Gold, antimony, marble, manganese, silver, zinc

Burkina's considerable mineral wealth includes large manganese and silver deposits. Falling world prices led to the closure in 1999 of the largest gold mine. Three hydroelectric dams will reduce dependence on thermal energy.

ENVIRONMENT

▷ Sustainability rank: 101st

 10% (8% partially protected)

0.1 tonnes per capita

Like other countries on the southern rim of the Sahara, desertification is the major ecological issue. The rate of tree cutting for fuel is on the increase.

MEDIA

▷ TV ownership low

 Daily newspaper circulation 1 per 1000 people

PUBLISHING AND BROADCAST MEDIA

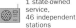 There are 6 daily newspapers, including *Sidwaya, Le Pays, Le Journal de Soir,* and *L'Observateur Paalga*

4 services: 1 state-owned, 3 independent

1 state-owned service, 46 independent stations

There are a number of small independent newspapers funded by opposition groups. A code of practice was introduced in 1999.

CRIME

▷ Death penalty not used in practice

6100 prisoners

⬇ Down 19% 1996–1998

Crime levels have traditionally been low. However, the urbanization of society and the increase in political violence have seen levels increase.

EDUCATION

▷ School leaving age: 16

 24%

8911 students

Education is based on the French system. Recently, practical subjects have received more emphasis.

HEALTH

▷ No welfare state health benefits

1 per 20,000 people

Malaria, diarrheal and respiratory diseases

Health spending focuses on primary health care and vaccination. More than 6% of adults were HIV positive in 2001.

SPENDING

▷ GDP/cap. increase

CONSUMPTION AND SPENDING

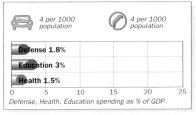

4 per 1000 population

4 per 1000 population

Defense 1.8%
Education 3%
Health 1.5%

Defense, Health, Education spending as % of GDP

Burkina is a country of extreme, almost universal, poverty. Displays of wealth are rare and ownership of high-tech items is limited to a small elite.

WORLD RANKING

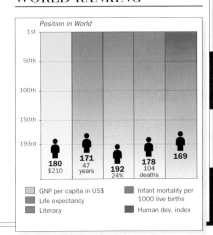

Position in World

180 $210
171 47 years
192 24%
178 104 deaths
169

GNP per capita in US$
Life expectancy
Literacy
Infant mortality per 1000 live births
Human dev. index

BURMA (MYANMAR)

OFFICIAL NAME: Union of Myanmar **CAPITAL:** Rangoon (Yangon)
POPULATION: 48.4 million **CURRENCY:** Kyat **OFFICIAL LANGUAGE:** Burmese (Myanmar)

SOUTHEAST ASIA

| 1948 | 1948 | Jan 4 | BUR | +6.5 | +95 | .mm |

A PREDOMINANTLY BUDDHIST country on the northeastern shores of the Indian Ocean, Burma is mountainous in the north and east, while the fertile Irrawaddy basin occupies most of the country. Rocked by ethnic conflict ever since gaining independence from the UK in 1948, Burma has been ruled by repressive military regimes since 1962. The National League for Democracy (NLD) gained a majority in free elections in 1990, but has been prevented from taking power by the military. Rich in natural resources, which include fisheries and teak forests, Burma remains a mostly agricultural economy.

Transporting timber on the Irrawaddy River near Mandalay. Burma once had the world's largest reserves of teak.

CLIMATE

▷ Tropical/mountain

WEATHER CHART FOR RANGOON

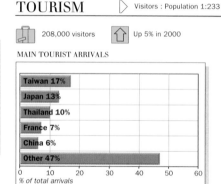

There are three seasons: the wet season, when rainfall in the far south Tenasserim region and Irrawaddy delta can reach 500 cm (197 in); summer, when northern Burma experiences 50°C (122°F) and 100% humidity; and winter, when it is rarely cooler than 15°C (59°F) except in the northern mountains.

TRANSPORTATION

▷ Drive on right

Mingaladon, Rangoon
580,000 passengers

124 ships
492,306 grt

THE TRANSPORTATION NETWORK

3440 km (2138 miles)		None	
3955 km (2458 miles)		12,800 km (7954 miles)	

Burma's main transportation corridors run north–south. Most traffic is concentrated between Rangoon and Mandalay. A daily express train runs between these two cities, although visitors are urged to take internal flights on the state-owned carriers. The Irrawaddy River and its tributaries also provide an important artery for travel.

Beyond the Irrawaddy basin transportation is limited and hazardous. The vast majority of roads remain unpaved, and access to neighboring countries is restricted. At certain points on the Chinese border, access is one-way only: out of China.

TOURISM

▷ Visitors : Population 1:233

208,000 visitors

Up 5% in 2000

MAIN TOURIST ARRIVALS

Taiwan 17%
Japan 13%
Thailand 10%
France 7%
China 6%
Other 47%

% of total arrivals

Since 1988 the military authorities have courted tourists for their economic value. The previous one-week restriction for visitors arriving to marvel at Burma's stunning Buddhist heritage was increased to 28 days. The state-run tourist agency is heavily promoted while independent travelers are required to spend a minimum of $200 during their stay. However, whole provinces remain entirely off-limits, and elsewhere official guides are often required. Opposition groups wish to discourage people from traveling to Burma at all.

PEOPLE

▷ Pop. density medium

Burmese, Karen, Shan, Chin, Kachin, Mon, Palaung, Wa

74/km² (191/mi²)

THE URBAN/RURAL POPULATION SPLIT

28%　　　72%

RELIGIOUS PERSUASION

Other 2%　Muslim 4%
Hindu 1%　Christian 6%
Buddhist 87%

ETHNIC MAKEUP

Rakhine 4%　Karen 6%
Shan 9%
Other 13%
Burman (Bamah) 68%

A savage history of ethnic repression at the hands of the majority Burman still plays a large part in the mistrust felt by the smaller minority communities. Each group maintains a distinct cultural identity. At independence the Chin, Kachin, Karen, Karenni, Mon, and Shan all unsuccessfully demanded their own state within a federation. Despite uniting against the military dictatorship in 1988, most factions had by 1996 signed peace agreements. Only the Shan and Karen remain militarily active to any significant degree, the former agreeing to a short-lived cease-fire in 2000. While the Burman claim racial purity, many of them are in fact of mixed blood or ethnically Chinese.

Accusations of forced labor lie at the heart of international criticism of the military regime, with ethnic minorities apparently at highest risk.

Domestic life in Burma is still based around the extended family. Women have a prominent role, and access to education. Many run or own businesses in their own right. However, top jobs in government are still held almost exclusively by men.

POPULATION AGE BREAKDOWN

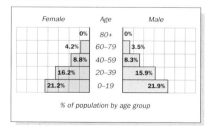

Female	Age	Male
0%	80+	0%
4.2%	60–79	3.5%
8.8%	40–59	8.3%
16.2%	20–39	15.9%
21.2%	0–19	21.9%

% of population by age group

B

POLITICS

▷ No legislative elections

1990/suspended | ▷ Chairman Than Shwe

AT THE LAST ELECTION

Constituent Assembly 485 seats

| | | | **2%** NUP | **1%** MNDF | **1%** NDP |

| **81%** NLD | | **5%** SNLD | **2%** RDL | **8%** Others |

NLD = National League for Democracy
SNLD = Shan National League for Democracy
NUP = National Unity Party
RDL = Rakhine Democracy League
MNDF = Mon National Democratic Front
NDP = National Democratic Party for Human Rights

A Constituent Assembly, responsible for the drafting of a new constitution and with no legislative power, was elected in 1990, but prevented from convening by the regime

Burma is ruled by the military-backed State Peace and Development Council (SPDC), under Gen. Than Shwe.

PROFILE

Seizing power amid mass pro-democracy protests in 1988, the military has suppressed all democratic opposition. The regime has never recognized the 1990 electoral victory of the NLD. Recent moves toward negotiation have produced few tangible results.

Ethnic rebellion in outer regions degenerated into cross-border guerrilla activity after a concerted government offensive in 1996.

MAIN POLITICAL ISSUES

Restoring democracy

In a nod to international pressure and the popularity of the prodemocracy movement, the junta has talked openly of steering Burma toward "disciplined democracy." In reality there has been little progress toward ending the dictatorship. The opposition NLD is led by the charismatic Aung San Suu Kyi who has been intermittently under house arrest since 1990. Her latest release, in 2002, quickly proved to have been a hollow gesture.

Aung San Suu Kyi, figurehead of the pro-democracy movement.

Gen. Than Shwe, leader of the military junta since 1992.

BURMA

Total Land Area : 678 500 sq. km (261 969 sq. miles)

POPULATION

- ⊡ over 1 000 000
- ◉ over 500 000
- ◎ over 100 000
- ○ over 50 000
- ● over 10 000
- · under 10 000

LAND HEIGHT

- 4000m/13 124ft
- 2000m/6562ft
- 1000m/3281ft
- 500m/1640ft
- 200m/656ft
- Sea Level

Forced labor

The practice of using slave labor in rural areas was officially banned in 2000. However, many groups, including the International Labor Organization, insist that the military authorities still regularly make use of "slave" laborers.

WORLD AFFAIRS

▷ Joined UN in 1948

CP | IAEA | ASEAN | NAM | WTO

Burma's key relationship is with China, which backs the SPDC military regime and is a major supplier of weapons to the Burmese army. The relationship allows China access to the Indian Ocean and gives it influence over a regime dependent on its support. While Burma's neighbors fear that the arrangement could destabilize the whole of the Asia–Pacific region, many favor a policy of "constructive engagement" with the SPDC. In 1997, Burma was admitted to ASEAN, despite continuing concerns about its human rights record.

The EU and Western members of the UN have strongly condemned the human rights violations in Burma and threatened to impose economic sanctions in response to the regime's policies. In practice, however, the West maintains an ambivalent position. Economic ties are expanding, particularly between SPDC-owned state enterprises and Western multinationals which have an interest in Burmese offshore oil and gas drilling sectors.

CHRONOLOGY

From the 11th century, Burma's many ethnic groups came under the rule of three Tibeto-Burman dynasties, interspersed with periods of rule by the Mongols and the Mon. The Third Dynasty came into conflict with the British in India, sparking the Anglo-Burmese wars of 1824, 1852, and 1885.

❏ **1886** Burma becomes a province of British India.
❏ **1930–1931** Economic depression triggers unrest.
❏ **1937** Separation from India.
❏ **1942** Japan invades.
❏ **1945** Antifascist People's Freedom League (AFPFL), led by Aung San, helps Allies reoccupy country.
❏ **1947** UK agrees to Burmese independence. Aung San wins elections, but is assassinated.
❏ **1948** Independence under new prime minister, U Nu, who initiates socialist policies. Revolts by ethnic separatists, notably Karen liberation struggle.
❏ **1958** Ruling AFPFL splits into two. Shan liberation struggle begins.
❏ **1960** U Nu's faction wins elections.
❏ **1961** Kachin rebellion begins. ➪

B

CHRONOLOGY *continued*

- ❏ **1962** Gen. Ne Win stages military coup. "New Order" policy of "Buddhist Socialism" deepens international isolation. Mining and other industries nationalized. Free trade prohibited.
- ❏ **1964** Socialist Program Party declared sole legal party.
- ❏ **1976** Social unrest. Attempted military coup. Ethnic liberation groups gain control of 40% of country.
- ❏ **1982** Nonindigenous people barred from public office.
- ❏ **1988** Thousands die in student riots. Ne Win resigns. Martial law. Aung San Suu Kyi, daughter of Aung San, and others form NLD. Gen. Saw Maung leads military coup. State Law and Order Restoration Council (SLORC) takes power. Ethnic resistance groups form Democratic Alliance of Burma.
- ❏ **1989** Army arrests NLD leaders and steps up antirebel activity. Officially renamed Union of Myanmar.
- ❏ **1990** Elections permitted. NLD wins landslide. SLORC remains in power, however. More NLD leaders arrested.
- ❏ **1991** Aung San Suu Kyi awarded Nobel Peace Prize.
- ❏ **1992** Gen. Than Shwe takes over as SLORC leader.
- ❏ **1996** Demonstrations against approval of Burma's membership of ASEAN.
- ❏ **1997** Ruling SLORC renamed State Peace and Development Council (SPDC). US imposes sanctions and bans further investment in Burma.
- ❏ **1998** NLD sets deadline for convening parliament; junta refuses.
- ❏ **1999** Aung San Suu Kyi rejects conditions set by SPDC for visiting the UK to see her husband, Michael Aris, who dies of cancer.
- ❏ **2000** Negotiations between junta and NLD begin.
- ❏ **2002** Aung San Suu Kyi released from house arrest.

AID

▷ Recipient

 $107m (receipts)   Up 32% in 2000

Moves in late 2000 to begin negotiations with the democratic opposition were welcomed by the UN, but were not enough to lift the sanctions on aid imposed in 1988. There is still some humanitarian assistance provided through agencies such as WHO. Japan has led the way in rewarding the regime's conciliatory gestures and has increased its contributions, making it the largest single donor.

DEFENSE

  No compulsory military service

$2.06bn   Up 3% in 2000

BURMESE ARMED FORCES

100 main battle tanks (PRC Type-69II)	325,000 personnel
68 patrol boats	10,000 personnel
113 combat aircraft (50 F-7, 10 FT-7, 22 A-5M)	9000 personnel
None	

The military authorities have steadily increased the country's military power, doubling the size of the army and obtaining modern weapons and military technology from around the world, primarily from China, which since 1989 has delivered arms worth over $1 billion to Burma, including tanks and jet fighters.

Burma's growing military capability is used mainly to control internal dissent, and the army has suppressed most ethnic insurgent campaigns by utilizing its military superiority and cutting numerous deals with rebel leaders. The remaining militant groups are now combated in cooperation with neighboring states. The army is accused of human rights abuses in rural areas.

ECONOMICS

▷ Inflation 26% p.a. (1990–1999)

$73.9bn   6.526–6.771 kyats

SCORE CARD

- ❏ WORLD GNP RANKING.........................43rd
- ❏ GNP PER CAPITA $1500
- ❏ BALANCE OF PAYMENTS..................−$293m
- ❏ INFLATION ....−0.1%
- ❏ UNEMPLOYMENT 7%

ECONOMIC PERFORMANCE INDICATOR

EXPORTS

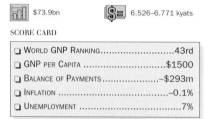

- Singapore 5%
- Japan 6%
- China 6%
- India 14%
- USA 24%
- Other 45%

IMPORTS

- Japan 9%
- Malaysia 10%
- South Korea 13%
- Singapore 19%
- China 22%
- Other 27%

STRENGTHS
Very rich in natural resources: fertile soil, rich fisheries, timber including diminishing teak reserves, gems, offshore natural gas, and oil.

WEAKNESSES
Shortage of skilled labor, managers, and technicians. Rudimentary financial systems and institutions. Nationwide black market. Huge external debt. Dependence on imported manufactures.

PROFILE
Burma's economy is agriculture-based and functions mainly on a cash and barter system. Its key industries are controlled by military-run state enterprises. Every aspect of economic life is permeated by a black market, where prices are rocketing – a reaction to official price controls.

Since 1989, the SPDC's open-door market-economy policy has brought a flood of foreign investment in oil and gas (by Western companies), and in forestry, tourism, and mining (by Asian companies). The resulting boom in trade with China has turned less developed Upper Burma into a thriving business center. A narcotics-eradication program has been initiated in the northeastern border states, which account for about 60% of the world's heroin, by encouraging farmers to grow food crops instead of poppies. Few plans exist for the manufacturing sector, however, and dependence on imports continues.

BURMA : MAJOR BUSINESSES

- 🏛 Real estate
- Defense
- Jade
- Teak industries
- Trading centre
- Gas
- Oil
- Fish processing
- Manufacturing
- Opium

B

RESOURCES

Electric power 1.4m kw

945,827 tonnes

7659 b/d (reserves 50m barrels)

11.2m cattle, 6.2m ducks, 4.14m pigs, 48.3m chickens

Oil, natural gas, tin, antimony, zinc, copper, tungsten, lead, coal, gemstones

ELECTRICITY GENERATION

Hydro 23% (0.9bn kwh)				
Combustion 77% (3.2bn kwh)				
Nuclear 0%				
Other 0%				

0   20   40   60   80   100

% of total generation by type

Burma is one of the world's largest teak exporters. It is also a producer of pearls, rubies, and other gems. Foreign capital is funding exploration for natural gas and oil in the Tenasserim strip, while offshore fields are already in full production.

SHAN PLATEAU

BURMA : LAND USE

Cropland
Pasture
Forest
Rice
Cattle

0   200 km
0   200 miles

ENVIRONMENT

Sustainability rank: 90th

0.3%

0.2 tonnes per capita

ENVIRONMENTAL TREATIES

No	No	Yes
No	No	No

Deforestation is a major problem, and it has increased since the 1988 coup. Chinese companies have been given unrestricted logging concessions.

MEDIA

TV ownership low

Daily newspaper circulation 10 per 1000 people

PUBLISHING AND BROADCAST MEDIA

There are 5 daily newspapers, including *Myanma Alin* and *New Light of Myanmar*

1 state-controlled service

1 state-controlled service

Political dissent of any kind is a criminal offense. An underground prodemocracy press produces antigovernment material.

CRIME

Death penalty in use

53,195 prisoners

Down 82% in 1999

CRIME RATES

Murders	
1	per 100,000 population

Rapes	
0.3	per 100,000 population

Thefts	
15	per 100,000 population

Compared with similar totalitarian regimes, levels of bribery, corruption, embezzlement, and black marketeering are high. The state is guilty of illegal activity. The UN reports regularly on human rights abuses against civilians, and the murder of innocent civilians including children, women, Buddhist monks, students, minorities, and political dissidents.

There is a nominal civilian judicial system in Burma, but in practice all judges and lawyers are appointed by the junta and all legal functions are executed by the SPDC. The most common charge is that of sedition against the state or the army under the 1975 "Law to Protect the State from Destructionists." Among the SPDC's frequent arbitrary "notices" is Order 2/88, prohibiting assemblies of more than five persons. Most detainees have no legal rights of representation and are either jailed, used as forced labor, or put under house arrest without public trial. Amnesty International is banned.

EDUCATION

School leaving age: 10

85%

385,300 students

THE EDUCATION SYSTEM

% of each age group in education

100% Primary
36% Secondary
5% Tertiary

The education system offers ten years of schooling. Ethnic-language schools are discouraged. A shortage of teachers, many of whom have left or are in jail, has disrupted education. All but two universities were closed in the late 1990s by the regime, but quietly reopened in 2000. The NLD has criticized the shortened and "sanitized" courses on offer.

HEALTH

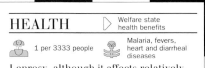

Welfare state health benefits

1 per 3333 people

Malaria, fevers, heart and diarrheal diseases

Leprosy, although it affects relatively few people compared with other diseases, has a higher prevalence in Burma than in the rest of Asia. In the last few years Burma has seen an increase in the incidence of malaria.

The growing number of AIDS cases is largely due to migrant prostitution across the Thai–Burmese border. They put an additional strain on health facilities, which are well developed but not comprehensive.

SPENDING

GDP/cap. increase

CONSUMPTION AND SPENDING

1 per 1000 population

6 per 1000 population

Defense 0.6%				
Education 1.2%				
Health 0.2%				

0   5   10   15   20   25

Defense, Health, Education spending as % of GDP

The state monopoly of the production and distribution of goods by rationing under Gen. Ne Win's administration led to an increase in corruption and the rise of a nationwide black market, with huge disparities between official and unofficial prices. Only the military elite and their supporters could afford to live well. The situation has not changed significantly since 1988. Giant military enterprises grouped under a Defense Services holding company, whose capital amounts to 10% of GDP, now reap wealth and distribute privileges for a minority. Nevertheless, traditional social and economic mobility still exists. Climbing the socioeconomic ladder is mainly a matter of loyalty to the military. Dissidents forced out of their jobs and hill tribes form the poorest groups.

WORLD RANKING

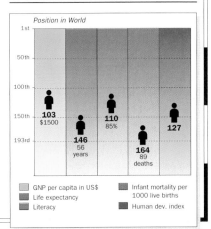

Position in World

1st
50th
100th
150th
193rd

103 $1500
146 56 years
110 85%
164 89 deaths
127

GNP per capita in US$
Life expectancy
Literacy
Infant mortality per 1000 live births
Human dev. index

BURUNDI

OFFICIAL NAME: Republic of Burundi   **CAPITAL:** Bujumbura
POPULATION: 6.5 million   **CURRENCY:** Burundi franc   **OFFICIAL LANGUAGES:** French and Kirundi

LANDLOCKED BURUNDI lies just south of the equator on the Nile–Congo watershed. Lake Tanganyika forms part of its border with the Democratic Republic of the Congo (DRC). Tension between the Hutu majority and the dominant Tutsi minority remains the main factor in politics. The current political unrest dates from October 1993, when the assassination of the first Hutu president in a coup by the Tutsi-dominated army sparked terrible violence.

Pig farming and fish ponds. The majority of Burundi's population depends on subsistence farming.

CLIMATE
▷ Tropical wet & dry

WEATHER CHART FOR BUJUMBURA

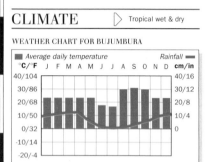

Burundi is temperate with high humidity, much cloud, and frequent heavy rain. The highlands have frost.

TRANSPORTATION
▷ Drive on right

Bujumbura International 65,472 passengers — Has no fleet

THE TRANSPORTATION NETWORK

1028 km (639 miles)	None
None	Lake Tanganyika is navigable

The dense road network has been rehabilitated. There are plans for a railroad to link Burundi with Rwanda, Uganda, and Tanzania.

TOURISM
▷ Visitors : Population 1:217

30,000 visitors — Up 15% in 2000

MAIN TOURIST ARRIVALS

Africa 48%
Europe 37%
Asia 8%
Americas 7%
% of total arrivals

A lack of basic infrastructure and violent political strife have deterred tourists. The industry has limited potential, since Burundi lacks its neighbors' spectacular scenery and game parks.

PEOPLE
▷ Pop. density high

Kirundi, French, Kiswahili
253/km² (656/mi²)

THE URBAN/RURAL POPULATION SPLIT
9% — 91%

ETHNIC MAKEUP
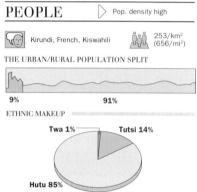
Twa 1%   Tutsi 14%   Hutu 85%

Burundi's history has been marked by violent conflict between the majority Hutu and the Tutsi, formerly the political elite, who still control the army. Large-scale massacres have occurred repeatedly over the past two decades. Hundreds of thousands of people, mostly Hutu, have been killed in political and ethnic conflict since 1993. The Twa pygmy minority has not been greatly affected.

Most Burundians are subsistence farmers, and the vast majority are Roman Catholic.

POLITICS
▷ In transition

Both houses 2002/Transitional   President Pierre Buyoya

AT THE LAST ELECTION
Transitional National Assembly 170 seats

38% Frodebu   10% UPRONA   52% Others

Frodebu = Front for Democracy in Burundi
UPRONA = Union for National Progress
The Transitional National Assembly consists of members of all parties that signed the Arusha peace accord, and members of "civil society."

Transitional Senate 51 seats

The indirectly elected Transitional Senate comprises 24 Hutus, 24 Tutsis, and 3 Twas, as formulated under the 2000 Arusha peace accord.

From 1966, Tutsi dominated all areas of life. An attempt by Tutsi president Buyoya to promote integration backfired in a bloody coup in 1993, when the first Hutu president, Melchior Ndadaye of Frodebu, was assassinated soon after his election and hundreds of thousands of Hutu were killed or fled; 1994 saw vicious civil war begin. Buyoya was returned to power in 1996 by a military coup, but despite his efforts to regain international acceptance, a regional economic boycott caused serious damage. A peace accord was forged in 2000, and in 2001 a joint Hutu–Tutsi transitional government was installed, but fighting continued, with the Hutu National Liberation Front refusing to sign a cease-fire or join the government.

BURUNDI

Total Land Area : 27 830 sq. km (10 745 sq. miles)

LAND HEIGHT
2000m/6562ft
1000m/3281ft
500m/1640ft

POPULATION
◎ over 100 000
○ over 50 000
● over 10 000
· under 10 000

WORLD AFFAIRS
▷ Joined UN in 1962

    ACP   CEPGL   COMESA   OIF   AU

Since 1995 Burundi has resisted proposals for international intervention to prevent further bloodshed.

AID
▷ Recipient

 $93m (receipts)   Up 26% in 2000

The flight of hundreds of thousands of people since 1993 has disrupted agriculture, and many people remain dependent on UN food aid.

DEFENSE
▷ No compulsory military service

 $65m   Down 6% in 2000

The 40,000-strong army is run by Tutsi. The attempt to bring Hutu into officer ranks was a cause of the 1993 coup. A state of virtual civil war now exists between it and rebel Hutu militias. A UNICEF program to demobilize over 14,000 child soldiers from these militias was announced in 2001.

ECONOMICS
▷ Inflation 12% p.a. (1990–2000)

 $732m   780.4–864.0 Burundi francs

SCORE CARD

- ❏ World GNP Ranking......................161st
- ❏ GNP per Capita $110
- ❏ Balance of Payments.......................–$49m
- ❏ Inflation 24.3%
- ❏ Unemployment.......Widespread underemployment

Strengths
Small quantities of gold and tungsten. Potential of massive nickel reserves and oil in Lake Tanganyika.

Weaknesses
Harsh regional sanctions since 1996 coup. Overwhelmingly agricultural economy under pressure from high birthrate and war damage. Little prospect of lasting political stability.

EXPORTS

IMPORTS

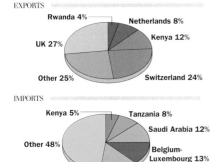

Rwanda 4%   Netherlands 8%   Kenya 12%   UK 27%   Other 25%   Switzerland 24%

Kenya 5%   Tanzania 8%   Saudi Arabia 12%   Other 48%   Belgium-Luxembourg 13%   France 14%

RESOURCES
▷ Electric power 43,000 kw

 9254 tonnes   Not an oil producer and has no refineries

 600,000 goats, 315,000 cattle, 4.7m chickens   Gold, tungsten, nickel, vanadium, uranium, oil

Burundi has around 5% of the world's nickel reserves. Extraction, however, is not economically viable. There are also deposits of gold and vanadium. Surveys in the 1980s detected oil reserves below Lake Tanganyika, but production has yet to begin. Burundi imports gasoline from Iran and electricity from the DRC. New HEP plants at Mugera and Rwegura, in the far north, are intended to meet most domestic electricity requirements.

ENVIRONMENT
▷ Sustainability rank: 115th

 6%   0.03 tonnes per capita

Only 2% of Burundi is forest and even this is now under pressure from one of Africa's highest birthrates. Burundi suffers from the problems associated with deforestation, particularly soil erosion. Some soils are also being exhausted from overuse. Several tree-planting programs have been introduced. UNESCO is also running ecological education initiatives at village level, aimed at women farmers.

MEDIA
▷ TV ownership low

 Daily newspaper circulation 3 per 1000 people

PUBLISHING AND BROADCAST MEDIA

Le Renouveau du Burundi is published three times a week by the government

1 state-controlled service

1 state-controlled service, some independent stations

Pro-Hutu/anti-Tutsi radio stations have been broadcasting since 1994. Radio Umwizero, an EU-funded station promoting peace, was launched in 1996.

CRIME
▷ Death penalty in use

 9411 prisoners   Up 78% 1990–1998

Burundi has an appalling human rights record. There have been frequent massacres of Hutu by the army. The worst pogroms occurred in 1972, 1988, 1993, and 1994.

EDUCATION
▷ School leaving age: 12

 48%   5037 students

Elementary schooling begins at seven, and is compulsory, though further schooling is not. There are around 70 elementary schoolchildren per teacher. There is one university.

CHRONOLOGY

From the 16th century, Burundi (formerly Urundi) was ruled by the minority Tutsi with the majority Hutu as their serfs. Merged with Rwanda, Burundi was controlled by Germany from 1884 and by Belgium from 1919.

- ❏ **1946** UN trust territory.
- ❏ **1959** Split from Rwanda.
- ❏ **1962** Independence.
- ❏ **1966** Army overthrows monarchy.
- ❏ **1972** 150,000 Hutu massacred.
- ❏ **1993** Ndadaye wins first free elections; killed four months later.
- ❏ **1996** Pierre Buyoya retakes power.
- ❏ **1999** Talks between warring groups.
- ❏ **2000** Renewed violence. Arusha peace accord signed by most groups.
- ❏ **2001** Multiethnic transitional government appointed.

HEALTH
▷ No welfare state health benefits

 1 per 10,000 people   Communicable infections, parasitic diseases

Over half of the population of Burundi are underfed. Over 30% of the people do not have access to health services, and, as at end-2001, 390,000 people were living with HIV/AIDS.

SPENDING
▷ GDP/cap. decrease

CONSUMPTION AND SPENDING

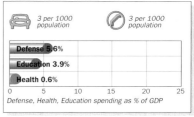

3 per 1000 population   3 per 1000 population

Defense 5.6%
Education 3.9%
Health 0.6%

0   5   10   15   20   25
Defense, Health, Education spending as % of GDP

Wealth is concentrated within the Tutsi political and business elite. Most of Burundi's people live at the level of subsistence farming.

WORLD RANKING

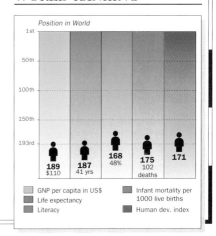

Position in World

1st   50th   100th   150th   193rd

189 $110   **187** 41 yrs   **168** 48%   **175** 102 deaths   **171**

GNP per capita in US$ Infant mortality per 1000 live births
Life expectancy   Human dev. index
Literacy

B

C

CAMBODIA

OFFICIAL NAME: Kingdom of Cambodia **CAPITAL:** Phnom Penh
POPULATION: 13.4 million **CURRENCY:** Riel **OFFICIAL LANGUAGE:** Khmer

 1953 1953 Nov 11 K +7 +855 .kh

THE ANCIENT KINGDOM of Cambodia emerged from French colonial rule in 1953, only to be plunged into violent civil conflict. Under the extremist Khmer Rouge, headed by the infamous Pol Pot, the country endured one of the world's most brutal totalitarian regimes. Since the withdrawal of Vietnamese troops in 1989 the country has gradually returned to relative stability. The dominating geographic feature is the Tônlé Sap, or Great Lake, which drains into the Mekong River. Over three-quarters of Cambodia is forested, with mangroves lining the coast.

CLIMATE
▷ Tropical monsoon

WEATHER CHART FOR PHNOM PENH

Cambodia has a varied climate. Low-lying regions have moderate rainfall and the most consistent year-round temperatures. The dry season from December to April is characterized by high temperatures and an average of eight hours of sunshine a day. From May to September, winds are southeasterly, while from October to April they are north or northeasterly. During the rainy season, Cambodia is sultry and humid. The monsoons in 2000 caused severe flooding of the Mekong River, which inundated Phnom Penh.

TRANSPORTATION
▷ Drive on right

 Pochentong, Phnom Penh
860,613 passengers

 405 ships
1.45m grt

THE TRANSPORTATION NETWORK

4165 km (2588 miles)		None
650 km (404 miles)		3700 km (2299 miles)

Years of war led to a near-collapse of Cambodia's rail and road systems. Although international aid has helped fund key projects, the state of the country's roads remains appalling. The government has pledged to make reconstruction a priority. The Mekong River is vital for accessing the interior. Taxi-mopeds, bicycles, and rickshaws dominate urban transportation.

Angkor Wat stands in the ruins of the ancient city of Angkor, once the capital of the Khmer empire. It is now one of Cambodia's leading tourist attractions.

TOURISM
▷ Visitors : Population
1:51

 265,000 visitors   ⬆ Up 1% in 2000

MAIN TOURIST ARRIVALS

USA	12%
China	10%
France	9%
Taiwan	8%
Japan	7%
Other	54%

% of total arrivals

Cambodia, the center of the Khmer empire between 800 and 1400 C.E., has some of the most impressive temples in southeast Asia. The most famous is Angkor Wat, near Siem Reap (Siemreab), which is now largely safe for tourists after the Khmer Rouge relinquished control of the area in 1998. Kidnappings and murders of tourists by the Khmer Rouge kept Cambodia off the backpacker circuit in the mid-1990s. Once the political situation is fully stabilized and land mines have been cleared, there is considerable potential, not just for adventurous independent travelers.

PEOPLE
▷ Pop. density medium

Khmer, French, Chinese, Vietnamese, Cham

76/km²
(197/mi²)

THE URBAN/RURAL POPULATION SPLIT

16%   84%

RELIGIOUS PERSUASION

Christian 1%   Muslim 6%
Buddhist 93%

ETHNIC MAKEUP

Chinese 1%   Vietnamese 4%
Other 5%
Khmer 90%

Cambodia underwent one of the 20th century's most horrific experiments in social transformation between 1975 and 1979 under Pol Pot's Khmer Rouge regime. Warfare, starvation, exhaustion, or execution killed one in eight of the population. Half a million more fled to Thailand. The Pol Pot regime's extreme radical beliefs led to the scrapping of money, possessions, and hierarchy. "Bourgeois" learning was despised, whereas peasants, soldiers of the revolution, and some industrial workers were officially given higher status. Boys and girls of 13 and 14 were taken from their homes, indoctrinated in the tenets of revolution, and allowed to kill those held guilty of bourgeois crimes. Violence at all levels was sanctioned in the name of revolution. The legacies of the regime are both the emigration of surviving professionals, and one of the world's highest rates of orphans and widows.

Religious and ethnic tensions are minimal, although there is a traditional hostility in Khmer culture toward ethnic Vietnamese.

POPULATION AGE BREAKDOWN

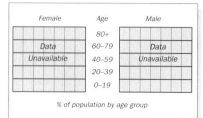

Female	Age	Male
	80+	
Data	60–79	Data
Unavailable	40–59	Unavailable
	20–39	
	0–19	

% of population by age group

POLITICS   ▷ Multiparty elections

L. House 1998/2003
U. House 1999/2004

H.M. King Norodom
Sihanouk

AT THE LAST ELECTION

National Assembly 122 seats

| 53% | 35% | 12% |
| CPP | Funcinpec | SRP |

CPP = Cambodian People's Party
Funcinpec = United National Front for an Independent
Neutral Peaceful and Cooperative Cambodia
SRP = Sam Rainsy Party

Senate 61 seats

The membership of the Senate, first established in March
1999, is determined in proportion to the results of the 1998
legislative elections

Cambodia is a constitutional monarchy.

PROFILE

In 1975, a US-installed government
was overthrown by the Maoist Khmer
Rouge under Pol Pot. That extremist
and murderous regime was ousted in
1979 thanks to a Vietnamese invasion.
The Khmer Rouge then joined a
Western-backed anti-Vietnamese
exile coalition with the supporters
of the then Prince Sihanouk and the
Khmer People's National Liberation
Front (KPNLF), gaining UN recognition
against the Vietnam-backed regime
in Phnom Penh. In 1989 Vietnam
withdrew its forces, paving the way
for UN-supervised elections in 1993.
The royalist Funcinpec emerged as
the main winners and King Sihanouk
formed a coalition government of
national reconciliation, but the Khmer
Rouge remained outside this coalition
and resumed armed resistance until
its surrender in 1998. The strife-torn
coalition meanwhile had degenerated
into open hostility in 1997, when Hun
Sen of the communist CPP ousted his
co-prime minister Prince Ranariddh. A
year later the CPP failed to win elections
outright, forcing it to seek a new
coalition agreement with Funcinpec.

MAIN POLITICAL ISSUES
Settling accounts with Khmer Rouge
The Khmer Rouge, which had resumed
its armed struggle in 1993, surrendered
in 1998. Mass defections and the
death of Pol Pot earlier that year had
weakened the group. The legislature
in 2001 approved plans for a tribunal
to try the surviving leaders for crimes
against humanity.

Royalist–CPP rivalry
Power struggles between the
Funcinpec and the CPP came to a head
with CPP leader Hun Sen's mid-1997
coup. However, inconclusive elections
in July 1998 led the two parties to
re-form their uneasy coalition in
November. Against expectations,
that compromise proved a
durable one.

***Prime Minister Hun
Sen,*** *who ousted his
co-prime minister
in 1997.*

***King Norodom
Sihanouk,*** *the pivotal
figure in Cambodian
society and politics.*

WORLD AFFAIRS   ▷ Joined UN in 1955

 ASEAN   IAEA   OIF   Mekong River   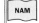 NAM

During the civil war that followed
the Vietnamese invasion of 1979,
the Phnom Penh government (along
with Vietnam) was reduced to an
international pariah. It was recognized
by few countries outside the Soviet bloc,
and its seat at the UN was allotted to
the exiled resistance coalition, despite
one of its components being the Khmer
Rouge, which had inflicted appalling
violence and suffering on Cambodians.

Although the 1993 constitution
aims to make Cambodia a nonaligned
"island of peace," a neutral foreign
policy was difficult to pursue so long
as China continued to advocate the
political rehabilitation of the Khmer
Rouge. Cambodia's relations with
Vietnam also remained problematic,
fueled in part by the historic animosity
between the two countries. The
situation has improved since the late
1990s, and Cambodia's membership
of ASEAN, which had been on hold
pending the consolidation of full
democratic government, was
confirmed in 1999.

AID   ▷ Recipient

💲 $398m (receipts)   ⬆ Up 43% in 2000

Aid is crucial to Cambodia's economy,
providing the bulk of government
revenues. Widespread corruption
and political instability prompted
some countries to withhold assistance
in the late 1990s, but NGOs continued
working in the country, and Western
donors made fresh pledges in 2000.

CAMBODIA

Total Land Area : 181 040 sq. km (69 900 sq. miles)

POPULATION

⊙ over 500 000
○ over 50 000
● over 10 000
• under 10 000

LAND HEIGHT

1000m/3281ft
500m/1640ft
200m/656ft
Sea Level

CHRONOLOGY

A former French protectorate, Cambodia gained independence in 1953 as a constitutional monarchy with Norodom Sihanouk as king.

❏ **1955** Sihanouk abdicates to pursue political career; takes title "prince."

❏ **1970** Right-wing coup led by Prime Minister Lon Nol deposes Sihanouk. Exiled Sihanouk forms Royal Government of National Union of Cambodia (GRUNC), backed by communist Khmer Rouge. Lon Nol proclaims Khmer Republic.

❏ **1975** GRUNC troops capture Phnom Penh. Prince Sihanouk head of state, Khmer Rouge assumes power. Huge numbers die under radical extremist regime.

❏ **1976** Country renamed Democratic Kampuchea. Elections. Sihanouk resigns; GRUNC dissolved. Khieu Samphan head of state; Pol Pot prime minister.

❏ **1978** December, Vietnam invades, supported by Cambodian communists opposed to Pol Pot.

❏ **1979** Vietnamese capture Phnom Penh. Khmer Rouge ousted by Kampuchean People's Revolutionary Party (KPRP), led by Pen Sovan. Khmer Rouge starts guerrilla war. Pol Pot held responsible for genocide and sentenced to death in absentia.

❏ **1982** Government-in-exile including Khmer Rouge and Khmer People's National Liberation Front, headed by Prince Sihanouk, is recognized by UN.

❏ **1989** Withdrawal of Vietnamese troops.

❏ **1990** UN Security Council approves plan for UN-monitored cease-fire and elections.

❏ **1991** Signing of Paris peace accords. Sihanouk reinstated as head of state of Cambodia.

❏ **1993** UN-supervised elections won by royalist Funcinpec. Sihanouk takes title of "king."

❏ **1994** Khmer Rouge refuses to join peace process.

❏ **1995** Former finance minister Sam Rainsy forms opposition party.

❏ **1996** Leading Khmer Rouge member Ieng Sary defects.

❏ **1997** Joint prime minister Hun Sen mounts coup against royalist co-premier Prince Ranariddh.

❏ **1998** April, death of Pol Pot; June, Khmer Rouge surrender; July, parliamentary elections; November, Hun Sen heads coalition government including Funcinpec.

❏ **1999** Cambodia admitted to ASEAN.

❏ **2001** Law approved on trials of Khmer Rouge leaders for atrocities committed by regime.

DEFENSE

 ▷ No compulsory military service

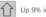 $192m   ⬆ Up 9% in 2000

CAMBODIAN ARMED FORCES

150 main battle tanks (100 T-54/55, 50 PRC Type-59)	90,000 personnel	
4 patrol boats	3000 personnel	
24 combat aircraft (MiG-21)	2000 personnel	
None		

The coalition government's initial defense priority was to unify the command structures of armies under the control of diverse parties. The surrender of Khmer Rouge forces in mid-1998 and the disintegration of remaining pockets of Khmer resistance later that year improved the prospects for a unified national army. Plans for the demobilization of 30,000 soldiers over a three-year period began in May 2000.

Under the nominal overall structure of the Royal Cambodian Armed Forces, there remain in existence three main armies – the CPP's Cambodian People's Armed Forces, Funcinpec's Armée Nationale Sihanoukiste, and the KPNLF's Khmer People's National Liberation Armed Forces. The rivalries between them remain intense, with the two first-named in open conflict as recently as 1997–1998. Although well equipped, their soldiers are poorly paid.

A system of conscription for five years between the ages of 18 and 35 has not been implemented since 1993.

ECONOMICS

▷ Inflation 25% p.a. (1990–2000)

 $3.15bn   3835 riels

SCORE CARD

❏ WORLD GNP RANKING.........................130th
❏ GNP PER CAPITA $260
❏ BALANCE OF PAYMENTS.....................–$19m
❏ INFLATION ....–0.8%
❏ UNEMPLOYMENT 3%

EXPORTS

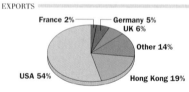

France 2% — Germany 5% — UK 6% — Other 14% — USA 54% — Hong Kong 19%

IMPORTS

Singapore 7% — China 8% — Other 39% — Taiwan 12% — Thailand 16% — Hong Kong 18%

STRENGTHS

Currently very few, as economy still recovering from long-running conflicts. Considerable future potential. Growth in tourism. Relatively unbureaucratic mentality. Self-sufficiency in rice achieved by 1999. Gems, especially sapphires. Possible offshore oil wealth. Export-oriented garment industry.

WEAKNESSES

Tiny tax base makes economic reform hard to implement. Dependence on overseas aid; corruption at most levels of government limits its effectiveness. Disputes over land ownership rights.

ECONOMIC PERFORMANCE INDICATOR

— Consumer Price Index   GDP ▇

(Consumer price index 1995=100; GDP 1996=100; years 1996–2000)

PROFILE

The economy was devastated during the Pol Pot years. The Vietnamese attempted some reconstruction based on central planning, then switched to encouraging the private sector. Investment in the 1990s was heavily aid-dependent. The Asian financial crisis and internal turmoil affected funding after 1997.

CAMBODIA : MAJOR BUSINESSES

Bătdâmbâng, Mémót, Ta Khmau, Kâmpôt, Phnom Penh

🐙 Rubber
✳ Textiles
⚗ Fertilizers
⛏ Gold mining

0   200 km
0   200 miles

RESOURCES Electric power 35,000 kw

 284,100 tonnes   Not an oil producer

4.6m ducks, 2.87m cattle, 2.12m pigs, 15.3m chickens   Salt, phosphates, gemstones

Few resources are currently exploited, apart from tropical rainforest timber, particularly teak and rosewood, much of which is felled illegally.

ELECTRICITY GENERATION

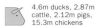

Type	
Hydro 37% (80m kwh)	
Combustion 63% (135m kwh)	
Nuclear 0%	
Other 0%	

% of total generation by type

CAMBODIA : LAND USE

Cropland	
Pasture	
Wetland	
Forest	
Cattle	
Rice	
Rubber – cash crop	

0   200 km
0   200 miles

ENVIRONMENT Sustainability rank: 97th

16%   0.1 tonnes per capita

ENVIRONMENTAL TREATIES

Yes	Yes	Yes	
Yes	No	No	

Deforestation is one of the most serious problems facing Cambodia. Illegal logging is the main culprit. Timber, one of the country's most valuable assets, was sold in huge quantities by all Cambodian factions to finance their war efforts. A moratorium on logging was declared at the end of 1992, but was largely ignored. Despite international pressure and efforts from 2000 to tighten controls, in many parts of the country logging is impossible to police. Tropical hardwoods extracted illegally from Cambodia find lucrative outlets through Thailand in particular. The environmental consequences – topsoil erosion and increased risk of flooding – are enormous and will hold back Cambodia's reconstruction.

MEDIA TV ownership low

Daily newspaper circulation 2 per 1000 people

Phnom Penh has several independent TV stations in addition to the national network. However, many are reliant on party political support, which compromises their independence. The government has used a 1995 press law to prosecute numerous newspapers for defamation and disinformation.

CRIME ▷ No death penalty

5502 prisoners   Civilian crime rates are now fairly stable

CRIME RATES

Statistics for murders, rapes, and thefts are unavailable

There are allegations that Cambodia is becoming Asia's new "narco-state." It is claimed that there has been a proliferation of narcotics trading, money laundering, and illegal banking operations. Mob killings go largely unremarked. Corruption in business is a major issue. Phnom Penh witnessed an increase in violent crime in the aftermath of the 1997 Hun Sen coup, owing to the spread of illegally owned firearms. Until the surrender of the Khmer Rouge in 1998, areas under its command, especially in the west around Pailin and Battambang, were particularly dangerous. Banditry remains rife and policing virtually nonexistent.

EDUCATION ▷ School leaving age: 12

37%   8901 students

THE EDUCATION SYSTEM

Primary	100%
Secondary	22%
Tertiary	1%

% of each age group in education

The government aims to put in place a nine-year period of education. Currently primary education is compulsory, and lasts for six years between the ages of six and 12. Only 5000 of Cambodia's 20,000 teachers survived the Pol Pot period; the Vietnamese-installed government trained or retrained about 40,000.

PUBLISHING AND BROADCAST MEDIA

There are 2 daily newspapers; only 10 newspapers and magazines publish regularly

6 services: 1 state-run, 5 independent

10 services; 2 state-run, 8 independent

HEALTH No welfare state health benefits

1 per 3333 people   Circulatory and infectious diseases, cancers

The Cambodian health system was effectively destroyed in the Pol Pot period; only 50 doctors survived, and Cambodia's health indicators were among the worst in the world.

Conditions have since improved, but AIDS is widespread, affecting even children in rural areas. Infant mortality remains high, and malaria and cholera are endemic. In 2000, UNICEF helped mount an immunization campaign against tetanus, a major cause of neonatal mortality.

SPENDING ▷ GDP/cap. increase

CONSUMPTION AND SPENDING

5 per 1000 population   2 per 1000 population

Defense 6.1%	
Education 5.5%	
Health 0.6%	

Defense, Health, Education spending as % of GDP

New industries such as textiles, in which female garment workers may earn $40 a month in vast workshops, help to attract migrants to the towns, although they risk unemployment and homelessness. Cambodians in rural areas face more severe poverty, exacerbated by land shortage.

WORLD RANKING

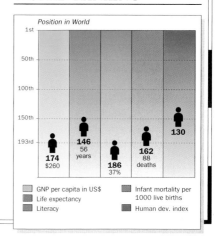

Position in World

174 $260	GNP per capita in US$
146 56 years	Life expectancy
186 37%	Literacy
162 88 deaths	Infant mortality per 1000 live births
130	Human dev. index

CAMEROON

OFFICIAL NAME: Republic of Cameroon **CAPITAL:** Yaoundé
POPULATION: 15.2 million **CURRENCY:** CFA franc **OFFICIAL LANGUAGES:** French and English

LOCATED ON THE CENTRAL west African coast, over half of Cameroon is forested, with equatorial rainforest to the south and evergreen forest and wooded savanna north of the Sanaga River. Most cities are located in the south, although there are densely populated areas around Mount Cameroon, a dormant volcano. For 30 years Cameroon was effectively a one-party state. Democratic elections in 1992 returned the former ruling party to power.

*Savanna landscape below **Mindif Pic** in Cameroon's far north. From here, the land slopes down to the hot, arid Lake Chad basin.*

CLIMATE ▷ Tropical equatorial

WEATHER CHART

Climate varies from the equatorial south, with 500 cm (200 in) of rain a year, to the drought-beset Sahelian north.

TRANSPORTATION ▷ Drive on right

Douala International
439,279 passengers

61 ships
13,600 grt

THE TRANSPORTATION NETWORK

4288 km (2664 miles)		Trans-African Highway
1016 km (631 miles)		2090 km (1299 miles)

Major projects are the east–west Trans-African Highway and the realigning of the Douala–Nkongsamba railroad.

TOURISM ▷ Visitors : Population 1:113

135,000 visitors

Up 32% in 1998

MAIN TOURIST ARRIVALS

Africa 36%	
France 27%	
USA 9%	
Other 28%	

0   10   20   30   40
% of total arrivals

The government and Commonwealth aim to boost visitor numbers, with a yearly target of 500,000. A new airport near Yaoundé will replace the present one. There are beach hotels near Kribi and package tours to northern game parks.

PEOPLE ▷ Pop. density low

Bamileke, Fang, Fulani, French, English

33/km² (85/mi²)

THE URBAN/RURAL POPULATION SPLIT

49%   51%

RELIGIOUS PERSUASION

Roman Catholic 35%
Protestant 18%
Muslim 22%
Traditional beliefs 25%

Cameroon is ethnically diverse – there are 230 groups, no single group being dominant. The largest is the Bamileke of the central southwest region, but this group has never held political power. When President Ahidjo, a northern Fulani, retired, he was replaced by Paul Biya of the southeastern Bulu-Beti group. The north–south enmity which affects many other west African states is also present in Cameroon, albeit diminished by the great diversity of peoples. There are tensions between the French- and English-speaking communities, with sections of the latter demanding independence.

POLITICS ▷ Multiparty elections

2002/2007

President Paul Biya

AT THE LAST ELECTION

National Assembly 180 seats

2% CDU

74% RDPC   12% SDF   12% Others

RDPC = Cameroon People's Democratic Rally **SDF** = Social Democratic Front **CDU** = Cameroon Democratic Union

A Senate is to be created under the 1995 constitution.

Incumbent president Paul Biya's RDPC narrowly won control of the new parliament in multiparty elections in 1992, which were boycotted by the main opposition SDF. It has held on to power in the face of SDF claims of corruption, steadily increasing its majority in 1997 and again in 2002. Similarly, Biya's own reelections in 1992 and 1997 were condemned as the products of fraud and intimidation by SDF candidate John Fru Ndi.

CAMEROON

Total Land Area : 475 400 sq. km (183 567 sq. miles)

POPULATION

over 1 000 000	▣
over 500 000	◉
over 100 000	◎
over 50 000	○
over 10 000	●
under 10 000	•

LAND HEIGHT

2000m/6562ft	
1000m/3281ft	
500m/1640ft	
200m/656ft	
Sea Level	

WORLD AFFAIRS

 Joined UN in 1960

Cameroon's most important relationship continues to be with France, although the country has attempted to diversify its international links, joining the Commonwealth in 1995. A territorial dispute with Nigeria of long duration concerns sovereignty over the oil-rich Bakassi peninsula, where there were clashes in 1996 and 1998.

AID

 Recipient

 $380m (receipts)   Down 12% in 2000

France is by far the most important donor, even having twice paid up Cameroon's back debts to the IMF to prevent its being blacklisted. Lack of funding has forced many development projects to be abandoned. Despite poor economic performance, relations with the IMF are improving.

DEFENSE

 No compulsory military service

 $154m   No change in 2000

The 11,500-strong army has been active in supporting the regime and maintaining order in the face of prodemocratic protests since before independence. Military equipment and training comes mainly from France. There is also a 9000-strong paramilitary gendarmerie.

ECONOMICS

Inflation 5.1% p.a. (1990–2000)

$8.64bn   698.7–736.7 CFA francs

SCORE CARD

❏ WORLD GNP RANKING	90th
❏ GNP PER CAPITA	$580
❏ BALANCE OF PAYMENTS	$–153m
❏ INFLATION	5.3%
❏ UNEMPLOYMENT	30%

STRENGTHS

French and US companies exploit moderate oil reserves. Very diversified agriculture includes timber, cocoa, bananas, and coffee. Self-sufficiency in food. Strong informal sector. Private sector in relatively good state. Electricity is 95% HEP.

WEAKNESSES

Massive fuel smuggling from Nigeria affects refinery profits. Inflated civil service. Widespread corruption.

EXPORTS

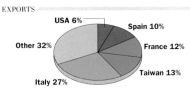

USA 6%
Spain 10%
Other 32%
France 12%
Taiwan 13%
Italy 27%

IMPORTS

Italy 4%
Belgium 7%
USA 4%
Nigeria 13%
Other 39%
France 33%

RESOURCES

 Electric power 627,000 kw

 95,067 tonnes   80,000 b/d (reserves 400m barrels)

 5.9m cattle, 4.4m goats, 3.8m sheep, 30m chickens   Oil, coal, tin, natural gas, bauxite, iron, uranium, gold

New oil discoveries may be able to bolster declining extraction rates. In spite of large bauxite deposits, much is imported for the Edea smelter, which takes 50% of national electricity output.

ENVIRONMENT

Sustainability rank: 93rd

 5% (2% partially protected)   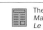 0.1 tonnes per capita

The rate of commercial logging and the planned Cameroon–Chad oil pipeline constitute major threats to Cameroon's environment.

MEDIA

TV ownership low

 Daily newspaper circulation 0.5 per 1000 people

PUBLISHING AND BROADCAST MEDIA

There are 3 daily newspapers, *Politiks Matinal*, *Le Tribune du Cameroun*, and *Le Quotidien*

1 state-owned service   1 state-owned service

There are frequent allegations of censorship and violence against journalists. English-language media are generally more outspoken.

CRIME

 Death penalty used

 15,903 prisoners   Up 238% 1996–1998

Armed robbery and burglary in Douala and Yaoundé are rising fast. The police are known to use torture.

EDUCATION

School leaving age: 12

 76%   66,902 students

The French-speaking majority has failed in its attempt to take over the bilingual system. Cameroon has a high literacy rate compared with much of the rest of Africa.

CHRONOLOGY

One of the great trading emporia of west Africa, Cameroon was divided between the French and British in 1919, after 30 years of German rule.

- ❏ **1955** Revolt; French kill 10,000.
- ❏ **1960** French sector independent.
- ❏ **1961** British south joins Cameroon (north joins Nigeria).
- ❏ **1982** Ahidjo dies; Biya president.
- ❏ **1983–1984** Coup attempts. Heavy casualties; 50 plotters executed.
- ❏ **1990** Demonstrations and strikes; declaration of multiparty state.
- ❏ **1992** Multiparty elections.
- ❏ **1997** President and ruling RDPC returned in disputed elections.
- ❏ **2000** World Bank funds for pipeline project, despite environmental fears.
- ❏ **2001** Over 80% of indigenous forests allocated for logging.
- ❏ **2002** RDPC increases its majority.

HEALTH

No welfare state health benefits

 1 per 10,000 people   Malaria, diarrheal and respiratory diseases

A sharp fall in government health provision means that more people are using the private health sector or traditional practitioners.

SPENDING

GDP/cap. increase

CONSUMPTION AND SPENDING

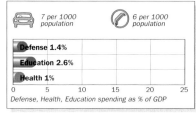

7 per 1000 population   6 per 1000 population

Defense 1.4%
Education 2.6%
Health 1%

Defense, Health, Education spending as % of GDP

Wealth is unevenly distributed and has been declining since the end of the oil boom. There is still a very wealthy, albeit small, sector of the population.

WORLD RANKING

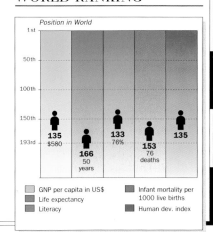

Position in World

135	$580
166	50 years
133	76%
153	76 deaths
135	

GNP per capita in US$
Life expectancy
Literacy
Infant mortality per 1000 live births
Human dev. index

C

CANADA

OFFICIAL NAME: Canada **CAPITAL:** Ottawa **POPULATION:** 31 million
CURRENCY: Canadian dollar **OFFICIAL LANGUAGES:** English and French

NORTH AMERICA

 1867   1949   July 1   CDN   -3.5 to -8   +1 .ca

C ANADA IS THE WORLD'S THIRD-LARGEST country, stretching north to Cape Columbia on Ellesmere Island, south to Lake Erie, and across six time zones from Newfoundland to the Pacific seaboard. The interior lowlands around Hudson Bay make up 80% of Canada's land area and include the vast Canadian Shield, with the plains of Saskatchewan and Manitoba and the Rocky Mountains to the west. The St. Lawrence, Yukon, Mackenzie, and Fraser Rivers are among the world's 40 largest. The Great Lakes–St. Lawrence River lowlands are the most populous areas. An Inuit homeland, Nunavut, formerly the eastern part of the Northwest Territories, was created in 1999, covering nearly a quarter of Canada's land area. French-speaking Québec's relationship with the rest of the country causes recurring constitutional arguments.

CANADA

Total Land Area : 9 976 140 sq. km (3 851 788 sq. miles)

POPULATION

- ▣ over 1 000 000
- ◉ over 500 000
- ◎ over 100 000
- ○ over 50 000
- ● over 10 000
- • under 10 000

LAND HEIGHT

- 3000m/9843ft
- 2000m/6562ft
- 1000m/3281ft
- 500m/1640ft
- 200m/656ft
- Sea Level

CLIMATE ▷ Continental/subarctic/ mountain

WEATHER CHART FOR OTTAWA

Canada's climate ranges from polar and subpolar in the north, to cool in the south. Summers in the interior are hotter, and winters colder and longer than on the coast, with temperatures well below freezing and deep snow. The Pacific coast around Vancouver has the warmest winters, where temperatures rarely fall below zero.

TRANSPORTATION ▷ Drive on right

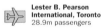 Lester B. Pearson International, Toronto
28.9m passengers

 861 ships
2.66m grt

THE TRANSPORTATION NETWORK

318,371 km (197,826 miles)		16,571 km (10,297 miles)	
14,000 km (8700 miles)		3769 km (2342 miles)	

The emergence of a national economy depended on the development of an efficient system of transportation. The Trans-Canada Highway and two transcontinental rail systems are the east–west backbones of the road and rail networks, which also reach into the far north. The Great Lakes–St. Lawrence Seaway system's cheap transportation helped Ontario and Québec dominate the economy for most of the 20th century. Air services were hit by the industry's crisis after 11 September 2001.

TOURISM

▷ Visitors : Population 1:1.6

🧳 19.7m visitors   ⬍ Little change in 2001

MAIN TOURIST ARRIVALS

USA	78%
UK	4%
Japan	3%
France	2%
Germany	2%
Other	11%

% of total arrivals
(scale: 0 10 20 30 40 50 60 70 80)

Most tourist visitors come from the US, often on short tours. Efforts to attract European visitors are based around campaigns emphasizing Canada's unpolluted natural beauty. Bizarrely, the fictional home of the eponymous heroine of *Anne of Green Gables* on Prince Edward Island is a magnet for tourists from Japan, where the novels about her by E. M. Montgomery enjoy enormous popularity.

PEOPLE

▷ Pop. density low

👤 English, French, Chinese, Italian, German, Ukrainian, Portuguese, Inuktitut, Cree

👥 3/km²
9/mi²

THE URBAN/RURAL POPULATION SPLIT

77%   23%

RELIGIOUS PERSUASION

- Nonreligious 12%
- Roman Catholic 47%
- Protestant 41%

ETHNIC MAKEUP

- Indigenous Amerindian and Inuit 4%
- British origin 44%
- French origin 25%
- Other European 20%
- Other 7%

Relations between French-speaking Québécois and the English-speaking majority in Canada have been the dominant ethnic issue of the past 40 years. Support for separatist parties increased mainly because of the failure of Canada's other provinces to deal with Québec's demand to be recognized as a "distinct society," with powers to preserve its culture and language from further anglicization. Québec's still controversial 1977 language law made French the province's official language. Two-thirds of Canada's population live in the 5% of its land area taken up by the Great Lakes–St. Lawrence lowlands. However, Canada's

*A dude ranch in British Columbia.
Many tourists are attracted by Canada's wide choice of outdoor pursuits.*

POPULATION AGE BREAKDOWN

Female	Age	Male
1.8%	80+	1%
7.2%	60–79	6.2%
12.6%	40–59	12.6%
15.8%	20–39	16.1%
13%	0–19	13.7%

% of population by age group

ethnic mix has changed significantly since the 1970s, due to a move from a restrictive immigration policy to one which welcomes those with money or skills. Significant numbers of Asians have moved to Canada. The government promotes a policy which encourages each group to maintain its own culture. Canada is now officially a "community of communities."

The largest element of the indigenous population is the 800,000 people of native Amerindian descent, known in Canada as First Nations. There are also 213,000 Métis (French–Amerindians) and an Inuit population of some 50,000 in the north. In 1992 the Inuit successfully settled their long-standing land claim, and in 1999 the Nunavut area, with only 25,000 mainly Inuit inhabitants, gained the status of a territory, the first part of Canada to be governed by indigenous Canadians in modern history. A Supreme Court land rights ruling in 1997, establishing the principle of "aboriginal title," opened the way for the return of ancestral lands claimed by native Amerindian nations, and in 1998 the federal government formally apologized for their past mistreatment.

Canada has a long tradition of state welfare more akin to Scandinavia than the US. Unemployment provision and health care, supported by high taxes, are still generous, despite recent cutbacks. The government has sought to end inequalities. Measures include the "pay-equity" laws, which aim to specify pay rates for jobs done mainly by women – such as receptionists – equivalent to similar skill jobs for men. Women are well represented at most levels of business and government.

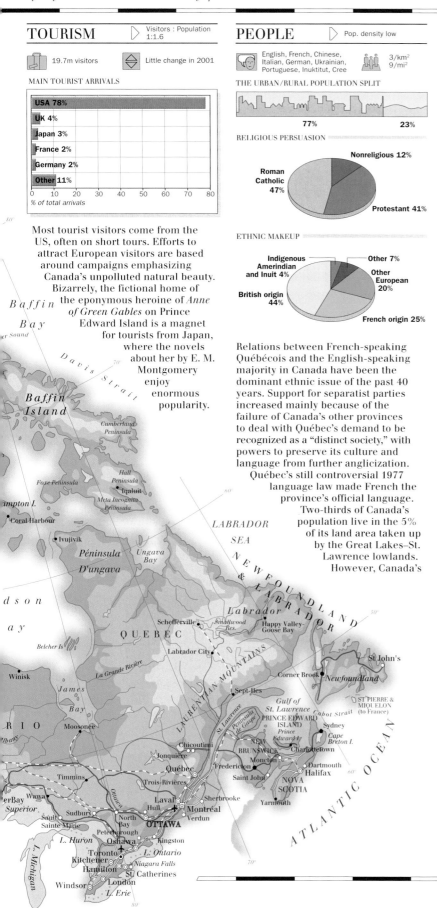

C

C

CHRONOLOGY

Peopled for centuries by indigenous Inuits and Amerindians, Canada began to experience extensive European settlement following the landing of the English expedition led by John Cabot in 1497 and the French landing of Jacques Cartier in 1534.

❏ **1754** British fight French and Indian War. France forced to relinquish St. Lawrence and Québec settlements to Britain.
❏ **1774** Act of Québec recognizes Roman Catholicism, French language, culture, and traditions.
❏ **1775–1783** American War of Independence. Canada becomes refuge for loyalists to British crown.
❏ **1867** Federation of Canada created under British North America Act.
❏ **1885** Transcontinental railroad completed.
❏ **1897** Klondike gold rush begins.
❏ **1914–1918, 1939–1945** Canada supports Allies in both world wars.
❏ **1931** Autonomy within Commonwealth.
❏ **1949** Founder member of NATO. Newfoundland joins Federation.
❏ **1968** Liberal Party under Pierre Trudeau in power. Separatist Parti Québécois (PQ) formed.
❏ **1970s** Québec secessionist movement grows, accompanied by terrorist attacks.
❏ **1976** In Québec, PQ wins elections.
❏ **1977** French made official language.
❏ **1980** Separation of Québec rejected at referendum. Trudeau prime minister again.
❏ **1982** UK transfers all powers relating to Canada in British law.
❏ **1984** Trudeau resigns. Elections won by PCP. Brian Mulroney prime minister until 1993.
❏ **1987** Meech Lake Accord.
❏ **1989** Canadian–US Free Trade Agreement.
❏ **1992** Charlottetown Agreement on provincial–federal issues rejected at referendum. Canada, Mexico, and US finalize terms for NAFTA.
❏ **1993** Crushing election defeat of PCP, rise of regional parties.
❏ **1994** PQ regains power in Québec. NAFTA takes effect.
❏ **1995** Narrow "no" vote in second Québec sovereignty referendum.
❏ **1995** Fishing dispute with EU.
❏ **1997** Regionalism dominates federal election; Liberals retain power based on support in Ontario.
❏ **1998** PQ only narrowly holds power in Québec.
❏ **2000** November, early elections. Liberals retain power.
❏ **2001** December, Smart Border Declaration and antiterrorist legislation.

POLITICS

 Multiparty elections

 L. House 2000/2005   H.M. Queen Elizabeth II

AT THE LAST ELECTION

House of Commons 301 seats

13% BQ
4% PCP
57% LP
22% CA
4% NDP

LP = Liberal Party **CA** = Canadian Reform Conservative Alliance **BQ** = Bloc Québécois **NDP** = New Democratic Party **PCP** = Progressive Conservative Party **Ind** = Independents

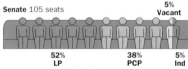

Senate 105 seats

5% Vacant
52% LP
38% PCP
5% Ind

Senators are appointed for life by the governor-general in Council, to a usual maximum of 105; the prime minister may appoint an extra 8 senators.

Canada is a federal multiparty democracy.

PROFILE

Canadian politics was traditionally dominated by two main parties, the PCP and LP, which had few major ideological differences. The third party, the NDP (which has never held federal power), advocated greater government intervention than the other two.

The 1990s brought a major political shift – the eclipse of the PCP, three successive victories for Jean Chrétien and the LP (1993, 1997, and 2000), and an eventual realignment on the right, based on the emergence of the populist Reform Party (RP) in the western provinces. The trend away from Canada-wide politics, toward parties representing strong regional interests, left the LP with few seats outside its strongholds in the east. The Bloc Québécois (BQ), espousing the separatist cause at federal level, was the second-largest party in the federal parliament between 1994 and 1997. Since then the role of official opposition has been held by the RP and then by its successor, the conservative Canadian Alliance, which was riven in 2001–2002 by a fractious leadership dispute.

MAIN POLITICAL ISSUES
The unity of the state

Opposition to federal government is not confined to Québec – the 1997 and 2000 federal elections confirmed support for greater autonomy for Canada's western provinces – but Canada has agonized over separatist tendencies in francophone Québec almost since the foundation of the state. Québec did not take part in the 1997 Calgary conference, where a Canadian unity framework was agreed by the other provinces, together with recognition of Québec's "unique character." A series of earlier proposals, to

The Niagara Falls are situated between Lakes Erie and Ontario on the Canada–US border. Horseshoe Falls, in Canada, are 49 m (160 ft.) high and 790 m (2591 ft.) across.

recognize Québec as a distinct society and strengthen the powers of all the federal provinces, had failed to gain ratification or had been rejected by the electorate. The Parti Québécois (PQ), back in power at provincial level since 1994, advocates another referendum on separatism, despite losing those held in 1980 and 1995. The Supreme Court has ruled, however, that secession would require federal approval and the agreement of at least seven of the ten provinces. The April 2000 Clarity Act set strict criteria for any pro-secession referendum result to be valid.

North American integration

The North American Free Trade Agreement (NAFTA), a hotly debated issue in Canada when it was being negotiated in the early 1990s, has produced a trade boom, especially for Ontario. However, Canadians have problems competing for foreign investment with Mexico, where labor costs, social welfare, and environmental standards are lower. Most Canadians oppose such ideas as a currency union and ever closer integration with the US.

Bernard Landry, the outspoken separatist premier of Québec.

Stephen Harper was elected leader of the CA in 2002.

Jean Chrétien, prime minister since 1993.

WORLD AFFAIRS　▷ Joined UN in 1945

Canada's most important relationship is with the USA, its main trading partner. Canada backed the US-led "war on terrorism" after the attacks on US targets in September 2001; security along the immense US–Canadian border was tightened in an agreement three months later. There are tensions, such as over Cuba – Canada opposes US sanctions – and transboundary pollution. Canadians greatly fear damage from oil transported between Alaska and the rest of the US. Canada has protested to the WTO over US duties on its softwood timber, but a fisheries dispute relating to Pacific salmon was settled in 1999.

In the forefront on debt relief for the poorest countries, Canada also led the world campaign against antipersonnel mines, hosting the conference which agreed the 1997 Ottawa treaty.

C

AID　▷ Donor

 US$1.74bn (donations)　 Up 2% in 2000

Canada's aid budget, earmarked for cuts in the 1990s, was given extra funding again at the end of the decade. NGOs supported by the Canadian International Development Agency (CIDA) are prominent on global development issues, and most Canadians support aid.

Aid now aims to provide know-how skills, rather than funding for large-scale development projects. CIDA has pioneered a theme-based approach, stressing human rights, basic needs, gender issues, and good governance. The regional focus of aid has gradually shifted, with less emphasis on Africa. Programs in the 1990s supported recovery and reform in former communist countries. More aid now goes toward the development needs of the Indian subcontinent.

DEFENSE　▷ No compulsory military service

 US$7.46bn　⬇ Down 11% in 2000

CANADIAN ARMED FORCES

🛡	114 main battle tanks (*Leopard* C-1/C-2)	18,600 personnel
🚢	1 submarine, 12 frigates, 4 destroyers, and 14 patrol boats	9000 personnel
✈	140 combat aircraft (122 CF-18)	13,500 personnel
	None	

Canada cooperates closely with the US on North American defense and security issues.

Defense spending was cut after the end of the Cold War, and many Canadians continue to advocate further reductions. Canada withdrew its forces stationed in Europe in 1992. The focus of defense planning is now the creation of rapid reaction forces. Canadian troops have served in many UN peacekeeping operations, most recently in Kosovo, East Timor, and Sierra Leone. Their involvement in Somalia, however, which ended in 1993, was tarnished by a scandal over racism, torture, and murder which shocked Canadian society.

ECONOMICS　▷ Inflation 1.4% p.a. (1990–2000)

📊 US$650bn　💲 1.502–1.596 Canadian dollars

SCORE CARD

❏ WORLD GNP RANKING	8th
❏ GNP PER CAPITA	US$21,130
❏ BALANCE OF PAYMENTS	US$12.7bn
❏ INFLATION	2.7%
❏ UNEMPLOYMENT	7%

EXPORTS

China 1%, UK 1%, Germany 1%, Japan 2%, Other 8%, USA 87%

IMPORTS

Mexico 3%, UK 4%, China 3%, Japan 5%, Other 21%, USA 64%

STRENGTHS

A broad and rich resource base. Provides exports, raw materials for manufacturing sector, and massive cheap energy, notably HEP; also large oil and gas reserves. Agriculture and forestry contribute 3% of GDP, mining 4%. Successful manufacturing sector, contributes 17% of GDP, especially forestry products, transportation equipment, and chemicals. Strong recovery and growth from mid-1990s to 2001. Access to huge US and Mexican markets through NAFTA. Low inflation.

WEAKNESSES

Problems of competitiveness; higher taxes, more regulations, lower productivity relative to NAFTA; other threats from globalization. Vulnerable to price fluctuations for raw material exports. Federal and provincial budget deficits are still high.

CANADA : MAJOR BUSINESSES

✈ Aerospace industry		🔌 Electronics	
🚗 Vehicle manufacture		⚙ Engineering	
🌲 Timber industries		△ Chemicals	
📦 Pulp & paper		△ Metallurgy	
🍴 Food processing		⛽ Oil & gas	
🐟 Fish processing			

ECONOMIC PERFORMANCE INDICATOR

PROFILE

Canada has an enormous resource base, and the UN rates it as having one of the highest standards of living in the world. Since the mid-1980s, however, its manufactured exports have faced increasing competition, while prices for its primary exports fluctuate. Real growth averaged 3.5% a year for most of the 1980s, but then stagnated for five years, while budget deficits rose, forcing restructuring at both federal and provincial levels. Many welfare programs were cut back, while the defense budget was sharply reduced. Growth resumed after 1993 and strengthened before the global economic downturn in 2001. Within NAFTA, Canadian firms have had to become more competitive to maintain exports. Most have been successful, with better productivity and a high-tech shift. Unemployment, at almost 10% in the mid-1990s, was down to just over 6% by 2000. By December 2001, it was back to 8%, but then fell again.

RESOURCES

 Electric power 115m kw

 1.14m tonnes

2.76m b/d (reserves 6.6bn barrels)

 13m cattle, 12.6m pigs, 5m turkeys, 158m chickens

 Coal, oil, gas, gold, zinc, uranium, nickel, potash, asbestos, gypsum

ELECTRICITY GENERATION

Hydro 59% (332bn kwh)	
Combustion 28% (158bn kwh)	
Nuclear 13% (72bn kwh)	
Other 0%	

0   20   40   60   80   100

% of total generation by type

Canada is a country of enormous natural resources. It is the world's largest exporter of forest products and a top exporter of fish, furs, and wheat. Minerals have played a key role in Canada's transformation into an urban–industrial economy. Alberta, British Columbia, Québec, and Saskatchewan are the principal mining regions. Ontario and the Northwest (NWT) and Yukon Territories are also significant producers. Canada is the world's largest producer of uranium, the second-largest of nickel, asbestos, potash, and gypsum, and the third-largest of zinc. Oil and gas are exploited in Alberta, off the Atlantic coast, and in the northwest – huge additional reserves are thought to exist in the high Arctic. Most exports go to the US. Canada is also one of the world's top hydroelectricity producers.

CANADA : LAND USE

Cropland	
Forest	
Pasture	
Tundra	
High mountain regions	
Wheat	
Cattle	

0   1000 km
0   1000 miles

ENVIRONMENT

 Sustainability rank: 4th

 10% (2% partially protected)

15.4 tonnes per capita

ENVIRONMENTAL TREATIES

Yes	Yes	Yes
Yes	Yes	No

With a population of only some 31 million living in the world's third-largest country, Canada is justly renowned for vast tracts of wilderness untroubled by pollution either from industry or from intensive farming methods. A major conservation issue is the battle to stop the logging of virgin forest in northern Ontario and on the west coast. Notable successes were achieved in the late 1990s, pressuring timber companies to adopt more sustainable policies, and a landmark agreement in early 2001 promised protection for British Columbia's coastal Great Bear Rainforest.

Canadians have tighter pollution controls than the neighboring US. Ontario, the most polluted province, has imposed stricter limits on oil refineries and (from 2001) on electricity-generating plants. Carbon dioxide emissions (mainly from cars) are among the highest in the world per capita. Canada has accepted a target of a 6% cut by 2010. Production of hazardous waste is also higher than the European average.

MEDIA

 TV ownership high

Daily newspaper circulation 158 per 1000 people

PUBLISHING AND BROADCAST MEDIA

	There are 101 daily newspapers; over 80 daily or weekly publications appear in 20 languages, catering to immigrant groups
	1 publicly owned service, many independent services
	1 publicly owned service, many independent services

The public Canadian Broadcasting Corporation (CBC) runs two national TV channels, in English and French; the Canadian Alliance wants it partially privatized. Local cable services often include multilingual or ethnic channels. Canadian TV is renowned for its news and sports coverage. *La Presse* is the leading French-language daily and the *Globe and Mail* the leading serious newspaper in English.

Autumn in the tundra in northern Canada. Trees such as the black spruce are subject to the effects of acid rain originating in the USA's northern industrial regions.

CRIME

 No death penalty

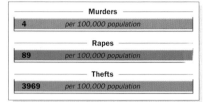 31,600 prisoners

Down 4% in 1999

CRIME RATES

Murders	
4	per 100,000 population

Rapes	
89	per 100,000 population

Thefts	
3969	per 100,000 population

Rates for serious crime are lower in Canada than in the US. Canadians ascribe this to their far stricter gun control laws, which were further tightened in the 1990s.

Newfoundland police began carrying guns routinely only in 1998, the last force in North America to do so. There have been careful efforts to maintain the inner cities as crime-free zones.

To address the narcotics problem, innovative "drug treatment courts" in Toronto and Vancouver link criminal justice with treatment and social care.

EDUCATION

 School leaving age: 16

99%

1.56m students

THE EDUCATION SYSTEM

% of each age group in education

Primary 100%   Secondary 100%   Tertiary 58%

Education is a responsibility of the individual provinces, rather than the federal government, and is accorded a very high priority. The period of free compulsory school attendance varies, but is a minimum of nine years.

The prime medium of instruction is English in all provinces except francophone Québec. In several other provinces, French-speaking students are entitled to be taught in French. Multicultural education also helps maintain the cultural identity of immigrant groups.

Canada has 76 universities and some 200 other higher education institutions. Nearly all high school graduates go on to some form of tertiary or further education. This represents the highest proportion in the industrialized world.

QUÉBEC'S DISTINCT SOCIETY AND SEPARATISM

QUÉBEC IS CANADA'S largest province, with an area of 1,667,926 sq. km (594,860 sq. miles). Its population, 7.3 million at the 1994 census, includes more than five million of Canada's 6.1 million French Canadians, most of whom are Catholics. Its capital is Québec City. The leading commercial center, Montréal, has suffered a decline in prestige compared with Toronto, in neighboring Ontario, which is overwhelmingly English-speaking. Québec has massive hydroelectric power resources and vast areas of forest, and its principal industries include timber, pulp, and paper, and mining, particularly for iron ore. Consequently, environmental protection legislation is relatively lax, and there is conflict with those seeking to protect First Nation lands from devastation by logging and massive dam schemes.

French shop signs above the streets of Québec City.

Conquered by the British in the 18th century, the Québécois retained the French civil code under the 1774 Québec Act, but French Canadians only gradually recovered minority language rights suppressed after an unsuccessful rebellion in the 1830s. A "quiet revolution" began in the 1960s, based initially on militant trade unionism. This brought far-reaching changes in the social, economic, and political balance in Québec. The wage gap closed, and francophones now slightly out-earn anglophones, partly because many of the best-educated anglophones tend to leave the province. Francophone-owned businesses were built up through the "Québec Inc." project. Anglophone dominance in government and the civil service was reversed, and higher education opportunities expanded for francophones. Francophone militancy was also channeled into party politics, fueled by a harsh security clampdown in the early 1970s against the guerrilla

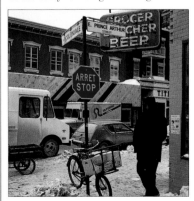

Bilingual street signs in Québec have now largely been replaced by French-only signs.

Québec Liberation Front. The Parti Québécois (PQ) unexpectedly gained control of the provincial assembly in 1976, and a French language charter, Bill 101, was enacted almost immediately. French became not only the official language but compulsory in government and on public signs. Non-francophones protested, and the Supreme Court ruled in 1988 that part of the language charter violated their human rights, but the provincial government managed to enforce these rules on a "temporary" basis.

The PQ's other initiative was sovereignty for Québec. After its defeat in a referendum in 1980, the idea was revived in the 1990s amid long-running disputes about the Canadian constitution and provincial–federal relationships. The PQ, promising a second referendum (and backed by its newly formed counterpart at federal level, the BQ), returned to power in Québec in 1994. The following year, 60% of Québec's francophone majority voted in favor of sovereignty in association with Canada, a high-water mark for separatism; the proposition was defeated by the narrowest of margins by the non-francophone vote. Although the PQ said that it would seek a third referendum, federal legislation limits the conditions under which it can do so, and the party's commitment to separatism has lost much of its impetus, with disappointing showings in the 1998 provincial and 2000 federal elections. Even francophones, still fiercely protective of their language and culture, appear increasingly disposed to settle for a federal compromise acknowledging Québec's "unique character."

HEALTH

 Welfare state health benefits

1 per 476 people

Heart and respiratory diseases, cancers, accidents

The comprehensive state health service is funded from national insurance.

Rising costs are the result of an aging population and the spread of more sophisticated and expensive treatments. Health care was the main issue of the 2000 election campaign. Popular backing for retaining the present publicly funded system has encouraged the LP government to restore spending to earlier levels, after a period of cuts made in an effort to reduce the budget deficit. About 25% of Canadians use private health facilities.

SPENDING

 GDP/cap. increase

CONSUMPTION AND SPENDING

459 per 1000 population

677 per 1000 population

Defense 1.2%

Education 5.6%

Health 6.6%

Defense, Health, Education spending as % of GDP

Despite strains caused by recession during the early 1990s – including a rise in unemployment to over 10% – life for most Canadians remains very good.

The UN ranks Canada as one of the best countries in the world in which to live. In its overall assessment of human development indicators such as income, education, and life expectancy, Canada consistently comes out near the top, and ahead of the USA.

However, disadvantaged groups do exist, in particular among indigenous Canadians. Unemployment, poor housing, and mortality rates for Amerindians and Inuits are well above those for other Canadians; the Inuit suicide rate is three times higher. Those Amerindians who live on reserves are the poorest group.

WORLD RANKING

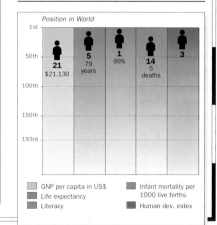

Position in World

21 $21,130	5 79 years	1 99%	14 5 deaths	3

GNP per capita in US$

Life expectancy

Literacy

Infant mortality per 1000 live births

Human dev. index

CAPE VERDE

OFFICIAL NAME: Republic of Cape Verde   **CAPITAL:** Praia
POPULATION: 437,000   **CURRENCY:** Cape Verde escudo   **OFFICIAL LANGUAGE:** Portuguese

C

The Cape Verde archipelago off the west coast of Africa became independent of Portugal in 1975. Most of the islands are mountainous and volcanic; the low-lying islands of Sal, Boa Vista, and Maio have agricultural potential, though they are prone to debilitating droughts. Around 50% of the population live on São Tiago. Following a period of single-party socialist rule, Cape Verde held its first multiparty elections in 1991.

CLIMATE ▷ Tropical oceanic

WEATHER CHART FOR PRAIA

■ Average daily temperature   Rainfall ■
°C/°F   J F M A M J J A S O N D   cm/in
40/104 .... 40/16
30/86 .... 30/12
20/68 .... 20/8
10/50 .... 10/4
0/32 0
-10/14
-20/-4

Cape Verde has a very dry climate, subject to droughts that sometimes last for years at a time.

TRANSPORTATION ▷ Drive on right

✈ **Amilcar Cabral, Sal Island**   🚢 42 ships
635,563 passengers   20,800 grt

THE TRANSPORTATION NETWORK

🛣 858 km (533 miles)	🛤 None
🚂 None	⚓ None

Cape Verde has a strategic position on international sea and air routes, which it is beginning to exploit.

TOURISM ▷ Visitors : Population 1:5.9

🧳 74,000 visitors   ⬆ Up 68% in 2000

MAIN TOURIST ARRIVALS

Portugal 31%
Italy 31%
Germany 12%
Other 26%

0   10   20   30   40   50   60
% of total arrivals

Tourism has not been a government priority and remains on a modest scale. The islands of São Tiago, Santo Antão, Fogo, and Brava have tourist potential, offering a combination of mountain scenery and extensive beaches.

PEOPLE ▷ Pop. density medium

Portuguese Creole, Portuguese   108/m² (281/mi²)

THE URBAN/RURAL POPULATION SPLIT

62%   38%

ETHNIC MAKEUP

Other 10%
African 30%
Mestiço 60%

The majority of the population is Portuguese–African *mestiço*; the remainder is largely African, descended either from slaves or from more recent immigrants from the mainland. The Creolization of the culture has led to a relative lack of ethnic tension, though there is some bad feeling between islands. African traditions of the extended family, as well as the Roman Catholic Church, have helped to ensure the vitality of family life. Women outnumber men in Cape Verde, with many single mothers acting as head of the household and main breadwinner.

POLITICS ▷ Multiparty elections

🗳 2001/2006   👤 President Pedro Pires

AT THE LAST ELECTION
National Assembly 72 seats

55% **PAICV**   42% **MPD**   3% **ADM**

PAICV = African Party for the Independence of Cape Verde
MPD = Movement for Democracy
ADM = Democratic Alliance for Change

Cape Verde experienced a peaceful transition to multipartyism in 1991, when elections brought the MPD to power. Although there had previously been a decade of single-party rule under the PAICV, it had in fact operated a liberal system in which opposition and dissent were tolerated. The large Cape Verdean diaspora had an important influence in effecting the transition to multiparty politics.

The MPD was defeated in legislative elections in January 2001, when the PAICV was returned to power with an absolute majority. Pedro Pires of the PAICV was elected president the following month, beating his MPD rival by just 17 votes. The main issue for the current government is that of economic survival, particularly during periods of drought.

WORLD AFFAIRS ▷ Joined UN in 1975

CPLP   ECOWAS   OIF   NAM   AU

Cape Verde wishes to diversify its international contacts in order to secure aid, while maintaining good relations with Portugal, the former colonial power. Within the region, Cape Verde seeks to restore normal relations with Guinea-Bissau, having withdrawn from a proposed union in 1980, and to improve contacts with other mainland states, such as Senegal.

CAPE VERDE

Total Land Area : 4033 sq. km (1557 sq. miles)

LAND HEIGHT

POPULATION
over 50 000   ○
over 10 000   ●
under 10 000   ·

2000m/6562ft
1000m/3281ft
500m/1640ft
200m/656ft
Sea Level

AID

 ▷ Recipient

 $94m (receipts)   ⬇ Down 31% in 2000

Portugal is the main source of aid. Other major donors include the World Bank, Germany, Japan, and the Netherlands. An important donor is the EU, which has provided substantial food aid in the wake of recent droughts, as well as funding aid programs. Aid finances almost all development in Cape Verde, which is one of the least industrialized countries in the world.

DEFENSE

 ▷ Compulsory military service

 $7m   ⬍ No change in 2000

After independence, small armed forces were established, now consisting of a 1000-strong army, a small air force, and a naval coastguard. They have never been called upon to play a political role; their main duties are to protect territorial waters against illegal fishing and to curb smuggling.

ECONOMICS

 ▷ Inflation 4.9% p.a. (1990–2000)

 $588m   118.2–119.8 Cape Verde escudos

SCORE CARD

❏ WORLD GNP RANKING	167th
❏ GNP PER CAPITA	$1330
❏ BALANCE OF PAYMENTS	–$67m
❏ INFLATION	4.4%
❏ UNEMPLOYMENT	24%

STRENGTHS

Strategic location, off the westernmost tip of Africa, close to the mid-Atlantic where Africa is nearest to Latin America. This has military and economic advantages, including shipping maintenance and air travel. Low debt-servicing costs.

WEAKNESSES

Permanent threat of drought and water supply problems, despite desalination plants. Lack of agricultural land and dependency on food aid. Difficulties of communications between islands.

EXPORTS

Other 9%
USA 9%
Portugal 82%

IMPORTS

France 5%
Japan 5%
USA 5%
Netherlands 6%
Portugal 48%
Other 31%

Portuguese colonial-style architecture on Fogo, one of the larger islands. The volcano in its center is the highest point in Cape Verde.

RESOURCES

 ▷ Electric power 7000 kw

 10,371 tonnes   Not an oil producer

200,000 pigs, 110,000 goats, 480,000 chickens   Salt, pozzolana

Cape Verde has no known strategic resources. With no oil or gas and no possibility of hydroelectric power, it depends on imported petroleum for energy. However, experimental projects have been carried out to investigate the potential of wave power, windmills, and biogas.

ENVIRONMENT

 ▷ Not available

 None   0.3 tonnes per capita

Cape Verde has recently suffered several years of persistent drought, which have affected food production and reduced livestock herds. It is a very active member of CILSS, which struggles against drought in the Sahel region. Environmental initiatives include reforestation, soil conservation, and a water resources program.

MEDIA

 ▷ TV ownership low

☰ There are no daily newspapers

PUBLISHING AND BROADCAST MEDIA

There are no daily newspapers. Independent publications suffer from financial pressures	
1 state-controlled service	1 state-controlled service

The government publishes three weeklies, but there are no daily newspapers. Press freedom is guaranteed by law. TV and radio broadcasting are in Portuguese and Creole, with the cooperation of the Portuguese service RTPI.

CRIME

 ▷ No death penalty

 775 prisoners   ⬍ Little change from year to year

Crime is not a serious problem, even in urban centers, though smuggling is fairly widespread.

EDUCATION

▷ School leaving age: 13

74%   Not available

At independence, education became a priority; 80% of children now attend elementary school, and more than 60% go on to secondary education.

HEALTH

▷ No welfare state health benefits

1 per 5882 people   Heart disease, tuberculosis, typhoid, and accidents

Health care has improved since the colonial period, and polio has been virtually eradicated.

SPENDING

▷ GDP/cap. increase

CONSUMPTION AND SPENDING

31 per 1000 population   126 per 1000 population

Defense 2.7%
Education 4.2%
Health 1.8%

Defense, Health, Education spending as % of GDP

In comparison with the 90% of the population engaged in primary production, the small business class in Praia is well off.

WORLD RANKING

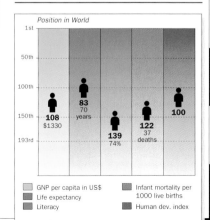

Position in World

108 $1330
83 70 years
139 74%
122 37 deaths
100

GNP per capita in US$ — Infant mortality per 1000 live births
Life expectancy
Literacy — Human dev. index

CENTRAL AFRICAN REPUBLIC

CENTRAL AFRICA • Africa

OFFICIAL NAME: Central African Republic **CAPITAL:** Bangui
POPULATION: 3.8 million **CURRENCY:** CFA franc **OFFICIAL LANGUAGE:** French

C

LANDLOCKED AT THE WESTERN end of the Sahel, the Central African Republic (CAR) is a low plateau stretching north from one of Africa's great rivers, the Ubangi, which forms its border with the Democratic Republic of the Congo. Almost all the population lives in the equatorial, rainforested south. "Emperor" Bokassa's eccentric rule from 1965 to 1979 was followed by military dictatorship. Democracy was restored in 1993.

CLIMATE ▷ Tropical equatorial

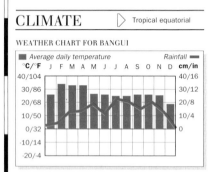

The south is equatorial, the north has a savanna-type climate, and the far north lies within the Sahel.

TRANSPORTATION ▷ Drive on right

The CAR has a limited transportation system, depending on the river link to Brazzaville, Congo, and rail from there to Pointe-Noire and the Congo River ports.

TOURISM ▷ Visitors : Population 1:380

10,000 visitors   Up 43% in 1999

MAIN TOURIST ARRIVALS

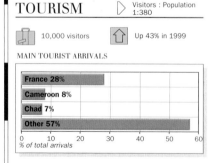

Tourist promotion is small-scale, but since 1979 there has been a modest increase in national park safaris. A new runway in Bangui will permit air charters, chiefly from France.

PEOPLE ▷ Pop. density low

Sango, Banda, Gbaya, French   6/km² (16/mi²)

THE URBAN/RURAL POPULATION SPLIT

41%   59%

ETHNIC MAKEUP

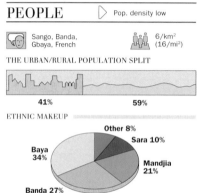

Other 8%
Sara 10%
Baya 34%
Mandjia 21%
Banda 27%

Although the Baya and the Banda are the largest ethnic groups, the lingua franca is Sango. This is spoken by the southern riverine minorities, who provided the political leaders from independence until 1993 (Presidents Dacko and Kolingba and "Emperor" Bokassa). President Patasse is from the interior. Resentment against the river peoples occasionally flares up, as happened after the coup attempt in 2001. As in other non-Muslim African countries, women have considerable power. Elizabeth Domitien was prime minister from 1975 to 1976 and Ruth Rolland ran for president in 1993.

POLITICS ▷ Multiparty elections

1998/2003   President Ange-Félix Patasse

AT THE LAST ELECTION

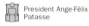

National Assembly 109 seats

			6%	6%	3%	4%
			Ind	FPP	PUN	Others
43%	18%	7%	6%	5%	2%	
MLPC	RDC	MDD	PSD	ADP	PLD	

MLPC = Central African People's Liberation Movement
RDC = Central African Democratic Rally **MDD** = Movement for Democracy and Development **Ind** = Independents
PSD = Social Democratic Party **FPP** = Patriotic Front for Progress **ADP** = Alliance for Democracy and Progress
PUN = National Unity Party **PLD** = Liberal Democratic Party

Democratic elections in 1993 ended four years of Gen. André Kolingba's single-party rule, bringing in Ange-Félix Patasse as president. Patasse,

WORLD AFFAIRS ▷ Joined UN in 1960

BDEAC CEMAC FZ LCBC AU

An improving image of political stability since 1996 was shattered by 2001. Libya emerged as the major ally of President Patasse, becoming directly involved in a series of coups and armed rebellions. Relations with Cameroon, which provides key access to the sea, were also threatened in heightened border tension in 2000. Financial help from France is relied upon and will be needed for some time.

AID ▷ Recipient

$76m (receipts)   Down 35% in 2000

Almost all development projects are funded from external aid. France, the former colonial power, provides one-third of the total. The EU and Japan are other major donors. The CAR also receives assistance from the World Bank and the IMF, which has provided funds to support the country's economic program. The Paris Club of creditor nations rescheduled and reduced the CAR's debt in 1998.

DEFENSE ▷ Compulsory military service

$43m   Down 4% in 2000

The well-equipped 3000-strong army is a heavy drain on the budget. Military service is selective. French officers fill senior army posts. In 1996, 1400 resident French troops intervened to quell army rebellions; they began to withdraw in 1997; UNMICAR forces were deployed from 1997 to 2000. Gen. Francois Bozize led the military rebellion in 2001.

Bokassa's prime minister during the 1970s, had been jailed for dissent and subsequently went into exile. Army mutinies in 1996 prompted the formation of a government of national unity, but fighting continued until in February 1997 a French-led multinational force intervened to keep the peace. It was replaced by a UN force (UNMICAR) in April 1998. Patasse's party, the MLPC, remained the most important in the new parliament after the 1998 elections but needs the support of opposition parties to maintain a workable coalition. In 2001, with assistance from Libyan troops, the government weathered a coup attempt and a military insurrection.

CENTRAL AFRICAN REPUBLIC

POPULATION   Total Land Area : 622 984 sq. km
                             (240 554 sq. miles)

- ⊙ over 500 000
- ○ over 50 000
- ● over 10 000
- · under 10 000

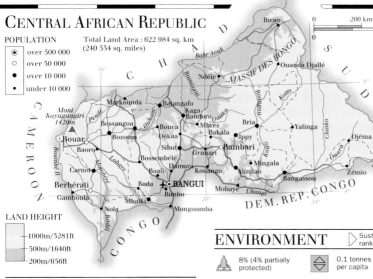

LAND HEIGHT

- 1000m/3281ft
- 500m/1640ft
- 200m/656ft

ECONOMICS

▷ Inflation 4.6% p.a.
(1990–2000)

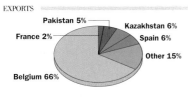 $1.03bn   698.7–736.7 CFA francs

SCORE CARD

❑ WORLD GNP RANKING	152nd
❑ GNP PER CAPITA	$280
❑ BALANCE OF PAYMENTS	Zero
❑ INFLATION	–1.5%
❑ UNEMPLOYMENT	6%

STRENGTHS

Self-sufficiency in food. Some diversity of export earnings (diamonds, cotton, timber, iron, coffee). Status as transit zone in central Africa. Trans-African Highway and waterways.

WEAKNESSES

Landlocked. Poor infrastructure. Not enough trained people to run economy.

EXPORTS

Pakistan 5%
France 2%
Kazakhstan 6%
Spain 6%
Other 15%
Belgium 66%

IMPORTS

Belgium 3%
Japan 3%
Germany 3%
Cameroon 14%
Other 45%
France 32%

RESOURCES

▷ Electric power 43,000 kw

15,117 tonnes   Not an oil producer and has no refineries

3.1m cattle, 2.6m goats, 680,000 pigs, 4.2m chickens   Diamonds, gold, uranium, iron, copper, manganese

Cotton and diamonds are the CAR's major exports. Other minerals are potentially important.

ENVIRONMENT

▷ Sustainability rank: 43rd

 8% (4% partially protected)   0.1 tonnes per capita

There has been an attempt to impose a conservationist forest policy. Hunting of elephants was banned in 1985, as numbers had fallen from 80,000 in the mid-1970s to just over 13,000.

MEDIA

▷ TV ownership low

 Daily newspaper circulation 2 per 1000 people

PUBLISHING AND BROADCAST MEDIA

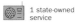

There are 3 daily newspapers, *E Le Songo*, *Le Citoyen*, and *Le Novateur*

1 state-owned service   1 state-owned service

The three weeklies and three daily newspapers have only limited circulation. A small opposition press has developed with multipartyism, but is inhibited by lack of resources.

CRIME

▷ Death penalty not used in practice

 The CAR does not publish prison figures   Crime is rising

Human rights abuses have been reduced dramatically since the excesses of the Bokassa years. The level of criminality is usually low. The major criminal problem appears to be the increase in urban robbery resulting from continued political instability from 1996.

***Baskets of cotton**, Meme village. Cotton is one of the Central African Republic's most significant export crops.*

CHRONOLOGY

The French established the colony of Ubangi-Chari in 1905 and gave it autonomy as the CAR in 1958.

- ❑ **1960** Independence under David Dacko; one-party state.
- ❑ **1965** Coup by Jean-Bédel Bokassa.
- ❑ **1977** Bokassa crowned "emperor."
- ❑ **1979** French help reinstate Dacko.
- ❑ **1981** Gen. Kolingba ousts Dacko.
- ❑ **1993** Democracy restored.
- ❑ **1996** Government of national unity formed following army rebellion.
- ❑ **2001** Coup attempt. Army rebellion.

EDUCATION

▷ School leaving age: 14

 47%   6229 students

Schooling, on the French model, is compulsory, but in practice is only received by 68% of 6–14 year olds.

HEALTH

▷ No welfare state health benefits

 1 per 20,000 people   Communicable and parasitic diseases, malnutrition

Colonial neglect and postcolonial maladministration have resulted in a poorly developed health system.

SPENDING

▷ GDP/cap. increase

CONSUMPTION AND SPENDING

1 per 1000 population   3 per 1000 population

Defense 3.7%
Education 1.9%
Health 2%

Defense, Health, Education spending as % of GDP

There is a small political–military elite in the CAR, which came into being only in postcolonial days. For its members, Paris is the chosen destination and style leader.

WORLD RANKING

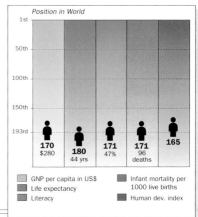

Position in World

170 $280	180 44 yrs	171 47%	171 96 deaths	165

- ▨ GNP per capita in US$
- ▨ Life expectancy
- ▨ Literacy
- ▨ Infant mortality per 1000 live births
- ▨ Human dev. index

CHAD

OFFICIAL NAME: Republic of Chad CAPITAL: N'Djamena POPULATION: 8.1 million
CURRENCY: CFA franc OFFICIAL LANGUAGES: Arabic and French

CENTRAL AFRICA

 1960 1960 Aug 11 TCH +1 +235 .td

LANDLOCKED IN NORTH central Africa, Chad has had a turbulent history since independence from France in 1960. Intermittent periods of civil war, involving French and Libyan troops, followed a coup in 1975. Another coup in 1990 preceded a transition to multipartyism, enshrined in a new constitution. The tropical, cotton-producing south is the most populous region. The discovery of large oil reserves could eventually have a dramatic impact on the economy.

CLIMATE
▷ Hot desert/steppe

WEATHER CHART FOR N'DJAMENA

There are three distinct zones: the tropical south, the central semiarid Sahelian belt, and the desert north.

TRANSPORTATION
▷ Drive on right

N'Djamena International
16,861 passengers

Has no fleet

THE TRANSPORTATION NETWORK

267 km (166 miles)		None
None		2000 km (1243 miles)

Chad has a limited transportation infrastructure. The nearest rail links are in Nigeria and Cameroon.

TOURISM
▷ Visitors : Population 1:203

40,000 visitors

Down 15% in 2000

MAIN TOURIST ARRIVALS

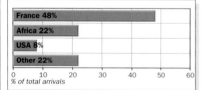

France 48%					
Africa 22%					
USA 8%					
Other 22%					

0 10 20 30 40 50 60
% of total arrivals

Tourism is virtually nonexistent. The national parks and game reserves are the main potential attractions. The prehistoric rock painting of the Tibesti plateau and the Muslim cities of central Chad attract the adventurous.

Watering hole at Oum Hadjer, *a village on the Batha watercourse in central Chad, 145 km (90 miles) east of Ati.*

PEOPLE
▷ Pop. density low

French, Sara, Arabic, Maba

6/km² (17/mi²)

THE URBAN/RURAL POPULATION SPLIT

24%   76%

RELIGIOUS PERSUASION

Christian 7%
Muslim 50%
Traditional beliefs 43%

About half the population, mainly the Sara-speaking and related peoples, is concentrated in the south in one-fifth of the national territory. Most of the rest are located in the central sultanates. The northern third of Chad has a population of only 100,000 people, mainly nomadic Muslim Toubou.

CHAD

Total Land Area : 1 284 000 sq. km
(495 752 sq. miles)

POPULATION
⊙ over 500 000
◎ over 100 000
○ over 50 000
● over 10 000
· under 10 000

LAND HEIGHT
3000m/9843ft
2000m/6562ft
1000m/3281ft
500m/1640ft
200m/656ft
100m/328ft

POLITICS
▷ Multiparty elections

2002/2006

President Idriss Déby

AT THE LAST ELECTION

National Assembly 155 seats

6% FAR
72% MPS
7% RDP
15% Others

MPS = Patriotic Salvation Movement
RDP = Rally for Democracy and Progress
FAR = Front of Action Forces for the Republic

Idriss Déby overthrew President Hissène Habré in 1990 after an armed invasion from Sudan. He promised multipartyism, and in 1992 – for the first time since the early 1960s – political parties were legalized. After many delays, the transitional process led to a successful referendum in 1996 on a new constitution based on the French model. President Déby was confirmed in office in elections in 1996 and again in 2001. His ruling MPS just achieved an overall majority in the 1997 elections, and expanded its share of seats further in 2002. A shaky Libyan-brokered peace deal in 2002 failed to end a rebellion in the north which had begun in 1999.

WORLD AFFAIRS

 Joined UN in 1960

 CILSS FZ LCBC AU OIC

Chad's most important relationship is with France. Libya occupied the uranium-rich Aozou strip in 1973–1994.

AID

 Recipient

 $131m (receipts) Down 30% in 2000

France is by far the major donor. Other sources include Libya, the EU, the World Bank, Germany, and the Arab members of OPEC. Without assistance to cover civil servants' pay over recent years, the administration would have collapsed.

DEFENSE

 Compulsory military service

 $47m ⇕ No change in 2000

On seizing power, Déby swelled the existing army with irregulars. This policy has now been reversed and the army reduced to 25,000, including former rebels. France provides military aid and personnel.

ECONOMICS

Inflation 7.1% p.a. (1990–2000)

 $1.54bn 698.7–736.7 CFA francs

SCORE CARD

- ❏ World GNP Ranking........................143rd
- ❏ GNP per Capita $200
- ❏ Balance of Payments...................–$158m
- ❏ Inflation 3.8%
- ❏ Unemployment ....Widespread underemployment

STRENGTHS

Discovery of large oil deposits and Chad–Cameroon pipeline could transform economy. Cotton industry; potential for other agriculture in south. Strategic trading location in heart of Africa. Natron and uranium deposits.

WEAKNESSES

Underdevelopment and poverty. Lack of transportation infrastructure. Political instability. Frequent droughts.

EXPORTS

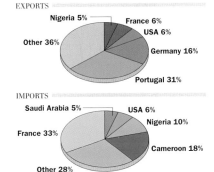

Nigeria 5%
France 6%
USA 6%
Germany 16%
Other 36%
Portugal 31%

IMPORTS

Saudi Arabia 5%
USA 6%
Nigeria 10%
France 33%
Cameroon 18%
Other 28%

RESOURCES

 Electric power 29,000 kw

 84,000 tonnes

Reserves currently unexploited

5.9m cattle, 5.25m goats, 2.4m sheep, 5m chickens

Natron, uranium, oil, kaolin, soda, rock salt

A consortium of Exxon, Shell, and Elf has discovered large oil reserves in the south, mostly near Doba, which could make Chad a major African producer. Natron, found north of Lake Chad, is the only mineral currently exploited. There is uranium in the Aozou strip.

ENVIRONMENT

 Sustainability rank: 96th

 9% ⇕ 0.02 tonnes per capita

President Déby's government has made protection of the environment a priority, with antidesertification measures such as tree-planting campaigns and aid-funded irrigation schemes. There is concern about potential environmental damage which may result from the planned Chad–Cameroon oil pipeline.

MEDIA

 TV ownership low

✕ Daily newspaper circulation 0.2 per 1000 people

PUBLISHING AND BROADCAST MEDIA

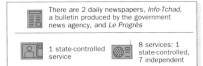

There are 2 daily newspapers, *Info-Tchad*, a bulletin produced by the government news agency, and *Le Progrès*

1 state-controlled service

8 services: 1 state-controlled, 7 independent

Broadcasting is controlled by the government, which sometimes allows the airing of opposition views. There are a few independent publications, of which the best known is the weekly *N'Djamena-Hebdo*.

CRIME

 Death penalty in use

 2521 prisoners

Crime is rising

The easy availability of weapons in the region in the past two decades has meant that local disputes, usually over water or grazing, often now lead to gun battles. Armed robbery, smuggling, and vandalism are widespread. In several areas, the activities of disaffected former rebels threaten security.

EDUCATION

 School leaving age: 12

43%

5901 students

Education is based on the French model, although there are Koranic schools in the north. Primary schooling is officially compulsory, but enrollment is only 46%. Recently, World Bank aid has been directed at elementary schooling.

HEALTH

 Welfare state health benefits

1 per 20,000 people

Diarrheal, parasitic, and communicable diseases

There are few city hospitals and fewer than 300 smaller health centers. One in five children die before the age of five.

SPENDING

GDP/cap. increase

CONSUMPTION AND SPENDING

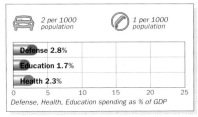

2 per 1000 population

1 per 1000 population

Defense 2.8%
Education 1.7%
Health 2.3%

Defense, Health, Education spending as % of GDP

Poverty is almost universal in Chad; the middle class is very small. There are few wealthy individuals. Habré looted the treasury when he was overthrown in 1990.

WORLD RANKING

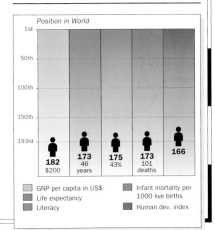

Position in World

| 182 | 173 | 175 | 173 | 166 |
| $200 | 46 years | 43% | 101 deaths | |

- ☐ GNP per capita in US$
- ☐ Life expectancy
- ☐ Literacy
- ☐ Infant mortality per 1000 live births
- ☐ Human dev. index

C

CHILE

OFFICIAL NAME: Republic of Chile   **CAPITAL:** Santiago
POPULATION: 15.4 million   **CURRENCY:** Chilean peso   **OFFICIAL LANGUAGE:** Spanish

C

 1818
 1883
 Sept 18
 RCH
 -4
 +56
.cl

CHILE EXTENDS IN A narrow ribbon 4350 km (2700 miles) down the Pacific coast of South America. Its extraordinary shape means that its physical geography ranges from the deserts of the High Andes in the north to fertile valleys in the center, while in the south are the fjords, lakes, and deep sea channels of the Southern Andes. In 1989, Chile returned to elected civilian rule, following a popular rejection of the Pinochet dictatorship. A collapse in copper prices, coupled with weaker export markets, has interrupted the high growth seen in the 1990s.

General Pinochet, *a dictatorial president rejected by popular referendum in 1989.*

Ricardo Lagos *was narrowly elected president in 2000.*

CLIMATE
▷ Desert/mountain/ maritime

WEATHER CHART FOR SANTIAGO

Chile has an immensely varied climate. The north, which includes the world's driest desert, the Atacama, is frequently cloudy and cool for its latitude. The central regions have an almost Mediterranean climate, with changeable winters and hot, dry summers. The higher reaches of the Andes have a typically alpine climate, with glaciers and year-round snow. The south is the wettest region.

TRANSPORTATION
▷ Drive on right

Comodoro Arturo Merino Benítez, Santiago
5.78m passengers

471 ships
842,300 grt

THE TRANSPORTATION NETWORK

11,012 km (6843 miles)	3455 km (2147 miles)
2084 km (1295 miles)	725 km (450 miles)

The state railroad is being upgraded with private capital assistance: suburban services into Santiago are being improved, and line repairs and better rolling stock will shorten journeys from Santiago to Temuco. Sections of the Pan-American Highway, the sole arterial road running from the Peruvian border to Puerto Montt, are being upgraded. The open skies agreement of the US reached with Chile in 1999, the first with a South American country, gives Chile strong regional and international links.

TOURISM
▷ Visitors : Population 1:8.8

1.74m visitors

Up 7% in 2000

MAIN TOURIST ARRIVALS

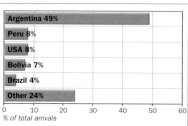

The Pinochet years saw a dramatic decline in tourists from the US and Europe, although the numbers from neighboring countries held up. Since the fall of the regime in 1989, visitors have returned, but over half of them still come from Argentina and Peru.

Investment in the sector totaled $2.17 billion in the first half of 2000, up 3.8% on the same period in 1999. Accordingly, Chile is making more of its stunning Andean scenery, its immensely long coastline, and a number of exceptional sites, including Chuquicamata, the world's largest copper mine, the Elqui Valley wine-growing region, and the spectacular glaciers and fjords of southern Chile. Easter Island in the Pacific is another major attraction.

Peaks in the Paine range, southern Chile. *Fjords, glaciers, and myriad islands typify Chile's very wet, wild, and stormy south.*

PEOPLE
▷ Pop. density low

Spanish, Amerindian languages

21/km² (53/mi²)

THE URBAN/RURAL POPULATION SPLIT

86%   14%

RELIGIOUS PERSUASION

Other and nonreligious 20%

Roman Catholic 80%

ETHNIC MAKEUP

Amerindian 10%

Mixed and European 90%

One-third of Chile's population lives in Santiago, where rapid growth has created large slums. Chile has relatively few immigrants; most people are of mixed Spanish–Amerindian descent. A Commission for Historic Truth was set up in 2000 to address the problems of Amerindians. There are some 80,000 Mapuche Amerindians around Temuco in the south, 20,000 Aymara in the northern Chilean Andes, and 2000 Rapa Nui on Easter Island.

Over 25% of working women are employed in domestic service.

POPULATION AGE BREAKDOWN

Female	Age	Male
0.8%	80+	0.4%
4.8%	60–79	3.8%
10.2%	40–59	9.8%
16.3%	20–39	16.4%
18.4%	0–19	19.1%

% of population by age group

CHILE

Total Land Area :
756 950 sq. km
(292 258 sq. miles)

POPULATION

▣	over 1 000 000
◉	over 100 000
○	over 50 000
•	over 10 000
·	under 10 000

LAND HEIGHT

	4000m/13124ft
	2000m/6562ft
	1000m/3281ft
	200m/656ft
	Sea Level

POLITICS

▷ Multiparty elections

   L. House 2001/2005   President Ricardo
U. House 2001/2005   Lagos Escobar

AT THE LAST ELECTION

Chamber of Deputies 118 seats

52%	47%	1%
CPD	APC	Ind

CPD = Concertación – Coalition of Parties for Democracy (Christian Democratic Party–**PDC**, Party for Democracy–**PPD**, and Socialist Party of Chile–**PS**) **APC** = Alliance for Chile (Independent Democratic Union–**UDI** and National Renewal Party–**RN**) **Ind** = Independents **App** = Appointed
SL = Senator-for-Life

Senate 48 seats

				13%	6%
				UDI	PPD
25%	18%	13%	13%	10%	2%
PDC	App	RN	Ind	PS	SL

There are 38 elected members and 9 appointed senators. Former president Frei is a senator-for-life

After 16 years of military rule under Gen. Augusto Pinochet, Chile returned to multiparty democracy in 1989.

PROFILE

Chilean politics is still strongly affected by the legacy of the military dictatorship of 1973–1989, which began when Pinochet's coup overthrew the elected Marxist government of Salvador Allende.

The CIA backed the Pinochet coup, anxious to halt Allende's program of nationalization of the largely US-owned copper mines. As a result thousands of Chileans were killed by the military or "disappeared," and a further 80,000 were taken as political prisoners.

Pinochet's nationalist politics drew on the example of Franco's Spain, while his economic policy was one of the first experiments in the free-market Chicago School of monetarism. Chile's business and middle classes prospered, while opposition, which was brutally suppressed by the DINA secret police, came most visibly from the Church and the urban poor.

In 1988 Pinochet, attempting to secure a popular mandate for continuing his regime, was surprised when the population emphatically voted for democracy. Pinochet stepped down, but remained head of the army. Patricio Aylwin won presidential elections held in 1989, heading Concertación, a center-left coalition.

Under Aylwin, politics became more stable, partly as a result of a cross-party consensus on economic policy. Continued growth and some progressive social measures attracted the support of the trade unions. These policies were continued under Eduardo Frei of the PS party, who was elected president in 1993.

When Pinochet retired as army chief in 1998, heated disagreements over his entry to the Senate as a senator-for-life split Concertación along broadly left–right lines. Disagreements over Pinochet's subsequent arrest and detention in Europe on human rights charges further complicated the picture. Ricardo Lagos of the PS emerged as front runner for the 1999 presidential elections, in which both the PS and the PDC presented themselves as the guarantors of peace and democracy. Right-wing opposition parties, however, also began to downplay their past links with Pinochet to broaden their electoral appeal. Lagos narrowly won a run-off poll in 2000. Facing a better organized opposition than hitherto, the ruling center-left Concertación performed reasonably well in the 2001 elections retaining its majority in the Chamber of Deputies but losing it in the Senate.

MAIN POLITICAL ISSUES
Political consolidation

President Lagos, still Chile's most charismatic and trusted politician, reshuffled what was seen as a weak cabinet after the 2001 legislative elections to reinvigorate his government for his remaining four years in office. Top priority was to guarantee a stable environment for economic growth and job creation. The strong electoral performance of the UDI, senior partner in the right-wing Alliance for Chile opposition coalition, boosted the presidential hopes of its leader, Joaquín Lavín, currently the mayor of Santiago.

Improved relations with military

Relations between government and military are improving. Lagos appointed Chile's first woman defense minister and installed as new army commander a moderate who had accepted that Pinochet should respect the authority of the courts, before the human rights charges against the former dictator were dropped on grounds of dementia.

CHRONOLOGY

The Spanish first attempted the conquest of Chile against the fierce indigenous Araucanian people in 1535. Santiago was founded in 1541. Chile was subject to Spanish rule until independence in 1818.

❏ **1817–1818** Bernardo O'Higgins leads republican Army of the Andes in victories against royalist forces.

❏ **1879–1883** War of the Pacific with Bolivia and Peru. Chile gains valuable nitrate regions.

❏ **1891–1924** Parliamentary republic ends with growing political chaos. ⇨

C

CHRONOLOGY *continued*

- ❏ **1936–1946** Communist, Radical, and Socialist parties form influential Popular Front coalition.
- ❏ **1943** Chile backs US in World War II.
- ❏ **1946–1964** Right-wing Chilean presidents follow US McCarthy policy and marginalize the left.
- ❏ **1970** Salvador Allende elected. Reforms provoke strong reaction from the right.
- ❏ **1973** Allende dies in army coup. Brutal dictatorship of Gen. Pinochet begins.
- ❏ **1988** Referendum votes "no" to Pinochet staying in power.
- ❏ **1989** Democracy peacefully restored; Pinochet steps down after Aylwin election victory.
- ❏ **1998** Pinochet detained in UK pending extradition to Spain on human rights charges.
- ❏ **2000** Ricardo Lagos (PS) sworn in as president. Pinochet, deemed unfit to face trial, returns to Chile, where charges are suspended in 2001.

WORLD AFFAIRS

▷ Joined UN in 1945

| APEC | G15 | NAM | OAS | RG |

Chile hopes to consolidate the key relationship with the US, its main trading partner, through a bilateral free trade agreement. During the Allende period in the early 1970s, the US had actively worked against the government, fearing that the spread of socialism would jeopardize its investments in Chile and the rest of Latin America. However, the human rights record of the Pinochet regime eventually became an embarrassment.

In 1999, in response to Pinochet's detention in the UK and planned extradition to Spain to stand trial on human rights charges, Chile argued that this amounted to an infringement of its sovereignty and the jurisdiction of its courts. Relations with Spain and the UK improved in 2000 following Pinochet's return to Chile.

An associate member of Mercosur, Chile also jealously guards its right to pursue its own bilateral economic agreements; a free trade agreement between Chile and the EU was initialled in June 2002. Border disputes with Bolivia and Peru are ongoing.

AID

▷ Recipient

💲 $49m (receipts) ⬇ Down 30% in 2000

The majority of aid is in the form of debts rescheduled by the World Bank at the instigation of the US.

DEFENSE

▷ Compulsory military service

💲 $2.89bn ⬆ Up 12% in 2000

CHILEAN ARMED FORCES

🛡	290 main battle tanks (250 *Leopard* 1, 40 AMX-30)	51,000 personnel
⚓	3 submarines, 3 destroyers, 3 frigates, and 27 patrol boats	24,000 personnel
✈	77 combat aircraft (16 F-5, 35 *Mirage*, 14 A-37B, 12 A-36)	12,500 personnel
	None	

Most of the officers linked to human rights abuses under the dictatorship (1973–1989) are now retired, but dozens still face investigation and possible trial. The army has a moderate commander, and for the first time a woman, also a member of the PS, has been appointed defense minister. Despite domestic concerns on pressing social needs, President Lagos signed up to reequip the air force with expensive, US-made F-16 fighters, still to be part-funded by 10% of copper exports under a controversial law inherited from the years of military rule. To dampen fears of arms escalation, Chile forewarned Argentina, Brazil, Peru, and Bolivia about the deal. Modernization of the navy, which was lobbying for four new frigates, is on hold.

ECONOMICS

▷ Inflation 7.3% p.a. (1990–2000)

📊 $69.9bn 💲 573.75–661.15 Chilean pesos

SCORE CARD

- ❏ WORLD GNP RANKING............................44th
- ❏ GNP PER CAPITA $4590
- ❏ BALANCE OF PAYMENTS.....................–$991m
- ❏ INFLATION 3.8%
- ❏ UNEMPLOYMENT......................................9%

EXPORTS

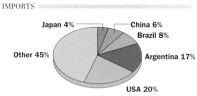

Brazil 5%
China 5%
UK 6%
Japan 14%
USA 17%
Other 53%

IMPORTS

Japan 4%
China 6%
Brazil 8%
Argentina 17%
USA 20%
Other 45%

STRENGTHS

World's largest copper producer. Fresh fruit exports. Strong investment inflows allowing rapid economic growth. Highest credit rating due to fiscal and monetary stability and highly liquid financial system. Development of nontraditional industries such as wine and fresh and prepared fish.

WEAKNESSES

Dependence on US as single largest trading partner. Prolonged weakness in Asian export markets. Vulnerability of copper revenues, representing 40% of exports, to low world market prices.

PROFILE

Competing ideologies have battled over Chile's economy. Allende's socialism brought huge corporations into the state sector. The Pinochet dictatorship

ECONOMIC PERFORMANCE INDICATOR

Consumer Price Index — GDP ▨

introduced radical monetarist policies. Drastic cutting of the state sector and the selling-off of state enterprises at below market value led to large profits for investors and speculators. Tough economic measures, irrespective of the social consequences, brought Chile's inflation rate down from 400%.

The Aylwin and Frei governments continued with neoliberal policies, including privatizing the pension system. However, some 30 companies, including the large Codelco copper company, remain in the state sector.

Though it avoided the worst of the Argentine crisis, the economy in 2002 suffered from sluggish international conditions and modest copper prices.

CHILE : MAJOR BUSINESSES

- 🛢 Oil
- ⬛ Oil refining
- ⛏ Copper mining
- 🏭 Manufacturing
- Pharmaceuticals
- ✿ Heavy engineering
- Fish processing
- Agribusiness

Iquique
Chuquicar
Vina del Mar
Santiago
Teniente
Talcahuano
Concepción
Punta Arenas
Straits of Magellan

0   500 km
0   300 miles

C

RESOURCES

 Electric power 8.4m kw

5.32m tonnes

4852 b/d (reserves 285m barrels)

4.2m sheep, 4.15m cattle, 2.5m pigs, 78m chickens

Coal, copper, gold, silver, iron, lithium, molybdenum, iodine, natural gas, oil

ELECTRICITY GENERATION

- Hydro 45% (16bn kwh)
- Combustion 55% (20bn kwh)
- Nuclear 0%
- Other 0%

% of total generation by type

Chile is the world's largest producer of copper, which accounts for a major proportion of its export revenues. There are important deposits of lithium, molybdenum, and especially of gold. Chile also has reserves of natural gas, oil, and coal, and plenty of hydroelectric potential. In addition, it is a leading producer of fishmeal, and has a flourishing wine industry.

CHILE : LAND USE

Cropland
Pasture
Forest
Desert
High mountain regions
Wheat
Fruits - cash crop
Sheep

0   300 km
0   300 miles

ENVIRONMENT

 Sustainability rank: 35th

19% (7% partially protected)

 4.1 tonnes per capita

ENVIRONMENTAL TREATIES

Yes | Yes | Yes
Yes | Yes | No

Environmental concerns do not rank highly on the political agenda. Severe smogs still cover Santiago, due in part to diesel fumes from the city's 14,500 buses. The chief concern is logging in the south by Japanese and other foreign companies. The huge growth of the salmon industry, which fences off sea lakes, is resulting in dolphins losing their natural habitats. Overfishing of swordfish has led to friction with the EU, particularly Spain.

MEDIA

 TV ownership medium

Daily newspaper circulation 98 per 1000 people

PUBLISHING AND BROADCAST MEDIA

There are 52 daily newspapers, including *El Mercurio, Las Ultimas Noticias,* and the best-selling *La Tercera*

1 state-owned service, many independent services

1 state-owned service, 1046 independent stations

A long-delayed liberalized press law was finally introduced in 2001. Military courts will no longer be able to try journalists, and political authorities will be denied the special procedures used to sue reporters for slander.

CRIME

 Death penalty not used in practice

 30,852 prisoners   Up 1% in 1999

CRIME RATES

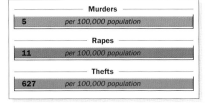

Murders	
5	per 100,000 population
Rapes	
11	per 100,000 population
Thefts	
627	per 100,000 population

The judiciary has been slow to pursue human rights cases from the Pinochet regime, despite the discoveries during the 1990s of mass graves of victims of the DINA (secret police). Mapuche leaders were among the "disappeared." Levels of child abuse are exceptionally high, although now starting to fall.

EDUCATION

 School leaving age: 13

96%   424,672 students

THE EDUCATION SYSTEM

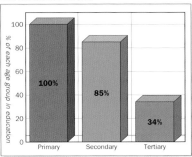

% of each age group in education

Primary 100%
Secondary 85%
Tertiary 34%

Economic growth has permitted public spending on education to increase substantially, but the sector suffers from budgetary cuts. Free primary education is officially compulsory for eight years. Human rights issues now appear in school curricula.

HEALTH

 Welfare state health benefits

1 per 909 people   Heart diseases, cancers, murders

Recent growth has meant increased public spending on health; however, 2002 saw cuts. The public health service covers 80% of people, but is mostly found in urban areas. There is private care for the rich. Infant mortality has fallen to one-third of the 1980 level.

SPENDING

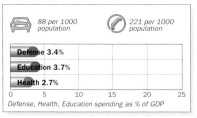 GDP/cap. increase

CONSUMPTION AND SPENDING

88 per 1000 population   221 per 1000 population

Defense 3.4%
Education 3.7%
Health 2.7%

Defense, Health, Education spending as % of GDP

Chile's traditionally large middle class did well under Pinochet and the economic policies of the Chicago School. The wealthiest sections benefited considerably from the sale of state assets at 40%–50% of their true market value. Five years into the regime, wealth had become highly concentrated, with just nine economic conglomerates controlling the assets of the top 250 businesses, 82% of banking, and 64% of all financial loans. The regime's artificially high domestic interest rates enabled those with access to international finance to earn an estimated $800 million between 1977 and 1980, simply by borrowing abroad and lending at home. These groups have retained their position.

The poor, by contrast, are over 15% worse off than in 1970, with an estimated four million people living just above the UN poverty line and one million below it.

WORLD RANKING

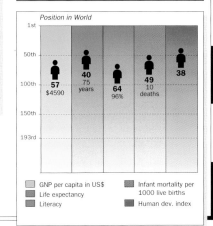

Position in World

	GNP per capita in US$		57 $4590
	Life expectancy		40 75 years
	Literacy		64 96%
	Infant mortality per 1000 live births		49 10 deaths
	Human dev. index		38

GNP per capita in US$
Life expectancy
Literacy
Infant mortality per 1000 live births
Human dev. index

CHINA

OFFICIAL NAME: People's Republic of China **CAPITAL:** Beijing
POPULATION: 1.29 billion **CURRENCY:** Renminbi (known as yuan) **OFFICIAL LANGUAGE:** Mandarin

COVERING A VAST AREA of eastern Asia, China is bordered by 14 countries; to the east it has a long Pacific coastline. Two-thirds of China is uplands: the southwestern mountains include the Tibetan Plateau; in the northwest, the Tien Shan Mountains separate the Tarim and Dzungarian basins. The low-lying east is home to two-thirds of the population. China was dominated by Mao Zedong from the founding of the Communist People's Republic in 1949 until his death in 1976. Despite the major disasters of the 1950s Great Leap Forward and the 1960s Cultural Revolution, it became an industrial and nuclear power. Today, China is rapidly developing a market-oriented economy. The current leadership remains set on accomplishing this without political liberalization, instead enforcing single-party rule as was advocated by "elder statesman" Deng Xiaoping, who died in 1997.

Li River (Xi Jiang), Guangxi, China's most beautiful region. Its spectacular scenery has encouraged large-scale tourist development.

CLIMATE

▷ Mountain/tropical/continental/steppe

WEATHER CHART FOR BEIJING

China is divided into two main climatic regions. The north and west are semiarid or arid, with extreme temperature variations. The south and east are warmer and more humid, with year-round rainfall.

Winter temperatures vary with latitude and are warmest on the subtropical southeast coast, where they average about 16°C (60°F). Summer temperatures are more uniform, rising above 21°C (70°F) throughout China; on the southeast coast, the July average is about 30°C (86°F). In the north and west, temperate summers contrast with harsh winters. In northern Manchuria, rivers freeze for five months and temperatures can fall to –25°C (–13°F). In the deserts of Xinjiang province, temperatures range from –11°C (12°F) in winter to 33°C (91°F) in summer.

Summer and autumn are China's wettest seasons. Winds from the Pacific during the summer monsoon bring rains to most of the country. The south and east also have wet winters, but elsewhere the winter monsoon brings cold, dry air from Siberia.

Floods are frequent and sometimes catastrophic, as in 2002. Droughts can be even more devastating; that of 1959–1962 contributed to a famine which killed millions.

CHINA

Total Land Area : 9 596 960 sq. km
(3 705 386 sq. miles)

POPULATION
- ▣ over 5 000 000
- ▣ over 1 000 000
- ◉ over 500 000
- ◎ over 100 000
- ○ over 50 000
- ● over 10 000

⎍⎍ Great Wall of China

LAND HEIGHT
6000m/19686ft
4000m/13124ft
3000m/9843ft
2000m/6562ft
1000m/3281ft
500m/1640ft
200m/656ft
Sea Level

0   400 km
0   400 miles

TOURISM

▷ Visitors : Population
1:39

🧳 33.2m visitors ⬆ Up 6% in 2001

MAIN TOURIST ARRIVALS

	% of total arrivals
Japan 22%	
South Korea 12%	
Russia 10%	
USA 9%	
Malaysia 4%	
Other 43%	

% of total arrivals

The easing of restrictions since the 1980s has led to the rapid growth of all kinds of tourism, from luxury tours to budget packages and backpacking. Hong Kong is a major entry point. Most of China is now open to visitors, and there have been some moves to allow tourists into Tibet, although access to Xinjiang in western China, and other areas, is sometimes impossible. The Great Wall, the Forbidden City in Beijing, and the terra-cotta warriors at Xi'an remain among the top attractions. The number of Chinese able to travel abroad now exceeds ten million.

TRANSPORTATION

▷ Drive on right

✈ **Chek Lap Kok, Hong Kong**
32.8m passengers

🚢 3319 ships
16.5m grt

THE TRANSPORTATION NETWORK

🛣 271,300 km (168,578 miles)	🛤 24,474 km (15,207 miles)
🚆 67,394 km (41,879 miles)	〰 110,263 km (68,514 miles)

Roads and railroads have been extended since 1949 to provide a basic national network. The transportation system is now being modernized and expanded to support the push for economic growth. The Ninth Five-Year Plan (1996–2000) provided for 8100 km (5000 miles) of new railroad. A high-speed rail link between Beijing and Shanghai is in prospect under the Tenth Plan. In 2001, plans were announced to connect the Tibetan capital, Lhasa, to the railroad system by extending the track from Golmud. If and when it is completed, this will present one of the world's highest railroads.

Container shipping is growing fast. Shanghai handled one-third of all Chinese container traffic before the reversion of Hong Kong (which is the world's biggest container port) to Chinese rule in 1997. The inland waterway system, which was hitherto in a state of disrepair, is being upgraded and now handles one-third of all internal freight. The Yangtze River (Chang Jiang) is navigable by ships of over 1000 tonnes for more than 1000 km (620 miles) from the coast. This capacity is planned to increase under the Three Gorges Dam project.

Nine small airlines which had sprung up since 1988 were consolidated into three carriers in 2002. Hong Kong's new airport opened in 1999. Air transportation is growing rapidly, like private car ownership, as wealth increases. However, bicycles still form the main mode of personal transportation.

Li River (Xi Jiang) valley. Irrigation helps Chinese farmers to feed 20% of the world's people, using only 7% of the world's farmland.

C

C

PEOPLE

 Pop. density medium

Mandarin, Wu, Cantonese, Hsiang, Min, Hakka, Kan

138/km² (357/mi²)

THE URBAN/RURAL POPULATION SPLIT

32%   68%

RELIGIOUS PERSUASION

Muslim 2%   Buddhist 6%
Other 13%
Traditional beliefs 20%
Nonreligious 59%

ETHNIC MAKEUP

Zhaung 1%   Hui 1%
Other 5%
Han 93%

The vast majority of China's population is Han Chinese. The rest belong to one of 55 minority nationalities, or recognized ethnic groups. The minorities have disproportionate political significance because many, like the Mongolians, Tibetans, or Muslim Uyghurs in Xinjiang, live in strategic border areas.

The deeply resented policy of resettling Han Chinese in remote regions has led to ruthlessly suppressed uprisings in Xinjiang and Tibet. Han Chinese are now a majority in Xinjiang and Nei Mongol Zizhiqu (Inner Mongolia). Tibetan calls for greater political and cultural autonomy get much more international attention.

A one-child policy was adopted in 1979. Most Han Chinese still face strict family-planning controls, though these are widely flouted. Cases of female infanticide at birth have produced a demographic imbalance, and rules were relaxed for minorities after some small groups came near to extinction.

Chinese society is patriarchal in practice, and generations tend to live together. However, economic change is putting pressure on family life, breaking down the social controls of the Mao era. Divorce and unemployment are rising; materialism has replaced the puritanism of the past. The Falun Gong spiritual movement, perceived as a rival to CCP authority, was banned in 1999.

POPULATION AGE BREAKDOWN

Female	Age	Male
0.6%	80+	0.3%
5%	60–79	4.7%
10.6%	40–59	10.9%
17.4%	20–39	17.5%
15.7%	0–19	17.3%

% of population by age group

POLITICS

 No multiparty elections

1998/2003   President Jiang Zemin

AT THE LAST ELECTION

National People's Congress 2979 seats

The Communist Party of China (CCP) is the only permitted party

China is a single-party state, dominated by the CCP, the world's largest political party. The National People's Congress, indirectly elected every five years, is theoretically the supreme organ of state power. It appoints the president and executive State Council, headed by the prime minister. The real focus of power, however, is the 22-member Politburo of the CCP and, in particular, its Standing Committee of seven.

PROFILE

The death in 1997 of Deng Xiaoping marked in effect the passing of the dominance of the "Immortals" – those who took part with Mao Zedong in the 1934–1935 Long March. Deng, the architect of China's economic reforms, had worked hard behind the scenes forming alliances to promote his reformist ideas and followers. Jiang Zemin consolidated his position as president and CCP general secretary after Deng's death, but a "fourth generation" leadership stands ready to emerge at the 2002 party congress, with Vice President Hu Jintao its most prominent figure.

MAIN POLITICAL ISSUES
Economic change and CCP authority
After the death of Mao Zedong in 1976, China embarked on economic reform, while seeking to secure the dominance of the CCP and avoid political upheaval. The "great helmsman" of this process for two decades was Deng Xiaoping, China's paramount leader, even after he had relinquished all official posts. Advocating a fast-track move to a "socialist market economy," Deng and his followers looked to South Korea and Taiwan as achieving high growth without political reform. At the 1997 party congress the reformers, led by

Nanjing Donglu (Nanking Road), in central Shanghai, is one of China's most famous shopping streets. A magnet for foreign investment, Shanghai is China's largest city.

Jiang Zemin, took their opportunity to realign formal party policy with their desire to privatize large areas of state-run industry. The transfer of much of the huge state economic system into private ownership has been a challenge to the CCP's ability to monopolize power. The 22 provinces, particularly those in the southeast, are acting increasingly independently of Beijing. At a popular level, there is growing rural discontent over widening wealth differentials. However, the party has allowed no political opposition to surface.

Shifting balances in the top leadership
The prodemocracy protests of 1989, culminating in the Tiananmen Square massacre, enabled conservatives within the party, under the then premier Li Peng, to gain the upper hand until Deng moved to restore the balance. His longevity shifted the advantage toward his heir apparent, President Jiang Zemin, who subsequently strengthened his own power base and international stature, while Premier Zhu Rongji spearheaded the economic reform and anticorruption campaigns. Li Peng, since 1998 president of the National People's Congress, still commands support, but all three are scheduled to stand down from the Politburo at the 2002 party congress. A major overhaul of party ideology in 2001, under Jiang's "Three Represents" doctrine, encouraged the equal promotion of business, culture, and the rural masses.

Deng Xiaoping was China's paramount leader until his death in 1997.

Jiang Zemin, CCP leader and China's president since Deng resigned the post.

Li Peng, president of the National People's Congress.

Vice President Hu Jintao, prominent among "fourth generation" leaders.

C

WORLD AFFAIRS ▷ Joined UN in 1945

The push for economic modernization, and concerns about regional stability, dominate Chinese foreign policy. Investment, technology, and trade considerations outweigh ideology.

Despite lingering concerns over human rights, relations with the West have rebounded from the low of the 1989 Tiananmen Square massacre. China was awarded Most Favored

Nation trading status by the US in 2000 and entered the WTO in 2001. Despite initial tensions in relations with the new US administration in 2001, the two countries had reached a "common understanding" by the end of the year on the issue of terrorism, which had shot to the top of the global agenda.

Relations with Russia have improved steadily. The two states condemned the UN bombing campaign in Yugoslavia in 1999, and are against the proposed US missile defense shield. Regionally, ties

have been normalized with Vietnam and strengthened with South Korea.

Taiwan remains the dark spot of Chinese external relations. Beijing strongly rejects any moves to recognize even de facto independence for the island, and cross-straits relations are frequently damaged by displays of military might and repetitive threats. Despite such alarming flare-ups, however, closer links have been forged since 2000, with an increase in tourist traffic and indirect trade.

AID ▷ Recipient

 $1.74bn (receipts) Down 27% in 2000

In the 1970s aid was an important part of Chinese diplomacy, going mostly to Africa, but other communist and southeast Asian states were also recipients. Outward aid flows almost ceased in the late 1970s, as the economic reform process turned China itself into a major aid recipient. Japan is the biggest bilateral donor to China, but the potential of the Chinese market means that most developed states provide aid. A significant portion of funding is linked to a donor country's interest in opportunities created by China's huge infrastructure problems, and used to finance high-tech imports.

DEFENSE ▷ Compulsory military service

$41.2bn Up 3% in 2000

CHINESE ARMED FORCES

8000 main battle tanks (T-59I/II, T-79, T-88B/C, T-98)	1.6m personnel
69 submarines, 21 destroyers, 41 frigates, and 368 patrol boats	250,000 personnel
2900 combat aircraft (300 Q-5, 1500 J-6, 624 J-7, 170 J-8, 40 Su-30)	420,000 personnel
ICBM (20 DF-5), IRBM (20 DF-4, 38 DF-3A, 8 DF-21), SLBM (12 CSS-N-3), SRBM (20 DF-15, 40 DF-11), 1 SSBN	

The People's Liberation Army (PLA) is being cut back as part of a

modernization process, reducing both its numbers and its involvement in the economy through army-run industries. Closely linked with the ruling CCP, it had appeared unassailable until the mid-1990s, when it numbered as many as three million personnel. A key instrument of ensuring the party's dominance, it was used in 1967 to restore order after the chaos of the Cultural Revolution, and in 1989 to suppress prodemocracy protests in Tiananmen Square, as well as to stamp out dissent in Tibet.

China has a large weapons industry, and has extended its nuclear weapons capability to include the neutron bomb. It is a significant arms exporter.

ECONOMICS ▷ Inflation 7.1% p.a. (1990–2000)

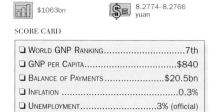 $1063bn 8.2774–8.2766 yuan

SCORE CARD

❏ WORLD GNP RANKING	7th
❏ GNP PER CAPITA	$840
❏ BALANCE OF PAYMENTS	$20.5bn
❏ INFLATION	0.3%
❏ UNEMPLOYMENT	3% (official)

STRENGTHS

Huge domestic market. Food self-sufficiency. Mineral reserves. Diversified industrial sector. Low wage costs. Rapid sustained growth. Growing export sector. Hong Kong as financial center. Heavy investment in communications and IT.

ECONOMIC PERFORMANCE INDICATOR

EXPORTS

Germany 4% South Korea 5% Japan 17% Other 35% Hong Kong 18% USA 21%

WEAKNESSES

Poor transportation. Debt-ridden state sector. Massive underemployment, rising unemployment (132 million admitted in 2002). Unevenly distributed resources.

CHINA : MAJOR BUSINESSES

Ürümqi Lanzhou Xi'an Beijing Tianjin Shanghai Wuhan Guangzhou Kunming Chongqing Hong Kong

Oil
Textiles
Chemicals
Computers
Electronics
Iron & steel
Engineering
Consumer goods
Research & development

0 1000 km
0 1000 miles

* significant multinational ownership

IMPORTS

Germany 5% South Korea 10% USA 10% Other 46% Taiwan 11% Japan 18%

PROFILE

China has demonstrably shifted from a centrally planned to a market-oriented economy since the 1980s; liberalization is greatest in the south. Growth had to be curbed on several occasions to control inflation. In the Ninth Five-Year Plan (1996–2000) the government retained strict controls, promoting intensive growth. The 1997 CCP congress boosted privatization of the huge state-owned sector. The Tenth Five-Year Plan (2001–2005) placed fresh emphasis on rapid development, reforms, and improving competitiveness.

Trade entered a new era in 2000–2001. A substantial growth in imports followed a deal with the EU and the normalization of US trade relations. Even Taiwan dropped its ban on direct trade with the mainland. China's accession to the WTO in 2001 paved the way for further economic expansion.

C

MINORITIES IN THE AUTONOMOUS REGIONS

AUTONOMOUS REGIONS OF CHINA

CHINA'S CONSTITUTION states that "regional autonomy is practiced in areas where people of minority nationalities live in compact communities." The Manchu, Miao, Yi, and many others among the 55 recognized ethnic minorities are not regarded as meeting this criterion, but there are five areas which do officially have the status of autonomous region or *zizhiqu*. The two most westerly such regions, sparsely populated Tibet and the home of the Uyghur people in Xinjiang, are both particularly sensitive border areas, and opposition to Chinese rule there has been suppressed.

TIBET
The vast mountainous region of Tibet has China's lowest population density, at only two inhabitants per sq. km (five per sq. mile), spread across an area of 1,221,600 sq. km (472,000 sq. miles). Lhasa is its capital. Tibetans were officially recorded as making up 95% of its total population of 2.2 million at the 1990 census. Han Chinese, whose immigration has aroused strong local resentment, were recorded as under 4%, though another census in 1993 gave a total population figure of nearly three million.

In 1950, Chinese forces invaded Tibet, which had been part of the Chinese empire from the 18th century until 1911, and ruthlessly crushed the 1959 independence uprising. The Dalai Lama, spiritual head of Tibetan Buddhism, fled to India amid an exodus of refugees and established a government in exile, but he has recently come close to accepting Chinese rule in an effort to win greater real self-government and cultural freedom for the Tibetan people. In 1965 Tibet was made a region of China, known as Xizang Zizhiqu. Opponents of Chinese rule were imprisoned or executed and many Buddhist monasteries were destroyed. Clashes between nationalists and Chinese troops in 1987 led to a renewed clampdown. In 1988, in an apparently more conciliatory spirit, the Chinese government accepted Tibetan as a "major official language."

Since 1995, the Chinese authorities have detained a young boy recognized by Tibetan Buddhists as the latest incarnation of their second-ranking spiritual leader, the Panchen Lama. The Chinese government named its own candidate to this office. Tibetans see this as unwarranted interference in their religious life. They are also alarmed by Chinese promotion of Tibet as a tourist destination, and the start of major restoration projects on the Potala and Norbulinka palaces in Lhasa.

XINJIANG UYGHUR
Consisting largely of desert, and bordering on Kyrgyzstan and Kazakhstan, the Xinjiang Uyghur Zizhiqu has a total area of 1,646,900 sq. km (636,000 sq. miles), or nearly 18% of China's total land mass. Han Chinese are narrowly in the majority in its

The Kumbum dagoba *at the Palkhor Tschöde monastery, Gyangze (Chiang-tzu), Tibet. At the tip of this 15th-century structure is a Buddhist chapel.*

population of 15.4 million, the total number of (mainly Muslim) Uyghurs being 7.2 million. Access is restricted and dissent is firmly suppressed, with a major clampdown on "terrorists" in 2002. The regional capital, Urumqi, is an industrial city noted for iron and steel, oil, and chemicals.

INNER MONGOLIA
Inner Mongolia, or Nei Mongol Zizhiqu, occupies most of the area of northern China along the long land border with Mongolia (i.e. Outer Mongolia, as seen from the Chinese perspective). This totals 1,177,500 sq. km (455,000 sq. miles). Han Chinese are the majority in a total population of 21.1 million, of whom 4.8 million, or less than a quarter, are ethnic Mongolian.

NINGXIA HUI
The smallest of all the autonomous regions, Ningxia Hui Zizhiqu occupies an area of 66,000 sq. km (25,000 sq. miles), just south of Nei Mongol Zizhiqu. Its capital is Yinchuan. The population was 4.7 million at the 1990 census. Many of the Hui people, who number some 8.6 million in all, live outside the region.

GUANGXI ZHUANG
This autonomous region lies in the south, close to China's booming economic heartland. Relatively small in terms of area (220,000 sq. km – 85,000 sq. miles), it is by far the largest in terms of population, with a density of 192 per sq. km (497 per sq. mile) and 42.5 million inhabitants at the 1990 census. The capital is Nanning. The Zhuang people, although China's largest ethnic minority, number only 15.5 million, and thus are substantially in the minority in the region.

Monasteries

SPECIAL ECONOMIC ZONES, OPEN CITIES, AND SPECIAL ADMINISTRATIVE REGIONS

ECONOMIC REFORMS FIRST INSTITUTED by Deng Xiaoping in 1978 began opening China to foreign business. This has spread from a small number of zones operating under special tax regimes to other cities all along the coastal belt and inland, and has fueled an extraordinary urban investment boom.

Special Economic Zones (SEZs)   ■ Special Administrative Regions (SARs)

SPECIAL ECONOMIC ZONES
The creation of five special economic zones (SEZs) on the south coast was a major early milestone. Shenzhen, the trailblazer for the concept, was established in 1980, adjoining Kowloon in Hong Kong. Also in Guangdong province are Zhuhai, adjoining Macao, and Shantou further east, while Xiamen SEZ is in Fujian province, facing Taiwan. These locations reflect the aim of attracting investment from the 30 million Chinese overseas, especially in Hong Kong and Taiwan, whose ancestral homes were in Guangdong and Fujian. The fifth and least dynamic of the SEZs comprises the southern island province of Hainan.

FOREIGN INVESTMENT
Foreign direct investment reached over $45 billion in 1998, more than for any other country except the US. The inflow then began falling, for the first time for a decade. Most investment goes into the Open Coastal Belt, based on 14 cities, from Dalian in the north to Zhanjiang and Beihai in the south, which were picked in the mid-1980s for the dual role of "windows," opening to the outside world, and "radiators," spreading the development of an export-oriented economy.

Among the most successful development areas is the great port of Shanghai, where the stock exchange reopened in 1990. Investors in the city's Pudong New Zone, on the east bank of the Huangpu River, enjoy more preferential conditions than do the SEZs, including the right to sell goods and financial services. Pudong has attracted major foreign companies keen to establish a foothold in the potentially massive Chinese market, including General Motors, NEC, Sharp, Hitachi, Siemens, Unilever, BASF, and Pilkington. Pudong also forms the "dragon head" for a chain of open cities, extending up the Yangtze River (Chang Jiang) valley, where foreign investment has been encouraged since 1990. China has also designated a new set of open cities since 1992, this time in areas near its land borders, adjoining Russia, Mongolia, Kazakhstan, Burma, and Vietnam, to develop infrastructure and promote trade and the growth of export-oriented industries.

INDUSTRIAL GROWTH IN GUANGDONG
Guangdong province, with Guangzhou and Shenzhen at its heart, promises to become Asia's largest industrial region. One in ten of the world's top 500 multinational companies has invested in the Guangzhou Economic and Technical Development District (GET). Shenzhen, meanwhile, was the first Chinese city to start selling state-owned apartments freehold. It is being revamped as the science and technology city of the future, complete with parks, pedestrian precincts, civic amenities, and a sophisticated communications infrastructure. The process of reform was extended to politics in 2002 in an experiment to tackle rampant corruption.

HONG KONG AND MACAO SARS
Hong Kong, recession-hit after its reversion to Chinese sovereignty in mid-1997, recovered to record impressively high growth in 2000, but then suffered the knock-on effects of the global economic slowdown. As a British colony, Hong Kong had flourished through its textile industry, subsequently expanding into electronics, but it was above all a trade and financial services center. This role is preserved under its status as a special administrative region (SAR). Hong Kong's population is estimated at 6.69 million, in an area of only 1098 sq. km (424 sq. miles). Its Basic Law guarantees a high level of autonomy for 50 years under the "one country, two systems" formula. Its Chinese-appointed chief executive, Tung Chee-hwa, has been unpopular, however – increasingly at odds with members of his administration (several of whom resigned in 2000–2001), and with the region's Legislative Council. The number of directly elected seats on the Council was increased somewhat (to 24 out of 60) for the September 2000 elections, half of which were won by the opposition Democratic Party, but public apathy produced a low turnout. English and Chinese are both official languages, and the freely convertible Hong Kong dollar remains as the SAR's currency, while Hong Kong has its own separate membership of key international organizations, notably the WTO.

China's other SAR, Macao, ceased to be a Portuguese colony at midnight on 19–20 December 1999. Macao too has free port status, its own currency, the pataca, two official languages (Portuguese and Chinese), a partly elective Legislative Assembly, and a guarantee under its Basic Law that the "one country, two systems" formula will apply for 50 years. Macao's area of only 25.4 sq. km (9.8 sq. miles) was due to increase by 20% on completion of the Nam Van lakes project. The population of 414,128 at the 1996 census included more than 100,000 people with Portuguese nationality. The economy, based on tourism and gambling, has recently been in decline.

Hong Kong's return to Chinese sovereignty after 157 years of British rule, took place at midnight on 30 June 1997.

C

CHRONOLOGY

China's recorded history begins 4000 years ago with the Shang dynasty, which was founded in the north in 1766 BCE. Succeeding dynasties expanded China's boundaries; it reached its greatest extent under the Manchu (Qing) dynasty in the 18th century. Chinese isolationism frustrated Europe's attempts to expand trade into the empire until the 19th century, when China had fallen behind the industrializing West. For the previous 3000 years, it had been one of the world's most advanced nations.

❏ **1839–1860** Opium Wars with Britain. China defeated; forced to open ports to foreigners.
❏ **1850–1873** Internal rebellions against Manchu Empire.
❏ **1895** Defeat by Japan in war over Korean peninsula.
❏ **1900** Boxer Rebellion to expel all foreigners suppressed.
❏ **1911** Manchu Empire overthrown by nationalists led by Sun Yat-sen. Republic of China declared.
❏ **1912** Sun Yat-sen forms National People's Party (Guomindang).
❏ **1916** Nationalists factionalize. Sun Yat-sen sets up government in Guangdong. Rest of China under control of rival warlords.
❏ **1921** CCP founded in Shanghai.
❏ **1923** CCP joins Soviet-backed Guomindang to fight warlords.
❏ **1925** Chiang Kai-shek becomes Guomindang leader on death of Sun Yat-sen.
❏ **1927** Chiang turns on CCP. CCP leaders escape to rural south.
❏ **1930–1934** Mao Zedong formulates strategy of peasant-led revolution.
❏ **1931** Japan invades Manchuria.
❏ **1934** Chiang forces CCP out of its southern bases. Start of 12,000-km (7450-mile) Long March.
❏ **1935** Long March ends in Yanan, Shaanxi province. Mao becomes CCP leader.
❏ **1937–1945** War against Japan: CCP Red Army in north, Guomindang in south. Japan defeated.
❏ **1945–1949** War between Red Army and Guomindang. US-backed Guomindang retreats to Taiwan.
❏ **1949** 1st October, Mao proclaims People's Republic of China.
❏ **1950** Invasion of Tibet. Mutual assistance treaty with USSR.
❏ **1950–1958** Land reform; culminates in setting up of communes. First Five-Year Plan (1953–1958) fails.
❏ **1958** "Great Leap Forward" to boost production fails; contributes to millions of deaths during 1959–1961 famine. Mao resigns ⇨

RESOURCES

  Electric power 231m kw

41.5m tonnes

636m ducks, 454m pigs, 208m geese, 3.77bn chickens

3.31m b/d (reserves 24bn barrels)

Coal, oil, natural gas, salt, iron, molybdenum, titanium, tungsten

ELECTRICITY GENERATION

Hydro 18% (208bn kwh)
Combustion 81% (944bn kwh)
Nuclear 1% (14bn kwh)
Other 0%

% of total generation by type

China dominates the world market in molybdenum, titanium, and tungsten; it has the world's largest deposits of more than a dozen minerals, and commercial deposits of most others.

China is the world's second-largest coal producer, with reserves of about 800 billion tonnes, primarily in the Shaanxi and Sichuan basins. Annual output (over one billion tonnes) considerably exceeds demand; this and appalling safety records are forcing the closure of many mines.

Power generation, previously well behind demand, expanded rapidly to create overcapacity by the late 1990s. Nuclear power capacity in 2002 was just over 2100 MW from three reactors,

with major expansion to 50,000 MW planned by 2020. The world's largest hydropower plant, the "Three Gorges" scheme on the Yangtze River, is due for completion in 2009, but faces sustained controversy over its proposed benefits, costs, and environmental consequences.

Crude oil production has risen only slightly since reaching 160 million tonnes in 1997. Eastern oil fields are depleted, and hopes now center on enormous reserves in the Tarim basin in the far west. A 4000-km (2500-mile) pipeline is due to come on line in 2004.

CHINA : LAND USE

Cropland
Forest
Pasture
Wetlands
Desert
High mountain regions
Sheep
Tea
Rice

0   200 km
0   200 miles

ENVIRONMENT

 Sustainability rank: 129th

6%

2.5 tonnes per capita

ENVIRONMENTAL TREATIES

Yes   Yes   Yes
Yes   No   No

Climate and geology mean that natural disasters are quite frequent in China, and their impact is often made worse by human actions. The economic policies of the 1950s turned drought into a devastating famine, while poor building standards helped push the death toll in the 1976 Tangshan earthquake to over 500,000.

Economic growth is the priority of China's leaders. Industrial pollution and environmental degradation, already widespread, are increasing. However, the environment is a growing concern among educated Chinese. The campaign, unsuccessful in 1992, to stop the Three Gorges hydroelectric scheme was revived with growing open criticism in the late 1990s. The government is becoming less suspicious of Western pressure for environmental controls, and is taking steps to respond to acute problems of urban air pollution, deforestation, and water quality in particular.

MEDIA

▷ TV ownership high

☒ Daily newspaper circulation 40 per 1000 people

PUBLISHING AND BROADCAST MEDIA

There are 39 major daily newspapers. *Renmin Ribao* is the CCP daily; *China Daily* is published in English

1 state-owned service

2 state-owned services

China's more open, market-oriented economy has created access to nonofficial sources of information. TV ownership is rising and satellite-dish owners can choose what to view, while growing Internet usage makes central control even more difficult. Since 2000, Internet sites have had to obtain official approval and have been held responsible for their content. Many Internet cafés were closed in 2001, and new rules in 2002 restricted the media's use of Internet sources.

The ideological influence of the huge-circulation party newspaper *Renmin Ribao* (*People's Daily*) and the trade union *Gongren Ribao* (*Workers' Daily*) is much diminished, but "undesirable" papers have their licenses removed in periodic cleanups. Beijing and the provinces have their own dailies.

C

CHRONOLOGY *continued*

as CCP chairman; succeeded by Liu Shaoqi.

- ❑ **1960** Sino-Soviet split.
- ❑ **1961–1965** More pragmatic economic approach led by Liu and Deng Xiaoping.
- ❑ **1966** Cultural Revolution initiated by Mao to restore his supreme power. Youthful Red Guards attack all authority. Mao rules, with Military Commission under Lin Biao and State Council under Zhou Enlai.
- ❑ **1967** Army intervenes to restore order amid countrywide chaos. Liu and Deng purged from party.
- ❑ **1969** Mao regains chair of CCP. Lin Biao designated his successor, but quickly attacked by Mao.
- ❑ **1971** Lin dies in plane crash.
- ❑ **1972** US president Nixon visits. More open foreign policy initiated.
- ❑ **1973** Mao's wife Jiang Qing, Zhang Chunquio, and other "Gang of Four" members elected to CCP Politburo. Deng Xiaoping rehabilitated.
- ❑ **1976** Death of Zhou Enlai. Mao strips Deng of posts. September, Mao dies. October, Gang of Four arrested.
- ❑ **1977** Deng regains party posts, begins to extend power base.
- ❑ **1978** Decade of economic modernization launched. Open door policy to foreign investment; farmers allowed to farm for profit.
- ❑ **1980** Deng emerges as China's paramount leader. Economic reform gathers pace, but hopes for political change suppressed.
- ❑ **1983–1984** Elderly conservative leaders attempt to slow reform.
- ❑ **1984** Industrial reforms announced.
- ❑ **1989** Prodemocracy demonstrations in Tiananmen Square. Crushed by army; 1000–5000 dead.
- ❑ **1992–1995** Trials of prodemocracy activists continue. Plans for market economy accelerated.
- ❑ **1993** Jiang Zemin president.
- ❑ **1997** February, Deng Xiaoping dies at 92. July, UK hands back Hong Kong. September, party congress confirms reformist policies.
- ❑ **1999** China develops neutron bomb. Portugal hands back Macao. Friction over Taiwanese claim of statehood. Clampdown on Falun Gong sect.
- ❑ **2000** Taiwanese presidential election causes tension. USA normalizes trade relations.
- ❑ **2001** Major diplomatic incident with USA when Chinese pilot is killed and US spy-plane is forced down on Hainan Island. December, accession to WTO.
- ❑ **2002** Crackdown on Uyghur separatism. Severe flooding: 750 killed.

CRIME

 Death penalty in use

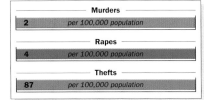

1.43m prisoners   **Down 33%** 1990–1996

CRIME RATES

Murders	
2	per 100,000 population

Rapes	
4	per 100,000 population

Thefts	
87	per 100,000 population

China's legal system is a mix of custom and statute; judges have only been required to hold law degrees since 2002.

Economic reform and social changes have been paralleled by a rise in violent crime and corruption. Many party officials were condemned in China's largest anticorruption trial in 2000. A crackdown on people trafficking followed the breaking of rings smuggling Chinese into Europe. The death penalty is used extensively – China carried out over 80% of the world's executions in 2001. Many detainees have been released since the clampdown on dissent after the 1989 Tiananmen Square massacre, but many more remain incarcerated.

EDUCATION

 School leaving age: 16

84%   **5.56m students**

THE EDUCATION SYSTEM

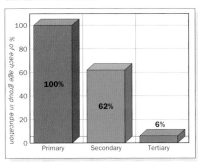

Despite the expansion of education since 1949, illiteracy and semiliteracy are still widespread. School attendance fell when fees at all levels were introduced in the 1980s, but now most children of secondary school age are in school – including those catching up with primary education.

Selection for higher education is now based on academic rather than political criteria. Internet-based distance learning degrees, first allowed in 1998, are especially popular in English, computing, and business studies. The government legalized private schools (tacitly permitted since the early 1980s) in 2001, in an effort to regulate and profit from them. About seven million pupils are thought to attend them.

HEALTH

 Welfare state health benefits

1 per 588 people   Cardiovascular and diarrheal diseases, cancers, tuberculosis

China's network of primary health care, combining traditional and Western medicine, used to extend to the remotest regions, and a free health system meant that the Chinese enjoyed a life expectancy on a par with many richer countries. The change to a market-oriented economy, however, has produced a two-tier system. A gaping divide exists between city and rural provision, fees for treatment are rising, and fewer people are covered by free health care as a benefit of state employment. The UN estimated in 2002 that there would be over ten million HIV/AIDS sufferers in China by 2010.

SPENDING

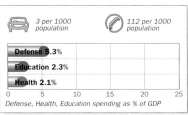 GDP/cap. increase

CONSUMPTION AND SPENDING

3 per 1000 population   **112 per 1000 population**

Defense 5.3%
Education 2.3%
Health 2.1%

Defense, Health, Education spending as % of GDP

The majority of Chinese are still farmers, whose living standards are threatened by rising production costs. Economic change has led to widening wealth disparities. The burgeoning small-business class and employees of companies with foreign investment have benefited most. They mainly live in the east, especially the southeast, where there are a number of dollar millionaires. The main losers are the 150 million "surplus" agricultural workers, many of whom have migrated to the cities in search of jobs. By July 2001, there were more mobile phones in China than in the US, 120.6 million in total, about one for every ten Chinese.

WORLD RANKING

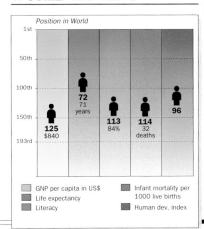

- GNP per capita in US$
- Life expectancy
- Literacy
- Infant mortality per 1000 live births
- Human dev. index

COLOMBIA

SOUTH AMERICA

OFFICIAL NAME: Republic of Colombia **CAPITAL:** Bogotá
POPULATION: 42.8 million **CURRENCY:** Colombian peso **OFFICIAL LANGUAGE:** Spanish

C

 1819   1903   July 20   CO   -5   +57   .co

LYING IN NORTHWEST South America, Colombia has coastlines on both the Caribbean and the Pacific. The east is densely forested and sparsely populated, and separated from the western coastal plains by the Andes mountains. The Andes divide into three ranges (cordilleras) in Colombia. The eastern range is divided from the two western ranges by the densely populated Magdalena River valley. The Colombian lowlands are very wet, hot, and fertile, supporting two harvests and allowing many crops to be planted at any time of year. A state plagued by instability and violence, Colombia is noted for its coffee, emeralds, gold, and narcotics trafficking.

CLIMATE

▷ Tropical/mountain

WEATHER CHART FOR BOGOTÁ

Most of Colombia is wet, and the hot Pacific coastal areas receive up to 500 cm (200 in) of rain a year. The Caribbean coast is a little drier. The Andes have three climatic regions: the *tierra caliente* (hot lowlands), *tierra templada* (temperate uplands), and *tierra fría* (cold highlands); the last has year-round springlike conditions such as those found in Bogotá. The equatorial east has two wet seasons.

TRANSPORTATION

▷ Drive on right

✈ **El Dorado, Bogotá**
4.66m passengers

⚓ 111 ships
81,400 grt

THE TRANSPORTATION NETWORK

🛣 13,868 km
(8617 miles)

🛤 Caribbean Trunk Highway

🚂 2113 km
(1313 miles)

⛰ 18,140 km
(11,272 miles)

Roads in the north are in reasonable condition. Those in the south and east tend to be rutted and badly affected by the frequent rains. Continuing instability means that roads are frequently blocked by the guerrillas and the military. Most of the railroad is closed.

Rivers are an important means of transportation; the Magdalena, Orinoco, Atrato, and Amazon river systems are all extensively navigable. Plans exist to connect Colombia to the Pan-American Highway.

TOURISM

▷ Visitors : Population
1:81

📦 530,000 visitors

📉 Down 3% in 2000

MAIN TOURIST ARRIVALS

	% of total arrivals
USA 30%	
Venezuela 14%	
Ecuador 11%	
Panama 9%	
Peru 5%	
Other 31%	

Tourism in Colombia is largely limited to the beaches of the Caribbean coast. Cartagena, Barranquilla, and Santa Marta are the main resorts. Cartagena has also been developed as a major Latin American conference center.

The expansion of tourism has been limited by Colombia's political instability and the prevalence of narcotics-related crime. The well-publicized activities of drugs cartels in Medellín and Cali, and instances of kidnappings in Bogotá, are major deterrents for travelers.

Limited infrastructure makes many regions of the country, particularly Amazonia to the east of the Andes, almost inaccessible. The Caribbean coast is also barely exploited.

Simón Bolívar and Cristóbal Colón, twin peaks with a height of 5775 m (18,947 ft) are the highest in the Colombian Andes.

PEOPLE

▷ Pop. density low

👤 Spanish, Amerindian languages, English Creole

👥 41/km²
(107/mi²)

THE URBAN/RURAL POPULATION SPLIT

74%   26%

RELIGIOUS PERSUASION

Other 5%
Roman Catholic 95%

ETHNIC MAKEUP

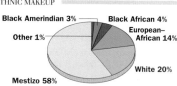

Black Amerindian 3%
Black African 4%
European–African 14%
Other 1%
White 20%
Mestizo 58%

The majority of Colombians are people of mixed blood. An estimated 450,000 indigenous Amerindians are largely concentrated in the southwest and Amazonia, although some communities are scattered throughout the country. A small black population lives along both coasts, particularly in Chocó, Colombia's poorest region. Blacks are the most unrepresented group.

Some progress has been made in giving Amerindians a greater political voice. In 1991, constitutional reforms reserved two seats in the Senate for indigenous representatives, and Amerindian pressure groups are increasingly active. Harassment by landowners and narcotics traffickers continues in Amazonia, and very few investigations into suspected human rights violations against Amerindians have led to prosecutions.

Women in Colombia have a higher profile than in much of the rest of Latin America. Many are prominent in the professions, though few reach the top in politics. The traditional extended Catholic family is still the norm. Regional identity is strong.

POPULATION AGE BREAKDOWN

Female	Age	Male
0.4%	80+	0.3%
3.1%	60–79	2.6%
8.1%	40–59	7.5%
17.3%	20–39	16.9%
21.6%	0–19	22.2%

% of population by age group

POLITICS ▷ Multiparty elections

 L. House 2002/2006
U. House 2002/2006

 President Alvaro
Uribe Velez

Colombia is a presidential democracy, with a bicameral Congress. Presidents may not serve two consecutive terms.

PROFILE

The two-party system which had held sway from the late 1950s appeared to have fractured by the 21st century. The dominance of the PSC and the PL, with few ideological differences, was undermined by electoral breakthroughs by new, smaller parties, culminating in the presidential victory of right-wing independent Alvaro Uribe Velez in 2002.

Pervasive official corruption and the violence associated with drugs cartels, guerrillas, paramilitaries, and the military have seriously weakened confidence in the state and the government, and have deterred foreign investors.

MAIN POLITICAL ISSUES
Elusive peace
Against a background of decades of civil conflict, former president Andres Pastrana Arango

AT THE LAST ELECTION
House of Representatives 166 seats

| 33% PL | 13% PCC | 10% C | 4% CR | 3% AL | 37% Others |

PL = Liberal Party **PCC** = Colombian Conservative Party
C = Coalition **CR** = Radical Change **AL** = Liberal Opening
MN = National Movement

Senate 102 seats

| | | | 6% MN | |
| 27% PL | 13% PCC | 6% C | | 48% Others |

Two special representatives of the Amerindian communities are appointed to the Senate, and one is elected.

(1998–2002) took political risks in pursuit of a lasting peace, granting "safe havens" in remote areas to the Armed Revolutionary Forces of Colombia (FARC) and National Liberation Army (ELN). Mutual suspicions undermined these peace efforts, which were abandoned in early 2002. A renewed military offensive focused mainly on the FARC.

Violence
At least 2856 people were kidnapped in 2001, including 35 foreigners, with guerrilla groups and right-wing paramilitaries blamed for 85% of cases. Unions said that 152 labor leaders were killed. The murder in 2002 of the Archbishop of Cali rocked the nation.

Alvaro Uribe Velez, *the right-leaning independent elected president in 2002.*

Pablo Escobar, *ex-leader of Medellín drugs cartel, gunned down in 1993.*

C

WORLD AFFAIRS ▷ Joined UN in 1945

 ACS   AP   AmCC   OAS   RG

Good relations with the US are conditional on tough measures to fight the narcotics trade. A $1.3 billion military aid package, approved in 2000, provided training, intelligence, and hardware, ostensibly to assist the "Plan Colombia" antidrugs program. The boundary between this aim and the targeting of guerrilla groups was always blurred.

The army's invasion in 2002 of the FARC's demilitarized haven reportedly was assisted by US military advisers. However, the US administration fought shy of describing events as a chapter in the US-led "war on terrorism," insisting that it was operating within legal limits set down by the Congress. Colombia's neighbors fear a "spillover" of violence and refugees.

AID ▷ Recipient

$187m (receipts) Down 38% in 2000

US military "antinarcotics aid" forms some 75% of the US total. The World Bank approved a loan of $400 million in 2002 to aid fiscal reform and improve delivery of public health services.

COLOMBIA

Total Land Area :
1 138 910 sq. km
(439 733 sq. miles)

LAND HEIGHT

 5000m/9843ft
2000m/6562ft
1000m/3281ft
500m/1640ft
Sea Level

POPULATION

◻	over 1 000 000
◉	over 500 000
◎	over 100 000
○	over 50 000
●	over 10 000
•	under 10 000

0   200 km
0   200 miles

CHRONOLOGY

In 1525, Spain began the conquest of Colombia, which became its chief source of gold.

❏ **1819** Simón Bolívar defeats the Spanish at Boyacá. Republic of Gran Colombia formed with Venezuela, Ecuador, and Panama.
❏ **1830** Venezuela and Ecuador split away during revolts and civil wars.
❏ **1849** The Centralist Conservative and federalist Liberal parties are established.
❏ **1861–1886** Liberals hold monopoly on power.
❏ **1886–1930** Conservative rule.
❏ **1899–1903** Liberal "War of 1000 Days" revolt fails; 120,000 die. ⇨

CHRONOLOGY *continued*

- ❏ **1903** Panama secedes, but is not recognized by Colombia until 1921.
- ❏ **1930** Liberal President Olaya Herrera elected by coalition in first peaceful change of power.
- ❏ **1946** Conservatives take over.
- ❏ **1948** Shooting of Liberal mayor of Bogotá and riot known as *El Bogotazo* spark civil war, *La Violencia*, to 1957; 300,000 killed.
- ❏ **1953–1957** Military dictatorship of Rojas Pinilla.
- ❏ **1958** Conservatives and Liberals agree to alternate government in a National Front until 1974. Other parties banned.
- ❏ **1965** Left-wing guerrilla National Liberation Army and Maoist Popular Liberation Army founded.
- ❏ **1966** Pro-Soviet FARC guerrilla group formed.
- ❏ **1968** Constitutional reform allows new parties, but two-party parity continues. Guerrilla groups proliferate from now on.
- ❏ **1984** Minister of justice assassinated for attempting to enforce antinarcotics campaign.
- ❏ **1985** M-19 guerrillas blast their way into ministry of justice; 11 judges and 90 others killed. Patriotic Union (UP) party formed.
- ❏ **1986** Liberal Virgilio Barco Vargas wins presidential elections, ending power-sharing. UP wins ten seats in parliament. Right-wing paramilitary start murder campaign against UP politicians. Violence by both left-wing groups and death squads run by narcotics cartels continues.
- ❏ **1989** M-19 reaches peace agreement with government, including the granting of a full pardon. Becomes legal party.
- ❏ **1990** Presidential candidates of UP and PL murdered during campaign. Liberal César Gaviria elected on antidrugs platform.
- ❏ **1991** New constitution legalizes divorce, prohibits extradition of Colombian nationals. Indigenous peoples' democratic rights guaranteed, but territorial claims are not addressed.
- ❏ **1992–1993** Medellín drugs cartel leader, Pablo Escobar, captured, escapes, and shot dead by police.
- ❏ **1995–1996** President Samper cleared of charges of receiving Cali cartel drug funds for elections.
- ❏ **1999** Earthquake kills thousands.
- ❏ **2001** US-backed spraying of coca plantations and destruction of food crops by herbicides provokes resentment.
- ❏ **2002** Peace talks abandoned. Renewed military offensive. Independent candidate Alvaro Uribe Velez elected president.

DEFENSE

 Compulsory military service

 $1.96bn   ⬇ Down 10% in 2000

COLOMBIAN ARMED FORCES

🛡	30 light tanks (M-3A1)	136,000 personnel
🚢	4 submarines, 4 corvettes, and 27 patrol boats	15,000 personnel
✈	58 combat aircraft (7 *Mirage* 5, 11 *Kfir*)	7000 personnel
	None	

The military is powerful, but rarely intervenes directly in politics. Human rights groups accuse the armed forces and their paramilitary allies of gross and systematic abuses, involving torture and murder, in their fight against guerrilla groups and the production of narcotics. Despite its restructuring in 1998, the army high command remained suspicious of peace negotiations and supported a tougher stance against the rebels, successfully exploiting tensions to expand substantially under President Pastrana, with numbers set to rise still more. Orders were given in February 2002 to retake the FARC "safe haven."

Colombia participates in the joint Latin American Defense Force. The US supplies most arms and training, especially through "Plan Colombia."

ECONOMICS

 Inflation 21% p.a. (1990–2000)

📊 $85.3bn   💲 2236–2278 Colombian pesos

SCORE CARD

- ❏ WORLD GNP RANKING..........................40th
- ❏ GNP PER CAPITA $2020
- ❏ BALANCE OF PAYMENTS $41m
- ❏ INFLATION 9.5%
- ❏ UNEMPLOYMENT 20%

EXPORTS

Germany 3% — Ecuador 4%
Peru 3% — Venezuela 10%
USA 51%
Other 29%

IMPORTS

Mexico 5% — Japan 5%
Brazil 4% — Venezuela 8%
Other 44%
USA 34%

ECONOMIC PERFORMANCE INDICATOR

Consumer Price Index — GDP

Consumer price index 1995=100 / GDP 1996=100
1996 1997 1998 1999 2000

Colombia has a successful private export sector. A program of privatization and macroeconomic stabilization was rewarded with IMF assistance of $1.5 billion in 2002.

Regional disparities remain marked. Most wealth is found in the Bogotá, Medellín, and Cali regions. Rural areas are largely underdeveloped. The main obstacle to growth is the instability caused by the narcotics business and protracted conflict. Given stability and investment, Colombia's potential for growth is considerable.

STRENGTHS

Substantial oil and coal deposits plus well-developed hydroelectric power makes Colombia almost self-sufficient in energy. Diversified export sector – especially coffee and coal. Light manufactures. Worldwide market for cocaine.

WEAKNESSES

Narcotics-related violence, corruption, and political instability discourage foreign investors. Domestic industry uncompetitive owing to protection. High unemployment. Coffee and oil subject to world price fluctuations.

PROFILE

Of all the Latin American economies, Colombia's is probably the closest to the US model. The state has traditionally played a relatively minor role, and

COLOMBIA : MAJOR BUSINESSES

Pulp and paper
Narcotics
Steel
Chemicals
Vehicle assembly
Food processing
Textiles
Oil

Barranquilla
Medellín
Cali
Bogotá
Ibagué
Orito

0   200 km
0   200 miles
* significant multinational ownership

RESOURCES

▷ Electric power 11.7m kw

170,896 tonnes

627,000 b/d (reserves 1.8bn barrels)

28.3m cattle, 2.75m pigs, 2.6m horses, 110m chickens

Oil, coal, natural gas, silver, emeralds, gold, platinum

ELECTRICITY GENERATION

Hydro 67% (31bn kwh)	
Combustion 33% (15bn kwh)	
Nuclear 0%	
Other 0%	

0   20   40   60   80   100

% of total generation by type

Colombia has substantial oil reserves but needs increasing investment to maintain production. Coal and gas are important, and it is a major producer of gold, platinum, silver, and emeralds.

COLOMBIA : LAND USE

Cropland
Pasture
Forest
High mountain regions
Coffee - cash crop
Cattle

0   200 km
0   200 miles

ENVIRONMENT

▷ Sustainability rank: 22nd

9%

1.7 tonnes per capita

ENVIRONMENTAL TREATIES

Yes	Yes	Yes
Yes	Yes	Yes

Cattle ranching, logging, coca growing, and spraying cause soil degradation and loss of natural habitats.

MEDIA

▷ TV ownership high

Daily newspaper circulation 26 per 1000 people

PUBLISHING AND BROADCAST MEDIA

	There are 37 daily newspapers. *El Tiempo* and *El Espectador* have the largest circulations
	3 services: 1 state-owned, 2 independent
	589 stations: 31 state-owned, 558 independent

The independent press is small. Journalists have been murdered by paramilitaries and held by guerrillas.

CRIME

▷ No death penalty

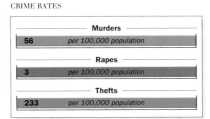

57,068 prisoners

Up 35% 1996–1998

CRIME RATES

Murders	
56	*per 100,000 population*

Rapes	
3	*per 100,000 population*

Thefts	
233	*per 100,000 population*

Colombia is one of the most violent countries in the world. Armed groups assassinated 20 mayoral candidates and 20 mayors, and kidnapped 200 other candidates, in regional and local elections in 2000. In late 2001, the local monitoring group Fundación País Libre reported 2856 cases of kidnapping that

EDUCATION

▷ School leaving age: 15

92%

644,188 students

THE EDUCATION SYSTEM

% of each age group in education

100% (Primary)
53% (Secondary)
21% (Tertiary)

Education in Colombia is free and compulsory, and is a mix of French and US models, with a *baccalauréat* examination taken at the end of secondary school. Provision in rural areas is poor and absenteeism high. Where provided, public and university education is generally of a high standard, but the resources available to public education have decreased due to budget cuts. The rich send their children to private schools and universities in the US.

HEALTH

▷ Welfare state health benefits

1 per 833 people

Heart diseases, cancers, violence, accidents

Budget cuts have reduced health spending. Private care is growing. Only 16% of Colombians benefit from any social security system, rather fewer than in most Latin American states. Rural areas have little health provision, since most doctors work in the larger cities. A polio vaccination campaign has largely eradicated the virus from Colombia, except in coastal regions.

year, mostly by paramilitaries and guerrillas, the rest blamed on other criminal groups. Anonymous gunmen murdered the Archbishop of Cali in 2002. NGOs estimate that two million people have been displaced over the last 14 years. Homicide is the main cause of death among young men in cities; overall, it rates third in the mortality stakes. Much of the violence is narcotics-related, and Bogotá, Cali, Barrancabermeja, and Medellín can all be dangerous; the army, the police, paramilitaries, and guerrillas are thought to be involved. Frequent armed robberies and kidnappings make wealthy residents extremely security conscious.

A relatively new phenomenon is that of "social cleansing" – the murder of street children and beggars by armed gangs, some in Bogotá funded by businesses.

SPENDING

▷ GDP/cap. increase

CONSUMPTION AND SPENDING

43 per 1000 population

169 per 1000 population

Defense 2.4%	
Education 4.1%	
Health 5.2%	

0   5   10   15   20   25

Defense, Health, Education spending as % of GDP

There is little social mobility; the historically wealthy Spanish families are still dominant in political and business life, but the entry of narcotics-related money has created new layers of rich in cities and among landowners. Drug money also finances the import of consumer goods such as TV sets, computers, and perfume. The wealthy go to the US for medical treatment and educate their children overseas. The rural poor are mostly landless. The inhabitants of shanty towns in Barranquilla, Buenaventura, Cali, and Cartagena form the poorest groups.

WORLD RANKING

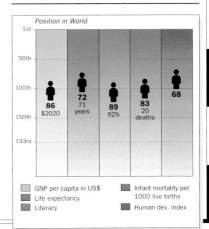

Position in World

1st
50th
100th
150th
193rd

86 $2020
72 71 years
89 92%
83 20 deaths
68

GNP per capita in US$
Life expectancy
Literacy

Infant mortality per 1000 live births
Human dev. index

COMOROS

OFFICIAL NAME: Federal Islamic Republic of the Comoros   **CAPITAL:** Moroni
POPULATION: 727,000   **CURRENCY:** Comoros franc   **OFFICIAL LANGUAGES:** Arabic, French, and Comoran

C

 1975   1975   July 6   COM   +3   +269   .km

THE ARCHIPELAGO republic of the Comoros lies off the east African coast, between Mozambique and Madagascar. It consists of three main islands and a number of islets. Most of the population are subsistence farmers. In 1975, the Comoros islands, except for Mayotte, became independent of France. Since then instability has plagued this poor region, with countless coups and countercoups, and repeated attempts at secession by smaller islands.

Moroni, the capital, on Grande Comore.
The Comoros islands are fertile and heavily forested. Many are ringed by coral reefs.

CLIMATE ▷ Tropical oceanic

WEATHER CHART FOR MORONI

The islands are tropical; it is hot and humid on the coasts and cooler higher up, notably on Mount Kartala.

TRANSPORTATION ▷ Drive on right

Moroni–Hahaya, Grande Comore
118,565 passengers

3 ships
2959 grt

THE TRANSPORTATION NETWORK

673 km (418 miles)	None
None	None

Recent projects have included the development of the port at Moroni and upgrading the international airport.

TOURISM ▷ Visitors : Population 1:30

24,000 visitors   ▽ Down 11% in 1999

MAIN TOURIST ARRIVALS

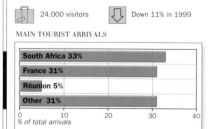

South Africa 33%	
France 31%	
Réunion 5%	
Other 31%	

% of total arrivals

In 1988, Sun International of South Africa joined a major project to build four hotels designed to attract 12,000 visitors a year. However, political instability has wrecked the islands' tourism prospects.

PEOPLE ▷ Pop. density high

Arabic, Comoran, French   326/km² (844/mi²)

THE URBAN/RURAL POPULATION SPLIT

33%   67%

RELIGIOUS PERSUASION

Other 1% — Roman Catholic 1%

Muslim (mainly Sunni) 98%

The Comoros has absorbed Polynesians, Africans, Indonesians, Persians, and Arabs over time, as well as immigrants from Portugal, Holland, France, and India. Some communities retain their individual character; for instance, Mwali (Mohéli) is still primarily African. Ethnic tension is rare, partly owing to the unifying force of Islam, the predominant religion. A more potent divisive factor, especially on Anjouan, is regionalism.

POLITICS ▷ In transition

1996/2000 (postponed)   President Assoumani Azzali

AT THE LAST ELECTION

Federal Assembly 43 seats   2% Ind

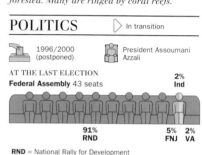

91% RND   5% FNJ   2% VA

RND = National Rally for Development
FNJ = National Front for Justice
Ind = Independent   **VA** = Votes annulled

The Senate was abolished under the 1996 constitution

Attempts to introduce democracy have been consistently undermined by repeated coups. Anjouan declared independence in 1997, since when rival militias have favored either cohabitation with Grande Comore or reattachment to France. Renewed violence in April 1999 provided Col. Assoumani Azzali with the pretext to install himself as Comoran president. In a new "Union of the Comoros" a loose confederation gives each island its own president and legislature, with a federal president having overall authority.

In 2002 Azzali was returned to power unopposed, although his victory was undermined by low voter turnout.

COMOROS

Total Land Area : 2170 sq. km (858 sq. miles)

POPULATION
over 10 000 ●
under 10 000 •

LAND HEIGHT
2000m/6562ft
1000m/3281ft
500m/1640ft
Sea Level

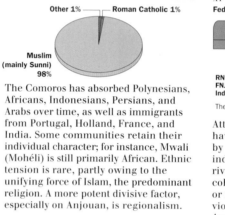

Mitsamiouli
Ntsaouéni
Mbéni
Grande Comore (Njazidja)
Koimbani
Itsandra
MORONI
Pidjani
La Kartala 2361m
Mitsoudjé
Foumbouni
Dembéni

Mohéli (Mwali)
Hoani   Fomboni
Itsamia
Ndréméani

0 20 km
0 20 miles

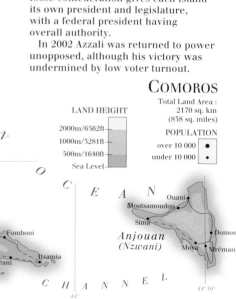

Ouani
Moutsamoudou
Sima
Anjouan (Nzwani)
Domoni
Moya   Nréméani

WORLD AFFAIRS

 Joined UN in 1975

France remains the main aid donor, although economic ties with South Africa are also strong. The turbulent

 situation on the war-torn islands necessitated the evacuation in 1999 of a visiting OAU assessment team from Anjouan. An army of European mercenaries attempted to take over Mohéli in December 2001.

AID

 Recipient

$19m (receipts) Down 10% in 2000

Foreign aid, mainly from France, the EU, the UN, and the World Bank, accounts for over 40% of GDP. Because of its Islamic links, the Comoros also gets aid from Arab states and OPEC. In 1998, major donors attacked the government for spending more than 70% on "political superstructure."

DEFENSE

No compulsory military service

$3m (estimate) No significant change

France and South Africa finance the small presidential guard, the principal security force. Mauritian aid was also sought after clashes on Anjouan.

ECONOMICS

 Inflation 4.9% p.a. (1990–2000)

$212m 528–558 Comoros francs

SCORE CARD

❏ World GNP Ranking	183rd
❏ GNP per Capita	$380
❏ Balance of Payments	–$1m
❏ Inflation	3.5%
❏ Unemployment	20%

STRENGTHS
Vanilla, ylang-ylang, and cloves are the main cash crops.

WEAKNESSES
Subsistence-level farming. Over 50% of food requirements are imported. Lack of basic infrastructure, especially electricity and transportation. Open to allegations of financial mismanagement. Persistent political instability is hindering growth of tourism.

EXPORTS

Germany 6%
Japan 6%
France 31%
Other 19%
Singapore 19%
USA 19%

IMPORTS

Belgium 5%
UAE 7%
Kenya 7%
Other 39%
South Africa 17%
France 25%

RESOURCES

 Electric power 5000 kw

12,200 tonnes Not an oil producer

172,000 goats, 52,000 cattle, 490,000 chickens None

There are few strategic resources. An HEP plant is under construction on Anjouan, but most fuel for energy is still imported. Fishing remains a neglected source of future growth.

ENVIRONMENT

 Not available

None 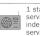 0.1 tonnes per capita

The environment is not a major priority in the Comoros; natural disasters, such as the volcanic eruption in 1977 which left 20,000 people homeless, are of more immediate concern. The government is promoting tourism and recognizes the long-term commercial value of imposing environmental controls on new developments.

MEDIA

 TV ownership low

There are no daily newspapers

PUBLISHING AND BROADCAST MEDIA

There are 2 weekly newspapers, the state-owned *Al Watwan* and the independent *La Gazette des Comores*

No TV service

1 state-controlled service, some independent services

Radio is strictly controlled, and there is no single national newspaper. Although there is criticism of the government, self-censorship is reportedly common.

CRIME

 Death penalty in use

200 prisoners Crime is rising

A climate of lawlessness has been created by the continuing power struggles between rival militias – particularly on the island of Anjouan since 1997.

EDUCATION

 School leaving age: 16

60% 649 students

There is a very limited education system beyond secondary level. Schools are equipped to teach only basic literacy, hygiene, and agricultural techniques. Pupil–teacher ratios are high.

CHRONOLOGY

The Comoros was ruled by matrilineally inherited sultanates until shortly before becoming a French protectorate in 1886.

- ❏ **1961** Internal self-government.
- ❏ **1975** Independence.
- ❏ **1978** Mercenaries restore Ahmed Abdallah to power.
- ❏ **1989** Abdallah assassinated.
- ❏ **1992** Chaotic first multiparty polls.
- ❏ **1996** Mohammed Taki Abdoulkarim elected president.
- ❏ **1997** Anjouan separatists beat off government troops.
- ❏ **1999** Col. Azzali seizes power. Anjouan militias clash.
- ❏ **2000** Fomboni declaration signed with Anjouan.
- ❏ **2001** Successive coups on Anjouan.
- ❏ **2002** New constitution. Azzali returned to office.

HEALTH

 No welfare state health benefits

1 per 14,000 people Malaria, infectious intestinal and bacterial diseases

Health care is rudimentary, except that there are two maternity clinics and 30 health centers.

SPENDING

GDP/cap. decrease

CONSUMPTION AND SPENDING

13 per 1000 population 10 per 1000 population

Defense	No data	
Education 4.2%		
Health 4.9%		

0   5   10   15   20   25
Defense, Health, Education spending as % of GDP

A political and business elite controls most of the wealth. Bridegrooms win social status according to the size of their weddings. Government workers often suffer from wage arrears.

WORLD RANKING

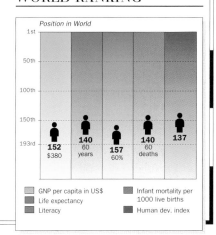

Position in World

152	140	157	140	137
$380	60 years	60%	60 deaths	

GNP per capita in US$
Life expectancy
Literacy
Infant mortality per 1000 live births
Human dev. index

CONGO

OFFICIAL NAME: Republic of the Congo **CAPITAL:** Brazzaville
POPULATION: 3.1 million **CURRENCY:** CFA franc **OFFICIAL LANGUAGE:** French

 1960   1960   Aug 15   RCB   +1   +242 .cg

S TRADDLING THE EQUATOR in west central Africa,
Congo achieved independence from France in 1960,
soon falling under a Marxist–Leninist form of government
which discouraged much foreign investment. Multiparty democracy was
achieved in 1991, but was soon overshadowed by years of violence.

CLIMATE ▷ Tropical equatorial

WEATHER CHART FOR BRAZZAVILLE

In most years there are two wet
seasons and two dry seasons
in Congo. The rainfall is heaviest
in the coastal regions.

TRANSPORTATION ▷ Drive on right

 Brazzaville International
411,678 passengers

 18 ships
3400 grt

THE TRANSPORTATION NETWORK

1242 km (772 miles)	None
895 km (556 miles)	4385 km (2725 miles)

Pointe-Noire is a major port, used by
the Central African Republic, Chad,
and Cameroon. The Congo Ocean
Railroad (to Brazzaville) reopened in
2000. There are plans for a second, and
larger, international airport near Ewo.

TOURISM ▷ Visitors : Population 1:119

 26,000 visitors Up 420% in 2000

MAIN TOURIST ARRIVALS

Africa 50%	
Europe 43%	
Americas 6%	
Other 1%	

% of total arrivals

The Marxist–Leninist regime did not
seek to develop tourism, and visitors,
mostly on safaris and business-related
trips, are still rare, though increasing.

The Loufoulakari Falls, *near Brazzaville.
The Congo River is a key transportation
artery for the region.*

PEOPLE ▷ Pop. density low

 Kongo, Teke, Lingala, French   9/km² (24/mi²)

THE URBAN/RURAL POPULATION SPLIT

63%   37%

ETHNIC MAKEUP

- Other 3%
- Mbochi 12%
- Teke 17%
- Sangha 20%
- Bakongo 48%

Congo is one of the most
tribally conscious countries
in Africa. The main tensions
are between the Bakongo, who
live in the north, and the Mbochi,
who are concentrated in the
more prosperous south.
Since the 1950s, women
have achieved considerable
freedom.

POLITICS ▷ Multiparty elections

L. House 2002/2007
U. House 2002/2008

President Denis
Sassou-Nguesso

AT THE LAST ELECTION

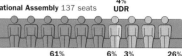

National Assembly 137 seats

- 4% UDR
- 61% PCT
- 6% Vacant
- 3% UPADS
- 26% Others

PCT = Congolese Labor Party and allies
UDR = Union for Democracy and the Republic
UPADS = Pan-African Union for Social Democracy

Senate 66 seats

- 9% Vacant
- 85% PCT
- 6% Others

Former Marxist dictator Denis Sassou-
Nguesso seized power in 1997, amid
intense fighting which left thousands
dead. Relative peace was secured in
1999. A new constitution giving greater
power to the presidency was approved
in 2002, and Sassou-Nguesso was easily
elected. Fighting broke out again later
that year, but legislative elections were
eventually held, and were won by
Sassou-Nguesso's supporters.

CONGO

Total Land Area :
342 000 sq. km
(132 046 sq. miles)

POPULATION

- ⊙ over 500 000
- ◎ over 50 000
- ○ over 100 000
- ● over 10 000
- · under 10 000

LAND HEIGHT

- 500m/1640ft
- 200m/656ft
- Sea Level

WORLD AFFAIRS

 Joined UN in 1960

 BDEAC   CEMAC   FZ   NAM ACP

Carefully balancing relations with France and the US is a priority, since both seek to extend their stake in the oil industry; nevertheless relations with old Eastern bloc allies remain strong. Congo has been susceptible in recent years to political instability and wars in the neighboring Democratic Republic of the Congo and Angola.

AID

 Recipient

 $33m (receipts)   Down 77% in 2000

Until 1990, the USSR, Cuba, and China were major donors. Most aid now comes from Italy and France. The IMF has begun moves to clear Congo's high level of debt built up during the 1970s and 1980s.

DEFENSE

No compulsory military service

$72m   Down 1% in 2000

The militias of the various political forces were being integrated into the 8000-strong army, until fighting broke out again between them in mid-1997. The air force numbers 1200, and is equipped with 12 MiG-21s.

ECONOMICS

 Inflation 11% p.a. (1990–2000)

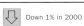 $1.74bn   698.7–736.7 CFA francs

SCORE CARD

- ❏ WORLD GNP RANKING.........................141st
- ❏ GNP PER CAPITA $570
- ❏ BALANCE OF PAYMENTS....................–$252m
- ❏ INFLATION ....–0.9%
- ❏ UNEMPLOYMENT.......Widespread underemployment

STRENGTHS
Increase in importance of oil, now providing 95% of export revenues. Significant timber supplies. Skilled and well-trained workforce helps sustain substantial industrial base in the capital and Pointe-Noire.

WEAKNESSES
Massive debt burden. Top-heavy bureaucracy. Overdependence on oil. Political instability. Large refugee population.

EXPORTS

Brazil 3%
China 11%
Taiwan 37%
Other 15%
USA 16%
South Korea 18%

IMPORTS

Netherlands 4%
Belgium 5%
Italy 7%
USA 10%
Other 54%
France 20%

RESOURCES

 Electric power 118,000 kw

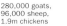 43,886 tonnes   271,000 b/d (reserves 1.5bn barrels)

280,000 goats, 96,000 sheep, 1.9m chickens   Oil, natural gas, zinc, gold, copper, potash, diamonds

Oil is by far the Congo's most important resource. Natural gas reserves have yet to be exploited. There are some deposits of diamonds and potash. Bauxite and iron ore reserves are not large enough to be profitably mined and phosphate production was abandoned in 1977. Chinese aid has helped build two hydroelectric dams, on the Bouenza and Djoué rivers. A third is currently being built on the Léfini at Imboulou.

ENVIRONMENT

 Sustainability rank: 40th

 5%   0.6 tonnes per capita

The 1999 Yaoundé Declaration should help control exploitation of tropical timber. Congo has been used in the past as a dumping ground for dangerous toxic waste from the West.

MEDIA

 TV ownership low

Daily newspaper circulation 8 per 1000 people

PUBLISHING AND BROADCAST MEDIA

 There are 64 daily newspapers, including *Mweti* and *Aujourd'hui*

 1 state-controlled service   4 state-controlled services

With only moderate media control, the press often takes an antigovernment stance. During World War II, Radio Brazzaville, still the official state radio, was vital to de Gaulle's French forces.

CRIME

 Death penalty not used in practice

918 prisoners   Crime is rising

Armed robbery and smuggling are widespread. Years of conflict and instability in neighboring countries mean that guns are easily available.

EDUCATION

 School leaving age: 11

 81%   16,602 students

Originally pioneered by French Roman Catholic missions, schools are still subject to inspection from Paris.

CHRONOLOGY

The kingdoms of Teke and Loango were incorporated as the Middle Congo (part of French Equatorial Africa) between 1880 and 1883.

- ❏ **1960** Independence.
- ❏ **1964** Marxist–Leninist National Revolution Movement (MNR) sole legal party.
- ❏ **1977** Yhompi-Opango head of state after President Ngoumbi's murder.
- ❏ **1979** Col. Denis Sassou-Nguesso president.
- ❏ **1991** Multiparty democracy.
- ❏ **1992** Pascal Lissouba president.
- ❏ **1993** Elections: Lissouba's UPADS party gains majority.
- ❏ **1997** Sassou-Nguesso ousts Lissouba.
- ❏ **1999** Cease-fire signed.
- ❏ **2001** IMF starts to clear debt.
- ❏ **2002** New constitution approved. Sassou-Nguesso wins elections.

HEALTH

Welfare state health benefits

1 per 3333 people   Diarrheal, parasitic and respiratory diseases, malaria

The health service, set up by French military doctors at the start of the 20th century, has been devastated by civil war.

SPENDING

GDP/cap. decrease

CONSUMPTION AND SPENDING

14 per 1000 population   7 per 1000 population

Defense 2.5%
Education 4.7%
Health 2%

0   5   10   15   20   25

Defense, Health, Education spending as % of GDP

Wealth generated from oil extraction has sustained an active and confident middle class. French-label products are considered to be status symbols.

WORLD RANKING

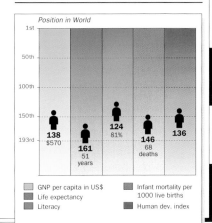

Position in World

1st
50th
100th
150th
193rd

138 $570
161 51 years
124 81%
146 68 deaths
136

- ▢ GNP per capita in US$
- ▢ Life expectancy
- ▢ Literacy
- ▢ Infant mortality per 1000 live births
- ▢ Human dev. index

CONGO (DEMOCRATIC REPUBLIC)

OFFICIAL NAME: Democratic Republic of the Congo   **CAPITAL:** Kinshasa
POPULATION: 52.5 million   **CURRENCY:** Congolese franc   **OFFICIAL LANGUAGE:** French

C

CENTRAL AFRICA   Africa

1960 | 1960 | June 6 | CGO | +1 | +243 | .zr

LYING IN EAST CENTRAL AFRICA, the Democratic Republic of the Congo (DRC), known as Zaire from 1971 to 1997, is one of Africa's largest countries. The rainforested basin of the Congo River occupies 60% of the land area. On independence in 1960, civil war broke out. The notoriously corrupt Gen. Mobutu ruled from 1965 until his overthrow in 1997 by rebel forces under Laurent-Désiré Kabila. A rebellion launched in 1998 plunged the country into renewed chaos, and spiraled into regional conflict. Peace initiatives were revitalized by Joseph Kabila's succession in January 2001.

The Congo River is navigable for 1357 km (848 miles) and provides one of the most convenient ways of traveling in the country.

CLIMATE
▷ Tropical equatorial/ wet & dry

WEATHER CHART FOR KINSHASA

The climate is tropical and humid. Temperatures average 25°C (77°F) and vary little through the year. Annual rainfall is around 150–200 cm (59–79 in); mountainous areas are wetter. The equator passes through the north of the country, causing marked regional variations. To its south, well-differentiated wet and dry seasons are October–May and June–September respectively. North of the equator, a short dry season lasts from December to February; the rest of the year is wet.

TRANSPORTATION
▷ Drive on right

✈ N'Djili, Kinshasa
273,563 passengers

⚓ 20 ships
12,900 grt

THE TRANSPORTATION NETWORK

🛣 157,000 km (97,555 miles)	30 km (19 miles)
🚆 3641 km (2263 miles)	15,000 km (9321 miles)

The Congo River and its many tributaries provide the main means of communication. The size of the country and the fact that most of it is covered by dense rainforest have severely limited the development of road and rail networks. Many forest settlements are inaccessible except by air. Road maintenance, always poor, has virtually ceased outside the main towns since 1990, isolating even more settlements away from the main rivers.

TOURISM
▷ Visitors : Population 1:991

🧳 53,000 visitors

⬇ Down in 1997–2001

MAIN TOURIST ARRIVALS

Africa 42%
Europe 15%
Americas 2%
Other 41%

% of total arrivals

Political turmoil and widespread anarchy since early 1997 ensure that the country remains off the itinerary for most travelers.

Potential tourist attractions consist mainly of scenery – mountains and lakes – and wildlife, but there are few facilities for tourists even in the capital. The Congo, 16 km (10 miles) wide in places, is Africa's second-longest river after the Nile. Visitors were formerly also attracted by the vibrant music of Kinshasa's many bands.

The once-large number of visitors on business has also collapsed as a consequence of the chronic instability of the 1990s.

PEOPLE
▷ Pop. density low

Kiswahili, Tshiluba, Kikongo, Lingala, French

23/km² (60/mi²)

THE URBAN/RURAL POPULATION SPLIT

30%   70%

RELIGIOUS PERSUASION

Protestant 13%
Traditional beliefs 50%
Roman Catholic 37%

ETHNIC MAKEUP

Bantu and Hamitic 45%
Other 55%

The Shaba mining area and major urban centers are densely populated, while the rainforests have a density of fewer than three people per sq. km, mostly subsisting on the margins of the cash economy. There is great ethnic diversity, with more than 12 main groups and around 190 smaller ones. The majority are of Bantu origin, but there are also large Hamitic and Nilotic populations, mainly in the north and northeast. The original inhabitants, the forest pygmies, today form a tiny and marginalized group.

Ethnic tensions inherited from the colonial period were contained under Mobutu until the 1990s. Ethnic violence in southeastern provinces in 1993 cost many thousands of lives, and a Hutu refugee influx from Rwanda the following year provoked serious tension among Tutsis in eastern areas; revenge killings on a large scale soon became commonplace. Regarded by Mobutu as foreigners, Tutsis provided the backbone of the 1996–1997 insurgency that overthrew him, and subsequently turned against Laurent Kabila also.

POPULATION AGE BREAKDOWN

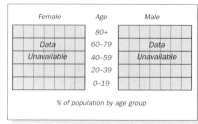

Female	Age	Male
	80+	
Data	60–79	Data
Unavailable	40–59	Unavailable
	20–39	
	0–19	

% of population by age group

CONGO, DEMOCRATIC REPUBLIC

Total Land Area : 2 345 410 sq. km
(905 563 sq. miles)

POPULATION

⊡	over 1 000 000
◉	over 500 000
◎	over 100 000
○	over 50 000
●	over 10 000
•	under 10 000

LAND HEIGHT

	2000m/6562ft
	1000m/3281ft
	500m/1640ft
	200m/656ft
	Sea Level

POLITICS ▷ No legislative elections

 1987/Transitional   President Joseph Kabila

AT THE LAST ELECTION

Constituent and Legislative Assembly 300 seats

The last legislative poll was in 1987 when members were chosen from Mobutu's Popular Revolutionary Movement (**MPR**). An interim legislature was appointed in July/August 2000 with 300 members chosen by Laurent Kabila.

The authoritarian regime of Laurent-Désiré Kabila continues under his son.

PROFILE

Kabila overthrew Mobutu's 32-year dictatorial regime in 1997, but his commitment to pluralist democracy soon came under question as he dissolved parliament and scrapped the constitution. A new Constituent and Legislative Assembly did not convene until August 2000. Meanwhile Kabila's ethnic Tutsi supporters rose against him in 1998, with backing from Uganda and Rwanda. As neighboring countries joined the fighting, the war became a regional crisis. The arrival of a UN peacekeeping mission was consistently delayed by Kabila in 2000.

Kabila's murder in January 2001 left a vacuum that exposed the extent of his grip on power. His son Joseph, head of the armed forces, was rapidly appointed

Joseph Kabila, who succeeded his father as president in 2001.

Mobutu, the ousted dictator, held power from 1965–1997.

as his successor and has promised to promote political pluralism and economic liberalization.

MAIN POLITICAL ISSUE
Securing peace and democracy
Positive change after Mobutu's overthrow proved elusive. The 1999 elections did not take place as fighting spiraled, with the government in true control of only half of the country. Kabila's death in 2001 brought all factions together for talks but fighting continued. Talks in South Africa in 2002 again failed to make any real progress, although a deal was signed with Rwanda to bring security to the east.

WORLD AFFAIRS ▷ Joined UN in 1960

The protracted civil war which has devastated this huge country has irrevocably muddied the DRC's international relations. While the West has grown frustrated with the slow pace of peace, neighboring countries have been drawn into one of the continent's largest conflicts.

The collapse of the autocratic Mobutu regime in 1997 did not bring the long-hoped for stability, and instead prompted a savage rebellion by the erstwhile Rwandan and Ugandan allies of new leader Laurent Kabila. The vast potential wealth of the Congolese forests and mineral deposits prompted a quick spiral into regional war. Angola, Burundi, Chad, Namibia, Sudan, and Zimbabwe all sent troops, leaving the DRC divided into government- and rebel-controlled areas.

Fighting continued despite a 1999 cease-fire signed in Lusaka. Laurent Kabila attracted intense international criticism in 2000 for suspending the accords, for taking an autocratic approach to a new transitional assembly, and for obstructing the arrival of the UN peacekeeping mission, MONUC. Talks after Kabila's death led to the withdrawal of some Ugandan, Rwandan, and Burundian troops, but fighting continued, and in early 2002 both Rwanda and Uganda reinforced troops. However, in July Rwanda agreed to recall its forces in return for Congolese assurances of security.

The involvement of foreign firms in the running of the long-neglected mines in the south of the country has drawn sharp internal criticism.

AID ▷ Recipient

💲 $184m (receipts)   ⬆ Up 39% in 2000

The regime's importance to the West during the Cold War brought in aid revenues on a large scale. Between 1970 and 1989, it received $8.3 billion in economic aid as well as large-scale military assistance.

By 1990, changing political priorities led the US to act on long-deferred problems of human rights abuses and misappropriation of aid. It suspended all but humanitarian aid; most other donors quickly followed suit, and the IMF declared the government "noncooperative" over its $10 billion foreign debt. Joseph Kabila's accession in early 2001 improved the country's international standing; aid was resumed and debt cancellations followed.

C

CHRONOLOGY

The modern Congo was the site of the Kongo and other powerful African kingdoms, and a focus of the slave trade. Belgium's King Leopold II claimed most of the Congo basin after 1876 as his personal possession.

❏ **1885** Brutal colonization of Congo Free State (CFS) as King Leopold's private fief.

❏ **1908** Belgium takes over CFS after international outcry.

❏ **1960** Independence of Republic of Congo. Katanga (Shaba) province secedes. UN intervenes.

❏ **1963** Katanga secession collapses.

❏ **1965** Gen. Joseph-Désiré Mobutu seizes power.

❏ **1970** Mobutu elected president; his MPR becomes sole legal party.

❏ **1971** Country renamed Zaire.

❏ **1977–1978** Two invasions by former Katanga separatists repulsed with Western help.

❏ **1982** Opposition parties set up Union for Democracy and Social Progress (UDPS).

❏ **1986–1990** Civil unrest and foreign criticism of human rights abuses.

❏ **1990** Belgium suspends aid after security forces kill prodemocracy demonstrators. Mobutu announces transition to multiparty rule.

❏ **1991** Opposition leader Etienne Tshisekedi heads short-lived "crisis government" formed by Mobutu.

❏ **1992–1993** Rival governments claim legitimacy.

❏ **1994** Combined High Council of the Republic–Transitional Parliament established.

❏ **1995** Regime demands international assistance to support a million Rwandan Hutu refugees.

❏ **1996** Major insurgency launched in east by Alliance of Democratic Forces for the Liberation of the Congo (AFDL) including Laurent Kabila's Popular Revolutionary Party (PRP), with disaffected ethnic Tutsi Banyamulenge.

❏ **1997** Forces led by Kabila sweep south and west. Kabila takes power. Country renamed DRC. Mobutu dies in exile.

❏ **1998** Banyamulenge join Kabila's opponents and launch rebellion in the east, backed by Rwanda and Uganda. Southern African states, give military backing to Kabila.

❏ **2000** UN approves peacekeeping mission; arrival stalled by Kabila.

❏ **2001** Laurent Kabila assassinated; succeeded by son Joseph. Peace talks start, with troop withdrawals from front line.

❏ **2002** April, peace talks inconclusive. May, 180 massacred in Kisangani by Rwandan-backed forces.

DEFENSE

 No compulsory military service

💲 $392m   ⬇ Down 5% in 2000

Although the military strongly backed Mobutu's regime, it offered no real resistance when Laurent Kabila's insurgents swept the country in 1996–1997. Government troops, poorly paid and undisciplined, but supported by foreign allies, have been fighting in the civil war since 1998, led by Kabila's son and successor Joseph Kabila. The UN peacekeeping force, MONUC, had established a small presence by 2001.

CONGOLESE ARMED FORCES

🚜	60 main battle tanks (20 PRC Type-59, 40 PRC Type-62)	79,000 personnel
🚢	2 patrol boats	900 personnel
✈	No combat aircraft	1500 personnel
⚔	None	

ECONOMICS

 Inflation 1423% p.a. (1990–1999)

📊 $5.4bn   💲 4.5–313 Congolese francs

SCORE CARD

❏ WORLD GNP RANKING	108th
❏ GNP PER CAPITA	$110
❏ BALANCE OF PAYMENTS	–$798m
❏ INFLATION	540%
❏ UNEMPLOYMENT	Very high

ECONOMIC PERFORMANCE INDICATOR

— Consumer Price Index   GDP ▨

EXPORTS

India 4%
Netherlands 3%
Finland 6%
Other 9%
USA 19%
Belgium 59%

IMPORTS

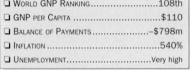

Kenya 5%
France 6%
Nigeria 11%
Other 41%
South Africa 18%
Belgium 19%

STRENGTHS

Rich resource base. Minerals – notably copper, cobalt, diamonds – provide 85% of export earnings. Energy: oil; possibly Africa's largest hydropower potential. Rich soil; much unutilized arable land.

WEAKNESSES

Decades of mismanagement and corruption: $12 billion foreign debt; lack of food self-sufficiency; inadequate, disintegrating infrastructure. Political instability. Hyperinflation. Loss of export income. Mineral resources plundered by foreign troops.

PROFILE

Political instability, systematic corruption, long-term mismanagement, and outright civil war have brought what is potentially a leading African economy to a state of collapse.

By the mid-1990s real GDP was falling by 10% or more each year. The government budget ran record deficits, and inflation ran virtually out of control from 1994 onward. Lack of spares and

power cuts have since closed many mines and halted most other industry. Strikes and riots over plummeting living standards hastened the flight of foreign capital. Subsistence farming and petty trade keep most people going.

The Kabila regime, despite its originally Marxist background, claims to want an effective free-market economy. Resumption of essential large-scale aid and debt relief will depend on difficult reforms and paying off arrears to the IMF and other creditors. Restructuring of state-owned enterprises in 2001 aimed to attract foreign investment back.

CONGO, DEM. REP. : MAJOR BUSINESSES

🛢	Oil	
🏭	Textiles	
⛏	Copper mining	
▦	Cobalt mining	
🗎	Food processing	
⊙	Industrial diamonds	
⚙	Light engineering	

0   200 km
0   200 miles

RESOURCES

 Electric power 3.2m kw

208,862 tonnes

23,158 b/d (reserves 194m barrels)

4.07m goats, 999,748 pigs, 20.6m chickens

Copper, diamonds, oil, coltan, cobalt, zinc, uranium, manganese

ELECTRICITY GENERATION

- Hydro 99% (5.4bn kwh)
- Combustion 1% (0.02bn kwh)
- Nuclear 0%
- Other 0%

0   20   40   60   80   100

% of total generation by type

What should be a prosperous country, with its rich resources, is instead one of the world's poorest states, exploited and mismanaged by its rulers for decades and plundered further by occupying forces during the civil war. In the 1980s, the country was the world's largest cobalt exporter and second-largest industrial diamond exporter. Since 1990, copper and cobalt output have collapsed and diamond smuggling is booming. There are oil reserves and hydroelectric installations with sufficient potential capacity to export power, but instead lack of maintenance has shut down many turbines and most urban areas face power cuts. Despite rich soils and the fact that 60% of people are involved in farming, the DRC is not even self-sufficient in food.

ENVIRONMENT

 Sustainability rank: 109th

5%

0.1 tonnes per capita

ENVIRONMENTAL TREATIES

Yes	Yes	Yes
Yes	Yes	No

Rainforests cover over 60% of the country, representing almost 6% of the world's and 50% of Africa's remaining woodlands. They are home to several endangered species. The poor transportation network has so far prevented large-scale commercial exploitation of timber, but clearance for fuelwood is a problem. The collapse of many urban refuse and sewage disposal systems has led to major health and pollution problems. Environmental damage caused by the civil war is estimated at $320 million.

MEDIA

 TV ownership low

Daily newspaper circulation 3 per 1000 people

PUBLISHING AND BROADCAST MEDIA

There are 9 daily newspapers, including *Le Palmarès*, *Elima*, *Boyoma*, and *Mjumbe*

1 state-controlled service, some independent services

2 state-controlled services, some independent services

Fighting makes services less accessible outside the capital. Unlike the broadcast media, press outlets are privately owned, and many newspapers openly criticize the authorities, although self-preservation requires a degree of self-censorship. A ban on foreign radio broadcasts was lifted in 2001. MONUC operates the only countrywide radio station, Radio Okapi.

CRIME

 Death penalty in use

The DRC does not publish prison figures

Violence and crime have risen rapidly since 1990

CRIME RATES

All types of crime are on the increase in the DRC

Political crisis and economic collapse exacerbate long-standing problems of corruption and human rights abuses. Violence and crime of all kinds, including extortion, robbery, rape, and murder, are widespread and on the increase, and war zones have seen the abandonment of the rule of law. Ethnic violence, suppressed after 1965, resurfaced in the 1990s in the south and between the Hema and Lendu tribes in the northeast.

EDUCATION

 School leaving age: 12

 61%

60,341 students

THE EDUCATION SYSTEM

Primary 46%   Secondary 18%   Tertiary 1%

% of each age group in education

In 1997 just over 37% of secondary age children were attending classes, but this figure has dropped sharply during the civil war. State provision, as with health care, is patchy and has faced sharp budget cuts since 1980. About 70% of schooling is now provided by the Roman Catholic Church.

CONGO, DEM. REP.: LAND USE

- Cropland
- Forest
- Pasture
- Wetlands
- Cattle
- Coffee
- Palm oil – cash crop

0   200 km
0   200 miles

HEALTH

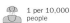 No welfare state health benefits

1 per 10,000 people

Malaria, respiratory and diarrheal diseases

State services have now virtually collapsed. Disease and death rates are rising, especially in rural areas. A new health insurance plan was announced in 2001, designed to enable greater access to health care. As at end-2001, 1.3 million people were estimated to be HIV/AIDS infected.

SPENDING

GDP/cap. decrease

CONSUMPTION AND SPENDING

17 per 1000 population

0.1 per 1000 population

- Defense 8.4%
- Education 0.1%
- Health 1.7%

0   5   10   15   20   25

Defense, Health, Education spending as % of GDP

Before his death in exile in 1997, ex-dictator Mobutu was one of the world's richest men, worth an estimated $4 billion. Most of his former subjects live in poverty, exacerbated by civil war.

WORLD RANKING

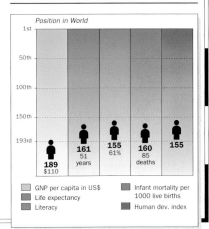

Position in World

1st
50th
100th
150th
193rd

189 $110
161 51 years
155 61%
160 85 deaths
155

- GNP per capita in US$
- Life expectancy
- Literacy
- Infant mortality per 1000 live births
- Human dev. index

COSTA RICA

OFFICIAL NAME: Republic of Costa Rica CAPITAL: San José
POPULATION: 4.1 million CURRENCY: Costa Rican colón OFFICIAL LANGUAGE: Spanish

SPANNING THE CENTRAL AMERICAN isthmus and wedged between Nicaragua and Panama, Costa Rica was under Spanish rule until 1821 and gained full independence in 1838. From 1948 until the end of the 1980s, it had the most developed welfare state in Central America. Costa Rica is nominally a multiparty democracy, but two parties dominate. Its army was abolished in 1948; the 1949 constitution then forbade national armies.

CLIMATE
▷ Tropical wet & dry

WEATHER CHART FOR SAN JOSÉ

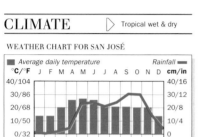

The Caribbean coast has heavy rainfall, while the Pacific coast is much drier. The central uplands are temperate.

TRANSPORTATION
▷ Drive on right

Juan Santamaria, San José
988,000 passengers

15 ships
5700 grt

THE TRANSPORTATION NETWORK

7827 km (4863 miles)

Pan-American Highway, 663 km (412 miles)

471 km (293 miles)

730 km (454 miles)

San José is the hub of a well-used bus network. Railroads, badly earthquake-damaged in 1991, are being reconstructed.

TOURISM
▷ Visitors : Population 1:3.8

1.09m visitors

Up 5% in 2000

MAIN TOURIST ARRIVALS

USA 38%
Nicaragua 16%
Panama 5%
Other 41%

0 10 20 30 40 50 60
% of total arrivals

Tourism brought in $1 billion in 1999 and is expanding, with the help of both domestic and foreign investment. A new 154-room hotel in San José and a comparable one on the Pacific coast attract wealthy visitors.

PEOPLE
▷ Pop. density medium

Spanish, English Creole, Bribri, Cabecar

80/km² (208/mi²)

THE URBAN/RURAL POPULATION SPLIT

52% 48%

RELIGIOUS PERSUASION

Other (including Protestant) 24%

Roman Catholic 76%

The majority of the population is *mestizo*, of partly Spanish origin. One-third of people in the Limón area are black and often English-speaking. There are only about 5000 indigenous Amerindians.

POLITICS
▷ Multiparty elections

2002/2006

President Abel Pacheco de la Espriella

AT THE LAST ELECTION

Legislative Assembly 57 seats

33% PUSC 30% PLN 25% PRC 10% ML 2% PAC

PUSC = Social Christian Unity Party PLN = National Liberation Party PRC = Costa Rican Renewal Party ML = Liberty Movement PAC = Citizens' Action Party

Politics has long been dominated by the PUSC and PLN, both of which have close ties to major banana- and coffee-growing families. Historically the US has exercised a very powerful influence on politics.

The PLN in 1994 promised reforms to its previous austerity policies, but soon came under pressure from international financial organizations to reduce the budget deficit. Harsh structural adjustment measures proved highly unpopular.

In 1998 the PUSC regained power. President Miguel Angel Rodríguez launched a three-year plan to reduce inflation and poverty, create thousands of jobs, and stimulate foreign investment in state companies. His chosen successor, Abel Pacheco, needed an unprecedented second round to clinch the presidency in 2002, when voter turnout hit an all-time low.

WORLD AFFAIRS
▷ Joined UN in 1945

ACS Geplac RG OAS San José

Trade ties with the US and protection of prices for coffee and bananas are priorities. Trade ties have been agreed with Canada and Chile. Tensions with Nicaragua over their mutual border were resolved in 2000, but illegal immigrants remain an issue.

Pineapple plantation near Buenos Aires, crossed by the Pan-American Highway which runs for 663 km (412 miles) through Costa Rica.

COSTA RICA

Total Land Area : 51 100 sq. km (19 730 sq. miles)

POPULATION
over 100 000
over 50 000
over 10 000
under 10 000

LAND HEIGHT
3000m/9843ft
2000m/6562ft
1000m/3281ft
500m/1640ft
200m/656ft
Sea level

0 50 km
0 50 miles

AID

 Recipient

 $12m (receipts)　　Down 60% in 1999–2000

During the 1980s Costa Rica was a large recipient of US aid designed to inoculate it against left-wing insurgencies such as those in El Salvador, Guatemala, and neighboring Nicaragua. Peace in the region has led to a sharp decline in such aid, especially given the country's relatively high per capita income. World Bank aid will help modernize Juan Santamaría international airport.

DEFENSE

 No compulsory military service

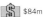 $84m　　Up 22% in 2000

Costa Rica emerged from the 1948 civil war as a neutral, demilitarized modern state. A 4400-strong Civil Guard is complemented by a largely military-trained police force. Spending on security has long been the lowest in the region. Lack of a common command structure hinders the influence of the security forces but also renders them less open to public control. Right-wing paramilitary groups are known to exist.

ECONOMICS

 Inflation 17% p.a. (1990–2000)

 $14.5bn　　317.5–341.3 Costa Rican colones

SCORE CARD

❏ World GNP Ranking	76th
❏ GNP per Capita	$3810
❏ Balance of Payments	–$650m
❏ Inflation	11%
❏ Unemployment	5%

Strengths

Major coffee, beef, and banana exports. Expanding tourism also fueling construction. Strong inward investment. Favorable WTO ruling on banana access to EU market.

Weaknesses

Coffee, beef, and bananas all vulnerable to falling prices. History of high inflation. Dependence on imported oil. Large

EXPORTS

Guatemala 3%
Germany 3%
USA 44%
UK 10%
Netherlands 11%
Other 29%

IMPORTS

Japan 4%
Guatemala 3%
Venezuela 5%
Mexico 5%
Other 42%
USA 41%

domestic debt. Competitiveness hindered by insufficient investment in infrastructure. State monopolies have deterred investment in energy, telecommunications, and insurance sectors. Inefficient management.

RESOURCES

 Electric power 1.5m kw

 35,003 tonnes　　Not an oil producer

 1.72m cattle, 430,000 pigs, 17m chickens　　Bauxite, gold, silver, manganese, mercury

Costa Rica has large bauxite deposits in the south – aluminum smelting is an important industry. Small quantities of gold, silver, manganese, and mercury are also mined. Self-sufficiency in energy is being pursued through the development of hydroelectric power. Forests cover 31% of the country.

ENVIRONMENT

 Sustainability rank: 9th

 14% (3% partially protected)　　1.4 tonnes per capita

Despite good environmental regulation, reckless economic development has contributed to extensive deforestation. Pasture land now covers some 46% of the territory and pesticide abuse by agribusiness has poisoned rivers and threatened species. Urban sprawl has degraded the fertile central valley.

MEDIA

 TV ownership medium

 Daily newspaper circulation 72 per 1000 people

PUBLISHING AND BROADCAST MEDIA

There are 8 daily newspapers, including *La Nación*, *La República*, *La Prensa Libre*, and *Diario Extra*	
8 stations: 1 state-owned, 7 independent	State-owned and independent stations

The media are free but dominated by conservative opinion. Entry into journalism is strictly licensed.

CRIME

No death penalty

8526 prisoners　　Up 71% 1989–1994

Costa Rica is the least violent Central American country. Attacks on and kidnappings of tourists are rare but have dented its image as a safe haven. Drug cartels use the country to transfer cocaine to the US and Europe. Police show hostility toward immigrants from neighboring countries.

CHRONOLOGY

Costa Rica, ruled since the 16th century by Spain, became an independent state in 1838.

- ❏ **1948** Disputed elections lead to civil war; ended by Social Democratic Party (later the PLN) forming provisional government under José Ferrer. Army abolished.
- ❏ **1949** New constitution promulgated.
- ❏ **1987** Central American Peace Plan initiated by President Arias.
- ❏ **1998** PUSC returns to power.

C

EDUCATION

 School leaving age: 14

96%　　59,947 students

Costa Rica has the highest literacy rate in the isthmus, and is home to the University of Central America.

HEALTH

 Welfare state health benefits

1 per 1111 people　　Heart diseases, accidents, cancers, perinatal deaths

The public health system is one of the most developed in Latin America. Health was allocated 29.3% of total public spending in 1999.

SPENDING

 GDP/cap. increase

CONSUMPTION AND SPENDING

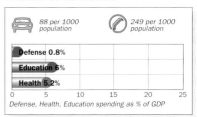

88 per 1000 population　　249 per 1000 population

Defense 0.8%
Education 6%
Health 5.2%

0　　5　　10　　15　　20　　25
Defense, Health, Education spending as % of GDP

The plantation-owning families are the wealthiest group; official figures claim that one-fifth of Costa Rica's population is living in poverty.

WORLD RANKING

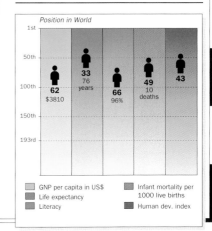

Position in World

62 $3810	33 76 years	66 96%	49 10 deaths	43

GNP per capita in US$
Life expectancy
Literacy
Infant mortality per 1000 live births
Human dev. index

CROATIA

EUROPE
Europe

OFFICIAL NAME: Republic of Croatia **CAPITAL:** Zagreb
POPULATION: 4.7 million **CURRENCY:** Kuna **OFFICIAL LANGUAGE:** Croatian

C

ALTHOUGH IT WAS CONTROLLED by Hungary from medieval times and was a part of the Yugoslav state for much of the 20th century, Croatia still has a strong national identity. It includes the historic provinces of Slavonia, Istria, and Dalmatia (Dinara). Actively involved in the conflicts which broke up Yugoslavia in the early 1990s, Croatia only regained full control of Serb-occupied Eastern Slavonia, around Vukovar, in 1998.

Dubrovnik, Dalmatia. *This historic city on the Adriatic coast was shelled and besieged by the Yugoslav federal army in 1991.*

CLIMATE
▷ Mediterranean/ continental

WEATHER CHART FOR ZAGREB

Northern Croatia has a temperate continental climate. Its Adriatic coast has a Mediterranean climate.

TRANSPORTATION
▷ Drive on right

Pleso International, Zagreb
1.15m passengers

246 ships
734,300 grt

THE TRANSPORTATION NETWORK

23,497 km (14,600 miles)	330 km (205 miles)
2726 km (1694 miles)	785 km (488 miles)

Zagreb has recovered from the effects of war, and of sanctions against Yugoslavia, and is once again an important regional road and rail hub. The Adriatic Highway affords fantastic views along the Dalmatian coast.

TOURISM
▷ Visitors : Population 1.2:1

5.83m visitors
Up 53% in 2000

MAIN TOURIST ARRIVALS

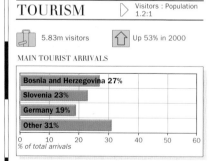

% of total arrivals

The Adriatic coast is regaining its popularity as a tourist destination. There are also many historical sites.

PEOPLE
▷ Pop. density medium

Croatian
83/km² (215/mi²)

THE URBAN/RURAL POPULATION SPLIT

58%   42%

RELIGIOUS PERSUASION

Muslim 1%   Orthodox 11%
Other 12%
Roman Catholic 76%

Before the breakup of Yugoslavia, Croats made up nearly 80% of the population, and Serbs 12%. In 1991, the Serbs, alienated by Croatian nationalism, proclaimed the Republic of Serbian Krajina in areas where they formed a majority. Croatian forces retook most Serb-held territory in offensives in 1995, precipitating a huge exodus of Serb refugees. The last rebel Serb enclave, Eastern Slavonia, was reintegrated into Croatia in 1998. Many of the 250,000 Bosnian refugees who fled to Croatia have now settled there. By 2001 Serbs comprised only 4% of the population.

POLITICS
▷ Multiparty elections

2000/2004
President Stipe Mesic

AT THE LAST ELECTION

House of Representatives 151 seats
3% HSP–HKDU

47% SDP–HSLS   30% HDZ   17% All   3% EM

SDP–HSLS = *SDP*–Social Democratic Party, *HSLS*–Croatian Social Liberal Party **HDZ** = Croatian Democratic Union **All** = Alliance (*HSS*–Croatian Peasant Party, *IDS*–Istrian Democratic Assembly, *HNS*–Croatian People's Party, *LS*–Liberal Party, *ASH*–Croatian Social Democrats' Action) **HSP–HKDU** = *HSP*–Croatian Party of Rights, *HKDU*–Croatian Christian Democratic Union **EM** = Ethnic minorities

Six seats are reserved for representatives of Croats living abroad (all are currently held by the HDZ). Five seats are reserved for ethnic minorities (one is currently held by the HSS)

Croatia left the former Yugoslavia in 1991 under the direction of the right-wing HDZ led by Franjo Tudjman. However, the country suffered growing international isolation because of its nationalist stance and ambiguous support for the 1995 Bosnian peace accord. The HDZ was left rudderless after Tudjman's death in December 1999 and, mired in corruption and spying scandals, was swept from power in elections in January 2000 by left-leaning parties. The eclipse of the HDZ was confirmed in presidential elections the following month, won by one-time Yugoslav federal president Stipe Mesic. The new government's priorities were to rehabilitate Croatia internationally and to reduce the authoritarian powers of the presidency.

WORLD AFFAIRS
▷ Joined UN in 1992

CE   WTO   PfP   CEI   OSCE

Tudjman's nationalist stance after independence worked against closer relations with the West. Croatia's military successes in 1995, hastening the end of the Bosnian war, were reflected in the Bosnian peace accord. Tudjman's successors seek closer cooperation with the EU. In 2000 Croatia joined NATO's Partnerships for Peace program.

AID
▷ Recipient

$66m (receipts)
Up 38% in 2000

EU states have spent over $1 billion on reconstruction in Croatia since 1991.

DEFENSE
▷ Compulsory military service

$509m
Down 23% in 2000

The army proved its effectiveness when in 1995 it recaptured territory held by Serb forces. Under 2002 proposals its size will be halved by 2005.

C

ECONOMICS

 Inflation 86% p.a. (1990–2000)

 $20.2bn

 8.087–8.254 kuna

SCORE CARD

❑ WORLD GNP RANKING	61st
❑ GNP PER CAPITA	$4620
❑ BALANCE OF PAYMENTS	–$399m
❑ INFLATION	5.4%
❑ UNEMPLOYMENT	20%

STRENGTHS
Moderate growth as economy made recovery from 1998–1999 recession. Progress in reducing government overspending, backed by IMF.

WEAKNESSES
War damage estimated at $50 billion. Slow privatization until 2001. High and persistent unemployment.

EXPORTS

Austria 6%
Other 34%
Slovenia 11%
Bosnia & Herzegovina 12%
Germany 15%
Italy 22%

IMPORTS

Austria 7%
Other 44%
Slovenia 8%
Russia 8%
Germany 16%
Italy 17%

RESOURCES

 Electric power 3.8m kw

 25,519 tonnes

1.23m pigs, 710,000 turkeys, 10.8m chickens

 31,859 b/d (reserves 47m barrels)

Coal, bauxite, iron, oil, china clay, natural gas

Croatia generates 40% of its energy needs from combustion and 60% from hydroelectric sources. It has very few minerals, although it does have oil and gas fields. The rich fishing grounds of the Adriatic are a major resource.

ENVIRONMENT

 Sustainability rank: 12th

8%

4.5 tonnes per capita

Croatia was the first Yugoslav republic to create reserves in order to protect endangered and unique wetlands.

MEDIA

TV ownership high

Daily newspaper circulation 115 per 1000 people

PUBLISHING AND BROADCAST MEDIA

There are 12 daily newspapers, published locally, including *Vecernji List* in Zagreb and *Slobodna Dalmacija* in Split

1 state-controlled service

4 stations: 1 state-controlled, 3 independent

The three TV channels are state-owned. Media freedoms, eroded under Tudjman, have gradually improved.

CRIME

No death penalty

2617 prisoners

Up 4% in 1999

Under Tudjman, former Croat HOS militiamen escaped prosecution for "ethnic cleansing" in Bosnia, but the post-Tudjman government ordered a number of arrests.

CHRONOLOGY
Between 1945 and 1991 Croatia was a republic of the Yugoslav federation.

❑ **1991** Independence. Rebel Croatian Serb republic proclaimed.
❑ **1992** Tudjman president. Involvement in Bosnian civil war.
❑ **1995** Krajina and Western Slavonia recaptured. Dayton agreement ends fighting.
❑ **1998** Eastern Slavonia reintegrated.
❑ **1999** Death of Tudjman.
❑ **2000** Center-left wins elections.

EDUCATION

 School leaving age: 15

98%

100,297 students

The education system is well developed. There are four universities, at Zagreb, Rijeka, Osijek, and Split.

HEALTH

Welfare state health benefits

1 per 435 people

Cerebrovascular and heart diseases, cancers

Most Croats are covered by a health insurance scheme. However, an extra strain on already scarce funds is created by the demands of refugees and disabled war veterans.

SPENDING

GDP/cap. increase

CONSUMPTION AND SPENDING

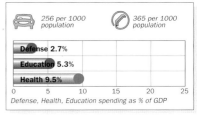

256 per 1000 population

365 per 1000 population

Defense 2.7%
Education 5.3%
Health 9.5%

Defense, Health, Education spending as % of GDP

Wage rises in the mid-1990s and again in 1999 led to spending booms. Consumers' high expectations were reined in by tighter wage policies.

WORLD RANKING

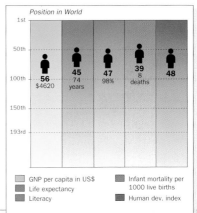

Position in World

1st				
50th				
56 $4620	**45** 74 years	**47** 98%	**39** 8 deaths	**48**
100th				
150th				
193rd				

GNP per capita in US$
Life expectancy
Literacy
Infant mortality per 1000 live births
Human dev. index

CROATIA
Total Land Area : 56 542 sq. km (21 831 sq. miles)

LAND HEIGHT
1000m/3281ft
500m/1640ft
200m/656ft
Sea Level

POPULATION
over 500 000
over 100 000
over 50 000
over 10 000
under 10 000

CUBA

OFFICIAL NAME: Republic of Cuba **CAPITAL:** Havana
POPULATION: 11.2 million **CURRENCY:** Cuban peso **OFFICIAL LANGUAGE:** Spanish

THE CARIBBEAN'S LARGEST ISLAND, Cuba has widely cultivated lowlands which fall between three mountainous areas. The fertile soil of the lowlands supports the sugarcane, rice, and coffee plantations. Sugar, the country's major export, suffers from underinvestment, low yields, and fluctuating world prices. A former Spanish colony, Cuba in 1959 became the only communist state in the Americas. In 1962, the deployment of Soviet nuclear missiles on the island shocked the US and brought the two superpowers close to war. Veteran president Fidel Castro is still very much in control, but, since the collapse of the USSR, the US sees Cuba as less of a threat.

Valle de Viñales, Pinar del Río province. Cuba's undulating countryside is ideal for growing the main export crop, sugar.

CLIMATE
▷ Tropical oceanic

WEATHER CHART FOR HAVANA

Cuba's subtropical climate is hot all year round and very hot in the summer. Rainfall is heaviest in the mountains, which receive up to 250 cm (98 in) a year. Generally, the north is wetter than the south; the Guantánamo area receives only 20 cm (8 in) of rainfall annually. In winter, the west is affected sometimes by cold air from the US, but only for a day or two at a time.

TRANSPORTATION
▷ Drive on right

José Martí, Havana
2.5m passengers

99 ships
120,000 grt

THE TRANSPORTATION NETWORK

29,820 km (18,529 miles)	638 km (396 miles)
14,331 km (8905 miles)	240 km (149 miles)

Public transportation in Cuba has been extremely cheap, although fuel shortages have made it increasingly erratic and unreliable. Cubans rely mostly on traditional black bicycles, imported by the thousand from China. Havana owes much of its charm to the number of 50-year-old Chevrolets and Oldsmobiles still being driven around. This is another result of sanctions, but keeps the many inventive local spare-parts workshops in business.

TOURISM
▷ Visitors : Population 1:6.4

1.74m visitors

No change in 2001

MAIN TOURIST ARRIVALS

Canada 17%	
Germany 11%	
Italy 10%	
Spain 9%	
France 8%	
Other 45%	

% of total arrivals

Tourism began to develop after 1977 (when the US relaxed some travel restrictions), and Cuba is now among the Caribbean's most popular tourist destinations. Tourism has supplanted sugar as the most important motor of the economy and largest generator of foreign exchange. Official estimates are that the number of arrivals will rise to more than five million by 2010. The government seeks to promote family tourism by cracking down on prostitutes who target Havana's main hotels.

Guanabo, 25 km east of Havana, is a low-key holiday resort favored by Cubans. The most modern cars in Cuba are imported, along with computers, in exchange for sugar in a special trading deal with Japan.

CUBA

Total Land Area : 110 860 sq. km (42 803 sq. miles)

POPULATION	LAND HEIGHT
▣ over 1 000 000	
◉ over 500 000	1000m/3281ft
◎ over 100 000	500m/1640ft
○ over 50 000	200m/656ft
● over 10 000	Sea Level
• under 10 000	

PEOPLE   ▷ Pop. density medium

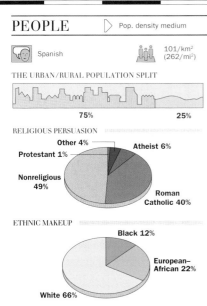

Spanish   101/km² (262/mi²)

THE URBAN/RURAL POPULATION SPLIT

75%   25%

RELIGIOUS PERSUASION

Other 4%
Atheist 6%
Protestant 1%
Nonreligious 49%
Roman Catholic 40%

ETHNIC MAKEUP

Black 12%
European–African 22%
White 66%

Ethnic tension in Cuba is minimal. About 70% of Cubans are of Spanish descent, mainly from the settlers, but also from the more recent influx of exiles from Franco's Spain. The black population is descended from the slaves and migrants from neighboring states, in particular Jamaica.

Living standards in Cuba fell dramatically in the early 1990s after the collapse of the east European communist bloc, previously its main trading partner, and rationing for most basic foodstuffs was subsequently introduced. The "dollarization" of the economy in recent years has led to great divisions between those who survive on pesos and the more than 50% of the population who have access to dollars. Since the early 1990s the number of those trying to leave, legally or otherwise, has risen markedly.

An increasing number of women are playing a prominent role in politics, the professions, and the armed forces. Child care facilities are widespread.

POPULATION AGE BREAKDOWN

Female	Age	Male
0.5%	80+	0.4%
5.9%	60–79	5.7%
11.1%	40–59	10.9%
18%	20–39	18.2%
14.3%	0–19	15%

% of population by age group

POLITICS   ▷ No multiparty elections

1998/2003   President Fidel Castro Ruz

AT THE LAST ELECTION

National Assembly of the People's Power 601 seats

100% PCC

PCC = Cuban Communist Party

Fidel Castro, who has led Cuba since 1959, founded its one-party communist system set out in the 1976 constitution.

PROFILE

The 1959 popular revolution, led by Castro, toppled the corrupt Batista dictatorship and launched a far-reaching program of social, economic, and political reforms.

In the 1990s the revolution seemed under siege in the wake of the collapse of the Soviet Union and tightened trade sanctions by the US. Supporters continue to see Cuba as living proof of the triumph of socialist development over adversity, but critics offer the view that the Castro administration is an intolerant dictatorship.

MAIN POLITICAL ISSUES
The succession

The aging Castro remains firmly in place, but his successor is a major issue. Some observers predict that a younger, reform-minded leadership would normalize relations with the US and steer Cuba toward Western-style democracy. Others warn that Castro's departure will lead to a power vacuum, vulnerable to social unrest and internal divisions between reformers and communist hard-liners.

Democracy

The 1999 trial of four moderate dissidents was part of a tough clampdown on internal opposition. This has reduced hopes for the type of opening seen in the economic sphere and threatens diplomatic and economic ties. It offers the US justification for its trade embargo.

***Raúl Castro,** brother of Fidel and the minister of defense.*

***Fidel Castro,** Cuba's leader since 1959. The US is keen to oust his regime.*

WORLD AFFAIRS   ▷ Joined UN in 1945

 ACP   IAEA   SELA   NAM   ACS

Since the 1959 revolution, and particularly after the 1962 stand-off over Soviet missiles, the US has considered Cuba a danger. The US trade blockade, first imposed in 1961, has left Cuba economically isolated despite regular votes in the UN condemning sanctions. The end of Soviet aid in the 1990s was another serious blow to Castro's embattled regime, but it has forced Cuba to soften its anti-Western stance in the search for alternative aid. Partly as a result, the US embargo has been progressively loosened since 1999 and now allows for more flights, direct mail, essential medicines, and more food imports.

Ties to Russia have been diluted further but relations with the US remain fraught, despite the arrival in 2001 of the first direct trade between the two countries' governments, in the form of emergency aid. In 2002 the US included Cuba in its "axis of evil" terrorist-sponsoring states. Relations with Europe are now improving, after being dented over the treatment of dissidents.

AID   ▷ Recipient

 $44m (receipts)   Down 25% in 2000

Spain, France, and UNICEF have given aid, and China loaned $400 million in 2001. A cash purchase of food and supplies from the US was made after Hurricane Michelle in late 2001.

CHRONOLOGY

Originally inhabited by the Arawak people, Cuba was claimed for Spain by Columbus in 1492. Development of the sugar industry from the 18th century, using imported slave labor, made Cuba the world's third-largest producer by 1860.

❑ **1868** End of the slave trade.
❑ **1868–1878** Ten Years' War for independence from Spain.
❑ **1895** Second war of independence. Thousands die in Spanish concentration camps.
❑ **1898** In support of Cuban rebels US declares war on Spain to protect strong American financial interests in Cuba.
❑ **1899** US takes Cuba and installs military interim government.
❑ **1901** US is granted intervention rights and military bases, including Guantánamo Bay naval base. ⇨

Moa
Baracoa
El Salvador
Guantánamo
GUANTÁNAMO BAY (to US)
Windward passage

C

C

CHRONOLOGY *continued*

- **1902** Tomás Estrada Palma takes over as first Cuban president. US leaves Cuba, but intervenes in 1906–1909 and 1919–1924.
- **1909** Liberal presidency of José Miguel Goméz. Economy prospers; US investment in tourism, gambling, and sugar.
- **1925–1933** Dictatorship of President Gerardo Machado.
- **1933** Years of guerrilla activity end in revolution. Sgt. Fulgencio Batista takes over; military dictatorship.
- **1955** Fidel Castro exiled after two years' imprisonment for subversion.
- **1956–1958** Castro returns to lead a guerrilla war in the Sierra Maestra.
- **1959** Batista flees. Castro takes over. Wholesale nationalizations; Cuba reorganized on Soviet model.
- **1960** US breaks off relations.
- **1961** US-backed invasion of Bay of Pigs by anti-Castro Cubans fails. Cuba declares itself Marxist–Leninist. US economic and political blockade.
- **1962** Missile crisis: Soviet deployment of nuclear weapons in Cuba leads to extreme Soviet–US tension; war averted by Khrushchev ordering withdrawal of weapons.
- **1965** One-party state formalized.
- **1972** Cuba joins COMECON (communist economic bloc).
- **1976** New socialist constitution. Cuban troops in Angola until 1991.
- **1977** Sends troops to Ethiopia.
- **1980** 125,000 Cubans, including "undesirables," flee to US.
- **1982** US tightens sanctions and bans flights and tourism to Cuba.
- **1983** US invasion of Grenada. Cuba involved in clashes with US forces.
- **1984** Agreement with US on Cuban emigration and repatriation of "undesirables" is short-lived.
- **1986** Soviet-style *glasnost* rejected.
- **1988** UN's second veto of US attempt to accuse Cuba of human rights violations. Diplomatic relations established with EC.
- **1989** Senior military executed for arms and narcotics smuggling.
- **1991** Preferential trade agreement with USSR ends. Severe rationing.
- **1992–1993** US tightens blockade. All former Soviet military leave.
- **1994–1995** Economic reforms to boost foreign trade and investment.
- **1996** US Helms-Burton Act tightens sanctions.
- **1998** Visit of Pope John Paul II.
- **1999** Leading moderate dissidents put on trial.
- **2001** Devastation by Hurricane Michelle.
- **2002** Guantánamo Bay used as high-security prison for captives from US "war on terrorism."

DEFENSE

 Compulsory military service

 $735m   Down 2% in 2000

CUBAN ARMED FORCES

900 main battle tanks (T-34, T-54/55, T-62)	35,000 personnel	
5 patrol boats	3000 personnel	
130 combat aircraft (MiG-21/23/29)	8000 personnel	
None		

From 1959 to the 1980s, Cuba's efficient military, well represented in the Council of Ministers and the Politburo, was one of the achievements of the revolution. Under Castro's brother Raúl, it succeeded in repelling the US-sponsored Bay of Pigs invasion in 1961, and saw effective action in Africa in the 1970s, preventing South Africa from taking control of Angola, and Somalia from occupying the Ogaden region in Ethiopia.

Today, with communist regimes collapsed around the world, it has lost much of its prestige. Russia is still the main source of arms. A siege mentality associated with the US economic embargo keeps the military on the alert for perceived internal and external threats.

ECONOMICS

Inflation 1.1% p.a. (1990–1999)

$18.3bn   21.00 Cuban pesos

SCORE CARD

- WORLD GNP RANKING...........................70th
- GNP PER CAPITA $1650
- BALANCE OF PAYMENTS ....In deficit
- INFLATION 0.3%
- UNEMPLOYMENT 6%

ECONOMIC PERFORMANCE INDICATOR

EXPORTS

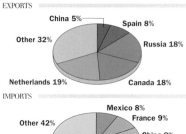

China 5% — Spain 8% — Russia 18% — Canada 18% — Netherlands 19% — Other 32%

IMPORTS

Mexico 8% — France 9% — China 9% — Italy 10% — Spain 22% — Other 42%

STRENGTHS

Buoyant tourism attracts strong foreign investment. Major exporter of sugar and nickel. Premium Cuban cigars. Strengthening banking sector.

WEAKNESSES

Denied a major market and investment capital by US trade embargo. Acute shortage of hard currency. Vulnerability of sugar and nickel to world price fluctuations. Difficult terms of trade and weak legal framework deter investment. Deficient infrastructure. Shortages of fuel, fertilizers, spare parts, and other inputs. Severe hurricane damage in 2001.

PROFILE

After a brief flirtation with the open market, total state control was reimposed in 1986. The collapse of the USSR meant the loss of some $5 billion in annual aid and led to a deep recession in the early 1990s. A cautious adoption of some capitalist-style reforms in the mid-1990s, including the free use of the US dollar, stimulated the growth of a dollarized sector centered on tourism, which has attracted strong foreign investment. Tourism rather than sugar now dominates the economy, and benefits some 160,000 self-employed and small businesses. Foreign companies are also involved in joint ventures in banking and the oil and gas sectors. There is a very large informal sector. The government lost significant revenue from 2001 when Russia terminated the lease of its information-gathering center.

CUBA : MAJOR BUSINESSES

Havana · Matahambre · Cardenas Bay · Ciego de Avila · Cienfuegos · Pinar del Rio · Isla de la Juventud · Santiago de Cuba

Oil refining · Manufacturing · Sugar cane refining · Pharmaceuticals · Nickel mining · Citrus fruits · Cigars · Oil · 0 100 km · 0 100 miles

RESOURCES

 Electric power 4m kw

122,425 tonnes

37,473 b/d (reserves 123m barrels)

4.4m cattle,
2.7m pigs,
13.3m chickens

Iron, nickel, cobalt,
chromite, gold,
manganese, oil

Cropland
Pasture
Forest
Wetlands
↓ Sugarcane – cash crop
🐄 Cattle

CUBA : LAND USE

0   100 km

0   100 miles

ELECTRICITY GENERATION

Hydro 1% (0.1bn kwh)	
Combustion 99% (15bn kwh)	
Nuclear 0%	
Other 0%	

% of total generation by type

ENVIRONMENT

 Sustainability rank: 58th

17% (2% partially protected)

2.2 tonnes per capita

ENVIRONMENTAL TREATIES

Yes		Yes		Yes	
Yes		Yes		Yes	

At the time of the revolution in 1959, only 14% of the country's forest cover remained, but a strong drive to replant has raised the tree cover level to over 20%. The intensive use of irrigation without adequate drainage has caused salinization and waterlogging. There is regional concern about the never-completed nuclear reactor at Juraguá.

MEDIA

TV ownership medium

Daily newspaper circulation 118 per 1000 people

PUBLISHING AND BROADCAST MEDIA

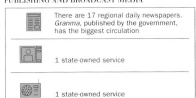

There are 17 regional daily newspapers. *Granma*, published by the government, has the biggest circulation

1 state-owned service

1 state-owned service

A catch-all anticrime law restricts and penalizes investigative reporting by independent journalists which is judged to be assisting the US foreign policy against Cuba.

The collapse of the USSR precipitated a steep decline in demand for sugar; production hit a 50-year low in 1998.

Cuba seeks to expand nickel and cobalt production, traditionally its biggest merchandise exports, assisted by private mining ventures. Several foreign companies are prospecting for gold, silver, and other metals, and for heavy crude oil and gas, through concessions. Work at Juraguá on a Russian-built nuclear reactor was abandoned in December 2000.

CRIME

 Death penalty in use

33,000 prisoners

⬆ Crime is rising

CRIME RATES

Cuba does not publish official statistics for murders, rapes, or thefts

Violent crime is officially viewed as a threat to national stability. In 1999 the penal code was amended to extend the death penalty to robbery involving firearms, certain narcotics offenses, attacks on security officers, and sexual corruption of minors. Widening inequality has led to growing crime, especially theft from state factories. The US base at Guantánamo Bay houses a prison for suspected members of the al-Qaida terrorist network.

EDUCATION

 School leaving age: 15

97%

116,700 students

THE EDUCATION SYSTEM

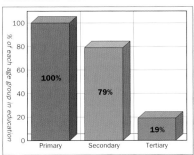

% of each age group in education

- Primary: 100%
- Secondary: 79%
- Tertiary: 19%

Education, which is universal and free at all levels, combines academic with manual work, in line with Marxist–Leninist principles. The importance given to education under Castro, which is reflected in the high literacy rate, is now being promoted to attract foreign investment in high-tech industries, particularly biotechnology. Spending on education was increased in the late 1990s to around 11.5% of the budget.

HEALTH

 Welfare state health benefits

1 per 189 people

Heart & cerebrovascular diseases, cancers, nutritional disorders

Spending on health accounted for around 10% of government expenditure in the late 1990s. Average life expectancy in Cuba is among the highest in Latin America, which is a reflection of its efficient, countrywide health service. The US trade embargo has led to shortages of hospital equipment and raw materials for drugs. The latter are normally supplied by Havana's sizable pharmaceuticals industry. Cuba's advanced surgery techniques attract patients from overseas.

SPENDING

 GDP/cap. increase

CONSUMPTION AND SPENDING

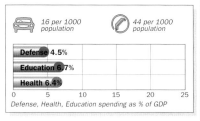

16 per 1000 population

44 per 1000 population

Defense 4.5%	
Education 6.7%	
Health 6.4%	

Defense, Health, Education spending as % of GDP

Under Batista there were huge wealth disparities, and Cuba was a playground for the rich. The 1959 revolution succeeded in reducing the disparities, partly by taking over all businesses, from oil companies to barbers' shops, and partly by prescribing not only minimum but also maximum wages. Economic regulations have varied since then; for a brief period in 1985, different wage rates were allowed in an attempt to provide incentives for hard workers, but this decision was reversed in 1986. Economic liberalization in the mid-1990s has created a large gulf between that half of the population with access to US dollars and those left in the peso economy who have to subsist on lower salaries.

WORLD RANKING

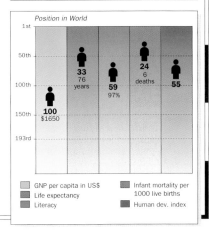

Position in World

- GNP per capita: 100 / $1650
- Life expectancy: 33 / 76 years
- Literacy: 59 / 97%
- Infant mortality: 24 / 6 deaths
- Human dev. index: 55

GNP per capita in US$
Life expectancy
Literacy
Infant mortality per 1000 live births
Human dev. index

CYPRUS

OFFICIAL NAME: Republic of Cyprus   CAPITAL: Nicosia   POPULATION: 790,000
CURRENCY: Cyprus pound (Turkish lira in TRNC)   OFFICIAL LANGUAGES: Greek and Turkish

C

THE ISLAND OF Cyprus, which rises from a central plateau to a high point at Mount Olympus, lies south of Turkey in the eastern Mediterranean. It was partitioned in 1974, following an invasion by Turkish troops. The south of the island is the Greek Cypriot Republic of Cyprus (Cyprus); the self-proclaimed Turkish Republic of Northern Cyprus (TRNC) is recognized only by Turkey.

CLIMATE   ▷ Mediterranean

WEATHER CHART FOR NICOSIA

The climate is typically Mediterranean: summers are hot and dry and winters mild, though there is mountain snow.

TRANSPORTATION   ▷ Drive on left

 Larnaca
4.88m passengers

 1475 ships
23.2m grt

THE TRANSPORTATION NETWORK

6249 km (3883 miles)	178 km (111 miles)
None	None

Travel between the two zones is impeded. The south regards the airport at Ercan as an illegal point of entry.

TOURISM   ▷ Visitors : Population 3.4:1

2.7m visitors   Up 10% in 2000

MAIN TOURIST ARRIVALS

UK 47%
Germany 10%
Russia 5%
Other 38%

% of total arrivals

Tourism expanded rapidly in the 1980s in the south, and more recently in the north. A ten-year plan aims to double the number of visitors by 2010. Tourists come for beaches, archeology, or the abundant wildlife, notably on the Akamas peninsula and the Troodos Mountains.

PEOPLE   ▷ Pop. density medium

Greek, Turkish   85/km²
(221/mi²)

THE URBAN/RURAL POPULATION SPLIT

57%   43%

ETHNIC MAKEUP

Other 5%
Turkish 18%
Greek 77%

Cyprus's Greek majority are Orthodox Christian. The Turkish minority are Muslim. Most Turkish Cypriots are the descendants of Turks who settled on the island from the 16th century, while it was under the rule of the Ottoman empire. Both Cypriot communities have suffered great upheavals: in 1974 the island was partitioned following the Turkish invasion, and 200,000 Greek Cypriots were forced to flee to the south, while 65,000 Turkish Cypriots fled in the other direction. Northern Cyprus is officially recognized as a political entity only by Turkey. Thousands of mainland Turks have settled there.

East European contract labor is brought in to staff hotels in the south, where wage levels are on average three times higher than in the north. Unemployment levels in the north, meanwhile, are rising.

The 2nd-century theater *at Curium, 14 km (19 miles) west of Limassol. Curium was the site of a flourishing Mycenaean colony before 1100 BCE.*

POLITICS   ▷ Multiparty elections

2001/2006 Cyprus
1998/2003 TRNC

President Glafcos Clerides (Cyprus)
President Rauf Denktash (TRNC)

AT THE LAST ELECTION

House of Representatives 80 seats

30% TC	25% AKEL	24% DISY	11% DIKO	5% Others	5% KISOS

TC = Reserved for Turkish Cypriots   AKEL = Progressive Party of the Working People   DISY = Democratic Rally   DIKO = Democratic Party   KISOS = Movement of Social Democrats

The 24 seats reserved for Turkish Cypriots have not been occupied since December 1963

Legislative Assembly (TRNC) 50 seats

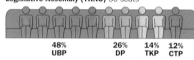

48% UBP	26% DP	14% TKP	12% CTP

UBP = National Unity Party   DP = Democrat Party
TKP = Social Welfare Party   CTP = Republican Turkish Party

The UN-backed proposal of a two-zone federation for Cyprus would give each community its own territory while sharing a number of government functions and ministries. TRNC president Rauf Denktash, mindful of the Greek Cypriots' repression of the Turks prior to 1974, is unwilling to accept a plan that does not ensure full sovereignty and political equality for Turks. Greek Cypriots, in turn, fear the plan would give too much influence over their affairs to the small Turkish minority, who would be able to veto all government decisions. Fresh impetus was given to talks in 2002 by the imminence of EU membership for Cyprus.

WORLD AFFAIRS   ▷ Joined UN in 1960

 CE   Comm   IBRD   NAM   OSCE

The presence of over 1000 UN troops since 1974 staffing the "Green Line" (only the Middle East and Kashmir have longer-standing UN peacekeeping forces) costs $43 million a year. Only Turkey recognizes the TRNC.

Cyprus applied to join the EU in 1990, and formal negotiations opened in 1998.

AID   ▷ Recipient

 $54m (receipts) (Cyprus)   Up 2% in 2000 (Cyprus)

Cyprus receives aid from international agencies, as well as from the EU and individual countries such as the UK. The TRNC is dependent on aid from Turkey of more than $60 million a year.

CYPRUS

Total Land Area :
9250 sq. km
(3571 sq. miles)

POPULATION

◎	over 100 000
○	over 50 000
●	over 10 000
•	under 10 000

LAND HEIGHT

1000m/3281ft
500m/1640ft
200m/656ft
Sea Level

Cease-fire line ×××××

0   25 km
0   25 miles

DEFENSE

▷ Compulsory military service

 $453m (Cyprus)

 Up 28% in 2000 (Cyprus)

In addition to two sovereign British bases and UN forces, there are 36,000 Turkish troops in northern Cyprus and 1250 Greek troops in the buffer zone. The 10,000-strong Greek Cypriot army and the 5000-strong Turkish Cypriot army both rely heavily on conscripts.

ECONOMICS

▷ Inflation 3.5% p.a. (1990–2000)

 $9.36bn

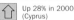 0.6116–0.6461 Cyprus pounds

SCORE CARD

❏ WORLD GNP RANKING	86th
❏ GNP PER CAPITA	$12,370
❏ BALANCE OF PAYMENTS	–$456m
❏ INFLATION	4.1%
❏ UNEMPLOYMENT	4%

STRENGTHS

Booming tourism industry, accounting for over 20% of GDP. Manufacturing sector and provision of services to Middle Eastern countries.

WEAKNESSES

Pressure for tighter supervision of offshore finance and crackdown on tax evasion. Limited liberalization. TRNC starved of foreign investment.

EXPORTS

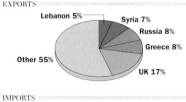

Lebanon 5%   Syria 7%   Russia 8%   Greece 8%   Other 55%   UK 17%

IMPORTS

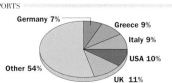

Germany 7%   Greece 9%   Italy 9%   USA 10%   Other 54%   UK 11%

RESOURCES

▷ Electric power 699,000 kw

 6695 tonnes

 Not an oil producer

418,500 pigs, 378,600 goats, 3.2m chickens

Asbestos, gypsum, iron, bentonite, copper

Cyprus continues to supply electricity to TRNC, but has not been paid for this. The possibility of offshore oil and gas to the south has attracted interest. Water is precious but new desalinization plants have reduced shortages.

ENVIRONMENT

▷ Not available

0.2% partially protected

7.9 tonnes per capita

Campaigners demand that the 155 sq. km (60 sq. miles) of the Akamas peninsula be fully protected from the threat of being sold for tourist development. Akamas is home to an unusual variety of plant and bird life, and contains breeding sites of the rare green turtle.

MEDIA

▷ TV ownership high

 Daily newspaper circulation 69 per 1000 people

PUBLISHING AND BROADCAST MEDIA

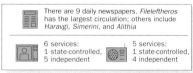

There are 9 daily newspapers. *Fileleftheros* has the largest circulation; others include *Haravgi*, *Simerini*, and *Alithia*.

6 services:
1 state-controlled,
5 independent

5 services:
1 state-controlled,
4 independent

Cyprus's press is lively and tends to be highly politicized. The radio and TV services for British troops based in Cyprus are also popular.

CRIME

▷ Death penalty not used in practice

 247 prisoners

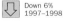 Down 6% 1997–1998

Crime rates are low and violence is rare. The unruly and sometimes violent behavior of foreign forces has on occasions led Cypriots to object to their presence. The arrest of a Cypriot politician during protests at a UK military base in 2001 provoked riots.

CHRONOLOGY

Cyprus was dominated, in turn, by Egypt, Greece, the Byzantines, the Ottomans, and the UK, before independence in 1960.

- ❏ **1963** Turkish Cypriots abandon parliament.
- ❏ **1974** President Makarios deposed by Greek military junta. Turkey invades. Partition.
- ❏ **1983** Self-proclamation of TRNC.
- ❏ **1998** Talks on EU membership start, following 1990 application to join.
- ❏ **2001** President Clerides visits TRNC.

EDUCATION

▷ School leaving age: 15

 97%

 25,163 students

Education is free and compulsory up to the age of 15. Many Greek Cypriots go abroad to university.

HEALTH

▷ Welfare state health benefits

1 per 392 people

Heart diseases, accidents, cancers

Health care is more advanced in the south; sophisticated surgery is carried out at Lefkosia General Hospital.

SPENDING

▷ GDP/cap. increase

CONSUMPTION AND SPENDING

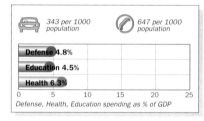

343 per 1000 population

647 per 1000 population

Defense	4.8%
Education	4.5%
Health	6.3%

Defense, Health, Education spending as % of GDP

The average income per capita in the southern part of Cyprus is higher than in Greece and Portugal, but slightly lower than that in Spain.

WORLD RANKING

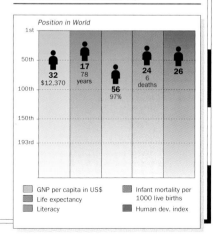

Position in World

1st
50th
100th
150th
193rd

| 32 | 17 | 56 | 24 | 26 |
| $12,370 | 78 years | 97% | 6 deaths | |

GNP per capita in US$
Life expectancy
Literacy

Infant mortality per 1000 live births
Human dev. index

CZECH REPUBLIC

OFFICIAL NAME: Czech Republic **CAPITAL:** Prague
POPULATION: 10.3 million **CURRENCY:** Czech koruna **OFFICIAL LANGUAGE:** Czech

EUROPE

LANDLOCKED IN CENTRAL Europe, the Czech Republic comprises the territories of Bohemia and Moravia, and for most of the 20th century it was part of Czechoslovakia. In 1989, the "Velvet Revolution" ended four decades of communist rule, and free elections followed in 1990. In 1993 the Czech Republic and Slovakia peacefully dissolved their federal union to become two independent states.

CLIMATE

▷ Continental

WEATHER CHART

The Czech climate is more moderate than that of Slovakia, though easterly winds bring low temperatures in winter.

TRANSPORTATION

▷ Drive on right

Ruzyné, Prague 5.55m passengers

18 ships 228,000 grt

THE TRANSPORTATION NETWORK

| 127,693 km (79,345 miles) | 499 km (310 miles) |
| 9365 km (5819 miles) | 677 km (421 miles) |

There are new expressways and rail links to Germany. Prague is a busy regional center for passenger air traffic.

TOURISM

▷ Visitors : Population 1:2.2

4.67m visitors

Down 17% in 2000

MAIN TOURIST ARRIVALS

Germany 41%
Poland 19%
Slovakia 11%
Other 29%

% of total arrivals

Revenues from tourism amount to nearly $4 billion a year and are an invaluable source of foreign earnings for the Czech economy. Germans are the most numerous among the millions of visiting tourists, mainly from Europe. Prague, which rivals Paris as the most beautiful capital in Europe, is still the main destination for visitors, although a growing proportion now seek other attractions such as spa towns and skiing in the Carpathian Mountains.

CZECH REPUBLIC

Total Land Area : 78 866 sq. km (30 450 sq. miles)

LAND HEIGHT	POPULATION
1000m/3281ft	over 1 000 000
500m/1640ft	over 500 000
200m/656ft	over 100 000
150m/492ft	over 50 000
	over 10 000
	under 10 000

PEOPLE

▷ Pop. density medium

Czech, Slovak, Hungarian

131/km² (338/mi²)

THE URBAN/RURAL POPULATION SPLIT

75%   25%

RELIGIOUS PERSUASION

Protestant 3%
Hussites 2%
Roman Catholic 39%
Other 18%
Atheist 38%

Czechs make up over 80% of the population; Moravians are the next largest group. Some 300,000 Slovaks were left in the country after partition, and dual citizenship is now permitted. Ethnic tensions are few, except that the Roma community faces discrimination. Divorce rates are high.

POLITICS

▷ Multiparty elections

L. House 2002/2006
U. House 2000/2002

President Vaclav Havel

AT THE LAST ELECTION

Chamber of Deputies 200 seats

| 35% CSSD | 29% ODS | 21% KSCM | 15% K |

CSSD = Czech Social Democratic Party **ODS** = Civic Democratic Party **KSCM** = Communist Party of Bohemia and Moravia **K** = Coalition of the Christian Democratic Union–Czech People's Party (**KDU–CSL**) and the Freedom Union (**US**) **QC** = Quad Coalition, comprising the KDU–CSL, the US, the Civic Democratic Alliance (ODA), and the Democratic Union (DEU) **Ind** = Independents

Senate 81 seats

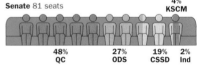

4% KSCM

| 48% QC | 27% ODS | 19% CSSD | 2% Ind |

The pro-democratic solidarity of 1989–1990, which saw the election of the Civic Forum and dissident playwright Vaclav Havel as president, soon gave way to a two-party system. The right-of-center ODS government pursued market economics, and oversaw the split with the Slovak Republic in 1993, but it lost its overall majority in 1996. It then gave tacit support to the social-democratic CSSD's minority government on the understanding that constitutional reforms would favor the two larger parties. The CSSD under Vladimir Spidla was able to form a majority government, without ODS backing, in 2002.

C

WORLD AFFAIRS
▷ Joined UN in 1993

 CE CEFTA NATO OECD OSCE

The Czech Republic began formal negotiations on EU membership in 1998. It joined NATO in 1999.

AID
▷ Recipient

 $438m (receipts)   Up 35% in 2000

Aid for economic restructuring has been crucial for modernizing infrastructure such as telecommunications.

DEFENSE
▷ Compulsory military service

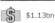 $1.13bn   Down 2% in 2000

The split with Slovakia left an oversized, expensive army. In 1994, plans to cut the military by 20,000 were approved. Professional soldiers with a communist past were the first to go. The Czech armaments industry has a long tradition based on precision engineering, and the Czech Republic is among the world's 20 largest arms exporters.

ECONOMICS
▷ Inflation 12% p.a. (1990–2000)

$53.9bn   37.63-35.56 Czech koruny

SCORE CARD
❏ World GNP Ranking	46th
❏ GNP per Capita	$5250
❏ Balance of Payments	–$2.34bn
❏ Inflation	3.9%
❏ Unemployment	9%

STRENGTHS
Skilled industrial labor force. Good industrial base. Speed of privatization of state industries. Attractive to German investors. Draw of Prague for tourists.

WEAKNESSES
Lack of diversification in sectors usually attractive to overseas investors. Limited restructuring, banking sector problems. Pressure to cut government expenditure to reduce serious budget deficit.

EXPORTS
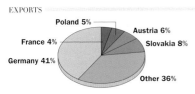

Poland 5%
France 4%
Germany 41%
Austria 6%
Slovakia 8%
Other 36%

IMPORTS

France 5%
Italy 5%
Other 44%
Slovakia 6%
Russia 7%
Germany 33%

Relations with Germany are a priority. The issue of the forced expulsion of Germans in 1945 remains a stumbling block. Austria and Germany strongly opposed the opening of the Temelín nuclear plant in 2000.

RESOURCES
▷ Electric power 13.9m kw

 22,965 tonnes
3.59m pigs, 1.58m cattle, 14.7m chickens
 3589 b/d (reserves 88m barrels)
Oil, natural gas, copper, lead, zinc, coal, uranium

Copper, lead, zinc, and coal are the Czech Republic's chief resources. The government is aiming to phase out the worst-polluting coal-fired power plants. Opposition to a planned 2000 MW Soviet-designed nuclear power station at Temelín delayed its completion until late 2000.

ENVIRONMENT
▷ Sustainability rank: 64th

 16%   11.5 tonnes per capita

Pollution from the power, chemical, and cement industries and the new Temelín nuclear plant are key concerns.

MEDIA
▷ TV ownership high

 Daily newspaper circulation 254 per 1000 people

PUBLISHING AND BROADCAST MEDIA

There are 21 daily newspapers. *Mladá Fronta Dnes* has the largest circulation

5 services: 2 state-owned, 3 independent

1 state-owned service, over 44 independent services

Government-influenced appointments of senior media officials provoked mass protests and a change in the law in 2001.

CRIME
▷ No death penalty

 21,538 prisoners   Little change in 1999

Prostitution is becoming a growing problem, especially in regions bordering Austria and Germany.

EDUCATION
▷ School leaving age: 15

 99%   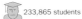 233,865 students

Schooling has reverted to the pre-1945 system. Charles University in Prague was founded in the 13th century.

HEALTH
▷ Welfare state health benefits

 1 per 333 people   Cancers, heart and cerebrovascular diseases, accidents

Health care expenditure as a share of GDP has increased by 30% since 1990. Wealthy Czechs travel to Germany for complex surgery.

The Vltava River *in Prague. Millions of tourists, mainly from Europe and the US, visit Prague every year.*

CHRONOLOGY
Formerly part of the Austro-Hungarian Empire, the Republic of Czechoslovakia was established in 1918. It was invaded by Hitler in 1939.

- ❏ **1968** "Prague Spring." Invasion by Warsaw Pact countries.
- ❏ **1989** "Velvet Revolution."
- ❏ **1990** Free elections won by Civic Forum; Vaclav Havel president.
- ❏ **1993** Split with Slovakia.
- ❏ **1998** Start of EU membership negotiations. Elections: CSSD forms minority government.
- ❏ **1999** Joins NATO.
- ❏ **2002** CSSD reelected.

SPENDING
▷ GDP/cap. increase

CONSUMPTION AND SPENDING

335 per 1000 population   378 per 1000 population

Defense 2.2%
Education 4.2%
Health 6.6%

0   5   10   15   20   25
Defense, Health, Education spending as % of GDP

A new entrepreneurial business class has emerged since the "Velvet Revolution" in 1989. Many Czechs have shares in privatized enterprises.

WORLD RANKING

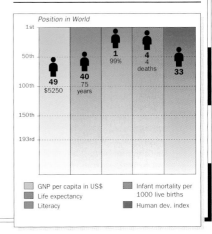

Position in World

1st
50th
100th
150th
193rd

49 $5250
40 75 years
1 99%
4 4 deaths
33

- ▢ GNP per capita in US$
- ▢ Life expectancy
- ▢ Literacy
- ▢ Infant mortality per 1000 live births
- ▢ Human dev. index

D

DENMARK

OFFICIAL NAME: Kingdom of Denmark **CAPITAL:** Copenhagen
POPULATION: 5.3 million **CURRENCY:** Danish krone **OFFICIAL LANGUAGE:** Danish

 950   1944   April 16   DK   +1   +45   .dk

THE MOST SOUTHERLY COUNTRY in Scandinavia, Denmark occupies the Jutland (Jylland) peninsula, the islands of Sjælland, Fyn, Lolland, and Falster, and more than 400 smaller islands. Its terrain is among the flattest in the world. The Faeroe Islands and Greenland in the North Atlantic are self-governing associated territories. Politically, Denmark is stable, despite a preponderance of minority governments since 1945. It possesses a long liberal tradition and was one of the first countries to establish a welfare system, in the 1930s.

TOURISM

Visitors : Population
1:2.5

2.1m visitors Up 3% in 2000

MAIN TOURIST ARRIVALS

Sweden 33%			
Norway 14%			
Germany 10%			
UK 6%			
USA 6%			
Other 31%			

0 ... 10 ... 20 ... 30 ... 40
% of total arrivals

CLIMATE

Maritime

WEATHER CHART FOR COPENHAGEN

Average daily temperature   Rainfall
°C/°F J F M A M J J A S O N D cm/in
40/104   40/16
30/86   30/12
20/68   20/8
10/50   10/4
0/32   0
-10/14
-20/4

Denmark's temperate, damp climate is one of the keys to its agricultural success. The Faeroes are windy, foggy, and cool. Greenland's climate ranges north–south from arctic to subarctic.

The island of Fyn, like the rest of Denmark, is flat and depends on coastal defenses to prevent flooding by the sea.

Principal attractions for tourists are Copenhagen (with its Tivoli Gardens and 18th-century architecture), Legoland, the countryside, and seaside resorts. Since 1959, only Danes have been allowed to own vacation homes.

TRANSPORTATION

Drive on right

Kastrup, Copenhagen
18.3m passengers

942 ships
6.82m grt

THE TRANSPORTATION NETWORK

71,681 km (44,540 miles)	953 km (592 miles)
2324 km (1444 miles)	417 km (259 miles)

There is an extensive, well-integrated transportation network of bus, rail, and ferry services. State-owned companies predominate, although privatization of some ferry and rail services has been mooted. Denmark wishes to reduce significant state transportation subsidies. A few private companies operate in the Faeroes and Greenland with state support.

Major new construction projects focus on bridge and tunnel links, such as the Storebælt project connecting the islands of Fyn and Sjælland. A 16-km (10-mile) Øresund road and rail link by bridge and tunnel, connecting Copenhagen with Malmö in Sweden, opened in July 2000. Copenhagen's new Metro light rail system has now been completed.

DENMARK

Total Land Area : 43 094 sq. km
(16 659 sq. miles)

POPULATION

over 1 000 000
over 100 000
over 10 000
under 10 000

LAND HEIGHT

175m/574ft
Sea Level

Ferry link

D

PEOPLE

 Pop. density medium

Danish

125/km²
(324/mi²)

THE URBAN/RURAL POPULATION SPLIT

85%   15%

RELIGIOUS PERSUASION

Roman Catholic 1%   Other 10%

Evangelical
Lutheran 89%

ETHNIC MAKEUP

Faeroe and
Inuit 1%   Other (including
Scandinavian
and Turkish)
3%

Danish 96%

Danish society is homogeneous, but the small population of foreign citizens doubled over 1984–1999, and the current right-leaning government has pledged to curb immigration. The most visible minority groups are the Inuit, Greenland's indigenous inhabitants, and the Turkish community. Rising unemployment has engendered some ethnic tension, although racially motivated attacks are still rare.

Denmark has undergone profound social changes over the last 20 years, and the role of women has been transformed. Helped by extensive social and educational provision, three-quarters of women now work in part-time or full-time jobs. Denmark provides the best state child support in Europe: almost half of children under two, and two-thirds of three- to six-year-olds, are in day nurseries.

Less than half the population lives in a nuclear family, in part reflecting the high divorce rate. Marriage is becoming less common; almost 40% of children are brought up by unmarried couples or single parents. Cohabiting couples now have the same legal rights as those who are married. In 1990, Denmark became the first country to allow registered partnerships between homosexual couples, effectively granting them the same legal status as heterosexual couples.

POPULATION AGE BREAKDOWN

Female	Age	Male
2.6%	80+	1.3%
8.6%	60–79	7.2%
13.4%	40–59	13.7%
14.5%	20–39	15.2%
11.5%	0–19	12%

% of population by age group

POLITICS

 Multiparty elections

2001/2005   H.M. Queen Margrethe II

AT THE LAST ELECTION

Parliament 179 seats

12% DFP	9% KF	7% SF

31% V	29% SD	9% Other CL	3% Other CR

V = Liberal Party (Venstre)   **SD** = Social Democrats
DFP = Danish People's Party   **KF** = Conservative People's
Party   **Other CL** = Other Center Left   **SF** = Socialist People's
Party   **Other CR** = Other Center Right
Greenland and the Faeroe Islands send 2 members each to
the Parliament

***Anders Fogh
Rasmussen,*** *heads
a right-wing,
minority coalition.*

Queen Margrethe II,
*who succeeded to the
throne in 1972.*

Denmark is a constitutional monarchy and a multiparty democracy. The associated territories of Greenland and the Faeroe Islands have home rule. The latter are divided over the issue of independence.

PROFILE

The intricate proportional electoral system ensures that Parliament truly reflects voters' wishes, but also tends to lead to minority governments. After a decade of Conservative–Liberal rule, the SD regained power in 1993, at the head of a center-left coalition under Poul Nyrup Rasmussen. The pendulum swung back with a victory in 2001 for the Liberals and Anders Fogh Rasmussen. Tax and immigration are important issues, but major policy differences between the two main political groups are few.

MAIN POLITICAL ISSUES
Relations with the EU

In recent years, Denmark's left-of-center parties have been suspicious of further EU integration. In 1992, ratification of the Maastricht Treaty, approved by Parliament, was rejected in a referendum, voters objecting to monetary union, a common defense force, and local election voting rights for European citizens living in Denmark. Later that year an EU summit allowed Denmark to opt out of monetary union, defense, and European citizenship. A referendum in 1993 finally approved the Maastricht Treaty, and voters approved the Amsterdam Treaty, its successor, in 1998, but a referendum in 2000 on joining the euro (which Denmark had decided not to introduce from its outset in 1999) produced a highly significant, and unexpected, "no" result.

Immigration

Despite Danish liberal traditions, the situation of immigrants and refugees – under 5% of the population – is a controversial issue. The election of a right-wing government in 2001 signaled a change to an overtly restrictive policy and a greater emphasis on integration for those immigrants already there. The ruling coalition relies for support on the far-right, anti-immigrant Danish People's Party, which had almost doubled its representation. Legislation passed in June 2002 – on the eve of Denmark's taking over the EU presidency – prevents the foreign (non-EU) spouse of a Danish citizen living in or emigrating to the country.

WORLD AFFAIRS

Joined UN in 1945

 CE EU   NATO   OECD OSCE

Relations with the rest of Europe are the major foreign policy concern, notably the issues of a common defense policy and monetary union. Denmark has decided against introducing the euro, but the krone is pegged to it and economic policies follow those of the participating states. Promoting economic ties with Norway, Sweden, and Finland is a priority, as are improving links with former Eastern bloc states, especially those on the Baltic – not least to assist pollution reduction, a serious concern. Denmark is also a strong supporter of Third World development, especially in Africa.

CHRONOLOGY

Founded in the 10th century, Denmark's monarchy is Europe's oldest. It was the dominant Baltic power until the 17th century, when it was eclipsed by Sweden.

❏ **1815** Denmark forced to cede Norway to Swedish rule.
❏ **1849** Creation of first democratic constitution.
❏ **1864** Denmark forced to cede provinces of Schleswig and Holstein after losing war with Prussia.
❏ **1914–1918** Denmark neutral in World War I.
❏ **1915** Universal adult suffrage introduced. Rise of SD.
❏ **1920** Northern Schleswig votes to return to Danish rule. ⇨

D

CHRONOLOGY *continued*

- ❏ **1929** First full SD government takes power under Prime Minister Thorvald Stauning.
- ❏ **1930s** Implementation of advanced social welfare legislation and other liberal reforms under SD.
- ❏ **1939** Outbreak of World War II; Denmark reaffirms neutrality.
- ❏ **1940** Nazi occupation. National coalition government formed.
- ❏ **1943** Danish Resistance successes lead Nazis to take full control.
- ❏ **1944** Iceland declares independence from Denmark.
- ❏ **1945** Denmark recognizes Icelandic independence. After defeat of Nazi Germany, SD leads postwar coalition governments.
- ❏ **1948** Faeroes granted home rule.
- ❏ **1952** Founder member of the Nordic Council.
- ❏ **1953** Constitution reformed; single-chamber, proportionally elected parliament created.
- ❏ **1959** Denmark joins EFTA.
- ❏ **1973** Denmark joins European Communities.
- ❏ **1979** Greenland granted home rule.
- ❏ **1975–1982** SD's Anker Jorgensen heads series of coalitions; elections in 1977, 1979, and 1981. Final coalition collapses over economic policy differences.
- ❏ **1982** Poul Schlüter first Conservative prime minister since 1894.
- ❏ **1992** Referendum rejects Maastricht Treaty on European Union.
- ❏ **1993** Schlüter resigns over "Tamilgate" scandal. Center-left government led by Poul Nyrup Rasmussen. Danish voters ratify revised Maastricht Treaty.
- ❏ **1994–1998** Elections: Rasmussen heads SD-led minority coalition.
- ❏ **2000** Referendum rejects joining eurozone.
- ❏ **2001** Elections: Liberals return to power. Anders Fogh Rasmussen appointed prime minister.

AID ▷ Donor

 $1.66bn (donations)   Down 4% in 2000

During the 1990s, Denmark was the world's leading aid donor in GNP terms, contributing an average 1% of national income. It supports both economic and social development projects and policy reforms. The Liberal government elected in 2001 is committed to reducing the foreign aid budget.

Denmark provides aid to Asia and Latin America, but its closest ties are with Africa. Tanzania is the largest single aid recipient. Denmark has also provided considerable support to the other SADC states.

DEFENSE ▷ Compulsory military service

 $2.4bn   Down 10% in 2000

Denmark was neutral until 1945. Apart from NATO commitments, defense has a low priority; spending is less than 2% of GDP (well below the NATO average). Denmark provides troops for the NATO-led forces in former Yugoslavia and observers for other UN peacekeeping operations. One-quarter of its armed forces are conscripts, and its reserves include a Home Guard. Denmark has observer status at the WEU.

DANISH ARMED FORCES

238 main battle tanks (220 *Leopard* 1A5, 18 *Leopard* 2)	12,900 personnel
4 submarines, 3 corvettes, and 27 patrol boats	4000 personnel
68 combat aircraft (F-16A/B)	4500 personnel
None	

ECONOMICS ▷ Inflation 2.2% p.a. (1990–2000)

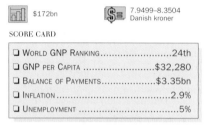 **$172bn**   7.9499–8.3504 Danish kroner

SCORE CARD

- ❏ WORLD GNP RANKING 24th
- ❏ GNP PER CAPITA $32,280
- ❏ BALANCE OF PAYMENTS..................$3.35bn
- ❏ INFLATION 2.9%
- ❏ UNEMPLOYMENT 5%

EXPORTS
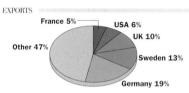
France 5% | USA 6% | UK 10% | Sweden 13% | Germany 19% | Other 47%

IMPORTS
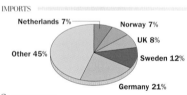
Netherlands 7% | Norway 7% | UK 8% | Sweden 12% | Germany 21% | Other 45%

ECONOMIC PERFORMANCE INDICATOR

Consumer Price Index — GDP

STRENGTHS
Low inflation and unemployment. Substantial balance-of-payments surplus. Gas and oil reserves. Strong high-tech, high-profit manufacturing sector. Skilled workforce.

WEAKNESSES
Heavy tax burden. Labor costs and historically strong currency affect competitiveness.

PROFILE
Denmark's mix of a large state sector and a private sector has been successful. GDP per capita is one of the highest among the OECD countries. However, total taxation, at about 50%, is among the world's highest. In November 2001 the Liberal government promised to fund increased welfare spending and lower taxation by extending privatization and reducing development aid.

In the 1980s the government stabilized the exchange rate and tightened budget controls in order to reduce inflation and reverse the balance of payments deficit. Growth slowed in the early 1990s, but an economic upturn between 1993 and 2000 was first led by private consumption and then buoyed by exports and business investment. Growth slowed significantly in 2001, but showed signs of regaining momentum by mid-2002.

Voters have refused to accept EU monetary union, most recently in a "no" vote in 2000 in a referendum on the euro. However, Denmark meets the EU's convergence criteria for monetary union, the krone is pegged to the euro, and economic policies follow those of the participating states.

DENMARK : MAJOR BUSINESSES

- 🔺 Oil & gas
- 🍺 Brewing
- Textiles
- Chemicals
- Agribusiness
- Electronics
- Transportation services
- Light engineering
- Trading center
- Fish processing

0   100 km
0   100 miles

D

RESOURCES

 Electric power 12.1m kw

 1.45m tonnes

342,000 b/d (reserves 1.1bn barrels)

12.1m pigs, 1.89m cattle, 21.5m chickens

 Natural gas, oil

ELECTRICITY GENERATION

Hydro 0%

Combustion 93% (43bn kwh)

Nuclear 0%

Other 7% (3.1bn kwh)

% of total generation by type

Despite expansion of North Sea oil and gas output, Denmark is still an overall importer of energy. Agriculture is highly efficient, and Denmark is the world's biggest exporter of pork. The use of wind power is expanding rapidly. Denmark is a world leader in this technology.

DENMARK : LAND USE

Cropland
Forest
Pasture
Pigs
Cereals

0   100 km
0   100 miles

ENVIRONMENT

 Sustainability rank: 31st

 33%

10.1 tonnes per capita

ENVIRONMENTAL TREATIES

Yes	Yes	Yes
Yes	Yes	Yes

Denmark has some of the strictest regulations in Europe, including those aimed at reducing ozone-destroying emissions and water pollution, and met its 2000 target – recycling 54% of all waste – a year early. There was a marked change in policy under the incoming Liberal government of 2001. The bans on house-building in state forests and on the sale of beer in cans were lifted, three planned windpower plants were shelved, and the environment ministry's budget was cut by one-third.

MEDIA

 TV ownership high

Daily newspaper circulation 311 per 1000 people

PUBLISHING AND BROADCAST MEDIA

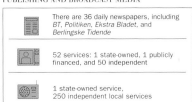

There are 36 daily newspapers, including *BT*, *Politiken*, *Ekstra Bladet*, and *Berlingske Tidende*

52 services: 1 state-owned, 1 publicly financed, and 50 independent

1 state-owned service, 250 independent local services

The media have a long history of political independence, and objectivity is prized. The tone of both TV and the press is serious; there is no scandal-mongering tabloid press as found in the US, the UK, and Germany. Invasion of privacy laws are strict. Legislation proposed in May 2002 would extend competition in broadcasting and prepare for the privatization of TV2.

CRIME

 No death penalty

3279 prisoners

Down 1% in 1999

CRIME RATES

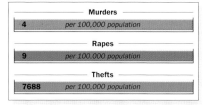

Murders

4 — per 100,000 population

Rapes

9 — per 100,000 population

Thefts

7688 — per 100,000 population

Problems are the potential importing of Mafia-style organized crime from eastern Europe, computer hacking, and drug trafficking. In 2001, stricter penalties for violence and rape were promised.

EDUCATION

 School leaving age: 16

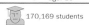 99%

170,169 students

THE EDUCATION SYSTEM

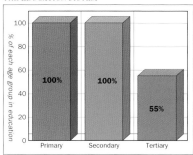

% of each age group in education

Primary	Secondary	Tertiary
100%	100%	55%

The educational level is generally high, in part reflecting the need for a skilled workforce. Formal schooling begins at age seven and is mandatory for nine years. However, most children receive preschool education, and over 90% of pupils go on at the age of 16 to further academic or vocational training. Traditional folk high schools offer a wide range of further education options, but no professional qualifications.

HEALTH

 Welfare state health benefits

1 per 294 people

Heart diseases, cancers, accidents

Denmark was one of the first countries to introduce a state social welfare system. The national health service, which still provides free treatment for almost everything, is the main reason for Denmark's high taxes – in 2000 almost a quarter of government spending was allocated to social services. Any attempts to reduce expenditure will meet with strong opposition. Repeated surveys show that most Danes prefer their system to those based on private health insurance. In the early 1980s, Denmark had the highest incidence of AIDS in Europe, but after peaking in 1993, it has dropped markedly, due to the free availability of drug therapy.

SPENDING

GDP/cap. increase

CONSUMPTION AND SPENDING

352 per 1000 population

720 per 1000 population

Defense 1.5%

Education 8.2%

Health 6.9%

Defense, Health, Education spending as % of GDP

Most Danes are comfortably off. Income distribution is more even than in many Western countries and social mobility is high. Free higher education means that access to the professions is more a question of ability than wealth or connections. Denmark forms one of the world's most egalitarian societies. The generous social security system means that Danes suffer little from social deprivation. The SD government of 1994–2001 has created more kindergarten places and increased time off for those with young children. Refugees and recent immigrants tend to be the most disadvantaged members of Danish society.

WORLD RANKING

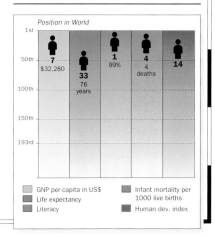

Position in World

7 — $32,280	
33 — 76 years	
1 — 99%	
4 — 4 deaths	
14	

GNP per capita in US$
Life expectancy
Literacy
Infant mortality per 1000 live births
Human dev. index

See also OVERSEAS TERRITORIES *p.640* 221

DJIBOUTI

EAST AFRICA · Africa

OFFICIAL NAME: Republic of Djibouti **CAPITAL:** Djibouti
POPULATION: 644,000 **CURRENCY:** Djibouti franc **OFFICIAL LANGUAGES:** Arabic and French

D

1977 · 1977 · June 27 · DJI · +3 · +253 · .dj

A CITY WITH A DESERT HINTERLAND, Djibouti lies in northeast Africa on the strait linking the Red Sea and the Indian Ocean. Known from 1967 as the French Territory of the Afars and Issas, Djibouti became independent in 1977. Its economy relies on the main port, the railroad to Addis Ababa, and French aid. A guerrilla war which erupted in 1991 as a result of tension between the Issas in the south and the Afars in the north has largely been resolved.

CLIMATE   ▷ Hot desert

WEATHER CHART FOR DJIBOUTI

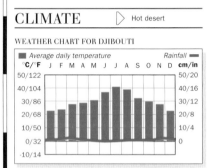

■ Average daily temperature   Rainfall ■
°C/°F J F M A M J J A S O N D cm/in
50/122 .... 50/20
40/104 .... 40/16
30/86 .... 30/12
20/68 .... 20/8
10/50 .... 10/4
0/32 0
-10/14

Despite extremely low rainfall, the monsoon season is characterized by very humid conditions. Even locals find the heat in June–August hard to bear.

TRANSPORTATION   ▷ Drive on right

Ambouli International, Djibouti   13 ships 4356 grt

THE TRANSPORTATION NETWORK

| 364 km (226 miles) | None |
| 121 km (75 miles) | None |

Djibouti's port, created by the French in the 19th century and now a modern container facility, is its key asset. Landlocked Ethiopia's vital link to the sea is the Addis Ababa–Djibouti railroad.

TOURISM   ▷ Visitors : Population 1:31

21,000 visitors   Little change 1995–1998

MAIN TOURIST ARRIVALS

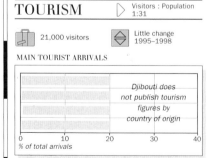

Djibouti does not publish tourism figures by country of origin

0   10   20   30   40
% of total arrivals

Most visitors are passing through on their way to Ethiopia, or coming to see relatives working in Djibouti port.

Nomadic Djiboutian village, close to Balho near the Ethiopian border.

PEOPLE   ▷ Pop. density low

Somali, Afar, French, Arabic   28/km² (72/mi²)

THE URBAN/RURAL POPULATION SPLIT

83%   17%

ETHNIC MAKEUP

Other 5%
Afar 35%
Issa 60%

The main ethnic groups are the Afars and Issas; tension between these groups developed into a guerrilla war in 1991. The population was swelled in 1992 by 20,000 Somali refugees. The rural people are mostly nomadic.

POPULATION
◎ over 100 000
• under 10 000

LAND HEIGHT
1000m/3281ft
500m/1640ft
200m/656ft
Sea Level
-200m/656ft

POLITICS   ▷ Multiparty elections

1997/2002   President Ismael Omar Guelleh

AT THE LAST ELECTION
National Assembly   65 seats

100% RPP–FRUD

RPP–FRUD = Alliance of the Popular Rally for Progress (RPP) and the Front for the Restoration of Unity and Democracy (FRUD)

President Hassan Gouled Aptidon, an Issa, backed by France, dominated politics from independence in 1977 until his retirement in 1999. Afar fears of Issa domination erupted in 1991, and the Afar guerrilla group FRUD took control of much of the country. The French intervened militarily to keep Gouled in power, but forced him to hold elections in 1992, which were won by the RPP. The FRUD became a legal political party following a 1996 peace agreement. An alliance of the RPP and FRUD made a clean sweep of elections in 1997. Presidential elections in 1999 were won by Ismael Omar Guelleh, a former close aide of Gouled, amid opposition claims of electoral fraud.

DJIBOUTI

Total Land Area : 22 000 sq. km (8494 sq. miles)

WORLD AFFAIRS
▷ Joined UN in 1977

France, with a key military presence, is pressing for greater democratization. Djibouti, Ethiopia, and Eritrea all seek to contain Afar secessionism. In 2000 the southern town of Arta hosted the Somali reconciliation conference.

AID
▷ Recipient

 $71m (receipts)   Down 5% in 2000

France is the major donor, effectively financing one-third of government expenditure. Djibouti has also received aid from Saudi Arabia and Kuwait.

DEFENSE
▷ No compulsory military service

 $23m   Up 5% in 2000

The size of the armed forces is a state secret, but is estimated at 9600 personnel; former FRUD guerrillas were integrated into the army. There is a 3200-strong French garrison.

ECONOMICS
▷ Inflation 3.7% p.a. (1990–2000)

$553m   174.8–170.0 Djibouti francs

SCORE CARD

- ❏ WORLD GNP RANKING.........................169th
- ❏ GNP PER CAPITA $880
- ❏ BALANCE OF PAYMENTS.....................–$14m
- ❏ INFLATION 2%
- ❏ UNEMPLOYMENT..................................50%

STRENGTHS
Free port in key Red Sea location; large profits from 1991 Gulf War and from 1992 US and UN intervention in Somalia. Continuing development of Djibouti and Tadjoura port facilities.

WEAKNESSES
Dependent on French aid and garrison. Planned Saudi investment held back by civil war in 1990s. Other ports on Red Sea now providing stiff competition.

EXPORTS

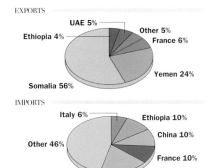

UAE 5%
Ethiopia 4%
Other 5%
France 6%
Yemen 24%
Somalia 56%

IMPORTS

Italy 6%
Ethiopia 10%
China 10%
Other 46%
France 10%
Saudi Arabia 18%

RESOURCES
▷ Electric power 85,000 kw

350 tonnes   Not an oil producer

513,000 goats, 465,000 sheep, 70,000 camels   Gypsum, mica, amethyst, sulfur, natural gas

The few mineral resources are scarcely exploited. Geothermal energy is being developed and natural gas has recently been found. The guerrilla war delayed attempts to develop underground water supplies for agriculture.

ENVIRONMENT
▷ Not available

0.4%   0.6 tonnes per capita

The concentration of business around Djibouti port means that inland desert areas are largely untouched. Ecological issues are not a national concern.

MEDIA
▷ TV ownership low

☒ There are no daily newspapers

PUBLISHING AND BROADCAST MEDIA

There are no daily newspapers. The weekly *La Nation de Djibouti* is published by the government

1 state-controlled service   1 state-controlled service

Djibouti is a member of the Arab Satellite Communications Organization. The media are largely state-controlled, but there is one opposition newspaper.

CRIME
▷ No death penalty

384 prisoners   Up 74% 1996–1998

The government accused the FRUD of atrocities, but the state's own human rights record has since been criticized by Amnesty International. While drug smuggling and prostitution are rife, petty crime, rather than violence, is the norm.

EDUCATION
▷ School leaving age: 12

65%   478 students

Schooling is mostly in French, although there has been a growing emphasis on Islamic teaching, particularly as Saudi Arabia has declared an interest in providing aid for education. Djibouti has no university.

HEALTH
▷ Welfare state health benefits

1 per 7143 people   Respiratory and heart diseases

AIDS is a growing problem in Djibouti port, with its large prostitute population. UN estimates for 1999 suggested that there were some 37,000 HIV/AIDS sufferers. Small French-financed hospitals cater for the urban elite.

CHRONOLOGY

The French set up a coaling station at Djibouti in the 1880s, to balance the British presence in Aden.

- ❏ **1917** Railroad from Addis Ababa reaches Djibouti port.
- ❏ **1977** Independence.
- ❏ **1981–1992** One-party state.
- ❏ **1989** Eruption of violence between Afars and Issas.
- ❏ **1991** FRUD launches armed insurrection.
- ❏ **1994** Peace agreement with FRUD.
- ❏ **1999** Ismael Omar Guelleh becomes president.
- ❏ **2000** Coup attempt by police officers fails.

SPENDING
▷ GDP/cap. decrease

CONSUMPTION AND SPENDING

15 per 1000 population   15 per 1000 population

Defense 5%
Education 2.9%
Health 5.4%

0   5   10   15   20   25
Defense, Health, Education spending as % of GDP

As happens in many African states, the wealth in Djibouti tends to be concentrated among those closest to government. Djiboutians working in the ports also do well, although much port labor is expatriate. The guerrilla war has had little effect on port life, since it is almost completely isolated from the rest of the country. The nomads of the interior are the poorest group.

Trade in the mild narcotic *qat*, or "green gold," which is grown in Ethiopia and shipped through Djibouti, is highly lucrative, to the extent that the state is now taking its share of the profits. In Djibouti, *qat* chewing is an age-old social ritual.

WORLD RANKING

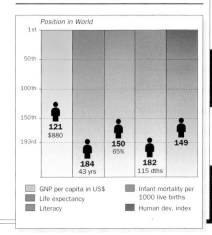

Position in World

1st
50th
100th
150th
193rd

121    $880
184   43 yrs
150   65%
182   115 dths
149

- ▢ GNP per capita in US$
- ▢ Life expectancy
- ▢ Literacy
- ▢ Infant mortality per 1000 live births
- ▢ Human dev. index

DOMINICA

OFFICIAL NAME: Commonwealth of Dominica **CAPITAL:** Roseau
POPULATION: 73,000 **CURRENCY:** Eastern Caribbean dollar **OFFICIAL LANGUAGE:** English

| 1978 | 1978 | Nov 3 | WD | -4 | +1767 | .dm |

DOMINICA IS RENOWNED as the Caribbean island that resisted European colonization until the 18th century, when it came under French control, passing to the UK in 1759. It is known as the "Nature Island" because of its spectacular, lush, and abundant flora and fauna, protected by extensive national parks. The most mountainous of the Lesser Antilles, Dominica is located between Guadeloupe and Martinique in the West Indian Windward Islands group. Its volcanic origin has given it very fertile soils and the second-largest boiling lake in the world.

CLIMATE ▷ Tropical oceanic

WEATHER CHART FOR ROSEAU

Like the other Windward Islands in the eastern Caribbean, Dominica is subject to constant trade winds. The rainy season is in the summer, and tropical depressions and hurricanes are likely between June and November. Short, thundery showers in the late afternoon and evening are common throughout the year.

TRANSPORTATION ▷ Drive on left

🛩 **Canefield, Roseau**
108,179 passengers

🚢 7 ships
2233 grt

THE TRANSPORTATION NETWORK

393 km
(244 miles)

None

None

None

Both airports take only small propeller aircraft. Roads are well maintained. There is no speed limit in rural areas.

TOURISM ▷ Visitors : Population 1:1

74,000 visitors

Up 12% in 1999

MAIN TOURIST ARRIVALS

Caribbean 58%	
USA 21%	
UK 9%	
Other 12%	

0 10 20 30 40 50 60
% of total arrivals

The national parks, with their rare indigenous birds, hot springs, and sulfur pools, are a major attraction for tourists. However, the lack of an airport able to take commercial jetliners (visitors use connecting flights from Barbados or Antigua) has made Dominica less accessible to mass-market tourism than its neighbors.

PEOPLE ▷ Pop. density medium

French Creole, English

97/km²
(252/mi²)

THE URBAN/RURAL POPULATION SPLIT

71%   29%

RELIGIOUS PERSUASION

Other 8%
Protestant 15%
Roman Catholic 77%

The majority of Dominicans are descendants of Africans brought over to work the banana plantations. The Carib Territory on the northeast of the island contains the only surviving Carib population in the Caribbean.

POLITICS ▷ Multiparty elections

2000/2005

President Vernon Shaw

AT THE LAST ELECTION
House of Assembly 30 seats

| 33% DLP | 30% DUWP | 30% App | 7% DFP |

DLP = Dominica Labour Party **DUWP** = Dominica United Workers' Party **App** = Appointed **DFP** = Dominica Freedom Party

9 senators are appointed to the House of Assembly by the head of state

Politicians tend to come from the professional classes – usually young lawyers and doctors. Occasionally the larger farmers, who provide most party funding, stand for election. The center-left DUWP narrowly won the 1995 elections, ending 15 consecutive years of rule by the right-wing DFP. A further swing to the left produced a DLP victory in January 2000, and its leader Rosie Douglas became prime minister, Pierre Charles succeeding him on his sudden death in October. The main political issue is how to cope with the ending of preferential banana exports to the EU.

WORLD AFFAIRS ▷ Joined UN in 1978

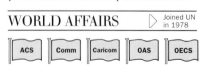

| ACS | Comm | Caricom | OAS | OECS |

Dominica has close links with France and the US. Preferential access to the EU for Caribbean bananas, crucial for Dominica's economy, was lost after a successful protest to the WTO in 1999 by the US.

DOMINICA

Total Land Area : 754 sq. km
(291 sq. miles)

LAND HEIGHT

1000m/3281ft
500m/1640ft
200m/656ft
Sea Level

POPULATION

over 10 000 ●
under 10 000 ·

0 10 km
0 10 miles

D

Inshore fishing boats, *which mostly supply the domestic market, on a typical Dominican beach.*

AID Recipient

 US$15m (receipts) ⬆ Up 50% in 2000

The European Development Fund allocated money in 1998 for sustainable tourism projects; Japan and Taiwan gave aid for firefighting and education.

DEFENSE 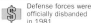 No compulsory military service

Defense forces were officially disbanded in 1981　　Not applicable

Dominica has no armed forces, but it does participate in the US-sponsored Regional Security System.

ECONOMICS Inflation 3% p.a. (1990–2000)

US$238m　　2.7 Eastern Caribbean dollars

SCORE CARD

❏ WORLD GNP RANKING.......................181st
❏ GNP PER CAPITA.........................US$3260
❏ BALANCE OF PAYMENTS....................−US69m
❏ INFLATION..0.8%
❏ UNEMPLOYMENT.....................................20%

STRENGTHS

Bananas, though this sector has declined since the loss of EU preferential access. Offshore business center and "economic citizenship" scheme. Growing services sector.

WEAKNESSES

Dependence on US and EU markets for its banana crop, threatened by WTO ruling. Low productivity in public sector. Poor infrastructure.

EXPORTS

Antigua & Barbuda 7%　France 7%
Other 31%　USA 7%
　　UK 24%
Jamaica 24%

IMPORTS

Japan 6%　UK 7%
Barbados 3%　Trinidad & Tobago 16%
USA 37%
　　Other 31%

RESOURCES Electric power 8000 kw

 1205 tonnes　　Not an oil producer

13,400 cattle, 9700 goats, 7600 sheep, 190,000 chickens　　None

Dominica has no natural resources. A hydroelectric power plant in the Morne Trois Pitons national park provides most of the island's power.

ENVIRONMENT Not available

9%　　1.2 tonnes per capita

Increased agriculture and timber harvesting is threatening Dominica's rainforest; already there is more land under cultivation than planned by the government. The current promotion of the rainforest as a tourist attraction poses a threat, as does a possible expansion in HEP generators. Two species of parrot – the imperial, or sisserou, and the red-necked – are threatened, despite conservation orders. Endangered hawksbill turtles, living on coral reefs off the island, are traditionally hunted.

MEDIA TV ownership medium

 There are no daily newspapers

PUBLISHING AND BROADCAST MEDIA

There are no daily newspapers. The dominant newspaper is the weekly *Chronicle*, which takes a center-left editorial stance

No TV service　　3 services: 1 state-owned, 2 independent

Local franchises, offering cable TV with selected US networks, serve one-third of the island. Broadcasts from other Caribbean states can also be received. There are five weekly newspapers.

CRIME Death penalty in use

298 prisoners　　Up 8% in 1999

Dominica has a lower crime rate than most of its Caribbean neighbors. Burglary and armed robbery are the major concerns; murders are rare. Justice is based on British common law and administered by the Eastern Caribbean Supreme Court, which is based on the island of St. Lucia.

EDUCATION School leaving age: 16

94%　　461 students

Dominican education is based on the British model, and retains the selective 11-plus exam for entrance into high school. Students go on to the University of the West Indies or, increasingly, to colleges in the US and the UK.

CHRONOLOGY

Colonized first by the French, Dominica came under British control in 1759.

❏ **1975** Morne Trois Pitons national park established.
❏ **1978** Independence from UK. Patrick John first prime minister.
❏ **1980** Eugenia Charles becomes Caribbean's first woman prime minister.
❏ **1981** Two coup attempts, backed by Patrick John, foiled.
❏ **1995** Opposition DUWP defeats DFP. Dame Eugenia Charles retires after 27 years in politics.
❏ **1999** WTO ruling on preferential access for bananas to EU market.
❏ **2000** DLP wins elections.

HEALTH Welfare state health benefits

1 per 2041 people　　Heart and respiratory diseases, cancers

There are numerous health centers. Difficult communications hamper emergency hospital access for people living in the interior.

SPENDING GDP/cap. increase

CONSUMPTION AND SPENDING

90 per 1000 population　　294 per 1000 population

Defense　None
Education　5.8%
Health　3.8%

0　5　10　15　20　25
Defense, Health, Education spending as % of GDP

Wealth disparities are not as marked in Dominica as they are on the larger Caribbean islands, but the alleviation of poverty has become a major plank of government policy. Measures taken include increased benefits and help for the country's pensioners.

WORLD RANKING

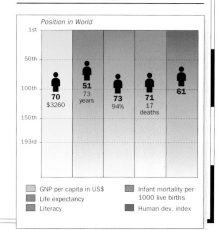

Position in World

1st
50th
100th
150th
193rd

70　$3260
51　73 years
73　94%
71　17 deaths
61

GNP per capita in US$　　Infant mortality per 1000 live births
Life expectancy　　Human dev. index
Literacy

DOMINICAN REPUBLIC

OFFICIAL NAME: Dominican Republic   **CAPITAL:** Santo Domingo
POPULATION: 8.5 million   **CURRENCY:** Dominican Republic peso   **OFFICIAL LANGUAGE:** Spanish

1865	1865	Feb 27	DOM	-4	+1809	.do

CARIBBEAN

THE MOST POPULAR tourist destination in the Caribbean, the Dominican Republic lies 970 km (600 miles) southeast of Florida. Once ruled by Spain, it occupies the eastern two-thirds of the island of Hispaniola and boasts both the region's highest point (Pico Duarte, 3088 m – 10,131 ft) and its lowest (Lake Enriquillo, 44 m – 144 ft – below sea level). Spanish-speaking, it seeks closer ties with the anglophone West Indies.

View south from Pico Duarte along the fertile banks of the Río Yaque del Norte.

CLIMATE
▷ Tropical equatorial/oceanic

WEATHER CHART FOR SANTO DOMINGO

The trade winds blow all year round, providing relief from the tropical heat and humidity. The hurricane season runs from June until November.

TRANSPORTATION
▷ Drive on right

Aeropuerto International de las Américas, Santo Domingo
4.65m passengers

24 ships
10,400 grt

THE TRANSPORTATION NETWORK

| 6224 km (3867 miles) | None |
| 1600 km (994 miles) | None |

Urban and rural transportation is poor; railroads are mainly for transporting sugarcane and ores. An international consortium in 1999 won a 30-year concession to operate four airports.

TOURISM
▷ Visitors : Population 1:2.9

2.97m visitors   Up 12% in 2000

MAIN TOURIST ARRIVALS

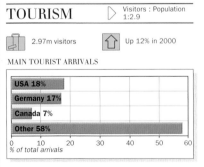

USA 18%
Germany 17%
Canada 7%
Other 58%

% of total arrivals

Ample accommodation and excellent beaches attract many tourists each year, mainly from Europe and North America.

PEOPLE
▷ Pop. density medium

 Spanish, French Creole   176/km² (455/mi²)

THE URBAN/RURAL POPULATION SPLIT

65%   35%

RELIGIOUS PERSUASION

Other and nonreligious 8%

Roman Catholic 92%

The white population, primarily the descendants of Spanish settlers, still owns most of the land. The mixed race majority – about 73% – controls much of the republic's commerce, and forms the bulk of the professional middle classes. Blacks, the descendants of Africans, are mainly small-scale farmers and often the victims of latent racism, especially in business. Women in the black community work the farms; in the white and mixed race communities women are starting to make professional careers.

DOMINICAN REPUBLIC

Total Land Area : 48 380 sq. km (18 679 sq. miles)

POLITICS
▷ Multiparty elections

L. House 2002/2006
U. House 2002/2006

President Hipolito Mejia

AT THE LAST ELECTION

Chamber of Deputies 150 seats

49% PRD   27% PLD   24% PRSC

PRD = Dominican Revolutionary Party   **PLD** = Dominican Liberation Party   **PRSC** = Christian Social Reform Party

Senate 32 seats

91% PRD   6% PRSC   3% PLD

Joaquín Balaguer of the PRSC, a political patriarch since the 1960s and representative of the white elite and the military, achieved a bogus victory in a 1994 poll. He was forced to agree to fresh elections in 1996 which were narrowly won by Leonel Fernández of the more moderate PLD. The PRSC and PLD later joined forces against the opposition center-left PRD, which won control of the Congress in 1998. The PRD called for a fair presidential election in 2000; it was won by the PRD candidate, Hipolito Mejia.

Map labels

POPULATION
- over 1 000 000
- over 100 000
- over 50 000
- over 10 000
- under 10 000

LAND HEIGHT
- 2000m/6562ft
- 1000m/3281ft
- 500m/1640ft
- 200m/656ft
- Sea Level

Monte Cristi, Puerto Plata, Dajabón, Esperanza, José E. Bisonó, Sabaneta, Mao, Moca, Santiago, Salcedo, Nagua, San Francisco de Macorís, La Vega, Samaná, Pedro Santana, Jarabacoa, Cotuí, Yuna, Sabana de la Mar, Constanza, Bonao, Comendador, San Juan, Villa Altagracia, Monte Plata, El Seibo, Malpasso, Lago Enriquillo, S. José de Ocoa, San Cristóbal, SANTO DOMINGO, Hato Mayor, San Pedro de Macorís, Higüey, Jimaní, Neiba, Azua, Bajos de Haina, Intl Airport Las Américas, Boca de Yuma, La Romana, Barahona, Baní, Vicente Noble, Pedernales, Isla Saona, Isla Beata

ATLANTIC OCEAN, CARIBBEAN SEA, Mona Passage, CORDILLERA CENTRAL, Pico Duarte 3088m

0   50 km
0   50 m

WORLD AFFAIRS

▷ Joined UN in 1945

 ACS Geplac IBRD OAS SELA

Relations with Haiti, with which it shares the island of Hispaniola, are important. The Dominican Republic favors a "strategic alliance" between the Caribbean and Central America.

AID

▷ Recipient

 $62m (receipts)   Down 68% in 2000

Multilateral and bilateral aid of some $235 million was granted in 1998 to repair severe hurricane damage.

DEFENSE

▷ No compulsory military service

 $112m   Down 2% in 2000

The military has economic and political interests, but no longer holds the defense portfolio. It focuses on illegal immigration from Haiti. The main arms supplier is the US.

ECONOMICS

▷ Inflation 9.4% p.a. (1990–2000)

$17.8bn   16.12–16.40 Dominican Republic pesos

SCORE CARD

- ❏ WORLD GNP RANKING.............................71st
- ❏ GNP PER CAPITA $2130
- ❏ BALANCE OF PAYMENTS.....................–$1.03bn
- ❏ INFLATION...6.5%
- ❏ UNEMPLOYMENT 14%

STRENGTHS

Sustained tourism growth. Mining – mainly of nickel and gold – and sugar major sectors. Hand-made cigars are biggest sellers in US. Large hidden economy based on transshipment of narcotics to US.

WEAKNESSES

Major sectors severely affected by fluctuating world prices and cutbacks in US import quotas. Poor creditworthiness. Failure to diversify. Stalled privatizations.

EXPORTS

Netherlands 1% — UK 1%
Canada 1% — Haiti 2%
Other 8%
USA 87%

IMPORTS

Japan 3% — Mexico 5%
Spain 3% — Venezuela 10%
USA 60% — Other 19%

RESOURCES

▷ Electric power 2.2m kw

 9269 tonnes   Not an oil producer

2.11m cattle, 565,529 pigs, 47.4m chickens   Ferro-nickel, bauxite, copper, gold, silver

The Dominican Republic is a net energy importer: hydroelectric generators are the only domestic source of power, and electricity blackouts can be a major problem. Oil prospecting has been unsuccessful, and oil is imported from Mexico and Venezuela on preferential terms under the San José Agreement. The Dominican Republic's quota from Venezuela was increased under the 2000 Caracas Accord.

ENVIRONMENT

▷ Sustainability rank: 79th

 32% (10% partially protected)   2.5 tonnes per capita

Forests are threatened by destructive agricultural practices and also by the use of wood as fuel by rural communities. Deforestation has accelerated soil erosion.

MEDIA

▷ TV ownership medium

Daily newspaper circulation 52 per 1000 people

PUBLISHING AND BROADCAST MEDIA

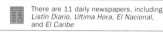

There are 11 daily newspapers, including *Listín Diario*, *Ultima Hora*, *El Nacional*, and *El Caribe*

7 services: 1 state-owned, 6 independent   131 services: 1 state-owned, 130 independent

Television broadcasts from both Mexico and the US can easily be received in the Dominican Republic.

CRIME

▷ No death penalty

 15,340 prisoners   Thefts decreased 1996–1998

The Dominican Republic is increasingly used by narcotics cartels as a transit point to the US. Narcotics trafficking and arms smuggling are linked to the high levels of violent crime.

EDUCATION

▷ School leaving age: 17

 84%   176,995 students

State schools are badly underfunded. The state university of Santo Domingo suffered a financial crisis in 1999. The rich send their children to study in the US and Spain.

HEALTH

▷ Welfare state health benefits

 1 per 455 people   Heart attacks, infectious and parasitic diseases

Wealthy Dominicans fly to Cuba and the US for treatment. The poor rely on a basic public service, inadequately provided by some 20 state hospitals.

CHRONOLOGY

The 1697 Franco-Spanish partition of Hispaniola left Spain with the eastern two-thirds of the island, now the Dominican Republic.

- ❏ **1865** Independence from Spain.
- ❏ **1930–1961** Gen. Molina dictator.
- ❏ **1965** Civil war. US intervention.
- ❏ **1966** Joaquín Balaguer begins first of seven presidential terms over next 30 years.
- ❏ **1996** Center-left PRD candidate Fernández succeeds Balaguer.
- ❏ **1998** Major hurricane damage.
- ❏ **2000** Hipolito Mejia of PRD wins presidency.

D

SPENDING

▷ GDP/cap. increase

CONSUMPTION AND SPENDING

28 per 1000 population   105 per 1000 population

Defense 0.8%
Education 2.3%
Health 1.9%

Defense, Health, Education spending as % of GDP

Great disparities exist between rich and poor. The government in 1998 announced a seven-year plan to relieve poverty and reduce the level of malnutrition affecting well over two million people. Black Dominicans remain at the bottom of the economic and social ladder, accounting for the major proportion of small farmers and unemployed. Haitian immigrants are poorly paid, badly treated, and liable to be deported at short notice. Mixed races have shown most upward mobility in recent years, but, nevertheless, the old Spanish families still form the wealthiest section of society and retain their grip on valuable estates.

WORLD RANKING

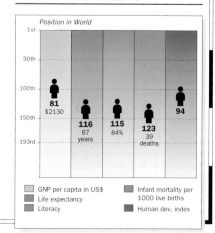

Position in World

GNP per capita in US$ — 81 $2130
Life expectancy — 116 67 years
Literacy — 115 84%
Infant mortality per 1000 live births — 123 39 deaths
Human dev. index — 94

EAST TIMOR

SOUTHEAST ASIA

Asia

OFFICIAL NAME: East Timor **CAPITAL:** Dili **POPULATION:** 737,811
CURRENCY: US dollar **OFFICIAL LANGUAGES:** Tetum (Portuguese/Austronesian) and Portuguese

E

  2002 2002 May 20 n/a +8 +670 .tp

LYING NORTH OF Australia across the Timor Sea, the island of Timor has a narrow coastal plain giving way to forested highlands. Its mountainous backbone rises to 2963 m (9715 ft). The eastern half was colonized for over 400 years by the Portuguese, then occupied from 1975 by Indonesia, whose forces hunted down all resistance. A referendum in 1999 launched a turbulent transition to independence in May 2002.

CLIMATE ▷ Tropical equatorial

WEATHER CHART FOR DILI

■ Average daily temperature   Rainfall ▬

The climate is tropical, with heavy rain from December to March, then dry and increasingly hot weather for the rest of the year, especially in the north.

TRANSPORTATION ▷ Not available

✈ Dili International   🚢 Not available

THE TRANSPORTATION NETWORK

🛣 1414 km (879 miles)	🛤 None	🛤 None	🌊 None

Roads are of poor quality and public transportation beyond Dili is unreliable and sparse.

TOURISM ▷ Visitors : Population 1:246

🧳 3000 visitors   ⬇ Down 70% in 1998

MAIN TOURIST ARRIVALS

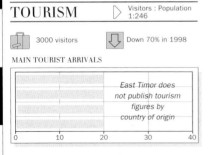

East Timor does not publish tourism figures by country of origin

The number of tourists fell dramatically after the preindependence violence in the region. Visitors are now generally discouraged by the high levels of crime and the almost complete lack of a tourist infrastructure.

PEOPLE ▷ Pop. density medium

Tetum (Portuguese/Austronesian), Bahasa Indonesia, and Portuguese

51/km² (131/mi²)

THE URBAN/RURAL POPULATION SPLIT

8%   92%

RELIGIOUS PERSUASION

Other (including Muslim and Protestant) 5%

Roman Catholic 95%

East Timor is almost entirely Roman Catholic. Preindependence violence was politically rather than ethnically motivated. The Timorese are a mix of Malay and Papuan peoples. Many indigenous Papuan tribes survive and there is an urban Chinese minority. Ethnic Indonesian settlers became numerous after annexation, and consisted 20% of the population by 1999.

Women do not enjoy a high profile in public life. The incidence of domestic violence is notably high.

EAST TIMOR

Total Land Area : 14,874 sq. km (5756 sq. miles)

POPULATION   LAND HEIGHT

over 10 000 ●
under 10 000 •

2000m/6562ft
1000m/3281ft
500m/1640ft
Sea Level

POLITICS ▷ Multiparty elections

2001/Transitional   President Xanana Gusmão

AT THE LAST ELECTION

Constituent Assembly 88 seats

7% PSD

63% Fretilin   8% PD   7% ASDT   15% Others

Fretilin = Revolutionary Front of an Independent East Timor
PD = Democratic Party   **PSD** = Social-Democratic Party
ASDT = Timorese Social-Democratic Association

East Timor is an emerging multiparty democracy. The Fretilin movement was the leading voice in the long struggle for independence. Turnout at recent elections was high.

After years of unrest and human rights abuses committed by the Indonesian army, the Indonesian government in 1999 conceded a referendum on East Timor's future. Pro-Indonesian militias went on the rampage, murdering hundreds in indiscriminate attacks, and forcing thousands of people into the Indonesian-controlled western half of the island. An international peacekeeping force, led by Australia, eventually secured relative calm and organized the promised vote on 30 August 1999. An overwhelming 80% of voters endorsed independence.

The UN Mission in East Timor was given full power over the territory in October 1999. Fretilin emerged as the outright victor in elections in late 2001 to the new Constituent Assembly. Its popular leader, Xanana Gusmão, reversed his decision to retire from politics and was duly elected president in 2002. Independence took effect on 20 May that year.

The new government, headed by Prime Minister Mari Alkatiri, has pledged to concentrate spending on health and education.

WORLD AFFAIRS

Joined UN in 2002

Relations with Indonesia remain strained over the issue of justice for past human rights abuses. Portugal, the former colonial power, had led international opposition to Indonesia's annexation of the territory. Australia, one of the few Western countries to recognize that annexation, later swung vital support behind the cause of independence. It is also a key player in providing future aid and assistance for reconstruction.

East Timor applied to join the UN and ASEAN shortly after it achieved independence in 2002.

AID

Recipient

 $233m (receipts)   Up 52% in 2000

International development aid provides the backbone of East Timor's GDP. Australia and Portugal are by far the biggest donors.

DEFENSE

No compulsory military service

 Not available   Not available

The East Timorese Defense Force was established in 2001 from the remains of pro-independence militia. Its role is largely as an extension of the police force, as external defense is still provided by UN troops.

ECONOMICS

Not available

 $403m   Currency is US dollar

SCORE CARD

❑ World GNP Ranking	172nd
❑ GNP per Capita	$458
❑ Balance of Payments	$8m
❑ Inflation	0%
❑ Unemployment	4%

Strengths
Potential from oil and natural gas reserves in Timor Sea. Traditional agricultural base: coffee the main export.

Weaknesses
Infrastructure devastated by 1999 violence. Insecurity deterred investment. Undeveloped industrial sector.

EXPORTS/IMPORTS

Export and import figures are not available for East Timor

Despite its young age, *East Timor has a strong national identity, based largely on the domination of Roman Catholicism.*

RESOURCES

Electric power 14,400 kw

 513 tonnes   Oil figures not available

378,000 pigs, 187,000 cattle, 1.02m chickens   Oil, natural gas, gold, manganese, marble

The country has few natural resources. In 2001 East Timor successfully negotiated with Australia for a greater share of the oil under the Timor Sea.

ENVIRONMENT

Not available

 Not available   Not available

Unrestricted logging under Indonesian rule has greatly diminished important species and contributed to erosion of the country's poor quality soil.

MEDIA

TV ownership low

Daily newspaper circulation figures are not available

PUBLISHING AND BROADCAST MEDIA

There is 1 daily newspaper, the *Timor Post*. The UN Mission publishes the biweekly *Tais Timor*

1 service, run by the UN   4 stations: 1 UN-run, 1 run by the Catholic Church

Official newsletters have the highest circulation. A number of private papers were established in 2000, including the *Timor Post*, which was formed from the remains of the main preindependence daily, *Suara Timor Timur*.

CRIME

No death penalty

East Timor does not publish prison figures   Crime levels are low but rising

Petty and violent crimes are common. Most of the perpetrators of violence in 1999 have yet to be brought to justice.

EDUCATION

Not available

 41%   7500 students

During Indonesian domination, classes were taught in Bahasa Indonesia. The number of students attending school and university has already begun to recover from the sharp decline in 1999.

CHRONOLOGY

The Portuguese arrived in Timor in the 1520s. It was formally divided by Portugal and the Netherlands in 1859.

- ❑ **1949** Dutch west Timor becomes part of Indonesia.
- ❑ **1975** Fretilin declares East Timor independent; Indonesia invades.
- ❑ **1991** Massacre of pro-independence demonstrators in Dili.
- ❑ **1996** Timorese leaders receive Nobel Peace Prize, raising global awareness.
- ❑ **1999** Indonesian government agrees to hold referendum; resulting violence quelled by UN force.
- ❑ **2001** Elections to new Constituent Assembly; Fretilin wins majority.
- ❑ **2002** Xanana Gusmão elected president. Independence.

HEALTH

Not available

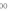 1 per 40,000 people   Not available

Life expectancy in East Timor has traditionally been lower than in the rest of the Indonesian archipelago. Most doctors left the province in 1999.

SPENDING

Not available

CONSUMPTION AND SPENDING

No data   7 per 1000 population

No data available

0   5   10   15   20   25
Defense, Health, Education spending as % of GDP

Living standards, already relatively low for the region, were made worse by the events of 1999. Thousands were left homeless. Well-paid UN staff enjoy a sharply contrasting lifestyle.

WORLD RANKING

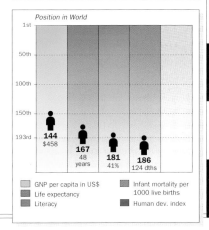

Position in World

1st
50th
100th
150th
193rd

144 $458
167 48 years
181 41%
186 124 dths

❑ GNP per capita in US$ ❑ Infant mortality per 1000 live births
❑ Life expectancy
❑ Literacy   ❑ Human dev. index

ECUADOR

SOUTH AMERICA

OFFICIAL NAME: Republic of Ecuador   **CAPITAL:** Quito
POPULATION: 12.9 million   **CURRENCY:** US dollar   **OFFICIAL LANGUAGE:** Spanish

E

  1830 1941 Aug 10 EC -5 +593   .ec

ONCE PART OF THE INCA heartland, Ecuador lies on the western coast of South America. It was ruled by Spain from 1533, when the last Inca emperor was executed, until independence in 1830. Most Ecuadorians live either in the lowland coastal region or in the Andean Sierra. The Amazonian Amerindians are now pressing for their land rights to be recognized. Massive depreciation of the sucre forced the government to dollarize the currency in 2000.

CLIMATE
▷ Tropical/mountain

WEATHER CHART FOR QUITO

Climate varies from hot equatorial in the Amazon forests, to dry heat in the south and "perpetual spring" in Quito.

TRANSPORTATION
▷ Drive on right

Mariscal Sucre, Quito
2.04m passengers

171 ships
300,900 grt

THE TRANSPORTATION NETWORK

8165 km (5073 miles)	Pan-American Highway
956 km (594 miles)	1500 km (932 miles)

The road network and railroad are grossly underfunded. Serious political unrest in 2000–2001 blocked roads.

TOURISM
▷ Visitors : Population 1:21

615,000 visitors

Up 21% in 2000

MAIN TOURIST ARRIVALS

Colombia 33%	
USA 22%	
Peru 8%	
Other 37%	

0   10   20   30   40
% of total arrivals

Tourism is growing. Quito, once the capital of the Inca empire, has restored many of its Spanish imperial buildings, including 86 churches. Access to the Galapagos Islands is restricted to 60,000 visitors a year.

PEOPLE
▷ Pop. density low

Spanish, Quechua, other Amerindian languages

47/km² (121/mi²)

THE URBAN/RURAL POPULATION SPLIT

65%   35%

RELIGIOUS PERSUASION

Protestant, Jewish and Other 7%

Roman Catholic 93%

Over half of the population is of Amerindian–Spanish extraction (*mestizo*). Black communities exist on the coast. The Amerindians, who make up about one-quarter of the population, are pressing for Ecuador to be described as a plurinational state, within which the different indigenous communities are recognized as distinct nationalities. The strong and largely unified Amerindian movement is at the forefront of social protests.

POLITICS
▷ Multiparty elections

1998/2002

President Gustavo Noboa

AT THE LAST ELECTION
National Congress 121 seats

2% 12%
FRA Others

27% DP   23% PSC   18% PRE   14% ID   4% NMN–PP

DP = Popular Democracy   **PSC** = Social Christian Party
PRE = Ecuadorian Roldosist Party   **ID** = Democratic Left
NMN–PP = New Country–Pachakutik Movement
FRA = Alfarist Radical Front

101 provincial members are elected for two-year terms, and 20 national members are directly elected for four-year terms

After the instability and corruption of the late 1990s, a new broad-based alliance was formed in 1999. Severe austerity reforms provoked widespread protests; the army intervened and Vice President Gustavo Noboa took over the presidency in 2000. The dollarization of the economy in that year led to popular protest, which resurfaced more widely and violently in response to IMF-backed austerity in 2001.

WORLD AFFAIRS
▷ Joined UN in 1945

AP   AmCC   NAM   OAS   RG

Access to US and EU markets for bananas and oil prices are major concerns. There are serious security problems on the border with Colombia.

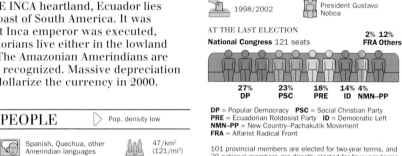

ECUADOR

Total Land Area : 283 560 sq. km
(109 485 sq. miles)

POPULATION

▣	over 1 000 000
◉	over 500 000
◎	over 100 000
○	over 50 000
●	over 10 000
•	under 10 000

Galapagos Islands

LAND HEIGHT

4000m/13124ft
2000m/6562ft
500m/1640ft
Sea Level

Quito is the highest capital in the world after La Paz in Bolivia. It lies in an Andean valley, lined with 30 volcanoes.

AID
 Recipient

 $147m (receipts)   Down 1% in 2000

Aid from the US, Japan, Spain, and the IDB alleviates the heavy foreign debt burden. The Galapagos Islands receive generous grants from UNESCO.

DEFENSE
 Compulsory military service

 $314m   Down 7% in 2000

The army kept out of politics from the mid-1970s until its intervention in 2000. Moves to reduce the military's 50% "royalty" share of oil revenues, its prime source of funding, have raised tensions.

ECONOMICS
 Inflation 37% p.a. (1990–2000)

 $15.3bn   Currency is US dollar

SCORE CARD

- ❏ World GNP Ranking..........................74th
- ❏ GNP per Capita $1210
- ❏ Balance of Payments $928m
- ❏ Inflation 96.1%
- ❏ Unemployment..................................13%

Strengths
Net oil exporter. World's second-biggest banana producer. Fishing industry.

Weaknesses
Poor infrastructure and land productivity. Energy crises. High inflation. Financial instability. Confusion over adoption of US dollar as official currency in 2000.

EXPORTS

IMPORTS

RESOURCES
 Electric power 3.9m kw

625,247 tonnes 416,000 b/d (reserves 2.1bn barrels)

5.57m cattle, 2.39m pigs, 1.98m sheep, 138m chickens Oil, natural gas, gold, silver, copper, zinc

The government is encouraging faster oil exploration and higher output. Ecuador left OPEC in 1992. Overfishing is threatening mackerel and squid stocks.

ENVIRONMENT
 Sustainability rank: 41st

 44% 2.2 tonnes per capita

Oil drilling in new areas of Amazonia threatens indigenous tribes. Tourism, some of it illegal, has upset the delicate ecosystems of the Galapagos Islands; the land iguana is endangered, and black coral is stolen in quantity for souvenirs. The breaching of the *Jessica* oil tanker just offshore in 2001 raised concerns about shipping oil through ecologically sensitive areas.

MEDIA
 TV ownership medium

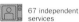 Daily newspaper circulation 70 per 1000 people

PUBLISHING AND BROADCAST MEDIA

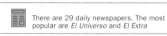

There are 29 daily newspapers. The most popular are *El Universo* and *El Extra*

67 independent services 1 state-owned, 320 independent stations

The press is largely independent. It is highly regionalized, based either in the Quito region or around Guayaquil on the coast. The latter is also a center for commercial radio stations. There are ten cultural and ten religious radio stations.

CRIME
 No death penalty

 8520 prisoners Up 1% in 1999

Right-wing paramilitaries were blamed for the murders of a trade union leader in 1998 and a left-wing congressman in 1999. The paramilitaries are rumored to be supported by Colombians. Left-wing urban guerrillas are also reported. Unprecedented numbers of citizens are applying for arms permits, while the illegal arms trade is thriving.

EDUCATION
 School leaving age: 15

 91% 206,541 students

Some 20% of Ecuadorians in the relevant age group receive higher education at 16 universities.
Programs have been launched to combat high levels of adult illiteracy in rural areas. Secondary schools are badly underfunded.

CHRONOLOGY
Alternating republican and military governments ruled Ecuador from independence in 1830 to 1978.

- ❏ **1941–1942** Loss of mineral-rich El Oro region to Peru.
- ❏ **1948–1960** Prosperity from bananas.
- ❏ **1972** Oil production starts.
- ❏ **1979** Return to democracy.
- ❏ **1992** Amerindians win land in Amazonia.
- ❏ **1996–1997** Abdalá Bucarám Ortíz removed from presidency on grounds of mental incapacity.
- ❏ **1998–1999** Jamil Mahuad of DP wins elections; forms new majority alliance. Economic crisis.
- ❏ **2000** Army sides with Amerindian protestors. Vice President Gustavo Noboa replaces Mahuad.

HEALTH
 Welfare state health benefits

1 per 588 people Malnutrition, intestinal infectious diseases, pneumonia, accidents

Health care is seriously underfunded. Some services exist in poor urban districts but are still unavailable in many rural areas. Severe budget cuts mean that any improvement will depend on more outside aid.

SPENDING
 GDP/cap. increase

CONSUMPTION AND SPENDING

41 per 1000 population 100 per 1000 population

Defense 1.6%
Education 3.5%
Health 1.7%

Defense, Health, Education spending as % of GDP

An estimated 60% of the population live in poverty, but disparity between rich and poor is not as great as in other South American countries.

WORLD RANKING
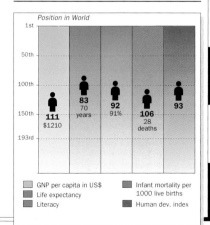

E

EGYPT

OFFICIAL NAME: Arab Republic of Egypt   CAPITAL: Cairo
POPULATION: 69.1 million   CURRENCY: Egyptian pound   OFFICIAL LANGUAGE: Arabic

NORTH AFRICA

 1936   1982   July 23   ET   +2   +20   .eg

OCCUPYING THE NORTHEAST corner of Africa, Egypt is bisected by the highly fertile Nile valley separating the arid western desert from the smaller semiarid eastern desert. Egypt's 1979 peace treaty with Israel brought security, the return of the Sinai, and large injections of US aid. Its essentially pro-Western military-backed regime is now being challenged by an increasingly influential Islamic fundamentalist movement.

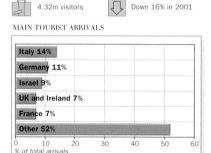

18th-Dynasty Temple of Queen Hatshepsut dating from the Middle Kingdom, c.1480 BCE. It is at Deir el-Bahri on the west bank of the Nile opposite Thebes, Egypt's capital at the time.

CLIMATE

▷ Hot desert/ Mediterranean

WEATHER CHART FOR CAIRO

■ Average daily temperature   Rainfall ■

Summers are very hot, especially in the south, but winters are cooler. The only significant rain falls in winter along the Mediterranean coast.

TRANSPORTATION

▷ Drive on right

Cairo International
8.94m passengers

372 ships
1.35m grt

THE TRANSPORTATION NETWORK

49,684 km (30,872 miles)	None	
5024 km (3122 miles)	3500 km (2175 miles)	

Cities are linked by adequate roads, but railroads are the main transportation arteries, and trains are frequently overcrowded. The Suez Canal is a vital international shipping lane.

TOURISM

▷ Visitors : Population 1:16

4.32m visitors

Down 16% in 2001

MAIN TOURIST ARRIVALS

Italy 14%
Germany 11%
Israel 9%
UK and Ireland 7%
France 7%
Other 52%

% of total arrivals

Egypt's wealth of antiquities from its ancient civilizations have made it a key tourist destination since the 1880s. Today, it also offers Nile cruises and some of the world's best scuba diving, notably on the coral reefs near Hurghada on the Red Sea.

The industry went into sharp decline, however, when militant Islamists began targeting Western tourists in the mid-1990s; in an attack in Luxor in November 1997, 58 tourists were killed. Heightened security measures allowed for a brief recovery, but the situation worsened following the September 2001 attacks in the US; the resultant contraction in the global tourism industry affected Islamic countries in particular.

EGYPT

Total Land Area : 1 001 450 sq. km
(386 660 sq. miles)

POPULATION

over 5 000 000
over 1 000 000
over 500 000
over 100 000
over 50 000
over 10 000
under 10 000

LAND HEIGHT

2000m/6562ft
1000m/3281ft
500m/1640ft
200m/656ft
Sea Level
-200m/-656ft

0   200 km
0   200 miles

E

PEOPLE ▷ Pop. density medium

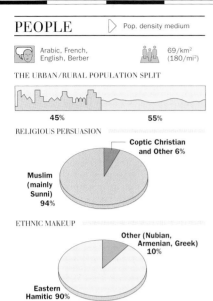

Arabic, French, English, Berber

69/km²
(180/mi²)

THE URBAN/RURAL POPULATION SPLIT

45%   55%

RELIGIOUS PERSUASION

Coptic Christian and Other 6%

Muslim (mainly Sunni) 94%

ETHNIC MAKEUP

Other (Nubian, Armenian, Greek) 10%

Eastern Hamitic 90%

Egypt has a long tradition of ethnic and religious tolerance, though the rise in Islamic fundamentalism has sparked sectarian clashes between Muslims and Coptic Christians. Most Egyptians speak Arabic, though many also have French or English as a second language. There are Berber-speaking communities in the western oases. Small colonies of Greeks and Armenians live in the larger towns. Islam is the dominant religion, followed by Coptic Christianity. Although many Jews left Egypt for Israel after 1948, a small community remains in Cairo.

Cairo is the second most populous city in Africa, and a key social question in Egypt is the high birthrate. In 1985 the government set up the National Population Council, which made birth control readily available. Since then, the birthrate has dropped from 39 to fewer than 25 per 1000 people, but population growth is still high. The population is predicted to reach almost 100 million by 2025. The growing influence of Islamic fundamentalists, who oppose contraception, could see the rate accelerate once more.

Egyptian women have been among the most liberated in the Arab world, and under a law passed in 2000 they now have the right to initiate divorce proceedings. The steady rise of Islamic fundamentalism, however, threatens their position, particularly in rural areas.

POPULATION AGE BREAKDOWN

Female	Age	Male
0.5%	80+	0.4%
2.6%	60–79	2.3%
7.7%	40–59	7.8%
14.3%	20–39	15.1%
24%	0–19	25.3%

% of population by age group

POLITICS ▷ Multiparty elections

 2000/2005

President Mohammed Hosni Mubarak

AT THE LAST ELECTION

People's Assembly 454 seats

2% NWP   1% Others

86% NDP   8% Ind   2% App   1% NPU

NDP = National Democratic Party   **Ind** = Independents
App = Appointed   **NWP** = New Wafd Party   **NPU** = National Progressive Unionist Party

444 members of the Assembly are elected and 10 are appointed by the head of state

Egypt is a multiparty system in theory. In practice, the ruling NDP, backed by the military, runs a one-party state.

PROFILE

Egypt has been politically stable since World War II, with just three leaders since 1954 when Nasser came to power. Anwar Sadat was assassinated in 1981, but was immediately replaced by Hosni Mubarak, a man in the same mold. The NDP retains its grip on the political process by means of the state of emergency, and has close links with the military. Elections in 2000 were more transparent than before and the Islamic opposition fared slightly better, but many candidates elected as independents then joined the NDP.

While Nasser promoted Arab socialism, influenced by the Soviet model, Sadat and Mubarak encouraged private enterprise and a liberalized economy. However, there has been no parallel liberalization in politics – one reason for the growing success of Islamist militants.

MAIN POLITICAL ISSUES
Islamic fundamentalism

The NDP government is engaged in a struggle against Islamist terrorist groups seeking to turn Egypt into a Muslim theocracy along Iranian lines. Extremists have been responsible for numerous attacks on police and tourists. The fundamentalist message, with promises of improved conditions, has proved attractive to both urban and rural poor. Mosques are often the main providers of education

*Hosni Mubarak,
president since the
assassination of
Anwar Sadat in 1981.*

*Gemal Abd
al-Nasser, pan-Arab
nationalist, president
from 1954 to 1970.*

and health services that parallel the state's. Although the government uses draconian measures to counter the terrorist threat, and banned the only legal Islamic party, the Labor Party, in May 2000, it continues to allow religious organizations to pursue their social programs.

The state of emergency

The ruling NDP have repeatedly extended the national state of emergency, in force since the assassination of President Sadat by Islamic terrorists in 1981. The most recent extension was in February 2002 for a further three years. Emergency laws have been invoked to justify the ban on religious parties, especially the Muslim Brotherhood. It is claimed by human rights groups that emergency powers are routinely applied to silence the NDP's political opponents.

WORLD AFFAIRS ▷ Joined UN in 1945

AL   Damasc   OAPEC   AU   OIC

Egypt has close relations with the West, particularly the US. Its support for the alliance against Iraq in the Gulf War was crucial, and Egypt received a massive economic reward from Saudi Arabia. It is also one of only two Arab countries to be technically at peace with Israel, for which it faces criticism from hard-line Islamic states. However, the Israeli military action in 2002 has weakened Egypt's support, and it continues to lobby for the creation of an autonomous Palestinian state.

Relations with Iran are particularly tense: Iran actively supports the Islamist groups operating against the NDP government, and characterizes Egypt as a corrupt state under US influence.

Egypt has expressed concern over international treatment of Iraq; President Mubarak calls for a diplomatic solution. He has opposed recent US-led air strikes against Iraq and urged restraint in the "war on terrorism."

Egypt's diplomatic service is the Arab world's largest, and many Egyptians, such as former UN Secretary General Boutros Boutros Ghali, have served on international bodies.

AID ▷ Recipient

$1.33bn (receipts)   Down 16% in 2000

Egypt has received massive levels of US military aid since the late 1970s, and currently heads the list of US aid recipients, having overtaken Israel. International aid was pledged in 2002 to make up for lost tourism revenue.

CHRONOLOGY

Egypt's centuries-long Ottoman occupation ended in 1914, when it came under direct British rule. It became fully independent in 1936. Army officers led by Lt. Col. Gemal Abd al-Nasser seized power in 1952.

- ❏ **1953** Political parties dissolved, monarchy abolished. Republic proclaimed with Gen. Mohammed Neguib as president.
- ❏ **1954** Nasser deposes Neguib to become president.
- ❏ **1956** Suez Crisis over nationalization of Suez Canal. Israeli, British, and French forces invade, but withdraw after pressure from UN and US.
- ❏ **1957** Suez Canal reopens after UN salvage fleet clears blockade.
- ❏ **1958** Egypt merges with Syria as United Arab Republic.
- ❏ **1960–1970** Aswan High Dam built.
- ❏ **1961** Syria breaks away from union with Egypt.
- ❏ **1967** Six-Day War with Israel; loss of Sinai.
- ❏ **1970** Nasser dies; succeeded by Anwar Sadat.
- ❏ **1971** Readopts the name Egypt. Islam becomes state religion.
- ❏ **1972** Soviet military advisers dismissed from Egypt.
- ❏ **1974–1975** US brokers partial Israeli withdrawal from Sinai.
- ❏ **1977** Sadat visits Jerusalem: first-ever meeting between Egyptian president and Israeli prime minister.
- ❏ **1978** Camp David accords, brokered by US, signed by Egypt and Israel.
- ❏ **1979** Egypt and Israel sign peace treaty, alienating most Arab states.
- ❏ **1981** Sadat assassinated; succeeded by Hosni Mubarak.
- ❏ **1982** Last Israeli troops leave Sinai.
- ❏ **1986** President Mubarak meets Israeli prime minister Shimon Peres to discuss Middle East peace.
- ❏ **1989** After 12-year rift, Egypt and Syria resume diplomatic relations.
- ❏ **1990–1991** Egypt participates in UN operation to liberate Kuwait.
- ❏ **1991** Damascus Declaration provides for a defense pact among Egypt, Syria, and GCC countries against Iraq.
- ❏ **1994–1998** Islamist extremists begin campaign of terrorism, killing civilians and tourists. Government steps up countermeasures.
- ❏ **1999** Banned Gamaat Islamiya ends campaign to overthrow government.
- ❏ **2000** Egypt recalls ambassador to Israel because of escalating Israeli aggression against Palestinians.
- ❏ **2001** Heavy decline in tourist numbers following September 2001 attack on US.

DEFENSE

 Compulsory military service

💲 $2.82bn   ⬇ Down 6% in 2000

EGYPTIAN ARMED FORCES

🛡	3860 main battle tanks (T-54/55, M1A1 *Abrams*, *Ramses* II, T-62, M-60)	320,000 personnel
🚢	4 submarines, 1 destroyer, 10 frigates, and 38 patrol boats	19,000 personnel
✈	580 combat aircraft (*Alpha Jet*, PRC J-6, F-4E, *Mirage* 5E2)	29,000 personnel
🚀	None	

Egypt's armed forces, the largest in the Arab world, are battle-hardened from successive wars with Israel and from participation in Operation Desert Storm to liberate Kuwait in 1991. More than 500,000 reservists augment the regular troops.

After the 1978 Camp David framework agreements were reached with Israel, Egypt stopped buying Soviet weapons and aircraft, and turned instead to Western suppliers. Cooperation with the US has reaped dividends in the form of access to more sophisticated defense equipment and improved training. Egypt has a small arms industry and sells light weapons, notably its version of the Soviet-developed AK-47 assault rifle, to other developing countries.

ECONOMICS

 Inflation 8.2% p.a. (1990–2000)

📊 $95.4bn   💱 3.890–4.575 Egyptian pounds

SCORE CARD

- ❏ WORLD GNP RANKING..........................38th
- ❏ GNP PER CAPITA.............................$1490
- ❏ BALANCE OF PAYMENTS....................–$971m
- ❏ INFLATION.....................................2.7%
- ❏ UNEMPLOYMENT................................12%

ECONOMIC PERFORMANCE INDICATOR

EXPORTS

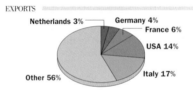

Netherlands 3% — Germany 4% — France 6% — USA 14% — Italy 17% — Other 56%

IMPORTS

France 6% — Chile 4% — Germany 7% — Italy 7% — USA 17% — Other 59%

STRENGTHS

Oil and gas revenues. Well-developed tourist infrastructure. Remittances from Egyptians working throughout the region. Suez Canal tolls. Agricultural produce, especially cotton. Light industry and manufacturing.

WEAKNESSES

Tourism revenues hit by "war on terrorism." Dependence on imported technology. High birthrate.

PROFILE

Under President Nasser, Egypt followed an economic policy inspired by the Soviet model. Rigid and highly centralized, it gave Egypt one of the largest public sectors of all developing countries. Economic restrictions were first relaxed in 1974. President Sadat's open-door policy allowed joint ventures with foreign partners for the first time, although the business classes were the only ones to profit. Most Egyptians suffered from new austerity measures.

Under President Mubarak, economic reform has quickened and there is more awareness of poverty and the high levels of unemployment. Priorities now are to encourage manufacturing, sustain economic growth, and reduce the gap between rich and poor.

EGYPT : MAJOR BUSINESSES

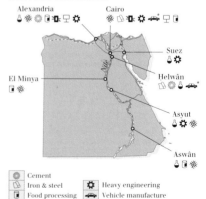

Cement, Iron & steel, Food processing, Consumer goods, Electronics, Heavy engineering, Vehicle manufacture, Chemicals, Textiles

* significant multinational ownership

0   200 km
0   200 miles

E

RESOURCES

 Electric power 17.5m kw

 606,780 tonnes

758,000 b/d (reserves 2.9bn barrels)

9.2m ducks, 9.1m geese, 4.6m sheep, 88m chickens

Natural gas, oil, phosphates, manganese, uranium

ELECTRICITY GENERATION

Hydro 21% (12bn kwh)

Combustion 79% (45bn kwh)

Nuclear 0%

Other 0%

% of total generation by type

Oil and gas are Egypt's most valuable resources. Oil multinationals are involved in new explorations, but more competitive oil-rich countries, such as Algeria and Yemen, are more profitable; 55% of Egypt's oil production is consumed locally.

Most electricity is derived from coal and hydroelectric power. The Aswan High Dam, built between 1960 and 1970 and with a maximum output of ten billion kWh, provides the bulk of hydroelectricity. Within four years, revenue from it had covered its construction costs. The US gave aid in 2000 to upgrade the power plant, with work going ahead in 2001.

EGYPT : LAND USE

Cropland
Pasture
Wetlands
Desert
Goats
Cotton – cash crop
Citrus fruits

Nile Delta
Suez Canal
SINAI
LIBYAN DESERT

0   200 km
0   200 miles

ENVIRONMENT

 Sustainability rank: 74th

0.8% (0.7% partially protected)

1.7 tonnes per capita

ENVIRONMENTAL TREATIES

Yes	Yes	Yes
Yes	Yes	No

Egypt suffers from a chronic lack of water. The Nile, the only perennial source, is increasingly saline because of its much-reduced flow, due to irrigation use and the Aswan High Dam. The main cities suffer heavy industrial pollution, and environmental controls are few. In Cairo a sewerage system has improved sanitary conditions.

MEDIA

 TV ownership medium

 Daily newspaper circulation 38 per 1000 people

Pressure from Islamists has resulted in more airtime for Islamic sermons. Severe restrictions were imposed in 1998, after criticism of the government's security clampdown. Egypt was the first Arab state to have its own satellite, Nilesat 101, and is now the center of a flourishing satellite TV industry.

CRIME

 Death penalty in use

80,000 prisoners

Crime is rising

CRIME RATES

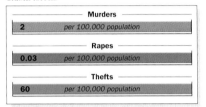

Murders
2   per 100,000 population

Rapes
0.03   per 100,000 population

Thefts
60   per 100,000 population

Terrorist attacks have tarnished Egypt's reputation as a law-abiding country; street crime and muggings were previously rare.

Intercommunity violence – in particular between Muslims and Christians – has become more common, as have attacks on Western tourists by Islamic extremists. Human rights groups have criticized the police for their abuse of current emergency laws, which results in the routine torture and death in police custody of scores of political prisoners.

EDUCATION

 School leaving age: 14

 55%   850,051 students

THE EDUCATION SYSTEM

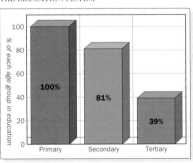

% of each age group in education

100% Primary
81% Secondary
39% Tertiary

Most Egyptians attend elementary school until the age of 11, but not many complete secondary education. A small majority of men, but a minority of women, are literate. A government initiative to improve girls' primary education was launched in 2000. The quality of the education given by Egyptian universities is widely respected in the Arab world.

PUBLISHING AND BROADCAST MEDIA

There are 17 daily newspapers. The serious, prestigious *Al Ahram* and the more populist *Al Akhbar* have the highest circulation

1 state-owned service

2 services: 1 state-owned, 1 independent

HEALTH

 Welfare state health benefits

 1 per 625 people

Digestive, respiratory, and heart diseases, perinatal deaths

Health care, although improved, remains basic – there is only one hospital bed for every 500 people. Islamic medical centers based on the mosque organization are spreading, and are replacing the state system. In 1996 the government banned female genital mutilation, a move upheld in 1997 by the Supreme Consitutional Court after being overturned by a lower court.

SPENDING

GDP/cap. increase

CONSUMPTION AND SPENDING

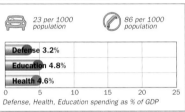

23 per 1000 population   86 per 1000 population

Defense 3.2%
Education 4.8%
Health 4.6%

Defense, Health, Education spending as % of GDP

Wealth disparities are highly marked in Egypt. The largely urban Coptic Christian community is the group with the country's highest standard of living. Most Egyptians remain subsistence farmers wih low incomes. The return of many unemployed workers from the Gulf states to their mainly rural homes has further depressed conditions in the countryside.

WORLD RANKING

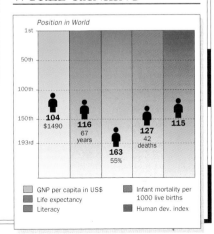

Position in World

1st
50th
100th
150th
193rd

104 $1490
116 67 years
163 55%
127 42 deaths
115

GNP per capita in US$
Life expectancy
Literacy

Infant mortality per 1000 live births
Human dev. index

EL SALVADOR

OFFICIAL NAME: Republic of El Salvador **CAPITAL:** San Salvador
POPULATION: 6.4 million **CURRENCIES:** Salvadorean colón & US dollar **OFFICIAL LANGUAGE:** Spanish

E

| 1841 | 1841 | Sept 15 | ES | -6 | +503 | .sv |

T HE SMALLEST AND MOST densely populated
Central American republic, El Salvador won full
independence in 1841. Located on the Pacific coast, it lies within a zone
of seismic activity. Between 1979 and 1991, El Salvador was engulfed in
a civil war between US-backed right-wing government forces and left-
wing FMLN guerrillas. Since the UN-brokered peace agreement, the
country has been concentrating on rebuilding its shattered economy.

*View over the capital, San Salvador. It lies
in a depression in the southern and higher of
El Salvador's two mountain ranges, which is
punctuated by more than 20 volcanoes.*

CLIMATE ▷ Tropical wet & dry

WEATHER CHART FOR SAN SALVADOR

The tropical coastal *tierra caliente* is
very hot, with seasonal rains. The low
hills are cooler at night; the higher
tierra templada is drier and also cooler.

TRANSPORTATION ▷ Drive on right

Cuscatlan, San Salvador
1.34m passengers

13 ships
1600 grt

THE TRANSPORTATION NETWORK

1986 km (1234 miles)	Pan-American Highway: 327 km (203 miles)
674 km (419 miles)	Rio Lempa

Earthquakes in 2001 further damaged
the already war-ravaged road and
rail networks. Reconstruction will
take many years.

TOURISM ▷ Visitors : Population 1:8.1

795,000 visitors

Up 21% in 2000

MAIN TOURIST ARRIVALS

Guatemala 31%	
USA 22%	
Honduras 17%	
Other 30%	

0   10   20   30   40
% of total arrivals

Peace has brought visitors back to
the unspoiled beach resorts, but crime,
earthquakes, and high prices for rooms
and air travel hinder tourist expansion.

PEOPLE ▷ Pop. density high

Spanish

309/km²
(800/mi²)

THE URBAN/RURAL POPULATION SPLIT

47%   53%

RELIGIOUS PERSUASION

Other 2%

Evangelical 18%

Roman
Catholic 80%

Salvadorans are largely *mestizo* (mixed
race); there are few ethnic tensions.
The civil war was fought over gross
economic disparities, which still exist.

POLITICS ▷ Multiparty elections

2000/2003

President Francisco
Flores

AT THE LAST ELECTION

Legislative Assembly 84 seats

6%   4%
PDC Others

37%   34%   17%   2%
FMLN   ARENA PCN   CDU

FMLN = Farabundo Martí National Liberation Front
ARENA = Nationalist Republican Alliance **PCN** = National
Conciliation Party **PDC** = Christian Democratic Party
CDU = United Democratic Center

El Salvador had been dominated by the
centrist PDC and right-wing ARENA.
The latter, however, now faces greater
opposition from the FMLN, leftist
former guerrillas, who in 1997 won
the mayorship of San Salvador and
half the state capitals.

In the 1999 presidential election
the FMLN, split between center-left
pragmatists and hard-liners, came a
poor second to ARENA's Francisco
Flores, who promised reduced poverty
and income redistribution. By March
2000, however, with the economy
in difficulties, voters punished ARENA
by returning the FMLN as the largest
party in parliament.

WORLD AFFAIRS ▷ Joined UN in 1945

| ACS | Geplac | IBRD | OAS | San José |

El Salvador was an international pariah
in the 1980s because of the human rights
abuses committed by military death
squads. Today it cooperates with its
neighbors in pressing the US on key
issues such as trade and immigration.
It relied heavily on US aid in 2001 after
three devastating earthquakes. In 2000
it signed a free trade treaty with Mexico,
Guatemala, and Honduras. A territorial
dispute with Honduras was finally
settled in 1998, when those affected
were permitted to choose between
Honduran and Salvadoran citizenship.

AID ▷ Recipient

$180m (receipts)

Down 2% in 2000

Post-civil war aid focused on efforts
to secure peace and achieve national
reconciliation by funding rebuilding
and refugee resettlement programs.
The current emphasis is on a shift
toward supporting growth.

The UN received a slow international
response in 2001 to its appeal for $34.8
million in emergency housing, medicine,
and disaster prevention programs after
El Salvador's devastating earthquakes.

DEFENSE ▷ Compulsory military service

$168m

Down 2% in 2000

Between 1979 and 1991, the role of
the US-backed military was to fight
an unrestricted war against the FMLN.
Human rights were in effect suspended
and governments that opposed the
military were overthrown. Under the
peace accords the military agreed to
withdraw from politics and internal
security matters, but it remains a
potent force capable of intervention.

ECONOMICS ▷ Inflation 7.4% p.a. (1990–2000)

📊 $12.6bn 💲 8.740–8.747 Salvadorean colones

SCORE CARD

- ❏ WORLD GNP RANKING..........................78th
- ❏ GNP PER CAPITA $2000
- ❏ BALANCE OF PAYMENTS.....................–$48m
- ❏ INFLATION 2.3%
- ❏ UNEMPLOYMENT.................................10%

STRENGTHS

Coffee. Foreign investment in *maquila* assembly plants. Sizable family remittances from US.

WEAKNESSES

Exports uncompetitive. High tax evasion and unemployment. Low savings. Vast reconstruction needed after earthquakes in 2001.

EXPORTS

Canada 3% | Costa Rica 5% | Honduras 6% | Guatemala 12% | USA 60% | Other 14%

IMPORTS

France 6% | Germany 4% | USA 50% | Mexico 7% | Guatemala 11% | Other 22%

RESOURCES ▷ Electric power 996,000 kw

 15,467 tonnes   Not an oil producer

1.22m cattle, 150,000 pigs, 8.1m chickens Salt, limestone, gypsum

No significant resources. Several volcanoes facilitate abundant and relatively cheap geothermal energy.

ENVIRONMENT ▷ Sustainability rank: 75th

 0.3% (0.2% partially protected)   1 tonne per capita

Deforestation has led to erosion and desertification – worsening landslides during the earthquakes of 2001. Overuse of pesticides is a major problem.

MEDIA ▷ TV ownership medium

 Daily newspaper circulation 48 per 1000 people

PUBLISHING AND BROADCAST MEDIA

There are 8 daily newspapers. *El Diario de Hoy* has the highest circulation

10 channels: 2 state-owned, 8 independent

66 stations: 1 state-owned, 65 independent

The media is owned by powerful groups, such as the Dutriz family. Intimidation and self-censorship exist.

CRIME ▷ Death penalty not used in practice

 6914 prisoners 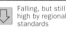 Falling, but still high by regional standards

A corrupt judiciary and police force have failed to stem a postwar crime wave fueled by readily available arms; armed robberies, kidnappings, and murders deter investment and tourism. Uncompleted elements of the peace accords, particularly land transfers, often lead to violence.

EDUCATION ▷ School leaving age: 15

 79%   114,675 students

Education is based on the US system and is limited in rural areas. During the civil war, state universities were closed by the military and replaced by private universities which continue to thrive despite their low standards. A 1995 reform bill tried to address the negative impact of deregulation.

EL SALVADOR

Total Land Area : 21 040 sq. km (8124 sq. miles)

POPULATION
- over 500 000 ◉
- over 100 000 ◎
- over 50 000 ○
- over 10 000 ●
- under 10 000 ·

LAND HEIGHT
- 2000m/6562ft
- 1000m/3281ft
- 500m/1640ft
- 200m/656ft
- Sea Level

CHRONOLOGY

El Salvador was a Spanish colony until 1821. Part of the United Provinces of Central America in 1823–1839, it became fully independent in 1841.

- ❏ **1932** Army crushes popular insurrection led by Farabundo Martí.
- ❏ **1944–1979** Army rules through PCN.
- ❏ **1979** Reformist officers overthrow PCN government.
- ❏ **1981** Left-wing FMLN launches civil war.
- ❏ **1991** UN-brokered peace. FMLN recognized as a political party.
- ❏ **1997** Leftist wins San Salvador mayoralty.
- ❏ **2001** Devastating earthquakes kill hundreds; dollarization of economy.

E

HEALTH ▷ Welfare state health benefits

1 per 909 people Accidents, violence, circulatory diseases, infections

Health spending, almost halved during the civil war, has been slow to recover. The wealthy go to the US for surgery.

SPENDING ▷ GDP/cap. increase

CONSUMPTION AND SPENDING

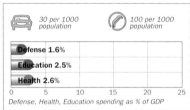

30 per 1000 population | 100 per 1000 population

Defense 1.6% | Education 2.5% | Health 2.6%

Defense, Health, Education spending as % of GDP

Gross wealth disparities see 20% of the population owning 70% of national wealth. Land distribution remains highly skewed, and some three million – nearly half the population – live in poverty.

WORLD RANKING

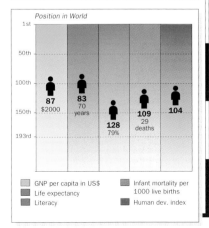

Position in World

87 $2000 | 83 70 years | 128 79% | 109 29 deaths | 104

- ▢ GNP per capita in US$
- ▢ Life expectancy
- ▢ Literacy
- ▢ Infant mortality per 1000 live births
- ▢ Human dev. index

EQUATORIAL GUINEA

OFFICIAL NAME: Republic of Equatorial Guinea **CAPITAL:** Malabo
POPULATION: 470,000 **CURRENCY:** CFA franc **OFFICIAL LANGUAGES:** Spanish and French

COMPRISING FIVE ISLANDS and the territory of Río Muni on the west coast of Africa, Equatorial Guinea lies just north of the equator. Mangrove swamps border the mainland coast. The republic gained its independence in 1968 after 190 years of Spanish rule. Multipartyism was accepted in 1991, but the fairness of subsequent general elections has been questioned.

Bioko, formerly Fernando Po. Although the volcanic land is very fertile, cocoa production fell by 90% during the Macías years.

CLIMATE
▷ Tropical equatorial

WEATHER CHART FOR MALABO

The island of Bioko is extremely wet and humid, with an annual rainfall of 200 cm (79 in), while the mainland is only marginally drier and cooler.

TRANSPORTATION
▷ Drive on right

Malabo International
187,474 passengers

64 ships
45,838 grt

THE TRANSPORTATION NETWORK

508 km (316 miles)	None
None	None

Apart from once- or twice-weekly flights to Madrid by the Spanish airline Iberia, all air links are through neighboring countries. There are six flights a week between Malabo and Bata.

TOURISM
▷ Not available

Tourism receipts totaled $2m in 1998

Numbers are increasing slowly

MAIN TOURIST ARRIVALS

Equatorial Guinea does not publish tourism figures by country of origin

% of total arrivals

Equatorial Guinea is only of interest to the adventurous, independent tourist, despite the potential attraction of its beaches and the island of Bioko's spectacular mountain scenery.

PEOPLE
▷ Pop. density low

Spanish, Fang, Bubi

17/km² (43/mi²)

THE URBAN/RURAL POPULATION SPLIT

48%   52%

RELIGIOUS PERSUASION

Other 10%

Roman Catholic 90%

The mainland has a majority of Fang, a people who also inhabit Cameroon and north Gabon. Bioko is populated by a majority of Bubi and a minority of Creoles, known as Fernandinos. The Macías dictatorship consolidated the power of the Fang, especially the Mongomo clan, from which both Macías and his successor Obiang come. The extended family has maintained its solidarity, despite disruptive social pressure during the Macías dictatorship.

EQUATORIAL GUINEA

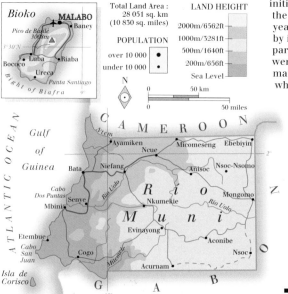

Bioko **MALABO** ·Baney
Pico de Basilé 3008m
Bococo ·Luba ·Riaba
Ureca ·Punta Santiago
Bight of Biafra

Total Land Area : 28 051 sq. km (10 850 sq. miles)

POPULATION
over 10 000 ●
under 10 000 ·

LAND HEIGHT
2000m/6562ft
1000m/3281ft
500m/1640ft
200m/656ft
Sea Level

0   50 km
0   50 miles

ATLANTIC OCEAN
Gulf of Guinea
Cabo Dos Puntas
CAMEROON
·Ayamiken ·Micomeseng ·Ebebiyin
·Ncue
Bata ·Niefang ·Anisoc ·Nsoc-Nsomo
Rio Muni
Senye ·Nkumekie ·Mongomo
Mbini
·Evinayong ·Aconibe
Etembue
Cabo San Juan
·Cogo ·Acurnam ·Nsoc
Isla de Corisco
GABON

POLITICS
▷ Multiparty elections

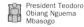

1999/2004

President Teodoro Obiang Nguema Mbasogo

AT THE LAST ELECTION

House of Representatives of the People 80 seats

94% PDGE   5% UP   1% CDS

PDGE = Equatorial Guinea Democratic Party **UP** = Popular Union **CDS** = Convergence for Social Democracy

Despite officially being a multiparty state since 1991, some of the several exiled political parties have not yet found it safe to return. The ruling PDGE was set up in 1987 by Teodoro Obiang Nguema Mbasogo, nephew of the dictator Francisco Macías Nguema, whom he overthrew in 1979. It replaced Macías' National Workers' Party (PUNT). The PDGE benefits from heavy government patronage, receiving 3% of all salaries.

The continuing movement toward multipartyism – which was initiated in 1988 following the first elections for 20 years – has been marked by instability. The 1993 parliamentary elections were boycotted by the main opposition parties, while the presidential poll in 1996, in which Obiang was the only candidate, was declared farcical by foreign observers. The 1999 legislative elections were won easily by the PDGE, but denounced by the opposition. In 2002, 68 opposition leaders were jailed in connection with a 1997 plot to overthrow the government.

E

WORLD AFFAIRS

▷ Joined UN in 1968

BDEAC ACP FZ NAM AU

After a period of extreme isolation at the time of the Macías dictatorship, Equatorial Guinea sought to rebuild links, especially with Spain, the former colonial power and traditionally a haven for political dissenters. A maritime border dispute with Nigeria was finally settled in 2000, paving the way for the exploitation of large oil reserves in the Gulf of Guinea. Wealth from these reserves is likely to affect regional relations greatly in the future.

AID

▷ Recipient

 $21m (receipts)   Up 5% in 2000

Equatorial Guinea is poorly developed and therefore heavily dependent on aid. Inefficiency, corruption, and a shortage of skilled people hinder the planning and implementation of projects, and the government's political record threatens funding. Spain and France are the main donors while aid from UN agencies is also important. An IMF program was suspended in 1997 after the government failed to implement reforms.

DEFENSE

▷ No compulsory military service

 $12m   Up 20% in 2000

The main concern for the military and paramilitary force is internal security. Cuba and North Korea provided Macías with a presidential guard, while Obiang has been protected by Moroccan troops. Nigeria, Cameroon, and Gabon have interests in maintaining the autonomy of the Malabo and Río Muni regions.

ECONOMICS

▷ Inflation 15% p.a. (1990–2000)

 $363m   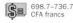 698.7–736.7 CFA francs

SCORE CARD

❏ WORLD GNP RANKING.......................175th
❏ GNP PER CAPITA $800
❏ BALANCE OF PAYMENTS....................–$344m
❏ INFLATION 6%
❏ UNEMPLOYMENT..................................30%

STRENGTHS

Fertile soils. Timber. Cocoa and coffee. Extensive territorial waters, with potential for fisheries. The economy is strengthening as oil and gas reserves are exploited.

WEAKNESSES

Lasting effects of economic regression under Macías dictatorship. Maladministration and ideological

RESOURCES

▷ Electric power 5000 kw

 7001 tonnes

 181,000 b/d (reserves 3.6m barrels)

 37,600 sheep, 30,000 ducks, 320,000 chickens

 Oil, natural gas, gold

Oil production levels doubled between 1998 and the end of 2001. President Obiang has pledged to use income from oil to promote development. The region around Bata is served by a 3.2 MW hydropower station.

ENVIRONMENT

▷ Not available

 None   0.6 tonnes per capita

The government has failed to take any serious measures to stop timber companies depleting the rainforest.

MEDIA

▷ TV ownership low

 Daily newspaper circulation 5 per 1000 people

PUBLISHING AND BROADCAST MEDIA

 There is no regular daily press. The formerly daily newspaper *Poto Poto* now appears irregularly

1 state-owned service

3 services: 1 state-owned, 2 independent

The press remains tightly controlled, with little liberalization, despite the state's adoption of multipartyism. The press association was closed down by the police in February 2001.

CRIME

▷ Death penalty in use

 Equatorial Guinea does not publish prison figures

 Little change from year to year

The level of recorded crime is relatively low, although many offenses do not get reported. Many human rights abuses still occur.

EXPORTS

France 2%
Other 2%
Japan 4%
USA 13%
Spain 53%
China 26%

IMPORTS

France 6%
Ivory Coast 7%
USA 33%
Spain 12%
UK 15%
Other 27%

attacks on the educated have restricted growth; under Macías, cocoa production slumped by 90%. Deterioration of rural economy under successive brutal regimes. Undeveloped natural resources.

CHRONOLOGY

Equatorial Guinea remained a backwater of Spanish colonialism until development began after 1939.

- ❏ **1968** Independence. President Macías begins reign of terror.
- ❏ **1979** Coup puts nephew in power.
- ❏ **1991** Multiparty constitution.
- ❏ **1999** Ruling party wins majority in election condemned as fraudulent.
- ❏ **2001** Government ministers resign over allegations of mismanagement and corruption.

E

EDUCATION

▷ School leaving age: 11

 83%   578 students

Education declined in the Macías years, when attendance rates fell from 90% to 55%. Although education is declared the state's first priority, funding is poor.

HEALTH

▷ No welfare state health benefits

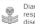 1 per 4000 people   Diarrheal and respiratory diseases, malaria

Life expectancy – just 37 years in 1960 – had risen substantially by 2000. There are 25 doctors to every 100,000 people.

SPENDING

▷ GDP/cap. increase

CONSUMPTION AND SPENDING

4 per 1000 population   13 per 1000 population

Defense 1.7%
Education 1.7%
Health 4.2%

0   5   10   15   20   25
Defense, Health, Education spending as % of GDP

What wealth there is in Equatorial Guinea tends to be concentrated in the ruling clan. There is also a remnant of the former Spanish plutocracy.

WORLD RANKING

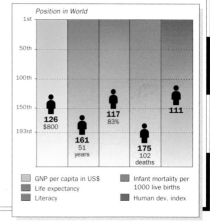

Position in World

1st		
50th		
100th		
150th	126 $800	117 83%	111
	161 51 years	175 102 deaths	
193rd		

- ▢ GNP per capita in US$
- ▢ Life expectancy
- ▢ Literacy
- ▢ Infant mortality per 1000 live births
- ▢ Human dev. index

ERITREA

OFFICIAL NAME: State of Eritrea **CAPITAL:** Asmara
POPULATION: 3.8 million **CURRENCY:** Nakfa **OFFICIAL LANGUAGE:** Tigrinya

LYING ON THE SHORES of the Red Sea, Eritrea's landscape is one of rugged mountains, bush, and desert. A former Italian colony later annexed by Ethiopia, Eritrea fought a long war to win independence in 1993. Like its southern neighbor, Eritrea is prone to recurring droughts and the threat of famine. War with Ethiopia in 1998–2000 brought heavy losses on both sides, until the signing of a comprehensive peace agreement in December 2000.

CLIMATE
▷ Hot desert/mountain

WEATHER CHART FOR ASMARA

Eritrea's harvest is dependent on mid-year rainfall in the highlands. Lowland temperatures may exceed 50°C (122°F).

TRANSPORTATION
▷ Drive on right

 Yohannes IV, Asmara
103,161 passengers

 10 ships
15,913 grt

THE TRANSPORTATION NETWORK

| 874 km (543 miles) | None |
| 117 km (73 miles) | None |

All transportation infrastructure requires massive investment. Ports have potential as transit points for Ethiopia.

TOURISM
▷ Visitors : Population 1:54

70,000 visitors ⬆ Up 23% in 2000

MAIN TOURIST ARRIVALS

Sudan 5%
Japan 2%
Italy 2%
Other 91%

0 10 20 30 40 50 60 70 80 90 100
% of total arrivals

There is currently very little tourism, but Eritrea has considerable long-term potential, especially along the Red Sea coast, with its underwater attractions, and in the spectacular Danakil depression. Guides are essential.

PEOPLE
▷ Pop. density low

Tigrinya, English, Tigre, Afar, Arabic, Bilen, Kunama, Nara, Saho, Hadareb

32/km² (84/mi²)

THE URBAN/RURAL POPULATION SPLIT

19%   81%

RELIGIOUS PERSUASION

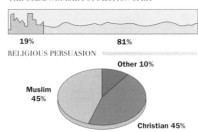

Other 10%
Muslim 45%
Christian 45%

Tigrinya-speakers, mainly Orthodox Christians, form the largest of Eritrea's nine main ethnic groups. A strong sense of nationhood has been forged by the 30-year struggle for independence. Women played an important role in the war; from 1973, 30,000 fought alongside men, some in positions of command. The nomadic peoples of the Danakil desert remain fiercely independent. Subsistence farmers account for 80% of the population.

ERITREA

Total Land Area : 121 320 sq. km
(46 842 sq. miles)

Mersa Teklay
Kerora
Algena
HAGAR NISH PLATEAU
Barisa
2576m Enghershatu
Saatta
Keren
Akordat
Sedberat
Adi Tekelezan
Barentu
Teseney
Biyagundi
Sitona
Om Hajer
Tekeze
Nakfa
DAHLAK ARCHIPELAGO
Dahlak I.
Massawa
Saati
Ginda
Nefasit
ASMARA
Dekemhare
Mendefera
Adi Kwala
Adi Keyih
Senafe
Dangal
Channel
Howakil Bay
Mersa Fatma
Anfile Bay
Suwa
Ed
Beylul
Aseb
DJIBOUTI
E T H I O P I A
S U D A N
R E D   S E A
DANAKIL

LAND HEIGHT

POPULATION
◎ over 100 000
○ over 50 000
● over 10 000
• under 10 000

2000m/6562ft
1000m/3281ft
500m/1640ft
200m/656ft
Sea Level
-200m/-656ft

N

0   100 km
0   100 miles

POLITICS
▷ In transition

 Elections not yet held
 President Issaias Afewerki

AT THE LAST ELECTION
National Assembly 150 seats

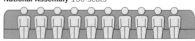

The National Assembly comprises 75 People's Front for Democracy and Justice (**PFDJ**) central committee members and 75 directly elected members, including 11 seats reserved for women. Elections expected in 1997 under the new constitution have not yet taken place

A former Italian colony, Eritrea was dominated by Ethiopia in a federation set up in 1952. Ethiopia annexed the country in 1961, prompting a long secessionist struggle. The Eritrean People's Liberation Front (EPLF) and its Tigrean allies helped defeat the Ethiopian regime in 1991. In 1993 a referendum gave overwhelming support to independence.

Pending multiparty elections, the country is run by a core leadership from the EPLF (now the PFDJ), with the National Assembly only convening sporadically. The 1997 constitution forbids parties based on religious or ethnic affiliations. Issaias Afewerki, a Christian, has been careful to include Muslims in his transitional cabinet. He lashed out at political opponents in 2001 and faced unprecedented criticism, even from within the PFDJ.

WORLD AFFAIRS
▷ Joined UN in 1993

COMESA iBRD IGAD NAM AU

Eritrea's secession represented the first major redrawing of borders established by Africa's colonizers. Attracting Western and Arab aid for reconstruction is a priority. A border conflict with Ethiopia erupted into open warfare in 1998. Under the 2000 UN-sponsored peace accord, a new border was to be established. In February 2001 Ethiopia completed its troop withdrawal. Work started on the border demarcation in spring 2002.

EAST AFRICA

E

E

AID
 Recipient

 $176m (receipts)   Up 18% in 2000

The economy is highly aid-dependent, and 75% of the population survive on food aid. This is an obvious and pressing need, given the country's vulnerability to famine, but Western donors have been less generous with aid for the $2 billion reconstruction costs. Emergency UN aid was requested in mid-2000 to assist over a million people displaced by the Ethiopian incursion. WHO supplied emergency medical aid.

DEFENSE
 Compulsory military service

 $206m   Down 33% in 2000

Defense expenditure is massive. The 50,000-strong permanent army (of whom about a third are women) is swelled by vast numbers of conscripts. Troops were being reintegrated into the economy on "food for work" schemes until the latest war with Ethiopia, which inflicted heavy losses. Mass demobilization was restarted in 2002.

ECONOMICS
 Inflation 9.4% p.a. (1993–2000)

 $696m   10.2–13.55 nakfa

SCORE CARD

- ❏ WORLD GNP RANKING 162nd
- ❏ GNP PER CAPITA $170
- ❏ BALANCE OF PAYMENTS...................–$208m
- ❏ INFLATION 14%
- ❏ UNEMPLOYMENT ....Widespread underemployment

STRENGTHS
Resourceful, hard-working population. Strategic position on Red Sea – tourism and transportation. Potential for mining and oil industry. Government committed to cutting dependence on food aid.

WEAKNESSES
Destruction of infrastructure and equipment; port of Massawa heavily bombed. Dependent on aid. Most of population living at subsistence level. Susceptibility to drought and famine. Return of some 750,000 refugees.

EXPORTS

USA 2%   Italy 5%
Other 27%   Japan 13%
Sudan 27%   Ethiopia 26%

IMPORTS

UK 4%   South Korea 4%
Germany 6%
Other 53%   UAE 16%
Italy 17%

RESOURCES
 Electric power: Not available

 7042 tonnes   Not an oil producer; oil refinery at Assab

2.2m cattle, 1.7m goats, 1.57m sheep, 1.3m chickens   Copper, potash, gold, iron, silver, zinc, oil, silica, granite, marble

Eritrea has substantial copper reserves, and lesser ones of silver, zinc, and gold. High-quality silica, granite, and marble deposits could be exploited. Onshore and offshore oil deposits are believed to exist, but exploration work is at an early stage. There is potential for power generation from geothermal sources.

ENVIRONMENT
 Not available

 5%   0.03 tonnes per capita

Deforestation and soil erosion are major problems. The Ethiopian army uprooted trees to destroy the cover they provided for Eritrean soldiers. Since 1991, 22 million seedlings have been grown in a replanting scheme. The Red Sea coast is a conservation priority.

MEDIA
 TV ownership low

 There are no daily newspapers

PUBLISHING AND BROADCAST MEDIA

New Eritrea, owned by the PFDJ, is published every 3 days in English, Tigrinya, and Arabic

1 state-controlled service   1 state-controlled service

The media are largely controlled by the PFDJ, which runs both the radio and TV services. Independent newspapers are not encouraged.

CRIME
 Death penalty in use

 Eritrea does not publish prison figures   Crime levels remain low

Crime has not been a major problem since independence. The judiciary and police answer to the PFDJ. There are a number of political prisoners.

EDUCATION
 School leaving age: 13

 56%   3994 students

Very few schools functioned during the war. There is one university. In an attempt to reduce potential ethnic tension, all children above the age of 11 are being taught in English.

HEALTH
No welfare state health benefits

1 per 20,000 people   Malaria, potential risk of famine

The risk of famine overrides normal health concerns. Eritreans built their own hospitals during the independence struggle. Health provision is basic.

Seasonal river beds carry rain from the Ethiopian highlands into Eritrea, providing essential irrigation for agriculture.

CHRONOLOGY
British military rule replaced Italian colonial authority in 1941.

- ❏ **1952** Ethiopia absorbs Eritrea.
- ❏ **1961** Beginning of armed struggle.
- ❏ **1987** EPLF refuses offer of autonomy; fighting intensifies.
- ❏ **1991** EPLF takes Asmara.
- ❏ **1993** Formal independence.
- ❏ **1998** Border war with Ethiopia.
- ❏ **2000** OAU peace treaty signed.
- ❏ **2001** Ethiopia completes troop withdrawal.
- ❏ **2002** Border demarcation begins.

SPENDING
GDP/cap. increase

CONSUMPTION AND SPENDING

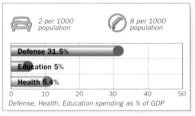
2 per 1000 population   8 per 1000 population

Defense 31.5%
Education 5%
Health 5.4%

0   10   20   30   40   50
Defense, Health, Education spending as % of GDP

Over 80% of Eritrea's population are subsistence farmers. A few of the 150,000 refugees who fled to Arab and Western countries have built up some personal savings.

WORLD RANKING

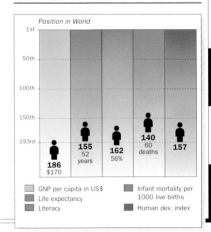

Position in World

	GNP per capita in US$	Life expectancy	Literacy	Infant mortality	Human dev. index
Value	186 $170	155 52 years	162 56%	140 60 deaths	157

- GNP per capita in US$
- Life expectancy
- Literacy
- Infant mortality per 1000 live births
- Human dev. index

ESTONIA

OFFICIAL NAME: Republic of Estonia **CAPITAL:** Tallinn
POPULATION: 1.4 million **CURRENCY:** Kroon **OFFICIAL LANGUAGE:** Estonian

EUROPE Europe

E

TRADITIONALLY THE MOST Western-oriented of the Baltic states, Estonia is bordered by Latvia and the Russian Federation. Its terrain is flat, boggy, and partly wooded, and includes more than 1500 islands. Estonia formally regained its independence as a multiparty democracy in 1991. In contrast to the peoples of Latvia and Lithuania, Estonians are Finno-Ugric, speaking a language related to Finnish.

CLIMATE ▷ Continental

WEATHER CHART FOR TALLINN

Estonia's coastal location gives it cool summers, and cold winters when the Baltic Sea freezes.

TRANSPORTATION ▷ Drive on right

Tallinn Ulemiste
559,800 passengers

209 ships
379,110 grt

THE TRANSPORTATION NETWORK

10,935 km
(6795 miles)

75 km
(47 miles)

968 km
(602 miles)

320 km
(199 miles)

Railroads have improved and buses are reliable. Baltic ferries link Tallinn with Finland, Sweden, and Germany.

TOURISM ▷ Visitors : Population 1:1.3

1.24m visitors

Up 31% in 2000

MAIN TOURIST ARRIVALS

Finland 60%	
Latvia 14%	
Russia 9%	
Other 17%	

% of total arrivals

Estonia is particularly popular with Finns. Water sports, winter sports, folk and architectural heritage, and nature tours, are the main attractions. Tallinn's medieval center is a major draw for tourists on short visits.

PEOPLE ▷ Pop. density low

Estonian, Russian

31/km²
(80/mi²)

THE URBAN/RURAL POPULATION SPLIT

69%   31%

ETHNIC MAKEUP

Other 8%
Estonian 62%
Russian 30%

After the decades of Soviet rule, the relationship between Estonians and the Russian minority was problematic. Rules on citizenship, introduced in 1992 and 1995, excluded many ethnic Russians who could not meet the Estonian language and minimum residency requirements. Over 100,000 took Russian rather than Estonian citizenship. A 2000 language law met international demands for an end to discrimination against the Russian-speaking minority. Estonians are predominantly Protestant. Families are small; divorce rates are high.

POLITICS ▷ Multiparty elections

1999/2003   President Arnold Ruutel

AT THE LAST ELECTION
Parliament 101 seats

7% CPP   6% UPP

27% K   18% PPU   18% R   17% M   7% KMU

K = Center Party **PPU** = Pro Patria Union **R** = Reform Party
M = Moderates **CPP** = Rural People's Party
KMU = Coalition Party **UPP** = United People's Party

Coalition government has been the norm since the end of communist rule. In elections in 1999 the Center Party won the most seats, on a platform of graduated income tax to help overcome wealth disparities. However, Mart Laar, who had led the government in power from 1992 to 1995, became prime minister, leading a center-right coalition of the PPU, the Reform Party, and the Moderates. Although hindered by a slim majority, Laar's efforts to pursue free-market reforms became easier as the economy grew strongly from the beginning of 2000. However, cracks in the coalition led to his resignation in January 2002. The new government led by Siim Kallas included the Center Party.

ESTONIA

Total Land Area :
45 226 sq. km
(17 462 sq. miles)

LAND HEIGHT

200m/565ft
Sea Level

POPULATION

⊙ over 500 000
◎ over 100 000
○ over 50 000
● over 10 000
• under 10 000

WORLD AFFAIRS

 Joined UN in 1991

Estonia's trade with the West has been growing, and "fast track" negotiations on full EU membership began in 1998. Ties with other Baltic countries and with Scandinavia have been particularly emphasized. Integration with the EU and NATO is the top priority.

Estonia has now accepted the de facto border with Russia, having effectively ceded a portion of its territory during the Soviet period.

AID

 Recipient

$64m (receipts)

Down 24% in 2000

Although an aid recipient, since 1997 Estonia has also been an aid donor, mainly through technical assistance.

DEFENSE

Compulsory military service

$79m

Up 11% in 2000

The government agreed in July 2000 to shorten compulsory military service from 12 to eight months. Initial US opposition to full membership of NATO has now been changed to support for the Baltic states' entry into the organization.

ECONOMICS

Inflation 53% p.a. (1990–2000)

$4.89bn

16.675–17.577 krooni

SCORE CARD

- ❏ WORLD GNP RANKING.......................112th
- ❏ GNP PER CAPITA $3580
- ❏ BALANCE OF PAYMENTS...................–$315m
- ❏ INFLATION 4%
- ❏ UNEMPLOYMENT...................................13%

STRENGTHS

Improved productivity and stable currency are pegged to the euro. Simplicity of tax regime. More advantage is being taken of natural resources, including timber and oil shale. Transportation infrastructure has been upgraded. Exports growing.

WEAKNESSES

Poor raw materials base. Dependence on imported energy supplies.

EXPORTS

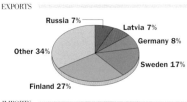

Russia 7%
Latvia 7%
Germany 8%
Other 34%
Sweden 17%
Finland 27%

IMPORTS

Chile 3%
Germany 9%
Sweden 9%
Other 41%
Russia 14%
Finland 24%

RESOURCES

 Electric power 2.6m kw

 111,993 tonnes

Oil figures not published

300,200 pigs, 252,800 cattle, 2.37m chickens

Oil shale, coal, peat, phosphorite

The chief energy resource is oil shale. Phosphorite mining has been stopped. Timber is processed to make paper.

ENVIRONMENT

 Sustainability rank: 18th

12%

12.1 tonnes per capita

Industrial pollution comes especially from power stations burning oil shale. Danger of radioactive leaks from former Soviet bases remains. Water supply and sewage treatment have improved.

MEDIA

 TV ownership high

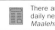 Daily newspaper circulation 174 per 1000 people

PUBLISHING AND BROADCAST MEDIA

There are 17 daily newspapers. The main daily newspapers are *Eesti Ekspress*, *Maaleht*, and *Postimees*

3 services: 1 state-owned, 2 independent

30 services: 1 state-owned, 29 independent

The media are mostly pro-government. The number of Russian-language programs is declining. Estonians have been able to receive Finnish satellite TV for some years.

CRIME

 No death penalty

4742 prisoners

Up 13% in 1999

Robbery and narcotics are the main crime problems. Generally, however, crime levels are still relatively low.

EDUCATION

 School leaving age: 16

 99%

50,814 students

Education is becoming increasingly Westernized. There are two main universities, in Tallinn and Tartu.

HEALTH

 Welfare state health benefits

1 per 333 people

Heart diseases, cancers, accidents, violence

The health system, improved since the collapse of communism, is better than that of most former Soviet republics.

***The Russian Orthodox convent** of Pühtitsa at Kuremäe in Estonia's marshy north. Most of the population is Evangelical Lutheran.*

E

CHRONOLOGY

After Swedish and then Russian rule, Estonia briefly enjoyed independence from 1921 until its incorporation into the Soviet Union in 1940.

- ❏ **1990** Unilateral declaration of independence; achieved in 1991.
- ❏ **1992** First multiparty elections: election of center-right government.
- ❏ **1996** President Lennart Meri wins second term of office.
- ❏ **1998** EU opens negotiations.
- ❏ **1999** Elections: new center-right government.
- ❏ **2001** Communist-era leader Arnold Ruutel elected president.

SPENDING

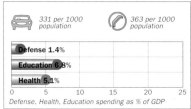 GDP/cap. decrease

CONSUMPTION AND SPENDING

331 per 1000 population

363 per 1000 population

Defense 1.4%
Education 6.8%
Health 5.1%

0   5   10   15   20   25
Defense, Health, Education spending as % of GDP

Market reforms have led to increased prosperity. A few have become very rich. Average wages are higher than in other Baltic states.

WORLD RANKING

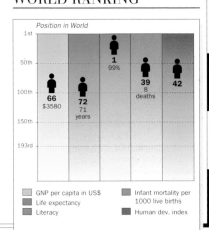

Position in World

1st
50th
100th
150th
193rd

66
$3580

72
71 years

1
99%

39
8 deaths

42

❏ GNP per capita in US$
❏ Life expectancy
❏ Literacy

❏ Infant mortality per 1000 live births
❏ Human dev. index

ETHIOPIA

OFFICIAL NAME: Federal Democratic Republic of Ethiopia **CAPITAL:** Addis Ababa
POPULATION: 64.5 million **CURRENCY:** Ethiopian birr **OFFICIAL LANGUAGE:** Amharic

E

LOCATED IN NORTHEAST Africa, the former empire of Ethiopia is the cradle of an ancient civilization which adopted Orthodox Christianity in the 4th century. It has been landlocked since 1993, when Eritrea, on the Red Sea, seceded. Ethiopia is mountainous except for desert lowlands in the northeast and southeast, and is prone to devastating drought and famine. A long civil war ended in 1991 with the defeat of the Stalinist military dictatorship that had ruled since 1974. A free-market, multiparty democratic system now provides substantial regional autonomy. War with Eritrea in 1998–2000 brought heavy losses on both sides before a peace agreement was signed in December 2000. Arbitrators began redefining the border in spring 2002.

CLIMATE

> Mountain/steppe

WEATHER CHART FOR ADDIS ABABA

In general, the climate is moderate, except in the lowlands of the Danakil and the Ogaden deserts, which are hot all year round and can suffer severe drought. The highlands are temperate, with night frost in the mountains. The single rainy season in the west brings twice as much rain as do the two wet seasons in the east. During these cloudy periods, thunderstorms occur almost daily.

TRANSPORTATION

> Drive on right

Bole International, Addis Ababa
1.04m passengers

12 ships
92,434 grt

THE TRANSPORTATION NETWORK

4275 km (2656 miles)		Trans-East Africa Highway	
681 km (423 miles)		None	

The single railroad linking Addis Ababa with Djibouti has grown in strategic importance due to the conflict with Eritrea. Ethiopia's main access to the sea by road has been through the Red Sea ports of Assab and Massawa, now part of an independent Eritrea. Inland, pack mules and donkeys are widely used. Ethiopian Airlines has good services to much of Africa, and to European and US cities.

TOURISM

> Visitors : Population 1:516

125,000 visitors

Up 36% in 2000

MAIN TOURIST ARRIVALS

USA	15%
Saudi Arabia	9%
UK	5%
Kenya	5%
Italy	4%
Other	62%

% of total arrivals

Despite Ethiopia's unique attractions, tourism is on a small scale, although since 1991 there has been a sizable increase in the number of visitors, mostly on organized tours. Several new hotels are being built. The Rift Valley lakes, Lake Tana, the Gonder castles, and the Blue Nile gorge, with its spectacular scenery, are popular tourism destinations, but guides are essential. Ancient churches and cities such as Aksum, the royal capital of the first Ethiopian kingdom, are now accessible. Visitors are also attracted to the five national parks.

Lalibela lies 120 km (75 miles) northwest of Desē in Ethiopia's plateau region, and is famous for the rock-hewn churches created by King Lalibela of the Zagwe dynasty.

PEOPLE

> Pop. density medium

Amharic, Tigrinya, Galla, Sidamo, Somali, English, Arabic

58/km² (151/mi²)

THE URBAN/RURAL POPULATION SPLIT

18%   82%

RELIGIOUS PERSUASION

Muslim 40%
Other 5%
Traditional beliefs 15%
Ethiopian Orthodox 40%

ETHNIC MAKEUP

Oromo 40%
Berta 6%
Somali 6%
Sidamo 9%
Other 14%
Amhara 25%

There are 76 ethnic nations in Ethiopia, speaking 286 languages. Oromos (or Gallas) form the largest group, whereas less than 5% of the population are Tigreans.

Civil war was sparked by fighting between different ethnic groups, but they later united in opposition to the Mengistu regime. Ethnic tensions are still near the surface, in spite of the new federal structure, and there have been reports of boundary disputes in several regions. The Oromos withdrew from the Tigrean-dominated government in 1992. Hostility to the government has also been voiced by disaffected Amharas, who had been dominant for several centuries, and by the Orthodox Church. The aspirations of ethnic Somalis in the southeast are another source of tension.

Most of the small Jewish community, which has lived in Ethiopia for 2000 years, was evacuated to Israel in 1991, but more than 20,000 remain, waiting for Israel to offer them citizenship.

The participation of women in rural organizations is increasing, reflecting the key role women played in the war.

POPULATION AGE BREAKDOWN

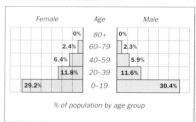

Female	Age	Male
0%	80+	0%
2.4%	60–79	2.3%
6.4%	40–59	5.9%
11.8%	20–39	11.6%
29.2%	0–19	30.4%

% of population by age group

ETHIOPIA

Total Land Area :
1 127 127 sq. km
(435 184 sq. miles)

POPULATION

▣	over 1 000 000
◉	over 100 000
○	over 50 000
●	over 10 000
•	under 10 000

LAND HEIGHT

	4000m/13 124ft
	3000m/9843ft
	2000m/6562ft
	1000m/3281ft
	500m/1640ft
	200m/656ft
	Sea Level
	-200m/656ft

0 ——— 200 km
0 ——— 200 miles

E

WORLD AFFAIRS

▷ Joined UN in 1945

COMESA · G24 · IGAD · NAM · AU

Landlocked since the secession of Eritrea, Ethiopia needs continued access to the Red Sea ports of Massawa and Assab, and maintained cordial relations with Eritrea until a border dispute escalated into armed conflict in 1998. A peace accord reached two years later allowed for the demarcation of a new border.

Addis Ababa is the headquarters of the AU (successor to the Organization of African Unity) and of the UN Economic Commission for Africa. Ethiopia is active in regional diplomacy, including numerous attempts at brokering peace in Somalia, although tension with Somali factions during 1998 and 1999 escalated into armed intervention by Ethiopia.

The government's official policy is one of noninterference in the affairs of neighboring countries, although Sudan and Somalia have accused Ethiopia of supporting rebels. Links with other African states have been strengthened, as have those with the US, the EU, and Israel.

POLITICS

▷ Multiparty elections

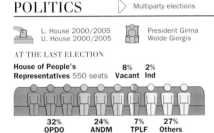

L. House 2000/2005
U. House 2000/2005

President Girma Wolde Giorgis

AT THE LAST ELECTION

House of People's Representatives 550 seats 8% Vacant 2% Ind

32% OPDO	24% ANDM	7% TPLF	27% Others

OPDO = Oromo People's Democratic Organization
ANDM = Amhara National Democratic Movement
TPLF = Tigre People's Liberation Front **Ind** = Independents

The Ethiopian People's Revolutionary Democratic Front (**EPRDF**) includes all the main parties and controls over 90% of the lower house

Federal Council 108 seats

The Federal Council is elected indirectly on a nonparty basis

The transitional period which followed the collapse of the Mengistu military dictatorship in 1991 ended in 1995 with multiparty elections.

PROFILE

The current government, reelected in 2000, succeeded one set up in 1991 by the EPRDF, the strongest of the groups that fought Mengistu's Marxist regime and chiefly responsible for winning the civil war. Prime Minister Meles Zenawi is the leader of the Tigrean People's Liberation Front, the largest group within the EPRDF. There is growing opposition from the Oromos and Amharas to the dominance of Tigreans. The nine states are largely governed by elected governments dominated by local liberation movements.

MAIN POLITICAL ISSUE
Ethnic representation

The 1994 constitution establishing a nine-state federation grants the states considerable autonomy, including the right to secede, as Eritrea did in 1993. The EPRDF government believes this to be the best way to prevent secessionist conflict and maintain national unity. The ruling broad-based coalition is ideologically dominated by Tigrean politicians.

Prime Minister Meles Zenawi, leader of the EPRDF, which ousted the Mengistu regime.

Mengistu Haile Mariam, who ran Ethiopia on Soviet lines from 1977–1991.

CHRONOLOGY

After repelling a devastating Muslim invasion in 1523, Ethiopia developed as an isolated empire until Egyptian and Sudanese incursions in the 1850s led to its renewed political power under Emperor Teodros. His successor, Menelik II, doubled the empire southward and eastward.

❏ **1896** Italian invasion of Tigre defeated. Europeans recognize Ethiopia's independence.
❏ **1913** Menelik II dies.
❏ **1916** His son, Lij Iyasu, is deposed for his conversion to Islam and a proposed alliance with Turkey. Menelik's daughter, Zauditu, becomes empress with Ras (Prince) Tafari as regent.
❏ **1923** Joins League of Nations.
❏ **1930** Zauditu dies. Ras Tafari crowned Emperor Haile Selassie.
❏ **1936** Italians occupy Ethiopia. League of Nations fails to react.
❏ **1941** British oust Italians and restore Haile Selassie, who sets up a constitution, parliament, and cabinet, but retains personal power and the feudal system.
❏ **1952** Eritrea, ruled by Italy until 1941, then under British mandate, federated with Ethiopia. ➪

E

CHRONOLOGY *continued*

- ❏ **1962** Unitary state created; Eritrea loses its autonomy despite demands of secessionists.
- ❏ **1972–1974** Famine kills 200,000.
- ❏ **1974** Strikes and army mutinies at Haile Selassie's autocratic rule and country's economic decline. Dergue (Military Committee) stages coup.
- ❏ **1975** Becomes socialist state: nationalizations, worker cooperatives, and health reforms.
- ❏ **1977** Col. Mengistu Haile Mariam takes over. Somali invasion of Ogaden defeated with Soviet and Cuban help.
- ❏ **1978–1979** Thousands of political opponents killed or imprisoned.
- ❏ **1984** Workers' Party of Ethiopia (WPE) set up on Soviet model. One million die in famine after drought and years of war. Live Aid concert raises funds for relief.
- ❏ **1986** Eritrean rebels now control the whole northeastern coast.
- ❏ **1987** Serious drought again threatens famine.
- ❏ **1988** Eritrean and Tigrean People's Liberation Fronts (EPLF and TPLF) begin new offensives. Mengistu's budget is for "Everything to the War Front." Diplomatic relations with Somalia restored.
- ❏ **1989** Military coup attempt fails. TPLF in control of most of Tigre. TPLF and Ethiopian People's Revolutionary Movement form alliance – EPRDF.
- ❏ **1990** Military gains by opponents of Mengistu regime. Moves toward market economy and restructuring of ruling party to include non-Marxists. Distribution of food aid for victims of new famine is hampered by government and rebel forces.
- ❏ **1991** Mengistu accepts military defeat and flees country. EPRDF enters Addis Ababa, sets up provisional government, promising representation for all ethnic groups. Outbreaks of fighting continue, between mainly Tigrean EPRDF and opposing groups.
- ❏ **1993** Eritrean independence recognized following referendum.
- ❏ **1995** Transitional rule ends. EPRDF wins landslide in multiparty elections, sets up first democratic government. New nine-state federation is formed.
- ❏ **1998–2000** Border war with Eritrea.
- ❏ **2000** OAU peace treaty signed. Haile Selassie's remains buried in Trinity Cathedral, Addis Ababa.
- ❏ **2001** Ethiopia completes troop withdrawal from Eritrea.
- ❏ **2002** Over 120 human rights demonstrators killed by police.

AID

 ▷ Recipient

 $693m (receipts)   Up 8% in 2000

The World Food Program and the EU are the largest sources of assistance, while the US has taken over from Italy and the former Soviet Union as the major bilateral donor. Aid per capita is low by regional standards. However, long-term development assistance and balance-of-payments support look set to continue their recent growth. Aid is now playing an increasingly important part in the economy. The emphasis was shifting from food aid toward credit for infrastructure development – until the 2000 drought disaster.

DEFENSE

▷ No compulsory military service

$448m   Up 1% in 2000

Ethiopia is one of the most heavily militarized states in Africa. Its sizable standing army is boosted by conscription at times of crisis. Heavy losses have been sustained in fighting with Eritrea, and a 12-month international arms embargo was imposed following the Ethiopian advance in May 2000. The government is trying to gain control of the many ethnic and clan-based militias throughout the country.

ETHIOPIAN ARMED FORCES

300 main battle tanks (T-54/55, T-62)	250,000 personnel	
None	None	
51 combat aircraft (24 MiG-21MF, 17 MiG-23BN, 4 Su-25, 6 Su-27)	2500 personnel	
None		

ECONOMICS

▷ Inflation 7% p.a. (1990–2000)

$6.74bn   8.225–8.450 Ethiopian birr

SCORE CARD

❏ WORLD GNP RANKING	99th
❏ GNP PER CAPITA	$100
❏ BALANCE OF PAYMENTS	$16m
❏ INFLATION	5.9%
❏ UNEMPLOYMENT	63%

ECONOMIC PERFORMANCE INDICATOR

EXPORTS

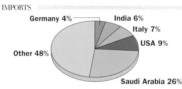

Italy 7%
Saudi Arabia 8%
Djibouti 11%
Japan 11%
Germany 18%
Other 45%

IMPORTS

Germany 4%
India 6%
Italy 7%
USA 9%
Saudi Arabia 26%
Other 48%

STRENGTHS

Increased economic aid in 1990s. End of total state control. Coffee production.

WEAKNESSES

Overwhelming dependence on agriculture. Periodic serious droughts. War-damaged infrastructure. Massive displacement of population by war and drought. Small industrial base. Lack of skilled workers. Legacy of Mengistu regime's centrally planned economy.

PROFILE

After the end of the civil war in 1991, Ethiopia began moving toward a market economy by encouraging foreign investment and reforming land tenure. Economic decline was reversed in 1993 as agricultural and industrial output grew, with foreign aid used to fund the purchase of parts and raw materials for manufacturing. These gains were undermined by war with Eritrea in 1998–2000 and the renewed danger of severe drought-related famine in 2000.

ETHIOPIA : MAJOR BUSINESSES

Light engineering
Coffee processing
Food processing
Publishing
Textiles

Gonder
Addis Ababa
Dirē Dawa
Nazrēt
Jima
Yirga 'Alem

0   250 km
0   250 miles

RESOURCES
  Electric power 456,000 kw

15,858 tonnes

34.5m cattle, 22.5m sheep, 17m goats, 55.8m chickens

Oil reserves currently unexploited

Oil, gold, platinum, copper, potash, iron, natural gas

ELECTRICITY GENERATION

Hydro 94% (1.6bn kwh)	
Combustion 6% (0.1bn kwh)	
Nuclear 0%	
Other 0%	

% of total generation by type

Manpower and financial constraints have prevented a systematic survey of mineral resources. At present, mining contributes less than 1% of GDP. Ethiopia has great potential for hydroelectric power which, in the long run, could offset a domestic reliance on fuelwood and slow massive deforestation and soil erosion. Current exploration for oil and gas has revealed reserves in the Ogaden, but exploitation has not begun. When Eritrea seceded in 1993, Ethiopia lost other substantial oil reserves and many oil concessions.

ETHIOPIA : LAND USE

Cropland
Forest
Pasture
Cattle
Coffee - cash crop
Cereals

0   250 km
0   250 miles

ENVIRONMENT
 Sustainability rank: 113th

6%   0.1 tonnes per capita

ENVIRONMENTAL TREATIES

No	Yes
Yes	No
Yes	No

Deforestation for fuelwood and the resultant rapid soil erosion, particularly in the highlands, are serious problems. Forest cover has fallen from 40% in 1900 to only 2% today. Dung is being used for fuel, instead of as a fertilizer. The amount used – with a fertilizing value of $123 million a year – would increase annual grain harvests by up to 1.5 million tonnes. Local projects include terracing hillsides to prevent soil and water run-off – 36,000 km (22,370 miles) of terraces were built in Tigre in 1992.

MEDIA
 TV ownership low

 Daily newspaper circulation 2 per 1000 people

The government remains uneasy about the post-Mengistu independent press, which has become prolific and critical, although circulation is small. Legal action has been taken to silence several publications. All main newspapers and the TV broadcasting station are government-owned and operated.

PUBLISHING AND BROADCAST MEDIA

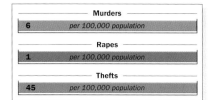

There are 3 daily newspapers, including *Addis Zemen* and *Ethiopian Herald* published by the government

1 state-owned service

4 services: 1 state-owned, 3 independent

CRIME
 Death penalty in use

 13,585 prisoners   Down 19% in 1999

CRIME RATES

Murders	
6	per 100,000 population

Rapes	
1	per 100,000 population

Thefts	
45	per 100,000 population

EDUCATION
 School leaving age: 13

38%   67,682 students

THE EDUCATION SYSTEM

% of each age group in education

Primary	Secondary	Tertiary
63%	17%	1%

Secondary education is in English and Amharic. Schools are basic and classes crowded; education was severely disrupted during the civil war. Addis Ababa University, a center of political activity (usually anti-EPRDF), suffers periodic closures and the dismissal of leading academics.

HEALTH
 No welfare state health benefits

1 per 20,000 people   Diarrheal and respiratory diseases, tuberculosis, malaria

Only about half of the population lives within 12 km (8 miles) of a health unit. Hospital building, distribution of resources to rural areas, outpatient visits, and referrals are all very slow. Skin and eye diseases are common. Mission hospitals are of a reasonably high standard, although the use of traditional remedies is widespread. The incidence of HIV/AIDS is falling.

A number of human rights abuses by the transitional government have been documented by the independent Ethiopian Human Rights Council. These include detention without trial, "disappearances," and extrajudicial killings. There is some concern over indiscipline among EPRDF forces, who provide a de facto police force in many regions. In many rural areas, the state system has yet to replace traditional forms of justice.

SPENDING
GDP/cap. increase

CONSUMPTION AND SPENDING

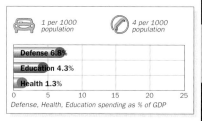

1 per 1000 population   4 per 1000 population

Defense 6.8%	
Education 4.3%	
Health 1.3%	

Defense, Health, Education spending as % of GDP

Most Ethiopians are extremely poor, many of the country's wealthier families having fled into exile in recent years. Ethiopian Christian culture places more value on maintaining traditional social structures than on realising individual ambition. Living at subsistence level and a reliance on traditional agriculture remain the general expectation.

WORLD RANKING

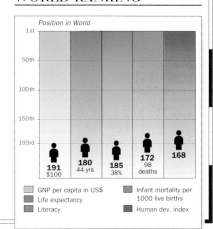

Position in World

191 $100	180 44 yrs	185 38%	172 98 deaths	168

GNP per capita in US$
Life expectancy
Literacy
Infant mortality per 1000 live births
Human dev. index

E

FIJI

SOUTH PACIFIC

Pacific Ocean

OFFICIAL NAME: Republic of the Fiji Islands **CAPITAL:** Suva
POPULATION: 823,000 **CURRENCY:** Fiji dollar **OFFICIAL LANGUAGE:** English

| 1970 | 1970 | Oct 10 | FJI | +12 | +679 | .fj |

FIJI IS A VOLCANIC archipelago in the southern Pacific Ocean, comprising two main islands and nearly 900 smaller islands and islets. The Melanesian Fijian population was outnumbered in the post-1945 period by ethnic Indians, descended from workers brought over by the British in 1879–1916. Coups led by Fijian supremacists between 1987 and 2000 led to a mass exodus of Indo-Fijians, reversing the ethnic balance and seriously damaging the economy.

F

CLIMATE
▷ Tropical oceanic

WEATHER CHART FOR SUVA

■ *Average daily temperature*   *Rainfall* ■
°C/°F J F M A M J J A S O N D cm/in
40/104   40/16
30/86   30/12
20/68   20/8
10/50   10/4
0/32   0
-10/14
-20/-4

The eastern sides of the main islands are wettest, having more than twice the annual rainfall of the western flanks. Fiji lies in a cyclone path.

TRANSPORTATION
▷ Drive on left

🛩 **Nadi International**
926,677 passengers

🚢 53 ships
29,500 grt

THE TRANSPORTATION NETWORK

🛣 1695 km (1053 miles)	🛤 None		
🚂 595 km (370 miles)		⚓ 203 km (126 miles)	

On the axis of Australian–US west coast air routes, Fiji is well served by international flights. An international airport is proposed for Vanua Levu.

TOURISM
▷ Visitors : Population 1:2.8

🧳 294,000 visitors   ⬇ Down 28% in 2000

MAIN TOURIST ARRIVALS

| Australia 29% |
| New Zealand 18% |
| USA 15% |
| Other 38% |

0   10   20   30   40   50   60
% of total arrivals

Tourism – once Fiji's greatest earner – is adversely affected by political instability: 7500 jobs in the tourism sector were lost in 2000 alone.

PEOPLE
▷ Pop. density low

👤 Fijian, English, Hindi, Urdu, Tamil, Telugu

👥 45/km² (117/mi²)

THE URBAN/RURAL POPULATION SPLIT

49%   **51%**

RELIGIOUS PERSUASION

Muslim 8%
Other 8%
Roman Catholic 9%
Hindu 38%
Methodist 37%

A delicate ethnic balance was shattered by the exodus of Indo-Fijians in 1987–1989 and again in 2000–2001. The lawlessness accompanying the recent upheavals exaggerated ethnic tensions and brought racist rhetoric back to the political mainstream. A substantial population of Polynesians live on Rotuma and have traditionally pressed for autonomy. Women are lobbying for more rights.

POLITICS
▷ Multiparty elections

🗳 L. House 2001/2006
U. House 2001/2006

👤 President Ratu Josefa Iloilo

AT THE LAST ELECTION

House of Representatives 71 seats

3% 3%
NLUP Others

45% SDL   **38%** FLP   **8%** MV   **3%** Ind

SDL = Fijian People's Party **FLP** = Fiji Labor Party
MV = Conservative Alliance (Matanitu Vanua) **NLUP** = New Labor Unity Party **Ind** = Independents

Senate 32 seats

The Senate is appointed by the president

The issue of political representation for Fiji's large Indian population has dominated politics since the 1980s. The 1987 coup, justified as defending ethnic Fijian land rights, was condemned internationally. The traditional Great Council of Chiefs emerged as the power broker after the 2000 coup, despite its lack of constitutional power. It ensured that the Indian-dominated FLP government, elected in 1999, was barred from returning to power and that the multiethnic 1997 constitution remained buried.

FIJI

Total Land Area : 18 270 sq. km (7054 sq. miles)

POPULATION

over 50 000 ○
over 10 000 ●
under 10 000 ·

LAND HEIGHT

1000m/3281ft
500m/1640ft
Sea Level

Rotuma

Cikobia
Vetauua

Navoalevu
Labasa
Naduri   Rabi
Yaqaga
Nabavatu   Kioa
Yadua   *Vanua Levu*   Buca   Qamea
Nacula   Savusavu   Somosomo   Bouma
Matacawa Levu   Nabouwalu   Kanacea   *Taveuni*   Naitaba
Yaqeta   *Bligh Water*
Viwa   Naviti   Koro   Vanua Balavu
Waya   Navaga · Nasau   Yacata   Munia
Tavua   Kakiraki   Makogai   Mago
Ba   Vatukoula   Wakaya
Mamanuca-i-ra   Tomaniivi   Ovalau   *Levuka*   Cicia   Tuvuca
Group   (Mt Victoria)   Korovou   Batiki   Nayau
Lautoka   1323m   Nairai   *K o r o*   Lakeba
Malolo   Nadi   Korovou   Lamiti   *S e a*
Viti Levu   Lami   Nausori   Gau   *Vanua Vatu*   Oneata
Sigatoka   Korolevu   Navua   **SUVA**   Moce
Beqa   Namuka-i-lau
Vatulele   Moala   Yagasa Cluster
Kadavu Passage   Tuaqava   Fulaga
Ono   Kabara   Ogea Levu
Totoya   Ogea Driki
Tavuki   Kadavu   Matuku
Vunisea   Vatoa

N

0   50 km
0   50 miles

Ono-i-lau
Tuvana-i-colo
Tuvana-i-ra

WORLD AFFAIRS ▷ Joined UN in 1970

 ACP CP Comm PC PIF

Fiji's international reputation has been severely damaged by its discrimination against Indo-Fijians and the recent coups. Fiji has been intermittently suspended from the Commonwealth.

AID ▷ Recipient

 US$29m (receipts)   Down 17% in 2000

Fiji traditionally received a lot of overseas aid, but international reaction to the 2000 coup prompted drastic cuts.

DEFENSE ▷ No compulsory military service

 US$32m   Down 9% in 2000

Of the almost entirely ethnic Fijian military, significant numbers – around 20% – are assigned to UN duties and have served in Lebanon and Egypt.

ECONOMICS ▷ Inflation 3.3% p.a. (1990–2000)

US$1.48bn   2.176–2.298 Fiji dollars

SCORE CARD

❏ World GNP Ranking	145th
❏ GNP per Capita	US$1820
❏ Balance of Payments	US$13m
❏ Inflation	1.1%
❏ Unemployment	5%

Strengths
Relatively well-diversified economy, with strong tourist infrastructure. Location on Pacific air routes. Many regional and international organizations located in Suva.

Weaknesses
2000 coup caused dramatic contraction in economy – 12.5%. Migration of many Indo-Fijian professionals. Sugar crops vulnerable to drought. Major exports – sugar, copra, and gold – subject to large fluctuations in world prices.

EXPORTS

IMPORTS

Cane field on the west side of Viti Levu, between Nadi and Lautoka. Sugar accounts for about one-third of Fiji's exports.

RESOURCES ▷ Electric power 200,000 kw

 38,471 tonnes   Not an oil producer

245,749 goats, 340,000 cattle, 3.7m chickens   Gold, silver

The varied terrain allows diversified agriculture. Gold and minerals are mined. A hydroelectric plant at Monasavu provides 95% of electricity.

ENVIRONMENT ▷ Not available

 0.3%   0.9 tonnes per capita

Governments are environmentally aware. Tourism is damaging coral reefs. Fiji was downwind of French Pacific nuclear tests. Fertilizers are overused.

MEDIA ▷ TV ownership low

 Daily newspaper circulation 51 per 1000 people

PUBLISHING AND BROADCAST MEDIA

 There are 2 English-language dailies, *Fiji Times* and *Fiji Daily Post*. *Nai Lalakai* and *Shanti Dut* are Fijian and Indian weeklies

 2 services: 1 state-owned, 1 independent   4 services: 1 state-controlled, 3 independent

Freedom of the press is championed by the government, and cases of corruption are often reported in the media. However, the police have blocked politically sensitive broadcasts.

CRIME ▷ Death penalty not used in practice

 1185 prisoners   Up in 2000

Usually theft and drink-related violence top the crime list. The 2000 coup spurred Fijians to settle old scores by force.

EDUCATION ▷ Schooling is not compulsory

 93%   9208 students

Education, originally modeled on the British system, is now mostly run by local committees and is increasingly racially segregated. The use of the birch in schools was banned in 2002.

CHRONOLOGY
The British decision to import Indian sugar workers in 1879–1916 dramatically changed Fijian society.

- ❏ **1970** Independence from Britain.
- ❏ **1987** Election win for Indo-Fijian coalition. Sitiveni Rabuka's coups secure minority ethnic Fijian rule. Ejected from Commonwealth.
- ❏ **1989** Mass Indo-Fijian emigration.
- ❏ **1990** Constitution discriminating against Indo-Fijians introduced.
- ❏ **1992** Rabuka wins legislative polls.
- ❏ **1997** Census shows ethnic Fijians outnumber Indo-Fijians. Fiji rejoins Commonwealth. New constitution.
- ❏ **1999** General election won by FLP. First Indo-Fijian prime minister.
- ❏ **2000** Civilian-led coup; new ethnic Fijian government.
- ❏ **2001** Nationalists win elections.

HEALTH ▷ Welfare state health benefits

1 per 1922 people   Cerebrovascular and heart diseases, cancers, accidents

Medical treatment is provided for all at a nominal charge. Fiji is free of almost all tropical diseases. Suicide is increasing, particularly among ethnic Indians.

SPENDING ▷ GDP/cap. increase

CONSUMPTION AND SPENDING

59 per 1000 population   106 per 1000 population

Defense 2.1%
Education 5.4%
Health 2.9%

Defense, Health, Education spending as % of GDP

Ostentatious displays of wealth are rare in Fiji; prestige derives from family and landholdings. The professional middle class, traditionally dominated by Indo-Fijians, is becoming more mixed.

WORLD RANKING

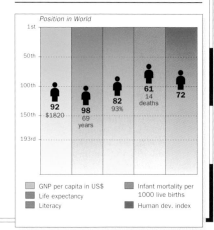

GNP per capita in US$
Life expectancy
Literacy
Infant mortality per 1000 live births
Human dev. index

FINLAND

OFFICIAL NAME: Republic of Finland **CAPITAL:** Helsinki **POPULATION:** 5.2 million
CURRENCY: Euro (markka until 2002) **OFFICIAL LANGUAGES:** Finnish and Swedish

BORDERED TO THE north and west by Norway and Sweden, and to the east by Russia, Finland is a low-lying country of forests and 187,888 lakes. Politics is based on consensus, and the country has been stable despite successive short-lived coalitions. Russia annexed Finland in 1809, ruling it until 1917, and subsequently Finland accepted a close relationship with the USSR as the price of maintaining its independence. It joined the EU in 1995 and, despite popular suspicion of Brussels bureaucracy, Finland was among the 12 EU states to adopt the euro from 2002.

CLIMATE
▷ Subarctic/continental

WEATHER CHART FOR HELSINKI

North of the Arctic Circle the climate is extreme. Temperatures fall to –30°C (–22°F) in the six-month winter and rise to 27°C (81°F) during the 73 days of summer midnight sun. In the south, summers are mild and short, winters are cold.

TRANSPORTATION
▷ Drive on right

Helsinki–Vantaa, 10m passengers

280 ships 1.62m grt

THE TRANSPORTATION NETWORK

| 49,853 km (30,977 miles) | 512 km (318 miles) |
| 5836 km (3626 miles) | 6715 km (4172 miles) |

The transportation system is well integrated. The railroad connects with the Swedish and Russian networks. There are frequent air services to most neighboring states, and links with Baltic states are being expanded. With one of the densest domestic networks in Europe, internal air travel is important, particularly north of the Arctic Circle.

With 187,888 lakes and a major river network, Finland has Europe's largest inland waterway system. It still carries freight, but is now used mainly for recreation. Finland's international ports handle around 70 million tonnes a year. Kotka is the chief export port. Helsinki, with five specialized harbors, handles most imports.

TOURISM
▷ Visitors : Population 1:1.9

2.71m visitors

Up 11% in 2000

MAIN TOURIST ARRIVALS

- Sweden 23%
- Russia 23%
- Germany 8%
- Estonia 8%
- UK 5%
- Other 33%

% of total arrivals

The scenery of the southern lakes and the vast forests of its Arctic north are Finland's main attractions. Helsinki is an important cultural center and hosts an annual arts festival. There are many first-class restaurants and its opera house has an international reputation. Most tourists try a sauna, a Finnish invention, and the local vodka, which is reputedly among the world's finest.

Visitors come largely from Sweden, Russia, the Baltic states, and Germany. By 2001 Finland emerged as the seventh most popular European destination for employers offering vacations as an incentive to their workforce.

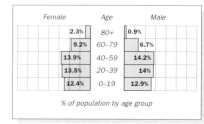

A summer's night at Kilpisjärvi, "The Way of the Four Winds," which lies at the point where Finland, Sweden, and Norway meet.

PEOPLE
▷ Pop. density low

Finnish, Swedish, Sami

17/km² (44/mi²)

THE URBAN/RURAL POPULATION SPLIT

67% 33%

RELIGIOUS PERSUASION

- Roman Catholic 1%
- Finnish Orthodox 1%
- Other 9%
- Evangelical Lutheran 89%

ETHNIC MAKEUP

- Other (including Sami) 7%
- Finnish 93%

Most Finns are of Scandinavian–Baltic extraction. Finnish belongs to the small Finno-Ugric linguistic group and is a legacy of the country's early Asian invaders. These tribes integrated with local and surrounding European peoples, but preserved their distinct language. Sami, also a Finno-Ugric language, is spoken by the small Sami population, who live above the Arctic Circle. Around 6% of the population speak Swedish, most of whom live in the southwestern coastal regions and on the Åland Islands.

More than 50% of Finns live in the five southernmost districts around Helsinki. Families tend to be close-knit, although divorce rates are high. The sauna is an integral part of everyday life; there are 1.5 million saunas among 5.2 million Finns.

Finnish women have a long tradition of political and economic participation. They were the first in Europe to get the vote, in 1906, and the first in the world able to stand for parliament. Almost 50% of women now work outside the home, and the president and one-third of the cabinet are female.

POPULATION AGE BREAKDOWN

Female	Age	Male
2.3%	80+	0.9%
9.2%	60–79	6.7%
13.9%	40–59	14.2%
13.5%	20–39	14%
12.4%	0–19	12.9%

% of population by age group

F

POLITICS ▷ Multiparty elections

1999/2003　　　President Tarja Halonen

Finland's constitution combines parliamentary government with a strong presidency. The external territory of the Åland Islands, inhabited by Swedish-speakers, has internal self-government.

PROFILE

Proportional representation has led to government by coalition, usually dominated by the SDP or KESK. The emphasis on consensus has favored stability but resulted in slow decision-making. The current "rainbow" coalition comprises parties from across the political spectrum.

AT THE LAST ELECTION
Parliament 200 seats

25% SDP	24% KESK	23% KOK	10% VL	6% SFP	6% G	6% Others

SDP = Social Democratic Party　KESK = Center Party
KOK = National Coalition Party　VL = Left-wing Alliance
SFP = Swedish People's Party　G = Greens

MAIN POLITICAL ISSUES
EU membership

Finland joined the EU in 1995, many Finns supporting entry as a way of identifying with western Europe. The small but influential farming community was hostile to membership. Others feared that welfare cuts would be more far-reaching if the economy were liberalized in line with EU expectations. In the event, after a dose of austerity, EU membership became associated with greater prosperity from the late 1990s. Finland's decision to join the eurozone in January 1999 contrasted with the nonmembership of its closest EU neighbor, Sweden.

Unemployment

The victory of SDP candidate Martti Ahtisaari in the 1994 presidential election was a sign of discontent with the conservative coalition then in power, whose handling of the recession resulted in record unemployment levels and welfare cuts. The 1995 general election led to the return of an SDP-led coalition which continued many of the previous government's austerity policies. Unemployment has since been substantially reduced.

Tarja Halonen,
who became Finland's
first female president
in 2000.

Prime Minister
Paavo Lipponen,
heads the SDP-led
coalition.

WORLD AFFAIRS ▷ Joined UN in 1955

CE	EU	OECD	OSCE	PfP

After carefully balancing its relations with the USSR and the West during the Cold War, Finland has now decided that its national interest lies with western Europe. In addition to joining the EU, it has observer status at the WEU. However, acknowledging historical and geographic realities, the government is also keen to maintain a special relationship with Russia.

AID ▷ Donor

$371m (donations)　　Down 11% in 2000

Finland's aid budget is still well below the UN target of 0.7% of GNP, despite vigorous campaigning. The main recipients are China, Bosnia and Herzegovina, and southern Africa.

CHRONOLOGY

Finland's history has been closely linked with the competing interests of Sweden and Russia.

❑ **1323** Treaty of Pähkinäsaari. Finland part of Swedish Kingdom.
❑ **1809** Treaty of Fredrikshamn, Sweden cedes Finland to Russia. Finland becomes a Grand Duchy enjoying considerable autonomy.
❑ **1812** Helsinki becomes capital.
❑ **1863** Finnish becomes an official language alongside Swedish.
❑ **1865** Grand Duchy acquires its own monetary system.
❑ **1879** Conscription law lays the foundation for a Finnish army.
❑ **1899** Czar Nicholas II begins process of Russification. Labor Party founded.
❑ **1900** Gradual imposition of Russian as the official language begins.
❑ **1901** Finnish army disbanded, Finns ordered into Russian units. Disobedience campaign prevents men being drafted into the army. ▷

FINLAND

Total Land Area :
357 050 sq. km
(130 127 sq. miles)

POPULATION
◎ over 100 000
○ over 50 000
● over 10 000

LAND HEIGHT
500m/1640ft
200m/656ft
Sea Level

F

251

CHRONOLOGY *continued*

- ❏ **1903** Labor Party becomes SDP.
- ❏ **1905** National strike forces restoration of 1899 status quo.
- ❏ **1906** Parliamentary reform. Universal suffrage introduced.
- ❏ **1910** Responsibility for important legislation passed to Russian Duma.
- ❏ **1917** Russian revolution allows Finland to declare independence.
- ❏ **1918** Civil war between Bolsheviks and right-wing government. Gen. Gustav Mannerheim leads government to victory at Battle of Tampere.
- ❏ **1919** Finland becomes republic. Kaarlo Ståhlberg elected president with wide political powers.
- ❏ **1920** Treaty of Tartu: USSR recognizes Finland's borders.
- ❏ **1921** London Convention. Åland Islands become part of Finland.
- ❏ **1939** August, Hitler–Stalin nonaggression pact gives USSR a free hand in Finland. November, Soviet invasion; strong Finnish resistance in ensuing Winter War.
- ❏ **1940** Treaty of Moscow. Finland cedes a tenth of national territory.
- ❏ **1941** Finnish troops join Germany in its invasion of USSR.
- ❏ **1944** June, Red Army invades. August, President Risto Ryti resigns. September, Finland, led by Marshal Mannerheim, signs armistice.
- ❏ **1946** President Mannerheim resigns, Juho Paasikivi president.
- ❏ **1948** Signs friendship treaty with USSR. Agrees to resist any attack on USSR made through Finland by Germany or its allies.
- ❏ **1952** Payment of $570 million in war reparations completed.
- ❏ **1956** Uhro Kekkonen, leader of the Agrarian Party, becomes president.
- ❏ **1956–1991** A series of coalition governments involving SDP and Agrarians, renamed KESK in 1965, hold power.
- ❏ **1981** President Kekkonen resigns.
- ❏ **1982** Mauno Koivisto president.
- ❏ **1989** USSR recognizes Finnish neutrality for first time.
- ❏ **1991** Non-SDP government elected. Austerity measures.
- ❏ **1992** January, signs ten-year agreement with Russia which, for first time since World War II, involves no military agreement.
- ❏ **1994** SDP candidate Martti Ahtisaari elected president.
- ❏ **1995** Finland joins EU. General election returns SDP-led coalition under Paavo Lipponen.
- ❏ **1999** General election returns Lipponen's coalition to power.
- ❏ **2000** Tarja Halonen elected as first woman president.
- ❏ **2002** Euro fully adopted. Fifth nuclear power plant approved.

DEFENSE

 ▷ Compulsory military service

$1.52bn | Down 7% in 2000

Finland is a neutral country. Its armed forces, the majority of whom are conscripts, are backed up by 500,000 active reservists and 3400 border guards. Russia's relative instability has reinforced concern about border security, the main defense issue. Finland participates in NATO's Partnerships for Peace program and has WEU observer status. Military service lasts for up to 12 months.

FINNISH ARMED FORCES

🛡	230 main battle tanks (70 T-55, 160 T-72)	24,550 personnel
🚢	10 patrol boats	5000 personnel
✈	64 combat aircraft (57 F/A-18C)	2700 personnel
	None	

ECONOMICS

 ▷ Inflation 1.9% p.a. (1990–2000)

$130bn | 1.0651–1.1231 euros

SCORE CARD

- ❏ World GNP Ranking..........................28th
- ❏ GNP per Capita $25,130
- ❏ Balance of Payments....................$8.89bn
- ❏ Inflation 3.4%
- ❏ Unemployment..................................10%

EXPORTS

France 5% | USA 7% | UK 9% | Sweden 9% | Germany 13% | Other 57%

IMPORTS

Netherlands 6% | UK 6% | Russia 9% | Sweden 14% | Germany 15% | Other 50%

STRENGTHS

Industry export- and quality-oriented. Large high-tech sector, especially Nokia mobile phones and Internet services. World leader in pulp and paper. Exports quick to recover from recession. Low inflation, averaging about 2% a year. Improved foreign investment incentives. Gateway to Russian and Baltic economies. Membership of eurozone. Strong recent economic growth. Finland named as world's most competitive economy by World Economic Forum in 2001.

WEAKNESSES

Severe recession in 1991–1993 when real GDP declined 15%. Rapidly aging population and low retirement age. High level of public and foreign debt. High unemployment. Small domestic market. Peripheral position in Europe.

PROFILE

Finland is a wealthy market economy. In the early 1990s it experienced the worst recession in 60 years, chiefly as

ECONOMIC PERFORMANCE INDICATOR

Consumer Price Index — GDP

a result of the collapse of the former Soviet Union. Russia took only 4% of Finland's exports in 1999, compared with over 25% to the Soviet Union before 1990.

A rapid rise in unemployment and business failures after 1990 pushed up government spending. The floating of the markka in 1992 and austerity measures improved competitiveness, overturned substantial fiscal deficits and allowed tax cuts in 2002.

Unemployment has fallen, but is still around 10%. Rapid growth in 2000 was followed by a slowdown in 2001. Finland was one of the 12 EU countries to adopt the euro in January 2002.

FINLAND : MAJOR BUSINESSES

- 🚢 Shipbuilding
- 🔌 Electronics
- ⚙ Light engineering
- ✱ Heavy engineering
- ⚡ Electrometallurgy
- 🌿 Pulp & paper
- 🐂 Agribusiness
- ⊗ Ceramics
- 🧪 Chemicals
- ✿ Textiles
- ■ Retail

Oulu
Vaasa
Tampere
Pori
Rauma
Turku
Jyväskylä
Lappeenranta
Lahti
Helsinki

0 — 100 km
0 — 100 miles

RESOURCES
 Electric power 14.9m kw

176,018 tonnes

Not an oil producer; refines 200,000 b/d

1.3m pigs, 1.09m cattle, 100,000 sheep, 6m chickens

Gold, copper, zinc, iron, lead, silver

ELECTRICITY GENERATION

Hydro 21% (15bn kwh)
Combustion 47% (33bn kwh)
Nuclear 31% (22bn kwh)
Other 1% (0.02bn kwh)

% of total generation by type

Finland's trees are its prime natural resource. Commercial forests cover 65% of the land, and wood products account for 30% of exports. Finland has no oil, but has significant hydroelectric resources. The high energy demands of industry are met chiefly by combustion and nuclear power. A fifth nuclear power plant was approved in 2002. Oil import costs have risen since 1990, when the collapse of the USSR ended a 42-year agreement on the exchange of Finnish manufactures for Soviet oil.

FINLAND : LAND USE

Cropland
Forest
Pasture
Reindeer
Barley

0 100 km
0 100 miles

ENVIRONMENT
 Sustainability rank: 1st

6%

10.3 tonnes per capita

ENVIRONMENTAL TREATIES

 Yes Yes Yes / Yes Yes Yes

Finland has strict laws on industrial emissions. Energy efficiency is a priority; nearly half of all homes are connected to district heating systems. Although there is opposition to the fifth nuclear plant, proponents argue that it will help the country meet its target emissions for greenhouse gases. The government is funding nuclear safety programs in Russia. Rising levels of pollution in the Baltic have given rise to concern.

MEDIA
 TV ownership high

Daily newspaper circulation 452 per 1000 people

PUBLISHING AND BROADCAST MEDIA

There are 200 daily newspapers. The most important are *Helsingin Sanomat*, *Aamulehti*, *Ilta-Sanomat*, *Turun Sanomat*, and *Kaleva*

3 services: 1 state-owned, 2 independent

5 services: 1 state-owned, 4 independent

Nine out of ten adult Finns read a daily paper, the world's third-highest per capita ratio. Regional papers dominate; the only national is the independent *Helsingin Sanomat*. There is no censorship, but the press shows restraint in criticizing the government.

CRIME
 No death penalty

2703 prisoners Up 7% in 1999

CRIME RATES

Murders
0.7 per 100,000 population

Rapes
10 per 100,000 population

Thefts
2619 per 100,000 population

The jump in unemployment in the early 1990s was seen as one of the causes of rising crime. There is concern about links with organized crime in Russia.

EDUCATION
 School leaving age: 16

99% 271,215 students

THE EDUCATION SYSTEM

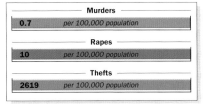
Primary 100%, Secondary 100%, Tertiary 83%

Compulsory education, introduced in 1921, lasts from seven to 16 years of age. By 1977/1978 the whole country had transferred to a new comprehensive system. Almost all children receive preschool education and also go on to three years of upper secondary education. Tough examinations mean that only 35% of entrants qualify to attend one of the 20 universities.

HEALTH
 Welfare state health benefits

1 per 323 people

Cerebrovascular and heart diseases, cancers, suicides

Of total government expenditure, 14% is spent on Finland's well-developed health system. Every Finn is legally guaranteed access to a local health center which is staffed by up to four doctors, as well as nurses and a midwife. Most non-hospital medical costs are covered by national health insurance; hospital fees are moderate. Diabetes and osteoporosis are increasing, and obesity is a growing health problem.

SPENDING
 GDP/cap. increase

CONSUMPTION AND SPENDING

403 per 1000 population

550 per 1000 population

Defense 1.3%
Education 7.5%
Health 5.2%

Defense, Health, Education spending as % of GDP

Income disparities are more marked in Finland than in other Scandinavian countries. However, the economic boom and labor shortages of the 1980s led to a sharp rise in all living standards. Personal consumption reached Swedish levels, and many families were able to take two vacations a year. Social security benefits were extended.

During the deep recession which began in 1990, this improvement was reversed. Wealth disparities widened and expenditure cuts led to lower social security benefits for the jobless. Those in work had to accept lower pay rises and higher taxes. Average real disposable incomes dropped sharply. The situation started to improve in 2000, with a temporary downturn in 2001.

Estonian immigrants form the poorest group in Finnish society.

WORLD RANKING

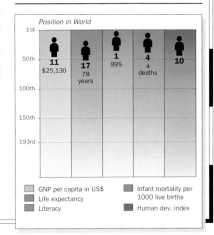

Position in World

11 $25,130
17 78 years
1 99%
4 4 deaths
10

GNP per capita in US$
Life expectancy
Literacy
Infant mortality per 1000 live births
Human dev. index

FRANCE

OFFICIAL NAME: French Republic   **CAPITAL:** Paris   **POPULATION:** 59.5 million
CURRENCY: Euro (French franc until 2002)   **OFFICIAL LANGUAGE:** French

 987 1919 July 14 F +1 +33 .fr

STRADDLING WESTERN EUROPE from the English Channel (La Manche) to the Mediterranean, France was Europe's first modern republic, and possessed a colonial empire second only to that of the UK. Today, it is one of the world's major industrial powers and its fourth-largest exporter. Industry is the leading economic sector, but the agricultural lobby remains powerful – French farmers will mount the barricades in defense of their interests. France's focus is very much on Europe. Together with Germany it was a founder member of the European Economic Community (EEC), and has supported successive steps to build a more closely integrated European Union. Paris, the French capital, is generally considered to be one of the world's most beautiful cities. Some of the most influential artists, writers, and filmmakers of the modern era have lived there.

Le Plessis-Bourré, Loire Valley. The region is famous for its many chateaux, which attract thousands of visitors every year.

CLIMATE ▷ Maritime/Mediterranean/mountain/continental

WEATHER CHART FOR PARIS

France's climate differs significantly from region to region. The northwest, in particular Brittany, is mild but damp. The east has hot summers and stormy winters, whereas summers in the south are dry and hot, and forest fires are a frequent occurrence.

TRANSPORTATION ▷ Drive on right

Charles de Gaulle, Paris
48.2m passengers

700 ships
1.5m grt

THE TRANSPORTATION NETWORK

893,300 km (555,070 miles)	9310 km (5785 miles)
31,589 km (19,629 miles)	14,932 km (9278 miles)

France led the world in high-speed train technology in 1983 with the TGV (*train à grande vitesse*) from Paris to Lyon. TGV lines have since been extended, and also link up with Belgium, Italy, Spain, the Channel Tunnel, and the Mediterranean; a Paris–Strasbourg line is planned for 2005. Traffic through Paris's two airports is set to double to 140 million by 2020, but plans for a third airport at Chaulnes, 130 km (80 miles) to the north, have become a political football.

TOURISM ▷ Visitors : Population 1.3:1

76.5m visitors

Up 1% in 2001

MAIN TOURIST ARRIVALS

Germany 21%	
UK & Ireland 16%	
Netherlands 16%	
Belgium & Luxembourg 12%	
Italy 9%	
Other 26%	

% of total arrivals (0 – 10 – 20 – 30 – 40)

France is the world's leading tourist destination, with over 76 million visitors a year. It ranks high as a destination for visitors from neighboring countries, particularly Germany and the UK. Most French people also prefer to take vacations in their own country, although many do visit Spain and Italy.

Paris is the most visited city in Europe. Its attractions include the Eiffel Tower, Nôtre Dame cathedral, Eurodisney, the Pompidou Center, and the Louvre, the world's most popular art museum.

The Côte d'Azur in the southeast became a byword for fashionable tourism when royalty and other notables flocked to resorts such as Nice at the end of the 19th century. Today Cannes hosts the world's leading film festival, and has a growing business convention trade. Other destinations throughout the country attract tourists for a variety of reasons such as wine production, historic and archaeological sites, and good beaches. There are resorts for skiing and hiking in the Alps and Pyrenees, and sailing off the varied coastline is also popular.

EUROPE

FRANCE

Total Land Area :
547 030 sq. km
(211 208 sq. miles)

POPULATION

◻ over 1 000 000
◉ over 100 000
○ over 50 000
● over 10 000

LAND HEIGHT

3000m/9843ft
2000m/6562ft
1000m/3281ft
500m/1640ft
200m/656ft
Sea Level

0 — 100 km
0 — 100 miles

F

PEOPLE   ▷ Pop. density medium

French, Provençal, German, Breton, Catalan, Basque

108/km²
(280/mi²)

There is a strong sense of national identity, and compulsory use of French has traditionally been promoted as a unifying force. The cultural traditions of Bretons, Flemings, Alsatians, Basques,

THE URBAN/RURAL POPULATION SPLIT

76%   24%

Occitans, Catalans, and Corsicans are now respected, but the Constitutional Court has struck down legislation on the use of regional language in government.

France encouraged immigration until the early 1970s, but laws have since become more restrictive. The 1999 census recorded 3.26 million resident foreigners (5.6% of the population) and a similar number of

POPULATION AGE BREAKDOWN

Female	Age	Male
2.7%	80+	1.2%
8.7%	60–79	7%
11.8%	40–59	11.8%
15%	20–39	15%
13.1%	0–19	13.7%

% of population by age group

RELIGIOUS PERSUASION

Buddhist 1%   Protestant 2%
Jewish 1%   Muslim 8%

Roman Catholic 88%

ETHNIC MAKEUP

Other (including Corsicans) 1%   German (Alsace) 2%
Breton 1%   North African (mainly Algerian) 6%

French 90%

foreign-born immigrants have been naturalized. Racist National Front (FN) propaganda has periodically whipped up anti-immigrant feeling, but there is also a strong current of youth solidarity among "black, blanc, beur" ("black, white, Arab"), spectacularly boosted in 1998 by the World Cup success of the multiracial national soccer team.

The Roman Catholic Church is still dominant, but there are sizable Muslim, Protestant, Buddhist, and Jewish minorities. Abortion and birth control were both legalized in the 1970s, despite strong Catholic opposition, and couples now commonly live together before marriage. Some two million unmarried couples of two or more years' standing, including gay couples, gained legal status with social and tax rights under 1998 legislation recognizing the civil solidarity pact.

Women did not get the vote until 1944. Although there was a woman prime minister, Edith Cresson, in 1991–1992, and women took five senior cabinet posts in the incoming Socialist-led government in 1997, the proportion of women in parliament was the second lowest in the EU until the introduction of a "parity" law in 2000, which requires an equal number of male and female candidates on party electoral lists.

F

F

CHRONOLOGY

The French Revolution of 1789–1794 overthrew a monarchy that had lasted for more than 800 years. It ushered in a period of alternating republicanism, Napoleonic imperialism, and monarchism. In 1870 the founding of the Third Republic established France firmly in the republican tradition.

❏ **1914–1918** 1.4 million Frenchmen killed in World War I.

❏ **1918–1939** Economic recession and political instability; 20 prime ministers and 44 governments.

❏ **1940** Capitulation to Germany. Puppet Vichy regime. Gen. de Gaulle leads "Free French" abroad.

❏ **1944** Liberation of France.

❏ **1946–1958** Fourth Republic. Political instability: 26 governments. Nationalizations. France takes leading role in EEC formation.

❏ **1958** Fifth Republic. De Gaulle president with strong powers.

❏ **1960** Most French colonies gain independence.

❏ **1962** Algerian independence after bitter war with France.

❏ **1966** France withdraws from NATO military command.

❏ **1968** General strike and riots over education policy and low wages. Assemblée Nationale dissolved; Gaullist victory in June elections.

❏ **1969** De Gaulle resigns after defeat in referendum on regional reform; replaced by Georges Pompidou.

❏ **1974** Valéry Giscard d'Estaing president. Center-right coalition.

❏ **1981** Left wins elections; François Mitterrand president.

❏ **1983–1986** Government U-turn on economic policy.

❏ **1986** Cohabitation between socialist president and new right-wing government led by Jacques Chirac. Privatization program introduced.

❏ **1988** Mitterrand wins second term. PS-led coalition returns.

❏ **1991** Edith Cresson becomes first woman prime minister.

❏ **1993** Center-right wins elections. Second period of cohabitation.

❏ **1995** Jacques Chirac president.

❏ **1995–1996** Controversial series of Pacific nuclear tests.

❏ **1996** Unpopular austerity measures to prepare economy for European monetary union.

❏ **1997** PS-led government takes office in reversed cohabitation.

❏ **1999** France introduces euro.

❏ **2000** 35-hour week becomes law.

❏ **2002** January, euro fully adopted. April–June, center-right victory in presidential and legislative elections. July, Chirac unharmed in assassination attempt.

POLITICS ▷ Multiparty elections

L. House 2002/2007
U. House 2001/2004

President Jacques
Chirac

AT THE LAST ELECTION

National Assembly 577 seats

4% PCF

62% UMP   24% PS   5% UDF   5% Others

UMP = Union for a Presidential Majority (Rally for the Republic – **RPR** and Liberal Democracy – **DL**)
PS = Socialist Party   **UDF** = Union for French Democracy
PCF = Communist Party of France   **UC** = Centrist Union
Rep = Republicans and Independents   **RCC** = Republicans, Communists, and Citizens   **RDSE** = European Democratic and Social Rally   **Ind** = Independents   **G** = Greens

Senate 321 seats

7% RCC   2% Ind

30% RPR   26% PS   16% UC   12% Rep   6% RDSE   1% G

PROFILE

France is a multiparty democracy. The constitution of the Fifth Republic, framed by Charles de Gaulle in 1958, ensures that the president has strong executive powers, but rules in tandem with a government and prime minister chosen by the Assemblée Nationale. Under changes agreed in 2000, the president and parliament are no longer elected according to separate timetables; the 2002 elections chose both for five-year terms. Traditionally the president attends to foreign policy and defense issues, while the government focuses on domestic and economic policy.

Apart from the two-term presidency of François Mitterrand of the PS (1981–1995), French presidents of the Fifth Republic have all been right-of-center. From 1986 onwards, however, both Mitterrand and his successor Jacques Chirac had to coexist for much of the time with a government and parliament dominated by political opponents. Thus Chirac, a conservative Gaullist, appointed Lionel Jospin of the PS as prime minister in 1997, heading a government coalition with the Communists, Greens, and Radical Socialists. In 2002, however, after Chirac won reelection as president in May, the sweeping victory of his UMP in legislative elections the following month marked a decisive rejection of so-called "cohabitation."

The far left has declined since 1945, when the PCF had 25% of the vote. The Greens suffered a serious setback in the 2002 elections, winning just three Assembly seats. The racist National Front (FN) failed to win any seats, but had created a major shock in the first round of the presidential poll when its leader Jean-Marie Le Pen finished second. Although the "democratic" parties united to ensure

that he was trounced by Chirac in the run-off, his score of almost 18% was still a high-water mark for the FN.

MAIN POLITICAL ISSUES
Liberal economic reform
The center-right program which brought electoral success in 2002 was built around promises of immediate income tax cuts and pro-business measures.

Racism and "exclusion"
Exploiting concerns about crime, urban violence, and unemployment, the racist right has several times turned such fears to its political advantage by blaming immigrants. Legislation on immigration has been tightened and the new government in June 2002 promised tough new measures on maintaining order. Inner-city deprivation and "exclusion" of the unemployed and homeless, although widely recognized as divisive, were issues on which the PS-led government of 1997–2002 failed to find effective policies.

European integration and globalization
Opposition to European integration has grown since the early 1990s, fueled by fears of losing French sovereignty, while opponents of "globalization" tap into similar concerns that French jobs and culture are under threat. With their country now part of the eurozone, the French have been encouraged to look for benefits in terms of stability and economic growth, but many still see giving up the franc as losing a symbolic part of their national identity. The most strident opponent of EU integration is the FN, but the center-right has also promised a stout defense of French interests, notably over agricultural reform.

Jacques Chirac, *president of France since 1995.*

Jean-Marie Le Pen, *leader of the far-right FN and presidential runner-up in 2002.*

Jean-Pierre Raffarin *was appointed prime minister by Chirac in 2002.*

WORLD AFFAIRS

▷ Joined UN in 1945

| EU | G8 | NATO | OECD | OSCE |

French foreign policy has followed two, apparently contradictory, strands since World War II – maintenance of a strongly independent line and furtherance of French interests within a united Europe. France's leading role within the EU is seen as a way of combining the two strands. Along with Germany, France called for greater European integration, and fully supports the EU's eastward expansion.

Relations with the UK have been troubled from time-to-time, notably over French health-related restrictions on the entry of UK farm products, and because illegal immigrants passing through France persistently use the Channel Tunnel to enter the UK.

France seeks to offset US dominance in both foreign affairs and culture. It left NATO's military command in 1966 and maintained an independent nuclear deterrent (which it insisted on testing in the Pacific in 1995–1996 despite a wave of international criticism). France has latterly not supported the US and the UK in a hard-line attitude to Saddam Hussein's regime in Iraq, but it has backed other recent US-led military campaigns, including the "war on terrorism." It also seeks to maintain its influence over its former empire and francophone regions in general.

AID

▷ Donor

$4.11bn (donations) — ⬇ Down 27% in 2000

France is one of the world's major aid donors. Its motives are not simply commercial; it also wishes to maintain the influence of the French language, particularly in west Africa, the main aid recipient. Médecins sans Frontières is part of a tradition of active involvement through NGO aid agencies.

DEFENSE

▷ No compulsory military service

$34.3bn — ⬇ Down 9% in 2000

FRENCH ARMED FORCES

809 main battle tanks (498 AMX-30B2, 311 *Leclerc*)	150,000 personnel	
1 carrier, 10 submarines, 1 cruiser, 3 destroyers, 30 frigates, & 39 patrol boats	45,600 personnel	
532 combat aircraft (352 *Mirage* F-1B/1C/1CR, and 120 *Jaguar*	63,000 personnel	
64 SLBM in 4 SSBN		

France was a founder member of NATO, but left its military command in

ECONOMICS

▷ Inflation 1.5% p.a. (1990–2000)

$1438bn — 1.0651–1.1231 euros

SCORE CARD

❏ WORLD GNP RANKING	5th
❏ GNP PER CAPITA	$24,090
❏ BALANCE OF PAYMENTS	$20.4bn
❏ INFLATION	1.7%
❏ UNEMPLOYMENT	10%

EXPORTS

USA 9%
Spain 9%
Italy 9%
UK 10%
Germany 14%
Other 49%

IMPORTS

Netherlands 7%
UK 8%
Italy 9%
Belgium 10%
Germany 17%
Other 49%

STRENGTHS

Engineering, reflected in the TGV and nuclear industries. Specializations such as cars (Citroën, Peugeot, and Renault) and telecommunications (Alcatel). Defense sector a major exporter, as are pharmaceutical and chemical industries. Success in attracting inward investment. Strong technocratic traditions: unlike in USA or UK, top graduates are attracted into engineering. Luxury goods, cosmetics, perfumes, and quality wines. Most agriculture well modernized; France is Europe's leading agricultural producer.

WEAKNESSES

High taxes, social charges, and labor costs. France is losing its positions in traditional industries such as iron and steel, metallurgy, and textiles. Some major high-tech industries, such as

ECONOMIC PERFORMANCE INDICATOR

— Consumer Price Index — GDP

telecommunications, run partly to further national pride, rather than on a strictly commercial basis.

PROFILE

At first slow to industrialize, protectionist France started competing in world markets and modernizing its industry in the 1950s and 1960s. Integration in western Europe, starting with the coal and steel industry in the 1950s, placed France at the heart of the EU. It was one of the 12 EU countries fully to adopt the euro in 2002. One of the world's top exporters, France's foreign trade balance runs a healthy surplus. It has a long tradition of state involvement in running the economy. Nationalization of key industries began in the late 1930s, with a fresh burst in 1981–1983, but since then both right-of-center and socialist governments have pursued privatization with vigor, reaching into the defense industry, aviation, banking, telecommunications, and insurance. Regional hubs are of growing economic significance. France is the EU's largest agricultural producer, and its farmers form a powerful political lobby. Active trade unions succeeded in getting a maximum 35-hour week introduced in 2000.

FRANCE : MAJOR BUSINESSES

✈	Aerospace
❋	Textiles
🍷	Chemicals
▣	Electronics
✿	Engineering
🍇	Wine
🚗	Vehicle assembly

0   200 km
0   200 miles

1966 in opposition to US domination. It maintained an independent nuclear deterrent through the Cold War, but went though a rapprochement with NATO in the 1990s. Joint participation between France and Germany in European army units is partly symbolic of reconciliation, as well as an expression of the need for an EU defense structure.

The influence of the army, which was once very strong, is now much diminished. Compulsory military service ended in 2001.

France has one of the world's largest and most export-oriented defense industries, producing its own tanks, jet fighter aircraft, and missiles.

F

RESOURCES

▷ Electric power 112m kw

845,649 tonnes

34,045 b/d (reserves 137m barrels)

42m turkeys, 23.5m ducks, 20.5m cattle, 230m chickens

Coal, oil, natural gas, iron, zinc

ELECTRICITY GENERATION

Hydro 13% (66bn kwh)	
Combustion 10% (49bn kwh)	
Nuclear 77% (388bn kwh)	
Other 0%	

% of total generation by type

F

France is the world's most committed user of nuclear energy, which provides over three-quarters of its electricity requirements. The policy reflects a desire for national energy self-sufficiency. Coal, once plentiful in the north and Lorraine, is now mostly exhausted, as are the gas fields off the southwest coast.

FRANCE : LAND USE

Forest
Pasture
Cropland
High mountain regions
Cattle
Sugarbeet
Vineyards

CORSICA

0	200 km
0	200 miles

ENVIRONMENT

▷ Sustainability rank: 33rd

14% (9% partially protected)

6.3 tonnes per capita

ENVIRONMENTAL TREATIES

	Yes		Yes		Yes
	Yes		Yes		Yes

Awareness of "green" issues has risen with a series of campaigns against major infrastructure projects. Nuclear power's importance, however, puts the environmentalist lobby in perspective. Transportation of oil by sea poses the threat of pollution of the Atlantic coast. Brittany's beaches and fisheries were severely affected by the wreck of the *Erika* in 1999. Severe storms devastated woodlands in 2000, focusing attention on the dangers of global warming.

THE IMPACT OF MODERN TRANSPORTATION

WHILE THE FRENCH REGIONS retain distinctive characteristics, the country's modern transportation and telecommunications network, with Paris at its hub, has brought the main provincial cities within easy reach of the capital. It also helps France to play a central role in European business.

THE AUTOROUTE NETWORK

The toll-charging *autoroutes*, or freeways, with a speed limit of 130 km/h (81 mph), now reach almost every corner of the map except the Cherbourg peninsula and Brittany in the northwest. They have relieved the previous generation of town-to-town highways of much of their former traffic congestion, but have also encouraged the upsurge in trans-European heavy freight vehicle movement, carrying through traffic to and from Spain, Italy, Switzerland, Germany, the Benelux countries, and the Channel ports. There has been much controversy about the environmental impact of several as yet uncompleted cross-country links, notably the E11 route south from Clermont-Ferrand to join the Mediterranean network.

FRENCH LEADERSHIP IN HIGH-SPEED TRAINS

French pride in leading-edge engineering, and a capacity for ambitious centralized planning and state-backed investment, have been apparent in its high speed train system, the *train à grande vitesse* (TGV). Rivaled only by the Japanese "bullet train" as the fastest in passenger service, the TGV can run at sustained speeds of over 300 km/h (186 mph). It has hit domestic airline traffic hard on some of the prime routes, by offering a combination of comfort and shorter door-to-door journey times in many instances. During the 1990s, some double-decker trains were introduced, to help meet high demand for seats, and a new generation of TGV trains was unveiled by the engineers Alstom in 1998–1999.

The first element was the TGV southeast. Construction began in 1975 and the Paris–Lyon section came into service in 1983, with subsequent extensions southward, culminating in the high-profile opening in 2001 of the direct service from Paris to Marseille. By the end of the 1980s a western route from Paris, initially to

FRANCE : ROAD AND RAIL

—— Autoroute
—— TGV route
—— Other rail

Le Mans, was also in service. The northern Paris–Lille route opened in 1993 with extensions to Brussels, Calais, and the Channel Tunnel.

EXTENDING THE TRAIN NETWORK

The existing TGV lines, which use mainly modernized but also some purpose-built track, are a highly profitable part of the French rail network. Seeing the likely economic benefits, many towns have lobbied hard to be included. The next planned line, however, from Paris east to Strasbourg, has faced more opposition as well as escalating cost estimates. Construction began in early 1999, aiming for completion in six to seven years. To make money, this line will need to provoke a big shift in existing traveling habits. The French government's commitment is also in part a strategic political decision, designed to reinforce Strasbourg's role "at the heart of European integration."

A link via Tours to Bordeaux is intended to be part of the next stage, along with further international links – from Lyon to Turin, from Marseille and Montpellier to Barcelona, and via Strasbourg to Stuttgart and Frankfurt and thus linking into the German high-speed ICE train network.

Modern office blocks in Montpellier, a city whose new dynamic image owes much to its transportation links and investment in communications technology.

MEDIA ▷ TV ownership high

📊 Daily newspaper circulation 218 per 1000 people

PUBLISHING AND BROADCAST MEDIA

There are 117 daily newspapers, including *Le Monde*, *Libération*, and *Le Figaro*. *Ouest-France* has the highest circulation

10 services: 5 state-controlled, 5 independent

7 services: 3 state-controlled, 4 independent

TV and radio were freed from direct state influence in the 1980s. Two of the main TV channels are still state-owned, but TF1 was privatized in 1987. Canal Plus mixes pay-per-view and advertising-backed services. A Breton-language station started up in 2000. Commercial channels have multiplied with the growth of satellite and cable.

The once innovative Minitel electronic communications system is now overshadowed by the Internet. Circulation of prestigious national newspapers has dwindled and regional papers too have suffered from a gradual shift to electronic media.

EDUCATION ▷ School leaving age: 16

👤 99% 🎓 2.09m students

THE EDUCATION SYSTEM

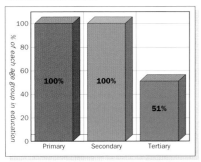

Education is highly centralized, a situation which is slowly generating a desire for greater flexibility. The education ministry organizes the curriculum, sets examinations, including the *baccalauréat*, and decides staffing issues. Roman Catholic schools, which take most of the 17% of privately educated children (but are not fee paying and receive large state subsidies), are the exception. However, they are still obliged to follow the national curriculum.

The focus in the classroom remains the acquisition of a broad range of knowledge. Pupils' academic records are impressive, despite frequent staff strikes.

CRIME ▷ No death penalty

🔲 46,376 prisoners ⬍ Little change in 1999

CRIME RATES

Murders
3 per 100,000 population

Rapes
14 per 100,000 population

Thefts
3849 per 100,000 population

The Code Napoléon, enacted in 1804 by Napoléon I, still forms the basis of French law. Criminal justice is based on inquisitorial rather than adversarial principles. The *juge d'instruction* has considerable powers to examine witnesses and assess evidence. The press are not restricted by *sub judice* rules in reporting trials. Political corruption cases, reaching into government, attract much attention.

Public concern about rising petty crime and violence has encouraged successive governments to promote tough policing. An "antiterrorism" bill was introduced in October 2001, giving police extra powers.

France has more than 70 universities – 13 in Paris – and higher education bodies with 1.2 million students. Entry is not competitive, but based on passing the secondary-level exam, the *baccalauréat*. Most students attend the university nearest to home. The universities have not been given the funds or staff to cope with the huge increase in student numbers in recent years. The 150 *Grandes Ecoles*, the most influential tertiary institutions, are outside the university system, and each takes just a few hundred carefully selected students. They groom the future governing elite, opening the way for their successful graduates to gain the top civil service and professional jobs.

Massif Central, Auvergne. *The Massif's lonely granite plateaus and extinct volcanoes are France's oldest rock formations.*

HEALTH ▷ Welfare state health benefits

🧑 1 per 333 people 💀 Heart and cerebro-vascular diseases, cancers, accidents

The French consume more medicines per capita than any other nation, and a significant number take medically approved, and prescribed, cures at health spas. French health care was rated the most efficient in the world by WHO in 2000. Under the national health system patients pay for treatment, and then get the majority of the cost reimbursed by an insurance company paid by the social services. Although health awareness has risen in recent years, a 1992 law banning smoking in public places is widely ignored and alcoholism remains a problem, with cirrhosis of the liver not uncommon as a cause of death.

SPENDING ▷ GDP/cap. increase

CONSUMPTION AND SPENDING

🚗 469 per 1000 population 📞 579 per 1000 population

Defense 2.6%
Education 5.9%
Health 7.3%

Defense, Health, Education spending as % of GDP

Wealth and income disparities in France are higher than in most OECD states. The Socialists narrowed the gap a little in the 1980s with the introduction of the legal minimum wage (*le SMIC*). Most tax is indirect – a result of a long French tradition of income-tax evasion. Major tax cuts announced in 2000 aimed to redress the imbalance of income tax on the rich and poor. The wealthy take exotic vacations to the Himalayas, the Andes, and Polynesia. The French lag behind their European neighbors in using the Internet; fewer than one-fifth of the population had done so by 2001.

WORLD RANKING

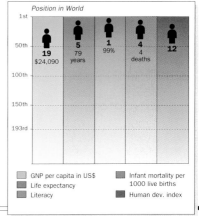

Position in World

19 $24,090 | 5 79 years | 1 99% | 4 4 deaths | 12

GNP per capita in US$
Life expectancy
Literacy
Infant mortality per 1000 live births
Human dev. index

F

GABON

OFFICIAL NAME: Gabonese Republic **CAPITAL:** Libreville
POPULATION: 1.3 million **CURRENCY:** CFA franc **OFFICIAL LANGUAGE:** French

WEST AFRICA

1960 1960 Aug 17 G +1 +241 .ga

AN EQUATORIAL COUNTRY on the west coast of Africa, Gabon's major economic activity is the production of oil. Only a small area of Gabon is cultivated, and more than two-thirds of it constitutes one of the world's finest virgin rainforests. Gabon became independent of France in 1960. A single-party state from 1968, it returned to multiparty democracy in 1990. Gabon's population is small, and the government is encouraging its increase.

CLIMATE
▷ Tropical equatorial

WEATHER CHART FOR LIBREVILLE

The climate is heavily equatorial – hot all year round with a long rainy season from October to May. The cold Benguela current lowers coastal temperatures.

TRANSPORTATION
▷ Drive on right

Léon M'Ba, Libreville
753,557 passengers

26,532 grt

THE TRANSPORTATION NETWORK

629 km (391 miles)	30 km (19 miles)
814 km (506 miles)	1600 km (994 miles)

The Trans-Gabon Railroad from Owendo port near Libreville to Massoukou is the key transportation link.

Air transportation is well developed, and most big companies have airstrips.

TOURISM
▷ Visitors : Population 1:8.4

155,000 visitors

Down 11% in 2000

MAIN TOURIST ARRIVALS

Europe 54%	
Africa 23%	
Americas 14%	
Other 9%	

% of total arrivals

Despite Libreville's many hotels, Gabon has little tourism, in part a reflection of its lack of good beaches.

PEOPLE
▷ Pop. density low

Fang, French, Punu, Sira, Nzebi, Mpongwe

5/km² (13/mi²)

THE URBAN/RURAL POPULATION SPLIT

81% 19%

ETHNIC MAKEUP

French 2%
European and other African 9%
Fang 35%
Eshira 25%
Other Bantu 29%

The largest ethnic group in Gabon is the Fang, who live mainly in the north. President Omar Bongo, from a subgroup of the Bateke in the southeast, has artfully united the common interests of other ethnic groups to keep the Fang from government. The Myene group around Port-Gentil consider themselves as the aristocrats of Gabonese society owing to their long-standing ex-colonial contacts. Oil wealth has led to the growth of a distinct bourgeoisie. Gabon is one of Africa's most urbanized countries.

POLITICS
▷ Multiparty elections

L. House 2001/2006
U. House 1997/2003

President Omar Bongo

AT THE LAST ELECTION

National Assembly 120 seats

5% RNB 2% PGP
71% PDG 19% Vacant 3% Others

PDG = Gabonese Democratic Party **RNB** = National Rally of Woodcutters **PGP** = Gabonese Progress Party

Senate 91 seats

The Senate is indirectly elected by regional councils

Gabon has had a constitution enshrining a multiparty system since 1990, when elections confirmed in power the former sole ruling party – Omar Bongo's PDG. Bongo, in power since 1967, won the first multiparty presidential poll in 1993, but its fairness was widely disputed. He was reelected president in 1998, this time for a seven-year term. In parliamentary elections in December 2001, the PDG won a majority in the National Assembly, as they had in 1996.

WORLD AFFAIRS
▷ Joined UN in 1960

FZ G24 AU OIC ACP

Gabon maintains close links with France, although US companies are also making inroads into Gabon's oil-rich economy. Gabon remains influential regionally (the Gulf of Guinea Commission was launched in Libreville in 2001), and relations further afield, particularly with OPEC, are also important.

GABON

Total Land Area : 267 667 sq. km (103 346 sq. miles)

POPULATION
⊙ over 100 000
● over 10 000
· under 10 000

LAND HEIGHT
500m/1640ft
200m/656ft
Sea Level

AID

 Recipient

 $12m (receipts)   Down 75% in 2000

France is by far the major aid donor, providing two-thirds of total receipts. For a middle-income country with one of the highest GNPs per capita in the developing world, Gabon has benefited from considerable aid. Its indebtedness is the result of excessive borrowing, which was encouraged by Western banks in the 1970s. Much aid goes to servicing this debt.

DEFENSE

 No compulsory military service

 $123m   Down 9% in 2000

President Bongo's background in the military is reflected in Gabon's large defense budget and prestige weaponry, which includes French *Mirage* jets. France guarantees Gabon's security and keeps a 750-strong garrison in Libreville. A recruitment drive was launched in April 2001 to recruit 1500 18- to 25-year-olds by 2006.

ECONOMICS

 Inflation 6.2% p.a. (1990–2000)

 $3.93bn   698.7–736.7 CFA francs

SCORE CARD

❏ World GNP Ranking	118th
❏ GNP per Capita	$3190
❏ Balance of Payments	$385m
❏ Inflation	1.5%
❏ Unemployment	21%

STRENGTHS
Oil and a relatively small population give Gabon a high per capita GNP. The country's other abundant resources – including some of the world's best tropical hardwoods – are just beginning to be tapped.

WEAKNESSES
Large debt burden incurred in the 1970s. Continuing dependence on French technical assistance.

EXPORTS

Netherlands Antilles 4%
China 8%
South Korea 4%
Other 17%
USA 50%
France 17%

IMPORTS

Netherlands 2%
Belgium 4%
Japan 2%
USA 5%
Other 22%
France 65%

RESOURCES

 Electric power 379,000 kw

 53,440 tonnes   301,000 b/d (reserves 2.5bn barrels)

213,000 pigs, 198,000 sheep, 3.2m chickens   Oil, manganese, uranium, gold, iron, natural gas

Oil is the major export earner. Gabon also has large deposits of uranium and over 100 years' reserves of manganese. The unexploited iron ore deposits at Bélinga are the world's largest.

ENVIRONMENT

 Sustainability rank: 36th

 3%   2.4 tonnes per capita

The Trans-Gabon Railroad has sliced through one of the world's finest virgin rainforests and has opened the interior to indiscriminate exploitation of rare woods such as oleoirme. Gabon abandoned plans for nuclear power following the 1986 Chernobyl disaster.

MEDIA

 TV ownership high

 Daily newspaper circulation 30 per 1000 people

PUBLISHING AND BROADCAST MEDIA

There are 2 daily newspapers, *L'Union* and *Gabon-Matin*

3 services: 1 state-owned, 2 independent

7 services: 2 state-controlled, 5 independent

The media are mostly government-controlled. There was a crackdown in 1998 on independent media – which in the 1990s had become quite diverse – raising concerns about freedom of expression.

CRIME

 Death penalty in use

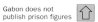

Gabon does not publish prison figures

Recorded crime rose sharply from 1992–1996

Urban crime rates are rising. A campaign to combat child trafficking was launched in 2001.

Gabon's human rights record has improved in recent years.

Albert Schweitzer Hospital, *Lambaréné, on the lower Ogooué River. Schweitzer won a Nobel Prize for his pioneering work in Africa.*

G

EDUCATION

 School leaving age: 16

 71%   7473 students

Education follows the French system. Université Omar Bongo in Libreville, founded in the 1970s, now has more than 4000 students.

HEALTH

 No welfare state health benefits

1 per 5000 people   Heart and diarrheal diseases, pneumonia, accidents

Oil revenues have allowed substantial investment in the health service, which is now among the best in Africa.

SPENDING

GDP/cap. increase

CONSUMPTION AND SPENDING

22 per 1000 population
32 per 1000 population

Defense 2.2%
Education 3.3%
Health 2.1%

0   5   10   15   20   25
Defense, Health, Education spending as % of GDP

Oil wealth has led to the growth of an affluent bourgeoisie. Menial and low-income jobs are done by immigrant workers.

WORLD RANKING

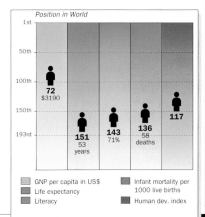

Position in World

1st
50th
100th
150th
193rd

72 $3190

151 53 years

143 71%

136 58 deaths

117

GNP per capita in US$
Life expectancy
Literacy
Infant mortality per 1000 live births
Human dev. index

GAMBIA

OFFICIAL NAME: Republic of the Gambia **CAPITAL:** Banjul
POPULATION: 1.34 million **CURRENCY:** Dalasi **OFFICIAL LANGUAGE:** English

WEST AFRICA

| 1965 | 1965 | Feb 18 | WAG | 0 | +220 | .gm |

A NARROW COUNTRY on the western coast of Africa, the Gambia was renowned as a stable democracy until an army coup in 1994. Agriculture accounts for 65% of GDP, yet many Gambians are leaving rural areas for the towns, where average incomes are four times higher. Its position as an enclave within Senegal seems likely to endure, following the failure of an experiment in federation in the 1980s.

CLIMATE ▷ Tropical wet & dry

WEATHER CHART FOR BANJUL

The subtropical and sunny dry season is punctuated by intermittent hot *harmattan* winds.

TRANSPORTATION ▷ Drive on right

Yundum International, Banjul
321,126 passengers

8 ships
1884 grt

THE TRANSPORTATION NETWORK

| 956 km (594 miles) | None |
| None | 400 km (249 miles) |

The Gambia River carries more traffic than the roads – ships of up to 3000 tonnes can reach Georgetown. Yundum airport was upgraded by NASA in 1989 for US space shuttle emergency landings.

TOURISM ▷ Visitors : Population 1:14

96,000 visitors

Up 5% in 1999

MAIN TOURIST ARRIVALS

UK 41%
Germany 24%
Netherlands 12%
Other 23%

% of total arrivals

Ecotourism is being developed, but most of those enjoying the beaches and resort hotel life are Europeans escaping winter, including many single women.

PEOPLE ▷ Pop. density medium

Mandinka, Fulani, Wolof, Jola, Soninke, English

134/km² (346/mi²)

THE URBAN/RURAL POPULATION SPLIT

33% 67%

ETHNIC MAKEUP

Other 5%
Mandinka 42%
Serahuli 9%
Jola 10%
Wolof 16%
Fulani 18%

Until the 1994 coup, led by Lt. Col. Yahya Jammeh, the 1962–1994 government of President Sir Dawda Jawara had tried to offset minority resentment of the Mandinka's domination of politics, distributing political offices fairly according to ethnic origins. Jammeh, a fervent Muslim, is from the minority Jola (or Diola) community, numerous across the border in Senegal, where they are active in a local rebellion. About 85% of Gambians follow Islam, although there is no official state religion. There is a yearly influx of migrants, who come from Senegal, Guinea, and Mali to trade in groundnuts. The Gambia is still a very poor country, with 80% of the labor force engaged in agriculture. Women are active as traders in an otherwise male-dominated society.

Fishing village. *Overfishing in the waters off the Gambia and Senegal, mainly by distant nations, is a growing problem.*

POLITICS ▷ Multiparty elections

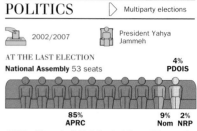

2002/2007

President Yahya Jammeh

AT THE LAST ELECTION

National Assembly 53 seats

4% PDOIS

85% APRC
9% Nom
2% NRP

APRC = Alliance for Patriotic Reorientation and Construction **Nom** = Nominated **PDOIS** = People's Democratic Organization for Independence and Socialism **NRP** = National Reconciliation Party

The People's Progressive Party (PPP) was in government from 1962 until 1994, for most of which time the Gambia was one of Africa's few democracies. A ban, imposed after the 1994 coup on the PPP and the three main parties which had been in opposition, was lifted in 2001 in time for elections.

During the army's coup, Sir Dawda Jawara took refuge aboard a visiting US warship, and he then went into exile in Britain. The coup's leaders claimed that it had been initiated in a bid to end corruption and pledged to preserve democracy. In the new government several portfolios went to civil servants who had served in the Jawara administration. Military leader Yahya Jammeh was elected president in controversial elections in September 1996, and the following January his APRC won the majority of seats in a parliamentary election. Jammeh and the APRC were reelected in 2001–2002; the PPP won no seats.

WORLD AFFAIRS ▷ Joined UN in 1965

| CILSS | Comm | ECOWAS | AU | OIC |

Commonwealth and Western criticism of the 1994 coup which brought President Jammeh to power has softened over recent years, but an apparent crackdown on opponents following Jammeh's reelection in 2001 reawoke international concern. Relations with Senegal have been troublesome since the end of federation in 1989.

AID ▷ Recipient

$49m (receipts)

Up 48% in 2000

Western aid flows, suspended after the 1994 coup, have largely resumed. The World Bank, the IMF, the AfDB, the UK, the US, Japan, Libya, Germany, the Gulf states, the UN, and the EU are all significant donors.

G

GAMBIA

Total Land Area : 11 300 sq. km
(4363 sq. miles)

LAND HEIGHT
100m/328ft
Sea Level

POPULATION
over 10 000 ●
under 10 000 •

DEFENSE

▷ No compulsory military service

💲 $15m   ⬇ Down 6% in 2000

The Gambia National Army, with two infantry battalions, takes about half of the defense budget; the rest finances the 600-strong gendarmerie. Most arms are bought from the UK, although supplies are now increasingly coming from Nigeria too. A defense pact with Senegal collapsed along with the federation in 1989.

ECONOMICS

▷ Inflation 4.1% p.a. (1990–2000)

📊 $440m   💱 15.28–17.54 dalasis

SCORE CARD

❑ WORLD GNP RANKING	172nd
❑ GNP PER CAPITA	$340
❑ BALANCE OF PAYMENTS	–$48m
❑ INFLATION	0.8%
❑ UNEMPLOYMENT	Widespread underemployment

STRENGTHS

Low tariffs make the Gambia a focus of regional trade. Natural deepwater harbor at Banjul, one of the finest on the west African coast. Well-managed economy, favorably viewed by donors.

WEAKNESSES

Small size of market can inhibit investment. Smuggling: deprives government of significant revenues. Lack of resources, little agricultural diversification; consequent over-reliance on groundnuts, the main crop.

EXPORTS

IMPORTS

RESOURCES

▷ Electric power 29,000 kw

🐟 30,004 tonnes   🛢 Not an oil producer

🐄 365,000 cattle, 145,000 goats, 780,000 chickens   💎 Ilmenite, zirconium, rutile, kaolin, tin, oil

The Gambia River is one of Africa's few good waterways, but it is underused owing to its separation from its natural hinterland by the Gambia–Senegal border. Irrigation is at present provided by a single dam; plans for further dams for power generation have met with opposition. Oil deposits are believed to exist offshore.

ENVIRONMENT

▷ Sustainability rank: 103rd

🌲 2%   ⬍ 0.2 tonnes per capita

The impact of tourism and of overfishing in Gambian waters are major concerns, as are desertification and deforestation.

MEDIA

▷ TV ownership low

❌ Daily newspaper circulation 2 per 1000 people

PUBLISHING AND BROADCAST MEDIA

📰	There are 2 daily newspapers, the independent *Daily Observer* and the government-owned *Gambia Daily*
📺 1 state-owned service	📻 9 services: 1 state-owned 8 independent

The independent media are severely restricted. Radio news broadcasts are dominated by state-controlled Radio Gambia, and the government runs the only national television station.

CRIME

▷ Death penalty not used in practice

🏛 478 prisoners   ⬆ General crime levels are low, but rising

Crime levels are relatively low in what is a peaceful society compared with many other states in the region.

EDUCATION

▷ Schooling is not compulsory

📖 37%   🎓 1702 students

The aims are to increase enrollment to 75% in primary and 20% in secondary schools, and to improve teacher quality. A university was established in 1998.

CHRONOLOGY

Mandinka traders brought Islam in the 13th century and were the main influence until the 18th century. The 1700s and 1800s saw colonial rivalry between Britain and France.

- ❑ **1888** British possession.
- ❑ **1959** Dawda Jawara founds PPP.
- ❑ **1965** Independence from Britain.
- ❑ **1970** Republic; Jawara president.
- ❑ **1982–1989** Federation with Senegal.
- ❑ **1994** Jawara ousted in army coup.
- ❑ **1996** Yahya Jammeh wins presidential election.
- ❑ **2000** Military coup foiled.
- ❑ **2001** $2 million antipoverty program launched by government.
- ❑ **2002** Jammeh's party sweeps parliamentary elections.

G

HEALTH

▷ No welfare state health benefits

👥 1 per 20,000 people   ☠ Malaria, tuberculosis, parasitic diseases

Most people have access to basic medicines, but these are no longer free. Advanced medical care in the public sector is limited. An HIV/AIDS awareness campaign was launched in 2002.

SPENDING

▷ GDP/cap. increase

CONSUMPTION AND SPENDING

🚗 8 per 1000 population   📞 26 per 1000 population

Defense 3.2%
Education 4.8%
Health 2.3%

0   5   10   15   20   25
Defense, Health, Education spending as % of GDP

Public service and the professions have created wealth and some people are comfortably off, but great wealth is not a feature of Gambian life. Unemployed young men in Banjul are regarded as the underclass.

WORLD RANKING

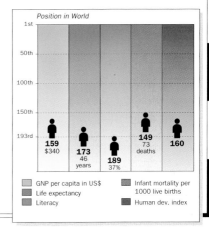

Position in World

	GNP per capita in US$	Life expectancy	Literacy	Infant mortality	Human dev. index
	159 $340	**173** 46 years	**189** 37%	**149** 73 deaths	**160**

GNP per capita in US$ Infant mortality per 1000 live births
Life expectancy
Literacy   Human dev. index

GEORGIA

OFFICIAL NAME: Georgia **CAPITAL:** Tbilisi
POPULATION: 5.2 million **CURRENCY:** Lari **OFFICIAL LANGUAGE:** Georgian; Abkhazian (in Abkhazia)

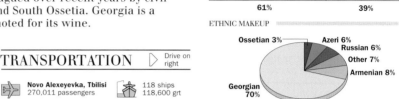

| 1991 | 1991 | May 26 | GE | +4 | +995 | .ge |

SANDWICHED BETWEEN the Greater and Lesser Caucasus, Georgia is a mountainous country, with a Black Sea coastline running north–south from Abkhazia to Ajaria. Georgia was one of the first republics to demand independence from the Soviet Union, but has been plagued over recent years by civil war and ethnic disputes in Abkhazia and South Ossetia. Georgia is a primarily agricultural country, and is noted for its wine.

CLIMATE
Mountain/subtropical

WEATHER CHART FOR TBILISI

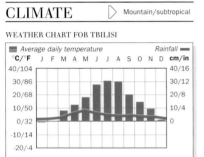

Georgia's climate is continental inland and subtropical along the coast, where grapes, citrus fruit, and tea are grown.

Tbilisi, Georgia's capital since the 5th century. Its buildings rise in steep terraces from both banks of the Kura River.

TRANSPORTATION
Drive on right

Novo Alexeyevka, Tbilisi
270,011 passengers

118 ships
118,600 grt

THE TRANSPORTATION NETWORK

19,354 km (12,026 miles)	None
1545 km (960 miles)	None

Civil war has seriously disrupted transportation. A new rail route and oil pipeline from Baku to the Black Sea ports of Poti and Supsa was opened in 1999.

TOURISM
Visitors : Population 1:14

384,000 visitors

Up 21% in 1999

MAIN TOURIST ARRIVALS

CIS 57%
Turkey 26%
USA 2%
Other 15%
% of total arrivals

The volatile political situation has discouraged tourism, but numbers are rising again. Most tourists still come from former Soviet states.

PEOPLE
Pop. density medium

Georgian, Russian, Abkhazian

75/km² (193/mi²)

THE URBAN/RURAL POPULATION SPLIT

61%   39%

ETHNIC MAKEUP

Ossetian 3%   Azeri 6%
Russian 6%
Other 7%
Armenian 8%
Georgian 70%

Georgia is a paternalistic society, with strong family and cultural traditions. Approximately 70% of the population is Georgian, with Armenian, Russian, Azeri, Ossetian, Greek, and Abkhaz minorities. More than 300,000 people were displaced by the violent internal conflicts of the 1990s.

POLITICS
Multiparty elections

1999/2003

President Eduard Shevardnadze

AT THE LAST ELECTION
Parliament 235 seats

| 56% CUG | 24% AGUR | 7% ISG | 5% AD | 8% Others |

CUG = Citizens' Union of Georgia **AGUR** = All Georgian Union of Revival **ISG** = Industry Will Save Georgia **AD** = Abkhazian Deputies

10 Abkhazian seats were not contested in 1999.

The political situation remains volatile. Eduard Shevardnadze, reelected as president in 2000, has been the target of several assassination attempts. He came to power in 1992 amid civil war with the "Zviadists," supporters of the ex-president Zviad Gamsakhurdia, who committed suicide while under fire in late 1993.

Fighting raged simultaneously in Abkhazia, where ethnic Abkhaz attempted to secede; ethnic Georgians were expelled. Fighting still flares up sporadically. A UN-brokered peace process begun in mid-2000 soon stalled over the basic issue of the future status of Abkhazia.

GEORGIA

Total Land Area : 69 700 sq. km (26 911 sq. miles)

POPULATION
☐ over 1 000 000
◎ over 100 000
○ over 50 000
● over 10 000
• under 10 000

LAND HEIGHT
3000m/9843ft
2000m/6562ft
1000m/3281ft
500m/1640ft
200m/656ft
Sea Level

G

WORLD AFFAIRS
▷ Joined UN in 1992

BSEC · CIS · CE · OSCE · PfP

Relations with Russia are strained over regional security and pipelines. Georgia is a focus of the US "war on terrorism."

AID
▷ Recipient

 $170m (receipts)   Down 31% in 2000

As well as aid for infrastructure projects, Georgia receives Western support for institutional and financial sector reform.

DEFENSE
▷ Compulsory military service

 $116m   Up 5% in 2000

The army's main focus remains Abkhazia. However, US concerns over the presence of Islamic terrorists prompted the arrival in 2002 of US forces to train the regular Georgian army; the security service is already CIA-trained. Russian troops, once numbering over 9000, have been steadily withdrawn since 2000.

ECONOMICS
▷ Inflation 389% p.a. (1990–2000)

$3.18bn   1.960–2.055 lari

SCORE CARD

❏ World GNP Ranking........................129th
❏ GNP per Capita $630
❏ Balance of Payment ....–$162m
❏ Inflation 19.1%
❏ Unemployment..................................15%

Strengths
Gateway to West for Azeri oil through pipelines across Georgia to Black Sea and Mediterranean ports. Hyperinflation brought under control in mid-1990s.

Weaknesses
War damage and severance of links with other ex-Soviet republics. Large black economy and influential Mafia. Drought and currency crisis in 1998. Serious budget deficit problems. Negative trade balance.

EXPORTS

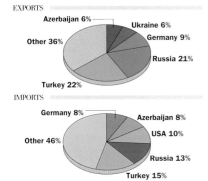

Azerbaijan 6%
Ukraine 6%
Germany 9%
Russia 21%
Other 36%
Turkey 22%

IMPORTS

Germany 8%
Azerbaijan 8%
USA 10%
Other 46%
Russia 13%
Turkey 15%

RESOURCES
▷ Electric power 4.6m kw

 1600 tonnes
 1.18m cattle, 545,000 sheep, 7.83m chickens

 2386 b/d (reserves 37m barrels)
 Manganese, coal, oil, natural gas, zinc, cobalt, vanadium

Known oil reserves are as yet barely developed. Georgia is dependent on Russia for much of its energy supply, although a new US–Georgian oil refinery was opened in eastern Georgia in 1998. Georgia is a predominantly agricultural country, and food processing and wine production continue to be the major industries. Manganese and small quantities of zinc, cobalt, and vanadium are mined.

ENVIRONMENT
▷ Not available

 3%   1 tonne per capita

Radiation from materials left by departing Russian soldiers is a growing problem, as is Black Sea pollution and the protection of upland pastures.

MEDIA
▷ TV ownership high

 Daily newspaper circulation figures not available

PUBLISHING AND BROADCAST MEDIA

 There are 3 daily newspapers, *Rezonansi*, published in Georgian, *Georgian Messenger* and *Georgian Times* published in English

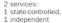 2 services: 1 state-controlled 1 independent   1 state-controlled service

Government newspapers are subsidized. The independent TV channel, Rustavi–2, was permitted to reopen in 1998, but has faced harassment.

CRIME
▷ No death penalty

 10,158 prisoners   Up 7% 1996–1998

Organized crime under the control of Mafia-style groups has flourished since independence in 1991. The judicial system currently favors Shevardnadze and his supporters.

EDUCATION
▷ School leaving age: 14

 99%   128,372 students

Since independence, education has stressed Georgian language and history. All levels of education are seriously underfunded. Tbilisi University was formerly of a high standard.

HEALTH
▷ Welfare state health benefits

1 per 227 people   Circulatory and respiratory diseases, cancers, accidents

The health system was limited under the control of the Soviet Union. Internal strife and a lack of resources have prevented any recent investment.

CHRONOLOGY
A Russian protectorate from 1763, Georgia was absorbed into the Russian Empire in 1801. It was established as an independent state under a Menshevik socialist government in 1918.

❏ **1920** Recognized by Soviet Russia as an independent state.
❏ **1921** Soviet Red Army invades. Effectively part of USSR.
❏ **1922–1936** Incorporated into Transcaucasian Soviet Federative Socialist Republic (TSFSR).
❏ **1989** Pro-independence riots in Tbilisi put down by Soviet troops.
❏ **1990** Declares sovereignty.
❏ **1991** Independence. Zviad Gamsakhurdia elected president.
❏ **1992** Gamsakhurdia flees Tbilisi. Shevardnadze elected chair of Supreme Soviet and State Council.
❏ **1992–1993** Abkhazia conflict.
❏ **1995** Shevardnadze narrowly survives assassination attempt, subsequently elected president.
❏ **1999** Opening of pipeline from Caspian to Black Sea.
❏ **2000** Shevardnadze reelected. Russian troop withdrawal begins.

G

SPENDING
▷ GDP/cap. decrease

CONSUMPTION AND SPENDING

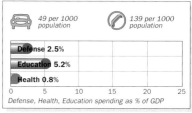

49 per 1000 population   139 per 1000 population

Defense 2.5%
Education 5.2%
Health 0.8%

0   5   10   15   20   25
Defense, Health, Education spending as % of GDP

There is a small wealthy and extravagant elite, but most Georgians live in poverty. Wages and welfare are often in arrears.

WORLD RANKING

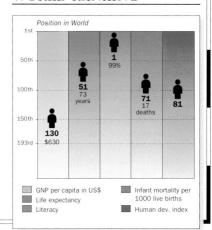

Position in World

1st
50th
100th
150th
193rd

130 $630
51 73 years
1 99%
71 17 deaths
81

GNP per capita in US$ Infant mortality per 1000 live births
Life expectancy   Human dev. index
Literacy

GERMANY

OFFICIAL NAME: Federal Republic of Germany **CAPITAL:** Berlin **POPULATION:** 82 million
CURRENCY: Euro (deutsche mark until 2002) **OFFICIAL LANGUAGE:** German

WITH COASTLINES on both the Baltic and North Seas, Germany is bordered by nine countries. Plains and rolling hills in the north give way to more mountainous terrain in the south. Europe's foremost industrial power, and its most populous country apart from Russia, Germany is the world's second-biggest exporter. Unified in the 1870s, it was divided after the defeat of the Nazi regime in 1945. The communist-ruled east was part of the Soviet bloc until the collapse of the East German regime in 1989, which paved the way for reunification in 1990. Tensions created by wealth differences between east and west were then exacerbated by record levels of unemployment. The government committed itself to European union and adopted the single currency, the euro, even though the stable deutsche mark had been a symbol of German pride.

GERMANY

Total Land Area : 357 021 sq. km
(137 846 sq. miles)

POPULATION

- ▣ over 1 000 000
- ◉ over 500 000
- ○ over 100 000
- ● over 10 000

LAND HEIGHT

- 2000m/6562ft
- 1000m/3281ft
- 500m/1640ft
- 200m/656ft
- Sea Level

CLIMATE ▷ Continental/maritime

WEATHER CHART FOR BERLIN

Germany has a broad climatic range. The upper Rhine valley is very mild and suitable for wine making. The Bavarian Alps, the Harz Mountains, and the Black Forest are by contrast cold, with heavy falls of snow in winter.

TRANSPORTATION ▷ Drive on right

Frankfurt/Main International
49.4m passengers

994 ships
6.55m grt

THE TRANSPORTATION NETWORK

650,891 km (404,444 miles)	11,515 km (7155 miles)
37,559 km (23,339 miles)	7467 km (4640 miles)

Germany virtually invented the modern highway with the first *Autobahnen* in the 1930s. These have since become Europe's most elaborate highway network; there are generally no tolls and few speed limits, despite protests from environmentalists. The efficient railroad system has been restructured as a first step toward privatization. Germany's high-speed ICE railroad opened its main north–south routes in 1991 and has expanded greatly since then. Urban transportation systems are highly efficient.

TOURISM ▷ Visitors : Population 1:4.6

17.9m visitors   Down 6% in 2001

MAIN TOURIST ARRIVALS

Northerly beaches and a colder climate make Germany less of a tourist destination than France or Italy. Skiing in the Bavarian Alps, the historic castles of the Rhine valley, the Black Forest, and Germany's excellent beer are all major attractions. Berlin, even before 1989, drew tourists with its rich cultural life and its Wall separating capitalist West and communist East. Now capital of the reunified Germany, and with a dynamic and vibrant atmosphere, it has undergone massive reconstruction.

The Stillach Valley, Allgäu Alps, Bavaria (Bayern). Alarm at acid rain damage to Germany's forests sparked off the rise of the Greens, who joined the government in 1998.

PEEPLE

▷ Pop. density high

German, Turkish

235/km² (608/mi²)

THE URBAN/RURAL POPULATION SPLIT

88% 12%

RELIGIOUS PERSUASION

- Muslim 3%
- Other 30%
- Protestant 34%
- Roman Catholic 33%

ETHNIC MAKEUP

- Turkish 2%
- Other European 3%
- Other 3%
- German 92%

The majority of German-speakers live in Germany itself, although Austria and the greater part of Switzerland are German-speaking, as are parts of eastern France and northern Italy. Germans share a common language, but they speak it in a variety of dialects, reflecting a strong sense of regionalism. The north is still largely Protestant, while the south and southwest, particularly Bavaria (Bayern), have strong Catholic traditions.

The large immigrant population now totals some 7.3 million, the 2.1 million Turks forming the largest single group. *Gastarbeiter* (guest workers) recruited from the mid-1950s to mid-1970s provided much of the labor on which the former West Germany's economic recovery was built. Legislation in 1999 improved the rights of their children to obtain German nationality.

Germany's once liberal asylum laws were tightened in 1993 in response to a new influx. Ethnic Germans enjoyed privileged access ahead of others fleeing westward after the collapse of communism. Extreme right-wing parties sought to exploit anti-immigrant feeling among disaffected groups including the young unemployed. In 2002 the government relaxed immigration quotas for skilled workers, recognizing the need for specific talents, but concurrently toughened the asylum procedure.

Family ties in Germany are little different from those in the US or the UK. Millions of couples live together in common-law arrangements, although this is frowned on by the Roman Catholic Church. In rural districts, notably in Bavaria (Bayern), more traditional habits are still observed. The birthrate is one of Europe's lowest, and the population would have fallen were it not for the influx of immigrants since the 1950s.

Germany has a tradition of strong feminism. Women have full rights under the law and play a bigger role in politics than in most other European countries. Over 30% of Bundestag (Federal Assembly) members elected in 2002 are women, and women ministers occupy several top cabinet posts. From 2001, women were permitted to take on combat roles in the armed forces. However, they are less well represented in top jobs in business and industry. Abortion remains a charged issue. Women in former East Germany once had the right to abortion on demand, but the Constitutional Court, after strong Catholic lobbying, overruled a relatively liberal 1992 compromise law for the whole country. The current regulations, dating from mid-1995, allow abortions (but only after counseling) within three months of conception.

POPULATION AGE BREAKDOWN

Female	Age	Male
2.8%	80+	1.0%
10.1%	60–79	7.5%
13.2%	40–59	13.5%
14.7%	20–39	15.6%
10.5%	0–19	11.1%

% of population by age group

CHRONOLOGY

German unification in the 19th century brought together a mosaic of states with a common linguistic, but varied cultural, heritage.

- ❑ **1815** German Confederation under nominal Austrian leadership.
- ❑ **1834** Zollverein Customs Union of 18 states, including Prussia.
- ❑ **1862** Otto von Bismarck appointed Prussian chancellor.
- ❑ **1864–1870** Prussia defeats Austrians, Danes, and French; north German states under Prussian control.
- ❑ **1871** Southern states join Prussian-led unified German Empire under Wilhelm I.
- ❑ **1870s** Rapid industrialization.
- ❑ **1890** Kaiser Wilhelm II accedes, with aspirations for German world role. Bismarck sacked.
- ❑ **1914–1918** World War I.
- ❑ **1918** Germany signs armistice; Weimar Republic created.
- ❑ **1919** Treaty of Versailles: colonies lost and reparations paid. Rhineland demilitarized.

G

G

CHRONOLOGY *continued*

- ❏ **1923** France occupies Ruhr; financial collapse, hyperinflation.
- ❏ **1933** Hitler chancellor after Nazis become largest party in elections. One-party rule; rearmament.
- ❏ **1935** Nuremberg Laws; official persecution of Jews begins.
- ❏ **1936** German entry into Rhineland. Axis alliance with Italy.
- ❏ **1938** Annexation of Austria and Sudetenland.
- ❏ **1939** Invasion of Poland starts World War II.
- ❏ **1940** France, Belgium, the Netherlands, and Norway invaded.
- ❏ **1941** USSR invaded.
- ❏ **1942–1943** Germans defeated by Red Army at Stalingrad.
- ❏ **1945** German surrender; Allies control four occupation zones.
- ❏ **1949** Germany divided: communist East led by Walter Ulbricht 1951–1971, Erich Honecker 1971–1989; liberal democratic West led by CDU's Konrad Adenauer, first Chancellor, 1949–1963.
- ❏ **1955** West Germany joins NATO.
- ❏ **1961** Berlin Wall built.
- ❏ **1966–1969** West German "grand coalition" of CDU and SPD.
- ❏ **1969–1982** SPD-led West German governments under Willy Brandt (1969–1974) and Helmut Schmidt (1974–1982).
- ❏ **1973** Both Germanies join UN.
- ❏ **1982** Helmut Kohl West German chancellor, CDU–FDP coalition.
- ❏ **1989** Fall of Berlin Wall.
- ❏ **1990** Reunification of Germany. First all-German elections since 1933; Kohl heads government.
- ❏ **1998** Gerhard Schröder heads coalition of SPD and Greens.
- ❏ **2000** Disgrace of Kohl in party funding scandal.
- ❏ **2001** Historic Berlin city government coalition of SPD and former communist PDS.
- ❏ **2002** Euro fully adopted. SPD–Green coalition reelected.

***The Messeturm**, Frankfurt, is the tallest office building in Europe. Frankfurt is Germany's financial services center, and many of its leading companies are located there.*

POLITICS ▷ Multiparty elections

 L. House 2002/2006
U. House varying

 President
Johannes Rau

AT THE LAST ELECTION
Federal Assembly 603 seats

| 42% SPD | 41% CDU/CSU | 9% A/G | 1% PDS | 7% FDP |

SPD = Social Democratic Party of Germany **CDU/CSU** = Christian Democratic Union/Christian Social Union
A/G = Alliance 90/Greens **FDP** = Free Democratic Party
PDS = Party of Democratic Socialism

Federal Council 69 seats

Each of the 16 states (*Länder*) is represented by between 3 and 6 members in the Federal Council (Bundesrat), who are appointed after the elections in each *Land*.

Germany is a federal democratic republic of 16 *Länder* (states). The government is led by the federal chancellor, elected by the Bundestag (Federal Assembly). The president's role is largely ceremonial. The "Basic Law" of West Germany, drawn up in 1948, became the 1990 federal constitution of reunified Germany.

PROFILE

Germany's politics are now strongly democratic, with a long tradition of federative association. Before 1871, Germany was a mass of separate principalities, kingdoms, and city-states, a situation in many ways respected by Bismarck's unification constitution. The 1933–1945 Nazi period, during which the federal system was abolished, was very much a hiatus. The Allies reestablished the federal system in West Germany in 1949; in the east, the *Länder* were restored after reunification in 1990. In many ways, the *Länder* are at the heart of German political life, each with its own elected parliament and largely controlling its finances. By general consensus the system delivers efficient and commercially astute government. There have been few major differences on domestic policy between the postwar ruling coalitions. All parties support the social market economy on which prosperity was built.

Germany has enjoyed stable governments, with coalitions of the center-left and center-right each holding sway since the "grand coalition" of 1966–1969. In 1998 the electorate chose moderate SPD leader Gerhard Schröder in a vote for change, ousting long-serving CDU chancellor Helmut Kohl. In opposition the CDU was beset by a party-funding scandal. Kohl was disgraced. Angela Merkel succeeded him, the first person from the former East Germany to become CDU leader. She accepted that Edmund Stoiber, leader of the CDU's Bavarian-based

sister party, the CSU, was the stronger candidate to contest the chancellorship in September 2002. In the poll, Stoiber's challenge fell just short of toppling Schröder's "red–green" coalition.

MAIN POLITICAL ISSUES
The economy
Recent recession has come as a shock to Germany, used to constant growth since the 1950s. Public spending has been reined in since the mid-1990s to meet targets for European monetary union, despite unrest as unemployment peaked at over four million. A brief return to growth enabled the SPD to start tackling pension reform, but the economy still dominates the political agenda.

East and west
Most Germans supported reunification after the fall of the Berlin Wall in 1989, but feelings soured as the true costs became clear. Over DM 1000 billion have been spent on reconstruction in the east, financed partly by a "solidarity surcharge" on income tax, but the tenth anniversary of reunification in November 2000 was less a celebration than a reflection on past mistakes. The east remains poorer, and people moving west, still seen as "*ossis*," may find it hard to fit in. Support for the former communist PDS is confined almost entirely to the east.

Far-right violence
Unemployment and resentment of "foreigners" led to a rise in support for far-right parties. Foreign workers, particularly Turks, and asylum seekers have been subject to shocking attacks. The problem of racism, even if no worse than in many other European states, is particularly sensitive, given Germany's history.

Helmut Kohl, *long-serving chancellor until 1998.*

Gerhard Schröder, *chancellor from 1998, reelected in 2002.*

Joschka Fischer, *leading Green politician and a popular foreign minister from 1998.*

WORLD AFFAIRS

▷ Joined UN in 1973

During the Cold War, a divided Germany was forced to play a subservient role in international affairs, the West closely adhering to US interests and the East taking its orders directly from the USSR. Since reunification in 1990, the emphasis has changed, and Germany has begun to voice a foreign policy which reflects its position as the most powerful country in Europe. Still firmly Western-oriented, Germany has championed European integration and has considerable economic power over the former Soviet bloc countries through large-scale investment there.

2001 proved to be a seminal year for German foreign policy. Germany was given command of the NATO peacekeeping mission in Macedonia and German troops were involved in the "war on terrorism" in Afghanistan.

AID

▷ Donor

$5.03bn (donations)   ⬇ Down 9% in 2000

Unlike the US, the UK, and France, Germany's aid programs are not directly motivated by its desire for political influence in the world's poorer regions. Most are multilateral, although there is also a strong tradition of direct aid. Much comes from Church organizations such as the Protestant Brot für die Welt. Many German volunteers and missionaries work overseas on aid programs.

DEFENSE

▷ Compulsory military service

$28.2bn   ⬇ Down 9% in 2000

The large German army, designed for national defense, is being slimmed down to focus more on mobility and providing support to allied states. The Constitutional Court ruled in 1994 that army units could take part in collective defense activities abroad, opening the way for German participation in NATO operations; the 1999 action against Serbia was a landmark.

GERMAN ARMED FORCES

🚜	2521 main battle tanks (751 *Leopard* 1A1/A3/A4/ A5, 1770 *Leopard* 2)	211,800 personnel
🚢	14 submarines, 2 destroyers, 12 frigates, and 28 patrol boats	26,050 personnel
✈	434 combat aircraft (131 F-4, 267 *Tornado*, 23 MiG-29)	70,550 personnel
🚀	None	

G

ECONOMICS

▷ Inflation 2% p.a. (1990–2000)

$2064bn   1.0651–1.1231 euros

SCORE CARD

❑ World GNP Ranking	3rd
❑ GNP per Capita	$25,120
❑ Balance of Payments	–$18.7bn
❑ Inflation	1.9%
❑ Unemployment	11%

EXPORTS

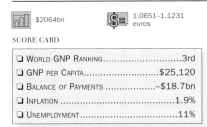

Netherlands 6%
Italy 8%
UK 8%
USA 10%
France 11%
Other 57%

IMPORTS

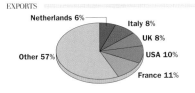

UK 7%
Italy 7%
Netherlands 9%
USA 9%
France 10%
Other 58%

ECONOMIC PERFORMANCE INDICATOR

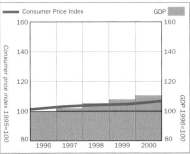

Consumer Price Index — GDP

economies. Relatively few small firms, a short working week in terms of hours, poorly developed service sector.

PROFILE

West Germany's remarkable postwar recovery, to become the world's third-strongest economy, was based on the concept of a social market economy, under which the state provided welfare and ensured workers' rights, while the economy was largely in private hands. Major banks and businesses are privately owned, except for the partly state-owned Volkswagen. After reunification in 1990, there was massive investment in the former East Germany, where state concerns were sold off.

Germany was one of the 12 EU states fully to adopt the euro in 2002. The SPD-led government elected in 1998 promised to tackle unemployment and to maintain growth, which reached a ten-year high of 3.1% in 2000. The $46 billion sale of "third generation" mobile phone operators' licenses reduced the budget deficit in 2001. Major tax reforms aimed to balance the budget by 2006. However,

GERMANY : MAJOR BUSINESSES

Kiel
Hamburg
Berlin
Ruhr Valley
Dresden
Nürnberg
Frankfurt am Main
München
Stuttgart

◊ Optics
🚢 Shipbuilding
🧪 Chemicals
🔌 Electronics
⚙ Engineering
🔬 Research & development
🚗 Vehicle assembly

0   200 km
0   200 miles

global economic slowdown brought Germany into recession at the end of 2001, with unemployment above 10% (meaning over four million out of work) and rising once again.

Potsdamer Platz, Berlin, *was rapidly reconstructed after reunification and is once again the commercial center of the capital.*

STRENGTHS

Europe's major industrial power. Efficient industry benefits from low inflation environment. Workers and managers live up to reputation for thoroughness and hard work. Strong sectors are cars, heavy engineering, electronics, and chemicals.

WEAKNESSES

Underestimation of costs of updating inefficient eastern German economy. High welfare costs; pension obligations (despite reforms in 2001) with an aging population. Rising unemployment. Competition from efficient Asian

RESODURCES

 Electric power 115m kw

312,492 tonnes

25.8m pigs, 14.6m cattle, 2.14m sheep, 108m chickens

58,827 b/d (reserves 215m barrels)

Coal, oil, natural gas, copper, salt, potash, tin, nickel

ELECTRICITY GENERATION

Hydro 4% (21bn kwh)

Combustion 65% (361bn kwh)

Nuclear 29% (161bn kwh)

Other 2% (13bn kwh)

| 0 | 20 | 40 | 60 | 80 | 100 |

% of total generation by type

With relatively few natural resources, Germany imports over 50% of its energy needs, mainly oil and gas. Coal, the basis of industrialization, now

ENVIRONMENT

 Sustainability rank: 50th

27%

10.1 tonnes per capita

ENVIRONMENTAL TREATIES

Yes — Yes — Yes

Yes — Yes — Yes

Germans are among the world's most environmentally conscious people. Environmental campaigns led by the Green Party, which emerged as a powerful political force in the 1980s, have influenced the policies of all major parties. At national level, the Greens are a significant force in the Bundestag; they joined the SPD-led federal government coalition in 1998, and are strongly represented in *Land* parliaments and local councils.

Germany has some of the strictest pollution controls in the world, with ambitious targets for reducing carbon dioxide emissions, compelling businesses to become more energy-efficient. Germans recycle half of their waste paper, and three-quarters of their used tires and glass.

The nuclear debate has been vigorously fought and won by the Greens; a gradual program of closing existing nuclear power plants was approved in 2001, although waste disposal is still an issue. Fears in the 1980s that up to 50% of trees were sick or dying because of car fumes and industrial pollution led to Germany becoming the first European country to insist that new cars be fitted with catalytic converters. The east had the highest per capita rate of sulfur emissions in the world, but these have been reduced by the closure of industrial plants and the elimination of the noxious Trabant cars.

accounts for under a quarter of energy consumption. West Germany invested less heavily than France in nuclear power, and Soviet-built plants in the east have been shut down. The "red–green" coalition government decided in mid-2000 to phase out nuclear power (which provides over 25% of electricity). Renewable resources account for 2% of primary energy consumption, but 6% of electricity production (with a target of 50% by 2050); Germany is the world's leading user of wind power.

GERMANY : LAND USE

Cropland
Forest
Pasture
Vineyards
Pigs
Cattle

| 0 | | 200 km |
| 0 | 200 miles | |

MEDIA

 TV ownership high

Daily newspaper circulation 300 per 1000 people

PUBLISHING AND BROADCAST MEDIA

There are 375 daily newspapers, including the *Frankfurter Allgemeine Zeitung*, the *Süddeutsche Zeitung*, and *Die Welt*

3 public service and several independent commercial channels

13 public service and several independent networks

TV is supervised by the political parties to ensure a balance of views. Satellite and cable TV have taken much of the audience once shared between the main public service channels, ARD and ZDF. Media conglomerates such as Kirch and Bertelsmann have expanded abroad. Newspapers are mostly regional and serious. An exception is *Bild Zeitung*, a right-wing, sensationalist tabloid, which sells 4.5 million copies daily.

Neuschwanstein Castle, *Bavaria (Bayern), one of Germany's major tourist attractions. It was built for the eccentric King Ludwig II.*

EDUCATION

 School leaving age: 18

99%

1.77m students

THE EDUCATION SYSTEM

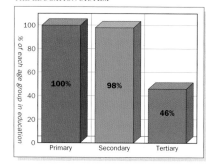

| Primary | Secondary | Tertiary |
| 100% | 98% | 46% |

% of each age group in education

Nearly one-tenth of total government expenditure goes on education, which is run by the *Länder*. They coordinate teaching policies, but have autonomy within their borders. The German approach to education stresses academic and vocational achievement. Sporting or cultural activities tend to be organized informally. Nearly all schools have Internet access.

Young people wanting to leave school must continue studying at least part-time until 18. Those who wish to go to university attend the upper-secondary *Gymnasien* to prepare for the *Abitur* exam. Students were taking an average of seven years to complete degrees, until new legislation added shorter bachelor's and master's degrees as in other countries. Research is done as much by major companies as by the universities.

CRIME

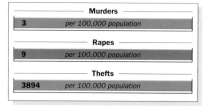 No death penalty

79,348 prisoners

Down 2% in 1999

CRIME RATES

| Murders | |
| 3 | per 100,000 population |

| Rapes | |
| 9 | per 100,000 population |

| Thefts | |
| 3894 | per 100,000 population |

Crime rates in Germany are lower than in most other European countries. This is largely the result of a genuine respect for the law, coupled with a strong police force. Recently, however, higher unemployment has led to an increase in petty theft and a wave of violence, notably against immigrants.

German politicians, once with an enviably clean reputation, have suffered several corruption scandals. Civil service corruption remains rare. People convicted under environmental laws can face ten-year jail sentences.

G

BERLIN, REUNIFICATION, AND CENTRAL EUROPE

THE SO-CALLED "BONN REPUBLIC," created in West Germany in the postwar period, ended symbolically in September 1999. Bonn could justifiably claim to have been home for some 50 years to Germany's only enduringly successful parliamentary democracy. The decision to move the capital to Berlin, however, symbolizing German unity, was enshrined in the 1990 unification treaty only months after the fall of the Berlin Wall in November 1989. It took time to confirm that both government and parliament would be transferred – the lower house of parliament, the Bundestag, only voted in 1994 to move – but five years after that vote, in the same month that Federal Chancellor Schröder moved into his office in Berlin, the Bundestag held its first full session there.

The dramatic glass dome of the new Reichstag building in Berlin.

PROSPECTS FOR BERLIN

The parliament's new home is the former Reichstag building, impressively redesigned by English architect Norman Foster. Topped with a transparent dome and lit up at night from within, it epitomizes the emphasis on architectural and engineering achievements in modern Berlin. New government, commercial, and tourist facilities have brought a lengthy construction boom in Berlin, which with a population of some 3.5 million is Germany's largest city. The proliferation of cultural and artistic activity has also contributed to a strong sense of excitement, boosting the city's image with the international media and the public. Berlin nevertheless faces many problems, including an unemployment rate well above the national average and the need to revitalize its declining industrial base. The outward movement of business and population, from central urban areas to surrounding regions, has been an established trend for decades in major western German cities, but is now happening much more rapidly in Berlin, whose western half was for years an enclave within East Germany. The city also has a long way to go in attracting major companies to make their headquarters there; only 12 of the largest German companies are located in Berlin, compared with over 40 each in Hamburg, Munich, and Frankfurt.

COSTLY REUNIFICATION

The costs of reconstruction of the former East Germany far outweighed initial expectations. Achievements stand out in telecommunications and rail transportation, but most infrastructure in the east is still well below the standard of the west, despite subsidies amounting to the transfer to the east of some 7% of GDP of western *Länder* per year for a decade. Unemployment is higher, labor productivity lower, and living conditions less attractive to the majority of Germans.

GERMANY'S NEW CENTER OF GRAVITY

The transfer of the capital reinforces the shift in Germany's center of gravity brought about by reunification. Coinciding with the collapse of communism across the whole former Soviet bloc came a revival of interest in Germany's role in central Europe or *Mitteleuropa*. In the former communist countries German economic influence is now particularly strong. West Germany was firmly anchored in western Europe, in economic terms by the European Economic Community (now the EU), and in a political– military sense by its membership of the NATO alliance. The imminent eastward expansion of the EU, however, suggests the emergence of France–Germany–Poland as a new and powerful axis in the Europe of the 21st century.

Unemployed Germans demonstrate for the right to work.

HEALTH

 Welfare state health benefits

1 per 286 people

 Heart and cerebro-vascular diseases, cancers, accidents

The German social security system, pioneered by Bismarck, is one of the most comprehensive in the world. Health insurance is compulsory, and employer and employee contributions are high. Although most hospitals are run by the *Länder*, some are still owned by Germany's wealthy churches. Almost one-third of health spending is now private.

Germans are increasingly health-conscious, paying great attention to diet. Nearly a million people go on cures every year to the country's 200-plus spas. In the east there is a higher incidence of lung diseases, the legacy of industrial pollution.

SPENDING

GDP/cap. increase

CONSUMPTION AND SPENDING

508 per 1000 population

611 per 1000 population

Defense	1.6%
Education	4.6%
Health	7.9%

0   5   10   15   20   25
Defense, Health, Education spending as % of GDP

The effects of the Nazi period, which discredited many of the ruling class, and the destruction of the property of millions of families in the war, explain the relatively classless nature of society. Status is now more closely linked to wealth than to birth. In the west, there are fewer disparities than in most of Europe; workers are generally well paid and social security is generous. Wages in the east, however, are 10% below western rates, and a disproportionate number of unemployed live on welfare benefit. Most Germans own a mobile phone, and almost a third of them had accessed the Internet by 2001.

WORLD RANKING

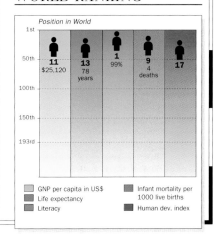

Position in World

1st
50th
100th
150th
193rd

11 $25,120
13 78 years
1 99%
9 4 deaths
17

GNP per capita in US$
Life expectancy
Literacy
Infant mortality per 1000 live births
Human dev. index

GHANA

OFFICIAL NAME: Republic of Ghana **CAPITAL:** Accra
POPULATION: 19.7 million **CURRENCY:** Cedi **OFFICIAL LANGUAGE:** English

WEST AFRICA

 1957   1957   March 6   GH   0   +233   .gh

THE HEARTLAND OF THE ancient Ashanti kingdom, modern Ghana is a union of the former British colony of the Gold Coast and the British-administered part of the UN Trust Territory of Togoland. Ghana gained independence in 1957, the first British colony to do so. Multiparty democracy was embraced in 1992, and the handover of power to the main opposition party in 2000 confirmed the shift away from Ghana's recent history of intermittent military rule.

G

CLIMATE
▷ Tropical wet & dry/ equatorial

WEATHER CHART FOR ACCRA

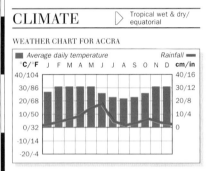

Southern Ghana has two rainy seasons: from April to July and September to November. The drier north has just one, from April to September.

TRANSPORTATION
▷ Drive on right

✈ **Kotoka International, Accra**
673,504 passengers

🚢 210 ships
118,700 grt

THE TRANSPORTATION NETWORK

🛣 11,653 km (7241 miles)	🛤 30 km (19 miles)
🚂 1300 km (808 miles)	⚓ 168 km (104 miles)

In 1983, work began to restore Ghana's roads, which had fallen into disrepair in the 1960s and 1970s; the network is now improving.

TOURISM
▷ Visitors : Population 1:53

🧳 373,000 visitors ⬆ Up 7% in 1999

MAIN TOURIST ARRIVALS

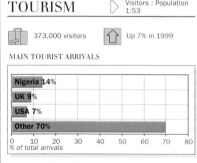

Nigeria	14%
UK	9%
USA	7%
Other	70%

0 10 20 30 40 50 60 70 80
% of total arrivals

Tourism is still small-scale; most visitors come from the rest of Africa, the UK, and the US. Good beaches and old coastal forts are major attractions.

PEOPLE
▷ Pop. density medium

Twi, Fanti, Ewe, Ga, Adangbe, Gurma, Dagomba (Dagbani)

86/km²
(222/mi²)

THE URBAN/RURAL POPULATION SPLIT

38%   62%

RELIGIOUS PERSUASION

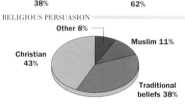

Other 8%
Muslim 11%
Christian 43%
Traditional beliefs 38%

Ghana contains various cultural-linguistic groups. The largest is the Akan, who include the Ashanti and Fanti peoples. Other important groups are the Mole-Dagbani in the north, Ga-Adangbe around Accra, and Ewe in the southeast. Tribal tensions are rare, but can flare up, most recently in 2002.

POLITICS
▷ Multiparty elections

2000/2004

President
John Kufuor

AT THE LAST ELECTION
Parliament 200 seats

1% PCP

50% NPP   46% NDC   2% Ind 1% PNC

NPP = New Patriotic Party **NDC** = National Democratic Congress **Ind** = Independents **PNC** = People's National Convention **PCP** = People's Convention Party

Ghana's return to multiparty rule in 1992 marked the legitimization of the military government of Jerry Rawlings. An air force flight-lieutenant of Ewe–Scottish descent and one of the great survivors of African politics, Rawlings staged coups in 1979 and 1981, and led the 1981–1992 Provisional National Defense Council (PNDC) military government. As the NDC candidate, Rawlings won 58% of the vote in the 1992 presidential election. Opposition parties boycotted the following parliamentary elections, which the NDC won easily. Elections in 1996 gave Rawlings a further and final term of office. In December 2000 the opposition NPP gained a historic victory when it stripped the NDC of its parliamentary majority and NPP candidate John Kufuor won the presidency.

WORLD AFFAIRS
▷ Joined UN in 1957

 Comm   ECOWAS G24 IAEA AU

Good relations with the West, which provides the bulk of Ghana's military and development aid, are a priority. Ghana has played a significant part in UN peacekeeping operations. It was also the main contributor, after Nigeria, to the ECOWAS forces (ECOMOG) in war-torn Liberia from 1990 to 1997. Ghana maintains good relations with its French-speaking neighbors, despite periods of strain with Togo.

GHANA
Total Land Area : 238 540 sq. km (92 100 sq. miles)

LAND HEIGHT
500m/1640ft
200m/656ft
Sea Level

POPULATION
● over 500 000
◉ over 100 000
◎ over 50 000
○ over 10 000
• under 10 000

0 100 km
0 100 miles

AID
 Recipient

 $609m (receipts)   Little change in 2000

Ghana is the third-largest recipient of World Bank aid (after India and China), supporting a largely successful economic recovery program which began in 1983. The most recent World Bank/IMF debt relief program started in April 2001.

DEFENSE
No compulsory military service

 $95m   Down 21% in 2000

In 1966, 1972, 1979, and 1981, the military mounted successful coups, and there have also been several unsuccessful coups. Outside Ghana, the 5000-strong army has been deployed mainly in UN and ECOWAS operations. Ghana's navy is small, with four patrol boats. The air force has 19 combat aircraft.

ECONOMICS
Inflation 27% p.a. (1990–2000)

$6.59bn 7275–7400 cedis

SCORE CARD

- ❏ WORLD GNP RANKING.......................101st
- ❏ GNP PER CAPITA $340
- ❏ BALANCE OF PAYMENTS...................–$413m
- ❏ INFLATION 25.2%
- ❏ UNEMPLOYMENT................................20%

STRENGTHS
GNP rose by 5% a year throughout the 1990s, following economic recovery policies begun in 1983. Second-largest gold producer in Africa. Since 1996, the Ashanti Goldfields Co. Ltd. has expanded into a multinational active in 12 African countries. Cocoa production accounts for 13% of world total.

WEAKNESSES
High budget deficits and debt repayments; the cedi was devalued in 1983 and has since tended to float downward. Foreign investors generally invest solely in gold mining. High inflation levels.

EXPORTS

Germany 6%
UK 8%
USA 11%
Netherlands 14%
Other 46%
Togo 15%

IMPORTS

Ivory Coast 5%
USA 7%
UK 9%
Other 48%
Italy 12%
Nigeria 19%

Dixcove harbor, *close to Ghana's most southerly cape. The majority of Ghanaians lead a traditional subsistence existence.*

RESOURCES
Electric power 1.2m kw

 493,206 tonnes Reserves of 500,000 barrels

 3.08m goats, 2.74m sheep, 1.3m cattle, 20.5m chickens   Gold, diamonds, bauxite, manganese

Gold production has expanded strongly since the mid-1980s; by 1993, gold had overtaken cocoa as the major export. Diamonds, bauxite, and manganese are also exported. Hydropower from the Volta Dam is exported to Togo and Benin, but is hit by periodic droughts.

ENVIRONMENT
Sustainability rank: 65th

 5%   0.2 tonnes per capita

Cutting of wood for fuel, timber, and farming has destroyed 70% of forests. Devastation caused by mining is now being tackled under a World Bank project.

MEDIA
TV ownership medium

 Daily newspaper circulation 14 per 1000 people

PUBLISHING AND BROADCAST MEDIA

 There are 2 daily newspapers, the *Ghanaian Times* and the *Daily Graphic*

 1 state-controlled service   1 state-controlled service

New independent weeklies reflect the increase in private press ownership. Radio and TV tend to follow government reporting guidelines.

CRIME
Death penalty in use

 10,444 prisoners   Down 20% 1997–1999

The judiciary has little independence and the government often resorts to ad hoc "people's tribunals." Corruption is now less of a problem.

EDUCATION
School leaving age: 15

 72%   38,305 students

All sectors of the education system are oversubscribed. There are a few high-quality boarding schools and four universities.

CHRONOLOGY
In 1874 Kumasi, capital of the Ashanti kingdom, was sacked by a British force to create the Gold Coast colony.

- ❏ **1957** Independence under Kwame Nkrumah.
- ❏ **1964** Single-party state.
- ❏ **1966** Army coup.
- ❏ **1972–1979** "Kleptocracy" of Gen. Acheampong. Executed 1979.
- ❏ **1979** Flt. Lt. Jerry Rawlings leads coup. Civilian Hilla Limann wins elections.
- ❏ **1981** Rawlings takes power again.
- ❏ **1992, 1996** Rawlings and NDC win multiparty elections.
- ❏ **2000** Opposition NPP wins elections; John Kufuor wins presidency.

G

HEALTH
No welfare state health benefits

 1 per 10,000 people   Malaria, diarrheal diseases, tuberculosis

The health of most of the population has benefited more from improvements in public hygiene than in medical care. Private health care is available.

SPENDING
GDP/cap. increase

CONSUMPTION AND SPENDING

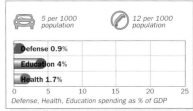
5 per 1000 population 12 per 1000 population
Defense 0.9%
Education 4%
Health 1.7%
0 5 10 15 20 25
Defense, Health, Education spending as % of GDP

Political uncertainty brought few opportunities for advancement, and many Ghanaians emigrated, but the situation is now improving. The main economic disparity is still between the poorer rural north and the richer, more urban, south.

WORLD RANKING

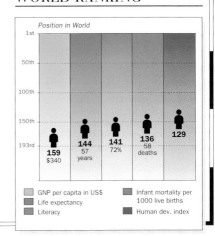
Position in World
1st
50th
100th
150th
193rd
159 $340 | 144 57 years | 141 72% | 136 58 deaths | 129

GNP per capita in US$ | Infant mortality per 1000 live births
Life expectancy |
Literacy | Human dev. index

GREECE

OFFICIAL NAME: Hellenic Republic **CAPITAL:** Athens **POPULATION:** 10.6 million
CURRENCY: Euro (drachma until 2002) **OFFICIAL LANGUAGE:** Greek

THE SOUTHERNMOST COUNTRY of the Balkans, Greece is embraced by the Aegean, Ionian, and Cretan Seas. Its mainly mountainous territory includes more than 2000 islands. Only one-third of the land is cultivated. There is a strong seafaring tradition, and some of the world's biggest shipowners are Greek. Greece is rich in minerals – including chromium, whose occurrence is rare. Relations with Turkey, marked by conflict and territorial disputes, have improved in recent years. To the north, however, upheavals in Albania and the conflicts in former Yugoslavia have made for greater instability.

CLIMATE

> Mediterranean

WEATHER CHART FOR ATHENS

The climate varies from region to region. The northwest is alpine, while parts of Crete border on the subtropical. The large central plain experiences high summer temperatures. Water is a problem, particularly on the many barren Aegean islands.

TRANSPORTATION

> Drive on right

Athinai, Athens
6.3m passengers

1529 ships
26.4m grt

THE TRANSPORTATION NETWORK

107,406 km (66,739 miles)	470 km (292 miles)
2299 km (1429 miles)	80 km (50 miles)

The easiest and cheapest method of transportation between the islands and the mainland is by boat or hovercraft. A major ferry disaster in 2000 prompted government moves to improve standards. Greece has 444 ports, of which Piraeus is the main one, and 123 are large enough to handle passenger or freight traffic. A new airport at Spata, 30 km east of Athens, opened in 2001. Greece has a good, if increasingly congested, road network. Two expressway routes are nearing completion, with the help of EU funds, as is the upgrading of the Athens metro. An interurban bus system and a fleet of air-conditioned tourist Pullmans offer an extensive service.

TOURISM

> Visitors : Population 1.2:1

12.5m visitors Up 3% in 2000

MAIN TOURIST ARRIVALS

UK	20%
Germany	20%
Italy	6%
Albania	6%
Netherlands	5%
Other	43%

% of total arrivals

Tourism is a mainstay of the Greek economy, with an annual turnover of some $10 billion, and is a major source of foreign exchange. Until recently, the state gave grants for hotel development and many third-grade hotels were built, especially on Crete and Rhodes. Smaller islands often lack sufficient water supplies or sandy beaches. To offset falling visitor numbers in the mid-1990s the industry has been encouraged to move upmarket, and is also promoting year-round activity vacations and conference tourism. The 2004 Olympics, to be held in Athens, are a stimulus to upgrade the city's facilities. A museum is planned to house the Parthenon Marbles, currently held in the British Museum (UK).

The theater at Dodona. *Classical sites, such as this amphitheater in northwestern Greece, have helped to make tourism one of Greece's most important industries.*

PEOPLE

> Pop. density medium

Greek, Turkish, Macedonian, Albanian

81/km² (210/mi²)

THE URBAN/RURAL POPULATION SPLIT

60% 40%

RELIGIOUS PERSUASION

Muslim 1% Other 1%
Greek Orthodox 98%

ETHNIC MAKEUP

Other 2%
Greek 98%

The Greeks were for many centuries a largely agrarian and seafaring nation. The German occupation during World War II, and the civil war that followed, destroyed much of the fabric of rural life and there was rapid urbanization after the 1950s. There was also extensive emigration in the 1950s and 1960s to northern Europe, Australia, the US, Canada, and southern Africa. However, many people returned to Greece in the 1980s, putting pressure on the labor market. The socialist PASOK governments of 1981–1989 spent large sums, mostly from EU sources, on developing the infrastructure and business life of the rural regions with a view to halting emigration to the cities. The policy was mostly successful, but a majority still lives in or near the capital, Athens, and Thessaloníki in the north.

Some 98% of the population belong to the Greek Orthodox Church. Civil marriage and divorce only became legal in 1982. There are minorities of Muslims, Roman Catholics, and Jews, and a recent influx of illegal immigrants, mainly from Albania.

POPULATION AGE BREAKDOWN

Female	Age	Male
2%	80+	1.4%
9.8%	60–79	8.4%
12.5%	40–59	12.2%
14.7%	20–39	14.9%
11.7%	0–19	12.4%

% of population by age group

EUROPE
Europe

POLITICS

 Multiparty elections

2000/2004

President Costas Stephanopoulos

AT THE LAST ELECTION
Parliament 300 seats

42% ND **3% KKE**

53% PASOK **2% Synaspismos**

PASOK = Pan-Hellenic Socialist Movement
ND = New Democracy **KKE** = Communist Party of Greece
Synaspismos = Left Coalition

Greece is a multiparty democracy. A military regime held power in 1967–1974.

PROFILE

On PASOK's election victory in 1981 its founder Andreas Papandreou headed the first socialist government in Greek history, which remained in power until 1989. Since 1993 PASOK has again held power continuously, but its economic policies have differed little from the previous conservative government; it only narrowly won the 2000 election. Kostas Simitis has led the party since Papandreou's resignation and death in 1996.

MAIN POLITICAL ISSUES
Closer European union
Greece joined the eurozone in 2001, but only after stringent austerity policies, which evoked widespread protests. Greece favors EU enlargement, and backs EU membership for Cyprus, to the annoyance of Turkey. Currently the poorest EU country, Greece could lose funding to new members, especially if Turkey joins.

Relations with FYRM
In 1995, Greece, with its northern province of Macedonia, finally agreed to recognize the sovereignty of the Former Yugoslav Republic of Macedonia. The crisis in that country in 2001 over ethnic Albanian separatism increased Greek regional security concerns.

Albanian refugees
Thousands of Albanians of Greek descent entered Greece illegally after 1990. Willing to work for very low wages, they swelled the thriving black economy.

Kostas Simitis, *prime minister since January 1996.*

Andreas Papandreou, *founded PASOK, prime minister 1981–1989, 1993–1996.*

A 1998 legalization program resulted in the registration of 375,000 Albanians.

G

WORLD AFFAIRS
 Joined UN in 1945

EU **NATO** **OECD** **OSCE** **CE**

Although closely allied to the West, Greece has sympathies with Russians and Serbs, who share its Orthodox heritage. Greece withdrew from NATO's military command in 1974, in protest at its failure to prevent the Turkish invasion of Cyprus, but has since rejoined. The island remains a sore point in Greece's otherwise improving ties with Turkey. Friction with FYRM over the use of the name "Macedonia" has eased, but relations remain tense. The current priority is Greece's role within an expanding EU.

AID
▷ Donor

$ $226m (donations) ⬆ Up 16% in 2000

Greece's contribution to overseas development assistance is the lowest in western Europe. However, Greek companies have invested elsewhere in the southern Balkans, and increasingly in Turkey since 1999. Greece receives regional development assistance from the EU, especially from the EU's structural and cohesion funds, its share of which was estimated to amount to around $3.5 billion over the period 1994–1999. Some of the money has been used to reverse the decline of northeast Greece – the EU's least developed region. Emergency humanitarian aid was given to Turkey following the severe earthquake there in 1999.

GREECE
Total Land Area : 131 940 sq. km
(50 942 sq. miles)

POPULATION
- ▣ over 1 000 000
- ◉ over 500 000
- ◎ over 100 000
- ○ over 50 000
- ● over 10 000

LAND HEIGHT
- 2000m/6562ft
- 1000m/3281ft
- 500m/1640ft
- 200m/656ft
- Sea Level

0   100 km
0   100 miles

G

CHRONOLOGY

Greece was occupied by Nazi Germany between 1941 and 1944. After liberation by the Allies, communists and royalists fought a five-year civil war. This ended with communist defeat, and King Paul became the constitutional monarch.

❏ **1964** King Constantine succeeds his father, King Paul.
❏ **1967** Military coup. King in exile. Col. Giorgios Papadopoulos premier.
❏ **1973** Greece declared a republic, with Papadopoulos as president. Papadopoulos overthrown in military coup. Lt. Gen. Ghizikis becomes president, Adamantios Androutsopoulos prime minister.
❏ **1974** Greece leaves NATO in protest at Turkish occupation of northern Cyprus. "Colonels' regime" falls. Constantinos Karamanlis becomes premier and his ND party wins subsequent elections.
❏ **1975** Konstantinos Tsatsou becomes president.
❏ **1977** Elections: ND reelected.
❏ **1980** Karamanlis president. Georgios Rallis prime minister. Greece rejoins NATO.
❏ **1981** PASOK wins elections. Andreas Papandreou first-ever socialist premier. Greece joins European Communities.
❏ **1985** Proposals to limit power of president. Karamanlis resigns. Christos Sartzetakis president. Greece and Albania reopen borders, closed since 1940.
❏ **1985–1989** Civil unrest caused by economic austerity program.
❏ **1988** Cabinet implicated in financial scandal. Leading members resign.
❏ **1989** Defense agreement with US. Two inconclusive elections lead to formation of all-party coalition.
❏ **1990** Coalition government collapses. ND wins elections. Konstantinos Mitsotakis prime minister, Karamanlis president.
❏ **1990–1992** Strikes against economic reform.
❏ **1992** EU persuaded not to recognize independent Macedonia.
❏ **1993** PASOK wins election, Andreas Papandreou premier.
❏ **1995** Costas Stephanopoulos elected president; recognition of sovereignty of Former Yugoslav Republic of Macedonia (FYRM).
❏ **1996** Andreas Papandreou resigns as prime minister; succeeded by Kostas Simitis.
❏ **1999** Earthquakes in Greece and Turkey. Sympathetic response shows improvement in relations.
❏ **2000** PASOK wins general election.
❏ **2001** Armed conflict in FYRM.
❏ **2002** Euro fully adopted.

DEFENSE

 Compulsory military service

 $5.46bn   ⬆ Up 5% in 2000

Greece spends a higher percentage of GDP on defense than any other NATO country except Turkey, whose perceived threat is its main concern, although tensions with that country are now less acute. In 1998 a law was passed on the conscription of women (for four days a year) for the defense of border regions. Greece has committed 3500 troops to the planned European rapid response force.

GREEK ARMED FORCES

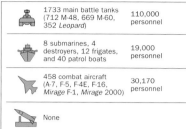

1733 main battle tanks (712 M-48, 669 M-60, 352 *Leopard*)	110,000 personnel
8 submarines, 4 destroyers, 12 frigates, and 40 patrol boats	19,000 personnel
458 combat aircraft (A-7, F-5, F-4E, F-16, *Mirage* F-1, *Mirage* 2000)	30,170 personnel
None	

ECONOMICS

 Inflation 9.2% p.a. (1990–2000)

▦ $126bn $⊟ 1.0651–1.1231 euros

SCORE CARD

❏ WORLD GNP RANKING...........................30th
❏ GNP PER CAPITA $11,960
❏ BALANCE OF PAYMENTS ....–$9.82bn
❏ INFLATION 3.2%
❏ UNEMPLOYMENT...................................11%

EXPORTS

Cyprus 5%   USA 5%
UK 6%
Italy 9%
Other 63%
Germany 12%

IMPORTS

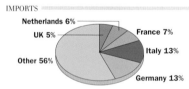

Netherlands 6%   France 7%
UK 5%
Italy 13%
Other 56%
Germany 13%

STRENGTHS

One of the major tourist destinations in Europe. Efficient agricultural exporter. Shipping: the world's largest beneficially owned fleet.

WEAKNESSES

High levels of public debt. Until recently, interest rates and bureaucratic banking system discouraged private initiative. State-owned sector, often poorly managed, remains large, as does the black economy.

PROFILE

Greece took longer than most other countries to recover from World War II. It was not until the 1960s that any substantial investment occurred. The Colonels' dictatorship curbed inflationary pressures with a wage freeze. When civilian government was restored in 1974, a spate of high wage settlements and the oil price shocks of 1973 and 1979 drove inflation over 20%. Greece's largest companies made substantial losses, until the socialists'

ECONOMIC PERFORMANCE INDICATOR

controversial austerity program of 1986–1987 reined in labor costs.

Greece failed in 1999 to meet the economic convergence criteria for introducing the euro. It then tackled the problems with determination, balancing the budget and bringing inflation under control, although public-sector debt remains high. In January 2001 Greece became the 12th member of the eurozone, and fully adopted the currency in 2002. Unemployment is still high and GDP per capita is the lowest in the EU.

GREECE : MAJOR BUSINESSES

Cement	
Textiles	
Chemicals	
Electronics	
Beverages	
Iron & steel	
Shipbuilding	
Pulp & paper	
Fruit processing	
Pharmaceuticals	
Tobacco processing	

0   200 km
0   200 miles

RESOURCES

 Electric power 10.1m kw

215,964 tonnes

5875 b/d (reserves 15m barrels)

9m sheep, 5.3m goats, 905,000 pigs, 28m chickens

Oil, gas, coal, iron, bauxite, marble, nickel, magnesite, chromium

ELECTRICITY GENERATION

Hydro 9% (4bn kwh)

Combustion 91% (42bn kwh)

Nuclear 0%

Other 0%

% of total generation by type

Greece has an oil and gas field off the coast of Thasos island. There may also be exploitable reserves in eastern waters, ownership of which is contested by Turkey. Coal, iron, and other mining contributes less than 2% to GDP. Greece is a leading producer of marble.

ENVIRONMENT

 Sustainability rank: 60th

4% (0.2% partially protected)

8.1 tonnes per capita

ENVIRONMENTAL TREATIES

Yes | Yes | Yes
Yes | Yes | Yes

Local fishing interests have formed a successful antipollution organization, HELMEPA. Smog in Athens is irritating to the eyes and throat and highly damaging to ancient monuments: the Parthenon in Athens has suffered more erosion in the last two decades than in the previous 2000 years. Forest fires regularly cause havoc, damaging flora and fauna.

MEDIA

 TV ownership high

Daily newspaper circulation 64 per 1000 people

PUBLISHING AND BROADCAST MEDIA

There are 122 daily newspapers. *Eleftheros Typos* has the highest circulation figure

18 services: 1 state-owned, 17 independent

2 services: 1 state-owned, 1 independent

After the state broadcasting monopoly ended in 1990, many private TV and radio networks emerged; the two national public TV channels account for only abut 10% of viewing. Commercial broadcasting has made politicians more answerable to the public, and has also had a cultural impact, with the import of more foreign, particularly US, programs.

GREECE : LAND USE

RHODOPE MTS

PINDVS

AEGEAN ISLANDS

Cropland
Forest
Pasture
High mountain regions
Sheep
Fruit

0   100 km
0   100 miles

CRIME

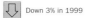 Death penalty not used in practice

8343 prisoners

Down 3% in 1999

CRIME RATES

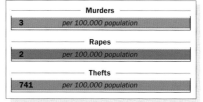

Murders

3 | per 100,000 population

Rapes

2 | per 100,000 population

Thefts

741 | per 100,000 population

An influx of migrants is blamed for an increase in violent crime. The terrorist group November 17, which had carried out high-profile assassinations, was finally tracked down in 2002.

EDUCATION

 School leaving age: 15

97%

363,150 students

THE EDUCATION SYSTEM

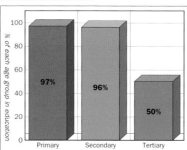

Primary 97% | Secondary 96% | Tertiary 50%

% of each age group in education

Some 9% of total government spending is on education, which is free and officially compulsory for nine years. Teachers are poorly paid and qualifications are low. University places are limited and many students go abroad for tertiary education. Technical courses, funded by the EU, have increased since the 1990s.

HEALTH

 Welfare state health benefits

1 per 244 people

Heart and cerebro-vascular diseases, cancers, accidents

The socialists (PASOK) introduced a national health service and a national pharmaceuticals industry. Some 12% of government expenditure goes on health, and every Greek is entitled to sickness benefit. Greece now has the second-highest number of doctors per head of population in the EU; however, primary care is poor, as is that in state hospitals. In the early 1990s the ND attempted to upgrade private medicine and to incorporate its activities with those in state hospitals. Many Greeks needing major surgery go abroad for treatment.

SPENDING

 GDP/cap. increase

CONSUMPTION AND SPENDING

254 per 1000 population

532 per 1000 population

Defense 4.9%

Education 3.1%

Health 4.7%

0   5   10   15   20   25

Defense, Health, Education spending as % of GDP

Greek society changed dramatically in the postwar period. Formerly a largely agricultural society living in isolated communities, it was rapidly urbanized in the 1950s. Former agricultural workers made fortunes, many by grabbing opportunities presented by the shipping industry. Among these were the now prominent Niarchos and Onassis families.

The advent of the republic in 1973 reflected the social changes which had occurred since the war. New wealth and success became more admired than aristocratic birth or prestige. Greece is now a socially mobile society. Living standards have improved universally since the 1950s.

WORLD RANKING

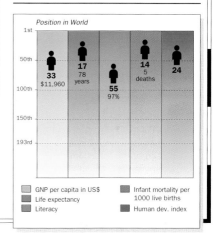

Position in World

1st

50th

100th

150th

193rd

33 $11,960 | 17 78 years | 55 97% | 14 5 deaths | 24

GNP per capita in US$
Life expectancy
Literacy

Infant mortality per 1000 live births
Human dev. index

G

GRENADA

OFFICIAL NAME: Grenada   CAPITAL: St. George's
POPULATION: 98,000   CURRENCY: Eastern Caribbean dollar   OFFICIAL LANGUAGE: English

 1974 1974 Feb 7 WG -4 +1473 .gd

 CARIBBEAN

THE MOST SOUTHERLY of the Windward Islands, Grenada also includes the southern Grenadine islands of Carriacou and Petite Martinique. It is the world's second-largest nutmeg producer. One of the seven members of the OECS, Grenada became a focus of international attention in 1983 when the US, with token backing from several Caribbean states, mounted an invasion to sever its growing links with Castro's Cuba.

CLIMATE
▷ Tropical oceanic

WEATHER CHART FOR ST. GEORGE'S

Rainfall totals 150 cm (60 in) on the coast, and twice that in the mountains. Hurricanes occur in the rainy season.

TRANSPORTATION
▷ Drive on left

Point Salines, St. George's
438,987 passengers

6 ships
1009 grt

THE TRANSPORTATION NETWORK

638 km (396 miles)	None
None	None

Roads in the interior are poor. Catamarans provide the fastest link between Grenada and Carriacou.

TOURISM
▷ Visitors : Population 1.3:1

129,000 visitors   Up 3% in 2000

MAIN TOURIST ARRIVALS

USA 25%	
UK 20%	
Trinidad & Tobago 8%	
Other 47%	

0 10 20 30 40 50 60
% of total arrivals

Tourism has developed since the 1984 completion of the international airport, though there has been a concurrent decline in the arrival of cruise ships; the Crown Dynasty company folded in 1999, with a major impact on tourist numbers.

PEOPLE
▷ Pop. density high

English, English Creole   288/km² (748/mi²)

THE URBAN/RURAL POPULATION SPLIT

38%   62%

RELIGIOUS PERSUASION

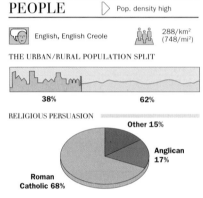

Other 15%
Anglican 17%
Roman Catholic 68%

Most Grenadians are descendants of Africans brought over to work sugar plantations in the 16th to 19th centuries. Intermarriage between this group and the small numbers of Europeans and indigenous Amerindians has meant that there is little racial tension. As in other Caribbean states, extended families with absentee fathers are not uncommon.

GRENADA

Total Land Area : 340 sq. km (131 sq. miles)

POPULATION
● over 10 000
· under 10 000

LAND HEIGHT
500m/1640ft
200m/656ft
Sea Level

POLITICS
▷ Multiparty elections

L. House 1999/2004
U. House 1999/2004

H.M. Queen Elizabeth II

AT THE LAST ELECTION

House of Representatives 15 seats

100% NNP

NNP = New National Party

Senate 13 seats

The members of the Senate are appointed by the prime minister and the leader of the opposition

The past 25 years have seen Grenada move from a position of political isolation toward integration with the rest of the region. The late Sir Eric Gairy, when prime minister, was as well known for his eccentric requests to the UN Security Council – he once asked it to investigate UFOs on the island – as for his intimidation of political opponents by means of organized gangs. Gairy was overthrown in 1979 by armed militants of the New Jewel Movement led by Maurice Bishop, a charismatic socialist who in turn was deposed, and executed by former allies, in 1983. This coup was the pretext for the US invasion, the primary motive of which was to end the perceived Cuban influence in Grenada. A new government was elected in 1984, and the US provided large amounts of aid. Politics has since been center-right, and there is little to choose ideologically between the parties. The NNP, led by Keith Mitchell, gained power in 1995, and went on to achieve an unparalleled victory over a divided opposition in the early general election held in 1999, taking all 15 seats. Mitchell promised "explosive" growth and lower unemployment.

G

WORLD AFFAIRS

 Joined UN in 1974

 ACS   Caricom   Comm   OAS   OECS

Priorities are the relations with the rest of the Windward Islands group, access to the EU for bananas, and strategies with Indonesia aimed at steadying world nutmeg prices. Since 1983, it has supported US policy in the Caribbean.

AID

 Recipient

 US$17m (receipts)   Up 70% in 2000

Main aid sources are Arab donors, Japan, the Caribbean Development Bank, and the EU. Cuba helped build the pre-1983 airport at Point Salines.

DEFENSE

No compulsory military service

Minimal expenditure Defense spending is falling

The People's Revolutionary Army, created by Maurice Bishop in the wake of his 1979 coup, was replaced in 1983 by a paramilitary defense unit trained by the US and the UK.

ECONOMICS

Inflation 2.5% p.a. (1990–2000)

 US$370m 2.7 Eastern Caribbean dollars

SCORE CARD

- ❑ WORLD GNP RANKING.........................174th
- ❑ GNP PER CAPITA ....US$3770
- ❑ BALANCE OF PAYMENTS.................–US$79m
- ❑ INFLATION 0.2%
- ❑ UNEMPLOYMENT.....................................15%

STRENGTHS

Second-largest producer of nutmeg after Indonesia. Important sectors are tourism, cocoa, bananas, construction, and financial services.

WEAKNESSES

Weak tax base, lack of diversification. Poor infrastructure. Low productivity. Large avoidance of customs duties. Smuggling.

EXPORTS

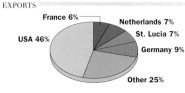

France 6%
Netherlands 7%
St. Lucia 7%
USA 46%
Germany 9%
Other 25%

IMPORTS

Barbados 4%
UK 7%
Japan 4%
USA 42%
Trinidad & Tobago 20%
Other 23%

RESOURCES

 Electric power 15,000 kw

1632 tonnes Not an oil producer

13,100 sheep, 7100 goats, 220,000 chickens None

Grenada has no strategic resources and has to import most of its energy. The major asset is Grenadian nutmeg, which is highly prized for its quality, but production levels fluctuate.

ENVIRONMENT

 Not available

None   1.9 tonnes per capita

Tourism threatens some key environmental sites, including a remnant of rainforest. Resort projects have caused serious beach erosion, in turn requiring costly coastal defenses. An environmental levy on visitors is opposed by cruise companies.

MEDIA

 TV ownership high

There are no daily newspapers

PUBLISHING AND BROADCAST MEDIA

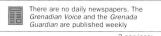
There are no daily newspapers. The *Grenadian Voice* and the *Grenada Guardian* are published weekly

1 state-owned service 3 services: 1 partly state-owned, 2 independent

The government sold a 60% share in the then Grenada Broadcasting Corporation (GBC) in 1999. There is an independent private press.

CRIME

 Death penalty not used in practice

327 prisoners Up 345% 1989–1996

The doubling of poverty during the 1990s and high unemployment have contributed to a rising crime rate. Narcotics trafficking is also a growing problem. However, while there is street crime, the level of violence is low.

EDUCATION

 School leaving age: 16

96% 651 students

Education follows the former British selective 11-plus system. Many students go on to the University of the West Indies, or to college in the US.

HEALTH

 Welfare state health benefits

1 per 2000 people Heart diseases, cancers, nutritional disorders

After 1979, Cuban physicians provided a basic health care system. There are free weekly clinics in each district, and treatment in subsidized state hospitals now matches the Caribbean average. In 1999 Cuba began its promised expansion of the general hospital.

St. George's harbor. *The newest hotel developments are on the beaches to the south of the capital.*

G

CHRONOLOGY

A French colony from 1650, Grenada was captured by the British in 1762.

- ❑ **1951** Universal suffrage introduced.
- ❑ **1967–1974** Internal self-government. Full independence from the UK.
- ❑ **1979** Coup. Maurice Bishop prime minister. Growing links with Cuba.
- ❑ **1983** US invasion establishes pro-US administration.
- ❑ **1999** NNP reelected, taking all 15 seats in House of Representatives.

SPENDING

GDP/cap. increase

CONSUMPTION AND SPENDING

83 per 1000 population 332 per 1000 population

Defense No data
Education 4.7%
Health 2.9%

0   5   10   15   20   25
Defense, Health, Education spending as % of GDP

Wealth disparities in Grenada are less marked than in most Caribbean states, but poverty is growing. The wealthiest groups are those in control of the nutmeg trade.

WORLD RANKING

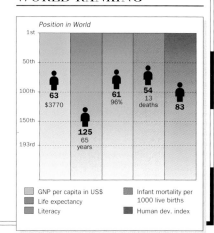

Position in World

63 $3770
125 65 years
61 96%
54 13 deaths
83

❑ GNP per capita in US$
❑ Life expectancy
❑ Literacy
❑ Infant mortality per 1000 live births
❑ Human dev. index

GUATEMALA

OFFICIAL NAME: Republic of Guatemala   **CAPITAL:** Guatemala City
POPULATION: 11.7 million   **CURRENCY:** Quetzal   **OFFICIAL LANGUAGE:** Spanish

THE LARGEST AND MOST POPULOUS of the states of the Central American isthmus, Guatemala was home to the ancient Mayan civilization. Its fertile Pacific and Caribbean coastal lowlands give way to the highlands which dominate the country. Independent since 1838, Guatemala fell under military rule in 1954. Civilian rule was not restored for over three decades, in 1986. Some 60% of people live below the poverty line.

G

CLIMATE
 Tropical equatorial/ wet & dry

WEATHER CHART FOR GUATEMALA CITY

The climate varies with altitude: daytime temperatures average 28°C (82°F) in tropical coast areas and 20°C (68°F) in the more temperate central highlands.

TRANSPORTATION
 Drive on right

 La Aurora, Guatemala City
939,000 passengers

 9 ships
4700 grt

THE TRANSPORTATION NETWORK

3616 km (2247 miles)	140 km (87 miles)
1994 km (1239 miles)	260 km (162 miles)

Good roads link the major towns. The railroad and two international airports are attracting foreign investment.

TOURISM
 Visitors : Population 1:14

826,000 visitors   No change in 2000

MAIN TOURIST ARRIVALS

El Salvador 33%
USA 22%
Honduras 7%
Other 38%

% of total arrivals

Tourism rapidly revived after the military excesses in the 1980s, but postwar crime, including an increase in mob violence, deters visitors. Mayan ruins are the top attractions.

PEOPLE
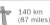 Pop. density medium

Quiché, Mam, Cakchiquel, Kekchí, Spanish   108/km² (279/mi²)

THE URBAN/RURAL POPULATION SPLIT

40%   60%

ETHNIC MAKEUP

Other 10%
Amerindian 60%
Mestizo 30%

More than half of Guatemalans are Amerindians, descendants of the original Mayans. Culture and language distinguish them from *ladino* groups. *Ladinos* include a white elite, a large mixed race group, and also Amerindians rejecting traditional dress and language to avoid discrimination. Political power and 65% of land are held by a few *ladino* families. Amerindians mainly live in the highlands, by subsistence farming. In a 1999 plebiscite, *ladinos* rejected proposed reforms recognizing 23 Amerindian languages and the right of Amerindians to have judicial hearings in their own languages.

GUATEMALA

Total Land Area : 108 890 sq. km (42 042 sq. miles)

POPULATION

⊡ over 1 000 000
◎ over 100 000
○ over 50 000
• over 10 000

LAND HEIGHT

3000m/9843ft
2000m/6562ft
1000m/3281ft
500m/1640ft
200m/656ft
Sea Level

POLITICS
 Multiparty elections

1999/2003   President Alfonso Portillo

AT THE LAST ELECTION
Congress of the Republic 113 seats

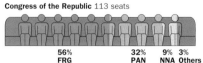

56% FRG	32% PAN	9% NNA	3% Others

FRG = Guatemalan Republican Front   **PAN** = National Advancement Party   **NNA** = New Nation Alliance

A military government which came to power in 1954 with US backing brutally suppressed opposition and persecuted the highland Amerindians until the return of democracy in 1986. Civil war effectively continued until President Arzú of the PAN concluded a peace agreement with the Guatemalan National Revolutionary Unity (URNG) guerrillas in 1996. The 36-year war had claimed 200,000 lives, mostly innocent civilians. In presidential elections in 1999 Alfonso Portillo of the right-wing FRG, running on a strong law and order platform, defeated the PAN candidate.

Efraín Ríos Montt, military ruler in 1982–1986, remains a force in politics and in 2001 was elected president of the Congress, despite being investigated by a Spanish court on charges of genocide. Constant tension exists between the legislative and executive branches of government.

WORLD AFFAIRS ▷ Joined UN in 1945

 ACS　 Geplac　 NAM　 OAS　San José

Economic relations with the US and neighboring states are priorities. A UN mission was deployed for six months to oversee the 1996 peace accord. A claim to half of Belize was revived in early 2000.

AID ▷ Recipient

 $264m (receipts)　 Down 10% in 2000

In 1998 the government agreed the disbursement of the second half of $1.9 billion pledged by the international donors for postwar reconstruction. In the same year $1.4 million were also loaned as hurricane relief.

DEFENSE ▷ Compulsory military service

 $115m　Down 23% in 2000

A damning "truth commission" report in 1999 found the armed forces and their allies guilty of 93% of human rights violations during the civil war. The army remains largely unreformed and a potent sociopolitical force.

ECONOMICS ▷ Inflation 10% p.a. (1990–2000)

 $19.2bn　 7.782–7.910 quetzales

SCORE CARD

❏ WORLD GNP RANKING..........................67th
❏ GNP PER CAPITA $1680
❏ BALANCE OF PAYMENTS ....–$1.05bn
❏ INFLATION 6%
❏ UNEMPLOYMENT 8%

STRENGTHS

Main exports are coffee, sugar, beef, bananas, and cardamom. Privatizations boost foreign investor confidence.

WEAKNESSES

Traditional exports vulnerable to world price changes. Shaky financial system. Inequalities in land and wealth limit domestic market. Tax evasion.

EXPORTS

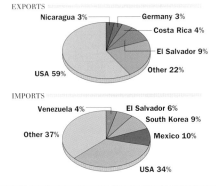

IMPORTS

CHRONOLOGY

The site of the Mayan civilization, Guatemala declared independence from Spain in 1821. It became a fully independent nation in 1838.

- ❏ **1954** US-backed coup topples reformist government.
- ❏ **1966–1984** Counterinsurgency war; highlands "pacification."
- ❏ **1986–1993** Return of civilian rule; President Serrano elected. Flees country after abortive "self-coup."
- ❏ **1996** President Arzú elected; peace deal with URNG guerrillas, ending 36 years of civil war.
- ❏ **1998** Bishop Juan Gerardi, human rights campaigner, murdered.
- ❏ **1999** "Truth Commission" blames army for most human rights abuses. Portillo and FRG win elections.

North Acropolis, Tikal, Petén. *One of the largest lowland Maya cities, Tikal was virtually abandoned by about 900 CE.*

RESOURCES ▷ Electric power 1.5m kw

 15,878 tonnes　 25,503 b/d (reserves 196m barrels)

 2.5m cattle, 1.45m pigs, 35m chickens　 Oil, antimony, lead, tungsten, nickel, copper

Agriculture provides over 20% of GDP and about 70% of export earnings. Guatemala is the world's largest producer of cardamom. The civil war hindered the exploitation of oil reserves and hydroelectric potential.

ENVIRONMENT ▷ Sustainability rank: 67th

 17% (0.5% partially protected)　 0.9 tonnes per capita

Forest cover has been halved to 35% since 1954 due to intensive agriculture. The excessive use of pesticides, many banned in the US, threatens health. Pollution and waste are serious problems, especially in the capital, Guatemala City.

MEDIA ▷ TV ownership medium

 Daily newspaper circulation 33 per 1000 people

PUBLISHING AND BROADCAST MEDIA

 There are 7 daily newspapers, including *Prensa Libre, Siglo Veintiuno, El Periódico,* and the state *Diario de Centroamérica*

 5 services: 1 state-owned, 4 independent　 85 stations: 5 state-owned, 80 independent

Powerful groups own the media, but newspapers can be hard-hitting. The four independent TV stations are controlled by a single owner.

CRIME ▷ Moratorium on death penalty

 8460 prisoners　 Crime is rising

Violence is high. In 2000 the interior ministry purged the police, which had been corrupted by drug trafficking.

EDUCATION ▷ School leaving age: 14

 69%　146,291 students

Education is only for the privileged. A literacy rate of 69% makes Guatemala one of the worst educated countries in Latin America.

HEALTH ▷ Welfare state health benefits

 1 per 1111 people　 Gastrointestinal infections, tuberculosis, heart disease, violence

Health spending is a budget priority as a result of pressure from the UN and multilateral lenders. Gastrointestinal and other infections directly linked to poverty remain the main causes of death.

SPENDING ▷ GDP/cap. increase

CONSUMPTION AND SPENDING

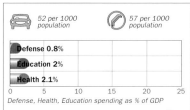

Poverty has risen since 1980: 60% of the population now live below the poverty line. The richest 10% control an estimated 46% of the national wealth.

WORLD RANKING

G

GUINEA

OFFICIAL NAME: Republic of Guinea   CAPITAL: Conakry
POPULATION: 8.3 million   CURRENCY: Guinea franc   OFFICIAL LANGUAGE: French

G

GUINEA LIES ON the western coast of Africa. Central highlands, either densely forested or savanna covered, slope down to coastal plains and swamps; the north is semidesert. Military rule from 1984 ended with disputed elections in 1995. Neighboring civil wars have spilled over into domestic conflict in Guinea.

CLIMATE ▷ Tropical monsoon

WEATHER CHART FOR CONAKRY

Conakry, Guinea's capital, receives particularly heavy rainfall, averaging 130cm in July alone

Guinea's climate is similar to Sierra Leone's. Conakry's six-month rainy season peaks dramatically in July.

TRANSPORTATION ▷ Drive on right

Conakry–Gbessia
293,484 passengers

33 ships
11,300 grt

THE TRANSPORTATION NETWORK

5033 km (3127 miles)		None	
1045 km (649 miles)		1295 km (805 miles)	

Major roads and rail lines are being rebuilt with World Bank and French aid. Much of the rail network is exclusively for the use of the bauxite industry.

A small mosque in Conakry. Muslims make up 85% of the population; 8% are Christian. The remainder follow traditional beliefs.

TOURISM ▷ Visitors : Population 1:252

33,000 visitors   Up 22% in 2000

MAIN TOURIST ARRIVALS

France 18%	
Sierra Leone 6%	
Belgium 6%	
Other 70%	

0   10   20   30   40   50   60   70   80
% of total arrivals

Limited infrastructure means that Guinea cannot exploit the tourist potential of its beaches, scenery, and rich culture.

PEOPLE ▷ Pop. density low

Fulani, Malinke, Soussou, French

34/km² (87/mi²)

THE URBAN/RURAL POPULATION SPLIT

33%   67%

ETHNIC MAKEUP

Other 5%
Malinke 30%
Kissi 10%
Other tribes 10%
Soussou 15%
Fila (Fulani) 30%

Since the death of Marxist dictator Sekou Touré in 1984, traditional rivalries have reemerged between ethnic groups. The two largest groups are the Fulani, based in the highland region of Fouta Djallon, and the Malinke, who lost the power they had held under Touré, and have suffered reprisals. Today, the coastal peoples, including the Soussou, are dominant, benefiting from renewed rivalry between the Malinke and Fulani.

The extended family system survived the climate of suspicion generated by paid informers under Sekou Touré. Women acquired influence within his Marxist party, but a Muslim revival since 1984 has reversed this trend.

Hundreds of thousands of refugees, fleeing from conflicts in neighboring countries, are now caught up in fighting in the southern border region.

POLITICS ▷ Multiparty elections

2002/2007

President Lansana Conté

AT THE LAST ELECTION

National Assembly 114 seats

3% PDG

74% PUP   17% UPR   3% UPG   3% Others

PUP = Party of Unity and Progress   UPR = Union for Progress and Renewal   UPG = Union for the Progress of Guinea   PDG = Democratic Party of Guinea

The death in 1984 of Sekou Touré, head of the Marxist single-party PDG regime from 1958, opened the way for the military to intervene, with promises of multiparty elections. A 1990 referendum overwhelmingly approved democratic changes, but the military appointed a Transitional Committee, delaying elections until 1993. Incumbent leader Gen. Lansana Conté heavily defeated the Malinke leader, Alpha Condé, who had been in exile until 1992. Amid claims that the election was rigged, serious violence erupted. A disputed victory for Conté's PUP in the 1995 legislative elections was followed by his own reelection in 1998. Fighting escalated into civil war in 2000, with cross-border incursions from rebels based in Sierra Leone and Liberia. A controversial constitutional referendum in 2001 allowed Conté to stand for a possible third term, but the legislative polls set for that December were postponed. When they were eventually held in July 2002, the PUP increased its majority.

WORLD AFFAIRS ▷ Joined UN in 1958

ECOWAS   OIF   AU   OIC   OMVG

Balancing the interests of the major aid donors, France and the US, is a growing concern. Relations with neighboring states, badly affected by conflicts in border regions, were patched up symbolically in February 2002.

AID ▷ Recipient

$153m (receipts)   Down 36% in 2000

In 1969, the World Bank funded the Boké bauxite project, then one of its most ambitious projects. Since 1986, Western aid has grown to finance over 85% of all development projects. The 1997–2000 World Bank/IMF structural reform program foresees an annual growth rate of 5%.

GUINEA

Total Land Area :
245 857 sq. km
(94 925 sq. miles)

POPULATION

- ⊙ over 500 000
- ○ over 50 000
- ● over 10 000
- · under 10 000

LAND HEIGHT

- 1000m/3281ft
- 500m/1640ft
- 200m/656ft
- Sea Level

DEFENSE

▷ Compulsory military service

$57m

Down 5% in 2000

Defense forces consist of an army and a militia – partly merged since the 1984 coup – a gendarmerie, and a tiny navy and air force. China, North Korea, and the Soviet bloc used to be the main arms procurement markets. Weaponry is now supplied by France and the US.

ECONOMICS

▷ Inflation 5.1% p.a. (1990–2000)

$3.3bn

1880–1970 Guinea francs

SCORE CARD

- ❑ WORLD GNP RANKING.......................128th
- ❑ GNP PER CAPITA $450
- ❑ BALANCE OF PAYMENTS...................–$165m
- ❑ INFLATION 6%
- ❑ UNEMPLOYMENT..........Widespread underemployment

STRENGTHS

Natural resources including bauxite, gold, and diamonds. Major iron ore deposits at Mount Nimba. Good soil and climate give high cash-crop yields. Relatively low inflation.

WEAKNESSES

Legacy of maladministration from Touré years. Poor infrastructure. 1990–1997 Liberian civil war set back major joint projects. Current refugee situation a drain on resources.

EXPORTS

Russia 9%
Spain 9%
Ireland 9%
USA 13%
Other 43%
Belgium 17%

IMPORTS

UK 6%
Belgium 7%
Ivory Coast 9%
USA 11%
Other 48%
France 19%

RESOURCES

▷ Electric power 186,000 kw

87,314 tonnes

Not an oil producer

2.68m cattle, 1.01m goats, 11.9m chickens

Bauxite, diamonds, gold, iron

Bauxite accounts for over 90% of export earnings. Guinea, with 30% of known world reserves, is the world's largest producer after Australia. Demand for electricity for bauxite processing is high. Gold production has grown steadily since 1995.

ENVIRONMENT

▷ Sustainability rank: 98th

 0.7%

0.2 tonnes per capita

Uncontrolled deforestation, particularly of large areas of rainforest, is the major long-term problem.

MEDIA

▷ TV ownership low

✕ Daily newspaper circulation 2 per 1000 people

PUBLISHING AND BROADCAST MEDIA

There is 1 daily newspaper, *Fonike*

1 state-owned service

1 state-owned service

Guinea's limited broadcast media are state-owned. The main newspaper, *Horoya*, is a weekly. There has been a slight relaxation in censorship.

CRIME

▷ Death penalty in use

 4014 prisoners

 Up 20% in 1992

The death penalty was reintroduced in 2001 in an attempt to crack down on spiraling crime. Cross-border diamond smuggling is at the root of anarchy in the south.

CHRONOLOGY

France colonized Guinea in 1890, strongly opposed by the Fulani Muslim empire of Fouta Djallon.

- ❑ **1958** Full independence under Sekou Touré.
- ❑ **1984** Touré dies. Army coup.
- ❑ **1993–1995** Disputed elections.
- ❑ **1998** Conté reelected president.
- ❑ **2000** Cross-border rebel attacks place Guinea in a state of civil war.
- ❑ **2002** PUP wins delayed elections.

EDUCATION

▷ School leaving age: 13

 41%

 8151 students

French was readopted as the main teaching language in 1984, after Sekou Touré's Marxist-inspired experiments.

HEALTH

▷ No welfare state health benefits

1 per 10,000 people

Malaria, diarrheal and respiratory diseases, tuberculosis

Health provision is very poor, reflected in Guinea's high infant mortality rate and low average life expectancy. Private health care was legalized in 1984.

SPENDING

▷ GDP/cap. increase

CONSUMPTION AND SPENDING

2 per 1000 population

8 per 1000 population

Defense 1.5%
Education 1.8%
Health 2.3%

Defense, Health, Education spending as % of GDP

Private enterprise has brought with it a new business class and Guinea now has some wealthy exiles, but much of the country remains poor and underdeveloped; GNP is below $500 per capita.

WORLD RANKING

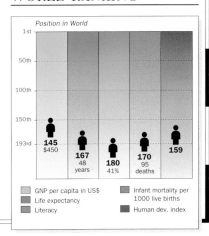

Position in World

1st
50th
100th
150th
193rd

145 $450
167 48 years
180 41%
170 95 deaths
159

- GNP per capita in US$
- Life expectancy
- Literacy
- Infant mortality per 1000 live births
- Human dev. index

GUINEA-BISSAU

OFFICIAL NAME: Republic of Guinea-Bissau   **CAPITAL:** Bissau
POPULATION: 1.2 million   **CURRENCY:** CFA franc   **OFFICIAL LANGUAGE:** Portuguese

L YING ON AFRICA'S west coast, impoverished Guinea-Bissau, a former Portuguese territory, is bordered by Senegal to the north and Guinea to the south and east. Apart from savanna highlands in the northeast, the country is low-lying. The PAIGC initiated a process of change to multiparty democracy in 1990, and elections were held in 1994. A military coup in 1999 followed army rebellion the previous year, but legislative and presidential elections have since been held.

G

CLIMATE ▷ Tropical monsoon

WEATHER CHART FOR BISSAU

■ *Average daily temperature*   *Rainfall* ■
°C/°F   J F M A M J J A S O N D   cm/in

60/140	60/24
50/122	50/20
40/104	40/16
30/86	30/12
20/68	20/8
10/50	10/4
0/32	0

90/35
80/31
70/28

The climate is tropical. The north is affected by the Sahel, the wetter south by the Atlantic. Droughts can occur.

TRANSPORTATION ▷ Drive on right

Bissalanca International, Bissau

25 ships
6685 grt

THE TRANSPORTATION NETWORK

453 km (281 miles)	None
None	Scattered stretches important to coastal commerce

The many waterways and islands make water transportation as vital as the roads. Both are being improved.

TOURISM ▷ Not available

A small number of visitors

No significant change from year to year

MAIN TOURIST ARRIVALS

Guinea-Bissau does not publish tourism figures by country of origin

0   10   20   30   40
% of total arrivals

Lack of tourist facilities means that the country remains a destination for only the most independent of travelers.

PEOPLE ▷ Pop. density low

Portuguese Creole, Balante, Fulani, Malinke, Portuguese

43/km²
(111/mi²)

THE URBAN/RURAL POPULATION SPLIT

24%   76%

RELIGIOUS PERSUASION

Christian 8%
Muslim 40%
Indigenous beliefs 52%

About 98% of Guinea-Bissau's people come from indigenous ethnic groups. The largest is the southern Balante, who form almost one-third of the population. Mixed-race *mestiço* and European minorities make up just 2% of the population. Although small in number, the *mestiços* – many of whom derive from Cape Verde, Portugal's other former west African colony – still dominate the bureaucracy. Resentment at this, especially among the Balante, who provided most of the PAIGC troops in the independence war, was one cause of the 1980 coup. The majority of the population live and work on small family farms, grouped in self-contained villages. The bulk of the urban population live in the capital, Bissau.

POLITICS ▷ Multiparty elections

1999/2003   President Kumba Yalla

AT THE LAST ELECTION
National People's Assembly 102 seats

37% PRS	27% RGB	24% PAIGC	4% AD	8% Others

PRS = Party for Social Renewal   **RGB** = Guinea-Bissau Resistance   **PAIGC** = African Party for the Independence of Guinea and Cape Verde   **AD** = Alliance for Democracy

Twenty years of one-party rule ended in 1994 with the holding of multiparty elections. However, opposition groups disputed the ruling PAIGC's victory in these polls.

A period of instability led to an army rebellion in 1998 and eight months of fighting between loyalists of President João Bernardo Vieira and the army chief, Gen. Ansumane Mane, which displaced about half the population. ECOWAS troops intervened and a national unity government was formed, only to be overthrown in May 1999.

Mane, the coup leader, convened fresh elections which were won by the PRS, whose candidate Kumba Yalla was elected president in January 2000. Mane was killed in another, failed, coup attempt, but Yalla now faces resistance from the Assembly itself over his increasingly erratic style of government.

WORLD AFFAIRS ▷ Joined UN in 1974

ECOWAS	CPLP	OIF	AU	OIC

Relations with neighboring states are extremely tense due to the activities of various rebel militias on the borders. Senegalese and Guinean troops intervened in mid-1998 to suppress the army rebellion.

GUINEA-BISSAU

Total Land Area :
36 120 sq. km
(13 946 sq. miles)

POPULATION

over 100 000   ◎
over 10 000   ●
under 10 000   ·

LAND HEIGHT

200m/656ft
Sea Level

AID

 Recipient

 $80m (receipts)   Up 54% in 2000

Portugal is Guinea-Bissau's most important single aid donor. Balance-of-payments support is critical to the economy. Export earnings are small, compared with the costs of imports and debt servicing. Donor support was frozen in 1991 beause of the country's World Bank arrears, but the government pushed ahead with economic reforms begun in the mid-1980s, and the World Bank and the IMF agreed a $790 million debt relief package in 2001. Education, the infrastructure, and health care are the main targets of project aid.

DEFENSE

 Compulsory military service

$6m   No change in 2000

There are around 9000 troops. The army led coups in 1980 and 1999, and suffered internal rebellions in 1998 and 2000, continuing a history of military interference in politics. ECOWAS soldiers have intervened to restore order on a number of occasions.

ECONOMICS

Inflation 33% p.a. (1990–2000)

$217m   698.7–736.7 CFA francs

SCORE CARD

❑ World GNP Ranking	183rd
❑ GNP per Capita	$180
❑ Balance of Payments	–$27m
❑ Inflation	8.6%
❑ Unemployment	Widespread underemployment

STRENGTHS
Minimal at present, but good potential in fisheries and timber. Offshore oil potential.

WEAKNESSES
Lack of sufficiency in rice staple. Fish stocks depleted by poaching. Few exports, mainly cashew nuts, groundnuts. Minimal industry. High illiteracy. Poor state economic management.

EXPORTS
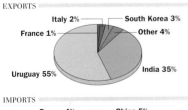

Italy 2%   South Korea 3%
France 1%   Other 4%
Uruguay 55%   India 35%

IMPORTS

France 4%   China 5%
   Thailand 9%
Other 37%   Senegal 15%
Portugal 30%

Bafatá, the chief town *in central Guinea-Bissau. It lies on the Gêba River and is also an important inland port.*

RESOURCES

 Electric power 11,000 kw

 5000 tonnes   Not an oil producer

515,000 cattle, 350,000 pigs, 1.4m chickens   Bauxite, phosphates

Fish and timber are the main natural resources, but local exploitation is only a tiny proportion of the sustainable levels. There is considerable potential for developing hydropower and for offshore oil production.

ENVIRONMENT

 Sustainability rank: 127th

None   0.2 tonnes per capita

Drought and locust plagues are serious natural hazards. A small population and minimal industry mean that there are few serious environmental problems.

MEDIA

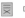 TV ownership low

Daily newspaper circulation 5 per 1000 people

PUBLISHING AND BROADCAST MEDIA

There is 1 daily newspaper, *Nô Printcha*, published by the government

1 state-owned service

3 services: 1 state-owned, 2 independent

There are three independently owned newspapers that publish intermittently due to the high costs of paper and distribution. Portugal helps to fund the TV service, started in 1989.

CRIME

 No death penalty

Guinea-Bissau does not publish prison figures   Up 66% in 1992

The death penalty was abolished in 1993. Reform of the legal system is in progress to make it more independent. Human rights abuses increased during the instability in the late 1990s.

EDUCATION

 School leaving age: 13

39%   463 students

Around 65% of children receive rudimentary education. Guinea-Bissau has no university.

CHRONOLOGY

Explored by the Portuguese in the 15th century, Portuguese Guinea was established in 1879. A war for independence began in the 1960s.

- ❑ **1974** Independence. PAIGC takes power.
- ❑ **1980** Military coup.
- ❑ **1990** Multiparty politics accepted.
- ❑ **1994** Multiparty elections.
- ❑ **1998** Army rebellion led by Gen. Mane. ECOWAS intervention.
- ❑ **1999** Transitional government. May, army seizes power. November, PRS defeats PAIGC in elections.
- ❑ **2000** Kumba Yalla president. Mane killed in failed coup attempt.

G

HEALTH

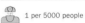 No welfare state health benefits

 1 per 5000 people   Parasitic, diarrheal, and communicable diseases, malaria

Guinea-Bissau's health statistics are among the world's worst, due partly to the minimal medical facilities. Average life expectancy is just 45 years; infant mortality is 126 per 1000 live births; the maternal death rate is high. In mid-2000 the AfDB provided $500,000 in funding for an emergency health program.

SPENDING

GDP/cap. increase

CONSUMPTION AND SPENDING

6 per 1000 population   9 per 1000 population

Defense 1.7%
Education No data
Health 4%

0   5   10   15   20   25
Defense, Health, Education spending as % of GDP

Living conditions for the majority of Guinea-Bissau's people are extremely poor; over 70% of the population are unable to meet their basic needs. The tiny elite is mainly *mestiço*.

WORLD RANKING

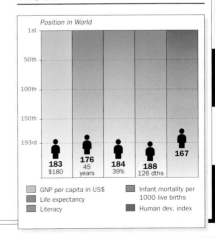

Position in World

183 $180	**176** 45 years	**184** 39%	**188** 126 dths	**167**

❑ GNP per capita in US$ ❑ Infant mortality per 1000 live births
❑ Life expectancy
❑ Literacy   ❑ Human dev. index

GUYANA

OFFICIAL NAME: Cooperative Republic of Guyana **CAPITAL:** Georgetown
POPULATION: 763,000 **CURRENCY:** Guyana dollar **OFFICIAL LANGUAGE:** English

 1966   1966   Feb 23   GUY   -4   +592 .gy

LYING ON THE NORTHERN EDGE of South America, Guyana stretches 600 km (375 miles) from dense tropical rainforests, through broad savanna and mountains dotted with waterfalls, to the narrow Atlantic coastal plain where most of the population lives. A British colony from 1814 until independence in 1966, Guyana has closer ties with the mostly anglophone Caribbean than with its Spanish-, Portuguese-, and Dutch-speaking neighbors.

CLIMATE
▷ Tropical equatorial

WEATHER CHART FOR GEORGETOWN

The lowlands are very humid with a constant temperature. The highlands are a little cooler, especially at night.

TRANSPORTATION
▷ Drive on left

Timehri International, Georgetown
270,500 passengers

61 ships
15,669 grt

THE TRANSPORTATION NETWORK

590 km (367 miles)	None
46 km (29 miles)	5900 km (3666 miles)

Reliable travel to the interior is by air or river; most paved roads are coastal. The only international airport is Timehri.

TOURISM
▷ Visitors : Population 1:10

75,000 visitors

Up 10% in 1999

MAIN TOURIST ARRIVALS

USA 47%
Canada 16%
Surinam 8%
Other 29%
% of total arrivals

The government promotes tourism, but the number of tourists is modest. Guyana means Land of Many Waters; the Kaieteur Falls are among the world's most impressive. Old Dutch wooden architecture characterizes Georgetown.

Modest homes, Georgetown. Most buildings are made of wood. The cathedral is one of the world's tallest freestanding wooden buildings.

PEOPLE
▷ Pop. density low

English Creole, Hindi, Tamil, Amerindian languages, English

4/km² (10/mi²)

THE URBAN/RURAL POPULATION SPLIT

38%   62%

ETHNIC MAKEUP

European and Chinese 2%
Amerindian 4%
Other 4%
East Indian 52%
Black African 38%

Tension exists between the Afro-Guyanese, descended from Africans brought over in the 17th to 19th centuries, and the Indo-Guyanese, descendants of south Asian laborers brought from India in the 19th century. This is currently displayed in the hostility existing between the opposition PNC, representing Afro-Guyanese, and the ruling PPP, traditionally representing the Indo-Guyanese.

GUYANA

Total Land Area : 214 970 sq. km (83 000 sq. miles)

POPULATION
◎ over 100 000
○ over 50 000
● over 10 000
• under 10 000

LAND HEIGHT
1000m/3281ft
500m/1640ft
200m/656ft
Sea Level

POLITICS
▷ Multiparty elections

2001/2006

President Bharrat Jagdeo

AT THE LAST ELECTION
National Assembly 65 seats

3% 2%
GAP–WPA TUF

52%
PPP–CIVIC

41%
PNC

2%
ROAR

PPP–CIVIC = People's Progressive Party–CIVIC
PNC = People's National Congress
GAP–WPA = Guyana Action Party–Working People's Alliance
ROAR = Rise, Organize, and Rebuild **TUF** = The United Force

The success of the PPP in the 1992 elections, widely seen as the first fair poll since independence, ended the dominance of the pro-Afro-Guyanese PNC. Veteran PPP leader Cheddi Jagan, a Marxist before adopting free-market policies, died in office in 1997. The PNC violently contested the succession of his wife Janet. A Caricom-brokered peace restored calm, though tension flared in early 2001 when the High Court condemned the 1997 poll. Violence erupted once again when elections in 2001 confirmed Janet Jagan's successor, Bharrat Jagdeo, as president.

WORLD AFFAIRS

▷ Joined UN in 1966

ACS | OIC | Caricom | Comm | OAS

Rescheduling debt with Western creditor states is paramount. Also important are the long-standing border dispute with Venezuela and closer integration with the Caribbean.

AID

▷ Recipient

 US$108m (receipts)　 Up 35% in 2000

Most aid comes from the US, the EU, and the UK. Recent grants covered public health projects, business development, and protection of the rainforest.

DEFENSE

▷ No compulsory military service

US$7m　　No change in 2000

The security forces, which include a small land army, benefit from financial support and training provided by the US and UK governments.

ECONOMICS

▷ Inflation 13% p.a. (1990–2000)

 US$652m　　 180.5 Guyana dollars

SCORE CARD

❏ WORLD GNP RANKING	163rd
❏ GNP PER CAPITA	US$860
❏ BALANCE OF PAYMENTS	–US$117m
❏ INFLATION	6.1%
❏ UNEMPLOYMENT	12%

STRENGTHS
Gold, rice, sugar, diamonds, bauxite, and timber production. Good tourism potential. Debt reduction plan agreed with multilateral agencies.

WEAKNESSES
High per capita foreign debt. Political instability dents investor confidence. Currency vulnerable to exchange rate pressure. Main exports vulnerable to fluctuations in international commodity prices. Weak manufacturing base.

EXPORTS

Portugal 6%
Netherlands Antilles 11%
Other 26%
UK 14%
USA 22%
Canada 21%

IMPORTS

Cuba 4%
UK 6%
USA 33%
Trinidad & Tobago 14%
Netherlands Antilles 19%
Other 24%

RESOURCES

▷ Electric power 202,000 kw

 54,450 tonnes　　Minimal oil production

220,000 cattle, 130,000 sheep, 12.5m chickens　　Gold, diamonds, bauxite, gemstones, oil, manganese, uranium

Gold, diamonds, bauxite, and timber are major resources. Offshore and onshore prospecting for oil has not reduced the need for petroleum imports for electricity generation. More hydroelectric power plants are being constructed.

ENVIRONMENT

▷ Not available

 0.05%　　2.2 tonnes per capita

The state of disrepair of the 18th-century sea defense system endangers the urbanized coastline that lies below sea level. Commercial logging threatens to deplete the rainforest. The pollution of rivers due to mining activities is now a serious problem.

MEDIA

▷ TV ownership medium

Daily newspaper circulation 50 per 1000 people

PUBLISHING AND BROADCAST MEDIA

There are 2 daily newspapers. The *Guyana Chronicle* is published by the government

16 services:
1 state-owned,
15 independent

1 state-owned service

The PPP and the PNC both publish periodicals. The government owns one TV service and the sole radio service as well as publishing a daily newspaper.

CRIME

▷ Death penalty in use

 1507 prisoners　　 Down 6% in 1999

The police are strongly criticized for corruption and ineffectiveness in the face of rising urban crime. Serious violence between PNC and PPP–CIVIC supporters erupted in 1998, 1999, and 2001.

EDUCATION

▷ School leaving age: 15

 98%　　 8965 students

Education is based on the former British system. Entry to high schools is by 11-plus examination. There is a state-financed university, though many students go to the US or the UK.

HEALTH

▷ Welfare state health benefits

 1 per 5556 people　　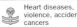 Heart diseases, violence, accidents, cancers

Nearly all of Guyana's population have access to the country's mainly state-run health service. The referral system is relatively good.

CHRONOLOGY

During the 17th and 18th centuries, the Dutch founded three colonies, Essequibo, Demerara, and Berbice, in the region. In 1814, these came under British control, and were later combined to form British Guiana.

❏ **1953** First universal elections won by PPP under Cheddi Jagan; parliament later suspended by UK.
❏ **1966** Independence from UK.
❏ **1973** PPP boycotts parliament, accusing PNC of electoral fraud.
❏ **1992** Fair elections won by PPP. Jagan president.
❏ **1997–1998** Jagan dies in office; PNC rejects his widow's election victory. Political crisis.
❏ **1999** Caricom-brokered peace deal. Janet Jagan resigns; Bharrat Jagdeo takes over as president.
❏ **2001** Jagdeo and PPP reelected. Political violence flares again.

SPENDING

▷ GDP/cap. increase

CONSUMPTION AND SPENDING

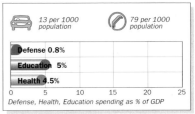

13 per 1000 population　　79 per 1000 population

Defense 0.8%
Education 5%
Health 4.5%

0　5　10　15　20　25
Defense, Health, Education spending as % of GDP

Significant urban and rural poverty in Guyana has forced the government to make provision in the budget for poverty alleviation. Redundancies in the public sector exacerbate the problem. The poorest group in society are Amerindian subsistence farmers. There are a few very affluent urban families who derive their wealth not only from business but also from rural farming interests.

WORLD RANKING

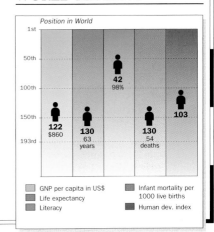

Position in World

1st
50th
100th
150th
193rd

42 98%

122 $860
130 63 years
130 54 deaths
103

☐ GNP per capita in US$
☐ Life expectancy
☐ Literacy
☐ Infant mortality per 1000 live births
☐ Human dev. index

HAITI

OFFICIAL NAME: Republic of Haiti   **CAPITAL:** Port-au-Prince   **POPULATION:** 8.3 million
CURRENCY: Gourde   **OFFICIAL LANGUAGES:** French and French Creole

HAITI OCCUPIES the western third of the Caribbean island of Hispaniola. A former Spanish colony, in 1804 it was the first Caribbean state to become independent, and has been in a state of political chaos virtually ever since. Democracy did not materialize with the exile of the dictator Jean-Claude Duvalier in 1986. Elections were held in 1990, but by 1991 the military were back in power and were ousted in 1994 only through US intervention.

H

CLIMATE   ▷ Tropical equatorial/ oceanic

WEATHER CHART FOR PORT-AU-PRINCE

Haiti lies mostly in the rain shadow of the central mountains, so humidity is lower than the Caribbean average.

TRANSPORTATION   ▷ Drive on right

Port-au-Prince
962,599 passengers

4 ships
1200 grt

THE TRANSPORTATION NETWORK

1011 km (628 miles)	None
None	100 km (62 miles)

Roads are poor, especially in the interior. Ferries provide the main transportation to the southern peninsula.

TOURISM   ▷ Visitors : Population 1:58

143,000 visitors

Down 3% in 1999

MAIN TOURIST ARRIVALS

USA 65%	
Canada 11%	
France 4%	
Other 20%	

0 10 20 30 40 50 60 70 80
% of total arrivals

Haiti's location, history, and culture provided much of its attraction for tourists in the 1960s and 1970s. Political instability and violence in the 1980s, however, led to the industry's near collapse and it has yet to recover.

PEOPLE   ▷ Pop. density high

French Creole, French

301/km² (780/mi²)

THE URBAN/RURAL POPULATION SPLIT

36%   64%

RELIGIOUS PERSUASION

Nonreligious 1%   Other 3%
Protestant 16%
Roman Catholic 80%

Most Haitians are the descendants of Africans; a few have European roots, primarily French. The majority of the population lives in extreme poverty: Haiti is the poorest country in the Americas, and Port-au-Prince has the worst slums in the Caribbean. Social tensions run high, and focus on class rather than race. In recent years, the combination of political repression and a collapsing economy led many to emigrate illegally to the US, or across the border to the neighboring Dominican Republic.

HAITI

Total Land Area : 27 750 sq. km
(10 714 sq. miles)

POPULATION	LAND HEIGHT
▣ over 1 000 000	1000m/3281ft
◎ over 500 000	500m/1640ft
● over 10 000	200m/656ft
• under 10 000	Sea Level

0   50 km
0   50 miles
N

POLITICS   ▷ Multiparty elections

L. House 2000/2004
U. House 2000/2002

President Jean-Bertrand Aristide

AT THE LAST ELECTION

Chamber of Deputies 83 seats

87%   2%   11%
Lavalas coalition   Vacant Others

Senate 27 seats

96%   4%
Lavalas coalition   Others

A wealthy elite, backed by the military, supported the Duvalier dictatorships and regularly financed coups after "Baby Doc" Duvalier's overthrow in 1986. UN sanctions followed the 1991 coup and US forces restored the elected president, Jean-Bertrand Aristide, in 1994. His left-wing Lavalas party won legislative elections in 1995 and René Préval was installed as the new president, with the party's backing, in 1996. A backlash against austerity policies insisted on by the US forced the prime minister to step down in 1997 and soured relations between the presidency and the legislature. New elections, finally held in May 2000, resulted in a strongly disputed but decisive victory for the Lavalas coalition. A similarly controversial presidential poll in November was won by Aristide. The opposition rejected his victory, leading to effective political stalemate.

WORLD AFFAIRS

▷ Joined UN in 1945

Continued political instability has cut off international aid and relations with major donors. Major issues are illegal immigration to the US and relations with the Dominican Republic.

AID

 ▷ Recipient

$208m (receipts)   ⬇ Down 21% in 2000

The IMF granted $21 million in emergency aid in 1998 for hurricane damage. The IDB approved loans for water and health problems, and Taiwan granted $60.4 million in aid.

DEFENSE

▷ No compulsory military service

$48m   ⬇ Down 4% in 2000

In 1994, the military were ousted and democracy was restored. The armed forces and police were disbanded and an interim public security force was created. A 5300-strong new national police force has now been formed, funded and trained by the US.

ECONOMICS

▷ Inflation 20% p.a. (1990–2000)

 $4.06bn   21.00-25.74 gourdes

SCORE CARD

❑ WORLD GNP RANKING	117th
❑ GNP PER CAPITA	$510
❑ BALANCE OF PAYMENTS	–$38m
❑ INFLATION	13.7%
❑ UNEMPLOYMENT	70%

STRENGTHS

Coffee exports. Remittances by Haitians living abroad. US demand for goods assembled in Haiti. Large profits from transshipment of narcotics to US.

WEAKNESSES

Huge tax avoidance. Foreign investment and promised aid deterred by political instability.

EXPORTS

Belgium 2%
France 2%
Canada 2%
Other 3%
Domican Republic 4%
USA 87%

IMPORTS

Colombia 2%
Japan 2%
France 3%
Dominican Republic 11%
USA 54%
Other 28%

Haiti: the poorest country in the Americas. In remote villages, most houses are made of earth and do not have glass in their windows.

RESOURCES

▷ Electric power 264,000 kw

 5000 tonnes   Not an oil producer

1.94m goats, 1.44m cattle, 1m pigs, 5.5m chickens   Marble, limestone, clay, silver, gold, natural asphalt

Haiti has no strategic resources. Under prolonged economic sanctions, it had to find unofficial sources of oil; much was imported from Europe.

ENVIRONMENT

▷ Sustainability rank: 137th

 0.4%   ⬆ 0.2 tonnes per capita

One-third of soil is seriously eroded and forest cover is now only 1.5% of total land area. The removal of 4000 tonnes of toxic waste, illegally dumped in 1988 near Gonaïves, finally began in 1998.

MEDIA

▷ TV ownership low

 Daily newspaper circulation 3 per 1000 people

PUBLISHING AND BROADCAST MEDIA

 There are 2 daily newspapers, *Le Nouvelliste* and *Le Matin*

 5 services: 1 state-owned, 4 independent   18 services: 1 state-owned, 17 independent

Under military rule, the media were largely controlled through intimidation. The transition to multiparty democracy has produced a more open press.

CRIME

▷ No death penalty

 4152 prisoners   ⬆ Crime is rising

Extrajudicial killings, torture, and brutality continue, despite the ending of military dictatorship. Narcotics trafficking is highly organized. Police are inexperienced and the judicial system is slow and open to corruption.

EDUCATION

▷ School leaving age: 15

 50%   6288 students

The run-down state system is based on the French model; the *baccalauréat* pass rate is consistently under 10%. The wealthy educate their children abroad.

CHRONOLOGY

In 1697, Spain ceded the west of Hispaniola to France. Ex-slave Toussaint Louverture's rebellion in 1791 led to independence in 1804.

- ❑ **1915–1934** US occupation.
- ❑ **1957–1971** François "Papa Doc" Duvalier's brutal dictatorship.
- ❑ **1971–1986** His son Jean-Claude, "Baby Doc," rules; eventually flees.
- ❑ **1986–1988** Military rule.
- ❑ **1990** Jean-Bertrand Aristide elected; exiled in 1991 coup.
- ❑ **1994–1995** US forces oust military. Aristide reinstated; elections.
- ❑ **1997–1999** Political deadlock.
- ❑ **2000** Lavalas coalition and Aristide reelected.

H

HEALTH

▷ No welfare state health benefits

 1 per 5000 people   Malaria, other parasitic diseases, tuberculosis

Most Haitians cannot afford health care. In rural areas, help is often sought from voodoo priests.

SPENDING

▷ GDP/cap. decrease

CONSUMPTION AND SPENDING

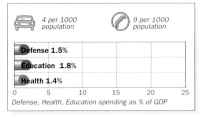

🚗 4 per 1000 population   💿 9 per 1000 population

Defense 1.5%
Education 1.8%
Health 1.4%

Defense, Health, Education spending as % of GDP

Haiti's rigid class structure maintains extreme disparities of wealth between a few affluent families and the mass of the population, who live in slums without running water or proper sanitation. According to the UN, around 80% of Haitians cannot meet their basic daily needs.

WORLD RANKING

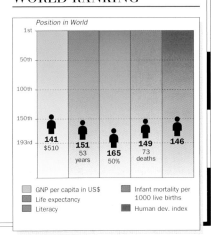

Position in World

141	151	165	149	146
$510	53 years	50%	73 deaths	

GNP per capita in US$
Life expectancy
Literacy
Infant mortality per 1000 live births
Human dev. index

HONDURAS

OFFICIAL NAME: Republic of Honduras **CAPITAL:** Tegucigalpa
POPULATION: 6.6 million **CURRENCY:** Lempira **OFFICIAL LANGUAGE:** Spanish

CENTRAL AMERICA

MOST OF HONDURAS is mountainous terrain, with a small sheltered Pacific coast to the south and a broad Caribbean shoreline to the north, including part of the virtually uninhabited Mosquito Coast. After a succession of military governments it returned to full civilian rule in 1984. In 1998 Honduras was devastated by Hurricane Mitch, which resulted in the death of at least 5600 people and damage estimated at some $3 billion.

H

CLIMATE ▷ Tropical equatorial

WEATHER CHART FOR TEGUCIGALPA

Honduras' Caribbean coastline is generally extremely hot. The rest of the country is much cooler.

TRANSPORTATION ▷ Drive on right

La Mesa, San Pedro Sula
635,859 passengers

1407 ships
1.11m grt

THE TRANSPORTATION NETWORK

3126 km (1942 miles)		None
996 km (619 miles)		465 km (289 miles)

In 1998 Hurricane Mitch destroyed roads and bridges across the country; reconstruction will take many years.

TOURISM ▷ Visitors : Population 1:16

408,000 visitors

Up 10% in 2000

MAIN TOURIST ARRIVALS

| USA 39% |
| El Salvador 15% |
| Guatemala 11% |
| Other 35% |

0   10   20   30   40
% of total arrivals

Caribbean coast resorts and the Bay Islands are popular, while exploring the remote region inland from the Mosquito Coast and jungle rafting appeal to the adventurous. The ruined Maya temples of Copán are a major draw.

PEOPLE ▷ Pop. density medium

Spanish, Black Carib, English Creole

59/km² (153/mi²)

THE URBAN/RURAL POPULATION SPLIT

53%   47%

ETHNIC MAKEUP

White 1% — Amerindian 4%
Black African 5%
Mestizo 90%

As in most of Central America, very few pure indigenous groups remain. The estimated 45,000 Miskito Amerindians, and an English-speaking *garífuna* (black) population on the Caribbean coast united in 1999 to oppose a constitutional amendment allowing foreigners to buy land in coastal areas, traditionally their communal lands. Poverty is at the root of social tension; whites still have the best opportunities.

Rural poverty and strong Roman Catholicism (97% are Roman Catholic) mean that the family is a powerful unifying force. The status of women is low; many work in domestic service.

POLITICS ▷ Multiparty elections

2001/2005

President Ricardo Maduro

AT THE LAST ELECTION

National Congress 128 seats

48% PNH
43% PLH
4% PDU
3% PINU–SD
2% PDCH

PNH = National Party of Honduras **PLH** = Liberal Party of Honduras **PDU** = Party of Democratic Unification
PINU–SD = Innovation and Unity Party–Social Democracy
PDCH = Honduran Christian Democratic Party

The traditional power brokers have been the military, the US embassy, and the United Fruit Company (now called Chiquita), the country's biggest banana producer.

The military held power intermittently from the mid-1950s, until pressure from the US government forced it to restore civilian rule in 1984. During the 1980s, US President Reagan effectively converted the country into a US "aircraft carrier" to counter a perceived communist threat from El Salvador and Nicaragua. Peace in the region then saw a cut in US aid.

The PNH and PLH have few real ideological differences. Presidents, able to serve only one four-year term, have tended to be weak. The PLH introduced unpopular austerity measures in 1994, but also began reducing the autonomy of the military by abolishing conscription. PLH president Carlos Flores, elected in 1997, continued this "demilitarization" process by naming a civilian defense minister in 1999. The presidency was won back by the PNH when Ricardo Maduro was elected in 2001.

Reconstruction after the devastation of Hurricane Mitch in 1998 will be a long-term undertaking.

HONDURAS

Total Land Area : 112 090 sq. km
(43 278 sq. miles)

LAND HEIGHT

| 2000m/6562ft |
| 1000m/3281ft |
| 500m/1640ft |
| 200m/656ft |
| Sea Level |

POPULATION

| over 500 000 |
| over 100 000 |
| over 50 000 |
| over 10 000 |
| under 10 000 |

WORLD AFFAIRS

▷ Joined UN in 1945

Hurricane aid, and trade with and immigration to the US are key issues. In 2001 free trade was agreed with El Salvador, Guatemala, and Mexico.

AID

▷ Recipient

💲 $449m (receipts)   ⬇ Down 45% in 2000

Aid from the IMF and the World Bank on favorable terms followed the 1998 earthquake. Western countries agreed debt relief of $1.2 billion.

DEFENSE

▷ No compulsory military service

 $93m   Down 2% in 2000

Until 1994 the military operated with virtual impunity. The first civilian defense minister was appointed in 1999, completing the "demilitarization" process begun with the return to civilian rule in 1984.

ECONOMICS

▷ Inflation 19% p.a. (1990–2000)

 $5.52bn   15.10–15.88 lempiras

SCORE CARD

- ❏ WORLD GNP RANKING........................107th
- ❏ GNP PER CAPITA $860
- ❏ BALANCE OF PAYMENTS....................–$204m
- ❏ INFLATION 11.1%
- ❏ UNEMPLOYMENT...................................28%

STRENGTHS

Coffee, flowers, fruit. Economic boost due to hurricane reconstruction. Barely exploited mineral deposits. Hardwoods.

WEAKNESSES

Servicing of foreign debt. Vulnerability of coffee exports. Banana industry yet to recover from hurricane damage. Slow rate of privatizations. Corruption. Lack of land reform. High unemployment and underemployment. Weak industrial base. Overdependence on hydroelectric power.

EXPORTS

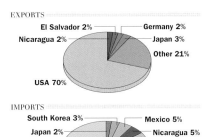

El Salvador 2%   Germany 2%
Nicaragua 2%   Japan 3%
Other 21%
USA 70%

IMPORTS

South Korea 3%   Mexico 5%
Japan 2%   Nicaragua 5%
Other 27%
USA 58%

Tobacco field. *As an export earner tobacco is overshadowed by coffee, shrimps, and melons.*

RESOURCES

▷ Electric power 518,000 kw

 15,395 tonnes   🛢 Not an oil producer

1.72m cattle, 480,000 pigs, 18m chickens   💎 Lead, zinc, silver, gold, copper, iron, tin, coal

Coffee exports dropped dramatically in 2001, due to low world prices. Oil and mineral deposits are being explored. Hydroelectric supply is erratic.

ENVIRONMENT

▷ Sustainability rank: 47th

 6%   0.8 tonnes per capita

The unregulated timber, cotton, and cattle industries, land colonization, and pesticides have led to ecological crisis.

MEDIA

▷ TV ownership medium

 Daily newspaper circulation 55 per 1000 people

PUBLISHING AND BROADCAST MEDIA

 There are 9 daily newspapers, including *La Prensa, El Heraldo,* and *La Tribuna*

 6 independent services   5 services: 1 state-owned, 4 independent

Self-censorship, dependence on US sources, corruption, and intimidation guarantee a largely compliant media.

CRIME

▷ No death penalty

 10,869 prisoners   ⬆ Violent crime is rising

Overzealous security forces are frequently accused of murdering child members of violent street gangs.

EDUCATION

▷ School leaving age: 13

 75%   76,573 students

State-run education follows the US system. The drop-out rate from secondary schools is high.

HEALTH

▷ Welfare state health benefits

 1 per 1250 people   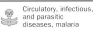 Circulatory, infectious, and parasitic diseases, malaria

Only 66% of people have easy access to health services. Plans were announced in 2000 to privatize the health service.

CHRONOLOGY

Honduras was a Spanish possession until 1821. In 1823, it formed the United Provinces of Central America with four neighboring nations.

- ❏ **1838** Declares full independence.
- ❏ **1890s** US banana plantations set up.
- ❏ **1932–1949** Dictatorship of General Tiburcio Carías Andino of PNH.
- ❏ **1954–1957** Elected PLH president Villeda Morales deposed, reelected.
- ❏ **1963** Military coup.
- ❏ **1969** 13-day Soccer War with El Salvador sparked by World Cup.
- ❏ **1980–1983** PLH wins elections but General Gustavo Alvarez holds real power. Military maneuvers with US. Trades unionists arrested; death squads operate.
- ❏ **1984** Return to democracy.
- ❏ **1988** 12,000 Contra rebels forced out of Nicaragua into Honduras.
- ❏ **1995** Military defies human rights charges.
- ❏ **1998** Hurricane Mitch wreaks havoc.
- ❏ **1999** Appointment of first civilian defense minister.

SPENDING

▷ GDP/cap. increase

CONSUMPTION AND SPENDING

🚗 52 per 1000 population   46 per 1000 population

Defense 1.6%
Education 4%
Health 3.9%

0   5   10   15   20   25
Defense, Health, Education spending as % of GDP

The social structure of Honduras is characterized by great inequalities: 4% of people own 60% of the land. Relief agencies estimate that 85% of people now live below the poverty line, compared with 80% before the devastation of Hurricane Mitch.

WORLD RANKING

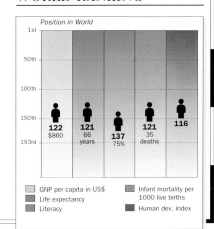

Position in World

	GNP per capita in US$	Life expectancy	Literacy	Infant mortality per 1000 live births	Human dev. index
	122 $860	121 66 years	137 75%	121 35 deaths	116

HUNGARY

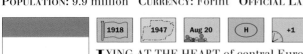

OFFICIAL NAME: Republic of Hungary **CAPITAL:** Budapest
POPULATION: 9.9 million **CURRENCY:** Forint **OFFICIAL LANGUAGE:** Hungarian

L YING AT THE HEART of central Europe, Hungary is landlocked and has borders with seven states. Historically, Hungary has been a cosmopolitan cultural center, and during its years of market socialism was more prosperous than the other Eastern bloc countries. Economic and political reforms have brought it closer to the EU, which it expects to join in the first "wave" of eastward enlargement; Hungary has also become a member of the NATO alliance. In foreign policy it is particularly sensitive about the treatment of Hungarian minorities in neighboring states.

CLIMATE ▷ Continental

WEATHER CHART FOR BUDAPEST

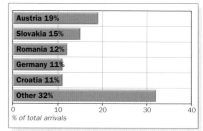

Hungary has a continental climate, with wet springs, late summers, and cold, cloudy winters. There are no great differences of weather and climate within the country. Conditions in summer and winter may, however, differ from one year to the next. The transition between seasons tends to be sudden.

TRANSPORTATION ▷ Drive on right

✈ **Budapest Ferihegy**
4.68m passengers

🚢 Has no fleet

THE TRANSPORTATION NETWORK

🛣 80,680 km (50,132 miles)	448 km (278 miles)
🚆 7988 km (4964 miles)	1373 km (853 miles)

Freight travels mainly via the rail link from Budapest to the Austrian border. Most foreign investment is located along this corridor. A direct link to Slovenia opened in mid-2001. The Budapest–Vienna expressway was the first of four big EU-backed road projects to be completed.

TOURISM ▷ Visitors : Population 1.5:1

🧳 15.3m visitors   ⬇ Down 2% in 2001

MAIN TOURIST ARRIVALS

Austria 19%	
Slovakia 15%	
Romania 12%	
Germany 11%	
Croatia 11%	
Other 32%	

% of total arrivals

Lake Balaton, the traditional destination for summer vacations, was a magnet for east European visitors during the communist period. Since then, Hungary has invested heavily in its tourist facilities, and the number of travel agents and hotels has risen dramatically. Austrians and Slovaks are most numerous among the new clientele the country now attracts. Budapest's baths, some of which date from the Ottoman period, are a distinctive feature, and the capital also promotes itself as an international business convention center.

HUNGARY

Total Land Area : 93 030 sq. km
(35 919 sq. miles)

POPULATION

over 1 000 000	▣
over 500 000	◉
over 100 000	◎
over 50 000	○
over 10 000	●

LAND HEIGHT

500m/1640ft	
200m/656ft	
80m/262ft	

PEOPLE ▷ Pop. density medium

 Hungarian   107/km²
                                           (278/mi²)

THE URBAN/RURAL POPULATION SPLIT

64%   **36%**

RELIGIOUS PERSUASION

Greek Orthodox 3% ———   Lutheran 4%
Other 2% ——   Nonreligious 7%
                             Calvinist 20%
Roman Catholic 64%

ETHNIC MAKEUP

Roma 1% ——   German 2%
Slovak 1% ——   Other 6%
Magyar 90%

Hungary is ethnically homogeneous, although there are small minorities of Germans, Slovaks, Roma, Serbs, Croats, Romanians, and Hungarian Jews. There is little ethnic tension at home, but considerable concern about the treatment of Hungarian minorities in Romanian Transylvania, Serbian Vojvodina, and Slovakia. New legislation in June 2001 gave them special status in Hungary, including the right to work there for three months a year. The 100,000-strong Jewish community in Hungary is the largest in the region, although a fraction of the pre-Holocaust numbers, and is the target of occasional antisemitic outbursts. Prejudice against Roma is widespread and there have been major discrimination rows.

A new bourgeoisie is emerging, but for the unskilled and unemployed life is tougher than under communism. Hungary's suicide rate was the highest in the world throughout almost the whole of the 20th century.

POPULATION AGE BREAKDOWN

Female		Age	Male	
	1.9%	80+	0.8%	
	9.9%	60–79	6.8%	
14.2%		40–59	13%	
13.7%		20–39	14.1%	
12.5%		0–19	13.1%	

% of population by age group

POLITICS ▷ Multiparty elections

2002/2006   President Ferenc Mádl

AT THE LAST ELECTION

National Assembly 386 seats

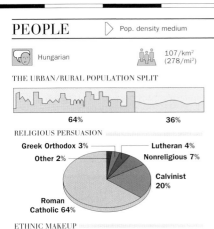

49%   **46%**   **5%**
Fidesz–MPP   **MSzP**   **SzDSz**

Fidesz–MPP = Young Democrats–Hungarian Civic Party
MSzP = Hungarian Socialist Party
SzDSz = Alliance of Free Democrats

Hungary has been a multiparty democracy since 1990.

PROFILE

Hungary's governments since the fall of communism have been relatively stable coalitions. The electoral pendulum has swung at four-yearly intervals between right and left.

József Antall, leader of the Christian democratic nationalist Hungarian Democratic Forum (MDF) and prime minister from 1990, was the dominant figure in Hungarian democratic politics until his death in 1993. However, party disintegrations and disappointing economic results increased apathy and disillusionment among voters, who returned the former communists to power in 1994. The victorious MSzP under Gyula Horn nevertheless preferred to work in coalition in order to ease the passage of economic and social reforms through parliament.

The right-of-center Fidesz–MPP coalition led by Viktor Orbán, which took office after the 1998 election, drove the country toward completing its transition to a market economy ready for EU membership. However, Orbán was narrowly defeated by a coalition of the MSzP and the SzDSz in the 2002 elections. Peter Medgyessy was appointed prime minister.

MAIN POLITICAL ISSUE
Social welfare versus free-market economics

Reforms to assist transition to a market economy have led to strong economic recovery in the Budapest area and the western part of the country. Widening income differentials between young, skilled workers and those in education, health, and other state sectors have provoked protests and strikes, as the new prosperity eludes others.

Ferenc Mádl, known as "Mr. Professor," was elected president in 2000.

Peter Medgyessy, leader of the center-left MSzP and prime minister since 2002.

The majestic Danube River divides the modern city of Budapest between the ancient towns of Buda (foreground) and Pest.

WORLD AFFAIRS ▷ Joined UN in 1955

 CE   CEFTA   NATO   OECD   OSCE

Hungary gained WEU associate status in 1994. In a 1997 referendum 85% of voters endorsed joining NATO, and in 1999 Hungary became a full NATO member. Joining the EU is a slower process. With an association agreement since early 1994, in 1998 Hungary was one of six applicant countries to open formal membership negotiations.

Hungary has a cooperation and friendship treaty with Russia, but relations have been strained by Hungary's open courting of the West. Difficult relations with Slovakia and Romania were eased by friendship treaties concluded in the mid-1990s, but were troubled again by the controversial status law in regard to Hungarians resident abroad.

CHRONOLOGY

The region today occupied by Hungary was first settled by the Finno-Ugrian Magyar peoples from the 8th century. In the 16th and 17th centuries, it came under Austrian domination, lasting until 1867, when Austria-Hungary was formed.

❑ **1918** Hungarian Republic created as successor state to Austria-Hungary.
❑ **1919** Béla Kún leads a short-lived communist government. Romania intervenes militarily and hands power to Adm. Horthy.
❑ **1938–1941** Hungary gains territory from Czechoslovakia, Yugoslavia, and Romania in return for supporting Nazi Germany.
❑ **1941** Hungary drawn into World War II on Axis side when Hitler attacks Soviet Union.
❑ **1944** Nazi Germany preempts Soviet advance on Hungary by invading. Deportation of Hungarian Jews and Roma to extermination camps begins. Soviet Red Army enters in October. Horthy forced to resign.
❑ **1945** Liberated by Red Army. Soviet-formed provisional government ⇨

H

CHRONOLOGY *continued*

installed. Imre Nagy introduces land reform.

❏ **1947** Communists emerge as largest party in second postwar election.

❏ **1948** Forcible merger of Social Democrats with communists; known as Hungarian Socialist Workers' Party (HSWP) from 1956.

❏ **1949** New constitution; formally becomes People's Republic.

❏ **1950–1951** First Secretary Mátyás Rákosi uses authoritarian powers to collectivize agriculture and industrialize the economy.

❏ **1953** Nagy, Rákosi's rival, becomes premier and reduces political terror.

❏ **1955** Nagy deposed by Rákosi.

❏ **1956** Rákosi out. Student demonstrations, demanding withdrawal of Soviet troops and Nagy's return, become popular uprising. Nagy appointed premier and János Kádár First Secretary. Nagy announces Hungary will leave Warsaw Pact. Three days later, Soviet forces suppress protests. About 25,000 killed. Kádár becomes premier.

❏ **1958** Nagy executed.

❏ **1968** Kádár introduces New Economic Mechanism to bring market elements to socialism.

❏ **1986** Police suppress commemoration of 1956 uprising. Democratic opposition demands Kádár resign.

❏ **1987** Party reformers establish MDF as a political movement.

❏ **1988** Kádár ousted. Protests force suspension of plans for Nagymaros Dam on the Danube.

❏ **1989** Parliament votes to allow independent parties. Posthumous rehabilitation of Nagy, who is given state funeral. Round table talks between HSWP and opposition.

❏ **1990** József Antall's MDF wins multiparty elections decisively. Speed of economic reform hotly debated. Árpád Göncz president.

❏ **1991** Warsaw Pact dissolved. Last Soviet troops leave.

❏ **1994** Hungary joins NATO's Partnerships for Peace program. Former communist MSzP wins general election. Austerity program prompts protests.

❏ **1998** EU entry negotiations open. Elections: Viktor Orbán (Fidesz–MPP) forms right-of-center coalition.

❏ **1999** Joins NATO. Airspace used in NATO bombing of Yugoslavia.

❏ **2000** Ferenc Mádl succeeds Göncz as president.

❏ **2002** Elections won by socialist and free democrat alliance. Peter Medgyessy prime minister.

AID

 Recipient

 $252m (receipts)   ⬆ Up 1% in 2000

Hungary received substantial Western aid in 1990–1996, but by the end of the decade was considered able to attract investment mainly on commercial terms. EU and World Bank assistance moved to focus on targeting disadvantaged social groups, raising environmental standards, and strengthening market institutions.

DEFENSE

 Compulsory military service

 $777m   ⬆ Up 1% in 2000

Troop numbers were more than halved, and conventional arms and the military hierarchy were updated in advance of NATO membership in 1999. The emphasis has switched toward more flexibility and rapid response. Almost immediately on joining NATO, Hungary permitted it to use its airspace to bomb Yugoslavia. Military service has been shortened to six months with effect from 2002.

HUNGARIAN ARMED FORCES

753 main battle tanks (515 T-55, 238 T-72)	13,160 personnel
None	None
46 combat aircraft (27 MiG-29)	7500 personnel
None	

ECONOMICS

 Inflation 19% p.a. (1990–2000)

 $47.2bn   282.3–274.8 forint

SCORE CARD

❏ World GNP Ranking 52nd
❏ GNP per Capita $4710
❏ Balance of Payments ....–$1.49bn
❏ Inflation 9.8%
❏ Unemployment 6%

EXPORTS

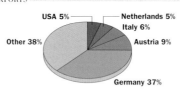

USA 5%
Netherlands 5%
Italy 6%
Other 38%
Austria 9%
Germany 37%

IMPORTS

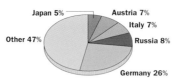

Japan 5%
Austria 7%
Italy 7%
Other 47%
Russia 8%
Germany 26%

STRENGTHS

Openness to foreign direct investment, especially since 1998. Favorable tax regime, streamlined bureaucracy. Strong export-led growth since late 1990s. High industrial production, especially at new, state-of-the-art factories. Currency fully convertible from mid-2001. Inflation dropping.

WEAKNESSES

Low energy efficiency. East–west split as development bypasses rural eastern areas. Widening income differentials. Money laundering a challenge to finance industry regulators; Hungary has featured on OECD blacklists.

ECONOMIC PERFORMANCE INDICATOR

Consumer Price Index — GDP

PROFILE

The collapse of COMECON (communist economic bloc) caused a reorientation of trade toward western Europe. Exports increased rapidly and competitiveness has improved. However, the economy did not recover to its pre-1989 level until 1999. Privatization has reduced the state-owned share of the economy, from 85% to 15%, and has helped cut external debt.

HUNGARY : MAJOR BUSINESSES

Agribusiness	Heavy engineering
Metallurgy △	Light engineering
Electronics	Pharmaceuticals
Chemicals	Vehicle assembly
Textiles	Computer software

0   100 km
0   100 miles

* significant multinational ownership

RESOURCES

 Electric power 7m kw

 19,461 tonnes

25,263 b/d (reserves 55m barrels)

4.83m pigs, 3.3m turkeys, 30.7m chickens

Bauxite, coal, oil, natural gas, lignite

ELECTRICITY GENERATION

- Hydro 0%
- Combustion 62% (23bn kwh)
- Nuclear 38% (14bn kwh)
- Other 0%

% of total generation by type

Hungary has bauxite, brown coal, lignite, and natural gas reserves. It depends for about 40% of its electricity on nuclear

HUNGARY : LAND USE

- Forest
- Pasture
- Cropland
- Pigs
- Cereals
- Vineyards – cash crop

energy from the Paks complex, north of Baja. Fertile farmlands provide grains, sugar beet, and potatoes. Wine production is also important.

ENVIRONMENT

 Sustainability rank: 11th

7% (5% partially protected)

5.8 tonnes per capita

ENVIRONMENTAL TREATIES

Yes	Yes	Yes
Yes	Yes	No

A high sulfur content in Hungary's fossil fuels exacerbates the serious air pollution in industrial zones. The energy industry must meet new standards by 2003. A "green card" system has been introduced to favor the use of cars with catalytic convertors and reduce serious pollution from older vehicles.

Industrial pollution of the Sajo, Tisza, and Danube river systems, sometimes originating in neighboring countries, is a major problem.

MEDIA

 TV ownership high

Daily newspaper circulation 186 per 1000 people

PUBLISHING AND BROADCAST MEDIA

There are 40 daily newspapers, including *Népszabadság* and *Népszava*

2 services: 1 state-owned, 1 independent

4 services: 1 state-owned, 3 independent

Newspapers and magazines are fiercely independent and critical of government policy. In 1994, the Constitutional Court declared that state interference in the media was unlawful, but allegations of interference persist. The boards controlling state TV and radio must have equal representation from government and opposition under a 1996 media law, but the Orbán government was accused of bending or ignoring the rules.

CRIME

 No death penalty

15,588 prisoners

Down 15% in 1999

CRIME RATES

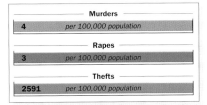

Murders
4 per 100,000 population

Rapes
3 per 100,000 population

Thefts
2591 per 100,000 population

An alarming trend in the late 1990s was the increase in murders of elderly people for financial gain. Organized crime, money laundering, and smuggling of illegal immigrants are rising.

EDUCATION

 School leaving age: 16

99%

171,516 students

THE EDUCATION SYSTEM

- Primary: 100%
- Secondary: 98%
- Tertiary: 34%

% of each age group in education

Education is free and compulsory from the age of six to 16. Bilingual schools have been established in southern Hungary to promote the languages of the national minorities. In 1999–2000 a major transformation of the education system took place, as a result of which, with effect from 2000, there are 30 universities and colleges run by the state, 26 run by the Church, and six colleges run by various foundations.

HEALTH

 Welfare state health benefits

1 per 313 people

Cerebrovascular and heart diseases, cancers, accidents

Medical treatment has traditionally been free to all, although there is a contribution to prescription costs. State sickness benefits remain relatively generous. Spending on the health service has fallen in recent years in real terms; at $600 per capita, it is only one-third the OECD average, and there is concern that Hungary's health care sector is among the least developed of OECD countries. The ratio of doctors to patients is high, but there is a shortage of nurses. Family physician services are being privatized rapidly under a law passed in 2000.

SPENDING

 GDP/cap. increase

CONSUMPTION AND SPENDING

238 per 1000 population

372 per 1000 population

- Defense 1.7%
- Education 4.6%
- Health 5.2%

Defense, Health, Education spending as % of GDP

Hungary enjoys a higher standard of living than other former communist countries except Slovenia, and demand for luxury goods is rising. Access to mobile phones and the Internet is relatively high. Real wages, which fell by 15% in the mid-1990s, had regained most of this ground by 2000. Hungarians still have to work longer hours to pay for basic consumer goods than workers in western Europe. Salaries in the public services have not kept pace with the rising cost of living, and as a result there is a growing disparity between those working in the state and private sectors. The Roma minority suffers particularly over access to housing, which is in short supply.

WORLD RANKING

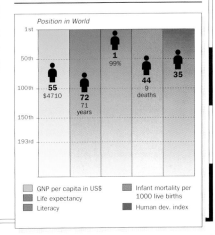

Position in World

55	72	1	44	35
$4710	71 years	99%	9 deaths	

- GNP per capita in US$
- Life expectancy
- Literacy
- Infant mortality per 1000 live births
- Human dev. index

H

ICELAND

EUROPE

OFFICIAL NAME: Republic of Iceland   CAPITAL: Reykjavík
POPULATION: 281,000   CURRENCY: Icelandic króna   OFFICIAL LANGUAGE: Icelandic

 1944   1944   June 17   IS   0   +354   .is

EUROPE'S WESTERNMOST country, Iceland has a strategic location in the North Atlantic, just south of the Arctic Circle. Its position, on the rift where the North American and European continental plates are pulling apart, accounts for its 200 volcanoes and its numerous geysers and solfataras. Previously a Danish possession, Iceland became fully independent in 1944. Most settlements are along the coast, where ports remain ice-free in winter.

CLIMATE ▷ Subarctic

WEATHER CHART FOR REYKJAVÍK

Iceland sits in the Gulf Stream. Winters are consequently mild. Summers are cool, with fine, long sunny days.

TRANSPORTATION ▷ Drive on right

Keflavík International, Reykjavík
1.46m passengers

1027 ships
210,864 grt

THE TRANSPORTATION NETWORK

| 3439 km (2137 miles) | None |
| None | None |

Icelanders rely entirely on cars and internal airplane and helicopter flights. Most freight moves by sea. The only main road circles the island.

TOURISM ▷ Visitors : Population 1.1:1

303,000 visitors   Up 15% in 2000

MAIN TOURIST ARRIVALS

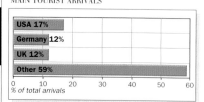

USA 17%		
Germany 12%		
UK 12%		
Other 59%		

0   10   20   30   40   50   60
% of total arrivals

Iceland is promoting itself, especially in Japan, as an upmarket destination for tourists who are attracted by its spectacular scenery, glaciers, green valleys, fjords, and hot springs.

PEOPLE ▷ Pop. density low

Icelandic   3/km² (7/mi²)

THE URBAN/RURAL POPULATION SPLIT

93%   7%

RELIGIOUS PERSUASION

Other Christian 1%   Nonreligious 6%
Evangelical Lutheran 93%

Descended from Norwegians and Celts, Icelanders form an ethnically homogeneous society. Almost all follow the Evangelical Lutheran Church. More than half the population live in or near Reykjavík. Living standards are high, and there are few social tensions.

POLITICS ▷ Multiparty elections

1999/2003   President Olafur Ragnar Grimsson

AT THE LAST ELECTION
Parliament 63 seats

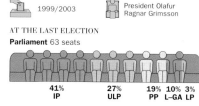

| 41% IP | 27% ULP | 19% L–GA | 10% PP | 3% LP |

IP = Independence Party   **ULP** = United Left Party
PP = Progressive Party   **L–GA** = Left–Green Alliance
LP = Liberal Party

From independence Iceland has been ruled by coalitions, but in the 1980s the traditional four-party system began to splinter. After the 1991 election, a new IP–Social Democratic coalition promoted market-led reforms. Arguments over whether or not to join the EU were defused in 1992 with the successful negotiation of the EEA, giving Iceland access to the key EU market.

The coalition collapsed after the 1995 general election, when both parties had lost support, and was replaced by a center-right government led by the IP, with David Oddsson as prime minister. He has successfully built on an economic recovery under way since 1994, and strengthened his position in the 1999 general election.

ICELAND

Total Land Area : 103 000 sq. km
(39 768 sq. miles)

POPULATION
○ over 50 000
● over 10 000
• under 10 000

LAND HEIGHT
1000m/3281ft
500m/1640ft
200m/656ft
Sea Level
Ice Cap

WORLD AFFAIRS ▷ Joined UN in 1946

CE NATO OECD OSCE EEA

Although a member of NATO and EFTA, Iceland has traditionally maintained arm's length relations with the EU and the US. Major disputes have concerned fishing rights. Iceland left the IWC in 1992 when a commercial whaling ban was extended, but rejoined in 2001. Links with other Nordic states are strong.

AID ▷ Donor

 $9m (donations)   Up in 2000

Aid donations are modest, and form a smaller proportion of the budget than in other Scandinavian states.

DEFENSE ▷ No compulsory military service

Coastguard is only military force   Not applicable

A NATO member and an associate WEU member, Iceland has no armed forces. US troops are based at Keflavík.

ECONOMICS ▷ Inflation 3.2% p.a. (1990-2000)

 $8.54bn   84.7–102.9 Icelandic krónur

SCORE CARD

- ❑ World GNP Ranking...........................91st
- ❑ GNP per Capita $30,390
- ❑ Balance of Payments....................–$851m
- ❑ Inflation 5.2%
- ❑ Unemployment 1%

STRENGTHS
High-tech fishing industry with exclusive access to prime fishing grounds. Strong economic recovery in late 1990s; low inflation and unemployment. Very cheap geothermal power.

WEAKNESSES
Over 70% of export earnings derived from single source: fish. State-owned banking sector restricts flexibility.

EXPORTS

Portugal 6%, Netherlands 7%, USA 12%, Other 40%, Germany 16%, UK 19%

IMPORTS
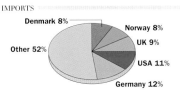
Denmark 8%, Norway 8%, UK 9%, Other 52%, USA 11%, Germany 12%

Lava towers, *near Lake Mývatn in northern Iceland – an area of grassy lowlands. Iceland's center consists of lava desert and glaciers.*

RESOURCES ▷ Electric power 1.2m kw

 1.74m tonnes   Not an oil producer

465,000 sheep, 78,000 horses, 180,000 chickens   Diatomite

Iceland has virtually no minerals. All energy needs are met by geothermal and hydroelectric sources. It has implemented measures to try to restore once abundant fish stocks.

ENVIRONMENT ▷ Sustainability rank: 8th

 9% (7% partially protected)   7.6 tonnes per capita

Iceland has no nuclear or coal-fired power plants. Pollution levels are low. Believing that minke whales eat valuable cod stocks, Iceland resumed whale hunting in 1992. The 1996 eruption of the Loki volcano under Vatna glacier caused extensive flooding and damage.

MEDIA ▷ TV ownership high

 Daily newspaper circulation 535 per 1000 people

PUBLISHING AND BROADCAST MEDIA

There are 4 daily newspapers, including *Dagbladid-Visir* and *Morgunbladid*, which has the largest circulation

11 services: 1 state-owned, 10 independent   17 services: 1 state-owned, 16 independent

Iceland is renowned for having one of the highest per capita newspaper circulations in the world.

CRIME ▷ No death penalty

 82 prisoners   Crime rates are rising

Crime rates are comparatively low. The rate of alcohol-related murders is higher than the European average.

EDUCATION ▷ School leaving age: 16

99%   10,543 students

Icelanders buy more books per capita than any other nation. Education is state-run; some 40% of school students go on to university at Reykjavík or Akureyri, or to colleges in the US.

HEALTH ▷ Welfare state health benefits

1 per 307 people   Heart disease, cancers, accidents

The state health system is free to all Icelanders. Iceland has one of the lowest infant mortality rates and one of the highest longevity rates in the world.

SPENDING ▷ GDP/cap. increase

CONSUMPTION AND SPENDING

572 per 1000 population   701 per 1000 population

Defense Not applicable
Education 5.4%
Health 7.4%
Defense, Health, Education spending as % of GDP

Wealth distribution in Iceland is comparatively even and social mobility is high. Domestic heating, extracted from geothermal sources, is provided at almost no cost.

WORLD RANKING

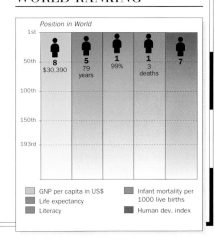
Position in World: 8 $30,390 | 5 79 years | 1 99% | 1 3 deaths | 7

GNP per capita in US$; Life expectancy; Literacy; Infant mortality per 1000 live births; Human dev. index

INDIA

SOUTH ASIA

OFFICIAL NAME: Republic of India CAPITAL: New Delhi
POPULATION: 1.03 billion CURRENCY: Indian rupee OFFICIAL LANGUAGES: Hindi and English

| 1947 | 1947 | Jan 26 | IND | +5.5 | +91 | .in |

SEPARATED FROM the rest of Asia by the Himalaya mountain range, India forms the bulk of a subcontinent. As well as the Himalayas, there are two other main geographic regions, the Indo-Gangetic plain, which lies between the foothills of the Himalayas and the Vindhya Mountains, and the central–southern plateau. India is the world's largest democracy and second most populous country after China. The birthrate has recently been falling, but even at its current level India's population will probably overtake China's by 2030. After years of protectionism, India is opening up its economy to the outside world. The hope is that the free market will go some way to alleviating one of the country's major problems, poverty.

CLIMATE

▷ Tropical/subtropical/desert/mountain/monsoon

WEATHER CHART FOR NEW DELHI

During the hot season, temperatures in the north can reach 40°C (105°F). The monsoon breaks in June and peters out in September or October. In the cool season, average temperatures are 5°C–15°C (40°F–60°F) in the north and the weather is mainly dry. The south has a less variable climate. Chennai (Madras) is always hot: average temperatures range from 24°C (75°F) in January to 32°C (90°F) in May and June.

TRANSPORTATION

▷ Drive on left

Mumbai International
12.1m passengers

987 ships
6.66m grt

THE TRANSPORTATION NETWORK

| 1.52m km (942,666 miles) | 33,500 km (20,816 miles) |
| 62,810 km (39,030 miles) | 16,180 km (10,054 miles) |

India's state-owned railroad system, the largest in Asia, spans all the major cities. Rail carries 40% of passenger traffic and 65% of freight. Strict controls on diesel emissions from cars and buses to check pollution were enforced in 2001. Cycle and scooter rickshaws are common in urban centers. Kolkata (Calcutta) still has rickshaws pulled by hand.

INDIA

Total Land Area : 3 287 590 sq. km
(1 269 338 sq. miles)

POPULATION

▣	over 5 000 000
▣	over 1 000 000
◉	over 500 000
◎	over 100 000
●	over 10 000

LAND HEIGHT

5000m/16 405ft
4000m/13 124ft
3000m/9843ft
2000m/6562ft
1000m/3281ft
500m/1640ft
200m/656ft
Sea Level

0   200 km
0   200 miles

*A **religious festival**. Such festivals are a frequent occurrence and form an important part of Hindu culture.*

PEOPLE

 Pop. density high

Hindi, English, Urdu, Bengali, Marathi, Telugu, Tamil, Bihari, Gujarati, Kanarese

345/km² (893/mi²)

THE URBAN/RURAL POPULATION SPLIT

28%   72%

RELIGIOUS PERSUASION

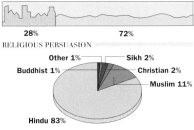

- Other 1%
- Buddhist 1%
- Sikh 2%
- Christian 2%
- Muslim 11%
- Hindu 83%

ETHNIC MAKEUP

- Mongoloid and Other 3%
- Dravidian 25%
- Indo-Aryan 72%

India is the world's second most populous country after China, officially passing the one billion mark in 2000. Despite a major birth control program, the decrease in population growth has been marginal. Nationwide awareness campaigns aim to promote the idea of smaller families. India's planners consider the rise in the population the most significant brake on development. Cultural and religious pressures encourage large families, however, and the extended family is seen as essential security for old age.

The fertile rice-growing areas of the Gangetic plain and delta are very densely populated. The northern state of Uttar Pradesh has the largest population, followed by neighboring Bihar and the western state of Maharashtra. Maharashtra is also the most urbanized state, with more than half of its people living in towns or cities. Elsewhere, most Indians live in rural areas, although poverty continues to drive many to the swelling cities.

The overwhelming majority of the population are Hindus, who belong to thousands of castes and subcastes, which largely determine status, occupation, and whom they marry. Tension between Hindus and Muslims has grown in recent years and escalated sharply in 2002 following violent clashes in Gujarat.

POPULATION AGE BREAKDOWN

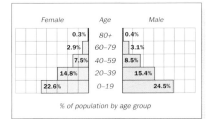

Female	Age	Male
0.3%	80+	0.4%
2.9%	60–79	3.1%
7.5%	40–59	8.5%
14.8%	20–39	15.4%
22.6%	0–19	24.5%

% of population by age group

TOURISM

Visitors : Population 1:388

2.64m visitors   Up 6% in 2000

MAIN TOURIST ARRIVALS

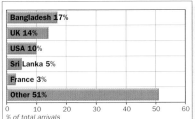

Bangladesh	17%
UK	14%
USA	10%
Sri Lanka	5%
France	3%
Other	51%

0   10   20   30   40   50   60
% of total arrivals

Tourism is India's sixth-largest foreign exchange earner. More luxury hotels are being built, and wildlife and adventure tourism are being promoted. However, India has only a small share of the world tourism market, and has suffered recently from security worries over the repercussions of the US-led "war on terrorism" and acute tensions with Pakistan, particularly over Kashmir.

CHRONOLOGY

The origins of an Indus Valley civilization may be traced back to the third millennium BCE. By the 3rd century BCE, the Mauryan kingdom under Ashoka encompassed most of modern India. Following the Battle of Plassey in 1757, British rule – through the East India Company – was consolidated.

- ❏ **1885** Formation of Indian National Congress.
- ❏ **1919** Act of Parliament introduces "responsible government."
- ❏ **1920–1922** Mahatma Gandhi's first civil disobedience campaign.
- ❏ **1935** Government of India Act grants autonomy to provinces.
- ❏ **1936** First elections under new constitution.
- ❏ **1942–1943** "Quit India" movement.
- ❏ **1947** August, independence and partition into India and Pakistan. Jawarhalal Nehru becomes first prime minister.
- ❏ **1948** Assassination of Mahatma Gandhi. War with Pakistan over Kashmir. India becomes a republic.
- ❏ **1951–1952** First general election won by Congress party.
- ❏ **1957** Congress party reelected. First elected communist state government installed in Kerala.
- ❏ **1960** Bombay divided into states of Gujarat and Maharashtra.
- ❏ **1962** Congress party reelected. Border war with China.
- ❏ **1964** Death of Nehru. Lal Bahadur Shastri becomes prime minister.
- ❏ **1965** Second war with Pakistan over Kashmir.
- ❏ **1966** Shastri dies; Indira Gandhi (daughter of Jawarhalal Nehru) becomes prime minister.
- ❏ **1969** Congress party splits into two factions; larger faction led by Indira Gandhi.
- ❏ **1971** Indira Gandhi's Congress party wins elections. Third war with Pakistan, over creation of Bangladesh.
- ❏ **1972** Simla (peace) Agreement signed with Pakistan.
- ❏ **1974** Explosion of first nuclear device in underground test.
- ❏ **1975–1977** Imposition of state of emergency.
- ❏ **1977** Congress loses general election. People's Party (JD) takes power at the center.
- ❏ **1978** New political group, Congress (Indira) – Congress (I) – formally established.
- ❏ **1980** Indira Gandhi's C(I) wins general election.
- ❏ **1984** Indian troops storm Sikh Golden Temple in Amritsar. Assassination of Indira Gandhi

I

I

CHRONOLOGY *continued*

by Sikh bodyguard; her son Rajiv becomes prime minister and C(I) leader. Gas explosion at US-owned Union Carbide Corporation plant in Bhopal kills 2000 people in India's worst industrial disaster.

❑ **1985** Peace accords with militant separatists in Assam and Punjab.

❑ **1987** Deployment of Indian peacekeeping force in Sri Lanka to combat Tamil Tigers.

❑ **1989** General election; National Front forms minority government with BJP support. C(I) implicated in Bofors scandal.

❑ **1990** Withdrawal of troops from Sri Lanka.

❑ **1991** Rajiv Gandhi assassinated. Narasimha Rao becomes prime minister of a C(I) minority government and initiates economic liberalization.

❑ **1992** Demolition of the Babri Masjid mosque at Ayodhya by Hindu extremists triggers widespread violence; 1200 people die.

❑ **1993** Resurgence of Hindu–Muslim riots. Bomb explosions in Bombay (Mumbai). Border troop agreement with China.

❑ **1994** Rupee made fully convertible. C(I) routed in key state elections amid increasing allegations of corruption in ruling party. Outbreak of pneumonic plague.

❑ **1995** Punjab chief minister assassinated by Sikh extremists.

❑ **1996** Corruption scandal triggers political crisis. C(I) suffers its worst electoral defeat. Leftist United Front coalition government takes office.

❑ **1997** Successive governments fall as C(I) withdraws support.

❑ **1998** General election; BJP led by Atal Bihari Vajpayee forms coalition government. Sonia Gandhi, widow of Rajiv Gandhi, becomes president of C(I). India tests nuclear bomb.

❑ **1999** Vajpayee travels to Pakistan to inaugurate bus service between India and Pakistan. India and Pakistan test nuclear missiles, and engage in violent confrontation in Kashmir. BJP returned to power after elections triggered by vote of no confidence.

❑ **2001** Earthquake kills more than 25,000 in Gujarat. BJP government implicated in major bribery scandal.

❑ **2001–2002** Terrorist attacks by Kashmiri separatists precipitate crisis with Pakistan.

❑ **2002** More than 2000, mainly Muslims, killed in Gujarat following worst intercommunal riots since independence. Heatwave kills more than 600.

POLITICS ▷ Multiparty elections

L. House 1999/2003
U. House 2002/2004

President A. P. J. Abdul Kalam

AT THE LAST ELECTION

House of the People (Lok Sabha) 545 seats

| 49% BJP | 31% C(I) | 18% UF | 2% Others |

BJP = Bharatiya Janata Party and Allies, including the All-India Anna Dravida Munnetra Kazhagam
C(I) = Congress (I) and Allies
UF = United Front and Allies, including the Communist Party of India (Marxist) – **CPI(M)**
Nom = Nominated

Others include the Rashtriya Janata Dal, independents, and two appointed seats reserved for Anglo-Indians

Council of States (Rajya Sabha) 245 seats

| 24% C(I) | 19% BJP | 5% CPI(M) | 5% Nom | 47% Others |

233 members are elected to the Rajya Sabha by State Legislative Assemblies, and 12 "distinguished citizens" are nominated by the head of state

India is a multiparty democracy. The Lok Sabha (lower house) is directly elected by universal adult suffrage, while the Rajya Sabha (upper house) is indirectly elected by the state assemblies. There are 28 self-governing states. Of the seven union territories, Delhi and Pondicherry have their own assemblies.

Profile

The ascendancy of the Hindu nationalist BJP, in government since 1998, has relegated the C(I) – founded in 1978 as successor to the historic Congress Party which led India to independence in 1947 – to an unprecedented period of opposition at union level. Under P. V. Narasimha Rao's leadership in 1991–1996, a bold program of economic liberalization broke with the C(I)'s traditionally left-of-center policies. However, allegations of corruption undermined the party, resulting in heavy electoral defeats in 1996 and 1998. After 1998 hopes for C(I)'s political revival rested on its president Sonia Gandhi, who, as the widow of the assassinated former prime minister Rajiv Gandhi, restored the influence of the Nehru family over the party.

In 1996 regional parties formed the core of a center-left United Front coalition government in power until 1997. However, the 1998 election established the BJP with a strong enough mandate to lead a coalition government headed by its leader, A. B. Vajpayee. Elections held in 1999 returned the coalition to power with an overall majority.

Main Political Issues
Hindu militancy

The right-wing Hindu BJP has emerged as a credible alternative to the C(I). In 1996 it won most parliamentary seats, despite being tainted by corruption allegations. Although it failed on that occasion to form a viable administration, the BJP was returned to power in 1998 and 1999. Its rise has encouraged the spread of Hindu nationalism and raised fears about the future of India's secular constitution. In 2002 the BJP state government in Gujarat was accused by its critics of sanctioning mob attacks against the minority Muslim community in Ahmadabad during communal riots in the state.

Political corruption

Allegations of political corruption have dominated Indian politics. In 1989, the C(I) prime minister Rajiv Gandhi was accused of accepting bribes from a Swedish arms company. The C(I) was also implicated in a financial scandal in 1992. In 1996 corruption forced the resignations of several C(I) government ministers and the leader of the opposition BJP. The issue resurfaced in 2001 when the BJP government was implicated in a bribery scandal over arms sales which led to the resignation of the defense minister.

The free market

The introduction of the free-market economy has been vigorously resisted. Critics contend that free trade would undermine local production and participation by foreign companies damage the national economy. In the mid-1990s governments increased spending on rural development programs in order to soften the impact of economic liberalization. The BJP-led coalition, although more sympathetic to urban interests, is generally opposed to competition from foreign businesses.

A. B. Vajpayee, BJP leader and prime minister since 1998.

A. P. J. Abdul Kalam, a Muslim nuclear scientist, elected president in 2002.

Sonia Gandhi has revived the fortunes of the opposition Congress (I).

WORLD AFFAIRS ▷ Joined UN in 1945

| Comm | G15 | G24 | NAM | SAARC |

The overriding preoccupation in foreign policy is the dispute with Pakistan over Kashmir, which has sparked two wars, in 1948 and 1965, and intensified the nuclear race between the two countries. High-level talks occasionally raise hopes of better relations, but border clashes stoke tensions. In 2002 a series of attacks across the Line of Control in Kashmir by Muslim militants, allegedly sponsored by Pakistan, pushed the two countries to the brink of war.

In 1998 India carried out nuclear weapons tests, prompting Pakistan to follow suit and sparking international condemnation. The US imposed sanctions on India for ignoring the ban on nuclear weapons testing, but India has still not signed the Comprehensive Test Ban Treaty. US sanctions were lifted in 2001 following India's support for the "war on terrorism."

AID ▷ Recipient

 $1.49bn (receipts) ⬍ Little change in 2000

India does not depend on aid. The US suspended aid following nuclear tests in 1998, but restored payments in late 2001. The World Bank has withdrawn funding for the Narmada Dam project. International relief aid helped victims of the 2001 Gujarat earthquake.

DEFENSE ▷ No compulsory military service

$14.5bn ⬆ Up 4% in 2000

INDIAN ARMED FORCES

🛡	3414 main battle tanks (700 T-55, 1500 T-72M1, 1200 *Vijayanta*, 14 *Arjun*)	1.1m personnel
🚢	16 submarines, 1 carrier, 8 destroyers, 11 frigates, 7 corvettes, and 39 patrol boats	53,000 personnel
✈	738 combat aircraft (84 *Jaguar* S(I), MiG-21/23/27/29)	110,000 personnel
	Capability undisclosed; weapons tested in 1998	

India considers the possession of a nuclear deterrent to be vital, and tested its own weapons in 1998. It has the world's fourth-largest army, and produces its own *Arjun* battle tank. The *Agni*-II intermediate-range missile, capable of carrying a nuclear warhead to all parts of Pakistan, was tested in 2001. Also that year a much-delayed jet fighter program reached test flight stages. However, the replacement of India's outdated foreign weaponry was disrupted by US sanctions between 1998 and 2001. Russia signed a $10 billion air defense deal in 2001.

ECONOMICS ▷ Inflation 8.5% p.a. (1990–2000)

📊 $455bn   💲 46.68–48.22 Indian rupees

SCORE CARD

❑ WORLD GNP RANKING	12th
❑ GNP PER CAPITA	$450
❑ BALANCE OF PAYMENTS	–$4.2bn
❑ INFLATION	4%
❑ UNEMPLOYMENT	Widespread underemployment

EXPORTS

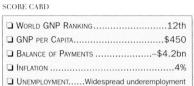

Germany 5% UK 5% Japan 5% Hong Kong 6% USA 23% Other 56%

IMPORTS

Singapore 6% Japan 6% Belgium 7% UK 7% USA 8% Other 66%

STRENGTHS

Massive home market of over one billion people. Cheap labor. Some of the workforce possess skills for new high-tech industries such as software programming, a major boom area of the economy. Vibrant film industry: "Bollywood." Highly efficient textile sector and garment manufacturers. Growing competitiveness in world market, reflected in strong export growth. Competition is encouraging firms to manufacture to international standards. Despite strong objections from opposition parties, India ratified the GATT world trade agreement in 1995.

Since the economy was opened up to foreign competition in 1991, foreign direct investment has risen massively. Much of this went into the power sector. Large multinationals, such as Coca-Cola and IBM, are expanding, despite protests from some sections of the ruling BJP hostile to the growing presence of foreign businesses.

WEAKNESSES

Large budget deficit dogs economy. Governments have found it politically difficult to move away from the old system of widespread subsidies. Value of rupee has declined sharply. Inflation compounded by rise in wholesale food prices. Poor communications systems and power shortages hinder growth.

PROFILE

India has the fastest growing economy in Asia after China. From a highly protectionist mixed economy, which succeeded in building the basis of a modern industrial state, India has to a large extent converted to a free-market economy and is entering the global marketplace. Wide-ranging reforms, from lowering trade barriers to attracting foreign investment, have been put in place. The United Front and BJP-led governments in power since 1996 have not undone these reforms, though they have been criticized for not being wholehearted about driving them forward. Meanwhile, in the rural economy millions of people grapple with the problems of subsistence farming.

ECONOMIC PERFORMANCE INDICATOR

Consumer Price Index ─── GDP ▨

INDIA : MAJOR BUSINESSES

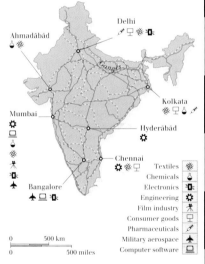

Textiles	※
Chemicals	⬦
Electronics	◈
Engineering	⚙
Film industry	✇
Consumer goods	🖵
Pharmaceuticals	◔
Military aerospace	✈
Computer software	▭

0   500 km
0   500 miles

Hillside monastery in Ladakh, *Kashmir, northern India. The Ladakhi Buddhists maintain their traditional farming existence and are known for their friendliness.*

RESOURCES ▷ Electric power 107m kw

 5.35m tonnes

782,000 b/d (reserves 4.8bn barrels)

 220m cattle, 124m goats, 413m chickens

Iron, diamonds, coal, limestone, zinc, lead

ELECTRICITY GENERATION

Hydro 17% (83bn kwh)
Combustion 81% (399bn kwh)
Nuclear 2% (12bn kwh)
Other 0%

0   20   40   60   80   100

% of total generation by type

ENVIRONMENT ▷ Sustainability rank: 116th

 5% (3% partially protected)

1.1 tonnes per capita

ENVIRONMENTAL TREATIES

🌿 Yes   🦏 Yes   🌐 Yes

Yes   No   No

Deforestation is one of India's most pressing environmental problems. Industrial and agricultural pressures have felled almost 90% of original forest cover. This results in major soil erosion, the silting up of dams, and landslides. Unusually serious flooding in eastern states in 2000 was largely attributed to deforestation. On the other hand, dealing with water scarcities, such as the drought which affected much of the northwest in 2000, has become a major public policy issue.

MEDIA ▷ TV ownership medium

Daily newspaper circulation 28 per 1000 people

PUBLISHING AND BROADCAST MEDIA

There are 5157 daily newspapers. The *Times of India*, the *Statesman*, and the *India Express* publish nationally

1 state-owned service

1 state-owned service

Satellite TV is increasingly popular in India. Services range from the BBC World Service to CNN, Hindi-language Zee TV to MTV, and one state-run channel. More than seven million households are estimated to have acquired dishes. State-run terrestrial TV has suffered as a result. Critics fear a Western onslaught on Indian values. Recent newspaper launches include the *Asian Age*, which is simultaneously published in London by satellite and claims to be India's first truly international paper.

India's most significant mineral exports are iron ore and cut diamonds. In addition, there are large coal reserves; India is the world's fourth-largest coal producer. The steel industry has been privatized. Steel imports are now subject to lower duties, but the industry has so far withstood external competition, and exports have increased. However, steel production, which consumes up to twice as much energy as the method used by some foreign competitors, is

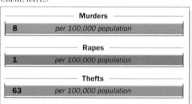

INDIA - LAND USE

Cropland
Forest
Pasture
Wetlands
Desert
High mountain regions
Cattle
Tea – cash crop
Cotton – cash crop
Rice

0   500 km
0   500 miles

CRIME ▷ Death penalty in use

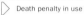 381,147 prisoners   Crime is rising

CRIME RATES

Murders	
8	*per 100,000 population*

Rapes	
1	*per 100,000 population*

Thefts	
63	*per 100,000 population*

Interreligious violence is sporadic but serious. Attacks on Christians drew particular attention, but were overshadowed by Hindu–Muslim violence in Gujarat in 2002. Security forces gained increased powers that year under the Prevention of Terrorism Ordinance, passed in response to Kashmiri separatist attacks in Delhi and other cities.

Violent crime is increasing, especially in the big cities. Gangs have made vast profits from smuggling, prostitution, narcotics, protection, and extortion rackets. Theft has risen sharply as consumer spending increases, and Internet crime is a problem.

Dacoits still operate in large areas of central India. Modeled on the *thugee* gangs of the 19th century, they are outlaws who live by highway robbery and terrorizing small rural communities.

inefficient by international standards.

The state is currently unable to meet the country's demand for electricity. Petroleum and coal are the main sources of energy generation, although these are imported. A scheme in Maharashtra to increase output by attracting investment from a US-led consortium was mired in controversy in the mid-1990s when the BJP state government temporarily suspended negotiations in response to nationalist groups opposed to foreign businesses.

EDUCATION ▷ School leaving age: 12

57%   6.06m students

THE EDUCATION SYSTEM

% of each age group in education

Primary 100%
Secondary 49%
Tertiary 7%

Education is primarily the responsibility of the individual state governments. There is now a primary school in every village across the subcontinent, but many children drop out of school to provide supplementary income for their families. There are more than 50 million students at secondary level, and an estimated 10 million graduates from more than 200 universities. Women make up almost 10% of those enrolled in higher education; this is a good percentage for a low-income economy. Although the high level of illiteracy is a brake on development, India has one of the largest pools of science graduates anywhere in the world.

Terraced fields in central India. In addition to rice, wheat, sorghum, maize, millet, and barley are also important cereal crops.

I

THE SANGH MOVEMENT FOR A HINDU NATIONAL IDENTITY

THE IDEA OF *HINDUTVA*, a specifically Hindu-based national identity, has since preindependence days existed in Indian politics. For decades, however, its supporters were no more than a minor strand of the opposition to the secularist Congress party, which ruled almost uninterruptedly for over 40 years. In the late 1980s, however, the rise of the BJP made *Hindutva* a national issue. When the party came to power, heading its first coalition government in 1998, Prime Minister A. B. Vajpayee kept its ideology in the background, but it is pressed by the nationalist Sangh movement to deliver on its pro-Hindu aims. The so-called "saffron brigade" calls in particular for a Hindu temple to be built in place of a Muslim mosque at Ayodhya, Uttar Pradesh, the supposed birthplace of the Hindu god Rama. In 1992 Hindu activists destroyed the mosque, setting off communal rioting in which 1200 people died. Another major flare-up came in 2002, sparked by plans to begin construction of the proposed temple. When Hindu activists returning from Ayodhya were massacred on a train, revenge attacks set off intercommunal violence in Gujarat in which some 2000 people died.

Multifaith demonstrators *condemned the intercommunal violence which left two thousand dead in Gujarat in early 2002.*

THE SANGH MOVEMENT
The BJP's ideological origins lay in the Sangh movement built around the Rashtriya Swayamsevak Sangh (RSS). The RSS, whose name translates as the National Union of Selfless Servants, was first formed in Nagpur in 1925 to recruit men and women as *swayemsevaks*, volunteers dedicated to the creation of a Hindu nation (Hindu *rashtra*). Banned briefly in 1948 over its association with the Hindu extremist assassin of Mahatma Gandhi, the RSS has no clear organizational structure or membership. It has educational and trade union-based arms, and a religious wing in the militant World Hindu Council (VHP). The Sangh's influence is still strong in the BJP, although in 2000 the BJP-led government was compelled by the strength of pro-secular parties to reaffirm a ban on RSS members in the civil service. The BJP also tried to distance itself from the Sangh leader K. S. Sudarshan's increasingly vocal campaign for the expulsion of Christian missionaries and the replacement of "foreign churches" with a Hindu national church. The movement is linked with fringe organizations blamed for attacks on Christians.

REGIONAL APPEAL
The appeal of *Hindutva* is strongest in the Hindi-speaking heartland, the "cow belt," centered on Uttar Pradesh and Madhya Pradesh, with Bihar and Orissa to the east and in the west Maharashtra and Gujarat. While seeking to be inclusive and to appeal to all Hindus, the movement is rooted in Brahmin culture, and faces rivalry in some states from parties combating the oppression of the lowest castes (*dalits*).

In Maharashtra, the RSS inspires a more extreme local rival Hindu chauvinist party, Shiv Sena (the "army of Shivaji," a 17th-century warrior king). Shiv Sena leader Bal Thackeray, an open admirer of Adolf Hitler, is the main force in local politics. Thackeray also led the campaign for Bombay's change of name to Mumbai in 1995.

The growth of regional parties has made coalition building a feature of government, and has been a major restraint on the BJP's more militant ideologues at national level. The party under Vajpayee's leadership at first downplayed controversial policies, but he provoked a storm in 2000 by describing the temple-building movement as embodying national sentiment and part of an unfinished agenda.

Young Hindu militants *on a pilgrimage to Ayodhya in Uttar Pradesh, northern India.*

HEALTH

▷ Welfare state health benefits

👥 1 per 2500 people

☠ Respiratory, nutritional, and diarrheal diseases, malaria

Malnutrition is extremely common, increasing infant mortality, much of which is due to preventable diseases. Air pollution from the domestic use of solid fuel kills half a million children every year. AIDS has accelerated and HIV infection rates are rising. State governments are responsible for most health programs, but there are various national health projects, including a massive polio eradication program. The government announced plans in 2000 to subsidize health insurance for the poorest of the poor.

SPENDING

▷ GDP/cap. increase

CONSUMPTION AND SPENDING

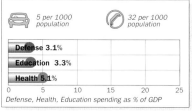

🚗 5 per 1000 population   💿 32 per 1000 population

Defense 3.1%		
Education 3.3%		
Health 5.1%		

0   5   10   15   20   25
Defense, Health, Education spending as % of GDP

According to the government, 240 million people (23% of the population at that time), mostly in rural areas, were living below the poverty line in the late 1980s. Recent studies dispute whether this figure is rising or falling. Extremes of wealth, particularly with the opening up of the economy, are frequently seen alongside extremes of poverty. The middle classes, who number some 150–200 million, have an exceedingly comfortable lifestyle, with servants and plush housing. Many of the slums in cities such as Mumbai (Bombay) and Kolkata (Calcutta) have five to nine people living in one room; few slum houses have sanitation. In Mumbai alone, over 100,000 people live on the sidewalks.

WORLD RANKING

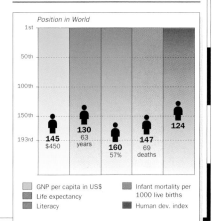

Position in World

1st
50th
100th
150th
193rd

145 $450
130 63 years
160 57%
147 69 deaths
124

▮ GNP per capita in US$ ▮ Infant mortality per 1000 live births
▮ Life expectancy
▮ Literacy   ▮ Human dev. index

INDONESIA

OFFICIAL NAME: Republic of Indonesia   **CAPITAL:** Jakarta
POPULATION: 214 million   **CURRENCY:** Rupiah   **OFFICIAL LANGUAGE:** Bahasa Indonesia

INDONESIA IS THE WORLD'S LARGEST archipelago. Its 13,677 islands stretch 5000 km (3100 miles) from the Indian Ocean to New Guinea. Java, Kalimantan, Papua, Sulawesi, and Sumatra are mountainous, volcanic, and densely forested. Politics after independence was dominated by the military for over three decades, until the fall of the Suharto regime in 1998, when a partial "civilianization" began. In outlying regions, the forcibly suppressed demands for greater autonomy have flared up, bringing renewed violence. East Timor, which Indonesia invaded in 1975 and then annexed, voted for independence in 1999 and became a fully sovereign state in 2002.

Rice terraces on Bali, one of Indonesia's 13,677 islands and its most popular tourist destination. Rice is the staple food crop.

CLIMATE

▷ Tropical equatorial/ monsoon

WEATHER CHART FOR JAKARTA

Indonesia's climate is predominantly tropical. Variations relate mainly to differences in latitude, but hilly areas are cooler overall. Rain falls throughout the year, often in thunderstorms, but there is a relatively dry season from June to September. December to March is the wettest period, except in the Moluccas, which receive the bulk of their rain between June and September.

TRANSPORTATION

▷ Drive on left

 Sukarno-Hatta, Jakarta
10.7m passengers

2480 ships
3.38m grt

THE TRANSPORTATION NETWORK

 158,670 km
(98,593 miles)

200 km
(124 miles)

 6485 km
(4030 miles)

21,579 km
(13,409 miles)

For a multi-island state spread across three time zones, communications are an obvious government priority. Indonesia was an early entrant into satellite communications, providing an international satellite-based telephone system as early as 1976.

However, provisions vary greatly in the different provinces. Road surfaces in Java and Sumatra are excellent. Rail services are restricted to these two islands. In contrast, roads in Kalimantan and Papua are poor, and most travel is by air or river.

TOURISM

▷ Visitors : Population 1:42

5.06m visitors

Up 7% in 2000

MAIN TOURIST ARRIVALS

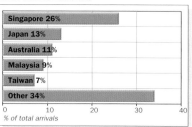

Singapore 26%	
Japan 13%	
Australia 11%	
Malaysia 9%	
Taiwan 7%	
Other 34%	

% of total arrivals

Tourism took off during the 1980s. In 1997, the number of tourists exceeded five million, although political unrest in 1998 and 1999 discouraged many visitors. Bali, Java, and Sumatra are the most popular destinations. Expansion has been encouraged by a major investment in hotels and the opening of Bali to airlines other than the national carrier, Garuda Indonesia.

INDONESIA

Total Land Area : 1 919 440 sq. km
(741 096 sq. miles)

LAND HEIGHT

4000m/13 124ft	
3000m/9843ft	
2000m/6562ft	
1000m/3281ft	
500m/1640ft	
Sea Level	

POPULATION

■ over 5 000 000
▣ over 1 000 000
◉ over 500 000
◎ over 100 000
○ over 50 000

PEEPLE ▷ Pop. density medium

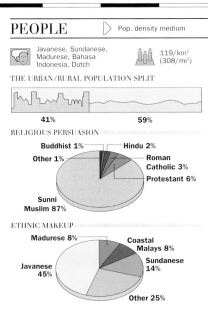

Javanese, Sundanese, Madurese, Bahasa Indonesia, Dutch

119/km² (308/mi²)

THE URBAN/RURAL POPULATION SPLIT

41%　　　　59%

RELIGIOUS PERSUASION

- Buddhist 1%
- Other 1%
- Hindu 2%
- Roman Catholic 3%
- Protestant 6%
- Sunni Muslim 87%

ETHNIC MAKEUP

- Madurese 8%
- Coastal Malays 8%
- Javanese 45%
- Sundanese 14%
- Other 25%

The basic Melanesian–Malay ethnic division disguises a diverse society. The national language, Bahasa Indonesia, coexists with at least 250 other spoken languages or dialects. Attempts by the Javanese political elite to suppress local cultures have been vigorously opposed, especially by the East Timorese, the Aceh of northern Sumatra, and the Papuans of New Guinea.

Religious and interethnic hostility is increasing. Since 1998 there have been violent clashes between Muslims and Christians on Sulawesi, and in the Moluccas. Similar clashes occurred in Kalimantan in 1999 and 2001 between indigenous Dayaks and ethnic Madurese immigrants. Aceh introduced *sharia* (Islamic law) in 2000.

Discrimination against ethnic Chinese has encouraged vicious attacks on their businesses, as in Jakarta in 1998. Gender equality is enshrined in law, and women are active in public life.

POPULATION AGE BREAKDOWN

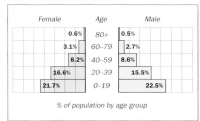

Female	Age	Male
0.6%	80+	0.5%
3.1%	60–79	2.7%
8.2%	40–59	8.6%
16.6%	20–39	15.5%
21.7%	0–19	22.5%

% of population by age group

POLITICS ▷ Multiparty elections

1999/2004

President Megawati Sukarnoputri

AT THE LAST ELECTION

House of Representatives 500 seats

| 31% PDI–P | 24% Gol | 11% PPP | 10% PKB | 7% PAN | 8% U | 9% Others |

PDI–P = Indonesian Democratic Party of Struggle
Gol = Golkar PPP = United Development Party
PKB = National Awaking Party U = Unelected (army) seats
PAN = National Mandate Party

Indonesia is a multiparty democracy. The president is indirectly elected by a 700-member People's Consultative Assembly.

PROFILE

Under Gen. Suharto, the autocratic ruler from 1968, Golkar – an amalgam of interest groups – and the army were dominant. Suharto was forced to resign in 1998 amid widespread protest over corruption, economic mismanagement, and denial of democratic rights. After his fall, the mainly Muslim PDI–P, led by Megawati Sukarnoputri, daughter of the first president, won elections, but Abdurrahman Wahid of the PKB was elected president; Megawati became vice president.

Disenchantment with Wahid pressured him into devolving many of his duties to her in 2000, and increasingly vehement opposition in parliament culminated in his removal in July 2001, whereupon Megawati became president herself.

MAIN POLITICAL ISSUES
The army
The army's involvement in politics remains a serious problem. In early 2000, the influential Gen. Wiranto left the government, but the army refused to relinquish its 38-seat entitlement in the legislature any earlier than 2009.

Separatist movements
Separatism – the legacy of Suharto's centralist policies – and violence between Muslims and religious minorities threaten national unity. The movements in Aceh (northern Sumatra) and Papua continue to gather momentum; both were granted greater autonomy in 2001.

Abdurrahman Wahid, dismissed as president in 2001.

President Megawati Sukarnoputri, Wahid's successor.

WORLD AFFAIRS ▷ Joined UN in 1950

APEC　ASEAN　G15　OIC　OPEC

Indonesia pursues a largely pro-Western foreign policy, although the government is under pressure to improve its human rights record. The scale and nature of the East Timor massacres in 1999 severely damaged its standing. Internal security in the far-flung provinces tests regional relations.

China remains a major foreign policy concern, despite the restoration of diplomatic ties in 1990. Indonesia and Australia signed a groundbreaking security cooperation agreement in late 1995 but the movement of illegal immigrants through Indonesia creates tensions.

AID ▷ Recipient

$1.73bn (receipts)　Down 22% in 2000

Japan accounts for the bulk of bilateral aid. Multilateral aid comes above all from the World Bank, although it called for urgent reform in late 2001 before loaning more money. Aid has been notoriously subject to "leakage."

CHRONOLOGY

Hindu, Buddhist, and Islamic interest preceded 16th-century European rivalries over valuable spices. The Dutch had won control by the 17th century, when colonization began on Java. By 1910, the Dutch East Indies encompassed present-day Indonesia.

- ❑ **1912** Sarekat Islam party formed.
- ❑ **1920** Indonesian Communist Party (PKI) formed; leads revolt in West Java (1926) and Sumatra (1927).
- ❑ **1927** Indonesian National Party formed under Dr. Sukarno.
- ❑ **1930s** Dutch repress nationalists.
- ❑ **1942–1945** Japanese occupation. Autonomy within "Greater East Asia" state promised. Sukarno works with Japanese while promoting independence.
- ❑ **1945** Sukarno declares Indonesian independence. Dutch forces attempt to reassert control.
- ❑ **1945–1949** Nationalist guerrilla war.
- ❑ **1949** Dutch grant independence to federal-style United States of Indonesia under Sukarno as president.
- ❑ **1950** Union dissolved, Sukarno grows increasingly authoritatian.
- ❑ **1950s** Moluccas declare independence from Indonesia, fight unsuccessful separatist war.
- ❑ **1957–1959** Sukarno introduces authoritarian Guided Democracy. ⇨

I

CHRONOLOGY *continued*

- ❑ **1959** Sukarno extends presidential powers. Civilian legislature replaced by military. Extreme nationalist and pro-Chinese policies.
- ❑ **1962** Dutch relinquish Western New Guinea.
- ❑ **1965** Communist PKI alliance with military ends. Army led by Gen. Suharto crushes abortive coup and acts to eliminate banned PKI; up to one million killed.
- ❑ **1966** Sukarno hands over power to Gen. Suharto. Temporary handover becomes permanent in following year.
- ❑ **1968** Suharto becomes president: declares "New order"; introduces pro-Western liberal economic policies while transferring real power from cabinet to small group of officers.
- ❑ **1971** First elections for 16 years. Government-sponsored Golkar wins landslide. Opposition parties now passive partners of government.
- ❑ **1975** Invasion of East Timor; its incorporation as 27th province in 1976 is not recognized by UN.
- ❑ **1984** Muslim protesters clash with troops in Jakarta. Start of resurgence of Islamic protest.
- ❑ **1985** Fretilin movement declares East Timor independent.
- ❑ **1989** Growing discontent with authoritarian government; student protests, unrest in Java and Sumbawa. Demands for Suharto to retire. Low-key official response.
- ❑ **1991** Indonesian troops massacre pro-independence demonstrators in East Timor. New organizations, including Democratic Forum and League for Restoration of Democracy, allowed to form in response to growing demands for "openness."
- ❑ **1993** Suharto wins sixth term in office.
- o **1996** Antigovernment demonstrations in Jakarta.
- o **1997** Economic recession. Smog across region from forest fires.
- o **1998** Suharto resigns amid unrest.
- o **1999** Election victory for opposition led by Megawati Sukarnoputri. East Timor referendum backing independence triggers violent backlash. Abdurrahman Wahid elected president, Megawati named vice president.
- o **2000** Aceh becomes first province to introduce Islamic *sharia* law. Violence erupts again in Moluccas.
- o **2001** Wahid removed, replaced by Megawati.
- o **2002** January, autonomy officially granted to Papua. May, East Timor independent.

DEFENSE

 Compulsory military service

💲 $1.49bn ⬇ Down 1% in 2000

INDONESIAN ARMED FORCES

🛡	355 light tanks (275 AMX-13, 30 PT-76, 50 *Scorpion*-90)	230,000 personnel
🚢	2 submarines, 17 frigates, 36 patrol boats	40,000 personnel
✈	108 combat aircraft (A-4, F-16, *Hawk* Mk 53, 109, and 209)	27,000 personnel
	None	

Defense spending is low by regional standards, despite the military's high profile. The constitution enshrines the military's political role, and even now it remains the key influence in Indonesia. The recent "civilianization" of political parties, the bureaucracy, and state companies has reduced the presence of the military in these areas, if not their influence. The government is under pressure from the IMF to audit "off-budget funds," which often supplement official defense allocations. Western arms sales are increasingly being made dependent on the improvement of Indonesia's human rights record. The main defense issues are internal security and the perceived Chinese threat.

Military service is selective and lasts for two years.

ECONOMICS

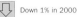 Inflation 16% p.a. (1990–2000)

📊 $120bn 💲 9675–10,400 rupiahs

SCORE CARD

o WORLD GNP RANKING	32nd
o GNP PER CAPITA	$570
o BALANCE OF PAYMENTS	$7.99bn
o INFLATION	3.7%
o UNEMPLOYMENT	5%

ECONOMIC PERFORMANCE INDICATOR

EXPORTS

China 4%
South Korea 7%
Singapore 11%
USA 14%
Japan 23%
Other 41%

IMPORTS

South Korea 6%
China 6%
USA 10%
Singapore 11%
Japan 16%
Other 51%

STRENGTHS

Varied resources, especially oil. Signs of return to high growth. Debt successfully rescheduled. International credit rating improved.

WEAKNESSES

High level of bureaucracy. Endemic corruption. Huge wealth disparities. Regional insecurity deters investment. High underemployment.

PROFILE

Under Suharto the economy grew rapidly, fueled largely by oil, until its collapse in 1997–1998. State-owned corporations played a significant role, protected from foreign competition. Non-oil exports were diversified, but the debt burden used up a third of export earnings. Reform was delayed by conflict between "technologists" who favored industrialization over profit for state concerns and advocates of deregulation. Further reforms have been stymied by the government's poor representation in parliament. Corruption remains rife, embroiling then president Wahid in 2001. Persistent weaknesses are likely to continue to hold back real growth. However, relations with the IMF and other donors were sufficiently improved by 2002 for Indonesia's international credit rating to be upgraded.

INDONESIA : MAJOR BUSINESSES

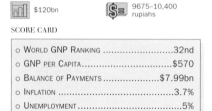

🐙	Rubber
⚙	Heavy engineering
◊	Gas
⚗	Chemicals
🌲	Timber industries
⛽	Oil
🛢	Oil refining
💻	Electronics
🚗	Vehicle assembly
✈	Aerospace industry

* significant multinational ownership

Medan
Banjarmasin
Balikpapan
Kendari
Sorong
Palembang
Jakarta
Bandung
Surabaya
Ujung Pandang

0 500 km
0 500 miles

RESOURCES

 Electric power 22.9m kw

4.8m tonnes

1.41m b/d (reserves 5bn barrels)

28.1m ducks, 12.5m goats, 751m chickens

Oil, natural gas, coal, bauxite, nickel, copper, gold, tin

ELECTRICITY GENERATION

Hydro 16% (14bn kwh)

Combustion 81% (73bn kwh)

Nuclear 0%

Other 3% (2.6bn kwh)

0   20   40   60   80   100

% of total generation by type

INDONESIA : LAND USE

Cropland
Forest
Pasture
Wetlands
Rice
Nutmeg - cash crop
Cattle

0   500 km
0   500 miles

Indonesia is rich in energy sources. The main export earners are oil and liquefied natural gas (LNG) – Indonesia is the world's largest LNG exporter. However, oil output has been falling. Combined with rapid growth in domestic energy demand, this could turn the country into an oil importer in the next decade. The government is therefore encouraging oil exploration in remote regions. It is also considering developing geothermal and nuclear energy sources. Indonesia's other main resources are coal, bauxite, and nickel, along with agricultural products such as rubber and palm oil. Processed wood products are a significant export commodity; the rapid depletion of the rainforests has given rise to attempts at control over log exports.

ENVIRONMENT

 Sustainability rank: 100th

11% (3% partially protected)

1.1 tonnes per capita

ENVIRONMENTAL TREATIES

Yes   Yes   Yes
Yes   Yes   No

Environment legislation is poorly enforced, which means that the rich tropical forests are threatened by excessive logging and that rare flora and fauna are disappearing. In 2002, the government asked Greenpeace to help monitor illegal logging in remote areas. Smog from forest fires is a serious problem, causing a regional health alert in 1997. Frequent oil spillages in the Malacca Strait are a hazard.

MEDIA

 TV ownership medium

Daily newspaper circulation 23 per 1000 people

PUBLISHING AND BROADCAST MEDIA

There are 69 daily newspapers. *Kompas* and *Pos Kota* have the largest circulations

6 services: 1 state-owned, 5 independent

1 state-owned service, 1 independent station

The 1999 press law prohibits censorship, but journalists can still be fined for violating "religious and moral norms." Independent press and broadcasting have flourished since 1998.

CRIME

 Death penalty in use

53,399 prisoners   Up 17% in 1999

CRIME RATES

Murders	
1	*per 100,000 population*

Rapes	
0.6	*per 100,000 population*

Thefts	
55	*per 100,000 population*

Suppression of secessionists is harsh. There is brutal ethnic violence in the provinces.

EDUCATION

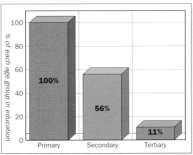 School leaving age: 12

87%   3.13m students

THE EDUCATION SYSTEM

100% Primary
56% Secondary
11% Tertiary

% of each age group in education

Primary education is subsidized by the state, compulsory, and often provided by Islamic schools. In contrast, good secondary education is hard to find in rural areas. University students come predominantly from the richer elites.

HEALTH

 Welfare state health benefits

1 per 5000 people   Lower respiratory and diarrheal diseases

An extensive network of clinics, down to village level, means that access to health care is reasonable. As a result, health indicators have improved significantly. The death rate declined from 2% in 1965 to 0.7% in 2000, thus helping to increase life expectancy, while infant mortality has more than halved over this period. However, malnutrition, particularly among children, continues to remain a real problem, although chronic levels of vitamin A deficiency have been largely eradicated. HIV infections among intravenous drug users, particularly in prisons, are rising.

SPENDING

GDP/cap. increase

CONSUMPTION AND SPENDING

14 per 1000 population   31 per 1000 population

Defense 1%
Education 1.4%
Health 0.8%

0   5   10   15   20   25

Defense, Health, Education spending as % of GDP

Many Indonesians live in relative poverty and those on the peripheral islands in real poverty; large wealth disparities exist between the Javanese middle classes and the subsistence farmers and tribesmen of Papua and Kalimantan. This reflects both a concentration of wealth in the hands of a limited number of key political and business figures, and a concentration of development and investment on the main islands, particularly Java. Since 1998 attempts have been made in the courts to tackle the issues of corruption and concentration of wealth in the hands of close associates and relatives of former President Suharto and his political successors.

WORLD RANKING

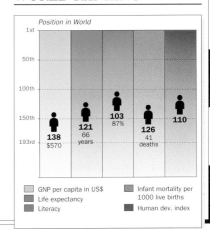

Position in World

1st
50th
100th
150th
193rd

138 $570
121 66 years
103 87%
126 41 deaths
110

GNP per capita in US$
Life expectancy
Literacy
Infant mortality per 1000 live births
Human dev. index

I

IRAN

MIDDLE EAST

OFFICIAL NAME: Islamic Republic of Iran **CAPITAL:** Tehran
POPULATION: 71.4 million **CURRENCY:** Iranian rial **OFFICIAL LANGUAGE:** Farsi

  1502 1990 Feb 11 IR +3.5 +98 .ir

IRAN IS SURROUNDED by turbulent neighbors, with republics of the former Soviet Union to the north, Afghanistan and Pakistan to the east, and Iraq and Turkey to the west. The south faces the Persian Gulf and the Gulf of Oman. Since 1979, when a revolution led by Ayatollah Khomeini deposed the shah, Iran has become the world's largest theocracy and the leading center for militant Shi'a Islam. Iran's active support for Islamic fundamentalist movements has led to strained relations with central Asian, Middle Eastern, and north African states, as well as the US and Europe.

The Reshteh-ye Kuhhā-ye Alborz (Elburz Mountains). Their Caspian Sea slopes are rainy and forested; the southern slopes are dry.

CLIMATE

▷ Mountain/cold desert

WEATHER CHART FOR TEHRAN

The area bordering the Caspian Sea is Iran's most temperate region. Most of the country has a desert climate.

TRANSPORTATION

▷ Drive on right

Mehrabad International, Tehran
8.47m passengers

395 ships
4.23m grt

THE TRANSPORTATION NETWORK

49,440 km (30,721 miles)	470 km (292 miles)
6398 km (3976 miles)	904 km (562 miles)

Adequate roads link main towns, but rural areas are less well served. Most freight travels by rail. A ferry runs from Bandar-e Abbas to the UAE.

TOURISM

▷ Visitors : Population 1:42

1.7m visitors

Up 29% in 2000

MAIN TOURIST ARRIVALS

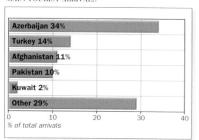

Azerbaijan 34%	
Turkey 14%	
Afghanistan 11%	
Pakistan 10%	
Kuwait 2%	
Other 29%	

% of total arrivals

Iran's historical heritage, mosques, and bazaars formerly attracted sizable numbers of tourists. This flow was cut off by the 1979 revolution, which deterred visitors, especially from the West. In the 1990s, however, there was a rise in the number of business people visiting Iran. Procedures at Tehran's Mehrabad airport have been simplified and the capital's hotels refurbished. In late 1998 President Khatami's more liberal regime welcomed a delegation of US tourists, despite opposition from conservative groups.

PEOPLE

▷ Pop. density low

Farsi, Azerbaijani, Gilaki, Mazanderani, Kurdish, Baluchi, Arabic, Turkmen

44/km² (113/mi²)

THE URBAN/RURAL POPULATION SPLIT

62% 38%

RELIGIOUS PERSUASION

Other 1% Sunni Muslim 4%
Shi'a Muslim 95%

ETHNIC MAKEUP

Kurd 8% Lur and Bakhtiari 8%
Persian 50% Other 10%
Azari 24%

The population comprises several ethnic groups. The people of the north and center – about half of all Iranians – speak Farsi (Persian), while about a quarter speak related languages, including Kurdish in the west and Baluchi in the southeast. Another quarter of the population speaks Turkic languages, primarily the Azaris and the

Turkmen in the northwest. Smaller groups, such as the Circassians and Georgians, are found in the northern provinces.

Until the 16th century, much of Iran followed the Sunni interpretation of Islam, but since then the Shi'a sect has been dominant. Religious minorities, accounting for just 1% of the population, include followers of the Bahai faith, who suffer discrimination, Zoroastrians, Christians, and Jews. The regime has a remarkably liberal attitude to refugees of the Muslim faith. Nearly three million Afghan refugees were received during the height of the Afghan civil war in the 1980s–1990s, although many have since returned. In Khorosan province in the east, refugees account for nearly a quarter of the population; near the Turkish border they constitute half the total population. Many are young, resulting in intense competition with Iranians for jobs and consequent ethnic tensions.

One of the main consequences of the 1979 Islamic revolution was to reverse the policy of female emancipation. The revolution restricted the public role of women and enforced a strict dress code, obliging women to cover themselves from head to foot in the *chadoor* or veil. More liberal attitudes have gradually emerged, and reform of the divorce laws has been proposed, under which it would become possible for proceedings to be initiated by the wife.

POPULATION AGE BREAKDOWN

Female	Age	Male
0.2%	80+	0.2%
2.5%	60–79	3%
5.9%	40–59	6.2%
14.4%	20–39	14.7%
26%	0–19	26.9%

% of population by age group

I

POLITICS ▷ Multiparty elections

2000/2004 | ▷ President Mohammad Khatami

AT THE LAST ELECTION

Consultative Council (Majlis) 290 seats

2% Rel

65% IIPF | **17% C** | **11% Ind** | **5% Vacant**

IIPF = Islamic Iran Participation Front (reformists)
C = Coalition of Followers of the Line of Imam (conservatives)
Ind = Independents **Rel** = Religious Minorities

Iran is a theocracy. Tension exists between the conservative mullahs and the reformist government.

PROFILE

Iran's religious revolution of 1979 was fueled by popular outrage at the corruption, repression, and inequalities of the shah's regime. Since the time of Ayatollah Khomeini, successive Iranian governments have maintained that the clergy have a religious duty to establish a just social system. Accordingly, the legislature, the executive, and the judiciary may, in theory, be

Ayatollah Khamenei, who became spiritual leader after the death of Ayatollah Khomeini.

Mohammad Khatami, reformist president, in office since 1997.

overruled by the religious leadership; former president Hashemi Rafsanjani's moderate policies were questioned by radical clergymen advocating "permanent revolution." However, the mullahs' failure to address Iran's economic problems has eroded their political standing. Reformists were encouraged by the election of President Mohammad Khatami in 1997, but the clergy remains powerful. Huge student demonstrations in 1999 and 2000 in favor of reform were offset by crackdowns on reformist politicians and newspapers. The mullahs, so far unbowed by reformist election victories, have continued to confront the modernizers.

MAIN POLITICAL ISSUE
Mosque versus secular state

A power struggle between the clergy and the secular state has arisen from the ill-defined division of power between the two. The conservative faction in parliament lost its overall majority in 1996 and has been steadily displaced by reformists, who in 2000 made sweeping gains in parliamentary elections. In 2001 reformist president Mohammad Khatami was reelected, this time with an overwhelming 77% of the vote.

Khatami is committed to modernizing the economy, but is strongly opposed by the mullahs, for whom adherence to religious values is more important than material welfare. Student proreform protests were heavily suppressed by hard-liners, who remain a force, despite the 2000 elections, through the powerful Council of Guardians.

WORLD AFFAIRS ▷ Joined UN in 1945

ECO | G24 | NAM | OIC | OPEC

Following the 1979 revolution, Iran assumed international significance as the voice of militant Shi'a Islam. Iran is accused of backing terrorist activity by Muslim extremists and of fostering unrest throughout the Middle East and central Asia.

Under President Khatami, Iran has tried to convey a less confrontational image. Improved relations with Saudi Arabia, troubled since Iran's seizure of the islands of Abu Musa and the Tunbs in 1970, resulted in the signing of a pact in 2001. Ties with the West have also improved, although relations with the US remain tense and suffered a sharp setback in early 2002 after US President George W. Bush cast Iran (with Iraq and North Korea) as part of an international "axis of evil." Iran's most pressing security preoccupation is with Iraq, which allows *mujahideen* guerrillas to mount attacks on Iran from its territory.

IRAN

Total Land Area : 1 648 000 sq. km
(636 295 sq. miles)

POPULATION

▣ over 1 000 000
◉ over 500 000
◎ over 100 000
○ over 50 000

LAND HEIGHT

3000m/9845ft
2000m/6562ft
1000m/3281ft
500m/1640ft
200m/656ft
Sea Level

0 200 km
0 200 miles

CHRONOLOGY

Persia was ruled by the shahs as an absolute monarchy until 1906, when the first constitution was approved. The Pahlavis took power in 1925 and changed the country's name to Iran in 1935.

❏ **1957** SAVAK, shah's secret police, established to control opposition.
❏ **1964** Ayatollah Khomeini is exiled to Iraq for criticizing secular state.
❏ **1971** Shah celebrates 2500th anniversary of Persian monarchy.
❏ **1975** Agreement with Iraq over Shatt al Arab waterway.
❏ **1977** Khomeini's son dies. Anti-shah demonstrations during mourning.
❏ **1978** Riots and strikes. Khomeini settles in Paris.
❏ **1979** Shah goes into exile. Ayatollah Khomeini returns, declares an Islamic republic. Students seize 63 hostages at US embassy in Tehran.
❏ **1980** Shah dies in exile. Start of eight-year Iran–Iraq war.
❏ **1981** US hostages released. Hojatoleslam Ali Khamenei elected president.
❏ **1985** Khamenei reelected.
❏ **1987** Around 275 Iranian pilgrims killed in riots in Mecca.
❏ **1988** USS *Vincennes* shoots down Iranian airliner; 290 killed. End of Iran–Iraq war.
❏ **1989** Khomeini issues *fatwa* condemning UK author Salman Rushdie to death for blasphemy. Khomeini dies. President Ali Khamenei appointed Supreme Religious Leader. Hashemi Rafsanjani elected president.
❏ **1990** Earthquake in northern Iran kills 45,000 people.
❏ **1992** Majlis elections.
❏ **1993** Rafsanjani reelected president.
❏ **1995** Imposition of US sanctions.
❏ **1996** Majlis elections. Society for Combatant Clergy loses ground to more liberal Servants of Iran's Construction.
❏ **1997** Earthquake south of Mashhad kills 1500 people. Mohammad Khatami elected president.
❏ **1998** Khatami government dissociates itself from *fatwa* against Salman Rushdie.
❏ **1999** First nationwide local elections since 1979. President Khatami visits Italy: first Iranian leader to be welcomed by a Western government since 1979.
❏ **2000** Sweeping election victory for reformists. Crackdown on reformist newspapers.
❏ **2001** Khatami reelected, winning 77% of vote.

AID Recipient

 $130m (receipts) Down 20% in 2000

As an oil exporter, Iran does not qualify for much aid, and hard-liners are opposed to money from the West. However, Iran receives some UN aid for its millions of refugees from Afghanistan and Iraq.

Concern that Iran supports Muslim terrorism has affected aid programs. In 1994, the World Bank suspended loans, and US sanctions imposed in 1995 ended bilateral assistance, although European oil companies have more recently announced new deals.

DEFENSE Compulsory military service

 $7.33bn Up 28% in 2000

Iran has more than 500,000 men under arms, including the 125,000-strong Revolutionary Guard Corps (*Pasdaran Inqilab*), and is regarded by neighboring states as a serious military threat. The testing of medium-range cruise and ballistic missiles in 1998 heightened concern over Iran's possible military objectives.

Before the 1979 revolution Iran was part of a pro-Western alliance structure. The long war with Iraq in the 1980s diminished the military power of the revolutionary regime.

IRANIAN ARMED FORCES

	1565 main battle tanks (M-47/48/60A1, *Chieftain* Mk3/5, T-54/55/62/72)	325,000 personnel
	6 submarines, 3 frigates, 53 patrol boats	18,000 personnel
	283 combat aircraft (F-4D/E, F-5E/F, Su-24, F-14, F-7, MiG-29)	45,000 personnel
	None	

A new defense agreement with Russia allowing for the sale of arms to Iran was reached in 2000.

ECONOMICS Inflation 26% p.a. (1990–2000)

 $107bn 1747.5–1750.0 Iranian rials

SCORE CARD

❏ WORLD GNP RANKING............................34th
❏ GNP PER CAPITA $1680
❏ BALANCE OF PAYMENTS....................$12.6bn
❏ INFLATION 14.5%
❏ UNEMPLOYMENT....................................25%

EXPORTS

China 6%
Saudi Arabia 7%
Italy 7%
South Korea 8%
Japan 18%
Other 54%

IMPORTS

France 5%
Italy 5%
UAE 6%
South Korea 9%
Germany 10%
Other 65%

STRENGTHS

OPEC's second-biggest oil producer; soaring world oil prices in 2000. Potential for related industries and increased production of traditional exports: carpets, pistachio nuts, and caviar.

WEAKNESSES

Theocratic authorities restrict contact with West and access to technology. High unemployment and inflation. Excessive foreign debts.

ECONOMIC PERFORMANCE INDICATOR

Consumer Price Index — GDP

PROFILE

With few industries other than oil, US sanctions and fluctuations in oil prices made foreign earnings volatile; higher prices in 2000 held out the prospect of being able to invest in diversification.

IRAN : MAJOR BUSINESSES

Cement
Textiles
Oil refining
Engineering
Carpet weaving
Food processing
Petrochemicals/chemicals

RESOURCES

 Electric power 30.6m kw

419,000 tonnes

3.69m b/d (reserves 89.7bn barrels)

53m sheep, 25.2m goats, 7m cattle, 260m chickens

Iron, copper, lead, oil, zinc, chromite, coal, manganese, gypsum

ELECTRICITY GENERATION

Hydro 7% (7bn kwh)
Combustion 93% (96bn kwh)
Nuclear 0%
Other 0%

0　20　40　60　80　100

% of total generation by type

Iran has substantial oil reserves. It also has metal, coal, and salt deposits, but these are relatively undeveloped. The agricultural sector is an important part of Iran's economy. Principal crops are wheat, barley, rice, sugar beet, tobacco, and pistachio nuts.

Iran was once an opium exporter, but its cultivation and use have since been banned. The vodka industry has also been closed down. Enough wool is produced to supply the carpet weaving industry. Iran has insufficient livestock to supply the domestic meat market and has to import large quantities. The Caspian Sea fisheries are controlled by the state, which sells caviar for export.

IRAN : LAND USE

PLATEAU OF IRAN

Cropland
Forest
Pasture
Wetlands
Desert
Sheep
Wheat
Tobacco

0　400 km
0　400 miles

ENVIRONMENT

 Sustainability rank: 104th

5% (3% partially protected)

4.7 tonnes per capita

ENVIRONMENTAL TREATIES

Yes　Yes　Yes
Yes　Yes　No

War damage to southern Iran, especially at Bandar Khomeini, the tanker terminal at Kharg Island, and the refinery at Abadan, has caused significant environmental damage. Environmental issues are not of concern to the religious leadership.

MEDIA

 TV ownership medium

Daily newspaper circulation 26 per 1000 people

PUBLISHING AND BROADCAST MEDIA

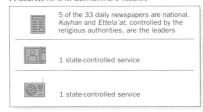

5 of the 33 daily newspapers are national. *Kayhan* and *Ettela'at*, controlled by the religious authorities, are the leaders

1 state-controlled service

1 state-controlled service

Radio and TV are state-controlled. Satellite dishes are banned. Closures of reformist newspapers by the conservative Council of Guardians, and prosecutions of their editors, continue.

CRIME

 Death penalty in use

158,000 prisoners

Little change from year to year

CRIME RATES

Iran does not publish crime statistics. However, general crime rates are relatively low.

Revolutionary guards enforce the law. More than 100 offenses carry the death sentence. Executions, of both men and women, are common for political "crimes." Western governments accuse Iran of supporting international terrorism carried out by Muslim extremists abroad. In 2000 the government finally admitted that narcotics addiction, prostitution, and the violent abuse of women were rife.

EDUCATION

 School leaving age: 14

77%　　679,000 students

THE EDUCATION SYSTEM

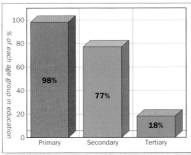

98% Primary
77% Secondary
18% Tertiary

% of each age group in education

Most of the population is literate. Primary education, which lasts for five years from the age of six, is free, as are universities, but a small fee is charged for secondary education. Most schools are single-sex. There are 36 universities. Students are strong supporters of liberalization and the reformist agenda of President Khatami.

HEALTH

 Welfare state health benefits

1 per 1111 people

Heart and respiratory diseases, injuries, neonatal deaths

Although an adequate system of primary health care exists in the cities, conditions in rural areas are basic. Under Khomeini, having children became a political and religious duty, but the high birthrate has now forced the introduction of birth control programs, and sterilization and contraception are now officially promoted. Growing drug addiction has resulted in rehabilitation programs and antidrugs propaganda. AIDS is spreading; some 20,000 adults were estimated to be living with HIV/AIDS in 2001.

SPENDING

 GDP/cap. increase

CONSUMPTION AND SPENDING

30 per 1000 population　　149 per 1000 population

Defense 7.5%
Education 4.6%
Health 1.7%

0　5　10　15　20　25

Defense, Health, Education spending as % of GDP

After the 1979 revolution, living standards in Iran declined markedly. A shortage of foreign exchange has stifled imports of consumer goods. Rationing, brought in during the war with Iraq, is still partly in force, and smuggling from the Arab Gulf states is rife. Unemployment is high, and few Iranians are able to gain access to modern technology such as telephones. Official figures for income per capita do not relate to conditions on the ground. In reality, oil wealth fails to reach the economically deprived. Private businesses have gradually emerged in Iran since the launch in 1994 of the country's first private savings and loans associations.

WORLD RANKING

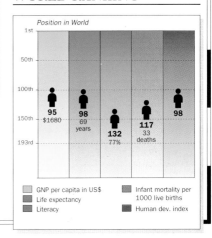

Position in World

1st
50th
100th
150th
193rd

95 $1680
98 69 years
132 77%
117 33 deaths
98

GNP per capita in US$
Life expectancy
Literacy

Infant mortality per 1000 live births
Human dev. index

I

IRAQ

OFFICIAL NAME: Republic of Iraq   CAPITAL: Baghdad
POPULATION: 23.6 million   CURRENCY: Iraqi dinar   OFFICIAL LANGUAGE: Arabic

OIL-RICH IRAQ, divided by the Euphrates and Tigris Rivers, shares borders with Iran, Turkey, Syria, Jordan, Saudi Arabia, and Kuwait. The Euphrates valley is fertile, but most of the country is desert or mountains. Iraq was the site of the ancient civilization of Babylon. Today, it encompasses Shi'a Muslim holy shrines. After the removal of the monarchy in 1958, it experienced domestic political turmoil. Despite Iraq's defeat in the 1991 Gulf War, the current regime (in place since 1968) retains power – through repression.

Golden Mosque at Sāmarrā' on the Tigris. Among the extensive remains of its ancient city are those of the Great Mosque built in 847 CE.

CLIMATE
Hot desert/steppe

WEATHER CHART FOR BAGHDAD

The weather is dry and rainfall is low and unreliable, except in the north. Iraq experiences a wide range of temperatures. The south has a desert climate, with hot, dry summers and mild winters. In mountainous Iranian and Turkish border regions winters can be harsh, with frost and heavy falls of snow. In the Mesopotamian plain huge dust storms are a regular feature of the summer.

TRANSPORTATION
Drive on right

 Saddam International, Baghdad
 99 ships 511,000 grt

THE TRANSPORTATION NETWORK

| 38,400 km (23,861 miles) | 1264 km (785 miles) |
| 2339 km (1453 miles) | 1015 km (631 miles) |

Some international air links have been reopened, but the main road access remains through Ramadi to Jordan.

TOURISM
Visitors : Population 1:303

 78,000 visitors   Up 53% 1999–2000

MAIN TOURIST ARRIVALS

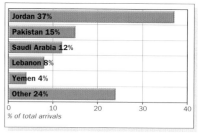

Jordan	37%
Pakistan	15%
Saudi Arabia	12%
Lebanon	8%
Yemen	4%
Other	24%

% of total arrivals

The Shi'a holy shrines attract thousands of pilgrims each year, but Iraq is effectively closed to the tourists who once visited its many archaeological sites. In particular, the ruins of Babylon, and its fabled hanging gardens, were once a major tourist attraction. Tourists also used to visit the marshlands close to the Shatt al Arab waterway. However, this area of great ecological importance continues to be drained as part of the effort to suppress the mainly Shi'a Marsh Arabs.

IRAQ

Total Land Area : 457 072 sq. km (168 755 sq. miles)

PEOPLE
 Pop. density medium

 Arabic, Kurdish, Turkic languages, Armenian, Assyrian

 54/km² (140/mi²)

THE URBAN/RURAL POPULATION SPLIT

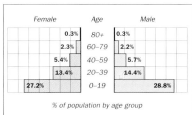

77% 23%

RELIGIOUS PERSUASION

Other (including Christian) 5%

Sunni Muslim 33%

Shi'a ithna Muslims 62%

ETHNIC MAKEUP

Turkman 2% — Persian 3%

Kurdish 16%

Arab 79%

In addition to the Arab and Kurdish populations, Iraq has a small number of minority groups, such as Turkmen and Persians. Over 90% of the population are Muslim, while the rest comprise a variety of Christian sects. Since the creation of Israel, most Iraqi Jews have emigrated. The Arab Muslims are divided into Sunni and Shi'a sects. The Shi'a form the largest single religious group; however, Shi'a divines do not have as intimate a connection with the people as they do in Iran and their influence on government is limited.

Since the mid-1970s, many Iraqis have moved, or been forced to move, to the cities, where some three-quarters of the population now live.

In the marshes of the southeast, communities of mainly Shi'a Marsh Arabs survive. After the 1991 Gulf War, some of these attempted a rebellion against the state, which continues to drain the marshes in order to destroy both the people and their culture.

POPULATION AGE BREAKDOWN

Female		Age	Male	
	0.3%	80+	0.3%	
	2.3%	60–79	2.2%	
	5.4%	40–59	5.7%	
	13.4%	20–39	14.4%	
27.2%		0–19		28.8%

% of population by age group

WORLD AFFAIRS
 Joined UN in 1945

 AL   NAM   OAPEC   OIC OPEC

The legacy of the Iran–Iraq war in the 1980s, the invasion of Kuwait in 1990, and the Gulf War in 1991 still dominate foreign relations.

A US-led military force, assisted by several Arab states, inflicted a crushing defeat on Iraq in 1991. Since then the UN has imposed a severe sanctions regime. Continued tensions still lead to punitive air strikes by US and UK forces. Russia, France, and several Arab states led the way in late 2000 in undermining sanctions on humanitarian grounds. Although some limited economic links have been resumed, no major Western states have yet restored diplomatic relations as the regime continues to deny access to UN weapons inspectors. By early 2002 Iraq was high on the list of possible targets in the US-led "war on terrorism," despite the government's pledge never to invade Kuwait again.

Though there have been some signs of improvement, relations with Iran remain tense. Iranian opposition groups continue to use Iraq as a base for their operations.

I

POLITICS
 No multiparty elections

2000/2004

President Saddam Hussein

AT THE LAST ELECTION
National Assembly 250 seats

The National Assembly is composed of Ba'athists and their allies. Thirty seats representing the Kurdish region were not elected.

President Saddam Hussein has dominated Iraqi politics since overthrowing his predecessor in 1979. In theory, the highest state authority rests with the nine-member Revolutionary Command Council, of which Saddam Hussein is the chairman.

PROFILE
Iraq's repressive regime is controlled by Saddam Hussein and trusted members of his Takriti tribe. The defection to Jordan in 1995 of Saddam Hussein's relative, senior government minister Gen. Hussein Kamil, suggested dissent within the ruling circle, but his assassination in 1996 effectively neutralized a potential threat.

Saddam Hussein has promoted a massive personality cult. His regime relies on terror and a ruthless intelligence network. Iraq has the highest rate of "disappearances" in the world. Legislative elections allow voters only to choose between approved candidates, and produce results such as the 99.99% vote for Saddam's son Uday. The formation of a new Communist Party was announced in 2000, but it forms a section of the Ba'ath Party.

MAIN POLITICAL ISSUES
Sanctions
Iraq's invasion of Kuwait in 1990 and its defeat in the 1991 Gulf War resulted in the imposition of UN sanctions. These remain in force pending Iraq's full compliance with UN Gulf War resolutions. In 1994, Iraq recognized Kuwait but continued to defy UN weapons inspection programs. Campaigns for the lifting of sanctions focus on the impact they have on innocent civilians, particularly children. Since 1996 Iraq has been authorized by the UN to sell limited quantities of oil to purchase humanitarian supplies. Despite the sanctions, Iraq pursues a program of reconstruction. From 2000 limited oil revenue was permitted to be reinvested in the oil industry. In 2002 the list of sanctioned items was restricted to items of potential military use.

Threats to the regime
There is little unity among the opposition groups. The most significant are the Tehran-based Supreme Council for the Islamic Revolution in Iraq and the Iraqi National Congress operating from London. The murder in 1996 of Gen. Hussein Kamil was a blow to the

***Tariq Aziz**, deputy prime minister and mediator between Iraq and the UN.*

***Saddam Hussein**, Iraq's dictatorial leader since he seized power in 1979.*

opposition and a further demonstration of the regime's ruthlessness.

The separatist Kurdish minority in the north suffers factional disputes which triggered a civil war and the intervention of Iraqi troops in 1996. Iraq's enemies continue to be reluctant to endorse Kurdish plans for Iraq's territorial dismemberment.

AID
 Recipient

 $101m (receipts) Up 33% in 2000

Before it invaded Kuwait, Iraq received economic aid from neighboring Gulf states. Under UN sanctions, Iraq has been entitled only to humanitarian aid, but there is mounting evidence of covert trade, especially through Jordan and Turkey.

I

CHRONOLOGY

Iraq became independent in 1932. In 1958, the Hashemite dynasty was overthrown when King Faisal died in a coup led by the military under Brig. Kassem. He was initially supported by the Iraqi Ba'ath Party.

- ❏ **1961** Start of Kurdish rebellion. Iraq claims sovereignty over Kuwait on the eve of Kuwait's independence.
- ❏ **1963** Kassem overthrown. Col. Abd as-Salem Muhammad Aref takes power. Kuwait's sovereignty recognized.
- ❏ **1964** Ayatollah Khomeini, future leader of Iran, takes refuge at Najaf in Iraq.
- ❏ **1966** Aref is succeeded by his brother, Abd ar-Rahman.
- ❏ **1968** Ba'athists under Ahmad Hassan al-Bakr take power.
- ❏ **1970** Revolutionary Command Council agrees manifesto on Kurdish autonomy.
- ❏ **1972** Nationalization of Western-controlled Iraq Petroleum Company.
- ❏ **1978** Iraq and Syria form economic and political union.
- ❏ **1979** Saddam Hussein replaces al-Bakr as president.
- ❏ **1980** Outbreak of Iraq–Iran war.
- ❏ **1982** Shi'a leader Mohammed Baqir al-Hakim, exiled in Tehran, forms Supreme Council of the Islamic Revolution in Iraq.
- ❏ **1988** Iraq and Iran agree cease-fire. Iraqi chemical weapons attack on Kurdish village of Halabja.
- ❏ **1990** British journalist Farzad Bazoft hanged for spying. Iraq and Iran restore diplomatic relations. Iraq invades Kuwait. UN imposes trade sanctions.
- ❏ **1991** Gulf War. US-led military coalition defeats Iraq and liberates Kuwait. Iraqi regime suppresses Shi'a rebellion. Northern no-fly zone enforced.
- ❏ **1992** Western powers proclaim air exclusion zone over south.
- ❏ **1994** Outbreak of Kurdish civil war. Iraq recognizes Kuwaiti sovereignty.
- ❏ **1995** Government minister Gen. Hussein Kamil defects to Jordan, and is murdered on his return to Iraq in January 1996.
- ❏ **1996** First legislative elections since 1989 won by ruling Ba'ath Party. UN supervises limited sales of Iraqi oil to purchase humanitarian supplies.
- ❏ **1998–1999** UN weapons inspection teams refused reentry into Iraq; US and UK mount punitive air strikes.
- ❏ **2002** Threat of US military attack to oust Saddam Hussein.

DEFENSE

 Compulsory military service

$1.47bn — Down 2% in 2000

IRAQI ARMED FORCES

🛡	2200 main battle tanks (1500 T-55/62, PRC Type-59, 700 T-72)	375,000 personnel
🚢	6 patrol boats	2000 personnel
✈	316 combat aircraft (*Mirage* F1EQ5, Su-20/22M/24MK/25, MiG-21/23/25/29)	30,000 personnel
🚀	None	

Iraq's military defeat by the US-led coalition in 1991 led to the destruction of much of its arsenal. Since then UN Security Council resolutions have required the elimination of the bulk of Iraq's weapons of mass destruction, and inspection teams have sought with some effect to enforce this. There is a shortage in Iraq of high-tech weaponry that could match the kind acquired by Kuwait and Saudi Arabia from US and other Western suppliers since the Gulf War. The army is large, but poorly trained and equipped; military service, lasting between 18 months and two years, is compulsory for all men at the age of 18. The military relies on tanks and aircraft from the former Soviet Union and China. The air force has some French *Mirage* fighters and US helicopters.

ECONOMICS

 Not available

📊 $20bn — 💲 0.3124–0.3110 Iraqi dinars

SCORE CARD

- ❏ World GNP Ranking.........................64th
- ❏ GNP per Capita $950
- ❏ Balance of Payments................Not available
- ❏ Inflation 100%
- ❏ Unemployment.........................Not available

ECONOMIC PERFORMANCE INDICATOR

EXPORTS

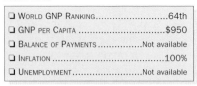

Jordan 4% · France 8% · Spain 8% · Italy 11% · Other 28% · USA 41%

IMPORTS

Jordan 6% · Italy 10% · China 13% · Australia 14% · France 14% · Other 43%

STRENGTHS
Second-largest crude oil and natural gas reserves in OPEC. Large labor force.

WEAKNESSES
Inability to sell oil on the international market; Iraq's GNP halved by UN sanctions. Once-thriving agricultural sector devastated by war.

PROFILE
Before 1990, Iraq was the world's third-largest oil supplier. Under sanctions, oil was produced only for domestic consumption. Limited oil exports under strict UN supervision were resumed for the first time in 1996, and in 2000 the UN Security Council approved a resolution permitting Iraq to buy parts and equipment for the oil industry.

The denial of Western assistance following the 1991 Gulf War has stifled Iraq's economy, although the resumption of some informal economic links and the revision of UN sanctions in 2002 may lead to some improvement. The once thriving agricultural sector was badly affected by the war. The manufacturing industry is at a standstill. The introduction of draconian penalties, including the death sentence, have failed to curb the black market or halt the sharp depreciation in the value of the dinar.

IRAQ : MAJOR BUSINESSES

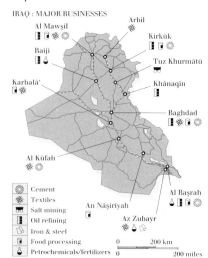

Arbil · Al Mawşil · Kirkūk · Baiji · Tuz Khurmātū · Karbalā' · Khānaqīn · Baghdad · Al Kūfah · Al Başrah · An Nāşirīyah · Az Zubayr

🗼 Cement
❉ Textiles
Salt mining
Oil refining
Iron & steel
Food processing
Petrochemicals/fertilizers

0   200 km
0   200 miles

RESOURCES

 Electric power 9.5m kw

26,789 tonnes

2.41m b/d (reserves 112.5bn barrels)

6.78m sheep, 1.6m goats, 1.35m cattle, 23m chickens

Oil, natural gas, phosphates, sulfur, gypsum

ELECTRICITY GENERATION

Hydro 2% (0.6bn kwh)						
Combustion 98% (30bn kwh)						
Nuclear 0%						
Other 0%						
0	20	40	60	80	100	

% of total generation by type

Iraq has huge reserves of oil and gas. The oil industry is controlled by the Iraqi National Oil Company. Total gas

reserves, many of which are associated with oil, are proven to be 3.11 trillion cu. m (110 trillion cu. ft), with estimates of a further 4.25 trillion cu. m (150 trillion cu. ft). Most electricity is generated from oil, although hydroelectric power also makes a small contribution. Reserves of phosphates, sulfur, gypsum, and salt are also exploited.

Before the invasion of Kuwait and subsequent war, Iraq supplied 80% of the world's trade in dates. Production is now sharply down; food is now produced simply for domestic consumption. Iraq has, however, achieved a degree of self-sufficiency in such crops as wheat, rice, and sugarcane.

ENVIRONMENT

 Sustainability rank: 139th

None

3.7 tonnes per capita

ENVIRONMENTAL TREATIES

No		No		No	
No		No		No	

Wars with Iran and with the US-led alliance over the Kuwait occupation led to massive environmental damage. Hundreds of thousands of land mines remain in the Kuwait border regions, posing lethal hazards to farmers, livestock, and wild animals. The north has been affected by chemical weapons, used by the regime against the Kurds. In the southeast, an entire wetland ecosystem is being destroyed by a program to drain the marshes for political reasons.

MEDIA

 TV ownership medium

Daily newspaper circulation 20 per 1000 people

PUBLISHING AND BROADCAST MEDIA

The 9 daily newspapers, including *Ath-Thawra* and the English-language *Baghdad Observer*, are all state-controlled

1 state-controlled service

1 state-controlled service

The media are strictly controlled, although rebel groups circulate clandestine newspapers. Saddam Hussein's son Uday controls two newspapers, one of which is the influential *Babil*, which opposes UN Gulf War resolutions. All foreign journalists are vetted. Opposition groups began satellite broadcasts with US assistance in 2001.

CRIME

 Death penalty in use

Iraq does not publish prison figures

Crime is rising

CRIME RATES

Murders	
7	per 100,000 population

Rapes	
Iraq does not publish rape statistics	

Thefts	
63	per 100,000 population

Economic collapse has sent crime rates soaring, especially in cities. Theft has been made a capital offense, encouraging thieves to murder their victims in order to escape detection.

EDUCATION

 School leaving age: 11

56%

209,818 students

THE EDUCATION SYSTEM

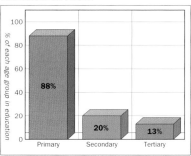

% of each age group in education

- Primary 88%
- Secondary 20%
- Tertiary 13%

Education is free and universal, except in remote rural areas. Primary education has been made compulsory in an effort to reduce illiteracy. There are six universities. Academics authorized the organized plunder of antiquities and university equipment from Kuwait during the 1990 occupation. University scientists work closely with the regime on weapons research programs.

IRAQ : LAND USE

- Cropland
- Forest
- Pasture
- Wetlands
- Desert
- Sheep
- Wheat
- Dates - cash crop

MESOPOTAMIA

SYRIAN DESERT

Tigris

Euphrates

| 0 | 200 km |
| 0 | 200 miles |

HEALTH

 Welfare state health benefits

1 per 2000 people

Pneumonia, influenza, cancers, heart diseases

An effect of UN sanctions has been to aggravate the shortage of medical supplies and equipment, and deaths among children and the elderly have spiraled sharply. The increase in the number of children born with birth defects since 1991 is attributed to the Allies' use of depleted uranium shells during the Gulf War.

SPENDING

GDP/cap. decrease

CONSUMPTION AND SPENDING

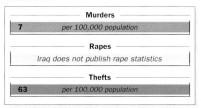

36 per 1000 population

29 per 1000 population

Defense 9.7%				
Education 5.1%				
Health 3.8%				
0	5	10	15	20	25

Defense, Health, Education spending as % of GDP

Vulnerable sections of society have been particularly affected by UN sanctions. Middle-class citizens and traders are able to benefit from Iraq's open border with Jordan.

WORLD RANKING

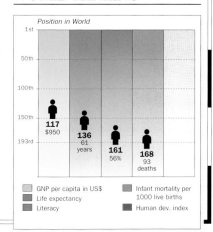

Position in World

- 1st
- 50th
- 100th
- 150th
- 193rd

117 $950

136 61 years

161 56%

168 93 deaths

- GNP per capita in US$
- Life expectancy
- Literacy
- Infant mortality per 1000 live births
- Human dev. index

IRELAND

OFFICIAL NAME: Ireland **CAPITAL:** Dublin **POPULATION:** 3.8 million
CURRENCY: Euro (punt until 2002) **OFFICIAL LANGUAGES:** Irish and English

EUROPE

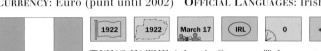

| 1922 | 1922 | March 17 | IRL | 0 | +353 | .ie |

L YING IN THE Atlantic Ocean, off the west coast of
Great Britain, the Irish republic occupies about 85%
of the island of Ireland. Low coastal ranges surround a central basin with
lakes, hills, and peat bogs. Centuries of struggle against English domination
led in 1922 to the formation of the Irish Free State and in 1937 full
sovereignty. Efforts to resolve the Northern Ireland conflict center on the
1998 Good Friday accord, under which Ireland gave up its territorial claim.

CLIMATE
▷ Maritime

WEATHER CHART FOR DUBLIN

Moderated by the Gulf Stream, the Irish
climate is mild, equable, and wet. The
mean annual temperature is 12°C (54°F).

TRANSPORTATION
▷ Drive on left

Dublin International
13.8m passengers

172 ships
248,236 grt

THE TRANSPORTATION NETWORK

| 87,043 km (54,086 miles) | 103 km (64 miles) |
| 1919 km (1192 miles) | 710 km (441 miles) |

EU funds have improved road networks,
especially around Dublin, which still,
however, suffers from congestion.

TOURISM
▷ Visitors : Population 1.8:1

6.75m visitors

Up 5% in 2000

MAIN TOURIST ARRIVALS

| UK 59% |
| USA 14% |
| Germany 5% |
| Other 22% |

% of total arrivals

Tourist numbers have increased steadily
in recent years and exceed six million
a year. Vibrant Dublin attracts many on
"city breaks." Other draws are scenery,
Ireland's "clean" environmental image,
and the relaxed lifestyle.

PEOPLE
▷ Pop. density medium

English, Irish Gaelic

55/km² (143/mi²)

THE URBAN/RURAL POPULATION SPLIT

59%   41%

RELIGIOUS PERSUASION

Jewish 1%
Anglican 3%
Other and nonreligious 8%
Roman Catholic 88%

The population is 95% ethnic Irish.
The influence of the Roman
Catholic Church is declining.
Ireland is now a country
of net immigration,
against the trend
of the past
150 years.

POLITICS
▷ Multiparty elections

L. House 2002/2007
U. House 2002/2007

President Mary
McAleese

AT THE LAST ELECTION

House of Representatives 166 seats

| 49% FF | 19% FG | 12% LP | 4% GP | 5% PD | 3% SF | 8% Others |

FF = Fianna Fail **FG** = Fine Gael **LP** = Labour Party
PD = Progressive Democrats **GP** = Green Party
SF = Sinn Fein **Ind** = Independents

Senate 60 seats

| 50% FF | 25% FG | 8% LP | 7% PD | 8% Ind | 2% Others |

In 1973, an FG–LP coalition took
power, marking the end of FF as the
traditional party of government – a
position which it had held since 1932.
Government leadership then alternated
between FG and FF. The latter
dominated an FF–PD coalition under
Bertie Ahern as *taoiseach* (prime
minister) from 1997, but
although it improved
its position in the April
2002 elections, it
still failed to win
an outright majority
of seats in the
legislature.

REPUBLIC OF IRELAND

Total Land Area :
70 280 sq. km
(27 135 sq. miles)

POPULATION

- ◉ over 500 000
- ◎ over 100 000
- ○ over 50 000
- ● over 10 000
- • under 10 000

LAND HEIGHT

- 1000m/3281ft
- 500m/1640ft
- 200m/656ft
- Sea Level

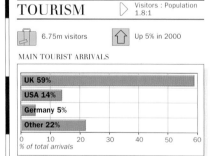

WORLD AFFAIRS
▷ Joined UN in 1955

CE   ESA   EU   OECD   OSCE

Ireland, the UK, and the US are involved in the Northern Ireland peace process. A 2001 referendum rejected the EU's Nice treaty, embarrassing the government.

AID
▷ Donor

 $235m (donations)   Down 4% in 2000

Africa is the main recipient of Irish aid. Ireland itself benefits from EU funding.

DEFENSE
▷ No compulsory military service

 $684m   Down 8% in 2000

Ireland is determined to maintain its neutrality, despite EU moves to establish a common European defense policy. It has observer status at the WEU.

ECONOMICS
▷ Inflation 3.5% p.a. (1990–2000)

 $86bn   1.0651–1.1231 euros

SCORE CARD

- ❏ WORLD GNP RANKING..........................39th
- ❏ GNP PER CAPITA $22,660
- ❏ BALANCE OF PAYMENTS...................–$593m
- ❏ INFLATION 5.6%
- ❏ UNEMPLOYMENT 4%

STRENGTHS
The "Celtic tiger" – one of Europe's fastest-growing economies: real GDP growth of 9% a year in the latter half of the 1990s. Trade surplus. Low inflation. Efficient agriculture and food processing. Expanding high-tech sector: electronics account for 25% of exports. Large recipient of EU infrastructure aid. Highly educated workforce.

WEAKNESSES
Many key sectors owned by overseas multinationals. Danger of economy overheating. Housing shortage. Rapid growth is now straining the country's infrastructure.

EXPORTS

Netherlands 6%   France 8%   Germany 11%   Other 38%   USA 17%   UK 20%

IMPORTS

France 4%   Japan 5%   Germany 6%   Other 36%   USA 16%   UK 33%

RESOURCES
▷ Electric power 4.4m kw

 329,777 tonnes   Not an oil producer; refines 56,000 b/d

 6.46m cattle, 5.13m sheep, 1.73m pigs, 11.3m chickens   Lead, zinc, natural gas, silver, coal, oil

Oil has been found off the south coast. Studies suggest that this may be in commercially exploitable quantities.

ENVIRONMENT
▷ Sustainability rank: 37th

 0.9%   10.3 tonnes per capita

The main environmental concerns are overexploitation of the country's peat bogs for fuel and the recent expansion of conifer plantations. While Ireland's levels of forest cover will increase in the next few years, most new planting is of conifers. In 1994 stringent new laws increased pollution controls. A small windpower industry is expanding.

MEDIA
▷ TV ownership high

 Daily newspaper circulation 149 per 1000 people

PUBLISHING AND BROADCAST MEDIA

There are 6 daily newspapers. These include the *Irish Times* and the *Irish Independent*

2 services: 1 state-owned, 1 independent   3 services: 1 state-owned, 2 independent

The national Radio Telefís Éireann (RTE) dominates broadcasting. Radió na Gaeltachta and TG4 broadcast in Irish. Access to UK media is widespread.

CRIME
▷ No death penalty

 2948 prisoners   Down 26% in 1999

Rural Ireland has one of the EU's lowest crime rates. Urban crime and narcotics are a problem in Dublin and Cork.

EDUCATION
▷ School leaving age: 16

 99%   113,338 students

The Roman Catholic Church runs many schools. Increased education spending has resulted in a skilled workforce.

Clew Bay in County Mayo, *on the western coast of Connaught, viewed from the slopes of neighboring Croagh Patrick.*

CHRONOLOGY
English colonization, begun in 1167, was reinforced after 1558 by anti-Catholic legislation and settlement of Scottish Protestants in the north.

- ❏ **1845–1855** Famine. One million die, 1.5 million emigrate.
- ❏ **1919–1921** Anglo-Irish war after Sinn Féin proclaims independence.
- ❏ **1922** Irish Free State established.
- ❏ **1973** FG/LP alliance wins elections, ending FF's ascendancy since 1932.
- ❏ **1990** Mary Robinson elected first woman president.
- ❏ **1995** Referendum favors divorce.
- ❏ **1998** Good Friday accord on Northern Ireland.
- ❏ **2001** Voters reject EU treaty on expansion.
- ❏ **2002** Euro fully adopted.

HEALTH
▷ Welfare state health benefits

 1 per 435 people   Heart diseases, cancers, accidents

Free care is means tested; about one-third of people qualify. Others pay to visit their doctor and for prescriptions, and there is a modest charge for hospital care.

SPENDING
▷ GDP/cap. increase

CONSUMPTION AND SPENDING

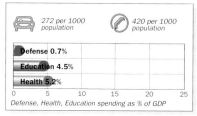
272 per 1000 population   420 per 1000 population

Defense 0.7%   Education 4.5%   Health 5.2%

0   5   10   15   20   25
Defense, Health, Education spending as % of GDP

Living standards for those with jobs are rising steadily. Welfare for those not in work is low by OECD standards.

WORLD RANKING

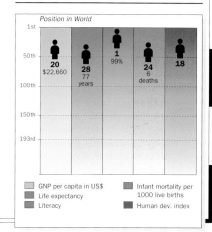
Position in World

1st   50th   100th   150th   193rd

20 $22,660   28 77 years   1 99%   24 6 deaths   18

GNP per capita in US$ Infant mortality per 1000 live births
Life expectancy
Literacy   Human dev. index

I

ISRAEL

MIDDLE EAST

OFFICIAL NAME: State of Israel **CAPITAL:** Jerusalem (not internationally recognized)
POPULATION: 6.2 million **CURRENCY:** Shekel **OFFICIAL LANGUAGES:** Hebrew and Arabic

THE CREATION OF ISRAEL in 1948 in the then UN-mandated territory of Palestine fulfilled the Zionist ambition for a Jewish state. Subsequent military victories over its Arab neighbors enabled Israel to annex or occupy additional territory, some of which it returned to Egypt under the 1978 Camp David agreement. Hopes in the 1990s for a "land for peace" deal to end the Israeli–Palestinian conflict were mired in a cycle of violence by 2002.

CLIMATE

▷ Hot desert/ Mediterranean

WEATHER CHART FOR JERUSALEM

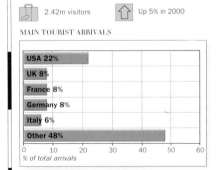

Summers are hot and dry. The wet season is between November and March, when the weather is mild.

TRANSPORTATION

▷ Drive on right

Ben-Gurion International, Tel Aviv–Yafo
9.88m passengers

49 ships
611,730 grt

THE TRANSPORTATION NETWORK

15,965 km (9920 miles)	56 km (35 miles)
663 km (412 miles)	None

Railroads are being extended. There are three commercial ports. Ben-Gurion international airport is being expanded.

TOURISM

▷ Visitors : Population 1:2.6

2.42m visitors

Up 5% in 2000

MAIN TOURIST ARRIVALS

USA 22%
UK 8%
France 8%
Germany 8%
Italy 6%
Other 48%

0   10   20   30   40   50   60
% of total arrivals

Tourism has been damaged by the sustained violence of the Israeli–Palestinian conflict.

PEOPLE

▷ Pop. density high

Hebrew, Arabic, Yiddish, German, Russian, Polish, Romanian, Persian

305/km² (790/mi²)

THE URBAN/RURAL POPULATION SPLIT

91%   9%

RELIGIOUS PERSUASION

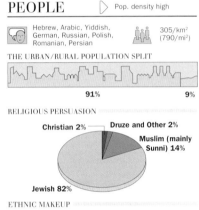

Christian 2% — Druze and Other 2%
Muslim (mainly Sunni) 14%
Jewish 82%

ETHNIC MAKEUP

Other (mostly Arab) 18%
Jewish 82%

Large numbers of Jewish immigrants settled in Palestine before Israel was founded. Sephardi Jews from the Middle East and Mediterranean are now probably in the majority, but Ashkenazi Jews, most of central European origin, still dominate society. Hundreds of thousands of Russian Jews have arrived since 1989. Israel's non-Jewish population – mostly Arab and predominantly Muslim (with Christian and Druze minorities) – totals more than one million, but remains sidelined in Israeli life. There are tensions between secular and Orthodox Jews and between left and right over the pursuit of peace with the Palestinians.

POPULATION AGE BREAKDOWN

Female	Age	Male
1.3%	80+	0.9%
5.9%	60–79	4.7%
9.9%	40–59	9.4%
14.5%	20–39	14.7%
18.8%	0–19	19.9%

% of population by age group

POLITICS

▷ Multiparty elections

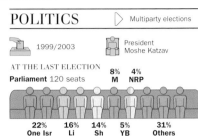

1999/2003

President Moshe Katzav

AT THE LAST ELECTION
Parliament 120 seats

| 22% One Isr | 16% Li | 14% Sh | 5% YB | 8% M | 4% NRP | 31% Others |

One Isr = One Israel (comprising Labor, Gesher, and Meymad) **Li** = Likud **Sh** = Shas
M = Meretz–Democratic Israel **YB** = Yisrael Ba'aliya
NRP = National Religious Party

Israel is a multiparty democracy.

PROFILE

Conflict with the Palestinians dominates the agenda. Peace initiatives of the 1990s have foundered. Labor prime minister Ehud Barak was replaced by hard-liner Ariel Sharon, elected in 2001. Fractious coalitions are endemic and Sharon's broad "government of national unity" is plagued by interparty disputes.

MAIN POLITICAL ISSUES
Peace with the Palestinians

Whether Israel's security should be based on strength, or on agreements with Palestinians and neighboring countries, is a central issue. Small but influential religious parties support Jewish settlement of the occupied territories, the main obstacle to any "land for peace" deal. The Oslo Peace Process, begun in 1993, seemingly heralded the gradual end of occupation, but Palestinian terror attacks and, latterly, suicide bombings of civilian targets, reinforced Israeli skepticism about prospects for peace.

Political unity

The national unity government merely papers over major differences. Sharon's Likud vehemently backs a hard-line, aggressive stance against the Palestinians and their Arab supporters. The centrist Labor party favors negotiation, which most religious groups reject.

Ariel Sharon, Israel's hard-line prime minister elected in 2001.

Yasser Arafat, the militant-turned-moderate leader of the PLO.

I

WORLD AFFAIRS

 Joined UN in 1949

Israel is technically at war with all Arab states except Egypt and Jordan. In 2000 it withdrew from southern Lebanon. Israel maintains close ties with the USA, but its harsh response to Palestinian militancy makes it increasingly isolated.

AID

 Recipient

$800m (receipts)   Down 12% in 2000

Israel receives massive military and economic aid from the US. Large ad hoc donations are also received from Jewish NGOs.

ISRAEL

Total Land Area : 20 770 sq. km (8019 sq. miles)

POPULATION
⊚ over 100 000
○ over 50 000
● over 10 000

LAND HEIGHT
1000m/3281ft
500m/1640ft
200m/656ft
Sea Level
-200m/-656ft

DEFENSE

 Compulsory military service

$9.37bn   Up 6% in 2000

Israel, the only known nuclear-armed power in the Middle East, has a small regular defense force, which can be boosted by nearly 600,000 reservists. Equipped with some of the latest US technology, it is vastly superior in firepower (and training) to the armies of its Arab neighbors. To counter the Palestinian *intifada* it uses punitive strikes and counterinsurgency methods rather than conventional tactics.

ISRAELI ARMED FORCES

	3930 main battle tanks (*Centurion*, M-48A5/60A3, *Magach 7, Merkava I/II/III*)	120,000 personnel
	3 submarines and 47 patrol boats	6500 personnel
	696 combat aircraft (50 F-4E-2000, 73 F-15, 237 F-16)	37,000 personnel
	Widely believed that Israel has a nuclear capacity with up to 100 warheads. Delivery via *Jericho* 1 and *Jericho* 2 missiles	

ECONOMICS

 Inflation 10% p.a. (1990–2000)

$104bn   4.044–4.412 sheqalim

STRENGTHS

Modern infrastructure; educated population. Huge potential of agriculture, manufacturing, and high-tech industry. Important banking sector.

WEAKNESSES

Violence and regional instability. Large defense budget. Financial burdens of absorbing immigrants and subsidizing settlement. Little trade with Arab neighbors. Corruption. Rising unemployment.

PROFILE

The government seeks ways to reduce massive state spending. The state owns most of the land and controls over 20% of all industries and services. Public companies are being privatized and there are plans to end restrictive labor practices. Agriculture, highly specialized and profitable, has been eclipsed by high-tech industries. The state aims to boost the service sector. Israel's economy expanded through the 1990s, benefiting from mass immigration of Jews, many highly educated, from the former Soviet Union. Although unemployment rose as a result, new skills and contacts also helped the Israeli economy toward

SCORE CARD

❏ WORLD GNP RANKING	35th
❏ GNP PER CAPITA	$16,710
❏ BALANCE OF PAYMENTS	–$1.42bn
❏ INFLATION	1.1%
❏ UNEMPLOYMENT	9%

EXPORTS

UK 4%   Germany 5%
Hong Kong 4%   Belgium 6%
Other 44%
USA 37%

IMPORTS

Switzerland 5%   Germany 7%
UK 7%
Other 53%   Belgium 10%
USA 18%

sustained export-led growth. Pockets of poverty remain in "development towns."

The Palestinian uprising has since 2000 seriously injured the economy, hitting trade, tourism, and investment. The Israeli military response has inflicted over $300 million of damage to Palestinian infrastructure.

ISRAEL : MAJOR BUSINESSES

Hefa
Nazerat
Tel Aviv-Yafo
Ashqelon
Jerusalem
Be'ér Sheva'

❋ Textiles
Chemicals
Computers
Food processing
Pharmaceuticals
Consumer goods

0   50 km
0   50 miles

ECONOMIC PERFORMANCE INDICATOR

Consumer Price Index   GDP

Consumer price index 1995=100
GDP 1996=100
1996 1997 1998 1999 2000

ISRAEL AND THE PALESTINIANS

THE CONFLICT between Israel and the Palestinians is crucial in Middle East politics. The creation of Israel in 1948 effectively turned the almost one million Arabs living in the former British Mandate of Palestine into refugees overnight. Some 300,000 more left territories later occupied by Israel in 1967. Not until the 1993 accords were the Palestinians granted limited self-rule in the West Bank and Gaza – inhabited by two million Palestinians and 300,000 Jewish settlers – under the jurisdiction of the Palestinian National Authority (PNA), headed by Palestine Liberation Organization (PLO) leader Yasser Arafat. From 2000, hopes of a peaceful resolution of "permanent status" issues, and accession to full statehood, were lost from sight amid a return to violent conflict.

BACKGROUND

In 1947 a UN plan partitioned Palestine into separate Jewish and Arab states. When in May 1948 the British Mandate ended, Arab states (which had all rejected the plan) invaded Palestine but were pushed back well beyond the UN partition lines by Israeli forces. The 1949 armistices left only East Jerusalem, the West Bank (5900 sq. km – 2300 sq. miles), and the Egyptian-administered Gaza Strip (1000 sq. km – 400 sq. miles) outside Israel. Jordan declared East

Poor slum housing in Gaza.

GAZA STRIP

Jerusalem and the West Bank to be part of its territory, only renouncing this claim formally in 1988. In the 1967 Six-Day War, Israel annexed the Palestinian territories. Jewish settlers began moving in, regarding the new borders as encompassing the biblical-era "Land of Israel."

Palestinian autonomy, part of the 1978 Camp David agreement, did not happen; there was diplomatic deadlock. The PLO did not recognize Israel's right to exist until 1988, while Israel refused to "negotiate with terrorists."

PALESTINIAN AUTONOMY

Turning away from armed struggle, the PLO concluded a historic "land for peace" deal with Israel in 1993, known as the "Oslo Accords." The two sides formally recognized one another, and Arafat and Israeli Prime Minister Yitzhak Rabin signed a Declaration of Principles in Washington D.C. A five-year timetable for "permanent status" negotiations would tackle the future of Jewish settlements and the Palestinian demand for East Jerusalem as their capital. Palestinians were to get interim self-rule, initially in Gaza and Jericho (achieved in 1994), and gradually in the whole West Bank. This meant Palestinian police taking over from the Israeli military, who had been struggling since 1987 to end an uprising (*intifada*) led by the radical Islamic organization Hamas. The PNA was established, based in Gaza, and Arafat made a triumphal return in July 1994. He was elected president of the newly created 88-member Palestinian Legislative Council in January 1996.

THE DERAILING OF THE PEACE PROCESS

The "Oslo B" accord in 1995 extended PNA rule to six more West Bank towns. After repeated delays Israel also gave up control of Hebron (but not of rural areas). Mutual mistrust and violence, however, threatened to derail the peace process before it reached the stage of "final status" talks and Arafat risked losing credibility among radical Palestinians. Attacks by rogue Hamas guerrillas, including suicide bombings, dominated mid-1996 elections in Israel. A new right-wing government took a tougher stance and did little to restrain Jewish settlers, fierce opponents of "land for peace."

A Labor-led government, in power from 2000, laid out an ambitious timetable for a permanent agreement with the PNA. Intensive US diplomacy helped keep this plan alive until the eleventh hour, but heavy-handed Israeli retaliation to

WEST BANK

☐ Town under Palestinian control

■ Israeli settlement

☐ Other major settlement

violent incidents in September–October provoked Palestinian rage and a return to the tactics of *intifada*. In a bitter climate of atrocity, retaliation, and counterstrike, a new Israeli government was formed by Ariel Sharon, elected prime minister in 2001. Denouncing Palestinian militants who mounted suicide bomb attacks on civilian targets, Sharon blamed Arafat for failing to control them, while instituting a brutal policy of military reprisals and "targeted assassinations." Determined to sideline Arafat completely, the Sharon government gained US support in mid-2002 for the idea that only a Palestinian leadership change could bring Israel back to the negotiating table.

Bethlehem, situated in the troubled West Bank.

RESOURCES

▷ Electric power 7.9m kw

 24,661 tonnes

 5m turkeys, 1.4m geese, 30m chickens

Low levels of oil production

Natural gas, oil, salt, potash, copper, gold, magnesium, bromine

ELECTRICITY GENERATION

- Hydro 0%
- Combustion 100% (38bn kwh)
- Nuclear 0%
- Other 0%

% of total generation by type

The country's most valuable deposits of minerals are potash, bromine (of which Israel is the world's largest exporter), and other salts mined near the Dead Sea. Reserves of copper ore and gold were discovered in 1988. In the coastal plain, mixed farming, vineyards, and citrus groves are plentiful. Former desert areas now have extensive irrigation systems supporting specialized agriculture.

Israel's most critical resource is water. The 1994 treaty with Jordan provided for sharing the water of the Jordan River.

ISRAEL : LAND USE

- Cropland
- Forest
- Pasture
- Desert
- Sheep
- Citrus fruit – cash crop

0 — 50 km
0 — 50 miles

ENVIRONMENT

▷ Sustainability rank: 63rd

 16%

 10.1 tonnes per capita

Since 1993–1994, designated Environment Year, the government has aimed to promote recycling schemes, the cleanup of rivers, and a healthier urban environment.

ENVIRONMENTAL TREATIES

Yes | Yes | Yes
Yes | Yes | No

MEDIA

▷ TV ownership high

 Daily newspaper circulation 288 per 1000 people

PUBLISHING AND BROADCAST MEDIA

There are 34 daily newspapers. The leading papers are the Hebrew *Ha'aretz*, and the English *Jerusalem Post*

2 services: 1 state-owned, 1 independent

2 state-owned services, many independent stations

The left-wing press favors the peace process. The number of private radio stations, many right-wing, is rising.

CRIME

▷ Death penalty not used in practice

 9421 prisoners

 Up 25% in 1999

CRIME RATES

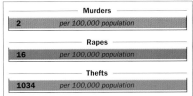

- Murders: 2 per 100,000 population
- Rapes: 16 per 100,000 population
- Thefts: 1034 per 100,000 population

Car theft is a rising concern. The vast majority of violent attacks are due to the Israeli–Palestinian conflict.

EDUCATION

▷ School leaving age: 16

 96%

 198,766 students

THE EDUCATION SYSTEM

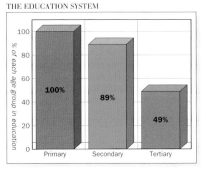

% of each age group in education

Primary 100%
Secondary 89%
Tertiary 49%

A highly educated population has been the engine of Israel's economic growth. State schools have religious (Jewish), secular, and Arab streams. Ultraorthodox and Sephardi Jews increasingly run their own private establishments.

HEALTH

▷ Welfare state health benefits

 1 per 256 people

 Heart and cerebrovascular diseases, cancers

The ratio of doctors to the total population in Israel is one of the highest in the world. Primary health care reaches all communities. Israel's hospitals have pioneered many innovative treatments.

War with the neighboring Arab states followed immediately upon the creation of the state of Israel in 1948.

- ❏ **1967** Israeli victory in Six-Day War.
- ❏ **1973** Egypt and Syria attack Israel.
- ❏ **1978** Camp David accords with Egypt.
- ❏ **1979** Formal peace treaty, Sinai returned to Egypt.
- ❏ **1982** Israel invades Lebanon.
- ❏ **1987** Palestinians launch *intifada*.
- ❏ **1993** Oslo Accords.
- ❏ **1994** Palestinian autonomy begins in Gaza and Jericho.
- ❏ **1995** Prime Minister Yitzhak Rabin assassinated.
- ❏ **1996** Palestinian elections.
- ❏ **1998** Government stalls on US-backed plan to revive peace process.
- ❏ **1999** Ehud Barak (Labor) prime minister. Renewed peace process with Palestinians and Syria.
- ❏ **2000** Israeli withdrawal from southern Lebanon. Large-scale Israeli–Palestinian clashes.
- ❏ **2001** Ariel Sharon (Likud) elected prime minister; forms unity government. Conflict worsens.
- ❏ **2002** Arafat briefly besieged in Ramallah. Suicide attacks and Israeli reprisals intensify.

SPENDING

▷ GDP/cap. increase

CONSUMPTION AND SPENDING

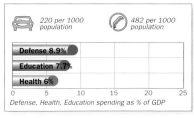

220 per 1000 population | 482 per 1000 population

- Defense 8.9%
- Education 7.7%
- Health 6%

Defense, Health, Education spending as % of GDP

Income per head is high, but taxation is heavy. In theory, those living in communes (*kibbutzim*) eschew personal material wealth.

WORLD RANKING

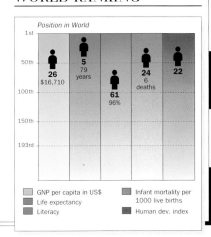

Position in World

26 $16,710 | 5 79 years | 61 96% | 24 6 deaths | 22

- GNP per capita in US$
- Life expectancy
- Literacy
- Infant mortality per 1000 live births
- Human dev. index

ITALY

OFFICIAL NAME: Italian Republic   **CAPITAL:** Rome   **POPULATION:** 57.5 million
CURRENCY: Euro (Italian lira until 2002)   **OFFICIAL LANGUAGE:** Italian

THE BOOT-SHAPED Italian peninsula stretches 800 km (500 miles) southward into the Mediterranean, while the Alps form a natural boundary to the north. Italy also includes Sicily, Sardinia, and several smaller islands. The south is an area of seismic activity, with two famous volcanoes, Vesuvius and Etna. Rival city-states flourished in Renaissance Italy, a unified country only in Roman times and since 1861. Fascist rule under Mussolini from 1922 ended with Italy's defeat in World War II. The Christian Democrats (DC) then dominated Italy's notoriously short-lived governments for decades, until in the 1990s the established parties and patronage systems were shaken up by corruption investigations. New groupings emerged, power alternating between a right-wing coalition and a broad center-left Olive Tree alliance.

CLIMATE ▷ Mediterranean/mountain

WEATHER CHART FOR ROME

■ Average daily temperature   Rainfall ■
°C/°F   J F M A M J J A S O N D   cm/in

A Mediterranean climate in the south contrasts with more temperate conditions in the north. Summers are hot and dry, especially in the south; Sardinia and Sicily have highs of more than 30°C (86°F). The Adriatic coast suffers from cold winds such as the *bora*. Southern winters are mild; northern ones are cooler and wetter, with heavy snow in the mountains.

TRANSPORTATION ▷ Drive on right

Leonardo da Vinci (Fiumicino), Rome
26.3m passengers

1457 ships
9.05m grt

THE TRANSPORTATION NETWORK

654,676 km
(406,796 miles)

6957 km
(4323 miles)

16,108 km
(10,010 miles)

2400 km
(1491 miles)

Italy's roads, which carry most its trade via Switzerland and Austria, are badly congested. The *autostrada* (expressway) network lacks key links, and serious bottlenecks affect the main north–south artery. Rail services are extensive and cheap. Luxurious Pendolino trains link Rome and Milan, but the high-speed Naples–Turin TAV project is behind schedule and over budget.

TOURISM ▷ Visitors : Population 1:1.5

39m visitors

Down 5% in 2001

MAIN TOURIST ARRIVALS

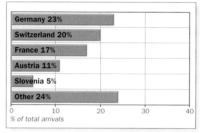

Germany 23%
Switzerland 20%
France 17%
Austria 11%
Slovenia 5%
Other 24%

0   10   20   30   40
% of total arrivals

Italy has been a tourist destination since the 16th century. Roman popes consciously aimed to make their city the most beautiful in the world to attract travelers. In the 18th century, Italy was the focus of any Grand Tour. Today, its many unspoiled centers of Renaissance and ancient culture continue to make Italy one of the world's major tourism destinations. The industry accounts for 3% of Italy's GDP, and employs a million people – 5% of the workforce.

Most visitors travel to the northern half of Italy, to cities such as Venice and Florence, and to Rome. Tourists are also drawn to the northern lakes, while beach resorts such as Rimini attract a large, youthful crowd in summer. Italy is also growing in popularity as a skiing destination. In the south the breathtaking ruins of Roman Pompeii are a particular magnet for visitors.

Fears have been expressed about the detrimental impact of tourism on Italy's environment. The pressure of visitors to Venice is such that in summer one-way systems for pedestrians have to be introduced and day trippers are often turned away.

Tuscan landscape. *Chianti wine is produced in this region, where many northern Europeans own holiday homes.*

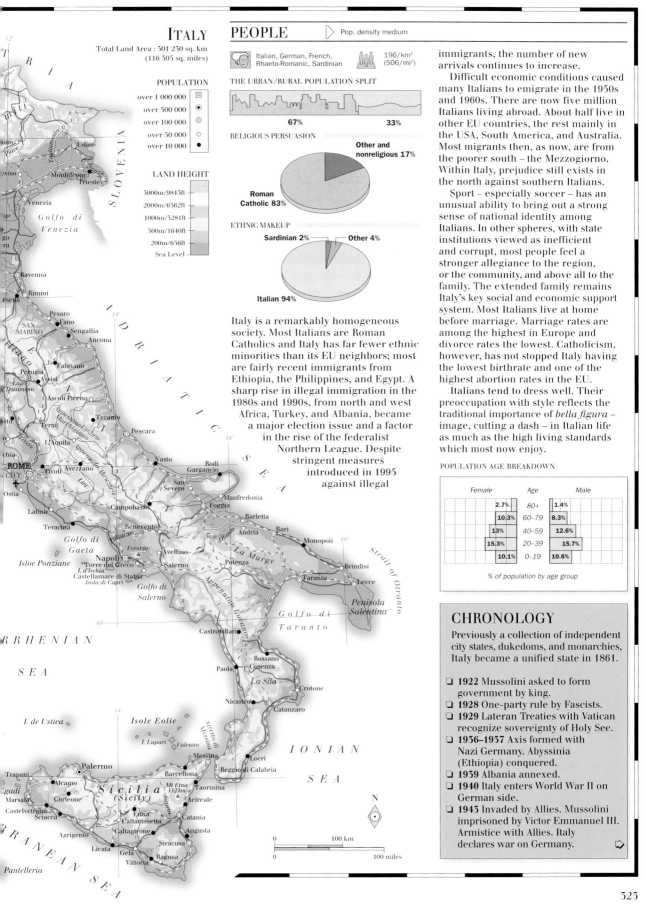

ITALY

Total Land Area : 301 230 sq. km
(116 305 sq. miles)

POPULATION

over 1 000 000
over 500 000
over 100 000
over 50 000
over 10 000

LAND HEIGHT

5000m/9843ft
2000m/6562ft
1000m/3281ft
500m/1640ft
200m/656ft
Sea Level

PEOPLE

Pop. density medium

Italian, German, French,
Rhaeto-Romanic, Sardinian

196/km²
(506/mi²)

THE URBAN/RURAL POPULATION SPLIT

67%   33%

RELIGIOUS PERSUASION

Other and
nonreligious 17%

Roman
Catholic 83%

ETHNIC MAKEUP

Sardinian 2%   Other 4%

Italian 94%

Italy is a remarkably homogeneous society. Most Italians are Roman Catholics and Italy has far fewer ethnic minorities than its EU neighbors; most are fairly recent immigrants from Ethiopia, the Philippines, and Egypt. A sharp rise in illegal immigration in the 1980s and 1990s, from north and west Africa, Turkey, and Albania, became a major election issue and a factor in the rise of the federalist Northern League. Despite stringent measures introduced in 1995 against illegal immigrants, the number of new arrivals continues to increase.

Difficult economic conditions caused many Italians to emigrate in the 1950s and 1960s. There are now five million Italians living abroad. About half live in other EU countries, the rest mainly in the USA, South America, and Australia. Most migrants then, as now, are from the poorer south – the Mezzogiorno. Within Italy, prejudice still exists in the north against southern Italians.

Sport – especially soccer – has an unusual ability to bring out a strong sense of national identity among Italians. In other spheres, with state institutions viewed as inefficient and corrupt, most people feel a stronger allegiance to the region, or the community, and above all to the family. The extended family remains Italy's key social and economic support system. Most Italians live at home before marriage. Marriage rates are among the highest in Europe and divorce rates the lowest. Catholicism, however, has not stopped Italy having the lowest birthrate and one of the highest abortion rates in the EU.

Italians tend to dress well. Their preoccupation with style reflects the traditional importance of *bella figura* – image, cutting a dash – in Italian life as much as the high living standards which most now enjoy.

POPULATION AGE BREAKDOWN

Female	Age	Male
2.7%	80+	1.4%
10.3%	60–79	8.3%
13%	40–59	12.6%
15.3%	20–39	15.7%
10.1%	0–19	10.6%

% of population by age group

CHRONOLOGY

Previously a collection of independent city states, dukedoms, and monarchies, Italy became a unified state in 1861.

❑ **1922** Mussolini asked to form government by king.
❑ **1928** One-party rule by Fascists.
❑ **1929** Lateran Treaties with Vatican recognize sovereignty of Holy See.
❑ **1936–1937** Axis formed with Nazi Germany. Abyssinia (Ethiopia) conquered.
❑ **1939** Albania annexed.
❑ **1940** Italy enters World War II on German side.
❑ **1943** Invaded by Allies. Mussolini imprisoned by Victor Emmanuel III. Armistice with Allies. Italy declares war on Germany.

CHRONOLOGY *continued*

- ❏ **1945** Mussolini released; establishes puppet regime in north; executed by Italian partisans.
- ❏ **1946** Referendum votes in favor of Italy becoming a republic.
- ❏ **1947** Italy signs peace treaty, ceding border areas to France and Yugoslavia, Dodecanese to Greece, and giving up colonies.
- ❏ **1948** Elections: DC under De Gaspieri heads coalition.
- ❏ **1949** Founder member of NATO.
- ❏ **1950** Agreement reached on US bases in Italy.
- ❏ **1951** Joins European Coal and Steel Community.
- ❏ **1957** Founder member of European Economic Community. Aided by funds from that organization and by Marshall Aid, industrial growth accelerates.
- ❏ **1964** DC government under Aldo Moro forms coalition with Socialist Party (PSI).
- ❏ **1969** Red Brigades, extreme left terrorist group, formed.
- ❏ **1972** Support for extreme right reaches postwar peak (9%). Rise in urban terrorism by both extreme left and right.
- ❏ **1976** Communist Party (PCI) support reaches a peak of 34% under Enrico Berlinguer's Eurocommunist philosophy.
- ❏ **1978** Aldo Moro abducted and murdered by Red Brigades.
- ❏ **1980** Extreme right bombing of Bologna station kills 84, wounds 200.
- ❏ **1983–1987** Center-left coalition formed under Bettino Craxi.
- ❏ **1990** LN attacks immigration policies and subsidies for the south.
- ❏ **1992** Corruption scandal, involving bribes for public contracts, uncovered in Milan. Government members accused.
- ❏ **1994** General election: DC support collapses; coalition government formed between Silvio Berlusconi's Forza Italia, LN, and neo-fascists.
- ❏ **1995–1996** Technocrat government tackles budget, pensions, media, and regional issues.
- ❏ **1996** Center-left Olive Tree alliance wins general election; Romano Prodi prime minister.
- ❏ **1998** May, Italy qualifies to join euro currency from January 1999. October, Prodi government falls, Massimo D'Alema prime minister.
- ❏ **1999** Carlo Ciampi president.
- ❏ **2000** D'Alema replaced by Giuliano Amato.
- ❏ **2001** May, Berlusconi victory in general election. June, right-wing government includes "post-fascist" National Alliance.
- ❏ **2002** Euro fully adopted.

POLITICS

  Multiparty elections

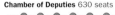 L. House 2001/2006
U. House 2001/2006
President Carlo
Azeglio Ciampi

AT THE LAST ELECTION

Chamber of Deputies 630 seats

| 58% PdL | 38% U | 2% PRC | 1% Others | 1% U–SVP |

PdL = Freedom Alliance (includes Forza Italia, National Alliance – **AN**, and Northern League – **LN**)
U = Olive Tree alliance (includes Democrats of the Left – **DS** and Party of Italian Communists – **PdCI**)
PRC = Communist Refoundation Party
U–SVP = Olive Tree–South Tyrolese People's Party
Nom = Nominated

Senate 324 seats

| 55% PdL | 38% U | 3% Nom | 1% PRC | 1% U–SVP | 2% Others |

The Senate comprises 315 elected members and several life senators

Italy is a multiparty democracy.

PROFILE

A coalition of Silvio Berlusconi's Forza Italia, Gianfranco Fini's "post-fascist" AN, and Umberto Bossi's xenophobic, secessionist LN won elections in 1994, but had collapsed by the end of that year. A technocratic government took over until 1996, when fresh elections resulted in a historic victory for the center-left Olive Tree alliance headed by Romano Prodi.

Bossi failed to rouse mass support for his declaration in September 1996 of an independent northern state of "Padania."

Success in qualifying for membership of the single European currency crowned the Prodi government's achievements in the economic sphere. After two years in office, it fell in October 1998 when the communists challenged its budget. The premiership

Romano Prodi, ex-premier (1996–1998) who heads European Commission.

Umberto Bossi, NL leader and minister in the Berlusconi government.

Silvio Berlusconi, conservative prime minister, elected for a second time in 2001.

The church of Santa Maria della Salute marks the entrance to Venice. The city-state managed to retain its independence until Napoleon Bonaparte's invasion of Italy.

passed to Massimo D'Alema of the DS (one of the successor groups to the Communist Party – PCI), then to Giuliano Amato, retaining a broadly similar coalition formula. Aware of the strength of the challenge from Berlusconi and the right, Amato then stood aside so that the popular mayor of Rome, Francesco Rutelli, could lead the center-left campaign in the May 2001 elections. The right-wing victory in that poll was essentially Berlusconi's personal triumph, although his new government repeated his 1994 formula by including the LN and the AN. It faced mass public protest in 2002 against proposed labor law reforms.

MAIN POLITICAL ISSUES
Corruption

The 1990s *mani pulite* (clean hands) investigations, initially concerned with financial scandals in Milan, went on to reveal a nationwide network of corruption permeating political and business life. The issue destroyed the old political order, and many public figures were disgraced. Berlusconi, who emerged as a leading figure in the country's new political makeup, was himself dogged by bribery-related charges. Although he was acquitted on several counts during 2000, his return to power the following year left the corruption issue still simmering, as he pushed through reforms which meant that outstanding charges against him could be quashed.

Institutional Reform

The old proportional representation (PR) electoral system, blamed for a lack of strong government, was much modified in the early 1990s, but the process lost impetus after the 1994 elections. Twice, in 1999 and in 2000, referendums failed to abolish the 25% of seats still elected by PR, not because voters opposed this, but merely due to inadequate turnout. Berlusconi favors a system with greater presidential powers, akin to that in France (and has made explicit his desire to go on to be head of state under such a system).

I

WORLD AFFAIRS ▷ Joined UN in 1955

EU · G8 · NATO · OECD · OSCE

Italy was one of the founders of the EU, but the return of a right-wing government in 2001 encouraged the opponents of further integration. Italy's strategic position in the Mediterranean makes it a central member of NATO, whose South European Command is based in Naples. Despite a pro-Western orientation, Italy often contributes to mediation in eastern Europe and the Middle East. Dependence on Libya for energy supplies made it especially keen to see UN sanctions lifted.

Major concerns in the 1990s were upheaval in Albania and conflict in the former Yugoslavia. NATO used Italian bases for air strikes against Yugoslavia in 1999, and Italy backed the US-led "war on terrorism" in Afghanistan in 2001.

AID ▷ Donor

💲 $1.38bn (donations) ⬇ Down 24% in 2000

A relatively small aid program is nearly all carried out through international organizations. Albania and the countries of the former Yugoslavia have received funding to ward off a feared influx of "economic" migrants, the Italian military running special relief operations.

DEFENSE ▷ Phasing out conscription

💲 $20.6bn ⬇ Down 9% in 2000

ITALIAN ARMED FORCES

🛡	1349 main battle tanks (819 *Leopard* 1, 368 *Centauro* B-1, 162 *Ariete*)	137,000 personnel
⚓	7 submarines, 1 carrier, 1 cruiser, 4 destroyers, 16 frigates, and 15 patrol boats	38,000 personnel
✈	405 combat aircraft (115 *Tornado*, 69 F-104, 74 AMX)	55,350 personnel
	None	

Since the ending of the Cold War, conflicts in former Yugoslavia have helped refocus defense priorities. A "New Model Defense" was announced in 1992, and in 2000 the legislature approved a law ending conscription by 2005 and also introducing women soldiers. The envisaged professional army is to play a rapid-intervention role on NATO's southern flank, while the navy fulfills Mediterranean coastal functions rather than retaining ocean-going capabilities. Defense spending remains low, despite pressures to modernize weapons systems. In 2001, 2700 troops were deployed in the "war on terrorism."

ECONOMICS ▷ Inflation 3.8% p.a. (1990–2000)

📊 $1163bn 💲 1.0651–1.1231 euros

SCORE CARD

❏ WORLD GNP RANKING	6th
❏ GNP PER CAPITA	$20,160
❏ BALANCE OF PAYMENTS	–$5.67bn
❏ INFLATION	2.5%
❏ UNEMPLOYMENT	11%

EXPORTS

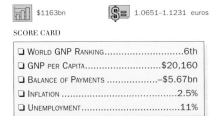

Spain 6% · UK 7% · USA 10% · France 13% · Germany 15% · Other 49%

IMPORTS

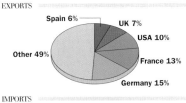

USA 5% · UK 5% · Netherlands 6% · France 11% · Germany 17% · Other 56%

STRENGTHS

Highly competitive, innovative small- to medium-size business sector. World leader in industrial and product design, textiles, and household appliances. Several highly innovative firms include Fiat (cars), Montedison (plastics), Olivetti (communications), and Benetton (clothes). Strong tourism and agriculture sectors, prestigious fashion houses.

WEAKNESSES

Public deficit and government debt remain high. Slow growth. Inefficient public sector undergoing major privatizations. Uneven wealth distribution: northern Italy far richer than the south, which suffers three times more unemployment. Poor record on tax collection, although now much improved. Relatively small companies facing foreign competition. Heavy dependence on imported energy.

PROFILE

Since World War II, Italy has developed from a mainly agricultural society into a world industrial power. The economy is characterized by a large state sector, a mass of family-owned businesses, relatively high levels of protectionism, and strong regional differences. Italy also has relatively few multinationals compared with other G7 economies.

The Institute for Industrial Reconstruction (IRI), a state-owned holding company dating from the Fascist era, progressively privatized its electronics, steel, engineering, shipbuilding, telecommunications, transportation, and aerospace companies, until closing down itself in

ECONOMIC PERFORMANCE INDICATOR

Consumer Price Index · GDP

Consumer price index 1995=100 / GDP 1996=100

1996 1997 1998 1999 2000

2000. The National Hydrocarbons Group (ENI), one of the world's top players in the energy and chemicals sectors, has been privatized, as has Telecom Italia, and the electricity corporation Enel. City and regional authorities own utilities, banks, and other businesses.

Family-owned businesses, which are the backbone of the private sector, include Fiat, whose interests include aero engines, telecommunications, and bioengineering, as well as cars. Similar businesses tend to congregate, encouraging local competition which has translated into national success.

The Mezzogiorno remains an exception. State attempts to attract new investment have met with success in areas immediately south of Rome, but elsewhere organized crime has deterred investors and siphoned off state funds. Anger at the misuse of state funds in the south was a powerful factor in the growth of the LN, with its demands for autonomy. One-third of Italian tax revenue is generated in Italy's industrial heartland of Milan.

ITALY : MAJOR BUSINESSES

Milano · Torino · Venezia · Bologna · Firenze · Genova · Rome · Napoli · Palermo

❊ Textiles
♨ Chemicals
👕 Garments
🔌 Electronics
💉 Pharmaceuticals
⚙ Light engineering
⬤ Defence industries
🚗 Vehicle manufacture
✈ Aerospace industries

0 200 km
0 200 miles

I

Remains of the Greek theater at Taormina, eastern Sicily. It was rebuilt by the Romans in the 2nd century CE. Today, the theater is the venue for an annual arts festival.

RESOURCES

Electric power
68.4m kw

540,523 tonnes

79,000 b/d
(reserves 600m
barrels)

25m turkeys, 11.1m
sheep, 8.33m pigs,
100m chickens

Coal, oil, lignite,
pyrites, fluorite,
barytes, bauxite

ELECTRICITY GENERATION

Hydro 18% (47bn kwh)	
Combustion 80% (209bn kwh)	
Nuclear 0%	
Other 2% (4bn kwh)	

0   20   40   60   80   100
% of total generation by type

Italy has very few natural resources. It produces just 1% of its oil needs and is highly vulnerable to both fluctuations in world prices and political instability in its north African suppliers. It has reduced its exposure since 1973, when oil accounted for 71% of its needs. Even so, oil still accounts for over 50% of energy consumption. Some power is generated from hydro and geothermal sources. Nuclear power was rejected in a 1987 referendum, and development has effectively been abandoned. Italy's mineral assets are small and the sector contributes little to national wealth.

ITALY : LAND USE

Cropland
Forest
Pasture
High mountain regions
Vineyards
Citrus fruits
Cattle

SARDEGNA

SICILIA

0   200 km
0   200 miles

ENVIRONMENT

Sustainability
rank: 84th

8%

7.2 tonnes
per capita

ENVIRONMENTAL TREATIES

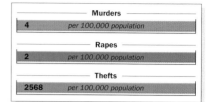

Yes   Yes   Yes
Yes   Yes   Yes

Italy has extensive environmental legislation, but has faced problems in enforcing directives. Wildlife successes include the return of the endangered lynx and brown bear, and growing numbers of wolves in the Appenines. The hunting of migrant birds, a popular sport in Italy, attracts international criticism. The use of drift nets, prone to catching dolphins and turtles as well as fish, has been made illegal under EU law. The right-wing government of the mid-1990s, returned to office in 2001, is suspicious of energy taxes and laws on waste recycling, not wanting to restrict business competitiveness. Green Party members in government in the Olive Tree alliance from 1996 to 2001 had insisted on a more active environmental stance.

Pollution in cities such as Naples and Rome is a major concern. Bans on traffic for up to seven hours during windless days are not uncommon. Acid rain has also damaged forests; 10% of trees are affected.

MEDIA

TV ownership high

Daily newspaper circulation 104 per 1000 people

PUBLISHING AND BROADCAST MEDIA

	There are 78 daily newspapers. The leading nationals are *Corriere della Serra* and *La Repubblica*
	1 publicly owned service, 16 independent national networks, over 900 independent stations
	1 publicly owned service, over 2100 independent stations

Italy's media are dominated by a few conglomerates. Mediaset, owned by Prime Minister Silvio Berlusconi, is the main commercial operator. The three state TV RAI channels have nearly 50% of the market. Until reforms in the 1990s, they were highly politicized, in that RAI 1 was apportioned to the Christian Democrats, RAI 2 to the Socialists, and RAI 3 to the Communists. News Corporation has brought out Vivendi to gain a near-monopoly on pay-TV. All the media reflect the Italian love of sport, especially soccer. *La Gazzetta dello Sport* has one of the largest circulations of the national dailies. The press is highly regionalized.

CRIME

No death penalty

53,481 prisoners

Down 1%
1996–1998

CRIME RATES

Murders	
4	*per 100,000 population*

Rapes	
2	*per 100,000 population*

Thefts	
2568	*per 100,000 population*

Over 25% of prisoners are foreigners, many held for narcotics offenses. There is a huge backlog of cases. Organized crime has been weakened by the anticorruption drive and a cleaned-up bureaucracy. The Sicilian Mafia, the Cosa Nostra, in particular, was hit hard by arrests and trials in which former members provided key evidence. However, the Cosa Nostra and its counterparts in Naples and Calabria – Camorra and 'ndrangheta – still control wholesale agricultural markets and much of the narcotics trade, bleed businesses of protection money, and manipulate public works contracts.

EDUCATION

School leaving
age: 16

98%

1.89m students

THE EDUCATION SYSTEM

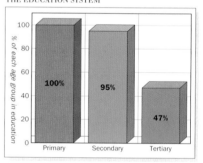

% of each age group in education

100%   95%   47%

Primary   Secondary   Tertiary

Schooling is state-run, apart from a few religious and elite private institutions. The pupil–teacher ratio in Italian schools is one of the best in Europe. In 1993, the minimum school leaving age was raised from 14 to 16 years, bringing Italy into line with most of Europe. However, the dropout rate in schools – as high as 50% in Sicily – remains a problem. An educational credit system aims to tackle shortcomings in information technology training.

Universities in Italy are oversubscribed. Rome has 180,000 students at any one time, only 30% of whom gain a degree. Many Italian educationalists wish to restrict entry. Another concern is the fact that Italy devotes only 1.4% of its GNP to research, compared with a European average of 2.5%.

REGIONALISM AND NORTHERN SEPARATISM

THE ECONOMIC DIVISION of Italy into the more dynamic and prosperous north and the impoverished south or Mezzogiorno is a source of continuing tension. It helps to explain why, in a country as homogeneous as Italy in terms of ethnicity, language, and religion, there is a strong demand for more devolution of power to regional level. In the mid-1990s, a political party calling for the breakup of the country was able to gain a significant following. The Northern League (LN), created in 1991 and built around the earlier Lombard League, won over 10% of the vote in the 1996 elections. Although it has lost momentum since then, regional feeling remained sufficient for the LN to be drafted into the right-wing government formed after the 2001 elections, and for a referendum in October to secure the first step toward limited devolution.

Run-down housing in the poor, rural area of Calabria in southern Italy.

REGIONS WITH MORE DYNAMIC ECONOMIES

GDP per capita is almost three times higher in the richest northern areas, such as Bologna in Emilia-Romagna and Milan in Lombardy, than in southern areas such as Reggio di Calabria. After the 1996 recession, the north achieved more rapid growth, creating more new jobs faster. Lombardy and Piedmont attract the lion's share of inward investment. Turin, the capital of Piedmont and home of Fiat, is boosting its industrial image by completing high-speed rail links and staging the Winter Olympics in 2006. Some southern areas try to attract investment by emphasizing that labor is cheaper there, and getting unions to accept lower wages locally in so-called "territorial pacts," but poor infrastructure remains a major obstacle. A road–rail bridge project linking Sicily to the Italian mainland was only revived in 2002.

PADANIA

The LN championed the idea of creating a "Republic of Padania" to secede from the rest of Italy (Rome,

its surrounding Lazio region, and the south). The LN leader Umberto Bossi declared Padania's "independence" in a ceremony in 1996 after a pilgrimage the length of the river Po, the symbolic artery of "Padania," from Piedmont in northwest Italy to the Adriatic south of Venice. The LN set up a self-styled government and parliament, a national guard, and a flag, while allowing a year to "negotiate" terms of separation with the Italian government. Mantua was described as the Padanian capital, but it remained unclear precisely how far its intended territory extended. Usually understood to refer to the nine most northerly regions, as far south as Florence and the rest of Tuscany, it was sometimes defined by the LN as also including the Umbria and Marche regions.

The Italian government refused to take the Padania idea seriously, dismissing it as a publicity stunt. LN leaders were subsequently persuaded, by a series of bad local election results including losing the mayorships of Mantua and Milan, that the time was not right to press forward with their project. Some of its impetus – the desire of northern businesses to be at the center of European Union integration, not held back by the south – also disappeared when Italy qualified to join the European single currency, the euro, in 1999.

Shopping center in Milan, Italy's center of fashion and commerce.

HEALTH

 Welfare state health benefits

1 per 169 people

 Heart and cerebro-vascular diseases, cancers, accidents

Italy's health care system is rated by WHO as the second most efficient in the world.

Standards of health care vary across the country, as the services are run by the regions. The state-run health system introduced in the 1970s initially provided services free at the point of use, but charges have been levied since 1988 for some dental and prescription costs; patients also have to pay a daily hospital charge and a yearly health fee. AIDS patients are exempt.

SPENDING

GDP/cap. increase

CONSUMPTION AND SPENDING

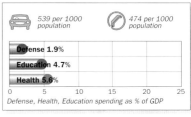

539 per 1000 population

474 per 1000 population

Defense 1.9%				
Education 4.7%				
Health 5.6%				
0	5	10	15	20	25

Defense, Health, Education spending as % of GDP

Italians, particularly in the north, are today among the world's wealthiest people in terms of disposable income. This is a result not only of economic growth, but also of the structure of Italian society.

Many Italians have more than one job. The extended families in which most people still live often have access to more than one income. Few people have mortgages, and savings and tax avoidance levels are high.

The main exceptions are in parts of the south. Although inward investment has been attracted to the Bari area, many people still live in poverty in other places, such as Naples and the Calabria region, where investment has been lowest, unemployment is highest, and even tourism is underdeveloped.

WORLD RANKING

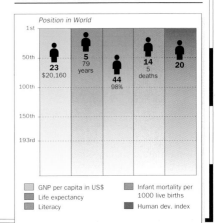

- GNP per capita in US$
- Life expectancy
- Literacy
- Infant mortality per 1000 live births
- Human dev. index

IVORY COAST

WEST AFRICA
Africa

OFFICIAL NAME: Republic of Côte d'Ivoire   **CAPITAL:** Yamoussoukro
POPULATION: 16.3 million   **CURRENCY:** CFA franc   **OFFICIAL LANGUAGE:** French

 1960   1960   Aug 7   CI   0 +225   .ci

ONE OF THE LARGER west African coastal countries, Ivory Coast – officially Côte d'Ivoire – is the world's biggest cocoa producer. The forested interior, apart from the capital, is more sparsely populated than the sandy coastal strip. Pro-Western President Houphouët-Boigny ruled from independence in 1960 until 1993. A period of turbulence followed his death. Military rule in 1999 gave way in 2000 to an elected president after a chaotic poll.

CLIMATE   ▷ Tropical wet & dry

WEATHER CHART FOR YAMOUSSOUKRO

The south's four seasons – two rainy and two dry – merge in the north into a single wet season with lower rainfall.

TRANSPORTATION   ▷ Drive on right

Abidjan (Port-Bouët)
1.08m passengers
33 ships
9200 grt

THE TRANSPORTATION NETWORK

4889 km (3038 miles)		None	
1955 km (1215 miles)		980 km (609 miles)	

The relatively good transportation system focuses on Abidjan, the premier port of francophone west Africa.

TOURISM   ▷ Visitors : Population 1:54

301,000 visitors   Up 10% in 1998

MAIN TOURIST ARRIVALS

| France 24% |
| USA 6% |
| Burkina 6% |
| Other 64% |

% of total arrivals

Ambitious plans for an "African Riviera" east of Abidjan and the opening of a hotel by the French Club Méditerranée have helped increase tourism. The giant Roman Catholic basilica at Yamoussoukro is an attraction.

PEOPLE   ▷ Pop. density medium

 Akan, French, Kru, Voltaic   51/km² (133/mi²)

THE URBAN/RURAL POPULATION SPLIT

46%   54%

RELIGIOUS PERSUASION

Muslim 25%
Protestant 6%
Other 23%
Roman Catholic 23%
Traditional beliefs 23%

There are more than 60 tribes in Ivory Coast. Larger groups among them can be recognized on the basis of cultural identity, the key ones being the Baoulé in the center, the Agri in the east, the Senufo in the north, the Dioula in the northwest and west, the Bété in the center-west, and the Dan-Yacouba in the west. President Houphouët-Boigny promoted his own group, the Baoulé, who account for 23% of the population.

The extended family is an important force, particularly in the shanty towns of Abidjan. Migrants from other west African countries account for up to 40% of the population.

IVORY COAST

Total Land Area : 322 460 sq. km (124 502 sq. miles)

POPULATION

⊡	over 1 000 000
◎	over 100 000
○	over 50 000
●	over 10 000
•	under 10 000

LAND HEIGHT

1000m/3281ft
500m/1640ft
200m/656ft
Sea Level

POLITICS   ▷ Multiparty elections

2000/2005   President Laurent Gbagbo

AT THE LAST ELECTION

National Assembly 225 seats

10% Ind   1% PIT   1% Vac
43% FPI   42% PDCI   2% RDR   1% Others

FPI = Ivorian Popular Front   **PDCI** = Democratic Party of Ivory Coast   **Ind** = Independents   **RDR** = Rally of the Republicans   **PIT** = Ivorian Labor Party   **Vac** = Vacant

A power struggle to replace President Houphouët-Boigny in 1993 was won by arch-conservative, and French-backed, National Assembly president Henri Konan Bédié. His main rival, the Muslim northerner Alassane Ouattara of the RDR, returned in 1999 to contest the presidency. Konan Bédié's attempts to prevent him from standing provoked a coup led by Gen. Robert Guei. Guei stood as a candidate in the 2000 presidential elections, while barring both Konan Bédié and Ouattara from the poll. A popular revolt was incited by Guei's attempts to falsify the results and he fled the country. Supporters of veteran socialist Laurent Gbagbo, who had led in the poll, then clashed violently with Ouattara's followers. Legislative elections in December were won by Gbagbo's FPI. A national reconciliation forum convened in October–December 2001 successfully promoted a dialog among the main political parties.

WORLD AFFAIRS   Joined UN in 1960

Instability in other west African countries, such as neighboring Liberia, is a major concern. Relations with the

main creditors, the World Bank group, and France, the chief source of private investment and bilateral loans, also form a major issue. Ivory Coast plays an important role in international cocoa and coffee organizations.

AID   Recipient

 $352m (receipts)   Down 21% in 2000

France is the largest source of bilateral aid. Structural adjustment loans from the World Bank were particularly important in easing the acute burden of a debt accumulated on the strength of overinflated oil hopes.

DEFENSE   Compulsory military service

$132m   Up 2% in 2000

France is the main supplier of equipment and trainer of officers for the 6800-strong army. It also maintains a garrison near Abidjan. Ivory Coast has a large paramilitary gendarmerie.

ECONOMICS Inflation 7.5% p.a. (1990–2000)

$9.59bn   698.7–736.7 CFA francs

SCORE CARD
- WORLD GNP RANKING ... 85th
- GNP PER CAPITA ... $600
- BALANCE OF PAYMENTS ... –$13m
- INFLATION ... 2.5%
- UNEMPLOYMENT ... 13%

STRENGTHS
Well-developed agriculture: major cocoa and coffee producer. Relatively good infrastructure. Expanding oil and gas industries. Healthy foreign investment levels. Successful debt rescheduling.

WEAKNESSES
Failure to invest adequately in education and professional training. Overdependence on cocoa and coffee. Slave labor on plantations.

EXPORTS
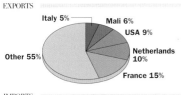
Italy 5%, Mali 6%, USA 9%, Netherlands 10%, France 15%, Other 55%

IMPORTS

Germany 3%, Italy 3%, Belgium-Luxembourg 4%, France 19%, Nigeria 25%, Other 46%

RESOURCES Electric power 1.2m kw

77,000 tonnes   25,704 b/d (reserves 103m barrels)

1.45m sheep, 1.41m cattle, 29.4m chickens   Oil, diamonds, cobalt, gold, iron, manganese, nickel

Close to self-sufficiency in energy. Substantial offshore oil and gas reserves. Forest resources badly depleted.

ENVIRONMENT Sustainability rank: 108th

6% (0.3% partially protected)   0.9 tonnes per capita

The government imposed a ban on unprocessed timber exports in 1995 to protect Ivorian forests.

MEDIA TV ownership medium

Daily newspaper circulation 17 per 1000 people

PUBLISHING AND BROADCAST MEDIA

There are 16 daily newspapers, including *Fraternité Matin* and *Ivoir Soir*, both published by the government

2 services: 1 state-owned, 1 independent   8 services: 1 state-owned, 7 independent

Although heavy censorship in Ivory Coast has eased since the early 1990s, there are still cases of official harassment of the media.

EDUCATION School leaving age: 13

47%   96,681 students

Baccalauréat pass rates are low. Cuts in spending have triggered student protests. Primary education fees ended in 2001.

CRIME No death penalty

13,670 prisoners   Up 32% 1992–1996

Foreign immigrants are often blamed for the widespread crime in Abidjan. Human rights abuses are common.

The basilica, Yamoussoukro. *Built in the new capital, Houphouët-Boigny's birthplace, it is modeled on St. Peter's, Rome.*

CHRONOLOGY
One of the great trading emporia of west Africa, the Ivory Coast was made a French colony in 1893. By 1918, the French had defeated the Malinke empire and the forest peoples of the interior.

- **1903–1935** Plantations developed.
- **1960** Félix Houphouët-Boigny declares independence.
- **1990** First contested polls: Houphouët-Boigny and PDCI win.
- **1993** Houphouët-Boigny dies.
- **1998** Power of president increased, Ouattara apparently barred from standing in presidential elections.
- **1999** Military coup by Gen. Guei.
- **2000** Guei ousted after false election victory claim. Gbagbo president.
- **2001** National Reconciliation Forum set up.

HEALTH No welfare state health benefits

1 per 10,000 people   Malaria, communicable diseases, neonatal deaths

The incidence of HIV/AIDS is high, affecting 9.7% of adults as at end-2001. In 2001 drugs companies agreed to cut the prices of treatments by 80–90%.

SPENDING GDP/cap. increase

CONSUMPTION AND SPENDING

20 per 1000 population   18 per 1000 population

Defense 0.9%, Education 4.2%, Health 1.2%
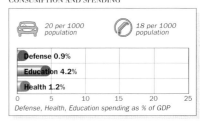
Defense, Health, Education spending as % of GDP

A large bourgeoisie grew rich in the boom years. Urban living standards in Ivory Coast are better than in many African countries.

WORLD RANKING

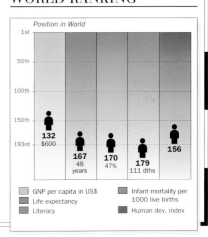
Position in World — 132 $600, 167 (48 years), 170 (47%), 179 (111 dths), 156

GNP per capita in US$, Life expectancy, Literacy, Infant mortality per 1000 live births, Human dev. index

JAMAICA

OFFICIAL NAME: Jamaica **CAPITAL:** Kingston
POPULATION: 2.6 million **CURRENCY:** Jamaican dollar **OFFICIAL LANGUAGE:** English

1962 | **1962** | **Aug 5** | **JA** | **-5** | **+1876** | **.jm**

FIRST COLONIZED BY the Spanish and then, from 1655, by the English, Jamaica is located in the Caribbean, 145 km (90 miles) south of Cuba. It was the first of the Caribbean island countries to become independent in the postwar years, and remains an active force in Caribbean politics. Jamaica is also influential on the world music scene; *reggae* and *ragga* (or dancehall) developed in the tough conditions of Kingston's poor districts.

CLIMATE ▷ Tropical oceanic

WEATHER CHART FOR KINGSTON

Tropical and humid conditions at sea level give way to temperate weather in mountain areas. Rainfall is seasonal, with marked regional variations.

TRANSPORTATION ▷ Drive on left

 Donald Sangster International, Montego Bay 2.99m passengers   9 ships 3600 grt

THE TRANSPORTATION NETWORK

13,100 km (8140 miles)	None
339 km (211 miles)	None

Kingston's harbor has been expanded and its airport improved. The road network is extensive. Private buses provide public transportation.

TOURISM ▷ Visitors : Population 1:2

1.32m visitors ⬆ Up 6% in 2000

MAIN TOURIST ARRIVALS

USA 70%	
UK 9%	
Canada 8%	
Other 13%	

0 10 20 30 40 50 60 70 80
% of total arrivals

Tourism is the major earner of foreign exchange. Most tourists stay in large, enclosed beach resorts. Recent bouts of social unrest have damaged the sector.

PEOPLE ▷ Pop. density high

 English Creole, English   240/km² (622/mi²)

THE URBAN/RURAL POPULATION SPLIT

56% 44%

RELIGIOUS PERSUASION

Other and nonreligious 39% — Other Protestant 20% — Other 6% — Anglican 7% — Baptist 10% — Church of God 18%

Jamaicans are a broad ethnic mix. Most are the descendants of Africans brought to the island between the 16th and 19th centuries, but there are minorities of Europeans, Indians, Chinese, and Arabs. Jamaica is also home to the Rastafarians, followers of Haile Selassie, the former emperor of Ethiopia.

Most social tension is the result of the marked disparities in wealth. The Caribbean women's rights movement originated in Jamaica, and today many Jamaican women hold senior positions in economic and political life.

Although life revolves around the family, absentee fathers are common. Many career women are single parents by choice. Life in the ghettos of Kingston is often violent and based largely on gun law. Kingston slums have their own patois.

Bauxite mine and terminal, *Runaway Bay. Bauxite – from which aluminum is extracted – is the main source of foreign income.*

POLITICS ▷ Multiparty elections

L. House 1997/2002
U. House 1998/2003 H.M. Queen Elizabeth II

AT THE LAST ELECTION

House of Representatives 60 seats

85% PNP 15% JLP

PNP = People's National Party **JLP** = Jamaica Labour Party

Senate 21 seats

The members of the Senate are appointed. Thirteen members are chosen by the prime minister and 8 by the leader of the opposition

The country's political complexion changed markedly in the late 1980s, as the ideologies of the once socialist PNP and the conservative JLP converged toward a moderate free-market economic approach. A general election in 1997 gave the PNP government a third consecutive term in office. The party subsequently swept the local elections in 1998.

Violent disturbances in 1998 and 1999 were in response to the government's attempts to deal with economic recession and a large fiscal deficit. The unrest, which led to several deaths, gave new life to the internally troubled JLP, as it then identified itself with opposition to fuel tax increases.

WORLD AFFAIRS ▷ Joined UN in 1962

ACS | **Caricom** | **Geplac** | **Comm** | **OAS**

Antidrugs cooperation with the US and future relations within Caricom and the Commonwealth predominate.

AID ▷ Recipient

$ US$10m (receipts) ⬇ Down 47% in 1999–2000

Most aid comes from the US, the EU, and Japan. It includes both project loans and balance-of-payments support.

DEFENSE ▷ No compulsory military service

$ US$49m ⬇ Down 4% in 2000

Jamaica's defense force buys its arms from the US, but is trained by the UK. Today, the defense force is used against narcotics smugglers and to assist the police to break up unrest, as in 1999.

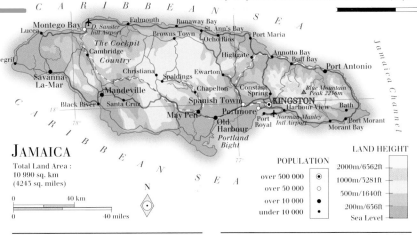

JAMAICA

Total Land Area :
10 990 sq. km
(4245 sq. miles)

0 — 40 km
0 — 40 miles

N

POPULATION
- over 500 000 ⊙
- over 50 000 ○
- over 10 000 ●
- under 10 000 ·

LAND HEIGHT
- 2000m/6562ft
- 1000m/3281ft
- 500m/1640ft
- 200m/656ft
- Sea Level

CHRONOLOGY

Spain occupied the island in 1510, wiping out the indigenous Arawak population. Britain seized it in 1655.

- ❏ **1958–1961** West Indies Federation.
- ❏ **1962** Independence under JLP.
- ❏ **1972** PNP elected. Reforms fail; street violence begins.
- ❏ **1980** Unpopular IMF austerity measures lead to JLP election win.
- ❏ **1989, 1993, 1997** PNP returned and austerity continues.
- ❏ **1999** Violent protests over fuel tax increases.

ECONOMICS

▷ Inflation 24% p.a. (1990–2000)

 US$6.88bn

 45.10–47.05 Jamaican dollars

SCORE CARD

❏ WORLD GNP RANKING	98th
❏ GNP PER CAPITA	US$2610
❏ BALANCE OF PAYMENTS	–US$275m
❏ INFLATION	8.2%
❏ UNEMPLOYMENT	16%

STRENGTHS

Relatively diversified economy. Tourism. Mining and refining of bauxite for aluminum. Agriculture, including sugar, bananas, rum, and coffee. Light manufacturing and data processing for US companies are growing sectors.

WEAKNESSES

Banking and insurance sectors. Financing of sugar production. Stagnant growth. High debt burden.

EXPORTS

Norway 9%
Canada 10%
USA 39%
UK 11%
Netherlands 11%
Other 20%

IMPORTS

Mexico 5%
Japan 6%
Venezuela 4%
Trinidad & Tobago 10%
USA 45%
Other 30%

RESOURCES

▷ Electric power 1.2m kw

 12,658 tonnes

Not an oil producer; refines 32,000 b/d

 440,000 goats, 400,000 cattle, 11m chickens

 Bauxite, marble, gypsum, silica, clay

Jamaica is the world's third-largest producer of bauxite. Sugar and bananas are major exports.

ENVIRONMENT

▷ Sustainability rank: 122nd

 0.1%

 4.3 tonnes per capita

Acidic dust from bauxite processing is a major problem, as is urban pollution in Kingston and its bay. Broad-leaved tropical forests have largely disappeared.

MEDIA

▷ TV ownership medium

 Daily newspaper circulation 63 per 1000 people

PUBLISHING AND BROADCAST MEDIA

There are 4 daily newspapers, the *Daily Gleaner*, the *Daily Star*, the *Jamaica Observer*, and the *Jamaica Herald*.

3 independent services

7 services:
1 public,
6 independent

The government has loosened its hold on broadcasting. The Jamaican press is one of the most influential in the Caribbean.

CRIME

▷ Death penalty in use

 4288 prisoners

 Down 18% 1996–1998

Armed crime is a major problem. Many murders are the result of armed robberies linked to narcotics gangs competing for territory. Much of the world crack trade is still controlled from Jamaica. Large areas of Kingston are ruled by violent gang leaders. The armed police are also frequently accused of the arbitrary shooting of suspects. An agreement to create a new Caribbean Court of Justice increased the likelihood of executions being carried out. The last hangings were in 1988 but sentences are still being imposed.

EDUCATION

▷ School leaving age: 12

 87%

 21,217 students

Education is based on the former British 11-plus selection system. Jamaica hosts the largest of the three campuses of the University of the West Indies.

HEALTH

▷ Welfare state health benefits

 1 per 714 people

 Cerebrovascular and heart diseases, cancers, diabetes

The once-efficient state health service is now seriously underfunded. There are fewer doctors and nurses than in the 1980s. Hospitals generally have a shortage of drugs and there is only rudimentary medical equipment.

SPENDING

▷ GDP/cap. increase

CONSUMPTION AND SPENDING

41 per 1000 population
199 per 1000 population

Defense 0.7%
Education 6.3%
Health 3.1%

0   5   10   15   20   25
Defense, Health, Education spending as % of GDP

Wealth disparities are highly marked in Jamaica, although better education has seen an increase in the number of Afro-Jamaicans taking more lucrative, white-collar jobs. The poorest in Jamaica, mostly migrants from rural areas, live in the slums of Kingston.

WORLD RANKING

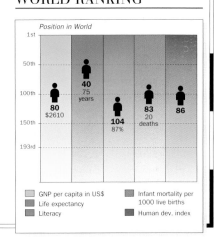

Position in World

1st
50th
100th
150th
193rd

80 $2610
40 75 years
104 87%
83 20 deaths
86

- GNP per capita in US$
- Life expectancy
- Literacy
- Infant mortality per 1000 live births
- Human dev. index

JAPAN

OFFICIAL NAME: Japan **CAPITAL:** Tokyo
POPULATION: 127.3 million **CURRENCY:** Yen **OFFICIAL LANGUAGE:** Japanese

 1590
 1972
 Dec 23
 J
 +9
 +81
.jp

EAST ASIA

A CONSTITUTIONAL MONARCHY, with an emperor as ceremonial head of state, Japan is located off the east Asian coast in the north Pacific. It comprises four principal islands and more than 3000 smaller islands. Sovereignty over the most southerly islands in the Kurile chain is disputed with the Russian Federation. The terrain is mostly mountainous, with fertile coastal plains; over two-thirds is woodland. The Pacific coast is vulnerable to tsunamis – tidal waves triggered by submarine earthquakes. Most cities are located by the sea; Tokyo, Kawasaki, and Yokohama together constitute the most populous and heavily industrialized area. Hokkaido is the most rural of the main islands. Japan's power in the global economy, with annual trade surpluses exceeding $100 billion and massive overseas investments, has been shaken since the early 1990s by a series of bad debt crises, bankruptcies in the financial sector, and two recessions.

The Shinkansen "bullet train" is the second fastest train in the world. Its speed is matched by its punctuality.

CLIMATE   ▷ Continental/subtropical

WEATHER CHART FOR TOKYO

The Sea of Japan has a moderating influence on the climate. Winters are less cold than on the Asian mainland, and rainfall is much higher. Spring is perhaps the most pleasant season, with warm, sunny days without the sultry, oppressive heat and rainfall of the summer. Recent freak storms and heavy floods have raised concern over the implications of global climate change.

TRANSPORTATION   ▷ Drive on left

 Haneda, Tokyo
56.4m passengers

 8012 ships
15.3m grt

THE TRANSPORTATION NETWORK

863,003 km (536,244 miles)	6617 km (4112 miles)
20,165 km (12,531 miles)	1770 km (1100 miles)

Railroads are the most important means of transportation in Japan. The Shinkansen, known in the West as the bullet train, is the second-fastest in the world. It is renowned as much for its reliability – timed to the second – as for its speed. The Tokyo–Sapporo air route is the busiest in the world.

JAPAN

Total Land Area : 377 835 sq. km
(145 882 sq. miles)

POPULATION

■	over 5 000 000
◨	over 1 000 000
◉	over 500 000
◎	over 100 000
○	over 50 000
●	over 10 000

LAND HEIGHT

1500m/4921ft
1000m/3281ft
500m/1640ft
Sea Level

J

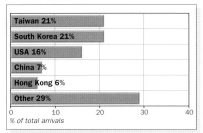

The Kurile islands are administered by the Russian Federation, but claimed by Japan

TOURISM ▷ Visitors : Population 1:27

🧳 4.76m visitors   ⬆ Up 7% in 2000

MAIN TOURIST ARRIVALS

	% of total arrivals
Taiwan 21%	
South Korea 21%	
USA 16%	
China 7%	
Hong Kong 6%	
Other 29%	

% of total arrivals (0 – 10 – 20 – 30 – 40)

Japan is expensive for foreign tourists, despite reductions in the yen exchange rate. Most come from Taiwan, South Korea, the US, and, increasingly, China. The ancient imperial capital, Kyoto, and the temples and gardens of Nara are popular tourist destinations. Other attractions include the extraordinary variety of energetic high-tech urban living in Tokyo and Osaka. Traditional agricultural life can be found in rural areas such as Tohoku in northern Honshu. Wilderness areas of Hokkaido attract mainly Japanese climbers and hikers.

High Street, Ginza District, Tokyo.
Japan's well-policed cities are among the safest in the world.

PEOPLE ▷ Pop. density high

 Japanese, Korean, Chinese   338/km² (876/mi²)

THE URBAN/RURAL POPULATION SPLIT

79%   **21%**

RELIGIOUS PERSUASION

Other (including Christian) 8%
Buddhist 16%
Shinto and Buddhist 76%

ETHNIC MAKEUP

Other (mainly Korean) 1%
Japanese 99%

Japan is racially one of the most homogeneous societies in the world. Its sense of order is reflected in the tradition of the lifetime employer. Many Japanese men define themselves by the company they work for rather than their job. An employer's influence stretches into employees' social time, and even to encouraging and approving marriages.

Traditionally women run the home and supervise the all-important education of their children. Many pursue careers until marriage, then continue to work part-time. However, trends are changing and some women are beginning to take on long-term careers, particularly in the medical and legal professions. Takako Doi became the country's first female political party leader in 1991.

Social form remains extremely important in Japanese society. Respect for elders and for social and business superiors is strongly ingrained. There is little tradition of generation rebellion, but the youth market is powerful and current fashions are geared toward teenagers. Many may still follow their parents' lifestyles, but established attitudes are under challenge. Working for the same company for life, and giving up evenings and weekends to entertain company clients, have become harder to justify since the economically turbulent 1990s.

POPULATION AGE BREAKDOWN

Female	Age	Male
2.2%	80+	1.1%
9.8%	60–79	8.1%
14.5%	40–59	14.3%
13.7%	20–39	14.1%
10.8%	0–19	11.4%

% of population by age group

CHRONOLOGY

Japan's tendency to limit its contacts with the outside world ended in 1853, when a US naval squadron coerced trading concessions from the last of the Tokugawa shoguns.

❑ **1868** Meiji Restoration; overthrow of Tokugawa regime and restoration of imperial power.
❑ **1872** Modernization along Western lines. Japan's strong military tradition becomes state-directed.
❑ **1889** Constitution modeled on Bismarck's Germany adopted.
❑ **1894–1895** War with China; ends in Japanese victory.
❑ **1904–1905** War with Russia; ends in Japanese victory. Formosa (Taiwan) and Korea later annexed.
❑ **1914** Joins World War I on Allied side. Sees limited naval action.
❑ **1919** Versailles peace conference gives Japan limited territorial gains in the Pacific.
❑ **1923** Yokohama earthquake kills 140,000.
❑ **1927** Japan enters period of radical nationalism, and introduces the notion of a "coprosperity sphere" in southeast Asia under Japanese control. Interpreted in the US as a threat to its Pacific interests.
❑ **1931** Chinese Manchuria invaded and renamed Manchukuo.
❑ **1937** Japan launches full-scale invasion of China proper.
❑ **1938** All political parties placed under one common banner; Japan effectively ruled by militarists.
❑ **1939** Undeclared border war with Soviet Union; Japan defeated.
❑ **1940** Fall of France in Europe; Japan occupies French Indo-China.
❑ **1941** US imposes total trade embargo, including oil, on Japan thereby threatening to stifle its military machine. Japan responds in December by launching attack on US fleet at Pearl Harbor and invading US, British, and Dutch possessions in the Pacific.
❑ **1942** Japan loses decisive naval battle of Midway.
❑ **1945** Huge US bombing campaign culminates in atomic bombing of Hiroshima and Nagasaki. Soviet Union declares war on Japan. Emperor Hirohito surrenders, gives up divine status. Japan placed under US military government with Gen. MacArthur installed in Japan as supreme commander of Allied Powers.
❑ **1947** New Japanese constitution: modeled on that of the US, but retains emperor in ceremonial role.
❑ **1950** Korean War. US army contracts lead to quick expansion of Japanese economy. ⇨

J

CHRONOLOGY *continued*

- ❑ **1952** Treaty of San Francisco. Japan regains independence. Industrial production recovers to 15% above 1936 levels.
- ❑ **1955** Formation of LDP, which governs for next 38 years.
- ❑ **1964** Tokyo Olympics. Bullet train (Shinkansen) inaugurated. Japan admitted to OECD.
- ❑ **1973** Oil crisis. Economic growth cut. Government-led economic reassessment decides to concentrate on high-tech industries.
- ❑ **1976** LDP shaken by Lockheed bribery scandal; in subsequent election it remains in power but loses outright majority for first time.
- ❑ **1979** Second oil crisis. Growth continues at 6% per year.
- ❑ **1980** Elections: restoration of LDP overall majority.
- ❑ **1982** Honda establishes first car factory in US.
- ❑ **1988** Japan becomes world's largest aid donor and overseas investor.
- ❑ **1989** Death of Emperor Hirohito. Recruit–Cosmos bribery scandal leads to resignation of Prime Minister Noburo Takeshita; replaced by Sosuke Uno, who is in turn forced to resign over sexual scandal.
- ❑ **1990** Tokyo stock market crash.
- ❑ **1991–1992** LDP torn by factional disputes, further financial scandals, and the issue of electoral reform.
- ❑ **1993** Reformists split from LDP and create new parties. Elections; LDP loses power. Morihiro Hosokawa becomes prime minister at head of seven-party coalition.
- ❑ **1994** Hosokawa resigns. Withdrawal of SDPJ causes collapse of coalition. New three-party coalition includes LDP and SDPJ. Opposition parties unified by creation of Shinshinto. Implementation of far-reaching political and electoral reforms designed to eradicate "money politics."
- ❑ **1995** Kobe earthquake kills more than 5000 people.
- ❑ **1996** Elections: LDP minority government. Copper trader Yasuo Yamanaka sentenced to eight years in prison for incurring losses of $2.6 billion while acting for the Sumitomo Corporation.
- ❑ **1997** Severe economic recession.
- ❑ **1998** Crisis over reform of banking and financial system.
- ❑ **2000** April, Prime Minister Keizo Obuchi falls into coma, replaced by Yoshiro Mori. June, LDP loses overall majority in general election.
- ❑ **2001** April, LDP turns to populist right-winger Junichiro Koizumi as prime minister; five women appointed to cabinet.

POLITICS Multiparty elections

 L. House 2000/2004
U. House 2001/2004

Emperor Tsegu no Miya Akihito

AT THE LAST ELECTION

House of Representatives 480 seats

			6% NK	4% JCP	6% Others
49% LDP	26% DPJ	5% LP	4% SDPJ		

LDP = Liberal Democratic Party **DPJ** = Democratic Party of Japan **NK** = New Komeito **LP** = Liberal Party **JCP** = Japan Communist Party **SDPJ** = Social Democratic Party of Japan

House of Councillors 247 seats

			9% NK	3% SDPJ	8% Others
45% LDP	24% DPJ	8% JCP	3% LP		

Japan is a multiparty democracy. The emperor has a purely ceremonial role.

PROFILE

Morihiro Hosokawa, who became the first non-LDP prime minister in 1993, was helped by not being associated with the tainted world of Tokyo politics. Four LDP prime ministers had resigned because they were implicated in scandals, or had failed to stamp out corruption. However, after becoming prime minister Hosokawa was also accused of financial irregularities, and resigned in 1994. His government, a fragile coalition, nevertheless laid the basis for electoral reform, apologized for Japan's war crimes, and began the process of institutional deregulation.

In 1996, the LDP returned to power in a coalition including the SDPJ, with Ryutaro Hashimoto as prime minister. Elections later that year were mainly a contest for the center-right vote. The LDP emerged as the largest party, ahead of Shinshinto, formed two years earlier

Junichiro Koizumi,
populist premier
who has promised
economic reform.

Yoshiro Mori,
prime minister
(2000–2001), but not
liked by the voters.

Emperor Akihito.
He acceded in 1989
on the death of his
father, Hirohito.

Traditional paddy field in Hokkaido.
Rice farming is among the most protected
sectors of the Japanese economy.

by Ichiro Ozawa as a merger of opposition groups. Hashimoto formed a minority LDP government, but was replaced as LDP leader in mid-1998 by Keizo Obuchi. His coalition was shattered in April 2000 when Ozawa withdrew his repeatedly sidelined Liberal Party. Obuchi himself fell into a coma after suffering a stroke and died. His successor as prime minister and LDP leader, Yoshiro Mori, led a lackluster campaign in a general election in 2000, from which the LDP emerged with a reduced representation but still as the largest party and the main force in government. Mori's unpopularity finally led to his replacement as LDP leader (and prime minister) in April 2001 by charismatic newcomer Junichiro Koizumi.

MAIN POLITICAL ISSUES

LDP strength and opposition weakness
Japanese postwar politics was until 1993 dominated by a system of patronage linking big business, the bureaucracy, and the ruling LDP. Numerous scandals and public disaffection with the LDP then saw the party briefly ousted from government. Far-reaching electoral reforms adopted in 1994 abolished multimember constituencies. This weakened the system of LDP factions (although these later returned), striking at the "money politics" by which they were financed. A shortlived unification of opposition groups, creating a Shinshinto alliance, then promised a realignment toward a two-party system. Adroit maneuvering, however, restored the LDP's position, which was bolstered by its hold on the centralized bureaucracy.

The military issue

The Japanese constitution enshrines pacifism. This prohibition of the use of force is a matter of hot debate within the country; critics argue that its economic power should be better reflected in foreign policy and the defense arena. Japan's role in the US-led "war on terrorism" from 2001 brought it a step closer to this aim.

WORLD AFFAIRS

▷ Joined UN in 1956

APEC | G8 | IAEA | WTO | OECD

After years of limiting its role on the world stage, Japan began to make its influence felt in the 1990s. Its eventual aim is a seat on the UN's Security Council, which would be commensurate with its economic influence. Tentative moves were made in 1993, when Japanese forces joined UN peacekeepers in Cambodia, and in 2001, when Japan assisted in the US-led "war on terrorism." However, the lobby fearing a resurgence of militarism in Japan is still strong. Relations with the West are seriously strained over Japan's continuing to carry out "scientific" whale hunts. In Asia, Japan remains burdened by the legacy of its wartime aggression, exacerbated by revision of school history texts downplaying its crimes.

AID

▷ Donor

$13.5bn (donations) | Down 12% in 2000

Japan's aid donations are the largest of any single country in the world. Most aid goes to Asia and the Pacific, particularly China. Polynesian islands are heavily dependent on Japanese aid in support of their main livelihood, fishing. In 2001, Japan admitted to "buying" support for whaling.

DEFENSE

▷ No compulsory military service

$44.4bn | Up 10% in 2000

JAPANESE ARMED FORCES

🛡	1050 main battle tanks (840 Type-74, 210 Type-90)	148,700 personnel
🚢	16 submarines, 12 frigates, 42 destroyers, 3 patrol boats	44,200 personnel
✈	297 combat aircraft (F-1, F-4EJ, F-15J/DJ)	45,400 personnel
	None	

Article 9 of the Japanese constitution renounces war as a means of settling international disputes, and in the postwar period any military activity in Japan has aroused fierce debate. Involvement even in UN peacekeeping duties, which Japan first undertook in Cambodia in 1993, is hotly contested by pacifists. Japan's Self-Defense Forces, however, have grown quite large. Since 1999 force has been used to deter intrusions by North Korean vessels, sparking fresh debate. A bill was passed in 2001 allowing noncombat assistance specifically to the "war on terrorism."

ECONOMICS

▷ Inflation 0.1% p.a. (1990–2000)

📊 $4519bn | 💱 114.20–131.06 yen

ECONOMIC PERFORMANCE INDICATOR

SCORE CARD

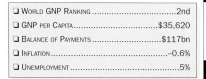

❑ WORLD GNP RANKING	2nd
❑ GNP PER CAPITA	$35,620
❑ BALANCE OF PAYMENTS	$117bn
❑ INFLATION	–0.6%
❑ UNEMPLOYMENT	5%

EXPORTS

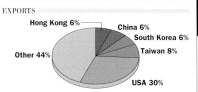

Hong Kong 6% | China 6% | South Korea 6% | Taiwan 8% | Other 44% | USA 30%

IMPORTS

South Korea 5% | Taiwan 5% | Indonesia 4% | China 15% | Other 52% | USA 19%

STRENGTHS

Established market leader in high-tech products and cars. Commitment to long-term research. Talent for capitalizing on imported ideas. Manufacturing plants already established in the West. Domestic economy heavily protected from outside competition. Weak yen has promoted exports.

WEAKNESSES

Recent recession. Dependence on oil imports. Secretive and debt-ridden financial system. Falling industrial production, high-profile bankruptcies, and record unemployment levels. Trade surplus damages international relations.

PROFILE

Once among the world's strongest performing economies, Japan's strengths have been overshadowed for the last decade by its growing weaknesses. The country entered a brief recession for the second time in five years in 2001.

The 1990 crash of the Tokyo stock market marked the end of a period of remarkable growth. The government managed to spend its way out of disaster, effectively delaying the full impact of the downturn.

In an attempt to appease Western discontent over Japan's huge trade surplus, the government encouraged a move away from Japan's dependence on export revenues through stimulation of the domestic economy. However, the contracting trade balance remains above $100 billion a year. Elsewhere the Koizumi government has attempted to promote greater flows of bilateral trade by steering Japan into its first free trade agreement: with Singapore from April 2002.

However, the domestic situation remains critical. The financial sector is in desperate need of reform. The prominent corporate collapses of 1997 were repeated in 2001, with record

losses reported across the high-tech industries in particular. Koizumi has promised radical change, rejecting the standard increase in government spending in favor of basic structural reform. Banks' bad loans have been cleared and the system of privileged "special public institutions" has been overhauled.

Nonetheless, critics have slated Koizumi's plans as overcautious, and have pointed to the increasing pressure from traditional economic power brokers which has put a brake on the pace of change.

JAPAN : MAJOR BUSINESSES

☢ Research & development | 🍶 Brewing
🚗 Vehicle manufacture | 🧵 Textiles
⚙ Heavy engineering | 💻 Computers
🖥 Consumer goods | 🏦 Banking & Finance
⚓ Shipbuilding
⛓ Iron & steel
⚡ Electronics
⚗ Chemicals

Sapporo
Toyama
Kóbe
Hitachi
Hiroshima
Tokyo
Kitakyúshu
Yokohama
Nagoya
Nagasaki
Kyóto
Osaka

0 — 500 km
0 — 500 miles

J

J

RESOURCES

▷ Electric power 245m kw

5.94m tonnes

9023 b/d (reserves 53m barrels)

9.79m pigs, 4.53m cattle, 297m chickens

Limestone, sulfur, coal

ELECTRICITY GENERATION

Hydro 10% (103bn kwh)

Combustion 58% (608bn kwh)

Nuclear 32% (332bn kwh)

Other 0%

% of total generation by type

Japan has few exploitable resources. High production costs have made it the world's largest coal importer. In

ENVIRONMENT

▷ Sustainability rank: 78th

7%

9 tonnes per capita

ENVIRONMENTAL TREATIES

| | Yes | | Yes | | Yes |
| Yes | | Yes | | Yes | |

Japan supports moves to establish a global foundation to aid sustainable development in the Third World. In 1997 it played host to the Kyoto climate conference, although it only agreed to a modest cut in its "greenhouse gas" emissions. It faces strong criticism for its consumption of tropical timber, overfishing, and continuing to catch whale species under the umbrella of "scientific research."

Traditional Japanese respect for nature has spawned a vigorous grassroots ecological movement which prevented a second runway at Tokyo's Narita airport, and opposes nuclear power expansion and waste processing. The most serious environmental disasters have been a nuclear accident at Tokaimura in 1999 and the breakup in early 1997 of a Russian oil tanker along Japan's western shoreline.

***Datsetsusan National Park**, Hokkaido. Japan's northerly island is the least populous of the main group.*

an attempt to reduce dependence on imported fuels, Japan has developed alternative energy sources. It is now the world's third-biggest generator of nuclear power. However, environmentalists strongly oppose any expansion of this sector. Nuclear safety became a priority issue after a serious accident at the Tokaimura plant in 1999.

HOKKAIDŌ

HONSHU

CHUGOKU-SANCHI

SHIKOKU

KYŪSHŪ

JAPAN : LAND USE

Cropland	
Forest	
Pasture	
Sheep	
Fruits	
Rice	

0   500 km
0   500 miles

MEDIA

▷ TV ownership high

Daily newspaper circulation 574 per 1000 people

PUBLISHING AND BROADCAST MEDIA

There are 122 daily newspapers. *Asahi Shimbun, Mainichi Shimbun,* and *Yomiuri Shimbun* are among the most popular

128 services: 1 publicly owned, 127 commercial

100 services: 1 publicly owned, 99 commercial

The Japanese are among the world's most avid newspaper readers. Major papers are issued in simultaneous editions in the main urban centers. Most dailies are owned by large media groups who also have TV and cable interests. Weekly newspapers carry more tabloid journalism. *Manga,* Japanese comics, are hugely popular, with their characteristic artwork influencing design and art across Japanese culture. They now account for 40% of all published material in Japan: the most popular title, *Jump,* sells over six million copies a week.

Japanese technology has defined the world's media. Along with the personal stereo, Japanese companies effectively created the huge international computer games market. Nintendo, a leading games company, is among the most profitable in Japan. Ironically, the Internet was slow to take off, although by 2001 Japan had the third-highest number of people "online." That year also saw Japan launch the world's first "third-generation" mobile phone service.

EDUCATION

▷ School leaving age: 15

99%

3.13m students

THE EDUCATION SYSTEM

% of each age group in education

| Primary | Secondary | Tertiary |
| 100% | 100% | 44% |

The Japanese education system is highly pressurized and competitive. One of the key dividing lines is between university graduates, who get the most coveted white-collar jobs, and nongraduates, who have difficulty reaching management level.

Competition for university places is intense, and starts with the choice of kindergarten, which the Japanese attend from the age of four. Academic pressure diminishes once at university. Graduates from Tokyo, Kyoto, Waseda, and Keio, which are the most prestigious universities, have access to top civil service and business jobs. The system succeeds in producing a uniformly well-educated workforce. However, it has also been criticized for not fostering individual responsibility, flexibility, or entrepreneurship.

CRIME

▷ Death penalty in use

56,133 prisoners

Up 6% in 1999

CRIME RATE

| Murders | |
| 1 | per 100,000 population |

| Rapes | |
| 1 | per 100,000 population |

| Thefts | |
| 1508 | per 100,000 population |

Japan has one of the Western world's lowest crime rates, despite petty crime levels being at a 50-year high. Cities are safe, with police kiosks at frequent intervals on street corners. However, crime is involving more young people, and narcotics abuse is increasing.

The major crime problem is fraud and the activities of the kumi, organized Mafia-style syndicates. The authorities have been reluctant to challenge these groups, seeking to contain rather than halt their activities. Kumi are suspected of having connections with the political extreme right.

THE 2002 WORLD CUP: A GAME OF TWO HALVES

SEEKING BOTH AN economic boost, and a way of improving rocky relations across the Korea Strait, Japan successfully bid to cohost the world's premier soccer tournament, the FIFA World Cup, with South Korea in 2002. It was the first World Cup to be held in Asia and the first to be staged by two countries. Japan's bid ignored the relative infancy of soccer in the country, where it only recently became a professional game. The Japanese Professional Football League itself (popularly known as the J League) has been in existence only since 1993, prompted by the return in 1986 of Japan's first professional player, Yasuhiko Okudera, who had been playing in the West German Bundesliga.

CAUGHT OFFSIDE

The principle behind sharing the monthlong competition was to spread both the cost and the expected economic rewards. However, the 2002 World Cup was by far the most costly ever staged in the tournament's 72-year history and the much-vaunted financial benefits failed to materialize. Under FIFA rules at least eight stadiums must be supplied for each tournament. Japan and its neighbor went well beyond this minimum requirement. Each built eight brand new arenas from scratch, refurbished a further two, and upgraded local facilities and infrastructure. The cost for Japan alone was estimated at $4.5 billion. To offset this outlay the authorities predicted rewards totaling somewhere near $30 billion from ticket sales, advertising, and "soccer tourists." In the event, although companies selling televisions saw a 38% increase in trade, and imports of foreign beer tripled, actual revenues were far less spectacular. An embarrassing mix-up with ticketing left each game being played before a significant number of empty seats. Advertisers declined to risk the early morning/midnight exposure which the time difference imposed on key Western audiences, and "soccer tourists"

A "Corean" fan at the World Cup in 2002.

The Sapporo Dome features a "floating" pitch, which is kept outside when not in use.

came in far fewer numbers than the extravagant 800,000 predicted. Those who did come spent little, and left quickly. Of the stadiums constructed in Japan, few are likely to play host to professional soccer again, leaving remote communities with annual bills of over $2 million in maintenance for impressive but now empty arenas.

THE FAILED FRIENDLY

The World Cup had also been expected to contribute to a warming of relations between its ambitious cohosts. However, in the year leading up to the opening match the intensified press attention seemed only to put greater strain on the two neighbors. Junichiro Koizumi's new populist administration in Japan stirred Korean ire by staging visits to the Yasakuni war shrine (which includes Japanese war criminals among those it commemorates) and by approving "revisionist" school textbooks which, in Korean eyes, glossed over the Imperial army's war crimes in occupied east Asia during World War II. Korean fans took the opportunity presented by the Cup to publicize the popular campaign to reinstate their country's English-language spelling as "Corea." The "K" spelling, they claim, came about through a conspiracy by their imperial overlords after 1910, designed to give Japan precedence in alphabetical ranking.

Although the ailing Japanese economy saw little overall benefit, and relations with Korea (or Corea) failed to transcend traditional enmities, the tournament itself was celebrated around the world. Expected underdogs, including Japan and South Korea, did well enough to challenge the supposed supremacy of European and South American teams. In the end, however, the competition had its most fitting finale as the world's most celebrated team, Brazil, won through to take the title for a record fifth time.

HEALTH

 Welfare state health benefits

1 per 526 people Heart and circulatory diseases, cancers, accidents

Japan's health care system, which is ranked by WHO as the best in the world, delivers some of the highest longevity and lowest infant mortality rates. The poorest in society receive free treatment; expensive high-tech hospital facilities can also offer the latest techniques. Contributory national health insurance is based on earnings-related premiums, and the cost of medical care for the elderly and the self-employed is subsidized, though the rapidly aging population presents a major future funding challenge.

SPENDING

▷ GDP/cap. increase

CONSUMPTION AND SPENDING

395 per 1000 population   586 per 1000 population

| Defense 1% |
| Education 3.5% |
| Health 5.7% |

Defense, Health, Education spending as % of GDP (0 – 25)

Measured in consumer goods, the Japanese are wealthy; car ownership is only low because city parking is so restricted. Most households have substantial savings, enabling them to withstand economic recession.

The country's wealthiest men remain extremely wealthy; the fortunes of the top ten averaged $7 billion in 2000. The richest, software king Masayoshi Son, tripled his fortune in one year, to $19.4 billion. Tokyo living costs are high and most who work there live outside the city center, facing a long, cramped commuter journey. Girls and young women still living in their parents' homes are one group with a high disposable income.

WORLD RANKING

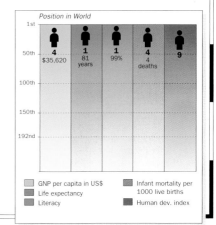

Position in World

| | 4 $35,620 | 1 81 years | 1 99% | 4 4 deaths | 9 |

☐ GNP per capita in US$
☐ Life expectancy
☐ Literacy
☐ Infant mortality per 1000 live births
☐ Human dev. index

J

JORDAN

OFFICIAL NAME: Hashemite Kingdom of Jordan **CAPITAL:** Amman
POPULATION: 5.1 million **CURRENCY:** Jordanian dinar **OFFICIAL LANGUAGE:** Arabic

MIDDLE EAST

SURROUNDED ON ALL SIDES by the deserts of the Middle East, Jordan has just 26 km (16 miles) of maritime coastline on the Gulf of Aqaba. The vast majority of the population lives in the northwest, on the east bank of the Jordan River. Jordan ceded its claim to the West Bank of the river to the aspiring Palestinian state in 1988. Tourism, associated with important historical sites such as Petra, and phosphates are the mainstays of the economy.

CLIMATE
▷ Hot desert/steppe/ Mediterranean

WEATHER CHART FOR AMMAN

Summers are hot and dry, winters cool and wet. Areas below sea level are very hot in summer and warm in winter.

TRANSPORTATION
▷ Drive on right

Queen Alia International, Amman
2.38m passengers

9 ships
41,800 grt

THE TRANSPORTATION NETWORK

8000 km (4971 miles)	None
293 km (182 miles)	None

Adequate roads link main cities. A railroad links the port of Aqaba with the Syrian capital, Damascus.

TOURISM
▷ Visitors : Population 1:3.5

1.48m visitors

Up 4% in 2001

MAIN TOURIST ARRIVALS

Syria 28%
Saudi Arabia 18%
Iraq 11%
Other 43%

0 10 20 30 40 50 60
% of total arrivals

Aqaba offers fine beaches, water sports, and scuba diving, while the ancient city of Petra attracts those visitors interested in Nabatean remains. Amman is developing as a center for Arab culture and the arts.

PEOPLE
▷ Pop. density medium

Arabic

57/km² (149/mi²)

THE URBAN/RURAL POPULATION SPLIT

74%   **26%**

ETHNIC MAKEUP

Circassian 1%   Armenian 1%

Arab 98%

Jordan is a predominantly Muslim country drawn from Bedouin roots, with a Christian minority. About half the population are Palestinian in origin. The monarchy's power base lies among the rural tribes, which also provide the backbone of the military. National identity is strong.

POLITICS
▷ Multiparty elections

L. House 1997/2002
U. House 2001/2005

H.M. King Abdullah II

AT THE LAST ELECTION
House of Deputies 80 seats

63% Ind	20% IAF	2% AAP	15% Others

Ind = Independents   **IAF** = Islamic Action Front
AAP = Al-Ahd Party

Senate 40 seats

The members of the Senate (Majlis Al-Aayan) are appointed by the king

King Abdullah II acceded to the throne in February 1999 upon the death of his father, King Hussein. Although lacking in political experience, he is respected by the army and enjoys the support of Jordan's tribal leaders. Multiparty elections, initiated in 1993, have benefited pro-government parties, despite a strong Islamist opposition lobby. The appointment in 2000 of Prime Minister Ali Abu al-Ragheb marked a shift toward a modernizing and pro-business government.

JORDAN

Total Land Area : 92 300 sq. km (35 637 sq. miles)

POPULATION

over 100 000	◉
over 50 000	◎
over 10 000	●
under 10 000	•

LAND HEIGHT

1000m/3281ft
500m/1640ft
200m/656ft
Sea Level
-200m/-656ft

J

WORLD AFFAIRS ▷ Joined UN in 1955

AL   AMF   WTO   NAM   OIC

Jordan's position as a key player in Middle East politics is under question. Policy toward the emerging Palestinian state is uncertain since Jordan's relations with Israel remain much less hostile than those of other Arab countries. The US signed a ten-year free trade agreement with Jordan in 2000.

Jordan has recently been at the forefront of the rehabilitation of Iraq, having refused to join the anti-Iraq coalition formed by the Gulf states in 1991.

AID ▷ Recipient

 $552m (receipts)   ⬆ Up 28% in 2000

The Gulf states undertook to restore aid to Jordan when King Hussein moved to distance himself from Iraq in 1995.

DEFENSE ▷ No compulsory military service

 $510m   ⬇ Down 13% in 2000

The armed forces are loyal to the monarchy. They have a reputation for thorough training and professionalism. The forces are dependent on Western support for credit in purchasing advanced arms and equipment, but Jordanian forces played no part in the 1991 Gulf War.

ECONOMICS ▷ Inflation 3.2% p.a. (1990–2000)

 $8.36bn   0.7100–0.7088 Jordanian dinars

SCORE CARD

- ❑ WORLD GNP RANKING 92nd
- ❑ GNP PER CAPITA $1710
- ❑ BALANCE OF PAYMENTS $59m
- ❑ INFLATION 0.7%
- ❑ UNEMPLOYMENT.................................15%

STRENGTHS
Major exporter of phosphates. Skilled workforce. Recovery of tourist industry after 1991 Gulf War. The port of Aqaba is a special economic zone.

WEAKNESSES
Reliant on imports of energy. Poor export to import ratio. Unemployment, exacerbated by influx of refugees from Kuwait after Gulf crisis. Little arable land.

EXPORTS

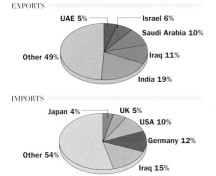

UAE 5%
Israel 6%
Saudi Arabia 10%
Iraq 11%
Other 49%
India 19%

IMPORTS

Japan 4%   UK 5%
USA 10%
Germany 12%
Other 54%
Iraq 15%

RESOURCES ▷ Electric power 1.3m kw

 1025 tonnes   40 b/d

1.85m sheep, 640,000 goats, 23.7m chickens   Oil, phosphates, potash

Oil deposits have been discovered. Phosphates, livestock, and crops such as tomatoes, wheat, olives, and vegetables are the main resources.

ENVIRONMENT ▷ Sustainability rank: 53rd

 3%   ⬆ 3 tonnes per capita

Conservation is a government priority. Rare animals are protected and species that became extinct in the wild in the 1950s are being reintroduced into controlled environments.

MEDIA ▷ TV ownership medium

 Daily newspaper circulation 42 per 1000 people

PUBLISHING AND BROADCAST MEDIA

There are 8 daily newspapers, including *Ad-Dustour* and *Ar-Rai*

1 state-controlled service   1 state-controlled service

A restrictive press and publications law was enacted in 1998. Radio and TV are controlled by the state.

CRIME ▷ Death penalty in use

 5600 prisoners   ⬆ Up 7% in 1999

Jordan is largely peaceful. Crime levels are generally low, although theft in urban areas is rising.

EDUCATION ▷ School leaving age: 16

 90%   142,190 students

Men and women receive the same education. Jordanian teachers work all over the Middle East.

HEALTH ▷ Welfare state health benefits

 1 per 588 people   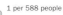 Heart, digestive, and respiratory diseases, accidents, cancers

Health care is subsidized by the government. Hospitals are well distributed throughout the country.

CHRONOLOGY

Jordan, previously the British-mandated territory of Transjordan, became independent in 1946.

- ❑ **1953** Hussein becomes king.
- ❑ **1967** Israel seizes West Bank territories.
- ❑ **1970** Massive crackdown on Palestine Liberation Organization.
- ❑ **1988** Jordan cedes claims to West Bank to PLO.
- ❑ **1994** Peace treaty with Israel.
- ❑ **1999** Death of King Hussein; succession of King Abdullah II.

The King's Highway, *seen from the castle at Al Karak, a strategic fortress built by Crusader knights in the 12th century.*

SPENDING ▷ GDP/cap. increase

CONSUMPTION AND SPENDING

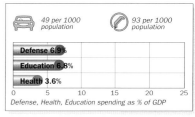

49 per 1000 population   93 per 1000 population

Defense 6.9%
Education 6.8%
Health 3.6%

0   5   10   15   20   25
Defense, Health, Education spending as % of GDP

Poverty is relatively rare, though refugee camps still exist and 25% unemployment damaged many family incomes in the late 1990s.

WORLD RANKING

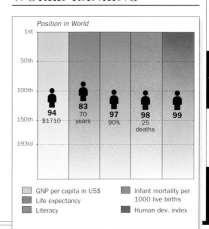

Position in World

1st
50th
100th
150th
193rd

94   83   97   98   99
$1710   70 years   90%   25 deaths

❑ GNP per capita in US$ ❑ Infant mortality per 1000 live births
❑ Life expectancy
❑ Literacy   ❑ Human dev. index

J

KAZAKHSTAN

OFFICIAL NAME: Republic of Kazakhstan **CAPITAL:** Astana
POPULATION: 16.1 million **CURRENCY:** Tenge **OFFICIAL LANGUAGE:** Kazakh

THE SECOND-LARGEST of the former Soviet republics, Kazakhstan extends almost 3000 km (1900 miles) from the Caspian Sea in the west to the Altai Mountains in the east and 1600 km (1000 miles) north to south. It borders Russia to the north and China to the east. Kazakhstan was the last Soviet republic to declare its independence, in 1991. In 1999, elections confirmed the former communist Nursultan Nazarbayev and his supporters in power. Kazakhstan has considerable economic potential, and many Western companies seek to exploit its mineral resources.

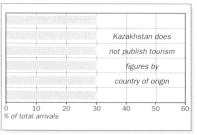

***The Altai Mountains**, eastern Kazakhstan. Subject to harsh continental winters, the Altai range is a cold, inhospitable place. Rivers carry meltwater down onto the vast steppe.*

CLIMATE

Cold desert/steppe

WEATHER CHART FOR ASTANA

Kazakhstan has a continental climate with large temperature variations: average January temperatures range from –18°C (0°F) on the northern Kazakh steppe to –3°C (27°F) in the deserts 1600 km (1000 miles) to the south; July temperatures average 19°C (66°F) and 30°C (86°F) respectively. As the Caspian Sea never freezes, winters are mildest on Kazakhstan's southwestern coast.

TRANSPORTATION

Drive on right

Astana
242,936 passengers

18 ships
9253 grt

THE TRANSPORTATION NETWORK

| 103,272 km (64,170 miles) | None |
| 13,601 km (8452 miles) | 3900 km (2423 miles) |

Transportation networks focus on the north and east as the key economic areas. Most of the roads in Kazakhstan are in urgent need of repair. In 1998, measures to restructure and privatize the railroad were announced. The railroad system links into that of the Russian Federation. Helicopters are frequently used to reach remote destinations.

TOURISM

Not available

Visitors are mainly on business

Gradually increasing

MAIN TOURIST ARRIVALS

Kazakhstan does not publish tourism figures by country of origin

% of total arrivals

The number of visitors to Kazakhstan is increasing, but very few come solely as tourists. The majority are business travelers, and a dense web of contacts with foreign companies has evolved. Of the central Asian states, Kazakhstan has cultivated the closest links with the West. Most foreign businesses are concentrated in Almaty.

KAZAKHSTAN

Total Land Area : 2 717 300 sq. km
(1 049 150 sq. miles)

POPULATION

over 500 000	⊙
over 100 000	◎
over 50 000	○
over 10 000	●
under 10 000	•

LAND HEIGHT

3000m/9843ft
2000m/6562ft
1000m/3281ft
500m/1640ft
200m/656ft
Sea Level
-200m/-656ft

PEOPLE ▷ Pop. density low

Kazakh, Russian,
Uighur, Korean, German

6/km²
(15/mi²)

THE URBAN/RURAL POPULATION SPLIT

56%   **44%**

RELIGIOUS PERSUASION

Protestant 1%
Russian Orthodox 13%
Other 36%
Muslim (mainly Sunni) 50%

ETHNIC MAKEUP

German 2%
Tatar 2%
Ukrainian 4%
Other 9%
Kazakh 53%
Russian 30%

POPULATION AGE BREAKDOWN

Female	Age	Male
0.9%	80+	0.3%
5.4%	60–79	3.3%
10.3%	40–59	9.2%
15.6%	20–39	16%
19.2%	0–19	19.8%

% of population by age group

Kazakhstan's ethnic diversity arose mainly from forced settlement of Tatars, Germans, and Russians during the Soviet era. By 1959, Kazakhs were outnumbered by ethnic Russians. This balance has been redressed by the immigration of ethnic Kazakhs from neighboring states and the departure in the 1990s of some 1.5 million ethnic Russians. In addition, a majority of ethnic Germans have opted to live in Germany, although in 2000 the government announced a campaign to try to lure some of them back.

In 1995, ethnic Russians criticized the country's new constitution for preventing dual citizenship with Russia and refusing to recognize Russian as an official language. Central control over ethnic Russians has been reinforced by shifting the capital to Astana (formerly Akmola) in the north, where the majority of ethnic Russians reside.

Few Kazakhs retain their traditional nomadic life. Commitment to Islam and loyalty to the clan remain strong.

POLITICS ▷ Multiparty elections

L. House 1999/2004
U. House 1999/2002

President Nursultan Nazarbayev

AT THE LAST ELECTION

Majlis 77 seats

4% AP
47% Others
31% Otan
14% CPK
4% CP

Otan = Fatherland Republican Party of Kazakhstan
CPK = Civil Party of Kazakhstan   AP = Agrarian Party
CP = Communist Party of Kazakhstan

Senate 39 seats

Two members are elected by each of 16 districts and 7 are nominated by the president

Legislative authority is vested in the bicameral Parliament. The president, who must be fluent in Kazakh, has supreme executive power.

PROFILE

Despite a democratic government, the president enjoys political dominance, and the patronage of the Kazakh clans is still important. Since coming to power in 1989, President Nazarbayev has concentrated on market reforms. There have been allegations of electoral fraud, and domestic and international criticism of Nazarbayev's attempts to expand the scope of presidential powers. In 2000 the Majlis granted him special powers to advise future presidents after his term expires in 2006, and in 2002 opposition parties were effectively neutered by reform of the party registration process.

MAIN POLITICAL ISSUE
Presidential powers

The increased powers of Nazarbayev are the focus of political controversy. Critics accuse him of developing a personality cult. The 1995 constitution strengthened presidential powers, conferring a veto over the decisions of the Constitutional Council. A referendum extended Nazarbayev's term of office until 2000, but in 1998 the legislature approved constitutional amendments forcing him to hold a presidential election in 1999. Although reelected, Nazarbayev was tarnished by allegations of voting irregularities. In April 2002 Nazarbayev's credibility was again shaken by the revelation that he had moved $1 billion of state oil revenues into a secret overseas bank account, without parliament's knowledge.

President Nursultan Nazarbayev, *who steered Kazakhstan to independence.*

WORLD AFFAIRS ▷ Joined UN in 1992

CIS   SCO   EAPC   OIC   OSCE

Relations with Russia, although strained at times by Moscow's concern over Kazakhstan's ethnic Russians, have been cemented by a 25-year cooperation treaty. However, Kazakhstan's earlier enthusiasm for greater integration of the former Soviet states has cooled.

Kazakhstan's location, bridging Europe and Asia, has led to close relations with a variety of partners, and its rich mineral resources have attracted investors from Europe, the US, and Asia. Relations with China have improved, with agreements in 1998 and 1999 on border issues, and it joins China, Russia, and three other central Asian republics in the Shanghai Cooperation Organization. However, it competes with Uzbekistan, with which its has a border dispute, to be the acknowledged central Asian regional power.

AID ▷ Recipient

$189m (receipts)   Up 8% in 2000

Kazakhstan joined the IMF and the World Bank in 1992, and is also a member of the EBRD. Most multilateral and bilateral aid is aimed at supporting economic reform and improving health care, transportation, and communications.

K

CHRONOLOGY

Once part of the Mongol empire, Kazakhstan was absorbed by the Russian Empire in the 19th century. Ethnic Russians began to settle on land used by nomadic Kazakhs. Russian settlement intensified after the 1917 Revolution and Kazakhstan was subjected to intensive industrial and agricultural development.

❏ **1916** Rebellion against Russian rule brutally suppressed.
❏ **1917** Russian Revolution inspires civil war in Kazakhstan between Bolsheviks, anti-Bolsheviks, and Kazakh nationalists.
❏ **1918** Kazakh nationalists set up autonomous republic.
❏ **1920** Bolsheviks take control. Kirghiz Autonomous Soviet Socialist Republic (ASSR) set up within Russian Soviet Federative Socialist Republic.
❏ **1925** Kirghiz ASSR renamed Kazakh ASSR.
❏ **1936** Kazakhstan becomes full union republic of the USSR as Kazakh SSR.

K

CHRONOLOGY *continued*

- ❏ **1930s** Stalin's collectivization program leads to increase in Russian settlement and the deaths of an estimated one million Kazakhs. Large penal settlements established for victims of Stalinist purges.
- ❏ **1941–1945** Large-scale deportations from Russia of Germans, Jews, Crimean Tatars, and others to Kazakhstan.
- ❏ **1949–1989** Nuclear test site at Semipalatinsk carries out nearly 500 nuclear explosions.
- ❏ **1954–1960** Khrushchev's policy to plow "Virgin Lands" for grain most vigorously followed in Kazakhstan. Russian settlement reaches peak.
- ❏ **1986** Riots in Almaty after ethnic Russian Gennadi Kolbin appointed head of Kazakhstan Communist Party (CPK) to replace Kazakh Dinmukhamed Kunyev.
- ❏ **1989** Kolbin replaced by Nursultan Nazarbayev, ethnic Kazakh and chair of Council of Ministers. Reform of political and administrative system.
- ❏ **1990** CPK wins elections to Supreme Soviet by overwhelming majority. Nazarbayev appointed first president of Kazakhstan. Kazakhstan declares sovereignty.
- ❏ **1991** Kazakhstan votes to preserve USSR as union of sovereign states. USSR authorities hands over control of enterprises in Kazakhstan to Kazakh government. CPK ordered to cease activities in official bodies following abortive August coup in Moscow. CPK restructures itself as Socialist Party of Kazakhstan (SPK). Independence of Republic of Kazakhstan declared; joins CIS. Announcement of closure of Semipalatinsk nuclear test site.
- ❏ **1992** Opposition demonstrations against dominance of reformed communists in Supreme Soviet, now Supreme Kenges. Nationalist groups form Republican Party, Azat.
- ❏ **1993** Adoption of new constitution. Introduction of new currency, the tenge.
- ❏ **1994** Legislative elections annulled after proof of widespread voting irregularities.
- ❏ **1995** Adoption of new constitution broadening presidential powers; referendum extends Nazarbayev's term until 2000; legislative elections.
- ❏ **1998** Legislature approves constitutional amendments, including the holding of early presidential election.
- ❏ **1999** Nazarbayev reelected president for seven more years.

DEFENSE

 Phasing out conscription

 $357m   Up 23% in 2000

KAZAKH ARMED FORCES

🛡	930 main battle tanks (650 T-72, 280 T-62)	45,000 personnel
	None	None
✈	164 combat aircraft (MiG-29, Su-24/25/27)	19,000 personnel
	None	

As a former nuclear power and the largest of the five former Soviet central Asian republics, Kazakhstan is a potential guarantor of regional peace and stability. Kazakhstan ratified the START-I nuclear reduction treaty in 1992 and the NPT in 1993. The US agreed in 1993 to grant Kazakhstan $84 million to enable it to dismantle its nuclear weapons armory. In 1995, Kazakhstan announced that all its nuclear weapons had been transferred to Russia or else had been destroyed.

A restructuring of Kazakhstan's armed forces was launched in 2001. This included moves to end conscription, whose duration had been halved, from two years to one in 1998. It is expected that the restructuring will take at least three years.

ECONOMICS

▷ Inflation 205% p.a. (1990–2000)

$18.8bn   145.56-150.90 tenge

SCORE CARD

- ❏ WORLD GNP RANKING...........................68th
- ❏ GNP PER CAPITA $1260
- ❏ BALANCE OF PAYMENTS...................$1.07bn
- ❏ INFLATION 13.2%
- ❏ UNEMPLOYMENT 4%

EXPORTS

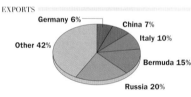

Germany 6%
China 7%
Italy 10%
Other 42%
Bermuda 15%
Russia 20%

IMPORTS

UK 4%
USA 5%
Italy 3%
Germany 7%
Russia 49%
Other 32%

STRENGTHS

Mineral resources, notably oil and gas, and bismuth and cadmium, used in electronics industry. Joint oil and gas ventures with Western companies. Mass privatization program launched in 1994.

WEAKNESSES

Collapse of former Soviet economic and trading system. Reliance on imported consumer goods. Rapid introduction of the tenge in 1993 increased instability and fueled sharp price rises. Inefficient industrial plants.

PROFILE

Kazakhstan has moved faster than other former Soviet republics to establish a market economy. Prices have been freed, foreign trade deregulated, and the tax system reformed. Petroleum revenues have been the main engine of

ECONOMIC PERFORMANCE INDICATOR

— Consumer Price Index   GDP

growth, and inflation, at one time rising sharply, has been brought under control.

Foreign direct investment is mainly in the energy sector. Outdated equipment and inadequate distribution networks mean that energy has to be imported, although Kazakhstan exports fossil fuel.

An oil price boom set the tone for the country's first "five-year plan" in 2000, part of Nazarbayev's "Kazakhstan 2030" program. It promised long-term land leases, although not full private ownership, from 2001.

KAZAKHSTAN : MAJOR BUSINESSES

🛢 Oil	
🏛 Steel	
❋ Textiles	
🧪 Chemicals	📦 Food processing
Oil refining	❋ Light engineering
Coal mining	Pharmaceuticals

0   500 km

0   500 miles   * significant multinational ownership

RESOURCES

 Electric power 19m kw

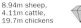 26,951 tonnes

828,000 b/d (reserves 8bn barrels)

8.94m sheep, 4.11m cattle, 19.7m chickens

Oil, gas, manganese, gold, silver, coal, iron, tungsten, chromite, bismuth, cadmium

ELECTRICITY GENERATION

Hydro 12% (6.1bn kwh)

Combustion 88% (43bn kwh)

Nuclear 0%

Other 0%

0 20 40 60 80 100

% of total generation by type

Mining is the single most important industry in Kazakhstan. The US company Chevron has been

ENVIRONMENT

 Sustainability rank: 88th

3%

8.2 tonnes per capita

ENVIRONMENTAL TREATIES

No

Yes

Yes

No

No

No

Environmental damage caused by intensive industrial and agricultural development is a major concern. The eastern cities are heavily polluted and farmlands are being eroded. The Aral Sea, polluted by overuse of fertilizers, has shrunk by 50% owing to the Soviet-era diversion of rivers for irrigation.

In 1991, environmental pressure groups succeeded in ending more than 40 years of nuclear testing at Semipalatinsk. The green lobby now presses for tighter pollution controls.

MEDIA

 TV ownership medium

Daily newspaper circulation 30 per 1000 people

PUBLISHING AND BROADCAST MEDIA

There are 5 principal daily newspapers and over 400 other registered newspapers

3 services: 1 state-owned, 2 independent

1 state-owned service, several private stations

The state-owned media operate alongside independent publications and privately owned radio and television stations. Amendments in 2001 to the Media Law strengthened state control over small broadcast outlets and Internet sites, and created more grounds for libel charges against editors and proprietors. Such state pressure has produced a drastic decline in the number of media outlets.

developing the huge Tengiz oil field since 1993. Joint ventures to exploit substantial oil and gas reserves in the Caspian Sea were agreed with Russia in 1995, and the US and Japan in 1998. Kazakhstan also possesses vast iron ore and gold reserves.

KAZAKHSTAN : LAND USE

0 500 km
0 500 miles

Cropland
Forest
Pasture
Desert
Sheep
Cereals

CRIME

 Death penalty in use

84,000 prisoners

Down 2% in 1999

CRIME RATES

Murders	
16	per 100,000 population

Rapes	
7	per 100,000 population

Thefts	
334	per 100,000 population

Narcotics smuggling is increasing. Corruption is rife. Kazakhstan retains and uses the death penalty, but does not publish statistics on its use.

EDUCATION

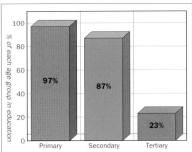 School leaving age: 15

99%

419,460 students

THE EDUCATION SYSTEM

% of each age group in education

97% Primary
87% Secondary
23% Tertiary

Education remains based on the Soviet model. Since its adoption as the state language in 1995, Kazakh is gradually replacing Russian as the main instruction medium in schools, but there is a shortage of Kazakh textbooks and Kazakh-speaking teachers. There are a large number of higher-education institutions and medical schools.

HEALTH

 Welfare state health benefits

1 per 286 people

Heart attacks, cancers, accidents, violence, tuberculosis

Kazakhstan's ill-equipped and poorly funded health system has produced the lowest average life expectancy in central Asia.

The health system is limited in terms of both facilities and coverage. Rural people have minimal access to clinics. The country's size means that extending coverage and improving the quality of care will be costly. Attempts are therefore being made to attract foreign investment into the health sector. Many doctors have emigrated to Russia.

SPENDING

GDP/cap. decrease

CONSUMPTION AND SPENDING

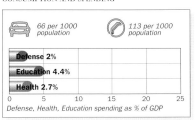

66 per 1000 population

113 per 1000 population

Defense 2%
Education 4.4%
Health 2.7%

0 5 10 15 20 25

Defense, Health, Education spending as % of GDP

Life for the majority of Kazakhs has always been hard, and has grown even more difficult since 1989. Unemployment has risen and living standards have deteriorated as a result of the market-oriented reforms within Kazakhstan. In addition, the liberalization of the economy has had the effect of fueling sharp price rises for essential commodities.

The rural population, the poorest group in Kazakhstan, has been badly affected. The small wealthy elite is made up mainly of former communist officials, many of whom have benefited from privatization, or belong to President Nazarbayev's clan.

WORLD RANKING

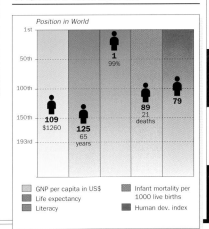

Position in World

1st
50th
100th
150th
193rd

1 99%
109 $1260
125 65 years
89 21 deaths
79

GNP per capita in US$
Life expectancy
Literacy

Infant mortality per 1000 live births
Human dev. index

K

343

KENYA

OFFICIAL NAME: Republic of Kenya **CAPITAL:** Nairobi
POPULATION: 31.3 million **CURRENCY:** Kenya shilling **OFFICIAL LANGUAGES:** Kiswahili and English

KENYA STRADDLES the equator on Africa's east coast, its central plateau bisected by the Great Rift Valley. The land to the north is desert, while to the east lies a fertile coastal belt. After independence from the UK in 1963, politics was dominated by Jomo Kenyatta. He was succeeded as president in 1978 by Daniel arap Moi, whose divide-and-rule policies have drawn accusations of favoritism and of fomenting ethnic hatreds. His KANU won elections easily in 1992 and 1997, amid accusations of fraud. Economic mainstays are tourism and agriculture. High population growth is a major problem.

Kenyatta Conference Center, *Nairobi. The modern skyline of the business center contrasts sharply with the slums on the city's outskirts.*

CLIMATE

Steppe/mountain/tropical

WEATHER CHART FOR NAIROBI

The coast and Great Rift Valley are hot and humid, the plateau interior is temperate, and the northeastern desert hot and dry. Rain generally falls from April to May and October to November.

TRANSPORTATION

Drive on left

Jomo Kenyatta, Nairobi
2.96m passengers

39 ships
20,900 grt

THE TRANSPORTATION NETWORK

8868 km (5510 miles)	None
2634 km (1637 miles)	Lake Victoria is navigable

Kenya's railroads, ports, and main airport are currently being upgraded, while in addition a five-year road improvement program was begun in 2000.

Great Rift Valley, *Kenya. This huge crack in the Earth's crust runs from the River Jordan right through Africa to the Zambezi River.*

TOURISM

Visitors : Population 1:33

943,000 visitors

Up 10% in 1999

Tourism is vital to the economy and a key foreign exchange earner. After a boom in package safaris and beach vacations during the 1980s, Kenya saw a decline in visitor numbers in the 1990s due to world recession, reports of instability, and the much-publicized murder of several tourists. A slight upturn in 1999–2000 was set back again by the global decline in the industry in 2001.

MAIN TOURIST ARRIVALS

Germany	15%
UK	15%
Tanzania	12%
Uganda	7%
USA	7%
Other	44%

% of total arrivals

KENYA

Total Land Area :
582 650 sq. km
(224 961 sq. miles)

POPULATION

▣	over 1 000 000
◉	over 500 000
◎	over 100 000
○	over 50 000
●	over 10 000
•	under 10 000

LAND HEIGHT

	3000m/9843ft
	2000m/6562ft
	1000m/3281ft
	500m/1640ft
	200m/656ft
	Sea Level

0   100 km
0   100 miles

K

PEOPLE Pop. density medium

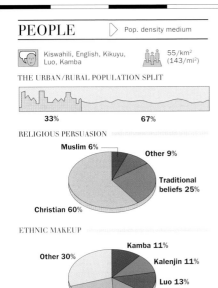

Kiswahili, English, Kikuyu, Luo, Kamba

55/km²
(143/mi²)

THE URBAN/RURAL POPULATION SPLIT

33%   67%

RELIGIOUS PERSUASION

Muslim 6%
Other 9%
Traditional beliefs 25%
Christian 60%

ETHNIC MAKEUP

Kamba 11%
Kalenjin 11%
Luo 13%
Luhya 14%
Kikuyu 21%
Other 30%

Kenya's ethnic diversity, with about 70 different groups, reflects its past as a focus of population movements. Asians, Europeans, and Arabs form 1% of the population. The rural majority retains strong clan and extended family links, although these are being weakened by urban migration. Poverty, severe drought, and a high population growth rate are the root causes of a land hunger which has fueled a surge in ethnic violence. There is much violence in western Kenya, where Kikuyu, the formerly dominant tribe, are the main targets of attacks by other groups. Several hundred thousand Kikuyu are believed to have been displaced from their villages.

Female genital mutilation, criminalized in 2001, remains widespread.

POPULATION AGE BREAKDOWN

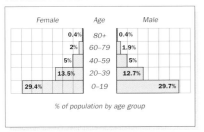

Female	Age	Male
0.4%	80+	0.4%
2%	60–79	1.9%
5%	40–59	5%
13.5%	20–39	12.7%
29.4%	0–19	29.7%

% of population by age group

WORLD AFFAIRS Joined UN in 1963

Comm   COMESA   EAC   IGAD   AU

Relations with neighboring states and with key Western donors are Kenya's priorities. The US embassy closed temporarily in 1998 after a terrorist bombing caused carnage there. In 1991, human rights concerns were partly responsible for a two-year suspension of aid. Talks were held in spring 2001 in an effort to end Kenya's border dispute with Sudan. Relations with Rwanda improved in 2000, when it was agreed to reopen the Kenyan embassy in Kigali and to establish a joint commission for bilateral relations. Kenya is greatly concerned by the situation in Burundi and the DRC, as well as by tensions in Sudan, Somalia, and Ethiopia.

POLITICS Multiparty elections

1997/2002

President Daniel arap Moi

AT THE LAST ELECTION

National Assembly 224 seats

51% KANU   18% DP   10% NDP   7% SDP   4% Others   8% FORD–K   2% Saf

KANU = Kenya African National Union
DP = Democratic Party   **NDP** = National Development Party
FORD–K = Forum for the Restoration of Democracy–Kenya
SDP = Social Democratic Party   **Saf** = Safina
Others include FORD–People, Kenya Social Congress, Shirikisho, FORD–Asili, and two ex-officio members.

The National Assembly comprises 210 elected members, 12 nominated by the president, and two ex-officio members

Kenya has been led by President Daniel arap Moi since 1978, and became a multiparty democracy in 1992.

PROFILE

After 1982, President Moi's efforts to entrench KANU in power stirred up demands at home for the introduction of multiparty politics and provoked condemnation abroad for the human rights abuses committed. Forced in 1992 to concede free elections, Moi helped ensure KANU's victory by curtailing the campaign period. His reelection in 1997 was marred by widespread allegations of intimidation and electoral fraud.

Moi has attempted to break down barriers with some opponents. In 1999 he appointed paleontologist Richard Leakey, a former political opponent, to head a drive against official corruption, which was completed in 2001. In 2002, KANU absorbed the "opposition" NDP, which had joined the government the previous year.

MAIN POLITICAL ISSUE
Ethnic violence

The ethnic polarization of political parties in Kenya and rising poverty are fueling ethnic violence. Determined to ensure KANU dominance, President Moi, a Kalenjin, is turning the party into an alliance of smaller ethnic groups opposed to the once dominant Kikuyu. The latter are the largest ethnic group, the main victims of violence, and the leading supporters of the opposition.

Daniel arap Moi,
succeeded Kenyatta as president in 1978.

Mwai Kibaki,
opposition leader.

AID Recipient

 $512m (receipts)   Up 65% in 2000

Kenya has been a major recipient of aid from donors including the UK, Japan, the EU, the World Bank, and the IMF. Little, however, has trickled down to the majority of the population. This is partly because of the high proportion of aid tied to construction projects and donor-country firms, and partly because of mismanagement and official corruption. In 1996 Western creditor governments decided to link aid disbursements to improvements in human rights, and in 2001 the IMF and the World Bank withheld all aid pending anticorruption reforms.

K

CHRONOLOGY

From the 10th century, Arab coastal settlers mixed with indigenous peoples in the region. Britain's need for a route to landlocked Uganda led to the formation in 1895 of the British East African Protectorate in the coastal region.

❏ **1900–1918** White settlement.
❏ **1920** Interior becomes British colony.
❏ **1930** Jomo Kenyatta goes to UK; stays 14 years.
❏ **1944** Kenyan African Union (KAU) formed; Kenyatta returns to lead it.
❏ **1952–1956** *Mau Mau*, Kikuyu-led violent campaign to restore African lands. State of emergency; 13,000 people killed.
❏ **1953** KAU banned. Kenyatta jailed.
❏ **1960** State of emergency ends. Tom Mboya and Oginga Odinga form KANU.
❏ **1961** Kenyatta freed; takes up presidency of KANU.
❏ **1963** KANU wins elections. Kenyatta prime minister. Full independence declared.
❏ **1964** Republic of Kenya formed with Kenyatta as president and Odinga as vice president.
❏ **1966** Odinga defects to form Kenya People's Union (KPU).

CHRONOLOGY *continued*

- ❏ **1969** KANU sole party to contest elections (also 1974). Tom Mboya of KANU assassinated. Unrest. KPU banned and Odinga arrested.
- ❏ **1978** Kenyatta dies. Vice President Daniel arap Moi succeeds him.
- ❏ **1982** Kenya declared a one-party state. Opposition to Moi. Abortive air force coup. Odinga rearrested.
- ❏ **1986** Open "queue-voting" replaces secret ballot in first stage of general elections. Other measures to extend Moi's powers incite opposition.
- ❏ **1988** Moi wins third term and extends his control over judiciary.
- ❏ **1990** Government implicated in deaths of Foreign Minister Robert Ouko and Anglican archbishop. Riots. Odinga and others form FORD, outlawed by government.
- ❏ **1991** Arrest of FORD leaders and attempts to stop prodemocracy demonstrations. Donors suspend aid. Moi agrees to introduce multiparty system. Ethnic violence increases.
- ❏ **1992** FORD splits into factions led by ex-minister Kenneth Matiba and Odinga. Opposition weakness helps Moi win December elections.
- ❏ **1994** Odinga dies.
- ❏ **1997** December, Moi wins further term in widely criticized elections.
- ❏ **1998** Bomb at US embassy kills 230.
- ❏ **1999** Moi appoints paleontologist Richard Leakey to lead government drive against corruption.
- ❏ **2000** Worst drought since 1947.
- ❏ **2001** Leakey resigns. Drought threatens starvation for millions.
- ❏ **2002** Thousands displaced by recurrence of major floods.

DEFENSE

▷ No compulsory military service

 $307m

 Down 6% in 2000

KENYAN ARMED FORCES

🚜	78 main battle tanks (*Vickers* Mk 3)	20,000 personnel
🚢	4 patrol boats	1400 personnel
✈	29 combat aircraft (9 F-5E/F)	3000 personnel
	None	

Destabilization of the northeastern border by insecurity in Somalia is the main defense issue. The army has recently been deployed to suppress tribal fighting in the Rift Valley. Military assistance is given by the UK and the US.

ECONOMICS

 ▷ Inflation 14% p.a. (1990–2000)

 $10.6bn

78.05–78.60 Kenya shillings

SCORE CARD

❏ WORLD GNP RANKING	82nd
❏ GNP PER CAPITA	$350
❏ BALANCE OF PAYMENTS	–$238m
❏ INFLATION	5.9%
❏ UNEMPLOYMENT	50%

EXPORTS

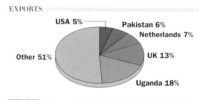

USA 5%
Pakistan 6%
Netherlands 7%
Other 51%
UK 13%
Uganda 18%

IMPORTS

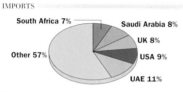

South Africa 7%
Saudi Arabia 8%
UK 8%
Other 57%
USA 9%
UAE 11%

STRENGTHS

Tourism – largest foreign exchange earner. Broad agricultural base, especially cash crops such as coffee and tea. East Africa's largest, most diversified manufacturing sector.

WEAKNESSES

Fluctuating world prices for coffee and tea. Corruption. Poor recent GDP growth. High population growth. Land shortage means uneconomic small units. Country's image problem affects tourism.

PROFILE

Kenya has been hailed as an example to the rest of Africa of the benefits of a mainly free-market economy. Government involvement has been relatively limited, and recently further reduced by privatization. Foreign investment has been encouraged, with some success. Tourism has become the leading foreign exchange earner over the past 20 years, despite suffering serious setbacks since the 1990s. Manufacturing now accounts for 21% of GDP, and is the most diversified sector in east Africa, but needs to expand rapidly so as to create more jobs.

Economic growth was good by African standards during the 1980s, averaging over 4% a year. However, it was barely sufficient to compensate for one of the world's highest population growth rates, approximately 3.5%, although the UN estimates that this rate will have almost halved by 2015. For the majority of Kenyans, farming ever-smaller landholdings or earning a living in the informal sector, life has

ECONOMIC PERFORMANCE INDICATOR

Consumer Price Index — GDP

become harsher. The situation was exacerbated in 2001 by a ban on all trade with Somalia, which affects exports of the mild narcotic, *qat*, widely grown in Kenya. Severe drought since 2000 has crippled the agricultural sector and put 20 million people at risk of starvation.

Other problems, including inflation, a heavy debt burden, and growing dependence on balance-of-payments support had come to a head in the early 1990s, when economic growth gave way to recession. Real GDP growth fell to 0.4% in 1992 and has remained low, falling to –0.2% in 2000. The rise in poverty-linked violence and political unrest hit tourism; earnings fell by 15% in the early 1990s, and the industry has yet to recover fully, a global downturn causing further problems in 2001.

Partly as a response to pressure from donors, the government has implemented some economic liberalization measures. These include floating the Kenya shilling, raising interest rates, and giving exporters direct access to their hard currency earnings. However, sustained growth is likely to remain elusive until Kenya overcomes the official corruption which drains vital resources, and the poor image affecting its tourist industry.

KENYA : MAJOR BUSINESSES

Eldoret
Kisumu
Lake Victoria
Nakuru
Nairobi
Nanyuki
Thika
Mombassa

Cement			
Tobacco			
Oil refining		Vehicle assembly	
Electronics		Food processing	
Steel		Agribusiness	
Textiles		Chemicals	

* significant multinational ownership

0   100 km
0   100 miles

RESOURCES

 Electric power 889,000 kw

 205,587 tonnes

Oil reserves not currently exploited

12.5m cattle, 9m goats, 6.5m sheep, 32m chickens

Soda ash, fluorite, limestone, rubies, gold, vermiculite, oil

ELECTRICITY GENERATION

Hydro 74% (3.3bn kwh)	
Combustion 14% (0.61bn kwh)	
Nuclear 0%	
Other 12% (0.55bn kwh)	

0 20 40 60 80 100

% of total generation by type

Agriculture is still the largest sector of the economy. Kenya's varied topography means that tropical, subtropical, and temperate crops may be grown. Coffee and tea, the main export crops, have been affected by falling world prices. Efforts to reduce dependence on these have led to the growth of a successful export-oriented horticultural industry.

Kenya has few mineral resources, though oil exploration has revealed deposits in Turkana District. Hydroelectric and geothermal sources are being developed to reduce energy imports – currently 70% of total requirements. However, droughts caused power shortages in 2000 and 2001, leading to the imposition of daily power cuts.

KENYA : LAND USE

Cropland
Forest
Pasture
Desert
Cattle
Maize
Coffee – cash crop

0 100 km
0 100 miles

ENVIRONMENT

Sustainability rank: 89th

6%

0.3 tonnes per capita

ENVIRONMENTAL TREATIES

	Yes		Yes		Yes
	Yes		Yes		No

The importance to tourism of wildlife conservation is recognized, and recent elephant protection schemes have been a success, but proposed national reserves compete with agriculture for land. Opposition to government plans to reallocate some national park land to squatters is growing.

MEDIA

 TV ownership low

Daily newspaper circulation 8 per 1000 people

PUBLISHING AND BROADCAST MEDIA

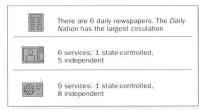

There are 6 daily newspapers. The *Daily Nation* has the largest circulation

6 services: 1 state-controlled, 5 independent

9 services: 1 state-controlled, 8 independent

Government intolerance of criticism is long-standing and includes plays and novels as well as the media. Ngugi wa Thiongo, Kenya's most famous novelist, was exiled for his criticism of KANU.

CRIME

Death penalty in use

 35,340 prisoners

Crime is rising

CRIME RATES

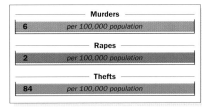

Murders
6 per 100,000 population

Rapes
2 per 100,000 population

Thefts
84 per 100,000 population

Nairobi's high crime levels are spreading countrywide, as a result of worsening poverty, ethnic violence, and rising banditry in the northeast. An increase in the use of guns underlies the rapid increase in violent crime.

EDUCATION

School leaving age: 14

 82%

44,411 students

THE EDUCATION SYSTEM

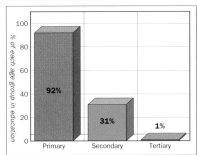

% of each age group in education

100
80
60
40
20
0

92% Primary
31% Secondary
1% Tertiary

The education system is loosely based on the British model. Primary education, from six to 14 years of age, is free and compulsory; some 92% of children attend. The drop-out rate at secondary level is high, with only about 31% attendance. In higher education, the emphasis is on vocational training. There are five state universities.

HEALTH

 Welfare state health benefits

1 per 10,000 people

Respiratory and diarrheal diseases, malaria, AIDS

The health system comprises a mixture of state and private facilities, the latter mainly run by charities and missions. The state system has been badly hit by recession, worsening the already limited access of the rural majority. Poverty-related illnesses, particularly among women and children, are increasing. HIV/AIDS reached epidemic proportions in some areas in the 1990s, and the level of infection with HIV/AIDS was estimated at 15% of the adult population as at end-2001. Kenya has ten doctors and 23 qualified nurses for every 100,000 people.

SPENDING

GDP/cap. increase

CONSUMPTION AND SPENDING

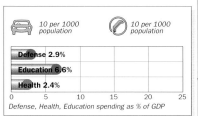

10 per 1000 population

10 per 1000 population

Defense 2.9%	
Education 6.6%	
Health 2.4%	

0 5 10 15 20 25

Defense, Health, Education spending as % of GDP

Wealth disparities in Kenya are large and growing, exacerbated by land hunger and migration to the cities, where jobs are few and existence depends on the informal economy. More than half of all town dwellers live in slums, and the slum dwellers of Nairobi's Amarthi Valley are among Africa's poorest, worst-nourished people. Their lives contrast sharply with those of the country's elite – top government officials with access to patronage; white Kenyans, who derive their wealth largely from agricultural estates; and the largely Asian business community. Wealthy Kenyans often send their children abroad for higher education.

WORLD RANKING

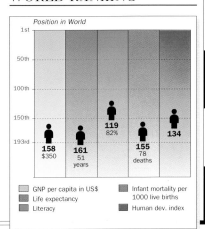

Position in World

1st
50th
100th
150th
193rd

158 $350

161 51 years

119 82%

155 78 deaths

134

GNP per capita in US$
Life expectancy
Literacy

Infant mortality per 1000 live births
Human dev. index

K

KIRIBATI

OFFICIAL NAME: Republic of Kiribati   **CAPITAL:** Bairiki (Tarawa Atoll)
POPULATION: 92,000   **CURRENCY:** Australian dollar   **OFFICIAL LANGUAGE:** English

  1979   1979   July 12   KIR   +12   +686 .ki

FORMERLY PART OF THE colony of the Gilbert and Ellice Islands, the Gilberts became independent from Britain in 1979 and took the name Kiribati (pronounced "Kir-ee-bahs"). British interest in the Gilbert Islands rested solely on the exploitation of the phosphate deposits on Banaba; these ran out in 1980. In 1981, Kiribati won damages (but not the costs of litigation) from the British for decades of phosphate exploitation.

Banreaba Island, Tarawa Atoll. *None of the atolls is more than 8 m (26 ft) high except Banaba, once the source of phosphates.*

CLIMATE
▷ Tropical oceanic

WEATHER CHART FOR BAIRIKI

Kiribati's small land area in the vast Pacific means that some atolls can often go for months without rain. In 1999, a nationwide drought emergency was declared.

TRANSPORTATION
▷ Drive on right

Bonriki International, Tarawa 51,000 passengers

4200 grt

THE TRANSPORTATION NETWORK

483 km (300 miles)		None	
None		5 km (3 miles)	

Kiribati has a limited air link with Fiji. Transportation around and between the atolls is provided mostly by small canoes.

TOURISM
▷ Visitors : Population 1:92

1000 visitors

No change in 2000

MAIN TOURIST ARRIVALS

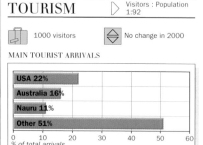

USA 22%	
Australia 16%	
Nauru 11%	
Other 51%	

0 10 20 30 40 50 60
% of total arrivals

A weekly flight from Honolulu to Kiritimati (Christmas Island) ensures a small but steady stream of visitors.

PEOPLE
▷ Pop. density medium

English, Micronesian dialect

130/km² (336/mi²)

THE URBAN/RURAL POPULATION SPLIT

36%   64%

RELIGIOUS PERSUASION

Other 8%

Kiribati Protestant Church 39%

Roman Catholic 53%

Locals still refer to themselves as Gilbertese. Almost all Gilbertese are Micronesian, although the inhabitants of Banaba employed anthropologists to establish their racial distinctness. Tension with the Banabans is intense, mostly fueled by the historic value of Banaba's phosphate deposits. Most Gilbertese are poor. Many go to Nauru as guest workers, living in barrack-room conditions, or work as merchant shipping crew. Those who stay at home go through a circular migration from the outlying islands to Tarawa, returning to see relatives. Women play a prominent role, especially on outlying islands, where they run most of the farms.

POLITICS
▷ Nonparty elections

1998/2002

President Teburoro Tito

AT THE LAST ELECTION
House of Assembly 42 seats

36% Ind	33% MTM	26% NPP	5% App

Ind = Independents   **MTM** = Maneaban Te Mauri
NPP = National Progressive Party   **App** = Appointed

The House of Assembly has one appointed member and one ex-officio member

The traditional chiefs still effectively rule Kiribati, through a party system on the British model. Victory for the MTM in the 1994 elections ended 15 years of rule by the NPP. The main concern is the economy, which is extremely vulnerable to any fluctuations in world demand for coconuts. The overpopulation of Tarawa is the other major issue. Possible restrictions on travel to the island have been discussed. In part, the problem of migration is caused by the poverty and lack of opportunity on the outer islands. A resettlement program, aiming to move people out of Tarawa, began in 1998.

KIRIBATI

Total Land Area : 717 sq. km (277 sq. miles)

Banaba (Ocean I.)

Tapiwa
Ooma

Tarawa

Bonriki
Betio
BAIRIKI

Makin
Butaritari
Abaiang Marakei
Tarawa
Maiana Abemama
Aranuka
Banaba (Ocean I.) Nonouti
Beru
Tabiteuea Nikunau
Onotoa
Tamana Arorae

PHOENIX Kanton
McKean I. Enderbury I.
Birnie I. Rawaki
Nikumaroro Manra
Orona
ISLANDS

Teraina
Tabuaeran
Kiritimati
Equator

Starbuck I.
Malden I.

Kiritimati

London Banana
Paris

Vostok I. Millenni
Flint I.

POPULATION
• under 10 000

LAND HEIGHT
under 100m

0 600 km
0 600 miles

K

WORLD AFFAIRS Joined UN in 1999

Kiribati has little international significance because of its tiny size and remote location, but is able to make its voice heard regionally through the Pacific Islands Forum. In 1986, Kiribati was a signatory to a deal between the US and a number of Pacific Island states that resulted in the US paying $60 million in return for access to Pacific fishing grounds. In the Cold War era Kiribati played the USSR off against the US, extracting a high price for fishing leases, which allowed boats to spy on US nuclear testing on the neighboring Kwajalein Atoll in the Marshall Islands.

AID Recipient

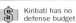 US$18m (receipts)   Down 14% in 2000

Aid is mostly offered for small projects to improve infrastructure. New Zealand is the most generous donor, granting US$1.7 million in 2000.

DEFENSE 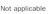 No compulsory military service

Kiribati has no defense budget   Not applicable

Australia and New Zealand provide de facto protection, with regular antisubmarine patrols.

ECONOMICS Inflation 3.8% p.a. (1990–2000)

US$86m   1.7997–1.9535 Australian dollars

SCORE CARD
- WORLD GNP RANKING.....................188th
- GNP PER CAPITA ....US$950
- BALANCE OF PAYMENTS ....US$1m
- INFLATION 2%
- UNEMPLOYMENT 2%

STRENGTHS
Subsistence economy; only Tarawa imports food. Coconuts provide some export income: the EU is the biggest market. Fisheries have limited potential. Upgraded port facilities at Betio.

WEAKNESSES
Lack of resources. High levels of poverty. Isolation, and large distances between islands. Heavy dependence on international aid. Almost no economic potential.

EXPORTS

IMPORTS

RESOURCES Electric power 2000 kw

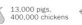 48,218 tonnes   Not an oil producer

13,000 pigs, 400,000 chickens   None

Phosphate deposits on Banaba ran out in 1980. All energy supplies have to be imported. An underwater agriculture scheme is under development.

ENVIRONMENT Not available

 39% (including marine and semi-protected areas)   0.3 tonnes per capita

Rising sea levels cause coastal erosion and ultimately threaten Kiribati's existence. Global warming is a critical issue both for this reason, and because of its damaging effects on the coral reef which protects Tarawa from the sea and holds important inshore fish stocks in the lagoon. The coral has also suffered from pollution by untreated effluent.

MEDIA TV ownership low

There are no daily newspapers

PUBLISHING AND BROADCAST MEDIA

There are no daily newspapers. The weekly newspapers are Butim'aea Manin te Euangkerio, Kiribati Newstar, and Te Uekera

1 independent service   1 independent service

The independent Newstar competes with the state-owned Te Uekera and the Protestant Church's paper. Churches also produce monthly newsletters.

CRIME No death penalty

57 prisoners   Crime is minimal

Crime, apart from brawls resulting from drunkenness, is minimal. The islands' judicial system is based on the British model.

EDUCATION School leaving age: 15

 98%   568 students

Education is British-inspired and compulsory from six to 15. The best students go on to university in Fiji.

CHRONOLOGY
The British established the phosphate-producing colony of the Gilbert and Ellice Islands in 1892.
- **1957** British nuclear tests take place near Kiritimati.
- **1979** Independence as two states, Kiribati and Tuvalu. NPP in power.
- **1981** Kiribati wins damages for phosphate mining from UK.
- **1986** Kiribati–US fishing deal.
- **1994** NPP loses election to MTM. Teburoro Tito elected president.
- **1999** National drought emergency.

K

HEALTH Welfare state health benefits

1 per 7600 people   Heart diseases, diabetes

Most Gilbertese are healthy despite the high levels of poverty. Free medical care is provided to all. Nutrition is becoming a problem on Tarawa where overpopulation limits agriculture.

SPENDING 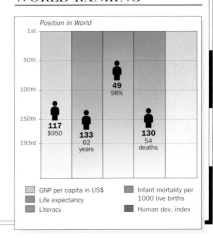 GDP/cap. increase

CONSUMPTION AND SPENDING

2 per 1000 population   43 per 1000 population

Defense Not applicable
Education 11.4%
Health 9.9%

Defense, Health, Education spending as % of GDP

Life in Kiribati is modest. Most Gilbertese live by subsistence farming and fishing. Civil servants in Bairiki form the wealthiest group. The cost of living in Tarawa is higher than in the outlying islands due to the need to import food, although fish is abundant and cheap everywhere.

WORLD RANKING

Position in World

117 $950 | 133 62 years | 49 98% | 130 54 deaths

GNP per capita in US$ / Life expectancy / Literacy / Infant mortality per 1000 live births / Human dev. index

349

NORTH KOREA

OFFICIAL NAME: Democratic People's Republic of Korea   **CAPITAL:** Pyongyang
POPULATION: 22.4 million   **CURRENCY:** North Korean won   **OFFICIAL LANGUAGE:** Korean

COMPRISING THE NORTHERN half of the Korean peninsula, North Korea is separated from the US-dominated South by an armistice line straddling the 38th parallel. Much of the country is mountainous; the Chaeryong and Pyongyang plains in the southwest are the most fertile regions. An independent communist republic from 1948, it remains largely isolated. With its economy starved of capital, it now faces a food crisis requiring large-scale international assistance.

CLIMATE ▷ Continental

WEATHER CHART FOR PYONGYANG

North Korea has a typically continental climate. Winters in the north can be extreme.

TRANSPORTATION ▷ Drive on right

 Sunan, Pyongyang

182 ships
631,000 grt

THE TRANSPORTATION NETWORK

1997 km (1241 miles)	524 km (326 miles)
5214 km (3240 miles)	2253 km (1400 miles)

The railroad network built by the occupying Japanese is heavily relied on. Highways are open only to very limited, officially approved traffic. Improving relations with South Korea in 2000 led to plans to construct cross-border links.

TOURISM ▷ Visitors : Population 1:172

130,000 visitors   ⬆ Up 2% 1995–1998

MAIN TOURIST ARRIVALS

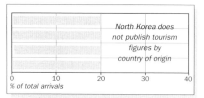

North Korea does not publish tourism figures by country of origin

Economic need has forced limited tourism. South Korean firms have developed resorts such as Mt. Kumgang.

Rice paddy field. The hot, wet summers are ideal for rice growing. Most farms are run as cooperatives.

PEOPLE ▷ Pop. density medium

Korean, Chinese   186/km² (482/mi²)

THE URBAN/RURAL POPULATION SPLIT

60%   **40%**

ETHNIC MAKEUP

Korean 100%

The Korean peninsula is unusual in having been inhabited by a single ethnic group for the last 2000 years. There is a tiny Chinese minority in North Korea.

The religions practiced under strict state control are Buddhism, Christianity, and Chondogyo, a combination of Confucianism, Buddhism, and Christianity which is peculiar to Korea.

North Koreans live highly regulated lives. Divorce is nonexistent and extramarital sex highly frowned upon. Women form more than 50% of the workforce, but are also expected to run the home; it is not uncommon for them to rise at 4 a.m., and end their working day at 7 p.m. From an early age, children are looked after by an extensive system of state-run crèches. The privileged lifestyle of the political elite – some 200,000 in number – is rumored to be a source of popular resentment.

POLITICS ▷ No multiparty elections

1998/2003   Eternal President Kim Il Sung

AT THE LAST ELECTION

Supreme People's Assembly 687 seats

100% DFRF

DFRF = Democratic Front for the Reunification of the Fatherland
The Democratic Front for the Reunification of the Fatherland (led by the Korean Workers' Party – **KWP**) was the sole party permitted to take part in the 1998 elections

The three million-strong KWP is the only legal party; membership is essential for individual advancement. Kim Il Sung, the subject of a lavish personality cult, died in 1994 after almost 50 years as leader. Since then the key question has been how Kim Jong Il, his son and chosen successor, would handle the leadership. He lacks his father's authority, and has yet to seal the succession; in 1998 Kim Il Sung, by then four years dead, was declared "Eternal President."

WORLD AFFAIRS ▷ Joined UN in 1991

NAM

The collapse of Soviet communism in 1991 destroyed North Korea's framework of foreign relations, leaving China as its closest ally. A breakthrough in North–South relations came in 2000, with their leaders' first direct encounter since the 1953 armistice. In 2001 Kim Jong Il paid a landmark visit to Russia. Despite North Korea's nuclear missile program, several countries have established diplomatic links, and many US trade sanctions were lifted; however, the US designated North Korea part of an "axis of evil" in 2002.

AID ▷ Recipient

$75m (receipts)   ⬇ Down 63% in 2000

Rice harvests were badly damaged by alternate drought and flood in the mid-1990s, precipitating a famine. Efforts to stave off starvation using international aid are threatened by lack of funds.

DEFENSE ▷ Compulsory military service

$2.05bn   ⬇ Down 2% in 2000

North Korea is thought to have manufactured nuclear weapons prior to the 1994 freeze on its nuclear program. It developed, and exports, missiles.

NORTH KOREA

Total Land Area : 120 540 sq. km (46 540 sq. miles)

ECONOMICS ▷ Not available

 $12.6bn

 2.2 North Korean won

SCORE CARD

❏ WORLD GNP RANKING..........................77th
❏ GNP PER CAPITA $573
❏ BALANCE OF PAYMENTS ....*Closed economy;*
❏ INFLATION ....*does not publish*
❏ UNEMPLOYMENT ....*any figures*

STRENGTHS

Other than minerals, North Korea's economy has few strengths.

WEAKNESSES

GNP has declined steadily since 1990. The acute shortage of foreign capital and technology has been catastrophic.

EXPORTS

Saudi Arabia 6%
Bangladesh 6%
Spain 6%
Brazil 14%
Other 43%
Japan 25%

IMPORTS

Hong Kong 4%
Brazil 8%
China 31%
Thailand 13%
Japan 14%
Other 30%

RESOURCES ▷ Electric power 9.5m kw

 278,500 tonnes

Not an oil producer; refines 42,000 b/d

3.16m ducks, 3.14m pigs, 2.57m goats, 16.9m chickens

Coal, iron, lead, gold, copper, zinc, tungsten, silver, tin, uranium

A shortage of electricity remains a major problem; blackouts are frequent. Under the 1994 agreement with the US, two new reactors are to be built with outside assistance. North Korea is relatively rich in metals including gold, silver, and tungsten.

ENVIRONMENT ▷ Sustainability rank: 140th

3% (0.1% partially protected)

10.3 tonnes per capita

Excessive use of fertilizers and unchecked pollution from heavy industry are the major problems.

MEDIA ▷ TV ownership medium

⊠ Daily newspaper circulation 199 per 1000 people

PUBLISHING AND BROADCAST MEDIA

There are 5 daily newspapers, including the leading *Rodong Sinmun*, the party newspaper, and *Minju Choson*

1 state-controlled service

1 state-controlled service

TV consists mostly of musical shows praising Kim Il Sung and Kim Jong Il, and anti-American tirades directed against the Korean War.

CRIME ▷ Death penalty in use

North Korea does not publish prison figures

Low level of violent street crime

At an individual level, crime is officially said hardly to exist. The criminal code is weighted to protect the state against "subversion," rather than the rights of the individual. North Korea has a very poor human rights record and there is a *gulag* of more than 100,000 "subversives," where whole families are sent along with those accused, and where torture is routine.

EDUCATION ▷ School leaving age: 16

95%   390,000 students

English is compulsory as a second language at the age of 14. Kim Il Sung, Pyongyang, is the only university.

HEALTH ▷ Welfare state health benefits

1 per 333 people

Heart disease, cancers, digestive diseases

Health care is free. Reasonable life expectancy is now threatened by malnutrition and outright starvation.

SPENDING ▷ GDP/cap. decrease

CONSUMPTION AND SPENDING

No data

46 per 1000 population

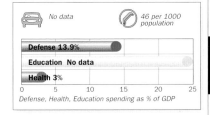

Defense 13.9%
Education No data
Health 3%

0   5   10   15   20   25

Defense, Health, Education spending as % of GDP

An elite within the KWP lives well, with access to specialist shops and consumer goods such as VCRs. Ownership of telephones, private cars, and, in many areas, bicycles, is forbidden.

WORLD RANKING

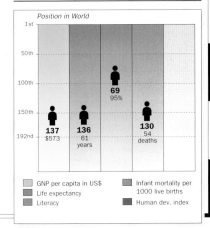

Position in World

1st
50th
100th
150th
192nd

137 $573
136 61 years
69 95%
130 54 deaths

GNP per capita in US$
Life expectancy
Literacy
Infant mortality per 1000 live births
Human dev. index

K

SOUTH KOREA

OFFICIAL NAME: Republic of Korea CAPITAL: Seoul
POPULATION: 47.1 million CURRENCY: South Korean won OFFICIAL LANGUAGE: Korean

O CCUPYING THE SOUTHERN half of the Korean peninsula in East Asia, over 80% of South Korea is mountainous and two-thirds is forested. Rice is the major agricultural product, grown by over 85% of South Korea's three million farmers. The whole peninsula was annexed by Japan from 1910 to 1945, and the split between South Korea and the communist North originated with the arrival of rival US and Soviet armies in 1945. Although the two states have discussed reunification, the legacy of hostility arising from the 1950–1953 Korean War remains a major obstacle.

TOURISM

Visitors : Population 1:9.2

5.15m visitors   Down 3% in 2001

MAIN TOURIST ARRIVALS

Japan 47%	
USA 9%	
China 7%	
Hong Kong 5%	
Philippines 4%	
Other 28%	

% of total arrivals (0, 10, 20, 30, 40, 50, 60)

CLIMATE

Continental

WEATHER CHART FOR SEOUL

South Korea has four distinct seasons. Winters are dry and can be bitterly cold. Summers are hot and humid, especially during July and August.

Overseas tourism to South Korea has increased ten-fold since 1969. Most visitors are Japanese, who come for the golf and Seoul's nightlife; Jeju-do (Cheju-do) is a favored honeymoon destination. Whereas visiting relations of US army personnel once made up 13% of all tourists, today Los Angeles-based Korean–Americans make up the greatest proportion of US visitors. However, despite the publicity generated by the 1988 Seoul Olympics and the 2002 soccer World Cup, South Korea is still not seen in the West as a prime tourist destination.

TRANSPORTATION

Drive on right

Kimpo International, Seoul
36.7m passengers

2381 ships
5.69m grt

THE TRANSPORTATION NETWORK

64,808 km (40,270 miles)	2477 km (1539 miles)
3098 km (1925 miles)	1609 km (1000 miles)

South Korea has an efficient and highly integrated public transportation system. A toll-based nationwide motor expressway network joins most major urban centers. Air travel has expanded rapidly as a convenient way to traverse the mountainous interior. Korean Air competes with Asiana. Buses, trains, boats, and planes are integrated in one timetable, and have a reputation for punctuality. A high-speed rail link will connect Seoul and Busan (Pusan).

Improving relations with North Korea in 2000 produced proposals for the construction of cross-border rail and road links and north–south connecting flights. Concrete projects have been slow to materialize, due to continuing political uncertainty, but the first flights since 1953 took place in mid-2002, albeit limited to teams of construction workers in the nuclear industry.

SOUTH KOREA

Total Land Area : 98 480 sq. km
(38 023 sq. miles)

POPULATION

over 5 000 000	▣
over 1 000 000	▣
over 500 000	◉
over 100 000	◎
over 50 000	○
over 10 000	●
under 10 000	•

LAND HEIGHT

1000m/3281ft	
500m/1640ft	
200m/656ft	
Sea Level	

K

PEOPLE ▷ Population density high

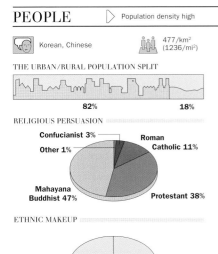

Korean, Chinese

477/km²
(1236/mi²)

THE URBAN/RURAL POPULATION SPLIT

82%   18%

RELIGIOUS PERSUASION

- Confucianist 3%
- Other 1%
- Roman Catholic 11%
- Mahayana Buddhist 47%
- Protestant 38%

ETHNIC MAKEUP

Korean 100%

The Korean peninsula is unusual in having been inhabited by one ethnic group for the last 2000 years. Family life is a central and clearly defined part of society, although the nuclear family model is becoming the norm rather than the old-style household of the extended family. Most Koreans can trace their ancestry back thousands of years. Regional origin is important in determining blood heritage, since there are only 270 Korean surnames and half the population is named Kim, Lee, Park, or Choi. Chondogyo, combining elements of Shamanism, Buddhism, and Christianity, is peculiar to the Koreas but has only a tiny following. Traditional values condition attitudes to women, and it is still not respectable for those who are married to have a job.

Economic growth has attracted illegal immigrants from the poorer Asian countries, who take menial jobs that South Koreans now refuse to do.

POPULATION AGE BREAKDOWN

Female	Age	Male
0.6%	80+	0.2%
5%	60–79	3.5%
10.6%	40–59	10.8%
18.5%	20–39	19.2%
15.1%	0–19	16.5%

% of population by age group

POLITICS ▷ Multiparty elections

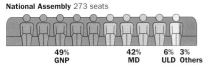

2000/2004

President Kim Dae Jung

AT THE LAST ELECTION

National Assembly 273 seats

49% GNP	42% MD	6% ULD	3% Others

GNP = Grand National Party **MD** = Millennium Democratic Party **ULD** = United Liberal Democrats

Officially a democracy since its inception, South Korea was in practice ruled by military dictators until 1987.

PROFILE

South Korea's politics changed radically in 1987 with the introduction of direct presidential elections and a parliament with enhanced powers.

In 1993, Kim Young Sam became president, the first nonmilitary leader in 30 years. He launched a popular anticorruption campaign, targeting former presidents. His New Korea Party (NKP) (formerly the Democratic Liberal Party – DLP), returned to power with a reduced majority in 1996, was brought down amid a steel scandal in 1997. Veteran opposition leader Kim Dae Jung was elected president later that year in the first peaceful transfer of power to an opposition politician in South Korea's history. Supporters of the new president gained ground in legislative elections in 2000, but political instability continued, with frequent changes of prime minister and opposition byelection victories. Chang Sang was appointed the first female prime minister in 2002, but was rejected by parliament.

MAIN POLITICAL ISSUES
The economy

South Korea's economy, once one of the most impressive in the world, was severely shaken by the regional crisis of late 1997. Severe financial austerity and retrenchment failed to prevent large-scale bankruptcies in 2000.

Relations with North Korea

The sudden flowering of North–South relations in 2000 soon lost popularity. Expensive cross-border projects, and promises of aid for the impoverished North, raised fears of the mounting cost for the embattled southern economy.

***President Kim Dae Jung.** A veteran activist, he has won the Nobel Peace Prize.*

***Chang Sang,** first woman appointed as premier, but rejected by parliament in 2002.*

WORLD AFFAIRS ▷ Joined UN in 1991

APEC   CP   IAEA   OECD   WTO

Since the division of Korea, relations with the North have dominated foreign policy. A historic summit meeting in 2000 in the North Korean capital, Pyongyang, opened a new phase, and cross-border diplomatic and economic cooperation briefly flourished. Reunification remains the ultimate goal of both Koreas, but with doubts about its social and economic costs. Military tensions persist, and a large detachment of US troops remains on the border – the most heavily defended in the world. Relations with China, the closest ally of North Korea, have improved. Japan is also a major trading partner, although South Koreans continue to harbor resentment over the 1910–1945 Japanese annexation.

AID ▷ Donor

$212m (donations)   Down 33% in 2000

Once a massive recipient of US aid, and then of Japanese war reparations, South Korea emerged in the 1970s and 1980s as a major aid donor. However, the economic crisis which hit South Korea in 1997–1998 forced it to seek international financial assistance to salvage key sectors of its threatened economy.

CHRONOLOGY

The Yi dynasty, founded in Seoul in 1392, ruled the kingdom of Korea until 1910. Korea became a vassal state of China in 1644.

❏ **1860** Korea reacts to French and British occupation of Peking by preventing Western influence: becomes the "Hermit Kingdom."
❏ **1904–1905** Russo-Japanese War. Japan conquers Korea.
❏ **1910** Japan annexes Korea.
❏ **1919** Independence protests violently suppressed.
❏ **1945** US and Soviet armies arrive. Korea split at 38°N. South comes under de facto US rule.
❏ **1948** Republic of South Korea created; Syngman Rhee becomes president at head of an increasingly authoritarian regime.
❏ **1950** Hostilities between North and South, each aspiring to rule a united Korea. North invades, sparking Korean War. US, with UN backing, enters on South's side; China unofficially assists North. In 1951 fighting stabilizes near 38th parallel.
❏ **1953** Armistice; de facto border at cease-fire line, close to the 38th parallel. ⇨

K

CHRONOLOGY *continued*

- **1960** Syngman Rhee resigns in face of popular revolt.
- **1961** Military coup leads to authoritarian junta led by Park Chung Hee.
- **1963** Pressure for civilian government. Park reelected as president (also in 1967 and 1971). Strong manufacturing base and exports drive massive economic development program.
- **1965** Links restored with Japan.
- **1966** 45,000 troops engaged in South Vietnam.
- **1972** Martial law stifles political opposition. New constitution with greater presidential powers.
- **1979** Park assassinated. Gen. Chun Doo Hwan, intelligence chief, leads coup. Kim Young Sam, opposition leader, expelled from parliament.
- **1980** Chun chosen as president. Kim Dae Jung and other opposition leaders arrested.
- **1986** Car exports start.
- **1987** Emergence of prodemocracy movement. Roh Tae Woo, Chun's chosen successor, elected president.
- **1988** Inauguration of Sixth Republic which includes genuine multiparty democracy. Restrictions on foreign travel lifted.
- **1990** Government party and two opposition parties, including Kim Young Sam's, merge to form DLP.
- **1991** South Korea joins UN.
- **1992** Diplomatic links with China established. December, Kim Young Sam elected president.
- **1996** Chun sentenced to death on charges of organizing 1979–1980 overthrow of civilian government; Roh given a lengthy prison term. Both sentences were rescinded.
- **1997** Violent protests against new labor laws. Steel scandal brings down government. Economic crisis.
- **1998** Kim Dae Jung president.
- **2000** Historic North–South summit in Pyongyang.
- **2002** Short-lived appointment of Chang Sang as first female premier.

Seoul lit up at night. *The city is home to more than ten million people – one-quarter of South Korea's population. Seoul means "capital."*

DEFENSE Compulsory military service

$12.5bn — Up 3% in 2000

SOUTH KOREAN ARMED FORCES

2330 main battle tanks (Type 88, M-47, M-48, T–80U)	560,000 personnel	
19 submarines, 6 destroyers, 9 frigates, 24 corvettes, 84 patrol boats	60,000 personnel	
555 combat aircraft (160 F-16C/D, 195 F-5E/F, 130 F-4D/E)	63,000 personnel	
None		

The main defense concern is North Korea. South Korea has fewer troops, tanks, artillery, and aircraft, but it claims parity through the permanent presence of 35,000 US troops on its territory and its superior technology. The manufacture of missiles capable of striking any target in North Korea was legalized in 2001. However, US computer simulations question South Korea's ability to resist an invasion by the North, since Seoul is only 55 km (35 miles) from the demilitarized zone.

In the mid-1990s, the army's role and standing in national politics was sharply downgraded after a vigorous government campaign to investigate corruption in the armed forces, especially pertaining to arms procurement and past military involvement in politics.

ECONOMICS Inflation 5% p.a (1990–2000)

$421bn — 1265–1314 South Korean won

SCORE CARD

- WORLD GNP RANKING...........................13th
- GNP PER CAPITA $8910
- BALANCE OF PAYMENTS....................$11.4bn
- INFLATION 2.3%
- UNEMPLOYMENT 4%

ECONOMIC PERFORMANCE INDICATOR

EXPORTS

Taiwan 5%, Hong Kong 6%, China 11%, Japan 12%, USA 22%, Other 44%

IMPORTS

Australia 4%, Saudi Arabia 6%, China 8%, USA 18%, Japan 20%, Other 44%

STRENGTHS
World's most successful shipbuilder, with 45% of the market. Strong yen makes South Korean exports more competitive than Japan's. Demand from China, particularly for cars.

WEAKNESSES
High level of indebtedness and vulnerability to international capital movements. Increasingly militant workforce since 1997. State sector a burden on the economy. Strong competition from Japan.

PROFILE
South Korea's economic miracle began with centralized planning. *Chaebol* (conglomerates) such as Samsung achieved impressive growth rates in strategic industries such as car making, shipbuilding, and semiconductors.

A well-educated workforce and cheap state credit gave Korea a competitive edge. The government then encouraged foreign investment and an emphasis on smaller industries to maintain growth. In 1996 South Korea joined the OECD. The following year, however, a major financial crisis and the threat of a debt implosion forced the government to turn to the IMF for a huge credit agreement. By 2001 the banking system showed signs of recovery, although global slowdown continued to affect the economy.

SOUTH KOREA : MAJOR BUSINESSES

Seoul, Inchŏn, Pohang, Ulsan, Gunsan, Busan, Changwŏn, Masan, Gwangju

Garments, Chemicals, Electronics, Iron & steel, Shipbuilding, Fish processing, Vehicle assembly, Telecommunications

0 50 km / 0 50 miles

RESOURCES

 Electric power 48m kw

2.42m tonnes

Not an oil producer; refines 1.15m b/d

8.72m pigs, 6m ducks, 102m chickens

Coal, iron, lead, zinc, tungsten, gold, graphite, fluorite

ELECTRICITY GENERATION

Hydro 3% (6.1bn kwh)	
Combustion 60% (145bn kwh)	
Nuclear 37% (90bn kwh)	
Other 0%	

0   20   40   60   80   100

% of total generation by type

South Korea has few natural resources. It has to import all of its oil and has built a series of nuclear reactors for generating electricity. Under the terms of the 1994 agreement between North Korea and the US, South Korea is constructing two reactors in the North which, in the event of reunification, will be connected to the national grid.

Agriculture remains a highly protected sector of the economy. Plans to open up the rice market have in the past provoked massive demonstrations in Seoul.

ENVIRONMENT

 Sustainability rank: 135th

 7%

 7.8 tonnes per capita

ENVIRONMENTAL TREATIES

Yes   Yes   Yes

Yes   Yes   No

Environmental groups remain hostile to South Korea's nuclear power program. Rapid industrialization has resulted in environmental problems. The government has signed but not yet ratified the Kyoto Protocol, which aims to reduce carbon dioxide emissions. More recently, steps have been taken to address the severe problem of air pollution in urban areas, particularly in Seoul. Rivers in rural areas have been polluted by fertilizers and chemicals.

MEDIA

 TV ownership high

Daily newspaper circulation 394 per 1000 people

PUBLISHING AND BROADCAST MEDIA

There are 60 daily newspapers. *Chosun Ilbo* has the highest circulation figures

7 services: 1 publicly owned, 6 independent

9 services: 1 publicly owned, 8 independent

South Korea's media have been freed of most restrictions since the advent of full multiparty democracy. However, criticisms of the armed forces are still frowned upon, and journalists tend to avoid altogether the subject of the role of the military in society. Caution also has to be exercised in reporting facts about North Korea. In the past, South Korean journalists who have made favorable mention of the North Korean communist regime have suffered harassment and intimidation.

CRIME

 Death penalty in use

64,038 prisoners   Up 159% in 1999

CRIME RATES

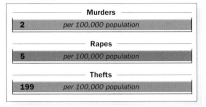

Murders

2   *per 100,000 population*

Rapes

5   *per 100,000 population*

Thefts

199   *per 100,000 population*

Crime rates are low but rising; violent crime is uncommon. The government has begun to treat corruption as a crime. Since 1987, the internal security forces' operations have been restricted, although left-wing activists are still harassed. Striking workers and student demonstrators encounter – and prepare for – forceful and confrontational crowd control.

EDUCATION

 School leaving age: 15

98%   1.79m students

THE EDUCATION SYSTEM

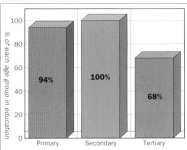

% of each age group in education

100
80
60
40
20
0

Primary 94%   Secondary 100%   Tertiary 68%

South Korea began a concentrated education program in the 1950s, and a well-educated workforce has been the foundation of impressive economic growth. Secondary education, which begins at 12 years of age, comprises two three-year cycles, of which one is compulsory. Tertiary enrollment is nearly 70% – one of the highest rates in the world.

SOUTH KOREA : LAND USE

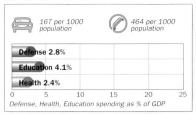

	Cropland
	Pasture
	Forest
∨	Poultry
	Rice
	Cereals

0   50 km
0   50 miles

Cheju-do

HEALTH

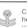 Welfare state health benefits

1 per 769 people

Cancers, accidents, heart and cerebro-vascular diseases, TB

The health service has improved in line with economic growth. Most hospitals are equipped with modern facilities, and many offer advanced treatments comparable with those in the US and western Europe. Health indicators such as infant mortality and longevity have likewise improved.

SPENDING

GDP/cap. increase

CONSUMPTION AND SPENDING

167 per 1000 population

464 per 1000 population

Defense 2.8%	
Education 4.1%	
Health 2.4%	

0   5   10   15   20   25

Defense, Health, Education spending as % of GDP

Most South Koreans have felt the effects of economic growth – and recession. The Jolla (Cholla) region in the southwest remains the poorest.

WORLD RANKING

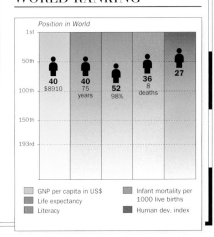

Position in World

1st
50th
100th
150th
193rd

40    $8910
40   75 years
52   98%
36   8 deaths
27

☐ GNP per capita in US$	☐ Infant mortality per 1000 live births	
☐ Life expectancy		
☐ Literacy	☐ Human dev. index	

K

KUWAIT

MIDDLE EAST

OFFICIAL NAME: State of Kuwait   **CAPITAL:** Kuwait City
POPULATION: 2 million   **CURRENCY:** Kuwaiti dinar   **OFFICIAL LANGUAGE:** Arabic

 1961   1961   Feb 25   KWT   +3   +965   .kw

AT THE NORTHWEST EXTREME of the Gulf, Kuwait is dwarfed by its neighbors. The flat, almost featureless landscape conceals huge oil and gas reserves which put Kuwait among the world's first oil-rich states. In 1990 Iraq invaded, claiming it as its 19th province. A US-led alliance, under the aegis of the UN, expelled Iraqi forces following a short war in 1991 and restored the rule of the al-Sabah dynasty. Kuwait has now built a wall separating its territory from Iraq.

Saffar Towers *in the business center of Kuwait City. Rebuilding Kuwait's postwar economy is estimated to have cost $25 billion.*

CLIMATE

▷ Hot desert

WEATHER CHART FOR KUWAIT CITY

Summer temperatures can soar to over 40°C (104°F), but winters can be cold, with frost at night.

TRANSPORTATION

▷ Drive on right

✈ Kuwait International, Kuwait City
3.87m passengers

🚢 202 ships
2.46m grt

THE TRANSPORTATION NETWORK

3590 km (2231 miles)	280 km (174 miles)
None	None

Kuwait has a system of radial expressways around the capital and good connecting roads to Saudi Arabia.

TOURISM

▷ Visitors : Population 1:26

🧳 77,000 visitors   ⬆ Up 12% 1995–1998

MAIN TOURIST ARRIVALS

Saudi Arabia	31%
Egpyt	12%
India	12%
Other	45%

% of total arrivals

The limited tourism from neighboring Arab states, notably Saudi Arabia, has not recovered since the 1990–1991 Gulf War. Most Western visitors to Kuwait go specifically to see relatives working in the oil industry.

PEOPLE

▷ Pop. density medium

Arabic, English   112/km² (291/mi²)

THE URBAN/RURAL POPULATION SPLIT

98%   2%

ETHNIC MAKEUP

Kuwaiti 45%
Other Arab 35%
South Asian 9%
Other 7%
Iranian 4%

Kuwait is a conservative Sunni Muslim society (27% of the population are Shi'a). Women have considerable freedom, although the amir's decree providing for female enfranchisement has been repeatedly rejected by the National Assembly.

Kuwait's oil wealth has drawn in thousands of workers from other Arab countries and south Asia. The Palestine Liberation Organization's support for the Iraqi invasion led to most Palestinians, hitherto more numerous in Kuwait than elsewhere in the Arabian peninsula, being driven out. Native Kuwaitis are outnumbered by resident foreign nationals.

KUWAIT

Total Land Area : 17 820 sq. km (6880 sq. miles)

POPULATION

◎ over 100 000
○ over 50 000
● over 10 000
• under 10 000

LAND HEIGHT

200m/656ft
Sea Level

0   25 km
0   25 miles

POLITICS

▷ No multiparty elections

🗳 1999/2003   Amir Shaikh Jabir al-Ahmad al-Jabir al-Sabah

AT THE LAST ELECTION
National Assembly 50 seats

The electorate comprises civilian men over 21 years of age whose families have been resident in Kuwait since before 1921. Elections on 3 July 1999 were contested by independents. The 50 seats were split evenly between Islamists, liberals, and government supporters.

In 1992 Amir Shaikh Jabir restored the National Assembly. There was then a government of "national unity" until 1999, when elections strengthened the amir's Islamist and liberal opponents. The Council of Ministers resigned in 2001 in the face of Assembly criticism. In the new cabinet the ruling al-Sabah family still held the top posts.

K

WORLD AFFAIRS
 Joined UN in 1963

AL | Damasc | GCC | OIC | OPEC

Strategically important as a major exporter of crude oil and natural gas, Kuwait has always maintained very close

AID
 Donor

 $165m (donations)   Up 12% in 2000

The Kuwait Fund for Arab Economic Development continued to give aid even during the invasion crisis.

DEFENSE
 Compulsory military service

 $3.21bn   Down 2% in 2000

In August 1990 Kuwait's 11,000-strong, partly volunteer army was easily overrun by vastly superior Iraqi forces. Since its liberation, defense pacts have been signed with the US, the UK, France, and Russia. Kuwait rearmed fast, with weapons purchased from major Western suppliers.

ECONOMICS
Inflation 3% p.a. (1990–1999)

$35.8bn   0.3055–0.3071 Kuwaiti dinars

SCORE CARD

- ❏ WORLD GNP RANKING..........................54th
- ❏ GNP PER CAPITA $18,030
- ❏ BALANCE OF PAYMENTS....................$14.9bn
- ❏ INFLATION 1.8%
- ❏ UNEMPLOYMENT 2%

STRENGTHS
Recovery of oil and gas production. Large overseas investments. Stable banking system.

WEAKNESSES
Overreliance of economy on oil and gas. Adverse consequences of Iraqi invasion. Strategic vulnerability deters investment. Reliance on imported skilled labor, food, and raw materials. Delays in enacting privatization package.

EXPORTS

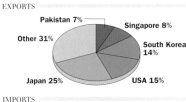

Pakistan 7%
Other 31%
Singapore 8%
South Korea 14%
Japan 25%
USA 15%

IMPORTS

China 6%
Germany 10%
Other 47%
UK 10%
Japan 11%
USA 16%

links with the West, which have strengthened since the war with Iraq, and on which it depends for its future security. In 2002 Iraq promised to respect the common border, but has yet to return POWs and detained Kuwaiti nationals.

RESOURCES
 Electric power 7m kw

6535 tonnes

2.14m b/d (reserves 96.5bn barrels)

630,000 sheep, 130,000 goats, 32.5m chickens

Oil, natural gas, salt

The oil industry is Kuwait's most profitable sector, accounting for over 80% of export earnings. Although badly hit by the Gulf War, when a number of wells were deliberately fired, it was quickly rehabilitated. Kuwait also possesses valuable reserves of natural gas. Other resources are dates, fish, ammonia, and chemicals.

ENVIRONMENT
 Sustainability rank: 142nd

 2%

26.3 tonnes per capita

The Iraqi invasion and the subsequent war caused an ecological disaster. Although the effects of this did not prove as grave as some observers first feared, marine life was damaged and many thousands of hectares of cultivated land were obliterated. Millions of land mines still litter border areas. Water is scarce.

MEDIA
 TV ownership high

Daily newspaper circulation 377 per 1000 people

PUBLISHING AND BROADCAST MEDIA

There are 7 daily newspapers, including *Al-Qabas* and *As-Seyassah*

1 state-controlled service

1 state-controlled service

Radio and TV are state-controlled, but satellite TV is freely available. Press freedom exists in theory.

CRIME
 Death penalty in use

 1735 prisoners   Up 8% 1996–1998

Isolated acts of terrorism occur. There have been complaints of human rights abuses.

EDUCATION
 School leaving age: 14

 83%   29,509 students

Kuwaiti citizens receive free education from nursery to university. Since the liberation, more emphasis has been placed on technology in the curriculum.

CHRONOLOGY
Kuwait traces its independence to 1710, but was under British rule from the late 18th century until 1961. The government denies any historical link with Iraq.

- ❏ **1961** Independence from UK. Iraq claims Kuwait.
- ❏ **1976** Amir suspends National Assembly.
- ❏ **1990** Iraqi invasion.
- ❏ **1991** Liberation following Gulf War.
- ❏ **1992** National Assembly elections.
- ❏ **1999** Elections; Islamists and liberals win most seats.

HEALTH
 Welfare state health benefits

1 per 526 people

Heart diseases, accidents, cancers, perinatal deaths

Despite theft of equipment during the Iraqi invasion, Kuwait has restored its Western-standard health care service. Nationals receive free treatment.

SPENDING
 GDP/cap. increase

CONSUMPTION AND SPENDING

317 per 1000 population   244 per 1000 population

Defense 9.8%
Education 6.5%
Health 3.4%

0 5 10 15 20 25
Defense, Health, Education spending as % of GDP

As well as the oil-rich elite, most Kuwaitis enjoy high incomes, and the government has repeatedly rescued citizens who have suffered stock market or other financial losses. School and university leavers are guaranteed jobs. Capital is easily transferred abroad and there are effectively no exchange controls.

WORLD RANKING

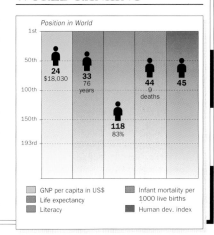

Position in World

1st
50th
100th
150th
193rd

24 $18,030
33 76 years
44 9 deaths
45
118 83%

▢ GNP per capita in US$	▢ Infant mortality per 1000 live births
▢ Life expectancy	
▢ Literacy	▢ Human dev. index

K

KYRGYZSTAN

CENTRAL ASIA

OFFICIAL NAME: Kyrgyz Republic **CAPITAL:** Bishkek
POPULATION: 5 million **CURRENCY:** Som **OFFICIAL LANGUAGES:** Kyrgyz and Russian

KYRGYZSTAN IS A SMALL and very mountainous state in central Asia. It is the least urbanized of the former Soviet republics (the rural population is growing faster than that in the towns) and was among the last to develop its own cultural nationalism. Its increasingly autocratic government tries to steer between Kyrgyz nationalist pressures and ensuring that the minority Russians are not alienated, since they tend to possess the skills necessary to run a market-based economy.

CLIMATE ▷ Mountain

WEATHER CHART FOR BISHKEK

Conditions vary from permanent snow and cold deserts at altitude to hot deserts in low regions. Intermediate slopes and valleys receive some rain.

TRANSPORTATION ▷ Drive on right

✈ **Bishkek International**   🚢 Has no fleet

THE TRANSPORTATION NETWORK

🛣 16,854 km (10,473 miles)	140 km (87 miles)
🚉 417 km (259 miles)	600 km (373 miles)

Kyrgyzstan does not have the financial budget to improve on its poor mountain road network.

TOURISM ▷ Visitors : Population 1:72

🧳 69,000 visitors   ⬆ Up 17% in 1999

MAIN TOURIST ARRIVALS

% of total arrivals

Tourism is undeveloped; most visitors are on business or working on multilateral aid projects. Tourism promotion centers on Kyrgyzstan's position on the Silk Road.

PEOPLE ▷ Pop. density low

Kyrgyz, Russian   👥 25/km² (65/mi²)

THE URBAN/RURAL POPULATION SPLIT

33% **67%**

ETHNIC MAKEUP

Ukrainian 2%   Other 7%
Tatar 2%   Uzbek 13%
Russian 19%
Kyrgyz 57%

Like other former Soviet republics, Kyrgyzstan has witnessed the rise of militant nationalism. Relations are most strained with the large Uzbek minority. The preference given to Kyrgyz in the political system and in particular in the land laws, which exclude all others from full title, has aggravated ethnic tensions. The trend in politics is toward greater Islamization, which is linking religion and race issues more closely and adding social pressure on "foreigners," particularly ethnic Russians, to leave.

Since 1989 a high birthrate has enabled the Kyrgyz to resume their position as the main ethnic group, replacing the Russian community which until recently controlled the economy. However, the government moved to stem the tide of Russian emigration by declaring Russian an official language with full equal status with effect from 2000.

Loess landscape, Naryn valley. Kyrgyzstan is dominated by the ice-capped Tien Shan mountains, but valleys are green and fertile.

POLITICS ▷ Multiparty elections

L. House 2000/2005
U. House 2000/2005 President Askar Akayev

AT THE LAST ELECTION
Legislative Assembly 60 seats

Election results by party were announced only for the 15 national list seats. The other 45 seats are elected on a constituency basis. The Party of Communists of Kyrgyzstan is the largest party and the Union of Democratic Forces the second-largest.

Assembly of People's Representatives 45 seats

The members represent Kyrgyzstan's different regional and ethnic communities

President Akayev's administration has become increasingly autocratic, ending Kyrgyzstan's reputation as one of the region's most liberal societies. Already damaged democratic credentials were destroyed by accusations of fraud during the 2000 legislative and presidential elecions, and the main opposition leader, Felix Kulov, was imprisoned only weeks before the presidential poll. The temporary imprisonment in March 2002 of his successor, Azimbek Beknazarov, prompted mass protests, when five demonstrators were shot dead by police. Spiraling protests forced the entire cabinet to resign in May.

Akayev himself has been accused of fostering a personality cult. Already in power under the Soviet regime before independence, he was reelected in 1995 and in 2000, after objections to a third term were overruled by the Constitutional Court.

WORLD AFFAIRS ▷ Joined UN in 1992

CIS   SCO   OIC   OSCE   EAPC

Kyrgyzstan is working to reduce its dependence on Russia. Turkey is developing close links based on ethnic similarities and aimed at restraining Iranian influence. Relations with Uzbekistan, which allegedly supports some of the antigovernment forces in Kyrgyzstan, are tense, although in 2000 both countries joined with Tajikistan to combat Islamist militants in the region.

AID ▷ Recipient

 $215m (receipts)   Down 24% in 2000

The US and Japan are the main aid donors. The World Bank is also an important source of financial assistance.

K

KYRGYZSTAN

Total Land Area : 198 500 sq. km
(76 641 sq. miles)

POPULATION		LAND HEIGHT
over 500 000	◉	4000m/13 124ft
over 100 000	◎	3000m/9843ft
over 50 000	○	2000m/6562ft
over 10 000	●	1000m/3281ft
under 10 000	•	500m/1640ft

DEFENSE

▷ Compulsory military service

💲 $31m   ⬇ Down 11% in 2000

In 1992, a national army was set up and a defense treaty was signed with five other CIS states. The army is weak and not influential in politics.

ECONOMICS

▷ Inflation 110% p.a. (1990–2000)

📊 $1.35bn   💲 48.24–47.72 soms

SCORE CARD

❑ WORLD GNP RANKING	148th
❑ GNP PER CAPITA	$270
❑ BALANCE OF PAYMENTS	–$77m
❑ INFLATION	18.7%
❑ UNEMPLOYMENT	6%

STRENGTHS

Agricultural self-sufficiency. Private land ownership since 2000. Gold and mercury exports. Hydropower potential.

WEAKNESSES

Dominant state and collective farming mentality. Sharp economic decline since breakup of USSR, on which it depended for trade and supplies. Chronic inflation.

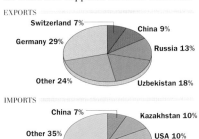

EXPORTS

Switzerland 7%   China 9%
Germany 29%   Russia 13%
Other 24%   Uzbekistan 18%

IMPORTS

China 7%   Kazakhstan 10%
Other 35%   USA 10%
Russia 24%   Uzbekistan 14%

RESOURCES

▷ Electric power 3.7m kw

🐟 198 tonnes   🛢 1544 b/d (reserves 37m barrels)

🐑 4.16m sheep, 985,000 cattle, 3.12m chickens   💎 Coal, antimony, gas, oil, tin, mercury, iron, uranium, zinc, gold

Kyrgyzstan has small quantities of commercially exploitable coal, oil, and gas, and great hydroelectric power potential. Energy policy, which relies on Western aid and technology, is primarily aimed at developing these further in order to reduce dependence on supplies from Russia, and eventually to achieve self-sufficiency in energy.

ENVIRONMENT

▷ Sustainability rank: 56th

🌲 4%   ⬇ 1.3 tonnes per capita

The major problem is the salination of the soil caused by excessive irrigation of cotton crops. Kyrgyzstan has a poor record in limiting industrial pollution.

MEDIA

▷ TV ownership low

 Daily newspaper circulation 9 per 1000 people

PUBLISHING AND BROADCAST MEDIA

There are 4 daily newspapers, including *Kyrgyz Tuusu*, *Slovo Kyrgyzstana*, and *Vechernii Bishkek*

1 state-owned service   4 services: 1 state-owned, 3 independent

The Kyrgyz press is increasingly under pressure from the government's efforts to silence opposition.

CRIME

▷ Moratorium on death penalty

 20,000 prisoners   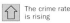 The crime rate is rising

Ethnic tension fuels violence. The narcotics trade flourishes. A 1998 moratorium on the death penalty was extended in 2002 for a further year.

CHRONOLOGY

The Kyrgyz first developed a recognizable ethnic consciousness in the late 18th century.

- ❑ **1860s** Expansion of Russian Empire into Kyrgyz lands.
- ❑ **1924–1991** Incorporated in USSR.
- ❑ **1995** New constitution adopted.
- ❑ **2000** Legislative and presidential elections; Akayev reelected.
- ❑ **2002** Government resigns after police shoot demonstrators.

EDUCATION

▷ School leaving age: 15

 97%   49,744 students

Replacing Russian as the main teaching language is proving an enormous task. Russian is likely to survive at tertiary level, as the Kyrgyz language lacks key technical and scientific terms.

HEALTH

▷ Welfare state health benefits

1 per 333 people   Heart diseases, cancers, accidents, violence, tuberculosis

Kyrgyzstan has one of the least developed public health systems in central Asia. Infant mortality is high.

SPENDING

▷ GDP/cap. decrease

CONSUMPTION AND SPENDING

🚗 39 per 1000 population   📞 77 per 1000 population

Defense 2.4%
Education 5.4%
Health 2.2%

0   5   10   15   20   25
Defense, Health, Education spending as % of GDP

In 2000, almost 90% of the population were estimated as living in poverty. The old Communist Party *nomenklatura* are the main beneficiaries of privatization.

WORLD RANKING

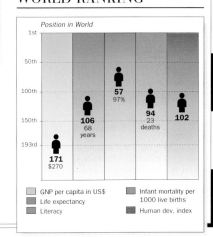

Position in World

57 97%	
106 68 years	94 23 deaths	102
171 $270	

1st, 50th, 100th, 150th, 193rd

- ▢ GNP per capita in US$
- ▢ Life expectancy
- ▢ Literacy
- ▢ Infant mortality per 1000 live births
- ▢ Human dev. index

LAOS

OFFICIAL NAME: Lao People's Democratic Republic **CAPITAL:** Vientiane
POPULATION: 5.4 million **CURRENCY:** New kip **OFFICIAL LANGUAGE:** Lao

THE MEKONG RIVER forms Laos's main thoroughfare and feeds the fertile lowlands of the Mekong valley. Two decades of civil war followed independence from France in 1953, and Laos was bombed heavily during the Vietnam War. The communist LPRP has held power since 1975. Market-oriented reforms began to be introduced in 1986. A transfer of power to a younger generation within the LPRP took place during the 1990s.

CLIMATE ▷ Tropical monsoon

WEATHER CHART FOR VIENTIANE

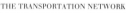

■ *Average daily temperature* Rainfall ■
°C/°F J F M A M J J A S O N D cm/in
40/104 — 40/16
30/86 — 30/12
20/68 — 20/8
10/50 — 10/4
0/32 — 0
-10/14
-20/-4

The tropical southerly monsoon brings heavy rains from May to September.

TRANSPORTATION ▷ Drive on right

✈ **Wattay, Vientiane**
165,000 passengers

🚢 1 ship
2400 grt

THE TRANSPORTATION NETWORK

| 9674 km (6011 miles) | None |
| None | 4600 km (2858 miles) |

A major new Thailand–Vietnam road is planned via Savannakhét. Freight goes mainly by river; there is no railroad and roads are poor and few.

TOURISM ▷ Visitors : Population 1:18

300,000 visitors ⬆ Up 16% in 2000

MAIN TOURIST ARRIVALS

Thailand 58%	
Vietnam 12%	
USA 4%	
Other 26%	

0   10   20   30   40   50   60
% of total arrivals

Tourists were first allowed into Laos in 1989; numbers have risen rapidly since then. Mass tourism is discouraged and preference given to small package tours. Hotels are few, and travel outside Vientiane is difficult.

PEOPLE ▷ Pop. density low

Lao, Mon-Khmer, Yao, Vietnamese, Chinese, French

👥 23/km² (61/mi²)

THE URBAN/RURAL POPULATION SPLIT

24%   76%

RELIGIOUS PERSUASION

Other (including animist) 15%

Buddhist 85%

There are more than 60 ethnic groups in Laos and this considerable diversity has hindered national integration. Society is broadly divided by geography, but by altitude rather than by region. The lowland Laotians (*Lao Loum*) make up some 60% of the population. The upland Laotians (*Lao Theung*), who live in the hills above the valleys, account for 30%.

The small minority of mountain-top Laotians (*Lao Soung*), who include the Hmong, Yao, and Man groups, have resisted government efforts to introduce substitutes for traditional cash crops such as opium. The government continues to face small pockets of Hmong resistance.

Two-thirds of the population speak Lao, and a large number of tribal dialects are also spoken. Buddhism is the main religion, but there are some Christians and animists.

POLITICS ▷ No multiparty elections

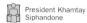

🗳 2002/2007

President Khamtay Siphandone

AT THE LAST ELECTION

National Assembly 109 seats

99%   1%
LPRP   Ind

LPRP = Lao People's Revolutionary Party (the sole legal political party) **Ind** = Independent

All candidates were approved by the LPRP

Laos has been ruled by the same circle of communist revolutionaries, one of the world's most durable and closely knit hierarchies, since 1975. The military, the LPRP, and the executive branch remain closely intertwined, with party chairman Gen. Khamtay Siphandone becoming the country's president in February 1998. Despite limited moves toward political reform, the LPRP, which is modeled on the Communist Party of Vietnam, continues to dominate political life at every level. The long-standing problem of corruption, sometimes at high levels, has become a matter of concern as Laos has opened to foreign investors. Economic reform has not been accompanied by political liberalization. Tensions continue to be felt between the government and the rural areas, where there is particular resistance to attempts to alter traditional farming methods.

LAOS

Total Land Area : 236 800 sq. km (91 428 sq. miles)

0   100 km
0   100 miles

LAND HEIGHT

2000m/6562ft
1000m/3281ft
500m/1640ft
75m/246ft

POPULATION

◎ over 100 000
○ over 50 000
● over 10 000
• under 10 000

WORLD AFFAIRS   ▷ Joined UN in 1955

Vietnam was Laos's most important ally from 1975 until the late 1980s, when the LPRP began to seek improved relations with Thailand and the West.

AID   ▷ Recipient

 $281m (receipts)   Down 5% in 2000

Laos has one of the highest per capita aid inflows in the developing world, Japan being a major donor.

DEFENSE   ▷ Compulsory military service

 $19m   Down 14% in 2000

The armed forces are estimated by the West to number around 30,000 personnel. This total is further swelled by a paramilitary militia. Eighteen months' military service is compulsory for all Laotian men.

ECONOMICS   ▷ Inflation 27% p.a. (1990–2000)

 $1.52bn   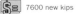 7600 new kips

SCORE CARD

❏ WORLD GNP RANKING	144th
❏ GNP PER CAPITA	$290
❏ BALANCE OF PAYMENTS	–$121m
❏ INFLATION	25.1%
❏ UNEMPLOYMENT	6%

EXPORTS

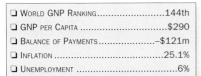

Belgium 3%
Germany 5%
France 6%
Thailand 16%
Other 37%
Vietnam 33%

IMPORTS

Singapore 5%
China 5%
France 4%
Other 13%
Thailand 59%
Vietnam 14%

STRENGTHS
Rising levels of investment from overseas. Potential of garment manufacturing, timber plantations, mining, wood processing, tourism, banking, and aviation. Minerals and possible oil and gas deposits.

WEAKNESSES
One of the world's least developed countries. Lack of technical expertise. Imbalance in sources of foreign investment – most is Thai. Problems in targeting aid efficiently.

The change was mainly due to the need for foreign aid. In 1992, Laos acceded to the Treaty of Amity and Concord of ASEAN, marking the beginning of a new relationship with former adversaries. Laos was admitted to full membership of ASEAN in 1997.

RESOURCES   ▷ Electric power 256,000 kw

 60,403 tonnes   Not an oil producer

 1.7m ducks, 1.5m pigs, 14m chickens   Tin, gypsum, iron, coal, copper, potash, lead, limestone, antimony

Laos's most important agricultural resources are timber and coffee. It has awarded foreign countries concessions to mine for gold and precious stones.

ENVIRONMENT   ▷ Sustainability rank: 32nd

 None   0.1 tonnes per capita

Bombing and the use of defoliants in the Vietnam War did serious ecological damage. Slash and burn farming and illegal logging are destroying forests.

MEDIA   ▷ TV ownership low

 Daily newspaper circulation 4 per 1000 people

PUBLISHING AND BROADCAST MEDIA

There are 3 daily newspapers, including the government-published *Vientiane Mai*

2 state-owned services

2 services: 1 fully state-owned, 1 part state-owned

Newspapers are owned and controlled by the LPRP; one is published by the Lao People's Army. Revelations of corruption by state officials are not uncommon, but criticism of the party and its leaders remains taboo.

CRIME   ▷ Death penalty in use

 Laos does not publish prison figures   Rising overall, particularly corruption

Laos is the world's third-largest opium producer. The US has provided funds to replace poppies with alternative cash crops in the northeast provinces.

Farm in northeastern Laos. *The only lowlands are along the Mekong River. Three-quarters of Laotians are subsistence farmers.*

EDUCATION   ▷ School leaving age: 11

 49%   12,732 students

Adult education is being expanded. Since 1990, private schools have been allowed, in order to help meet demand.

HEALTH   ▷ Welfare state health benefits

1 per 5000 people   Diarrheal, respiratory, and parasitic diseases, malaria, influenza

Since 1975, public health care has developed steadily. Malaria and hemorrhagic fever are on the increase.

SPENDING   ▷ GDP/cap. increase

CONSUMPTION AND SPENDING

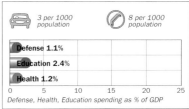

3 per 1000 population   8 per 1000 population

Defense 1.1%
Education 2.4%
Health 1.2%

Defense, Health, Education spending as % of GDP

While a rapidly expanding group of entrepreneurs profits from the gradual liberalization of the country's economy, many in highland and mountainous regions lead a subsistence existence.

WORLD RANKING

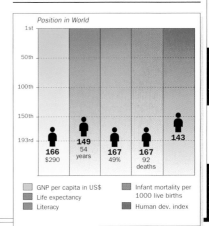

Position in World

166 $290	GNP per capita in US$
149 54 years	Life expectancy
167 49%	Literacy
167 92 deaths	Infant mortality per 1000 live births
143	Human dev. index

L

LATVIA

OFFICIAL NAME: Republic of Latvia **CAPITAL:** Riga
POPULATION: 2.4 million **CURRENCY:** Lats **OFFICIAL LANGUAGE:** Latvian

LATVIA IS ONE of the three Baltic states (with Estonia to the north and Lithuania to the south) which regained independence from Soviet rule in 1991. It lies on a low plain, which nowhere rises above 300 m (975 ft). Almost one-third of the population lives in the capital, Riga. Defense-related industries and agriculture play an important role in the economy. Only just over half of the population are ethnic Latvians.

CLIMATE ▷ Continental

WEATHER CHART FOR RIGA

Latvia's coastal position moderates its continental-type climate, but winters are cold and summers are cool.

TRANSPORTATION ▷ Drive on right

Riga International
574,870 passengers

186 ships
118,000 grt

THE TRANSPORTATION NETWORK

22,843 km
(14,194 miles)

None

2413 km
(1499 miles)

300 km
(186 miles)

Riga has the busiest container port in the Baltic. The EU-backed Via Baltica highway, linking Poland and Finland, runs north–south through Latvia. An east–west link of similar standard is a priority.

LATVIA

Total Land Area :
64 589 sq. km
(24 938 sq. miles)

POPULATION

⊙ over 500 000
◎ over 100 000
○ over 50 000
● over 10 000
• under 10 000

LAND HEIGHT

200m/656ft

Sea Level

The Russian Orthodox Cathedral in Riga. Used as a planetarium during the Soviet era, its interior has now been restored.

TOURISM ▷ Visitors : Population 1:4.9

 490,000 visitors

 Down 14% in 1999

MAIN TOURIST ARRIVALS

| Lithuania 30% |
| Estonia 29% |
| Russia 8% |
| Other 33% |

% of total arrivals

Riga is the main tourist destination, with many hotels and restaurants. Its medieval center is being restored.

PEOPLE ▷ Pop. density low

Latvian, Russian

37/km²
(96/mi²)

THE URBAN/RURAL POPULATION SPLIT

69%   31%

ETHNIC MAKEUP

Polish 2%   Ukrainian 3%
Other 2%   Belarussian 4%
Latvian 57%   Russian 32%

Latvians make up just over half the population, but are a minority in Riga. Naturalization procedures were simplified in 1998, easing tension with the country's large minority population of ethnic Russians, but a restrictive language law reopened controversy. In mid-2000 Latvian was proclaimed the only official language for the public and private sectors. The divorce rate is high.

POLITICS ▷ Multiparty elections

1998/2002

President Vaira
Vike-Freiberga

AT THE LAST ELECTION
Parliament 100 seats

24%	21%	17%	16%	14%	8%
TP	LC	TB/LNNK	TSP	LSDA	JP

TP = People's Party **LC** = Latvia's Way **TB/LNNK** = Fatherland and Freedom **TSP** = National Harmony Party
LSDA = Social Democratic Party **JP** = New Party

The 1998 general election boosted center-right parties, all in favor of EU membership and continuing market reforms. The largest single party was the recently formed TP, led by former premier Andris Skele, whose 1995–1997 government prepared the way for economic recovery. The TP was excluded from the new government, but this coalition was short-lived, the populist Skele taking over again in 1999. In May 2000, however, Riga's high-profile mayor, Andris Berzins, was appointed prime minister as Skele's coalition disintegrated.

L

WORLD AFFAIRS Joined UN in 1991

Latvia is one of six "second wave" candidates for admission to the EU which began membership negotiations in 2000. It also has US backing for entry to NATO. Discrimination against Russian-speakers, exemplified by laws passed in 2000 making Latvian the only official language, have strained relations with Russia.

AID ▷ Recipient

 $91m (receipts)   Down 9% in 2000

Aid to Latvia comes mainly from the World Bank, the IMF, and the EU. Most goes toward improving the country's infrastructure.

DEFENSE ▷ Compulsory military service

🏦 $70m   ⬆ Up 21% in 2000

Building up the military is a priority, and there is now US backing for entry into NATO. Latvia has been participating in NATO's Partnership for Peace program. In February 2000, Russian forces finished dismantling their last military installation in Latvia, the Skrunda radar station.

ECONOMICS ▷ Inflation 49% p.a. (1990–2000)

 $6.93bn   📈 0.6178–0.6292 lats

SCORE CARD

❑ WORLD GNP RANKING	97th
❑ GNP PER CAPITA	$2920
❑ BALANCE OF PAYMENTS	–$485m
❑ INFLATION	2.7%
❑ UNEMPLOYMENT	8%

STRENGTHS
Industrial production improving after slump. Service sector now providing more than half of GDP. Inflation under control. Foreign investment rising.

WEAKNESSES
Dependence on imported oil and natural gas for energy. Lack of raw materials. Farming is technically backward after the dismantling of the collective farms.

EXPORTS
IMPORTS

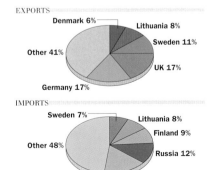

RESOURCES ▷ Electric power 2.1m kw

 125,857 tonnes   🛢 Not an oil producer

🐷 437,000 turkeys, 393,500 pigs, 3.11m chickens   💎 Amber, dolomite, gravel, gypsum, limestone, peat, sand

Latvia has very limited natural resources, and is dependent on imports (mainly from Russia) to meet its energy needs. Peat is still burned for energy. Electricity comes chiefly from hydroelectric power and imports from Lithuania and Estonia. Offshore oil exploration is planned. Ventspils deepwater port is being promoted as an oil terminal.

ENVIRONMENT ▷ Sustainability rank: 10th

🌲 13%   ⬇ 3.2 tonnes per capita

Peat extraction has damaged valuable bog habitat. Pollution of the Baltic Sea and air and water quality in industrial centers are also of concern.
Environmental issues featured in pre-independence protests and have retained a high profile.

MEDIA ▷ TV ownership high

📄 Daily newspaper circulation 247 per 1000 people

PUBLISHING AND BROADCAST MEDIA

There are 23 daily newspapers, including *Diena* and *Neatkariga Rita Avize*

2 services:
1 state-owned,
1 independent

16 services:
1 state-owned,
15 independent

The press is now relatively free from state interference. Previously, the media were predominantly in Russian. Since 1991 the state, aiming to broaden the use of the official language, has actively promoted Latvian publications.

CRIME ▷ Death penalty not used in practice

🔒 8603 prisoners   ⬆ Up 41% in 1999

Organized crime is a growing problem, but general crime levels are lower than in other former Soviet states.

EDUCATION ▷ School leaving age: 16

📖 99%   🎓 101,278 students

Nearly three-quarters of pupils receive teaching solely in Latvian. Schools have opened for many minority ethnic groups.

CHRONOLOGY

Latvia was dominated by Germany and, briefly, Sweden before Russia completed its conquest in 1795.

- ❑ **1917** Opposes Russian Bolshevik revolution. Declares independence.
- ❑ **1918–1920** Invaded by Bolsheviks and Germany.
- ❑ **1920** Gains independence.
- ❑ **1944** Incorporated into USSR.
- ❑ **1989** Popular Front wins elections; declares independence.
- ❑ **1991** Independence recognized.
- ❑ **1995** TP-led coalition formed.
- ❑ **1998** Elections; LC-led coalition. Naturalization procedure eased.
- ❑ **1999** First woman president. Andris Skele of TP returns as premier.
- ❑ **2000** Skele resigns; replaced by Andris Berzins.

HEALTH ▷ Welfare state health benefits

 1 per 357 people   Heart diseases, cancers, accidents, respiratory diseases

The state-run system suffers shortages of medicines and equipment. Some improvements have been made, but it is still seriously underfunded.

SPENDING ▷ GDP/cap. decrease

CONSUMPTION AND SPENDING

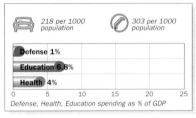
218 per 1000 population   303 per 1000 population

Defense 1%
Education 6.8%
Health 4%

Defense, Health, Education spending as % of GDP

The old bureaucracy has retained its privileged status and contacts, and remains the wealthiest group. Farmers are among the poorest.

WORLD RANKING

L

LEBANON

MIDDLE EAST

Cyprus · Syria

LEBANON

Israel · Jordan

Asia

Africa

OFFICIAL NAME: Republic of Lebanon **CAPITAL:** Beirut
POPULATION: 3.6 million **CURRENCY:** Lebanese pound **OFFICIAL LANGUAGE:** Arabic

1941 · 1941 · Nov 22 · RL · +2 · +961 · .lb

LEBANON LIVES in the shadow of its powerful neighbors, Syria and Israel. The coastal strip is fertile and the hinterland mountainous. The minority Maronite Christians have traditionally dominated the government. Civil war between Muslim and Christian factions from 1975, complicated by an Israeli invasion in 1982, threatened a breakup of the state, until Saudi Arabia brokered a peace deal in 1989. Greater political stability and reconstruction have ensued.

CLIMATE

> Mediterranean/ mountain

WEATHER CHART FOR BEIRUT

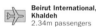

Winters are mild and summers hot, with high humidity on the coast. Snow falls on high ground in the winter.

TRANSPORTATION

> Drive on right

Beirut International, Khaldeh
2.34m passengers

106 ships
263,500 grt

THE TRANSPORTATION NETWORK

| 6200 km (3852 miles) | None |
| 40 km (25 miles) | None |

The redevelopment of Beirut could see it regain its position as one of the Middle East's major entrepôts.

TOURISM

> Visitors : Population 1:4.9

742,000 visitors

Up 10% in 2000

MAIN TOURIST ARRIVALS

| Saudi Arabia 12% |
| Jordan 10% |
| France 9% |
| Other 69% |

0 · 10 · 20 · 30 · 40 · 50 · 60 · 70 · 80
% of total arrivals

Tourists have gradually returned since the devastation of the civil war. Beirut remains the main destination, its battle scars even adding to its attraction. In 1998 the US lifted its restrictions on travel by US citizens to Lebanon.

PEOPLE

> Pop. density high

Arabic, French, Armenian, Assyrian

352/km² (911/mi²)

THE URBAN/RURAL POPULATION SPLIT

90%   10%

RELIGIOUS PERSUASION

Christian 30%

Muslim 70%

The Lebanese population is fragmented in religious terms into subsects of Christians and Muslims, but retains a strong sense of national identity. There has been a large Palestinian refugee population in the country since 1948. Islamic fundamentalism is influential among poorer Shi'a Muslims, who constitute the largest single group.

POLITICS

> Multiparty elections

2000/2004

President Emile Lahoud

AT THE LAST ELECTION

National Assembly 128 seats

| 6% MLU |
| 18% RD | 16% Ind | 14% D | 8% BH | 6% NS | 32% Others |

RD = Resistance and Development List **Ind** = Independents
D = Dignity **BH** = Baalbek–Hermel List **NS** = National Struggle List **MLU** = Mount Lebanon Unity

The Arab-brokered 1989 Taif peace agreement ending the civil war redressed the constitutional balance between Christians and Muslims and guaranteed power-sharing. Relative stability has been maintained under Rafiq al-Hariri, who won the first postwar legislative elections in 1992. After a brief term as premier by Salim al-Hoss, Hariri was elected to a third term by the National Assembly in 2000. Gen. Emile Lahoud was elected president in 1998. Syria remains the main power broker in Lebanon, especially following the withdrawal of Israeli troops in 2000.

LEBANON

Total Land Area : 10 400 sq. km
(4015 sq. miles)

0 · 20 km
0 · 20 miles

El Mina
Tripoli · Zgharta
Amioun · 3087m · Hermel
Batroûn · Bcharré
Jbaïl · Qartaba
Qoubaïyât
Joûnié · Baalbek
BEIRUT
Baabda · Zahlé
Aaley · Rayak
Bhamdoun
Damoûr · Aanjar
Saïda · Lac de Qaraoûn
Jezzine
Rachaïya
Nabatiyé · Marjayoun
Soûr · El Khiyam
Tibnine · GOLAN HEIGHTS
En Nâqoûra
Bent Jbaïl

S Y R I A
MEDITERRANEAN SEA
ISRAEL

LAND HEIGHT

3000m/9843ft
2000m/6562ft
1000m/3281ft
500m/1640ft
200m/656ft
Sea Level

POPULATION

over 1 000 000
over 100 000
over 10 000
under 10 000

WORLD AFFAIRS

 Joined UN in 1945

The 1989 Taif Agreement ending the civil war has left Syria with enormous influence in domestic politics. Anti-Israeli rhetoric remains the political lingua franca. The Hezbollah militia frequently skirmished with Israeli occupying forces (and their proxy militias) until the Israeli withdrawal in 2000, and both sides continue to exchange fire across the border, despite the presence of a UN patrol force.

AID

 Recipient

 $197m (receipts)   Up 2% in 2000

The government has sought billions of dollars for reconstruction. The World Bank prefers working through NGOs.

DEFENSE

 Compulsory military service

 $553m   Down 2% in 2000

The army has over 70,000 troops. Hezbollah guerrillas rapidly regained control of southern Lebanon after the Israeli withdrawal in 2000. There is a UN peacekeeping force on the Israeli border. Syrian forces, whose dominant security role had been formalized in 1991, withdrew from Beirut in 2001.

ECONOMICS

 Inflation 17% p.a. (1990–2000)

 $17.4bn   1507.0–1513.8 Lebanese pounds

SCORE CARD

- ❏ WORLD GNP RANKING 72nd
- ❏ GNP PER CAPITA $4010
- ❏ BALANCE OF PAYMENTS ....–$3.07bn
- ❏ INFLATION 0%
- ❏ UNEMPLOYMENT..................................18%

STRENGTHS

Peace will allow Lebanon to regain its position as an Arab center for banking and services. Potentially a major producer of wine and fruit. Tight fiscal policy has kept inflation down.

WEAKNESSES

Dependent on imported oil and gas. Agriculture still at 40% of prewar levels. High public debt. Alleged Syrian "dumping" of cheap products.

EXPORTS

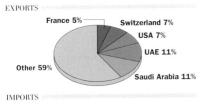

France 5%
Switzerland 7%
USA 7%
UAE 11%
Other 59%
Saudi Arabia 11%

IMPORTS

Switzerland 7%
USA 7%
Germany 8%
France 8%
Other 59%
Italy 11%

RESOURCES

 Electric power 1.4m kw

3860 tonnes

445,000 goats, 380,000 sheep, 32m chickens

Not an oil producer; refines 37,500 b/d

Lignite, iron ore

Wine, cotton, fruit, and vegetables are the main crops. Power plants are fueled by imported petroleum.

ENVIRONMENT

 Sustainability rank: 106th

0.5%   3.9 tonnes per capita

Lack of central authority during the civil war allowed unregulated building, logging, and quarrying to flourish.

MEDIA

 TV ownership high

❌ Daily newspaper circulation 74 per 1000 people

PUBLISHING AND BROADCAST MEDIA

There are 41 daily newspapers, including *Al-Anwar*, *An-Nahar*, and its French companion, *L'Orient-Le Jour*

5 services:
1 state-controlled,
4 independent

1 state-owned service

In the late 1990s the government banned news and political programs on private satellite television channels.

CRIME

Death penalty in use

7296 prisoners   Up 13% in 1999

The kidnapping of hostages and the breakdown of law during the civil war made Beirut a dangerous city for visitors.

Politically motivated violence has recently declined, though the risk of urban terrorism remains. Rural areas untouched by the conflict have low levels of crime.

The Corniche, Beirut, was rebuilt after the civil war by US consultant engineers and architects in a privately financed scheme.

CHRONOLOGY

Under French mandate from 1920, Lebanon declared independence in 1941, achieving full autonomy in 1946.

- ❏ **1975** Civil war erupts.
- ❏ **1982** Israeli invasion.
- ❏ **1989** Taif Agreement ends civil war.
- ❏ **1992** First election in 20 years. Rafiq al-Hariri prime minister.
- ❏ **1996** Israeli attack kills over 100 civilians at UN base in Qana.
- ❏ **1998** Emile Lahoud president.
- ❏ **2000** Israeli forces withdraw. Hariri reelected by a landslide.

EDUCATION

 School leaving age: 11

86%   81,588 students

Lebanon has one of the highest literacy rates in the Arab world. Education was severely disrupted by the war.

HEALTH

 Welfare state health benefits

1 per 476 people   Heart disease, infectious and parasitic diseases

An adequate system of primary health care exists. Hospital staffing is returning to prewar levels.

SPENDING

GDP/cap. increase

CONSUMPTION AND SPENDING

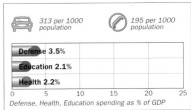

313 per 1000 population   195 per 1000 population

Defense 3.5%
Education 2.1%
Health 2.2%

Defense, Health, Education spending as % of GDP

Average income per capita statistics conceal the fact that a huge gulf exists between the poor and a small, massively rich elite.

WORLD RANKING

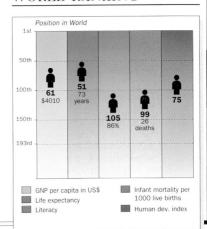

Position in World

61 $4010
51 73 years
105 86%
99 26 deaths
75

GNP per capita in US$
Life expectancy
Literacy
Infant mortality per 1000 live births
Human dev. index

LESOTHO

OFFICIAL NAME: Kingdom of Lesotho **CAPITAL:** Maseru
POPULATION: 2.1 million **CURRENCY:** Loti **OFFICIAL LANGUAGES:** English and Sesotho

SOUTHERN AFRICA

 1966   1966   Oct 4   LS +2   +266   .ls

A MOUNTAINOUS AND landlocked country entirely
surrounded by South Africa, Lesotho is economically
dependent on its larger neighbor. However, Lesotho
is beginning to benefit from the export of energy from the recently
completed Highlands Water Scheme. Elections in 1993 ended a period
of military rule, but South Africa had to send in its troops when
serious political unrest erupted in 1998.

CLIMATE

▷ Mountain

WEATHER CHART FOR MASERU

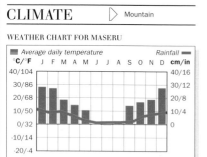

Drought is often followed by torrential
rain storms. Snow is frequent in winter
in the mountains.

TRANSPORTATION

▷ Drive on left

Moshoeshoe International, Maseru
43,000 passengers

Has no fleet

THE TRANSPORTATION NETWORK

887 km (551 miles)		None	
3 km (2 miles)		None	

Lesotho has to rely on South African
road and rail outlets. New roads
have been constructed to service
the Highlands Water Scheme.

TOURISM

▷ Visitors : Population 1:11

186,000 visitors

Up 24% in 1999

MAIN TOURIST ARRIVALS

South Africa 97%	
Other 3%	

0 10 20 30 40 50 60 70 80 90 100
% of total arrivals

Tourists, mainly South African, are
attracted to the dramatic mountain
scenery and watersports on artificially
created lakes. Another draw is Thaba-
Bosiu, King Moshoeshoe the Great's
mountain stronghold. Political violence
in 1998 deterred many visitors.

PEOPLE

▷ Pop. density medium

English, Sesotho, Zulu

69/km² (179/mi²)

THE URBAN/RURAL POPULATION SPLIT

28%   72%

ETHNIC MAKEUP

European and Asian 3%

Sotho 97%

The overwhelming majority of the
population are Sotho, though there is
a small community of European origin,
as well as south Asian and Chinese
minorities, which are active in
the retailing business. Ethnic
homogeneity and a strong sense
of national identity have tended
to minimize ethnic tension.

The export of male contract labor to
South African mines means that women
head 72% of households; they also run
farming, regarded by
Lesotho men as
"women's work."

POLITICS

▷ Multiparty elections

L. House 2002/2007
U. House 1998/2003

H.M. King Letsie III

AT THE LAST ELECTION

National Assembly 120 seats

4% LPC

64% LCD   18% 4% 10%
           BNP NIP Others

LCD = Lesotho Congress for Democracy Party
BNP = Basotho National Party **NIP** = National Independent
Party **LPC** = Lesotho People's Congress

Senate 33 seats

The Senate comprises 22 principal chiefs and 11 other
members named by the king

The armed forces have been key
political players in Lesotho since a
bloodless coup in 1986. Although direct
military rule ended in 1993, and a free
and peaceful general election resulted
in a sweeping victory for the Basotho
Congress Party (BCP), the army
maintained its powers over national
security. Tensions between the army
and the civilian administration
escalated in 1994, when mutinous
troops killed the deputy prime
minister. King Moshoeshoe II
was restored to the throne, and was
succeeded by his son Letsie III in 1996.

Accusations of vote rigging and mass
protests greeted a general election win
in 1998 by the LCD. After an attempted
coup in September, the South African
military intervened to restore
democracy, brokering an
agreement between
the king and Lesotho's 12
parties. When elections
were finally held in
2002, the LCD
retained its
majority.

LESOTHO

Total Land Area : 30 355 sq. km
(11 720 sq. miles)

POPULATION

⊙ over 100 000
• under 10 000

LAND HEIGHT

5000m/9843ft
2000m/6562ft
1000m/3281ft

L

WORLD AFFAIRS

Joined UN in 1966

 Comm ACP NAM AU SADC

Foreign policy is dominated by the nature of Lesotho's relationship with South Africa. Lesotho currently has duty-free access to the EU for most manufactured goods, and also has preferential access to US and Scandinavian markets.

AID

Recipient

 $41m (receipts)   Up 32% in 2000

Aid, mostly devoted to agricultural development, accounts for 26% of Lesotho's GNP; about half comes from SACU. A "national famine" was declared in 2002 to prompt donations of emergency assistance.

DEFENSE

No compulsory military service

 $29m   Down 15% in 2000

Lesotho's 2000-strong army relied on South African assistance to quell political violence in 1998.

ECONOMICS

Inflation 9.9% p.a. (1990–2000)

 $1.18bn   7.570–11.995 maloti

SCORE CARD

- ❏ WORLD GNP RANKING......................150th
- ❏ GNP PER CAPITA $580
- ❏ BALANCE OF PAYMENTS....................–$151m
- ❏ INFLATION 6.1%
- ❏ UNEMPLOYMENT.................................45%

STRENGTHS
Potential of educated workforce. Boom in textiles and other manufacturing. Membership of SACU. Future revenues from water sales.

WEAKNESSES
Dependent on South Africa. Loss of workers to mining in South Africa. Weak agricultural sector. Retail sector affected by sporadic unrest. High unemployment.

EXPORTS

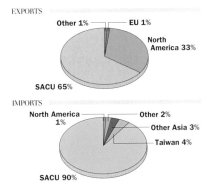

Other 1%   EU 1%
North America 33%
SACU 65%

IMPORTS

North America 1%   Other 2%
Other Asia 3%
Taiwan 4%
SACU 90%

Landscape near Mohales Hoek in Lesotho's lowest lands – over 1300 m (4260 ft) above sea level.

RESOURCES

Electric power: Included in South African total

 34 tonnes   Not an oil producer

 730,000 sheep, 570,000 goats, 1.7m chickens   Diamonds

The Highlands Water hydroelectric scheme has the capacity to supply all of Lesotho's energy requirements, as well as 62 cu. m (2200 cu. ft) of water per second for South African use. Diamonds are mined in the northeast.

ENVIRONMENT

Not available

 0.2%   Negligible emissions per capita

Climate and overgrazing have seriously eroded the land. The Highlands Water Scheme has flooded acres of peasant farmland. Supporters of this massive dam project stress encouragement for wildlife in reservoirs and on bird-friendly pylons.

MEDIA

TV ownership low

Daily newspaper circulation 8 per 1000 people

PUBLISHING AND BROADCAST MEDIA

There are 2 daily newspapers. *Leselinyana la Lesotho* is a popular religious periodical

1 state-owned service   1 state-owned service

State control of the media was used for political ends in 1998. The *Mirror* is the only independent paper.

CRIME

Death penalty in use

 2552 prisoners   Up 25% 1993–1999

The 1998 political crisis increased crime levels, which previously were much lower than in South Africa.

EDUCATION

School leaving age: 13

 83%   4046 students

Lesotho's schools have very high enrollment levels and one of the highest literacy rates in Africa.

CHRONOLOGY

As Basutoland, Lesotho became a British Crown colony in 1884.

- ❏ **1966** Independent kingdom.
- ❏ **1986** Military coup.
- ❏ **1990** King Moshoeshoe II exiled. Son installed as Letsie III.
- ❏ **1993** Free elections.
- ❏ **1994** Return of Moshoeshoe II.
- ❏ **1996** Letsie III succeeds to throne.
- ❏ **1998** New LCD wins polls. South Africa intervenes after coup attempt, and reconciles king and parties.
- ❏ **2002** Food emergency follows successive poor harvests. LCD wins long-postponed elections.

HEALTH

Welfare state health benefits

 1 per 10,000 people   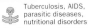 Tuberculosis, AIDS, parasitic diseases, nutritional disorders

Private health organizations and NGOs account for half of all health services. A government-operated flying doctor service covers the highlands. An estimated 25% of adults are HIV positive and life expectancy is declining. Food shortages in 2002 exacerbated the country's health problems.

SPENDING

GDP/cap. increase

CONSUMPTION AND SPENDING

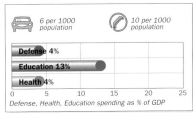

6 per 1000 population   10 per 1000 population

Defense 4%
Education 13%
Health 4%

0   5   10   15   20   25
Defense, Health, Education spending as % of GDP

Social mobility is limited in Lesotho; the ruling elite keeps a tight control on power and wealth. Over 90% of the population live below the poverty line and many are migrant laborers.

WORLD RANKING

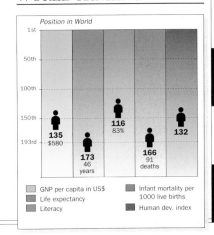

Position in World

135 $580
173 46 years
116 83%
166 91 deaths
132

- GNP per capita in US$
- Life expectancy
- Literacy
- Infant mortality per 1000 live births
- Human dev. index

L

LIBERIA

OFFICIAL NAME: Republic of Liberia **CAPITAL:** Monrovia
POPULATION: 3.1 million **CURRENCY:** Liberian dollar **OFFICIAL LANGUAGE:** English

FOUNDED IN 1847 by freed US slaves, Liberia today is struggling to recover from a civil war which reduced it to anarchy in the 1990s but new rebel factions have relaunched the fighting. Facing the Atlantic in equatorial west Africa, most of its coastline is characterized by lagoons and mangrove swamps. Inland, a grassland plateau supports the limited agriculture (just 1% of land is arable). Liberia has the world's largest flag of convenience merchant fleet.

CLIMATE   ▷ Tropical equatorial

WEATHER CHART FOR MONROVIA

There is one long rainy season from May to October, with a brief interlude in most of the country of about two weeks in August.

Temperatures are consistently high. During the dry season, when the dust-laden *harmattan* wind blows, they rise even higher inland.

TRANSPORTATION   ▷ Drive on right

Roberts Field International, Monrovia

1717 ships
60.5m grt

THE TRANSPORTATION NETWORK

657 km (408 miles)

None

480 km (298 miles)

None

Most roads in Liberia are unpaved. The 480-km (298-mile) railroad was built to transport iron ore and carries little other traffic. Roberts Field airport was built by the US during World War II.

TOURISM   ▷ Not available

Tourists deterred by civil war

Little change from year to year

MAIN TOURIST ARRIVALS

Liberia does not publish tourism figures by country of origin

% of total arrivals

As a result of the civil war and continuing insecurity, tourism, never a significant activity in Liberia, is now nonexistent.

PEOPLE   ▷ Pop. density low

Kpelle, Vai, Bassa, Kru, Grebo, Kissi, Gola, Loma, English

33/km² (83/mi²)

THE URBAN/RURAL POPULATION SPLIT

45%   55%

ETHNIC MAKEUP

Americo-Liberians 5%

Indigenous tribes (16 main groups) 95%

A key distinction in Liberia has been between Americo-Liberians, the descendants of those freed from slavery (known as "civilized persons"), and the majority indigenous "tribals." The latter were long held in contempt by the Americos, but intermarriage and political assimilation since 1944 have softened attitudes. Intertribal tension in Liberia is now a much more serious problem, and was the main cause of the civil war which erupted in 1990.

POLITICS   ▷ Multiparty elections

L. House 1997/2003
U. House 1997/2006

President Charles Taylor

AT THE LAST ELECTION

House of Representatives 64 seats

76% NPP
5% ALCP
11% UP
8% Others

NPP = National Patriotic Party **UP** = Unity Party
ALCP = All Liberia Coalition Party

Senate 26 seats

81% NPP
11% UP
8% ALCP

Liberian politics collapsed after 1990 into a chaotic, bloody, and many-sided conflict. A peace agreement finally signed in 1996 provided for presidential and legislative elections in 1997, won by Charles Taylor and his NPP, formerly the National Patriotic Front of Liberia (NPFL), the predominant armed faction. Some 700,000 refugees began returning, but instability continued, and intense fighting broke out in the north in mid-2000 following the formation of a new rebel faction, Liberians United for Reconciliation and Democracy. Thousands of civilians fled the area in 2001 and 2002 as fighting escalated.

LIBERIA

Total Land Area : 111 370 sq. km
43 000 sq. miles

POPULATION

over 100 000
over 10 000
under 10 000

LAND HEIGHT

1000m/3281ft
500m/1640ft
200m/656ft
Sea Level

WORLD AFFAIRS
▷ Joined UN in 1945

 ACP ECOWAS IAEA NAM AU

The ECOWAS peacekeeping force ECOMOG, backed chiefly by Nigeria and Ghana, arrived in 1990, and it was ECOWAS that kept the UN-led peace process on track in the mid-1990s. The UN imposed sanctions in 2001, accusing Liberia of fomenting war in west Africa. Relations with neighboring states, badly affected by these conflicts, were patched up symbolically in February 2002.

AID
▷ Recipient

 US$68m (receipts) ⬇ Down 28% in 2000

Liberia remains heavily dependent on foreign aid. In mid-2000 the EU cut an aid package, accusing the regime of selling arms to rebels in Sierra Leone.

DEFENSE
▷ No compulsory military service

 US$25m ⬍ No change in 2000

Peace agreements in 1995 and 1996 provided for the demobilization of the various warring factions and the formation of a single national army.

ECONOMICS
▷ Not available

 US$1.03bn 1 Liberian dollar

SCORE CARD
- ❏ WORLD GNP RANKING......................153rd
- ❏ GNP PER CAPITA ....US$330
- ❏ BALANCE OF PAYMENTS.............–US$145m
- ❏ INFLATION 5%
- ❏ UNEMPLOYMENT.................................70%

STRENGTHS
Potential for reviving the Firestone rubber plantation and huge LAMCO iron ore mine. Tropical timber, but reserves are declining.

WEAKNESSES
Little commercial activity, and low confidence. Political instability. UN sanctions on diamond trade from 2001.

EXPORTS
USA 7%, Italy 8%, Norway 9%, Belgium 19%, Germany 26%, Other 31%
IMPORTS
Germany 7%, Singapore 9%, France 29%, Japan 15%, South Korea 22%, Other 18%

Village near Gbarnga. *The Kpelle, the largest of Liberia's 16 indigenous ethnic groups, are concentrated in this part of Liberia.*

RESOURCES
▷ Electric power 334,000 kw

15,472 tonnes Not an oil producer
220,000 goats, 210,000 sheep, 4m chickens Iron ore, diamonds, gold, barytes, kyanite, columbite, manganese

Liberia has an estimated billion tonnes of iron ore reserves at Mount Nimba, but even when peaceful conditions return, the current state of world demand would barely justify exploitation.

ENVIRONMENT
▷ Sustainability rank: 130th

1% 0.1 tonnes per capita

The NPFL and other armed groups cut down tropical forests to help finance their armies.

MEDIA
▷ TV ownership low

Daily newspaper circulation 16 per 1000 people

PUBLISHING AND BROADCAST MEDIA

There are 6 daily newspapers, including the independent *Daily Observer*

1 partly state-owned service

10 services: 4 state-owned, 6 independent

The Liberian Communication Network, owned by President Charles Taylor and his party, effectively controls the print and broadcast media.

CRIME
▷ Death penalty in use

Liberia does not publish prison figures Crime is rampant. There are no enforcing agencies

In the 1990s warring factions regularly massacred civilians, press-ganged armies, and displaced thousands into seeking refuge in neighboring states. As fighting erupted again in 2001, it was government forces that were accused of human rights abuses.

EDUCATION
▷ School leaving age: 16

54% 20,804 students

Originally based on the US model, the education system effectively collapsed during the civil war.

CHRONOLOGY
Between 1816 and 1892, 22,000 liberated slaves, most from the US, settled in Liberia, established as a republic in 1847.

- ❏ **1980** Coup. President assassinated by Samuel Doe.
- ❏ **1990** Outbreak of civil war.
- ❏ **1991** Doe assassinated.
- ❏ **1996** Second peace agreement.
- ❏ **1997** Charles Taylor president.
- ❏ **2001** Borders with Guinea and Sierra Leone closed. Conflict with rebels escalates.
- ❏ **2002** State of emergency declared.

HEALTH
▷ No welfare state health benefits

1 per 20,000 people Communicable, diarrheal, parasitic, and heart diseases

Very few people have access to basic health care. Liberia's infant mortality rate remains among the highest in the world.

SPENDING
▷ GDP/cap. increase

CONSUMPTION AND SPENDING

3 per 1000 population 2 per 1000 population
Defense 5.6%
Education 5.7%
Health 0.8%
Defense, Health, Education spending as % of GDP

Power and wealth have a very direct connection in Liberia. Both the Americo-Liberian regimes and the Doe regime which replaced them, saw the state as a source of plunder in the form of well-paid jobs and kickbacks from contracts. The factions in the 1990–1996 civil war sought similar power. Most ordinary Liberians live in rural poverty.

WORLD RANKING

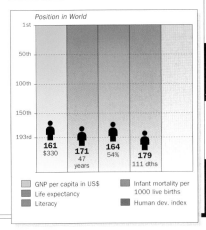
Position in World
161 $330 — GNP per capita in US$
171 47 years — Life expectancy
164 54% — Literacy
179 111 dths — Infant mortality per 1000 live births
Human dev. index

L

LIBYA

OFFICIAL NAME: Great Socialist People's Libyan Arab Jamahiriyah
CAPITAL: Tripoli   POPULATION: 5.4 million   CURRENCY: Libyan dinar   OFFICIAL LANGUAGE: Arabic

NORTH AFRICA

LIBYA IS SITUATED between Egypt and Algeria on the Mediterranean coast of north Africa, with Chad and Niger on its southern borders. Apart from the coastal strip and the mountains in the south, it is desert or semidesert. Libya's strategic position in north Africa and its abundant oil and gas resources made it an important trading partner for European states. It has for many years been politically marginalized by the West for its links with terrorist groups, but UN sanctions were suspended in 1999, when it handed over the two men suspected of the 1988 Lockerbie bombing.

Roman amphitheater, Sabrata. *Libya's impressive classical heritage testifies to its importance in ancient times.*

CLIMATE ▷ Hot desert

WEATHER CHART FOR TRIPOLI

  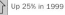

The coastal region has a warm, temperate climate, with mild, wet winters and hot, dry summers.

TRANSPORTATION ▷ Drive on right

 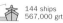
Tripoli International   144 ships 567,000 grt

THE TRANSPORTATION NETWORK

| 47,590 km (29,571 miles) | None |
| None | None |

The National Coast Road runs 1825 km (1135 miles) between the Tunisian and Egyptian borders, linking the principal urban centers. There are no railroads, but some are planned. Since sanctions were lifted in 1999, international airlines have resumed flights to Libya.

Al Kufrah Oasis. *As 90% of Libya is arid rock and sand, oases provide essential agricultural land, besides being tourist attractions.*

TOURISM ▷ Visitors : Population 1:135

40,000 visitors   Up 25% in 1999

Libya possesses a rich Roman and Greek heritage, centered on the ancient Roman coastal towns of Labdah (Leptis Magna) and Sabrata near Tripoli, and Shahhat (Cyrene) further east. There are fine beaches at Tripoli. A $2–3 billion investment program launched in 2000 aims to attract thousands of visitors. Western tourists have begun to return since sanctions were lifted in 1999.

MAIN TOURIST ARRIVALS

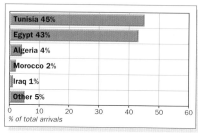

Tunisia 45%
Egypt 43%
Algeria 4%
Morocco 2%
Iraq 1%
Other 5%
% of total arrivals

PEOPLE ▷ Pop. density low

Arabic, Tuareg   3/km² (8/mi²)

THE URBAN/RURAL POPULATION SPLIT

88%   12%

RELIGIOUS PERSUASION

Other 3%
Muslim (mainly Sunni) 97%

ETHNIC MAKEUP

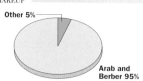

Other 5%
Arab and Berber 95%

Arabs and Berbers, split into many tribal groupings, form 95% of the population. They were artificially brought together when Libya was created in 1951 by the unification of three historic Ottoman provinces. The pro-Western monarchy then perpetuated the dominance of Cyrenaican tribes and the Sanusi religious order.

The 1969 revolution brought to the fore Arab nationalist Col. Muammar al-Gaddafi, who embodied the character

POPULATION AGE BREAKDOWN

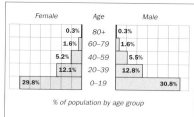

Female	Age	Male
0.3%	80+	0.3%
1.6%	60–79	1.6%
5.2%	40–59	5.5%
12.1%	20–39	12.8%
29.8%	0–19	30.8%

% of population by age group

and aspirations of the rural Sirtica tribes from Fazzan: fierce independence, deep Islamic convictions, belief in a communal lifestyle, and hatred for the urban rich. His revolution wiped out private enterprise and the middle class, banished European settlers and Jews, undermined the religious Muslim establishment, and imposed a form of popular democracy through the *jamahiriyah* (state of the masses). However, resentment of the regime grew as it became clear that power now lay mainly with the Sirtica tribes, especially Gaddafi's own clan, the Qadhadhfa.

Since the revolution, Libya has become a society where most are city dwellers. Jews have been invited to return as investors, and immigrants from sub-Saharan Africa have been drawn in to provide low-cost labor. However, clashes in 2000, in which 100 died, highlighted unresolved social issues.

L

POLITICS

 No multiparty elections

 Not applicable

Leader of the Revolution
Col. Muammar al-Gaddafi

LEGISLATIVE OR ADVISORY BODIES

General People's Congress 750 seats

The constitution makes no provision for direct elections. Last renewal May 2000

Executive power is exercised by the General People's Committee. The General People's Congress elects the head of state, the Leader of the Revolution.

PROFILE

In 1977, a new form of direct democracy was promulgated, through which some 2000 People's Congresses sought to involve every adult in policy-making. In theory, their wishes are carried out by popular committees. In practice, ultimate control rests with Col. Gaddafi and his collaborators, many of whom date from the 1969 revolution. In recent years some are thought to have been alienated from Gaddafi, including his deputy, Maj. Abdessalem Jalloud, who in 1994 was reportedly marginalized after expressing differences with him. In 1995, another of Gaddafi's close associates, Khoueldi Hamidi, a defense commander, was also said to have become disillusioned with Gaddafi. In 2000, Gaddafi embraced African unity – an unpopular concept among most Libyans, increasing Gaddafi's alienation from his fomer associates. He is now believed to rely on members of his own clan, particularly his five sons.

MAIN POLITICAL ISSUES
Repression
Political dissidents, including Islamist militants, have been violently suppressed. Libyan dissidents have been murdered abroad, allegedly by government agents. Political parties were banned in 1971, but opposition groups are active in Egypt and Sudan.

The regime's public image
In the past few years, the regime has made an effort to improve its image. Measures have included freeing political prisoners, welcoming back exiles, permitting foreign travel, and accepting responsibility for past acts of terrorism.

Col. Gaddafi,
Libya's leader since 1969, shies from official titles.

Ex-king Idris
was deposed by Col. Gaddafi in 1969.

WORLD AFFAIRS

Joined UN in 1955

 AL   AMU   NAM   OIC   OPEC

Gaddafi has attempted to style himself as the champion of African integration and regional stability. He was a chief architect of the African Union, and has hosted various peace negotiations in the trans-Saharan area. This transformation has also involved a less confrontational stance toward the West, dropping key policies which had left Libya internationally isolated in the past – support for various terrorist groups and strong opposition to Israel – and pursuing a crackdown on internal corruption. UN sanctions imposed in 1992 were eased, and relations with the UK were resumed in 1999. Gaddafi tacitly approved the US-led "war on terrorism" in 2001. However, the US still added Libya to the "axis of evil" in 2002.

L

CHRONOLOGY

Italy occupied Libya and expelled the Turks in 1911. Britain and France agreed to a UN plan for an independent monarchy in 1951.

❑ **1969** King Idris deposed in coup by Revolutionary Command Council led by Col. Gaddafi. Tripoli Charter sets up revolutionary alliance with Egypt and Sudan.

❑ **1970** UK and US military ordered out. Property belonging to Italians and Jews confiscated. Western oil company assets nationalized, a process completed in 1973.

❑ **1973** Libya forms abortive union with Egypt. Gaddafi launches Cultural Revolution. Libya occupies Aozou Strip in Chad.

❑ **1974** Libya forms union of Libya and Tunisia.

❑ **1977** Official name changed to the Great Socialist People's Libyan Arab Jamahiriyah.

❑ **1979** Members of Revolution Command Council replaced by elected officials. Gaddafi remains Leader of the Revolution. ➪

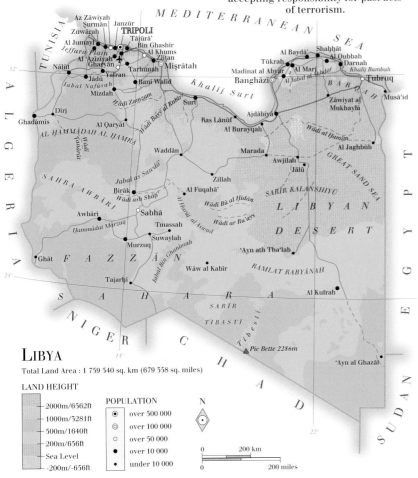

LIBYA

Total Land Area : 1 759 540 sq. km (679 358 sq. miles)

LAND HEIGHT

2000m/6562ft
1000m/3281ft
500m/1640ft
200m/656ft
Sea Level
-200m/-656ft

POPULATION

⊙ over 500 000
◎ over 100 000
○ over 50 000
● over 10 000
• under 10 000

N

0   200 km

0   200 miles

CHRONOLOGY *continued*

- ❑ **1981** US shoots down two Libyan aircraft over Gulf of Sirte.
- ❑ **1984** Gunman at Libyan embassy in UK kills British policewoman; UK severs diplomatic relations (until 1999). Oudja Accord signed with Morocco for Arab Africa Federation.
- ❑ **1985** Libya expels 30,000 foreign workers. Tunisia cuts diplomatic links.
- ❑ **1986** US aircraft bomb Libya, killing 101 people and destroying Gaddafi's residence.
- ❑ **1988** Pan-Am airliner explodes over Lockerbie, Scotland; allegations of Libyan complicity.
- ❑ **1988–1989** Army restructured.
- ❑ **1989** Arab Maghreb Union established with Algeria, Morocco, Mauritania, and Tunisia. Cease-fire in Aozou Strip.
- ❑ **1990** Libya expels Palestinian splinter group led by Abu Abbas.
- ❑ **1991** Opening of first branch of Great Man-Made River project.
- ❑ **1992–1993** UN sanctions imposed as Libya fails to hand over Lockerbie suspects; sanctions made stricter.
- ❑ **1994** Religious leaders obtain right to issue religious decrees (*fatwas*) for first time since 1969. Return of Aozou strip to Chad.
- ❑ **1996** US legislation imposes penalties on foreign companies investing in Libya's energy sector.
- ❑ **1999** Lockerbie suspects handed over for trial in the Netherlands under Scottish law; UN sanctions eased.
- ❑ **2001** Lockerbie trial verdict: one suspect convicted. Sanctions eased further.
- ❑ **2002** US–Libya talks aim to mend relations. Lockerbie bomber begins life sentence.

AID

$15m (receipts)   Up 114% in 2000

As an oil-exporting state, Libya fails to qualify for much international aid, despite being a developing country. During the 1970s, Col. Gaddafi aided several African liberation movements, such as the ANC in South Africa. He backed Hissène Habré's forces in Chad, and helped dissidents by training them in his Pan-African legion. He has also financed or supplied arms to the PLO in the Middle East, Irish republicans in Northern Ireland, the Moros in the southern Philippines, and the Basques, Corsicans, and other separatist causes in Europe. In 1993, Libya granted aid totaling $27 million, despite UN sanctions and a lack of surplus resources.

DEFENSE

 Compulsory military service

$1.18bn   Down 10% in 2000

LIBYAN ARMED FORCES

🪖	2210 main battle tanks (1040 T-54/55, 280 T-62, 145 T-72)	45,000 personnel
🚢	1 submarine, 2 frigates, and 16 patrol boats	8000 personnel
✈	372 combat aircraft (40 MiG-23BN, 15 MiG-23U, 58 *Mirage*, 53 Su-20/22)	23,000 personnel
	None	

The armed forces suffered a blow in 1987 with the loss of thousands of men and equipment worth $1.4 billion when Libya became embroiled in the Chad civil war; in 1994 it agreed to hand back to Chad the Aozou Strip, which it had occupied in 1973. In 1988–1989 the armed forces were replaced by "the Armed People." Conscription is selective, and can last up to two years. In addition, there is a People's Militia of 40,000.

Attempts to depoliticize the army received a setback following confirmation of an abortive military coup in 1993. UN sanctions resulted in military hardware becoming outdated. Despite the suspension of the sanctions in 1999, fresh arms contracts would still be too controversial for most potential suppliers.

ECONOMICS

 Not available

$29.2bn   0.5423–0.6416 Libyan dinars

SCORE CARD

❑ World GNP Ranking	59th
❑ GNP per Capita	$5220
❑ Balance of Payments	$1.48bn
❑ Inflation	18.5%
❑ Unemployment	30%

EXPORTS

IMPORTS

ECONOMIC PERFORMANCE INDICATOR
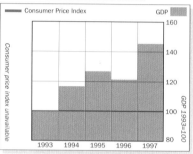

In the 1970s, an ambitious program of industrialization was launched. Gaddafi's most controversial economic project has been the Great Man-Made River. Started in 1984 and engineered by European and Korean companies, this scheme was designed to bring underground water from the Sahara to the coast, but the pipes are already corroding, with water leaking into the sand.

LIBYA : MAJOR BUSINESSES

STRENGTHS

Oil and gas production. High investment in downstream industries – petrochemicals, refineries, fertilizers, and aluminum smelting.

WEAKNESSES

Single-resource economy subject to oil-market fluctuations. Most food is imported. Reliance on foreign labor. Lack of water for agriculture. History of international unreliability.

PROFILE

Western oil companies had close business ties with Libya until the imposition in 1992 of UN sanctions over the Lockerbie affair. In 1993, Gaddafi called for the program of privatization, authorized by the General People's Congress in 1992, to be revived, but there have been few tangible results.

L

RESURCES

 Electric power 4.6m kw

32,550 tonnes

1.43m b/d (reserves 29.5bn barrels)

5.1m sheep, 1.95m goats, 25m chickens

Oil, natural gas, iron, potassium, sulfur, magnesium, gypsum

ELECTRICITY GENERATION

Hydro 0%

Combustion 100% (19bn kwh)

Nuclear 0%

Other 0%

% of total generation by type

Libya's economy depends almost entirely on petroleum and natural gas resources. It has considerable crude oil reserves and is likely to remain an oil-exporting country well into the century. Natural gas potential is more limited but, provided links are developed with other north African states, the future is assured. Libya also has reserves of iron ore, potassium, sulfur, magnesium, and gypsum. The Great Man-Made River project means that the area of irrigated land has grown, but 90% of Libya is desert. Animal husbandry is the basis of farming, but some cereal crops are grown, as well as dates, olives, and citrus fruits. Cement production is sufficient to meet national demand; raw materials are local, but most other manufacturing inputs are imported.

LIBYA : LAND USE

Cropland
Pasture
Desert
Sheep
Dates

0   400 km
0   400 miles

ENVIRONMENT

 Sustainability rank: 124th

 0.1%   7.2 tonnes per capita

ENVIRONMENTAL TREATIES

Yes   No   Yes

Yes   No   No

The UN Development Program has described Libya as more than 90% "wasteland." Both nature and man have conspired against the environment. Apart from two coastal strips – the Jeffara Plain and the Jabal al Akhdar in Cyrenaica – together with the Fazzan Oasis, most of Libya is desert. Much of the irrigated area is saline because of unwise use of naturally occurring water from artesian wells. Near Tripoli, seawater has penetrated the water table as far as 20 km (12 miles) inland.

MEDIA

 TV ownership medium

Daily newspaper circulation 14 per 1000 people

PUBLISHING AND BROADCAST MEDIA

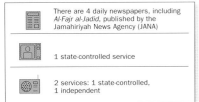

There are 4 daily newspapers, including *Al-Fajr al-Jadid*, published by the Jamahiriyah News Agency (JANA)

1 state-controlled service

2 services: 1 state-controlled, 1 independent

Libya's press and TV are a mouthpiece for the leadership. Satellite TV and the Internet are widely available, but heavily censored. The main daily newspaper is published in Arabic and has a circulation of 40,000 readers. The TV station broadcasts mostly in Arabic. Radio broadcasts in Kiswahili, Hausa, Fulani, and Amharic were due to begin in 1999, but were postponed.

CRIME

 Death penalty in use

6750 prisoners   Up 10% in 1999

CRIME RATES

Murders
2   *per 100,000 population*

Rapes
5   *per 100,000 population*

Thefts
346   *per 100,000 population*

Policing is often in the hands of gangs appointed by Gaddafi's lieutenants to root out student protestors and other dissidents. Hit squads allegedly operate abroad against Libyan exiles.

EDUCATION

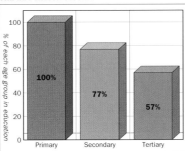 School leaving age: 15

80%   126,348 students

THE EDUCATION SYSTEM

% of each age group in education

100% — Primary
77% — Secondary
57% — Tertiary

Some one million Libyans are in formal education. It is compulsory between the ages of six and 15, and rates of attendance are very high, but it varies in quality and can be rudimentary in rural areas. Secondary education, from the age of 15, lasts for three years. There are 13 universities, and institutes for vocational training. The literacy rate has more than doubled from a level of 39% in 1970.

HEALTH

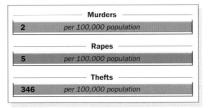 Welfare state health benefits

1 per 769 people   Pneumonia, diarrheal diseases, accidents, cancers

An adequate system of free primary health care exists except in remote areas, and there are two big hospitals, in Benghazi and Tripoli. However, hospitals lack equipment, and there is a shortage of medical supplies.

SPENDING

GDP/cap. decrease

CONSUMPTION AND SPENDING

159 per 1000 population   108 per 1000 population

Defense 3.2%
Education 7.1%
Health 3.4%

0   5   10   15   20   25
Defense, Health, Education spending as % of GDP

There is widespread poverty after years of import constraints; UN sanctions worsened the situation. Gaddafi refuses to use oil revenues for basic expenses, such as salaries – teachers earn about $1200 a year.

WORLD RANKING

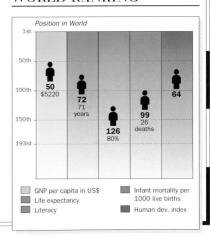

Position in World

1st
50th
100th
150th
193rd

50
$5220

72
71 years

126
80%

99
26 deaths

64

GNP per capita in US$
Life expectancy
Literacy

Infant mortality per 1000 live births
Human dev. index

L

LIECHTENSTEIN

OFFICIAL NAME: Principality of Liechtenstein **CAPITAL:** Vaduz
POPULATION: 32,200 **CURRENCY:** Swiss franc **OFFICIAL LANGUAGE:** German

PERCHED IN THE ALPS, Liechtenstein is rare among small states in having both a thriving banking sector and a well-diversified manufacturing economy. It is closely allied to Switzerland, which handles its foreign relations and defense. Life in Liechtenstein is stable and conservative. The traditional secrecy surrounding the country's financial industry, and low taxes, means that many overseas trusts, banks, and investment companies are located there.

CLIMATE ▷ Mountain

WEATHER CHART FOR VADUZ

Climate varies with altitude. Excellent skiing conditions are the result of heavy settling snow from December to March. Summers are warm and wet.

TRANSPORTATION ▷ Drive on right

 None   Has no fleet

THE TRANSPORTATION NETWORK

| 250 km (155 miles) | None |
| 19 km (12 miles) | 26 km (16 miles) |

Public transportation in Liechtenstein is mostly by the postal bus network. The single-track railroad has few stops. Zürich, a two-hour drive away, is the nearest airport.

TOURISM ▷ Visitors : Population 1.9:1

61,000 visitors   Up 2% in 2000

MAIN TOURIST ARRIVALS

Germany 36%
Switzerland 22%
USA 7%
Other 35%

% of total arrivals

Liechtenstein's alpine scenery attracts skiers in the winter, and climbers and hikers in the summer.

PEOPLE ▷ Pop. density high

German, Alemannish dialect, Italian   201/km² (521/mi²)

THE URBAN/RURAL POPULATION SPLIT

21%   79%

RELIGIOUS PERSUASION

Protestant 7%
Other 12%
Roman Catholic 81%

Liechtenstein's role as a financial center accounts for the many foreign residents (over 35% of the population), of whom half are Swiss and the rest mostly German. The high standard of living results in few ethnic or social tensions. Family life is very traditional; women received the vote only in 1984, after much controversy. A proposal to enshrine equal rights for women in the constitution was rejected in a referendum in 1985 by a large majority, and was only finally passed in 1992.

POLITICS ▷ Multiparty elections

2001/2005   Prince Hans Adam II von und zu Liechtenstein

AT THE LAST ELECTION
Parliament 25 seats

52% FBP   44% VU   4% FL

FBP = Progressive Citizens' Party VU = Fatherland Union
FL = Free List

Between 1938 and 1997 the VU and the FBP alternated as coalition leaders, except briefly in 1993. This partnership ended in 1997 when Mario Frick, at the age of 28 Europe's youngest premier, formed a VU-only government. In elections in 2001 the FBP under Otmar Hasler overtook the VU. Referendums have increasingly been used to decide policy issues.

WORLD AFFAIRS ▷ Joined UN in 1990

 CE   EEA   IAEA   OSCE   WTO

Liechtenstein effectively gave up control of its external relations in 1924 when it signed a Customs Union Treaty with Switzerland. This agreement requires Swiss approval for any treaty arrangements between Liechtenstein and a third state. Liechtenstein became a member of the UN only in 1990. It joined EFTA and the EBRD in 1991, and has been a participant in the EEA since 1995. However, Swiss rejection of EU membership in 1992 effectively ended any prospect of Liechtenstein joining the EU in the foreseeable future.

AID ▷ Donor

Donor, but does not publish figures   Not available

Although overseas aid donations are small and aid issues have little political importance, Liechtenstein has helped to fund shelter and reconstruction projects in former Yugoslavia and local development projects in Bulgaria.

LIECHTENSTEIN

Total Land Area : 160 sq. km (62 sq. miles)

POPULATION
under 10 000 •

LAND HEIGHT
2000m/6562ft
1500m/4921ft
1000m/3281ft
500m/1640ft
400m/1312ft

***Alpine scenery near Vaduz.** The state budget includes 2% allocated to restoring mountain vegetation and coordinating land use.*

DEFENSE

 No compulsory military service

 No defense force

 Not applicable

There has been no standing army since 1868, and there is only a small police force. De facto protection is provided by Switzerland. In theory, any male under 60 is liable for military service during a national emergency, although this law has never been invoked.

ECONOMICS

 Inflation 2.9% p.a. (1985–1996)

$1.6bn

1.6205–1.6603 Swiss francs

SCORE CARD

- ❑ WORLD GNP RANKING 142nd
- ❑ GNP PER CAPITA $50,000
- ❑ BALANCE OF PAYMENTS.....Included in Swiss total
- ❑ INFLATION 0.5%
- ❑ UNEMPLOYMENT 2%

STRENGTHS

Stability and customs union with Switzerland make Liechtenstein a favored tax haven; its lack of EU membership makes the banking sector less vulnerable to future changes in EU banking laws. The economy is well diversified; chemicals, furniture, coatings for the electro-optical industry, construction services, and precision instruments are all thriving sectors.

WEAKNESSES

Very few. Need to balance integration with other countries with safeguarding economic independence.

EXPORTS

Switzerland 13%

EU 49%

Other 38%

IMPORTS

With a limited domestic market, Liechtenstein's industry is export-oriented. Liechtenstein has a customs union with Switzerland and does not publish separate import figures.

RESOURCES

 Electric power: Included in Swiss total

 None

Not an oil producer

6000 cattle, 3000 pigs, 2900 sheep, 280 goats

 None

Liechtenstein has to import most of its energy. Almost all of its electricity comes from German power plants.

ENVIRONMENT

 Not available

38% partially protected

Not available

Protection of Liechtenstein's alpine scenery is high enough on the political agenda for one of the five councillors, or ministers, to have responsibility for the environment. As in Switzerland, the greatest worry is the effect of high rates of car use and of through traffic. However, a 1988 trial in providing free public bus transportation proved a failure, as Liechtensteiners remained firmly wedded to their automobiles.

MEDIA

 TV ownership high

Daily newspaper circulation 602 per 1000 people

PUBLISHING AND BROADCAST MEDIA

There are 2 daily newspapers, *Liechtensteiner Vaterland* and *Liechtensteiner Volksblatt*

No TV service

1 radio service

The two daily newspapers, although free of formal state control, are both run by political parties: the *Vaterland* by the VU, the *Volksblatt* by the FBP. Both have circulations of around 10,000.

CRIME

 No death penalty

24 prisoners

Crime does not pose any great problems

Crime is a minor problem, a result of the relatively even distribution of wealth and high average living standard. Liechtenstein has also taken great care to protect its tax-haven status by careful regulation of its financial sector. It has avoided major scandals, and took steps in 2000 to tighten precautions against the growing problem of money laundering.

EDUCATION

School leaving age: 16

99%

Not available

Education, modeled on the German system, includes two types of school at secondary level – the more academic *Gymnasium* and the *Realschule*. Liechtenstein has no university; students go on to colleges in Austria, Switzerland, or Germany, or to business schools in the US.

CHRONOLOGY

In 1719 Liechtenstein became an independent principality of the Holy Roman Empire.

- ❑ **1924** Customs union with Switzerland.
- ❑ **1990** Joins UN.
- ❑ **1995** Joins EEA.
- ❑ **1997** End of VU–FBP coalition dominant since 1938. Mario Frick heads VU government.
- ❑ **2001** FBP wins majority in elections; Otmar Hasler premier.

HEALTH

 Welfare state health benefits

1 per 948 people

Heart and respiratory diseases, cancers

Although clinics and hospitals are few, the health system provides advanced care. Many Liechtensteiners have private health insurance arrangements, so that they have access to Swiss medical expertise and facilities as well as their own.

SPENDING

 GDP/cap. increase

CONSUMPTION AND SPENDING

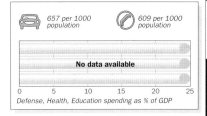

657 per 1000 population

609 per 1000 population

No data available

0   5   10   15   20   25

Defense, Health, Education spending as % of GDP

Unlike other tax havens, Liechtenstein displays a more conservative prosperity. Private deposit accounts are not a key part of its banking business, but an increase in money-laundering activities and the country's appearance on a blacklist of financial centers led to a ban on anonymous accounts in mid-2000.

WORLD RANKING

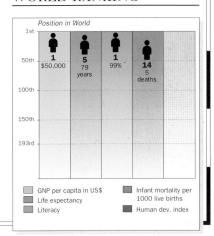

Position in World			
1st			
1 $50,000	**5** 79 years	**1** 99%	**14** 5 deaths

- ■ GNP per capita in US$
- ■ Life expectancy
- ■ Literacy
- ■ Infant mortality per 1000 live births
- ■ Human dev. index

 L

LITHUANIA

OFFICIAL NAME: Republic of Lithuania **CAPITAL:** Vilnius
POPULATION: 3.7 million **CURRENCY:** Litas **OFFICIAL LANGUAGE:** Lithuanian

 EUROPE

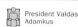

LYING ON THE EASTERN COAST of the Baltic Sea, Lithuania was the last European country formally to embrace Christianity, around 1400. Its terrain is mostly flat, with many lakes, moors, and bogs. Now a multiparty democracy, Lithuania regained independence from the former USSR in 1991. Industrial production and agriculture are the mainstays of the economy. Russia finally withdrew all its troops from Lithuania in 1993.

CLIMATE ▷ Continental

WEATHER CHART FOR VILNIUS

Lithuania's coastal position moderates an otherwise continental-type climate. Summers are cool.

TRANSPORTATION ▷ Drive on right

Vilnius International
521,529 passengers

206 ships
481,100 grt

THE TRANSPORTATION NETWORK

64,951 km (40,359 miles)	417 km (259 miles)
1905 km (1184 miles)	600 km (373 miles)

Lithuania is crossed by international rail routes, and Klaipeda provides extensive connections to other Baltic ports.

TOURISM ▷ Visitors : Population 1:3.4

 1.08m visitors

 Down 24% in 2000

MAIN TOURIST ARRIVALS

Russia 33%
Latvia 27%
Belarus 17%
Other 23%
% of total arrivals

The tourist industry boomed after independence. Vilnius is well preserved; its historic center survived German and Russian occupation. Trakai, the capital of the Grand Duchy in the 16th century, is also popular with visitors.

PEOPLE ▷ Pop. density medium

Lithuanian, Russian

57/km² (147/mi²)

THE URBAN/RURAL POPULATION SPLIT

68%   32%

ETHNIC MAKEUP

Belarussian 2%   Polish 7%
Other 2%   Russian 9%
Lithuanian 80%

With a mainly Catholic population, Lithuania has strong historical links with Poland, with which it was once united, although there is tension between ethnic Lithuanians and Poles. Relations with the Jewish minority are strained. More than 90% of nonethnic Lithuanians have been granted citizenship. Of all the Baltic states, Lithuania has the best relations with ethnic Russians, but there are fewer of them who live there.

POLITICS ▷ Multiparty elections

 2000/2004

President Valdas Adamkus

AT THE LAST ELECTION

Parliament 141 seats

36% ABSD	24% LLS	21% NS(SL)	6% TS(LK)	13% Others

ABSD = A. Brazauskas Social Democratic Coalition
LLS = Lithuanian Liberal Union
NS(SL) = New Union (Social Liberals)
TS(LK) = Homeland Union (Lithuanian Conservatives)

Lithuania was the most politically stable of the three Baltic republics in the 1990s. Strongly nationalist when it reestablished independence from the disintegrating Soviet Union in 1991, it unexpectedly voted the former communists back into office in 1992 and 1993. Former communist leader Algirdas Brazauskas held the presidency for five years until returning emigré businessman Valdas Adamkus was elected in 1998. Meanwhile the conservative TS(LK) had won the 1996 legislative elections in the wake of a banking scandal, but their popularity dwindled as successive prime ministers were forced to resign. The TS(LK) suffered a massive defeat in the 2000 legislative elections. Adamkus controversially bypassed Brazauskas' ABSD, the largest bloc in the new Parliament, in favor of a pro-market coalition. This collapsed in mid-2001, and Brazauskas became prime minister in a government dominated by the ABSD.

LITHUANIA

Total Land Area :
65 200 sq. km
(25 174 sq. miles)

POPULATION
⊙ over 500 000
◎ over 100 000
○ over 50 000
● over 10 000
· under 10 000

LAND HEIGHT
200m/656ft
Sea Level

0   50 km
0   50 miles

WORLD AFFAIRS

 Joined UN in 1991

 CE CBSS WTO OSCE | PfP

As a "second wave" applicant, Lithuania began EU membership negotiations in 2000. It currently has the best relations with Russia of all the Baltic states, but is keen for all three to join NATO.

AID

 Recipient

$99m (receipts) Down 26% in 2000

Aid, mostly from the IMF and the EU, is used for infrastructure projects and to promote private enterprise.

DEFENSE

Compulsory military service

$195m Up 82% in 2000

Lithuania's security is in the hands of its army, the small navy and air force, and a large National Guard formed to patrol its frontiers. Twelve months' military service is compulsory. The US is now supporting Lithuania's entry into NATO.

ECONOMICS

 Inflation 75% p.a. (1990–2000)

 $10.8bn | 3.9990–3.9984 litai

SCORE CARD

❑ WORLD GNP RANKING	81st
❑ GNP PER CAPITA	$2930
❑ BALANCE OF PAYMENTS	–$675m
❑ INFLATION	1%
❑ UNEMPLOYMENT	11%

STRENGTHS
Privatization has stimulated economy to some extent. Inflation under control. Successful liberal reforms. Litas pegged to euro.

WEAKNESSES
Agriculture in the doldrums following decollectivization. Exports dependent on health of Russian economy. Poor raw materials base. Difficulty in attracting significant foreign investment.

EXPORTS

Poland 5%
Russia 7%
UK 8%
Germany 14%
Other 51%
Latvia 15%

IMPORTS

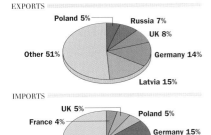

UK 5%
France 4%
Poland 5%
Germany 15%
Other 43%
Russia 28%

One of Lithuania's 3000 lakes. *The entire country is low-lying. Its coast, fringed by sand dunes and pine forests, is famous for amber.*

RESOURCES

 Electric power 5.8m kw

 35,244 tonnes | 5554 b/d (reserves 14m barrels)

855,600 pigs, 748,300 cattle, 5.58m chickens | Sand, gravel, clay, limestone, gypsum

Lithuania has significant reserves of peat and materials used in construction industry. The Ignalina nuclear plant provides more than 80% of the country's electricity. Oil is mostly imported from Russia.

ENVIRONMENT

 Sustainability rank: 27th

 12% | 4.2 tonnes per capita

Radioactive leaks and the risk of accident at the giant Chernobyl-type nuclear plant at Ignalina cause much concern. In 2002 the government agreed to decommission it by 2009, after pressure from the EU, which will share the $2.6 billion cost.

MEDIA

 TV ownership high

 Daily newspaper circulation 93 per 1000 people

PUBLISHING AND BROADCAST MEDIA

There are 19 daily newspapers, including *Lietuvos Rytas* and *Respublika*

10 services: 1 state-owned, 9 independent | 24 services: 1 state-owned, 23 independent

The mainstream media, Russian under communism, now publish and broadcast mainly in Lithuanian.

CRIME

 No death penalty

8960 prisoners | Down 1% in 1999

Levels of crime are low compared with other parts of the former USSR. Robbery is a growing problem.

EDUCATION

 School leaving age: 15

 99% | 84,300 students

Teaching at all levels is in Lithuanian, making access to higher education harder for minorities; 8% of the population are graduates.

CHRONOLOGY

Russia annexed Lithuania in 1795. The suppression of rebellions in 1831 and 1863 failed to undermine its nationalist movement.

- ❑ **1915** Occupied by German troops.
- ❑ **1918** Independence declared.
- ❑ **1926** Military coup; one-party rule.
- ❑ **1940** Annexed by Soviet Union.
- ❑ **1941–1944** Nazi occupation.
- ❑ **1945** Incorporated into USSR.
- ❑ **1991** Achieves full independence.
- ❑ **1992** First multiparty elections.
- ❑ **1993** Russian troops withdraw.
- ❑ **1996** Prime minister forced from office by banking scandal. General election; conservative TS(LK) wins.
- ❑ **1998** Valdas Adamkus president.
- ❑ **2000** Brief center-left coalition.
- ❑ **2001** Ex-president Brazauskas becomes prime minister.

HEALTH

 Welfare state health benefits

 1 per 250 people | Heart disease, cancers, accidents, tuberculosis

The 1997 reorganization of the health service involves replacing state funding with finance from insurance funds.

SPENDING

GDP/cap. decrease

CONSUMPTION AND SPENDING

334 per 1000 population | 321 per 1000 population

Defense 1.8%
Education 6.4%
Health 4.7%

0   5   10   15   20   25
Defense, Health, Education spending as % of GDP

Lithuanians are on average poorer than their neighbors in the other Baltic states. Since 1991 a large gap has opened between the incomes of rich and poor.

WORLD RANKING

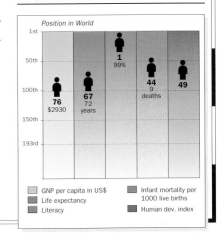

Position in World

1st
50th
100th
150th
193rd

1 99%
44 9 deaths
49
76 $2930
67 72 years

❑ GNP per capita in US$
❑ Life expectancy
❑ Literacy
❑ Infant mortality per 1000 live births
❑ Human dev. index

L

LUXEMBOURG

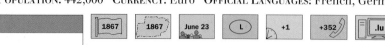

OFFICIAL NAME: Grand Duchy of Luxembourg **CAPITAL:** Luxembourg-Ville
POPULATION: 442,000 **CURRENCY:** Euro **OFFICIAL LANGUAGES:** French, German, & Luxemburgish

1867	1867	June 23	L	+1	+352	.lu

LUXEMBOURG SHARES BORDERS with the industrial regions of Germany, France, and Belgium, and has the highest per capita income in the EU. Making up part of the plateau of the Ardennes, its countryside is undulating and forested. Its prosperity was once based on steel; before World War II it produced more per capita than the US. Today, it is known as a tax haven and banking center, and as the headquarters of key EU institutions.

CLIMATE ▷ Maritime

WEATHER CHART FOR LUXEMBOURG-VILLE

- Average daily temperature
- Rainfall

The south, where vines grow, is the warmest area. Winter is cold and snowy, especially in the Ardennes.

TRANSPORTATION ▷ Drive on right

Findel, Luxembourg-Ville
1.66m passengers

Has no fleet

THE TRANSPORTATION NETWORK

5166 km (3210 miles)	115 km (71 miles)
274 km (170 miles)	37 km (23 miles)

There is an excellent road network, though congestion is a problem. Rail and bus services are integrated.

TOURISM ▷ Visitors : Population 1.8:1

807,000 visitors

Down 3% in 2000

MAIN TOURIST ARRIVALS

| Belgium 28% |
| Germany 15% |
| Netherlands 14% |
| Other 43% |

% of total arrivals

Key attractions are the mountains and forests, 76 castles, and the Benedictine abbey at Echternach. Foreign hotel workers learn about the history, language, and culture of the duchy under a government initiative.

PEOPLE ▷ Pop. density medium

Luxembourgish, German, French

171/km² (443/mi²)

THE URBAN/RURAL POPULATION SPLIT

92%   8%

RELIGIOUS PERSUASION

Protestant, Greek Orthodox, and Jewish 3%

Roman Catholic 97%

Nearly a third of its residents and half of Luxembourg's workers are foreigners. Integration has been straightforward; most are fellow western Europeans and Roman Catholics, mainly from Italy and Portugal. Life in Luxembourg is comfortable. Salaries are high, unemployment very low, and social tensions few.

POLITICS ▷ Multiparty elections

L. House 1999/2004

Grand Duke Henri

AT THE LAST ELECTION

Chamber of Deputies 60 seats

12% ACDJ	2% L

| 32% CSV/PCS | 25% DP/PD | 21% LSAP/POSL | 8% G |

CSV/PCS = Christian Social Party **DP/PD** = Democratic Party **LSAP/POSL** = Luxembourg Socialist Workers' Party **ACDJ** = Action Committee for Democracy and Justice **G** = Greens **L** = The Left

Council of State 21 seats

The members of the Council of State are appointed for life by the grand duke

There is remarkable political consensus, and governments are characterized by coalitions and long-serving prime ministers; the grand duke's role is mostly ceremonial. The main issues relate to European integration.

WORLD AFFAIRS ▷ Joined UN in 1945

Benelux	EU	NATO	OECD	OSCE

Luxembourg has long been the keenest member of the EU. It was during its EU presidency that the Maastricht agreement for closer European union was brokered; Luxembourg was not only the first member state to meet all the economic, financial, and legal requirements of union under Maastricht, but did so a year early. This commitment to the EU reflects the tremendous benefits Luxembourg has gained from membership. It is home to both the Secretariat of the European Parliament and the Court of Justice. In 1995, Prime Minister Jacques Santer left office to become president of the European Commission, but had to resign in 1999 amid allegations of corruption in the Commission.

LUXEMBOURG

Total Land Area : 2586 sq. km
(998 sq. miles)

LAND HEIGHT

500m/1640ft
200m/656ft
Sea Level

POPULATION

over 50 000
over 10 000
under 10 000

Charlotte Bridge, Luxembourg. *The modern road system provides excellent links with the rest of Europe.*

AID

 Donor

$127m (donations)   Up 7% in 2000

Aid has been increased substantially and equaled 0.69% of GNP in 2000. Most goes to sub-Saharan Africa.

DEFENSE

No compulsory military service

$126m Down 7% in 2000

The army numbers around 900 full-time soldiers. Spending has fallen slightly in recent years.

ECONOMICS

Inflation 2.2% p.a. (1990–2000)

$18.4bn 1.0651–1.1231 euros

SCORE CARD

❏ WORLD GNP RANKING	69th
❏ GNP PER CAPITA	$42,060
❏ BALANCE OF PAYMENTS	$1.59bn
❏ INFLATION	3.1%
❏ UNEMPLOYMENT	3%

STRENGTHS

Location for some EU institutions. Banking secrecy and expertise make the capital home to around 1000 investment funds and over 200 banks – more than in any other city in the world.

WEAKNESSES

International service industries account for 65% of GDP, making Luxembourg vulnerable to changing conditions overseas. Downturn in steel market.

EXPORTS

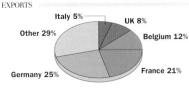

Italy 5%
UK 8%
Other 29%
Belgium 12%
France 21%
Germany 25%

IMPORTS

Netherlands 4%
Taiwan 4%
France 12%
Belgium 35%
Other 21%
Germany 24%

RESOURCES

Electric power 1.3m kw

Not available   Not an oil producer

207,862 cattle, 85,830 pigs, 8220 sheep, 2818 horses   Iron

Most energy is imported; Luxembourg produces only a small amount of hydro-electricity. Arbed is the world's third-largest steel producing company.

ENVIRONMENT

Not available

None   18 tonnes per capita

Acid rain from European industry has affected 20% of Luxembourg's trees and, in the worst cases, 30% of trees in mature stands. Luxembourg is a member of an international committee on reducing pollution of the Rhine.

MEDIA

TV ownership high

Daily newspaper circulation 328 per 1000 people

PUBLISHING AND BROADCAST MEDIA

There are 5 daily newspapers. The leading newspaper, in terms of both circulation and influence, is the *Luxemburger Wort*

2 independent services   11 independent services

Broadcasting is dominated by RTL (Radio–Television Luxembourg), one of the largest media groups in Europe, which exports its programs in a variety of languages.

CRIME

No death penalty

394 prisoners   Down 2% in 1999

Luxembourg's banking secrecy rules have provoked international criticism, as they can provide a cover for both tax evasion and fraud. Violent crime remains uncommon.

EDUCATION

School leaving age: 15

99%   2662 students

Teaching is mainly in German at primary and French at secondary level. Higher education is limited and many students go to universities in other European countries. Training given by Luxembourg banks is reputed to be the best in Europe.

HEALTH

Welfare state health benefits

1 per 368 people   Cerebrovascular and heart diseases, cancers, accidents

There are no private commercial hospitals in Luxembourg; they are run either by the state or by nuns. Patients' fees are refunded from the state sickness fund.

CHRONOLOGY

Until 1867, Luxembourg was ruled by a succession of neighboring European powers.

- ❏ **1890** Link with Dutch throne ends.
- ❏ **1921** Economic union with Belgium. End of German ties.
- ❏ **1940–1944** German occupation.
- ❏ **1948** Benelux treaty (1944) creating a customs union comes into effect.
- ❏ **1957** One of six signatories of Treaty of Rome, the principal foundation of what develops into the EU.
- ❏ **1995** Premier Jacques Santer is president of European Commission.
- ❏ **1999** Santer resigns amid corruption allegations. Socialist election losses.
- ❏ **2000** Grand Duke Jean abdicates in favor of his son, Henri.
- ❏ **2002** Euro fully adopted.

SPENDING

GDP/cap. increase

CONSUMPTION AND SPENDING

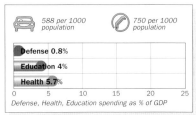

588 per 1000 population   750 per 1000 population

Defense 0.8%
Education 4%
Health 5.7%

0   5   10   15   20   25
Defense, Health, Education spending as % of GDP

With the world's second-highest per capita income, Luxembourgers enjoy a comfortable lifestyle. Recent strong economic performance has allowed them to benefit both from lower taxes and increased social security spending. Low unemployment has led to the recruitment of foreign workers, mainly from neighboring countries or from other EU countries such as Portugal and Italy, to take less well-paid jobs. As elsewhere in western Europe, financing care of the aging population is likely to be a burden in the future.

WORLD RANKING

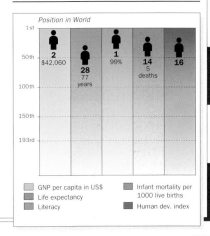

Position in World

1st

2 — $42,060
28 — 77 years
1 — 99%
14 — 5 deaths
16

50th

100th

150th

193rd

GNP per capita in US$ Infant mortality per 1000 live births
Life expectancy   Human dev. index
Literacy

MACEDONIA

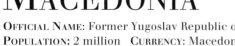

OFFICIAL NAME: Former Yugoslav Republic of Macedonia   **CAPITAL:** Skopje
POPULATION: 2 million   **CURRENCY:** Macedonian denar   **OFFICIAL LANGUAGES:** Macedonian and Albanian

  1991 1991 Aug 2 MK +1 +389 .mk

THE FORMER YUGOSLAV REPUBLIC of Macedonia (FYRM) is landlocked in southeastern Europe. Despite the signing of an accord in 1995, Greece remains suspicious that it harbors ambitions about absorbing northern Greece – also called Macedonia – in a "Greater Macedonia." A militant movement among ethnic Albanians erupted into violent conflict in March–September 2001, but a peace agreement was reached after the involvement of a NATO force.

A fisherman's hut on Lake Dojran. The lake lies on the border with Greece in southeastern Macedonia and is shared by the two countries.

CLIMATE ▷ Continental

WEATHER CHART FOR SKOPJE

The FYRM has a continental climate. Winter snow supports skiing.

TRANSPORTATION ▷ Drive on right

Skopje International
1.01m passengers

Has no fleet

THE TRANSPORTATION NETWORK

5540 km (3442 miles)	133 km (83 miles)
699 km (434 miles)	None

An east–west road and rail route from Tirana in Albania through Macedonia to Sofia in Bulgaria is planned, to reduce reliance on routes through Serbia.

TOURISM ▷ Visitors : Population 1:8.9

224,000 visitors   Up 24% in 2000

MAIN TOURIST ARRIVALS

Serbia & Montenegro 15%
Albania 11%
Bulgaria 10%
Other 64%

0 10 20 30 40 50 60 70 80
% of total arrivals

The major attraction is the ecclesiastical center of Ohrid, situated on Europe's deepest lake, with Roman and Byzantine ruins. Other lake resorts and skiing in the Sara mountains in the northwest, have potential once stability is restored.

PEOPLE ▷ Pop. density medium

Macedonian, Albanian, Serbo-Croat

78/km² (201/mi²)

THE URBAN/RURAL POPULATION SPLIT

62%   38%

ETHNIC MAKEUP

Roma 2%   Serb 2%
Other 2%   Turkish 4%
Albanian 23%
Macedonian 67%

Slav Macedonians, speaking a language akin to Bulgarian, are in the majority. The large Albanian minority, claiming to amount to over one-third of the population, maintains strong links with Albanians in neighboring states. Months of violent conflict sparked by ethnic Albanian insurgents seeking greater rights resulted in a new constitution in 2001 guaranteeing equality.

Macedonians are mostly Orthodox Christians, but there are a substantial number of Slavic Muslims (Pomaks), whose ancestors converted during the Ottoman occupation. Ethnic Albanians are mostly Muslim.

POLITICS ▷ Multiparty elections

2002/2006

President Boris Trajkovski

AT THE LAST ELECTION

Assembly of the Republic 120 seats

49% ZMZ   28% VMRO–DPMNE   13% BDI   2% PPD   1% SPM   6% PDS   1% PDK

ZMZ = Together for Macedonia, headed by the Social Democratic Alliance of Macedonia (**SDSM**) **VMRO–DPMNE** = Internal Macedonian Revolutionary Organization–Democratic Party for Macedonian National Unity **BDI** = Democratic Union for Integration **PDS** = Democratic Party of Albanians **PPD** = Party of Democratic Prosperity **PDK** = National Democratic Party **SPM** = Socialist Party of Macedonia

Political parties divide along ethnic lines. The Social Democrats and the right-wing VMRO–DPMNE are

WORLD AFFAIRS ▷ Joined UN in 1993

CE   IAEA   EAPC   PfP   OSCE

Macedonia's hosting of NATO troops in the Kosovo conflict in 1999 placed it firmly in the Western fold. Nationalists resented Western pressure in 2001 to compromise with Albanian militants.

AID ▷ Recipient

$252m (receipts)   Down 9% in 2000

The World Bank and the EU are the main channels for economic development assistance. The EU and US pledged $515 million for reconstruction in 2002. Regional security fears limit foreign investment.

DEFENSE ▷ Compulsory military service

$76m   Up 13% in 2000

The army relies heavily on officer training in NATO countries. A major overhaul has been announced, with the aim being to tackle weaknesses revealed by the 2001 conflict.

predominantly Slav, while Albanians vote mainly for the BDI, the PDS, and the smaller PPD. The September 2002 elections gave the Social Democrat-led ZMZ alliance a slight ascendancy.

Previously, VMRO–DPMNE Prime Minister Ljubco Georgievski had led a panethnic right-wing coalition (1998–2001), then a fragile "national unity" government formed in the face of armed insurrection by Albanian rebels (UCK). A peace deal in August 2001 ushered in a new, more equal constitution and guarantees on minority rights. With peace secured at the end of 2001, the "unity" government fractured, but Georgievski remained in power until the 2002 polls.

ECONOMICS

▷ Inflation 79% p.a. (1990–2000)

$3.7bn   66.26–69.17 Macedonian denari

SCORE CARD

- ❏ WORLD GNP RANKING........................124th
- ❏ GNP PER CAPITA $1820
- ❏ BALANCE OF PAYMENTS....................–$107m
- ❏ INFLATION ....–1.3%
- ❏ UNEMPLOYMENT................................32%

EXPORTS

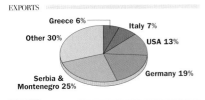

Greece 6%
Italy 7%
Other 30%
USA 13%
Germany 19%
Serbia & Montenegro 25%

IMPORTS

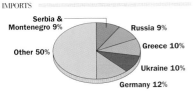

Serbia & Montenegro 9%
Other 50%
Russia 9%
Greece 10%
Ukraine 10%
Germany 12%

STRENGTHS

Growth in private sector and foreign investment. Mineral resources.

WEAKNESSES

Poorest of former Yugoslav republics. Loss of trade in mid-1990s due to Greek embargo and sanctions on Yugoslavia. Dependence on oil, gas, and machinery imports. Disruption caused by Kosovo conflict and 2001 violence.

FORMER YUGOSLAV REPUBLIC OF MACEDONIA

Total Land Area : 25 333 sq. km (9781 sq. miles)

LAND HEIGHT
- 2000m/6562ft
- 1000m/3281ft
- 500m/1640ft
- 50m/164ft

POPULATION
- ◉ over 500 000
- ◎ over 100 000
- ○ over 50 000
- ● over 10 000
- • under 10 000

RESOURCES

▷ Electric power 1.5m kw

1804 tonnes
1.25m sheep, 265,000 cattle, 3.35m chickens

Not an oil producer
Coal, copper, bauxite, iron, antimony, chromium, lead, zinc

Minerals remain underexploited. South-facing fertile plains produce early fruit and vegetables for EU markets.

ENVIRONMENT

▷ Sustainability rank: 83rd

7%   6.1 tonnes per capita

Industrial pollution affects water quality. The Titov Veles lead and zinc smelter is the worst culprit for toxic waste.

MEDIA

▷ TV ownership high

Daily newspaper circulation 21 per 1000 people

PUBLISHING AND BROADCAST MEDIA

There are 4 daily newspapers, including the government-funded Albanian *Flaka e Vellazerimit*

3 services: 1 state-owned, 2 independent

1 state-owned, also independent services

Newspaper sales have expanded rapidly. In 2001, two independent newspapers "voluntarily" ceased publication.

CRIME

▷ No death penalty

1394 prisoners   Down 11% 1996–1998

Cigarette smuggling is dominated by Albanian gangs also involved in the illegal arms trade and heroin trafficking.

EDUCATION

▷ School leaving age: 15

94%   36,922 students

A new university at Tetovo, offering teaching in Albanian as well as Macedonian, officially opened in 2001.

CHRONOLOGY

The end of Ottoman rule saw historic Macedonia divided between Serbia, Bulgaria, and Greece in 1912–1913. What is now the FYRM was incorporated into Serbia.

- ❏ **1944** Tito establishes republic, stressing Macedonian identity.
- ❏ **1989–1990** Multiparty elections.
- ❏ **1991** Independence declared. EU recognition delayed by Greeks.
- ❏ **1995** Accord with Greece.
- ❏ **1998–1999** Right-wing VMRO–DPMNE coalition wins elections.
- ❏ **1999** Upheaval over Kosovo conflict.
- ❏ **2001** New constitution after NATO involvement to end conflict with ethnic Albanian militants.
- ❏ **2002** Left-of-center ZMZ alliance gains upper hand in elections.

HEALTH

▷ Welfare state health benefits

1 per 455 people   Cerebrovascular and heart diseases, cancers

In theory, the state guarantees universal health care, but effective and speedy treatment is increasingly only available in the private sector.

SPENDING

▷ GDP/cap. decrease

CONSUMPTION AND SPENDING

139 per 1000 population   255 per 1000 population

Defense 2.1%
Education 5.1%
Health 5.3%

0   5   10   15   20   25
Defense, Health, Education spending as % of GDP

Incomes have fallen by more than two-thirds since 1990, although smuggling and organized crime have made a few people conspicuously wealthy.

WORLD RANKING

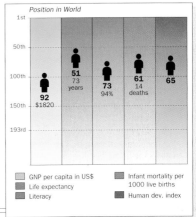

Position in World

1st				
50th		51		
100th	92	73 years	73	61	65
	$1820		94%	14 deaths	
150th					
193rd					

- ▇ GNP per capita in US$
- ▇ Life expectancy
- ▇ Literacy
- ▇ Infant mortality per 1000 live births
- ▇ Human dev. index

M

MADAGASCAR

INDIAN OCEAN — Africa

OFFICIAL NAME: Republic of Madagascar **CAPITAL:** Antananarivo
POPULATION: 16.4 million **CURRENCY:** Malagasy franc **OFFICIAL LANGUAGES:** French and Malagasy

1960 1960 June 26 RM +3 +261 .mg

LYING IN THE INDIAN Ocean, Madagascar is the world's fourth-largest island. Its isolation means that there is a host of unique wildlife and plants. To the east, the large central plateau drops precipitously through forested cliffs to the coast; in the west, gentler gradients give way to fertile plains. It became independent from France in 1960, and after 18 years of radical socialism under Didier Ratsiraka, became a multiparty democracy in 1993.

CLIMATE ▷ Tropical

WEATHER CHART FOR ANTANANARIVO

Tropical Madagascar often has cyclones. The coastal lowlands are humid, but the central plateau is cooler.

TRANSPORTATION ▷ Drive on right

 Ivato, Antananarivo
707,304 passengers

 101 ships
41,700 grt

THE TRANSPORTATION NETWORK

5781 km (3592 miles)	None
883 km (549 miles)	600 km (373 miles)

An extensive domestic air network – due for privatization – compensates for a very limited rail network and roads that are impassable during the rains.

TOURISM ▷ Visitors : Population 1:103

160,000 visitors   Up 16% in 2000

MAIN TOURIST ARRIVALS

France 54%
Réunion 8%
Italy 6%
Other 32%

% of total arrivals

Extensive tropical beaches and unique flora and fauna offer great tourism potential. Political stability after 1993 led to a marked increase in arrivals which was threatened by the chaos in 2002.

PEOPLE ▷ Pop. density low

 Malagasy, French   28/km² (73/mi²)

THE URBAN/RURAL POPULATION SPLIT

30%   70%

RELIGIOUS PERSUASION

Muslim 7%
Traditional beliefs 52%
Christian (mainly Roman Catholic) 41%

Madagascans are essentially Malay–Indonesian in origin. Their ancestors migrated across the Indian Ocean from the 1st century. Later migrants from the African mainland intermixed. Arab traders provided another ingredient to the mix. The main ethnic division is between the central plateau and *côtier* peoples. Of more pronounced Malay extraction, the plateau Merina were Madagascar's historic rulers. They remain the social elite – to the resentment of the poorer *côtiers*, the group to which long-term ruler Didier Ratsiraka belongs. The extended family remains the focus of social life for the rural majority.

MADAGASCAR

Total Land Area : 587 040 sq. km
(226 656 sq. miles)

POPULATION
⊙ over 500 000
◉ over 100 000
○ over 50 000
● over 10 000
• under 10 000

LAND HEIGHT
2000m/6562ft
1000m/3281ft
500m/1640ft
200m/656ft
Sea Level

POLITICS ▷ Multiparty elections

L. House 1998/2003
U. House 2001/2007

President Marc Ravalomanana

AT THE LAST ELECTION
National Assembly 150 seats

9% AVI 2% MFM
42% Arema 22% Ind 11% LF 7% RPSD 7% Others

Arema = Association for the rebirth of Madagascar
Ind = Independents **LF** = Leader–Torch (Fanilo)
AVI = "People are judged by the work they do"
RPSD = Rally for Socialism and Democracy
MFM = Militant Party for the Development of Madagascar

Senate 90 seats

Two-thirds of Senate members are elected by regional governments; the remainder are nominated by the president

From 1975 until 2002 Madagascan politics was dominated by radical socialist president Didier Ratsiraka. A return to multipartyism resulted in his removal in 1993, but in 1997 he was returned to power. Continuing economic gloom saw him defeated in elections in late 2001. Refusing to accept the victory of the liberal opposition leader Marc Ravalomanana, Ratsiraka precipitated an eight-month power struggle which violently divided the island. Eventually support for Ravalomanana proved overwhelming, and he was recognized internationally as president in June 2002.

M

WORLD AFFAIRS

▷ Joined UN in 1960

Once-close ties with Russia and North Korea waned as Madagascar cemented relations with its main Western trading partners, especially with France and the US. Since 1997 cooperation with the IMF has improved as well. Regionally, Madagascar has reestablished ties with South Africa and in 1994 joined COMESA.

AID

▷ Recipient

 $322m (receipts)   Down 10% in 2000

International donors pledged $2.3 billion in July 2002 in a four-year emergency aid package to support the new government and fund reconstruction.

DEFENSE

▷ Compulsory military service

 $41m   Down 5% in 2000

The army was briefly divided during the 2002 turmoil. It intervened against federalist *côtiers* in 1992.

ECONOMICS

▷ Inflation 19% p.a. (1990–2000)

 $3.87bn   6220–6370 Malagasy francs

SCORE CARD

❏ World GNP Ranking	120th
❏ GNP per Capita	$250
❏ Balance of Payments	–$260m
❏ Inflation	12%
❏ Unemployment	Widespread underemployment

STRENGTHS

Varied agricultural base; vanilla, coffee, and clove exports. Offshore oil and gas. Prawns. Tourism. Literate workforce.

WEAKNESSES

Losing out to cheaper vanilla exporters. Vulnerability to drought and cyclone damage, severe in early 2000. Economic reforms yet to bear fruit. Not self-sufficient in rice, the food staple.

EXPORTS

Indonesia 5%
Singapore 5%
France 35%
Germany 7%
USA 18%
Other 30%

IMPORTS

Singapore 4%
Taiwan 8%
Mauritius 9%
Other 41%
China 9%
France 29%

Tôlañaro (also known as Fort Dauphin), a port on the southeast coast. This was the area first settled by the French in the 16th century.

RESOURCES

▷ Electric power 220,000 kw

 141,057 tonnes   Not an oil producer

10.3m cattle, 3.8m ducks, 3m geese, 19m chickens   Chromite, graphite, oil, mica, iron, bitumen, gemstones, marble, gas

Electricity is hydrogenerated. There are underexploited mineral reserves and offshore oil and gas. There were large finds of high-quality sapphires in 1998.

ENVIRONMENT

▷ Sustainability rank: 128th

 2% (1% partially protected)   0.1 tonnes per capita

Madagascar's environment is a unique resource; 80% of its plant species and many animal species, such as the lemur, are found nowhere else. Aid helps to combat deforestation and soil erosion.

MEDIA

▷ TV ownership low

 Daily newspaper circulation 5 per 1000 people

PUBLISHING AND BROADCAST MEDIA

There are 5 daily newspapers, including the *Madagascar Tribune* and *Midi-Madagasikara*

1 state-owned service   1 state-owned service, many independent stations

Even before the return of multiparty democracy in 1993, there was a flourishing opposition press. There are 127 local radio stations.

CRIME

▷ Death penalty not used in practice

20,109 prisoners   Crime is rising

Urban crime levels are rising, with theft a particular concern. The army faces accusations of abusing human rights and of shooting federalists in 1993.

EDUCATION

▷ School leaving age: 11

 66%   31,013 students

Primary education will soon be based on French, not Malagasy. Attendance at secondary level is rising, as is the number of teachers. The OPEC Fund is backing a government literacy drive.

HEALTH

▷ Welfare state health benefits

1 per 10,000 people   Malaria, enteric, and respiratory diseases

Private health care was legalized in 1993. State care is free but inadequate. Malaria is at epidemic levels. There are outbreaks of bubonic plague.

SPENDING

▷ GDP/cap. increase

CONSUMPTION AND SPENDING

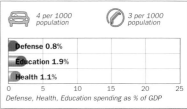

4 per 1000 population   3 per 1000 population

Defense 0.8%
Education 1.9%
Health 1.1%

Defense, Health, Education spending as % of GDP

Most of Madagascar's people are terribly poor, although those who live on the central plateau are richer than the *côtier* farmers and fishermen.

WORLD RANKING

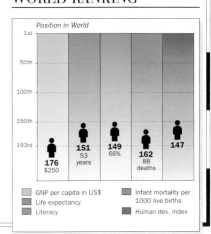

Position in World

	GNP per capita in US$	Life expectancy	Literacy	Infant mortality	Human dev. index
	176 $250	151 53 years	149 66%	162 88 deaths	147

❏ GNP per capita in US$
❏ Life expectancy
❏ Literacy
❏ Infant mortality per 1000 live births
❏ Human dev. index

M

MALAWI

OFFICIAL NAME: Republic of Malawi   **CAPITAL:** Lilongwe
POPULATION: 11.6 million   **CURRENCY:** Malawi kwacha   **OFFICIAL LANGUAGE:** English

LANDLOCKED IN SOUTHEAST Africa, Malawi lies on the Great Rift Valley and is dominated by Lake Nyasa, Africa's third-largest expanse of water. In the 1980s Malawi hosted large numbers of Mozambican refugees, at some cost to its fragile economy; food shortages are a recurrent and serious threat. A former British colony, Malawi established democracy in 1994 after three decades of one-party rule under Hastings Banda.

CLIMATE
▷ Tropical wet & dry

The south is hot and humid. The rest of Malawi is warm and very sunny in the dry season, but cooler in the highlands.

TRANSPORTATION
▷ Drive on left

Kamuzu International, Lilongwe
202,618 passengers

Has no fleet

THE TRANSPORTATION NETWORK

5254 km (3265 miles)

None

797 km (495 miles)

144 km (89 miles)

The Kamuzu Highway has been upgraded, and the Nacala Rail Corridor, a vital link to the distant sea, has attracted private investment.

TOURISM
▷ Visitors : Population 1:51

228,000 visitors

Down 10% in 2000

MAIN TOURIST ARRIVALS

Zambia 15%
UK & Ireland 9%
Mozambique 8%
Other 68%
0 10 20 30 40 50 60 70 80
% of total arrivals

The national parks and Lake Nyasa's fishing and water sports are the main tourist attractions. The opening of international airports at Blantyre and Lilongwe has increased accessibility.

PEOPLE
▷ Pop. density medium

Chewa, Lomwe, Yao, Ngoni, English

123/km² (319/mi²)

THE URBAN/RURAL POPULATION SPLIT

25%   75%

RELIGIOUS PERSUASION

Traditional beliefs 5%
Muslim 20%
Protestant 55%
Roman Catholic 20%

Ethnicity has not been exploited for political ends as has been the case in neighboring states. Most Malawians share a common Bantu origin. Of the various groups, the Chewa are dominant in the central region, Nyanja in the south, Tumbuka in the north, the mostly Muslim Yao in the southeast, and the Ngoni, a Zulu offshoot, in the lowlands. Other groups include the Chieoka and Tonga. Northerners felt ignored by Banda and his MCP, but the UDF government has largely succeeded in reducing the resulting tensions.

The election of President Muluzi, a member of Malawi's 20% Muslim minority, arguably signals the failure of Banda's plan to enforce Protestant domination in Malawi. Many of Malawi's Muslim Asians work in the retail sector.

Fruit and vegetable sellers offering their wares on the Mozambican border. The south of the country is intensively cultivated.

POLITICS
▷ Multiparty elections

1999/2004

President Bakili Muluzi

AT THE LAST ELECTION
National Assembly 193 seats

2% Ind   1% Others

48% UDF   34% MCP   15% AFORD

UDF = United Democratic Front **MCP** = Malawi Congress Party **AFORD** = Alliance for Democracy **Ind** = Independents

For 30 years from independence in 1964 Malawi was ruled by the autocratic Hastings Banda. His single-party regime outlawed dissent; torture and imprisonment without trial were common. In 1992 international aid was suspended because of the regime's human rights record. A referendum forced Banda to introduce multiparty politics in 1994, when the mainly southern-based UDF scored a dramatic victory, and its leader Bakili Muluzi won the presidency. He shrewdly recruited several prominent MCP politicians to his team, and vowed to restore personal and religious freedom, liberalize and revive the shattered economy, and improve Malawi's regional standing. Muluzi was narrowly reelected in 1999, but opposition leader Gwanda Chakuamba contested the results, and violence against Muslims and UDF supporters erupted in the north. A new opposition National Democratic Alliance emerged in 2001. Its leader Brown Mpinganjira was arrested for treason later that year.

WORLD AFFAIRS
▷ Joined UN in 1964

Comm   COMESA   NAM   AU   SADC

Malawi aims to protect its restored status as a recipient of Western aid. In 1998 the UK pledged to support President Muluzi's fiscal policies and antipoverty drive. Malawi also wants to preserve ties with South Africa, unbroken since 1967, a unique record in black Africa. One in ten Mozambicans fled to Malawi as refugees in the 1980s.

AID
▷ Recipient

$445m (receipts)

Little change in 2000

Nonhumanitarian aid resumed with the advent of democracy; international donors have pledged a total of $1.2 billion. The World Food Program launched an emergency aid program in April 2001 to help flood victims.

DEFENSE

▷ No compulsory military service

 $26m | ⬇ Down 4% in 2000

Muluzi's government has the backing of the 5300-strong army. In the last days of Banda's rule, the military lost confidence in the ruling MCP, forcing the pace of democratization. In 1993, it disarmed the Young Pioneers, a militarized section of the MCP.

ECONOMICS

▷ Inflation 34% p.a. (1990–2000)

$1.74bn | 80.40–67.01 Malawi kwacha

SCORE CARD

- ❏ World GNP Ranking.......................140th
- ❏ GNP per Capita $170
- ❏ Balance of Payments...................–$523m
- ❏ Inflation 29.5%
- ❏ Unemployment 1%

STRENGTHS

Tobacco, earning 76% of foreign exchange. Tea and sugar production. Unexploited bauxite, asbestos, and coal reserves. Much tourism potential.

WEAKNESSES

Agriculture vulnerable to drought and price fluctuations. Only 14% of GDP derived from industry. Small domestic market, few skilled workers. Strain of housing Mozambican refugees.

EXPORTS

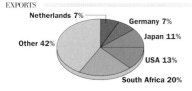

Netherlands 7%
Germany 7%
Japan 11%
USA 13%
South Africa 20%
Other 42%

IMPORTS

India 3%
UK 3%
South Africa 42%
Zambia 10%
Zimbabwe 15%
Other 27%

RESOURCES

▷ Electric power 185,000 kw

 45,982 tonnes | Not an oil producer

 1.45m goats, 750,000 cattle, 15.2m chickens | Coal, limestone, gemstones, bauxite, graphite, uranium

One 215 MW hydropower plant on Shire River accounts for nearly 85% of generating capacity, but only 3% of total energy use. Most rely on fuelwood for their energy needs. Malawi now encourages privatization, crop diversification, improved irrigation, and regional economic integration via the SADC to exploit its naturally limited resources. A deep-seam coal mine currently operates at Rumphi.

MALAWI

Total Land Area : 118 480 sq. km (45 745 sq. miles)

POPULATION
- ◎ over 100 000
- ● over 10 000
- • under 10 000

LAND HEIGHT
- 2000m/6562ft
- 1000m/3281ft
- 500m/1640ft
- 200m/656ft
- Sea Level

0   100 km
0   100 miles

ENVIRONMENT

▷ Sustainability rank: 82nd

11% | 0.1 tonnes per capita

Drought, with its devastating effects on agriculture, eclipses all other problems. Ecological husbandry attracts tourism.

MEDIA

▷ TV ownership low

 Daily newspaper circulation 3 per 1000 people

PUBLISHING AND BROADCAST MEDIA

There are 5 daily newspapers, including the *Daily Times*	
1 service	1 state-owned service

Violence against the staff of privately owned newspapers causes concern among human rights groups.

CRIME

▷ Death penalty in use

7024 prisoners | ⬆ Crime is rising

Urban crime is on the increase. The proliferation of weapons, particularly guns, is contributing to a rise in cases of armed robbery.

CHRONOLOGY

After strong Scottish missionary activity, Malawi came under British rule as Nyasaland in 1891.

- ❏ **1964** Independence under Hastings Banda.
- ❏ **1966** One-party state.
- ❏ **1992** Antigovernment riots. Illegal prodemocracy groups unite.
- ❏ **1993** Referendum for multipartyism.
- ❏ **1994** Muluzi's UDF wins elections.
- ❏ **2002** Worst-ever cholera epidemic exacerbated by food shortages.

EDUCATION

▷ School leaving age: 13

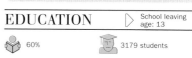 60% | 3179 students

Primary-level education is widespread, with 73% of boys and 60% of girls attending school regularly.

HEALTH

▷ Welfare state health benefits

 1 per 20,000 people | Infectious, parasitic, and respiratory diseases, AIDS

Life expectancy is around only 40 years, owing to the high incidence of HIV/AIDS.

SPENDING

▷ GDP/cap. increase

CONSUMPTION AND SPENDING

 3 per 1000 population | 4 per 1000 population

Defense 1.8%
Education 4.6%
Health 2.8%

0   5   10   15   20   25
Defense, Health, Education spending as % of GDP

The ousted MCP elite grew wealthy, allegedly through embezzlement. However, 80% of Malawians remain mired in poverty, and are forced to survive on less than $1 a day.

WORLD RANKING

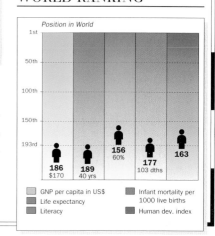

Position in World

186 $170	
189 40 yrs	
156 60%	
177 103 dths	
163	

- ▢ GNP per capita in US$
- ▢ Life expectancy
- ▢ Literacy
- ▢ Infant mortality per 1000 live births
- ▢ Human dev. index

MALAYSIA

OFFICIAL NAME: Federation of Malaysia **POPULATION:** 22.6 million **CURRENCY:** Ringgit
CAPITALS: Kuala Lumpur; Putrajaya (administrative) **OFFICIAL LANGUAGE:** Bahasa Malaysia

 1963 1965 Aug 31 MAL +8 +60 .my

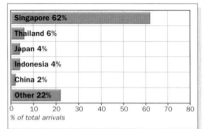
SOUTHEAST ASIA
Asia

MALAYSIA COMPRISES the three territories of Peninsular Malaysia, Sarawak, and Sabah, stretching over 2000 km (1240 miles) from the edge of the Indian Ocean to the northeastern end of the island of Borneo. A central mountain chain separates the coastal lowlands of Peninsular Malaysia; Sarawak and Sabah have swampy coastal plains rising to mountains on the border with Indonesia. Putrajaya, just south of Kuala Lumpur, is a high-tech new development intended as the home of government. The United Malays National Organization has dominated politics since independence.

TOURISM

 Visitors : Population 1:2.2

10.2m visitors   Up 29% in 2000

MAIN TOURIST ARRIVALS

Singapore 62%	
Thailand 6%	
Japan 4%	
Indonesia 4%	
China 2%	
Other 22%	

0 10 20 30 40 50 60 70 80
% of total arrivals

CLIMATE

 Tropical equatorial

WEATHER CHART FOR KUALA LUMPUR

Average daily temperature   Rainfall
°C/°F J F M A M J J A S O N D cm/in
40/104   40/16
30/86   30/12
20/68   20/8
10/50   10/4
0/32   0
-10/14
-20/-4

The whole of Malaysia has an equatorial climate. While there are two distinct rainy seasons, from March to May and from September to November, there is rain throughout the year: almost everywhere has rain on between 150 and 200 days a year. Coastal areas are also subject to monsoon winds, which alternate in direction between the southwest and northeast.

Tea plantations and colonial-style houses and gardens make the Cameron Highlands, in Peninsular Malaysia, one of Asia's most popular mountain resorts.

TRANSPORTATION

 Drive on left

 Subang International, Kuala Lumpur 14.7m passengers   828 ships 5.21m grt

THE TRANSPORTATION NETWORK

70,970 km (44,099 miles)		580 km (360 miles)	
1622 km (1008 miles)		7296 km (4534 miles)	

Transportation in Peninsular Malaysia is well developed. A major north–south highway connects the urban centers of the west coast. Roads in Sabah are also good, with an efficient bus network linking the towns. Travel in Sarawak, on the other hand, is hindered by poorly maintained roads and a lack of public transportation. East Malaysia is most effectively traversed by air.

Malaysia is southeast Asia's major tourist destination. Most tourists come for the excellent tropical beaches on the peninsula's east coast, to hike in the Cameron Highlands, or to trek in the world's oldest rainforests in Borneo. There has recently been an increase in the international business convention trade, and hotel capacity has been growing at 10% a year.

By 1990, when the government ran the Visit Malaysia Year campaign, tourism had become Malaysia's third-biggest foreign exchange earner. Two other such campaigns were launched in 1994 and 1998. However, the resurgence since 1999 of pro-Islamic parties, which favor stricter dress codes for women and a ban on alcohol, has deterred some Western tourists. In 2000 Malaysia backed an integrated tourism package with Thailand, Indonesia, and Singapore to enable tourists to visit the four countries under a common program.

MALAYSIA

Total Land Area : 329 750 sq. km (127 316 sq. miles)

POPULATION
⊙ over 500 000
◎ over 100 000
○ over 50 000
● over 10 000
• under 10 000

LAND HEIGHT
2000m/6562ft
1000m/3281ft
500m/1640ft
200m/656ft
Sea Level

M

PEOPLE ▷ Pop. density medium

 Bahasa Malaysia, Malay, Chinese, Tamil, English

 69/km² (178/mi²)

THE URBAN/RURAL POPULATION SPLIT

57%   43%

The key distinction in Malaysian society is between the indigenous Malays, termed the Bumiputras ("sons of the soil"), and the Chinese. The Malays form the largest group, accounting for just under half of the population. However, the Chinese have traditionally controlled most business activity. The New Economic Policy (NEP), introduced in the 1970s, was designed to address this imbalance by offering positive opportunities to the Malays through the education system and by making jobs available to them in both the state

RELIGIOUS PERSUASION

Traditional beliefs 2%
Other 7%
Christian 7%
Chinese faiths 12%
Muslim (mainly Sunni) 53%
Buddhist 19%

ETHNIC MAKEUP

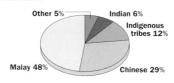

Other 5%
Indian 6%
Indigenous tribes 12%
Malay 48%
Chinese 29%

and private sectors. There are estimated to be more than one million Indonesian and Filipino immigrants in Malaysia, a dearth of employment in their own countries giving additional attraction to

POPULATION AGE BREAKDOWN

Female	Age	Male
0.6%	80+	0.5%
2.4%	60–79	2.2%
8%	40–59	8.4%
16%	20–39	16.9%
21.8%	0–19	23.2%

% of population by age group

Malaysia's need for labor. In addition, nearly 255,000 Vietnamese refugees were offered temporary refuge in Malaysia between 1975 and 1997; most have now been resettled in third countries, but around 6000 remain. Gender discrimination was only outlawed in 2001. Muslim women are encouraged to wear a veil.

POLITICS ▷ Multiparty elections

L. House 1999/2004
U. House varying

Raja Tuanku Syed Sirajuddin ibni al-Marhum Syed Putra Jamalullail

AT THE LAST ELECTION

House of Representatives 193 seats

5% 2%
DAP PBS

76% BN   14% PAS   3% PKN

BN = National Front (dominated by the United Malays National Organization – **UMNO**) **PAS** = Pan-Malaysian Islamic Party **DAP** = Democratic Action Party **PKN** = National Justice Party (Keadilan) **PBS** = United Sabah Party

The DAP, the PAS, and the PKN form the Alternative Front

Senate 69 seats

The Senate comprises 26 members indirectly elected by the State Legislative Assemblies, and 43 appointed by the head of state

Supreme power rests in theory with the monarch, acting on the advice of parliament. In practice, the prime minister wields executive authority. Opposition parties, while legal, are under tight control.

PROFILE

Malaysia has been dominated by UMNO, part of the ruling BN coalition, since Malay independence in 1957. It controls a huge network of patronage. In 2002, Prime Minister Mahathir Mohamed shocked the country by tearfully announcing that he would retire in 2003, after 22 years at the helm. He had appeared unassailable, although his authority had been shaken by the economic crisis of 1997–1998 and dissent within the ruling coalition. In 1998, Anwar Ibrahim, deputy prime minister and once Mahathir's chosen successor, was dismissed after challenging the government's economic policy and calling for political reform. He was convicted in 1999 on corruption charges; his prison sentence was lengthened to 15 years in 2000 after his conviction for sodomy. The initial verdict sparked riots and gained support for the new opposition PKN headed by Anwar's wife, Wan Azizah.

Anwar Ibrahim, *in 1998 controversially dismissed by Mahathir.*

Mahathir Mohamed, *prime minister since 1981.*

PKN activists have been hounded by the government. In the 1999 elections, the BN coalition retained its large majority but Mahathir's own UMNO lost ground.

MAIN POLITICAL ISSUE
Malay dominance of government
Mahathir's administration has declared that it no longer wishes to discriminate positively in favor of Malays, but the Chinese community accuses the government of corruption and uncompetitive practices, declaring that Malays are still favored for the placing of government contracts. The Chinese are further alienated by the more restrictive nature of Islamic society.

WORLD AFFAIRS ▷ Joined UN in 1957

 APEC ASEAN Comm G15 OIC

Mahathir sees himself as one of the developing world's leading voices. He maintains a strongly anti-US line in his public speeches and has chastised the West for singling out Islamic countries in its campaign against international terrorism. Mahathir's pro-Malay policies have caused tensions with Singapore, exacerbated by the latter's dependence on Malaysia for water.

AID ▷ Recipient

$45m (receipts)   ⬇ Down 69% in 2000

Most Western aid to Malaysia was used until recently to finance large infrastructure projects. The economic crisis which affected southeast Asia in 1997–1998 forced Malaysia to seek foreign assistance to support an economic recovery program.

M

CHRONOLOGY

The former British protectorate of Malaya gained independence in 1957. The federation of Malaysia, incorporating Singapore, Sarawak, and Sabah, was founded in 1963.

- ❏ **1965** Singapore leaves federation, reducing Malaysian states to 13.
- ❏ **1970** Malay–Chinese ethnic tension forces resignation of Prime Minister Tunku Abdul Rahman. New prime minister, Tun Abdul Razak, creates the BN coalition.
- ❏ **1976-1978** Guerrilla attacks by banned Communist Party of Malaya (CPM), based in southern Thailand.
- ❏ **1976** Death of Tun Abdul Razak.
- ❏ **1977** Unrest in Kelantan following expulsion of its chief minister from Pan-Malaysian Islamic Party (PAS). National emergency declared. PAS expelled from BN.
- ❏ **1978** Elections consolidate BN power. PAS marginalized. Government rejects plans for Chinese university.
- ❏ **1978-1989** Unrestricted asylum given to Vietnamese refugees.
- ❏ **1981** Mahathir Mohamed becomes prime minister.
- ❏ **1982** General election returns BN with increased majority.
- ❏ **1985** BN defeated by PBS in Sabah state elections.
- ❏ **1986** PBS joins BN coalition. Dispute between Mahathir and his deputy, Dakuk Musa, triggers general election, won by BN.
- ❏ **1987** Detention without trial of 106 politicians from all parties suspected of Chinese sympathies. Media censored.
- ❏ **1989** Disaffected UMNO members join PAS. Screening of Vietnamese refugees introduced. CPM signs peace agreement with Malaysian and Thai governments.
- ❏ **1990** General election. BN returned to power with reduced majority.
- ❏ **1993** Sultans lose powers, including legal immunity.
- ❏ **1995** BN wins landslide victory in the country's ninth general election.
- ❏ **1997** Major financial crisis ends decade of spectacular economic growth.
- ❏ **1998–1999** Deputy Prime Minister Anwar Ibrahim dismissed from office. Launches Reformasi (Reform) movement. Found guilty of corruption, later convicted of sodomy and six-year sentence extended to 15 years; his wife, Wan Azizah, forms National Justice Party (PKN). November, UMNO loses ground in general election.
- ❏ **2002** Mahathir announces that he will step down in 2003.

DEFENSE

▷ No compulsory military service

 $2.71bn

 Down 14% in 2000

MALAYSIAN ARMED FORCES

🛡	26 light tanks (*Scorpion*)	80,000 personnel
🚢	4 frigates and 41 patrol boats	12,500 personnel
✈	71 combat aircraft (25 *Hawk*, 15 MiG-29, 9 MB-339, 8 F/A-18D)	8000 personnel
⚓	None	

Malaysia's armed forces are predominantly composed of Malays. Main defense concerns are Singapore, with its large and highly mechanized army, and more recently, though to a lesser extent, Indonesia. Also important to Malaysia is the growing Chinese influence in the South China Sea. Patrolling east and west Malaysia is a key function of the navy, which is large by regional standards.

Malaysia is an important market for Western arms suppliers. However, since 1994 Malaysia has bought a number of Russian MiG-29 fighter aircraft.

ECONOMICS

▷ Inflation 3.9% p.a. (1990–2000)

 $78.7bn

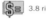 3.8 ringgits

SCORE CARD

- ❏ WORLD GNP RANKING 42nd
- ❏ GNP PER CAPITA $3380
- ❏ BALANCE OF PAYMENTS....................$12.6bn
- ❏ INFLATION 1.5%
- ❏ UNEMPLOYMENT 3%

EXPORTS

- Netherlands 4%
- China 5%
- Japan 13%
- Other 39%
- Singapore 18%
- USA 21%

IMPORTS

- South Korea 4%
- Taiwan 6%
- Singapore 14%
- Other 38%
- USA 17%
- Japan 21%

STRENGTHS

Electronics, computer hardware, and electrical appliances. Tourism. Heavy industries such as steel. Palm oil. Latex, rubber, chemical products. Success of "national car," the Proton. Return to growth from 2000.

MALAYSIA : MAJOR BUSINESSES

- Perai
- Kuala Terengganu
- George Town
- Ipoh
- Tanah Rata
- Kuala Lumpur
- Maran
- Port Dickson
- Melaka
- Johor Bahru

| 0 | 200 km |
| 0 | 200 miles |

ECONOMIC PERFORMANCE INDICATOR

— Consumer Price Index   GDP

WEAKNESSES

High level of debt. Shortage of skilled labor. High interest rates deter private investors. High government budget spending. Competition from NICs.

PROFILE

From 1987, for almost a decade, Malaysia expanded faster than any other southeast Asian nation, at an average yearly rate of 8%, with much of the growth state-directed. However, plans for full industrialization, named "Vision 2020," were revised after the 1997 financial crisis. The construction of Putrajaya provided a renewed stimulus to growth at the end of the 1990s. A project for a Multimedia Super Corridor (MSC), located south of Kuala Lumpur and aimed at attracting world-class companies, is expected to be completed by 2003.

- Lutong

Palm oil	🌴
Tin mining	⛏
Electronics	💻
Oil refining	⛽
Petrochemicals	🛢
Vehicle assembly	🚗

* significant multinational ownership

RESOURCES

 Electric power 13.6m kw

1.41m tonnes

788,000 b/d (reserves 3bn barrels)

13m ducks, 1.83m pigs, 125m chickens

Natural gas, oil, tin, bauxite, copper, iron, coal

ELECTRICITY GENERATION

Hydro 8% (4.8bn kwh)	
Combustion 92% (56bn kwh)	
Nuclear 0%	
Other 0%	

% of total generation by type

(scale: 0, 20, 40, 60, 80, 100)

Palm oil, of which Malaysia is the world's largest producer, is now the major export product, whereas rubber production is now overshadowed by that of neighboring Thailand. Malaysia is a significant exporter of petroleum and natural gas. Reserves lie offshore from Sabah and Sarawak. The petroleum is high grade, so that most is exported, while crude imports are refined.

PENINSULAR MALAYSIA

MALAYSIA : LAND USE

Cropland
Forest
Pigs
Rubber
Palm oil

SABAH
SARAWAK
KALIMANTAN

0 200 km
0 200 miles

Malaysia accounts for nearly half of world timber exports, most of which come from Sarawak.

ENVIRONMENT

 Sustainability rank: 68th

5%

5.4 tonnes per capita

ENVIRONMENTAL TREATIES

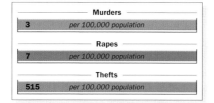

Yes	Yes	Yes
Yes	Yes	No

Logging is the overwhelming concern. World Bank estimates suggest that trees are being cut down at four times the sustainable rate. Indigenous forest communities are being destroyed, and some species of wood are near extinction. The government pledged in 1997 to plant 20 million trees across the country by 2000.

Smog caused by burning forests and scrub in Indonesia creates a pollution and health threat to which the whole region is alerted.

Traditional lifestyles are threatened by grandiose modernization schemes, and the Bakun Dam project, shelved in 1997 due to a lack of investment confidence, was restarted in 2000.

MEDIA

 TV ownership medium

Daily newspaper circulation 112 per 1000 people

PUBLISHING AND BROADCAST MEDIA

	There are 42 daily newspapers. The most influential are the *New Straits Times*, *Utusan Malaysia*, and *Sin Chew Jit Poh*
	7 services: 3 state-controlled, 4 independent
	5 services: 3 state-controlled, 2 independent

Almost all newspapers, TV stations, and radio broadcasts are strictly controlled by the state and UMNO, which claim a need to protect the country from un-Islamic influences. A series of laws passed in the 1980s regulate content. The Internet provides a medium for free reporting. The leading website is Malaysiakini.

CRIME

 Death penalty in use

25,994 prisoners

Up 80% 1996–1998

CRIME RATES

Murders	
3	*per 100,000 population*

Rapes	
7	*per 100,000 population*

Thefts	
515	*per 100,000 population*

The death sentence for possession of narcotics is mandatory. The Internal Security Act allows detention without trial. Some Islamic-dominated states have attempted to implement the harsh punishments of the Islamic penal code, including stoning and amputation.

EDUCATION

 School leaving age: 15

88%

230,000 students

THE EDUCATION SYSTEM

% of each age group in education

- Primary: 99%
- Secondary: 98%
- Tertiary: 12%

Racial integration in multiethnic schools is encouraged, but teaching is to remain in separate languages.

At tertiary level a quota system gives Malays preference for places. The Chinese community has its own schools, but plans for a private Chinese university were vetoed by the government. Many students, particularly the Chinese, complete their studies in the UK or the US. From 2002, university students and staff have had to swear allegiance to the state.

HEALTH

 Welfare state health benefits

1 per 1429 people

Heart diseases, cancers

There is growing disparity between the modern facilities available in cities and the traditional medicine practiced in rural and outlying areas. Traditional practices such as acupuncture and herbal medicine continue to be used by the Chinese community.

SPENDING

 GDP/cap. increase

CONSUMPTION AND SPENDING

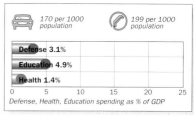

170 per 1000 population

199 per 1000 population

Defense 3.1%	
Education 4.9%	
Health 1.4%	

Defense, Health, Education spending as % of GDP

(scale: 0, 5, 10, 15, 20, 25)

The Chinese remain the wealthiest community in Malaysia. However, following riots in 1970, the UMNO government embarked on a deliberate program of achieving 30% Malay ownership of the corporate sector. The "extremely rich" were barred from government service in 2001 in an effort to stamp out "money politics."

WORLD RANKING

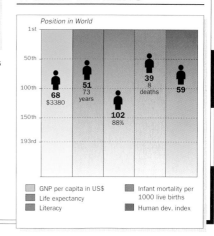

Position in World

- 68 $3380
- 51 / 73 years
- 102 / 88%
- 39 / 8 deaths
- 59

GNP per capita in US$
Life expectancy
Literacy
Infant mortality per 1000 live births
Human dev. index

M

389

MALDIVES

OFFICIAL NAME: Republic of Maldives **CAPITAL:** Male'
POPULATION: 300,000 **CURRENCY:** Rufiyaa **OFFICIAL LANGUAGE:** Dhivehi

 1965   1965   July 26   MV   +5   +960   .mv

AN ISLAMIC SULTANATE until 1968, the Maldives is an archipelago of 1190 small coral islands or atolls (a word derived from the local Dhivehi language), set in the Indian Ocean southwest of India. The islands, none of which rise above 1.8 m (6 ft), are protected by encircling reefs or faros. Only 200 are inhabited. Tourism has grown in recent years, though vacation islands are separate from settled islands.

Traditional Maldivian trading yacht.
The 1190 coral islands are grouped in natural atolls, derived from the Dhivehi word "atolu."

CLIMATE ▷ Tropical oceanic

WEATHER CHART FOR MALE'

The Maldives has a tropical climate, with abundant rainfall and high temperatures throughout the year. The northern islands are occasionally affected by violent storms caused by tropical cyclones. Most rain falls in the southern islands.

TRANSPORTATION ▷ Drive on left

✈ **Male' International, Hulule Island** 1.71m passengers ⚓ 68 ships 101,100 grt

THE TRANSPORTATION NETWORK

🛣 10 km (6 miles)	🛣 None
🚃 None	〰 None

It is possible to walk across Male' island in 20 minutes. Interisland travel is mostly by ferry and traditional *dhoni*.

TOURISM ▷ Visitors : Population 1.6:1

🧳 467,000 visitors ⬆ Up 9% in 2000

MAIN TOURIST ARRIVALS

Italy 21%	
Germany 20%	
UK 15%	
Other 44%	

% of total arrivals

Tourism is the largest source of foreign exchange, accounting for almost 20% of GDP. The first resort was opened in 1972, and hotels financed by local and foreign capital have since been built on the uninhabited islands. There are now nearly half a million visitors a year.

POLITICS ▷ Nonparty elections

 1999/2004   President Maumoon Abdul Gayoom

AT THE LAST ELECTION

Citizens' Assembly 50 seats

There are no political parties. 42 members of the Majlis (Assembly) are elected, and 8 appointed by the president

Politics in the Maldives is the preserve of a small group of influential families. Most were already dominant under the sultanate. Formal parties with ideological objectives are virtually nonexistent, politics being organized around family and clan loyalties.

Former president Ibrahim Nasir abolished the premiership in 1975 and substantially strengthened the presidency. The main figure now is Maumoon Abdul Gayoom, a wealthy businessman who has been president since 1978. His brother-in-law, Ilyas Ibrahim, is regarded as his main rival.

A young Westernized elite has increased the pressure for political reform. Under a constitution effective since 1998, rival candidates may seek to be parliament's presidential nominee; only one name then goes forward for popular endorsement in a referendum.

PEOPLE ▷ Pop. density high

👤 Dhivehi (Maldivian) 👥 1000/km² (2590/mi²)

THE URBAN/RURAL POPULATION SPLIT

26%   74%

RELIGIOUS PERSUASION

Sunni Muslim 100%

It is believed that the islands were inhabited as early as 1500 BCE. Aryan immigrants arrived around 500 BCE. The islands were then discovered by Arab traders. The people, who are all Sunni Muslims, live on only 200 of the 1190 islands. About 25% of the total population live on the island capital of Male'. It is estimated that 12,000 guest workers from neighboring Sri Lanka and India work in the Maldives. The country's newfound prosperity has seen the emergence of a commercial elite.

Map

INDIAN OCEAN
MALDIVES · India · Asia · Sri Lanka

M

Eight Degree Channel

Ihavandippolhu Atoll
Thiladhunmathi Atoll
Makunudhoo Atoll
North Miladummadulu Atoll
South Miladummadulu Atoll
North Maalhosmadulu Atoll
Faadhippolhu Atoll
South Maalhosmadulu Atoll
Horsburgh Atoll
Rasdu Atoll
Male' Atoll
Ari Atoll
✈ **MALE'**
Felidhu Atoll
North Nilandhe Atoll
Mulaku Atoll
South Nilandhe Atoll
Kolhumadulu Atoll
Hadhdhunmathi Atoll

One and Half Degree Channel

North Huvadhu Atoll
South Huvadhu Atoll

Equatorial Channel — Equator

Fuammulah
Addu Atoll
Gan

INDIAN OCEAN

MALDIVES

Total Land Area : 300 sq. km (116 sq. miles)

POPULATION
over 10 000 ●
under 10 000 ·

LAND HEIGHT
100m/328ft
Sea Level

N

0 — 100 km
0 — 100 miles

WORLD AFFAIRS
 Joined UN in 1965

The Maldives is a long-standing member of the NAM. The government continues to support it and rejects the view that the organization does not have a role to play in the post-Cold War world. The Maldives' international standing was enhanced in 1990, when it hosted the fifth SAARC summit meeting, held in Male'.

AID
 Recipient

$19m (receipts)　　Down 39% in 2000

Aid has helped to finance the development of port and airport facilities. Japan is the most important bilateral aid donor. The Maldives is classed by the UN as a Least Developed Country, which grants it access to special financial programs; it hopes to graduate from this listing.

DEFENSE
No compulsory military service

$44m　　Up 7% in 2000

The British military presence ended in 1975, when troops were withdrawn from the staging post on Gan, in the Addu atoll. The Maldives follows a policy of nonalignment, but in 1988 called on India for military assistance to help suppress a coup attempt.

ECONOMICS
Inflation 5.8% p.a. (1990–2000)

$541m　　11.77 rufiyaa

SCORE CARD

- ❑ WORLD GNP RANKING........................170th
- ❑ GNP PER CAPITA $1960
- ❑ BALANCE OF PAYMENTS.....................–$60m
- ❑ INFLATION ....–1.1%
- ❑ UNEMPLOYMENT 1%

STRENGTHS
Boom in tourism. Thriving fishing industry, especially tuna. Shipping. Clothing. Coconut production. Economic reforms since 1989 have eased import restrictions and encouraged foreign investment.

WEAKNESSES
Too dependent on fluctuating tourist industry. Growing trade deficit. Skilled labor shortage. Small manufacturing base. Cottage industries employ 25% of workforce; little scope for expansion.

EXPORTS

IMPORTS

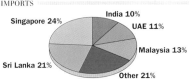

RESOURCES
Electric power 25,000 kw

133,547 tonnes　　Not an oil producer

31,000 cattle, 20,000 goats, 11,000 sheep　　None

Natural resources include abundant stocks of fish, particularly tuna. Fishing, still carried out by the traditional pole and line method to help conserve stocks, employs over 20% of the working population. Coconut production is also important. All oil products and virtually all staple foods are imported.

ENVIRONMENT
Not available

None　　1.3 tonnes per capita

Rising sea levels due to global warming and climate change threaten the islands, which have an average height of just 1.5 m (5 ft). A sea wall has been built around the capital island.

Other environmental concerns are sewerage, waste disposal, and the mining of coral for building.

MEDIA
TV ownership low

Daily newspaper circulation 19 per 1000 people

PUBLISHING AND BROADCAST MEDIA

There are 3 daily newspapers, including *Haveeru Daily* and *Aafathis*, published in Dhivehi and English

1 state-owned service　　2 services

There is a marked degree of press self-censorship; in the past, journalists have been imprisoned. An Internet café opened in Male' in 1998.

CRIME
Death penalty not used in practice

1098 prisoners　　Up 90% 1992–1996

The Maldives is a strict Islamic society. Narcotics crimes are heavily punished. Political prisoners are banished to outer islands. The judiciary and executive are closely linked.

EDUCATION
Schooling is not compulsory

96%　　Not available

Primary education has been improved. Secondary education is less developed in the outer islands; the first school outside Male' was opened in 1992.

HEALTH
Welfare state health benefits

1 per 1328 people　　Infectious and parasitic diseases, tuberculosis, perinatal deaths

There is a lack of general equipment and facilities. Health care is less developed on the outlying islands.

SPENDING
GDP/cap. increase

CONSUMPTION AND SPENDING

7 per 1000 population　　91 per 1000 population

Defense 9.5%
Education 6.4%
Health 3.7%

Defense, Health, Education spending as % of GDP

Great disparities of wealth exist between the people who live in Male' and those who live on the more distant outer islands.

WORLD RANKING

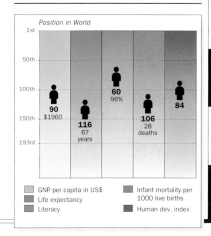

Position in World

GNP per capita in US$ — 90 $1960
Life expectancy — 116 67 years
Literacy — 60 96%
Infant mortality per 1000 live births — 106 28 deaths
Human dev. index — 84

M

MALI

WEST AFRICA

OFFICIAL NAME: Republic of Mali CAPITAL: Bamako
POPULATION: 11.7 million CURRENCY: CFA franc OFFICIAL LANGUAGE: French

MALI IS LANDLOCKED in the heart of west Africa. Its mostly flat terrain comprises virtually uninhabited Saharan plains in the north and more fertile savanna land in the south, where most of the population lives. The Niger River irrigates the central and southwestern regions. Mali achieved independence from France in 1960. Multiparty democratic elections under a new constitution, in 1992 and then in 1997, provoked accusations of severe irregularities.

POLITICS

 Multiparty elections

 2002/2007   President Amadou Toumani Touré

AT THE LAST ELECTION
National Assembly 147 seats

45%	35%	7%	5%	4%	4%
Espoir 2002	ARD	ACC	Vacant	SADI	Ind

Espoir 2002 = Hope 2002 (led by the Rally for Mali – **RPM**)
ARD = Alliance for the Republic and Democracy (led by the Alliance for Democracy in Mali – **ADEMA**)
ACC = Convergence for Rotation and Change
Ind = Independents   **SADI** = Party for African Solidarity, Democracy, and Integration

The successful transition to multiparty politics in 1992 followed the overthrow the previous year of Moussa Traoré, Mali's dictator for 23 years. The army's role was crucial in leading the coup, while Col. Amadou Toumani Touré, who acted as interim president, was responsible for the swift return to civilian rule in less than a year. For a decade President Alpha Oumar Konaré's ADEMA government attempted to alleviate poverty while placating the opposition. However, its economic austerity measures proved unpopular. Col. Touré was returned to power in the 2002 presidential elections with the support of all main opposition parties, and in the legislative polls ADEMA saw its majority disappear. It sought to form a coalition with smaller parties. Maintaining good relations with the Tuareg remains a key issue.

CLIMATE

Hot desert/steppe

WEATHER CHART FOR BAMAKO

In the south, intensely hot, dry weather precedes the westerly rains. Mali's northern half is almost rainless.

TRANSPORTATION

Drive on right

Bamako–Senou
417,974 passengers

Has no fleet

THE TRANSPORTATION NETWORK

1827 km (1135 miles)	None
729 km (453 miles)	1815 km (1128 miles)

Mali is linked by rail with the port of Dakar in Senegal, and by good roads to the port of Abidjan in Ivory Coast.

TOURISM

Visitors : Population 1:129

91,000 visitors   Up 5% in 2000

MAIN TOURIST ARRIVALS

France 27%
Africa 20%
USA 6%
Other 47%

0 10 20 30 40 50 60
% of total arrivals

Tourism is largely safari-oriented, though the historic cities of Djénné, Gao, and Mopti, lying on the banks of the Niger River, also attract visitors. A national domestic airline began operating in 1990.

PEOPLE

Pop. density low

Bambara, Fulani, Senufo, Soninke, French

10/km² (25/mi²)

THE URBAN/RURAL POPULATION SPLIT

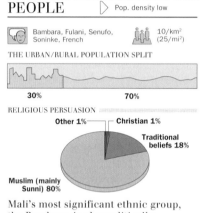

30%   70%

RELIGIOUS PERSUASION

Other 1%   Christian 1%
Traditional beliefs 18%
Muslim (mainly Sunni) 80%

Mali's most significant ethnic group, the Bambara, is also politically dominant. The Bambara speak the *lingua franca* of the Niger River, which is shared with other groups including the Malinke. The relationship between the Bambara–Malinke majority and the Tuareg nomads of the Saharan north is tense and sometimes violent. The extended family is a vital social security system and link between the urban and rural poor. There are a few powerful women in Mali but, in general, women have little status.

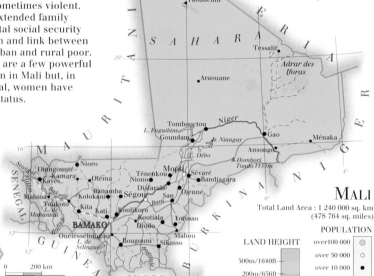

MALI

Total Land Area : 1 240 000 sq. km (478 764 sq. miles)

POPULATION

LAND HEIGHT	
500m/1640ft	over 100 000 ⊚
200m/656ft	over 50 000 ○
over 100m/328ft	over 10 000 ●
	under 10 000 ·

M

WORLD AFFAIRS

 Joined UN in 1960

 ECOWAS FZ AU OIC OIF

Mali concentrates on maintaining good relations with the ECOWAS countries to its south and northern neighbors such as Algeria. Relations with Libya, which is suspected of fomenting Tuareg revolt, are tense. There are good relations with the US and other Western aid providers.

AID

 Recipient

 $360m (receipts)   Up 2% in 2000

Mali is highly dependent on foreign aid, which principally comes from the World Bank, France, the US, Germany, the Netherlands, Japan, and the EU.

DEFENSE

 Compulsory military service

$29m   Down 15% in 2000

Mali's 7800-strong armed forces have stayed out of politics since the overthrow of President Traoré in 1991.

ECONOMICS

 Inflation 7.1% p.a. (1990–2000)

 $2.55bn   698.7–736.7 CFA francs

SCORE CARD

❏ World GNP Ranking	.......................132nd
❏ GNP per Capita	.............................$240
❏ Balance of Payments	...................–$178m
❏ Inflation	....–0.7%
❏ Unemployment	........Widespread underemployment

Strengths
Producer of high-quality cotton. Irrigation potential from the Niger and Senegal rivers. Rapid expansion of gold production now under way.

Weaknesses
Serious poverty and underdevelopment. Communications difficulties of vast landlocked country. Drought-prone climate. Cotton exports vulnerable to low world prices.

EXPORTS

Germany 7%
Portugal 7%
Canada 9%
Other 56%
Italy 9%
Brazil 12%

IMPORTS

Germany 4%
Senegal 4%
South Africa 3%
France 13%
Other 59%
Ivory Coast 17%

Village near Bandiagara. *The low, broken hills typical of the east and southeast of Mali are the homeland of the Dogon people.*

RESOURCES

 Electric power 114,000 kw

 98,766 tonnes   Not an oil producer

9.9m goats, 6.82m cattle, 6.4m sheep, 25m chickens   Gold, salt, marble, phosphates, tungsten, diamonds, oil

Gold deposits are now being mined, and prospecting is under way for tungsten, diamonds, and oil. Exploitation of natural resources is hampered by Mali's poor infrastructure and landlocked situation. Electric power comes from the Selingue Dam on the Niger and the Manantali Dam on the Senegal. The latter produced its first electricity in 2001 – 13 years after it was completed.

ENVIRONMENT

 Sustainability rank: 85th

 4%   0.05 tonnes per capita

Severe drought in 1983 destroyed herds and accelerated desertification and deforestation. The Selingue Dam seriously affects the levels of the Niger, even in years of good rainfall.

MEDIA

 TV ownership low

Daily newspaper circulation 1 per 1000 people

PUBLISHING AND BROADCAST MEDIA

There are 5 daily newspapers, including the pro-government *L'Essor – La Voix du Peuple*

3 services: 1 state-owned, 2 independent

15 services: 1 state-owned, 14 independent

Even before the 1991 coup, previously rigid controls were being relaxed. The 1992 constitution guarantees the freedom of the press, and Mali's broadcast and print media are now among the freest in Africa.

CRIME

 Death penalty not used in practice

 3135 prisoners   Crime is rising slowly

Crime is not particularly prevalent compared with some other countries in the region, owing at least in part to the relative lack of urbanization. In towns, robbery, juvenile delinquency, and smuggling are problems.

CHRONOLOGY

Mali was a major trans-Saharan trading empire. The French colonized the area between 1881 and 1895.

- ❏ **1960** Independence.
- ❏ **1968** Coup by Gen. Moussa Traoré.
- ❏ **1990** Pro-democracy demonstrations.
- ❏ **1991** Traoré arrested.
- ❏ **1992** Free multiparty elections.
- ❏ **1997** President Konaré and ADEMA party reelected in disputed polls.
- ❏ **2002** Elections: Col. Touré president; ADEMA loses majority.

EDUCATION

School leaving age: 13

41%   18,662 students

Only 25% of children go to primary school and just 7% to secondary school. A ten-year program to raise education levels for girls was launched in 2001.

HEALTH

No welfare state health benefits

1 per 10,000 people   Malaria, pneumonia, parasitic and diarrheal diseases

A health program began in 1998, with the aim of higher immunization rates for children and more health care access.

SPENDING

GDP/cap. increase

CONSUMPTION AND SPENDING

3 per 1000 population   3 per 1000 population

Defense 1%
Education 3%
Health 2.1%

0   5   10   15   20   25
Defense, Health, Education spending as % of GDP

Poverty is widespread, and wealth is limited to a very small group; Malians disapprove of flaunted wealth and public ostentation is rare.

M

WORLD RANKING

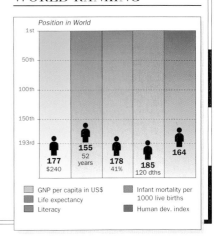

Position in World

1st
50th
100th
150th
193rd

177 $240
155 52 years
178 41%
185 120 dths
164

GNP per capita in US$
Infant mortality per 1000 live births
Life expectancy
Literacy
Human dev. index

MALTA

OFFICIAL NAME: Republic of Malta **CAPITAL:** Valletta
POPULATION: 392,000 **CURRENCY:** Maltese lira **OFFICIAL LANGUAGES:** Maltese and English

 1964   1964   Sept 21   M   +1   +356   .mt

THE MALTESE ARCHIPELAGO is strategically located, lying between Europe and north Africa. Controlled throughout its history by successive colonial powers, Malta finally gained independence from the UK in 1964. The islands are mainly low-lying, with rocky coastlines; only Malta, Gozo (Ghawdex), and Kemmuna are inhabited. Tourism is Malta's chief source of income, with an influx of tourists each year of over three times the islands' population.

CLIMATE ▷ Mediterranean

WEATHER CHART FOR VALLETTA

The climate is typical of the southern Mediterranean – with at least six hours of sunshine a day, even in winter.

TRANSPORTATION ▷ Drive on left

 Luqa International, Valletta
3m passengers

 3189 ships
28.6m grt

THE TRANSPORTATION NETWORK

1677 km (1042 miles)	None
None	None

Malta Freeport at Marsaxlokk exploits Malta's strategic shipping location in the Mediterranean. In summer, a five-minute helicopter flight from the international airport links the islands of Malta and Gozo. There is a well-developed public transportation system, with ferry and hovercraft services and buses on both islands.

Traditionally painted **luzzus** *at St. Julian's harbor. The fish caught are now only for domestic and tourist consumption.*

TOURISM ▷ Visitors : Population 3.1:1

 1.2m visitors ⬍ No change in 2000

MAIN TOURIST ARRIVALS

UK 35%	
Germany 18%	
Italy 8%	
Other 39%	

0 10 20 30 40
% of total arrivals

Tourism is vital to the economy and accounts for more than 30% of GDP, even though most visitors are budget vacationers. In addition to beaches and scenery, there are the historical attractions of Mdina and Valletta. Development on the quieter island of Gozo is limited to luxury-grade hotels.

PEOPLE ▷ Pop. density high

 Maltese, English 1225/km² (3173/mi²)

THE URBAN/RURAL POPULATION SPLIT

91% 9%

RELIGIOUS PERSUASION

Other and nonreligious 2%

Roman Catholic 98%

Malta's population has been subject over the centuries to diverse Arabic, Sicilian, Norman, Spanish, English, and Italian influences. Today, much of the younger Maltese population goes abroad to find work, especially to the US or Australia; opportunities for them on the islands are few.

The Maltese are staunch Roman Catholics, on a percentage basis more so than virtually any other nation. The remainder are mainly Anglicans, who are included within the diocese of Gibraltar. Divorce is illegal.

POLITICS ▷ Multiparty elections

1998/2003 President Guido de Marco

AT THE LAST ELECTION

House of Representatives 65 seats

54%
NP

46%
MLP

NP = Nationalist Party **MLP** = Malta Labour Party

Maltese politics is strongly adversarial and evenly split between the right-wing NP and the left-wing MLP. The latter dominated the government in the 1970s and 1980s, ensuring state control of industry and pursuing a nonaligned foreign policy.

The 1990s, however, saw a switch in favor of the NP, with Prime Minister Edward Fenech Adami at the helm in 1987–1996 and since 1998. Under Fenech Adami, Malta moved toward ever closer ties with Europe, and favors a free-market approach to its economy. The NP secured reelection in 1992, largely due to a rise in living standards. A modernized MLP ended the NP's nine-year reign in 1996. Under Alfred Sant, a leading writer and Harvard MBA, it diluted traditional links with the unions and "froze" Malta's EU application. However, the MLP's small parliamentary majority undermined the government, and the NP won early elections in 1998. Fenech Adami has now reset Malta on its course for membership of the EU.

WORLD AFFAIRS ▷ Joined UN in 1964

 CE   Comm   IBRD   NAM   OSCE

Malta has made the most of its location on the fringe of Europe, with a staunchly nonaligned foreign policy. Ties are traditionally strong with the Arab world and north Africa, and relations with Libya remain good. There are also close commercial links with Russia and China.

However, it is the island's relationship with Europe that has dominated recent policy. Malta's bid for EU membership, launched in 1990, was derailed by the anti-EU MLP government in 1996. With the application frozen, Malta was denied a place in the "first wave" of potential EU members. However, the return to power of the pro-EU NP in 1998 restarted the bid and Malta joined talks in March 2000 as part of the "second wave."

M

MALTA

Total Land Area : 316 sq. km
(122 sq. miles)

POPULATION

- over 10 000
- under 10 000

LAND HEIGHT

200m/656ft
Sea Level

AID

 Recipient

 $21m (receipts)   Down 16% in 2000

Malta receives economic assistance under an agreement with the EU. Italy is the main bilateral source of aid.

DEFENSE

No compulsory military service

$26m   Down 4% in 2000

The Maltese army, advised by the Libyans in the 1980s, now receives training and equipment from Italy, Germany, and the UK.

ECONOMICS

Inflation 2.8% p.a. (1990–2000)

$3.56bn   0.4357–0.4503 Maltese liri

SCORE CARD

- ❏ World GNP Ranking..........................127th
- ❏ GNP per Capita $9120
- ❏ Balance of Payments.....................–$515m
- ❏ Inflation 2.4%
- ❏ Unemployment 5%

STRENGTHS
Tourism and naval dockyards. Schemes to attract foreign high-tech industry. Malta Freeport container distribution center. Offshore banking. Strategic position between Europe and Africa, on main Mediterranean shipping lines.

WEAKNESSES
Cut-rate competition from Africa and Asia in traditional textile industry. Need to import almost all requirements.

EXPORTS

UK 7%
France 8%
Germany 10%
Singapore 16%
USA 27%
Other 32%

IMPORTS

Germany 8%
USA 11%
Singapore 15%
Italy 17%
France 19%
Other 30%

RESOURCES

Electric power 250,000 kw

 3035 tonnes   Reserves under exploration

80,074 pigs, 19,200 cattle, 820,000 chickens   Stone, sand, oil

Malta is dependent on desalination plants for most of its water supply. All oil has to be imported, mostly from Libya. However, there are petroleum reserves currently under exploration in Maltese waters.

ENVIRONMENT

Not available

 None   4.7 tonnes per capita

The main environmental concern is linked to the tourist industry. A lack of planning controls in the 1970s was responsible for unsightly beach developments. These are now tightly controlled, particularly on Gozo.

MEDIA

TV ownership high

 Daily newspaper circulation 127 per 1000 people

PUBLISHING AND BROADCAST MEDIA

There are 4 daily newspapers, *In-Nazzjon*, *L-Orizzont*, *The Times*, and *The Malta Independent*

7 services: 1 state-owned, 6 independent

12 services: 1 state-owned, 11 independent

The Maltese press is largely party politically oriented. Two of the three main press groups are affiliated to the NP or MLP; one is independent.

CRIME

No death penalty

 260 prisoners   Down 19% 1992–1996

Crime rates are low compared with those on the European mainland. There has been an increase in narcotics transshipment and associated crimes.

EDUCATION

School leaving age: 16

 92%   8260 students

One-third of pupils attend non-state schools, including heavily subsidized church-run institutions. There is a state university in Valletta.

HEALTH

Welfare state health benefits

1 per 383 people   Cerebrovascular and heart diseases, cancers, diabetes

Malta has five state-run and a couple of private hospitals. Diabetes is prevalent, as on other Mediterranean islands.

SPENDING

GDP/cap. increase

CONSUMPTION AND SPENDING

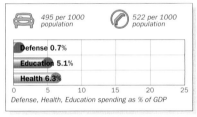

495 per 1000 population   522 per 1000 population

Defense 0.7%
Education 5.1%
Health 6.3%

Defense, Health, Education spending as % of GDP

Remittances from Maltese working abroad are an important source of income for many island families.

WORLD RANKING

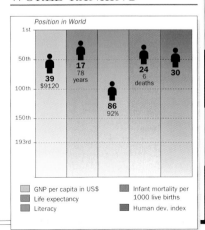

Position in World

39 $9120
17 78 years
86 92%
24 6 deaths
30

GNP per capita in US$
Life expectancy
Literacy
Infant mortality per 1000 live births
Human dev. index

M

MARSHALL ISLANDS

PACIFIC OCEAN

OFFICIAL NAME: Republic of the Marshall Islands   **CAPITAL:** Majuro
POPULATION: 68,100   **CURRENCY:** US dollar   **OFFICIAL LANGUAGES:** English and Marshallese

   1986   1986   May 1   MH   +12   +692 .mh

THE MARSHALL ISLANDS comprise a group of 34 widely scattered atolls in the central Pacific Ocean, formerly under US rule as part of the UN Trust Territory of the Pacific Islands. An agreement which granted internal sovereignty in free association with the US became operational in 1986, and the Trust was formally dissolved in 1990. The economy is almost entirely dependent on US aid and rent for the US missile base on Kwajalein Atoll.

Ebeye District on Kwajalein Atoll.
Population pressures on the island have led to the disappearance of most tree and grass cover.

CLIMATE ▷ Tropical oceanic

WEATHER CHART FOR MAJURO

The climate is tropical oceanic with little seasonal variation; temperatures average around 30°C (86°F).

TRANSPORTATION ▷ Drive on right

Majuro International

207 ships
6.44m grt

THE TRANSPORTATION NETWORK

| Surfaced roads only on larger islands | None |
| None | None |

The transportation system is limited, although there is some interisland shipping. State carrier Air Marshalls has experienced economic difficulties.

TOURISM ▷ Visitors : Population 1:14

5000 visitors   No change in 2000

MAIN TOURIST ARRIVALS

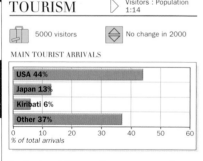

USA 44%
Japan 13%
Kiribati 6%
Other 37%

% of total arrivals

In the late 1990s major resort complexes were established on Majuro and on Mili Atoll. Attractions include diving, game fishing, and exploring the sites and relics of World War II battles.

PEOPLE ▷ Pop. density high

 Marshallese, English, Japanese, German   376/km² (975/mi²)

THE URBAN/RURAL POPULATION SPLIT

69%   31%

ETHNIC MAKEUP

Other 3%
Micronesian 97%

Of the 34 atolls making up the Marshall Islands, 24 are inhabited. Majuro, the capital and commercial center, is home to almost half of the population, many of whom live in its overcrowded slums. The other main center of population is Ebeye Island in the Kwajalein Atoll, where tensions are high due to poor living conditions. Most of Kwajalein Atoll's inhabitants were forcibly relocated to Ebeye in 1947 to make way for a US missile tracking, testing, and interception base; many still travel daily to work at the base. Life on the outlying islands is still centered on subsistence agriculture and fishing. Society is traditionally matrilineal.

MARSHALL ISLANDS

Total Land Area : 181 sq. km (70 sq. miles)

LAND HEIGHT
100m/328ft
Sea Level

POLITICS ▷ Multiparty elections

 1999/2003   President Kessai Note

AT THE LAST ELECTION

Parliament 33 seats

55% UDP   45% K

UDP = United Democratic Party   K = Pro-Kabua Grouping
The 33 members are elected from 25 districts

Council of Chiefs 12 seats

All 12 members are high chiefs

Politics is traditionally dominated by chiefs. Amata Kabua, the islands' high chief and first president until his death in 1996, was succeeded in early 1997 by his cousin Imata Kabua. However, the 1999 elections were won by the United Democratic Party, whose presidential candidate, commoner and former parliamentary speaker Kessai Note, was elected in early January 2000. Just over a year later Imata Kabua instigated an unsuccessful vote of no confidence in Note's administration – only the second in the islands' history. The vote was motivated by criticism of the government approach to the crucial renegotiation of the Compact of Free Association with the US. The original treaty, which provided most of the islands' revenue and defense, expired in 2001; a new treaty is under negotiation.

WORLD AFFAIRS
▷ Joined UN in 1991

The Compact of Free Association has made ties to the USA of key importance. From 1986 the US provided $1 billion in return for the use of Kwajalein Atoll as a missile range and has determined the islands' foreign and defense policies. A new Compact is under discussion. Taiwan has become a source of funding for development, provoking controversy over diplomatic recognition.

AID
▷ Recipient

 $57m (receipts)   Down 10% in 2000

US aid accounts for around two-thirds of the islands' revenue. Australia and Japan also provide some assistance.

DEFENSE
▷ No compulsory military service

 USA is responsible for defense   Not applicable

There is no defense force. All defense is provided by the US under the Compact of Free Association. The US does not have offensive weapons sited in the Marshalls, but its navy patrols regularly.

ECONOMICS
▷ Inflation 5.6% p.a. (1990–2000)

$102m   Currency is US dollar

SCORE CARD
- ❑ WORLD GNP RANKING.........................187th
- ❑ GNP PER CAPITA $1970
- ❑ BALANCE OF PAYMENTS $21m
- ❑ INFLATION 5%
- ❑ UNEMPLOYMENT................................31%

STRENGTHS
US guarantee against economic collapse to preserve strategic influence. Aid from the US, on which islands almost totally depend. Copra. Huge tourism potential.

WEAKNESSES
High unemployment. Dependence on imports, which are twice as large as exports. All fuel has to be imported. Vulnerability to storm damage. Large state sector employs 75% of workers.

EXPORTS

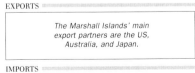

The Marshall Islands' main export partners are the US, Australia, and Japan.

IMPORTS

Singapore 3%   Australia 4%
Japan 3%   Guam 5%
USA 47%   Other 38%

RESOURCES
▷ Not available

 400 tonnes   Not an oil producer

Not available   Phosphates

There are few known strategic resources. Exploratory tests have revealed some high-grade phosphate deposits, but not in economically viable quantities. Small diesel generators are used for electricity production.

ENVIRONMENT
▷ Not available

 None   Not available

Between 1946 and 1958, Bikini, Enewetak, and neighboring atolls were rendered uninhabitable by a series of US nuclear military tests. Enewetak residents were allowed to return in 1980, and Rongelap was declared habitable in 2001. A 1999 tribunal adopted stringent standards for further decontamination. The US has now paid out over $101 million to victims of nuclear testing. Nuclear waste imports were banned in 1999. The effects of rising sea levels are a major concern. Erosion affects beaches and soil is being lost and also contaminated by brackish water.

MEDIA
▷ TV ownership low

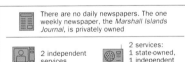 There are no daily newspapers

PUBLISHING AND BROADCAST MEDIA

There are no daily newspapers. The one weekly newspaper, the *Marshall Islands Journal*, is privately owned

2 independent services

2 services:
1 state-owned,
1 independent

Radio is the major source of information in the Marshalls. The main TV service is subscription-only. The US personnel stationed on Kwajalein have their own TV and radio stations.

CRIME
▷ No death penalty

 23 prisoners   Little change from year to year

Crime levels are generally low; however, the rate is up in Ebeye. Outlying islands are crime-free.

EDUCATION
▷ School leaving age: 14

 91%   251 students

Education, compulsory between the ages of six and 14 years, is based on the US model. The number of secondary school graduates exceeds the availability of suitable employment in the Marshall Islands. Many go on to university in the US.

CHRONOLOGY
After a period under Spanish rule, the Marshall Islands became a German protectorate in 1885; Japan took possession at the start of World War I. The islands were transferred to US control in 1945.

- ❑ **1946** US nuclear testing begins.
- ❑ **1947** UN Trust Territory of the Pacific Islands established.
- ❑ **1961** Kwajalein becomes US army missile range.
- ❑ **1979** Constitution approved in referendum. Government set up.
- ❑ **1986** Compact of Free Association with US operational.
- ❑ **1990** Trust terminated by UN.
- ❑ **1997** Imata Kabua elected president after death in office of Amata Kabua, his cousin.
- ❑ **2000** Kessai Note president after opposition election victory.

HEALTH
▷ No welfare state health benefits

1 per 3294 people   Respiratory, heart, and diarrheal diseases

Medical facilities are rudimentary. Complex operations are performed in Hawaii. Levels of malnutrition and vitamin A deficiency are high.

SPENDING
▷ Not available

CONSUMPTION AND SPENDING

No data   62 per 1000 population

Defense None
Education No data
Health 9%

0   5   10   15   20   25
Defense, Health, Education spending as % of GDP

Wealth disparities are small. Very few citizens can afford luxuries such as air conditioning and cars.

WORLD RANKING

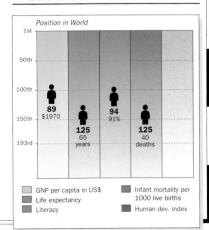

Position in World

1st
50th
100th
150th
193rd

89 $1970
94 91%
125 65 years
125 40 deaths

- ◻ GNP per capita in US$
- ◻ Life expectancy
- ◻ Literacy
- ◻ Infant mortality per 1000 live births
- ◻ Human dev. index

M

MAURITANIA

OFFICIAL NAME: Islamic Republic of Mauritania **CAPITAL:** Nouakchott
POPULATION: 2.7 million **CURRENCY:** Ouguiya **OFFICIAL LANGUAGE:** Arabic

LOCATED IN NORTHWEST AFRICA, Mauritania is a member of both the AU and the Arab League. Formerly a French colony, the country has taken a strongly Arab direction since 1964; today, it is the Maures who control political life and dominate the minority black population. The Sahara extends across two-thirds of Mauritania's territory; the only productive land is that drained by the Senegal River in the south and southwest.

CLIMATE
▷ Hot desert

WEATHER CHART FOR NOUAKCHOTT

The dusty Saharan *harmattan* wind often aggravates the very hot, dry conditions. Some rain falls in the south.

TRANSPORTATION
▷ Drive on right

Nouakchott
226,096 passengers

140 ships
47,959 grt

THE TRANSPORTATION NETWORK

866 km (538 miles)	None
740 km (460 miles)	Senegal River is navigable by small craft

The transportation system is limited and unevenly developed. There are two major roads, but shifting sands mean that they require constant maintenance.

TOURISM
▷ Visitors : Population 1:113

24,000 visitors

Little change from year to year

MAIN TOURIST ARRIVALS

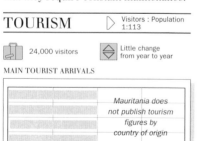

Mauritania does not publish tourism figures by country of origin

% of total arrivals

There are few tourists apart from desert safari enthusiasts. The more mountainous areas are especially dramatic, but access is difficult. Nouakchott has some hotels.

PEOPLE
▷ Pop. density low

Hassaniyah Arabic, Wolof, French

3/km² (7/mi²)

THE URBAN/RURAL POPULATION SPLIT

58%   42%

RELIGIOUS PERSUASION

Muslim (Sunni) 100%

The politically dominant Maures make up the majority of the population. The black population is composed of the Havalin, the Senegalese, and the Peulh, Tukolor, and Wolof groups. Ethnic tension centers on the oppression of blacks by Maures. The old black bourgeoisie has now been superseded by a Maurish class; tens of thousands of blacks are estimated to be in slavery. The arrival of 200,000 Maures from Senegal in 1989 caused ethnic tension to come to a head. There were attacks on Senegalese in Mauritania and many fled or were deported to refugee camps along the Senegal River.

Family solidarity among nomads is particularly strong.

MAURITANIA

Total Land Area : 1 030 700 sq. km
(397 955 sq. miles)

POPULATION
⊙ over 500 000
● over 10 000
• under 10 000

LAND HEIGHT
500m/1640ft
200m/656ft
Sea Level

POLITICS
▷ Multiparty elections

L. House 2001/2006
U. House 2002/2004

President Maaouya ould Sid Ahmed Taya

AT THE LAST ELECTION

National Assembly 81 seats

79%
PRDS

5%
AC

16%
Others

PRDS = Democratic and Social Republican Party
AC = Action for Change **Rep** = Representatives of Mauritanians living abroad **Ind** = Independents

Senate 56 seats

93%
PRDS

5%
Rep

2%
Ind

The Senate is indirectly elected

Mauritania officially adopted multiparty democracy in 1991. However, the 1992 and 1997 presidential elections simply returned to power the incumbent military ruler, President Maaouya ould Sid Ahmed Taya, with around 90% of the vote. Opposition parties have accused the government of electoral fraud; initially boycotting legislative elections, they increased their representation in the 2001 poll, but the AC was subsequently banned. The opposition parties are mainly Maure-led. The blacks of the south support exiled parties, such as the Senegal-based African Liberation Forces of Mauritania (FLAM).

WORLD AFFAIRS ▷ Joined UN in 1961

AL　CILSS　AU　OIF　OIC

Mauritania seeks to maintain a balance between sub-Saharan Africa and the Arab world, but has had tensions with all its neighbors. It has now effectively withdrawn from the Western Sahara dispute. Relations with Senegal have improved since the conflicts of 1989.

AID ▷ Recipient

 $212m (receipts)　Down 3% in 2000

France, Germany, the IMF, OPEC, and Iraq are all donors. Most aid is used for development projects, such as the EU-funded Trans-Mauritanian Highway.

DEFENSE ▷ Compulsory military service

$23m　Down 4% in 2000

The 15,000-strong army is a strain on Mauritania's budget. Troops are used increasingly in public works projects. France is the main arms supplier.

ECONOMICS ▷ Inflation 5.9% p.a. (1990–2000)

 $978m　250.97–264.01 ouguiyas

SCORE CARD

❏ World GNP Ranking	155th
❏ GNP per Capita	$370
❏ Balance of Payments	$90m
❏ Inflation	3.3%
❏ Unemployment	23%

Strengths
Iron from the Cominor mine at Zouérat. Largest gypsum deposits in the world. Copper, yet to be properly exploited. Offshore fishing among the best in West Africa. Significant debt cancellations in 2002.

Weaknesses
Poor land. Drought, locust attacks. Fluctuating commodity prices. Very hot, dry desert climate.

EXPORTS

Belgium 9%
Other 33%
Spain 11%
Italy 14%
France 18%
Japan 15%

IMPORTS

Italy 6%
Germany 5%
Spain 6%
Belgium 9%
Other 47%
France 27%

***Mauritania's extreme aridity** means that only 1% of the land is arable. Two-thirds of the country are part of the Sahara Desert; sparse vegetation over the rest supports some livestock.*

RESOURCES ▷ Electric power 105,000 kw

 47,811 tonnes　Not an oil producer

 7.6m sheep, 5.1m goats, 1.5m cattle, 4.1m chickens　 Iron, gypsum, copper, gold, phosphates, yttrium, diamonds

Iron continues to be exploited, despite low world prices. There are some gold and diamond deposits. Mining and fisheries represent 99.7% of exports. Electricity generation expanded by 40% between 1989 and 1996, and further expansion is expected to come from the Manantali Dam. Phosphates have been found near the Senegal River. Offshore oil exploration started in April 2001.

ENVIRONMENT ▷ Sustainability rank: 126th

 2% (0.2% partially protected)　 1.2 tonnes per capita

The chief environmental problem in Mauritania is that of the encroaching Sahara Desert, a situation worsened by the droughts of 1973 and 1983, which caused widespread loss of grazing land. The consequent exodus of people away from the land has raised Nouakchott's population from 20,000 in 1960 to almost a million today.

MEDIA ▷ TV ownership medium

 Daily newspaper circulation 0.5 per 1000 people

PUBLISHING AND BROADCAST MEDIA

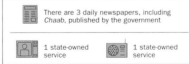

There are 3 daily newspapers, including *Chaab*, published by the government

1 state-owned service　1 state-owned service

The press is heavily censored, and the broadcast media are state-owned. *Chaab*, the government newspaper, is also published in French (*Horizons*).

CRIME ▷ Death penalty in use

 1400 prisoners　 Down 45% 1997–1999

Key issues are smuggling, robbery, and a growing number of abandoned children in Nouakchott and other towns.

EDUCATION ▷ School leaving age: 16

 42%　 12,912 students

Despite improvements in education, over half the population continues to be illiterate. Arabic has been compulsory in all schools since 1988.

HEALTH ▷ No welfare state health benefits

1 per 10,000 people　Diarrheal and respiratory diseases, influenza, tuberculosis

Historic regional inequalities persist and the best facilities are in the capital. The overall level of care is on a par with neighboring states.

SPENDING ▷ GDP/cap. increase

CONSUMPTION AND SPENDING

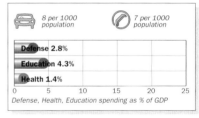

8 per 1000 population　7 per 1000 population

Defense 2.8%
Education 4.3%
Health 1.4%

0　5　10　15　20　25
Defense, Health, Education spending as % of GDP

The small ruling Maurish elite forms the richest sector. Wealthy Maures travel to Mecca, Saudi Arabia, to perform the *haj* (Muslim pilgrimage).

WORLD RANKING

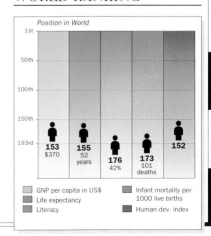

Position in World

1st
50th
100th
150th
193rd

| 153 $370 | 155 52 years | 176 42% | 173 101 deaths | 152 |

◼ GNP per capita in US$　◼ Infant mortality per 1000 live births
◼ Life expectancy
◼ Literacy　◼ Human dev. index

M

MAURITIUS

INDIAN OCEAN — Africa — Madagascar — MAURITIUS — Réunion (Fr.)

OFFICIAL NAME: Republic of Mauritius   **CAPITAL:** Port Louis
POPULATION: 1.2 million   **CURRENCY:** Mauritian rupee   **OFFICIAL LANGUAGE:** English

1968   1968   March 12   MS   +4   +230   .mu

THE ISLANDS THAT MAKE UP Mauritius lie in the Indian Ocean east of Madagascar. The main island, from which the country takes its name, is of volcanic origin and surrounded by coral reefs. Along with Rodrigues to the east, the country includes the Agalega Islands and the Cargados Carajos Shoals (500 km – 300 miles – to the north). Mauritius has enjoyed considerable economic success following recent industrial diversification and the expansion of tourism.

CLIMATE
▷ Tropical oceanic

WEATHER CHART FOR PORT LOUIS

The climate is subtropical and humid. December to March are the hottest and wettest months. Tropical cyclones are an occasional threat during this time.

TRANSPORTATION
▷ Drive on left

Sir Seewoosagur Ramgoolam International
1.76m passengers

51 ships
206,000 grt

THE TRANSPORTATION NETWORK

1834 km (1140 miles)	36 km (22 miles)
None	None

Roads are extensive, but often congested. Plans exist for a monorail link between Port Louis and Curepipe.

TOURISM
▷ Visitors : Population 1:1.8

656,000 visitors   Up 14% in 2000

MAIN TOURIST ARRIVALS

France 30%	
Réunion 14%	
UK 10%	
Other 46%	

% of total arrivals

Tourism expanded rapidly in the 1990s. Spectacular beaches, water sports, and big game fishing are major attractions. Around 30% of visitors each year come from France.

PEOPLE
▷ Pop. density high

French Creole, Hindi, Urdu, Tamil, Chinese, English, French

645/km² (1671/mi²)

THE URBAN/RURAL POPULATION SPLIT

41%   59%

RELIGIOUS PERSUASION

Protestant 2%   Other 3%
Muslim 17%
Hindu 52%
Roman Catholic 26%

Mauritius is one of the world's most densely populated countries. The majority of the population descend from indentured Indian laborers brought over in the 19th century. Creoles make up 27% of the population, while 3% are of Chinese origin. Clashes between Hindus, Muslims, and Creoles no longer occur, although Creoles complain of discrimination.

POLITICS
▷ Multiparty elections

2000/2005   President Karl Offmann

AT THE LAST ELECTION
National Assembly 70 seats

11% 3%
PTr/PMXD MR

83% MSM/MMM

3% OPR

MSM/MMM = Mauritian Socialist Movement/Mauritian Militant Movement   **PTr/PMXD** = Labour Party/Mauritian Social Democratic Party of Xavier Duval   **OPR** = Organization of the People of Rodrigues   **MR** = Mouvement Rodriguais

62 members of the National Assembly are directly elected, and up to eight are appointed from the highest losers

Mauritius became a republic in 1992. Navin Ramgoolam of the PTr became prime minister in 1995, promoting regional integration and economic liberalization to attract investment. However, corruption scandals led to early elections in 2000. The PTr was defeated and Sir Aneerood Jugnauth, Ramgoolam's predecessor, now heads a new coalition between his MSM and the MMM.

WORLD AFFAIRS
▷ Joined UN in 1968

Comm   COMESA   COI   AU   SADC

Mauritius hosted a francophone nations summit in 1995, and the first OAU human rights conference in 1999. Links with South Africa and India are important. Disputes persist over UK-administered Diego Garcia and the French-ruled island of Tromelin.

Round Island
Flat Island
Îlot Gabriel
Gunner's Quoin
Grand Baie
Île d'Ambre
Triolet   Goodlands
Mapou   Poudre d'Or
Pamplemousses   River Citrons
Belle Vue Maurel
Amaury
l'Aventure
PORT LOUIS
Centre de Flacq
Beau Bassin
Rose Hill   Quartier Militaire
Quatre Bornes   Île de l'Est
Phoenix   Île aux Cerfs
Vacoas   Floréal
Tamarin   Curepipe   Grande River South East
Piton de la Petite Rivière Noire 828m
Nouvelle France
Île aux Bénitiers
Rose Belle   Mahebourg
Sir Seewoosagur Ramgoolam Intl
Chamouny
Chemin Grenier
Rivière des Anguilles
Bel Ombre   Surinam   Souillac

RODRIGUES
Port Mathurin
Mont Limon 396m
Grand Montagne
Petite Butte   19°45'
(continuation on same scale)   63°25'

MAURITIUS

Total Land Area : 1860 sq. km (718 sq. miles)

LAND HEIGHT
500m/1640ft
200m/656ft
Sea Level

POPULATION
over 100 000
over 50 000
over 10 000
under 10 000

0   10 km
0   10 miles

M

AID

 Recipient

 $20m (receipts)   Down 52% in 2000

Aid is predominantly bilateral, with the EU and France as the main donors. Mauritius also receives aid from Arab aid agencies and Japan, and from the UN and other international organizations. The World Bank assisted a five-year conservation program, starting in 1990, and promised $53 million toward transforming Port Louis into a free port.

DEFENSE

 No compulsory military service

 $87m   Down 4% in 2000

Mauritius has no standing defense forces. There is, however, a 1100-strong special police mobile unit to ensure internal security. There is also a coastguard numbering 500.

ECONOMICS

Inflation 5.9% p.a. (1990–2000)

$4.45bn   27.82–30.25 Mauritian rupees

SCORE CARD

❏ WORLD GNP RANKING	116th
❏ GNP PER CAPITA	$3750
❏ BALANCE OF PAYMENTS	–$33m
❏ INFLATION	4.2%
❏ UNEMPLOYMENT	6%

STRENGTHS

Strong economic growth. The sugar industry accounts for 30% of export earnings. Export processing zone (EPZ), especially for clothing manufacture. Tourism. Highly educated workforce. Ranked as the most competitive economy in Africa by the World Economic Forum in 1999. Development as offshore financial center.

WEAKNESSES

Vulnerability to fluctuating world sugar price and droughts. 75% of food requirements are imported. Few crops other than sugar can be grown. Lack of strategic resources. Remoteness.

EXPORTS

Germany 4%
Madagascar 5%
UK 29%
Other 20%
France 22%
USA 20%

IMPORTS

UK 4%
China 8%
India 9%
Other 54%
France 10%
South Africa 15%

Villagers at a water source in the center of *Mauritius island. Mauritius' main rivers are used for hydropower generation.*

RESOURCES

 Electric power 364,000 kw

 12,089 tonnes   Not an oil producer

95,000 goats, 28,000 cattle, 4.5m chickens   None

Mauritius has to import oil, so the government has invested heavily in alternative indigenous energy schemes, including HEP generation, and power plants fueled by bagasse (a by-product of the sugar industry). Industrialization and agricultural diversification make up for limited natural resources.

ENVIRONMENT

 Not available

 8% (0.1% partially protected)   1.5 tonnes per capita

Rapid industrialization as well as unchecked hotel building have caused environmental problems. Coral reefs are under threat from both coral sand mining and the discharging of untreated sewage into the sea.

MEDIA

 TV ownership high

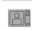 Daily newspaper circulation 76 per 1000 people

PUBLISHING AND BROADCAST MEDIA

There are 10 daily newspapers. *Le Quotidien, L'Express,* and *Le Mauricien* have the largest circulations

1 independent service   1 independent service

Mauritius has an active press, subject to few regulations and with a wide readership. Newspapers are published in English, French, Creole, Hindi, Chinese, and Tamil. Opposition parties complain that TV and radio broadcasts are consistently biased toward the government.

CRIME

 No death penalty

 1985 prisoners   Down 19% in 1999

Crime rates on the main island are fairly low. There has been a small increase in thefts and narcotics smuggling. Outlying islands are virtually crime-free.

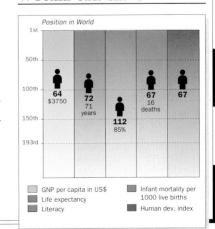

CHRONOLOGY

Mauritius was colonized and ruled by the Dutch in the 17th century, the French (1710–1810), and the British.

- ❏ **1959** First full elections.
- ❏ **1968** Independence. Riots between Creoles and Muslims.
- ❏ **1982–1995** Sir Aneerood Jugnauth prime minister; forms MSM.
- ❏ **1992** Becomes a republic.
- ❏ **1995** Elections won by PTr–MMM.
- ❏ **2000** Return of Jugnauth.

EDUCATION

 School leaving age: 12

 85%   7559 students

Educational provision is good, and 91% of Mauritians under 30 are literate. The University of Mauritius has about 2000 students.

HEALTH

Welfare state health benefits

1 per 1111 people   Circulatory and heart diseases, cancers, accidents, malnutrition

In Mauritius free health care is universally available. There are 14 state hospitals and six private clinics.

SPENDING

GDP/cap. increase

CONSUMPTION AND SPENDING

73 per 1000 population   235 per 1000 population

Defense 1.8%
Education 4%
Health 1.8%

Defense, Health, Education spending as % of GDP

French-descended hotel and plantation owners form the country's wealthiest social group. Government employees are well paid.

WORLD RANKING

Position in World

64 $3750
72 71 years
112 85%
67 16 deaths
67

GNP per capita in US$
Life expectancy
Literacy
Infant mortality per 1000 live births
Human dev. index

M

MEXICO

OFFICIAL NAME: United Mexican States **CAPITAL:** Mexico City
POPULATION: 100.4 million **CURRENCY:** Mexican peso **OFFICIAL LANGUAGE:** Spanish

 1836
 1848
 Sept 16
 MEX
 -6
 +52
.mx

INCREASINGLY CONSIDERED a part of North rather than Central America, Mexico separates the US from the rest of Latin America. Coastal plains along its Pacific and Caribbean seaboards rise into an arid central plateau, which includes one of the world's biggest conurbations, Mexico City, built on the site of the Aztec capital, Tenochtitlán. Colonized by the Spanish for its silver mines, Mexico achieved independence in 1836. In the "Epic Revolution" of 1910–1920, in which 250,000 died, much of modern Mexico's structure was established. In 1994, Mexico signed the North American Free Trade Agreement (NAFTA).

The cathedral of Santa Prisca at Taxco near Cuernavaca. It was built in Spanish Churriguera style between 1748 and 1758.

CLIMATE

▷ Tropical/mountain/desert

WEATHER CHART FOR MEXICO CITY

■ Average daily temperature   Rainfall ■

°C/°F J F M A M J J A S O N D cm/in
40/104 .... 40/16
30/86 .... 30/12
20/68 .... 20/8
10/50 .... 10/4
0/32 0
-10/14
-20/-4

The plateau and high mountains are warm for much of the year. The Pacific coast has a tropical climate.

TRANSPORTATION

▷ Drive on right

✈ **Benito Juárez International, Mexico City**
21m passengers

⚓ 626 ships
1.09m grt

THE TRANSPORTATION NETWORK

🛣 96,221 km (59,789 miles)		🛣 6335 km (3936 miles)	
🚂 26,595 km (16,526 miles)		⚓ 2900 km (1802 miles)	

A privately financed $14 billion road network, some 6000 km (3750 miles) of toll roads, is seriously underused and a commercial failure. Regional travel is mainly by bus; the unreliable railroad is largely for freight. Plans to construct a new international airport serving Mexico City were abandoned in 2002 as local farmers staged violent protests against the implied acquisition of their lands.

TOURISM

▷ Visitors : Population
1:5.1

🧳 19.8m visitors   ⬇ Down 4% in 2001

MAIN TOURIST ARRIVALS

	0	10	20	30	40	50	60	70	80	90	100
USA 92 %											
Europe 3%											
Canada 3%											
South America 1%											
Other 1%											

% of total arrivals

Tourism is one of the largest employment sectors in Mexico, and a major source of foreign exchange. Attractions include excellent beach resorts such as Acapulco on the Pacific coast, and the new resorts of the Peninsula de Yucatán on the Caribbean coast. Impressive coastal scenery, volcanoes, the Sierra Madre, and archaeological remains of Aztec and Mayan civilizations, designated as World Heritage sites, are major draws, as are the many Spanish colonial cities, such as Morelia and Guadalajara, which have remained virtually intact since their construction after the conquest.

MEXICO

Total Land Area : 1 972 550 sq. km
(761 602 sq. miles)

LAND HEIGHT

- 3000m/9843ft
- 2000m/6562ft
- 1000m/3281ft
- 500m/1640ft
- 200m/656ft
- Sea Level

POPULATION

- ▣ over 5 000 000
- ▣ over 1 000 000
- ◉ over 500 000
- ◎ over 100 000
- ○ over 50 000

0 ___ 200 km
0 ___ 200 miles

M

PEOPLE ▷ Pop. density medium

Spanish, Nahuatl, Mayan, Zapotec, Mixtec, Otomi, Totonac, Tzotzil, Tzeltal

53/km² (136/mi²)

THE URBAN/RURAL POPULATION SPLIT

74%   26%

RELIGIOUS PERSUASION

Protestant 1%   Other 4%
Roman Catholic 95%

ETHNIC MAKEUP

Other 9%
European 16%
Mestizo 55%
Indigenous Amerindian 20%

POPULATION AGE BREAKDOWN

Female	Age	Male
0.5%	80+	0.4%
2.9%	60–79	2.7%
7.4%	40–59	7.1%
16.7%	20–39	15.5%
23.2%	0–19	23.6%

% of population by age group

While most Mexicans are *mestizo* (mixed race), it is Mexico's Amerindian culture which is promoted by the state. This obscures the fact that rural Amerindians are largely segregated from Hispanic society, a situation that dates back to the Spanish colonial period and

which has only recently been seriously challenged. The 1994 Chiapas Zapatista (EZLN) guerrilla uprising was on behalf of Amerindian rights, and in protest against the poverty of landless Amerindians. President Vicente Fox promised to act, but the Indigenous Rights and Culture Bill, watered down by a hostile Congress and enacted in 2001, was rejected by the EZLN and all the main indigenous groups.

The small black community, which is concentrated along the eastern coast, is well integrated.

As in much of Latin America, men retain their dominance in business and relatively few women take part in the political process.

POLITICS ▷ Multiparty elections

L. House 2000/2003
U. House 2000/2006

President Vicente Fox

AT THE LAST ELECTION

Chamber of Deputies 500 seats

45% PAN–PVEM   42% PRI   13% PRD–PT

PAN–PVEM = Alliance for Change (National Action Party and Green Party)   **PRI** = Institutional Revolutionary Party
PRD–PT = Alliance for Mexico (Party of the Democratic Revolution and Labor Party)

Senate 128 seats

45% PRI   42% PAN–PVEM   13% PRD–PT

Mexico was a multiparty democracy in name only until electoral reform in 1997.

PROFILE

The PRI dominated Mexico from 1929. Regional elections in the 1990s showed the strength of opposition parties, and after grudging electoral reform, the PRI lost its monopoly on power in 1997. President Vicente Fox's PAN became the largest party in the Chamber of Deputies in the 2000 elections, but it lacks an overall majority. Fox promised a broadly conceived "government of transition."

MAIN POLITICAL ISSUES
President Fox's administration
Strong opposition in the Congress has

frustrated President Fox's program of structural reform: by relying on advisers and cabinet rule, he neglected to consult the opposition and failed to build a legislative consensus.

Indigenous rights
Fox initially honored some conditions of the Zapatista National Liberation Army (EZLN) for new peace talks, by releasing prisoners and withdrawing troops from Chiapas. Serious revisions by the Congress to the Indigenous Rights and Culture Bill were rejected by the EZLN and indigenous groups, setting the peace process back.

Future of the PRI
Election defeats left the PRI faction-ridden and rudderless. Roberto Madrazo, a hard-liner elected party leader in 2002, has to unite the PRI to win the 2006 presidential elections.

Vicente Fox, elected president in 2000, ending 70 years of PRI dominance.

Subcomandante Marcos, leader of the Zapatista National Liberation Army.

WORLD AFFAIRS ▷ Joined UN in 1945

G15   NAFTA   OECD   OAS   RG

NAFTA has bonded the economies of Mexico and the US. Mexican imports have undercut US jobs, as has the southward relocation of multinationals to benefit from cheap Mexican labor. Heightened US border security post-11 September complicated a bilateral agreement to deal with high flows of illegal immigrants but also hindered powerful narcotics cartels supplying the US market.

Mexico has free trade agreements with 32 countries, as well as with the EU and EFTA, and competes with Brazil to play a leading negotiating role for Latin America in the formation of a Free Trade Area of the Americas (FTAA) due to be signed in 2005.

Mexico's swing vote in the UN in 2002, which triggered a resolution on human rights in Cuba, inflamed relations with the Castro regime and set President Fox against the Senate on the issue of who decides foreign policy.

AID ▷ Recipient

  $37m (receipts)

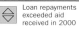 Loan repayments exceeded aid received in 2000

In 2002 the IDB made its biggest loan – $1 billion – to support a six-year antipoverty project in Mexico. European and US NGOs provide assistance.

CHRONOLOGY

The Aztec kingdom of Montezuma II was defeated in war by the Spaniard, Hernán Cortés, in 1521. By 1546, the Spaniards had discovered large silver mines at Zacatecas. Mexico, then known as New Spain, became a key part of the Spanish colonial empire.

❏ **1810** Fr. Miguel Hidalgo leads abortive rising against Spanish.
❏ **1821** Spanish viceroy forced to leave by Agustín de Iturbide.
❏ **1822** Federal Republic established.
❏ **1823** Texas opened to US immigration.
❏ **1829** Spanish military expedition fails to regain control.
❏ **1836** US is first country to recognize Mexico's independence. Spain follows suit. Texas declares its independence from Mexico.
❏ **1846** War breaks out with US.
❏ **1848** Loses modern-day New Mexico, Arizona, Nevada, Utah, California, and part of Colorado.
❏ **1858–1861** War of Reform won by anticlerical Liberals.
❏ **1862** France, Britain, and Spain launch military expedition.

M

CHRONOLOGY *continued*

- ❏ **1863** French troops capture Mexico City. Maximilian of Austria established as Mexican emperor.
- ❏ **1867** Mexico recaptured by Benito Juárez. Maximilian shot.
- ❏ **1876** Porfirio Díaz president. Economic growth; rail system built.
- ❏ **1901** First year of oil production.
- ❏ **1910–1920** Epic Revolution provoked by excessive exploitation by foreign companies and desire for land reform. 250,000 killed.
- ❏ **1911** Díaz overthrown by Francisco Madero. Guerrilla war breaks out in north. Emilio Zapata leads peasant revolt in the south.
- ❏ **1913** Madero murdered.
- ❏ **1917** New constitution limits power of Church. Minerals and subsoil rights reserved for the nation.
- ❏ **1926–1929** Cristero rebellion led by militant Catholic priests.
- ❏ **1929** National Revolutionary Party (later PRI) formed.
- ❏ **1934** Gen. Cárdenas president. Land reform accelerated, cooperative farms established, railroads nationalized, and US and UK oil companies expelled.
- ❏ **1940s** US war effort helps Mexican economy to grow.
- ❏ **1970** Accelerating population growth reaches 3% a year.
- ❏ **1982** Mexico declares it cannot repay its foreign debt of over $800 billion. IMF insists on economic reforms to reschedule the debt.
- ❏ **1984** Government contravenes constitution by relaxing laws on foreign investment.
- ❏ **1985** Earthquake in Mexico City. Official death toll 7000. Economic cost estimated at $425 million.
- ❏ **1988** Carlos Salinas de Gortari, minister of planning during the earthquake, elected president.
- ❏ **1990** Privatization program begun.
- ❏ **1994–1995** Guerrilla rebellion in southern Chiapas state brutally suppressed by army: 100 dead. Mexico joins NAFTA. PRI presidential candidate Luis Colosio murdered. Ernesto Zedillo replaces him and is elected. Economic crisis.
- ❏ **1997–1999** PRI's monopoly on power in Congress ended. Banks bailed out.
- ❏ **2000** July, PAN wins presidency and elections, ending 70 years of PRI rule. December, President Vicente Fox takes office.
- ❏ **2001** EZLN guerrillas and supporters make 16-day motorcade from Chiapas to Mexico City to push for indigenous rights law.
- ❏ **2002** Former governor of Tabasco, Roberto Madrazo, traditionalist hard-liner, elected new PRI leader.

DEFENSE

▷ Compulsory military service

 $5.23bn

 Up 22% in 2000

Mexico has no ambitions beyond its borders, and the army acts to defend internal security. The military has, on the whole, avoided direct interference in politics. Most arms procurement is from the US and France. In 1994, the role of controlling the border with the US was passed to the police.

The Zapatista rebellion in Chiapas in 1994 elicited a brutal response from the army, acting on PRI orders. The increasing militarization of the state over the next six years hindered the peace process and led to a proliferation of paramilitaries, with the tacit blessing of the local PRI, who were blamed by

MEXICAN ARMED FORCES

No main battle tanks	144,000 personnel
3 destroyers, 8 frigates, and 109 patrol boats	37,000 personnel
107 combat aircraft (8 F-5E, 2 F-5F, 70 PC-7, 17 AT-33)	11,770 personnel
None	

human rights groups for the massacre of Amerindians. The PAN government has withdrawn some forces from key areas of the state.

ECONOMICS

▷ Inflation 19% p.a. (1990–2000)

 $497bn

9.609–9.169 Mexican pesos

SCORE CARD

- ❏ WORLD GNP RANKING...........................11th
- ❏ GNP PER CAPITA $5070
- ❏ BALANCE OF PAYMENTS..................–$17.8bn
- ❏ INFLATION 9.5%
- ❏ UNEMPLOYMENT 2%

EXPORTS

Japan 1%
Spain 1%
Germany 1%
Canada 2%
Other 6%
USA 89%

IMPORTS

South Korea 2%
Canada 2%
Germany 3%
Japan 4%
Other 16%
USA 73%

STRENGTHS
Global oil producer, with substantial reserves. Extensive mineral resources. Strong foreign direct investment. Diversification of exports. NAFTA membership. Low overheads.

WEAKNESSES
Debt burden. Vulnerable currency. Corruption. Affected by oil price changes and US slowdown. Weak tax system.

PROFILE
While in power, the PRI effectively ran the economy. The debt crisis of the 1980s, however, forced privatizations. The 1994 peso crisis needed a US-led $20 billion international bailout and resulted in a severe slump. The Zedillo government launched tough reforms, but a global loss of confidence in emerging markets affected growth.

ECONOMIC PERFORMANCE INDICATOR

— Consumer Price Index   GDP ▨

Tighter fiscal management was rewarded by 2000, when investor confidence improved. The Fox government's tight fiscal and monetary stance secured the approval of the IMF and major credit agencies, giving the country greater access to cheaper foreign capital. However, pledges on social spending and poverty relief suffered, and promises of large job creation projects also went unfulfilled, as the export-led economy echoed the slowdown in the US in late 2001.

MEXICO : MAJOR BUSINESSES

Food processing   Petrochemicals
Vehicle assembly   Oil refining
Computers
Silver mining
Electronics
Brewing
Textiles

Tijuana
Ciudad Juárez
Monterrey
Reynosa
Tampico
Minatitlán
Durango
Guadalajara
Salamanca
Mexico City

0   400 km
0   400 miles
* significant multinational ownership

RESOURCES

 Electric power 45.6m kw

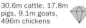

1.25m tonnes

3.56m b/d (reserves 26.9bn barrels)

30.6m cattle, 17.8m pigs, 9.1m goats, 496m chickens

Oil, gas, gold, silver, copper, coal, fluorite, mercury, antimony

ELECTRICITY GENERATION

Hydro 14% (25bn kwh)	
Combustion 78% (144bn kwh)	
Nuclear 5% (9.3bn kwh)	
Other 3% (5.7bn kwh)	

% of total generation by type

Mexico is one of the largest oil exporters outside OPEC. Most oil production comes from offshore drilling platforms in the Gulf of Mexico. The industry was state-owned and state-run by PEMEX, the world's fifth-largest oil company, employing 120,000 people. The decision to privatize petrochemical

MEXICO : LAND USE

Cropland	
Forest	
Pasture	
Wetlands	
Desert	
Cotton – cash crop	
Wheat	
Cattle	

0   400 km
0   400 miles

plants has provoked serious social unrest, and further sell-offs and deregulation remain politically highly sensitive. Despite its oil reserves, Mexico has embarked on a nuclear power program and projects to modernize the national electricity grid and boost natural gas production to overcome an energy crisis.

ENVIRONMENT

 Sustainability rank: 92nd

 4% (4% partially protected)   3.9 tonnes per capita

ENVIRONMENTAL TREATIES

Yes	Yes	Yes
Yes	Yes	Yes

Mexico City, largely unplanned, struggles to accommodate around 20 million inhabitants as the absence of environmental controls contributes to perhaps the world's worst air quality and waste problems. PEMEX (the state petroleum company) stands accused of massive pollution. *Maquiladoras* – assembly plants on the Mexico–US border – have no effective environmental controls and are usually surrounded by slums. Environmentalists oppose the intense development of tourism along the coast and are concerned about high rates of deforestation.

MEDIA

 TV ownership high

Daily newspaper circulation 97 per 1000 people

PUBLISHING AND BROADCAST MEDIA

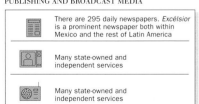

	There are 295 daily newspapers. *Excélsior* is a prominent newspaper both within Mexico and the rest of Latin America
	Many state-owned and independent services
	Many state-owned and independent services

The PRI in particular has historically manipulated the media, being accused of denying electoral opponents airtime.

CRIME

 Death penalty not used in practice

 144,261 prisoners   Little change from year to year

CRIME RATES

> *Mexico does not publish official statistics for murders, rapes, or thefts*

Northern Mexico is a major center for narcotics shipments to the US. Antidrugs police are accused by the US of corruption. Guns are rife and minor incidents may end in shootings. The high crime rate in Mexico City is a major political issue. Reforms of the corrupt judiciary and the whole police force are perennial issues.

EDUCATION

School leaving age: 14

91%   2.07m students

THE EDUCATION SYSTEM

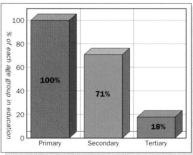

% of each age group in education

- Primary: 100%
- Secondary: 71%
- Tertiary: 18%

Public education, officially compulsory for the first nine years, is underfunded and rural provision is poor. The system is a mixture of the French and US models. There is a well-developed public university system.

HEALTH

 Welfare state health benefits

 1 per 588 people   Heart disease, accidents, cancers, violence

The national health care system is basic and badly underfunded, although an ambitious scheme was launched in 1996 to improve access to health services in marginalized areas. Mexico has a good reputation for surgery and dentistry, but this is mostly in the private sector. The rich also go to the US for treatment.

SPENDING

GDP/cap. increase

CONSUMPTION AND SPENDING

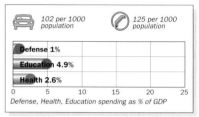

102 per 1000 population   125 per 1000 population

Defense 1%	
Education 4.9%	
Health 2.6%	

Defense, Health, Education spending as % of GDP

Mexico has enormous wealth disparities. Preliminary official figures in 2002 had six million families, or 26 million people, in extreme poverty. There is little social mobility; the old Spanish families retain their hold on institutions. In the past, the wealthy did not generally pay taxes and often benefited from the large state machine. Tax evasion remains a serious problem.

Rural Amerindians are probably the most disadvantaged group. In the last decade, poverty has forced them into city slums to work in factories or *maquiladoras*, where conditions and pay are poor. The 1994 Chiapas rebellion was fed by demands for more land and more assistance in farming it. The flow of poor rural migrants to the US stems largely from the need to subsidize families back home.

WORLD RANKING

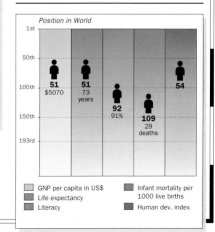

Position in World

- 51 — GNP per capita in US$ $5070
- 51 — Life expectancy 73 years
- 92 — Literacy 91%
- 109 — Infant mortality 29 deaths
- 54 — Human dev. index

GNP per capita in US$	Infant mortality per 1000 live births
Life expectancy	
Literacy	Human dev. index

M

MICRONESIA

OFFICIAL NAME: Federated States of Micronesia **CAPITAL:** Palikir (Pohnpei Island)
POPULATION: 133,000 **CURRENCY:** US dollar **OFFICIAL LANGUAGE:** English

THE FEDERATED STATES of Micronesia (FSM), situated in the Pacific Ocean, encompasses all the Caroline Islands except Palau. It is composed of four island cluster states: Pohnpei, Kosrae, Chuuk, and Yap. The FSM was formerly under US rule as part of the UN Trust Territory of the Pacific Islands. An agreement which granted internal sovereignty in free association with the US became operational in 1986, and the Trust was formally dissolved in 1990. The islands continue to receive considerable aid from the US.

CLIMATE ▷ Tropical oceanic

WEATHER CHART FOR PALIKIR

The islands are humid and fairly hot all year round, and the daily temperature range is small. Rainfall is abundant.

TRANSPORTATION ▷ Drive on right

Pohnpei 58,029 passengers — 19 ships 10,400 grt

THE TRANSPORTATION NETWORK

| 42 km (26 miles) | None |
| None | None |

The inauguration in 2000 of flights by a Boeing 737 opened the way for greater air traffic between the islands. Shipping is mainly used for bulk cargoes and copra. Some island roads are surfaced with coral.

Micronesia, aerial view of rock islands.
Like many Pacific states, Micronesia fears rising sea levels as a result of global warming.

TOURISM ▷ Visitors : Population 1:4

33,000 visitors — Up 65% 1997–2000

MAIN TOURIST ARRIVALS

Japan 40%	
USA 34%	
Europe 9%	
Other 17%	

% of total arrivals

Outlying islands remain untouched and unspoiled. Chuuk's underwater war wreckage and Kosrae's beaches attract visitors. Lack of infrastructure tends to hinder the growth of tourism.

PEOPLE ▷ Pop. density medium

Trukese, Pohnpeian, Mortlockese, Losrean, English — 190/km² (491/mi²)

THE URBAN/RURAL POPULATION SPLIT

28%   72%

RELIGIOUS PERSUASION

Other 2%
Roman Catholic 50%
Protestant 48%

The Micronesians are physically, linguistically, and culturally diverse. Melanesians live on Yap, and Polynesians occupy southwestern atolls in Pohnpei state. Most islanders live without electricity or running water, and many are effectively recipients of US welfare. Society is traditionally matrilineal.

POLITICS ▷ Nonparty elections

2001/2003 — President Leo Falcam

AT THE LAST ELECTION
Congress 14 seats

There are no political parties. Congress has 14 members, 10 senators directly elected for a two-year term and four "at-large" senators (one from each state) who are elected for a four-year term

Under the federal structure, the president and vice president are elected from among the four "at-large" senators (one from each state) by the federal legislature. However, the power of the traditional chiefs in politics remains very strong. Increasing Micronesia's economic independence is the key political issue and has dominated negotiations since 2000 on the outline of the next Compact of Free Association with the US.

MICRONESIA

Total Land Area : 702 sq. km (271 sq. miles)

POPULATION
• under 10 000

LAND HEIGHT
100m/328ft
Sea Level

PHILIPPINE SEA
Ulithi
Yap • Fais
Gaferut   Namonuito   East Fayu I.   Murilo   Minto Reef
STATE OF YAP
Faraulep   West Fayu   Pikelot   STATE OF CHUUK
Ngulu   Sorol   Olimarao   Lamotrek   Pulap   Nomwin   Oroluk   POHNPEI ISLANDS
Woleai   Elato   Satawal   Puluwat   CHUUK IS.   Losap   Pakin
Ifalik   Kuop • Weno   UPPER MORTLOCKS   ✈ PALIKIR
Eauripik   Pulusuk   Pohnpei   Mokil
WEST CAROLINE ISLANDS   STATE OF POHNPEI   Pingelap
Namoluk   LOWER MORTLOCKS   Ngatik
Satawan
Kosrae
KOSRAE STATE
EAST CAROLINE ISLANDS
Nukuoro
PACIFIC OCEAN
N
0 — 200 km
0 — 200 miles
140°   145°   150°   155°   160°
Kapingamarangi

M

WORLD AFFAIRS
 Joined UN in 1991

Micronesia's most important relationship is with the US, which administered the islands from 1947 as part of the UN Trust Territory of the Pacific Islands. Under the Compact of Free Association, the US has exclusive control over the FSM's foreign and defense policies. Alternative funds were sought after 2001 to compensate for the reduction in annual aid likely under a revised Compact. Japan is also important, with the Tokyo government providing aid, and the FSM has recently cultivated strong links with China.

AID
 Recipient

 $102m (receipts)　　 Down 6% in 2000

The US is the principal donor of aid, which funds hospitals, schools, food stamps, and construction projects.

DEFENSE
No compulsory military service

 US is responsible for defense　　Not applicable

Defense is entirely in the hands of the US. Airstrips in the FSM were used by the US in the Vietnam War.

ECONOMICS
 Inflation 2.6% p.a. (1990–2000)

$250m　　Currency is US dollar

SCORE CARD

❏ WORLD GNP RANKING	179th
❏ GNP PER CAPITA	$2110
❏ BALANCE OF PAYMENTS	$67m
❏ INFLATION	2.6%
❏ UNEMPLOYMENT	16%

STRENGTHS
Access to US economy, especially for garment manufacture, through preferential trading rights. Construction industry largest private-sector activity. Tourism, fishing, and copra production. US strategic interest in Micronesia, and US budget subsidies.

WEAKNESSES
Dependence on US for imports, especially for fuel. Heavy indebtedness. Acute shortage of water limits development potential. High levels of underemployment.

EXPORTS

Marshall Islands 1% — Other 15% — Japan 84%

IMPORTS

Japan 12% — Australia 20% — USA 44% — Other 24%

RESOURCES
 Not available

 11,886 tonnes　　Not an oil producer

32,000 pigs, 13,900 cattle, 4000 goats, 185,000 chickens　　None

The FSM is entirely dependent on external sources for its energy supply. Almost all electricity is produced by small diesel generators. The main resources are copra and valuable fish stocks, especially tuna.

ENVIRONMENT
 Not available

 None　　Not available

The FSM does not face pollution on the scale of that in the neighboring Marshall Islands. However, Chuuk suffers serious droughts; occasionally water rationing has had to be introduced for short periods. The growth of marine-based tourism is monitored by the South Pacific Environment Program, which aims to promote sustainable development.

MEDIA
 TV ownership medium

There are no daily newspapers

PUBLISHING AND BROADCAST MEDIA

There are no daily newspapers. *The National Union* is a popular biweekly

4 services:
1 state-owned,
3 independent

1 state-owned service

Press freedoms have not been infringed since the strongly criticized expulsion of a Canadian journalist in 1997.

CRIME
 No death penalty

39 prisoners　　Little change from year to year

Crime is rare and the outlying islands are crime-free. Some alcohol-related assault occurs on Chuuk.

EDUCATION
 School leaving age: 14

 89%　　 1461 students

Education is compulsory between the ages of six and 14 years. Most university students are supported by US grants and a large number attend US colleges.

CHRONOLOGY
The Caroline Islands were first colonized by the Spanish. Sold to Germany in 1899, the islands were occupied by Japan from 1914 and served as an important base in World War II. US control of the islands began in 1945.

- ❏ **1947** UN Trust Territory of the Pacific Islands established.
- ❏ **1979** Becomes independent.
- ❏ **1986** Compact of Free Association with US operational.
- ❏ **1990** Official termination of trusteeship agreement.
- ❏ **1991** Joins UN.
- ❏ **1999** Leo Falcam elected president.
- ❏ **2001** Submerged remains of USS *Mississinewa* causes oil leak in Yap.

HEALTH
 Welfare state health benefits

1 per 1640 people　　Cerebrovascular, heart, and intestinal diseases

Basic health care is accessible to all. Diabetes and drug abuse are growing problems. An increase in imported food has led to dietary problems.

SPENDING
GDP/cap. decrease

CONSUMPTION AND SPENDING

No data　　80 per 1000 population

Defense	None
Education	No data
Health	7.4%

Defense, Health, Education spending as % of GDP

The gap between rich and poor is increasing as Micronesia's businessmen and local officials exploit US aid donations.

WORLD RANKING

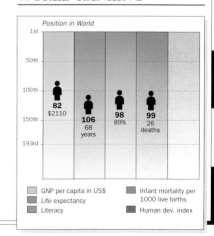
Position in World

82 $2110
106 68 years
98 89%
99 26 deaths

- GNP per capita in US$
- Life expectancy
- Literacy
- Infant mortality per 1000 live births
- Human dev. index

M

MOLDOVA

OFFICIAL NAME: Republic of Moldova **CAPITAL:** Chişinău
POPULATION: 4.3 million **CURRENCY:** Moldovan Leu **OFFICIAL LANGUAGE:** Moldovan

MOSTLY UNDULATING steppe country, Moldova is the most densely populated of the former Soviet republics. Once a part of Romania, it was incorporated into the Soviet Union in 1940. Independence in 1991 brought with it the expectation that Moldova would be reunited with Romania. In a 1994 plebiscite, however, Moldovans voted against the proposal. Most of Moldova's population is engaged in intensive agriculture.

Agricultural landscape. Warm summers and even rainfall are ideal for cereal and fruit farming. Moldova is famous for its wine.

CLIMATE ▷ Continental

WEATHER CHART FOR CHISINAU

Warm summers, mild winters, and moderate rainfall give Moldova an ideal climate for cultivation.

TRANSPORTATION ▷ Drive on right

Chisinau International
254,283 passengers

Small Black Sea fleet

THE TRANSPORTATION NETWORK

10,738 km (6672 miles)	None
1140 km (708 miles)	424 km (263 miles)

The transportation infrastructure is to be part of a planned "Transport Corridor Europe–Caucasus–Asia" (TRACECA).

TOURISM ▷ Visitors : Population 1:253

17,000 visitors   Up 21% in 2000

MAIN TOURIST ARRIVALS

Russia 18%
Romania 15%
Ukraine 14%
Other 53%

% of total arrivals

Few tourists go to Moldova, although some visitors to Romania do combine the two. Hopes for expansion of tourism focus on vineyards and underground wine vault "streets" as the main attractions.

PEOPLE ▷ Pop. density medium

Moldovan, Romanian, Russian

128/km² (330/mi²)

THE URBAN/RURAL POPULATION SPLIT

46%   54%

ETHNIC MAKEUP

Gagauz 4%
Other 4%
Russian 13%
Ukrainian 14%
Moldovan 65%

Moldovans are ethnically identical to Romanians. There are 153,000 Gagauz (Orthodox Christian Turks) in the south, and a population of mixed Russian–Moldovan–Ukrainian parentage on the eastern bank of the Dniester.

POLITICS ▷ Multiparty elections

2001/2005   President Vladimir Voronin

AT THE LAST ELECTION

Parliament 101 seats

70% CPM   19% BEAB   11% PPCD

CPM = Communist Party of Moldova
BEAB = Electoral Bloc Braghis Alliance
PPCD = Christian-Democratic People's Party

Moldova declared its independence in 1991. Reformist Petru Lucinschi was elected president in 1996 but faced stiff opposition from the increasingly powerful left: the revived CPM won most seats in the 1998 elections. Parliament ended direct presidential elections in 2000, but deadlock ensued over the appointment of Lucinschi's successor, forcing Parliament's dissolution in 2001. The new Parliament, with a big CPM majority, chose CPM leader Vladimir Voronin as president. However, the left's popularity has faltered, with mass discontent over its apparent eagerness to align Moldova with Russia rather than the West. Transdniestria (on the eastern bank of the river Dniester) and Gagauzia (in the south) declared themselves as republics in 1990. While Gagauzia accepted autonomous status as provided for in the 1994 constitution, Transdniestria still seeks independence.

MOLDOVA

Total Land Area : 33 843 sq. km (13 067 sq. miles)

POPULATION
- ⊙ over 500 000
- ◎ over 100 000
- ○ over 50 000
- ● over 10 000
- • under 10 000

LAND HEIGHT
200m/656ft
80m/262ft

N

WORLD AFFAIRS
 Joined UN in 1992

BSEC　CE　CIS　EAPC　OSCE

Moldova has not sought NATO membership, and in 2001 showed interest in joining a Union State with Russia and Belarus. Ties with countries in the Black Sea Economic Zone, including Romania and Ukraine, are being developed. The creation of a free economic zone near the mouth of the Danube is under discussion.

AID
 Recipient

 $123m (receipts)　　Up 15% in 2000

The World Bank resumed lending in 2002, granting $30 million. The US, the EU states, and the IMF are also important sources of aid.

DEFENSE
 Compulsory military service

$21m　　Down 22% in 2000

In 1999 plans were announced to cut army personnel by 30%. Military service has been reduced from 18 months to 12. Russian forces were withdrawn from Transdniestria by the end of 2001.

ECONOMICS
Inflation 120% p.a. (1990–2000)

$1.43bn　　12.395–13.085 Moldovan lei

SCORE CARD

❏ WORLD GNP RANKING	147th
❏ GNP PER CAPITA	$400
❏ BALANCE OF PAYMENTS	–$121m
❏ INFLATION	31.3%
❏ UNEMPLOYMENT	11%

STRENGTHS

Agriculture – notably wine, tobacco, and cotton – and food processing. Light manufacturing.

WEAKNESSES

Dependent on Russia as source of raw materials and fuel, and main market for exports. Dramatic shrinking of economy since independence. Isolated location; weak transportation network. Slow pace of reform. Cumbersome bureaucracy. Strong black economy. Foreign debt – over 50% of GDP – costs 20% of export earnings to service.

EXPORTS

Romania 5%　Italy 6%
Germany 7%
Russia 37%　USA 14%
Other 31%

IMPORTS

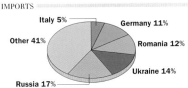

Italy 5%　Germany 11%
Other 41%　Romania 12%
Ukraine 14%
Russia 17%

RESOURCES
 Electric power 1m kw

1630 tonnes　　Oil and gas reserves not exploited

866,000 sheep, 543,000 pigs, 13m chickens　　Lignite, phosphates, gypsum, oil, natural gas

Moldova has few mineral resources. It has to import all its fuel and most of its electricity.

ENVIRONMENT
 Sustainability rank: 39th

1%　　2.2 tonnes per capita

Overuse of agricultural chemicals and pesticides on tobacco farms is a problem, as is soil erosion. There is little spending on environmental improvement.

MEDIA
 TV ownership high

Daily newspaper circulation 60 per 1000 people

PUBLISHING AND BROADCAST MEDIA

There are 4 leading daily newspapers, including the independent publication *Nezavisimaya Moldova*

1 state-controlled service　　1 state-controlled service

The many new publications represent widely differing interest groups. Russian-language broadcasting is restricted.

CRIME
 No death penalty

9837 prisoners　　Up 5% 1996–1998

Economic decline has caused crime to increase. The unstable situation in Transdniestria has encouraged smuggling, particularly of Russian arms. The Council of Europe has accused the police of routinely using torture.

EDUCATION
 School leaving age: 15

99%　　93,759 students

Education has followed a Romanian (French-inspired) system since 1991. Mass protests met plans in 2002 to make Russian compulsory in schools.

HEALTH
 Welfare state health benefits

1 per 286 people　　Circulatory diseases, cancers, accidents

The centralized health service is poor by regional standards. There are serious shortages of medical supplies.

CHRONOLOGY

Modern Moldova corresponds roughly to the eastern part of the Romanian principality of Moldavia, which existed for 500 years from 1359. Most of it was annexed by Russia in 1812 as Bessarabia.

- ❏ **1918** Bessarabia joins Romania.
- ❏ **1924** Moldovan Autonomous Soviet Republic formed within USSR.
- ❏ **1940** Romania cedes Bessarabia to Ukrainian and Moldovan SSRs.
- ❏ **1941–1945** Bessarabia again under Romanian control.
- ❏ **1945** Returns to Soviet control.
- ❏ **1990** Declares sovereignty.
- ❏ **1991** Independence.
- ❏ **1993–1994** Pro-unification parties' election defeat; referendum rejects Romanian unification. Rejoins CIS.
- ❏ **1998** Elections: communist revival.
- ❏ **2001** CPM wins big majority. Voronin becomes president.
- ❏ **2002** Mass protests over education plans.

SPENDING
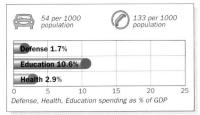 GDP/cap. decrease

CONSUMPTION AND SPENDING

54 per 1000 population　　133 per 1000 population

Defense 1.7%
Education 10.6%
Health 2.9%

0　5　10　15　20　25
Defense, Health, Education spending as % of GDP

Former communist officials have been well placed to benefit from the sale of state-owned businesses. Car ownership is low but rising. However, pensions and wages are often months in arrears. In 1998 the benefits for low-income families and veterans were scrapped. Ethnic Gagauz (Orthodox Christian Turks) are the poorest group.

WORLD RANKING

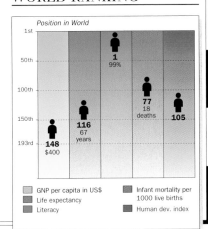

Position in World

1st
50th
100th
150th
193rd

1 — 99%
77 — 18 deaths
105
116 — 67 years
148 — $400

- GNP per capita in US$
- Life expectancy
- Literacy
- Infant mortality per 1000 live births
- Human dev. index

M

MONACO

EUROPE

OFFICIAL NAME: Principality of Monaco   **CAPITAL:** Monaco   **POPULATION:** 31,700
CURRENCY: Euro (French franc until 2002)   **OFFICIAL LANGUAGE:** French

M ONACO IS A TINY ENCLAVE on the French Côte d'Azur. Its destiny changed radically in 1863, when Prince Charles III, after whom Monte Carlo is named, opened the casino. Today, Monaco is a lucrative banking and services center, as well as a tourist destination. Prince Rainier's marriage to film star Grace Kelly, and some astute management of the economy, successfully transformed Monaco into a center for the international jet set. In 1962, the prince's absolute authority was abolished by a new, democratic constitution.

M

CLIMATE ▷ Mediterranean

Summers are hot and dry; days with 12 hours of sunshine are not uncommon. Winters are mild and sunny.

TRANSPORTATION ▷ Drive on right

An underground railroad system connecting to Cap d'Ail in France and opened in 1999, meant that 2% of Monaco's area could be reclaimed. Access to Monaco is mainly by land, although Air Monaco provides helicopter flights from Nice airport.

TOURISM ▷ Visitors : Population 9.5:1

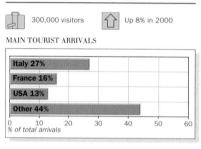

Huge numbers of tourists, greatly outnumbering the inhabitants, are attracted to Monaco, most coming from Italy and France. Almost all are day trippers drawn by the casinos and Monaco's conspicuous high society. Around 75% of hotel rooms are classed as "four-star deluxe," and the principality is a particular favorite of wealthy Italians. The Grimaldi Forum conference center, which opened in 2000, hopes to attract more business travelers.

A number of social and sporting events draw particularly large crowds each spring, including the Rose Ball (March), the Tennis Open (April), and the Grand Prix (May).

PEOPLE ▷ Pop. density high

French, Italian, Monégasque, English   16,253/km² (42,095/mi²)

THE URBAN/RURAL POPULATION SPLIT

100%

RELIGIOUS PERSUASION

Other 5%
Protestant 6%
Roman Catholic 89%

Less than a fifth of Monaco's residents are Monégasque. Around half are French, the rest Italian, American, British, and Belgian. Monégasques enjoy considerable privileges, including housing subsidies to protect them from Monaco's high property prices, and the right of first refusal before a job can be offered to a foreigner. Women have equal status, but only acquired the vote in the constitutional changes of 1962.

POLITICS ▷ Multiparty elections

1998/2003   H.S.H. Prince Rainier III

AT THE LAST ELECTION
National Council 18 seats

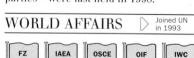

100%
NDU

NDU = National and Democratic Union
There are no formal political parties

The Grimaldi princes have been hereditary rulers of Monaco for more than 700 years. Prince Rainier III renounced absolute rule in 1962 but retains considerable power. The executive minister of state is appointed by the prince from a list of French diplomats. National Council elections – based on personalities rather than parties – were last held in 1998.

WORLD AFFAIRS ▷ Joined UN in 1993

FZ   IAEA   OSCE   OIF   IWC

A key concern is to protect both banking secrecy and the liberal tax regime from EU regulation, although the principality has adopted the euro. France is particularly critical, and French citizens have been banned from banking in Monaco since 1962.

MONACO

Total Land Area : 1.95 sq. km (0.75 sq. miles)

Places of Interest
Parks and Gardens
Grand Prix Circuit

0   500 m
0   1000 yds

Monte Carlo with its luxury hotels and yacht harbor. The only space for new development is on land reclaimed from the sea.

AID

 Not applicable

 Monaco has no aid receipts or donations — Not applicable

Monaco neither receives nor gives aid, and the issue is not of concern to Monégasques.

DEFENSE

No compulsory military service

France responsible for defense — Not applicable

Monaco has no armed forces and no defense budget. France, as the protecting power, bears responsibility for the defense of the principality.

ECONOMICS

Inflation 2.6% p.a. (1985–1996)

 $800m

 1.0651–1.1231 euros

SCORE CARD

- ❏ WORLD GNP RANKING.........................158th
- ❏ GNP PER CAPITA $25,000
- ❏ BALANCE OF PAYMENTS.....Included in French total
- ❏ INFLATION ....Included in French total
- ❏ UNEMPLOYMENT 3%

STRENGTHS

Strict banking confidentiality and low taxes attract billions of dollars of overseas deposits. Strong tourism sector. Assets managed by Monaco banks increased by 18% a year in the late 1990s. No formal debt and reserves of over 2.3 billion euros. Very low unemployment.

WEAKNESSES

Continuing vulnerability to money laundering despite revised banking secrecy laws under the 1994 accord with France obliging banks to furnish details of suspicious accounts. Subject to fluctuations of French and Italian economies. Dependence on VAT for 55% of revenues. Pressure from EU states to end privileged banking and tax laws. Total dependence on imports because of lack of natural resources.

EXPORTS/IMPORTS

Monaco has a full customs union with France

RESOURCES

Electric power: Included within French total

 3 tonnes

 Not an oil producer

Included within French total

None

Monaco has no strategic resources and imports all its energy from France. It has no agricultural land.

ENVIRONMENT

Not available

 None

 4.4 tonnes per capita

Monaco has built the most extensive underground car parking facilities in the world to tackle congestion. The quality of the built environment around the harbor occasionally arouses local passions. Important populations of red coral are under threat from land reclamation and pollution.

MEDIA

TV ownership high

Daily newspaper circulation 251 per 1000 people

PUBLISHING AND BROADCAST MEDIA

There is 1 daily newspaper. *Nice-Matin*, a regional French newspaper, publishes a Monaco edition

2 services

4 services: 1 part-owned by French state, 3 independent

In addition to its domestic radio and TV, Monaco receives all the mainstream French and Italian channels.

CRIME

No death penalty

 13 prisoners

 Up 3% in 1999

Low crime rates make it safe for the rich to sport their furs and jewelry in public. In late 1998 the appeals court upheld Monaco's first conviction of an individual for money laundering.

EDUCATION

School leaving age: 16

 99%

 Not available

The education system is essentially the same as that of France, with students studying for the *baccalauréat* exam. Most go on to university in France, but then return to claim good jobs in Monaco. The Catholic Church exerts considerable influence and is still responsible for primary schooling.

HEALTH

Welfare state health benefits

 1 per 333 people

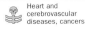 Heart and cerebrovascular diseases, cancers

Most medical care is provided by private health insurance. Doctors train in France. The Princess Grace Hospital can serve 60,000 people, also catering for patients from outside Monaco.

CHRONOLOGY

In 1297, the Grimaldis established themselves as the principality's hereditary rulers.

- ❏ **1861** Independent under French protection.
- ❏ **1949** Rainier III accedes to throne.
- ❏ **1962** Constitution rewritten: end of absolute authority of the prince.
- ❏ **1963** Democratic legislative elections held for first time.
- ❏ **1982** Princess Grace dies following car accident.
- ❏ **2002** Euro introduced.

SPENDING

GDP/cap. increase

CONSUMPTION AND SPENDING

680 per 1000 population

1030 per 1000 population

Defense	None	
Education	No data	
Health	8%	

0   5   10   15   20   25
Defense, Health, Education spending as % of GDP

Monaco's image abroad has changed dramatically since Prince Rainier acceded in 1949. From being considered a simple gambling spot, it is now ranked as one of the world's most glamorous international jet set destinations. In part, this was the result of Prince Rainier's wedding to Grace Kelly, then a leading Hollywood star, which brought Monaco to the attention of US high society. More important was the prince's work in turning Monaco into a major tax haven and an upmarket resort, by making the most of its Mediterranean coastal location. Many tax exiles have taken up residence, among them Luciano Pavarotti and Wall Street investment guru Bob Beckman.

WORLD RANKING

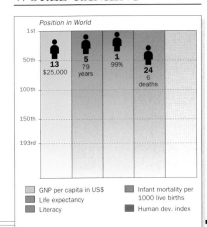

Position in World

13	5	1	24
$25,000	79 years	99%	6 deaths

- GNP per capita in US$
- Life expectancy
- Literacy
- Infant mortality per 1000 live births
- Human dev. index

M

MONGOLIA

OFFICIAL NAME: Mongolia   **CAPITAL:** Ulan Bator   **POPULATION:** 2.6 million
CURRENCY: Tugrik (tögrög)   **OFFICIAL LANGUAGE:** Khalkha Mongolian

LANDLOCKED BETWEEN Siberia and China's Mongolian provinces, Mongolia rises from the semiarid Gobi Desert to mountainous steppe. The traditionally nomadic Mongols were first unified by Genghis Khan in 1206. "Outer" Mongolia achieved independence from China as a communist state in 1924 and was officially aligned with the USSR from 1936. In 1990, it abandoned communist rule; widespread poverty ensued. Extremely harsh winters in 1999–2001 devastated the rural economy.

CLIMATE

▷ Mountain/cold desert/steppe

WEATHER CHART FOR ULAN BATOR

Temperature variations are extreme. Dry summers combine with severe winters, known as *zud*, to devastate livestock, as happened in 1999 and 2000.

TRANSPORTATION

▷ Drive on right

🛩 **Buyant–Ukhaa, Ulan Bator**   ⚓ Has no fleet

THE TRANSPORTATION NETWORK

🛣 1674 km (1040 miles)	🛤 None
🚂 1810 km (1125 miles)	〰 397 km (247 miles)

Lack of investment has left Mongolia's infrastructure to decay, increasing transportation and distribution costs. Links to China and the Pacific are priorities. Gasoline shortages have meant a large increase in the use of animal labor.

Traditional gers in the Gobi Desert.
Many Mongolians still choose to pursue a nomadic lifestyle, living in felt tents called gers.

TOURISM

▷ Visitors : Population 1:16

🧳 158,000 visitors   ⬇ Down 1% in 2000

MAIN TOURIST ARRIVALS

China	37%
Russia	35%
Japan	7%
Other	21%

% of total arrivals

Tourism has expanded overall since the easing of visa restrictions in 1991. Under communism, all travel was arranged through the state agency, Zhuuichin, but private companies are now entering the market.

PEOPLE

▷ Pop. density low

Khalkha Mongolian, Kazakh, Chinese, Russian   2/km² (4/mi²)

THE URBAN/RURAL POPULATION SPLIT

64%   36%

ETHNIC MAKEUP

Russian 2%   Chinese 2%
Other 2%   Kazakh 4%
Mongol 90%

Khalkh Mongols, who adhere to Tibetan Buddhism, are the main ethnic group. Although economic pressures keep many people near urban centers, most remain nomadic. One-third live in Ulan Bator. Turkic Kazakhs in the west form the largest minority, but emigration to Kazakhstan since 1990 has reduced their numbers. Tensions exist with the Chinese and Russian minorities.

POLITICS

▷ Multiparty elections

2000/2004   President Natsagyn Bagabandi

AT THE LAST ELECTION

State Great Hural 76 seats

95% MPRP   4% Ind   1% MNDP

MPRP = Mongolian People's Revolutionary Party
Ind = Independents
MNDP = Mongolian National Democratic Party

The end of communism and the advent of democracy in 1990 revolutionized Mongolian politics. The shock of economic reform led many Mongolians to regret the lost certainties of the communist era. In 1992 the democrats lost power to the renamed communists (MPRP), but their failure to revive the economy swung the pendulum back in favor of a democratic coalition in 1996. An uneasy cohabitation between President Natsagyn Bagabandi of the MPRP, who took office in 1997, and an MNDP-led government existed until the sweeping MPRP election victory in 2000.

MONGOLIA

Total Land Area : 1 565 000 sq. km (604 247 sq. miles)

POPULATION
⊙ over 500 000
○ over 50 000
● over 10 000
• under 10 000

LAND HEIGHT
3000m/9843ft
2000m/6562ft
1000m/3281ft
above 500m

M

WORLD AFFAIRS

 Joined UN in 1961

IAEA · IBRD · NAM · ADB · WTO

Closer relations with Japan and other east Asian states have failed to weaken Mongolia's ties with Russia and China. There are residual tensions with China, since the majority of ethnic Mongols actually resides in the adjoining Chinese province of Inner Mongolia, but there is no longer a fear of Chinese designs on Mongolian sovereignty.

AID

Recipient

$217m (receipts)   Down 2% in 2000

A large balance-of-payments deficit and severe weather make aid vital. The main donors are Japan and the ADB.

DEFENSE

Compulsory military service

$19m   No change in 2000

The last Soviet forces left in 1992. However, ties are still strong, and under agreements reached in 2000 and 2001 Russia is helping to reform the greatly reduced and poorly equipped Mongolian forces.

ECONOMICS

Inflation 58% p.a. (1990–2000)

$947m   1097–1102 tugriks

SCORE CARD

- ❏ WORLD GNP RANKING........................156th
- ❏ GNP PER CAPITA $390
- ❏ BALANCE OF PAYMENTS...................–$112m
- ❏ INFLATION 7.6%
- ❏ UNEMPLOYMENT 5%

STRENGTHS
Copper and cashmere. Largely untapped coal and oil reserves. Traditional and efficient rural economy.

WEAKNESSES
Harsh winters ravaged livestock between 1999 and 2001. Decaying infrastructure. Rising poverty.

EXPORTS

UK 3% — Italy 4%
Russia 9%
China 47%
Other 10%
USA 27%

IMPORTS

Germany 5% — Japan 5%
South Korea 10%
Russia 34%
China 21%
Other 25%

RESOURCES

 Electric power 901,000 kw

524 tonnes

Contracts have been signed with oil prospectors

15.7m sheep, 11.8m goats, 2.75m horses,

Oil, coal, copper, lead, fluorite, tungsten, tin, gold, uranium

Under communism, Mongolia's vast mineral resources were barely exploited, and prospecting has only recently begun. A uranium mining joint venture with Russia has been established. Mongolia is rich in oil, with sufficient reserves to meet future domestic needs. In 1999 an oil extraction agreement was signed with China.

ENVIRONMENT

 Sustainability rank: 42nd

12%   3.3 tonnes per capita

Industrial pollution around Ulan Bator is a health hazard. Lake Hövsgöl is seriously polluted. Renewed efforts were launched in 2001 to preserve the Bogd Khan, the world's oldest protected area, dating from 1788, from illegal logging, hunting, and air pollution.

MEDIA

 TV ownership medium

Daily newspaper circulation 17 per 1000 people

PUBLISHING AND BROADCAST MEDIA

There are 3 daily newspapers, *Önöödör, Zuuny Medee,* and *Ödriin Sonin*

5 services:
1 state-owned,
4 independent

7 services:
1 state-owned,
6 independent

Since 1990, Mongolia's press has enjoyed unlimited freedom; there are no slander or libel laws. Legislation enacted in 1999 eased remaining curbs on the media. However, the shortage of paper and fuel supplies has restricted both the number of publications and their distribution.

CRIME

 Death penalty in use

6414 prisoners   Up 8% in 1999

Crime rose rapidly in the early 1990s, particularly organized crime and muggings by knife gangs. Ulan Bator is the most dangerous area, especially for foreigners; Russians, Chinese, and dollar-carrying US tourists are the main targets.

EDUCATION

 School leaving age: 16

99%   84,970 students

Education is modeled on the former Soviet system. The majority of teachers are women on low salaries. Private-sector schools emphasizing Mongol culture are beginning to open.

CHRONOLOGY

In the 17th century, the Manchus took control of Mongolia. It stayed in Chinese hands until 1911.

- ❏ **1919** China reoccupies Mongolia.
- ❏ **1924** Independent communist state.
- ❏ **1989–1990** Prodemocracy protests; communist election defeat.
- ❏ **1992** Former communists, renamed MPRP, returned to power.
- ❏ **1996** Democratic Union coalition wins general election.
- ❏ **1997** MPRP wins presidency.
- ❏ **1999–2001** Severe winters.
- ❏ **2000** Landslide electoral victory for MPRP.

HEALTH

 Welfare state health benefits

1 per 417 people   Heart, parasitic, and respiratory diseases

Shortages of drugs and equipment have renewed interest in traditional Mongolian herbal medicine. As well as the state-run system, some Buddhist monasteries provide health care.

SPENDING

GDP/cap. increase

CONSUMPTION AND SPENDING

17 per 1000 population   56 per 1000 population

Defense 2%
Education 5.7%
Health 4.3%

0   5   10   15   20   25
Defense, Health, Education spending as % of GDP

Economic liberalization has fueled great disparities in wealth. An estimated 40% of the population now live below the poverty line; the poorest cannot even afford to buy bread. Starvation threatened after the severe winters of 1999–2001.

WORLD RANKING

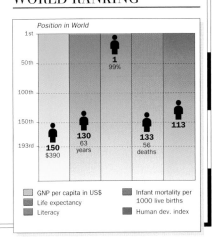

Position in World

1st
50th — 1 / 99%
100th
150th — 113
193rd — 150 / $390 · 130 / 63 years · 133 / 56 deaths

- GNP per capita in US$
- Life expectancy
- Literacy
- Infant mortality per 1000 live births
- Human dev. index

M

Morocco

NORTH AFRICA

OFFICIAL NAME: Kingdom of Morocco **CAPITAL:** Rabat
POPULATION: 30.4 million **CURRENCY:** Moroccan dirham **OFFICIAL LANGUAGE:** Arabic

1956 1956 July 30 MA 0 +212 .ma

MOROCCO IS SITUATED in northern Africa, but at its northernmost point lies only 12 km (8 miles) from mainland Europe, across the Strait of Gibraltar. The northern regions have a Mediterranean climate, while the south comprises semiarid desert. The late King Hassan's international prestige gave Morocco status out of proportion to its wealth. The main issues facing the country are the internal threat of Islamic militancy and the stalled political process in Western Sahara, the former Spanish colony occupied by Morocco since 1975. Key economic strengths are tourism, phosphates, and agriculture.

TOURISM

Visitors : Population
1:7.2

4.22m visitors   Up 3% in 2001

MAIN TOURIST ARRIVALS

France 20%
Spain 6%
Germany 6%
UK 3%
Italy 3%
Other 62%

% of total arrivals   0 10 20 30 40 50 60 70 80

CLIMATE

Hot desert/mountain/
Mediterranean

WEATHER CHART FOR RABAT

The climate ranges from warm and temperate in the north to semiarid in the south, but temperatures are cooler in the mountains, especially in the high Atlas. During the summer, the effects of the *sirocco* and *chergui*, hot winds from the Sahara, are felt.

TRANSPORTATION

Drive on
right

Mohammed V, Casablanca
3.57m passengers

496 ships
444,000 grt

THE TRANSPORTATION NETWORK

30,254 km
(18,799 miles)

327 km
(203 miles)

1907 km
(1185 miles)

None

Morocco has six international airports. A highway links Rabat and Casablanca, and plans for a new trans-Sahara highway from Tangiers to Lagos, Nigeria, were announced in mid-2000. In rural areas, however, roads tend to peter out.

Tourism is vital to the Moroccan economy. Good beaches abound; Agadir has 300 days of sunshine a year. Fès and Marrakech offer cultural interest, while the Atlas Mountains attract walkers and skiers. Desert safaris are offered in the Sahara. Most Western tourists come from France, Germany, and Spain.

M

WESTERN SAHARA

MOROCCO

Morocco

Total Land Area : 446 300 sq. km
(172 316 sq. miles)

POPULATION

over 1 000 000
over 500 000
over 100 000
over 50 000
over 10 000
under 10 000

LAND HEIGHT

3000m/9845ft
2000m/6562ft
1000m/3281ft
500m/1640ft
200m/656ft
Sea Level

PEOPLE ▷ Pop. density medium

 Arabic, Tamazight, French, Spanish   68/km² (176/mi²)

THE URBAN/RURAL POPULATION SPLIT

56%   44%

RELIGIOUS PERSUASION

Other (mostly Christian) 1%

Muslim (mainly Sunni) 99%

ETHNIC MAKEUP

European 1%
Berber 29%
Arab 70%

Morocco, the westernmost of the Maghreb states, is the main refuge for descendants of the original Berber inhabitants of northwest Africa. About 35% of Moroccans speak Berber (Tamazight is the main dialect). They live mainly in mountain villages, while the Arab majority inhabit the lowlands. Before independence from France, 450,000 Europeans lived in Morocco; numbers have since greatly diminished. Some 45,000 Jews enjoy religious freedom and full civil rights – a position in society unique among Arab countries. Most people speak Arabic, and French is also spoken in urban areas.

Sunni Islam is the religion of most of the population. The king is the spiritual leader through his position as Commander of the Faithful.

Female emancipation has been slow to take root in Morocco, but women are starting to take a more prominent role in society.

POPULATION AGE BREAKDOWN

Female		Age	Male	
	0.8%	80+	0.7%	
	3%	60–79	2.5%	
	7.3%	40–59	6.4%	
16.4%		20–39	15.1%	
23.4%		0–19	24.4%	

% of population by age group

POLITICS ▷ Multiparty elections

L. House 1997/2002
U. House 2000/2003   H.M. King Mohammed VI

AT THE LAST ELECTION
House of Representatives 325 seats   3% MPCD   1% PSD

31% K   31% W   30% C   3% FFD   1% Others

K = Koutla bloc (includes Socialist Union of Popular Forces and Istiqlal)   **W** = Wifaq bloc (includes Constitutional Union & Popular Movement)   **C** = Center bloc (includes National Rally of Independents and Democratic Popular Movement)   **MPCD** = Constitutional & Democratic Popular Movement   **FFD** = Democratic Forces Front   **PSD** = Democratic Socialist Party

House of Councillors 270 seats

An indirectly elected House of Councillors was formed in December 1997

Morocco is a constitutional monarchy with a bicameral legislature.

PROFILE

Although the majority parliamentary party now chooses the government, the king reserved the right to appoint or dismiss the prime minister. Legislative elections in 1997 left parliament split three ways, with only nine seats going to Islamists. The eventual formation of a socialist-led government was seen as the clearest expression to date of the increasing role of the party system.

MAIN POLITICAL ISSUES
The post-Hassan monarchy

King Mohammed VI is seen as a less dominating figure than his late father King Hassan II. His accession encouraged expectations of change. The king's marriage in 2002 to computer engineer Salma Bennani was presented as evidence of a more modern and open outlook. The queen is expected to provide a role model for Moroccan women.

Islamic militancy

Mohammed VI and the government in office since 1998 have ushered in a more liberal stance toward Islamist activists. In 2000, Abdessalam Yassine, spiritual leader of the banned Islamic movement Justice and Good Deeds, was released after ten years' imprisonment without trial. Popular support for Islamic fundamentalism is fueled by the fear that Morocco is losing its Islamic identity. Pro-Islamist rallies have far outnumbered those by supporters of greater rights for women.

Abderrahmane el Youssoufi, prime minister since 1998.

King Mohammed VI, who succeeded his father, Hassan II.

The town of Boumaine-Dadès lies in the southern foothills of the Atlas Mountains. The region's outstanding scenery makes it one of Morocco's major tourist attractions.

WORLD AFFAIRS ▷ Joined UN in 1956

 AL   AMU   IBRD   NAM   OIC

Morocco's important role in the quest for lasting peace in the Middle East was underlined by Israeli prime minister Yitzhak Rabin's visit to Rabat following the signing in Washington D.C. of the 1993 peace accord with the Palestine Liberation Organization. King Hassan's foreign policy was ambiguous, for while he negotiated with Israel he also headed the Jerusalem Committee of the Organization of the Islamic Conference (OIC). Generally more pro-Western than other Arab states, Morocco has also earned respect by protecting its Jewish minority.

International disapproval has focused on Morocco's occupation since 1975 of the former Spanish colony of Western Sahara. Resistance by Polisario Front guerrillas, who are fighting for an independent Western Sahara, commenced in 1983 and has continued, despite a UN-brokered peace plan in 1991. Since then, UN proposals have included a referendum on self-determination and autonomy for Western Sahara under Moroccan sovereignty. The process remained stalled in 2002, with the UN extending its mandate to the end of July.

Relations with the EU were strengthened with the signing of an association agreement in late 1995, envisaging free trade in industrial goods within 12 years.

Sovereignty disputes with Spain over rocky islets in the Mediterranean were at the heart of a high-profile confrontation in 2002.

AID ▷ Recipient

 $419m (receipts)   Down 38% in 2000

Saudi Arabia wrote off $2.7 billion of Moroccan debt after the Gulf War. The World Bank has given help to Morocco, but the country receives little aid.

M

CHRONOLOGY

Independence from France in 1956 was only the first step in ending colonial rule for the oldest kingdom in the Arab world, even though the present Alaoui dynasty has been in power for three centuries.

- ❑ **1956** France recognizes Moroccan independence under Sultan Mohammed ibn Yousif. Morocco joins UN. Spain renounces control over most of its territories.
- ❑ **1957** Sultan Mohammed king.
- ❑ **1961** Hassan succeeds his father.
- ❑ **1967** Morocco backs Arab cause in Six-Day War with Israel.
- ❑ **1969** Spain returns Ifni to Morocco.
- ❑ **1972** King Hassan survives assassination attempt.
- ❑ **1975** International Court of Justice grants right of self-determination to Western Saharan people. Moroccan forces seize Saharan capital.
- ❑ **1976** Morocco and Mauritania partition Western Sahara.
- ❑ **1979** Mauritania renounces claim to part of Western Sahara, which is added to Morocco's territory.
- ❑ **1984** King Hassan signs Oujda Treaty with Col. Gaddafi of Libya as first step toward Maghreb union. Morocco leaves OAU after criticism of its role in Western Sahara.
- ❑ **1986** Morocco abrogates Oujda Treaty.
- ❑ **1987** Defensive wall built around Western Sahara.
- ❑ **1989** Arab Maghreb Union (AMU) creates no-tariff zone between Morocco, Algeria, Tunisia, Libya, and Mauritania.
- ❑ **1991** Morocco accepts UN plan for referendum in Western Sahara.
- ❑ **1992** New constitution grants majority party in parliament right to choose the government.
- ❑ **1993** First general election for nine years. After major parties refuse his invitation, king appoints nonparty government.
- ❑ **1994** King Hassan replaces veteran prime minister Karim Lamrani with Abdellatif Filali.
- ❑ **1995** Islamist opposition leader Mohamed Basri returns to Morocco after 28 years of exile.
- ❑ **1998** Socialists enter government with Abderrahmane el Youssoufi as prime minister.
- ❑ **1999** Death of King Hassan. Mohammed VI enthroned. Liberalization program announced.
- ❑ **2001** UN special representative proposes a ten-year trial period for Western Sahara as part of Morocco.
- ❑ **2002** UN extends mandate in Western Sahara as political process stalls.

M

DEFENSE

 Compulsory military service

 $1.68bn — Down 5% in 2000

MOROCCAN ARMED FORCES

744 main battle tanks (224 M-48A5, 420 M-60, 100 T-72)	175,000 personnel	
1 frigate, 27 patrol boats	10,000 personnel	
95 combat aircraft (24 F-5E, 14 *Mirage* F-1EH, 15 *Mirage* F-1CH)	13,500 personnel	
None		

Morocco's long struggle in Western Sahara against Polisario Front guerrillas has given the kingdom's forces a formidable reputation. Moroccans have also fought as mercenaries in the Gulf. In the 1980s, Moroccan sappers constructed a 2500-km (1550-mile) defensive wall to cordon off Western Sahara in an attempt to prevent incursions from Polisario guerrillas based in Algeria. The Polisario forces themselves number some 3000–6000.

Morocco's pro-Western stance has allowed its forces access to sophisticated weapons and training from the West, particularly the US – unlike neighboring north African states, which are dependent on the former Soviet bloc.

The air force was formed in 1956 and flies US and European aircraft, notably *Mirage* interceptors. The navy uses Western-supplied ships, but is insignificant in regional terms. In addition, there are 42,000 paramilitaries.

Around 5% of national income is spent on defense – a relatively high figure for a developing country. Military service, lasting 18 months, is compulsory.

ECONOMICS

 Inflation 2.8% p.a. (1990–2000)

 $33.9bn — 10.563–11.519 Moroccan dirhams

SCORE CARD

❑ WORLD GNP RANKING	56th
❑ GNP PER CAPITA	$1180
❑ BALANCE OF PAYMENTS	–$475m
❑ INFLATION	1.9%
❑ UNEMPLOYMENT	23%

EXPORTS

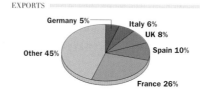

Germany 5%
Italy 6%
UK 8%
Spain 10%
Other 45%
France 26%

IMPORTS

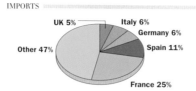

UK 5%
Italy 6%
Germany 6%
Spain 11%
Other 47%
France 25%

ECONOMIC PERFORMANCE INDICATOR

Consumer Price Index — GDP

STRENGTHS

Probusiness policies and abundant labor attract foreign investment. Low inflation. Great potential for tourist industry (already important), phosphates, and agriculture.

WEAKNESSES

High unemployment and population growth. Droughts have hit agriculture. Cannabis production (providing Europe's main source of resin) complicates closer EU links.

PROFILE

The government's large-scale privatization program, which began in 1992, was designed to attract investment, particularly from Europe. Severe drought in 1995 made austerity measures necessary. The new socialist-led government has given social policy a higher priority. Expected revenue from oil reserves will be channeled into the development of rural areas.

MOROCCO : MAJOR BUSINESSES

Tanger
Casablanca
Fès
Khouribga
Safi
Youssoufia
Agadir
Boudnib
Marrakech
Tan-Tan

Textiles
Chemicals
Phosphates
Fish processing
Food processing
Light engineering
Heavy engineering
Vehicle manufacture

0   200 km
0   200 miles

RESOURCES

 Electric power 4m kw

748,183 tonnes

241 b/d (reserves 659m barrels)

17.3m sheep, 5.2m goats, 2.7m cattle, 137m chickens

Phosphates, oil, gas, coal, iron, barytes, lead, copper, zinc

ELECTRICITY GENERATION

Hydro 13% (1.8bn kwh)

Combustion 87% (12bn kwh)

Nuclear 0%

Other 0%

% of total generation by type

Morocco possesses 75% of the world's phosphate reserves. The discovery of large oil and gas deposits in the northeastern desert in mid-2000 could yield an annual revenue of $400 million.

MOROCCO : LAND USE

Cropland
High mountain regions
Pasture
Forest
Desert
Sheep
Fruit
Wheat

0   200 km
0   200 miles

ENVIRONMENT

 Sustainability rank: 73rd

0.7%

1.2 tonnes per capita

ENVIRONMENTAL TREATIES

Yes   Yes   Yes

Yes   Yes   Yes

Morocco's wealth of plant and animal life has suffered severely from long periods of drought, most recently in the early 1980s and early 1990s. The unplanned development of tourist resorts is posing a threat to fragile coastal ecosystems.

CRIME

 Death penalty in use

48,600 prisoners

Crime has risen sharply

CRIME RATES

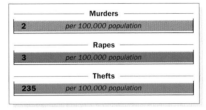

Murders
2   per 100,000 population

Rapes
3   per 100,000 population

Thefts
235   per 100,000 population

Urban crime is increasing, but muggings are rare. Apart from a 1990 strike that led to 40 deaths in Fès, there has been little civil unrest. Prisons are overcrowded, and conditions are poor.

MEDIA

 TV ownership medium

Daily newspaper circulation 27 per 1000 people

PUBLISHING AND BROADCAST MEDIA

There are 22 daily newspapers, including *Le Matin du Sahara et du Maghreb*, *Rissalat al-Oumma*, *al-Alam*, and *L'Opinion*

2 services: 1 state-owned, 1 independent

3 services: 1 state-owned, 2 independent

The succession of Mohammed VI fueled hopes of a more liberal climate, but the media remain strictly controlled, particularly over reporting of the Western Sahara issue. In 2000, the outspoken French-language weekly *Demain* was banned. The sports pages, especially the soccer reports, are the most dynamic sections of the press. State-owned TV began transmissions in Arabic and French in 1962. Radio broadcasts are in Arabic, Berber, French, Spanish, and English from Rabat and Tangier. Morocco can receive broadcasts from Spanish radio and television stations.

EDUCATION

 School leaving age: 14

49%

311,743 students

THE EDUCATION SYSTEM

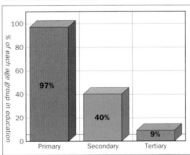

% of each age group in education

97%   Primary
40%   Secondary
9%   Tertiary

Literacy in rural areas – at only 14% – is much lower than in the cities. The literacy level and elementary school enrollment rates are well below average for countries with similar living standards; child labor is widely used. There are both state-controlled and private schools. In 1988 the secondary school graduation examination, the *baccalauréat*, was replaced by a system of continuous assessment.

HEALTH

 Welfare state health benefits

1 per 2000 people

 Neonatal causes, cerebrovascular and heart diseases

There is one hospital bed for every 1000 people. Despite recent progress, child mortality and nutritional standards for the poorest Moroccans remain substantially below the average. Outside the cities, primary health care is virtually nonexistent, with the result that people depend on traditional remedies for illnesses. All employees are required to contribute to a social welfare fund, which operates a system of benefits in the event of illness, occupational accidents, and old age.

SPENDING

GDP/cap. increase

CONSUMPTION AND SPENDING

41 per 1000 population

50 per 1000 population

Defense 5.1%
Education 5%
Health 1.2%

Defense, Health, Education spending as % of GDP

Average income per head is much lower than in the neighboring countries of Algeria and Tunisia. Almost one in five Moroccans live below the poverty line, and the rural–urban gap in wealth is considerable; the majority of the population lives in rural areas. A period of drought in the 1990s encouraged urban drift.

Unrest has largely been avoided owing to the fact that Morocco has a thriving informal sector. This provides jobs in food processing, clothes manufacturing, goods transportation, and the hotel and building trades. In addition, there is work to be found in the illegal hashish trade and the smuggling of alcohol and Western goods.

WORLD RANKING

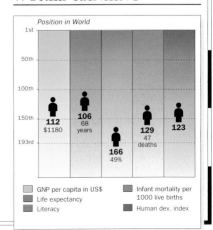

Position in World

1st
50th
100th
150th
193rd

112   $1180
106   68 years
166   49%
129   47 deaths
123

GNP per capita in US$
Life expectancy
Literacy

Infant mortality per 1000 live births
Human dev. index

M

MOZAMBIQUE

OFFICIAL NAME: Republic of Mozambique   **CAPITAL:** Maputo
POPULATION: 18.6 million   **CURRENCY:** Metical   **OFFICIAL LANGUAGE:** Portuguese

SOUTHERN AFRICA

 1975   1975   June 25   MOC   +2   +258 .mz

SITUATED ON THE SOUTHEAST African coast, Mozambique is bisected by the Zambezi River. South of the Zambezi lies a semiarid savanna lowland. The more fertile north-central delta provinces around Tete are home to most of Mozambique's ethnically diverse population. Following independence from Portugal in 1975, Mozambique was torn apart by civil war between the (then Marxist) FRELIMO government and the South African-backed Mozambique National Resistance (RENAMO). The conflict finally ended in 1992 after UN arbitration. Multiparty elections in 1994 returned FRELIMO to power. Devastating floods in 2000 and 2001 created a desperate situation for this impoverished country, with famine a major threat.

CLIMATE

> Tropical wet & dry

WEATHER CHART FOR MAPUTO

In theory, Mozambique has a rainy and a dry season. The coast at Beira and at Quelimane and the highlands west of Nampula are the wettest areas. The Zambezi valley is the hottest region.

Mozambique is prone to extremes of rainfall. In the 1980s, frequent failure of the rains contributed to two disastrous famines. Devastating floods occurred in 2000 and 2001.

TRANSPORTATION

> Drive on left

 Mavalane International, Maputo 436,849 passengers

 124 ships 35,300 grt

THE TRANSPORTATION NETWORK

5685 km (3532 miles)		None	
3114 km (1935 miles)		3750 km (2330 miles)	

The billion-dollar Maputo Corridor, launched in 1995, reconnects South African industrial centers with the Mozambican coast, and should also facilitate port modernization. CFM, the state-owned railroad company, is cooperating with other neighboring states. The national airline is returning profits. Even so, millions of land mines still hamper access, damaged bridges have yet to be rebuilt, and remote communities remain isolated.

Tea picking. *Other important cash crops are cashew nuts, cotton, sugar, copra, and citrus fruits. Agriculture employs 85% of workers.*

TOURISM

> Not available

 Tourism has still not recovered after war
 Little change from year to year

MAIN TOURIST ARRIVALS

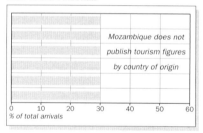

Mozambique does not publish tourism figures by country of origin

% of total arrivals

Mozambique used to attract around 300,000 South Africans and Rhodesians a year in the 1970s, but the tourist industry was destroyed by the civil war and is only slowly being rebuilt. Land mines still render travel outside the capital hazardous, while food shortages, poor infrastructure, and costly international flights are added obstacles. Further setbacks followed the floods in 2000 and 2001.

Given political stability, though, Mozambique could yet exploit its excellent beaches and game reserves, which include the Gorongosa Game Park. Some hotel groups are once more targeting Maputo as a luxury tourist and conference venue.

PEOPLE

> Pop. density low

 Makua, Tsonga, Sena, Lomwe, Portuguese

24/km² (61/mi²)

THE URBAN/RURAL POPULATION SPLIT

40%   60%

RELIGIOUS PERSUASION

- Muslim 10%
- Christian 30%
- Traditional beliefs 60%

ETHNIC MAKEUP

Other 3%
Yao 4%
Shona 11%
Makua Lomwe 47%
Malawi 12%
Tsonga 23%

Mozambique is ethnically diverse, with the overwhelming black African majority divided into numerous groups, and tiny minorities of whites, mixed-race groups, and Asians. However, the predominant social tensions are regional: RENAMO, strong in the north and central regions, accuses the FRELIMO government of consistently favoring the south. Antiwhite feelings are growing too, as "Africanists" claim that whites enjoy excessive political influence – incendiary charges in a country where average life expectancy is well below 50 years and poverty is endemic.

Mozambican society centers on the extended family. In some provinces, notably Zambezia, Cabo Delgado, and Tete, this is matriarchal. Polygamy is fairly widespread among men who can support second wives. FRELIMO pays special attention to women's rights. Many women served in FRELIMO armies, and are now protected by divorce, child-custody, and husband-desertion laws. The Mozambican Women's Organization encourages participation in political life.

POPULATION AGE BREAKDOWN

Female	Age	Male
0.1%	80+	0.1%
2.1%	60–79	1.7%
6.7%	40–59	6.1%
14.2%	20–39	12.5%
28.3%	0–19	28.2%

% of population by age group

M

POLITICS   ▷ Multiparty elections

1999/2004   President Joaquim Alberto Chissano

AT THE LAST ELECTION

Assembly of the Republic 250 seats

53% FRELIMO	47% RENAMO

FRELIMO = Front for the Liberation of Mozambique
RENAMO = Mozambique National Resistance

Mozambique held its first multiparty elections in 1994.

PROFILE

Between 1977 and 1990, Mozambique was a one-party state run by the Soviet-backed FRELIMO. It had campaigned for independence since the 1960s. White South Africa and Southern Rhodesia (now Zimbabwe) both backed anti-Marxist RENAMO rebels.

Changing international realities persuaded FRELIMO to adopt a democratic constitution in 1990. Meanwhile, RENAMO lost its external sponsors. Today little distinguishes the two ideologically. Although FRELIMO is the biggest party in parliament, RENAMO is clearly popular, and demands recognition for its 15 years of struggle. New groups, such as the antiwhite PALMO, COINMO, and UNAMO, have recently emerged; FRELIMO is pushing ahead with plans to decentralize power. Continuing disputes since 1999 between RENAMO and the government about provincial representation have rekindled fears for the future of multiparty democracy.

MAIN POLITICAL ISSUES
The move to democracy

In 1993, the UN secured, with difficulty, the $260 million and the 7500-strong multinational forces required to demobilize the warring factions and to stage the first democratic elections.

Despite last-minute hitches, elections were held in 1994 which returned FRELIMO to power. However, support for RENAMO was stronger than had been expected. Their leader Afonso Dhlakama polled strongly in the presidential elections in both 1994 and 1999, contesting Joaquim Chissano's claim of victory.

Reconstruction

The government faces a huge task in rebuilding a country where devastating floods came on top of the ravages of civil war. The fighting had left 900,000 dead, one million refugees, and an estimated 90% of the remaining population living below the poverty line.

Joaquim Chissano, president since 1986, has pushed toward political pluralism.

Afonso Dhlakama, RENAMO leader, has turned from militarism to politics.

INDIAN OCEAN

MOZAMBIQUE

Total Land Area : 801 590 sq. km
(309 494 sq. miles)

POPULATION

over 1 000 000	▣
over 100 000	◉
over 50 000	○
over 10 000	●
under 10 000	•

LAND HEIGHT

2000m/6562ft	
1000m/3281ft	
500m/1640ft	
200m/656ft	
Sea Level	

M

WORLD AFFAIRS   ▷ Joined UN in 1975

 Comm   CPLP   AU   OIC   SADC

Mozambique was a key Cold War battleground between Soviet-backed Marxism, and capitalism sponsored by the US and South Africa. The resulting civil war devastated the country between 1977 and 1992.

In the early 1980s, however, the FRELIMO government's position began to shift as Soviet aid became erratic. Responding to President Samora Machel's overtures, the US lifted its ban on economic assistance in 1984. Britain agreed to train FRELIMO's forces in 1987. South Africa continued tacitly to support RENAMO until at least 1990. Zimbabwean troops helped Mozambique guard the strategically important Beira and Limpopo corridors, but left in 1993.

In 1995, the UN withdrew its peacekeepers and a democratic Mozambique joined the Commonwealth, despite having no formal links with the old British Empire. President Chissano became deputy head of the SADC, but regional tensions persisted, with Mozambique accusing South Africans of gun-running, and Swaziland claiming Maputo Province as its own.

CHRONOLOGY

The Portuguese tapped the local trade in slaves, gold, and ivory in the 16th century and made Mozambique a colony in 1752. Large areas were run by private companies until 1929.

❏ **1964** FRELIMO starts war of liberation.
❏ **1975** Independence. FRELIMO leader Samora Machel is president.
❏ **1976** Resistance movement RENAMO set up inside Mozambique by Rhodesians.
❏ **1976–1980** Mozambique closes Rhodesian border and supports Zimbabwean freedom fighters. Reprisals by RENAMO.
❏ **1977** FRELIMO constitutes itself as Marxist–Leninist party.
❏ **1980** South Africa takes over backing of RENAMO.   ⇨

CHRONOLOGY *continued*

- **1982** Zimbabwean troops arrive to guard Mutare–Beira corridor.
- **1984** Nkomati Accord: South Africa agrees to stop support for RENAMO, and Mozambique for ANC, but fighting continues.
- **1986** RENAMO declares war on Zimbabwe. Tanzanian troops reinforce FRELIMO. Machel dies in mysterious air crash in South Africa. Joaquim Chissano replaces him.
- **1988** Nkomati Accord reactivated. Mozambicans allowed back to work in South African mines.
- **1989** War and malnutrition said to claim one million lives. FRELIMO drops Marxism–Leninism.
- **1990** Multipartyism and free-market economy in new constitution. RENAMO breaches cease-fire.
- **1992** Chissano signs peace agreement with RENAMO.
- **1994** Democratic elections return FRELIMO to power.
- **1995** Joins Commonwealth. Economic reforms begun.
- **1999** G7 chooses Mozambique as flagship for international debt relief initiative. RENAMO disputes results of December elections.
- **2000–2001** Thousands displaced by devastating floods.

M

AID   ▷ Recipient

 $876m (receipts)   ⬆ Up 9% in 2000

Mozambique is the world's second most aid-dependent country. Aid accounts for fully 60% of national earnings, and pays for the food needs of some seven million citizens. In 1999, Mozambique became one of only four countries to receive the G7 debt relief scheme for HIPCs, which is worth nearly $3 billion. The main donors are the World Bank, Portugal, the US, the EU and its member states, and the IMF. Debts from earlier Soviet aid have been written off.

DEFENSE   ▷ Compulsory military service

 $85m   ⬇ Down 10% in 2000

About 2.5 million men were deemed "fit for military service" in 1998, but since the civil war ended in 1992, the military's once dominant role in society has greatly diminished. Military figures, once prominent in the FRELIMO government, have been largely stripped of political influence.

Mozambique's new post-peace, British-trained permanent army was formally inaugurated in 1994. Truly national in character, and only around 10,000 strong, it contains both former government and RENAMO troops.

However, one by-product of reorganization was the demobilization of some 75,000 battle-hardened soldiers. Their severance pay ended in mid-1996, and it has not been easy to retrain them, or reintegrate them into civilian life. Some have turned to banditry.

The war's end also saw the departure of external forces, such as UN peacekeepers and the Zimbabwean troops who once guarded strategic railroads against RENAMO attack.

MOZAMBICAN ARMED FORCES

🚜	80 main battle tanks (T-54/55)	10,000 personnel
🚢	3 patrol boats	600 personnel
✈	No combat aircraft	1000 personnel
🚀	None	

ECONOMICS   ▷ Inflation 33% p.a. (1990–2000)

 $3.75bn   💲 17,175–22,885 meticais

SCORE CARD

- ❏ World GNP Ranking......................123rd
- ❏ GNP per Capita $210
- ❏ Balance of Payments...................–$764m
- ❏ Inflation...2%
- ❏ Unemployment...................................21%

EXPORTS

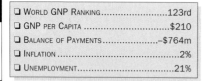
India 5%
Spain 11%
Other 39%
Portugal 12%
South Africa 15%
Zimbabwe 18%

IMPORTS

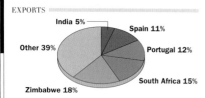
USA 3%
Japan 4%
France 2%
Portugal 8%
Other 46%
South Africa 37%

ECONOMIC PERFORMANCE INDICATOR

— Consumer Price Index   ▨ GDP
Consumer price index 1995=100 / GDP 1995=100
250, 200, 150, 100, 50
1995 1996 1997 1998 1999

Strengths
Following IMF advice, the government has adopted privatization, exchange rate reforms, and trade liberalization. This has enabled Mozambique to attract aid and increase exports.

Massive rural development programs target agriculture, which employs 85% of the workforce. Fisheries industry has great potential. Improved transportation links with Maputo, Africa's second-largest harbor, will help it to service southern Africa's landlocked regions. Go-ahead for a new port at Dobela, in the south, was given in 2002.

Weaknesses
Overseas aid is essential to prevent at least half the population starving. Overdependence on foreign donors and companies is another long-term concern. The country is susceptible to drought, floods, and cyclones. Skilled workers often choose to work in other countries; their absence has delayed the return to normal economic activity.

Profile
Although Mozambique has enormous problems, the government in 1995 produced an optimistic plan, based on World Bank recommendations, to eradicate poverty and raise annual GDP growth to 8–9% by 2000. Devastation resulting from the floods of 2000 and 2001 has effectively destroyed such hopes, and GDP growth in 2000 was only 2.1%.

MOZAMBIQUE : MAJOR BUSINESSES
Pemba
Nacala
Tete
Quelimane
Beira
Maputo
Textiles
Fertilizers
Coal mining
Agribusiness
Fish processing
Food processing
Manufacturing
0   200 km
0   200 miles
* significant multinational ownership

RESOURCES

 Electric power 2.4m kw

 35,560 tonnes

 Not an oil producer

1.32m cattle, 670,000 ducks, 28m chickens

Coal, iron, tantalite, uranium, gold, bauxite titanium, copper, gas

ELECTRICITY GENERATION

Hydro 93% (6.8bn kwh)	
Combustion 7% (0.5bn kwh)	
Nuclear 0%	
Other 0%	

% of total generation by type

Reserves of coal, iron, bauxite, uranium, and gas are underexploited, but a gas pipeline to South Africa is to be completed by 2004. Cotton vies with cashew nuts as the chief crop. Fishing is a vital sector. Electricity supplies are being restored, and the Mozal aluminum smelter opened in 2000.

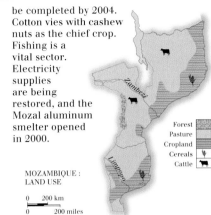

MOZAMBIQUE : LAND USE

Forest
Pasture
Cropland
Cereals
Cattle

0   200 km
0   200 miles

ENVIRONMENT

 Sustainability rank: 59th

 6%

0.1 tonnes per capita

ENVIRONMENTAL TREATIES

No	Yes	Yes
Yes	Yes	No

Floods followed by droughts are often devastating. The floods of early 2000, resulting from a combination of cyclones and torrential rain, affected an estimated one million people. More than 200,000 were displaced by the floods of 2001. The worst drought in living memory was in 1982–1984; it killed 100,000 and left four million people close to starvation. Civil war pushed rural populations toward the cities and coasts, resulting in overcrowding, disease, pollution, and desertification on abandoned farms. However, ecological concerns are still low on the agenda.

MEDIA

 TV ownership low

 Daily newspaper circulation 3 per 1000 people

PUBLISHING AND BROADCAST MEDIA

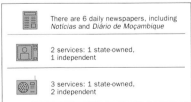

There are 6 daily newspapers, including *Notícias* and *Diário de Moçambique*	
2 services: 1 state-owned, 1 independent	
3 services: 1 state-owned, 2 independent	

The press, hitherto a FRELIMO publicity machine, has enjoyed greater freedom in the 1990s. The killing in 2000 of a popular and outspoken editor, Carlos Cardoso, shocked the country. TV sets are still a rarity. The state-owned radio station broadcasts in Portuguese, English, and vernacular languages.

CRIME

 No death penalty

10,800 prisoners

Crime is rising sharply

CRIME RATES

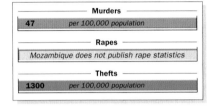

Murders	
47	per 100,000 population

Rapes	
Mozambique does not publish rape statistics	

Thefts	
1300	per 100,000 population

Weapons are easily obtainable. In rural areas there are many bandits, often former soldiers; road travel is unsafe. Senior officials stand accused of misappropriating food aid money.

EDUCATION

School leaving age: 13

44%

7143 students

THE EDUCATION SYSTEM

% of each age group in education

- Primary: 71%
- Secondary: 9%
- Tertiary: 1%

At independence, between 85% and 95% of the adult population were illiterate, and school closures during the civil war created a lost generation of uneducated people. A target of educating at least 86% of children was set for 2000, concentrating on rural areas. The government used World Bank/IMF debt relief to strengthen the education budget in 2001.

HEALTH

 Welfare state health benefits

1 per 20,000 people

Tuberculosis, gastroenteric infections, pneumonia, AIDS

Thousands of people lost limbs from land mines, or suffered other appalling injuries and psychological trauma, during Mozambique's savage civil war. Health services have improved since the war's end, and preventive medicines and antenatal care are provided free. Doctors serve a mandatory two years in rural areas. Many private clinics have been established since 1987. However, cholera, a lingering by-product of war, is still a serious issue in the far northern province of Cabo Delgado. An estimated 1.1 million Mozambicans have contracted HIV.

SPENDING

 GDP/cap. increase

CONSUMPTION AND SPENDING

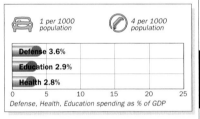

1 per 1000 population

4 per 1000 population

Defense 3.6%	
Education 2.9%	
Health 2.8%	

Defense, Health, Education spending as % of GDP

Mozambique is one of the world's poorest countries, with over 90% of the people living below the breadline even before the floods of 2000 and 2001, which left thousands homeless. Measures adopted in the 1990s to attract Western aid made conditions tougher, raising the price of rice by 600%. The recent export boom has generally bypassed the traditional subsistence farmer. Only the higher echelons of FRELIMO, RENAMO, and other political parties have cars, air-conditioning, and brick-built apartments. Free-market reforms, however, are gradually increasing access to consumer goods.

WORLD RANKING

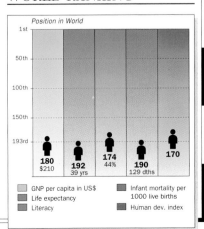

Position in World

1st				
50th				
100th				
150th				
193rd				

180
$210

192
39 yrs

174
44%

190
129 dths

170

GNP per capita in US$	Infant mortality per 1000 live births
Life expectancy	
Literacy	Human dev. index

M

NAMIBIA

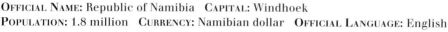

OFFICIAL NAME: Republic of Namibia **CAPITAL:** Windhoek
POPULATION: 1.8 million **CURRENCY:** Namibian dollar **OFFICIAL LANGUAGE:** English

LOCATED IN SOUTHWESTERN Africa, Namibia has an arid coastal strip formed by the Namib Desert. After many years of guerrilla warfare, Namibia won independence from South Africa in 1990. Despite the move away from apartheid, Namibia's economy remains reliant on the expertise of the small white population, a legacy of the previously poor education for blacks. Namibia is Africa's fourth-largest minerals producer.

CLIMATE
▷ Hot desert/steppe

WEATHER CHART FOR WINDHOEK

Namibia is almost rainless. The coast is usually shrouded in thick, cold fog unless the hot, very dry *berg* blows.

TRANSPORTATION
▷ Drive on left

Windhoek International
481,419 passengers

105 ships
54,794 grt

THE TRANSPORTATION NETWORK

5250 km
(3262 miles)

None

2382 km
(1480 miles)

None

Large-scale industry is well served by road and rail. Plans exist to build a new harbor at Walvis Bay.

TOURISM
▷ Visitors : Population 1:2.9

614,000 visitors

Up 54% 1995–1998

MAIN TOURIST ARRIVALS

South Africa 37%	
Angola 32%	
Germany 11%	
Other 20%	

0 10 20 30 40
% of total arrivals

Tourists, largely from South Africa, make a very limited contribution to GDP. German tourists come to see Windhoek's German sector. There are plans to limit tourists to 300,000 a year to preserve Namibia's fragile desert ecology.

Spitzkoppe, west of Karibib. *Unique scenery such as this is attracting increasing numbers of tourists to Namibia.*

PEOPLE
▷ Pop. density low

Ovambo, Kavango, English, Bergdama, German, Afrikaans

2/km²
(6/mi²)

THE URBAN/RURAL POPULATION SPLIT

31%   69%

ETHNIC MAKEUP

Ovambo 50%
Damara 8%
Herero 8%
Other 9%
Kavango 9%
Other tribes 16%

The largest ethnic group, the Ovambo, tend to live in the sparsely populated north of the country. Whites – 60% of whom speak Afrikaans – are concentrated in Windhoek. The capital is also home to a wealthy century-old German community. Namibia's original inhabitants, the San and Khoi (once called Bushmen) now constitute a tiny, marginalized minority.

The ethnic strife predicted in 1990 has not materialized. For the most part, black Namibians, predominantly subsistence farmers, have accepted the greater wealth of the white community. Families are large in Namibia; many black women have six or more children. The constitution supports gender equality and discriminates in favor of women; few, however, have official jobs or own property. Homosexuality is not tolerated.

POLITICS
▷ Multiparty elections

L. House 1999/2004
U. House 1998/2004

President Sam Nujoma

AT THE LAST ELECTION

National Assembly 72 seats

76% SWAPO   10% DTA   1% MAG   10% CoD   3% UDF

SWAPO = South West Africa People's Organization
CoD = Congress of Democrats **DTA** = Democratic Turnhalle Alliance **UDF** = United Democratic Front **MAG** = Monitor Action Group
Six additional nonvoting members may be appointed to the National Assembly by the president

National Council 26 seats

Two members are elected by each of the 13 Regional Councils to the National Council

SWAPO guerrillas fought for and won independence from South Africa in 1990. Namibia switched from a system of apartheid to a state-wide, multiparty democracy. The center-left SWAPO has dominated politically ever since. In 1998 a constitutional amendment allowed President Nujoma to run for a third term, which he duly won in late 1999. SWAPO's main opposition comes from the center-right DTA, a coalition of 11 parties favoring free-market practices.

WORLD AFFAIRS
▷ Joined UN in 1990

Comm COMESA NAM AU SADC

In 1992, South Africa settled its border dispute with Namibia, and in 1994 relinquished control of the enclave of Walvis Bay – Namibia's only deepwater port. South Africa has also written off Namibia's earlier debts. Withdrawal of Namibian troops from the war-torn DRC began in 2001.

AID
▷ Recipient

US$152m (receipts)

Down 15% in 2000

The EU provides most aid; Germany is the main unilateral donor. Around one-third of aid is spent on education.

DEFENSE
▷ No compulsory military service

US$103m

Down 14% in 2000

Fishing grounds are patrolled to prevent raids by foreign trawlers. Namibian soldiers performed a peacekeeping role in the DRC.

N

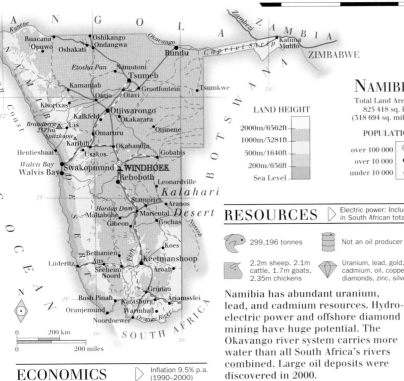

NAMIBIA

Total Land Area :
825 418 sq. km
(318 694 sq. miles)

LAND HEIGHT

2000m/6562ft
1000m/5281ft
500m/1640ft
200m/656ft
Sea Level

POPULATION

over 100 000 ◎
over 10 000 ●
under 10 000 ·

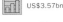

CHRONOLOGY

In 1915, South Africa took over the former German colony as a League of Nations' mandate known as South West Africa.

❑ **1966** Apartheid laws imposed. SWAPO begins armed struggle.
❑ **1968** Renamed Namibia.
❑ **1973** UN recognizes SWAPO.
❑ **1990** Independence.
❑ **1994** South Africa relinquishes Walvis Bay.
❑ **1999** President Sam Nujoma wins third term.

RESOURCES ▷ Electric power: Included in South African total

299,196 tonnes   Not an oil producer

2.2m sheep, 2.1m cattle, 1.7m goats, 2.35m chickens   Uranium, lead, gold, cadmium, oil, copper, diamonds, zinc, silver,

Namibia has abundant uranium, lead, and cadmium resources. Hydro-electric power and offshore diamond mining have huge potential. The Okavango river system carries more water than all South Africa's rivers combined. Large oil deposits were discovered in 2000.

ENVIRONMENT ▷ Sustainability rank: 26th

13%   Negligible emissions per capita

Illegal poaching and the presence of anthrax threaten the unique Namibian desert-adapted elephant (fewer than 50 remain) and the black rhino. Vast expanses of the fragile, unspoiled Namib and Kalahari Desert ecosystems are protected. The government is generally sensitive to environmental issues (the annual seal cull to protect fish stocks is an exception) and wishes to attract ecotourists rather than invest in mass-market developments.

MEDIA ▷ TV ownership low

Daily newspaper circulation 19 per 1000 people

PUBLISHING AND BROADCAST MEDIA

There are 4 daily newspapers, including *The Namibian* and *Die Republikein*

2 services:
1 state-owned,
1 independent

5 services:
1 state-owned,
4 independent

State radio transmits in 11 languages, including German and English. An active press targets corrupt politicians.

CRIME ▷ No death penalty

4397 prisoners   Down 51% in 1999

Burglary and theft are rising, particularly in urban areas. Ostrich smuggling to the US is common.

ECONOMICS ▷ Inflation 9.5% p.a. (1990–2000)

US$3.57bn   7.570–11.995 Namibian dollars

SCORE CARD

❑ WORLD GNP RANKING	126th
❑ GNP PER CAPITA	US$2030
❑ BALANCE OF PAYMENTS	US$162m
❑ INFLATION	8.6%
❑ UNEMPLOYMENT	40%

STRENGTHS

Varied mineral resources. Market conditions attractive to private investors. Rich fishing grounds. Potential of Walvis Bay as conduit for landlocked neighbors. Low external debt.

WEAKNESSES

Most goods imported. Fluctuations in mineral prices. Recessionary ripple effect of currency pegged to South African rand. Lack of skilled labor; high unemployment. Severe drought in 1996.

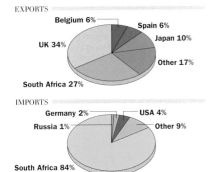

EXPORTS

Belgium 6%
Spain 6%
UK 34%
Japan 10%
Other 17%
South Africa 27%

IMPORTS

Germany 2%
Russia 1%
USA 4%
Other 9%
South Africa 84%

EDUCATION ▷ School leaving age: 16

82%   11,209 students

Nearly 90% of children attend primary school, but illiteracy among black adults remains a legacy of apartheid.

HEALTH ▷ Welfare state health benefits

1 per 3333 people   AIDS, respiratory, heart, and intestinal diseases

Preventive care and rural health care have top priority. Most areas lack safe water. AIDS is the leading cause of death.

SPENDING ▷ GDP/cap. increase

CONSUMPTION AND SPENDING

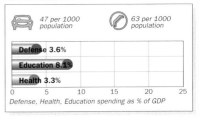

47 per 1000 population   63 per 1000 population

Defense 3.6%
Education 8.1%
Health 3.3%

0   5   10   15   20   25
Defense, Health, Education spending as % of GDP

Gross disparities in wealth persist throughout Namibia: the top 1% of households consumes as much as the poorest 50%.

WORLD RANKING

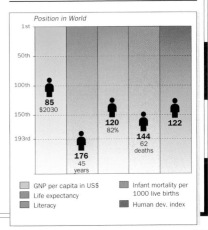

Position in World

1st
50th
100th
150th
193rd

85 $2030
120 82%
144 62 deaths
122
176 45 years

☐ GNP per capita in US$
☐ Life expectancy
☐ Literacy
☐ Infant mortality per 1000 live births
☐ Human dev. index

N

NAURU

OFFICIAL NAME: Republic of Nauru **CAPITAL:** *No official capital*
POPULATION: 11,800 **CURRENCY:** Australian dollar **OFFICIAL LANGUAGE:** Nauruan

1968 1968 Jan 31 NAU +12 +674 .nr

NAURU, THE WORLD'S smallest republic, lies in the Pacific Ocean, 4000 km (2480 miles) northeast of Australia. Once a British colony exploited for its phosphates by the UK, Australia, and New Zealand, it became independent in 1968. The phosphates industry made Nauruans among the wealthiest people in the world, but economic mismanagement and the imminent end of phosphate reserves has left Nauru facing financial ruin, prompting economic reform.

CLIMATE ▷ Tropical oceanic

WEATHER CHART FOR NAURU

Nauru's tiny size means that rain clouds often miss the island; years can pass without rain.

TRANSPORTATION ▷ Drive on left

Nauru Island International

3 ships 1000 grt

THE TRANSPORTATION NETWORK

24 km (15 miles)		None	
5 km (3 miles)		None	

Nauru operates its own airline with a Boeing 737 flown by Australian pilots. The Nauru Pacific Line provides infrequent commercial services to Australia. However, all external travel is very expensive. Nauru has no harbor: to load cargoes of phosphates, ships have to dock, engines still running, with huge concrete caissons floating out at sea. The circular ring road is often littered with abandoned cars, as it has been much cheaper for Nauruans to import new vehicles than to attempt to repair existing ones. The number of car accident fatalities is one of the highest in the South Pacific.

TOURISM ▷ Not available

Minimal tourist arrivals

Little variation from year to year

MAIN TOURIST ARRIVALS

Nauru does not publish tourism figures by country of origin

0   10   20   30   40
% of total arrivals

Even if Nauru had any conventional tourist attractions, the enormous cost of getting there would dissuade most tourists from making the journey. The main feature of interest on the island is the bizarre lunar landscape created by over 80 years of phosphate extraction. There are no beaches and only a few basic hotels.

NAURU

Total Land Area : 21 sq. km (8.1 sq. miles)

LAND HEIGHT
200m/565ft
Sea Level

Urban area
Phosphate mineworks

0   1 km
0   1 mile

PEOPLE ▷ Pop. density high

Nauruan, Kiribati, Chinese, Tuvaluan, English

564/km² (1461/mi²)

THE URBAN/RURAL POPULATION SPLIT

Nauru is 100% semi-urban

ETHNIC MAKEUP

European 5% Chinese and Vietnamese 8%

Other Pacific islanders 25%

Nauruan 62%

Indigenous Nauruans are a homogeneous blend of Melanesian, Micronesian, and Polynesian strands. They have traditionally held posts in government service, while a large imported workforce – mainly from Kiribati – mines the phosphates.

A society of just over 10,000 people, Nauru is mostly self-regulating. There is some tension between younger Nauruans, who go to Australia to study but have little incentive to do well, and their parents, who fought hard for independence. As the phosphates run out, an increasing feeling of futility is gripping the young. Many see their future in Australia or New Zealand, but fear a drop in living standards and the loss of the luxury of sovereignty. These fears led Nauruans to reject the offer of resettlement on an island off the Queensland coast of Australia.

POLITICS ▷ Nonparty elections

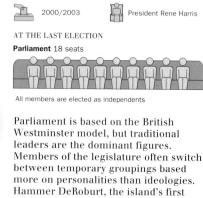

2000/2003 President Rene Harris

AT THE LAST ELECTION

Parliament 18 seats

All members are elected as independents

Parliament is based on the British Westminster model, but traditional leaders are the dominant figures. Members of the legislature often switch between temporary groupings based more on personalities than ideologies. Hammer DeRoburt, the island's first post-independence president, dominated the political landscape for two decades until a 1989 vote of no confidence.

Since DeRoburt's downfall, the presidency (elected by parliament) has changed hands frequently in a more volatile political atmosphere.

N

WORLD AFFAIRS

 Joined UN in 1999

| Comm | PC | PIF | ADB | ACP |

The case for compensation for phosphate exploitation brought by Nauru against the UK government was rejected in 1992 after the longest suit in British legal history. However, an Australian settlement that year brought payments eventually totaling US$79 million. Nauru's main concern is participation in the Pacific Islands Forum and the management of trust funds to support Nauruans when phosphate deposits run out. Nauru joined the UN in 1999. It seeks a voice on environmental issues.

AID

 Recipient

 US$4m (receipts)   ⬇ Down 43% in 2000

Nauru receives or donates little aid, except as a member of the Pacific Islands Forum.

DEFENSE

 No compulsory military service

$ Australia responsible for defense   ⬍ Not applicable

Nauru has no defense force. Australia, under a de facto arrangement, is responsible for the island's security.

ECONOMICS

 Not available

📊 US$80m   💲 1.7997–1.9535 Australian dollars

SCORE CARD

❏ WORLD GNP RANKING	190th
❏ GNP PER CAPITA	US$7270
❏ BALANCE OF PAYMENTS	Not available
❏ INFLATION	–3.6%
❏ UNEMPLOYMENT	Minimal unemployment

STRENGTHS

Considerable investments overseas. Tax haven. Offshore banking potential. Trust funds for post-phosphate era. Strong Australian dollar.

WEAKNESSES

Phosphate revenues due to end in 2005. High cost of rehabilitating the 80% of the island where mining occurred. Virtually no other resources. Bad investment decisions in 1990s.

EXPORTS

Nauru's only export commodity is phosphates, in which it trades with Australia and New Zealand.

IMPORTS

New Zealand 4%
Other 36%
Australia 60%

RESOURCES

 Electric power 10,000 kw

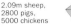 250 tonnes   Not an oil producer

2.09m sheep, 2800 pigs, 5000 chickens   Guano (phosphates)

Since 1888 Nauru has been exploited by the Germans, British, Australians, New Zealanders, and recently by Nauruans themselves, for its valuable phosphate reserves. Extraction has destroyed 80% of the island, and the deposits have been virtually exhausted. Nauru has no other resources. It is entirely dependent on outside energy supplies, and the cost of oil is 50% higher than the Pacific average, since Nauru does not lie on any shipping routes. Most electricity is produced by small diesel generators.

ENVIRONMENT

 Not available

 None   ⬍ Not available

Nauru is an environmental disaster area. Mining has destroyed 80% of its ecosystem and, like other Pacific islands, it faces the increasing threat of rising sea levels. Also of concern is contamination from the nearby former French nuclear test sites in the Pacific.

MEDIA

 TV ownership low

☰ There are no daily newspapers

PUBLISHING AND BROADCAST MEDIA

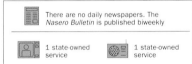

There are no daily newspapers. The *Nasero Bulletin* is published biweekly

1 state-owned service   1 state-owned service

Nauru has one national TV broadcasting service and one radio station. Both are state-run.

CRIME

 Death penalty not used in practice

 Nauru does not publish prison figures   Crime levels are rising slightly

Theft is almost nonexistent. Assaults and dangerous driving as a result of drunkenness are the major problems.

Nauru is almost circular with a 16-km (10-mile) ring road. The overcrowded coastal strip is the sole habitable land.

CHRONOLOGY

Colonized by Germany in 1888, from 1919 the island was administered jointly by the UK, Australia, and New Zealand.

- ❏ **1968** Independence.
- ❏ **1970** Gains phosphate control.
- ❏ **1992** Australia agrees compensation for phosphate extraction.

EDUCATION

 School leaving age: 16

 99%   Not available

Many Nauruans attend boarding school in Australia from a young age. Few go on to university.

HEALTH

 Welfare state health benefits

 1 per 700 people   Tuberculosis, vitamin deficiencies, diabetes

A diet of processed imported foods and widespread obesity are the major problems. Over one-third of the population suffers from non-insulin-dependent diabetes. Industrial accidents are treated in Australia.

SPENDING

GDP/cap. increase

CONSUMPTION AND SPENDING

🚗 No data   ☎ 185 per 1000 population

Defense	None
Education	No data
Health	5%

0 5 10 15 20 25
Defense, Health, Education spending as % of GDP

Nauru is carrying out a major economic adjustment program, to be funded by the ADB. The program is intended to allow it to adjust to the loss of income when its phosphate reserves are exhausted.

WORLD RANKING

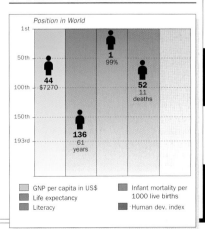

Position in World

1st
50th
100th
150th
193rd

44 $7270
1 99%
52 11 deaths
136 61 years

GNP per capita in US$ — Infant mortality per 1000 live births
Life expectancy
Literacy — Human dev. index

NEPAL

SOUTH ASIA

OFFICIAL NAME: Kingdom of Nepal CAPITAL: Kathmandu
POPULATION: 23.6 million CURRENCY: Nepalese rupee OFFICIAL LANGUAGE: Nepali

 1769   1769   Dec 28   NEP   +5.75   +977 .np

LYING ALONG the southern Himalayas, Nepal was an absolute monarchy until 1990, since when its politics became increasingly turbulent. The mainly agricultural economy depends heavily on the prompt arrival of the monsoon. Hopes for development have been invested in hydropower, despite the adverse impact of large dams. A Maoist insurgency, begun in 1999, threw the country into chaos in 2001.

CLIMATE ▷ Mountain/subtropical

WEATHER CHART FOR KATHMANDU

The warm July to October monsoon affects the whole country, causing flooding in the hot Terai plain, but generally decreases northward and westward. The rest of the year is dry, sunny, and mild, except in the Himalayas, where valley temperatures in winter may average –10°C (14°F).

TRANSPORTATION ▷ Drive on left

Tribhuvan International, Kathmandu
1.91m passengers

Has no fleet

THE TRANSPORTATION NETWORK

| 4073 km (2531 miles) | None |
| 101 km (63 miles) | None |

Domestic flights link the main towns. There are paved roads in the south and in the Kathmandu valley; only one runs north to China. Two short stretches of railroad cross into India.

Himalayan harvest. Steep mountainsides and easily eroded soils mean that most fields are terraced. A majority of Nepalese are farmers.

TOURISM ▷ Visitors : Population 1:51

464,000 visitors Down 6% in 2000

MAIN TOURIST ARRIVALS

The wish to preserve the environment conflicts with the desire for tourist revenue. Backpackers have been a major source of income, but recent terrorist activity, such as the 1999 hijacking of an Air India aircraft bound for Nepal and the increased violence in 2001 of the Maoist insurgency, threaten to deter visitors.

Child labor was banned in the tourism industry from 2000.

PEOPLE ▷ Pop. density medium

Nepali, Maithilli, Bhojpuri 173/km² (447/mi²)

THE URBAN/RURAL POPULATION SPLIT

12% 88%

RELIGIOUS PERSUASION

Other 2% Muslim 3%
Christian 1% Buddhist 4%
Hindu 90%

There are few tensions among diverse ethnic groups such as the Sherpas in the north, "Hill Hindu" Brahmins and Chhettris, Newars, and others in the Kathmandu valley, and Terai in the south. The Sherpa and other Buddhist women are less restricted than Hindus. Polygamy is practiced in the hills. Since 1990 many ethnic Nepalese refugees from Bhutan have settled in Nepal.

POLITICS ▷ Multiparty elections

L. House 1999/2002
U. House 2001/2003

H.M. King Gyanendra
Bir Bikram Shah Dev

AT THE LAST ELECTION

House of Representatives 205 seats

5% NDP
54% NCP
35% CPN–UML
6% Others

NCP = Nepali Congress Party CPN–UML = Communist Party of Nepal–United Marxist–Leninist NDP = National Democratic Party App = Appointed by the king

National Council 60 seats

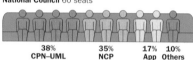

38% CPN–UML
35% NCP
17% App
10% Others

A multiparty system introduced in 1990 produced a short-lived communist government in 1994, then a series of unstable coalitions until the NCP won elections in 1999. Instability returned in 2001: the royal family was murdered and the Maoist rebel insurgency intensified. To tackle the Maoists, Prime Minister Sher Bahadur Deuba imposed a state of emergency, and its unilateral extension in 2002, supported by unpopular new King Gyanendra, sparked a political crisis which saw Deuba expelled from his own NCP.

WORLD AFFAIRS ▷ Joined UN in 1955

 CP   IBRD   NAM SAARC ADB

The government has revived links with India, and security relations are questioned by the UML, which is more pro-Chinese. Relations with Bhutan are strained over ethnic Nepali Bhutanese refugees in Nepal. The Maoist insurgency has increased regional tension.

AID ▷ Recipient

$390m (receipts) Up 11% in 2000

Nepal's strategic position has made it a focus for powerful donors, including the US, China, India, Japan, and member states of the CIS.

DEFENSE ▷ No compulsory military service

$49m Up 17% in 2000

The army, at 46,000 men, is small and has no tanks or combat aircraft. Weapons come from India and the UK, in whose own army the Nepalese Gurkhas serve.

N

NEPAL
Total Land Area : 140 800 sq. km
(54 363 sq. miles)

POPULATION
over 100 000
over 10 000
under 10 000

LAND HEIGHT
6000m/19 686ft
4000m/13 124ft
2000m/6562ft
1000m/3281ft
500m/1640ft
200m/656ft
50m/164ft

CHRONOLOGY

The foundations of the Nepalese state were laid in 1769, when King Prithvi Narayan Shah conquered the region.

- ❏ **1816–1923** Quasi-British protectorate.
- ❏ **1959** First multiparty constitution.
- ❏ **1960** Constitution suspended.
- ❏ **1962–1990** *Panchayat* nonparty system.
- ❏ **1972** Birendra succeeds to throne.
- ❏ **1991** NCP victory in elections.
- ❏ **1994** First communist government.
- ❏ **1995–1998** Succession of weak coalition governments.
- ❏ **1999** NCP election victory. Maoist insurgency in rural areas.
- ❏ **2001** King and family shot; Gyanendra crowned amid unrest. Upsurge in Maoist violence; state of emergency declared.

ECONOMICS

▷ Inflation 8.2% p.a. (1990–2000)

$5.58bn

74.58–76.59 Nepalese rupees

SCORE CARD

❏ WORLD GNP RANKING	106th
❏ GNP PER CAPITA	$240
❏ BALANCE OF PAYMENTS	–$293m
❏ INFLATION	1.5%
❏ UNEMPLOYMENT	1%

STRENGTHS
Self-sufficiency in grain most years. Economic liberalization under NCP government. Potential for hydroelectric power generation. Low debt level.

WEAKNESSES
Agricultural dependency: only 10% of GDP from manufacturing. Landlocked status. Low savings rate. Absence of active entrepreneurial class.

EXPORTS

Japan 4%
Argentina 6%
USA 32%
Germany 13%
Other 19%
India 26%

IMPORTS

Hong Kong 7%
Singapore 7%
India 34%
China 14%
Argentina 14%
Other 24%

RESOURCES

▷ Electric power 327,000 kw

25,780 tonnes

Not an oil producer

7m cattle, 6.5m goats, 3.6m buffaloes, 19.8m chickens

Mica, lignite, copper, cobalt, iron

The first privately owned power plant, situated near Ramechhap, opened in mid-2000.

ENVIRONMENT

▷ Sustainability rank: 99th

8% (1% partially protected)

0.1 tonnes per capita

Kathmandu has chronic traffic and pollution problems. Deforestation and soil erosion are serious. The native tiger is fast disappearing. Approval of the controversial Arun III hydroelectric project was granted in mid-2000.

MEDIA

▷ TV ownership low

Daily newspaper circulation 11 per 1000 people

PUBLISHING AND BROADCAST MEDIA

There are 29 daily newspapers, including the leading *Gorkhapatra*, *Nepali Hindi Daily*, and *Rising Nepal*

1 limited state-owned service

2 services: 1 state-owned, 1 independent

The Nepal TV service began in 1986; under 25% of the population receives it. The press is mainly Kathmandu-based with low circulations. Press watchdogs warned of censorship under the state of emergency imposed in 2001.

CRIME

▷ No death penalty

5878 prisoners

Up 5% in 1999

Petty theft and smuggling are the main problems. The legal provision for detention without trial is used, and police suppression of demonstrations is often brutal.

EDUCATION

▷ School leaving age: 11

42%

115,938 students

Over 80% of boys attend school in Nepal, but still only a minority of girls. Nepal's literacy rate is among the lowest in the world.

HEALTH

▷ Welfare state health benefits

1 per 20,000 people

Respiratory and diarrheal diseases, maternal deaths

There are about 100 *dharmi-jhankri* (faith healers) for every health worker. Maternal mortality is high, the result of harmful traditional birth practices; a re-education program for midwives has been established.

SPENDING

▷ GDP/cap. increase

CONSUMPTION AND SPENDING

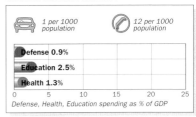

1 per 1000 population

12 per 1000 population

Defense 0.9%
Education 2.5%
Health 1.3%

0   5   10   15   20   25
Defense, Health, Education spending as % of GDP

Nepal is one of the poorest countries in the world. Bonded labor was abolished in mid-2000, releasing 36,000 people.

WORLD RANKING

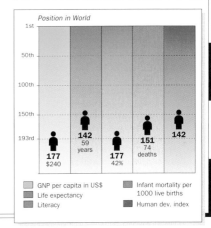

Position in World

1st
50th
100th
150th
193rd

177 $240
142 59 years
177 42%
151 74 deaths
142

GNP per capita in US$
Life expectancy
Literacy
Infant mortality per 1000 live births
Human dev. index

NETHERLANDS

OFFICIAL NAME: Kingdom of the Netherlands **POPULATION:** 16.2 million
CAPITALS: Amsterdam; The Hague (administrative) **CURRENCY:** Euro **OFFICIAL LANGUAGE:** Dutch

THE NETHERLANDS IS LOCATED at the delta of four major rivers in northwest Europe. The few hills in the eastern and southern part of the country fall into a flat coastal area, bordered by the North Sea to the north and west. This is protected by a giant infrastructure of dunes, dikes, and canals, as 27% of the coast is below sea level. The Netherlands became one of the world's first confederate republics after Spain recognized its independence in 1648. Its highly successful economy has a long trading tradition, and Rotterdam is the world's largest port.

PEOPLE

Dutch, Frisian — 478/km² (1237/mi²)

THE URBAN/RURAL POPULATION SPLIT
89% / 11%

RELIGIOUS PERSUASION

Muslim 3%, Protestant 27%, Roman Catholic 36%, Other 34%

ETHNIC MAKEUP

Surinamese 2%, Turkish 2%, Moroccan 2%, Other 12%, Dutch 82%

CLIMATE
Maritime

WEATHER CHART FOR AMSTERDAM

The Netherlands has a temperate climate, with mild winters which rarely fall much below freezing, and cool summers with a mean temperature of 20°C (68°F). The country's coastal areas have the mildest climate, though northerly gales are fairly frequent, particularly in autumn and winter.

TRANSPORTATION
Drive on right

Schiphol, Amsterdam 39.6m passengers — 1214 ships 4.26m grt

THE TRANSPORTATION NETWORK
113,018 km (70,226 miles) — 2200 km (1367 miles)
2808 km (1745 miles) — 5043 km (3134 miles)

Rotterdam, the key transshipment port for northern Europe, is also the world's largest. Schiphol airport is one of the air transportation hubs of Europe. A high-speed passenger rail line is due to link Amsterdam and Rotterdam with Brussels and Paris in 2003, and a high-speed freight line from Rotterdam to Germany should be completed in 2004.

The Dutch see their country as the most tolerant in Europe, and it has a long history of welcoming refugees seeking religious and political asylum. In the 20th century, immigrants from former colonies settled in the Netherlands and became fully accepted as citizens. They came first from Indonesia and then from Suriname, the Netherlands Antilles, and Morocco. The small Turkish community, however, does not enjoy full citizenship.

The tradition of tolerance is reflected in liberal attitudes to sexuality. In 2001 same-sex marriages were legalized, giving gay couples full equality, including the right of adoption (after three years of marriage).

The state does not try to impose a particular morality on its citizens. Drug taking is seen as a matter of personal choice, and in 2001 the Netherlands became the first country in the world to legalize euthanasia, albeit under strict conditions.

Women enjoy equal rights and hold 37% of seats in the Second Chamber of the States-General, but are not well represented in boardrooms.

TOURISM
Visitors : Population 1:1.6

 10m visitors   Up 1% in 2000

MAIN TOURIST ARRIVALS
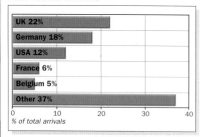
UK 22%, Germany 18%, USA 12%, France 6%, Belgium 5%, Other 37%

Tourism is a major business in the Netherlands. Visitors go mainly to Amsterdam, although cities such as Groningen and Maastricht are growing in popularity. Amsterdam caters for a diverse tourism market. Its world-famous museums include the Rijksmuseum, with its collection of Vermeers and Rembrandts, while its network of canals is popular. Amsterdam is also renowned for its liberal attitude to sex; its red-light district draws millions every year. In the past decade, the city has become a center for the European gay community, with celebrations on 30 April (Queen's Day – the monarch's official birthday) and in August (Amsterdam Pride). A thriving club scene and liberal drug laws draw enthusiasts from neighboring countries. In spring and summer, the tulip fields and North Sea beaches attract large numbers of visitors.

Windmill at Baambrugge, near Amsterdam. A century ago there were 10,000 in the country compared with today's 1000. A protective ring of 900 mills kept Amsterdam from flooding.

POPULATION AGE BREAKDOWN

Female	Age	Male
2.2%	80+	0.9%
8.1%	60–79	6.6%
12.9%	40–59	13.3%
15.6%	20–39	16.1%
11.9%	0–19	12.4%

% of population by age group

POLITICS ▷ Multiparty elections

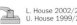

L. House 2002/2006
U. House 1999/2003   H.M. Queen Beatrix

AT THE LAST ELECTION
Second Chamber of the States-General 150 seats

29%	17%	16%	15%	7%	6%	10%
CDA	LPF	VVD	PvdA	GL	SP	Others

CDA = Christian Democratic Appeal   **LPF** = List Pim Fortuyn
VVD = People's Party for Freedom and Democracy
PvdA = Labor Party   **GL** = Green Left   **SP** = Socialist Party
D66 = Democrats 66

First Chamber of the States-General 75 seats

31%	25%	19%	9%	5%	11%
VVD	CDA	PvdA	D66	GL	Others

The First Chamber of the States-General is indirectly elected

The Netherlands is a constitutional monarchy. Legislative power is vested in parliament, and the monarch has only nominal power.

PROFILE
Dutch politics show a high degree of consensus. Since the early 1980s, governments have employed the "polder model," which focuses on pay moderation, job creation, economic deregulation, and generous social protection.

The CDA has traditionally led two-party coalitions, either with

NETHERLANDS

Total Land Area :
41 526 sq. km
(16 033 sq. miles)

Queen Beatrix, who acceded in 1980 and rebuilt support for the Dutch monarchy.

CDA leader Jan Peter Balkenende was appointed prime minister in 2002.

the left-of-center PvdA or with the right-wing VVD. However, after the 1994 election the PvdA under Wim Kok led the government, in coalition with the VVD and the left-liberal D66.

Reelected in 1998, this administration resigned in April 2002 after a report criticized Dutch troops serving with the UN in Bosnia in 1995 for failing to stop the Srebrenica massacre. In 2002 the

emergence of the ultranationalist Pim Fortuyn challenged current policy on immigration and integration. Fortuyn's influence, articulating a sense of widespread public alienation from a complacent political establishment, was expected to be lasting, despite his assassination only days before the May election. The strong sympathy vote boosted a swing to the right, resulting in victory for the CDA under Jan Peter Balkenende and second place for the LPF.

MAIN POLITICAL ISSUES
The future of social welfare
Despite cutbacks in the 1980s, the Dutch still have one of Europe's most generous welfare systems. Most parties accepted that levels of welfare could not be maintained indefinitely. The debate thus focuses on how much and in which areas cuts should be made.

Refugees and asylum seekers
Responding to rising numbers of asylum seekers in the Netherlands, asylum laws have been tightened since 1994. First- and second-generation immigrants made up 18% of the national population in 2002, but 40% (30% non-European) in Rotterdam, where the LPF won 35% of the vote in the March 2002 local elections. Fortuyn opposed multiculturalism and advocated amending the constitutional ban on discrimination.

WORLD AFFAIRS ▷ Joined UN in 1945

Benelux   EU   NATO   OECD   OSCE

Political and monetary integration within the EU have strong popular support. In 1995, internal border controls were lifted under the Schengen Convention and in 2002 the Netherlands fully adopted the euro. Traditionally the Netherlands favored enlargement of the EU, but the CDA government expressed reservations when it came to power in May 2002. The International Court of Justice, the International Criminal Tribunal for the former Yugoslavia, and the International Criminal Court sit in The Hague.

AID ▷ Donor

$3.14bn (donations)   Little change in 2000

The Netherlands continues to be one of the few countries which exceeds the UN target of devoting 0.7% of GNP to development aid. The government actively pursues a policy of linking foreign aid and human rights. It also gives priority to projects which link longer-term development goals with efforts to manage and reduce intergroup conflict.

POPULATION

over 1 000 000
over 500 000
over 100 000
over 50 000
over 10 000

LAND HEIGHT
100m/328ft
Sea Level
-100m/-328ft

CHRONOLOGY

Suppression of Protestantism by the ruling Spanish Habsburgs led to the revolt of the Netherlands and the declaration of independence of the northern provinces as a republic in 1581, recognized by Spain in 1648.

- ❏ **1813** Dutch oust French after 18 years of French rule and choose to become a constitutional monarchy.
- ❏ **1815** United Kingdom of Netherlands formed to include Belgium and Luxembourg.
- ❏ **1839** Recognition of 1830 secession of Catholic southern provinces as Belgium.
- ❏ **1848** New constitution – ministers to be accountable to parliament.
- ❏ **1897–1901** Wide-ranging social legislation enacted. Development of strong trade unions.
- ❏ **1898** Wilhelmina succeeds to throne, ending Luxembourg union, where male hereditary Salic Law is in force.
- ❏ **1914–1918** Dutch neutrality respected in World War I.
- ❏ **1922** Women fully enfranchised.
- ❏ **1940** Dutch assert neutrality in World War II, but Germany invades. Fierce resistance.
- ❏ **1942** Japan invades Dutch East Indies.
- ❏ **1944–1945** "Winter of starvation" in German-occupied western provinces.
- ❏ **1945** Liberation. International Court of Justice set up in The Hague.
- ❏ **1946–1958** PvdA leads center-left coalitions.
- ❏ **1948** Juliana becomes queen.
- ❏ **1949** Joins NATO. Most of East Indies colonies gain independence as Indonesia.
- ❏ **1957** Founder member of EEC.
- ❏ **1960** Economic union with Belgium and Luxembourg comes into effect.
- ❏ **1973** PvdA wins power after 15 years spent mainly in opposition. Center-left coalition until 1977.
- ❏ **1980** CDA alliance of the "confessional" parties forms a single party. Beatrix becomes queen.
- ❏ **1982–1994** CDA-led coalitions under Ruud Lubbers.
- ❏ **1990** 20-year National Environment Policy (NEP) introduced.
- ❏ **1992** Licensed brothels legalized.
- ❏ **1994** Elections: Wim Kok of PvdA heads coalition with VVD and D66.
- ❏ **2001** Euthanasia and gay marriage legalized.
- ❏ **2002** January, euro fully adopted. April, government resigns after publication of report blaming Dutch military in Bosnia for Srebrenica massacre in 1995. May, ultra-nationalist leader Pim Fortuyn assassinated. Elections result in CDA-led coalition under Jan Peter Balkenende.

DEFENSE

 ▷ No compulsory military service

💲 $6.39bn   ⬆ Up 3% in 2000

DUTCH ARMED FORCES

320 main battle tanks (*Leopard* 2)	23,100 personnel
4 submarines, 2 destroyers, and 10 frigates	12,130 personnel
157 combat aircraft (F-16A/B)	10,000 personnel
None	

The Dutch military has undergone major restructuring since the end of the Cold War with the aim of making it a rapidly deployable, more flexible military force as befits a NATO member state. Compulsory military service was abolished in 1996 and personnel cut by 44%, with the number of army divisions reduced from three to two. In 1995, a joint Dutch–German army corps numbering 28,000 was inaugurated. The Netherlands sent troops to the international peacekeeping effort in Bosnia; a 2002 report criticized them for failing to prevent the massacre at Srebrenica in 1995.

The Netherlands also has a large defense industry, which specializes in submarines, weapons systems, and aircraft.

ECONOMICS

▷ Inflation 1.9% p.a. (1990–2000)

📊 $398bn   💱 1.0651–1.1231 euros

SCORE CARD

- ❏ WORLD GNP RANKING...........................14th
- ❏ GNP PER CAPITA $24,970
- ❏ BALANCE OF PAYMENTS....................$16.3bn
- ❏ INFLATION 2.5%
- ❏ UNEMPLOYMENT 3%

EXPORTS

- UK 11%
- France 11%
- Belgium 12%
- Germany 27%
- Other 33%
- Italy 6%

IMPORTS

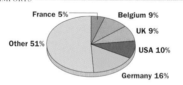

- Belgium 9%
- UK 9%
- USA 10%
- Germany 16%
- France 5%
- Other 51%

STRENGTHS

Highly skilled, educated, multilingual workforce. Sophisticated infrastructure. Many blue-chip multinationals, including Philips and Shell. Strong consensus between employers and employees. Low inflation and unemployment. Tradition of high-tech innovation, including development of music cassette and CD.

WEAKNESSES

Costly welfare system, resulting in high taxes and social insurance premiums; one-third of national income spent on social security. High labor costs.

PROFILE

Trade has been central to the success of the economy of the Netherlands since the 16th century. Most goods travel through Rotterdam, the world's biggest port. As well as high-tech industries such as electronics, telecommunications, and chemicals, there is a successful agricultural sector. Dependence on trade makes the economy vulnerable to world economic fluctuations; thus high growth rates in 1997–2000 were followed by a marked downturn in 2001.

ECONOMIC PERFORMANCE INDICATOR

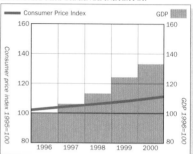

NETHERLANDS : MAJOR BUSINESSES

- Electronics
- Pharmaceuticals
- Light engineering
- Heavy engineering
- Aerospace industry
- Gas refining
- Oil refining
- Chemicals
- Textiles

0   50 km
0   50 miles

N

RESOURCES

 Electric power 20.2m kw

 623,396 tonnes

 34,366 b/d (reserves 113m barrels)

12.8m pigs, 4.05m cattle, 1.5m turkeys, 107m chickens

Natural gas, oil

ELECTRICITY GENERATION

Hydro 0%	
Combustion 95% (86bn kwh)	
Nuclear 4% (3.8bn kwh)	
Other 1% (0.6bn kwh)	

0   20   40   60   80   100

% of total generation by type

There are large natural gas reserves in the north. There is some oil production from offshore drilling in the North Sea.

ENVIRONMENT

 Sustainability rank: 34th

7%   10.4 tonnes per capita

ENVIRONMENTAL TREATIES

	Yes		Yes		Yes
	Yes		Yes		Yes

There is a strong environmental tradition, a legacy in part of living in one of the most densely populated states in the world. NGOs such as Greenpeace are well supported and the Green Left party is well represented in parliament.

The Dutch recycle domestic trash, have a good record on energy efficiency, and have developed innovative projects in housing and local transportation. An eco-tax on energy users was introduced in 1996 – the first of its kind in the West – though big businesses are exempt.

Serious flooding of the rivers Maas and Waal (an arm of the Rhine) in 1993 and 1995 raised concern about the state of the country's flood defenses and the use of floodplains for development.

MEDIA

 TV ownership high

Daily newspaper circulation 305 per 1000 people

PUBLISHING AND BROADCAST MEDIA

	There are 38 national dailies. The right-wing *De Telegraaf* has the largest circulation
	3 services: 1 public, 2 independent
	5 privately owned national stations, many regional and local stations

Newspaper circulation is high. While editorially independent, broadcasting is strongly regulated. Dutch law does not recognize a right of reply or a right to protect information sources.

NETHERLANDS : LAND USE

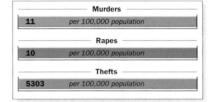

	Cropland
	Forest
	Pasture
	Wetlands
	Cattle
	Pigs
	Sugarbeet
	Bulbs & flowers

0   50 km
0   50 miles

CRIME

 No death penalty

13,847 prisoners   Down 16% 1996–1998

CRIME RATES

Murders	
11	*per 100,000 population*

Rapes	
10	*per 100,000 population*

Thefts	
5303	*per 100,000 population*

The Netherlands treats the use of hard drugs more as a medical and social than a criminal issue. Other member states of Europe's Schengen Convention, particularly France, fear that this makes Dutch ports a soft point of entry for narcotics. Possessing cannabis for personal use has been decriminalized – stopping short of actual legalization.

EDUCATION

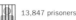 School leaving age: 16

99%   468,970 students

THE EDUCATION SYSTEM

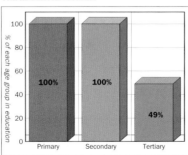

% of each age group in education

Primary	Secondary	Tertiary
100%	100%	49%

Corporate funding plays an important part in university research.

Public, municipally run schools attract 35% of pupils, and private (mostly denominational) schools are attended by 65%. Both types are fully funded by the state.

HEALTH

 Welfare state health benefits

1 per 323 people   Heart and respiratory diseases, cancers

Health care is largely funded by the state, though around 25% of funding comes from private sources. High spending ensures that the care that is offered is among the best in the world, but it is threatened by a rapidly aging population. The Netherlands was the first country to legalize abortion, but has the lowest rate of terminations in the world. Major health problems are similar to those in the rest of western Europe. Incidence of AIDS is higher than in Sweden or the UK but lower than in Switzerland, France, or Spain.

SPENDING

GDP/cap. increase

CONSUMPTION AND SPENDING

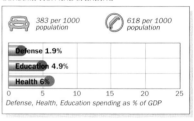

383 per 1000 population   618 per 1000 population

Defense 1.9%	
Education 4.9%	
Health 6%	

0   5   10   15   20   25

Defense, Health, Education spending as % of GDP

The Netherlands is, per capita, one of the richest countries in the world. Oil executives, stock market traders, and businessmen are among the wealthiest sector of the population. A progressive taxation system and extensive social welfare mean that wealth is quite evenly distributed. A small elite have considerable inherited wealth, but extravagant displays of affluence are rare.

Class does not play a big part in Dutch society. Most citizens would consider themselves middle class. Immigrant communities are the exception; they often live on the edges of towns in deprived areas. The poorest group of all are the illegal immigrants.

WORLD RANKING

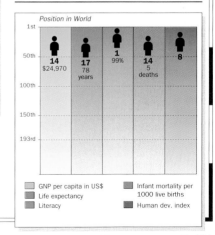

Position in World

14 $24,970	17 78 years	1 99%	14 5 deaths	8

1st — 50th — 100th — 150th — 193rd

GNP per capita in US$		Infant mortality per 1000 live births
Life expectancy		
Literacy		Human dev. index

NEW ZEALAND

OFFICIAL NAME: New Zealand   **CAPITAL:** Wellington   **POPULATION:** 3.8 million
CURRENCY: New Zealand dollar   **OFFICIAL LANGUAGES:** English and Maori

LYING IN THE SOUTH PACIFIC, 1600 km (992 miles) southeast of Australia, New Zealand comprises the main North and South Islands, separated by the Cook Strait, and a number of smaller islands. South Island is the more mountainous; North Island contains hot springs and geysers, and the bulk of the population. The political tradition is liberal and egalitarian, and has been dominated by the National and Labour Parties. Radical, and often unpopular, reforms since 1984 have restored economic growth, speeded up economic diversification, and strengthened New Zealand's position within the Pacific Rim countries.

CLIMATE   ▷ Maritime/subtropical

WEATHER CHART FOR WELLINGTON

New Zealand's climate is generally temperate and damp, with an average temperature of 12°C (54°F). There are differences between the islands, which extend north–south nearly 2000 km (1240 miles). The extreme north is almost subtropical; southern winters are cold. It is windy: Wellington, in particular, is known for bouts of blustery weather that can last for days.

TRANSPORTATION   ▷ Drive on left

✈ **Auckland International** 7.95m passengers	🚢 173 ships 336,278 grt

THE TRANSPORTATION NETWORK

53,568 km (33,286 miles)	226 km (140 miles)
3900 km (2423 miles)	1609 km (1000 miles)

Although both the main islands are well served by transportation services, the more populous North Island's road and rail network is more extensive than the South's. Air and ferry services complement the land networks and provide links between the North and South Islands, as well as with the numerous smaller islands. Cargo ferry services are particularly important for Antarctic bases in the Ross Dependency. Links with New Zealand's other associated territories – the Cook Islands, Niue, and the Tokelau atolls – are underdeveloped.

TOURISM   ▷ Visitors : Population 1:2.1

🧳 1.8m visitors   ⬆ Up 11% in 2000

MAIN TOURIST ARRIVALS

	% of total arrivals
Australia	33%
UK	11%
USA	11%
Japan	9%
South Korea	3%
Other	33%

New Zealand's prime attraction is its scenery. Unspoiled and, relative to the country's size, the most varied in the world, it offers mountains, fjords and lakes, glaciers, rainforests, beaches, boiling mud pools, and geysers. Other attractions are the Maori culture and outdoor activities such as river rafting, fishing, skiing, whale watching, and bungee jumping – a local invention.

Tourists come mainly from Australia, the US, the UK, and Japan. Tourism, the largest single foreign-exchange earner, continues to grow, although the 1997–1998 Asian economic crisis saw the number of Asian tourists drop by 10%. An increase in visitor numbers has followed the huge success of the first *Lord of the Rings* film: the whole series was shot in New Zealand.

***Mount Egmont**, an extinct volcano, is one of the numerous popular natural attractions of New Zealand's North Island.*

PEOPLE   ▷ Pop. density low

English, Maori   14/km² (37/mi²)

THE URBAN/RURAL POPULATION SPLIT

86%   14%

RELIGIOUS PERSUASION

- Methodist 5%
- Roman Catholic 15%
- Anglican 24%
- Nonreligious 16%
- Other 22%
- Presbyterian 18%

ETHNIC MAKEUP

- Pacific islanders 5%
- Other immigrant 6%
- Maori 12%
- European 77%

New Zealand is a country of migrants. The first settlers, the Maoris, migrated from Polynesia about 1200 years ago. Today's majority European population is mainly descended from British migrants who settled after 1840. Newer migrants include Asians from Hong Kong and Malaysia, and Polynesians. The government is keen to attract skilled South Americans, Russians, Chinese, and Africans to revitalize the economy.

The living standards and unemployment rates of the Maoris compare adversely with those of the European-descended majority, and relations can be tense. The Waikato Raupatu Claims Settlement Act was signed in 1995 and an official apology to the Maoris was made. In 1998 the Waitangi Tribunal ordered the return of confiscated land.

New Zealand became the first country in the world to give women the vote – in 1893. In 2001, the posts of prime minister, leader of the opposition, and governor-general were all filled by women.

POPULATION AGE BREAKDOWN

Female	Age	Male
1.7%	80+	0.9%
6.8%	60–79	6%
11.7%	40–59	11.5%
15.9%	20–39	15.2%
14.8%	0–19	15.5%

% of population by age group

POLITICS ▷ Multiparty elections

2002/2005   H.M. Queen Elizabeth II

AT THE LAST ELECTION
House of Representatives 120 seats   8% GP   2% PC

43% LP   **22%** NP   **11%** NZF   **8%** ACT   **6%** UFNZ

LP = Labour Party   **NP** = National Party   **NZF** = New Zealand First Party   **ACT** = ACT New Zealand (Association of Consumers and Taxpayers)   **GP** = Green Party **UFNZ** = United Future   **PC** = Progressive Coalition

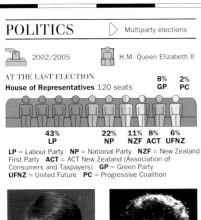

Helen Clark,
LP leader and prime
minister since 1999.

Jenny Shipley,
the first woman
prime minister
(NP, 1997–1999).

New Zealand is a parliamentary democracy. The Cook Islands and Niue are self-governing territories.

PROFILE

Since 1984 the economy has undergone massive reforms; cuts to the welfare system and privatization of public assets have been unpopular. The NP and the LP dominated politics until proportional representation (PR) was introduced in 1996. Former NP leader Jenny Shipley became the country's first woman prime minister in 1997 at the head of a coalition with the small NZF, and from 1998 led a minority administration when the NZF withdrew. A new minority coalition of the LP and the Alliance assumed power under LP leader Helen Clark in 1999. The LP remained in power after the 2002 elections, forming a minority coalition with the PC.

MAIN POLITICAL ISSUE
Electoral reform

New Zealand shifted from a first-past-the-post electoral system to PR for the 1996 general election. Endorsement of this reform in a referendum in 1993 reflected widespread disillusionment with the NP and LP. The new German-style system strengthened the role of smaller parties. As predicted, the first election to use the system in 1996 produced a coalition government, led by the NP. Forced into opposition in 1998, the NP has, unsuccessfully, called for a review of the PR system.

WORLD AFFAIRS ▷ Joined UN in 1945

Comm   APEC   OECD   PIF   PC

Many New Zealanders are strongly committed to the British monarchy and the Commonwealth, but the UK's EU involvement has forced New Zealand to reorient its trade and foreign policy toward its Pacific Rim neighbors, especially Australia, now New Zealand's single largest trading partner. Their 1983 Closer Economic Relationship (CER) treaty was strengthened in 1996 by the signing of a mutual recognition agreement.

Relations with Asia are growing in importance. The 1997–1998 Asian economic crisis significantly affected trade, particularly tourism.

Relations with the US are improving after a low point when New Zealand's antinuclear stance led to its exclusion from the ANZUS pact. Official ties with France, cut in 1985 after French agents bombed Greenpeace's *Rainbow Warrior* in Auckland harbor, were restored in 1997.

AID ▷ Donor

US$113m (donations)   Down 16% in 2000

Over half of New Zealand's overseas aid is bilateral. Particular areas of focus are the Pacific states and Pacific-wide organizations. New Zealand is a major supporter of the Pacific Islands Forum, the University of the South Pacific, and the Pacific Environment Program. It also offers scholarships to overseas students for study or training in New Zealand.

NEW ZEALAND

Total Land Area : 268 680 sq. km
(103 737 sq. miles)

LAND HEIGHT	POPULATION	
2000m/6562ft	over 500 000	⊙
1000m/3281ft	over 100 000	◎
500m/1640ft	over 50 000	○
200m/656ft	over 10 000	•
Sea Level	under 10 000	·

CHRONOLOGY

A former British colony, New Zealand became a dominion in 1907, self-governing from 1926, and fully independent in 1947.

- ❏ **1962** Western Samoa (now Samoa) gains independence.
- ❏ **1965** Cook Islands gain autonomy.
- ❏ **1975** Conservative NP wins elections. Economic austerity program introduced.
- ❏ **1976** Immigration cut by over 80%.
- ❏ **1984** LP elected; David Lange prime minister. Auckland harbor headland restored to Maoris.
- ❏ **1985** New Zealand prohibits nuclear vessels from ports and waters. French agents sink Greenpeace ship *Rainbow Warrior* in Auckland harbor.
- ❏ **1986** US suspends military obligations under ANZUS Treaty.
- ❏ **1987** LP wins elections. Introduction of controversial privatization plan. Nuclear ban enshrined in legislation.
- ❏ **1990** LP defeated by NP in elections. James Bolger prime minister.
- ❏ **1991** Widespread protests at spending cuts.
- ❏ **1992** Maoris win South Island fishing rights. Majority vote for electoral reform in referendum.
- ❏ **1993** Docking of first French naval ship for eight years. NP returned with single-seat majority in election. Proportional representation introduced by referendum.
- ❏ **1994** Senior-level US contacts restored; agrees not to send nuclear-armed ships to New Zealand ports. Maoris reject government ten-year land claims settlement of US$660 million.
- ❏ **1995** Waitangi Day celebrations abandoned after Maori protests. Crown apologizes to Maoris and signs Waikato Raupatu Claims Act. UK warship visits resume.
- ❏ **1996** NP forms coalition to preserve overall legislative majority. First general election under new proportional representation system.
- ❏ **1997** NP forms coalition with New Zealand First (NZF) party. Bolger resigns. Jenny Shipley becomes first woman prime minister.
- ❏ **1998** Shipley sacks NZF leader Winston Peters as deputy prime minister. Waitangi Tribunal orders government to return to Maoris US$3.3 million of confiscated land.
- ❏ **1999** LP led by Helen Clark wins general election.
- ❏ **2001** Air New Zealand renationalized.
- ❏ **2002** Combat wing of air force taken out of service. July, elections: LP reelected

DEFENSE

 No compulsory military service

 US$788m ⬇ Down 4% in 2000

Military cuts announced in May 2001 emphasized the aim to refocus defense policy on small-scale peacekeeping. The move put renewed stress on the 1951 security pact with Australia and the US (ANZUS). The 1984 decision to refuse access to nuclear warships from 1985 damaged defense cooperation with the US and other Western powers for ten years, forcing New Zealand to seek closer links with Australia. Senior-level contacts were resumed in 1994, the US announcing that it would not send nuclear-armed warships to New Zealand ports. Since then, the UK has also resumed naval visits.

NEW ZEALAND ARMED FORCES

🛡	8 light tanks (*Scorpion*)	4450 personnel
⚓	3 frigates and 4 patrol boats	1980 personnel
✈	No combat aircraft	2800 personnel
🚀	None	

New Zealand's defense strategy centers on protection against low-level economic threats, terrorism, regional security, and peacekeeping.

ECONOMICS

 Inflation 1.5% p.a. (1990–2000)

US$49.8bn   2.26–2.40 New Zealand dollars

SCORE CARD

- ❏ WORLD GNP RANKING..........................48th
- ❏ GNP PER CAPITA ....US$12,990
- ❏ BALANCE OF PAYMENTS..............–US$2.76bn
- ❏ INFLATION 2.6%
- ❏ UNEMPLOYMENT 6%

EXPORTS

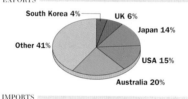

South Korea 4%   UK 6%
Japan 14%
Other 41%
USA 15%
Australia 20%

IMPORTS

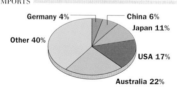

Germany 4%   China 6%
Japan 11%
Other 40%
USA 17%
Australia 22%

ECONOMIC PERFORMANCE INDICATOR

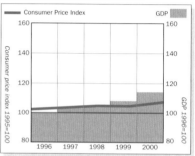

Consumer Price Index — GDP

government spending helped to restore growth and cut inflation to a minimum. Diversification into new markets and products recovered after the 1997–1998 Asian economic crisis. Prime Minister Clark dropped objections in 2001 to the idea of a unified Australia–New Zealand dollar. High public debt and poor levels of private investment remain a problem.

STRENGTHS

Modern agricultural sector; world's biggest exporter of butter and (per capita) of wool. Rapidly expanding tourist sector. Manufacturing, with emphasis on high tech. One of world's most open economies. Strong trade links within Pacific Rim.

WEAKNESSES

One of the highest levels of public debt outside developing world. Continuing reliance on imported manufactured goods and foreign investment.

PROFILE

Since 1984, New Zealand has changed from being one of the most regulated to one of the most open economies in the world. Radical reforms and drastic cuts in social security and related

NEW ZEALAND : MAJOR BUSINESSES

Auckland
Nelson
Napier
Wellington
Christchurch
Dunedin

Brewing
Chemicals
Pulp & paper
Meat packing
Dairy products
Fruit processing
Wool processing
Light engineering
Telecommunications

0   200 km
0   200 miles

RESOURCES

▷ Electric power 8m kw

685,734 tonnes

41,183 b/d (reserves 105m barrels)

44m sheep, 9.37m cattle, 354,489 pigs, 13m chickens

Coal, oil, natural gas, iron, gold, silica sand

ELECTRICITY GENERATION

Hydro 65% (24bn kwh)

Combustion 28% (11bn kwh)

Nuclear 0%

Other 7% (2.5bn kwh)

% of total generation by type

New Zealand's rich pastures, a result of even rainfall throughout the year, have traditionally been its key resource. The sheep, wool, and dairy products on which the country's wealth was built are still important. Newer export industries include products such as fruit, vegetables, fish, cork, wood, wine, and textile fibers.

New Zealand is well endowed with energy resources. It has coal, oil, and natural gas reserves, but most energy is generated by hydroelectric plants.

NORTH ISLAND

NEW ZEALAND : LAND USE

- Cropland
- Forest
- Pasture
- High mountain regions
- Cattle
- Sheep
- Cereals

SOUTH ISLAND

0   200 km
0   200 miles

ENVIRONMENT

▷ Sustainability rank: 19th

24%

7.9 tonnes per capita

ENVIRONMENTAL TREATIES

Yes	Yes	Yes
Yes	Yes	No

New Zealand's isolation, small population, and limited industry have helped to keep it one of the world's most pollution-free countries. It was a leading opponent of French nuclear testing in the Pacific and has banned nuclear vessels from its ports. Ozone depletion over Antarctica, deforestation, and protection of native flora and fauna are major issues.

MEDIA

▷ TV ownership high

Daily newspaper circulation 216 per 1000 people

PUBLISHING AND BROADCAST MEDIA

There are 29 daily newspapers. The leading newspaper is the *New Zealand Herald*

7 services: 1 state-owned, 6 independent

3 services: 1 state-owned, 2 independent

Deregulated in 1988, New Zealand television is one of the most liberal media in the world. Ruia Mai, the first Maori-language radio station, began broadcasting in 1996.

CRIME

▷ No death penalty

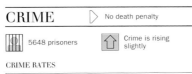

5648 prisoners

Crime is rising slightly

CRIME RATES

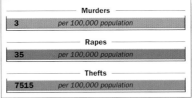

Murders	
3	per 100,000 population

Rapes	
35	per 100,000 population

Thefts	
7515	per 100,000 population

Crime rates in New Zealand's urban areas have increased in recent years. However, overall, the country remains one of the world's safest and most peaceful places in which to live.

EDUCATION

▷ School leaving age: 16

99%

237,781 students

THE EDUCATION SYSTEM

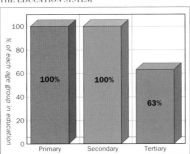

% of each age group in education

Primary 100%   Secondary 100%   Tertiary 63%

Education is free, and compulsory between six and 16. A number of schools are composite, providing both primary and secondary education. New Zealand has one of the highest proportions of the population with tertiary qualifications in the OECD. Nearly all adults are literate, but the level of literacy is not always very high: a government initiative, More than Words, was launched in 2000.

HEALTH

▷ Welfare state health benefits

1 per 435 people

Heart disease, cancers, accidents

In 1936 New Zealand was the first country to introduce a full welfare state. Government efforts since 1991 to impose UK-style market systems on the health service have been very unpopular. While life expectancy continues to improve, the nation's OECD health ranking is falling. In comparison with other OECD countries New Zealand has high mortality rates for heart disease, respiratory disease, breast and bowel cancer, motor vehicle accidents, and suicide.

SPENDING

▷ GDP/cap. increase

CONSUMPTION AND SPENDING

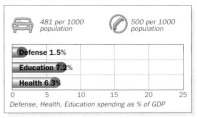

481 per 1000 population

500 per 1000 population

Defense 1.5%

Education 7.2%

Health 6.3%

Defense, Health, Education spending as % of GDP

The years since 1984 have been very difficult for New Zealanders, who are used to affluence within a generous welfare state. A rash of economic and social reforms has held back wages, raised unemployment, and cut welfare benefits. Even so, average living standards are still high, and a strong egalitarian tradition means that wealth remains quite evenly distributed.

The quality of life in New Zealand is among the world's highest, in terms of access to basic necessities, and a pure, healthy, urban and rural environment. Social mobility is fairly high. Wealthier people tend to spend their money on houses close to the water. Yachts are a major status symbol.

WORLD RANKING

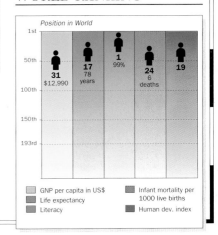

Position in World

1st					
50th	31 $12,990	17 78 years	1 99%	24 6 deaths	19
100th					
150th					
193rd					

- GNP per capita in US$
- Life expectancy
- Literacy
- Infant mortality per 1000 live births
- Human dev. index

N

NICARAGUA

CENTRAL AMERICA / North America

OFFICIAL NAME: Republic of Nicaragua   **CAPITAL:** Managua
POPULATION: 5.2 million   **CURRENCY:** Córdoba oro   **OFFICIAL LANGUAGE:** Spanish

 1838   1838   Sept 15   NIC   -6   +505   .ni

BOUNDED BY the Pacific Ocean to the west and the Caribbean Sea to the east, Nicaragua lies at the heart of Central America. After more than 40 years of dictatorship, the Sandinista revolution in 1978 led to social reforms, but also to 11 years of civil war, which almost destroyed the economy. Right-wing parties have held power since the Sandinistas unexpectedly lost the 1990 elections. Despite the devastation of Hurricane Mitch in 1998, the economy is slowly strengthening.

Oil refinery at Bluefields, on the Caribbean coast. Under the Sandinistas, most crude oil came from the Soviet Union, via Cuba.

CLIMATE

▷ Tropical equatorial/ wet & dry

WEATHER CHART FOR MANAGUA

■ Average daily temperature   Rainfall
°C/°F   J F M A M J J A S O N D   cm/in
40/104 .... 40/16
30/86 .... 30/12
20/68 .... 20/8
10/50 .... 10/4
0/32 0
-10/14
-20/-4

The climate is tropical and often violent, as evidenced by seasonal hurricanes, such as Hurricane Mitch in 1998.

TRANSPORTATION

▷ Drive on right

Augusto C. Sandino International, Managua
816,054 passengers

28 ships
4300 grt

THE TRANSPORTATION NETWORK

1818 km (1130 miles)	Pan-American Highway: 384 km (239 miles)
None	2220 km (1379 miles)

Nicaragua lacks a Caribbean deepwater port. Hurricane Mitch damaged major roads and destroyed 35 key bridges.

TOURISM

▷ Visitors : Population 1:11

486,000 visitors   Up 4% in 2000

MAIN TOURIST ARRIVALS

Honduras 29%	
USA 16%	
Costa Rica 15%	
Other 40%	

0 10 20 30 40
% of total arrivals

The civil war caused the near-collapse of tourism, and slow recovery has been interrupted by the devastation caused by Hurricane Mitch. Foreign direct investment in the sector grew in 1997–1998.

PEOPLE

▷ Pop. density low

Spanish, English Creole, Miskito

44/km²
(113/mi²)

THE URBAN/RURAL POPULATION SPLIT

65%   35%

ETHNIC MAKEUP

Zambos 4%   Amerindian 5%
Black 8%
White 14%
Mestizo 69%

The Caribbean regions, which in 1987 achieved limited autonomy, are isolated from the more populous Pacific regions. The indigenous Miskito tribes and the descendants of Africans, brought over by Spanish colonists in the 18th century to work the plantations, are concentrated along the Caribbean coast, where English Creole is widely spoken. The Sandinista revolution improved the status of women through changes in the legal system and the incorporation of women into economic and political life. However, poverty and lack of permanent employment have since forced many women into prostitution.

POLITICS

▷ Multiparty elections

2001/2006   President Enrique Bolanos

AT THE LAST ELECTION

National Assembly 92 seats

51%   47%   2%
PLC   FSLN   PCN

PLC = Liberal Constitutionalist Party
FSLN = Sandinista National Liberation Front
PCN = Conservative Party of Nicaragua

Defeated in the 1990 and 1996 polls by right-wing parties, the FSLN underwent an internal crisis. A PLC-dominated alliance took office in 1997, promising to unite the country but it quickly became unpopular as a result of its austerity measures and of allegations of corruption. Recent two-party domination, strengthened by a controversial pact between the ruling PLC and FSLN in 2000, has weakened democracy. Enrique Bolanos of the PLC was elected president in 2001.

NICARAGUA

Total Land Area : 129 494 sq. km
(49 998 sq. miles)

POPULATION

⊙ over 500 000
◎ over 100 000
○ over 50 000
● over 10 000
· under 10 000

LAND HEIGHT

1000m/3281ft
500m/1640ft
200m/656ft
Sea Level

0   100 km
0   100 miles

N

WORLD AFFAIRS

 Joined UN in 1945

Main issues are debt relief in the wake of Hurricane Mitch, cooperation with neighboring countries for increased US trade access, and the treatment of over

300,000 Nicaraguan immigrants in Costa Rica. A free trade agreement with Mexico is important. A dispute over the common border with Costa Rica was resolved in 2000. Ongoing border and navigation rights disputes exist with Honduras and Colombia.

AID

 Recipient

 $562m (receipts)   ⬇ Down 17% in 2000

Hurricane Mitch damage produced new World Bank and IDB loans. Cuba, France, Finland, and Spain pardoned

all or part of Nicaragua's debt. The US and other Western creditors canceled outstanding obligations when the IMF included Nicaragua in the Highly Indebted Poor Countries initiative.

DEFENSE

▷ Compulsory military service

💲 $26m   ⬆ Up 4% in 2000

FSLN forces once formed the basis of the army, which was cut from a civil war peak of 134,000 to 10,000 by 1995. Senior Sandinistas were among officers

retired in 1998. The army is to be involved in more community-based roles focused on the defense of natural resources and mine clearance. Russia agreed in 2001 to help upgrade the military.

ECONOMICS

▷ Inflation 34% p.a. (1990–2000)

 $2.05bn   💲 12.90–13.76 córdobas oro

SCORE CARD

❏ World GNP Ranking	136th
❏ GNP per Capita	$400
❏ Balance of Payments	–$505m
❏ Inflation	11.2%
❏ Unemployment	11%

STRENGTHS

Coffee, sugar, and grain exports. Foreign aid and public and private reconstruction work after Hurricane Mitch will benefit tourism, energy, services, and construction.

WEAKNESSES

Heavy debt burden. Main exports subject to commodity price fluctuations. High unemployment. Poor energy supply and infrastructure. Lack of investment and diversification. Weak banks. Delays in privatization. Skewed land ownership and protracted disputes over property. Corruption.

EXPORTS

Honduras 3% — Canada 4%
Costa Rica 3% — Germany 5%
USA 58% — Other 27%

IMPORTS

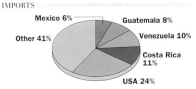

Mexico 6% — Guatemala 8%
Other 41% — Venezuela 10%
— Costa Rica 11%
— USA 24%

RESOURCES

▷ Electric power 614,000 kw

  24,767 tonnes   🛢 Not an oil producer

 2.28m cattle, 400,000 pigs, 17m chickens   ◆ Gold, silver, lead, zinc, copper, tungsten, salt

Nicaragua has small quantities of gold and silver. New thermal generation projects are planned to overcome energy deficits. There is possible offshore oil.

ENVIRONMENT

▷ Sustainability rank: 52nd

🌲 8%   ⬆ 0.7 tonnes per capita

Deforestation over large areas and the widespread use of pesticides are major problems.

MEDIA

▷ TV ownership medium

▤ Daily newspaper circulation 30 per 1000 people

PUBLISHING AND BROADCAST MEDIA

▤ There are 8 daily newspapers, including *La Prensa* and *El Nuevo Diario*	
7 services	62 stations: 1 state-owned, 61 independent

Since the civil war, radio, TV, and newspapers have tended to ally themselves with the government or the opposition; there is little room for political neutrality.

CRIME

▷ No death penalty

▦ 7198 prisoners   ⬆ Crime is rising

Former combatants have menaced parts of central and northern regions. Violent crime is rising, as is drug trafficking.

CHRONOLOGY

Nicaragua became an independent state in 1838. Guerrilla forces, led by Gen. Sandino, opposed the US marine presence in the early 1930s.

- ❏ **1978** FSLN, formed in 1961, ends 44-year Somoza dictatorship; ensuing conflict between FSLN and Contras.
- ❏ **1984** Daniel Ortega, FSLN leader, elected president.
- ❏ **1990** End of civil war. Center-right National Opposition Union (UNO) unexpectedly wins elections.
- ❏ **1998** Hurricane Mitch causes havoc.
- ❏ **2002** Bolanos elected president.

EDUCATION

▷ School leaving age: 12

👤 69%   🎓 56,558 students

The Sandinista "Literacy Crusade" achieved dramatic results in the 1980s, but has long since died away.

HEALTH

▷ Welfare state health benefits

👤 1 per 1111 people   ♥ Heart and intestinal infectious diseases, accidents, tuberculosis

Real spending on health fell by 71% between 1988 and 1993, and still has to recover.

SPENDING

▷ GDP/cap. increase

CONSUMPTION AND SPENDING

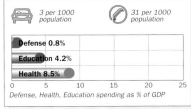

🚗 3 per 1000 population   📞 31 per 1000 population

Defense 0.8%
Education 4.2%
Health 8.5%

Defense, Health, Education spending as % of GDP

A 1998 UNDP study revealed that 44% of Nicaragua's population survive on less than $1 a day.

WORLD RANKING

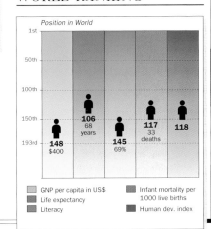

Position in World

148 $400
106 68 years
145 69%
117 33 deaths
118

❏ GNP per capita in US$
❏ Life expectancy
❏ Literacy
❏ Infant mortality per 1000 live births
❏ Human dev. index

N

NIGER

OFFICIAL NAME: Republic of Niger   **CAPITAL:** Niamey
POPULATION: 11.2 million   **CURRENCY:** CFA franc   **OFFICIAL LANGUAGE:** French

LANDLOCKED IN THE WEST of Africa, Niger is linked to the sea by the Niger River. Saharan conditions prevail in the northern regions, in the area around the Aïr Mountains, and, particularly, in the vast uninhabited northeast. Niger was ruled by one-party or military regimes until 1992. A much-troubled democratic process was then disrupted by military coups in 1996 and 1999. Niger is one of the poorest countries in Africa.

CLIMATE ▷ Hot desert/steppe

WEATHER CHART FOR NIAMEY

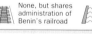

The Saharan north is virtually rainless. The south, in the Sahel belt, has an unreliable rainy season, preceded by a period of extreme daytime heat.

TRANSPORTATION ▷ Drive on right

Niamey International 84,096 passengers

Has no fleet

THE TRANSPORTATION NETWORK

798 km (496 miles)	Trans-Sahara Highway: 428 km (266 miles)
None, but shares administration of Benin's railroad	300 km (186 miles)

A very small proportion of Niger's road network is paved. There are international airports at Niamey and Agadez. There is no railroad.

TOURISM ▷ Visitors : Population 1:224

50,000 visitors   ⬆ Up 16% in 2000

MAIN TOURIST ARRIVALS

Africa 62%	
France 13%	
Germany 2%	
Other 23%	

0 10 20 30 40 50 60 70 80
% of total arrivals

The Aïr Mountains, southern Hausa cities, and Saharan Tuareg culture attract some tourists in spite of Niger's limited infrastructure and its instability.

PEOPLE ▷ Pop. density low

Hausa, Djerma, Fulani, Tuareg, Teda, French

9/km² (23/mi²)

THE URBAN/RURAL POPULATION SPLIT

21%   79%

ETHNIC MAKEUP

Other 6%
Tuareg 9%
Fulani 10%
Djerma and Songhai 21%
Hausa 54%

Considerable tensions exist in Niger between the Tuaregs in the north and the southern groups. The Tuaregs' sense of alienation from mainstream Nigerien politics has increased since the 1973 and 1984 droughts, which disrupted the Tuaregs' nomadic way of life. A five-year rebellion by northern Tuaregs ended in 1995 with a peace agreement. In eastern Niger, Toubou and Arab groups have also been in revolt.

A more subtle antagonism exists between the Djerma and Hausa groups. The Djerma elite from the southwest dominated politics for many years until 1993, when control passed to the Hausa majority.

Niger is an overwhelmingly Islamic society. Women have, on the whole, only limited rights and restricted access to education.

Testing boating poles in the market at Ayorou on the Niger River, the country's only major permanent watercourse.

POLITICS ▷ Multiparty elections

1999/2004   President Mamadou Tandja

AT THE LAST ELECTION

National Assembly 83 seats

5% ANDP

46% MNSD   20% CDS   19% PNDS   10% RDP

MNSD = National Movement for the Development of Society **CDS** = Democratic and Social Convention **PNDS** = Niger Party for Democracy and Socialism **RDP** = Rally for Democracy and Progress **ANDP** = Niger Alliance for Democracy and Progress

The death in 1987 of the military dictator, President Seyni Kountché, paved the way for prodemocracy demonstrations and eventually led to multiparty elections in 1993. An ensuing power struggle between President Mahamane Ousmane and his political opponents provoked a military coup in 1996. Gen. Ibrahim Barre Mainassara promulgated a new constitution and won a presidential election condemned as fraudulent by the opposition. Mainassara was assassinated by his presidential guard in early 1999. The new military leadership drew up yet another constitution. MNSD leader Mamadou Tandja won the presidential poll later that year. His party, allied with the CDS of former president Ousmane, dominates the new legislature.

WORLD AFFAIRS ▷ Joined UN in 1960

CILSS   ECOWAS   FZ   AU   OIC

Relations with Libya and Algeria have improved since the end of the Tuareg rebellion in 1995. ECOWAS members and the OAU condemned the 1999 coup, as did all key donors, led by France.

AID ▷ Recipient

$211m (receipts)   ⬆ Up 13% in 2000

France is the principal donor, followed by the World Bank and Japan. Most aid was frozen immediately following the 1999 coup, but in late 2000 the IMF approved a three-year loan under its Poverty Reduction and Growth Facility.

DEFENSE ▷ Compulsory military service

US$26m   ⬇ Down 7% in 2000

Niger's armed forces and paramilitary elements total 10,700. Politics has been dominated by the military since 1974.

N

NIGER

Total Land Area : 1 267 000 sq. km
(489 188 sq. miles)

POPULATION

- ◎ over 100 000
- ○ over 50 000
- ● over 10 000
- · under 10 000

LAND HEIGHT

- 1000m/3281ft
- 500m/1640ft
- 200m/656ft
- 150m/492ft

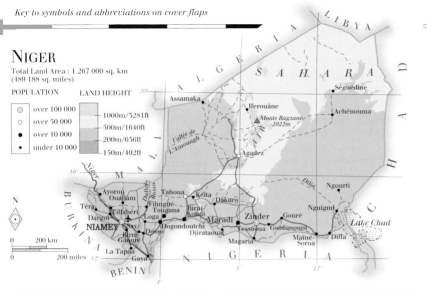

ECONOMICS ▷ Inflation 6% p.a. (1990–2000)

$1.94bn   698.7–736.7 CFA francs

SCORE CARD

- ❏ WORLD GNP RANKING.........................139th
- ❏ GNP PER CAPITA $180
- ❏ BALANCE OF PAYMENTS....................–$168m
- ❏ INFLATION 2.9%
- ❏ UNEMPLOYMENT 3%

STRENGTHS

Vast uranium deposits; gold and oil discoveries in late 1990s revived hopes for economic viability.

WEAKNESSES

Aid-dependent. Collapse of uranium prices in 1980s created large debt burden. Only 3% of land cultivable. Weak infrastructure. Frequent droughts. Political instability.

EXPORTS

- Benin 2%
- USA 3%
- Ghana 2%
- Other 4%
- France 48%
- Nigeria 41%

IMPORTS

- Japan 6%
- China 8%
- Nigeria 10%
- Other 43%
- Ivory Coast 13%
- France 20%

RESOURCES ▷ Electric power 105,000 kw

11,014 tonnes   Not an oil producer

6.9m goats, 4.5m sheep, 2.26m cattle, 24m chickens   Uranium, tin, gypsum, coal, salt, tungsten, oil, iron, phosphates, gold

During the 1970s, Niger's uranium mines boomed, but output collapsed in the 1980s when world prices slumped. Other mining is small-scale and oil reserves, discovered in the Lake Chad area, are not yet commercially viable. Salt is a traditionally exploited resource, as are such plant resources as the doum and palmyra palms.

ENVIRONMENT ▷ Sustainability rank: 123rd

8%   0.1 tonnes per capita

Serious droughts intensify desertification. Hunting was banned in 2001, in an effort to preserve wildlife numbers.

MEDIA ▷ TV ownership low

Daily newspaper circulation 0.2 per 1000 people

PUBLISHING AND BROADCAST MEDIA

There is 1 daily newspaper, *Le Sahel*, published by the government

3 services: 2 state-owned, 1 independent   3 services: 1 state-owned, 2 independent

The government controls most broadcasting. The BBC World Service's Hausa programming is influential.

CRIME ▷ Death penalty not used in practice

5263 prisoners   Down 36% 1996–1998

Rural banditry is common, often involving access to grazing and water. Urban crime levels are low, but in border areas smuggling is a way of life.

EDUCATION ▷ School leaving age: 12

16%   4513 students

Local languages are emphasized more strongly than in most francophone states. School attendance is only 30%.

CHRONOLOGY

The powerful Islamic Sokoto empire dissolved as the French took over Niger between 1883 and 1901.

- ❏ **1960** Independence.
- ❏ **1968** French open uranium mines.
- ❏ **1973** Drought; 60% of livestock die.
- ❏ **1974** Military coup. Gen. Seyni Kountché bans political parties.
- ❏ **1984** New drought; Niger River dries up. Uranium boom ends.
- ❏ **1987** Kountché dies. Gen. Ali Saibou eases transition to democracy.
- ❏ **1990–1995** Tuareg rebellion.
- ❏ **1992** Multiparty constitution.
- ❏ **1993** Democratic elections.
- ❏ **1996** Military coup. Staged elections.
- ❏ **1999** New constitution. Gen. Mainassara assassinated. Multiparty elections won by Mamadou Tandja.
- ❏ **2001** Hunting banned in effort to save wildlife.

HEALTH ▷ No welfare state health benefits

1 per 20,000 people   Malaria, tuberculosis, meningitis, measles, malnutrition

In spite of progress in rural health care, immunization, malaria control, and child nutrition are still limited.

SPENDING ▷ GDP/cap. increase

CONSUMPTION AND SPENDING

4 per 1000 population   2 per 1000 population

- Defense 1.5%
- Education 2.7%
- Health 1.2%

Defense, Health, Education spending as % of GDP

A small circle of secretive trading families controls much of Niger's wealth. They are successful in evading taxation.

WORLD RANKING

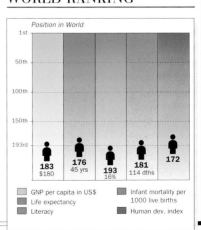

Position in World

183 $180	176 45 yrs	193 16%	181 114 dths	172

- GNP per capita in US$
- Life expectancy
- Literacy
- Infant mortality per 1000 live births
- Human dev. index

NIGERIA

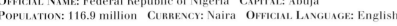

OFFICIAL NAME: Federal Republic of Nigeria **CAPITAL:** Abuja
POPULATION: 116.9 million **CURRENCY:** Naira **OFFICIAL LANGUAGE:** English

AFRICA'S MOST POPULOUS state, Nigeria gained its independence from Britain in 1960. Bordered by Benin, Niger, Chad, and Cameroon, its terrain varies from tropical rainforest and swamps in the south to savanna in the north. Nigeria has been dominated by military governments since 1966. After many delays, a promised return to civilian rule came about in 1999, with the election as president of Olusegun Obasanjo, a former general who had been head of state from 1976 to 1979. Nigeria is a major OPEC oil producer, but it has experienced a fall in living standards since the 1970s oil boom.

Village beneath Tengele Peak in Bauchi State. A large proportion of Nigerians live from subsistence agriculture.

CLIMATE

▷ Tropical/steppe

WEATHER CHART FOR ABUJA

■ Average daily temperature   Rainfall ■

The south is hot, rainy, and humid for most of the year. The arid north experiences only one, uncomfortably humid, rainy season from May to September. Its very hot dry season is marked by the *harmattan* wind. The Jos Plateau and the eastern highlands are cooler than the rest of Nigeria. Forcados, in the Niger delta, gets most rain, with 380 cm (150 in.) a year.

TRANSPORTATION

▷ Drive on right

Murtala Muhammad, Lagos
2.7m passengers

293 ships
451,900 grt

THE TRANSPORTATION NETWORK

| 60,068 km (37,324 miles) | 1194 km (742 miles) |
| 3557 km (2210 miles) | 8575 km (5328 miles) |

Nigeria relies almost entirely on road transportation. During the oil-boom years of the 1970s, new long-distance roads and stretches of freeway were built. The road network is now badly maintained and in urgent need of repair. The road accident rate is among the world's highest and there is severe and chronic traffic congestion in Lagos. In mid-2000 plans for a new trans-Sahara highway from Lagos to Tangiers, Morocco, were announced. Work started in 2001 on a $40 million ports project that will link the five southeastern states.

TOURISM

▷ Visitors : Population
1:144

813,000 visitors   Up 5% in 2000

Nigeria has attempted to build a tourist industry, but numbers remain low. Year-round tropical temperatures and poor infrastructure have limited its growth. The major deterrent to visitors, however, is crime. Travel can be hazardous, and Lagos has one of the world's highest crime rates.

MAIN TOURIST ARRIVALS

Niger 21%	
Benin 13%	
Ghana 7%	
Liberia 4%	
Cameroon 4%	
Other 51%	

% of total arrivals

NIGERIA

Total Land Area : 923 768 sq. km (356 667 sq. miles)

POPULATION

▣	over 1 000 000
◉	over 500 000
◎	over 100 000
○	over 50 000
●	over 10 000
•	under 10 000

LAND HEIGHT

	2000m/6562ft
	1000m/3281ft
	500m/1640ft
	200m/656ft
	Sea Level

PEOPLE ▷ Pop. density medium

Hausa, English, Yoruba, Ibo

128/km² (332/mi²)

THE URBAN/RURAL POPULATION SPLIT

44%   56%

RELIGIOUS PERSUASION

Traditional beliefs 10%

Muslim 50%

Christian 40%

ETHNIC MAKEUP

Fulani 11%

Other 29%

Ibo 18%

Yoruba 21%   Hausa 21%

The Hausa, Yoruba, and Ibo are the three main ethnic groups, and there are 245 smaller ones. There is intense ethnic rivalry, and religion is also a source of tension. Outbreaks of violence, particularly in the north, are often caused by clashes between Islamist militants and Christian proselytizers. When in 1999 Zamfara state introduced *sharia* – Islamic law – President Obasanjo, a Christian, would not attend the ceremony. The introduction of *sharia* in other states in the north has sparked violence, leaving over 400 dead in Kaduna in 2000. Religious tensions were exacerbated by the September 2001 attacks on the US, and 500 died in subsequent riots in Jos.

Traditionally, except in the Islamic north, women have had independent economic status. In recent years they have, however, been subjected to some prejudice in professional circles.

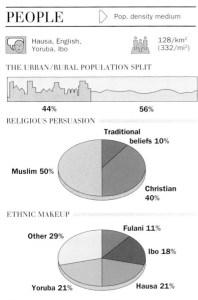

Lake Chad

asak Baza

Ngala

POPULATION AGE BREAKDOWN

Female		Age	Male	
	0.5%	80+	0.5%	
	1.9%	60–79	2.2%	
	5.2%	40–59	6%	
	15.1%	20–39	13.1%	
27.3%		0–19		28.2%

% of population by age group

POLITICS ▷ Multiparty elections

L. House 1999/2003
U. House 1999/2003

President Olusegun Obasanjo

AT THE LAST ELECTION
House of Representatives 360 seats

60% PDP   19% APP   18% AD   3% Vacant

PDP = People's Democratic Party **APP** = All People's Party
AD = Alliance for Democracy

Senate 109 seats

61% PDP   21% APP   17% AD   1% Vacant

Since May 1999 Nigeria has had a civilian constitution, after 16 years of military dictatorships. President Olusegun Obasanjo and his PDP promised national reconciliation.

PROFILE
The sudden death of military ruler Sani Abacha in 1998, followed by that of the imprisoned Chief Moshood Abiola, the presumed winner of the annulled 1993 presidential elections, left Gen. Abdulsalam Aboubakar to usher in civilian rule. Olusegun Obasanjo, a popular general who was head of state in 1976–1979, won elections held in 1999.

MAIN POLITICAL ISSUES
Corruption
Corruption has been identified as a major cause of Nigeria's debt levels. Accusations at all levels of government peaked with the impeachment in mid-2000 of the Senate president. President Obasanjo has promised to stamp out corruption.

Ethnic and religious tensions
The Obasanjo regime faces difficulties in reversing the rivalries among the Hausa, Yoruba, and Ibo. The situation deteriorated from early 2000, with ethnic violence involving Yoruba separatists. The adoption of *sharia* law, with strict punishments meted out for crimes such as theft and adultery, by many northern states has provoked rioting between Christians and the majority Muslims. Clashes in Jos in May 2002 left 500 dead.

Olusegun Obasanjo, elected president in 1999.

Gen. Abacha. Head of state from 1993 until 1998.

CHRONOLOGY
Before formal colonization by the British, begun only in 1861, Nigeria was a collection of African states owing their considerable wealth to trans-Saharan and transatlantic trade. During the 18th century the principal commodity was slaves: over 15,000 people a year were exported from the Bight of Benin and another 15,000 from the Bight of Biafra.

❏ **1885** Royal Niger Company given official responsibility for British sphere of influence along Niger and Benue Rivers. British armed forces coerce local rulers into accepting British rule.
❏ **1897** West Africa Frontier Force (WAFF) established; subjugation of the north begins.
❏ **1898** The Royal Niger Company's charter revoked.
❏ **1900** British Protectorate of Northern Nigeria established.
❏ **1906** Lagos incorporated into the Protectorate of Southern Nigeria.
❏ **1914** Protectorates of Northern and Southern Nigeria joined to form colony of Nigeria.
❏ **1960** Independence. Nigeria established as a federation.
❏ **1961** Northern part of UK-administered UN Trust Territory of the Cameroons incorporated as part of Nigeria's Northern Region.
❏ **1966** January, first military coup, led by Maj. Gen. Ironsi. July, countercoup mounted by group of northern army officers. Ironsi murdered. Thousands of Ibo in Northern Region massacred. Gen. Gowon in control of north and west.
❏ **1967–1970** Civil war. Lt. Col. Ojukwu calls for secession of oil-rich east under the new name Biafra. Over one million Nigerians die before secessionists defeated by federal forces.
❏ **1970** Gen. Gowon in power.
❏ **1975** Gowon toppled in bloodless coup. Brig. Mohammed takes power.
❏ **1976** Mohammed murdered in abortive coup. Succeeded by Gen. Olusegun Obasanjo.
❏ **1978** Political parties legalized, on condition they represent national, not tribal, interests.
❏ **1979** Elections won by Alhaji Shehu Shagari and the National Party of Nigeria (NPN), marking return to civilian government.
❏ **1983** Military coup. Maj. Gen. Mohammed Buhari heads Supreme Military Council.
❏ **1985** Maj. Gen. Ibrahim Babangida heads bloodless coup, promising a return to democracy.
❏ **1993** August, elections annulled; Babangida resigns; military sets

N

CHRONOLOGY *continued*

up Interim National Government (ING). November, ING dissolved. Military, headed by Gen. Sani Abacha, takes over.

❏ **1994** Moshood Abiola arrested, opposition harassed.

❏ **1995** Ban on parties lifted. Obasanjo and 39 others convicted of plotting coup. Execution of Ken Saro-Wiwa and eight other Ogoni activists: EU sanctions, suspension of Commonwealth membership.

❏ **1998** Abacha dies; Abiola dies; return to civilian rule timetabled.

❏ **1999** Elections: presidency won by Obasanjo. Sanctions lifted, Commonwealth membership restored. Zamfara becomes first state to introduce *sharia*.

❏ **2000** Ethnic violence escalates, threatens national unity.

❏ **2001** 200 villagers massacred by army in apparent revenge killing.

❏ **2002** 1000 killed in Lagos munitions dump explosion and ensuing chaos. 500 killed in clashes between Muslims and Christians in Jos.

WORLD AFFAIRS   ▷ Joined UN in 1960

Comm   ECOWAS   AU   OIC   OPEC

Liking to be regarded as Africa's leading voice, Nigeria is a keen sponsor of ECOWAS and of the AU, and has been the main contributor to ECOMOG forces. However, the military regime's violations of human rights (most notoriously the execution of Ken Saro-Wiwa and eight other Ogoni activists in 1995) prompted UN condemnation and suspension from the Commonwealth. Membership was restored only after Obasanjo came to power in 1999.

Nigeria was an implacable opponent of apartheid, and relations with South Africa were restored only after democratic elections there in 1994. A maritime border dispute with Equatorial Guinea was settled in 2001, opening up the further exploitation of undersea oil reserves.

AID   ▷ Recipient

 $185m (receipts)   Up 22% in 2000

The 1981 drop in world oil prices turned Nigeria from an aid donor into a major receiver of World Bank assistance. Aid flows interrupted in late 1995 were resumed in 2000. Aid agreed under the G8 summit in June 2002 was welcomed by President Obasanjo – one of the architects of the New Partnership for Africa's Development (NEPAD) plan.

DEFENSE   No compulsory military service

 $2.34bn   ⬆ Up 5% in 2000

The Nigerian government has contributed a significant number of forces to ECOMOG, the regional peacekeeping force set up in 1990 by ECOWAS, initially to restore order in Liberia. Nigerian forces served there until 1999, and in Sierra Leone in 1993–2000 and from 2002.

Since 2001 the armed forces have also been deployed to restore internal control after outbreaks of religious unrest in northern and central states, in which hundreds of civilians have been killed. However, government troops themselves were reported to have massacred some 200 villagers

NIGERIAN ARMED FORCES

200 main battle tanks (50 T-55, 150 *Vickers* Mk 3)	62,000 personnel	
1 frigate, 3 patrol boats	7000 personnel	
86 combat aircraft (19 *Alpha Jet*, 17 MiG-21, 15 *Jaguar*)	9500 personnel	
None		

in the eastern state of Benue in October 2001, in retaliation for the abduction of 19 members of the armed forces who had been deployed to quell unrest.

ECONOMICS   ▷ Inflation 29% p.a. (1990–2000)

 $32.7bn   110–120 naira

SCORE CARD

❏ WORLD GNP RANKING	57th
❏ GNP PER CAPITA	$260
❏ BALANCE OF PAYMENTS	$6.98bn
❏ INFLATION	6.9%
❏ UNEMPLOYMENT	28%

EXPORTS

France 5%
Brazil 3%
USA 44%
India 6%
Spain 10%
Other 32%

IMPORTS

China 7%
Other 59%
Germany 7%
France 8%
USA 9%
UK 10%

STRENGTHS

One of world's top oil producers at 2.15 million b/d. Vast reserves of natural gas, still only partly exploited. Soaring world oil prices in 2000 signified recovery from 1986 collapse. Almost self-sufficient in food. Strong entrepreneurial class. Large domestic market.

WEAKNESSES

Overdependence since the 1970s on oil, which encourages massive state inefficiency. Advantages of a large domestic market mitigated by low per capita purchasing power and high unit transportation costs. Entrepreneurs focus on trade rather than production. Of Nigeria's traditional agricultural exports only cocoa remains. Notorious corruption and maladministration undermine investors' confidence.

ECONOMIC PERFORMANCE INDICATOR

PROFILE

The economy has been characterized by massive government spending and the running up of debts which could not be serviced after the 1981 oil price fall. Led by the IMF, creditors want major cuts in spending – especially on loss-making public-sector companies – and subsidies. Gasoline subsidies alone are estimated to have cost $2.4 billion a year. Such changes are politically fraught. In May 1999 the government began an anticorruption drive.

NIGERIA : MAJOR BUSINESSES

Oil
Textiles
Brewing
Manufacturing
Pharmaceuticals
Oil refining
Chemicals
Palm oil
Cement

0   200 km
0   200 miles   * significant multinational ownership

N

RESOURCES Electric power 5.9m kw

477,365 tonnes

2.15m b/d (reserves 24bn barrels)

24.3m goats, 20.5m sheep, 19.8m cattle, 126m chickens

Oil, natural gas, coal, tin, iron, bauxite, columbite, lead

ELECTRICITY GENERATION

Hydro 35% (5.5bn kwh)	
Combustion 65% (10bn kwh)	
Nuclear 0%	
Other 0%	

% of total generation by type

(scale: 0 20 40 60 80 100)

Oil has been Nigeria's main resource since the 1970s. Government policy is to increase output to 3 million b/d by 2003. Domestic demand is 300,000 b/d, much of it smuggled to neighboring countries. Nigeria's vast gas deposits are still underexploited. The state retains 60% control of the oil and gas industry. Shell is the main foreign shareholder, but most oil multinationals are represented.

Nigeria has sizable iron ore deposits. These are not yet utilized in the state-run steel industry; imported ore is used instead. Bauxite deposits are also currently underexploited. There are, however, plans for establishing an aluminum industry. Nigeria also has deposits of coal and tin.

NIGERIA : LAND USE

Cropland	
Pasture	
Forest	
Wetlands	
Cacao – cash crop	
Goats	

0   200 km
0   200 miles

ENVIRONMENT Sustainability rank: 133rd

3% (1% partially protected)

0.6 tonnes per capita

ENVIRONMENTAL TREATIES

Yes		Yes		Yes	
Yes		Yes		No	

Oil industry pollution in the Niger delta, a major local concern, came to international attention in 1995. Shell has been particularly condemned. Before the discovery of a highly toxic cargo in 1988, Nigeria was a dumping ground for European chemical waste.

MEDIA TV ownership medium

Daily newspaper circulation 27 per 1000 people

The Nigerian press is traditionally one of Africa's liveliest. Media freedom has improved since the return to civilian government. There are about 90 TV stations; the state-run Nigerian Television Authority runs 32.

CRIME Death penalty in use

44,450 prisoners   Crime is rising

CRIME RATES

Murders	
94	per 100,000 population

Rapes	
Nigeria does not publish rape statistics	

Thefts	
1256	per 100,000 population

Nigeria has one of the highest crime rates in the world. Murder often accompanies even minor burglaries. Rich Nigerians live in high-security compounds. A special armed police unit was formed to restore order in Lagos in 2002. It is claimed that brutal *sharia* punishments have reduced crime in northern states. Vigilante gangs are used to fight crime in the southeastern states.

EDUCATION School leaving age: 15

64%   207,982 students

THE EDUCATION SYSTEM

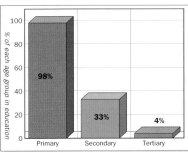

% of each age group in education

Primary 98% · Secondary 33% · Tertiary 4%

Responsibility for education is shared between the federal and the state governments. Education has suffered from the government's massive debt repayment burden. During the oil-boom years, Nigeria concentrated on creating 31 universities with prestigious medical and scientific schools. However, standards in primary education, which has not received the same level of investment, have fallen since the 1970s. Almost all children attend primary school, but only about 30% receive secondary education.

PUBLISHING AND BROADCAST MEDIA

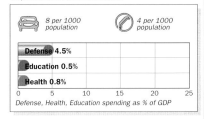

There are 25 daily newspapers. The *Daily Times*, published by the government, has the highest circulation

1 state-controlled service, about 60 private stations

5 services: 2 state-controlled, 3 independent

HEALTH Welfare state health benefits

1 per 5000 people   Yellow fever, malaria, trachoma, yaws

The health service functions mainly in urban areas and has suffered from the crisis in government revenues; a 1999 government report stated that only 49% of the population had access to water and health services. More than 3.5 million Nigerians were living with HIV/AIDS as at end-2001. Free generic versions of AIDS treatments became available in 2001 and the government supports a successful AIDS awareness strategy.

SPENDING ▷ GDP/cap. increase

CONSUMPTION AND SPENDING

8 per 1000 population   4 per 1000 population

Defense 4.5%	
Education 0.5%	
Health 0.8%	

Defense, Health, Education spending as % of GDP

(scale: 0 5 10 15 20 25)

Nigerians with access to the rich pickings of political office spent on a massive scale during the country's oil boom – on expensive cars and on overseas education for their children. Much was financed by government loans. Habits have not changed with the fall in oil revenues: borrowing has simply grown.

WORLD RANKING

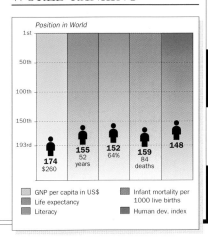

Position in World

1st	50th	100th	150th	193rd

174 $260 · 155 52 years · 152 64% · 159 84 deaths · 148

▢ GNP per capita in US$	▢ Infant mortality per 1000 live births
▢ Life expectancy	▢ Human dev. index
▢ Literacy	

N

NORWAY

OFFICIAL NAME: Kingdom of Norway   **CAPITAL:** Oslo
POPULATION: 4.5 million   **CURRENCY:** Norwegian krone   **OFFICIAL LANGUAGE:** Norwegian

| 1905 | 1905 | May 17 | N | +1 | +47 | .no |

OCCUPYING THE WESTERN PART of Scandinavia, Norway's western coastline is characterized by numerous fjords and islands. Large oil and gas revenues have brought prosperity. Gro Harlem Brundtland, Norway's first woman prime minister, went on to take top UN posts. Despite the Europe-wide recession in the early 1990s, Norway was able to contain rising unemployment, which peaked at 6% in 1993. A constitutional requirement is that government creates conditions that enable every person to find work.

The village of Reine on Moskenesøya, deep inside the Arctic Circle in the Lofoten Islands. It is a popular destination for summer visitors.

CLIMATE ▷ Maritime/subarctic

WEATHER CHART FOR OSLO

The whole of Norway's west coast is kept ice-free by the warm Gulf Stream. It receives much more precipitation than the rest of the country; Bergen has a yearly average of 225 cm (89 in). Norway enjoys the highest mean temperatures in Scandinavia, but in winter the temperature in Oslo can drop to –25°C (–13°F).

NORWAY

Total Land Area : 324 220 sq. km
(125 181 sq. miles)

LAND HEIGHT

2000m/6562ft
1000m/3281ft
500m/1640ft
200m/656ft
Sea Level

POPULATION

over 100 000 ◎
over 50 000 ○
over 10 000 ●
under 10 000 ·

TRANSPORTATION ▷ Drive on right

Fornebu International, Oslo
14.2m passengers

2325 ships
23.1m grt

THE TRANSPORTATION NETWORK

| 67,602 km (42,006 miles) | 128 km (80 miles) |
| 4179 km (2597 miles) | 1577 km (980 miles) |

It has been impossible to extend rail links further north than Bodø, inside the Arctic Circle. To reach the Lofotens or Narvik and beyond, the most common form of transportation is air. Some 5% of the world's total shipping tonnage is controlled by Norway, making it the sixth-largest shipping nation.

The royal palace, Oslo. This is situated near the national theater, at one end of the Karl Johanisgate, the city's main thoroughfare.

TOURISM

 Visitors : Population 1:1

 4.35m visitors   Down 3% in 2000

Norway is a popular destination, receiving visitors from Germany, Sweden, the UK, Denmark, and the US. Its winter tourism industry is based on skiing and was boosted by the 1994 Winter Olympics taking place in Lillehammer. Cruising along the fjords is popular with summer visitors. Areas within the Arctic Circle are a particular attraction in June, when tourists go in search of the midnight sun. Oslo has a reputation for good classical music and jazz. However,

MAIN TOURIST ARRIVALS

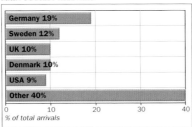

	% of total arrivals
Germany	19%
Sweden	12%
UK	10%
Denmark	10%
USA	9%
Other	40%

the strength of the krone and the high cost of living make Norway expensive.

PEOPLE

 Pop. density low

Norwegian (*Bokmål* "book language" and *Nynorsk* "new Norsk"), Sami   15/km² (38/mi²)

THE URBAN/RURAL POPULATION SPLIT

76%   24%

RELIGIOUS PERSUASION

- Roman Catholic 1%
- Other and nonreligious 10%
- Evangelical Lutheran 89%

ETHNIC MAKEUP

- Sami 1%
- Other 6%
- Norwegian 93%

Norway has a small but growing immigrant population, forming 5.6% of the population in 2001. In the early 1990s the number of asylum seekers increased, especially as a result of the Bosnian conflict. Some refugees have been attacked by right-wing groups.

The family is traditionally close and nuclear. Men are expected to share responsibility for raising children, who frequently attend day schools from under the age of two years. Women enjoy considerable power and freedom, and comprise at least 40% of members of the government. Over half of marriages end in divorce.

POPULATION AGE BREAKDOWN

Female	Age	Male
2.7%	80+	1.3%
8.6%	60–79	7.2%
12.1%	40–59	12.6%
14.6%	20–39	15.2%
12.5%	0–19	13.2%

% of population by age group

POLITICS

 Multiparty elections

2001/2005   H.M. King Harald V

AT THE LAST ELECTION

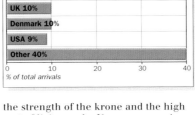

Parliament 165 seats

26% DNA	23% H	16% FrP	14% SV	13% KrF	6% SP	1% V	1% Others

DNA = Norwegian Labor Party   **H** = Hoeyre (Conservative Party)   **FrP** = Progress Party   **SV** = Socialist Left Party   **KrF** = Christian Democratic Party   **SP** = Center Party   **V** = Venstre (Liberal Party)
The Parliament (Storting) is elected as one body but divides itself for most legislative purposes into an upper chamber (Lagting, with 42 members) and a lower chamber (Odelsting, with 123 members)

Norway is a constitutional monarchy, with a king as head of state and an elected parliament.

PROFILE

Political decisions are based on consensus building between the government, parliament, and the strong trade unions. In 1993 opposition to EU membership boosted support for the SP but the DNA remained the largest party.

In 1997 the DNA lost ground to the center-right, who formed a coalition led by Kjell Magne Bondevik of the KrF. The DNA returned to office in 2000, but received heavy losses in the 2001 polls. Bondevik, who had campaigned on a platform of using oil revenues to fund tax cuts and public service improvements, returned as prime minister. A notable feature of the election was the capture of 26 seats by the anti-immigration FrP, which had emerged during the 1990s.

MAIN POLITICAL ISSUE
EU membership

In a 1994 referendum, 52% voted against EU membership, previously rejected in 1972. Terms for accession, supported by government and industry, had been agreed, but opponents argued successfully that it would lead to a loss of control of national resources, notably fisheries and offshore oil. Membership was not an electoral issue in 2001.

WORLD AFFAIRS

Joined UN in 1945

 CE   NATO   OECD   OSCE   EEA

A founder member of NATO, Norway continues to offer it strong support. Norway also became an associate member of the WEU in 1992.

The 1994 referendum decision to reject EU membership means that the European Economic Area (EEA) offers Norway its chief access to the European single market. As a Nordic Council member it is also associated with the Schengen Convention. In January 2002 Bondevik stated that there would be no referendum on joining the EU. However, a concurrent opinion poll suggested that a majority now favored membership.

Norway has played peacemaker in a number of major international conflicts, notably the Israeli–Palestinian conflict during the mid-1990s, and in Sri Lanka in 2002.

The government has been unable to control the ecological effects of acid rain, which is destroying its forests. Representatives of 25 European countries and Canada met in Oslo in 1994 and signed a UN protocol on reducing sulfur emissions.

King Harald V, who succeeded his father King Olaf V in 1991.   *Prime Minister Kjell Magne Bondevik, reelected in 2002.*

CHRONOLOGY

Norway gained independence from the Swedish crown in 1905 and elected its own king, Håkon VII.

- ❏ **1935** DNA forms government.
- ❏ **1940–1945** Nazi occupation. Puppet regime led by Vidkun Quisling.
- ❏ **1945** DNA resumes power.
- ❏ **1949** Founder member of NATO.
- ❏ **1957** King Håkon dies. Succeeded by son, Olaf V.
- ❏ **1960** Becomes member of EFTA.
- ❏ **1962** Unsuccessfully applies to join the European Communities (EC).
- ❏ **1965** DNA electoral defeat by SP coalition led by Per Borten.
- ❏ **1967** Second bid for EC membership.
- ❏ **1971** Prime Minister Per Borten resigns following disclosure of secret negotiations to join EC. DNA government, led by Trygve Bratteli.

N

CHRONOLOGY *continued*

- ❏ **1972** EC membership rejected in popular referendum by 3% majority. Bratteli resigns. Center coalition government takes power. Lars Korvald prime minister.
- ❏ **1973** Elections. Bratteli returns to power as prime minister.
- ❏ **1976** Bratteli succeeded by Odvar Nordli.
- ❏ **1981** Nordli resigns owing to ill health. Gro Harlem Brundtland becomes first woman prime minister. Elections bring to power Norway's first Conservative Party (H) government for 53 years. Kare Willoch prime minister.
- ❏ **1983** Conservatives form coalition with SP and KrF.
- ❏ **1985** Election. Willoch's H–SP–KrF coalition returned. Norway agrees to suspend commercial whaling.
- ❏ **1986** 100,000 demonstrate for better working conditions. Brundtland forms minority DNA government. Currency devalued by 12%.
- ❏ **1989** Brundtland resigns. H–KrF coalition in power. Soviet Union agrees exchange of information after fires on Soviet nuclear submarines off Norwegian coast.
- ❏ **1990** H–KrF coalition breaks up over closer ties with EU (formerly EC). Brundtland and DNA in power.
- ❏ **1991** Olaf V dies; succeeded by son, King Harald V.
- ❏ **1994** EEA comes into effect. Referendum rejects EU membership.
- ❏ **1996** Brundtland resigns; replaced by Thorbjørn Jagland (also DNA).
- ❏ **1997** Kjell Magne Bondevik forms center-right coalition.
- ❏ **2000** Jens Stoltenberg (DNA) heads three-party coalition.
- ❏ **2001** Right-wing victory in elections. Bondevik heads coalition government.

AID ▷ Donor

💲 $1.26bn (donations) ⬇ Down 8% in 2000

Norway has been granting more than the UN development target of 0.7% of GNP in aid every year since 1975. Although Norway's ratio of aid to GNP has declined somewhat, to 0.92% in 1999, it remains one of the highest in the world. The vast majority of Norway's bilateral aid goes to the least developed countries of southeast Africa, south Asia, and Central America, though Palestine and Bosnia are also important recipients. The Norwegian government allocates funds to various multilateral assistance programs, as well. The 1999 budget included a debt relief program to help reduce developing country indebtedness.

DEFENSE ▷ Compulsory military service

💲 $2.86bn ⬇ Down 12% in 2000

Plans have been announced to almost halve Norway's conscript army, which traditionally has absorbed most of the defense budget. Norway joined NATO in 1949. The overriding defense issue is the stability of Russia and the security of their common border. Five Russian diplomats were expelled in 1998 after a double agent revealed that Russia had extensive information on Norwegian defenses and oil industry.

NORWEGIAN ARMED FORCES

🛡	170 main battle tanks (*Leopard*)	14,700 personnel
🚢	6 submarines, 3 frigates, and 14 patrol boats	6100 personnel
✈	61 combat aircraft (57 F-16A/B)	5000 personnel
	None	

ECONOMICS ▷ Inflation 2.8% p.a. (1990–2000)

📊 $155bn 💲 8.819–8.969 Norwegian kroner

SCORE CARD

❏ WORLD GNP RANKING	26th
❏ GNP PER CAPITA	$34,530
❏ BALANCE OF PAYMENTS	$23bn
❏ INFLATION	3.1%
❏ UNEMPLOYMENT	3%

EXPORTS

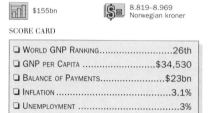

Germany 9%
Sweden 9%
France 10%
Netherlands 12%
UK 21%
Other 39%

IMPORTS

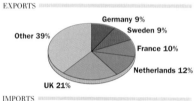

Denmark 7%
USA 7%
UK 9%
Germany 13%
Sweden 16%
Other 48%

ECONOMIC PERFORMANCE INDICATOR

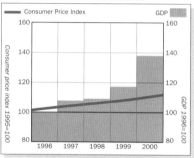

STRENGTHS

Western Europe's biggest producer and exporter of oil and natural gas. Mineral reserves. Hydroelectric power satisfies much of country's energy demands, allowing most oil to be exported. Soaring world oil prices in 2000. Petroleum fund for current profits to provide for future generations. Large merchant shipping fleet. Low inflation and unemployment compared with rest of Europe.

WEAKNESSES

Overdependence on oil revenue. Small home market and relatively remote location. Shortage of skilled labor. Harsh climate limits agriculture.

PROFILE

The state is interventionist by nature. In 1991, it stepped in to rescue most of the main commercial banks, which had been hit by bad loans. It began returning them to the private sector in 1994. The state also manages the distribution of offshore oil and gas licenses, and owns 45% of the Norsk Hydro conglomerate. In 2000 the state sold a 20% share in its oil and telecommunications companies.

Norway's immediate future prosperity is guaranteed by its offshore sector. There is a shortage of skilled labor, partly eased by the arrival of workers from other Scandinavian countries, but creating upward pressure on wages. Continuing the strong regional policy of redirecting resources from the more prosperous south to the isolated north is likely to remain a priority, both for social and strategic reasons.

NORWAY : MAJOR BUSINESSES

Hammerfest
Tromso
Bodø
Trondheim
Höyanger
Bergen
Stavanger
Kristiansand
Oslo
Larvik

Aluminum smelting
Electrometallurgy
Pulp & paper
Fish processing
Copper mining
Shipbuilding
Textiles
Oil refining

0   200 km
0   200 miles

RESOURCES
 Electric power 29.3m kw

- 3.09m tonnes
- 3.41m b/d (reserves 9.4bn barrels)
- 2.4m sheep, 979,800 cattle, 3.2m chickens
- Oil, natural gas, iron, coal, copper, lead, zinc

ELECTRICITY GENERATION
- Hydro 99% (116bn kwh)
- Combustion 1% (0.6bn kwh)
- Nuclear 0%
- Other 0%

% of total generation by type

Norway is Europe's largest oil producer; it also has sizable gas reserves. Most of Norway's electricity is produced by hydropower. In summer, the HEP surplus is exported. Fish and forestry are traditionally significant sectors – salmon farming, managed with particular efficiency, has grown rapidly – although with agriculture, they account for only 2.5% of GDP and 5% of the workforce.

A March 2002 report warned that cod stocks in the North Sea were under threat from oil-drilling waste products.

NORWAY : LAND USE
- Cropland
- Pasture
- Forest
- High mountain regions
- Tundra
- Cereals
- Sheep

0   200 km
0   200 miles

ENVIRONMENT
 Sustainability rank: 2nd

- 31%
- 7.6 tonnes per capita

ENVIRONMENTAL TREATIES
Yes / Yes / Yes / Yes / Yes / Yes

In 1986 northern Norway suffered radioactive contamination after the Chernobyl nuclear disaster. Norway has a tax on carbon dioxide emissions and was instrumental in securing agreement on the 1997 Kyoto Protocol on greenhouse gas emissions. In 1993, it lifted a ban on fishing minke whales, and in 2001 it allowed the export of whale products. It was also criticized in 2001 for a cull of (endangered) gray wolves and for plans to develop a coalfield on Svalbard.

MEDIA
 TV ownership high

Daily newspaper circulation 583 per 1000 people

PUBLISHING AND BROADCAST MEDIA
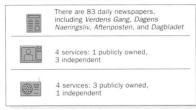
There are 83 daily newspapers, including *Verdens Gang*, *Dagens Naeringsliv*, *Aftenposten*, and *Dagbladet*

4 services: 1 publicly owned, 3 independent

4 services: 3 publicly owned, 1 independent

The state broadcaster NRK enjoys comparatively generous funding and has a 38% share of viewers. Newspapers have the second-highest proportion of readers in the world.

CRIME
No death penalty

2643 prisoners   Down 3% in 1999

CRIME RATES
Murders
2 per 100,000 population

Rapes
12 per 100,000 population

Thefts
4577 per 100,000 population

Norway has low levels of crime, even by Scandinavian standards. Violent crime barely exists – the murder rate is one-fifth of that of Sweden, and there are considerably fewer assaults and robberies.

EDUCATION
School leaving age: 16

99%   184,063 students

THE EDUCATION SYSTEM
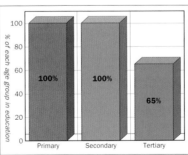
Primary 100%, Secondary 100%, Tertiary 65%

The period of compulsory schooling was increased from nine to ten years, with effect from the school year 1997/1998. Most schools are run by municipalities. There are four universities; specialized colleges include the Nordic College of Fisheries. Promotion of continuing education kept youth unemployment down during the early 1990s recession.

HEALTH
Welfare state health benefits

1 per 357 people   Heart and cerebro-vascular diseases, cancers, accidents

WHO ranks Norway as the world's third-best provider of health care. The country's infant mortality rate is among the world's lowest, and life expectancy at birth is one of the highest. Spending is among the highest of OECD states.

A new health plan announced in 2002 will place the high-spending hospitals under central control.

Telemedicine (online remote audio and image diagnosis) allows remote northern hospitals to obtain specialist consultations without having to send patients to the regional hospital.

SPENDING
GDP/cap. increase

CONSUMPTION AND SPENDING
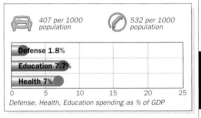
407 per 1000 population   532 per 1000 population
Defense 1.8%
Education 7.7%
Health 7%
Defense, Health, Education spending as % of GDP

In terms of income distribution, the Scandinavian countries are the most egalitarian in the world, and the richest 10% of Norway's population owns much less of the country's wealth than is the case in other developed countries. Homelessness and social deprivation are very rare. Recent refugees from the Bosnian conflict are the most disadvantaged group.

The discrepancy between men's and women's pay is greater than in either Sweden or Finland, although still well below the European average. Social provision has been maintained even through economic recession. Benefits are generous.

WORLD RANKING
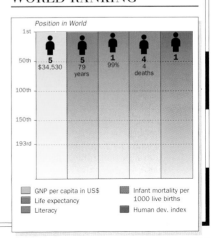
Position in World
- 5 $34,530
- 5 79 years
- 1 99%
- 4 4 deaths
- 1

- GNP per capita in US$
- Life expectancy
- Literacy
- Infant mortality per 1000 live births
- Human dev. index

N

OMAN

OFFICIAL NAME: Sultanate of Oman   **CAPITAL:** Muscat
POPULATION: 2.6 million   **CURRENCY:** Omani rial   **OFFICIAL LANGUAGE:** Arabic

  1951   1951   Nov 18   OM   +4   +968   .om

O CCUPYING THE EASTERN corner of the Arabian peninsula, Oman commands a strategic position at the entrance to the Persian Gulf. It is the least developed of the Gulf states. The most densely populated areas are the northern coast and the southern Salalah plain. Oil exports have given Oman modest prosperity under a paternalistic sultan. A Marxist-led insurgency supported by southern Dhofaris was defeated in the 1970s.

CLIMATE

▷ Hot desert

WEATHER CHART FOR OMAN

In the north temperatures often climb above 45°C (113°F) in summer. The south has a monsoon climate.

TRANSPORTATION

▷ Drive on right

Seeb International, Muscat
2.72m passengers

20 ships
15,000 grt

THE TRANSPORTATION NETWORK

9840 km (6114 miles)	550 km (342 miles)
None	None

Northern cities are well served by good roads, but some places, in the south particularly, are best reached by air.

TOURISM

▷ Visitors : Population 1:5.2

502,000 visitors

Down 19% in 1999

MAIN TOURIST ARRIVALS

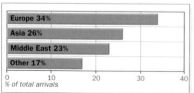

Europe 34%			
Asia 26%			
Middle East 23%			
Other 17%			

0   10   20   30   40
% of total arrivals

Until the late 1980s, Oman was closed to all but business or official visitors. The sultanate's rich cultural heritage, fine beaches, and luxury hotels are now enjoyed by thousands of Western visitors each year.

PEOPLE

▷ Pop. density low

Arabic, Baluchi

12/km² (32/mi²)

THE URBAN/RURAL POPULATION SPLIT

84%   16%

RELIGIOUS PERSUASION

Other Muslim and Hindu 25%

Ibadi Muslim 75%

Native Omanis, who include Arab refugees who fled Zanzibar in the 1960s, make up three-quarters of the population. Baluchis are the largest foreign grouping. Expatriates pose no threat to the regime and Westerners enjoy considerable freedom. Urban drift has taken place, and most Omanis now live in cities. Oman has a number of distinct minorities; the most numerous are the Jebalis in Dhofar – nomadic herdsmen who speak a language resembling Ethiopian. Many Dhofaris supported the Marxist-led insurgents in the 1970s, but they are now considered loyal. Most Omanis are Ibadi Muslims who follow an appointed leader, the imam. Ibadism does not oppose freedom for women, and a few enjoy positions of authority; two were elected to the Consultative Council in September 2000.

POLITICS

▷ No legislative elections

2000/2003

Sultan Qaboos bin Said

LEGISLATIVE OR ADVISORY BODIES

Consultative Council 82 seats

There are no political parties. The members of the Consultative Council (Majlis ash-shoura) were directly elected for the first time in 2000 by electoral committees in each province, and included two women.

Sultan Qaboos is an authoritarian but paternalistic monarch, whose dynasty traces its roots to the 18th century. As well as being head of state, the sultan is prime minister and minister for foreign affairs, defense, and finance. Family members hold other key positions. The regime faces no serious challenge, although Qaboos keeps a careful eye on the religious right wing. In 1991, he created the Consultative Council (Majlis ash-shoura), which gives a semblance of democracy. From 2000 its members were directly elected by provincial committees, rather than being appointed. Major political issues include the planned privatization of medium-sized government projects, and improving Oman's self-defense capability.

OMAN

Total Land Area : 212 460 sq. km
(82 031 sq. miles)

POPULATION

- over 50 000 ○
- over 10 000 ●
- under 10 000 ·

LAND HEIGHT

2000m/6562ft
1000m/3281ft
500m/1640ft
200m/656ft
Sea Level

WORLD AFFAIRS

 Joined UN in 1971

 AL   AMF   Damasc   GCC   OIC

Relations with Israel were cut off after the renewal of the Palestinian *intifada* in 2000. Although still basically pro-Western, and supportive of the US-led "war on terrorism," Oman has ties with Iran and calls for an easing of sanctions against Iraq.

A watchtower above an oasis. Most of Oman is gravelly desert. The only large area of cultivation is the 20-km-wide Al Batinah plain.

CHRONOLOGY

The present Albusaidi dynasty has ruled in Oman since 1749.

- ❑ **1932** Sultan bin Taimur in power.
- ❑ **1951** Sovereignty recognized by UK.
- ❑ **1970** Sultan Qaboos bin Said seizes power from his father.
- ❑ **1975** Suppression of Dhofar revolt.
- ❑ **1991** Consultative Council set up.
- ❑ **2000** Consultative Council members elected for first time.

AID

 Recipient

 $46m (receipts)   Up 15% in 2000

Aid used to come mainly from the West, and particularly the US, but Arab donors are now of greater significance. Japan is also a major contributor. Oman makes occasional donations of its own to Arab and Muslim causes.

EDUCATION

 Schooling is not compulsory

72%   16,032 students

Education has improved, but rural illiteracy is still high. Between 1996 and 2000, over 200 new schools were built.

DEFENSE

No compulsory military service

$1.73bn   Up 6% in 2000

The UK is the main supplier of equipment. In the 1991 Gulf War, Oman provided communications and services to US and UK forces. Oman's Defense Council, established in 1996, has replenished tanks, ships, and aircraft in recent years. Baluchi mercenaries supplement army strength.

HEALTH

Welfare state health benefits

1 per 769 people   Heart and cerebrovascular diseases, accidents

There is a policy of replacing expatriate medical staff with Omani nationals. Rural areas are served by clinics.

ECONOMICS

 Inflation 0.4% p.a. (1991–2000)

$11bn   0.3851 Omani rials

SCORE CARD

- ❑ WORLD GNP RANKING...........................80th
- ❑ GNP PER CAPITA $4778
- ❑ BALANCE OF PAYMENTS....................$3.35bn
- ❑ INFLATION ....−1.1%
- ❑ UNEMPLOYMENT 5%

STRENGTHS

Oil industry, led by Royal Dutch/Shell. Has benefited from staying out of OPEC and selling oil at spot prices without quotas. Soaring world oil prices in 2000 signified recovery from 1986 collapse. Rich Indian Ocean coastal waters have potential for sizable fishing industry.

WEAKNESSES

Overdependence on oil (90% of GNP), with less than 20 years' known reserves. Services sector could be better developed. Foreign workers needed in all economic sectors.

EXPORTS

China 28%
Thailand 9%
UAE 9%
South Korea 15%
Japan 18%
Other 21%

IMPORTS

Saudi Arabia 4%
USA 4%
UK 9%
Other 36%
Japan 15%
UAE 32%

RESOURCES

 Electric power 2.1m kw

 108,819 tonnes

959,000 b/d (reserves 5.5bn barrels)

 980,000 goats, 290,000 cattle, 3.4m chickens

Oil, natural gas, copper, chromite, marble, gypsum

Limiting oil production to conserve resources ended in 1993 following a number of exploration successes, but reserve levels are now falling again.

ENVIRONMENT

 Sustainability rank: 120th

 16%   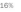 8.8 tonnes per capita

Overpumping of groundwater causes seawater to seep into traditional irrigation areas. Nature reserves and antihunting laws protect rich wildlife.

MEDIA

 TV ownership high

Daily newspaper circulation 28 per 1000 people

PUBLISHING AND BROADCAST MEDIA

There are 5 daily newspapers, including *Al-Watan*, *Oman Daily Newspaper*, and the English-language *Oman Daily Observer*

1 state-controlled service   2 state-controlled services

Nothing critical of the government may be published in Oman, despite a 1984 law allowing for "freedom of opinion."

CRIME

 Death penalty in use

 Oman does not publish prison figures   Up 18% in 1999

Reckless driving by young Omani males is a problem. A "flying court" serves remote communities.

SPENDING

GDP/cap. increase

CONSUMPTION AND SPENDING

97 per 1000 population   89 per 1000 population

Defense 10%
Education 3.9%
Health 2.9%

0   5   10   15   20   25

Defense, Health, Education spending as % of GDP

Omanis in urban areas enjoy the same high living standards that are to be found in other Gulf states. Hunting trips to Pakistan are popular among the rich Omani elite, and a *khanjar*, a curved dagger, is seen as a status symbol.

WORLD RANKING

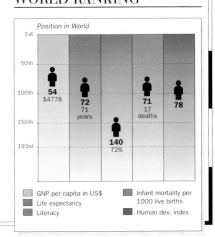

Position in World

1st				
50th				
100th	54 $4778	72 71 years		71 17 deaths	78
150th			140 72%	
193rd				

GNP per capita in US$
Life expectancy
Literacy
Infant mortality per 1000 live births
Human dev. index

O

PAKISTAN

OFFICIAL NAME: Islamic Republic of Pakistan **CAPITAL:** Islamabad
POPULATION: 145 million **CURRENCY:** Pakistani rupee **OFFICIAL LANGUAGE:** Urdu

  1947 1971 March 23 PK +5 +92 .pk

ONCE A PART OF BRITISH INDIA, Pakistan was created in 1947 in response to the demand for an independent and predominantly Muslim Indian state. Initially the new nation included East Pakistan, present-day Bangladesh, which seceded from Pakistan in 1971. Eastern and southern Pakistan, the flood plain of the Indus River, is highly fertile and produces cotton, the basis of the large textile industry.

Barren landscape in Kachhi, Baluchistan.
This area of Pakistan has some of the highest May-to-September temperatures in the world.

CLIMATE ▷ Mountain/steppe/ hot desert

WEATHER CHART FOR ISLAMABAD

Temperatures can soar to 50°C (122°F) in Sindh and Baluchistan and fall to –20°C (–4°F) in the northern mountains.

TRANSPORTATION ▷ Drive on left

Karachi International
4.94m passengers

57 ships
401,200 grt

THE TRANSPORTATION NETWORK

141,252 km (87,770 miles)	339 km (211 miles)
7792 km (4842 miles)	None

Most roads are poorly maintained. A modern highway linking Islamabad and Lahore was inaugurated in 1997.

TOURISM ▷ Visitors : Population 1:267

543,000 visitors
Up 26% in 2000

MAIN TOURIST ARRIVALS

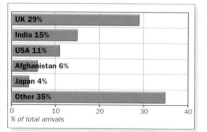

UK 29%	
India 15%	
USA 11%	
Afghanistan 6%	
Japan 4%	
Other 35%	

0   10   20   30   40
% of total arrivals

Relatively few tourists visit Pakistan, despite its rich cultural heritage and unspoiled natural beauty.

PEOPLE ▷ Pop. density medium

Punjabi, Sindhi, Pashtu, Urdu, Baluchi, Brahui
188/km² (487/mi²)

THE URBAN/RURAL POPULATION SPLIT

 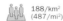

37%   63%

RELIGIOUS PERSUASION

Christian 1% — Hindu 2%
Shi'a Muslim 20%
Sunni Muslim 77%

ETHNIC MAKEUP

Baluch 5% — Other 7%
Mohajir 8%
Pashtun (Pathan) 15%
Punjabi 50%
Sindhi 15%

Punjabis account for 50% of the population, while Sindhis, Pathans (Pashtun), and Baluch are also prominent. *Mohajirs* – Urdu-speaking immigrants from pre-partition India – predominate in Karachi and Hyderabad. Punjabi political and military dominance of the centralized state has spawned many separatist and autonomy movements. Pathans have frequently threatened to establish a homeland with ethnic kin in Afghanistan. Tensions between the Baluch and Pashtun refugees from Afghanistan sporadically erupt into violence, as do those between native Sindhis and immigrant *mohajirs*.

The gap between rich and poor, for example between the feudal landowning class and their serfs, is considerable. There is an expanding middle class of small-scale traders and manufacturers.

Recent years have witnessed a marked increase in Islamist militancy, accompanied by growing discrimination against religious minorities. After the 1999 coup, the Musharraf regime trod a fine line in trying to avoid conflict

POPULATION AGE BREAKDOWN

Female	Age	Male
0.2%	80+	0.2%
2.7%	60–79	2.9%
7.4%	40–59	8.2%
13.2%	20–39	14.4%
24.9%	0–19	25.9%

% of population by age group

with Islamic militants, both over issues such as the strict application of Islamic *sharia*, and over foreign policy.

The extended family is an enduring institution, and ties between its members are strong, reflected in the dynastic and nepotistic nature of the political system. Although some women hold prominent positions, and Benazir Bhutto has twice been prime minister, relatively few are allowed to work by their religiously conservative menfolk. Pakistan has one of the world's lowest ratios of females to males, implying widespread neglect and some female infanticide.

Amnesty International criticized Pakistan in 2000 for its failure to give women's rights sufficient protection. Women's rights groups are mainly based in cities, and have made little overall impact.

POLITICS

No legislative elections

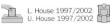
L. House 1997/2002　President Pervez
U. House 1997/2002　Musharraf

AT THE LAST ELECTION
National Assembly (dissolved)　　5%　4%　13%
237 seats　　　　　　　　　MQM(A)　N-M　Others

58%　　　　8%　8%　4%
PML　　　　W　PPP　ANP

PML = Pakistan Muslim League　**W** = Women
PPP = Pakistan People's Party　**MQM(A)** = Mohajir Qaumi
Movement (A) Haq Parast Group　**ANP** = Awami National
Party　**N-M** = Non-Muslim minorities
Others include Baluchistan National Party, Jamiat-Ulema-e-
Pakistan, Pakistan People's Party Jamhoori Watan Party
(Republican Nation Party), National People's Party, and
independents

Senate (dissolved)　87 seats

The Senate is indirectly elected

The military suspended multiparty democracy in 1999, and the National Assembly was dissolved in June 2001.

PROFILE
Throughout the 1990s fragile coalitions had to rule in cooperation with the president and the army, and were hampered by a large bureaucracy. The current military regime, which removed Nawaz Sharif in 1999, remains committed to restoring democracy by 2003, and faces pressure from all opposition parties to accelerate the transition. In June 2001 Gen. Pervez Musharraf appointed himself president, and in May 2002 he was endorsed for a further five years following a controversial referendum.

The National Security Council (NSC) acts as a cabinet under Musharraf. The popularity of the 1999 coup indicated the loss of respect for corrupt politicians and the country's much-abused democratic institutions. The ousted prime minister Nawaz Sharif was found guilty of treason in 2000.

MAIN POLITICAL ISSUE
Militant Islam
Under pressure to curb the ambitions of radical Islamists in Kashmir, as in Afghanistan, the Musharraf regime treads a tightrope in domestic politics, where militant Islam has a powerful grip on the urban poor. There was a fresh upsurge in 2002 in Sunni extremist violence against the Shi'a minority.

Benazir Bhutto, PPP leader and former prime minister.　*Pervez Musharraf, military ruler and self-styled president.*

WORLD AFFAIRS

Joined UN in 1947

 IAEA　 ECO　 NAM　 OIC　 SAARC

In 1998, Pakistan carried out a series of nuclear tests in response to similar tests by India, provoking international condemnation and three years of US sanctions. The last of three wars between India and Pakistan was three decades ago, but their dispute over Kashmir has repeatedly stoked tensions. In late 2001–early 2002, a series of attacks by Muslim Kashmiri separatists, allegedly supported by Pakistan, pushed the two countries again to the brink of war.

One of only three countries to recognize the Afghan *taliban* regime, Pakistan in September 2001 urged the *taliban* to hand over terrorism suspect Osama bin Laden, and then backed US-led action against the regime.

Pakistan's membership of the Commonwealth was suspended after the 1999 coup.

P

PAKISTAN

Total Land Area :
803 940 sq. km
(310 401 sq. miles)

LAND HEIGHT

6000m/19 686ft
4000m/13 124ft
3000m/9843ft
2000m/6562ft
1000m/3281ft
500m/1640ft
200m/656ft
Sea Level

POPULATION

over 5 000 000
over 1 000 000
over 500 000
over 100 000
over 50 000
over 10 000

0　　200 km
0　　200 miles

Rice paddy fields, with monsoon rains threatening from the Himalaya mountains. Rice is the second most valuable agricultural export after cotton.

CHRONOLOGY
From the 8th to the 16th centuries, Islamic rule extended to northwest and northeast India. Punjab and Sindh, annexed by the British East India Company in the 1850s, were ceded to the British Raj in 1857.

❏ **1906** Muslim League founded as organ of Indian Muslim separatism.
❏ **1947** Partition of India. Pakistan divided by 1600 km (994 miles) of Indian territory into East and West Pakistan. Millions displaced by large-scale migration. Muhammad Ali Jinnah first governor-general.
❏ **1948** First India-Pakistan war over Kashmir.
❏ **1949** New Awami League (AL) demands East Pakistan's autonomy.
❏ **1956** Constitution establishes Pakistan as an Islamic republic.
❏ **1958** Martial law. Gen. Muhammad Ayub Khan takes over; elected president two years later.

CHRONOLOGY *continued*

- ❏ **1965** Second India–Pakistan war over Kashmir.
- ❏ **1970** Ayub Khan resigns. Gen. Agha Yahya Khan takes over. First direct elections won by AL; West Pakistani parties reject results. War with India over East Pakistan.
- ❏ **1971** East Pakistan secedes as Bangladesh. PPP leader Zulfikar Ali Bhutto president.
- ❏ **1972** Simla (peace) Agreement with India.
- ❏ **1973** Bhutto, now prime minister, initiates Islamic socialism.
- ❏ **1977** General election. Riots over allegations of vote rigging. Gen. Zia ul-Haq stages military coup.
- ❏ **1979** Bhutto executed.
- ❏ **1986** Bhutto's daughter Benazir returns from exile to lead PPP.
- ❏ **1988** Zia killed in air crash. PPP wins general election.
- ❏ **1990** Ethnic violence in Sindh. President dismisses Benazir Bhutto. Nawaz Sharif becomes premier.
- ❏ **1991** Muslim *sharia* incorporated in legal code.
- ❏ **1992** Violence between Sindhis and *mohajirs* escalates in Sindh.
- ❏ **1993** President Khan and Prime Minister Sharif resign. Elections; Bhutto returns to power.
- ❏ **1996** President dismisses Bhutto.
- ❏ **1997** PML wins landslide election victory; Sharif prime minister.
- ❏ **1998** Nuclear tests.
- ❏ **1999–2000** Military coup. Sharif found guilty of treason.
- ❏ **2001** Parliament suspended, Gen. Musharraf appoints himself president. Pakistan key ally in US-led "war on terrorism."
- ❏ **2002** Controversial referendum endorses Musharraf as president. US and French nationals killed in terrorist attacks in Islamabad and Karachi. Threat of war with India over Kashmir.

AID

▷ Recipient

 $703m (receipts)   Down 4% in 2000

Pakistan is heavily dependent on aid, although the government has a long history of misdirecting aid payments. Aid intended for major projects has regularly been used to fund the current-account deficit. In mid-1998 the IMF agreed to help Pakistan meet its international debt obligations after the US and other Western aid donors cut off aid in protest against Pakistan's nuclear tests. The US subsequently agreed to resume aid, and boosted its support in late 2001 to reward the regime for supporting the "war on terrorism." Japan and Germany are among other main bilateral donors.

DEFENSE

▷ No compulsory military service

 $3.58bn   Up 2% in 2000

Pakistan has emerged as a significant regional arms trader. It established itself as a nuclear power by conducting a number of successful nuclear tests in 1998. Defense spending has a high priority, accounting for about a quarter of all government expenditure. The US, once the most important arms supplier, in 2001 lifted the sanctions imposed in 1990 and 1998. Pakistan's other main defense procurements are from France, the UK, and China. The army has been highly significant in politics throughout the period since independence, even when not actually in power in a military regime. Two years prior to the

PAKISTANI ARMED FORCES

	2300 main battle tanks (PRC Type-59/69/85, M-47/48A5, T-54/55/80)	555,000 personnel
	10 submarines, 8 frigates, and 9 patrol boats	25,000 personnel
	353 combat aircraft (52 *Mirage* 5, 42 Q-5)	45,000 personnel
	Capability undisclosed; weapons tested in 1998	

1999 coup, it had already assumed a formal role in civilian decision-making by its inclusion in a National Security Council.

ECONOMICS

▷ Inflation 10% p.a. (1990–2000)

$61bn 57.6–59.9 Pakistani rupees

SCORE CARD

- ❏ WORLD GNP RANKING...........................45th
- ❏ GNP PER CAPITA $440
- ❏ BALANCE OF PAYMENTS.................–$2.21bn
- ❏ INFLATION 4.4%
- ❏ UNEMPLOYMENT 6%

ECONOMIC PERFORMANCE INDICATOR

EXPORTS

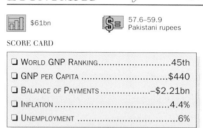

UAE 6%
Germany 6%
Hong Kong 6%
UK 7%
Other 50%
USA 25%

IMPORTS

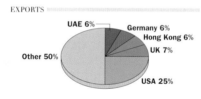

USA 6%
Japan 6%
UAE 11%
Saudi Arabia 11%
Other 54%
Kuwait 12%

STRENGTHS

Gas, water, coal, oil. Substantial untapped natural resources. Low labor costs. Potentially huge market. One of the world's leading producers of cotton and a major exporter of rice.

WEAKNESSES

Production and sales of cotton and rice vulnerable to weather conditions. History of inefficient and haphazard government economic policies. Weak and overstretched infrastructure.

PROFILE

Pakistan has yet to show progress in tackling its considerable economic problems. Although successive governments have reversed the nationalization policies instituted in the 1970s, private enterprise has been stifled by the rules of a massive

bureaucracy. There is some foreign investment in previously state-only sectors such as banking, and water and other utilities. However, corruption at all levels of government undermined economic confidence throughout the 1990s, and it was particularly acute under the administration of Benazir Bhutto. Efforts by the military government to tackle corruption and poverty were praised by the World Bank in 2001. Defense spending remains high.

PAKISTAN : MAJOR BUSINESSES

Light engineering Carpet weaving
Chemicals Electronics
Vehicle assembly Textiles
Shipbuilding Leather tanning
Food processing
Tobacco
Steel

* significant multinational ownership

RESOURCES

▷ Electric power 15.7m kw

674,606 tonnes

55,258 b/d (reserves 222m barrels)

49.1m goats, 24.2m sheep, 155m chickens

Oil, natural gas, coal, limestone, salt, gypsum, silica sand

ELECTRICITY GENERATION

- Hydro 35% (22bn kwh)
- Combustion 64% (40bn kwh)
- Nuclear 1% (0.4bn kwh)
- Other 0%

% of total generation by type

Apart from cotton and rice, Pakistan's major resources are oil, coal, gas, and water. The state hopes that the privatization of the utilities industries will reduce energy imports and shortages – peak electricity demand, for example, exceeds supply by 20%. Refining capacity is also well below present demand. Steps are being taken to attract more foreign investment in oil and gas exploration, extraction, and distribution, but privatization plans went on hold in October 2001 because of the region's acute instability.

PAKISTAN : LAND USE

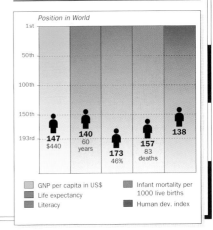

- Cropland
- Pasture
- Forest
- Desert
- Wetlands
- High mountain regions
- ↓ Sugarcane
- ⚥ Wheat
- 🐄 Cattle

ENVIRONMENT

▷ Sustainability rank: 112th

5% (3% partially protected)

0.7 tonnes per capita

ENVIRONMENTAL TREATIES

Yes	Yes	Yes
Yes	Yes	No

Tough measures are in force to curb illegal logging. Urban pollution affects many cities. Local groups increasingly voice environmental concerns.

MEDIA

▷ TV ownership medium

 Daily newspaper circulation 21 per 1000 people

PUBLISHING AND BROADCAST MEDIA

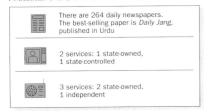

	There are 264 daily newspapers. The best-selling paper is *Daily Jang*, published in Urdu
	2 services: 1 state-owned, 1 state-controlled
	3 services: 2 state-owned, 1 independent

State-run services dominate the mass media. Journalists who challenge official views are systematically harassed.

CRIME

▷ Death penalty in use

78,938 prisoners

Down 4% in 1999

CRIME RATES

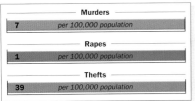

Murders	
7	*per 100,000 population*

Rapes	
1	*per 100,000 population*

Thefts	
39	*per 100,000 population*

Compared with similar Islamic states, rates of murder, kidnapping, narcotics trafficking, rape, and robbery are high. Corruption and the abuse of women are

EDUCATION

▷ Schooling is not compulsory

 46%

1.14m students

THE EDUCATION SYSTEM

% of each age group in education

- Primary: 86%
- Secondary: 37%
- Tertiary: 3%

Although universal free primary education is a constitutional right, it is not compulsory. Literacy rates are among the lowest in the world. The education system is heavily Islamized, and weighted toward educating males; a large majority of children enrolled in primary schools are boys.

There are 23 universities, 675 arts and sciences colleges, and 99 professional colleges, all of which have a heavy preponderance of arts students. Wealthy parents frequently choose to send their children abroad for higher education, mainly to colleges in the UK or the US.

major causes for concern; reports of deaths or death threats for refusing to accept arranged marriages are rising. Torture and rape of prisoners and deaths in custody are frequent. The most dangerous area is Sindh: Karachi is terrorized by severe factional violence. Militant sectarian groups are also blamed for a recent rise in crime in Punjab. Pressure from Islamic parties has forced the government of North West Frontier Province to replace British-based civil law with the rulings of *sharia* courts. Special part-military courts were established in 2002 to combat terrorism and other "serious" crimes.

HEALTH

▷ No welfare state health benefits

 1 per 1667 people

 Malaria, tuberculosis, diarrheal diseases

Availability of doctors and hospital beds is low, and there is a shortage of equipment and medicines. Uncontrolled counterfeit drugs are common. A specialized cancer hospital in Lahore, opened in 1995, offers modern facilities and advanced treatment. Pakistan has a high incidence of heroin addicts, due largely to its proximity to Afghanistan.

SPENDING

▷ GDP/cap. increase

CONSUMPTION AND SPENDING

5 per 1000 population

22 per 1000 population

- Defense 5.8%
- Education 2.7%
- Health 0.7%

Defense, Health, Education spending as % of GDP

Members of the bureaucratic and political elite tend to be extremely rich, as are some of the top military. Despite Pakistan's considerable economic potential, many people live below the poverty line.

WORLD RANKING

Position in World

- 147 $440
- 140 60 years
- 173 46%
- 157 83 deaths
- 138

Legend:
- GNP per capita in US$
- Life expectancy
- Literacy
- Infant mortality per 1000 live births
- Human dev. index

P

PALAU

PACIFIC OCEAN

OFFICIAL NAME: Republic of Palau CAPITAL: Koror
POPULATION: 19,100 CURRENCY: US dollar OFFICIAL LANGUAGES: Palauan and English

| 1994 | 1994 | Oct 1 | PAL | +9 | +680 | .pw |

THE REPUBLIC OF PALAU (locally known as Belau) is situated in the western Pacific and comprises more than 300 islands in the Caroline Islands archipelago, only nine of which are inhabited. Formerly a part of the US-administered Trust Territory of the Pacific Islands, Palau became independent in association with the US in 1994, but continues to be heavily dependent on US aid.

CLIMATE ▷ Tropical oceanic

WEATHER CHART FOR KOROR

The islands are humid, with fairly constant temperatures and heavy rainfall all year round. The mean temperature is 27°C (81°F).

TRANSPORTATION ▷ Drive on right

Palau International, Koror   Has no fleet

THE TRANSPORTATION NETWORK

| 36 km (22 miles) | None |
| None | None |

There are limited air and sea links between islands. A new bridge, opened in 2002, connects Koror to Babelthuap.

TOURISM ▷ Visitors : Population 2.9:1

55,000 visitors   Down 14% in 1999

MAIN TOURIST ARRIVALS

Japan	40%
Taiwan	20%
Guam	10%
Other	30%

0   10   20   30   40
% of total arrivals

Tourism is becoming more important. Improved transportation access is planned, amid concerns about impact on traditional culture. Several islands have battle sites from the Pacific War. The outlying islands remain unspoiled.

PEOPLE ▷ Pop. density low

Palauan, English, Japanese, Angaur, Tobi, Sonsorolese   38/km² (98/mi²)

THE URBAN/RURAL POPULATION SPLIT

70%   30%

RELIGIOUS PERSUASION

Modekngei 34%
Christian 66%

Palauans can claim a diverse genetic heritage. Palau was first colonized by southeast Asian peoples some 3000 years ago. There has also been mixing with other Melanesian and Polynesian peoples. More recently, low-skilled Filipinos have added to the population. Immigration from south Asia was banned in 2001 due to rising tensions.

Some 70% of Palauans live on the island city of Koror (a new capital is being constructed on neighboring Babelthuap island). The thinly scattered remaining population are linguistically diverse, with distinct separate languages in the most southern islands. Cultural influence from the US and Japan has been strong, though in the more remote islands a more traditional way of life has been maintained. Society remains largely matrilineal. The indigenous Modekngei religion is a unique blend of traditional beliefs and Christianity.

Palau's islands have many idyllic beaches, but a lack of resources means that tourism remains underdeveloped.

POLITICS ▷ Nonparty elections

L. House 2000/2004
U. House 2000/2004
President Tommy Remengesau

AT THE LAST ELECTION
House of Delegates 16 seats

One member is elected to the House of Delegates to represent each of the 16 states

Senate 14 seats

The 14 senators represent geographical districts, according to population

The Compact of Free Association with the US formed the basis of Palau's independence. One clause, requiring the transit and storage of nuclear materials, was the subject of several referendums until a simple majority in 1993 brought sovereignty the following year. Vice President Tommy Remengesau was elected president in November 2000.

WORLD AFFAIRS ▷ Joined UN in 1994

| PC | PIF | IBRD | ACP | IWC |

The US has exclusive control over Palau's foreign affairs and defense policies under the 1994 conditions of the Compact of Free Association. Cordial relations with the Pacific Islands Forum were restored in 1999 after tensions caused by Palau's bid to give Japan the right to veto the establishment of a whale sanctuary in the South Pacific.

AID ▷ Recipient

$39m (receipts)   Up 34% in 2000

Palau is heavily dependent on aid, which accounts for around 30% of GDP. Under the 15-year Compact of Free Association, signed with the US in 1994, Palau is set to receive up to $700 million in return for the use of military facilities. However, in recent years Japan has overtaken the US as the country's major donor.

DEFENSE ▷ No compulsory military service

There are no armed forces   Not applicable

Under the 1994 Compact of Free Association, the US is responsible for Palau's defense.

ECONOMICS

 Inflation 4% p.a. (1994–2000)

 $82m

$≣ Currency is US dollar

SCORE CARD

- ❏ WORLD GNP RANKING.........................189th
- ❏ GNP PER CAPITA $5000
- ❏ BALANCE OF PAYMENTS.....................$17.2bn
- ❏ INFLATION 3%
- ❏ UNEMPLOYMENT 2%

STRENGTHS

Relationship with aid donors – US and Japan – provides economic stability and access to lucrative markets. Improved transportation infrastructure. Tourism industry growing. Fishing and copra production important. Increasing regional trade. Trust funds established from Compact money.

RESOURCES

 Electric power 62,000 kw

 1801 tonnes

Not an oil producer

Not available

Gold

On some islands the soil is highly fertile, although the terrain of the larger islands makes farming difficult. Some islands are densely forested. Palau has copra and some gold deposits. There is also the possibility of exploitation of reserves of minerals on the seabed. Palau has a small fishing industry with the potential for development.

PALAU

Total Land Area : 458 sq. km
(177 sq. miles)

EXPORTS

Palau does not publish export figures by country

IMPORTS

Taiwan 5%
Other 11%
USA 40%
Japan 13%
Singapore 13%
Guam 18%

WEAKNESSES

Heavy dependence on aid. Remote location. Underemployment. Poor transportation links between islands and to other countries. Few resources.

ENVIRONMENT

 Not available

None

13.9 tonnes per capita

Palau suffers from inadequate facilities for the disposal of solid waste. Sand and coral dredging and illegal fishing practices pose a significant threat to the marine ecosystem. Palau and its surrounding waters are Micronesia's richest habitat; however, there is concern about the commercial export of fruit bats to neighboring islands as a delicacy. Typhoons sometimes cause severe damage to infrastructure.

MEDIA

TV ownership high

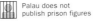 There are no daily newspapers

PUBLISHING AND BROADCAST MEDIA

There are no daily newspapers. *Tia Belau* is published biweekly in English and Palauan.

2 services:
1 state-owned,
1 limited
independent

3 radio stations

The country's TV and radio stations tend to deal in material which is largely derived from the US.

CRIME

No death penalty

Palau does not publish prison figures

Little change from year to year

There is a little alcohol-related crime, but much of the country, particularly the outlying islands, is crime-free.

EDUCATION

School leaving age: 14

92%

305 students

Elementary education is compulsory between the ages of six and 14. The Micronesian Occupational College, based in Palau, provides two-year training programs.

CHRONOLOGY

The Caroline Islands were colonized in turn by Spain, Germany, and Japan, coming under US control in 1945.

- ❏ **1947** UN Trust Territory of the Pacific Islands established.
- ❏ **1982** Compact of Free Association with US signed.
- ❏ **1993** Compact approved.
- ❏ **1994** Independence in free association with US.
- ❏ **2001** Tommy Remengesau president.

HEALTH

 Welfare state health benefits

1 per 83 people

Heart, cerebro-vascular, and intestinal diseases

Basic health care is available. Many outlying islands do not have easy access to qualified doctors and therefore often rely on nurses or traditional health remedies. An epidemic of mosquito-borne dengue fever hit Palau in 2000.

SPENDING

Not available

CONSUMPTION AND SPENDING

No data

19 per 1000 population

Defense No data

Education No data

Health 6%

0   5   10   15   20   25
Defense, Health, Education spending as % of GDP

The gap between rich and poor is growing steadily, as entrepreneurs and government officials exploit aid and develop the tourist industry. In 2001, a program providing cheap rental housing for low-income families received a US grant of $200,000.

WORLD RANKING

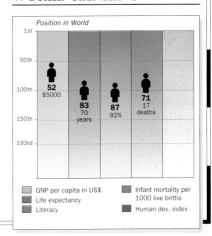

Position in World

52 $5000	83 70 years	87 92%	71 17 deaths

- GNP per capita in US$
- Life expectancy
- Literacy
- Infant mortality per 1000 live births
- Human dev. index

P

PANAMA

OFFICIAL NAME: Republic of Panama **CAPITAL:** Panama City
POPULATION: 2.9 million **CURRENCY:** Balboa **OFFICIAL LANGUAGE:** Spanish

PANAMA IS THE SOUTHERNMOST of the seven countries occupying the isthmus that joins North and South America. The rainforests of the southeastern Darien Peninsula are some of the wildest areas left in the Americas. Elected governments have held power since the US invasion of 1989. Panama's traditional economic strength is its banking sector. The US returned control of the Panama Canal Zone to Panama on 31 December 1999.

Cruise liner on the Panama Canal. The canal shortens the sea route between the east coast of the US and Japan by 4800 km (3000 miles).

CLIMATE ▷ Tropical wet & dry

WEATHER CHART FOR PANAMA CITY

Panama has a humid tropical climate; rainfall is twice as heavy on the Caribbean coast as on the Pacific coast.

TRANSPORTATION ▷ Drive on right

Tocumen International, Panama City
898,000 passengers

6143 ships
98.2m grt

THE TRANSPORTATION NETWORK

3783 km (2351 miles)	Pan-American Highway: 30 km (19 miles)
485 km (301 miles)	800 km (497 miles)

The 80-km (50-mile) Panama Canal cuts around two weeks off the voyage between the Atlantic and Pacific Oceans. Boats form the most convenient means of transportation for the coastal regions. Many roads are in disrepair.

TOURISM ▷ Visitors : Population 1:6.1

479,000 visitors

Up 5% in 2000

MAIN TOURIST ARRIVALS

USA 27%
Colombia 14%
Costa Rica 14%
Other 45%

0 10 20 30 40 50 60
% of total arrivals

Portobelo and Panama City have old Spanish colonial buildings. In 2000, new cruise-ship facilities opened in Colón.

PEOPLE ▷ Pop. density low

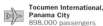 English Creole, Spanish, Amerindian languages, Chibchan

38/km² (99/mi²)

THE URBAN/RURAL POPULATION SPLIT

56% 44%

ETHNIC MAKEUP

Asian 4%
Other 2%
Amerindian 8%
Black 12%
White 14%
Mestizo 60%

The northwest coast has a large black community, mostly descended from African immigrants who worked the plantations. The majority speak English Creole rather than Spanish. About 8% of the population are Amerindians mainly from the Guaymies, Chocoes, Kunas, and Ngobe-Buglé tribes. Roman Catholicism and the extended family remain strong, although the canal and the former US military bases have given society a cosmopolitan outlook.

PANAMA

Total Land Area : 78 200 sq. km (30 193 sq. miles)

POPULATION
⊙ over 500 000
◎ over 100 000
○ over 50 000
● over 10 000
• under 10 000

LAND HEIGHT
2000m/6562ft
1000m/3281ft
500m/1640ft
200m/656ft
Sea Level

POLITICS ▷ Multiparty elections

 1999/2004

President Mireya Moscoso

AT THE LAST ELECTION

Legislative Assembly 71 seats

58% NN   34% UP   8% AO

NN = New Nation (comprises the Democratic Revolutionary Party (**PRD**), the Solidarity Party, the National Liberal Party, and the Papa Egoro Movement)
UP = Union for Panama (comprises the Arnulfisto Party (**PA**), MOLIRENA, the Party for Democratic Change, and MORENA)
AO = Action for the Opposition (comprises the Christian Democratic Party, the Liberal Party, the Popular Nationalist Party, and the Civil Renovation Party)

In 1989, the US invaded Panama and arrested its ruler, Gen. Manuel Noriega, for narcotics smuggling. US forces installed the compliant Endara government, criticized for corruption. The 1994 elections were won by Ernesto Pérez Balladares and the PRD, Noriega's old party, but the new government was largely pro-US, and its economic reforms attracted widespread discontent. A 1998 referendum denied Pérez Balladares a second consecutive term. In 1999 opposition leader Mireya Moscoso of the PA was elected as Panama's first woman president. She promised social justice for the poor, but has followed an IMF-backed program and seen her popularity decline.

P

WORLD AFFAIRS
▷ Joined UN in 1945

The Canal Zone reverted to Panama on December 31, 1999. After failing to reach agreement to allow some troops to stay on in Panama, the US, which had maintained 14 military bases there, has re-based its antinarcotics activities in Florida.

Panama has had observer status with the Andean Community since 1995.

AID
▷ Recipient

 $17m (receipts)   Up 13% in 2000

The IDB pledged $3.3 million in 2000 to rehabilitate a former US base; Japan has loaned $1 million for the project.

DEFENSE
▷ No compulsory military service

$127m   ⬇ Down 1% in 2000

The National Guard and defense forces were disbanded in 1990 following the 1989 US invasion. They were replaced by the Panamanian Public Force, numbering some 11,800 and comprising the National Police, the National Air Service, and the National Maritime Service.

ECONOMICS
▷ Inflation 1.9% p.a. (1990–2000)

 $9.31bn   1 balboa

SCORE CARD
- ❑ WORLD GNP RANKING..........................87th
- ❑ GNP PER CAPITA $3260
- ❑ BALANCE OF PAYMENTS...................–$927m
- ❑ INFLATION 1.4%
- ❑ UNEMPLOYMENT..................................13%

STRENGTHS
Colón Free Trade Zone second-largest in world. Strong banking, financial, insurance, and other allied services. Banana, shrimp exports. Merchant shipping payments for sailing under the Panamanian flag.

WEAKNESSES
History of political instability and corruption. Large foreign debt. High unemployment, underemployment. Poor infrastructure.

EXPORTS

Belgium–Luxembourg 5%
Italy 3%
Costa Rica 5%
Sweden 8%
USA 46%
Other 33%

IMPORTS

Costa Rica 4%
Mexico 4%
Japan 6%
Ecuador 7%
Other 46%
USA 33%

RESOURCES
▷ Electric power 1m kw

 123,734 tonnes   Not an oil producer; refines 100,000 b/d

 1.34m cattle, 277,900 pigs, 14.5m chickens   Copper, coal, gold, silver, manganese, salt, clay

The Petaquilla area, west of the canal, has great copper and gold potential. To reduce the country's dependence on oil imports, the government has stepped up hydroelectric production; four state energy plants were privatized in 1999. Tropical hardwoods are being cut down at an alarming rate.

ENVIRONMENT
▷ Sustainability rank: 17th

 19%   2.1 tonnes per capita

The destruction of rainforests is proceeding at an increasingly rapid rate, resulting in widespread soil erosion. Large numbers of rare bird and animal species are threatened. Sewage from Panama City and Colón is discharged directly into coastal waters, canals, and ditches. Stretches of mangrove swamps are cut down for urban development, shrimp farms, and resorts.

MEDIA
▷ TV ownership medium

 Daily newspaper circulation 62 per 1000 people

PUBLISHING AND BROADCAST MEDIA

 There are 8 daily newspapers, including *La Prensa* and *La Estrella de Panamá*

5 independent services   1 state-owned service, over 200 independent stations

A more independent press has flourished since Noriega's overthrow. Radio reaches the greatest number.

CRIME
▷ No death penalty

8517 prisoners   ⬇ Down 44% 1996–1998

Panama City and Colón have high crime levels. Money laundering, narcotics trafficking, and corruption are rife.

EDUCATION
▷ School leaving age: 15

 92%   84,452 students

Schooling is based on the US model. Provision for the urban poor, blacks, and indigenous people is limited.

CHRONOLOGY
On independence from Spain in 1821, Panama was incorporated into Gran Colombia. Panama gained independence from Colombia with US support in 1903.

- ❑ **1903** US buys concession for Panama Canal.
- ❑ **1914–1939** Canal opens to traffic. US protectorate status ended.
- ❑ **1968–1981** Rule of Col. Torrijos Herrera.
- ❑ **1989** Indicted narcotics trafficker Gen. Noriega annuls elections to retain power. US invasion.
- ❑ **1994** PRD wins presidency and is largest party in parliament.
- ❑ **1999** PA's Mireya Moscoso elected first woman president. December, control of Canal Zone reverts from US to Panama.

HEALTH
▷ Welfare state health benefits

 1 per 688 people   Heart disease, cancers, violence, accidents, tuberculosis

Primary health care is accessible to some two-thirds of the rural population. The isolation of many villages hinders efforts to improve the system.

SPENDING
▷ GDP/cap. increase

CONSUMPTION AND SPENDING

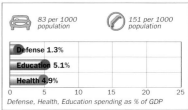

83 per 1000 population   151 per 1000 population

Defense 1.3%
Education 5.1%
Health 4.9%

0   5   10   15   20   25
Defense, Health, Education spending as % of GDP

Wealth disparities are large. Almost 40% of the population are estimated to live below the poverty line – clustered in the cities rather than in rural areas.

WORLD RANKING

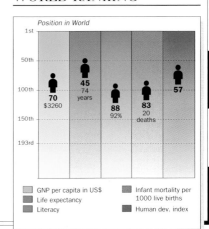

Position in World

1st
50th
100th
150th
193rd

70 $3260
45 74 years
88 92%
83 20 deaths
57

- GNP per capita in US$
- Life expectancy
- Literacy
- Infant mortality per 1000 live births
- Human dev. index

P

PAPUA NEW GUINEA

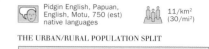

OFFICIAL NAME: Independent State of Papua New Guinea **CAPITAL:** Port Moresby
POPULATION: 5.2 million **CURRENCY:** Kina **OFFICIAL LANGUAGE:** English

 1975 1975 Sept 16 PNG +10 +675 .pg

THE MOST LINGUISTICALLY diverse country in the world, with approximately 750 languages, Papua New Guinea (PNG) achieved independence from Australia in 1975. The country occupies the eastern end of New Guinea, the world's third-largest island, and several other groups of islands. Much of the country is still isolated and much of the rural population experiences basic living conditions.

CLIMATE

▷ Tropical equatorial/monsoon

WEATHER CHART FOR PORT MORESBY

- Average daily temperature — Rainfall —

Unvaryingly hot lowlands contrast with snow on Mount Victoria. Severe weather followed El Niño of 1997–1998.

TRANSPORTATION

▷ Drive on left

Jacksons, Port Moresby
745,000 passengers

102 ships
61,000 grt

THE TRANSPORTATION NETWORK

686 km (426 miles)	None
None	10,940 km (6798 miles)

Infrastructure is improving with the construction and upgrading of major link roads, airports, and port facilities.

Papua New Guinea's 600 or so outer islands are mainly mountainous and volcanic, with lush vegetation and fringing coral reefs.

TOURISM

▷ Visitors : Population 1:90

 58,000 visitors Down 13% in 2000

MAIN TOURIST ARRIVALS

Australia 50%	
USA 8%	
New Zealand 6%	
Other 36%	

% of total arrivals

Tourism has great potential. However, it is hampered by the high rates of poverty-related violent crime, particularly in urban centers.

PEOPLE

▷ Pop. density low

Pidgin English, Papuan, English, Motu, 750 (est) native languages

11/km² (30/mi²)

THE URBAN/RURAL POPULATION SPLIT

17%   83%

RELIGIOUS PERSUASION

Other 3%
Roman Catholic 37%
Protestant 60%

PNG has an extraordinary diversity of peoples, with around 750 different languages and even more tribes. The key distinction is between the lowlanders, who have frequent contact with the outside world, and the very isolated highlanders. Highland tribes see all strangers as potentially hostile. Vendettas can last for generations and tribal battles are not infrequent. A majority of people are nominally Christian, but indigenous beliefs and practices are widespread.

POLITICS

▷ Multiparty elections

2002/2007   H.M. Queen Elizabeth II

AT THE LAST ELECTION
National Parliament 109 seats

5% PAP

17% NAP	16% Ind	11% PDM	7% PPP	6% V	6% PP	32% Others

NAP = National Alliance Party **Ind** = Independents
PDM = People's Democratic Movement
PPP = People's Progress Party **V** = Vacant
PP = Pangu Pati **PAP** = People's Action Party

The many parties lack clear ideological foundations, creating long-term political instability. The patronage required to maintain coalitions bred corruption. A PDM-led government worked to overhaul the system, and claimed the first simple majority in 2001. However, the party was overshadowed in the 2002 elections by the NAP. Veteran NAP leader Sir Michael Somare was appointed prime minister. A ten-year insurgency by separatists on Bougainville ended with the 1998 cease-fire. Autonomy has been promised, along with a future referendum on independence. Elsewhere, strong local traditions and communications problems have made centralization difficult.

PAPUA NEW GUINEA

Total Land Area :
462 840 sq. km
(178 703 sq. miles)

POPULATION
- ◎ over 100 000
- ○ over 50 000
- ● over 10 000
- • under 10 000

LAND HEIGHT
- 3000m/9843ft
- 2000m/6562ft
- 1000m/3281ft
- 500m/1640ft
- 200m/656ft
- Sea Level

0   200 km
0   200 miles

WORLD AFFAIRS

▷ Joined UN in 1975

APEC | Comm | NAM | PC | PIF

Accusations of PNG support for separatists in Papua, the neighboring Indonesian province, have strained relations. A pro-Taiwan policy, in return for aid funding, provoked Chinese anger in 1999.

AID

▷ Recipient

$275m (receipts) — Up 27% in 2000

Australia is the major aid donor. The World Bank in 2000 endorsed government reforms with a $90 million loan.

DEFENSE

▷ No compulsory military service

$55m — Up 20% in 2000

The army has expressed its doubts over political and economic reforms, most recently in a mutiny in March 2001.

ECONOMICS

▷ Inflation 7.9% p.a. (1990–2000)

 $3.61bn — 3.035–3.767 kina

SCORE CARD

❑ WORLD GNP RANKING	125th
❑ GNP PER CAPITA	$700
BALANCE OF PAYMENTS	–$8m
❑ INFLATION	15.6%
❑ UNEMPLOYMENT	8%

STRENGTHS

Significant mineral reserves of copper, gold, nickel, cobalt, oil, and natural gas. A proposed gas pipeline between the highlands and Australia is expected to net $219 million a year. Agriculture sustains the population.

WEAKNESSES

Agricultural production and mining were significantly disrupted by severe drought caused by El Niño of 1997–1998. Poor transportation and banking infrastructures. Political instability. Foreign exploitation of resources.

EXPORTS

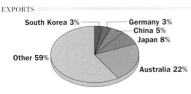

South Korea 3% — Germany 3% — China 5% — Japan 8%
Other 59%
Australia 22%

IMPORTS

New Zealand 4% — Japan 4%
Malaysia 3% — Other 19%
Australia 50%
Singapore 20%

RESOURCES

▷ Electric power 490,000 kw

53,763 tonnes — 57,000 b/d (reserves 200m barrels)

1.6m pigs, 88,000 cattle, 9000 ducks, 3.7m chickens — Copper, gold, silver, gas, oil, nickel, chromite, cobalt

PNG is rich in minerals. The Ok Tedi gold/copper mine in the Star Mountains is the most productive in the country; the Porgera gold mine is one of the world's largest. The severity of the 1998 drought significantly affected production at both. Prospecting has revealed extensive oil and natural gas reserves.

ENVIRONMENT

▷ Sustainability rank: 51st

None — 0.5 tonnes per capita

Deforestation and heavy-metal pollution are major issues. Cyanide poisoning from an Australian-owned mine in 2000 caused serious water pollution. Subduction of continental plates has forced the relocation of thousands of people from the low-lying islands.

MEDIA

▷ TV ownership low

Daily newspaper circulation 15 per 1000 people

PUBLISHING AND BROADCAST MEDIA

There are 3 daily newspapers, the *National*, *Papua New Guinea Post-Courier*, and the *Niugini Nius*.

2 independent services — 3 services: 1 state-owned, 2 independent

The dismissal of the head of the National Broadcasting Commission in 2001 raised concerns over the freedom of the media.

CRIME

▷ Death penalty not used in practice

3566 prisoners — Up 22% in 1992

Violent crime by gangs of "Rascals" is very common. A cultural tradition of vendettas persists in rural communities.

EDUCATION

▷ School leaving age: 16

64% — 14,755 students

Education is not compulsory. Equipment charges and fees have been introduced. Universities are suffering funding cuts.

HEALTH

▷ Welfare state health benefits

1 per 10,000 people — Malaria, pneumonia, diarrheal diseases, tuberculosis

The health system has suffered from recent cuts. HIV and tuberculosis co-infections are at crisis level. Life expectancy rates are among the lowest in the Pacific. Access to clean water and sanitation are major issues.

CHRONOLOGY

The British annexed the southeast and the Germans the northeast of the island of New Guinea in 1884.

- ❑ **1904** Australia takes over British sector; renamed Papua in 1906.
- ❑ **1914** German sector occupied by Australia.
- ❑ **1942–1945** Japanese occupation.
- ❑ **1964** National Parliament created.
- ❑ **1971** Renamed Papua New Guinea.
- ❑ **1975** Independence under Michael Somare, leader since 1972.
- ❑ **1988** Bougainville Revolutionary Army begins guerrilla campaign.
- ❑ **1997** El Niño causes severe drought and tsunamis. Prime Minister Julius Chan resigns over use of Western-led mercenaries in Bougainville.
- ❑ **2000** Loloata Understanding promises autonomy for Bougainville.
- ❑ **2001** PDM claims parliamentary majority. Final peace accord with Bougainville ratified after three-year cease-fire.
- ❑ **2002** NAP wins elections. Somare returns as prime minister.

SPENDING

▷ GDP/cap. increase

CONSUMPTION AND SPENDING

7 per 1000 population — 13 per 1000 population

Defense 1.2%
Education 4.7%
Health 2.5%

0 — 5 — 10 — 15 — 20 — 25
Defense, Health, Education spending as % of GDP

There is a growing gap between rich and poor, particularly in urban areas. Spending on education and health was cut in 1998, when the kina dropped in value, but it was increased again in 2000, and the 2002 budget proposed the reintroduction of free schooling.

WORLD RANKING

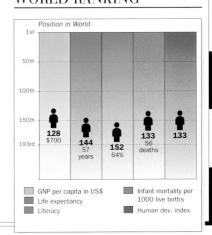

Position in World

1st	
50th	
100th	
150th	
193rd	

128 $700 — 144 57 years — 152 64% — 133 56 deaths — 133

GNP per capita in US$ — Infant mortality per 1000 live births
Life expectancy — Human dev. index
Literacy

P

PARAGUAY

SOUTH AMERICA

OFFICIAL NAME: Republic of Paraguay   **CAPITAL:** Asunción
POPULATION: 5.6 million   **CURRENCY:** Guaraní   **OFFICIAL LANGUAGE:** Spanish

| 1811 | 1938 | May 14 | PY | -4 | +595 | .py |

LANDLOCKED IN SOUTH America and a Spanish possession until 1811, Paraguay gained large tracts of land from Bolivia in 1938. From then until the overthrow in 1989 of Gen. Alfredo Stroessner, South America's longest-surviving dictator, it experienced periods of anarchy and military rule. The Paraguay River divides the eastern hills and fertile plains, where 90% of people live, from the almost uninhabited Chaco in the west. Paraguay's economy is largely agricultural.

CLIMATE ▷ Tropical/subtropical

WEATHER CHART FOR ASUNCIÓN

Paraguay is subtropical, with all parts experiencing floods and droughts, but the Chaco is generally drier and hotter.

TRANSPORTATION ▷ Drive on right

Silvio Pettirossi
International, Asunción
465,664 passengers

47 ships
44,900 grt

THE TRANSPORTATION NETWORK

| 15,000 km (9321 miles) | Pan-American Highway: 700 km (435 miles) |
| 370 km (230 miles) | 3100 km (1926 miles) |

Roads badly need upgrading. The government wants to part-privatize the near-paralyzed FCCAL (railroad).

TOURISM ▷ Visitors : Population 1:25

221,000 visitors   Down 18% in 2000

MAIN TOURIST ARRIVALS

Argentina 37%	
Brazil 23%	
Chile 6%	
Other 34%	

% of total arrivals

Tourist numbers are small. Most visitors are cross-border day trippers from Brazil and Argentina, who flock to Ciudad del Este to buy cheap, mainly Far Eastern, electrical goods. The Chaco attracts tourists for safaris.

PEOPLE ▷ Pop. density low

Guaraní, Spanish   14/km² (37/mi²)

THE URBAN/RURAL POPULATION SPLIT

56%   44%

ETHNIC MAKEUP

Amerindian 2%   Other 8%
Mestizo 90%

Most Paraguayans are of combined Spanish and native Guaraní origin. The majority are bilingual, although outside the large cities Guaraní is spoken almost exclusively. Rural Guaraní, deprived of ancestral lands, have been forced into marginal labor and prostitution. While most immigrants historically have been European, Japanese, Koreans, and South Africans are recent arrivals.

POLITICS ▷ Multiparty elections

L. House 1998/2003
U. House 1998/2003
President Luis Gonzalez Macchi

AT THE LAST ELECTION

Chamber of Deputies 80 seats

56% ANR–PC   44% DA

ANR–PC = National Republican Association–Colorado Party
DA = Democratic Alliance (led by the Authentic Radical Liberal Party – **PLRA**)

Senate 45 seats

53% ANR–PC   45% DA   2% Other

A 1989 coup ended Gen. Stroessner's 34-year dictatorship. In 1993, his PC won the first free elections in 60 years, but with continued reliance on the military. The PC again won in 1998, despite the last-minute annulment of former army chief Gen. Lino Oviedo's presidential candidacy. Unrest in 1999 after the assassination of Vice President Luis Argaña, Oviedo's main opponent in the PC, forced the resignation of President Raúl Cubas, a supporter of Oviedo. A fragile coalition government survived a coup in May 2000 led by Oviedo – his third such attempt since 1993. Vice-presidential elections in August 2000 led to an unprecedented cohabitation between a PC president and his PLRA deputy.

PARAGUAY

Total Land Area : 406 750 sq. km
(157 046 sq. miles)

POPULATION
- ◎ over 100 000
- ○ over 50 000
- ● over 10 000
- • under 10 000

LAND HEIGHT
- 1000m/3281ft
- 500m/1640ft
- 200m/656ft
- Sea Level

WORLD AFFAIRS

 Joined UN in 1945

 IBRD　 IAEA　 Geplac　 Mercosr　 RG

The main aims are fairer integration in the Mercosur common market and good relations with the US.

AID

 Recipient

 $82m (receipts)　 Up 5% in 2000

The World Bank offers development aid, the IMF conditional loans. NGO charities run small programs in rural areas.

DEFENSE

Compulsory military service

$121m　Down 5% in 2000

Under Stroessner, the military controlled political and economic life. In 1994–1995, Congress tried to limit its powers, but President Juan Carlos Wasmosy endorsed its political and institutional role. The pact between the military and the PC, in power since 1947, has been weakened by factionalism.

ECONOMICS

Inflation 13% p.a. (1990–2000)

 $7.93bn　 3545–4635 guaranies

SCORE CARD

❏ World GNP Ranking	94th
❏ GNP per Capita	$1440
❏ Balance of Payments	–$137m
❏ Inflation	9%
❏ Unemployment	16%

STRENGTHS

Electricity exporter – earnings obtain foreign exchange. Self-sufficiency in wheat and other staple foodstuffs. Cotton, oilseeds, notably soybeans.

WEAKNESSES

Reliance on agriculture and shaky Brazilian and Argentine markets. No hydrocarbons produced. Weak banking and financial sectors. High unemployment. Political instability deters foreign investment.

EXPORTS

Bermuda 4%　Spain 4%
Brazil 32%　Chile 6%
Other 25%
Argentina 29%

IMPORTS

China 3%　Hong Kong 7%
Brazil 30%　USA 16%
Argentina 23%　Other 21%

The Iguaçu Falls, on the border with Brazil and Argentina, are composed of over 20 cataracts, separated by rocks and tree-covered islands.

RESOURCES

 Electric power 7.1m kw

 25,095 tonnes　 Not an oil producer

9.74m cattle, 2.7m pigs, 710,000 ducks, 15.2m chickens

Gypsum, marble, clay, kaolin, iron, manganese, uranium

The joint Paraguay–Brazil Itaipú hydroelectric dam is the world's largest. The massive Yacyretá Dam is operated with Argentina.

ENVIRONMENT

 Sustainability rank: 25th

 4%　0.9 tonnes per capita

Apart from the destruction of forests for farming and for dams, a major ecological worry is the smuggling abroad of endangered species.

MEDIA

 TV ownership medium

Daily newspaper circulation 43 per 1000 people

PUBLISHING AND BROADCAST MEDIA

There are 7 daily newspapers, including *ABC Color*, *Noticias*, *Popular*, and *Ultima Hora*

4 independent services　21 independent services

The media, historically sponsored by political parties, flourished after the fall of Stroessner, publishing details of corruption and abuses of human rights. The constitution nominally protects the rights of columnists to air their views.

CRIME

 No death penalty

4088 prisoners　 Down 8% in 1999

Paraguay is the contraband capital of Latin America, with trade in everything from cars to cocaine. Jungle airstrips near Brazil provide a route for narcotics.

EDUCATION

 School leaving age: 12

 93%　42,302 students

Education is compulsory only to the age of 12, and fewer than half of children go on to secondary school. Provision is limited in remote rural areas.

CHRONOLOGY

Paraguay was controlled by Spain from 1536 until 1811.

- ❏ **1864–1870** Loses War of the Triple Alliance against Argentina, Brazil, and Uruguay.
- ❏ **1928–1935** Two Chaco Wars against Bolivia over disputed territory.
- ❏ **1938** Boundary with Bolivia fixed; Paraguay awarded large tracts.
- ❏ **1954–1989** Rule of Gen. Stroessner; repressive military regime.
- ❏ **1993** First democratic elections.
- ❏ **1996** Coup attempt by Gen. Lino Oviedo.
- ❏ **1998–1999** Raúl Cubas elected president; resigns after assassination of vice president; Cubas and Oviedo leave country.
- ❏ **2000** Further failed coup attempt by Oviedo.

HEALTH

Welfare state health benefits

1 per 909 people　Heart disease, cancers, obstetric causes, tuberculosis

Hepatitis, typhoid, dysentery, and tuberculosis are endemic and leprosy is common. Medical care is expensive.

SPENDING

GDP/cap. increase

CONSUMPTION AND SPENDING

14 per 1000 population　50 per 1000 population

Defense 1.3%
Education 4.5%
Health 1.7%

Defense, Health, Education spending as % of GDP

Income inequality is great and rural poverty serious. Top military ranks, business leaders, and landed elites control wealth.

WORLD RANKING

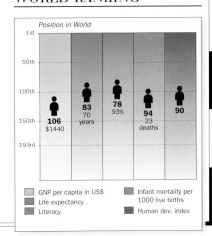

Position in World

106 $1440	83 70 years	78 93%	94 23 deaths	90

- ▪ GNP per capita in US$
- ▪ Life expectancy
- ▪ Literacy
- ▪ Infant mortality per 1000 live births
- ▪ Human dev. index

P

PERU

OFFICIAL NAME: Republic of Peru **CAPITAL:** Lima **POPULATION:** 26.1 million
CURRENCY: Nuevo sol **OFFICIAL LANGUAGES:** Spanish and Quechua

SOUTH AMERICA

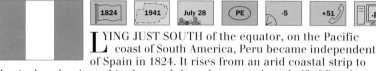

1824 **1941** **July 28** **PE** **-5** **+51** **.pe**

L YING JUST SOUTH of the equator, on the Pacific
coast of South America, Peru became independent
of Spain in 1824. It rises from an arid coastal strip to
the Andes, dominated in the south by volcanoes; about half of Peru's
population live in mountain regions. Peru's border with Bolivia to the
south runs through Lake Titicaca, the highest navigable lake in the
world. In 1995, Peru was involved in a brief border war with Ecuador,
its northern neighbor, and the issue was finally settled in 1998.

CLIMATE

▷ Tropical/mountain/desert

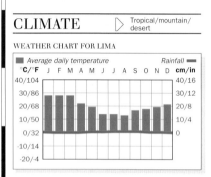
WEATHER CHART FOR LIMA

Peru has several distinct climatic
regions. The arid or desert coastal
region experiences the *garúa*,
persistent low cloud and fog, giving
Lima cool "winters" even though it is
close to the equator. The temperate
slopes of the Andes have large daily
temperature ranges and one rainy
season, while the tropical Amazon
Basin receives year-round rains.

TRANSPORTATION

▷ Drive on right

✈ **Jorge Chávez International, Lima**
4.51m passengers

🚢 719 ships
269,700 grt

THE TRANSPORTATION NETWORK

8700 km (5406 miles)	Pan-American Highway: 2495 km (1550 miles)
1639 km (1018 miles)	8600 km (5344 miles)

The World Bank in 2001 estimated that
Peru had a public infrastructure deficit
of $7 billion. Most roads remain unpaved.
Work on a transcontinental highway
from Ilo, a free port on the Pacific, via
Puerto Suárez in Bolivia, to the port of
Portos in Brazil is ongoing. The two rail
networks, the Central and Southern,
are as yet unconnected. The La Oroya–
Huancayo line is the world's highest
stretch of standard-gauge railroad.
River transportation provides major
access to Iquitos in Amazonia. As well
as four important international airports,
there are more than 130 airstrips
scattered throughout the country.

*Spanish colonial church near Urubamba.
The Urubamba River with its deep gorges was
known as the Sacred Valley to the Incas.*

TOURISM

▷ Visitors : Population 1:25

🧳 1.03m visitors   ⬆ Up 9% in 2000

MAIN TOURIST ARRIVALS

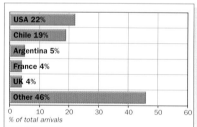

USA 22%	
Chile 19%	
Argentina 5%	
France 4%	
UK 4%	
Other 46%	

% of total arrivals

Tourism is gradually recovering, after
being plunged into crisis in the early
1990s by guerrilla activity, crime, and
cholera fears. The heavily indebted
industry has been unable to take
full advantage of new investment
opportunities, but privatization
programs have seen the sale of
state hotels. Visitors face poor
infrastructure and accommodation to
see incomparable sites such as the Inca
ruins at Machu Picchu in the Andes.
Tourism to the Amazon is also growing,
but environmentalists are concerned
about the impact on indigenous people.
The patterns in the desert made by the
Nazca people (known as the Nazca
lines), dating from the 2nd century
BCE, are also a major attraction. Lake
Titicaca and the Spanish colonial
architecture of Lima are other draws.

PEOPLE

▷ Pop. density low

Spanish, Quechua, Aymara   20/km² (53/mi²)

THE URBAN/RURAL POPULATION SPLIT
73%   27%

RELIGIOUS PERSUASION

Other 5%
Roman Catholic 95%

ETHNIC MAKEUP

Other 2%   White 12%
Mestizo 32%
Amerindian 54%

Most Peruvians are Amerindian or
mestizo (mixed race). The small elite
of Spanish descendants retain a strong
hold on the economy, power, and social
standing. A few Chinese and Japanese
live in the northern cities.

Previously remote Andean
Amerindians are increasingly informed
of events in Lima and the coastal strip
by radio and by relatives in cities.
This has compensated for problems
associated with the marginalization
of their native Quechua and Aymara
languages in a Spanish-speaking
culture. A further 250,000 Amazonian
Amerindians live in the eastern
lowlands. Together with the small
community of Africans (descendants of
plantation workers), they tend to suffer
the worst discrimination in towns.

The extended family remains
strong. A part of traditional native
Amerindian traditions, its role as
a social bond was strengthened by
Roman Catholicism. In recent years,
economic difficulties have raised its
profile as the key social support
system for most Peruvians.

POPULATION AGE BREAKDOWN

Female	Age	Male
0.4%	80+	0.3%
3.2%	60–79	2.9%
7.9%	40–59	7.5%
16%	20–39	15.5%
22.8%	0–19	23.5%

% of population by age group

P

POLITICS ▷ Multiparty elections

 2001/2006

 President Alejandro Toledo

AT THE LAST ELECTION

Congress of the Republic 120 seats

| 38% PP | 23% APRA | 14% NU | 9% FIM | 3% SP | 5% UPP | 8% Others |

PP = Peru Possible APRA = American Popular Revolutionary Alliance NU = National Unity FIM = Independent Moralizing Front UPP = Union for Peru SP = Somos Peru

Alberto Fujimori, *president 1990–2000, ran an increasingly autocratic regime.*

President Alejandro Toledo, *Peru's first Amerindian head of state, elected in 2001.*

Peru is a multiparty democracy with an executive presidency.

PROFILE

The long tradition of large parties dominating politics ended with President Alberto Fujimori's election in 1990. His "self-coup" created a compliant legislature and judiciary, and the approval of a new constitution permitted his reelection in 1995. His popularity was greatly boosted by successes against hyperinflation and the Sendero Luminoso (Shining Path) guerrillas. It faded in the late 1990s as he tightened his personal control of the government, and appeared increasingly reliant on the army. Few checks on the executive remained, and Fujimori was able to obtain his third term in 2000. However, blatant electoral fraud and the corrupt use of power were exposed when his security service chief Vladimiro Montesinos was videoed bribing opposition legislators. Fujimori's position was irrevocably damaged, and he resigned in November, having fled to Japan. Fresh presidential and legislative elections in April 2001 were won by populist Alejandro Toledo and his Peru Posible party.

MAIN POLITICAL ISSUES
Disenchantment with Toledo

President Toledo promised a new start for Peru, but his low ratings in opinion polls testified to early impatience with his administration. Respondents expressed disappointment that he had failed to fulfill election promises, especially on job creation, as the government itself committed itself to IMF goals on structural reform and fiscal restraint. Polls also pointed to public unease at the continuation of nepotism in high places, a much resented aspect of the Fujimori years. The dismissal of three cabinet ministers revealed discontent in the ruling PP congressional bloc with the defection of the highly popular Cecilia Tait, a former sporting star.

Corruption

The public endorses investigations into widespread corruption during the Fujimori era, but is growing fatigued by endless rounds of legislative commissions and probes. Many would prefer politicians to refocus on sorting out the economy.

Reform of military

Senior officers loyal to Fujimori and Montesinos have been sacked, but the military, and the militarized police, need reforming to make them professional forces. The likelihood of this happening soon is slim.

CHRONOLOGY

Francisco Pízarro's arrival in 1532 during a war of succession between two Inca rulers marked the start of the Spanish colonization of Peru, and the end of the Inca empire.

❑ **1821** Independence proclaimed in Lima after its capture by Argentine liberator, José de San Martín, who had just freed Chile.

❑ **1824** Spain suffers final defeats at battles of Junín and Ayacucho by Simón Bolívar and Gen. Sucre, liberators of Venezuela and Colombia.

❑ **1836–1839** Peru and Bolivia joined in short-lived confederation.

❑ **1866** Peruvian–Spanish War.

❑ **1879–1884** War of the Pacific. Chile defeats Peru and Bolivia. Peru loses territory in south.

❑ **1908** Augusto Leguía y Salcedo's dictatorial rule begins.

❑ **1924** Dr. Víctor Raúl Haya de la Torre founds nationalist APRA in exile in Mexico.

❑ **1930** Leguía ousted. APRA moves to Peru as first political party.

❑ **1931–1945** APRA banned.

❑ **1939–1945** Moderate, pro-US civilian government.

❑ **1948** Gen. Manuel Odría takes power. APRA banned again.

❑ **1956** Civilian government restored.

❑ **1962–1963** Two military coups.

❑ **1963** Election of Fernando Belaúnde Terry. Land reform, but military used to suppress communist-inspired insurgency. ⇨

PERU

Total Land Area : 1 285 200 sq. km (496 223 sq. miles)

POPULATION

⊡	over 1 000 000
⊙	over 500 000
◉	over 100 000
○	over 50 000
•	under 50 000

LAND HEIGHT

4000m/13124ft
2000m/6562ft
500m/1640ft
Sea Level

Gulf of Guayaquil
Equator
COLOMBIA
ECUADOR
BRAZIL
BOLIVIA
CHILE
PACIFIC OCEAN

Arica
Napo
Putumayo
Tigre
Pastaza
Marañón
Amazon
Iquitos
Tumbes
Sullana
Castilla
Piura
Bernal
Catacaos
Chulucanos
Moyobamba
Tarapoto
Chachapoyas
Chiclayo
Saña
Cajamarca
San Pedro de Lloc
Trujillo
Chimbote
Huaraz
Nevado Huascarán 6768m
Huánuco
Pucallpa
Cerro de Pasco
Huacho
La Oroya
Tarma
Comas
Huaral
Callao
Chaclacayo
LIMA
Huancavelica
Huancayo
Puerto Maldonado
Madre de Dios
Urubamba
Machu Picchu
Ayacucho
Abancay
Cusco
Chincha Alta
Pisco
Ica
Nazca
San Juan
Juliaca
Puno
Lake Titicaca
Socabaya
Arequipa
Matarani
Moquegua
Ilo
Tacna
las Piedras
Ucayali
Huallaga
Cordillera Occidental
N

0   200 km
0   200 miles

CHRONOLOGY *continued*

- ❏ **1968** Military junta takes over. Attempts to alleviate poverty. Large-scale nationalizations.
- ❏ **1975–1978** New right-wing junta.
- ❏ **1980** Belaúnde reelected. Maoist Sendero Luminoso (Shining Path) begins armed struggle.
- ❏ **1981–1998** Border war with Ecuador over Cordillera del Cóndor, given to Peru by a 1942 protocol. Ecuador wants access to Amazon.
- ❏ **1982** Deaths and "disappearances" start to escalate as army cracks down on guerrillas and narcotics.
- ❏ **1985** Electoral win for left-wing APRA under Alán García Pérez.
- ❏ **1987** Peru bankrupt. Plans to nationalize banks blocked by new Libertad movement led by writer Mario Vargas Llosa.
- ❏ **1990** Over 3000 political murders. Alberto Fujimori, an independent, elected president on anticorruption platform. Severe austerity program.
- ❏ **1992–95** Fujimori "self-coup." New constitution. Fujimori reelected.
- ❏ **1996–1997** Left-wing Tupac Amarú guerrillas seize hundreds of hostages at Japanese ambassador's residence in four-month siege.
- ❏ **2000** November, Fujimori seeks refuge in Japan and resigns amid corruption scandal despite having won controversial third term in May.
- ❏ **2001** Fresh presidential elections, won by Alejandro Toledo in run-off against García.

P

WORLD AFFAIRS ▷ Joined UN in 1945

Cooperation with the US, which is the main source of aid, extends to the war on cocaine, although Peru remains one of the world's largest coca producers.

Reported incursions by guerrillas, paramilitaries, and narcotics traffickers highlight the security of the border with Colombia as a key problem.

Peru competes with Chile for an alliance with Bolivia to process and pipe natural gas.

AID ▷ Recipient

💲 $401m (receipts) ⬇ Down 11% in 2000

Aid from the US is mostly directed at antinarcotics activity. Loans worth some $1.3 billion in recent years from the IDB, the World Bank, and Japan were conditional on Peru's meeting specific health and educational targets and on making progress on privatizations.

DEFENSE ▷ Compulsory military service

💲 $861m ⬇ Down 3% in 2000

The military, in power from 1968 to 1980, supported President Fujimori's 1992 presidential coup. A quarter of national territory remained under states of emergency until early 2000, despite the apparent defeat of the Sendero Luminoso guerrillas. Fujimori's control over promotions and the National Intelligence Service (SIN) guaranteed a loyal armed forces leadership. After his resignation and the dissolution of the SIN in late 2000, the interim government moved to cut back military influence. Bomb attacks in Lima prior to US president Bush's 2002 visit were widely blamed on guerrillas, but

PERUVIAN ARMED FORCES

🛡	275 main battle tanks (T-54/T55)	60,000 personnel
🚢	6 submarines, 1 cruiser, 4 frigates, 10 patrol boats	25,000 personnel
✈	116 combat aircraft (*Canberra*, Su-22/25, MiG-29, *Mirage*)	15,000 personnel
🚀	None	

an option studied by the Interior Ministry was that elements within the military, intelligence, or police still sympathetic to Montesinos were responsible.

ECONOMICS ▷ Inflation 27% p.a. (1990–2000)

📊 $53.4bn 💲 3.527–3.444 new soles

SCORE CARD

❏ WORLD GNP RANKING	47th
❏ GNP PER CAPITA	$2080
❏ BALANCE OF PAYMENTS	–$1.63bn
❏ INFLATION	3.8%
❏ UNEMPLOYMENT	8%

EXPORTS

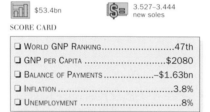

China 6% — Japan 6% — Switzerland 8% — UK 8% — USA 28% — Other 44%

IMPORTS

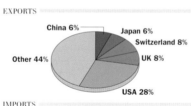

Colombia 5% — Venezuela 5% — Chile 7% — Spain 9% — USA 30% — Other 44%

STRENGTHS

Abundant mineral resources, including oil. Rich Pacific fish stocks. Wide climatic variation allows diverse and productive agriculture; cotton and coffee are important. Well-developed textile industry.

WEAKNESSES

Overdependence on metals and commodities whose fluctuating prices undermine trade and investment. Stalled privatization. Corruption and poor infrastructure deterring investment. Weak banks.

PROFILE

Economic activity is largely confined to the cities of the coastal plain. The inhabitants of the Andean uplands are subsistence farmers or coca producers. Peru's strict fiscal and monetary policy

ECONOMIC PERFORMANCE INDICATOR

— Consumer Price Index   GDP ▨

Consumer price index 1995=100 / GDP 1996=100 — 1996, 1997, 1998, 1999, 2000

continued under Fujimori. Growth was hit hard in 1998 by the disruption of fishing by El Niño-generated storms, the Asian economic crises, Russian-provoked turmoil in emerging markets, and depressed world commodity prices. In 1999 the IMF granted a three-year loan package, extended in 2002, to support comprehensive structural reform and fiscal restraint, with the aim of gradually reducing the state-sector deficit. Meanwhile public belief in Toledo's promises on jobs and poverty waned.

PERU : MAJOR BUSINESSES

Arica, Talara, Sechura, Trujillo, Pucallpa, Cerro de Pasco, Lima, Ica, Arequipa

Oil / Oil refining / Textiles / Mining / Fish processing / Food processing / Vehicle assembly

0 — 400 km / 0 — 400 miles

* significant multinational ownership

RESOURCES

 Electric power
5.2m kw

 8.44m tonnes

14.5m sheep,
4.93m cattle, 2.8m
pigs, 90m chickens

102,000 b/d
(reserves 300m
barrels)

Oil, coal, lead,
zinc, silver, iron,
gold, copper

ELECTRICITY GENERATION

Hydro 74% (14bn kwh)	
Combustion 26% (4.8bn kwh)	
Nuclear 0%	
Other 0%	

% of total generation by type

Peru is an important exporter of copper
and lead. Development of the huge
Antamina copper and zinc deposit is
under way. Development of the $3
billion Camisea hydrocarbon project is
ongoing. The full extent of the large oil
reserves still needs to be explored. A
priority is the further
development of
hydroelectric
power.

PERU : LAND USE

Cropland
Pasture
Forest
Desert
High mountain regions
Sugarcane - cash crop
Sheep

0   400 km
0   400 miles

ENVIRONMENT

 Sustainability
rank: 29th

 3%

 1.1 tonnes per capita

ENVIRONMENTAL TREATIES

	Yes		Yes		Yes
	Yes		Yes		No

Environmentalists have long been
concerned about coastal industrial
pollution and the activities of the
fishing industry. Overfishing of
anchovies almost resulted in their
extinction in the 1970s. Today,
attention has switched to the rising
number of dolphins being caught
in drift nets. Unchecked urban and
industrial pollution, especially in
Lima, is a major problem.

Environmentalists fear that Peru's
policy of using powerful air-sprayed
herbicides to destroy coca crops is
adding to river pollution in the Andes,
where mining also causes severe
environmental problems.

MEDIA

 TV ownership medium

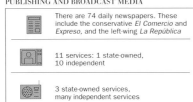 Daily newspaper circulation 4 per 1000 people

PUBLISHING AND BROADCAST MEDIA

	There are 74 daily newspapers. These include the conservative *El Comercio* and *Expreso*, and the left-wing *La República*
	11 services: 1 state-owned, 10 independent
	3 state-owned services, many independent services

Main cities have their own newspapers;
Lima's papers cater to diverse political
readerships. State TV is not popular.

CRIME

 Death penalty not used
in practice

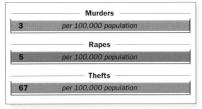 27,452 prisoners

Down 81%
1992–1998

CRIME RATES

Murders	
3	*per 100,000 population*

Rapes	
5	*per 100,000 population*

Thefts	
67	*per 100,000 population*

Kidnappings, murders, armed
robberies, and drugs-related crime
remain serious problems, especially
in Lima. Corruption is deep-seated in
the police and security forces. Despite
the near-destruction of the left-wing
Sendero Luminoso guerrillas, main
cities frequently have curfews, and
those who can afford it protect
themselves with high-security
homes and armed guards.

EDUCATION

 School leaving
age: 16

90%

657,586 students

THE EDUCATION SYSTEM

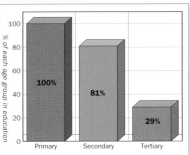

Education is based on the US system;
spending has been declining. The
provision of state education, especially
for the poor, remains a major challenge.
State and private universities are
accessible to a small minority.

HEALTH

 Welfare state
health benefits

1 per 1111 people

Respiratory, heart,
infectious, and
parasitic diseases

The poor public health system almost
collapsed in the 1980s. In many areas
primary care is nonexistent. Advanced
treatment is available only to private
patients in city clinics. Goiter, a thyroid
abnormality, is widespread, especially
in mountain areas. Infant mortality is
rising due to social deprivation,
diarrheal diseases, and tuberculosis.
Malaria is again widespread, and cholera
reached epidemic proportions in 1994.
Thousands of poor women were
forcibly sterilized in the late 1990s
as part of a government program to
lower the birthrate. Social welfare
is compulsory, and benefits cover
sickness, disability, and old age.

SPENDING

GDP/cap. increase

CONSUMPTION AND SPENDING

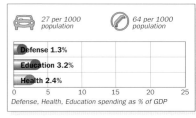

27 per 1000
population

64 per 1000
population

Defense 1.3%		
Education 3.2%		
Health 2.4%		

Defense, Health, Education spending as % of GDP

Most wealth and power in Peru is
still retained by old Spanish families.
Indigenous peoples remain excluded
from both. The rich live in a state of
siege; a key status symbol is the
number of armed guards and security
cameras protecting family property.
Overpopulation and rural migration
accentuate poverty in Lima, where
some 2.7 million people live in shanty
towns, many of them lacking such
basic utilities as running water and
electricity. The UN estimates that
over 30% of Peruvians live below
the poverty line.

WORLD RANKING

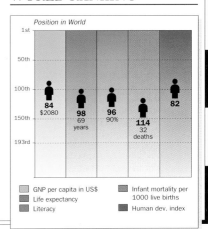

Position in World

84 $2080	98 69 years	96 90%	114 32 deaths	82

GNP per capita in US$
Life expectancy
Literacy

Infant mortality per
1000 live births
Human dev. index

P

PHILIPPINES

OFFICIAL NAME: Republic of the Philippines **CAPITAL:** Manila
POPULATION: 77.1 million **CURRENCY:** Philippine peso **OFFICIAL LANGUAGES:** English and Filipino

L YING ON THE WESTERN RIM of the Pacific Ocean, the Philippines is the world's second-largest archipelago-state. Of its 7107 islands, 4600 are named and 1000 inhabited. There are three main island groupings: Luzon, Visayan, and the Mindanao and Sulu islands. Located on the Pacific "ring of fire," it is thus subject to frequent earthquakes and volcanic activity. Economic growth outstripped population increase in the 1990s, until the 1997–1998 "Asian crisis," but efforts to build a stable democracy have been compromised by high-level corruption, leading to the ouster of President Estrada in 2001.

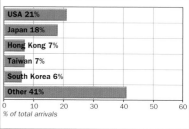
Bohol Island has over 1000 of these famous mounds, known as "the chocolate hills."

CLIMATE ▷ Tropical monsoon/equatorial

WEATHER CHART FOR MANILA

The Philippines is warm and humid all year. The rainy season lasts from June to October. Humidity falls from 85% in September to 71% in March.

TRANSPORTATION ▷ Drive on right

✈ **Ninoy Aquino International, Manila**
12.8m passengers

🚢 1726 ships
8.51m grt

THE TRANSPORTATION NETWORK

39,590 km (24,600 miles)	None
485 km (301 miles)	3219 km (2000 miles)

Basic infrastructure lacks investment and many main roads are in desperate need of repair. Chronic traffic congestion in Manila holds back economic growth.

Air travel is the only means of getting around the islands quickly. Philippines Airlines, privatized in 1992, has invested heavily in new aircraft and in expanding its regional route network. Work began in 2000 on a spectacular new terminal at Manila's international airport, nearly trebling its capacity.

Subic Bay, a massive US naval base until 1992, is now being exploited as a commercial asset, thanks to its prime location. Opening on to the South China Sea, its deep natural harbor has been developed as a free port and enterprise zone. The Taiwanese are the biggest investors in this project.

TOURISM ▷ Visitors : Population 1:36

2.17m visitors Down 8% in 2000

MAIN TOURIST ARRIVALS

USA 21%	
Japan 18%	
Hong Kong 7%	
Taiwan 7%	
South Korea 6%	
Other 41%	

% of total arrivals

There is less tourism in the Philippines than in other regional NICs. Dubious images conveyed by sex-tourism have become a liability. International pressure to end this abuse has intensified. Muslim secessionists have seized tourists on neighboring Malaysian islands as hostages. The tiny island of Boracay, off Panay, is a popular resort, and Palawan retains most of its tropical rainforest and coral lagoons, although coral reefs elsewhere are badly damaged. The rice terraces of northern Luzon are another attraction.

PHILIPPINES

Total Land Area : 300 000 sq. km (115 830 sq. miles)

POPULATION

▣	over 1 000 000
◉	over 500 000
◎	over 100 000
○	over 50 000

LAND HEIGHT

2000m/6562ft
1000m/3281ft
500m/1640ft
200m/656ft
Sea Level

PEOPLE   Pop. density high

 Filipino, Tagalog, Cebuano, Hiligaynon, Samaran, Ilocano, Bikol, English

259/km² (670/mi²)

RELIGIOUS PERSUASION

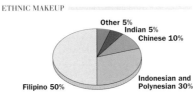

Other (including Buddhist) 3%
Muslim 5%
Protestant 9%
Roman Catholic 83%

POPULATION AGE BREAKDOWN

Female	Age	Male
0.3%	80+	0.2%
2.6%	60–79	2.3%
7.2%	40–59	7.3%
15.4%	20–39	15.5%
24.1%	0–19	25.1%

% of population by age group

THE URBAN/RURAL POPULATION SPLIT

59%   41%

The Philippines encompasses more than 100 distinct ethnic groups, many of Malay origin. The national language, Filipino, is based on Tagalog, spoken by the largest of the various groups. Other groups include Cebuano, Ilocan, Longgo, Bicolano, Waray, Pampangan, and Pangasinan. They are concentrated on the main island, Luzon, and are also a majority on Mindanao. Most Muslims live on Mindanao, but many are also found in the Sulu archipelago. The Chinese minority, which was well established by 1603, has remained significant in business and trade. More than 120 Chinese schools have ensured that it has retained a distinct identity.

There are also a number of cultural minorities who practice animist

ETHNIC MAKEUP

Other 5%
Indian 5%
Chinese 10%
Indonesian and Polynesian 30%
Filipino 50%

religions. They include the Ifugaos, Bontocks, Kalingas, and Ibalois on Luzon, the Manobo and Bukidnon on Mindanao, and the Mangyans on Palawan. Many of these groups speak Malayo-Polynesian dialects. Limited intermarriage with other peoples has meant that groups in the more remote regions have managed to retain their traditional ways of life.

The Philippines is one of only two Christian states in Asia. Over 80% of Filipinos are Roman Catholics, and the Church is the dominant cultural force. It opposes state-sponsored preventive family planning programs aimed at curbing accelerating population growth. Abortion is illegal but widespread.

Women have traditionally played a prominent part in Philippine public and professional life. Inheritance laws give them equal rights to men. Many go into politics, banking, and business, and in several professional sectors they form a majority.

POLITICS   Multiparty elections

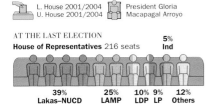 L. House 2001/2004
U. House 2001/2004
President Gloria Macapagal Arroyo

AT THE LAST ELECTION

House of Representatives 216 seats

39% Lakas–NUCD	25% LAMP	10% LDP	9% LP	12% Others	5% Ind

Lakas–NUCD = Lakas — National Union of Christian Democrats   **LAMP** = Party of the Filipino Masses
LDP = Fight of Democratic Filipinos   **LP** = Liberal Party
Ind = Independents   **PP** = People's Power (coalition led by Lakas–NUCD and including the LP)
PnM = Strength of the Masses (coalition of LAMP and LDP)

Senate 24 seats

50% PP	46% PnM	4% Ind

The Philippines is a multiparty democracy.

PROFILE

The 21-year dictatorship of Ferdinand Marcos was shown by 1986 to have no popular legitimacy as "people power" massed in support of his opponent Corazon Aquino, who was declared the true winner of the presidential elections. Losing the backing of the US, Marcos was forced into exile. Aquino was able to hand over power to Fidel Ramos through fair elections in 1992. His dependence on loose coalition arrangements in Congress slowed the economic liberalization program. His successor in 1998, the populist flamboyant former film star Joseph Estrada of LAMP, was accused of links

to a gambling syndicate, and in late 2000 impeachment proceedings began. A repeat of the mass "people power" demonstrations of 1986 toppled him in 2001. Vice President Gloria Macapagal Arroyo of the NUCD, who led the united political opposition to Estrada, was appointed president on his downfall.

MAIN POLITICAL ISSUES
Political stability

The downfall of the Marcos regime through popular demonstration ensured a return to democracy. The subsequent two presidents passed on power smoothly to an elected successor, but Estrada was quickly mired in corruption scandals. Moves for his impeachment were overtaken by mass rallies in Manila. Out of office he retained widespread support among the rural poor, questioning the legitimacy of Vice President Arroyo's assumption of the presidency.

Insurgency and separatism

Communist and Muslim separatists have been fighting Manila-based governments for over 30 years, with more than 10,000 armed confrontations with rebels recorded by the army. Much of the support for secession has been fueled by the failure of successive governments to curb poverty.

Since 1992, the government has been seeking peace with all armed groups. The communist New People's Army (NPA), once regarded as a heroic army of the oppressed and as an alternative to traditional politics, has declined in

Joseph Estrada, film-star-turned-president, ousted in 2001.

Gloria Arroyo, succeeded Estrada in a popular uprising.

significance, but launched new offensives in 2000. The Moro National Liberation Front (MNLF), representing secessionist Muslim rebels on Mindanao, signed a peace agreement in 1996. A militant breakaway Muslim Islamic Liberation Front (MILF) continued fighting, forcing the government to abandon a massive irrigation project. Fighting peaked in 2000, and the MILF joined the peace process in mid-2001. The smaller but internationally backed Abu Sayyaf continues low-level fighting in the Sulu region.

WORLD AFFAIRS   Joined UN in 1945

 APEC   ASEAN   G24   NAM   WTO

Regionally, there are sporadic frictions with Malaysia. Manila's claim to the Spratly Islands conflicts notably with China's. The resumption of US naval visits since 1999 indicate improved security cooperation. US troops provided logistical support against Muslim separatists from 2001.

P

CHRONOLOGY

Ceded to the US by Spain in 1898, the Philippines became self-governing in 1935, and an independent republic in 1946.

- ❑ **1965** Ferdinand Marcos president.
- ❑ **1972** Marcos declares martial law. Opposition leaders arrested, parliament suspended, press censored.
- ❑ **1977** Ex-Liberal Party leader Benigno Aquino sentenced to death. Criticism forces Marcos to delay execution.
- ❑ **1978** Elections won by Marcos's New Society (KBL). He is named president and prime minister.
- ❑ **1980** Aquino allowed to travel to US for medical treatment.
- ❑ **1981** Martial law ends. Marcos reelected president by referendum.
- ❑ **1983** Aquino shot dead on return from US. Inquiry blames military conspiracy.
- ❑ **1986** US compels presidential election. Result disputed. Army rebels led by Gen. Fidel Ramos, and public demonstrations, bring Aquino's widow, Corazon, to power. Marcos exiled to US.
- ❑ **1987** New constitution. Aquino-led coalition wins Congress elections.
- ❑ **1988** Marcos and wife Imelda indicted for massive racketeering.
- ❑ **1989** Marcos dies in US.
- ❑ **1990** Imelda Marcos acquitted of fraud charges in US. Earthquake in Baguio City leaves 1600 dead.
- ❑ **1991** Mt. Pinatubo erupts. US leaves Clark Air Base.
- ❑ **1992** Ramos wins presidential election. US withdrawal from Subic Bay base.
- ❑ **1996** Peace agreement with Muslim MNLF secessionists.
- ❑ **1998** Joseph Estrada president.
- ❑ **1999** First execution in 22 years.
- ❑ **2000** Tourists kidnapped by Islamic extremists.
- ❑ **2001** Estrada overthrown by popular protest. Gloria Macapagal Arroyo assumes presidency. August, Muslim MILF joins peace process.
- ❑ **2002** Local elections, described as "peaceful" despite 86 deaths.

AID

▷ Recipient

 $578m (receipts)　 Down 17% in 2000

The Philippines' main bilateral aid donors are Japan and the US. Many NGOs operate in the outlying islands. Large remittances are also received from Filipinos working overseas. Funds received from these hundreds of thousands of emigrant workers exceeded $6 billion in 2000.

DEFENSE

▷ No compulsory military service

 $1.5bn　 Down 8% in 2000

The military retains political influence. The armed forces are undergoing modernization with US assistance.

The historic tie to the US has been translated in modern times into a close strategic relationship. President Arroyo quickly pledged her country's support for the US-led "war on terrorism" in 2001, and US forces were deployed to help the government combat Muslim rebels in the south.

PHILIPPINE ARMED FORCES

🛡	40 light tanks (*Scorpion*)	67,000 personnel
🚢	1 frigate and 58 patrol boats	24,000 personnel
✈	44 combat aircraft (8 F-5A/B)	16,000 personnel
🚀	None	

ECONOMICS

▷ Inflation 8.4% p.a. (1990–2000)

 $78.8bn　50.0–51.6 Philippine pesos

SCORE CARD

❑ WORLD GNP RANKING	41st
❑ GNP PER CAPITA	$1040
❑ BALANCE OF PAYMENTS	$9.08bn
❑ INFLATION	4.4%
❑ UNEMPLOYMENT	10%

EXPORTS

Taiwan 7% | Netherlands 8% | Singapore 8% | Japan 15% | USA 30% | Other 32%

IMPORTS

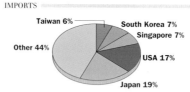

Taiwan 6% | South Korea 7% | Singapore 7% | USA 17% | Japan 19% | Other 44%

STRENGTHS

Now fully open to outside investment. Agricultural productivity rising. Strong pineapple and banana export industries. Substantial remittances from Filipinos working overseas.

WEAKNESSES

Power failures limit scope for expansion. Rudimentary infrastructure. Low domestic savings rates mean reliance on foreign finance. Low productivity in subsistence agriculture.

PROFILE

Once one of Asia's strongest economies, the Philippines has now fallen behind once much poorer countries such as Thailand, Malaysia, and South Korea. Around 50% of the rural population live on the poverty line, fueling many of the secessionist movements undermining the stability of successive governments.

The financial successes of the early 1990s – opening up to foreign investment, cutting back private monopolies – were effectively nullified by the crippling 1997–1998 Asian economic and financial crises. The government of Joseph Estrada failed to shore up the economy and ran up a crippling budget deficit, and investor confidence vanished amid political scandals. His successor, President Arroyo, has done much to regain the approval of the IMF and the international community, mostly through further deregulation and privatization. Although growth is back on track, despite the global slowdown in 2001, many public works projects have simply been put on hold.

ECONOMIC PERFORMANCE INDICATOR

Consumer Price Index — GDP

PHILIPPINES : MAJOR BUSINESSES

🍶 Brewing　👕 Garments　⚗ Chemicals　🔌 Electronics　⛏ Copper mining　🥫 Food processing　🚗 Vehicle assembly　💊 Pharmaceuticals　📞 Telecommunications

Baguio / San Fernando / Manila / Legaspi / Cebu / Cagayan de Oro / Davao / Lupon

0 — 200 km
0 — 200 miles

P

RESOURCES

▷ Electric power 11.6m kw

2.2m tonnes

822 b/d (reserves 205m barrels)

12.5m ducks, 6.95m goats, 115m chickens

Coal, copper, nickel, chromium, silver, manganese, gold, oil

ELECTRICITY GENERATION

Hydro 18% (7.3bn kwh)

Combustion 65% (27bn kwh)

Nuclear 0%

Other 17% (6.9bn kwh)

% of total generation by type

The Philippines is the world's biggest supplier of refractory chrome. Copper is also a significant export. Substantial gold reserves have been mined since 1996. However, more than 90% of mineral potential remains undeveloped. Oil production off Palawan began in 1979. The Philippines is the world's second-biggest user of geothermal power after the US. Almost 25% of electricity on Luzon is provided by this method. Although timber exports were halted in 1989, illegal logging continues to cause deforestation.

ENVIRONMENT

▷ Sustainability rank: 117th

5% (1% partially protected)

1 tonne per capita

ENVIRONMENTAL TREATIES

Yes   Yes   Yes

Yes   Yes   No

The environment has become a major issue. Most of the tropical rainforest has been destroyed, except for pockets such as the island of Palawan. Unique coral habitats have been dynamited, and fishermen continue to use cyanide and muro-ami (reef-hunting) techniques to increase the size of their catches.

The government recognizes the costs of environmental damage, as soil run-off silts rivers and reduces the power generated by hydroelectric dams, and fast-depleting coral habitats reduce the attraction of the Philippines for tourists.

Enforcement of a ban on logging is difficult; many loggers have their own private armies. In addition, continued use of slash and burn farming has contributed to deforestation.

MEDIA

▷ TV ownership medium

Daily newspaper circulation 63 per 1000 people

PUBLISHING AND BROADCAST MEDIA

There are 47 daily newspapers. The most influential are the *Philippine Star* and the *Philippine Daily Globe*

7 services: 1 state-owned, 6 independent

17 services

The lifting of censorship following Corazon Aquino's election in 1986 led to a burgeoning of the media. As well as the national press, there are more than 250 regional newspapers in local dialects. State TV and radio broadcast in English and Filipino. Four independent television stations serve Metro Manila.

CRIME

▷ Moratorium on death penalty

70,383 prisoners   ↑ Crime is rising

CRIME RATES

Murders

13 | per 100,000 population

Rapes

4 | per 100,000 population

Thefts

9 | per 100,000 population

The death penalty, reinstated in 1993, was suspended again in 2000, but President Arroyo made an exception for 95 kidnappers in 2001.

EDUCATION

▷ School leaving age: 12

95%   2.02m students

THE EDUCATION SYSTEM

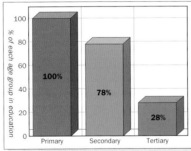

% of each age group in education

Primary 100%
Secondary 78%
Tertiary 28%

The Philippines has one of the highest literacy rates among developing countries. The education system is based on the US model, but with a higher proportion of private schools. The main teaching languages are English and Filipino/Tagalog.

Although there is a national curriculum up to age 15, sectarianism is common; the Chinese community has its own schools. Most colleges and universities are also run privately. The universities of San Carlos in Cebu and Santo Tomas in Manila are Spanish colonial foundations, dating from 1595 and 1611 respectively.

PHILIPPINES : LAND USE

Cropland
Forest
Pigs
Sugarcane
Coconuts

SIERRA MADRE

LUZON

0   200 km
0   200 miles

MINDANAO

HEALTH

▷ No welfare state health benefits

1 per 833 people

Pneumonia, tuberculosis, violence, accidents, malaria, typhoid

Most general hospitals are privately run. Malaria, which was once a major problem, has been eradicated in all but remote areas. Poor sanitation and disease are common in the sprawling slums around Manila.

SPENDING

▷ GDP/cap. increase

CONSUMPTION AND SPENDING

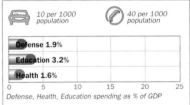

10 per 1000 population   40 per 1000 population

Defense 1.9%

Education 3.2%

Health 1.6%

Defense, Health, Education spending as % of GDP

The contrast between extremes of wealth and poverty is particularly marked. Wealth remains highly concentrated in a few select business families which are based in Manila.

WORLD RANKING

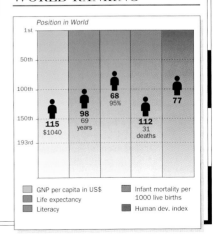

Position in World

1st
50th
100th
150th
193rd

115 $1040
98 69 years
68 95%
112 31 deaths
77

GNP per capita in US$
Life expectancy
Literacy
Infant mortality per 1000 live births
Human dev. index

P

POLAND

OFFICIAL NAME: Republic of Poland   **CAPITAL:** Warsaw
POPULATION: 38.6 million   **CURRENCY:** Zloty   **OFFICIAL LANGUAGE:** Polish

 1918   1945   May 3   PL   +1   +48   .pl

LOCATED IN THE HEART of Europe, Poland's low-lying plains extend from the Baltic shore in the north to the Tatra Mountains on its southern border with Slovakia. Since the collapse of communism, Poland has undergone massive social, economic, and political change. Opting for a radical form of economic "shock therapy" in the early 1990s to kick-start the switch to a market economy, it has experienced rapid growth, is one of the front-runners in negotiations to join the EU, and has already been accepted as a member of NATO.

CLIMATE   ▷ Continental

WEATHER CHART FOR WARSAW

Most of the country experiences a similar climate. Summers are hot, with heavy rainfall often accompanied by thunder. Winters are severe, with snow covering the ground on the southern mountains, for as much as 60–70 days in the east.

TRANSPORTATION   ▷ Drive on right

Okecie International, Warsaw
4.33m passengers

447 ships
1.42m grt

THE TRANSPORTATION NETWORK

249,966 km (155,321 miles)	358 km (222 miles)
22,891 km (14,224 miles)	3812 km (2369 miles)

The national airline LOT has increased its charter business as more middle-class Poles vacation abroad. Russian aircraft have all been replaced with Western models. A 15-year roads expansion program was begun in 1997. "Fast tram" systems for cities and long-distance high-speed rail links need major investment.

Telecommunications have been affected by the advent of mobile phones, and by the privatization of Telekomunikacja Polska and the end of its monopoly on long-distance calls. The government plans to slim down the rail workforce and to commercialize and part-privatize Polish State Railways (PKP) by 2003.

The medieval administrative center of Lublin lies in Poland's southeastern agricultural heartland.

TOURISM   ▷ Visitors : Population 1:2.6

15m visitors   ▽ Down 14% in 2001

MAIN TOURIST ARRIVALS

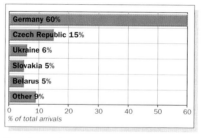

Germany	60%
Czech Republic	15%
Ukraine	6%
Slovakia	5%
Belarus	5%
Other	9%

% of total arrivals

Despite environmental problems, Poland is renowned for its skiing and hiking, especially in the Tatra Mountains. Kraków's medieval core has been preserved, while Toruń has restored its historic German Hanseatic buildings.

Warsaw's historic center has been reconstructed following the destruction of 80% of it by the German army in 1944. More hotels and restaurants are being opened.

Poznań has exploited its location between Warsaw and Berlin to create an international exhibition and business convention industry.

Airlines have increased their flights from the West to take advantage of the country's tourist potential.

PEOPLE   ▷ Pop. density medium

Polish

127/km² (328/mi²)

THE URBAN/RURAL POPULATION SPLIT

66%   34%

RELIGIOUS PERSUASION

Eastern Orthodox 2%
Other and nonreligious 5%
Roman Catholic 93%

ETHNIC MAKEUP

German 1%   Other 1%
Polish 98%

Poland has a strongly Roman Catholic population, and in addition there is little ethnic diversity. The Church believes that stronger links with the West, especially through joining the EU, will weaken its influence. Abortion is still a major issue, and attempts to liberalize the law in 1996 were overturned by the Constitutional Tribunal.

Some small ethnic groups have opened schools and cultural and religious centers. Others, particularly the Germans in Silesia, are becoming more assertive. Jews are still resentful of past discrimination, and there is some evidence of residual antisemitism at a high level. Disputes over the special significance of the site of the World War II Auschwitz concentration camp, near Kraków, have caused conflict between Jews and Catholics.

Wealth disparities are small, although the growing wealth of the entrepreneurial class is causing tension. The major political parties on left and right agree on continuing economic reform.

POPULATION AGE BREAKDOWN

Female	Age	Male
1.4%	80+	0.6%
8.1%	60–79	5.8%
12.7%	40–59	12.1%
14.2%	20–39	14.6%
14.9%	0–19	15.6%

% of population by age group

POLITICS

▷ Multiparty elections

L. House 2001/2005
U. House 2001/2005

President Aleksander Kwasniewski

AT THE LAST ELECTION
Diet 460 seats

| 47% SLD–UP | 14% PO | 11% S | 10% PiS | 8% LPR | 9% PSL | 1% GM |

SLD–UP = Democratic Left Alliance–Labor Union **PO** = Civic Platform **S** = Self Defense **PiS** = Law and Justice **PSL** = Polish Peasant Party **LPR** = League of Polish Families **GM** = German Minority of Lower Silesia **Bloc** = Senate 2001 Bloc **Ind** = Independents

Senate 100 seats

| 75% SLD–UP | 15% Bloc | 2% LPR | 2% S | 4% PSL | 2% Ind |

Since 1989, Poland has been a multiparty parliamentary democracy.

PROFILE
From 1993 until 1997, successive governments were formed by the reformed communists of the SLD and the PSL, which pursued a policy of market reforms. Aleksander Kwasniewski, leader of the SLD, was elected president in 1995, and reelected in 2000. More right-wing groups held sway for the 1997–2001 parliamentary term, with Jerzy Buzek, a member of Solidarity since its formation in 1980, as prime minister. His Solidarity Electoral Action (AWS) alliance, a right-wing grouping with vocal Catholic and nationalist elements, formed a coalition with the liberal Freedom Union (UW). A threatened split in early 1999 over controversial health service reforms was averted, but further contentious issues led to the withdrawal of the UW in May 2000. The AWS remained in office thereafter as a minority government. Buzek, who resisted pressure for early elections, came under attack from within the AWS itself.

In May 2001 the Solidarity trade union wing voted to withdraw from politics, and AWS lost all its seats in the legislative elections that September. A new left-of-center coalition was brought together under former communist Leszek Miller.

MAIN POLITICAL ISSUES
Coalition rule
Poland's emerging party system has been hindered by a superfluity of political factions, and sustaining coalitions has proved difficult. Parties are

Former president Lech Walesa. *He was awarded the Nobel Peace Prize in 1983.*

President Aleksander Kwasniewski, *leader of the SLD, who was elected in 1995.*

required to have at least 5% of the vote in order to gain a seat and 8% to be eligible to join a coalition government.

Church–state relations
The Roman Catholic Church has been outspoken in its views on social and political policy. Debates over abortion, worship in schools, and values in the media have fueled a heated dialogue over the proper role of the Church. Abortion laws were eased in 1996 until the Constitutional Tribunal ruled that this move was unconstitutional.

WORLD AFFAIRS

▷ Joined UN in 1945

CE   CEFTA   OECD   NATO   OSCE

Poland has good relations with the Baltic states and other countries of central Europe. It was admitted to NATO in 1999, and is among the region's "first wave" of applicants to join the EU, hoping for accession in 2004. A treaty has been signed with Germany recognizing the postwar border. The EU wants tighter controls along the eastern border with Belarus and Ukraine.

POLAND

Total Land Area : 312 685 sq. km (120 728 sq. miles)

POPULATION
- ▢ over 1 000 000
- ⊙ over 500 000
- ◎ over 100 000
- ○ over 50 000

LAND HEIGHT
- 1000m/3281ft
- 500m/1640ft
- 200m/656ft
- Sea Level

0   100 km
0   100 miles

CHRONOLOGY

Poland has Europe's second-oldest written constitution. In 1795, it was partitioned between Austria-Hungary, Prussia, and Russia.

- ❑ **1918** Polish state recreated.
- ❑ **1921** Democratic constitution.
- ❑ **1926–1935** Marshal Jozef Pilsudski heads military coup. Nine years of authoritarian rule.
- ❑ **1939** Germany invades and divides Poland with Russia.
- ❑ **1941** First concentration camps built on Polish soil.
- ❑ **1944** Warsaw Uprising.
- ❑ **1945** Potsdam and Yalta Conferences set present borders and determine political allegiance to Soviet Union.
- ❑ **1947** Communists manipulate elections to gain power. ⇨

P

CHRONOLOGY *continued*

- ❑ **1956** More than 50 killed in rioting in Poznań.
- ❑ **1970** Food price increases lead to strikes and riots in the Baltic port cities. Hundreds are killed.
- ❑ **1979** Cardinal Karol Wojtyla of Kraków is elected pope and takes the name of John Paul II.
- ❑ **1980** Strikes force the government to negotiate with the Solidarity union. Resulting Gdańsk Accords grant the right to strike and to form free trade unions.
- ❑ **1981** Gen. Wojciech Jaruzelski becomes prime minister.
- ❑ **1981–1983** Martial law. Solidarity forced into underground existence. Many of its leaders, including Lech Walesa, are interned.
- ❑ **1983** Walesa awarded Nobel Peace Prize.
- ❑ **1986** Amnesty for political prisoners.
- ❑ **1987** Referendum rejects government austerity program.
- ❑ **1988** Renewed industrial unrest.
- ❑ **1989** Ruling party holds talks with Solidarity, which is relegalized. Partially free elections are held. First postwar noncommunist government formed.
- ❑ **1990** Launch of market reforms. Walesa elected president.
- ❑ **1991** Free elections lead to fragmented parliament.
- ❑ **1992** Last Russian troops leave.
- ❑ **1993** Elections: reformed communists head coalition government.
- ❑ **1994** Launch of mass privatization.
- ❑ **1995** Leader of reformed communists Aleksander Kwasniewski is elected president.
- ❑ **1996** Historic Gdańsk shipyard declared bankrupt and closed down.
- ❑ **1997** April, parliament finally adopts new postcommunist constitution. September, legislative elections end former communist majority with big swing to AWS coalition. December, EU agrees to open membership negotiations.
- ❑ **1999** Joins NATO.
- ❑ **2001** Elections: AWS routed, left-of-center coalition formed under Leszek Miller.

AID ▷ Recipient

💲 $1.4bn (receipts)   ⬆ Up 18% in 2000

Large-scale aid for economic transformation was a phenomenon of the early 1990s. The IMF, the EBRD, and the EU all supported Poland's stabilization and reform program. EU aid now focuses on helping Poland to prepare to meet the environmental and other standards required of its members.

DEFENSE ▷ Compulsory military service

💲 $3.19bn   ⬇ Down 1% in 2000

Poland joined NATO in March 1999. Its standing army is among the largest in Europe, and there are also large paramilitary units, including border guards. A 15-year program to modernize the armed forces was introduced by the government in 1997. A civilian alternative to military service was first offered in 1998, and military service was reduced from 18 months to a year from 1999.

POLISH ARMED FORCES

🛡	1677 main battle tanks (786 T-55, 685 T-72, 206 PT-91)	120,300 personnel
⚓	3 submarines, 2 frigates, 1 destroyer, and 23 patrol boats	16,760 personnel
✈	212 combat aircraft (91 MiG-21, 22 MiG-29, 99 Su-22)	43,735 personnel
🚀	None	

ECONOMICS ▷ Inflation 23.4% p.a. (1990–2000)

📊 $162bn   💲 4.133–3.952 zlotys

SCORE CARD

❑ WORLD GNP RANKING	25th
❑ GNP PER CAPITA	$4190
❑ BALANCE OF PAYMENTS	–$10bn
❑ INFLATION	10.1%
❑ UNEMPLOYMENT	16%

EXPORTS

France 5%
UK 4%
Netherlands 5%
Italy 6%
Other 45%
Germany 35%

IMPORTS

UK 4%
France 6%
Italy 8%
Russia 9%
Other 49%
Germany 24%

ECONOMIC PERFORMANCE INDICATOR

— Consumer Price Index   GDP ▨

STRENGTHS
Restructuring of loss-making coal industry began in 1998. Successful privatizations accelerated again in late 1990s. High rates of foreign investment reflect status as largest market in central Europe. Modernization of electricity supply and oil refining. Booming construction industry.

WEAKNESSES
Agriculture suffers from overmanning, tiny farms, and lack of investment. Compensation for communist-era property expropriations unresolved. Heavy industries not competitive.

PROFILE
After a decade of economic crisis, the postcommunist government in 1990 drove through the most determined plan in the whole region to make the transition to a market economy. Most prices were freed, trade was opened and the zloty was made convertible. Economic growth and foreign investment soared, especially after

Western creditors agreed to cancel half of the country's foreign debt in 1994. Poland now attracts the most foreign capital in central and eastern Europe.

There are still large-scale heavy industrial plants left over from the communist era, but some have been converted or reorganized successfully. Many state farms have been liquidated, but agricultural efficiency is improving only slowly. Some 26% of the workforce is employed in farming.

Economic growth slowed at the end of the 1990s, but was back to 4% in 2000. Inflation has remained high and unemployment hit a postcommunist record level in 2002.

POLAND : MAJOR BUSINESSES

Gdańsk
Warsaw
Szczecin
Białystok
Poznań
Wrocław
Łódź
Kraków

🏭 Iron & steel		◊ Optics	
⛏ Coal mining		🚗 Vehicle assembly	
🚢 Shipbuilding		💉 Pharmaceuticals	
Electronics			
Textiles			
Engineering	0 200 km		
Chemicals	0 200 miles		

P

RESOURCES

 Electric power 29.5m kw

 268,822 tonnes

7218 b/d (reserves 37m barrels)

17m pigs, 5.72m cattle, 3.57m ducks, 48.3m chickens

Coal, copper, silver, sulfur, natural gas, lead, salt, iron

ELECTRICITY GENERATION

Hydro 3% (4.3bn kwh)
Combustion 97% (138bn kwh)
Nuclear 0%
Other 0%

0   20   40   60   80   100

% of total generation by type

Poland has significant quantities of coal, copper, silver, sulfur, natural gas, lead, and salt.

The government's aim is to achieve self-sufficiency in energy resources and eventually to be able to export them; plans are in place to privatize both the fuel and energy industries. Coal supplies two-thirds of the country's electricity generation.

POLAND : LAND USE

Cropland
Pasture
Forest
Pigs
Cattle
Cereals

0   200 km
0   200 miles

ENVIRONMENT

 Sustainability rank: 87th

10%

8.3 tonnes per capita

ENVIRONMENTAL TREATIES

Yes   Yes   Yes
Yes   Yes   No

Pollution problems are serious, but improving. Upper Silesia and the Kraków area are still badly affected, but industry there only emits a third of the pollutants it emitted in 1990. Now that much heavy industry has been cleaned up or closed down, there is more concern about small factories, domestic coal fires, and the increased use of private cars.

Water pollution is a major problem, mainly from untreated sewage and industrial discharges. Rivers flowing into the Baltic are badly affected by nitrates and phosphates used in farming. Polish standards, themselves widely disregarded, need to be raised to meet EU minimum requirements.

MEDIA

 TV ownership high

Daily newspaper circulation 37 per 1000 people

The constitution guarantees media freedom. *Gazeta Wyborcza*, set up by Solidarity in 1989, is still the leading daily, and its owners are expanding into other media. The head of election coverage at Polish Public Television (TVP) resigned in 2000, accused of favoring President Kwasniewski.

CRIME

 No death penalty

70,544 prisoners

Up 5% in 1999

CRIME RATES

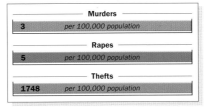

Murders
3   per 100,000 population

Rapes
5   per 100,000 population

Thefts
1748   per 100,000 population

Smuggling is the most significant problem, and Warsaw is a main center for this. Narcotics are transferred westward to Germany and expensive cars eastward to Russia. A National Remembrance Institute was set up in mid-2000 to investigate and prosecute the Nazi- and communist-era crimes of 1939–1989.

EDUCATION

 School leaving age: 18

99%   1.43m students

THE EDUCATION SYSTEM

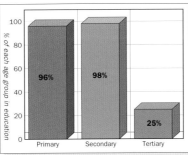

% of each age group in education

100
80
60   96%   98%
40
20   25%
0   Primary   Secondary   Tertiary

Primary education lasts from the age of seven to 13; lower secondary level follows until 16. At upper secondary level, exam-based selection separates the academic, technical, and vocational schools. A standard curriculum is followed in all schools. Despite the high official literacy figures, a relatively large proportion of school-leavers still lack basic skills. Public spending on education fell in real terms in the 1990s. Since 1989 the Roman Catholic Church has been allowed to operate schools. Most of the higher education institutions offer business-related courses.

PUBLISHING AND BROADCAST MEDIA

There are 52 daily newspapers, including *Gazeta Wyborcza*, *Rzeczpospolita*, *Kurier Polski*, and *Zycie Warszawy*

3 independent services

6 independent services

HEALTH

 Welfare state health benefits

1 per 435 people

Arteriosclerosis, heart disease, cancers, accidents

Fundamental reforms introduced in 1999 created a "market" health system, giving patients the right to choose where to go for treatment. Intended to be decentralized and less bureaucratic, the new system was confusing for some patients, with hospitals and doctors competing for business. Medical care is free for most people, but there are now a number of private health clinics.

SPENDING

 GDP/cap. increase

CONSUMPTION AND SPENDING

240 per 1000 population

282 per 1000 population

Defense 2%
Education 5.4%
Health 4.7%

0   5   10   15   20   25

Defense, Health, Education spending as % of GDP

Market reforms have led to some structural unemployment, and the inevitable hardship that this represents. More restructuring of heavy industry is planned. Pensioners have enjoyed benefits amounting to a higher percentage of GDP than in most countries, but state cutbacks are making private pensions more necessary.

WORLD RANKING

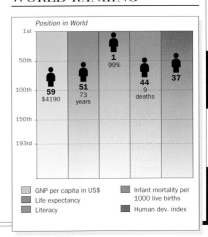

Position in World

1st
50th
100th
150th
193rd

59    $4190
51   73 years
1   99%
44   9 deaths
37

GNP per capita in US$
Life expectancy
Literacy

Infant mortality per 1000 live births
Human dev. index

P

PORTUGAL

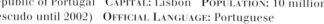

OFFICIAL NAME: Republic of Portugal   **CAPITAL:** Lisbon   **POPULATION:** 10 million
CURRENCY: Euro (escudo until 2002)   **OFFICIAL LANGUAGE:** Portuguese

 1139
 1640
 June 10
 P
 0
 +351
.pt

PORTUGAL, WITH ITS long Atlantic coast, lies on
the western side of the Iberian peninsula. The
river Tagus divides the more mountainous north
from the lower, undulating terrain to the south. In 1974, a bloodless
military coup overthrew a long-standing conservative dictatorship.
A constituent assembly was elected in 1975 and the armed forces
withdrew from politics thereafter. Portugal then began a substantial
program of economic modernization and accompanying social change.
Membership of the EU has helped underpin this process.

Santa Marta de Penaguiao, *a small village
in the heart of Portugal's wine-producing
region, which is centered on the Douro valley.*

CLIMATE   ▷ Mediterranean/ maritime

WEATHER CHART FOR LISBON

Portugal has a mild, Mediterranean
climate, which is moderated by the
influence of the Atlantic. Summers
can be hot and sultry, while winters
are relatively mild. Inland areas have
more variable weather than coastal
regions. Rainfall is generally higher
in the mountainous north, while the
central areas are more temperate.
The southern Algarve region is
predominantly dry and sunny.

TRANSPORTATION   ▷ Drive on right

 Portela de Sacavém, Lisbon
9.4m passengers

 442 ships
1.13m grt

THE TRANSPORTATION NETWORK

59,110 km (36,729 miles)	883 km (549 miles)
2813 km (1748 miles)	820 km (510 miles)

Road links with Spain remain limited,
despite modernization schemes and the
new southern Guadiana bridge. The
Lisbon–Madrid expressway was finally
completed in 1999, the year after the
18-km Vasco da Gama bridge in Lisbon
opened. Poor road construction, heavy
traffic, and dangerous driving mean
that Portugal has Europe's highest rate
of road deaths. Lisbon's small, efficient
metro complements its trams, but
Porto's metro remains uncompleted.
A high speed rail link with Spain is
due for completion in 2008.

TOURISM   ▷ Visitors : Population 1.2:1

 12m visitors   ⬆ Up 4% in 2000

MAIN TOURIST ARRIVALS

Spain 76%	
UK 7%	
Germany 4%	
France 3%	
Netherlands 2%	
Other 8%	

0   10   20   30   40   50   60   70   80
% of total arrivals

From the 1960s, Portugal's popularity
as a tourist destination has been linked
in part to qualities which reflected its
relatively poor economic development,
such as low prices and little crime.
Thus some of the consequences
of its substantial economic growth
may have eroded part of Portugal's
appeal, but since it now has around
12 million visitors a year, tourism
remains a major income-earner.
The most popular destination is the
Algarve, the southernmost province,
followed by the western resorts
of Figueira da Foz and the Tróia
Peninsula. Visitors are also attracted
by Portugal's architecture, notably
that dating from the Manueline period
(1490–1520), and by its handicrafts,
such as ceramics, lace, and tapestries.
In addition, Portugal is noted for being
the location of some of Europe's finest
golf courses.

PORTUGAL

Total Land Area : 92 391 sq. km (35 672 sq. miles)

Azores

Madeira Is

POPULATION

over 500 000 ⊙
over 100 000 ◎
over 50 000 ○
over 10 000 •

LAND HEIGHT

1000m/3281ft
500m/1640ft
200m/656ft
Sea Level

PEOPLE ▷ Population density medium

👤 Portuguese　　　　　🏃 109/km² (282/mi²)

THE URBAN/RURAL POPULATION SPLIT

64%　　　　36%

RELIGIOUS PERSUASION

Protestant 1%　Other 2%

Roman Catholic 97%

ETHNIC MAKEUP

African and Other 2%

Portuguese 98%

Portuguese society, once regarded as rather inward-looking, has become much more egalitarian since the 1974 revolution. It is increasingly integrated into the rest of western Europe.

The Roman Catholic Church has lost some of its social influence, as shown by falling birthrates and more liberal attitudes to abortion, divorce, and unmarried mothers (almost one in five children are born outside marriage). Apart from urban areas, the north is still devoutly Catholic.

Ethnic and religious tensions are limited. Immigration increased after 1974, and foreigners now constitute 2% of the population. Early arrivals came from the former African colonies, but recently there has been an influx of east European workers.

Family ties remain all-important. Women got the vote only in 1976; now 60% of university students are women, and 63% of women of working age have jobs.

POPULATION AGE BREAKDOWN

Female		Age	Male	
	1.9%	80+	1%	
	9.7%	60–79		7.6%
	12.6%	40–59		11.5%
	15.4%	20–39		15.2%
	12.3%	0–19		12.8%

% of population by age group

POLITICS ▷ Multiparty elections

2002/2006　　　　President Jorge Sampãio

AT THE LAST ELECTION

Assembly of the Republic 230 seats

5% CDU

46% PSD　　42% PS　　6% PP　1% BE

PSD = Social Democratic Party
PS = Socialist Party　**PP** = People's Party
CDU = United Democratic Coalition　**BE** = Left Bloc

Portugal is a multiparty democracy.

PROFILE

A decade of center-right government ended in the 1995 elections, when the PSD lost to the PS. Having campaigned on a platform of social reform, the PS, under António Guterres, gave priority to fiscal control. Despite a privatization program, however, the size of the public sector actually increased in 1995–1999. The PS was reelected in 1999, but soon lost popularity, and was ousted by the PSD in early elections in March 2002. PSD leader José Manuel Durão Barroso was appointed prime minister. Lacking an overall majority, the PSD allied with the right-wing PP. The new government pledged to cut public spending and accelerate privatization.

MAIN POLITICAL ISSUES
Transformation
Portugal has been transformed in recent decades. The consolidation of democracy since the 1974 "carnation revolution" and EU membership since 1986 have brought Portugal into the European mainstream. There are high expectations about completing the "catching-up" process. Expo '98 in Lisbon showed off the new confidence. In 2002 Portugal was among the first 12 EU countries fully to adopt the euro.

Presidency and parliament
For ten years up to 1995, the presidency and the government were controlled by opposing parties, a situation which encouraged conflict and obstruction. The relationship then began to function more smoothly, with PS presidents and minority governments. Former PS leader Jorge Sampãio, who succeeded Mário Soares as president in 1996, was reelected comfortably in early 2001, only to see a PSD government reclaim power a year later.

President Jorge Sampãio, *socialist president since 1996.*

José Manuel Durão Barroso, *PSD prime minister since 2002.*

WORLD AFFAIRS ▷ Joined UN in 1955

EU　CE　NATO　OECD　OSCE

Since 1986, Portugal's foreign policy has dealt mainly with the consequences of EU membership. It is a committed NATO member, though its relative strategic importance declined after Spain joined. Relations with its former African colonies, occasionally turbulent, are a high priority, as are those with Brazil. Portugal backed East Timor's struggle to reverse its annexation by Indonesia. Relations with China were cordial enough to ensure the smooth return of Macao to the latter at the end of 1999.

AID ▷ Donor

💲 $271m (donations)　⬇ Down 2% in 2000

Portugal is a major beneficiary of EU aid from the so-called structural funds. It currently earmarks 0.25% of its GDP for aid to developing countries. More than 60% goes to former colonies in Africa, especially Mozambique, where Portuguese funding helped rebuild the massive war-damaged Cahora Bassa Dam and power plant.

CHRONOLOGY

Portugal has existed as a nation state since 1139, although it was frequently challenged by Spain. It reached its zenith in the 16th century, before being annexed by Spain in 1580.

- ❏ **1640** Independence from Spain.
- ❏ **1755** Earthquake destroys Lisbon.
- ❏ **1793** Joins coalition against revolutionary France.
- ❏ **1807** France invades; royal family flees to Brazil.
- ❏ **1808** British troops arrive under Wellington. Start of Peninsular War.
- ❏ **1810** French leave Portugal.
- ❏ **1820** Liberal revolution.
- ❏ **1822** King John VI returns and accepts first Portuguese constitution. His son Dom Pedro declares independence of Brazil.
- ❏ **1834** Dom Pedro returns to Portugal to end civil war and installs his daughter as Queen Mary II.
- ❏ **1875–1876** Republican and Socialist parties founded.
- ❏ **1891** Republican uprising in Porto.
- ❏ **1908** Assassination of King Carlos I and heir to the throne.
- ❏ **1910** Abdication of Manuel II and proclamation of the Republic. Church and state separated.
- ❏ **1916** Portugal joins Allied side in World War I.
- ❏ **1917–1918** New Republic led by Sidónio Pais. ⇨

P

CHRONOLOGY *continued*

- ❏ **1926** Army overturns republic.
- ❏ **1928** António Salazar joins government as finance minister. Economy improves significantly.
- ❏ **1932** Salazar prime minister.
- ❏ **1933** Promulgation of the constitution of the "New State," instituting right-wing dictatorship.
- ❏ **1936–1939** Salazar assists Franco in Spanish Civil War.
- ❏ **1939–1945** Portugal neutral during World War II, but lets UK use air bases in Azores.
- ❏ **1949** Founder member of NATO.
- ❏ **1955** Joins UN.
- ❏ **1958** Américo Thómas appointed president, following fraudulent defeat of Gen. Humberto Delgado.
- ❏ **1961** India annexes Goa. Guerrilla warfare breaks out in Angola, Mozambique, and Guinea.
- ❏ **1970** Death of Salazar, incapacitated since 1968; succeeded by Marcelo Caetano.
- ❏ **1971** Caetano attempts liberalization.
- ❏ **1974** Carnation Revolution – left-wing Armed Forces Movement overthrows Caetano.
- ❏ **1974–1975** Portuguese possessions in Africa attain independence. Some 750,000 Portuguese expatriates return to Portugal.
- ❏ **1975** Communist takeover foiled by moderates and Mário Soares' PS.
- ❏ **1975–1976** Indonesia seizes former Portuguese East Timor unopposed.
- ❏ **1976** Gen. António Eanes elected president. New constitution. Soares appointed prime minister.
- ❏ **1978** Period of nonparty technocratic government instituted.
- ❏ **1980** Center-right wins elections. Gen. Eanes reelected.
- ❏ **1982** Full civilian government formally restored.
- ❏ **1983** Soares becomes caretaker prime minister; PS is majority party.
- ❏ **1985** Anibal Cavaco Silva becomes prime minister. Minority PSD government.
- ❏ **1986** Soares elected president. Portugal joins EU, which funds major infrastructure and construction projects.
- ❏ **1987** Cavaco Silva wins absolute majority in parliament.
- ❏ **1991** Soares reelected president.
- ❏ **1995** PS wins elections; António Guterres becomes prime minister.
- ❏ **1996** Former PS leader Jorge Sampãio elected president.
- ❏ **1999** PS strengthens its position in general election. December, Macao returned to China.
- ❏ **2001** Sampãio reelected.
- ❏ **2002** Euro fully adopted. PSD wins early elections, forms coalition with PP. José Manuel Durão Barroso becomes prime minister.

DEFENSE

 ▷ Phasing out conscription

 $2.2bn ⬇ Down 5% in 2000

Portugal has been a member of NATO since 1949. It has a small but relatively modern navy. The army and air force are less efficient. The government announced in 1999 that it planned to abolish compulsory miltary service (currently of four months, with a seven-month civilian alternative). The US, which is the major arms supplier, has a strategic air base in the Azores.

PORTUGUESE ARMED FORCES

🛡	187 main battle tanks (86 M-48A5, 101 M-60)	25,400 personnel
🚢	2 submarines, 6 frigates, and 31 patrol boats	10,800 personnel
✈	66 combat aircraft (40 *Alpha Jet*, 20 F-16A/B)	7400 personnel
🚀	None	

ECONOMICS

 ▷ Inflation 5.3% p.a. (1990–2000)

📊 $111bn 💲 1.0651–1.1231 euros

SCORE CARD

❏ WORLD GNP RANKING	33rd
❏ GNP PER CAPITA	$11,120
❏ BALANCE OF PAYMENTS	–$10.6bn
❏ INFLATION	2.9%
❏ UNEMPLOYMENT	4%

ECONOMIC PERFORMANCE INDICATOR

EXPORTS

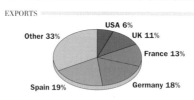

Other 33% | USA 6% | UK 11% | France 13% | Germany 18% | Spain 19%

IMPORTS

UK 6% | Italy 7% | Other 37% | France 11% | Germany 14% | Spain 25%

STRENGTHS

Relatively low, if rapidly rising, labor costs. Flexible labor market. High domestic and direct foreign investment. Strong banking sector. Potential for growth of tourism, now earning 6% of GDP – highest ratio in EU. Clothing and shoe manufacturing now joined by cars (notably Volkswagens) and machinery as major exports. Fast-track improvement of transportation infrastructure. Good deepwater port at Lisbon. Wine, especially port. Tomatoes, citrus fruit, cork, sardines.

WEAKNESSES

High dependence on imported oil. Reliance on public works to drive economic growth. Large agricultural sector (4% of GDP, 13% of workforce) most inefficient in EU. Outdated farming methods, small landholdings, low crop yields; product prices undercut by Spain.

PROFILE

EU membership in 1986 brought a sharp increase in foreign investment to largely rural Portugal. Exports rose dramatically. The economy recorded growth of over 3% a year in the second half of the 1990s. Maintaining growth thereafter required a big emphasis on raising labor productivity.

At one time, participation in EU monetary union seemed a distant prospect. In the event the socialist government of António Guterres (1995–2001) was able to meet the required economic criteria with relative ease, and in January 2002 Portugal was among the 12 EU countries which fully adopted the euro. While Portuguese wages are about 70% of the EU average, the unemployment rate is among the lowest in the EU.

PORTUGAL : MAJOR BUSINESSES

Steel	🛠
Wine	🍇
Textiles	✳
Cement	⚙
Ceramics	🏺
Chemicals	🧪
Vehicle manufacture	🚗
Light engineering	⚙
Fish processing	🐟
Shipbuilding	⚓

0   100 km
0   100 miles

Braga
Matosinhos
Porto
Aveiro
Lisbon
Setúbal
Faro

RESOURCES

 Electric power 9.8m kw

215,230 tonnes

Not an oil producer; refines 294,000 b/d

7m turkeys, 5.9m sheep, 2.35m pigs, 35m chickens

Coal, limestone, granite, marble, tin, copper, tungsten

ELECTRICITY GENERATION

Hydro 34% (13bn kwh)

Combustion 66% (26bn kwh)

Nuclear 0%

Other 0%

% of total generation by type

Portugal is disadvantaged by a lack of natural resources, including water. Mining has historically been important,

notably for tungsten, copper, and tin. The last coal mine closed in the mid-1990s. The fish catch, once central to the economy, has been declining in recent years.

PORTUGAL : LAND USE

Cropland
Pasture
Forest
Vineyards
Cereals
Sheep

0   100 km

0   100 miles

ENVIRONMENT

 Sustainability rank: 28th

7%

5.5 tonnes per capita

ENVIRONMENTAL TREATIES

Yes   Yes   Yes

Yes   Yes   Yes

The unrestricted development of tourist resorts in the Algarve and major infrastructure projects are having detrimental effects on natural habitats. EU agricultural grants for projects such as draining meadows, and monoculture afforestation, notably of eucalyptus and pine, are degrading biodiversity. Much toxic waste is dumped on any available land, as few official controls or infill sites exist. New waste management regulations are being introduced.

MEDIA

 TV ownership high

Daily newspaper circulation 73 per 1000 people

PUBLISHING AND BROADCAST MEDIA

There are 31 daily newspapers. The most prestigious is the *Diário de Notícias*, but the weekly *Expresso* has a wider circulation

1 state-owned service, several independent services

1 state-owned service, several independent national and regional services

Most newspapers have only regional distribution. TV is the dominant medium. In 1992–1993 two independent TV stations were launched. The Catholic TVI was sold in 1998 to Media Capital, one of four groups that control most of the press and broadcasting. The government increased its control over state-owned RTP in 2002, raising fears over its political independence.

CRIME

 No death penalty

13,106 prisoners

Up 1% in 1999

CRIME RATES

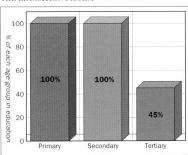

Murders

3   *per 100,000 population*

Rapes

1   *per 100,000 population*

Thefts

406   *per 100,000 population*

Portugal enjoys a low crime rate. Consumption and possession of small quantities of narcotics were decriminalized in mid-2000 and legalized a year later.

EDUCATION

 School leaving age: 15

92%

364,034 students

THE EDUCATION SYSTEM

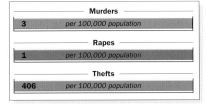

100%   100%   45%

Primary   Secondary   Tertiary

% of each age group in education

Portuguese is the sixth most widely spoken language in the world.

Free state education is available to all pupils between the ages of three and 15. Nursery provision has been greatly expanded, although the preschool stage is not compulsory. Middle-class parents rely heavily on the private sector. State universities have been expanded to ease the pressure on places. There are several prestigious private universities.

HEALTH

 Welfare state health benefits

1 per 313 people

Heart and cerebro-vascular diseases, cancers, accidents

Of total government expenditure nearly 10% is spent on health.

Portugal has had a publicly funded, free national health service since 1979; Spending on health has increased markedly in recent years, but care remains below the EU average. There are strong regional differences in facilities. Larger urban hospitals are modern and well equipped. Private health care schemes, which are allowed to coexist, are both affordable and good value for money; over 40% of the population use the private system.

SPENDING

GDP/cap. increase

CONSUMPTION AND SPENDING

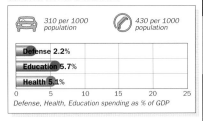

310 per 1000 population

430 per 1000 population

Defense 2.2%

Education 5.7%

Health 5.1%

0   5   10   15   20   25

Defense, Health, Education spending as % of GDP

Wealth differentials in Portugal are smaller than in most EU countries. The 1976 constitution committed Portugal to making the transition to socialism, and since then governments have introduced limited wealth redistribution measures.

Internal investment is directed chiefly through the property market, and there was a surge of purchases in the late 1990s. External investment goes to the EU and Brazil. In 2000, 14.5% of the population held shares directly. Average incomes, which were just over half the EU average in 1986, are now three-quarters of the EU average. Mobile phones, mainly on prepayment systems, pioneered in Portugal, are very widespread.

WORLD RANKING

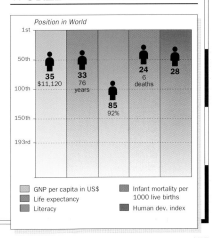

Position in World

1st
50th
100th
150th
193rd

35   $11,120
33   76 years
85   92%
24   6 deaths
28

GNP per capita in US$
Life expectancy
Literacy

Infant mortality per 1000 live births
Human dev. index

QATAR

OFFICIAL NAME: State of Qatar   **CAPITAL:** Doha
POPULATION: 575,000   **CURRENCY:** Qatar riyal   **OFFICIAL LANGUAGE:** Arabic

PROJECTING NORTH FROM the Arabian peninsula into the Gulf, Qatar is mostly flat, semiarid desert. Oil production began in the late 1940s and quickly transformed Qatar from an impoverished pearl producer into a prosperous shaikhdom and a founder member of OPEC. Plentiful oil and gas reserves have made it one of the wealthiest states in the region. Politics is being democratized gradually under the ruling al-Thani clan.

CLIMATE   ▷ Hot desert

WEATHER CHART FOR DOHA

The climate is hot and sultry, with midsummer temperatures reaching 44°C (111°F). Rainfall is rare.

TRANSPORTATION   ▷ Drive on right

Doha International
2.83m passengers

64 ships
744,131 grt

THE TRANSPORTATION NETWORK

| 1107 km (688 miles) | None |
| None | None |

There is no public transportation system to speak of outside Doha, and the rate of car ownership is high.

TOURISM   ▷ Visitors : Population 1:1.3

451,000 visitors

Up 6% in 1998

MAIN TOURIST ARRIVALS

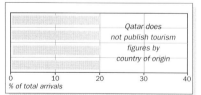

Qatar does not publish tourism figures by country of origin

0   10   20   30   40
% of total arrivals

Tourism is expanding. A government drive to improve Qatar's image as a tourist destination aims to net more than 1.5 million tourists a year by 2010. Attractions include unspoiled beaches, duty-free shopping, and modern hotels.

PEOPLE   ▷ Pop. density medium

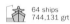 Arabic

52/km² (135/mi²)

THE URBAN/RURAL POPULATION SPLIT

93%   7%

ETHNIC MAKEUP

Arab 40%
Iranian 10%
Other 14%
Indian 18%
Pakistani 18%

Only one in five inhabitants is native-born. Most are guest workers from the Indian subcontinent, Iran, and north African countries. Western expatriates enjoy a high standard of living and take no part in politics.

Most Qataris are followers of the Wahhabi interpretation of Sunni Islam and espouse conservative religious views. However, women are not obliged to wear a veil and can hold a driving license. Expatriate Christians are allowed freedom to worship but not to promote Christianity.

Since the advent of oil wealth, the Qataris, who were formerly nomadic Bedouins, have become a nation of city dwellers. Almost 90% of the population now inhabit the capital Doha and its suburbs. As a result, northern Qatar is dotted with depopulated and abandoned villages.

Doha, the capital. Although desert covers the whole country, Qatar now grows most of its own vegetables by tapping groundwater.

POLITICS   ▷ No legislative elections

Not applicable

Amir Shaikh Hamad bin Khalifa al-Thani

LEGISLATIVE OR ADVISORY BODIES
Advisory Council 35 seats

Qatar is an absolute monarchy and has no legislature. The amir rules with the assistance of the Council of Ministers and the Advisory Council.

Qatar is a traditional emirate. The government and religious establishment is dominated by the amir, Shaikh Hamad, who took power from his father, Shaikh Khalifa, in 1995. A failed coup against Hamad in early 1996 was linked with efforts to regain power by Khalifa. The prodemocracy movement has called for reform of the 35-member Advisory Council. Shaikh Hamad responded by authorizing Qatar's first elections, to a new municipal council for Doha, in 1999, in which all adults, including women, were able to vote and stand as candidates. In early 2001 he promised that a directly elected parliament would be established before 2003.

WORLD AFFAIRS   ▷ Joined UN in 1971

 AL   OIC   GCC   OAPEC   OPEC

Although Qatar was a founder member of the GCC, Shaikh Hamad has adopted a somewhat ambivalent stance toward it. However, Qatar entered into the region's first mutual defense pact, under the aegis of the GCC, in 2000. In 2001 Qatar reached an agreement on the border with Saudi Arabia, but lost its claim to the Hawar Islands when the International Court of Justice ruled in Bahrain's favor.

Although keen to retain strong links with the West, the amir has criticized the US and the UK for their bombing of Iraq and over the campaign in Afghanistan. Qatar supplied liquified natural gas (LNG) to Israel in the late 1990s, but relations, previously the most cordial of the Gulf states, have suffered greatly since the renewal of the Palestinian *intifada* in 2000.

AID   ▷ Recipient

 No net receipts

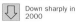 Down sharply in 2000

Qatar was a generous aid donor to developing countries during the 1970s and early 1980s. The small amounts of aid received come mainly from France.

Q

QATAR

Total Land Area : 11 437 sq. km
(4416 sq. miles)

POPULATION

over 100 000	◎
under 10 000	•

LAND HEIGHT

200m/1640ft

Sea Level

DEFENSE

▷ No compulsory military service

$1.43bn ⬇ Down 3% in 2000

The estimated 12,000-strong armed forces are too small to play a significant role in Qatari affairs, even in the event of political turmoil. A defense agreement with the USA has provided for joint exercises, the stockpiling of US equipment, and US access to bases.

ECONOMICS

▷ Inflation 2.5% p.a. (1990–2000)

$7.4bn   3.6393–3.6395 Qatar riyals

SCORE CARD

- ❏ WORLD GNP RANKING..........................96th
- ❏ GNP PER CAPITA $11,570
- ❏ BALANCE OF PAYMENTS..................–$1.66bn
- ❏ INFLATION ....–1%
- ❏ UNEMPLOYMENT.......................Not available

STRENGTHS

A steady supply of crude oil and huge gas reserves, plus related industries. Soaring world oil prices in 2000. Modern infrastructure. Budget surplus.

WEAKNESSES

Dependence on foreign workforce. All raw materials imported. Virtually all water has to be desalinated. Large

RESOURCES

▷ Electric power 1.9m kw

4207 tonnes

214,531 sheep, 179,000 goats, 4m chickens

783,000 b/d (reserves 15.2bn barrels)

Oil, natural gas

Qatar has the third-smallest reserves of crude oil within OPEC but abundant reserves of gas (the third-largest in the world), including the world's largest field of gas unassociated with oil.

ENVIRONMENT

▷ Not available

None

85.7 tonnes per capita

The desert hinterland supports little plant or animal life. Most native species are extinct in the wild. Oil pollution has damaged marine life. There are salt flats in the south.

MEDIA

▷ TV ownership high

Daily newspaper circulation 161 per 1000 people

PUBLISHING AND BROADCAST MEDIA

There are 6 daily newspapers, including Ar-Rayah and its English-language companion Gulf Times, Al-'Arab, and Ash-Sharq

2 services: 1 state-controlled

1 state-controlled service

Qatari TV is the most independent in the region; Al-Jazeera offers the leading Arab perspective internationally, especially since the events of 2001.

CRIME

▷ Death penalty in use

 527 prisoners

 Up 19% in 1999

Traditional Islamic punishments have deterred crime. However, narcotics trafficking is on the increase. The incidence of street crime is low.

EXPORTS

USA 3%
UAE 4%
Singapore 7%
South Korea 17%
Japan 45%
Other 24%

IMPORTS

UAE 7%
Germany 8%
UK 8%
USA 10%
Other 56%
Japan 11%

foreign reserves, but new industries depend on cementing agreements with foreign partners. Potential threats to security from Iraq and Iran makes some multinationals wary of investment.

EDUCATION

▷ Schooling is free but not compulsory

 81%

 8475 students

Education is free from primary to university level. The government finances students to study overseas.

HEALTH

▷ Welfare state health benefits

 1 per 794 people

Heart, circulatory, and infectious diseases, cancers

Primary health care is free to Qataris. Hospitals operate to Western standards of care and the government also funds treatment abroad.

SPENDING

▷ GDP/cap. increase

CONSUMPTION AND SPENDING

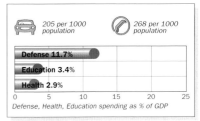

205 per 1000 population

268 per 1000 population

Defense 11.7%
Education 3.4%
Health 2.9%

Defense, Health, Education spending as % of GDP

Qataris have a high per capita income. There is no income tax, public services are free, and the government guarantees jobs for school-leavers. There are no exchange controls.

WORLD RANKING

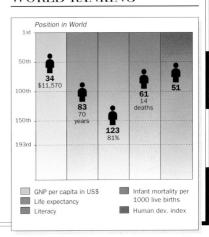

Position in World

34 $11,570
83 70 years
123 81%
61 14 deaths
51

▢ GNP per capita in US$		▢ Infant mortality per 1000 live births
▢ Life expectancy		
▢ Literacy		▢ Human dev. index

Q

ROMANIA

OFFICIAL NAME: Romania   **CAPITAL:** Bucharest
POPULATION: 21.7 million   **CURRENCY:** Romanian leu   **OFFICIAL LANGUAGE:** Romanian

ROMANIA LIES ON THE Black Sea coast, with the Danube as its southern border. The Carpathian Mountains form an arc across the country, curving around the upland basin of Transylvania. Long dominated by Poles, Hungarians, and Ottomans, Romania became an independent monarchy in 1878. After World War II, this was supplanted by a communist People's Republic, headed from 1965 by Nicolae Ceauşescu. A coup in 1989 resulted in his execution and a limited democracy under Ion Iliescu. Defeated in elections in 1996, Iliescu was returned to office in 2000.

Village in northeastern Romania, in the foothills of the Carpathian Mountains, close to the border with Ukraine. Corn and wheat are Romania's main crops.

CLIMATE

▷ Continental

WEATHER CHART FOR BUCHAREST

■ *Average daily temperature*   Rainfall ■

Romania has a continental climate with two growing seasons. Rainfall is generally moderate, with most rain falling in spring and early summer. Very heavy spring rains occasionally destroy new crops. Snow is frequent in winter, which can be bitterly cold.

TRANSPORTATION

▷ Drive on right

✈ **Bucharest–Otopeni International**
1.9m passengers

🚢 389 ships
2.09m grt

THE TRANSPORTATION NETWORK

103,671 km (64,418 miles)

133 km (83 miles)

11,364 km (7062 miles)

1724 km (1071 miles)

The road network is inadequate, and traffic levels are rising. EBRD, EU, World Bank, and Japanese funding has focused on the expressway from Bucharest to Hungary, and on improving major roads. Modernization of the port of Constanţa to include a container port and new grain silo is also under way.

TOURISM

▷ Visitors : Population 1:6.6

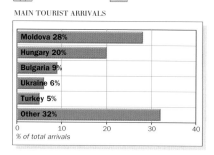

3.27m visitors

Up 2% in 2000

MAIN TOURIST ARRIVALS

| Moldova 28% |
| Hungary 20% |
| Bulgaria 9% |
| Ukraine 6% |
| Turkey 5% |
| Other 32% |

% of total arrivals

The Black Sea, Danube delta, and Carpathian Mountains are the primary natural attractions, while Transylvania has a rich historical heritage. A proposal to build a theme park to exploit the Dracula legend was scrapped in 2002 after an international outcry over its location. However, tourist facilities are generally poor. Under Ceauşescu, the need for foreign currency meant that tourist facilities came before Romanians' own housing needs. Today, privatization of property and an acute housing shortage have reduced the accommodation available to visitors.

ROMANIA

Total Land Area: 237 500 sq. km
(91 699 sq. miles)

POPULATION

over 1 000 000 ▣
over 100 000 ◎
over 50 000 ○

LAND HEIGHT

2000m/6562ft
1000m/3281ft
500m/1640ft
200m/656ft
Sea Level

R

PEOPLE
▷ Pop. density medium

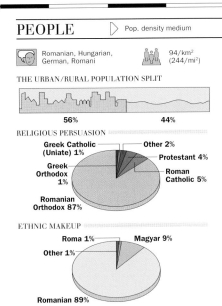

Romanian, Hungarian, German, Romani

94/km² (244/mi²)

THE URBAN/RURAL POPULATION SPLIT

56%   44%

RELIGIOUS PERSUASION

Greek Catholic (Uniate) 1% — Other 2%
Protestant 4%
Greek Orthodox 1%
Roman Catholic 5%
Romanian Orthodox 87%

ETHNIC MAKEUP

Roma 1% — Magyar 9%
Other 1%
Romanian 89%

Since 1989, Romanian nationalism has increased, aggravated by economic austerity measures. The incidence of ethnic violence has also risen, toward Roma and Hungarians in particular. Ethnic Hungarians, or Magyar, who form the largest minority group, are partly protected by the influence of Hungary, whereas the Roma do not have any similar support and tend to suffer greater discrimination.

The population is currently decreasing, due to rising emigration since 1989, mainly for economic reasons, and to a falling birthrate since the early 1990s. The latter trend is in sharp contrast to the 1980s, when the Ceauşescu regime enforced a "pronatalist" policy, banning abortion and contraception; the birthrate rose, but the population as a whole did not grow significantly due to an increase in the mortality rate. Abortion was legalized in 1989; maternal death rates have recently declined. Adoptions by foreigners, spurred partly by shocking conditions in orphanages, were banned for a year in 2001, amid concerns about a "trade in children." Romania, the last country in Europe to lift its ban on homosexuality, did so in 1996, although public acts by homosexuals still receive harsh sentences.

POPULATION AGE BREAKDOWN

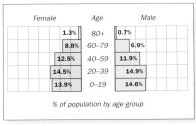

Female		Age	Male	
	1.3%	80+	0.7%	
	8.8%	60–79	6.9%	
	12.5%	40–59	11.9%	
	14.5%	20–39	14.9%	
	13.9%	0–19	14.6%	

% of population by age group

WORLD AFFAIRS
▷ Joined UN in 1955

 BSEC   CE   EBRD   OSCE   CEFTA

Romania's priority is building closer links with western Europe. In 1993, it signed an association agreement with the EU, and in 1995 formally applied for membership. Although not one of the front-runners with which the EU opened negotiations in 1998, it was among six "second wave" candidates which began membership negotiations in 2000.

In 1996 Romania signed a treaty of reconciliation and friendship with Hungary, although relations remain tense as Romania has resisted the demands of the Hungarian minority in Transylvania for greater autonomy. In 1997 Romania also signed a treaty with Ukraine recognizing the latter's sovereignty over parts of Bessarabia and Northern Bukovina.

AID
▷ Recipient

 $432m (receipts)   Up 12% in 2000

Western aid declined somewhat after the mid-1990s, reflecting uncertainty about the implementation of reform. A World Bank strategy in mid-2001 envisaged a significant increase in annual aid if its suggested reforms were given high priority.

POLITICS
▷ Multiparty elections

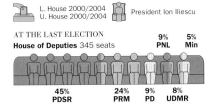

L. House 2000/2004
U. House 2000/2004

President Ion Iliescu

AT THE LAST ELECTION
House of Deputies 345 seats

45% PDSR   24% PRM   9% PD   8% UDMR   9% PNL   5% Min

PDSR = Social Democratic Pole of Romania (led by the Social Democracy Party of Romania) **PRM** = Greater Romania Party
PD = Democratic Party **PNL** = National Liberal Party
UDMR = Hungarian Democratic Union of Romania
Min = Minority representatives: 18 seats in the House of Deputies are reserved for national minorities

Senate 140 seats

46% PDSR   27% PRM   9% PD   9% PNL   9% UDMR

Romania is a multiparty democracy led by a directly elected president.

PROFILE
The 1989 "revolution" left an old communist elite in power, with no group ready to introduce reform. Only the victory of the center-right in 1996 brought more far-reaching change.

Many of the state assets privatized under the first regime of Ion Iliescu remained closely linked to the ruling clique. Now a declared social democrat,

Iliescu retains the support of such conservative groups as miners and rural workers. The coalition parties in power after 1996 came a poor third in the 2000 elections behind Iliescu's PDSR and the extreme nationalist PRM. In mid-2001 the PDSR electoral coalition was formalized in a merger of its main two constituent parties to form the Social Democrat Party (PSD).

MAIN POLITICAL ISSUES
Economic performance
Poverty has grown, while halfhearted attempts at economic reform failed to achieve structural change. Strikes are frequent and protests by miners have been especially forceful, notably against the center-right coalition in power in 1996–2000.

Ethnic tensions
Economic difficulties have led to increased ethnic tensions. The far right has made political gains and nationalism is increasingly accepted. Roma are victims of violent, racially motivated attacks. Benefits given to ethnic Hungarians by the Hungarian government have provoked an outcry from both the Romanian authorities and ordinary Romanians in mixed communities.

Ion Iliescu,
Romania's first post-communist president, reelected in 2000.

Adrian Nastase,
prime minister since December 2000.

R

CHRONOLOGY
Many foreign policy tensions stem from Romania's continually redrawn borders. It retains a Hungarian minority in Transylvania. Post-Soviet Moldova opted not to rejoin Romania.

❑ **1859** Unification of Moldavia and Wallachia forms basis of future Romania.
❑ **1878** Independence, but at cost of losing eastern Moldavia to Russia.
❑ **1916–1918** Enters World War I on Allied side. At end of war, gains substantial territory, including Transylvania from Hungary. ⇨

CHRONOLOGY *continued*

- **1924** Communists banned in unstable political arena. Rise of fascist "Iron Guard."
- **1938** King Carol establishes royal autocracy.
- **1940** Territory forcibly ceded to Soviet Union, Bulgaria, and Hungary. Coup by Iron Guard. King Carol abdicates in favor of son, Michael. Tripartite Pact with Germany.
- **1941** Enters war on Axis side, hoping to recover Moldavian lands.
- **1944** Romania switches sides as Soviet troops reach border.
- **1945** Soviet-backed regime installed. Romanian Communist Party plays an increasing role.
- **1946** Romania regains Transylvania. Eastern Moldavia reverts to USSR, which also demands huge reparations. Communist-led National Democratic Front wins majority in disputed elections.
- **1947** Michael forced to abdicate.
- **1948–1953** Centrally planned economy put in place.
- **1953** Leaders of Jewish community prosecuted for Zionism.
- **1958** Soviet troops withdraw.
- **1964** Prime Minister Gheorghiu-Dej declares national sovereignty. Proposes joint planning by all communist countries to lessen Soviet economic control.
- **1965** Ceauşescu party secretary after death of Gheorghiu-Dej.
- **1968–1980** Condemns Soviet invasion of Czechoslovakia; courts US and European Communities.
- **1982** Ceauşescu vows to pay off foreign debt.
- **1989** Demonstrations; many killed by military. Armed forces join with opposition in National Salvation Front (NSF) to form government. Ion Iliescu declared president. Ceauşescu summarily tried and shot.
- **1990** NSF election victory. Political prisoners freed.
- **1991** New constitution, providing for market reform, approved.
- **1992** Second free elections. NSF splits. Nicolae Vacaroiu forms minority government.
- **1994** General strike demands faster economic reform.
- **1996** Reconciliation treaty with Hungary. Center right wins elections, breaking with communist past; Emil Constantinescu president.
- **1997** Treaty recognizes Ukraine's sovereignty over territory ruled by Romania in 1919–1940.
- **1998** Coalition differences, Prime Minister Victor Ciorbea resigns.
- **2000** December, Ion Iliescu and social democrats win elections.

DEFENSE

 Compulsory military service

 $809m

Up 33% in 2000

The military received limited funding under the Ceauşescu regime, and troops were routinely deployed as cheap labor. Romania was the first country to join NATO's Partnerships for Peace program in 1994. Since 1996 the government has actively sought membership of NATO itself. Romanian soldiers were deployed in the US-led "war on terrorism" in Afghanistan from 2002.

ROMANIAN ARMED FORCES

1373 main battle tanks (821 T-55, 30 T-72, 314 TR-85, 208 TR-580)	52,900 personnel
1 submarine, 6 frigates, 1 destroyer, and 61 patrol boats	10,200 personnel
307 combat aircraft (72 IAR-93, 180 MiG-21, 27 MiG-23, 18 MiG-29)	18,900 personnel
None	

ECONOMICS

 Inflation 98% p.a. (1990–2000)

 $37.4bn

25,925–31,745 Romanian lei

SCORE CARD

- ❏ WORLD GNP RANKING..........................53rd
- ❏ GNP PER CAPITA $1670
- ❏ BALANCE OF PAYMENTS.................–$1.36bn
- ❏ INFLATION 45.7%
- ❏ UNEMPLOYMENT 7%

EXPORTS

UK 5% | Turkey 6% | France 7% | Germany 16% | Italy 22% | Other 44%

IMPORTS

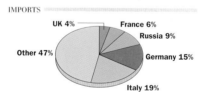

UK 4% | France 6% | Russia 9% | Germany 15% | Italy 19% | Other 47%

STRENGTHS
Many foreign joint ventures. Oil reserves. Tourism potential. Fall in inflation.

WEAKNESSES
Slow transition from centrally planned to market economy. Delays in economic reform. Low foreign investment levels. Large bureaucracy.

PROFILE
Despite being the first east European country to open its economy to foreign investment, Romania was relatively slow to launch economic reforms, and suffered severe recession for most of the 1990s. The chemical, petrochemical, metal, transportation, and food industries are the priority areas for liberalization and structural overhaul. The reform-oriented government elected in 1996 took tough measures to curb inflation and the budget deficit. Output fell sharply, worsened by disruption to trade during the 1999 Kosovo conflict, and not until 2000 did an export-led recovery appear likely.

ECONOMIC PERFORMANCE INDICATOR

Consumer Price Index — GDP. Consumer price index 1995=100. GDP 1996=100. Years: 1996, 1997, 1998, 1999, 2000.

Most farmland has been restored to private hands, and a program was launched in mid-2000 to sell off the remaining state farms by the end of 2002. Agriculture, severely undermechanized, still employs over 35% of the workforce.

From 1990 100% foreign ownership of ventures was permitted. Joint ventures, while now numerous, are small in scale; larger investors are put off by bureaucracy and doubts about stability. The privatization of state-owned enterprises, a priority for the center-right government of 1996–2000, was delayed by problems in the sale of larger companies, but by 2000 the private sector accounted for over 60% of GDP.

ROMANIA : MAJOR BUSINESSES

Cluj-Napoca, Braşov, Timişoara, Galaţi, Hunedoara, Craiova, Constanţa, Ploieşti, Bucharest

Gas, Oil refining, Chemicals, Metallurgy, Iron & steel, Textiles, Electronics, Pharmaceuticals, Heavy engineering, Vehicle manufacture

0   100 km

0   100 miles

R

RESOURCES
 Electric power 22.6m kw

 16,841 tonnes

130,000 b/d (reserves 1bn barrels)

 7.8m sheep, 5.08m pigs, 4m geese, 73.5m chickens

 Oil, coal, salt, iron, gas, methane, bauxite, copper, lead, zinc

ELECTRICITY GENERATION

Hydro 35% (19bn kwh)
Combustion 55% (29bn kwh)
Nuclear 10% (5.3bn kwh)
Other 0%

% of total generation by type

Romania has large reserves of oil, but there is little left of proven gas reserves. Oil and gas production from onshore fields has fallen for 25 years and meets barely 40% of domestic demand. Since the mid-1990s efforts have been concentrated on developing offshore reserves in the Black Sea, opening up exploration and processing to foreign investors. Deposits of other minerals are small and contribute little to export earnings. Many coal mines have been shut down.

The electricity industry is outdated but does produce a surplus for export. An agreement in 2000 connected the national grid with that of Bulgaria. The removal of price ceilings since 1997, with a consequent doubling of prices, provides strong incentives to users to improve poor efficiency.

ENVIRONMENT
 Sustainability rank: 66th

 5% (4% partially protected)

4.1 tonnes per capita

ENVIRONMENTAL TREATIES

Yes　Yes　Yes
Yes　Yes　Yes

Saddled with the disastrous legacy of its communist-era industry, Romania needs help with a major cleanup. Air pollution, mainly from emissions from cement and power plants but also from exhaust fumes and low-quality coal, is most serious in the south. Cyanide and heavy metal leaked from a gold mine in Baia Mare in 2000 creating a transboundary pollution catastrophe in the Tisza River in Hungary. The Danube delta, despite serious pollution, was designated a UNESCO biosphere reserve in 1998.

MEDIA
 TV ownership high

Daily newspaper circulation 298 per 1000 people

PUBLISHING AND BROADCAST MEDIA
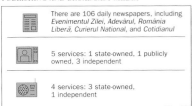

There are 106 daily newspapers, including *Evenimentul Zilei*, *Adevărul*, *România Liberă*, *Curierul National*, and *Cotidianul*

5 services: 1 state-owned, 1 publicly owned, 3 independent

4 services: 3 state-owned, 1 independent

Many of the newspapers which proliferated after 1989 are now closing, since rising prices mean that fewer readers choose to buy them. In practice the government controls the national TV service. The first satellite TV channel was launched in 1994, and the first exclusively Hungarian radio station began broadcasting in 1999.

CRIME
 No death penalty

 507,710 prisoners　Up 28% in 1999

CRIME RATES
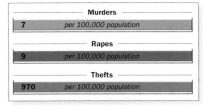

Murders
7　per 100,000 population
Rapes
9　per 100,000 population
Thefts
970　per 100,000 population

The black economy is the main source of income for a third of the population. Levels of tax evasion are extremely high. Romania is a source and transit country for people trafficking.

EDUCATION
 School leaving age: 15

 98%　452,621 students

THE EDUCATION SYSTEM

Primary 100% | Secondary 80% | Tertiary 23%

Attendance at secondary schools is below the European average. As university enrollment is no longer restricted, the number of tertiary students has risen rapidly. The government increased education expenditure in early 2000, pledging to devote at least 4% of GDP in future years and going some way to meet criticisms over the chronic underfunding of the school system.

ROMANIA : LAND USE

Cropland, Pasture, Forest, Wetlands, Potatoes, Cereals, Sheep

0 100 km
0 100 miles

HEALTH
Welfare state health benefits

1 per 556 people

Heart & cerebrovascular diseases, cancers, accidents, tuberculosis

Average life expectancy is among the lowest in Europe; in the worst-polluted parts of Transylvania it is as low as 61 years. The incidence of tuberculosis is the highest in Europe. After years of chronic state underfunding, there was a shift in 1999–2001 toward an insurance-based system.

SPENDING
GDP/cap. decrease

CONSUMPTION AND SPENDING
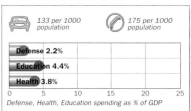

133 per 1000 population　175 per 1000 population

Defense 2.2%
Education 4.4%
Health 3.8%

Defense, Health, Education spending as % of GDP

Real incomes have been hit hard by a decade of economic decline; 40% live below the national poverty line. Most families own their own homes (often overcrowded) and many have small plots of land. Few rural homes have running water or sewerage.

WORLD RANKING
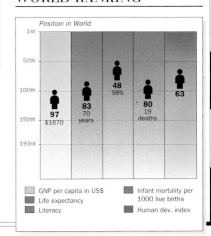

Position in World

97 $1670 | 83 70 years | 48 98% | 80 19 deaths | 63

GNP per capita in US$ | Infant mortality per 1000 live births
Life expectancy | Human dev. index
Literacy

R

RUSSIAN FEDERATION

OFFICIAL NAME: Russian Federation **CAPITAL:** Moscow
POPULATION: 144.7 million **CURRENCY:** Russian rouble **OFFICIAL LANGUAGE:** Russian

 1991   1991   June 12   RUS   +1 to +11   +7 .ru

ASIA
RUSSIAN FEDERATION
Asia

Russia's TERRITORY, bounded by the Arctic and Pacific Oceans to the north and east, extends over 16 million sq. km (6 million sq. miles). This makes it by far the world's largest state, almost twice as big as either the US or China. With borders established for the federative state in 1954, the Russian Federation's sovereign status dates from 1991 – the dissolution of the USSR. Within the CIS, it maintains a traditionally dominant role in central Asia and the Caucasus. Ethnic Russians make up 82% of the population, but there are around 150 smaller ethnic groups, many with their own national territories within Russia's borders. Regionalism and separatism are major political issues. The situation is complicated by the fact that many of these territories are rich in key resources such as oil, gas, gold, and diamonds.

The Kremlin, Moscow. *Rebuilt in 1475 by Ivan the Great, who commissioned architects from Pskov and Italy, it is enclosed by walls 4 km (1.5 miles) long and lies on the Moscow River.*

CLIMATE ▷ Subarctic/continental/ mountain/steppe

WEATHER CHART FOR MOSCOW

Russia has a cold continental climate, characterized by two widely divergent main seasons. Spring and autumn are very brief periods of transition between warm summers and freezing winters. The country is open to the influences of the Arctic and Atlantic to the north and west. However, mountains to the south and east prevent any warming effects from the Indian and Pacific Oceans filtering across. Severe winters affect most regions. Winter temperatures vary surprisingly little from north to south, but fall sharply in eastern regions. The January temperature of –70°C (–94°F) recorded at Verkhoyansk in Siberia is the world record low outside Antarctica.

Housing in Moscow. *Living conditions in major cities can be cramped, with families often sharing their small apartments.*

RUSSIAN FEDERATION

Total Land Area :
17 075 200 sq. km
(6 592 755 sq. miles)

POPULATION

- ▣ over 5 000 000
- ▣ over 1 000 000
- ◉ over 500 000
- ◎ over 100 000
- ○ over 50 000
- ● over 10 000

LAND HEIGHT

- 3000m/9843ft
- 2000m/6562ft
- 1000m/3281ft
- 500m/1640ft
- 200m/656ft
- Sea Level
- –200m/-656ft

TRANSPORTATION

▷ Drive on right

Sheremetyevo-2, Moscow
10.8m passengers

4723 ships
11.1m grt

THE TRANSPORTATION NETWORK

336,000 km (208,780 miles)		None	
86,031 km (53,460 miles)		101,000 km (62,758 miles)	

Russia has a comprehensive transportation network. However, since 1991, all systems have seen some decline due to lack of funding. Cities are still served by good trolley and bus systems and Moscow has one of the most impressive subway systems in the world. In rural areas, car ownership is low and the population relies on an extensive bus service.

About 20% of the railroad track should be renewed annually owing to frost and other damage. Shortage of funds means that this is no longer done. The railroads are heavily used but seriously overburdened and liable to accidents and delays. New track has been laid for the Sokol (Falcon) high-speed rail link between Moscow and St. Petersburg; the first trains to use it in 2000 cut over an hour off the previous minimum journey time and further dramatic reductions are expected.

Roads in major cities are deteriorating, as are interurban highways. Crime is a problem on railroads – notably the Trans-Siberian – and roads.

The former Aeroflot monopoly of air transportation has been broken up. Aeroflot now competes as Aeroflot Russian Airlines, but hundreds of regional "babyflot" airlines run mainly domestic routes, some with alarming accident records.

TOURISM ▷ Visitors : Population 1:6.8

21.2m visitors | Up 15% in 2000

MAIN TOURIST ARRIVALS

Ukraine 42%
Finland 8%
Georgia 6%
Poland 5%
Lithuania 5%
Other 34%

% of total arrivals
0 10 20 30 40 50 60

The breakup of the previous monopoly tourist agency, Intourist, has led to a vast expansion of tourism opportunities: each region is keen to earn hard currency and to attract rich visitors. By 2000 the total number of tourists a year had reached over 21 million.

At the luxury end of the market, trips from St. Petersburg to Tashkent on former president Brezhnev's official train are now available. River trips down the Volga and visits to medieval monasteries are increasingly popular. Tourists can also explore forests, or fish for salmon in the Kola Peninsula. The defense sector has opened up to tourism and now offers flights in MiG jets, or rides in T-84 Russian tanks. Even the space industry has branched into tourism.

Moscow and St. Petersburg remain favorite destinations, where hotels tend either to be for the well-off or of a basic standard. Near St. Petersburg, Novgorod has many fine churches, and the Pskov area is celebrated as the setting for many of Pushkin's works, including *Eugene Onegin* and *Boris Godunov*.

Many parts of Russia remain inaccessible to most tourists. The communist-era ban on foreigners visiting the Urals has been lifted, but the area still has very few facilities. Resorts such as the subtropical Sochi on the Black Sea, where powerful Russians have *dachas* (country houses), have experienced a building boom.

PEOPLE ▷ Pop. density low

Russian, Tatar, Ukrainian, Chavash, various other national languages | 9/km² (22/mi²)

THE URBAN/RURAL POPULATION SPLIT

78% | 22%

RELIGIOUS PERSUASION

Other 25%
Russian Orthodox 75%

ETHNIC MAKEUP

Ukrainian 3% | Tatar 4%
Chavash 1% | Other 10%
Russian 82%

In the former Soviet Union, Russians were just over 50% of the population, but in Russia they are an overwhelming majority. Although significant numbers of Russians still live in some of the neighboring former Soviet republics, notably Ukraine and Latvia, a rise in nationalism has persuaded many to return to Russia.

Within Russia there has also been an increase in ethnic tension, especially in the Caucasus. Hostility toward the Muslim Chechens is particularly acute. There are 57 nationalities with their own territories within the federation and 95 nationalities (but only 6% of the population) without a territory.

Social life in Russia has not changed significantly since the demise of communism. With the lifting of censorship, there has been a greater expression of sexuality and of political and religious views. The expensive rebuilding of Moscow's Church of Christ the Savior symbolized this change. The strong revival of the Russian Orthodox Church is boosted by legal recognition of its "special role" in Russia's history. All religious organizations were required to re-register by 1 January 2000, under conditions which many small minority faiths could not meet, and whereby a church deemed to "violate social order" would be banned.

One marked change of which Russians speak is the growing importance given to money. Mutual support systems of extended friendships are now in decline.

The position of women has changed little since the fall of communism. Many have suffered from the rise in unemployment, but this reflects the demise of part-time or badly paid jobs, rather than gender-motivated social change. Most Russians have very modest living standards and were further impoverished by the collapse of the economy in the late 1990s.

POPULATION AGE BREAKDOWN

Female	Age	Male
1.8%	80+	0.4%
9.3%	60–79	5.1%
13.1%	40–59	11.7%
14.8%	20–39	15.2%
14%	0–19	14.6%

% of population by age group

POLITICS ▷ Multiparty elections

L. House 1999/2003
U. House varying | President Vladimir Putin

AT THE LAST ELECTION

State Duma 450 seats

| 15% FAR | 5% Y | 2% Vac |

| 25% CP | 23% Ind | 16% U | 6% URF | 4% Z | 4% Others |

CP = Communist Party **Ind** = Independents **U** = Unity
FAR = Fatherland–All Russia **URF** = Union of Right Forces
Y = Yabloko **Z** = Zhirinovsky's Bloc **Vac** = Vacant

Council of the Federation 178 seats

Each of 89 regions is represented in the Council of the Federation (Soviet Federatsii) by two members chosen by the regional legislature

The government is responsible to the elected parliament (Duma), but executive power lies firmly with the president.

PROFILE

President Boris Yeltsin's second term (1996–1999) was overshadowed by his health problems, economic crisis, and corruption. Dramatic changes of government personnel were a characteristic of this period, as Yeltsin confronted the Duma in both 1998 and 1999 over his choice of prime minister. The confrontation ended with the appointment of Vladimir Putin, a little-known former head of the Federal Security Service (FSB), who became Yeltsin's favorite to succeed him.

Putin has greatly consolidated his position since coming to power as acting president from the end of 1999. Having stormed to victory in the first round of the 2000 presidential elections, he has tackled the power of the business "oligarchs," and of Russia's 89 regional governors, with his program of centralization. His support within the Duma has been consolidated by the transformation of the Unity bloc (which Putin had formed for the December 1999 legislative elections) into an official party. A merger with the Fatherland

Former president Boris Yeltsin, *renowned for his erratic behavior, stepped down in 1999.*

Mikhail Gorbachev, *whose restructuring ultimately led to the breakup of the Soviet Union.*

R

POLITICS *continued*

bloc in April 2001 made Unity the largest single party in parliament.

Although the conflict in Chechnya damaged Putin's image internationally, it has by contrast been a key element in his domestic appeal as a strong leader. This was shaken by criticism of his handling of the *Kursk* submarine disaster in August 2000, in which 118 sailors died. However, the marked improvement in the economy during his first year in office stood greatly to his credit in popular opinion, as it also did for Prime Minister Mikhail Kasyanov, appointed in May 2000 with revitalizing the economy as his central task.

The Communist Party, which until April 2001 was the single largest party in the Duma, remains powerful because of its effective organization and its ability to appeal to those who have suffered from the upheavals of the post-Soviet period. However, even with parliamentary allies, it has been unable to take power. Its leader Gennady Zyuganov has three times been beaten to the presidency.

MAIN POLITICAL ISSUES
Living standards
Russians are disillusioned at the failure of politicians to improve their living standards. The fall of communism swept away the securities which used to underpin life – long-term employment, guaranteed housing, and a basic diet – hitting the old particularly hard. Uneven and crisis-prone efforts at introducing a market economy under former president Yeltsin created much insecurity.

Crime and corruption
Crime levels rose alarmingly under the post-Soviet regime, and visitors began to be warned against walking the streets of St. Petersburg or Moscow after dark. Widespread bureaucratic corruption was countered by the power acquired by business tycoons, the so-called "oligarchs," who snapped up privatized industries at bargain prices. President Putin has launched a crusade against these oligarchs, but has shied away from a full-scale review of the privatization process.

Vladimir Putin
was handed power by Yeltsin in 1999, and was elected president in 2000.

Mikhail Kasyanov,
economic reformist appointed prime minister in 2000.

Regionalism and separatism
Nation-based separatism is brutally suppressed. Nowhere has this been made clearer than by the ferocious military campaign in Chechnya.

Influence accumulated under the Yeltsin regime by Russia's 89 regional governors was reversed by Putin's efforts to concentrate power in the presidency. Control of police and taxation has been centralized in seven huge federal districts, responsible to Putin, and the governors have been stripped of their seats in the upper house, the Federation Council.

Russia's loss of great power status
The collapse of the USSR and Russia's subsequent economic crisis seriously dented its importance on the global stage, where it once counterbalanced the US. Since 2001, however, Putin has had some success in reclaiming Russia's international influence.

WORLD AFFAIRS ▷ Joined UN in 1945

The September 11, 2001 terrorist attacks on the US laid the foundations for a massive change in Russia's relations with the West.

Before then, Russia had been pursuing an independent foreign policy with a renewed skepticism about Western intentions, especially NATO's eastward expansion. The 1999 NATO bombardment of Yugoslavia, Russia's natural coethnic and coreligionist ally, had pushed relations to a low ebb. Russia also felt threatened by US plans for a new missile defense system (NMD). On top of these underlying tensions was heaped Western criticism of the long-running and often bloody Russian campaign against Muslim separatists in Chechnya.

After September 11, by pledging immediate support to the US-led "war on terrorism," President Putin secured unqualified backing for the Chechnya campaign. This was now viewed by the West as a legitimate battle against "terrorism." By agreeing to share intelligence information as part of the new global war, Russia and NATO appeared to bury their grievances, and in 2002 a new NATO–Russia Council was established, effectively making Russia a junior partner in NATO. In return Russia dropped its fundamental objection to the NMD and agreed to a new nuclear disarmament treaty which was widely seen as favoring US interests.

The successor states of the USSR, the "near abroad," retain close links with Russia, particularly Belarus, which is keen on a full reunification.

St. Basil's Cathedral, Moscow.
Commissioned by Ivan the Terrible, it was built in 1555–1561. The exterior domes were decorated in the 1670s.

AID ▷ Recipient

$ $1.57bn (receipts) ⬇ Down 20% in 2000

Russia has received billions of dollars in aid from Western countries on several occasions to stave off government debt and to promote economic restructuring. Large-scale IMF credits were obtained during the economic crises of the mid- and late 1990s.

DEFENSE ▷ Phasing out conscription

$ $58.8bn ⬆ Up 4% in 2000

RUSSIAN ARMED FORCES

🚜	21,820 main battle tanks (T-34, T-55, T-62, T-64, T-72, T-80, T-90)	321,000 personnel
🚢	56 submarines, 1 carrier, 17 destroyers, 7 cruisers, 10 frigates, and 108 patrol boats	171,500 personnel
✈	1538 combat aircraft (MiG-25/29/31, Su-24/25/27)	184,600 personnel
🚀	776 ICBM, 19 SSBN, 100 ABM	

The loss of the 118-man *Kursk* nuclear submarine in August 2000 symbolized the long-term decline of Russia's military might. Maintaining and using the enormous former communist war machine has proved too expensive. There are plans to axe hundreds of thousands of troops by 2003, and conscription is set to be abolished in 2005. Public anger at conditions within the services has led to pay increases.

Spending on nuclear forces is limited to physical protection of warheads. In 2002, Russia agreed to cut further its nuclear arsenal, expanding on the 2000 Start II Treaty, and slashing the number of warheads by 60%. It also dropped previous objections to the proposed US national missile defense system.

The Northern and Pacific navy fleets are inactive and deteriorating fast. Fourteen admirals were disciplined in 2001 for the "failings" of the Northern fleet.

R

ECONOMIC UPHEAVAL, REGIONALISM, AND CENTRALISM

THE RUSSIAN ECONOMY surprised many by its recovery from dire crisis in 1998, to grow by 9% in 2000 and by 5% even in the adverse global climate of 2001. Massive problems remain, however, many of them relating to the sprawling and decaying industrial infrastructure, particularly in some remoter regions.

The Yeltsin regime in the 1990s had failed to manage an effective reform and restructuring process. When matters came to a head in 1998, collapse was only narrowly averted. The crisis spotlighted the chronic weakness of central finances. Since then, however, the government has overseen a real improvement in its financial position. The combination of devaluation, an international rescue package, and default on domestic debt was further helped by increased tax revenue and the introduction of a uniform rate of income tax in 2001.

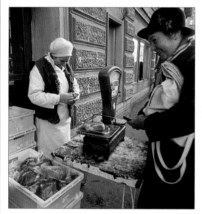

Residents of St. Petersburg *had to queue in the streets for basic foodstuffs.*

Imports became much more expensive after the 1998 devaluation, and Russian-made goods started appearing in Moscow and St. Petersburg again, beginning with big increases in supplies of such things as bicycles, electric kettles, refrigerators, and washing machines. Middle-class Russians in the cities, who had found ways to survive the crisis, were eager buyers. Gazprom, the giant company monopolizing the gas industry, benefited as an exporter from the low rouble and the improvement in world prices, while such industries as aluminum in Sakha (Yakutia) became even more powerful in the local economy.

Although wages and pensions may now be paid with greater regularity, most Russians are still far from catching up with the consequences of years of inflation. Teachers, for example, went five years without pay increases up to 1999. Pensioners could often be seen in the late 1990s attempting to sell possessions from makeshift stalls on street corners. Even by April 2002, as President Putin conceded, 40 million Russians – 28% of the population – were still living below the official poverty line.

VICTIMS OF CHANGE

For many people in the provinces, the dismantling of the old command economy meant the disappearance of what was often the sole source of employment. Far-flung towns had been built exclusively around a particular heavy industry, such as steel, coal, shipbuilding, machine tools, or arms production. Many of these concerns, no longer with a market, were impossible to privatize. Their workers, put on short time or laid off, have been reduced to living on promises, state handouts, and barter arrangements. In cities in the Arctic north, people leave if they can; Murmansk, the largest, has lost a fifth of its population in a decade. In the far east, the collapse of heating and power systems around Vladivostok left residents freezing; the 2000–2001 winter was the bitterest for 70 years.

The disappearance of proper state welfare provisions in areas such as health care makes the situation even more catastrophic. Rates of alcoholism, drug abuse, and suicide have all risen. Most hard pressed are people with neither work nor land, living with children in small urban settlements. Wherever possible, people grow food to be as self-sufficient as possible, and use what they can for barter. Great reliance is also placed on family support networks. The relevance of the formal economy is thereby much reduced, while rents, transportation, and utility bills are still heavily subsidized. Even when cash changes hands, a large proportion of all purchases involve the black market in one way or another.

Heavy industries *have declined dramatically, and many factories now lie idle.*

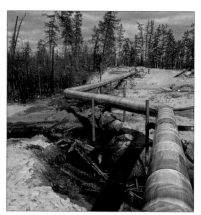

Poor maintenance *has resulted in cracks and leaks along Russia's network of oil pipelines.*

The running down of heavy industries has left some areas with massive environmental damage. One notorious instance is the pollution of rivers by toxic waste from the metallurgy industry in the Kola peninsula in the north. Problems of disposal of radioactive wastes affect northern coastal areas in particular, although the worst such legacy of the former Soviet Union – the effects of the nuclear accident at Chernobyl – has been to Belarus and Ukraine. Western countries agreed in June 2002 to spend $20 billion over the next decade on destroying discarded nuclear material, on environmental and global security grounds.

REPUBLICS AND CENTRAL CONTROL

Some of the 22 republics within the Russian Federation pressed hard for control over their mineral resources, believing that they could be managed more profitably at regional level, and resenting the outflow of their earnings. Regionalism based on dissatisfaction with central economic management began emerging as a key political issue. Tatarstan, lying on the Volga some 1000 km (625 miles) east of Moscow, notably avoided rushing into privatization, conducted elsewhere in such a way as to allow former managers effectively to appropriate huge assets. Its continuing control of the Tatneft oil company provides both a source of revenue and a means of foreign borrowing on relatively favorable terms.

Tatarstan, along with Chechnya, refused to sign the 1992 treaty between the Russian Federation and its constituent parts. Instead, its president, Mintimer Shaimiev, negotiated better autonomy terms, with detailed revenue-sharing provisions, in a 1994 agreement. Extended for a further five years from 1999, this treaty was to be replaced by a

R

new Tatarstan constitution in 2002. However, in spite of compromises expensively won by the allocation of Russian development funding, this constitution went far enough in its claims on Tatar sovereignty to provoke a crisis over federal acceptance.

President Putin has made no secret of his determination to reverse the fragmentation of the federation. Before becoming president he was already closely identified with "firmness" in relaunching a brutal war in 1999 (ostensibly as an antiterrorist operation) against Chechnya, the only republic actively seeking to secede. After his election in 2000 Putin's first major policy action was to group the federation's 89 regional administrative entities under seven federal districts headed by newly appointed presidential envoys. He asserted powers to disband regional legislatures and dismiss regional governors, and removed the governors from the upper house of parliament, the Federation Council, although they do now sit on a new advisory State Council and have been given the right to stand for more than just two terms of office.

Even in Tatarstan, recognizing the need to compromise, Shaimiev (reelected for a third term in 2001) reduced the emphasis on "Tatarization." Federal control over tax collection and spending became more evident again, as did the presence of federal agencies in Tatarstan's capital, Kazan. A push to reintroduce the Latin alphabet in place of Cyrillic in Tatarstan's schools, launched in 2000, was deferred the following year. The population has at least as many ethnic Russians as Tatars, who – like the Chechens – are mainly Muslims and fiercely proud of their distinctive language and culture.

New gas pipelines *being laid across the frozen expanses of Siberia.*

ECONOMICS

▷ Inflation 162% p.a. (1990–2000)

📊 $241bn

💲 28.65–30.50 Russian roubles (official rate)

SCORE CARD

- ❏ WORLD GNP RANKING............................20th
- ❏ GNP PER CAPITA..................................$1660
- ❏ BALANCE OF PAYMENTS....................$46.3bn
- ❏ INFLATION 20.8%
- ❏ UNEMPLOYMENT....................................11%

EXPORTS

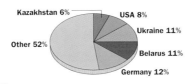

China 5% | Belarus 5% | Italy 7% | USA 8% | Germany 9% | Other 66%

IMPORTS

Kazakhstan 6% | USA 8% | Ukraine 11% | Belarus 11% | Germany 12% | Other 52%

ECONOMIC PERFORMANCE INDICATOR

Consumer Price Index — GDP ▨

Consumer price index 1995=100 / *GDP 1996=100*

550 / 220
400 / 180
250 / 140
100 / 100
／ 60

1996　1997　1998　1999　2000

STRENGTHS

Huge natural resources, in particular hydrocarbons, precious metals, fuel, timber. Potential from future international oil pipelines. Enormous engineering and scientific base. Small enterprises benefited from devaluation of rouble in 1998. Tax reforms increasing government revenue. Lucrative privatizations. Recognized as market economy in 2002, encouraging foreign investment.

WEAKNESSES

Oil profits vulnerable to fluctuating world prices. Crumbling infrastructure. Slow transition to market economy. Privatized companies asset-stripped by former managers. Organized crime controls huge areas of the economy. Regional investment hindered by uneven implementation of federal laws.

Tax evasion and corruption remain widespread. Rising wages and strengthening rouble have promoted boom in imports, narrowing trade gap, harming local production, and stifling small enterprises.

PROFILE

The few gains made in the early post-communist era were swept aside in the 1998 economic crisis. Powerful financial "oligarchs" emerged and organized crime moved into most areas of the economy. However, the devaluation of the rouble ironically served to promote a mini-boom in the last years of the 20th century. Real wages fell, encouraging small enterprises to expand, while the devaluation made imports too expensive for the average consumer – promoting local production. Industrial production increased and GDP grew by 7.6% in 2000. Putin has also moved to dismantle the power of the economic elites. The private sale of land was permitted from 2001, and a new 13% flat rate of income tax promised to help reduce widespread tax evasion. However, while the economy weathered the 2001 global slowdown, the rise in real wages and the strengthening of the rouble have begun to offset the previous years' gains.

R

RUSSIAN FEDERATION : MAJOR BUSINESSES

Nizhniy Novgorod
St. Petersburg
Kazan'
Krasnoyarsk
Moscow
Rostov-na-Donu
Volgograd
Samara
Novosibirsk
Perm'　Yekaterinburg
Vladivostok
Ufa　Chelyabinsk　Irkutsk

West Siberian Plain

Textiles
Chemicals
Metallurgy
Computers
Electronics
Engineering
Vehicle assembly
Defense industry
Oil refining
Oil & gas

0　1000 km
0　1000 miles

CHRONOLOGY

The first Russian state (Rus) was in present-day Ukraine. Occupation by the Tatars (1240–1480) marked the Russian language and character. From the 17th century, the Romanovs ruled an expanding empire.

❏ **1904–1905** Russian war against Japan; ends in defeat for Russia.
❏ **1905** "Bloody Sunday" revolution.
❏ **1909–1914** Rapid economic expansion.
❏ **1914** Enters World War I against Germany.
❏ **1917** February Revolution; abdication of Nicholas II. October Revolution; Bolsheviks take over with Lenin as leader.
❏ **1918** Nicholas II and family shot.
❏ **1918–1920** Civil war.
❏ **1921** New Economic Policy; retreat from socialism.
❏ **1922** USSR established.
❏ **1924** Lenin dies. Leadership struggle eventually won by Stalin.
❏ **1928** First Five-Year Plan: forced industrialization and collectivization.
❏ **1936–1938** Show trials and campaigns against actual and suspected members of opposition. Millions sent to gulags in Siberia and elsewhere. Purges widespread.
❏ **1939** Hitler–Stalin pact gives USSR Baltic states, eastern Poland, and Bessarabia (Moldova).
❏ **1941** Germany attacks USSR.
❏ **1943** February, tide of war turns with lifting of siege of Stalingrad.
❏ **1944–1945** Soviet offensive penetrates Balkans.
❏ **1945** Germany defeated. Under Yalta and Potsdam agreements eastern and southeastern Europe are Soviet zone of influence.
❏ **1947** Cold War begins; Stalin on defensive and fears penetration of Western capitalist values.
❏ **1953** Stalin dies.
❏ **1956** Hungarian uprising crushed. Krushchev's "secret speech" attacking Stalin at party congress.
❏ **1957** Krushchev consolidates power. *Sputnik* launched.
❏ **1961** Yuri Gagarin first man in space.
❏ **1962** Cuban missile crisis.
❏ **1964** Krushchev ousted in coup, replaced by Leonid Brezhnev.
❏ **1975** Helsinki Final Act; confirms European frontiers as at end of World War II. Soviets agree human rights are concern of international community.
❏ **1979** Invades Afghanistan. New intensification of Cold War.
❏ **1982** Brezhnev dies.
❏ **1985** Gorbachev in power. Start of *perestroika*, "restructuring." First of three US–USSR summits resulting in arms ⇨

RESOURCES

 Electric power 211m kw

 4.21m tonnes

7.06m b/d (reserves 48.6bn barrels)

27.3m cattle, 15.7m pigs, 14m sheep, 325m chickens

Coal, oil, gas, gold, diamonds, iron, aluminum, manganese

ELECTRICITY GENERATION

Hydro 19% (159bn kwh)	
Combustion 68% (564bn kwh)	
Nuclear 13% (104bn kwh)	
Other 0%	

0   20   40   60   80   100

% of total generation by type

RUSSIAN FEDERATION : LAND USE

SREDNE SIBIRSKOYE PLOSKOGORIE

▤	Cropland
	Forest
	Pasture
	Wetlands
	Tundra
	High mountain regions
🐂	Cattle
	Cereals
	Potatoes

0   1000 km
0   1000 miles

Russia is a leading world producer of oil, natural gas, and electricity, among other resources. Confirmed reserves make Russia the world's leading country in terms of hydrocarbons, gold, other precious metals, diamonds, and timber.

Unlike some of the other republics of the ex-USSR, Russia has been reluctant to open its resources up to foreign concerns, fearing the loss of control to Western multinationals. The lack of investment and technology is a reason for underexploitation of many of the country's major resources. Geographic remoteness is another.

Also, some of the richest energy and mineral deposits are located in national territories such as Tatarstan and Sakha (Yakutia) in Siberia. The desire of these regions for greater autonomy has turned the ownership of these resources into a delicate political issue.

ENVIRONMENT

 Sustainability rank: 72nd

 3%

9.8 tonnes per capita

ENVIRONMENTAL TREATIES

	Yes		Yes		Yes
	Yes		No		No

Although awareness of Russia's environmental problems has risen sharply, the resources, political will, and know-how to tackle them are still lacking. While Russia now has an active green movement, it has not as yet won significant support in general elections.

Each region has its own particular problems. The northwest risks contamination from the neglected Soviet-era nuclear submarine fleet and from nuclear waste containers dumped in the Barents Sea. Thousands of tonnes of chemical weapons have been dumped in the Baltic, although their exact location has not been revealed. Many fish species are now extinct in the Volga River in central Russia. In the Urals and the cities of European Russia, many chemical and heavy industrial plants do not treat their effluents at all. In 2001 parliament approved a bill to allow the atomic energy ministry Minatom to earn $2 billion a year from storing and reprocessing foreign nuclear waste.

MEDIA

 TV ownership high

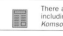 Daily newspaper circulation 105 per 1000 people

PUBLISHING AND BROADCAST MEDIA

There are 285 daily newspapers, including *Izvestiya*, *Rossiiskaya Gazeta*, *Komsomolskaya Pravda*, and *Trud*

2 main national and regional services, partly state-owned, several independent channels

1 main state-run service, broadcasting 2 channels, 1 foreign broadcast service, several independents

Argumenty i Fakty, founded in 1987, is the best-selling weekly paper, with a circulation of nearly three million. The trade union paper *Trud* is the top-selling daily, whereas the once dominant Communist Party daily *Pravda* now sells barely 70,000 copies. The former Soviet state organ *Izvestiya* survives as an independent paper.

Public Russian Television (ORT), established in 1995 under part private ownership, runs two main TV channels. Bias in TV reporting, rife under Yeltsin, continues under Putin. Reporting events in Chechnya is subject to particular pressure. The NTV network, flagship of Media-Most, the media empire of exiled magnate Vladimir Gusinsky, was taken over in 2001 by the state-owned Gazprom's media division, while critical editorial staff were dismissed from the daily *Sevodnya* and weekly *Itogi*. Many Russians have satellite dishes and tune in to CNN and other Western channels.

CHRONOLOGY *continued*

reduction treaties. Nationality conflicts surface.

❏ **1988** Law of State Enterprises gives more power to enterprises; inflation and dislocation of economy.

❏ **1990** Gorbachev becomes Soviet president. First partly freely elected parliament (Supreme Soviet) meets.

❏ **1991** Boris Yeltsin elected president of Russia. Yeltsin and Muscovites resist hard-line communist coup. Gorbachev sidelined. CIS established; demise of USSR.

❏ **1992** Economic shock therapy.

❏ **1993** Yeltsin decrees dissolution of Supreme Soviet and uses force to disband parliament. Elections return conservative State Duma.

❏ **1994** First Russian military offensive against Chechnya.

❏ **1995** Communists win elections.

❏ **1996** Yeltsin reelected despite strong Communist challenge. Peace accord in Chechnya.

❏ **1998** Economic turmoil forces devaluation of rouble. Severe recession, rampant inflation.

❏ **1998–1999** Yeltsin repeatedly changes prime minister in successive political crises.

❏ **1999** December, parliamentary elections. Yeltsin resigns; Prime Minister Putin is acting president.

❏ **1999–2000** Terrorist violence blamed on Islamic separatists in Dagestan and Chechnya. Military offensive against Chechnya; fall of Chechen capital Grozny to Russian forces.

❏ **2000** Putin wins presidential election, consolidates power. Attack on "oligarchs" in big business. Improvement in economy. *Kursk* nuclear submarine disaster.

❏ **2001** April, party mergers make Putin's Unity party the largest grouping in parliament. July, Russian–Chinese friendship treaty.

❏ **2002** April, Putin declares Chechen war "over." May, Cam Ranh Bay base in Vietnam, last Russian outpost beyond former USSR, closed. Nuclear arsenal cut by 60%. NATO–Russia Council formed.

Tundra in Russia's far east. Russia has some of the largest uninhabited tracts of land in the world.

CRIME

 Death penalty not used in practice

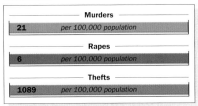

962,700 prisoners ⬆ Up sharply in 1999

CRIME RATES

Murders	
21	*per 100,000 population*

Rapes	
6	*per 100,000 population*

Thefts	
1089	*per 100,000 population*

Policing cannot keep pace with the formidable problems of rising crime in Russia. Intergang violence accounts for a sharp rise in the murder rate. Street crime has also increased in the larger cities. Corruption is rife, particularly in the regions. The Russian mafia profits from protection rackets, prostitution, smuggling operations, and narcotics, and is also active in western Europe.

Public fear resulting from the rise in crime contributed to a temporary rise in popularity for authoritarian political platforms.

With around one million inmates, overcrowding, poor conditions, and disease are major problems in prisons.

EDUCATION

School leaving age: 15

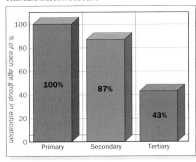 99%   🎓 4.64m students

THE EDUCATION SYSTEM

% of each age group in education

- Primary: **100%**
- Secondary: **87%**
- Tertiary: **43%**

Schooling is free, and is compulsory for nine years up to age 15.

Attempts to change the Soviet-based curriculum, still widely in use, are hampered by lack of funds. Hundreds of private lycées, such as those run by the Orthodox Church, offer courses in west European languages. German in particular has made a comeback as a key language for international commerce. The state-subsidized higher education system is seriously underfunded, and some institutions have begun charging students. Prestigious institutions such as the Academy of Sciences have been forced to cut staff and research. Most academics have to rely on extramural earnings.

HEALTH

Welfare state health benefits

 1 per 238 people   Heart disease, cancers, accidents, violence, tuberculosis

The health care system is in crisis and medicines are often in short supply.

Until 1991, state enterprises provided considerable health care for employees. Employers should now make payments through the Medical Insurance Fund, but many privatized concerns seek to cut costs. Local authorities lack the resources to take over these responsibilities. Bribing medical staff to obtain treatment is commonplace, and there is a lack of pharmaceutical products and drugs. Hospital patients are normally fed by their relatives.

SPENDING

GDP/cap. decrease

CONSUMPTION AND SPENDING

120 per 1000 population   218 per 1000 population

- Defense 5%
- Education 3.5%
- Health 4.6%

Defense, Health, Education spending as % of GDP

Wealth disparities in Russia have increased sharply. A small minority of the population made huge profits from the dismantling of the old Soviet command economy. About 10% are thought to have benefited in some way.

A growing number of dollar millionaires flaunt their wealth, especially in Moscow. The bosses of organized crime are Russian society's wealthiest group. Russia is now the biggest buyer of Rolls Royces, while BMWs, Mercedes, and Volvos are relatively common in Moscow and St. Petersburg. A considerable amount of wealth is deposited abroad. There are thousands of Russian offshore bank accounts; Northern Cyprus is a favorite location.

WORLD RANKING

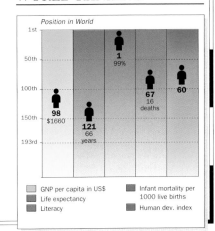

Position in World

- **98** $1660
- **121** 66 years
- **1** 99%
- **67** 16 deaths
- **60**

- GNP per capita in US$
- Life expectancy
- Literacy
- Infant mortality per 1000 live births
- Human dev. index

R

RWANDA

OFFICIAL NAME: Republic of Rwanda   **CAPITAL:** Kigali   **POPULATION:** 7.9 million
CURRENCY: Rwanda franc   **OFFICIAL LANGUAGES:** French, English, and Kinyarwanda

LANDLOCKED RWANDA lies just south of the equator in east central Africa. Since independence in 1962, ethnic tensions have dominated politics. In 1994, the violent death of the president led to appalling political and ethnic violence. Over half of the surviving population were displaced. The perpetrators of the genocide held sway in desperately overcrowded refugee camps in adjacent countries, greatly complicating the process of eventual repatriation and reintegration.

CLIMATE
▷ Tropical wet & dry

WEATHER CHART FOR KIGALI

Rwanda's climate is tropical, tempered by altitude. Two wet seasons allow for two harvests each year.

TRANSPORTATION
▷ Drive on right

Kanombe International, Kigali
121,323 passengers

Has no fleet

THE TRANSPORTATION NETWORK

1000 km (621 miles)		None	
None		Lake Kivu	

The road network is well developed. The international airport near Kigali was completed in 1986.

TOURISM
▷ Visitors : Population 1:3950

2000 visitors

Little change in 1998

MAIN TOURIST ARRIVALS

	Tourism		
	has been minimal		
	since the start of		
	civil war		

0	10	20	30	40

% of total arrivals

Tourism has effectively ceased as a result of the civil war. When peace is secured, Rwanda may be able to regain its status as a destination for wealthy wildlife enthusiasts. Top attractions are the mountain gorillas and Lake Kivu.

PEOPLE
▷ Pop. density high

Kinyarwanda, French, Kiswahili, English

317/km² (820/mi²)

THE URBAN/RURAL POPULATION SPLIT

6%   94%

ETHNIC MAKEUP

Other (including Twa) 1%   Tutsi 9%
Hutu 90%

The Hutu and Tutsi are the main groups; few of the Twa pygmies, the original inhabitants, remain. For more than 500 years, the cattle-owning Tutsi were politically dominant, oppressing the landowning Hutu majority. In 1959, violent revolt led to a reversal of the roles. The two groups have since been waging a spasmodic war. It is estimated that 800,000 people were killed in the violence of the mid-1990s, the majority of them Tutsi victims massacred by Hutus.

POLITICS
▷ In transition

1988/2003

President Paul Kagame

AT THE LAST ELECTION

Transitional National Assembly 74 seats

The last legislative election, to the National Development Council, took place in December 1988 when candidates were chosen from members of the National Republican Movement for Development and Democracy (MRND) list.
A Transitional National Assembly was set up in 1994.
Polls due in 1999 were postponed until 2003.

In 1993 a peace accord to end the rebellion launched in 1990 by the Tutsi-dominated Rwandan Patriotic Front (FPR) was signed. However, the fragile peace process was halted in 1994 by the death of the president in a plane crash. Genocidal violence was unleashed between the mainly Hutu supporters of the old regime and its mainly, but not exclusively, Tutsi opponents. An estimated 800,000 died and millions fled the conflict, in which the FPR eventually gained control. Hutu were allocated key government posts, including the presidency, but when, in March 2000, the balance was shifted to increase Tutsi representation, President Pasteur Bizimungu resigned. Vice President Paul Kagame, the regime's dominant figure and the FPR leader, was formally elected president in April. Presidential and parliamentary elections are planned for end-2003, following the approval of a constitution.

RWANDA

Total Land Area : 26 338 sq. km
(10 169 sq. miles)

POPULATION
over 100 000 ◎
over 10 000 ●
under 10 000 •

LAND HEIGHT
5000m/9843ft
2000m/6562ft
1000m/3281ft

0   40 km
0   40 miles

WORLD AFFAIRS
▷ Joined UN in 1962

 COMESA CEPGL OIF NAM AU

Accused of abandoning Rwanda during 1994, the UN in 1995 set up a war crimes tribunal on the genocide.

Relations with neighboring Uganda are particularly tense, especially over the conflict in the DRC.

AID
▷ Recipient

 $322m (receipts)   Down 14% in 2000

Large amounts of aid are required, particularly for the agricultural sector, which was severely disrupted by the war. Aid donors in November 2000 urged Rwanda to withdraw its troops from the DRC.

DEFENSE
▷ No compulsory military service

 $109m   Down 19% in 2000

The Rwandan army is sufficiently powerful to make it a strong influence in the war-torn region. A national police force was established in 1999.

ECONOMICS
▷ Inflation 15% p.a. (1990–2000)

 $1.99bn   359.0–452.9 Rwanda francs

SCORE CARD

- ❏ World GNP Ranking........................138th
- ❏ GNP per Capita $230
- ❏ Balance of Payments.......................–$7m
- ❏ Inflation 4.3%
- ❏ Unemployment ....Few have formal employment

Strengths
Currently none. Assuming stability, Rwanda could produce coffee and tea. Possible oil and gas reserves. Tourism potential.

Weaknesses
Economic activity completely disrupted by 1994 violence. Lengthy journey to Kenyan and Tanzanian ports means high transportation costs. Few resources.

EXPORTS

- Belgium 4%
- USA 5%
- Netherlands 8%
- Pakistan 8%
- Germany 18%
- Other 57%

IMPORTS

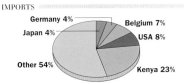

- Germany 4%
- Belgium 7%
- Japan 4%
- USA 8%
- Kenya 23%
- Other 54%

Terraced hillside. *Before the war, Rwanda was the most densely populated country in Africa and its land was intensively cultivated.*

RESOURCES
▷ Electric power 34,000 kw

 6733 tonnes   Not an oil producer

800,000 cattle, 700,000 goats, 1.2m chickens   Tin, tungsten, gold, columbo-tantalite, methane gas

Gas deposits in Lake Kivu are likely to be explored with the DRC. Only 20% of urban homes are on the national power grid.

ENVIRONMENT
▷ Sustainability rank: 119th

 15%   0.1 tonnes per capita

Apart from the effects of war, soil erosion and forest loss are the major environmental problems. The tourist industry underpinned the preservation of the mountain gorilla.

MEDIA
▷ TV ownership low

☒ Daily newspaper circulation 0.1 per 1000 people

PUBLISHING AND BROADCAST MEDIA

There is 1 daily newspaper. The monthly *Inkingi* and *La Relève* are published in Kinyarwanda and French respectively

1 state-controlled service   2 services: 1 state-controlled

The media have been used as an important propaganda tool by both sides in the political conflict.

CRIME
▷ Death penalty in use

 143,021 prisoners   Crime is rising

In 2001, 125,000 people were awaiting trial on genocide-related charges. A traditional system of participatory justice to address the backlog of cases became operational in 2002.

EDUCATION
▷ School leaving age: 13

 67%   5678 students

Schools are run by the state and by Christian missions. Primary education is officially compulsory, but only 78% of children attended in 1997; just 8% go on to secondary schooling.

CHRONOLOGY
The Hutu majority began to arrive in the 14th century, the warrior Tutsi in the 15th. From 1890, German and then Belgian colonizers acted to reinforce Tutsi dominance.

- ❏ **1962** Independence. Hutu-led government.
- ❏ **1960s** Tutsi revolt; massacres by Hutu; thousands of Tutsis in exile.
- ❏ **1973** Coup by Gen. Habyarimana.
- ❏ **1994** Habyarimana dies in plane crash. Genocidal violence unleashed by Hutu extremist regime, ousted by Tutsi-led FPR. Hutu refugee exodus.
- ❏ **1995** Start of war crimes tribunal.
- ❏ **1997** Refugees forcibly repatriated.
- ❏ **2000** Prominent Hutus leave office.
- ❏ **2001–2002** Limited troop withdrawal from DRC.

HEALTH
▷ Welfare state health benefits

 1 per 20,000 people   Malaria, measles, diarrheal diseases, violence

Rwanda has a network of 34 hospitals and 188 health centers. At the end of 2001, 9% of the population was estimated to be HIV-positive.

SPENDING
▷ GDP/cap. decrease

CONSUMPTION AND SPENDING

2 per 1000 population   2 per 1000 population

- Defense 4.7%
- Education 1.8%
- Health 2%

Defense, Health, Education spending as % of GDP

Wealth is limited to the country's political elite. Most Rwandans are poor farmers; Twa pygmies and refugees are poorer still.

WORLD RANKING

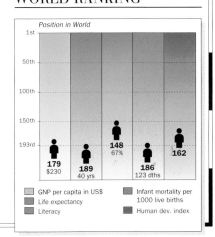

Position in World

- 179 $230
- 189 40 yrs
- 148 67%
- 186 123 dths
- 162

- GNP per capita in US$
- Life expectancy
- Literacy
- Infant mortality per 1000 live births
- Human dev. index

R

St. Kitts & Nevis

OFFICIAL NAME: Federation of Saint Christopher and Nevis   **CAPITAL:** Basseterre
POPULATION: 41,000   **CURRENCY:** Eastern Caribbean dollar   **OFFICIAL LANGUAGE:** English

 1983   1983   Sept 19   KN   -4   +1869   .kn

ONE OF THE CARIBBEAN'S most popular tourist destinations, St. Kitts and Nevis, a former British colony, lies at the northern end of the Leeward Islands chain. St. Kitts is of volcanic origin; Mount Liamuiga, a dormant volcano with a crater 227 m (745 ft) deep, is the highest point on the island. Nevis, separated from St. Kitts by a channel 3 km (2 miles) wide, is the lusher but less developed of the two islands. In the 18th century, its famed hot and cold springs gained Nevis the title "the Spa of the Caribbean."

CLIMATE   ▷ Tropical oceanic

WEATHER CHART FOR BASSETERRE

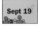

A combination of high temperatures, trade breezes, and moderate rainfall in summer constitute St. Kitts' typically Caribbean climate.

TRANSPORTATION   ▷ Drive on left

🛪 **Golden Rock International, Basseterre**   🚢 1 ship 600 dwt

THE TRANSPORTATION NETWORK

136 km (85 miles)	None
58 km (36 miles)	None

Most roads on the islands follow the coast; just a few cross the interior. Access to the remote southeast peninsula of St. Kitts has been improved. The airport on St. Kitts takes large jets; Nevis airport accepts only light aircraft. Regular ferries connect the islands.

The southeastern peninsula of St. Kitts, looking across to Nevis in the background, on a typical December evening.

TOURISM   ▷ Visitors : Population 2:1

🧳 84,000 visitors   ⬇ Down 10% in 1999

MAIN TOURIST ARRIVALS

USA 41%	
UK 16%	
Canada 7%	
Other 36%	

% of total arrivals

St. Kitts has long targeted the mass US tourist market. Improvements to boost tourism include the opening up of the St. Kitts southern peninsula to large-scale tourist developments, and the expansion of the main port to accommodate two cruise ships simultaneously. Most visitors come for sand, sun, and the Caribbean mood, although in recent years safaris inland to see local wildlife and mineral springs have become more popular. On St. Kitts, the old Brimstone Hill fortress has been converted into a museum, as has the Nevis birthplace of Alexander Hamilton, one of the architects of the US constitution.

St. Kitts & Nevis

Total Land Area : 261 sq. km (101 sq. miles)

LAND HEIGHT	POPULATION
1000m/3281ft	• over 10 000
500m/1640ft	• under 10 000
200m/656ft	
Sea Level	

PEOPLE   ▷ Pop. density medium

English, English Creole   👥 114/km² (295/mi²)

THE URBAN/RURAL POPULATION SPLIT

34%	66%

RELIGIOUS PERSUASION

Anglican 33%
Roman Catholic 7%
Moravian 9%
Other 22%
Methodist 29%

Most of the population is descended from Africans brought over in the 17th century; intermarriage has blurred other racial lines. There are small numbers of Europeans and South Asians. High levels of emigration ensure a stable population size and provide a source of foreign currency.

POLITICS   ▷ Multiparty elections

🗳 2000/2005   H.M. Queen Elizabeth II

AT THE LAST ELECTION

National Assembly 15 seats

53% SKLP	27% App	13% CCM	7% NRP

SKLP = St. Kitts Labour Party   **CCM** = Concerned Citizens' Movement   **NRP** = Nevis Reformation Party   **App** = Appointed

Nevis has its own executive and legislature, the Nevis Island Assembly, which exercise local power

The center-left SKLP ended 15 years of rule by the right-wing People's Action Movement (PAM) by winning the 1995 general election, and won a second term in 2000. A plan for secession by the Nevis government was narrowly defeated in a referendum held in 1998.

S

WORLD AFFAIRS
▷ Joined UN in 1983

ACS · Comm · Caricom · OAS · OECS

Resumption of hanging in 1998 after a 13-year moratorium drew strong international criticism. Suspected money laundering earned St. Kitts condemnation from G7 in 2000.

AID
▷ Recipient

 US$4m (receipts)   Down 20% in 2000

International aid in 1999 supported an economic recovery and relief program, following hurricanes Georges and Lenny, which significantly damaged buildings and infrastructure. Most aid is through the Caribbean Development Bank.

DEFENSE
▷ No compulsory military service

Army duties under-taken by Volunteer Defense Force   Not applicable

An army existed for six years before it was disbanded to cut government expenditure in 1981. A small paramilitary unit remains within the police; it made a token appearance with US forces during the 1983 invasion of Grenada.

ECONOMICS
▷ Inflation 3% p.a. (1990–2000)

 US$269m   2.7 Eastern Caribbean dollars

SCORE CARD

- ❏ World GNP Ranking........................178th
- ❏ GNP per Capita ....US$6570
- ❏ Balance of Payments..................−US$26m
- ❏ Inflation 3.9%
- ❏ Unemployment 5%

Strengths
Growth in tourism industry, light manufacturing, and financial services. Diversifying agricultural sector.

Weaknesses
Tourism and related industries prone to hurricane damage, most recently in 1998–1999. Agricultural exports sensitive to fluctuating world market.

EXPORTS
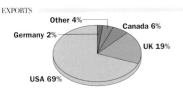
Other 4% · Canada 6% · Germany 2% · UK 19% · USA 69%

IMPORTS
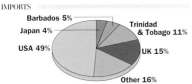
Barbados 5% · Japan 4% · Trinidad & Tobago 11% · USA 49% · UK 15% · Other 16%

RESOURCES
▷ Electric power 16,000 kw

 357 tonnes   Not an oil producer

14,500 goats, 7400 sheep, 3600 cattle, 60,000 chickens   None

St. Kitts has no strategic resources. Almost all energy has to be imported, mainly oil from Venezuela and Mexico. Sugar output declined rapidly in the 1990s. New crops, such as Sea Island cotton on Nevis, have been introduced. Offshore fishing has potential.

ENVIRONMENT
▷ Not available

 10%   2.5 tonnes per capita

Hurricanes are the greatest environmental threat. Hurricane Georges alone caused damage in 1998 estimated at US$400 million. It was followed in 1999 by Hurricane Lenny. As in the rest of the Caribbean, benefits from encouraging tourism must be set against potential ecological damage. The government has shown sensitivity, with strict preservation orders on the remaining rainforest and on indigenous monkeys.

MEDIA
▷ TV ownership high

 There are no daily newspapers

PUBLISHING AND BROADCAST MEDIA

There are no daily newspapers. The two main newspapers are the weekly *The Democrat* and the twice-weekly *Labour Spokesman*

1 state-owned service

4 services: 1 state-owned, 3 independent

Opposition parties have difficulty broadcasting their views on state-owned TV and radio. However, dissent is expressed in the printed press. The two main papers are run by political parties.

CRIME
▷ Death penalty in use

 135 prisoners   Down 31% in 1999

The judicial system is based on British common law. The police are UK-trained. Hanging was resumed in 1998, and plans agreed in 2001 to replace the role of the UK Privy Council with a Caribbean Court of Justice raised fears of more executions. Narcotics-related crimes are increasing.

EDUCATION
▷ School leaving age: 12

 90%   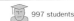 997 students

Education is free and the government is keen to promote information technology subjects. Students attend the regional University of the West Indies, or go on to colleges in the US and the UK.

CHRONOLOGY
A British colony since 1783, the islands were a part of the Leeward Islands Federation until 1956.

- ❏ **1932** Pro-independence St. Kitts–Nevis–Anguilla Labour Party set up.
- ❏ **1967** Internal self-government.
- ❏ **1980** Anguilla formally separates from St. Kitts and Nevis.
- ❏ **1983** Independence.
- ❏ **1995** Opposition SKLP wins election.
- ❏ **1998** Nevis referendum narrowly rejects secession.
- ❏ **2000** Prime Minister Denzil Douglas and SKLP reelected. Blacklisted by G7 for financial dealings.

HEALTH
▷ Welfare state health benefits

 1 per 855 people   Heart and respiratory diseases, cancers

The government-run health service now provides rudimentary care on both St. Kitts and Nevis. The EU and France provided EC$8 million in 1998 for repairs to the main hospital at Basseterre, which was badly damaged by Hurricane Georges.

SPENDING
▷ GDP/cap. increase

CONSUMPTION AND SPENDING

277 per 1000 population   569 per 1000 population

	0	5	10	15	20	25
Defense None						
Education 3.8%						
Health 3.1%						

Defense, Health, Education spending as % of GDP

Native professionals and civil servants have replaced expatriates over the past 20 years. They are now the best-paid group, but there are no great extremes of income.

WORLD RANKING

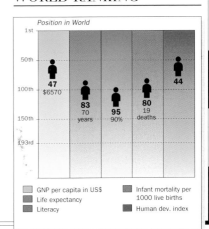
Position in World

47 $6570 · 83 70 years · 95 90% · 80 19 deaths · 44

- ▨ GNP per capita in US$
- ▨ Life expectancy
- ▨ Literacy
- ▨ Infant mortality per 1000 live births
- ▨ Human dev. index

S

ST. LUCIA

OFFICIAL NAME: Saint Lucia **CAPITAL:** Castries **POPULATION:** 156,300
CURRENCY: Eastern Caribbean dollar **OFFICIAL LANGUAGE:** English

ST. LUCIA IS ONE OF THE MOST beautiful islands of the Windward group of the Antilles. The twin Pitons, south of Soufrière, are one of the most striking natural features in the Caribbean. Ruled by the French and the British at different times in its past, St. Lucia retains the character of both. A multiparty democracy, it lives by banana-growing and tourism, with enticing beaches and a rich variety of wildlife in the rainforest.

CLIMATE ▷ Tropical oceanic

WEATHER CHART FOR CASTRIES

The dry season, from January to April, brings intense heat to sheltered parts of St. Lucia. During the rainy season, short warm showers can be expected daily. Rainfall is highest in the mountains.

TRANSPORTATION ▷ Drive on left

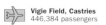 **Vigie Field, Castries** 446,384 passengers   3 ships 911 grt

THE TRANSPORTATION NETWORK

63 km (39 miles)	None
None	None

Roads are confined to the west and southeast coasts; only half are paved. Flights arrive from major European and North American cities, and other Caribbean locations. Direct passage to South America is largely by sea.

***One of the twin Pitons** south of Soufrière, marking the entrance to the Jalousie Plantation harbor.*

TOURISM ▷ Visitors : Population 1.7:1

 259,000 visitors Down 1% in 2000

MAIN TOURIST ARRIVALS

USA 32%	
UK 27%	
Canada 5%	
Other 36%	

% of total arrivals (0–40)

Tropical beaches and typical Caribbean towns make St. Lucia a favorite destination for cruise ships and stay-over tourists. The number of hotel rooms continues to rise. The pristine rainforest has become the focus of nature tourism, with tours often organized by the National Trust.

PEOPLE ▷ Pop. density high

 English, French Creole   256/km² (663/mi²)

THE URBAN/RURAL POPULATION SPLIT

38%   62%

RELIGIOUS PERSUASION

Other 10%
Roman Catholic 90%

St. Lucia has a rich, tension-free racial mix of descendants of Africans, Caribs, and European settlers. Despite relaxed attitudes, family life is central to most St. Lucians, many of whom are practicing Roman Catholics. Small families are the norm. In rural districts, where women run many of the farms, absentee fathers are fairly common. In recent years, women have had greater access to higher education and have moved into professions. A bill to permit the occasional use of Creole in parliament was passed in 1998.

POLITICS ▷ Multiparty elections

 L. House 2001/2006 U. House 2001/2006 H.M. Queen Elizabeth II

AT THE LAST ELECTION

House of Assembly 18 seats

78% SLP	17% UWP	5% App

SLP = St. Lucia Labour Party **UWP** = United Workers' Party
App = Appointed

Senate 11 seats

6 Senate members are nominated by the government, 3 by the opposition, and 2 by the governor-general on a nonparty basis

Politics revolved around John Compton of the UWP and Julian Hunte of the SLP, until from 1996 a less personalized climate prevailed with the emergence of new political figures. After losing heavily to the SLP in the 1997 general election, the UWP changed its leader three times in as many years, but failed to prevent the SLP's reelection in 2001.

WORLD AFFAIRS ▷ Joined UN in 1979

 ACS   Comm   Caricom   OECS OAS

Relations with the US were strained by the long-running, and ultimately successful, US campaign to end preferential access for Caribbean bananas to the EU. It is feared that from 2006 St. Lucia will be unable to compete with cheaper fruit from Central and South America. St. Lucia supports Japan, an aid donor, in its bid for a permanent seat on the UN Security Council. With the agreement in 2001 to form a Caribbean Court of Justice, one of the last constitutional ties to the UK was set to be cut.

AID ▷ Recipient

 US$11m (receipts)   Down 58% in 2000

The US, the EU, and Japan are the main donors. China has also given aid and grant loans in recent years.

DEFENSE ▷ No compulsory military service

 US$5m 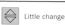 Little change

The police force is supported by a small paramilitary unit. Training is provided by the US and the UK.

S

ST. LUCIA

Total Land Area : 620 sq. km (239 sq. miles)

POPULATION

- ● over 10 000
- • under 10 000

LAND HEIGHT

- 500m/1640ft
- 200m/656ft
- Sea Level

ECONOMICS

▷ Inflation 2.9% p.a. (1990–2000)

US$642m

2.7 Eastern Caribbean dollars

SCORE CARD

- ❏ World GNP Ranking.......................165th
- ❏ GNP per Capita ....US$4120
- ❏ Balance of Payments..................–US$80m
- ❏ Inflation 5.4%
- ❏ Unemployment.................................16%

STRENGTHS

Banana industry (privatized in 1998). Tourism, fisheries potential.

WEAKNESSES

Preferential banana trade with EU to end by 2006. Most resorts foreign-owned.

EXPORTS

Dominica 2% — Barbados 6%
France 1% — Other 8%
UK 57%
USA 26%

IMPORTS

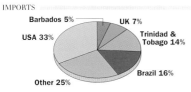

Barbados 5% — UK 7%
USA 33% — Trinidad & Tobago 14%
Brazil 16%
Other 25%

RESOURCES

▷ Electric power 22,000 kw

1719 tonnes

Not an oil producer

14,750 pigs, 12,500 sheep, 210,000 chickens

None

St. Lucia has no mineral resources and imports most of its energy. Plans exist to develop geothermal energy from the hot springs in the volcanic interior.

ENVIRONMENT

▷ Not available

2% partially protected

1.3 tonnes per capita

St. Lucians are proud of their island, and environmental questions arouse fierce debate. In recent years the greatest controversy surrounded the decision to allow a luxury hotel development on the ecologically important Jalousie Plantation, which encompasses the extraordinary twin Pitons and includes an important Amerindian archaeological site. The issue illustrates a key problem in St. Lucia, where business pressures to develop tourism can outweigh vital environmental concerns. One notable conservation success has been the St. Lucia parrot. In 1978, there were 150 birds; strict laws against the trade in parrots ensured that by 2000 numbers had risen to over 800.

MEDIA

▷ TV ownership medium

There are no daily newspapers

PUBLISHING AND BROADCAST MEDIA

There are no daily newspapers. *The Star* and *The Mirror* are published weekly

4 independent services

4 services: 1 state-owned, 3 independent

The privately owned press is free from government intervention. It is possible to receive TV programs from US, Mexican, and some Caribbean stations.

CRIME

▷ Death penalty in use

365 prisoners

Crime is rising

Murder is rare, but narcotics-related deaths are increasing, as is violence in schools. The government has strengthened the police force to combat rising urban crime.

EDUCATION

▷ School leaving age: 15

82%

3881 students

Education is based on the British system. Nobel prizewinners Sir Arthur Lewis (economics) and Derek Walcott (literature) give St. Lucia the world's highest per capita ratio of laureates.

CHRONOLOGY

An excellent naval raiding base in the Caribbean in the 17th and 18th centuries, St. Lucia was fought over by France and Britain. Ownership alternated before it was finally ceded to Britain in 1814. French influence survives in St. Lucian patois and the local cuisine.

- ❏ **1958** Joins West Indies Federation.
- ❏ **1964** Sugar growing ceases.
- ❏ **1979** Gains independence and joins Commonwealth.
- ❏ **1990** Establishes body with Dominica, Grenada, and St. Vincent to discuss forming a Windward Islands Federation.
- ❏ **1997** Hitherto ruling UWP reduced to one seat in general election.
- ❏ **2000** Blacklisted by OECD as international tax haven.

HEALTH

▷ Welfare state health benefits

1 per 2128 people

Heart and respiratory diseases, cancers

The National Insurance Scheme provides free medical cover for all. Work on a new general hospital began in 2001.

SPENDING

▷ GDP/cap. increase

CONSUMPTION AND SPENDING

95 per 1000 population

313 per 1000 population

Defense No data
Education 9.8%
Health 2.4%

0   5   10   15   20   25
Defense, Health, Education spending as % of GDP

The island's large-scale banana growers and hotel owners form the richest section of society. Nearly one-fifth of households are considered to be poor.

WORLD RANKING

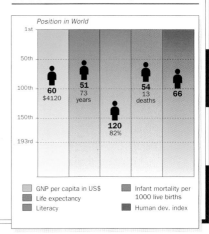

Position in World

1st
50th
100th
150th
193rd

60 $4120
51 73 years
120 82%
54 13 deaths
66

- ▧ GNP per capita in US$
- ▧ Life expectancy
- ▧ Literacy
- ▧ Infant mortality per 1000 live births
- ▧ Human dev. index

S

St. Vincent & the Grenadines

Official Name: Saint Vincent and the Grenadines **Capital:** Kingstown
Population: 115,500 **Currency:** Eastern Caribbean dollar **Official Language:** English

PART OF THE WINDWARD ISLANDS group, and bounded by submerged coral reefs, St. Vincent and the Grenadines is the Caribbean playground of the international celebrity circuit. Tourism and bananas are the economic mainstays; St. Vincent is also the world's largest arrowroot producer. It is mostly volcanic; the one remaining active volcano, La Soufrière, last erupted in 1979. The Grenadines are flat, mainly bare, coral islands.

CLIMATE
▷ Tropical oceanic

WEATHER CHART FOR KINGSTOWN

Constant trade winds moderate St. Vincent's tropical climate. Rainfall is heaviest during the summer months. Deep depressions and hurricanes are likely between June and November.

TRANSPORTATION
▷ Drive on left

Arnos Vale, Kingstown | 1317 ships 7.88m grt

THE TRANSPORTATION NETWORK

| 320 km (199 miles) | None |
| None | None |

Access by air is via neighboring islands. Paved roads encompass most of St. Vincent's coast. Port improvements have been completed in recent years. In 1992, an airport capable of taking executive jets was completed on the island of Bequia.

Aerial view of Union Island in the Grenadines chain. The government is developing the island as a major yachting center.

TOURISM
▷ Visitors : Population 1:1.6

73,000 visitors | Up 7% in 2000

MAIN TOURIST ARRIVALS

| USA 28% |
| UK 17% |
| France 10% |
| Other 45% |

% of total arrivals

Tourism is targeted at celebrities and cruise ships rather than the mass market, and is concentrated on the Grenadines. Mustique, long associated with Princess Margaret, has a rock music clientele. Union Island draws the yachting rich, and luxury villas, apartments, a golf course, and a casino have been built on Canouan. Layou, on St. Vincent, is the site of pre-Columbian Amerindian petroglyphs.

PEOPLE
▷ Pop. density high

English, English Creole | 340/km² (880/mi²)

THE URBAN/RURAL POPULATION SPLIT

55% | 45%

RELIGIOUS PERSUASION

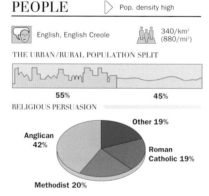

Other 19%
Anglican 42%
Roman Catholic 19%
Methodist 20%

Family life on St. Vincent is heavily influenced by the Anglican Church. Racial tensions are few, and intermarriage has meant that the original communities of descendants of African slaves, Europeans, and the few indigenous Caribs can no longer be distinguished. Many locals fear that traditional island life is being threatened by the expanding tourist industry.

POLITICS
▷ Multiparty elections

2001/2006 | H.M. Queen Elizabeth II

AT THE LAST ELECTION
House of Assembly 21 seats

| 57% ULP | 29% App | 14% NDP |

ULP = Unity Labour Party **App** = Appointed **NDP** = New Democratic Party
Six senators are appointed to the House of Assembly by the governor-general

In March 2001, 17 years of rule by the NDP ended with a crushing electoral defeat. The leader of the long-term opposition ULP, Ralph Gonsalves, was appointed prime minister. A new Caribbean Court of Justice was agreed in 2001, but constitutional ties to the British monarchy remain, despite a strong republican movement.

St. Vincent & the Grenadines

Total Land Area : 389 sq. km (150 sq. miles)

POPULATION
● over 10 000
• under 10 000

LAND HEIGHT
1000m/3281ft
500m/1640ft
200m/656ft
Sea Level

S

WORLD AFFAIRS
▷ Joined UN in 1980

The successful US bid to end the EU's preferential treatment of Caribbean banana imports has strained relations. St. Vincent supports a united move by Caricom members to promote a presidential system of government in place of the British monarchy.

AID
▷ Recipient

 US$6m (receipts)   Down 62% in 2000

The EU and Japan are the major sources of development aid. Significant funds also come from the Caribbean Development Bank and Arab countries.

DEFENSE
▷ No compulsory military service

US$3m (estimate)   No significant change from year to year

St. Vincent has no army. Its small police force, trained by the US and the UK, is part of the Windward and Leeward Islands' Regional Security System.

ECONOMICS
▷ Inflation 2% p.a. (1990–2000)

US$313m   2.7 Eastern Caribbean dollars

SCORE CARD
- ❏ World GNP Ranking.......................176th
- ❏ GNP per Capita ....US$2720
- ❏ Balance of Payments..................–US$66m
- ❏ Inflation 0.2%
- ❏ Unemployment..................................22%

STRENGTHS
Bananas, but preferential access to EU markets will end in 2006. Great tourist potential. Currency stability. Leading producer of arrowroot starch. Improving infrastructure.

WEAKNESSES
Little diversification. Strong potential competition from Central and South American banana producers.

EXPORTS

IMPORTS

RESOURCES
▷ Electric power 16,000 kw

15,573 tonnes

13,000 sheep, 9500 pigs, 200,000 chickens

Not an oil producer

None

There is a hydroelectric plant on the Cumberland River. Virtually all other energy requirements have to be imported. Some of the Grenadines have no fresh water sources.

ENVIRONMENT
▷ Not available

21% (including marine and semi-protected areas)   1.4 tonnes per capita

Hurricanes are the main environmental threat, sometimes destroying as much as 70% of the banana crop. The former inaccessibility of St. Vincent and the Grenadines meant that tourism was a minor environmental threat, and the untouched, idyllic landscape of islands such as Mustique was their attraction. Mustique is reasonably well protected – buildings have been restricted and further development is limited, since fresh water has to be shipped in. On Bequia, the new airport and consequent increase in visitors are seen as a mixed blessing. Schemes to develop Canouan have been opposed by locals.

MEDIA
▷ TV ownership medium

Daily newspaper circulation 9 per 1000 people

PUBLISHING AND BROADCAST MEDIA

There is 1 daily newspaper, *The Herald*. The main weekly newspaper is the independent *The Vincentian*

1 state-owned service   1 state-owned service

Of the many periodicals, three are published by the political parties; the rest are independent. Freedom of the press is written into the constitution.

CRIME
▷ Death penalty in use

405 prisoners   Crime is rising

The incidence of rape and robbery causes most concern, although on the outlying islands both crimes are very rare. St. Vincent is used for narcotics transshipment to the US.

EDUCATION
▷ School leaving age: 15

82%   677 students

State schools follow the former British 11-plus selective system. There are a few private schools. University students go on to the regional University of the West Indies in Jamaica, although increasing numbers are also studying in the US and the UK.

HEALTH
▷ Welfare state health beniefts

1 per 1136 people   Heart and respiratory diseases, cancers

Doctors train at the University of the West Indies. The system is a mixture of state and private hospitals and clinics; facilities are scarcer on the Grenadines.

SPENDING
▷ GDP/cap. increase

CONSUMPTION AND SPENDING

 44 per 1000 population   220 per 1000 population

Defense	No data
Education	6.7%
Health	4.2%

Defense, Health, Education spending as % of GDP

Jet-set wealth in the islands coexists with the low wages paid to most local workers. Union Island and Mustique in particular attract the wealthy, with their motor yachts and jeeps.

WORLD RANKING

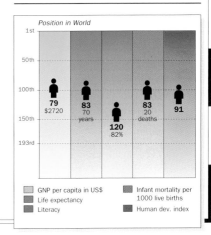

S

SAMOA

OFFICIAL NAME: Independent State of Samoa   **CAPITAL:** Apia
POPULATION: 159,000   **CURRENCY:** Tala   **OFFICIAL LANGUAGES:** Samoan and English

SAMOA, LYING IN THE HEART of the South Pacific, 2400 km (1500 miles) north of New Zealand, comprises nine volcanic islands. Four are inhabited – Apolima, Manono, Savai'i, and Upolu (where 72% of the population live). Rainforests cloak the mountains; vegetable gardens and coconut plantations thrive around the coasts. A relative boom in the economy in recent years has not lifted Samoa from the ranks of the UN's Least Developed Countries.

CLIMATE   ▷ Tropical oceanic

WEATHER CHART FOR APIA

The climate is humid and temperatures rarely drop below 25°C (77°F). December to March is the hurricane season.

TRANSPORTATION   ▷ Drive on right

Faleolo Apia
160,366 passengers

7 ships
3300 grt

THE TRANSPORTATION NETWORK

332 km (206 miles)		None
None		None

Apia port has been improved with Japanese aid. International links are mainly by air. Ferries provide interisland connections.

TOURISM   ▷ Visitors : Population 1:1.8

88,000 visitors

Up 4% in 2000

MAIN TOURIST ARRIVALS

| American Samoa 37% |
| New Zealand 27% |
| Australia 11% |
| Other 25% |

0   10   20   30   40
% of total arrivals

Tourism is a rapidly growing industry. Small-scale village-based tourism is encouraged. Tourists are attracted by the climate and the easygoing *fa'a Samoa* (Samoan way of life).

PEOPLE   ▷ Pop. density medium

Samoan, English

56/km²
(146/mi²)

THE URBAN/RURAL POPULATION SPLIT

22%   78%

ETHNIC MAKEUP

Other 1%   Euronesian 9%

Polynesian 90%

Ethnic Samoans are the world's second-largest Polynesian group, after Maoris. The *fa'a Samoa* – Samoan way of life – is communal and conservative. Extended family groups, in which most people live, own 80% of the land, and cannot sell it. They are each headed by a *matai*, or elected chief, who looks after its social and political interests. Large-scale migration to New Zealand and the US reflects a lack of jobs and the attractions of Western life. Conflict between the *fa'a Samoa* and modern life is strongest among the young, who have a high suicide rate. Euronesians are those of mixed European/Polynesian descent.

SAMOA

Total Land Area : 2860 sq. km (1104 sq. miles)

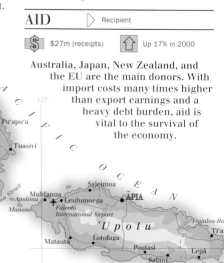

POPULATION

● over 10 000
• under 10 000

LAND HEIGHT

1000m/3281ft
500m/1640ft
200m/656ft
Sea Level

0   15 km
0   15 miles

POLITICS   ▷ Multiparty elections

2001/2006

H.H. Susuga Malietoa Tanumafili II

AT THE LAST ELECTION

Legislative Assembly 49 seats

47% HRPP   27% SNDP   26% Ind

HRPP = Human Rights Protection Party   **SNDP** = Samoan National Development Party   **Ind** = Independents

The conservatism of the *fa'a Samoa* and the Church underpins Samoa's political stability. Allegiance to the two main parties is quite fluid. Until 1990, only the 1800 elected chiefs, or *matai*, could vote for the 47 ethnic Samoan seats in the Assembly; the other two seats are elected by non-Samoans. Universal suffrage was introduced at the 1991 elections. Tofilau Eti Alesana of the HRPP, prime minister twice since 1988, resigned in 1998 amid widespread protest against the government's autocratic style. His successor, Tuilaepa Sailele Malielegaoi, also of the HRPP, was reappointed in 2001.

WORLD AFFAIRS   ▷ Joined UN in 1976

ACP   Comm   IBRD   PC   PIF

Australia is Samoa's main trading partner. New Zealand, the US, Japan, Fiji, American Samoa, and the EU are also important. Relations with China are well established. Samoa has trade links with the Cook Islands, and supports a Polynesian free trade agreement.

AID   ▷ Recipient

$27m (receipts)

Up 17% in 2000

Australia, Japan, New Zealand, and the EU are the main donors. With import costs many times higher than export earnings and a heavy debt burden, aid is vital to the survival of the economy.

DEFENSE
 No compulsory military service

 Samoa has no army and few police
 Not applicable

New Zealand looks after defense under a 1962 treaty. Internal order is mostly maintained by the *matai* (chiefs).

ECONOMICS
Inflation 3.3% p.a. (1990–2000)

$246m   3.339–3.558 tala

SCORE CARD

- ❏ WORLD GNP RANKING.........................180th
- ❏ GNP PER CAPITA $1450
- ❏ BALANCE OF PAYMENTS.......................–$19m
- ❏ INFLATION 1%
- ❏ UNEMPLOYMENT.......Widespread underemployment

STRENGTHS
Light manufacturing expanding, attracting foreign, especially Japanese, firms. Tourism growing rapidly with improved infrastructures. Services expanding rapidly since 1989 launch of offshore banking. Tropical agriculture: taro, coconut cream, cocoa, copra are main exports. Large fishing potential.

WEAKNESSES
Development adversely affected by cyclones. Fluctuating international markets for copra and cocoa. Dependence on aid and expatriate remittances.

EXPORTS

New Zealand 3%, American Samoa 3%, Australia 62%, Other 8%, USA 11%, Indonesia 13%

IMPORTS

Australia 27%, Japan 9%, Fiji 11%, Other 13%, New Zealand 14%, USA 26%

RESOURCES
Electric power 19,000 kw

 9750 tonnes   Not an oil producer

 170,000 pigs, 28,000 cattle, 450,000 chickens   None

With no minerals, Samoa's main resources are its forests and tropical agriculture. The rainforests in lower-lying areas are increasingly exploited for timber. Mahogany and teak plantations are being developed. The volcanic soils, particularly on Upolu, support a wide range of staple and export crops. Two-thirds of the population work in agriculture.

Apia, the capital, on Upolu, Samoa's second-largest island. It has a central volcanic range of mountains and many rivers.

ENVIRONMENT
Not available

 None   0.8 tonnes per capita

Strict logging regulations have been introduced to halt irreparable damage to the environment; over 80% of forests have been replaced by plantations. Overhunting and loss of habitat have endangered rare species of fruit bat and pigeon. Samoa is concerned about its marine resources and has taken a firm stance against driftnet fishing.

MEDIA
TV ownership medium

 There are no daily newspapers

PUBLISHING AND BROADCAST MEDIA

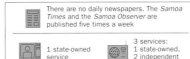 There are no daily newspapers. The *Samoa Times* and the *Samoa Observer* are published five times a week. 1 state-owned service. 3 services: 1 state-owned, 2 independent

The independent media, notably the *Samoa Observer*, can face strong governmental opposition.

CRIME
Death penalty not used in practice

 176 prisoners   Crime is rising slowly

Alcohol-related violence is a problem at weekends; otherwise, violent crime is almost unknown. Theft is increasing in urban areas.

EDUCATION
School leaving age: 14

 80%   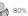 810 students

Education is based on the New Zealand system. School attendance is universal. A university was established in Samoa in 1988. There is widespread use of corporal punishment.

HEALTH
No welfare state health benefits

 1 per 2941 people   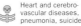 Heart and cerebro-vascular diseases, pneumonia, suicide

The Samoan preference for being big went well with traditional diets. Diabetes and heart disease are rising as people change to Western-style foods.

CHRONOLOGY
Polynesians settled Samoa in about 1000 BCE. Western rivalry after 1830 led to the 1899 division of the islands into German Western and American Eastern Samoa.

- ❏ **1914** New Zealand occupies Western Samoa.
- ❏ **1962** Becomes first independent Polynesian nation.
- ❏ **1990** Cyclone Ofa leaves 10,000 people homeless.
- ❏ **1991** HRPP retains power in first election under universal adult suffrage.
- ❏ **1996, 2001** HRPP returned to power in elections.
- ❏ **1997** Country's name changed from Western Samoa to Samoa.

SPENDING
GDP/cap. increase

CONSUMPTION AND SPENDING

30 per 1000 population   47 per 1000 population

Defense None
Education 4.2%
Health 4.8%

Defense, Health, Education spending as % of GDP

Most of the population depend on subsistence farming and the remittances of relatives for their livelihood. Samoa is classified by the UN as a Least Developed Country. Two-thirds of those with a permanent job work for the government. The prospect of earning higher wages by working in other, wealthier countries in the Pacific region, notably in the tuna canneries of the neighboring US dependency of American Samoa, leads thousands of young Samoans to emigrate every year.

WORLD RANKING

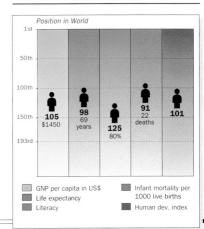
Position in World: 105 $1450, 98 69 years, 125 80%, 91 22 deaths, 101

GNP per capita in US$ / Life expectancy / Literacy / Infant mortality per 1000 live births / Human dev. index

S

SAN MARINO

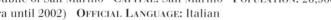

OFFICIAL NAME: Republic of San Marino **CAPITAL:** San Marino **POPULATION:** 26,900
CURRENCY: Euro (lira until 2002) **OFFICIAL LANGUAGE:** Italian

PERCHED ON THE SLOPES of Mount Titano in the Italian Appennines, tiny San Marino is the world's oldest republic and claims a lineage dating back to the 4th century. The territory is divided into nine castles, or districts. One-third of Sammarinesi live in the northern town of Serravalle. Today San Marino makes its living through agriculture, tourism, and limited industry. Italy effectively controls most of its affairs.

San Marino's second fortress, la Cesta, built in the 13th century, dominates the republic from its pinnacle, 755 m (2477 ft) above sea level.

CLIMATE ▷ Mediterranean

WEATHER CHART FOR SAN MARINO

San Marino's Mediterranean climate is moderated by cool sea breezes and its height above sea level. In summer, temperatures can reach 27°C (81°F), while in winter they fall to 7°C (45°F). There is rarely ever any snow.

TRANSPORTATION ▷ Drive on right

None — Has no fleet

THE TRANSPORTATION NETWORK

237 km (147 miles) — Not available

2 km (1 mile) — None

The 24-km (15-mile) highway to Rimini, which has the nearest airport, is San Marino's most important link. Congestion is a major problem, especially during the annual Mille Miglia car rally. A funicular railroad climbs the east side of Mount Titano. The railroad to Rimini has been closed since World War II.

PEOPLE ▷ Pop. density high

Italian — 442/km² (1144/mi²)

THE URBAN/RURAL POPULATION SPLIT

94% 6%

RELIGIOUS PERSUASION

Other and nonreligious 7%
Roman Catholic 93%

Citizenship requires 30 years' residence; it is no longer transmissible by marriage. Women gained the vote in 1960, but could not stand for public office until 1973. Around 20,000 live abroad, mainly in Italy. Some Sammarinesi speak a distinct regional dialect.

TOURISM ▷ Visitors : Population 20:1

532,000 visitors — Little change in 1999

Tourism is the mainstay of San Marino's economy, contributing more than half of government revenue and employment for almost 20% of the workforce. Earnings from tourism are the largest share of GDP. Every year half a million overnight visitors, and a further 2.5 million day visitors, are drawn by its mild climate and contrasting scenery, and come to sample its folklore and museums. The fortresses of Mount Titano – la Rocca, la Cesta, and Montale – built during the Middle Ages, command superb views and are the main attractions,

MAIN TOURIST ARRIVALS

Italy 73%
Other 27%

0 10 20 30 40 50 60 70 80
% of total arrivals

along with the medieval city of San Marino itself. Many visitors to San Marino are day trippers from Italy, although tourism is also boosted by the close proximity of the international airport at Rimini.

The San Marino tourist bureau also attracts thousands of sports enthusiasts to the republic by hosting a series of top international sporting events. In March, both the Rimini–San Marino marathon and the Mille Miglia veteran car meeting are held. May heralds the San Marino Grand Prix, when thousands of Formula One fans descend on the country. June, meanwhile, attracts more motor-racing fans for the World Motocross Championships. A renowned crossbow competition is held to mark San Marino's national day, September 3.

Efforts have been made to attract business meetings and conferences by means of extensive publicity in the Italian media.

Religious procession. The official state religion of San Marino is Roman Catholicism, in contrast to Italy, which has no state religion.

SAN MARINO

Total Land Area : 61 sq. km (23.6 sq. miles)

LAND HEIGHT
500m/1640ft
200m/656ft
above 175m/574ft

POPULATION
• under 10 000

Falciano
Dogana
Serravalle
Fiorina
Ventoso
Cailungo
Gualdicciolo
Acquaviva
Domagnano
Borgo Maggiore
SAN MARINO
Monte Titano 739 m
Faetano
Murata
Chiesanuova
Montegiardino

N

0 4 km
0 4 miles

POLITICS Multiparty elections

 2001/2006   Captains-Regent Antonio Volpinari and Giovanni Ugolini

San Marino is a parliamentary democracy headed by two captains-regent elected every six months. Although the PDCS is the largest single party, it is not guaranteed a place in the ever-shifting coalitions. The most recent government, formed in 2002, is headed by the PSS.

AT THE LAST ELECTION
Great and General Council 60 seats

| 42% PDCS | 25% PSS | 20% PPDS | 3% RC | 8% APDS | 2% AN |

PDCS = San Marino Christian Democratic Party
PSS = Socialist Party of San Marino   **PPDS** = Progressive Democratic Party   **APDS** = Popular Democratic Alliance
RC = Communist Refoundation   **AN** = National Alliance

WORLD AFFAIRS Joined UN in 1992

 CE   OSCE   IBRD   IMF   IWC

Foreign affairs are effectively decided by Italy, on which San Marino is entirely dependent. In 1992, San Marino acquired a seat at the UN.

AID Neither

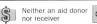 Neither an aid donor nor receiver   Not applicable

San Marino does not receive aid. However, annual subsidies from Italy and free access to the Italian market are essential to the economy.

DEFENSE No compulsory military service

 $1m   Little change in 1998

San Marino has a small territorial army and fortification guards. There is no compulsory military service, but males aged 16–55 may be called up in a national emergency.

ECONOMICS Inflation 5.9% p.a. (1985–1996)

 $190m   1.0651–1.1231 euros

SCORE CARD

- ❏ World GNP Ranking.........................185th
- ❏ GNP per Capita $7830
- ❏ Balance of Payments $11m
- ❏ Inflation 2.2%
- ❏ Unemployment 3%

Strengths
Tourism, providing 60% of government revenue. Light industry, notably mechanical engineering and clothing, with emphasis on sportswear and high-quality prestige lines. Philately.

Weaknesses
Need to import all raw materials.

Exports/Imports

San Marino does not publish independent trade statistics; trade movements are included in the Italian totals.

RESOURCES Electric power: Included in Italian total

 None   Not an oil producer

Small numbers of cattle, pigs, sheep, and horses   None

San Marino has to import all its energy from Italy. It has no exploitable mineral resources now that the stone quarry on Mount Titano has been exhausted.

ENVIRONMENT Not available

None   Not available

Mount Titano is a unique limestone outcrop in the surrounding Italian plain and so has a very localized ecosystem.

MEDIA TV ownership high

Daily newspaper circulation 71 per 1000 people

PUBLISHING AND BROADCAST MEDIA

 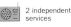

There are 4 daily newspapers. Regional Italian newspapers, especially *Il Resto del Carlino*, include coverage of San Marino

1 independent service   2 independent services

In 1993, a local TV station, San Marino RTV, began broadcasting. Sammarinesi can also receive Italian TV.

CRIME No death penalty

San Marino does not publish prison figures   Little change from year to year

San Marino has a low crime rate. Justice is mainly administered in Italy. Until mid-1997 homosexuality was illegal.

EDUCATION School leaving age: 16

 99%   Not available

All teachers are trained abroad, mostly in Italy. Secondary school pupils can go on to Italian universities.

HEALTH Welfare state health benefits

 1 per 375 people   Heart diseases, cancers, accidents

Health care is free and available to all. There is a hospital, but those requiring difficult operations normally go to Rimini for treatment.

CHRONOLOGY

Purportedly founded in the 4th century, the Republic of San Marino became one of many medieval Italian city-states. It was recognized by the papalcy in 1631.

- ❏ **1797** San Marino rejects expansion offered by Napoléon.
- ❏ **1861** Refuses to join unified Italy.
- ❏ **1914–1918** Fights for Italy in World War I.
- ❏ **1940** Supports Axis powers and declares war on the Allies.
- ❏ **1943** Declares neutrality shortly before Italy surrenders.
- ❏ **1960** Women obtain vote.
- ❏ **1978** Coalition of San Marino Communist Party (PCS) and PSS – sole communist-led government in Western Europe.
- ❏ **1986** Financial scandals lead to new PDCS–PCS government.
- ❏ **1988** Joins Council of Europe.
- ❏ **1990** PCS renames itself PPDS.
- ❏ **1992** Joins UN. Collapse of communism in Europe sees PDCS–PPDS alliance replaced by PDCS–PSS coalition.
- ❏ **2002** Adoption of euro.

SPENDING GDP/cap. increase

CONSUMPTION AND SPENDING

No data   689 per 1000 population

Defense	No data
Education	No data
Health	7.5%

| 0 | 5 | 10 | 15 | 20 | 25 |

Defense, Health, Education spending as % of GDP

Living standards are similar to those of northern Italy, while the unemployment rate is well below the Italian average.

WORLD RANKING

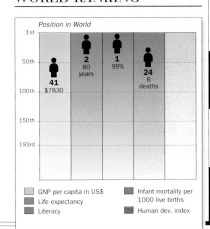

Position in World

1st				
50th	2 / 80 years	1 / 99%		
			24 / 6 deaths	
100th	41 / $7830			
150th				
193rd				

- ☐ GNP per capita in US$
- ☐ Life expectancy
- ☐ Literacy
- ☐ Infant mortality per 1000 live births
- ☐ Human dev. index

S

SÃO TOMÉ & PRÍNCIPE

WEST AFRICA

OFFICIAL NAME: Democratic Republic of São Tomé and Príncipe **CAPITAL:** São Tomé
POPULATION: 159,900 **CURRENCY:** Dobra **OFFICIAL LANGUAGE:** Portuguese

 1975 1975 July 12 STP 0 +239 .st

C OMPOSED OF the main islands of São Tomé and Príncipe and surrounding islets, the republic is situated off the western coast of Africa. In 1975, a classic Marxist single-party regime was established following independence from Portugal, but a referendum in 1990 resulted in a 72% vote in favor of democracy. São Tomé's main concerns are relations with Portugal and seeking closer ties with the EU and the US.

CLIMATE ▷ Tropical equatorial

WEATHER CHART FOR SÃO TOMÉ

The humid islands straddle the equator. Rainfall averages 500 cm (197 in) on São Tomé and 100 cm (39 in) on Príncipe.

TRANSPORTATION ▷ Drive on right

✈ São Tomé International
23,000 passengers

⚓ 10 ships
10,242 grt

THE TRANSPORTATION NETWORK

218 km (135 miles)	None
None	None

There are plans to construct a deepwater port and free trade zone at Agulhas Bay on Príncipe.

TOURISM ▷ Visitors : Population 1:32

5000 visitors

⬇ Down 17% 1995–1998

MAIN TOURIST ARRIVALS

	% of total arrivals
Portugal 35%	
France 20%	
Angola 7%	
Other 38%	

Tourism is still small-scale, attracting wealthy Africans and Europeans. Despite recent foreign investment, the islands attract relatively few tourists annually. The first modern hotel opened in 1986.

PEOPLE ▷ Pop. density medium

Portuguese Creole, Portuguese

167/km² (431/mi²)

THE URBAN/RURAL POPULATION SPLIT

47%   53%

ETHNIC MAKEUP

Portuguese and Creole 10%

Black 90%

The population is entirely descended from immigrants, since the islands were uninhabited when the Portuguese arrived in 1470. As the Portuguese settled, they imported Africans as slaves to work the sugar and cocoa plantations. The abolition of slavery in the 19th century, and the departure of 4000 Portuguese at independence, has resulted in a population which is 10% Portuguese and Creole and 90% black African, although Portuguese culture predominates. Blacks run the political parties. Society is well integrated and free of racial tensions. The main conflicts relate to class or differing ideologies. The extended family still offers the best, if not the only, form of social security. Women have a higher status than in most other African states; many have succeeded in attaining prominent positions in the professions.

Lush vegetation on São Tomé. *The tropical climate is slightly moderated by the cool Benguela current.*

POLITICS ▷ Multiparty elections

2002/2006

President Fradique de Menezes

AT THE LAST ELECTION

National Assembly 55 seats

44% MLSTP–PSD   42% MDFM–PCD   14% UK

MLSTP–PSD = São Tomé and Príncipe Liberation Movement –Social Democratic Party **MDFM–PCD** = Force for Change Democratic Movement–Democratic Convergence Party **UK** = Ue Kedadji coalition

In 1990, a new multiparty constitution did away with the Marxist single-party state in existence since independence in 1975. The opposition PCD was swept to victory in 1991, and later that year Miguel Trovoada returned from 11 years' exile to be elected as an independent to the presidency.

Elections in 1994 saw the MLSTP, renamed the MLSTP–PSD, return to power. In the 2001 presidential elections, businessman Fradique de Menezes defeated former Marxist president Manuel Pinto da Costa. He was forced to negotiate the formation of a coalition government with the MLSTP–PSD after it won elections in 2002.

Main political concerns are upholding the multiparty system and developing the economy in the light of undersea oil reserves, ownership of which was guaranteed when São Tomé and Nigeria reached agreement in 2000 over their common maritime border.

WORLD AFFAIRS ▷ Joined UN in 1975

CPLP ACP OIF NAM AU

São Tomé has achieved rapprochement with Portugal and seeks to maintain links with other former Portuguese colonies, notably Angola. It has always had close ties with Gabon and, while not dropping its ex-communist links, seeks closer relations with other central African states, France, and the US.

AID ▷ Recipient

$35m (receipts)

⬆ Up 25% in 2000

São Tomé has one of the highest aid-to-population ratios in Africa. Joining the Lomé Convention in the 1970s has meant that São Tomé has found new sources of aid fairly easily since the demise of communism worldwide. Portugal, France, and the EU are the main donors.

S

DEFENSE

 No compulsory military service

 $400,000   Little change

Since independence, the armed forces have figured prominently in national life. They have put down several attempted coups, notably in 1978, after which 2000 Angolan troops plus Soviet and Cuban advisers were invited in, and in 1988. In 1995, a group of army officers seized temporary control of the country. The national armed forces are believed to number 2000. With the collapse of the Eastern bloc, São Tomé now receives military assistance from the West.

ECONOMICS

Inflation 51% p.a. (1990–2000)

 $43m   2390–8937 dobras

SCORE CARD

❏ WORLD GNP RANKING	191st
❏ GNP PER CAPITA	$290
❏ BALANCE OF PAYMENTS	−$9m
❏ INFLATION	5%
❏ UNEMPLOYMENT	50%

EXPORTS

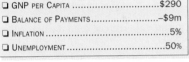

Norway 6%
Italy 6%
Portugal 39%
Netherlands 11%
Turkey 11%
Other 27%

SÃO TOMÉ & PRÍNCIPE

Total Land Area : 1001 sq. km (386 sq. miles)

POPULATION

- ● over 10 000
- • under 10 000

LAND HEIGHT

- 1000m/3281ft
- 500m/1640ft
- 200m/656ft
- Sea Level

0   10 km
0   10 miles

Príncipe
Ilha Bombom
Santo António
Infante Dom Henrique
Ilha Caroço
Tinhosa Pequena
Tinhosa Grande
(continuation on same scale)

N

Ilha das Cabras
✠ SÃO TOMÉ
Santana
Pico de São Tomé ▲ 2024m
São Tomé
Santa Cruz
Gulf of Guinea
Porto Alegre
Equator
Ilha das Rôlas

RESOURCES

 Electric power 6000 kw

 3756 tonnes   Reserves currently unexploited

 30,000 ducks, 4800 goats, 350,000 chickens

 None

An offshore oil exploration agreement with Nigeria was signed in 2001. There are no mineral resources on the islands. São Tomé is very fertile; cocoa estates are finally back to pre-1975 productivity, and diversification of crops is now a priority. Príncipe has better ports, but its wild scenery makes it more suitable for tourism than farming.

IMPORTS

Belgium–Luxembourg 5%
France 5%
Germany 8%
UK 8%
Portugal 48%
Other 26%

STRENGTHS

Legacy of Portuguese-built infrastructure. Potential for development of fisheries, agriculture, tourism, and oil. Able to attract substantial aid.

WEAKNESSES

Cocoa accounts for 90% of export earnings. Skillful diplomacy has attracted high levels of aid, but mismanagement of these funds has resulted in severe debt. Weak currency.

ENVIRONMENT

 Not available

 None   0.5 tonnes per capita

Fish conservation, deforestation for fuelwood, and potential tourism expansion are the major issues.

MEDIA

TV ownership medium

 There are no daily newspapers

PUBLISHING AND BROADCAST MEDIA

There are no daily newspapers. *Diário da República* and *Notícias* are published weekly by the government

1 state-controlled service   1 state-controlled service

Freedom of expression is respected. The state controls radio and TV stations. Radio ownership is high for Africa.

CRIME

No death penalty

 99 prisoners   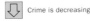 Crime is decreasing

Crime levels are fairly low owing to the tightly knit nature of the community. Urban robberies are a problem.

CHRONOLOGY

The entire preindependence history of the islands was as a Portuguese colony exploited by plantation owners.

- ❏ **1972–1973** Strikes by plantation workers.
- ❏ **1975** Independence as Marxist state. Plantations nationalized.
- ❏ **1978** Abortive coup.
- ❏ **1990** New democratic constitution.
- ❏ **1991–2000** Miguel Trovoada president for two terms.
- ❏ **1995** Príncipe granted autonomy.
- ❏ **2001** De Menezes wins presidency.

EDUCATION

 School leaving age: 14

 75%   Not available

Education is compulsory for 7–14-year-olds. All staff at the one technical and three secondary schools are foreigners.

HEALTH

No welfare state health benefits

 1 per 2128 people   Malaria, other parasitic diseases, respiratory and diarrheal diseases

Although health care is not free, São Tomé has a better system of basic care than other African countries.

SPENDING

GDP/cap. decrease

CONSUMPTION AND SPENDING

28 per 1000 population   31 per 1000 population

Defense	1%
Education	3.8%
Health	6.2%

0   5   10   15   20   25
Defense, Health, Education spending as % of GDP

Wealth disparities are not conspicuous. There is a growing business class. Cocoa workers form the country's poorest group.

WORLD RANKING

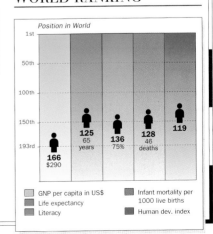

Position in World

1st
50th
100th
150th
193rd

166 $290
125 65 years
136 75%
128 46 deaths
119

■ GNP per capita in US$
■ Life expectancy
■ Literacy
■ Infant mortality per 1000 live births
■ Human dev. index

S

SAUDI ARABIA

OFFICIAL NAME: Kingdom of Saudi Arabia **CAPITAL:** Riyadh; Jiddah (administrative)
POPULATION: 21 million **CURRENCY:** Saudi riyal **OFFICIAL LANGUAGE:** Arabic

OCCUPYING MOST OF THE Arabian peninsula, Saudi Arabia covers an area as large as western Europe. Over 95% of its land is desert, with the most arid part, known as the Empty Quarter or Rub al Khali, being in the southeast. Saudi Arabia has the world's largest oil and gas reserves. It includes Islam's holiest cities, Medina and Mecca, visited each year by two million Muslims performing the pilgrimage known as the *haj*. The al-Sa'ud family have been Saudi Arabia's absolutist rulers since 1932. In theory, Islamic *sharia* underpins the constitution.

TOURISM

▷ Visitors : Population 1:5.7

3.7m visitors   Up 3% in 1998

MAIN TOURIST ARRIVALS

	% of total arrivals
Iran 16%	
Pakistan 10%	
Turkey 10%	
Egypt 10%	
Yemen 6%	
Other 48%	

0 10 20 30 40 50 60
% of total arrivals

Foreign tourism is discouraged. Until a limited relaxation in 2000, only Muslim pilgrims, business people, and foreign workers were permitted entry. Non-Muslims are banned from the holy cities of Mecca and Medina. Although strict quotas have been imposed to avoid overcrowding, stampedes of *haj* pilgrims in 1990, 1997, and 2001 killed or injured thousands. Many choose the port of Jiddah as a base from which to begin the pilgrimage. Also popular is the *umra*, or little pilgrimage, since it can be made at any time of year. An estimated $2.5 billion has been spent on improving *haj* facilities in recent years. Jizan on the Red Sea offers superb scuba diving. The Hejaz railroad and the Nabatean ruins at Medain Salih are of archaeological interest. To escape the summer heat, the government relocates to mountainous Taif, used as a resort by the Saudis.

CLIMATE

▷ Hot desert

WEATHER CHART FOR RIYADH

■ Average daily temperature   Rainfall ■
°C/°F J F M A M J J A S O N D cm/in
60/140 .... 60/24
50/122 .... 50/20
40/104 .... 40/16
30/86 .... 30/12
20/68 .... 20/8
10/50 .... 10/4
0/32 0

The kingdom's only reliable rainfall is in the southern Asir province, making agriculture viable there. The central plateau requires deep artesian wells to water crops. Inland, summer temperatures often soar above 48°C (118°F), but in winter, especially in the northwest, they may fall to freezing point.

SAUDI ARABIA

Total Land Area : 1 960 582 sq. km
(756 981 sq. miles)

POPULATION   LAND HEIGHT

▣ over 1 000 000	
◉ over 500 000	3000m/9843ft
◎ over 100 000	2000m/6562ft
○ over 50 000	1000m/3281ft
• over 10 000	500m/1640ft
· under 10 000	Sea Level

TRANSPORTATION

▷ Drive on right

King Abd al-Aziz International, Jiddah
10.6m passengers

279 ships
1.3m grt

THE TRANSPORTATION NETWORK

44,104 km (27,405 miles)		Trans-Arabian Highway
1392 km (865 miles)		None

A modern transportation infrastructure links the main population centers to the Gulf states, Jordan, and Egypt. Saudi Arabian Airlines, founded in 1945, is one of the largest airlines in the world.

Network of modern road junctions spread out across the landscape near Mecca.

Map labels

JORDAN
EGYPT
IRAQ
KUWAIT
The Gulf
BAHRAIN
QATAR
UNITED ARAB EMIRATES
OMAN
YEMEN
RED SEA
AN NAFŪD
AD DAHNĀ
RUB' AL KHALĪ
JABAL TUWAYQ

Ţurayf
Sakākah
Al Jawf
Rafḥā'
Hafar al Bāţin
Tabūk
Taymā'
Ḥā'il
Al Jubayl
Ras Tannūrah
Al Khubar
Ad Dammām
Az Zahrān
Al Wajh
Al 'Ula
Buraydah
Az Zilfī
Al Mubarraz
Al Hufūf
Unayzah
Al Majma'ah
Shaqrā'
RIYADH
Yanbu' al Baḥr
Al Madīnah (Medina)
Ḥaraḍ
Rābigh
Zalim
Laylá
Makkah (Mecca)
Jiddah
At Ţā'if
As Sulayyil
Qal 'at Bishah
Tathlīth
Jabal Sawda' 3133m
Khamis-Mushayt
Al Birk
Abhā
Zahrān
Najrān
Jizān
Wuday'ah

0   200 km
0   200 miles

PEOPLE ▷ Pop. density low

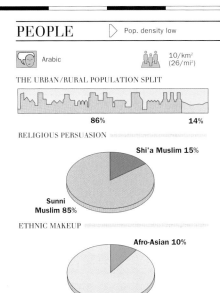

Arabic   10/km² (26/mi²)

THE URBAN/RURAL POPULATION SPLIT

86%   14%

RELIGIOUS PERSUASION

Shi'a Muslim 15%

Sunni Muslim 85%

ETHNIC MAKEUP

Afro-Asian 10%

Arab 90%

The Saudis, who take their name from the ruling al-Sa'ud family, were united by conquest between 1902 and 1932 by King Abd al-Aziz al-Sa'ud. The vast majority of Saudis are Sunni Muslims who follow the *wahhabi* (puritan) interpretation of Islam and embrace *sharia* (Islamic law). The politically dominant Nejdi tribes from the central plateau around Riyadh are Bedouin in origin. The Hejazi tribes, from the south and west, have a more cosmopolitan, mercantile background, but are largely displaced from politics. In the eastern Al-Hasa province there is a Shi'a minority of some 300,000, many of whom are employed in the oil fields.

Women have to wear the veil, cannot hold a driving license, and have no role in public life. In 2000, however, Saudi Arabia decided to sign the UN convention on women's rights – provided it did not contradict *sharia* – and issued women with identity cards from 2001.

POPULATION AGE BREAKDOWN

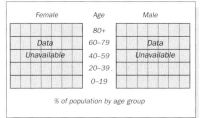

Female	Age	Male
	80+	
Data Unavailable	60–79	Data Unavailable
	40–59	
	20–39	
	0–19	

% of population by age group

POLITICS ▷ No legislative elections

 Not applicable   H.M. King Fahd ibn Abd al-Aziz

LEGISLATIVE OR ADVISORY BODIES
Consultative Council 120 seats

Saudi Arabia is an absolute monarchy. The king rules with the assistance of an appointed Council of Ministers and the Consultative Council

Saudi Arabia is an absolute monarchy. Since 1993 a Consultative Council (Majlis ash-Shoura) has been appointed by the king.

PROFILE

The royal family, the House of Sa'ud, rules by carefully manipulating appointments in all sectors of government. Frequent changes of personnel within the armed forces ensure that officers do not build personal followings. All influential cabinet portfolios, apart from those of oil and religious affairs, are held by members of the royal family.

Absolutist rule means that domestic politics are virtually nonexistent. The regime retains feudal elements: weekly *majlis*, or councils, are held where citizens can present petitions or grievances to leading members of the royal family. Large cash sums are often dispensed at these meetings.

The legitimacy of the regime is built on its adherence to Islamic values, and the backing of the *ulema* (scholars). It is the emphasis on Islam that most colors Saudi life. The 5000-strong *mutawa* (religious police) enforce the five-times-a-day call to prayer, when businesses must close. During Ramadan the *mutawa* are especially active.

MAIN POLITICAL ISSUES
Questioning the ruling family

Following the 1991 Gulf War, a civil rights campaign emerged to challenge the authority of the ruling family, demanding closer adherence to Islamic values. The movement objected to the presence of US troops on Saudi territory and the consequent exposure to corrupt Western culture. The al-Sa'uds moved swiftly to quash the protest but exiled opponents have continued their activities, using faxes and e-mail. The most vociferous denunciations of the royal family have come from Osama bin Laden, formerly part of the inner circle, operating from bases abroad.

The succession issue

The question of succession and the possibility of a future power struggle, rooted in rivalries endemic to the House of Sa'ud, emerged in 1996, when King Fahd suffered a stroke. The management of day-to-day affairs passed briefly to his half-brother, Crown Prince Abdullah, who remains in effective, if no longer formal, control.

WORLD AFFAIRS ▷ Joined UN in 1945

 AL   Damasc   GCC   OIC   OPEC

Saudi Arabia's strategic importance is derived largely from its oil reserves and worldwide investments. Relations with the US are particularly close, but frayed when Saudi Arabia refused to back war in Afghanistan in 2001. After Iraq's invasion of Kuwait in 1990, the Saudis had taken a leading role in consolidating the Arab coalition against Iraq. Saudi Arabia sheltered the Kuwaiti royal family, provided military bases to the Western allies, and supplied more troops than any other Arab country. The continued presence of foreign forces then provoked hostility from Islamic militants. Before this gained world attention through statements by Osama bin Laden at the time of the September 2001 terrorist attacks on the US, a bomb attack at a US military complex near Az Zahran in 1996 had killed 19 US personnel.

A pact signed with Yemen in June 2000 ended a simmering border dispute. In early 2002 Crown Prince Abdullah's new peace plan for Israel/Palestine won praise from Arabs and the West – and hinted at a major change in Saudi foreign policy.

AID ▷ Donor

 $295m (donations)   Up 59% in 2000

Generous loans and grants from the Saudi Fund for Development are made to other Arab and developing countries, mainly for infrastructure projects. Saudi Arabia promotes Islam through charitable foundations, especially in Africa, Asia, and the former Soviet Union. The royal purse also supports scientific and medical research. Since the liberation of Kuwait in 1991, Saudi Arabia has given large sums to countries that supported the US-led alliance, notably Egypt, Syria, Morocco, and Turkey. In addition, the Saudi government substantially reimbursed the US and the UK for the cost of their expeditionary forces, as well as favoring companies from the allied countries for reconstruction contracts.

King Fahd ibn Abd al-Aziz *acceded to the Saudi throne in 1982.*

Crown Prince Abdullah, *commander of the National Guard.*

S

CHRONOLOGY

The unification of Saudi Arabia under King Abd al-Aziz (ibn Sa'ud) was achieved in 1932. The kingdom remains the only country in the world which is named after its royal family.

- ❏ **1937** Oil reserves discovered near Riyadh.
- ❏ **1939** Ceremonial start of oil production at Az Zahran.
- ❏ **1953** King Sa'ud succeeds on the death of his father Abd al-Aziz.
- ❏ **1964** King Sa'ud abdicates in favor of his brother Faisal.
- ❏ **1973** Saudi Arabia imposes oil embargo on Western supporters of Israel.
- ❏ **1975** King Faisal assassinated by a deranged nephew; succeeded by his brother Khalid.
- ❏ **1979** Muslim fundamentalists led by Juhaiman ibn Seif al-Otaibi seize Grand Mosque in Mecca, proclaim a *mahdi* (savior) on first day of Islamic year 1400.
- ❏ **1981** Formation of GCC, with its secretariat in Riyadh.
- ❏ **1982** King Fahd succeeds on the death of his brother King Khalid. Promises to create consultative assembly.
- ❏ **1986** Opening of King Fahd Causeway to Bahrain. Shaikh Yamani sacked as oil minister.
- ❏ **1987** Diplomatic relations with Iran deteriorate after 402 people die in riots involving Islamic fundamentalists at Mecca during the *haj* (pilgrimage).
- ❏ **1989** Saudi Arabia signs nonaggression pact with Iraq. Saudi Arabia brokers political settlement to Lebanese civil war.
- ❏ **1990** Kuwaiti royal family seeks sanctuary in Taif after Iraqi invasion. Many allegedly pro-Iraqi Jordanians and Yemenis expelled.
- ❏ **1990–1991** US, UK, French, Egyptian, and Syrian forces assemble in Saudi Arabia for Operation Desert Storm. Public executions are halted.
- ❏ **1991** Iraqis seize border town of Al Khafji, but are repulsed by Saudi, US, and Qatari forces.
- ❏ **1993** King Fahd appoints Consultative Council (Majlis ash-Shoura).
- ❏ **1996** King Fahd briefly relinquishes control to Crown Prince Abdullah. Bomb attack at US military complex in Az Zahran kills 19 US citizens.
- ❏ **1997, 2001** Consultative Council expanded, first to 90 then to 120 members.
- ❏ **2002** Crown Prince Abdullah unveils Middle East peace plan; endorsed by AL summit in Beirut.

DEFENSE

 No compulsory military service

 $18.3bn

⬇ Down 16% in 2000

Saudi Arabia's substantial military contribution to the 1991 Gulf War, at a cost of $55 billion, enhanced its image as a major regional power. Military equipment is purchased mostly from the US, the UK, and France. Weapons systems are advanced and include Patriot missiles and AWACS early warning radar. However, skilled foreign personnel operate many of these: 1000 US air force troops are employed to keep AWACS flying.

The air force is the elite branch of the military. It had one brief period of politicization in 1969 when officers attempted a coup. The paramilitary

SAUDI ARABIAN ARMED FORCES

🛡	1055 main battle tanks (315 M-1A2 *Abrams*, 290 AMX-30, 450 M60A3)	75,000 personnel
🚢	4 frigates, 4 corvettes, and 26 patrol boats	15,500 personnel
✈	348 combat aircraft (68 F-5, 159 F-15, 85 *Tornado* IDS, 22 *Tornado* ADV)	20,000 personnel
🚀	None	

National Guard is drawn from tribal supporters of the al-Sa'ud regime. Its commander-in-chief is the crown prince rather than the defense minister.

ECONOMICS

 Inflation 2.2% p.a (1990–2000)

 $150bn

3.7505–3.7504 Saudi riyals

SCORE CARD

❏ World GNP Ranking	27th
❏ GNP per Capita	$7230
❏ Balance of Payments	$15.6bn
❏ Inflation	–0.8%
❏ Unemployment	6%

EXPORTS

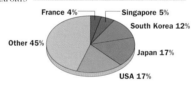

France 4% — Singapore 5% — South Korea 12% — Japan 17% — USA 17% — Other 45%

IMPORTS

France 4% — Germany 7% — UK 7% — Japan 9% — USA 21% — Other 52%

ECONOMIC PERFORMANCE INDICATOR

Consumer Price Index — GDP

Consumer price index 1995=100 / GDP 1996=100

(1996, 1997, 1998, 1999, 2000)

STRENGTHS

Vast oil and gas reserves. Soaring world oil prices in 2000 signified recovery from 1986 collapse. World-class associated industries. Accumulated surpluses and steady current income. Large income from two million pilgrims to Mecca annually.

WEAKNESSES

Lack of indigenous skilled workers. Food production requires heavy subsidy. Most consumer items and industrial raw materials imported. Up to 20% youth unemployment. Large national debt. Royal family technically owns most of national wealth.

PROFILE

Since the 1970s, great efforts have been made to reduce dependence on oil exports and to provide employment for young Saudis. By late 2000, Saudi Arabia had steered OPEC toward controlling oil prices to avoid recession in industrialized countries or spurring a drive to develop alternative energy. Since April 2000, approved foreigners have been allowed complete ownership of Saudi businesses and rights to property. Large sums have been spent on creating a US-standard infrastructure to provide the basis for a manufacturing economy. It remains, however, dependent on foreign workers.

SAUDI ARABIA : MAJOR BUSINESSES

Al Jubayl
Buraydah
Ras Tanūrah
Ad Dammān
Al Madinah
Al Hufūf
Yanbu' al Baḥr
Rabigh
Jiddah
Ḥaraḍ
Makkah
Riyadh

Cement	Oil
Steel	Oil refining
Food processing	Gas refining
Light engineering	
Petrochemicals/fertilizers	

0   400 km
0   400 miles

RESOURCES   ▷ Electric power 22.6m kw

51,949 tonnes

8.77m b/d (reserves 261.8bn barrels)

7.58m sheep, 4.31m goats, 130m chickens

Oil, natural gas, limestone, gypsum, marble, clay, salt

ELECTRICITY GENERATION

Hydro 0%	
Combustion 100% (113bn kwh)	
Nuclear 0%	
Other 0%	

% of total generation by type

With the world's biggest oil and gas reserves, Saudi Arabia plays a key role in the global economy and is among the

SAUDI ARABIA : LAND USE

Cropland
Desert
Pasture
Wheat
Dates
Sheep

RUB ' AL KHALI

0   400 km
0   400 miles

top ten traders of all the world's major industrialized nations.

ENVIRONMENT   ▷ Sustainability rank: 138th

2%

14.4 tonnes per capita

ENVIRONMENTAL TREATIES

No	Yes	Yes
Yes	Yes	No

Pollution in the Gulf and Red Sea has threatened some wildlife and their habitats, as have hunters using high-velocity rifles and off-road vehicles. The government has taken steps to confine manufacturing to industrial estates. Environmental legislation is, nevertheless, poorly developed, although planning controls apply in the major cities.

MEDIA   ▷ TV ownership high

☒ Daily newspaper circulation 59 per 1000 people

PUBLISHING AND BROADCAST MEDIA

There are 13 daily newspapers, in Arabic and English. The leading papers are *Ar-Riyadh*, *Sharq Al Awsat*, *Al-Jazirah*, and *Riyadh Daily*

2 state-owned services

2 services: 1 state-owned, 1 owned by a private oil company

The government imposes total press censorship and insists on strict morality. In 1994, private citizens were banned from owning satellite dishes, but the authorities turn a blind eye to their use. No allowance is made for Arab satellite broadcasts, which have been criticized for covering anti-Islamic views. The international *Sharq Al Awsat* is a leading Arabic daily. In 2001 the government announced strict rules regarding state and religion on the Internet.

CRIME   ▷ Death penalty in use

18,000 prisoners

Down 13% 1996–1998

CRIME RATES

Murders	
1	per 100,000 population

Rapes	
1	per 100,000 population

Thefts	
80	per 100,000 population

Strict Islamic punishments – stoning, amputation, and beheading – are enforced. Criticism for human rights abuses has increased, with an Amnesty International campaign in 2000.

EDUCATION   ▷ Schooling is free but not compulsory

77%

273,992 students

THE EDUCATION SYSTEM

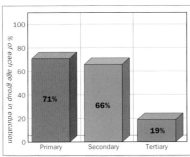

% of each age group in education

Primary	Secondary	Tertiary
71%	66%	19%

Growing numbers of Western-educated Saudis have intensified pressure for social and political change. In the 1950s, the religious establishment was persuaded to give women equal opportunities in education. Economic analysts have criticized Saudi universities for turning out many theology graduates, but not enough engineers and technocrats. Many Saudis still travel abroad to complete their studies.

HEALTH   ▷ Welfare state health benefits

1 per 588 people

Diarrheal, respiratory, heart, metabolic, and parasitic diseases

Infant mortality has dropped and endemic disease has been nearly eradicated. Health care outside major centers such as Riyadh and Jiddah still remains relatively undeveloped, given Saudi Arabia's huge economic resources. However, large sums have been spent on employing Western expertise. Many Saudis are still sent overseas by the government for treatment, especially for transplant operations, which pose some ethical problems for religious leaders. The private sector has also been encouraged.

SPENDING   ▷ GDP/cap. increase

CONSUMPTION AND SPENDING

93 per 1000 population

137 per 1000 population

Defense 10.1%	
Education 7.5%	
Health 6.4%	

Defense, Health, Education spending as % of GDP

Saudi citizens are among the most prosperous in the world. Non-Saudi citizens, especially the guest workers from the Indian subcontinent and the Philippines, are much poorer. The al-Sa'uds have used their wealth to create a cradle-to-grave welfare system. Ownership of TVs, telephones, and VCRs is among the region's highest. The distribution of wealth is carefully controlled by the royal family through the *majlis* system. There is no stock market, although shares in public companies are traded privately. Many Saudis refuse for religious reasons to accept interest on deposits with banks, but Islamic banks offer profit-sharing investment schemes as an alternative.

WORLD RANKING

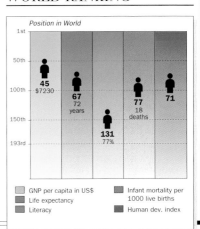

Position in World

45 $7230	
67 72 years	
131 77%	
77 18 deaths	
71	

GNP per capita in US$
Life expectancy
Literacy

Infant mortality per 1000 live births
Human dev. index

S

SENEGAL

OFFICIAL NAME: Republic of Senegal **CAPITAL:** Dakar
POPULATION: 9.7 million **CURRENCY:** CFA franc **OFFICIAL LANGUAGE:** French

| 1960 | 1960 | April 4 | SN | 0 | +221 | .sn |

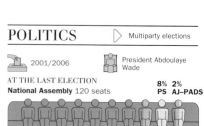

WEST AFRICA

SENEGAL'S CAPITAL, Dakar, lies on the westernmost cape of Africa. The country is mostly low, with open savanna and semidesert in the north and thicker savanna in the south. After independence from France in 1960, Senegal was ruled until 1981 by President Léopold Senghor. He was succeeded by his prime minister, Abdou Diouf, who held power for almost 20 years until his election defeat in March 2000.

POLITICS ▷ Multiparty elections

2001/2006 President Abdoulaye Wade

AT THE LAST ELECTION
National Assembly 120 seats

8% PS 2% AJ–PADS

74% SC 9% AFP 3% URD 4% Other

SC = Sopi (Change) coalition (led by the Senegalese Democratic Party – **PDS**) **AFP** = Alliance of Progressive Forces **PS** = Senegalese Socialist Party **URD** = Union for Democratic Renewal **AJ–PADS** = And Jëf – African Party for Democracy and Socialism

Senegal has been a multiparty democracy since 1981, when, under the then new president Abdou Diouf, the constitution was amended to allow more than four political parties. However, the PS held power from the 1950s until 2000, and its influence has been pervasive. Presidential elections in 2000 marked a political watershed. Diouf was defeated by Abdoulaye Wade of the liberal democratic PDS, the dominant party in the "Sopi" (Change) coalition which went on to win a landslide victory in the 2001 legislative elections.

A new constitution, approved in 2001 by referendum, abolished the Senate and restricts the president to two terms.

CLIMATE ▷ Steppe/tropical

WEATHER CHART FOR DAKAR

The coastal regions, which project into the path of the northern trade winds, are remarkably cool given their latitude.

TRANSPORTATION ▷ Drive on right

Dakar–Yoff International
1.1m passengers

198 ships
51,000 grt

THE TRANSPORTATION NETWORK

| 4271 km (2654 miles) | None |
| 751 km (467 miles) | 897 km (557 miles) |

Dakar is an important west African port, serving Senegal itself, Guinea, the hinterland of Mali, and southern Mauritania. The key rail link to Bamako, Mali's capital, was built in the 1920s.

PEOPLE ▷ Pop. density medium

Wolof, Fulani, Serer, Diola, Malinke, Soninke, Arabic, French

50/km² (130/mi²)

THE URBAN/RURAL POPULATION SPLIT

47% 53%

RELIGIOUS PERSUASION

Traditional beliefs 5%
Christian (mainly Roman Catholic) 5%
Sunni Muslim 90%

Senegal has a fairly well-developed sense of nationhood, and intermarriage between groups has reduced ethnic tensions. Groups can still be identified regionally, however. Dakar is a Wolof area, the Senegal River is dominated by the Toucouleur, the Malinke mostly live in the east, and the Diola in Casamance. The Diola have felt excluded from politics, and this has led to a long-running rebellion in Casamance. A French-influenced class system is still prevalent. The 2001 constitution gave women property rights for the first time.

WORLD AFFAIRS ▷ Joined UN in 1960

| CILSS | ECOWAS | FZ | OIC | OMVG |

Maintaining good relations with France, Senegal's main ally and aid donor, is the major foreign affairs concern. Relations with neighboring Gambia, Mauritania, and Guinea-Bissau continue to be a constant preoccupation.

TOURISM ▷ Visitors : Population 1:26

369,000 visitors Up 5% in 1999

MAIN TOURIST ARRIVALS

| France 51% |
| Africa 22% |
| Germany 5% |
| Other 22% |

0 10 20 30 40 50 60
% of total arrivals

In addition to French package tours to coastal resorts, tours for African–Americans to Gorée, an old slave island, are increasingly popular.

SENEGAL

Total Land Area :
196 190 sq. km
(75 749 sq. miles)

0 100 km
0 100 miles

POPULATION
over 1 000 000
over 100 000
over 50 000
over 10 000
under 10 000

LAND HEIGHT
200m/656ft
Sea Level

S

AID

 ▷ Recipient

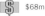 $423m (receipts)   ⬇ Down 21% in 2000

Senegal is one of the highest recipients of aid per capita in Africa, mostly from France, the World Bank, and Japan. Aid is used to import 400,000 tonnes of rice annually, but also helps to finance a sizable civil service, now being cut back. An important structural adjustment program, backed by the IMF, was renewed for a further year in April 2001.

The mosque in Touba, *religious capital of the Muslim Mouride sect, which was founded in 1887 in Senegal's groundnut-growing district.*

DEFENSE

 ▷ Compulsory military service

 $68m   ⬇ Down 16% in 2000

France maintains an important naval base at Dakar. The armed forces total 9400, plus a paramilitary force of 5800, but the military has never intervened in politics. Senegalese troops took part in Operation Desert Storm in 1991, and intervened in conflicts in Liberia, Rwanda, and the Central African Republic. They also helped to quell revolts in Gambia and Guinea-Bissau.

ECONOMICS

▷ Inflation 4.6% p.a. (1990–2000)

$4.71bn   698.7–736.7 CFA francs

SCORE CARD

- ❏ WORLD GNP RANKING........................114th
- ❏ GNP PER CAPITA $490
- ❏ BALANCE OF PAYMENTS...................–$310m
- ❏ INFLATION 0.7%
- ❏ UNEMPLOYMENT....................................4%

STRENGTHS

Good infrastructure. Relatively strong industrial sector. First west African country with international credit rating. Revenue from sale of fishing rights.

WEAKNESSES

Few natural resources exploited, other than groundnuts, phosphates, and fish.

EXPORTS

Spain 5% | Mali 6% | Italy 9% | France 16% | India 18% | Other 46%

IMPORTS

Germany 4% | USA 4% | Thailand 5% | Nigeria 19% | France 27% | Other 41%

Access to oil potential of Casamance region hampered by rebellion and poor transportation links.

RESOURCES

▷ Electric power 235,000 kw

418,280 tonnes   Not an oil producer

4.82m sheep, 4m goats, 45m chickens   Phosphates, bauxite, salt, natural gas, marble, iron, copper

Senegal's electricity capacity is largely dependent on imported fuel; cheaper supplies are expected to become available soon from the Manantali Dam in Mali. Initial explorations suggest that oil reserves may exist off Casamance.

ENVIRONMENT

▷ Sustainability rank: 81st

11% (6% partially protected)   0.4 tonnes per capita

The damming of the Senegal River has caused concern that traditional farming practices, which rely on seasonal floods, may be disrupted. Two major droughts in 1973 and 1983 led to the advance of the Sahara Desert in the west of the country.

MEDIA

▷ TV ownership low

 Daily newspaper circulation 5 per 1000 people

PUBLISHING AND BROADCAST MEDIA

There are 8 daily newspapers, including *Le Soleil*, *Wal Fadjiri*, and *Sud Quotidien*

2 services: 1 state-owned, 1 private   5 services: 1 state-owned, 4 independent

The independent media flourished with multipartyism. Senegal had the first satirical journal in Africa with the founding of *Le Politicien* in 1978.

CRIME

▷ Death penalty not used in practice

 4653 prisoners   Up 8% in 1999

Senegal has comparatively low crime rates, though levels are now rising in Dakar and the surrounding shanty towns, where gangs are based.

CHRONOLOGY

France colonized Senegal, a major entrepôt from the 15th century, in 1890. Dakar was the capital of French West Africa.

- ❏ **1885** Gambia split off from Senegal.
- ❏ **1960** Independence under Senghor.
- ❏ **1966–1976** One-party state.
- ❏ **1981** Full multipartyism restored.
- ❏ **2000** Presidency won by Abdoulaye Wade in first ever defeat for PS.
- ❏ **2001** Referendum approves new constitution.

EDUCATION

▷ School leaving age: 13

37%   29,303 students

Illiteracy is Senegal's major educational challenge. There are universities at Dakar and St.-Louis.

HEALTH

▷ No welfare state health benefits

1 per 10,000 people   Malaria, diarrheal diseases

The state health system is rudimentary. A successful education campaign helps to contain the incidence of HIV/AIDS.

SPENDING

▷ GDP/cap. increase

CONSUMPTION AND SPENDING

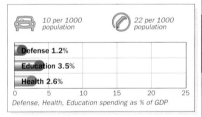

10 per 1000 population   22 per 1000 population

Defense 1.2% | Education 3.5% | Health 2.6%

Defense, Health, Education spending as % of GDP

Wealth disparities are considerable in Senegal, and poverty is widespread. Members of the former ruling PS are the wealthiest group.

WORLD RANKING

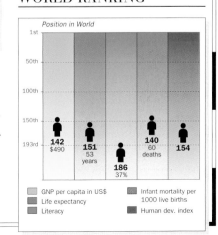

Position in World

142 $490 | 151 53 years | 186 37% | 140 60 deaths | 154

- GNP per capita in US$
- Life expectancy
- Literacy
- Infant mortality per 1000 live births
- Human dev. index

S

SERBIA & MONTENEGRO

OFFICIAL NAME: Serbia and Montengro   **CAPITAL:** Belgrade   **POPULATION:** 10.5 million
CURRENCY: Dinar (euro widely used in Montenegro)   **OFFICIAL LANGUAGE:** Serbo-Croat

 1992   1992   Nov 29   YU   +1   +381   .yu

THE LOOSE CONFEDERATION of Serbia and Montenegro emerged in 2002, when the troubled Federal Republic of Yugoslavia was finally dissolved after 84 years in existence. Literally a "union of south Slavs," Yugoslavia was created after World War I. A maverick communist state after 1945, it was torn apart by nationalist tensions in a violent civil war in 1991–1995, spawning four new Balkan republics. After the ousting in 2000 of the discredited Serbian nationalist Slobodan Milošević as federal president, a more democratic successor regime ended the country's international pariah status.

TOURISM

▷ Visitors : Population
1:69

🧳 152,000 visitors   ⬇ Down 46% in 1999

MAIN TOURIST ARRIVALS

Bosnia & Herzegovina 38%	
Macedonia 8%	
Italy 6%	
Greece 6%	
Russia 5%	
Other 37%	

0   10   20   30   40   50   60
% of total arrivals

CLIMATE

▷ Continental

WEATHER CHART FOR BELGRADE

There are three climate zones. The northern plains are characteristically continental: rainy springs, warm summers, and cold winters. While the southern highlands have colder winters with heavy snowfalls, the Adriatic coast boasts hot summers and milder winters.

The country's stunning scenery is a potential draw for tourists. Before the 1990s it attracted millions every year.

Instability, the impact of UN sanctions, and the Kosovo conflict meant that foreign tourism ceased in the 1990s. Serbia has never been a center of tourism. The Montenegrin coast, however, has renowned beaches. In the 1990s they were monopolized by Serbians, particularly by political and criminal elements of the Serbian elite. Hyperinflation and recession kept the average vacationer away.

SERBIA & MONTENEGRO
(YUGOSLAVIA)

Total Land Area : 102 350 sq. km (39 517 sq. miles)

POPULATION

over 1 000 000	▣
over 100 000	◉
over 50 000	○

LAND HEIGHT

2000m/6562ft	
1000m/3281ft	
500m/1640ft	
200m/656ft	
Sea Level	

TRANSPORTATION

▷ Drive on right

✈ Surcin, Belgrade
1.28m passengers

🚢 9 ships
4700 grt

THE TRANSPORTATION NETWORK

🛣 28,822 km (17,909 miles)		🛣 560 km (348 miles)	
🚆 4059 km (2522 miles)		587 km (365 miles)	

About one-third of railroads are electrified. The rail link to Greece is one of Serbia's main trading routes, and lines through Serbia remain the best link between Budapest and Sofia. For internal travel, trains are cheaper, but slower, than buses. Two flights each day link Belgrade with the airports in Montenegro.

Bridges and railroads were specifically targeted during the NATO bombing in 1999. The bombing of the bridge over the Danube at Novi Sad closed the river as a major regional trading artery, and its reopening was made a priority for international assistance once the Milošević regime had fallen.

S

PEOPLE ▷ Pop. density medium

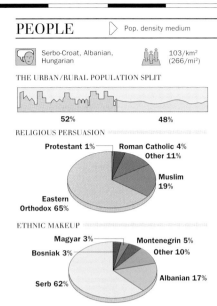

Serbo-Croat, Albanian, Hungarian   103/km² (266/mi²)

THE URBAN/RURAL POPULATION SPLIT

52%   48%

RELIGIOUS PERSUASION

Protestant 1% — Roman Catholic 4%
Other 11%
Muslim 19%
Eastern Orthodox 65%

ETHNIC MAKEUP

Magyar 3% — Montenegrin 5%
Bosniak 3% — Other 10%
Serb 62%
Albanian 17%

The social order has been devastated by the decade of conflict in the region. There is a depressingly high suicide rate among the old living in cities. In Kosovo, 900,000 ethnic Albanians fled (mainly to Albania and Macedonia) during the 1999 conflict. The 1999 agreement opened the way for most to return, since which time many of the province's embattled Serb minority have gone (or been driven out) themselves, fearing both vengeance and the longer-term prospect of Kosovan independence.

The Hungarian (and mainly Roman Catholic) minority is concentrated in, but does not dominate, the relatively prosperous northern Serbian province of Vojvodina. Limited autonomy was restored in 2002, after economically motivated pressure from within the region. Orthodox Christian Serbs who converted to Islam during Ottoman rule are considered to be an ethnically separate group.

POPULATION AGE BREAKDOWN

Female	Age	Male
1.1%	80+	0.8%
8.8%	60–79	7.2%
12.4%	40–59	12.2%
13.9%	20–39	14.3%
14.3%	0–19	15%

% of population by age group

POLITICS ▷ Multiparty elections

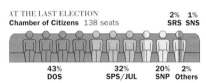

L. House 2000/2004   President Vojislav
U. House 2000/2004   Kostunica

AT THE LAST ELECTION
Chamber of Citizens 138 seats   2%   1%
SRS   SNS

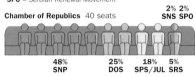

43% DOS   32% SPS/JUL   20% SNP   2% Others

DOS = Democratic Opposition of Serbia
SPS/JUL = Socialist Party of Serbia/Yugoslav Left
SNP = Socialist People's Party   **SRS** = Serbian Radical Party
SNS = Serbian People's Party
SPO = Serbian Renewal Movement

Chamber of Republics 40 seats   2% 2%
SNS SPO

48% SNP   25% DOS   18% SPS/JUL   5% SRS

20 delegates are elected to the Chamber of Republics by each of the two republics of Serbia and Montenegro

Serbia and Montenegro, each with a separate parliament and president, are represented in the Federal Assembly.

PROFILE

Slobodan Milošević, Serbian president from 1989 and federal president from 1997, dominated politics for a decade. Conflict from 1991 to 1995, then again over Kosovo, created a permanent sense of siege. The tightening of UN sanctions in 1999, rather than the NATO bombing, sealed Milošević's fate. After he was ousted in the October 2000 popular uprising, the reformists consolidated their position, dominating elections in December. However, Milošević's extradition in 2001 affronted nationalists in the DOS coalition, including President Vojislav Kostunica; Kostunica has since withdrawn his party.

MAIN POLITICAL ISSUES
Kosovo

The withdrawal from Kosovo of the federal army ended NATO bombing in 1999. The ethnic Albanian majority awaits a promised referendum on its future. International administrators in 2001 oversaw the election of a Kosovo assembly. Local Serbs took part, but are wary of Albanian calls for independence.

Montenegro

Although less subservient to Serbia from 1997, calls for Montenegro's independence were subdued in 2002 when President Milo Djukanović agreed to a trial period in a loose confederation.

Vojislav Kostunica,
federal president since the ousting of Milošević in 2000.

Slobodan Milošević,
disgraced former president charged with war crimes.

WORLD AFFAIRS ▷ Joined UN in 1945

 CEI   IAEA   EBRD   IMF   OSCE

In 1995, mutual recognition among the countries which had constituted the Socialist Federal Republic of Yugoslavia paved the way for the normalization of relations.

The pariah status of the Milošević regime in the late 1990s was underlined by his indictment by the International Criminal Tribunal for the former Yugoslavia (ICTY) in 1999, and the military action by NATO forces over the situation in Kosovo. Russia, long an ally of Serbia, shares its Orthodox Christianity and its Slavic ethnicity, and strongly opposed the NATO military action over Kosovo, but backed the settlement proposals required to end it. The inauguration of the Kostunica government in 2000 was welcomed enthusiastically by the West and Russia. The new regime was quickly invited to take up its vacant seat in the UN. Agreement to cooperate with the ICTY has reopened aid channels.

AID ▷ Recipient

$1.14bn (receipts)   Up 77% in 2000

Milošević's removal was an explicit condition of large-scale Western aid, urgently needed to rebuild the damaged economy. The country rejoined the World Bank in May 2001 and received pledges of $1.3 billion in aid immediately upon the extradition of Milošević.

CHRONOLOGY

The Serbs were defeated by the Turks at the Battle of Kosovo in 1389. Parts of the region were later ruled by the Austro-Hungarian Empire.

❏ **1878** Independence gained by Serbia and Montenegro at Congress of Berlin.
❏ **1918** Joint Kingdom of Serbs, Croats, and Slovenes created.
❏ **1929** King Alexander of Serbia assumes absolute powers over state, which changes name to Yugoslavia.
❏ **1941** Germans launch surprise attack. Rival resistance groups: Chetniks (Serb royalist) and Partisans (communist, under Tito).
❏ **1945** Federal People's Republic of Yugoslavia founded with Tito as prime minister (and president from 1953).
❏ **1948** Tito breaks with Stalin.
❏ **1951** Farmers permitted to sell products on free market.
❏ **1955** Yugoslav–Soviet détente. ⇨

S

CHRONOLOGY *continued*

- ❑ **1963** Third postwar constitution adopts name Socialist Federal Republic of Yugoslavia (SFRY).
- ❑ **1973** Economic cooperation agreement with West Germany. Agreement of noninterference signed with USSR.
- ❑ **1974** New constitution decentralizes government. Vojvodina and Kosovo given greater autonomy.
- ❑ **1980** Tito dies. Succeeded by collective presidency.
- ❑ **1981** Unrest among Kosovo Albanians; state of emergency.
- ❑ **1985** Serbian intellectuals publish memorandum listing Serb grievances within Yugoslavia.
- ❑ **1986** Slobodan Milošević becomes leader of Communist (later Socialist) Party of Serbia.
- ❑ **1987** Wage freeze to combat inflation. Banking system crisis.
- ❑ **1988** Belgrade protests against economic austerity. Government brought down over budget failure.
- ❑ **1989** 600th anniversary of Battle of Kosovo. Kosovo Albanians protest against Serb police unit; crackdown ends Kosovo's autonomy. Milošević elected president of Serbia.
- ❑ **1990** SPS wins elections in Serbia. Communists win presidency and dominate Montenegro elections.
- ❑ **1992** EU recognizes breakaway republics of Croatia, Slovenia, and Bosnia and Herzegovina. Bosnian war begins. UN sanctions imposed. Ibrahim Rugova elected president of self-declared republic of Kosovo. Milošević reelected president of Serbia, but SPS loses majority.
- ❑ **1995** Milošević signs Bosnian peace accord.
- ❑ **1996** UN sanctions formally lifted.
- ❑ **1997** Concessions after big protests, acknowledging malpractice in municipal elections. Milošević becomes federal president.
- ❑ **1998** Conflict in Kosovo escalates.
- ❑ **1999** March, Kosovo talks break down; "ethnic cleansing" precipitates mass exodus. NATO aerial bombing. June, withdrawal of Serbian forces and police from Kosovo, and entry of international force, KFOR.
- ❑ **2000** September, defeat of Milošević in first round of presidential election. October, opposition candidate Vojislav Kostunica swept to power by massive anti-Milošević protests. December, DOS dominates elections for Serbian parliament.
- ❑ **2001** April, arrest of Milošević. June, Milošević extradited to face war crimes tribunal in The Hague.
- ❑ **2002** Serbia and Montenegro agree to dissolve Yugoslavia.

DEFENSE Compulsory military service

 $1.79bn   Up 8% in 2000

SERBIA & MONTENEGRO ARMED FORCES

1016 main battle tanks (721 T-55, 230 M-84, 65 T-72)	79,000 personnel
4 submarines, 3 frigates, and 31 patrol boats	7000 personnel
111 combat aircraft (29 MiG-21, 14 *Orao* 2, 36 *Super Galeb* G-4)	19,500 personnel
None	

Military capability was specifically targeted for "degrading" by the NATO air strikes in 1999. The impact on antiaircraft defenses, heavy weaponry, logistics capacity, and infrastructure was less severe than had first been claimed, however.

The Serbian military had previously played a major role in the conflicts in the former Yugoslavia in the early 1990s. Traditionally the center of Yugoslav armaments manufacture, Serbia was able to arm itself – although the need to create money to pay for domestically produced weapons was a major factor in the crippling hyperinflation of that era.

In 2002 the country made its first (modest) contribution to a UN peacekeeping force, in East Timor.

ECONOMICS ▷ Not available

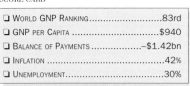 $10bn   13.65–66.48 dinars

SCORE CARD

- ❑ WORLD GNP RANKING.........................83rd
- ❑ GNP PER CAPITA $940
- ❑ BALANCE OF PAYMENTS ....–$1.42bn
- ❑ INFLATION 42%
- ❑ UNEMPLOYMENT..................................30%

EXPORTS

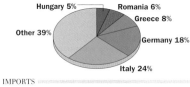

Hungary 5%   Romania 6%   Greece 8%   Germany 18%   Other 39%   Italy 24%

IMPORTS

Romania 5%   Austria 6%   Bulgaria 13%   Other 45%   Germany 15%   Italy 16%

ECONOMIC PERFORMANCE INDICATOR

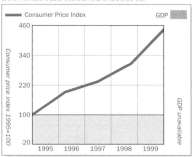

Consumer Price Index   GDP

Consumer price index 1995=100   GDP unavailable

and the state sectors. Hyperinflation had already pushed the economy to virtual collapse. NATO bombing in 1999 caused extensive damage to infrastructure; the EBRD estimated reconstruction costs at $20 billion over three years. Sanctions were lifted shortly after Milošević's downfall, and investment prospects were boosted by his extradition and the agreement to cooperate with the ICTY.

STRENGTHS

Return of international aid and investment in 2000–2001. Economic potential of Danube.

WEAKNESSES

Severe damage caused by sanctions and 1999 bombings. Low hard-currency reserves. Outflow of skilled professionals.

PROFILE

The living standards in the former Yugoslavia were among the most advanced of the socialist countries. It has now all but collapsed, although as much as 50% of all activity goes on within the resilient informal sector. The conflicts in the 1990s effectively stalled much-needed economic reform. Sanctions, maintained until 1996 and reimposed more fully in 1999, stifled trade and decimated both the emerging private

SERBIA & MONTENEGRO : MAJOR BUSINESSES

Novi Sad   Pančevo   Belgrade   Kragujevac   Paraćin   Kruševac   Niš

🖋 Pharmaceuticals   ⚙ Light engineering
🕸 Textiles   ✿ Heavy engineering
⚗ Chemicals   ▣ Food processing
△ Metallurgy
⚡ Electronics

0   100 km
0   100 miles

RESOURCES

 Electric power 11.8m kw

 9940 tonnes

18,305 b/d (reserves 80m barrels)

4.37m pigs, 1.92m sheep, 1.83m cattle, 21.1m chickens

Coal, bauxite, iron, lead, copper, zinc

ELECTRICITY GENERATION

Hydro 30% (12bn kwh)				
Combustion 70% (28bn kwh)				
Nuclear 0%				
Other 0%				

% of total generation by type

The country is self-sufficient in coal and electricity production. Vojvodina's oil industry could cater for one-third of its needs, but was badly hit by NATO bombing in March–June 1999.

ENVIRONMENT

 Not available

 3%

Not available

ENVIRONMENTAL TREATIES

Yes   Yes   Yes

Yes   No   No

Ecological awareness peaked in the late 1980s, when the Ecological Forum was active. NATO bombing of Serbia in 1999 caused extensive pollution of the Danube and raised fears of contamination from dioxins. Depleted uranium from NATO munitions has also aroused serious concerns both in Serbia proper and in Kosovo.

MEDIA

 TV ownership high

Daily newspaper circulation 106 per 1000 people

PUBLISHING AND BROADCAST MEDIA

	There are 18 daily newspapers. *Politika* has the largest circulation
	2 state-controlled, several independent services
	3 state-controlled, several independent services

The state broadcasting center was a prime target in the October 2000 popular uprising against the Milošević regime; it had used the broadcast media to control public opinion, while the press was held in check by government control of newsprint. Many journalists supported the opposition. The B92 radio station, when not forced off air, was a beacon of independent reporting. Free media have flourished under the post-Milošević authorities.

Cropland
Forest
Pasture
High mountain regions
Pigs
Cereals

SERBIA & MONTENEGRO: LAND USE

| 0 | | 100 km |
| 0 | | 100 miles |

CRIME

 Death penalty not used in practice

5566 prisoners

High crime levels

CRIME RATES

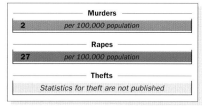

Murders	
2	*per 100,000 population*
Rapes	
27	*per 100,000 population*
Thefts	
Statistics for theft are not published	

Crime was rife under Milošević – from currency trading and black marketing to narcotics and extortion. The country has fully cooperated with the ICTY since 2001, and began its own war crimes trials in Serbia in 2002. The death penalty is set to be abolished.

EDUCATION

 School leaving age: 15

93%

172,313 students

THE EDUCATION SYSTEM

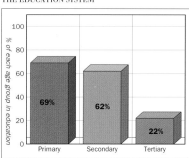

% of each age group in education

Primary 69%   Secondary 62%   Tertiary 22%

Schooling was totally disrupted by the Kosovo conflict in 1999, leaving the education system in crisis; the wealthy go abroad for their education. Literacy rates in Kosovo, where ethnic Albanian schools were closed in 1990, were low even before the conflict. Rebuilding the basic education system is key to reconstruction and reconciliation.

HEALTH

 Welfare state health benefits

 1 per 500 people

Cerebrovascular and heart diseases, cancers, accidents

Isolation from former trading partners has affected the quality of the health service, despite the exemption of medicines and medical supplies from sanctions. Social insurance is obligatory for those in employment, but medicines are scarce and costly, and death rates among infants and the elderly have risen dramatically. Health problems can be aggravated by bitingly cold winters.

SPENDING

GDP/cap. increase

CONSUMPTION AND SPENDING

176 per 1000 population

226 per 1000 population

Defense 10%				
Education 4.2%				
Health 4.5%				

Defense, Health, Education spending as % of GDP

The country as a whole was seriously impoverished by sanctions; real incomes fell dramatically, yet food prices remained higher than in much of western Europe. Bank collapses in 1992 and continuing hyperinflation in 1992–1994 wiped out dinar savings. One business that did expand was the illegal import of sanctions-busting goods for the few who could afford them – black marketeers and those close to Milošević.

The lifting of sanctions in 1995–1996 had hardly begun to be reflected in improvements in living conditions when the Kosovo conflict brought further dislocation and hardship in 1999. Even before it erupted, an estimated two-thirds of the population were living below subsistence level. Apart from the desperate situation of refugees and internally displaced families, unsupported pensioners fare worst.

WORLD RANKING

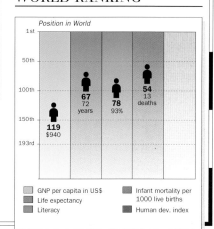

Position in World

119 $940   67 72 years   78 93%   54 13 deaths

	GNP per capita in US$		Infant mortality per 1000 live births
	Life expectancy		
	Literacy		Human dev. index

S

SEYCHELLES

INDIAN OCEAN

OFFICIAL NAME: Republic of the Seychelles **CAPITAL:** Victoria
POPULATION: 79,300 **CURRENCY:** Seychelles rupee **OFFICIAL LANGUAGE:** Seselwa (French Creole)

  1976   1976   June 18   SY   +4   +248   .sc

THE 115 ISLANDS of the Seychelles, lying in the Indian Ocean, support unique flora and fauna, including the giant tortoise and the world's largest seed, the *coco-de-mer*. Formerly a UK colony and then under one-party rule for 16 years, the Seychelles became a multiparty democracy in 1993. The economy relies on tourism.

CLIMATE

▷ Tropical oceanic

WEATHER CHART FOR VICTORIA

The islands have a tropical oceanic climate, with only small temperature variations throughout the year.

TRANSPORTATION

▷ Drive on left

Pointe Larue International, Mahé
155,000 passengers

16 ships
17,699 grt

THE TRANSPORTATION NETWORK

176 km (109 miles)		None	
None		None	

Nine islands have airstrips. The public transportation fleet and roads are being renewed. Victoria's deep-sea harbor is one of the best run in the region.

TOURISM

▷ Visitors : Population 1.6:1

130,000 visitors

Up 4% in 2000

MAIN TOURIST ARRIVALS

France 19%
Italy 16%
Germany 15%
Other 50%

% of total arrivals

Since the international airport was opened at Mahé in 1971, tourism has become the mainstay of the economy, and employs 30% of the workforce. New hotels must comply with laws to protect the islands' beauty and unique wildlife. There is substantial foreign investment.

PEOPLE

▷ Pop. density high

French Creole, English, French

294/km² (761/mi²)

THE URBAN/RURAL POPULATION SPLIT

64%   36%

RELIGIOUS PERSUASION

Other (including Muslim) 2%
Anglican 8%
Roman Catholic 90%

The Seychelles islands were uninhabited before French settlers arrived in the 1770s. Today, the population is markedly homogeneous as a result of intermarriage between different ethnic groups. The Creoles are the descendants of the French settlers and of the Africans who were settled on the islands by British administrators. There are small Chinese and Indian minorities.

Almost 90% of Seychellois live on Mahé. Population growth has been very low, as about 1000 people a year have been emigrating. The transition to democracy may yet prove to reverse this trend.

SEYCHELLES

Total Land Area : 455 sq. km (176 sq. miles)

POPULATION
• over 10 000
• under 10 000

LAND HEIGHT
500m/1640ft
200m/656ft
Sea Level

N

0   100 km
0   100 miles

INDIAN OCEAN

INNER ISLANDS
Île aux Vaches
Île Denis
Île Aride
Grand'Anse
Praslin
La Digue
Silhouette
Mahé
Frégate

Inset area

AMIRANTE ISLANDS
Île Desroches
Poivre Atoll
Île Platte

Coëtivy

Alphonse Group

OUTER

ISLANDS

Aldabra Atoll
Grande Terre
Assomption
Cosmolédo Atoll
St. Pierre
Providence Atoll
FARQUHAR GROUP
ALDABRA GROUP
Astove
Farquhar Atoll

Mahé
Bel Ombre
Morne Seychellois 905m
Anse Boileau
VICTORIA
Anse Étoile
Cascade
Anse Royale
0   50 km Takamaka Quatre Bornes
0   50 miles

POLITICS

▷ Multiparty elections

1998/2003

President
France-Albert René

AT THE LAST ELECTION

National Assembly 34 seats

88% SPPF   9% UO   3% DP

SPPF = Seychelles People's Progressive Front **UO** = United Opposition **DP** = Democratic Party

In 1993, the Seychelles returned to democracy after 16 years of one-party socialist rule under President France-Albert René, who had seized power soon after independence. Opposition divisions in the 1993 elections allowed René to retain the presidency. His SPPF kept its majority in the 1998 polls – having dramatically abandoned its leftist ideology, and adopted wide-ranging reforms, encouraging privatization and legalizing trade unions. The highlight is a scheme to develop Seychelles as an International Trading Zone, with free-port facilities and new industry.

WORLD AFFAIRS

▷ Joined UN in 1976

| OIF | Comm | COI | NAM | AU |

The Seychelles is nonaligned, but its strategic location encourages competing world powers to seek its friendship. It claims the Chagos Archipelago from the UK. Trade accords exist with other Indian Ocean states.

S

AID
 Recipient

 $18m (receipts) | ⬆ Up 38% in 2000

Multilateral agencies, notably the EU and the Arab Development Fund, support a range of development projects. In 1996, $13 million was spent on improving transportation links, protecting the environment, and rehabilitating Victoria Market. Bilateral aid comes from the EU, Arab donors, France, and Japan.

DEFENSE
 No compulsory military service

 $10m | ⬇ Down 9% in 2000

The Seychelles has a 200-strong army, and a paramilitary guard. The latter includes the coast guard made up of air and sea forces. The army, set up in 1977, was initially trained by Tanzania, and Tanzanian troops were brought in for three years after a coup attempt in 1981. North Korea provided advisers until 1989.

ECONOMICS
Inflation 3.3% p.a. (1990–2000)

📊 $573m | 6.261–5.618 Seychelles rupees

SCORE CARD

❑ WORLD GNP RANKING	168th
❑ GNP PER CAPITA	$7050
❑ BALANCE OF PAYMENTS	–$114m
❑ INFLATION	6.3%
❑ UNEMPLOYMENT	9%

STRENGTHS
Tourism. Fish exports, especially shrimp and tuna. Profitable re-export trade. International Trading Zone attracting foreign industrial interest. Copra, cinnamon, tea.

WEAKNESSES
Growing deficits in early 1990s, caused by drop in tourism following 1991 Gulf War, spending on hosting 1993 Indian Ocean Games, and cost of four elections. High debt-servicing costs. Reliance on food imports, especially for tourist industry. Copra production declining. Reliance on expatriate labor.

EXPORTS

IMPORTS

One of the Inner Islands, *which are home to most of the population. Unlike any other mid-ocean islands, all but two are granitic.*

RESOURCES
 Electric power 28,000 kw

 37,992 tonnes | Not an oil producer

18,400 pigs, 5250 goats, 560,000 chickens | Phosphates (guano), salt, granite, natural gas

There are virtually no mineral resources. All fuel is imported; only three islands have electricity. Offshore discoveries of natural gas have spurred a search for oil. Natural habitat and free trade environment are tremendous assets.

ENVIRONMENT
 Not available

🌳 95% | ⬆ 2.5 tonnes per capita

The Seychelles has been praised for its commitment to conservation. It has two natural World Heritage sites, and helped promote the idea of whale sanctuaries.

MEDIA
 TV ownership medium

 Daily newspaper circulation 46 per 1000 people

PUBLISHING AND BROADCAST MEDIA

There is 1 daily newspaper, the government-owned *Seychelles Nation*	
2 independent services	3 independent services

The state broadcasting company has been reorganized and is now ostensibly free of government control. Privately owned periodicals are now permitted.

CRIME
 No death penalty

 157 prisoners | ⬆ Up 72% 1997–1999

Violent crime is rare in the Seychelles. The main concern is the increasing rate of petty theft.

EDUCATION
School leaving age: 16

84% | 🎓 1682 students

The 1995–2008 Educational and Training Plan places special emphasis on increasing levels of female enrollment. National Youth Service is mandatory for entry to higher education.

HEALTH
 Welfare state health benefits

👤 1 per 758 people | Heart and cerebrovascular diseases, cancers

State health care is free. Private medicine is allowed under new social legislation. Life expectancy is over 70 years.

SPENDING
GDP/cap. increase

CONSUMPTION AND SPENDING

🚗 94 per 1000 population | ☎ 235 per 1000 population

Defense 1.8%
Education 7.9%
Health 4.8%

Defense, Health, Education spending as % of GDP

Living standards are the highest among AU states. There are no slums in the Seychelles, and the state welfare system caters for all.

WORLD RANKING

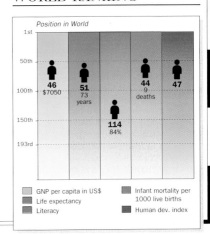

Position in World

GNP per capita in US$
Life expectancy
Literacy
Infant mortality per 1000 live births
Human dev. index

S

SIERRA LEONE

OFFICIAL NAME: Republic of Sierra Leone **CAPITAL:** Freetown
POPULATION: 4.6 million **CURRENCY:** Leone **OFFICIAL LANGUAGE:** English

THE WEST AFRICAN state of Sierra Leone was founded by the British in 1787 for Africans freed from slavery. The terrain rises from coastal lowlands to mountains in the northeast. A democratic government took office in 1996 against a background of bloody rebellion. Sierra Leone soon plunged back into a savage civil war. Although a 1999 peace agreement was short-lived, an ECOWAS-brokered accord signed in late 2000 seems to be holding.

CLIMATE

▷ Tropical equatorial/monsoon

WEATHER CHART FOR FREETOWN

Coastal rainfall can be as high as 500 cm (197 in) a year, making Sierra Leone one of the wettest places in coastal west Africa. Humidity is consistently high – about 80% – during the rainy season. The dusty, northeasterly *harmattan* wind often blows during the hotter dry season from November to April. The northeastern savannas are drier, with 190–250 cm (75–98 in) of rain, and are one of the hottest areas.

TRANSPORTATION

▷ Drive on right

Lungi International, Freetown
84,547 passengers

52 ships
18,792 grt

THE TRANSPORTATION NETWORK

904 km (562 miles)

None

84 km (52 miles)

800 km (497 miles)

Little progress has been made in improving Sierra Leone's roads. The 300-km (190-mile) narrow-gauge railroad was abandoned in 1971 as uneconomic, although 84 km (52 miles) of track still runs to the closed iron ore mines at Marampa. Having failed in 1987, Sierra Leone's national airline resumed flights – to Paris only – in 1991. A limited ferry service across the estuary is the only link between Freetown and the airport.

TOURISM

▷ Visitors : Population 1:460

10,000 visitors

Up 67% in 2000

MAIN TOURIST ARRIVALS

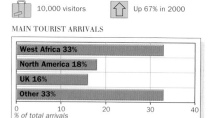

% of total arrivals

Sierra Leone has never attracted many tourists, apart from occasional cruise ship calls. Years of civil war have prevented the development of tourism. Among the chief potential attractions are the beaches along the Freetown peninsula, at present virtually undeveloped.

SIERRA LEONE

Total Land Area : 71 740 sq. km
(27 698 sq. miles)

LAND HEIGHT

POPULATION
◎ over 100 000
● over 10 000
• under 10 000

1000m/3281ft
500m/1640ft
200m/656ft
Sea Level

PEOPLE

▷ Pop. density medium

Mende, Temne, Krio, English

64/km² (166/mi²)

THE URBAN/RURAL POPULATION SPLIT

37%   63%

ETHNIC MAKEUP

Kuranko 4%
Limba 8%
Mende 35%
Other 21%
Temne 32%

Freetown was founded as a settlement for people freed from slavery. Its citizens' British and North American origins account for Sierra Leone's strongly anglicized Creole culture. An estimated two million people were displaced by the civil war.

POLITICS

▷ Multiparty elections

2002/2007

President Ahmad Tejan Kabbah

AT THE LAST ELECTION

National Assembly 124 seats

67% SLPP   22% APC   10% App   1% PLP

SLPP = Sierra Leone People's Party **APC** = All People's Congress **App** = Appointed: 12 paramount chiefs are indirectly elected to represent each province **PLP** = Peace and Liberation Party

A rebellion by the Revolutionary United Front (RUF) in 1991 sparked a decade of particularly savage civil war. President Ahmad Kabbah was popularly elected at the head of a civilian government in 1996. A peace and power-sharing agreement, reached in 1999, collapsed in 2000. A large UN and British force helped to secure a new cease-fire later that year. Remodeled as a political party, the RUF failed to win a single seat in the 2002 elections, which resulted in a convincing victory for Kabbah and his SLPP.

S

WORLD AFFAIRS

▷ Joined UN in 1961

| Comm | ECOWAS | MRU | AU | OIC |

UN peacekeepers and British forces assisted government forces during renewed fighting in 2000.

AID

▷ Recipient

 $182m (receipts)   Up sharply in 2000

The IMF and the World Bank agreed in 2002 to drop 80% of Sierra Leone's debt in return for key reforms.
Aid funds efforts to cope with the humanitarian needs of refugees from Liberia, internal migrants displaced by the civil war, and the near-collapse of public services.

DEFENSE

▷ No compulsory military service

 $9m   Down 18% in 2000

The army is an ineffectual fighting force, prompting assistance from the UK with training in 2000. Both sides in the civil war exploited child fighters.

ECONOMICS

▷ Inflation 29% p.a. (1990–2000)

 $647m   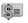 1899–2091 leones

SCORE CARD

❏ World GNP Ranking	........................164th
❏ GNP per Capita	..............................$130
❏ Balance of Payments	....................–$127m
❏ Inflation	....–0.8%
❏ Unemployment	..........................Widespread

STRENGTHS
Diamonds, although much of the output is smuggled; official exports resumed in late 2000 under a UN certification scheme. Some bauxite and rutile production.

WEAKNESSES
Years of instability affected the most productive areas, including diamond fields, with severe disruption of agricultural and mining sectors.

EXPORTS

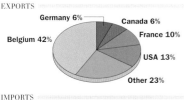

Germany 6%
Canada 6%
France 10%
USA 13%
Other 23%
Belgium 42%

IMPORTS

Germany 4%
Netherlands 6%
USA 6%
Nigeria 4%
Other 40%
UK 40%

RESOURCES

▷ Electric power 126,000 kw

 59,437 tonnes   Not an oil producer

 420,000 cattle, 365,000 sheep, 6m chickens   Diamonds, rutile, bauxite, gold, titanium

The large diamond deposits need fresh investment as areas currently being mined become depleted. The southeast is the most fertile region.

ENVIRONMENT

▷ Sustainability rank: 134th

 1%   0.1 tonnes per capita

Population pressures and the neglect resulting from years of civil war have depleted the land's productivity.

MEDIA

▷ TV ownership low

 Daily newspaper circulation 5 per 1000 people

PUBLISHING AND BROADCAST MEDIA

There is 1 daily newspaper, the *Daily Mail*, published by the government

1 state-controlled service   1 state-controlled service

A broad range of periodicals is available. The government has promised press freedom, but the Internet is heavily censored.

CRIME

▷ Death penalty in use

 Sierra Leone does not publish prison figures   Crime is rising

The civil war resulted in savage atrocities and the mass looting of resources. As a consequence, the UN has instigated a war crimes tribunal. International restrictions on trade in diamonds from war-torn areas cut RUF revenues, but illegal diamond mining and smuggling remain lucrative crimes.

EDUCATION

▷ Schooling is not compulsory

 36%   6744 students

Freetown has a long tradition of education, and its university, Fourahbay College, became affiliated with Durham University in the UK in 1876. In recent times, its students have often been active in political dissent. Educational provision has inevitably deteriorated over the past decade.

HEALTH

▷ No welfare state health benefits

1 per 10,000 people   Communicable diseases, malaria, malnutrition

Only traditional care is available outside the capital. WHO has ranked Sierra Leone's health care bottom in the world in terms of attainment and efficiency.

The main street, Kabala. *Sierra Leone is consistently near the bottom of the UN's Human Development Index.*

CHRONOLOGY

Freetown was founded in 1787 and became a British colony in 1808; the interior was annexed in 1896.

- ❏ **1961** Independence.
- ❏ **1978** Single-party republic.
- ❏ **1991** RUF rebellion starts.
- ❏ **1996** Civilian rule restored after 1992 army coup; Kabbah president.
- ❏ **1997** Coup ousts Kabbah for a year.
- ❏ **1999–2000** Power-sharing attempt.
- ❏ **2001** RUF ends insurgency.
- ❏ **2002** Government and UN agree to set up war crimes court. Kabbah and SLPP reelected.

SPENDING

▷ GDP/cap. decrease

CONSUMPTION AND SPENDING

2 per 1000 population   4 per 1000 population

Defense 1.2%
Education 1%
Health 0.9%

0   5   10   15   20   25
Defense, Health, Education spending as % of GDP

In terms of quality of life, the UN has repeatedly ranked Sierra Leoneans as the world's poorest people. Any wealth is associated with political power.

WORLD RANKING

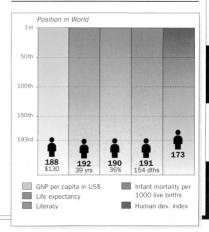

Position in World

1st
50th
100th
150th
193rd

188 $130
192 39 yrs
190 36%
191 154 dths
173

GNP per capita in US$	Infant mortality per 1000 live births
Life expectancy	
Literacy	Human dev. index

S

SINGAPORE

OFFICIAL NAME: Republic of Singapore   CAPITAL: Singapore   POPULATION: 4.1 million
CURRENCY: Singapore dollar   OFFICIAL LANGUAGES: Malay, English, Mandarin, and Tamil

1965   1965   Aug 9   SGP   +8   +65   .sg

SOUTHEAST ASIA

 Asia

AN ISLAND STATE now linked to the southernmost tip of the Malay Peninsula by a causeway, Singapore ("lion city") was largely uninhabited until the 19th century. In 1819, an official of the British East India Company, Stamford Raffles, recognized the island's strategic position on key trade routes, and established a trading settlement. Today, Singapore remains one of the most important entrepôts in Asia.

CLIMATE ▷ Tropical equatorial

WEATHER CHART FOR SINGAPORE

The only variations in the hot, wet, and humid climate are the airless months of September and March, when the trade winds change direction.

TRANSPORTATION ▷ Drive on left

Changi International
28.6m passengers

1677 ships
20.4m grt

THE TRANSPORTATION NETWORK

3038 km
(1888 miles)

150 km
(93 miles)

26 km
(16 miles)

None

The Mass Rapid Transit System (subway), completed in 1991, is among the world's most efficient. Space for new roads has run out and monthly auctions are held to sell certificates entitling people to buy from a quota of new cars. The massive port at Pasir Panjang is being expanded on reclaimed land.

The financial center. More than a quarter of Singapore's GDP is generated by financial and business services.

TOURISM ▷ Visitors : Population 1.5:1

6.26m visitors

Up 11% in 2000

MAIN TOURIST ARRIVALS

Indonesia 17%
Japan 13%
Australia 6%
Other 64%

0   10   20   30   40   50   60   70   80
% of total arrivals

The buildings of Chinatown, recognized as a picturesque tourist asset, are being restored. A Singaporean consortium is involved in developing a resort on Indonesia's Bintan island, some 45 km (28 miles) across the Strait of Singapore.

PEOPLE ▷ Pop. density high

Mandarin, Malay, Tamil, English

6721/km²
(17,408/mi²)

THE URBAN/RURAL POPULATION SPLIT

100%

ETHNIC MAKEUP

Other 1%
Indian 8%
Malay 14%
Chinese 77%

Singapore is dominated by the Chinese – the old-established English-speaking Straits Chinese and newer Mandarin speakers – who make up almost 80% of the community. Indigenous Malays are generally the poorest group, but today there is little overt ethnic tension. There is also a significant foreign workforce in Singapore.

Long-term plans to stabilize the population structure included the announcement in mid-2000 of cash bonuses for families with more than one child. Society is highly regulated and government campaigns to improve public behavior are frequent.

POLITICS ▷ Multiparty elections

2001/2006   President S. R. Nathan

AT THE LAST ELECTION
Parliament 94 seats

1% SPP   1% WP

87% PAP   10% Nom   1% NC

PAP = People's Action Party   Nom = Nominated
SPP = Singapore People's Party   NC = Nonconstituency
member   WP = Workers' Party

In addition to the 84 elected members, up to six "nonconstituency" members may be nominated from the losers with the most votes and nine members may be nominated to ensure a wider representation in Parliament

Singapore is a multiparty democracy, although the ruling PAP effectively controls all parts of the political process and much of the economy. There are plans to create a national ideology ("shared values") based on Confucian traditions. In 1993, Ong Teng Cheong became the first president to be directly elected. The veteran politician, Lee Kuan Yew, prime minister for more than 30 years until his resignation in 1990, still exercises influence.

The PAP retains its grip on power, having given Singapore one of the highest living standards in the world, based on a free-market economy. The first antigovernment rally was permitted in 2001, but even before the ballot the PAP was sure of reelection, fewer than half of the seats being contested. To improve parliamentary debate, the PAP formed an internal opposition faction.

WORLD AFFAIRS ▷ Joined UN in 1965

 APEC   ASEAN   Comm   NAM   WTO

Singapore has established diplomatic relations with China, while continuing to maintain close economic ties with Taiwan. It became the beneficiary in 2001 of Japan's first free trade agreement.

AID ▷ Recipient

US$1m (receipts)

Down 50% in 1999–2000

Aid is not an important issue in Singapore. The state does not provide aid to any states in southeast Asia.

DEFENSE ▷ Compulsory military service

US$4.71bn

Slight increase in 2000

Despite Singapore's small size, its armed forces have a total strength of over 60,000.

S

Key to symbols and abbreviations on cover flaps

ECONOMICS

 Inflation 1.3% p.a. (1990–2000)

US$99.4bn

1.734–1.847
Singapore dollars

SCORE CARD

❏ WORLD GNP RANKING	37th
❏ GNP PER CAPITA	US$24,740
❏ BALANCE OF PAYMENTS	US$21.8bn
❏ INFLATION	1.4%
❏ UNEMPLOYMENT	3%

STRENGTHS

Massive accumulated wealth is derived from success as an entrepôt and as a center of high-tech industries. Singapore is a major producer of computer disk drives. Huge state enterprises, such as TAMESEK, with over 450 companies, have proved highly flexible in responding to market conditions. World leader in new biotechnologies.

RESOURCES

 Electric power 5.7m kw

 9081 tonnes

Not an oil producer; refines 1.03m b/d

600,000 ducks, 190,000 pigs, 2m chickens

None

Singapore has no strategic resources and has to import almost all the energy and food it needs. Its main resources, on which its wealth as a center of commerce has been built, are its strategic position and its people.

ENVIRONMENT

Not available

 5%

21 tonnes per capita

There is a small green belt around the causeway. Singapore sees itself as a world leader in providing the perfect urban environment. There is no litter, thanks to instant heavy fines; chewing gum is banned by law.

SINGAPORE

Total Land Area : 648 sq. km (250 sq. miles)

Urban Areas
Open Areas
Nature Reserve

EXPORTS

Taiwan 6%
Japan 8%
Hong Kong 8%
Other 43%
USA 17%
Malaysia 18%

IMPORTS

Taiwan 4%
China 5%
USA 15%
Other 42%
Malaysia 17%
Japan 17%

WEAKNESSES

Dependence on Malaysia for water. Almost all food and energy imported. Skills shortages, notably in engineering. Fluctuations in world electronics market caused recession in 2001. Lack of land.

MEDIA

 TV ownership high

Daily newspaper circulation 324 per 1000 people

PUBLISHING AND BROADCAST MEDIA

There are 8 daily newspapers, including the *Straits Times* and *Lianhe Zaobao* in English and Chinese respectively

5 services:
4 independent,
1 US-controlled

6 privately owned
services

The government is very sensitive to any criticism. However, it has declared its intention partially to liberalize the local media, as a part of "constructive competition." Foreigners may not own newspapers or broadcasting stations.

CRIME

Death penalty in use

13,791 prisoners

Down 21% in 1999

Crime is limited and punishment can be severe. The Triads are no longer a problem; the main issue is intellectual piracy.

CHRONOLOGY

In 1819, Sir Stamford Raffles set up a trading post in the village of Singapore; by 1826 it was Britain's colonial center in southeast Asia.

- ❏ **1959** PAP becomes ruling party.
- ❏ **1965** Independence.
- ❏ **1990** Lee Kuan Yew resigns as prime minister.
- ❏ **1993** Ong Teng Cheong first directly elected president.
- ❏ **2001** PAP, in power since independence, wins elections.

EDUCATION

Schooling is not compulsory

92%

124,180 students

Schooling is not compulsory, but attendance is high. Education is seen as the key to a good salary, especially among the Chinese community.

HEALTH

Welfare state health benefits

1 per 625 people

Heart and cerebrovascular diseases, cancers

Singapore has an efficient modern health system. There are incentives to ensure the continuation of the extended family, so that the elderly are cared for at home.

SPENDING

GDP/cap. increase

CONSUMPTION AND SPENDING

97 per 1000 population

484 per 1000 population

Defense 4.9%
Education 3%
Health 1.2%

0   5   10   15   20   25
Defense, Health, Education spending as % of GDP

The 2001 "Singapore Share" scheme promises to give Singaporeans a share in the country's economy.

WORLD RANKING

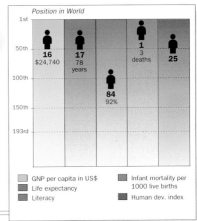

Position in World

1st
50th
100th
150th
193rd

16 $24,740
17 78 years
3 deaths
25
84 92%

- GNP per capita in US$
- Infant mortality per 1000 live births
- Life expectancy
- Literacy
- Human dev. index

S

SLOVAKIA

OFFICAL NAME: Slovak Republic **CAPITAL:** Bratislava
POPULATION: 5.4 million **CURRENCY:** Slovak Koruna **OFFICAL LANGUAGE:** Slovak

DOMINATED FOR 900 YEARS by neighboring Hungary, Slovakia spent much of the 20th century as the less developed half of communist Czechoslovakia. An independent democracy since 1993, Slovakia has struggled to create a modern market-led economy, but is now among the "second wave" of states conducting membership negotiations with the EU.

Levoča, in northeastern Slovakia, dates from the 13th century and still retains its medieval street plan and town walls.

CLIMATE ▷ Continental

WEATHER CHART FOR BRATISLAVA

Slovakia has a continental climate. Snowfalls are heavy in winter, while summers are moderately warm.

TRANSPORTATION ▷ Drive on right

Milan Rastislav Stefanik, Bratislava
292,515 passengers

Has no fleet

THE TRANSPORTATION NETWORK

17,533 km (10,894 miles)	288 km (179 miles)
3662 km (2276 miles)	172 km (107 miles)

The Danube River is a vital artery. Trains are cheap and efficient. Buses and trams are the mainstay of urban transport.

TOURISM ▷ Visitors : Population 1:5.1

1.05m visitors   Up 8% in 2000

MAIN TOURIST ARRIVALS

Czech Republic 28%
Poland 18%
Germany 14%
Other 40%
% of total arrivals

The High Tatras, one of the smallest high-mountain ranges in the world, and the Vrátna valley in the Little Tatras draw hikers and skiers. Tourists also visit Bratislava's castle and old city, and the many thermal-spring health spas.

PEOPLE ▷ Pop. density medium

Slovak, Hungarian, Czech   110/km² (285/mi²)

THE URBAN/RURAL POPULATION SPLIT

57%   43%

RELIGIOUS PERSUASION

Orthodox 4%
Protestant 8%
Atheist 10%
Other 18%
Roman Catholic 60%

Slovaks dominate society, but 11% of the population is Hungarian, and there is a significant Roma minority which faces discrimination. The Hungarian community, backed by Hungary, seeks protection for its language and culture. Tensions lessened in 1998 when its main political voice, the Hungarian Coalition Pary, joined Dzurinda's government. There were 300,000 Slovaks living in Czech lands in 1993. Dual citizenship is now permitted.

POLITICS ▷ Multiparty elections

2002/2006   President Rudolf Schuster

AT THE LAST ELECTION

National Council 150 seats

24% HZDS	19% SDKU	17% Smer	13% SMK	10% KDH	10% ANO	7% KSS

HZDS = Movement for a Democratic Slovakia
SDKU = Slovak Democratic and Christian Union
Smer = Direction **SMK** = Hungarian Coalition Party
KDH = Christian Democratic Movement **ANO** = New Civic Alliance **KSS** = Slovak Communist Party

The move which led to the separation of the two halves of the former Czechoslovakia in 1993 gathered momentum when Slovak leader Vladimir Meciar was tempted by independence as a way of enhancing his power base. The populist Meciar dominated Slovak politics until 1998, but clashed repeatedly with President Michal Kovac, both of them members of the HZDS.

A broad coalition led by center-right politician Mikulas Dzurinda won power in elections in 1998 and retained its position in 2002. Slovakia's first direct presidential election in May 1999 was won by the pro-Western Rudolf Schuster, frustrating Meciar's bid for this office.

The Hungarian minority has its own parties, whereas the Roma have no official representation.

WORLD AFFAIRS ▷ Joined UN in 1993

CE   CEFTA   OECD   EAPC   OSCE

The 1998 change of government reversed the pro-Russian Meciar years. Slovakia was among six "second wave" candidates which began EU membership talks in March 2000. It also became a member of the OECD in 2000.

AID ▷ Recipient

$113m (receipts)   Down 65% in 2000

Foreign aid fell after the mid-1990s, but EU programs are now in place in preparation for EU membership.

DEFENSE ▷ Compulsory military service

$340m   Up 11% in 2000

The Slovak armed forces include some 13,600 conscripts at any one time. Prime Minister Dzurinda, in office since late 1998, reversed Meciar's pro-Russian defense policies.

RESOURCES ▷ Electric power 7.8m kw

2263 tonnes   1203 b/d (reserves 7.5m barrels)

1.49m pigs, 646,100 cattle, 13.6m chickens   Coal, lignite, gas, oil, antimony, copper, iron, mercury, zinc

44% of electricity was nuclear-generated, even before the Mochovce nuclear plant began operations in 1998.

S

Slovakia

Total Land Area : 48 845 sq. km
(18 859 sq. miles)

CHRONOLOGY

Once part of the Austro-Hungarian Empire, Slovakia and the Czech Lands formed the Republic of Czechoslovakia in 1918.

- ❏ **1939–1945** Separate Slovak state under pro-Nazi Jozef Tiso.
- ❏ **1945** Czechoslovak state restored.
- ❏ **1947** Communists seize power.
- ❏ **1968** "Prague Spring" ended by Warsaw pact invasion.
- ❏ **1989** "Velvet Revolution."
- ❏ **1990** Free multiparty elections.
- ❏ **1993** 1st January, separate Slovak and Czech states established.
- ❏ **1994** HZDS election victory.
- ❏ **1998** Broad-based coalition under Dzurinda wins general election.
- ❏ **1999** Rudolf Schuster defeats Meciar in direct presidential poll.
- ❏ **2002** Dzurinda coalition reelected.

POPULATION
over 100 000 ◎
over 50 000 ○
over 10 000 ●

LAND HEIGHT
2000m/6562ft
1000m/3281ft
500m/1640ft
200m/656ft
Sea Level

ECONOMICS   ▷ Inflation 11% p.a. (1990–2000)

$20bn   46.88–48.01 Slovak koruny

SCORE CARD

❏ World GNP Ranking	66th
❏ GNP Per Capita	$3700
❏ Balance of Payments	–$694m
❏ Inflation	12%
❏ Unemployment	19%

Strengths

Increase in manufacturing, especially in Bratislava and surrounding area. Recent progress in cutting budget deficits and restructuring public and private sector. Growth in exports to EU. Potential for tourism, particularly skiing in the Tatra Mountains.

Weaknesses

High foreign indebtedness. Dependence on foreign trade makes the economy vulnerable to global recession. Heavy industry has found some new markets in the West, but struggles with poor productivity. Slow to attract foreign investment until 2000. Much poorer eastern region. Growing unemployment.

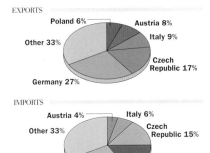

EXPORTS

Poland 6%
Austria 8%
Italy 9%
Other 33%
Czech Republic 17%
Germany 27%

IMPORTS

Austria 4%
Italy 6%
Other 33%
Czech Republic 15%
Russia 17%
Germany 25%

ENVIRONMENT   ▷ Sustainability rank: 14th

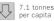

23%   7.1 tonnes per capita

The Gabcikovo Dam and the Bohunice nuclear reactors, now scheduled for partial closure, have provoked criticism.

MEDIA   ▷ TV ownership high

Daily newspaper circulation 184 per 1000 people

PUBLISHING AND BROADCAST MEDIA

There are 19 daily newspapers. *Nový Čas* and *Pravda* have the highest circulation figures

3 services: 1 state-controlled, 2 independent

1 state-controlled service and many privately owned stations

The state news agency TASR, accused of lacking objectivity and depending on government funding, resisted the emergence of independent rival SITA.

CRIME   ▷ No death penalty

7285 prisoners   Down 6% 1996–1998

Organized crime has increased rapidly in recent years, as has "white collar crime" such as business fraud. A new law to control money laundering took effect in 2001. A former economics minister accused of embezzlement was murdered in 1999.

EDUCATION   ▷ School leaving age: 15

99%   101,764 students

Schooling now draws on pre-1939 Slovak traditions but it is not adequately resourced, especially in rural areas. There is a modern university in Bratislava.

HEALTH   ▷ Welfare state health benefits

1 per 286 people   Cancers, heart and cerebrovascular diseases, accidents

Rising demand and costs are straining the health service severely. Restoring viability is now a government priority.

SPENDING   ▷ GDP/cap. increase

CONSUMPTION AND SPENDING

229 per 1000 population   314 per 1000 population

Defense 1.8%
Education 4.3%
Health 5.7%

Defense, Health, Education spending as % of GDP

A new elite is increasing demand for Western goods. Rural workers, Roma, and those living in the east are the poorest.

WORLD RANKING

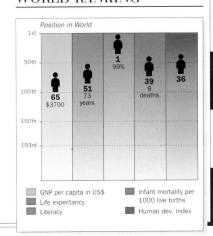

Position in World

65 $3700	
51 73 years	
1 99%	
39 8 deaths	
36	

- GNP per capita in US$
- Life expectancy
- Literacy
- Infant mortality per 1000 live births
- Human dev. index

S

SLOVENIA

OFFICIAL NAME: Republic of Slovenia   **CAPITAL:** Ljubljana
POPULATION: 2 million   **CURRENCY:** Tolar   **OFFICIAL LANGUAGE:** Slovene

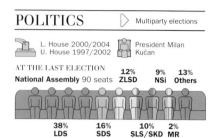

EUROPE

O F ALL THE FORMER Yugoslav republics, Slovenia has the closest links with western Europe. Located at the northeastern end of the Adriatic Sea, this small, Alpine country controls some of Europe's major transit routes. Slovenia's transition to independence in 1991 avoided the violence of the breakup of Yugoslavia. The most prosperous of the former communist European states, it is the only former Yugoslav republic on the "fast track" to EU membership.

CLIMATE

▷ Continental/ Mediterranean

WEATHER CHART FOR LJUBLJANA

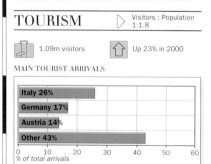

Slovenia's interior has a continental climate. Its small coastal region has a mild Mediterranean climate.

TRANSPORTATION

▷ Drive on right

Brnik International, Ljubljana
991,693 passengers

10 ships
1767 grt

THE TRANSPORTATION NETWORK

17,745 km (11,026 miles)	249 km (155 miles)
1202 km (747 miles)	None

Slovenia is strategically situated at some of Europe's major crossroads. In addition, its Adriatic ports provide Austria with its main maritime outlet.

TOURISM

▷ Visitors : Population 1:1.8

1.09m visitors

Up 23% in 2000

MAIN TOURIST ARRIVALS

Italy 26%
Germany 17%
Austria 14%
Other 43%

0   10   20   30   40   50   60
% of total arrivals

A revival in tourism has been helped by Slovenia's political stability. Particular attractions include skiing in the Julian Alps, picturesque Ljubljana, and the wine-growing region around Ptuj.

PEOPLE

▷ Pop. density medium

Slovene, Serbo-Croat

99/km² (256/mi²)

THE URBAN/RURAL POPULATION SPLIT

50%   50%

ETHNIC MAKEUP

Serb 2%   Croat 3%
Muslim 1%   Other 6%
Slovene 88%

Slovenes are ethnically very similar to the neighboring Croats and, like them, are predominantly Roman Catholic. However, Slovenia's long historical association with western Europe, and particularly with Austria, created a distinct Slovene identity. This enabled a smooth transition to independence in 1991. The major non-Slavic minorities are small communities of Hungarians in the east and Italians in the southwestern Istrian region; tensions are few.

Women are not heavily disadvantaged in Slovenian society.

POLITICS

▷ Multiparty elections

L. House 2000/2004
U. House 1997/2002

President Milan Kučan

AT THE LAST ELECTION

National Assembly 90 seats

38% LDS   16% SDS   12% ZLSD   10% SLS/SKD   9% NSi   2% MR   13% Others

LDS = Liberal Democracy of Slovenia
SDS = Social Democratic Party of Slovenia
ZLSD = United List of Social Democrats
SLS/SKD = Slovene People's Party/Christian Democrats of Slovenia   **NSi** = New Slovenia – Christian People's Party
MR = Two seats are reserved for Italian and Hungarian minority representatives

National Council 40 seats

22 members of the National Council, which has an advisory role, are indirectly elected, and 18 are chosen by an electoral college to represent various interests

Slovenia has been strikingly stable since independence, with Milan Kučan as president from 1990. Fragmented party politics makes coalitions essential. Janez Drnovsek, leader of the center-left LDS, has been prime minister almost continuously since 1992, apart from a six-month period in 2000 when the more right-wing SLS pulled out of the coalition. Andrej Bajuk, as SLS leader, then became prime minister, leaving his party after a failed attempt at electoral reform. Drnovsek returned to office following elections in 2000.

Many former communist officials still occupy top posts. Ownership of denationalized property remains an issue and there is still bitterness over the killing of opponents of Tito's partisans in the 1940s.

SLOVENIA

Total Land Area : 20 253 sq. km
(7820 sq. miles)

POPULATION

over 100 000
over 50 000
over 10 000
under 10 000

LAND HEIGHT

1000m/3281ft
500m/1640ft
200m/656ft
Sea Level

S

WORLD AFFAIRS ▷ Joined UN in 1992

CE · EAPC · OSCE · PfP · WTO

Slovenia is on the "fast track" to EU membership, probably in 2004. It also aims to be part of the next round of NATO enlargement.

AID ▷ Recipient

 $61m (receipts)　　 Up 97% in 2000

EU and World Bank aid focuses on infrastructure, the environment, and agricultural reform, in preparation for EU membership.

DEFENSE ▷ Phasing out conscription

 $223m　　 Down 34% in 2000

Troops staved off Yugoslav forces after secession in 1991. It was announced in 2002 that military conscription would be phased out by 2004.

ECONOMICS ▷ Inflation 20% p.a. (1993–2000)

 $20bn　　 227.5–245.6 tolars

SCORE CARD

- ❏ World GNP Ranking..........................65th
- ❏ GNP per Capita $10,050
- ❏ Balance of Payments..................–$594m
- ❏ Inflation 10.8%
- ❏ Unemployment 7%

EXPORTS

France 7%, Austria 8%, Croatia 8%, Italy 14%, Germany 27%, Other 36%

IMPORTS

Croatia 4%, Austria 8%, France 10%, Italy 17%, Germany 19%, Other 42%

STRENGTHS
Stability. Competitive manufacturing industry. EU and former Yugoslav export markets. Revoz car plant very productive, making Renault Clios. Competitive port at Koper. Free trade pact with Bosnia from 2001. Least indebted of central and eastern European states.

WEAKNESSES
Economy only partly liberalized. Capital market controls. Slower growth rates than other emerging markets. Slow pace of privatization.

Lake Bled in the Julian Alps, which lie astride the Slovenian–Italian border. The lake is a popular tourist destination.

RESOURCES ▷ Electric power 2.6m kw

 3215 tonnes　　 20 b/d

603,594 pigs, 493,670 cattle, 7.15m chickens　　Coal, lignite, lead, zinc, uranium, silver, mercury, oil

Slovenia has come under pressure from Austria to close the nuclear plant at Krško, which provides one-third of Slovenia's power. There are deposits of brown coal and lignite, but they are difficult to extract and of poor quality.

ENVIRONMENT ▷ Sustainability rank: 23rd

 6%　　 7.4 tonnes per capita

Protecting the country's alpine ecology is a priority. Pollution comes mainly from smelting, the chemicals industry, and burning brown coal and lignite.

MEDIA ▷ TV ownership high

 Daily newspaper circulation 199 per 1000 people

PUBLISHING AND BROADCAST MEDIA

There are 7 daily newspapers. *Dnevnik* is independently owned

4 services: 1 state-controlled, 3 independent　　4 services and many regional stations

A free and critical press has developed. State broadcasters have a new ethical code, protecting journalists' sources. POP TV is a commercial success.

CRIME ▷ No death penalty

 1136 prisoners　　 Up 11% in 1999

Slovenia's prison population is proportionally among the lowest in Europe. Smuggling people into western Europe is overtaking narcotics smuggling as the focus of organized crime.

EDUCATION ▷ School leaving age: 15

 99%　　 74,148 students

School is compulsory from seven to 15 years of age, and standards are high. The university at Ljubljana was founded in 1595.

CHRONOLOGY

Slovenia was part of the Austro-Hungarian Empire until 1918, when it joined the Kingdom of the Serbs, Croats, and Slovenes (Yugoslavia).

- ❏ **1949** Tito's break with Moscow.
- ❏ **1989** Parliament confirms right to secede. Calls multiparty elections.
- ❏ **1990** Control over army asserted, referendum approves secession.
- ❏ **1991** Independence declared; first republic to secede. Yugoslav federal army repelled.
- ❏ **1992** First multiparty elections. Milan Kučan president, Janez Drnovsek prime minister.
- ❏ **1993** Joins IMF and IBRD.
- ❏ **1998** EU membership talks begin.
- ❏ **2000** Drnovsek briefly ousted; returns to office after elections.

HEALTH ▷ Welfare state health benefits

 1 per 435 people　　 Cerebrovascular and heart diseases, cancers, accidents,

National health care in Slovenia uses health centers and outpatient clinics to increase accessibility for patients.

SPENDING ▷ GDP/cap. increase

CONSUMPTION AND SPENDING

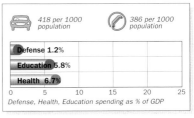

418 per 1000 population　　386 per 1000 population

Defense 1.2%　Education 5.8%　Health 6.7%

Defense, Health, Education spending as % of GDP

Slovenia has the highest standard of living of all the central and eastern European states of the former Soviet bloc.

WORLD RANKING

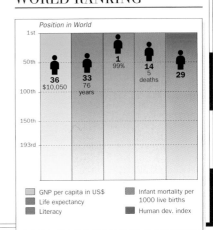

Position in World: 36 $10,050 | 33 76 years | 1 99% | 14 5 deaths | 29

- GNP per capita in US$
- Life expectancy
- Literacy
- Infant mortality per 1000 live births
- Human dev. index

S

SOLOMON ISLANDS

OFFICIAL NAME: Solomon Islands **CAPITAL:** Honiara
POPULATION: 463,000 **CURRENCY:** Solomon Islands dollar **OFFICIAL LANGUAGE:** English

 1978   1978   July 7   SLB   +11   +677 .sb

SCATTERED OVER 645,000 sq. km (250,000 sq. miles), the Solomons archipelago has several hundred islands, but most people live on the six largest – Guadalcanal, Malaita, New Georgia, Makira, Santa Isabel, and Choiseul. The Solomons have been settled since at least 1000 BCE; the Spanish arrived in 1568. Ethnic conflict between rival islanders ravaged the country from 1998 to 2000. Most of the Solomons are coral reefs. Just 1% of the land area is cultivable.

CLIMATE

▷ Tropical equatorial

WEATHER CHART FOR HONIARA

There is little temperature variation in the humid climate, but ferocious cyclones can occur in the rainy season.

TRANSPORTATION

▷ Drive on left

Henderson, Honiara 22,000 passengers

31 ships 1400 grt

THE TRANSPORTATION NETWORK

34 km (21 miles)	None		
None	None		

International flights from the principal airport, 13 km (8 miles) outside Honiara, were resumed in late 2000 after the ending of open hostilities.

Unloading seed coconuts near Munda on New Georgia in the Solomons' northern chain of islands. Coconuts are by far the largest and most commercially important crop.

TOURISM

▷ Visitors : Population 1:22

21,000 visitors   Up 62% in 1999

MAIN TOURIST ARRIVALS

Australia 23%	
Philippines 10%	
USA 7%	
Other 60%	

0 10 20 30 40 50 60
% of total arrivals

The importance of Guadalcanal during World War II and the tranquility of the outer islands used to attract tourists. However, ethnic conflict all but destroyed tourism in 1998 and again when fighting intensified in 2000. Lack of funding hampers recovery.

PEOPLE

▷ Pop. density low

English, Pidgin English, Melanesian Pidgin

17/km² (43/mi²)

THE URBAN/RURAL POPULATION SPLIT

20%   80%

RELIGIOUS PERSUASION

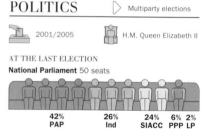

Church of Melanesia (Anglican) 34%
Other 9%
Seventh-day Adventist 10%
Methodist 11%
South Seas Evangelical Church 17%
Roman Catholic 19%

Almost all Solomon Islanders are Melanesian; relations between islands are tense. During the 1998–2000 conflict, 20,000 Malaitans were forced from their homes on Guadalcanal by native (Isatabu) militias. Authorities in outlying islands have pressed for greater autonomy. There are small communities of Micronesians who are descended from I-Kiribati temporarily relocated in 1957. More than 50 dialects are spoken. Although the islanders are nominally Christian, animist beliefs are widespread.

POLITICS

▷ Multiparty elections

2001/2005   H.M. Queen Elizabeth II

AT THE LAST ELECTION

National Parliament 50 seats

42% PAP   26% Ind   24% SIACC   6% PPP   2% LP

PAP = People's Alliance Party **Ind** = Independents
SIACC = Solomon Islands Alliance for Change Coalition
PPP = People's Progressive Party **LP** = Labour Party

Parliament is based on the Westminster model, but with no dominant political class, prominent local figures – known as "big men" – stand as candidates and turnover of members is high, leading to fluid and unstable coalitions. The civil conflict on Guadalcanal briefly ousted the government in 2000. A semblance of stability has since been restored with a new devolved "state system," which gives greater regional autonomy. Rebuilding the economy is now the major issue: "compensation" claims from the recent conflict quickly outstripped aid funds.

SOLOMON ISLANDS
Total Land Area : 28 450 sq. km (10 985 sq. miles)

POPULATION
over 10 000 ●
under 10 000 ·

LAND HEIGHT
1000m/3280ft
500m/1640ft
Sea Level

WORLD AFFAIRS
▷ Joined UN in 1978

   PC

The intensification of violence in 2000 caused great concern around the Pacific, and increased mediation efforts, in which

Australia was particularly involved. An Australian warship was the scene of key events during the resolution of the conflict, including the prime ministerial election and ratification of the Townsville peace accord.

AID
▷ Recipient

 US$68m (receipts)   Up 70% in 2000

Aid has focused very specifically on restoring stability and rebuilding infrastructure after two years of brutal

conflict. Regional powers Australia, New Zealand, and Taiwan are key in aiding recovery. Improved relations with Papua New Guinea brought aid of US$23 million between 1998 and 2001.

DEFENSE
▷ No compulsory military service

 Australia responsible for defense   Not applicable

The Peace Plan 2000 includes the creation of a panethnic security force. Under the Townsville peace accord

security was overseen by unarmed peacekeepers from neighboring Pacific states. The rival militias are effectively in control on Guadalcanal and Malaita.

ECONOMICS
▷ Inflation 8.5% p.a. (1990–2000)

US$278m   5.163–5.640 Solomon Islands dollars

SCORE CARD
- ❏ WORLD GNP RANKING........................177th
- ❏ GNP PER CAPITA ....US$620
- ❏ BALANCE OF PAYMENTS ....US$21m
- ❏ INFLATION 8.3%
- ❏ UNEMPLOYMENT............Some underemployment

EXPORTS

UK 7%
Other 30%
Philippines 11%
China 13%
South Korea 17%
Japan 22%

IMPORTS
USA 5%
Australia 27%
New Zealand 5%
Japan 5%
Singapore 25%
Other 33%

STRENGTHS
Good mineral and agricultural resources. Influx of international aid.

WEAKNESSES
Economy near collapse after ethnic conflict. Destruction of infrastructure. Key gold mine shut by militias. Grossly inflated compensation claims from conflict. Revenue from copra, gold, fish, and palm oil dried up. Social insecurity deters investment.

RESOURCES
▷ Electric power 12,000 kw

 82,347 tonnes   Not an oil producer

 67,000 pigs, 12,500 cattle, 200,000 chickens   Gold, copper, bauxite, lead, zinc, silver, cobalt, phosphates

Bauxite deposits have been discovered on Rennell Island. In addition, there are traces of gold and copper on Guadalcanal. There is increasing concern over the exploitation of forest and marine resources.

ENVIRONMENT
▷ Not available

 None   0.4 tonnes per capita

The environmental movement is strong. Depletion of forest and marine resources are a major concern. In 1998 a sustainable forest-harvesting policy was introduced, but the need to restore the economy puts pressure on environmentally sensitive areas.

MEDIA
▷ TV ownership low

 Daily newspaper circulation 16 per 1000 people

PUBLISHING AND BROADCAST MEDIA

There are two daily newspapers, *The Solomon Star* and *Solomon Voice*

 1 US-based Christian service   1 independent service

The one radio station broadcasts in English and Pidgin. The first TV service, a Christian channel, began in 2002.

CRIME
▷ No death penalty

 134 prisoners   Crime is rising

Under the armed militias, extortion and gang-related violence have flourished, notably on Guadalcanal and Malaita.

EDUCATION
▷ School leaving age: 14

 62%   Not available

Education is modeled on the British system. Tertiary students go to the University of the South Pacific in Fiji.

HEALTH
▷ Welfare state health benefits

 1 per 7143 people   Not available

The main hospital has seriously reduced services; local patients are now expected to provide their own food. Gang violence has spread to hospital wards.

SPENDING
▷ GDP/cap. decrease

CONSUMPTION AND SPENDING

3 per 1000 population   18 per 1000 population

Defense None		
Education 4.2%		
Health 4.2%		

0   5   10   15   20   25
Defense, Health, Education spending as % of GDP

Solomon Islanders in government jobs are the wealthiest group. Inhabitants of the outlying islands are extremely poor.

WORLD RANKING

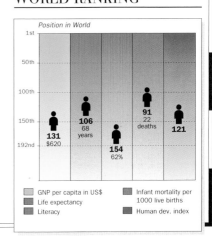

Position in World

1st
50th
100th
150th
192nd

131 $620
106 68 years
154 62%
91 22 deaths
121

- ◻ GNP per capita in US$
- ◻ Life expectancy
- ◻ Literacy
- ◻ Infant mortality per 1000 live births
- ◻ Human dev. index

S

527

SOMALIA

OFFICIAL NAME: Somalia   **CAPITAL:** Mogadishu   **POPULATION:** 9.2 million
CURRENCY: Somali shilling   **OFFICIAL LANGUAGES:** Somali and Arabic

OCCUPYING THE HORN of Africa, Italian Somaliland and British Somaliland joined in 1960 to form an independent Somalia. Except in the more fertile south, the land is semiarid. Years of clan-based civil war have resulted in the collapse of central government, the frustration of US and UN intervention initiatives aimed at easing a huge refugee crisis, and mass starvation.

CLIMATE
▷ Hot desert/steppe

WEATHER CHART FOR MOGADISHU

Somalia is very dry. The northern coast is very hot and humid, the eastern less so. The interior has some of the world's highest mean yearly temperatures.

TRANSPORTATION
▷ Drive on left

Mogadishu International   22 ships 11,400 grt

THE TRANSPORTATION NETWORK

2608 km (1621 miles)	None
None	None

About 50% of Somalis are nomads for whom the camel is the principal means of transportation. In 1990, the IDA agreed to repair the road network, but work on the seven-year project has not yet begun.

TOURISM
▷ Visitors : Population 1:920

10,000 visitors   No change in 1998

MAIN TOURIST ARRIVALS

Somalia does not publish tourism figures by country of origin

Aid workers and foreign journalists are the only visitors. Land mines are a major hazard.

Baydhabo market. Subsistence farming supports most people, despite chaos created by the fighting.

PEOPLE
▷ Pop. density low

 Somali, Arabic, English, Italian   15/km² (38/mi²)

THE URBAN/RURAL POPULATION SPLIT

28%   72%

RELIGIOUS PERSUASION

Christian 2%

Sunni Muslim 98%

The clan system is fundamental to Somalia. Shifting allegiances characterize its structure – a tendency stifled by Siad Barre's dictatorship but revived after his fall in 1991. His undermining of the traditional brokers of justice, the elders, contributed to the power vacuum that resulted in civil war, and his persecution of the Issaqs led to Somaliland's declaration of secession in 1991. However, the entire population is ethnic Somali, and national identity remains strong, shown by the widespread opposition to the UN's peacekeeping force.

POLITICS
▷ No legislative elections

1984/Uncertain   No internationally recognized head of state

AT THE LAST ELECTION
National Assembly (suspended)

There has been no prospect of organizing new elections since the overthrow of Siad Barre. A transitional assembly was formed in 2000 on a nonparty basis.

Somalia has remained in anarchy since the former dictator President Siad Barre fled in 1991. The unified state dissolved amid conflict in the south and separatism in the north.

The US led a UN peacekeeping force to the south in 1992, but failed to loosen the grip of the warring factions. Throughout the 1990s rival warlords, including the powerful Gen. Aideed, contended for supreme power. A National Salvation Council sank without trace in 1997.

A conference of businessmen and influential figures, held in neighboring Djibouti in 2000, established a transitional assembly and appointed former Barre minister Abdulkassim Salat Hassan as president. The new government, although warmly received in Mogadishu, was immediately rejected by most of the warlords and by the northern separatist authorities in "Somaliland" and "Puntland."

SOMALIA

Total Land Area : 637 657 sq. km (246 199 sq. miles)

POPULATION

over 1 000 000	▣
over 100 000	◉
over 50 000	○
over 10 000	•
under 10 000	·

LAND HEIGHT

2000m/6562ft
1000m/3281ft
500m/1640ft
200m/656ft
Sea Level

S

WORLD AFFAIRS
 Joined UN in 1960

After the UN force withdrew in 1995, the international community appeared to abandon Somalia until it gave support to the transitional parliament from 2000. Relations with Ethiopia are particularly tense; the government accuses Addis Ababa of sending troops to assist opposition warlords. Belief in the existence of terrorist training camps in Somalia has eroded relations with the US since September 11, 2001.

AID
 Recipient

 $104m (receipts)   Down 10% in 2000

Mass starvation among the Somali population in 1991 finally prompted the UN to launch a large-scale humanitarian aid effort. In this the UN was largely effective, averting widescale starvation and restoring food security.

DEFENSE
 No compulsory military service

 $39m   Down 2% in 2000

Former soldiers have been urged to reenlist. Efforts to demobilize the estimated 75,000 militia began in 2000.

ECONOMICS
 Not available

 $835m   2620 Somali shillings

SCORE CARD

- ❏ WORLD GNP RANKING........................157th
- ❏ GNP PER CAPITA $100
- ❏ BALANCE OF PAYMENTS...................–$157m
- ❏ INFLATION................................Over 100%
- ❏ UNEMPLOYMENT..........Widespread underemployment

STRENGTHS
Very few. Export of livestock to Arabian peninsula resumed in the north. Inflow of money from Somalis abroad. Growing market in stolen food aid.

WEAKNESSES
Every commodity, except arms, in extremely short supply. Little economic potential in the south. Destruction by drought of livestock. Mogadishu port closed by civil unrest until October 2000.

EXPORTS

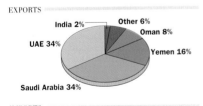

India 2%
Other 6%
Oman 8%
UAE 34%
Yemen 16%
Saudi Arabia 34%

IMPORTS

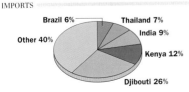

Brazil 6%
Thailand 7%
India 9%
Other 40%
Kenya 12%
Djibouti 26%

RESOURCES
 Electric power 79,000 kw

 20,250 tonnes   Not an oil producer

13.2m sheep, 12.5m goats, 6.2m camels   Salt, tin, zinc, copper, gypsum, manganese, uranium, iron

Commercially exploitable minerals remain untapped. An oil exploration agreement was signed with a French oil group in February 2001.

ENVIRONMENT
 Sustainability rank: 132nd

 0.3%   Negligible emissions per capita

Human deprivation and starvation caused by the effects of drought and war on land and livestock outweigh all other ecological considerations.

MEDIA
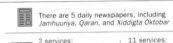 TV ownership low

Daily newspaper circulation 1 per 1000 people

PUBLISHING AND BROADCAST MEDIA

There are 5 daily newspapers, including *Jamhuuriya*, *Qaran*, and *Xiddigta Oktobar*

2 services: limited to the Mogadishu area   11 services: mostly political or religious

In Mogadishu there are three faction-run radio stations. Somali Television Network, an independent multichannel, multilingual service, began broadcasting in 1999. There are few newspapers, paper being in very short supply.

CRIME
 Death penalty in use

 Somalia does not publish prison figures   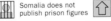 Widespread breakdown of law and order since 1991

Armed clan factions (some, in remoter regions, engaged in family feuds rather than the war) and bandits rule large areas. In Mogadishu a "national" police force has been established, and possession of firearms outlawed. *Sharia* (Islamic law), now the de facto system, is run in a makeshift fashion by elders.

EDUCATION
 School leaving age: 14

 24%   10,400 students

The system collapsed during the civil war. There were reports of improvised open-air schools starting up again in urban areas in 1993. Somali has been a written language only since 1972.

HEALTH
No welfare state health benefits

1 per 20,000 people   Diarrheal, communicable, and parasitic diseases

The state-run system has collapsed entirely. A few very rudimentary facilities are run by foreign workers.

SPENDING
GDP/cap. decrease

CONSUMPTION AND SPENDING

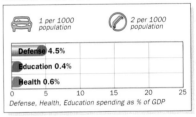

1 per 1000 population   2 per 1000 population

Defense 4.5%
Education 0.4%
Health 0.6%

0   5   10   15   20   25

Defense, Health, Education spending as % of GDP

Bandits and warlords gained rich pickings. Money sent by relatives living overseas is the main income for some people.

S

WORLD RANKING

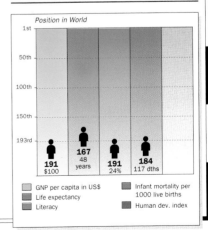

Position in World

1st
50th
100th
150th
193rd

	GNP per capita in US$	Life expectancy	Literacy	Infant mortality per 1000 live births / Human dev. index
	191 $100	167 48 years	191 24%	184 117 dths

Legend:
- GNP per capita in US$
- Life expectancy
- Literacy
- Infant mortality per 1000 live births
- Human dev. index

SOUTH AFRICA

OFFICIAL NAME: Republic of South Africa   CAPITALS: Pretoria; Cape Town; Bloemfontein
POPULATION: 43.8 million   CURRENCY: Rand   OFFICIAL LANGUAGES: Afrikaans, English, and 9 African languages

RICH IN NATURAL RESOURCES, South Africa comprises a central plateau, or *veld*, bordered to the south and east by the Drakensberg Mountains. After eight decades of white minority rule, with racial segregation under the apartheid policy since 1948, South Africa held its first multiracial elections in 1994. The revolution in South Africa's politics began in 1990, when black freedom groups were legalized and the dismantling of apartheid began. The African National Congress (ANC), under Nelson Mandela and his successor Thabo Mbeki, is now the leading political movement.

Nelson Mandela, who became president of South Africa in April 1994.

Thabo Mbeki, elected president in 1999 to succeed Mandela.

CLIMATE
▷ Desert/subtropical/ Mediterranean

WEATHER CHART FOR PRETORIA

Despite the moderating effects of oceans on three sides, the warm temperate climate is dry; 65% of the country has less than 50 cm (20 in) of rain a year. Drought is a periodic hazard.

TRANSPORTATION
▷ Drive on left

Jan Smuts International, Johannesburg
11.7m passengers

192 ships
383,700 grt

THE TRANSPORTATION NETWORK

63,027 km (39,163 miles)	2032 km (1263 miles)
22,686 km (14,097 miles)	None

Priorities include expanding port capacity and cross-border rail networks. Public transportation is limited and expensive, but there is an extensive informal network of minibuses and taxis.

Cape Town, set on a peninsula ending at the Cape of Good Hope, where the Indian and Atlantic oceans meet.

TOURISM
▷ Visitors : Population 1:7.3

6m visitors   No change in 2000

MAIN TOURIST ARRIVALS

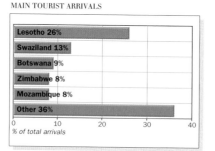

- Lesotho 26%
- Swaziland 13%
- Botswana 9%
- Zimbabwe 8%
- Mozambique 8%
- Other 36%

% of total arrivals

South Africa has huge tourist potential, with attractions ranging from beaches to mountains, from prizewinning vineyards to world renowned wildlife reserves. The enormous Kruger National Park boasts 137 mammal species and 450 bird species. Visitor numbers increased throughout the 1990s, but tourism is still recovering from the country's isolation during the apartheid era. Today, the key constraint on growth is rising crime. Studies suggest that by 2005 tourism could create an additional 450,000 jobs and contribute 10% toward GDP (compared with 4% in 1995).

PEOPLE
▷ Pop. density low

English, Afrikaans, Zulu, Xhosa, Ndebele, Setswana, Siswati, North Sotho, South Sotho, Tsongo, Venda

36/km² (93/mi²)

THE URBAN/RURAL POPULATION SPLIT

55%   45%

RELIGIOUS PERSUASION

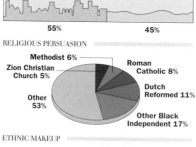

- Methodist 6%
- Zion Christian Church 5%
- Other 53%
- Roman Catholic 8%
- Dutch Reformed 11%
- Other Black Independent 17%

ETHNIC MAKEUP

- Other 4%
- Xhosa 9%
- Mixed 10%
- White 16%
- Zulu 23%
- Other Black 38%

Under apartheid, South Africans were divided into racial categories: whites (Afrikaners and English-speakers), and three black groups (Coloreds, people whose descent was deemed mixed; Asians, mainly Indians; and Africans). Each category had different political, economic, and social rights, with whites enjoying the most privileges and Africans the fewest. While blacks now dominate politics, English-speaking whites continue to control the economy.

The extended family has been undermined by regulations forcing men to migrate for work, leaving their wives and children in the rural areas. A small black middle class has developed, but most black South Africans are underemployed.

The expected postapartheid ethnic conflict failed to materialize, although Inkatha has exploited feelings of Zulu identity in its quest for greater political power. An area of the Kalahari Desert was returned to a Khomani San tribe (Bushmen) in 1999.

Many women are now prominent in public life. The new constitution guarantees equality of the sexes.

POPULATION AGE BREAKDOWN

Female	Age	Male
0.5%	80+	0.2%
3.2%	60–79	2.5%
8.1%	40–59	8.1%
16%	20–39	16.5%
22.2%	0–19	22.7%

% of population by age group

POLITICS ▷ Multiparty elections

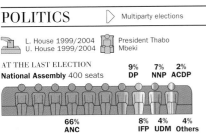

L. House 1999/2004
U. House 1999/2004

President Thabo Mbeki

AT THE LAST ELECTION
National Assembly 400 seats

9% DP	7% NNP	2% ACDP

66% ANC	8% IFP	4% UDM	4% Others

ANC = African National Congress **DP** = Democratic Party
IFP = Inkatha Freedom Party **NNP** = New National Party
UDM = United Democratic Movement **ACDP** = African
Christian Democratic Party

National Council of Provinces 90 seats

10 members are elected to the National Council of Provinces
by each of the nine provincial legislatures

South Africa became a multiracial
democracy following elections in 1994.

PROFILE
The 1994 elections ended 45 years
of apartheid, and Nelson Mandela
was elected president. A new liberal
constitution was adopted in 1996. The
1999 elections saw the ANC slightly
increase its majority; Thabo Mbeki
succeeded Mandela as president. The
DP overtook the strife-ridden NNP to
become the official opposition, and
in 2000 the two organizations
merged briefly to form an
opposition Democratic

Alliance (DA). However, in 2001, the
NNP allied itself with the ANC.

MAIN POLITICAL ISSUES
Maintaining unity
In April 1994, South Africa confounded
the proponents of violence and ethnic
division by holding peaceful elections.
The ANC-dominated government, while
heeding the aspirations of the black
majority, has tried to avoid
marginalizing minorities. To deal
with the apartheid past, it mandated
an innovative Truth and Reconciliation
Commission (TRC). Two years of
painful and sometimes controversial
hearings culminated in a final report
in 1998. Initial praise for the report
has been modified more recently
by allegations of corruption and
mishandling of compensation payments
for the victims of apartheid.

Reconstruction and development
The costly Reconstruction and
Development Program (RDP) aims to
improve health, housing, and education,
and boost employment. Implementation
is too slow for many blacks, who had
hoped for immediate benefits from
democratic rule, while some
members of the minorities
see the RDP's affirmative
action as reverse
discrimination.

SOUTH AFRICA

Total Land Area : 1 219 912 sq. km
(471 008 sq. miles)

WORLD AFFAIRS ▷ Joined UN in 1945

Comm WTO G24 AU SADC

After several decades of political
isolation and economic sanctions,
South Africa has been welcomed
back to the international fold,
and has rejoined the UN and the
Commonwealth. It now hopes to
reattract international investors.
South Africa has joined the SADC
and leads continental opinion on
regional issues. Former President
Mandela often intervened to help
resolve foreign conflicts, and
continues to act as a regional
mediator, most notably in Burundi.
His successor, Thabo Mbeki, has
continued this role, notably in pushing
forward the peace process in the
Democratic Republic of the Congo.
However, Mbeki has been criticized
internationally for his support of Robert
Mugabe's regime in neighboring
Zimbabwe, and for his controversial
opinions about HIV and AIDS.

SOUTH AFRICA's THREE CAPITALS

Pretoria - *administrative*
Cape Town - *legislative*
Bloemfontein - *judicial*

POPULATION

over 1 000 000
over 500 000
over 100 000
over 50 000
over 10 000

LAND HEIGHT

2000m/6562ft
1000m/3281ft
500m/1640ft
Sea Level

S

AID ▷ Recipient

 $488m (receipts)   ⬇ Down 10% in 2000

Apartheid-era South Africa was denied aid, particularly from the World Bank and the IMF. It now seeks financial assistance for massive reconstruction programs. As part of NEPAD – Africa's "Marshall Plan" launched in 2002 – President Mbeki stressed the importance of ending reliance on foreign aid.

CHRONOLOGY

Until 1652, what is now South Africa was peopled by Bantu-speaking groups and Bushmen. Then Dutch settlers arrived. British colonizers followed in the 18th century.

❏ **1910** Union of South Africa set up as British dominion; white monopoly of power formalized.
❏ **1912** ANC formed.
❏ **1934** Independence.
❏ **1948** NP takes power; apartheid segregationist policy introduced.
❏ **1958–1966** Hendrik Verwoerd prime minister. "Grand Apartheid" policy implemented.
❏ **1959** Pan-Africanist Congress (PAC) formed.
❏ **1960** Sharpeville massacre. ANC, PAC banned.
❏ **1961** South Africa becomes republic; leaves Commonwealth.
❏ **1964** Senior ANC leader Nelson Mandela jailed.
❏ **1976** Soweto uprisings by black students; hundreds killed.
❏ **1978** P. W. Botha in office.
❏ **1984** New constitution: Indians and Coloreds get some representation. Growing black opposition.
❏ **1985** State of emergency introduced. International sanctions.
❏ **1989** F. W. De Klerk replaces Botha as president.
❏ **1990** De Klerk legalizes ANC and PAC; frees Nelson Mandela.
❏ **1990–1993** International sanctions gradually withdrawn.
❏ **1991** Convention for a Democratic South Africa (CODESA) starts work.
❏ **1993** Mandela and De Klerk win Nobel Peace Prize.
❏ **1994** Multiracial elections won by ANC; Mandela president.
❏ **1996** TRC begins work.
❏ **1997** New constitution takes effect.
❏ **1998** TRC report condemns both apartheid crimes and ANC excesses.
❏ **1999** ANC election victory; Thabo Mbeki succeeds Mandela.
❏ **2000** DA wins nearly 25% of votes in local elections.
❏ **2001** Rand hits record low.
❏ **2002** August, world summit on sustainable development held in Johannesburg.

DEFENSE ▷ No compulsory military service

 $1.91bn   ⬆ Up 9% in 2000

SOUTH AFRICAN ARMED FORCES

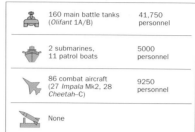

🛡	160 main battle tanks (*Olifant* 1A/B)	41,750 personnel
	2 submarines, 11 patrol boats	5000 personnel
✈	86 combat aircraft (27 *Impala* Mk2, 28 *Cheetah*–C)	9250 personnel
	None	

The creation by postapartheid South Africa of a truly national defense force seems almost miraculous, as it fuses together once bitter enemies: soldiers from the old white-run army, and guerrillas from the liberation groups.

However, doubts have been raised over the army's ability to operate effectively. A freeze on recruitment since 1994 has raised the average age of troops and created a glut of higher-ranking officers, while the incidence of AIDS is increasing. A large arms procurement program, announced in late 1998, failed to overcome the effects of previous swingeing cuts in spending. Few tanks are operational and the air force tends to run out of fuel toward the end of each financial year.

The legacy of years of sanctions is a major arms industry.

ECONOMICS ▷ Inflation 9.6% p.a. (1990–2000)

📊 $129bn   💲 7.570–11.995 rand

SCORE CARD

❏ WORLD GNP RANKING...........................29th
❏ GNP PER CAPITA $3020
❏ BALANCE OF PAYMENTS....................–$469m
❏ INFLATION 5.3%
❏ UNEMPLOYMENT...............................30%

EXPORTS

Netherlands 5%
Japan 6%
Germany 9%
UK 10%
USA 11%
Other 59%

IMPORTS

France 5%
Japan 8%
UK 9%
USA 12%
Germany 14%
Other 52%

STRENGTHS

Africa's largest and most developed economy; highly diversified with modern infrastructure. Strong financial sector for mobilizing investment. Growing manufacturing sector. Varied resource base, particularly of strategically important minerals.

WEAKNESSES

Foreign investors deterred by political fears. Growth too low to overcome deprivation. Black unemployment growing by 2.5% a year. Emigration of skilled workers. Population boom. Rand at record low in 2001. Falling gold price undermines many sectors.

PROFILE

South Africa has a large and diverse private sector, much of it controlled by multinationals. International sanctions

ECONOMIC PERFORMANCE INDICATOR

— Consumer Price Index   GDP

forced the government to play a central economic role through state corporations in the 1980s. This is now being reduced in a series of privatizations. The ANC has declared its intention to work with big business in order to revivify the economy and develop the townships.

SOUTH AFRICA : MAJOR BUSINESSES

Johannesburg
Pretoria
Potchefstroom
Kroonstad
Kimberley
Port Nolloth
Durban
Cape Town
Port Elizabeth

🍴 Food processing		🏦 Banking & finance	
🐟 Fish processing		⚙ Light engineering	
📖 Publishing		⚙ Heavy engineering	
⛽ Oil refining		🚗 Vehicle manufacture	
Gold mining		💻 Hi-tech	
Diamond mining		Textiles	

* significant multinational ownership

0   500 km
0   500 miles

S

RESOURCES

▷ Electric power 35.9m kw

592,144 tonnes

146,365 b/d (reserves 107m barrels)

28.8m sheep, 13.7m cattle, 119m chickens

Gold, coal, vanadium, vermiciline, diamonds, chromium, manganese, uranium, nickel

ELECTRICITY GENERATION

Hydro 1% (1.6bn kwh)
Combustion 92% (179bn kwh)
Nuclear 7% (14bn kwh)
Other 0%

0 20 40 60 80 100

% of total generation by type

South Africa has some of the continent's richest natural resources, in particular minerals. Its dominance of the world market in gold and diamonds helped it survive sanctions during apartheid. The falling price of gold in 2000 meant that for the first time sales of platinum group metals outstripped those of gold. South Africa is the largest single producer of manganese, chrome ore, vanadium, and vermiciline. It also produces uranium.

With no oil, South Africa pioneered the transformation of coal into oil, and otherwise uses its huge coal reserves to generate electricity. About 80% of black homes lack electricity, and the government is considering non-grid options. Agriculture is varied and provides lucrative export earnings.

SOUTH AFRICA : LAND USE

Cropland
Forest
Pasture
Desert
High mountain regions
Sheep
Corn
Fruit - cash crop

0   500 km
0   500 miles

ENVIRONMENT

▷ Sustainability rank: 77th

5% (4% partially protected)

8.3 tonnes per capita

ENVIRONMENTAL TREATIES

Yes   Yes   Yes
Yes   Yes   No

Floods and drought are familiar hazards. The main concern is protecting animal species. The creation of the world's largest game park, in cooperation with the governments of Zimbabwe and Mozambique, began in 2001.

MEDIA

▷ TV ownership medium

Daily newspaper circulation 29 per 1000 people

PUBLISHING AND BROADCAST MEDIA

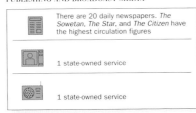

There are 20 daily newspapers. *The Sowetan, The Star,* and *The Citizen* have the highest circulation figures

1 state-owned service

1 state-owned service

A drive to combat racial stereotyping in the media was launched following a report on the subject to the Human Rights Commission in early 2000.

CRIME

▷ No death penalty

166,334 prisoners

Up 6% in 1999

CRIME RATES

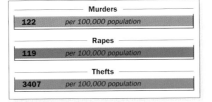

Murders
122   per 100,000 population

Rapes
119   per 100,000 population

Thefts
3407   per 100,000 population

South Africa is a dangerous country and crime rates are rising: murders occur with extreme frequency, and rape, armed robberies, and muggings are rife. Vigilantism is a huge problem in the Cape. The government introduced new gun laws in 2000. The death penalty was abolished in 1997.

EDUCATION

▷ School leaving age: 15

85%

633,918 students

THE EDUCATION SYSTEM

% of each age group in education

Primary 100%
Secondary 100%
Tertiary 17%

Education reform is a central task of the postapartheid government. Progress has been made in improving national literacy, and access to education has been widened through the Tirisano (working together) education program, launched in 2000. Long-established universities continue to be white-dominated.

HEALTH

▷ Welfare state health benefits

1 per 1667 people

Heart, respiratory, and diarrheal diseases, road deaths, cancers, AIDS

Health services were desegregated formally in 1990, but equal access to care is still a distant goal. Statistics on medical provision hide a strong bias toward whites and urban areas, where 80% of doctors work; 20% of children in rural areas die before the age of five, which is a rate considerably higher than the sub-Saharan average. South Africa has five million AIDS sufferers, more than any other country. The government has won the right to buy cheaper generic drugs for AIDS sufferers and has increased spending, but balked at the potential cost of widespread provision.

SPENDING

▷ GDP/cap. increase

CONSUMPTION AND SPENDING

94 per 1000 population

114 per 1000 population

Defense 1.6%
Education 6.1%
Health 3.3%

0   5   10   15   20   25
Defense, Health, Education spending as % of GDP

In South Africa, the black majority forms the poorest group in society. Wealth disparities are marked. At the top, the white elite enjoys living standards similar to those of Californians. In contrast, black living conditions are among Africa's poorest. Nearly half of black adults are unemployed. In between are the mixed race and Asian communities, who enjoyed more privileges under apartheid's strict racial hierarchy. However, a small black middle class is growing slowly, with some black-owned firms doing well on the stock market.

S

WORLD RANKING

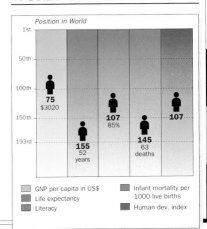

Position in World

1st
50th
100th
150th
193rd

75 $3020
107 85%
155 52 years
145 63 deaths
107

GNP per capita in US$
Life expectancy
Literacy
Infant mortality per 1000 live births
Human dev. index

SPAIN

OFFICIAL NAME: Kingdom of Spain **CAPITAL:** Madrid **POPULATION:** 39.9 million
CURRENCY: Euro (peseta until 2002) **OFFICIAL LANGUAGES:** Spanish, Galician, Basque, and Catalan

OCCUPYING THE MAJOR PART of the Iberian peninsula in southwest Europe, Spain has both an Atlantic and a Mediterranean coast, and is dominated by a central plateau. After the death of Gen. Franco in 1975, the country managed a rapid and relatively peaceful transition to democracy under the supervision of King Juan Carlos I. Since EU membership in 1986, there has been an increasing devolution of power to the regions. For just over 13 years from 1982, Spain had a center-left government, but the right-of-center Popular Party has dominated since 1996.

Alcaudete, Jaén Province, in the Andalusian mountains between the Guadalquivir River and Granada. The ruined castle is Moorish.

CLIMATE ▷ Mediterranean/maritime/mountain

WEATHER CHART FOR MADRID

The central plateau, or *meseta*, endures an extreme climate. Coastal areas are milder, and wetter in the north than in the south.

TRANSPORTATION ▷ Drive on right

✈ **Barajas, Madrid** 32.9m passengers

🚢 1570 ships 1.8m grt

THE TRANSPORTATION NETWORK

343,389 km (213,372 miles)	9063 km (5631 miles)
13,878 km (8624 miles)	1045 km (649 miles)

The AVE high-speed train links Madrid and Seville; more routes are planned. The state-run rail company RENFE is to be privatized.

TOURISM ▷ Visitors : Population 1.2:1

49.5m visitors

⬆ Up 3% in 2001

MAIN TOURIST ARRIVALS

UK 26%	
Germany 25%	
France 12%	
Netherlands 4%	
Italy 4%	
Other 29%	

% of total arrivals

PEOPLE ▷ Pop. density medium

Spanish, Catalan, Galician, Basque

80/km² (207/mi²)

THE URBAN/RURAL POPULATION SPLIT

78%   22%

RELIGIOUS PERSUASION

Other 4%
Roman Catholic 96%

ETHNIC MAKEUP

Other 2%   Basque 2%
Roma 1%   Galician 6%
Catalan 17%
Castilian Spanish 72%

A vigorous regionalism, suppressed under Franco, now flourishes. Catalonia is an example, with Barcelona its vibrant capital. In the Basque region, the ETA separatists who fight for independence by waging a high-profile terror campaign remain in a minority.

Spain today has one of the lowest birthrates in Europe, just half that of 1975. The influence of the Roman

POPULATION AGE BREAKDOWN

Female	Age	Male
2.3%	80+	1.2%
9.8%	60–79	8%
11.8%	40–59	11.5%
15.9%	20–39	16.3%
11.3%	0–19	11.9%

% of population by age group

Catholic Church on personal behavior has declined, and attitudes to sexuality are now relaxed. The divorce rate is very low, and family ties remain strong; men often live at home until their late 20s.

Economic growth from the 1970s led to a change in the composition of society. Migration from poor rural regions to the coast was associated with the arrival of job-seeking immigrants from Latin America and – especially – north Africa. A rise in racial tensions and racism has resulted from the subsequent economic downturn and competition for scarce jobs.

Spanish women are increasingly emancipated and more influential in public life, making up 27% of the deputies and senators in the Spanish parliament, and heading 30% of businesses.

Tourism earnings in 2001 topped $33 billion, with Germany and the UK still accounting for around 50% of all arrivals. Long dominant in the vacation-package sector, Spain has recently adopted marketing strategies to boost additional cultural, historical, and environmental tourism. Several areas began levying an environmental tax on tourist arrivals in 2001. The cut-price package industry has benefited from political turbulence in potential competitor countries in the Mediterranean.

POLITICS ▷ Multiparty elections

L. House 2000/2004
U. House 2000/2004

H.R.H. King Juan
Carlos I de Borbón

AT THE LAST ELECTION

Congress of Deputies 350 seats

| 52% PP | 36% PSOE | 5% CiU | 2% Others | 2% IU | 2% PNV | 1% CC |

PP = Popular Party **PSOE** = Spanish Socialist Workers' Party
CiU = Convergence and Union **IU** = United Left **PNV** =
Basque Nationalist Party **CC** = Canary Islands Coalition
App = Appointed

Senate 259 seats

| 49% PP | 24% PSOE | 20% App | 2% CC | 3% CiU | 2% PNV |

208 members are directly elected to the Senate, and 51
appointed by autonomous communities

Since 1978, Spain has been a semifederal
multiparty parliamentary monarchy.
Each region has a legislative assembly.

PROFILE

The PSOE's long period in power blurred
the boundaries between party and state.
The Cortes (parliament) failed to check
executive power, and political disputes
were often left to the judiciary, while
political corruption undermined voters'
faith in Spain's political system. The PP
government, led by José
María Aznar
from 1996,
benefited from being seen as a fresh
start. Its relative success in running the
economy helped it win a second term in
2000, despite an electoral pact between
the PSOE and the United Left. Ideological
issues no longer sharply divide the main
parties, which hold similar views on
economic policy and EU membership.

MAIN POLITICAL ISSUES
Increasing regionalism

Spain's 17 autonomous regions all vie
for greater funds or independence from
Madrid. Many have bypassed central
government to borrow funds on the
international money markets, and have
come close to breaching their legal debt
limits. In 1996 the PP government
approved a new model of financing for
the regions which gave them new powers
for raising tax revenue. The Basque
separatist movement ETA has, with
intermittent cease-fire announcements,
waged a protracted violent struggle for
independence, prompting large-scale
demonstrations against violence. The
Basque country, Catalonia, and Galicia
each use their own language alongside
Castilian Spanish.

Clean government

Initially the PP government suffered less
from the corruption scandals which
dogged the last years of its PSOE
predecessor, but in 2001 its reputation
was tarnished by major allegations
relating to the Telefonica telecoms group
and the Gescarta stockbroking house.

King Juan Carlos,
*who became head of
state on the death of
Franco in 1975.*

José María Aznar,
*of the PP, was
appointed prime
minister in 1996.*

WORLD AFFAIRS ▷ Joined UN in 1955

 CE   NATO   OECD OSCE   EU

Spain, an enthusiastic EU member, has
been chary of enlargement to include
central Europe, which it sees as a threat
to its direct financial benefit. Relations
with Morocco, strained over illegal
immigrants, were jeopardized in 2002
over sovereignty of rocky islets in the
Mediterranean. There is a long-standing
dispute with the UK over the status
of Gibraltar; Spain and the UK have
recently been willing to negotiate, but
the colony itself opposes any change.
Elsewhere, Spain has sponsored an
Ibero-American Community of Nations
(a Hispanic Commonwealth). Spain's
first contribution to a UN peacekeeping
force was for operations in the Balkans.

SPAIN

Total Land Area : 504 782 sq. km
(194 896 sq. miles)

POPULATION

over 1 000 000	▣
over 500 000	◉
over 100 000	◎
over 50 000	○
over 10 000	●

LAND HEIGHT

3000m/9843ft
2000m/6562ft
1000m/3281ft
500m/1640ft
Sea Level

100 km
100 miles

Islas Canarias

100 km
100 miles

CHRONOLOGY

United under Ferdinand and Isabella
in 1492, Spain became a dominant
force. A long period of economic and
political decline followed, however,
and by the mid-19th century, Spain
lagged behind many other European
countries in stability and prosperity.

❑ **1874** Constitutional monarchy
restored under Alfonso XII.
❑ **1879** PSOE founded.
❑ **1881** Trade unions legalized.
❑ **1885** Death of Alfonso XII.
❑ **1898** Defeat in war with US
results in loss of Cuba, Puerto
Rico, and the Philippines.
❑ **1914–1918** Spain neutral in
World War I.
❑ **1921** Spanish army routed by
Berbers in Spanish Morocco.
❑ **1923** Coup by Gen. Primo de Rivera
accepted by King Alfonso XIII.
Military dictatorship.
❑ **1930** Primo de Rivera dismissed
by monarchy.
❑ **1931** Second Republic proclaimed.
Alfonso XIII flees Spain.
❑ **1933** Center-right coalition
wins general election. ⇨

S

CHRONOLOGY *continued*

- ❏ **1934** Asturias uprising quashed by army. Failure of attempt to form Catalan state.
- ❏ **1936** Popular Front wins elections. Right-wing military uprising against Republic. Gen. Francisco Franco subsequently appointed leader.
- ❏ **1939** Franco wins civil war which claims 300,000 lives.
- ❏ **1940** Franco meets Hitler, but does not enter World War II.
- ❏ **1946** UN condemns Franco regime.
- ❏ **1948** Spain excluded from the Marshall Plan.
- ❏ **1950** UN lifts veto.
- ❏ **1953** Concordat with Vatican. Spain grants US military bases.
- ❏ **1955** Spain joins UN.
- ❏ **1959** Stabilization Plan is basis for 1960s rapid economic growth.
- ❏ **1962** Franco government applies for eventual membership of EEC.
- ❏ **1969** Gen. Franco names Juan Carlos, grandson of Alfonso XIII, his successor.
- ❏ **1970** Spain signs preferential trade agreement with EEC.
- ❏ **1973** Basque separatists assassinate Prime Minister Carrero Blanco; replaced by Arias Navarro.
- ❏ **1975** Death of Franco. Proclamation of King Juan Carlos I.
- ❏ **1976** King appoints Adolfo Suárez as prime minister.
- ❏ **1977** First democratic elections since 1936 won by Suárez's Democratic Center Union.
- ❏ **1978** New constitution declares Spain a parliamentary monarchy.
- ❏ **1981** Leopoldo Calvo Sotelo replaces Suárez. King foils military coup. Calvo takes Spain into NATO.
- ❏ **1982** Felipe González wins landslide victory for PSOE.
- ❏ **1986** Joins European Communities. González wins referendum on keeping Spain in NATO.
- ❏ **1992** Olympic Games held in Barcelona, Expo '92 in Seville.
- ❏ **1996** PSOE loses election; José María Aznar of PP prime minister.
- ❏ **1998** Former PSOE minister found guilty of involvement in Basque kidnappings. September, ETA cease-fire; holds until December 1999.
- ❏ **2000** Aznar and PP win elections.
- ❏ **2002** Euro fully adopted.

AID

 Donor

 $1.2bn (donations)   Down 12% in 2000

Spain has taken steps to increase grant aid after criticism that Spanish aid was of poor quality and tied to the acquisition of goods and services. Aid in 2000 represented 0.2% of GNP.

DEFENSE

 No compulsory military service

$7.05bn   Down 2% in 2000

A substantial, largely state-owned and commercially nonviable defense industry is subsidized for strategic reasons. Full integration of NATO military structures was approved in 1997. Defense spending has fallen in recent years, and is now well below the NATO average. National service has been abolished; the last conscripts worked out their terms of duty in the course of 2002.

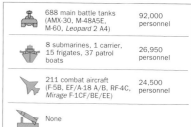

SPANISH ARMED FORCES

688 main battle tanks (AMX-30, M-48A5E, M-60, *Leopard* 2 A4)	92,000 personnel
8 submarines, 1 carrier, 15 frigates, 37 patrol boats	26,950 personnel
211 combat aircraft (F-5B, EF/A-18 A/B, RF-4C, *Mirage* F-1CF/BE/EE)	24,500 personnel
None	

ECONOMICS

Inflation 3.9% p.a. (1990–2000)

$595bn   1.0651–1.1231 euros

SCORE CARD

- ❏ WORLD GNP RANKING.....................10th
- ❏ GNP PER CAPITA $15,080
- ❏ BALANCE OF PAYMENTS ....–$17.3bn
- ❏ INFLATION 3.4%
- ❏ UNEMPLOYMENT...............................14%

EXPORTS

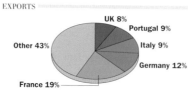

UK 8%
Portugal 9%
Italy 9%
Germany 12%
France 19%
Other 43%

IMPORTS

Netherlands 5%
UK 7%
Italy 9%
Germany 15%
France 18%
Other 46%

STRENGTHS

One of the fastest-growing OECD economies. Well-qualified labor force with relatively low labor costs. Privatization has introduced greater competition into gas, oil-refining, electricity, and telecommunications sectors. Recent major investment and expansion into Latin American market.

WEAKNESSES

Massive foreign penetration of economy and absence of Spanish multinationals. Low investment in research and development, concentration in declining industries, and low productivity – notably in agriculture. Persistent high unemployment.

PROFILE

Real convergence with the major European economies first became a realistic objective in the late 1980s, as Spain posted the highest investment-led output growth in the OECD. By

ECONOMIC PERFORMANCE INDICATOR

Consumer Price Index   GDP

Consumer price index 1995=100
GDP 1996=100

1996 1997 1998 1999 2000

1991, GDP per capita stood at almost 80% of the EU average. Recession in the early 1990s was turned around in mid-decade, and growth averaged 4% over 1997–2000, with public debt brought below 60% of GDP by 2001. The economy slowed markedly through 2001, and the annualized growth rate dropped to around 2%. Spain succeeded in meeting the economic convergence criteria necessary for European economic and monetary union and was among the 12 EU countries to adopt the euro fully in January 2002.

SPAIN : MAJOR BUSINESSES

La Coruña
Bilbao
Zaragoza
Barcelona
Vigo
Madrid
Huelva
Valencia
Sevilla
Cartagena
Málaga

❋ Textiles
Agribusiness
Chemicals
Shipbuilding
Vehicle manufacture
❂ Heavy engineering
❀ Light engineering
Fish processing

0   200 km
0   200 miles

* significant multinational ownership

RESURCES

 Electric power 50.1m kw

 1.49m tonnes

10,606 b/d (reserves 7.7m barrels)

24.4m sheep, 23.3m pigs, 128m chickens

Coal, oil, iron, uranium, mercury, fluorite, gypsum

ELECTRICITY GENERATION

Hydro 18% (36bn kwh)
Combustion 51% (99bn kwh)
Nuclear 30% (59bn kwh)
Other 1% (1.4bn kwh)

% of total generation by type

Spain lacks natural resources, especially water, and is heavily dependent on imported oil and gas. Coal, mined mainly to generate industry, is a declining but still subsidized sector, concentrated in the Asturias region. Spain has one of the world's largest fishing fleets, but EU restrictions have forced cuts in catches since the 1990s.

SPAIN : LAND USE

Forest
Pasture
Cropland
Wetlands
High mountain regions
Sheep
Olives - cash crop
Citrus fruits
Vineyards

0   200 km
0   200 miles

ENVIRONMENT

 Sustainability rank: 44th

 9%

6.3 tonnes per capita

ENVIRONMENTAL TREATIES

Yes   Yes   Yes
Yes   Yes   Yes

Public opinion is becoming increasingly vocal about environmental matters. Renewable energy, although still tiny in extent, is becoming more visible, particularly with the growth of wind farms. The benefits of a national tree-planting scheme to reduce soil erosion have been offset by increasingly frequent intentional forest fires. More land has national park status than in any other country in Europe. While dams and canals are necessary to counter desertification in the south, environmentalists oppose their construction. A project to bring water from the Ebro River to the parched south is causing controversy, particularly in areas further downstream.

MEDIA

 TV ownership high

Daily newspaper circulation 99 per 1000 people

PUBLISHING AND BROADCAST MEDIA

There are 87 daily newspapers. *ABC*, *El País*, *Marca*, and *El Mundo* are national

16 state-owned services, also independent services

363 services: 13 state-owned, 350 independent

Despite the large number of daily newspapers, readership is among the lowest in Europe. Both public and private TV are popular. Radio is of a generally high standard.

CRIME

 No death penalty

45,633 prisoners

Up 92% in 1999

CRIME RATES

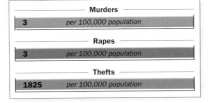

Murders
3   per 100,000 population

Rapes
3   per 100,000 population

Thefts
1825   per 100,000 population

Spain is a major crossroads in the world narcotics trade, and drugs-related crime is rising. Illegal immigration soared in 2000, with authorities in the south unable to cope with the influx.

EDUCATION

 School leaving age: 16

 98%

1.79m students

THE EDUCATION SYSTEM

Primary 100%
Secondary 100%
Tertiary 56%

% of each age group in education

The school leaving age has risen since 1990 from 14 to 16. The latest secondary education reforms, announced in 2000, offer a number of additional subjects, improvements in mathematics, philosophy, and languages, and increasing attention to information technology. Autonomous regions regulate by decree the teaching of languages other than Castilian Spanish, such as Basque or Catalan.

HEALTH

 Welfare state health benefits

1 per 323 people

Heart and circulatory diseases, cancers, accidents

Public health care is of high quality and readily available. Public hospitals, although widely considered to be superior, are outnumbered by private ones. In spite of very high tobacco and alcohol consumption, Spain has a healthy population, possibly due to its Mediterranean diet. The incidence of AIDS has risen alarmingly, however, to become one of the highest in western Europe.

SPENDING

 GDP/cap. increase

CONSUMPTION AND SPENDING

389 per 1000 population   421 per 1000 population

Defense 1.3%
Education 4.5%
Health 5.4%

Defense, Health, Education spending as % of GDP

In the late 1980s, it became fashionable in Spain to compete openly, make money, and acquire consumer goods. Rapid economic growth at this time greatly enriched the professional and managerial classes. The latter became the best-paid, in real terms, in Europe, and Spain quickly became an important market for luxury cars and yachts. In the early 1990s this ostentatious affluence waned in the face of recession and as the unemployment rate soared to become one of the highest in Europe. The boom of 1997–2000 boosted the number of jobs, but also gave rise to anxiety over inflation. By 2001 this boom too had subsided, and there was less financial security, with about one-third of employees on only temporary contracts.

S

WORLD RANKING

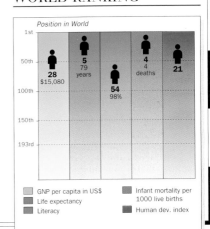

Position in World

28 $15,080
5 79 years
54 98%
4 4 deaths
21

GNP per capita in US$
Life expectancy
Literacy
Infant mortality per 1000 live births
Human dev. index

SRI LANKA

OFFICIAL NAME: Democratic Socialist Republic of Sri Lanka CAPITAL: Colombo
POPULATION: 19.1 million CURRENCY: Sri Lanka rupee OFFICIAL LANGUAGES: Sinhala, Tamil, and English

SOUTH ASIA

THE TEARDROP-SHAPED island of Sri Lanka is separated from India by the Palk Strait. Rugged central uplands give way to fertile plains in the north. The majority Sinhalese, an Indo-Aryan people originating in northern India, have a Buddhism-based identity. Independent since 1948, Sri Lanka suffered from 1983 from a protracted civil war involving the attempted secession of the minority (and mainly Hindu) Tamils in the north and east.

CLIMATE

▷ Tropical monsoon/ equatorial

WEATHER CHART FOR COLOMBO

The climate is tropical, with afternoon breezes on the coast and cooler air in the highlands. The northeast is driest.

TRANSPORTATION

▷ Drive on left

Bandaranaike, Katunayake
2.88m passengers

60 ships
189,200 grt

THE TRANSPORTATION NETWORK

10,721 km (6662 miles)		None
1463 km (909 miles)		430 km (267 miles)

Main roads are crowded and slow. Travel to the Tamil-dominated north and east is being reopened.

TOURISM

▷ Visitors : Population 1:48

400,000 visitors

Down 8% in 2000

MAIN TOURIST ARRIVALS

UK 19%
Germany 18%
India 10%
Other 53%

0 10 20 30 40 50 60
% of total arrivals

Sri Lanka has remained a popular destination, despite the civil war. A Tamil attack in 2001 on the military air base next to Colombo's Bandaranaike international airport threatened the lives of dozens of tourists.

PEOPLE

▷ Pop. density high

Sinhalese, Tamil, Sinhalese-Tamil, English

295/km² (764/mi²)

THE URBAN/RURAL POPULATION SPLIT

24%   76%

ETHNIC MAKEUP

Burgher, Malay, and Veddha 1%
Moor 7%
Tamil 18%
Sinhalese 74%

Ethnic tensions focus on the 19-year conflict between the minority, Hindu, Tamils and majority, Buddhist, Sinhalese. Favored by the British colonial administration, after independence the Tamils were subject to attempts by the Sinhalese to redress the balance. The Tamil Tigers seek a Tamil homeland in the north. A tentative peace was reached in 2002.

The Moors are the Muslim descendants of Arab traders. A few indigenous forest-dwelling Veddhas survive in the remote east of the island.

SRI LANKA

Total Land Area : 65 610 sq. km (25 332 sq. miles)

POPULATION
⊙ over 500 000
◉ over 100 000
◎ over 50 000
● over 10 000
• under 10 000

LAND HEIGHT
2000m/6562ft
1000m/3281ft
500m/1640ft
200m/656ft
Sea Level

0   100 km
0   100 miles

N

POLITICS

▷ Multiparty elections

2001/2007

President Chandrika Bandaranaike Kumaratunga

AT THE LAST ELECTION

Parliament 225 seats

7% 1%
TULF EPDP

48% UNP
34% PA
7% JVP
2% SLMC
1% DPLF

UNP = United National Party PA = People's Alliance, dominated by the Sri Lanka Freedom Party – SLFP
JVP = People's Liberation Front TULF = Tamil United Liberation Front SLMC = Sri Lanka Muslim Congress
EPDP = Eelam People's Democratic Party
DPLF = Democratic People's Liberation Front

Politics has been indelibly marked by the 19-year civil war which has claimed more than 50,000 lives. Breakthrough came in 2001 when the ruling left-wing PA was soundly beaten in elections by the right-of-center UNP. Under Prime Minister Ranil Wickremasinghe, the UNP sought to bring a swift end to the conflict. A permanent cease-fire was agreed in 2002, and the Liberation Tigers of Tamil Eelam (LTTE or Tamil Tigers) hinted that they were willing to drop their long-standing demand for an independent state. The conservative Buddhist clergy remain influential, helping to preserve Sinhalese nationalism, and opposing compromise with the Tamils.

S

WORLD AFFAIRS ▷ Joined UN in 1955

Comm | G24 | NAM | SAARC | WTO

Relations with India are paramount. However, India's role as peacemaker under the 1987 Indo-Sri Lankan accords was fiercely resisted by the Tamil Tigers, and India was forced to withdraw its peacekeeping troops. In recent years Norway has taken on the role of mediator in the conflict.

AID ▷ Recipient

 $276m (receipts) ⬆ Up 5% in 2000

The president has responded positively to Western aid donors seeking improvements in Sri Lanka's human rights record.

DEFENSE ▷ No compulsory military service

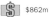 $862m ⬆ Up 7% in 2000

Recent drives to recruit 5000 more soldiers have shown a marked shift in policy, from simply enlarging the army to modernizing it.

ECONOMICS ▷ Inflation 9.1% p.a. (1990–2000)

 $16.4bn 82.70–93.16 Sri Lanka rupees

SCORE CARD

- ❑ WORLD GNP RANKING..........................73rd
- ❑ GNP PER CAPITA $850
- ❑ BALANCE OF PAYMENTS.................–$1.04bn
- ❑ INFLATION 6.2%
- ❑ UNEMPLOYMENT 9%

STRENGTHS

World's largest tea exporter. Foreign investment attracted by privatization, despite left-wing stance of President Kumaratunga.

WEAKNESSES

Civil war has severely drained government funds, and lingering tensions continue to deter investors and many tourists. High unemployment.

EXPORTS

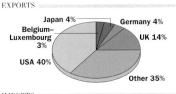

Japan 4% | Germany 4%
Belgium–Luxembourg 3% | UK 14%
USA 40% | Other 35%

IMPORTS

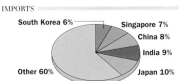

South Korea 6% | Singapore 7%
China 8%
India 9%
Other 60% | Japan 10%

Adam's Peak in mountainous central Sri Lanka is a famous religious site with a Buddhist shrine at the summit.

RESOURCES ▷ Electric power 1.7m kw

 279,900 tonnes

Not an oil producer; refines 50,000 b/d

1.56m cattle, 690,000 buffaloes, 11m chickens

Gemstones, graphite, iron, monazite, uranium, ilmenite, clay

Sri Lanka has to import all its oil. Hydropower supplies around 70% of electricity; droughts are frequent and supplies can be erratic. Sri Lanka is keen to diversify power sources and is turning to coal-powered generation.

ENVIRONMENT ▷ Sustainability rank: 55th

 14% (4% partially protected) ⬆ 0.4 tonnes per capita

Sri Lanka has successfully promoted national parks. Their development is opposed by the Veddha people, who have traditionally occupied such land.

MEDIA ▷ TV ownership medium

 Daily newspaper circulation 29 per 1000 people

PUBLISHING AND BROADCAST MEDIA

There are 13 daily newspapers, including the *Daily News, Dawasa, Observer*, and *Dinamina*

4 independent services

5 services: 1 state-owned, 4 independent

Press censorship was imposed in 1995 to control war reporting by local news agencies. Since 1998 foreign journalists have faced similar restrictions.

CRIME ▷ Death penalty not used in practice

 16, 850 prisoners ⬆ Up 52% 1992–1996

Extrajudicial killings and other human rights abuses increased the toll of deaths and disappearances during the civil war. New laws are being introduced to combat sex tourism.

EDUCATION ▷ School leaving age: 14

 92% 86,931 students

Sri Lanka has a high literacy rate for a developing nation. Many Sri Lankans attend US universities.

CHRONOLOGY

Sri Lanka was inhabited by Tamils and Sinhalese since before the 6th century. Named Ceylon by the British, it became independent in 1948.

- ❑ **1948** Indian Tamil workers stripped of suffrage and citizenship rights.
- ❑ **1956** SLFP wins election, promotes Sinhalese language.
- ❑ **1972** Renamed Sri Lanka.
- ❑ **1983** Tamil Tigers begin civil war.
- ❑ **1993** President Premadasa killed.
- ❑ **1994** Left-wing PA wins election; Chandrika Kumaratunga president.
- ❑ **1995–1996** Collapse of peace talks.
- ❑ **1999** Kumaratunga reelected.
- ❑ **2000** Sirimavo Bandaranaike, world's first woman prime minister, dies.
- ❑ **2001** UNP wins early elections.
- ❑ **2002** Comprehensive cease-fire promises end to civil war.

HEALTH ▷ Welfare state health benefits

 1 per 2500 people

Heart attacks, cancers, pneumonia, strokes

Years of high spending on health have resulted in an accessible, fee-free system. Ayurvedic medicine is popular.

SPENDING ▷ GDP/cap. increase

CONSUMPTION AND SPENDING

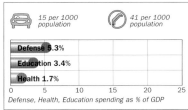

15 per 1000 population | 41 per 1000 population

Defense 5.3%
Education 3.4%
Health 1.7%

0　5　10　15　20　25
Defense, Health, Education spending as % of GDP

Economic growth has created a new class of wealthy Sinhalese. Tamil tea workers are the poorest group.

WORLD RANKING

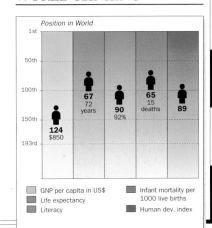

Position in World

1st
50th
100th
150th
193rd

124 $850
67 72 years
90 92%
65 15 deaths
89

GNP per capita in US$
Life expectancy
Literacy
Infant mortality per 1000 live births
Human dev. index

S

SUDAN

EAST AFRICA

OFFICIAL NAME: Republic of the Sudan   **CAPITAL:** Khartoum   **POPULATION:** 31.8 million
CURRENCY: Sudanese pound or dinar   **OFFICIAL LANGUAGE:** Arabic

| 1956 | 1956 | Jan 1 | SUD | +2 | +249 | .sd |

BORDERED BY THE RED SEA, Sudan is the largest country in Africa. Its landscape changes from desert in the north to lush tropical in the south, with grassy plains and swamps in the center. Since independence from British and Egyptian rule in 1956, tensions between the Arab north and African south have led to two civil wars. The second of these conflicts remains unresolved. In 1989, an army coup installed a military Islamic fundamentalist regime.

Camel caravan in the dry north. Periodic drought coupled with war disruption mean that Sudan requires large amounts of food aid.

CLIMATE

▷ Hot desert/steppe/tropical

WEATHER CHART FOR KHARTOUM

■ *Average daily temperature*   *Rainfall* ■

Sudan's northern half is hot arid desert with constant dry winds. The rest has a rainy season varying from two months in the center to eight in the south.

TRANSPORTATION

▷ Drive on right

✈ Khartoum International   🚢 19 ships / 43,078 grt

THE TRANSPORTATION NETWORK

| 4320 km (2684 miles) | None |
| 4595 km (2855 miles) | 5310 km (3299 miles) |

The Port Sudan–Khartoum railroad and road are Sudan's most important links. There are few other roads, but Iran is financing a north–south highway. Civil war has stopped all Nile shipping.

TOURISM

▷ Visitors : Population 1:636

🧳 50,000 visitors   ⬆ Up 28% in 2000

MAIN TOURIST ARRIVALS

| UK 7% |
| Egypt 7% |
| Germany 5% |
| Other 81% |

0 10 20 30 40 50 60 70 80 90 100
% of total arrivals

The civil war means that Sudan has very few tourists. Visitors are mostly aid workers or on business.

PEOPLE

▷ Pop. density low

Arabic, Dinka, Nuer, Nubian, Beja, Zande, Bari, Fur, Shilluk, Lotuko   13/km² (35/mi²)

THE URBAN/RURAL POPULATION SPLIT

36%   64%

RELIGIOUS PERSUASION

Other 1%   Christian 9%
Traditional beliefs 20%
Muslim (mainly Sunni) 70%

Sudan has a large number of ethnic and linguistic groups. About two million Sudanese are nomads. The major social division, however, is between the Arabized Muslims in the north and the mostly African, largely animist or Christian population in the south. Attempts to impose Arab and Islamic values throughout Sudan have been the root cause of the civil war that has ravaged the south since 1983. There are some non-Arab groups in the north and in the densely populated Darfur region. Women not wearing Islamic dress can suffer harassment or even public flogging. In 2002 the UN reported that thousands of people have been abducted into slavery within Sudan over the last 20 years.

SUDAN

Total Land Area : 2 505 810 sq. km (967 493 sq. miles)

0   400 km
0   400 miles

LAND HEIGHT

2000m/6562ft
1000m/3281ft
500m/1640ft
200m/656ft
Sea Level

POPULATION

◉ over 500 000
◎ over 100 000
○ over 50 000
• over 10 000
· under 10 000

POLITICS

▷ No multiparty elections

2000/2004   President Omar Hassan Ahmad al-Bashir

AT THE LAST ELECTION

National Assembly 400 seats

89% NC   10% Vac   1% Ind

NC = National Congress supporters   **Vac** = Vacant
Ind = Independents

Elections in 2000, under the 1999 constitution allowing "political associations," were boycotted by the opposition, and returned Gen. Omar al-Bashir and his NC bloc to power. The willingness of the Sudanese People's Liberation Army (SPLA) and its former rival, the Sudan People's Defense Force (SPDF), to enter peace talks raised the possibility of an end to nearly 20 years of civil conflict between Muslim north and Christian south. However, increasing hostilities in the oil-rich Upper Nile region in spring 2002 put the peace process at risk.

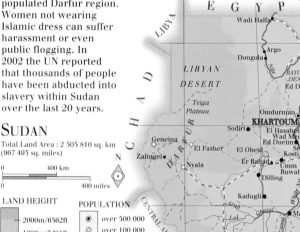

Map labels: EGYPT, LIBYA, CHAD, RED SEA, ERITREA, ETHIOPIA, CENTRAL AFRICAN REPUBLIC, DEM. REP. CONGO, UGANDA, KENYA, NUBIAN DESERT, LIBYAN DESERT, BAYUDHA DESERT, DARFUR, SUDD, Teiga Plateau, Red Sea Hills, Wadi Halfa, Argo, Dongola, Port Sudan, Tokar, Atbara, Ed Damer, Shendi, Omdurman, Khartoum North, KHARTOUM, Halfa el Gadida, Kassala, Khashm el Girba, Gedaref, Qala'en Nahl, Suki, Singa, Sennar, Kosti, Wad Medani, Ed Dueim, El Obeid, Er Rahad, Umm Ruwaba, Dilling, Ed Damazin, Kurmuk, Kadugli, Malakal, Wau, Rumbek, Juba, Yambio, Sodiri, El Hasaheisa, Geneina, Zalingei, Nyala, El Fasher, Imatong Mountains, Kinyeti 3187m

S

WORLD AFFAIRS ▷ Joined UN in 1956

 AL OIC COMESA IGAD AU

Sudan's support for Iraq in the Gulf War and suspicion that it sponsors terrorism have led to its increasing isolation from the West and the Arab world. Only Iran, Yemen, and Libya maintain friendly relations.

AID ▷ Recipient

$225m (receipts) — Down 7% in 2000

The only substantial bilateral aid comes from Iran. IMF funding ceased in 1990. Sudan depends on food aid.

DEFENSE ▷ Compulsory military service

$568m — Up 34% in 2000

The NC controls the military and police and has its own paramilitary militia. Sudan's 112,500-strong army was for many years engaged in fighting the two southern factions – the SPLA and SPDF – with forces of a similar size.

ECONOMICS ▷ Inflation 61% p.a. (1990–2000)

$9.6bn — 258.70 Sudanese dinars

SCORE CARD

❑ WORLD GNP RANKING	84th
❑ GNP PER CAPITA	$310
❑ BALANCE OF PAYMENTS	–$557m
❑ INFLATION	16%
❑ UNEMPLOYMENT	4%

STRENGTHS

Oil, gas, cotton, gum arabic, sesame, sugar. Some gold mining.

WEAKNESSES

Low industrialization. Lack of foreign exchange for importing energy and spare parts for industry. Drought. Little transportation infrastructure. Huge distances between towns. Civil war delayed exploitation of oil reserves. Alienation of Arab donors and investors.

EXPORTS

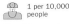

Taiwan 6% — South Korea 7% — Saudi Arabia 9% — China 38% — Japan 16% — Other 24%

IMPORTS

France 5% — Germany 7% — UK 7% — Saudi Arabia 10% — Other 59% — China 12%

RESOURCES ▷ Electric power 606,000 kw

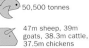

50,500 tonnes — 200,000 b/d (reserves 300m barrels)

47m sheep, 39m goats, 38.3m cattle, 37.5m chickens — Oil, gas, gold, copper, gypsum, marble, mica, silver, chromium, zinc

Large oil and gas reserves were found in the south in the 1980s; oil exports started in 1999. The half-thermal, half-hydroelectric generating capacity is insufficient, and weeklong power cuts are frequent. Gold mining has the potential for expansion.

ENVIRONMENT ▷ Sustainability rank: 102nd

 4% (0.3% partially protected) — 0.1 tonnes per capita

The Jonglei canal project, halted by rebel attacks when only 70% complete, could still devastate the Sudd, the world's largest swamp and a rich wetland habitat, fed by the White Nile.

MEDIA ▷ TV ownership high

Daily newspaper circulation 27 per 1000 people

PUBLISHING AND BROADCAST MEDIA

There are 7 daily newspapers, including *Al-Anbaa*, *Ar-Rai al-Amm*, and *Al-Nasr*

1 state-controlled service — 2 services: 1 state-controlled, 1 rebel-controlled

The media are controlled either by the government or the army, and are heavily censored.

CRIME ▷ Death penalty in use

32,000 prisoners — Down 7% 1992–1994

Antigovernment dissent is often suppressed by violence, and torture by the security forces is widespread. The UN has condemned Sudan's poor human rights record.

EDUCATION ▷ School leaving age: 14

 58% — 200,538 students

In 1991, measures were introduced to Islamize education. Primary school children must have two years of Islamic religious instruction, and men wishing to enter university must first serve for a year in the People's Militia.

HEALTH ▷ Welfare state health benefits

1 per 10,000 people — Infectious and parasitic diseases, malnutrition

Health service standards in rural areas are basic. The civil war has led to an increase in communicable diseases. The parasitic infection leishmaniasis is prevalent.

CHRONOLOGY

Northern Sudan was taken by Egypt in 1821, the south by Britain in 1877.

❑ **1882** British invade Egypt.
❑ **1883** Muslim revolt in Sudan led by Muhammad Ahmed, the Mahdi.
❑ **1898** Mahdists defeated. Anglo-Egyptian condominium set up.
❑ **1954** Becomes self-governing.
❑ **1955** Rebellion in south starts 17 years of civil war.
❑ **1956** Independence as republic.
❑ **1958–1964** Military rule.
❑ **1965** Civilian revolution, elections.
❑ **1969** Coup led by Col. Jaafar Nimeiri.
❑ **1972** South gets limited autonomy.
❑ **1973** Socialist Union sole party.
❑ **1983** Southern rebellion resumes. *Sharia* (Islamic law) imposed.
❑ **1984** Devastating drought.
❑ **1986** Army coup.
❑ **1989** Gen. Bashir takes over.
❑ **1991** *Sharia* penal code instituted. Pro-Iraq stance in Gulf War.
❑ **2000** Bashir ousts fundamentalist al-Turabi from leadership of NC.
❑ **2002** SPLA signs cease-fire. Fighting between rival factions over southern oil reserves escalates.

SPENDING ▷ GDP/cap. increase

CONSUMPTION AND SPENDING

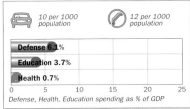

10 per 1000 population — 12 per 1000 population

Defense 6.1%
Education 3.7%
Health 0.7%

0 5 10 15 20 25
Defense, Health, Education spending as % of GDP

There are large disparities between rich and poor. Wealth is limited to the NC and southern rebel elites. Most of the population struggles to survive.

WORLD RANKING

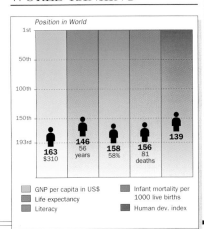

Position in World

1st
50th
100th
150th
193rd

163 $310
146 56 years
158 58%
156 81 deaths
139

GNP per capita in US$ — Infant mortality per 1000 live births
Life expectancy — Human dev. index
Literacy

S

SURINAME

OFFICIAL NAME: Republic of Suriname   **CAPITAL:** Paramaribo
POPULATION: 419,000   **CURRENCY:** Suriname guilder or florin   **OFFICIAL LANGUAGE:** Dutch

SOUTH AMERICA

 1975   1975   Nov 25   SME   -3   +597   .sr

BOUNDED EAST AND WEST by rivers, Suriname sits on the north coast of South America in the center of the "Guyana Plateau." The interior is rainforested highlands; most people live near the coast. In 1975, after almost 300 years of Dutch rule, Suriname became independent. The Netherlands is still its main aid supplier, and is home to one-third of Surinamese. Multiparty democracy was restored in 1991, after almost 11 years of military rule.

Congested street in Paramaribo. It boasts 18th- and 19th-century Dutch architecture. The large mosque is next to a Jewish synagogue.

CLIMATE   ▷ Tropical equatorial

WEATHER CHART FOR PARAMARIBO

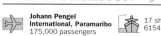

Suriname's tropical climate is cooled by the trade winds. Annual rainfall varies from 150 to 300 cm (60 to 120 in) between coast and interior.

TRANSPORTATION   ▷ Drive on left

 Johann Pengel International, Paramaribo
175,000 passengers

 17 ships
6154 grt

THE TRANSPORTATION NETWORK

1178 km (732 miles)		None
157 km (98 miles)		1200 km (746 miles)

Rivers provide the main north–south links, and the vast interior relies on water or air transportation. The road network runs east–west and focuses on the coast and its immediate hinterland.

TOURISM   ▷ Visitors : Population 1:7.4

57,000 visitors   Up 4% in 1999

MAIN TOURIST ARRIVALS

Netherlands 83%	
Guyana 6%	
China 3%	
Other 8%	

0 10 20 30 40 50 60 70 80 90 100
% of total arrivals

Tourism is undeveloped. Travelers outside Paramaribo are advised to carry their own hammock and food.

PEOPLE   ▷ Pop. density low

Sranan, Dutch, Javanese, Sarnami Hindi, Saramaccan, Chinese, Carib

3/km² (7/mi²)

THE URBAN/RURAL POPULATION SPLIT

74%   **26%**

ETHNIC MAKEUP

Other 5%
Black 9%
South Asian 34%
Javanese 18%
Creole 34%

About 250,000 Surinamese have emigrated since 1975. Of those who remain, 90% live near the coast, while the rest live in scattered rainforest communities. Tension between the Creole-dominated government, *bosnegers* (the descendants of runaway slaves), and Amerindians spilled over into armed rebellion in the 1980s. Many South Asians and Javanese work in farming. Christianity, Hinduism, and Islam are the dominant religions.

POLITICS   ▷ Multiparty elections

2000/2005   President Ronald Venetiaan

AT THE LAST ELECTION
National Assembly 51 seats   **19% MC**   **10% Others**

65% NF   **6% DNP 2000**

NF = New Front for Democracy and Development (includes the Suriname National Party, the Progressive Reform Party, the Suriname Labor Party, and Pertjajah Luhur)
MC = Millennium Combination (includes the National Democratic Party (**NDP**), the Democratic Alternative, and the Party for Unity and Harmony)
DNP 2000 = Democratic National Platform 2000

A coalition government representing Creoles, South Asians, and Javanese took power under Ronald Venetiaan in 1991. Five years later it was defeated by the NDP, controlled by Desi Bouterse, the military dictator from 1980 to 1988 and the man behind the 1990 coup which ended Suriname's first attempt to return to democracy. Between 1996 and 2000 President Jules Wijdenbosch of the NDP withstood the efforts of opponents in the National Assembly to replace him. In the 2000 legislative elections, however, the NDP was massively defeated by the opposition NF. The new Assembly went on to elect NF leader Venetiaan as president.

SURINAME

Total Land Area : 163 270 sq. km
(63 039 sq. miles)

LAND HEIGHT

1000m/3281ft
500m/1640ft
200m/1640ft
Sea Level

POPULATION

over
over
under

WORLD AFFAIRS ▷ Joined UN in 1975

 ACS AmCC Caricom OAS ▷ OIC

Relations with the Netherlands and the US, Suriname's key aid and trading partners, have been weakened over

charges of official connivance in narcotics trafficking. Although ties with Suriname's immediate neighbors are damaged by border disputes, greater regional integration, particularly with the Caribbean, is a priority.

AID ▷ Recipient

 $34m (receipts)   Down 6% in 2000

The Netherlands is the largest donor, but it has on occasion suspended aid amid deteriorating relations. The IDB and European Investment Bank have granted loans for agricultural and industrial development.

DEFENSE ▷ No compulsory military service

 $11m   No change in 2000

The army was politically dominant in the 1980s under Lt. Col. Desi Bouterse. A six-year war with *bosneger* rebels ended in 1992. Aid and training have been provided in recent years by both the US and China.

ECONOMICS ▷ Inflation 88% p.a. (1990–2000)

 $788m   981–2179 Suriname guilders

SCORE CARD

❑ World GNP Ranking	159th
❑ GNP per Capita	$1890
❑ Balance of Payments	$32m
❑ Inflation	64.3%
❑ Unemployment	11%

Strengths
Bauxite. Gold. Timber potential. Oil. Agricultural exports: rice, bananas, citrus fruits. Shrimp exports.

Weaknesses
Overdependence on declining bauxite reserves and decreased foreign aid. Weak currency. Severe shortage of foreign exchange. Banana industry damaged by ending of preferential EU access. Net food importer.

EXPORTS

- Canada 7%
- France 7%
- Other 33%
- Netherlands 11%
- Norway 18%
- USA 24%

IMPORTS

- Japan 7%
- Netherlands Antilles 8%
- USA 31%
- Trinidad & Tobago 12%
- Other 25%
- Netherlands 17%

RESOURCES ▷ Electric power 425,000 kw

 13,066 tonnes   4972 b/d (reserves 87m barrels)

 135,000 cattle, 66,000 ducks, 2.9m chickens   Bauxite, iron, gold, manganese, copper, nickel, platinum, oil

Suriname is a major exporter of aluminum and bauxite, but the minerals sector is affected by poor world prices, as is raw gold production. Oil consumption is almost double the level of oil production. Exploitation of the rainforests has begun. Rice and fruit are Suriname's key agricultural products.

ENVIRONMENT ▷ Not available

 10% partially protected   5.2 tonnes per capita

In 1998 the government declared some 16,000 sq. km (6150 sq. miles) of rainforest – almost 10% of the country – to be a natural reserve barred to logging but its exploitation for economic gain is still of real concern to environmentalists.

MEDIA ▷ TV ownership medium

 Daily newspaper circulation 122 per 1000 people

PUBLISHING AND BROADCAST MEDIA

 There are 2 daily newspapers, *De Ware Tijd* and *De West*

 2 state-owned services   10 services: 1 state-owned, 9 independent

There are radio broadcasts in a number of languages. Dutch is used by the daily newspapers and for most TV programs.

CRIME ▷ Death penalty not used in practice

 1933 prisoners   Relatively high crime levels

Human rights abuses associated with the former military regime have largely ended. Rival armed factions remain in some regions in the interior. Narcotics trafficking and money laundering are a problem, as is urban street crime.

EDUCATION ▷ School leaving age: 12

  94%   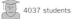 4037 students

Education is free and includes adult literacy programs. There is a long tradition of higher education, but most graduates now live in the Netherlands.

HEALTH ▷ Welfare state health benefits

 1 per 4000 people   Heart attacks, cancers, malaria, malnutrition, tuberculosis

Urban medical facilities in Suriname are relatively good; Paramaribo has several hospitals. However, provision in the interior is basic.

SPENDING ▷ GDP/cap. increase

CONSUMPTION AND SPENDING

- 144 per 1000 population
- 174 per 1000 population
- Defense 2.7%
- Education 3.5%
- Health 2%

Defense, Health, Education spending as % of GDP

Living standards have fallen since 1982, due to the effects of civil war and to aid and loan suspension. Urban Creoles dominate the rich elite. Amerindians and *bosnegers* are the poorest groups.

WORLD RANKING

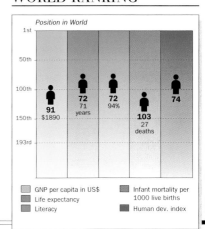

- *Position in World*
- 91 $1890
- 72 71 years
- 72 94%
- 103 27 deaths
- 74

Legend:
- GNP per capita in US$
- Life expectancy
- Literacy
- Infant mortality per 1000 live births
- Human dev. index

S

SWAZILAND

OFFICIAL NAME: Kingdom of Swaziland   **CAPITAL:** Mbabane
POPULATION: 938,000   **CURRENCY:** Lilangeni   **OFFICIAL LANGUAGES:** English and siSwati

THE TINY SOUTHERN African kingdom of Swaziland, bordered on three sides by South Africa and to the east by Mozambique, comprises mainly upland plateaus and mountains. Governed by a strong hereditary monarchy, Swaziland is a country in which tradition is being challenged by demands for modern multiparty government. King Mswati III, crowned in 1986, has overhauled the electoral process, but has still to legalize party politics.

CLIMATE   ▷ Subtropical

WEATHER CHART FOR MBABANE

Swaziland is temperate. Temperatures rise and rainfall declines as the land descends eastward, from high to low *veld*. The Low Veld is prone to drought.

TRANSPORTATION   ▷ Drive on left

Matsapha, Manzini 93,000 passengers   Has no fleet

THE TRANSPORTATION NETWORK

814 km (506 miles)	None
297 km (185 miles)	None

A sharp rise in road traffic has necessitated road improvements. The railroad, running to Mozambique and South Africa, mainly carries freight.

TOURISM   ▷ Visitors : Population 1:2.9

319,000 visitors   Down 1% in 1998

MAIN TOURIST ARRIVALS

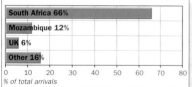

South Africa 66%	
Mozambique 12%	
UK 6%	
Other 16%	

% of total arrivals

Swaziland's attractions are its game reserves, mountain scenery, and, for the South Africans who make up two-thirds of tourists, its casinos.

The outskirts of Mbabane. *It lies on the High Veld, where traditional cattle farming has become more difficult owing to overgrazing.*

PEOPLE   ▷ Pop. density medium

English, siSwati, Zulu, Tsonga   55/km² (141/mi²)

THE URBAN/RURAL POPULATION SPLIT

26%   74%

RELIGIOUS PERSUASION

Traditional beliefs 40%
Christian 60%

Over 95% of the population belong to the Swazi ethnic group, making Swaziland one of Africa's most homogeneous states. It is also very conservative, but is now facing pressure from urban-based modernizers. The powerful monarchy dominates politics. Ancient traditions, such as *incwala*, the rainy season's annual movable feast, remain popular. Society is patriarchal and focused around the clan. Chiefs own much "national land," and wield authority through local consultations, called *tindkhundla*. Polygamy is tolerated. Women farm and may vote, but lack economic or political power. The exception is the Queen Mother, the "Great She Elephant," whose power as regent was clear during the mid-1980s.

POLITICS   ▷ No multiparty elections

L. House 1998/2003
U. House 1998/2003
H.M. King Mswati III

AT THE LAST ELECTION

House of Assembly 65 seats

There are no political parties. 10 members of the House of Assembly are appointed by the king

Senate 30 seats

20 members of the Senate are appointed by the king and 10 elected by the House of Assembly

Politics is dominated by a strong executive monarchy and rivalries within the ruling Dlamini clan. Royal advisers complement a nominated cabinet. Direct elections were held in 1993, but parties remain banned. Responding to mounting popular unrest, in 1996 the king appointed Sibusiso Dlamini as prime minister and set up a commission to review the political system. Mass antigovernment protests were staged by prodemocracy activists in 2000.

SWAZILAND

Total Land Area : 17 363 sq. km (6704 sq. miles)

POPULATION
○ over 50 000
● over 10 000
· under 10 000

LAND HEIGHT
1000m/3281ft
500m/1640ft
200m/656ft
Sea Level

WORLD AFFAIRS ▷ Joined UN in 1968

Swaziland's membership of the SACU reinforces its traditional dependence on its giant neighbor, South Africa. Having welcomed the election of an ANC-led government there, King Mswati has expressed concern over its support for Swazi prodemocracy campaigners. Peace in Mozambique has meant the return there of 134,000 refugees.

AID ▷ Recipient

 $13m (receipts)   Down 55% in 2000

Aid helps the balance of payments, and funds the development of the Matsapha industrial estate, roads, and social projects. Donors include Germany, Japan, the UK, and the ADB. EU aid mainly targets "microprojects," such as schools, and supports constitutional reform.

DEFENSE ▷ No compulsory military service

 $23m   Up 15% in 2000

The Swaziland Defense Force numbers just 3000 troops. Although it does not play an overt political role, its loyalty is to the monarch and the status quo.

ECONOMICS ▷ Inflation 13% p.a. (1990–2000)

 $1.45bn   7.570–11.995 emalangeni

SCORE CARD

❏ WORLD GNP RANKING	146th
❏ GNP PER CAPITA	$1390
❏ BALANCE OF PAYMENTS	–$40m
❏ INFLATION	16.7%
❏ UNEMPLOYMENT	22%

STRENGTHS
Economy quite diversified and buoyant. Manufacturing 32% of GDP. Investment rules attractive. Sugar 33% of export earnings. Wood pulp. Debt service low: only 3.8% of export earnings in 1993. Renewed regional stability has reduced risk to exports.

WEAKNESSES
Sugar is vulnerable to world price fluctuations. Dependence on South Africa for jobs, revenue, investment, electricity, and imported goods. Small plots of land and lack of land title hinder farm modernization. High population growth.

EXPORTS

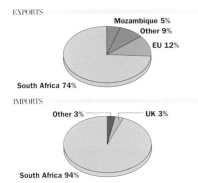

- Mozambique 5%
- Other 9%
- EU 12%
- South Africa 74%

IMPORTS

- Other 3%
- UK 3%
- South Africa 94%

RESOURCES ▷ Electric power: Included in South African total

 131 tonnes   Not an oil producer

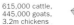 615,000 cattle, 445,000 goats, 3.2m chickens   Coal, diamonds, gold, asbestos, cassiterite, iron, tin

Swaziland's main export is sugarcane, followed by wood pulp, coal, and asbestos. The development of hydroelectric power plants has cut energy imports from South Africa.

ENVIRONMENT ▷ Not available

 2% (3% partially protected)   0.4 tonnes per capita

In 1998 Swaziland, Mozambique, and South Africa began an ecological project on the world's largest wetlands – the foothills of the Lebombo Mountains.

MEDIA ▷ TV ownership medium

 Daily newspaper circulation 27 per 1000 people

PUBLISHING AND BROADCAST MEDIA

 There is currently only one daily newspaper, *The Times of Swaziland*

 1 state-owned service   3 services: 1 state-owned, 2 independent

The king, responding to pressure, rapidly reversed his 2001 decree which had ensured the continued closure of the independent *Swaziland Observer*.

CRIME ▷ Death penalty in use

 2221 prisoners   Down 1% in 1999

The crime rate is low. The numbers of illegal weapons brought in by refugees have boosted armed crime.

EDUCATION ▷ School leaving age: 13

 80%   4880 students

Education is compulsory. Parents pay fees at all levels; even so, primary enrollment is about 93%. Drop-out rates at secondary level are high.

CHRONOLOGY

Swaziland became a British protectorate in 1903.

- ❏ **1968** Independence.
- ❏ **1973** King bans political activity, repeals constitution.
- ❏ **1978** New constitution confirms king's executive, legislative control.
- ❏ **1982** King Sobhuza dies. Queen Mother becomes regent for Prince Makhosetive. Power struggle between modernists and traditionalists in royal Dlamini clan.
- ❏ **1986** Makhosetive crowned King Mswati III at the age of 18.
- ❏ **1992** Limited electoral reforms; parties still banned.
- ❏ **1993** Elections under new system.
- ❏ **1996** Review of political system.
- ❏ **1998** Poor turnout at elections.
- ❏ **2000** Mass prodemocracy protests.

HEALTH ▷ No welfare state health benefits

 1 per 5000 people   Diarrheal and respiratory diseases, AIDS

Health facilities are rudimentary. About a third of the population aged 15 to 49 is estimated to be living with HIV/AIDS.

SPENDING ▷ GDP/cap. increase

CONSUMPTION AND SPENDING

34 per 1000 population   32 per 1000 population

Defense 1.6%	
Education 6.1%	
Health 2.5%	

Defense, Health, Education spending as % of GDP

About 50% of Swazis live below the UN poverty line. The royal Dlamini clan enjoys Western luxuries and travel.

WORLD RANKING

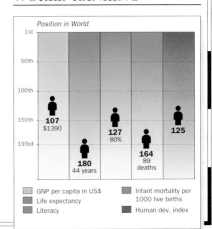

Position in World

- **107** $1390
- **180** 44 years
- **127** 80%
- **164** 89 deaths
- **125**

- GNP per capita in US$
- Life expectancy
- Literacy
- Infant mortality per 1000 live births
- Human dev. index

S

SWEDEN

EUROPE

OFFICIAL NAME: Kingdom of Sweden **CAPITAL:** Stockholm
POPULATION: 8.8 million **CURRENCY:** Swedish krona **OFFICIAL LANGUAGE:** Swedish

 1523   1905   June 6   S   +1   +46   .se

SITUATED ON THE SCANDINAVIAN peninsula with Norway to its west, Sweden is a densely forested country with numerous lakes. The north of Sweden falls within the Arctic Circle; much of the south is fertile and widely cultivated. Sweden has one of the most extensive welfare systems in the world, and is among the world's leading proponents of equal rights for women. Its economic strengths include high-tech industries, such as Ericsson, and car production, most notably Volvo and Saab. Unlike neighboring Norway, it is an EU member, having joined in January 1995.

CLIMATE
▷ Subarctic/continental

WEATHER CHART FOR STOCKHOLM

Sweden has a largely continental climate. The Baltic Sea often freezes in winter, making the east coast much colder than western regions. Summers are mild everywhere, with temperatures varying surprisingly little between northern and southern regions.

TRANSPORTATION
▷ Drive on right

 Arlanda, Stockholm
18.4m passengers

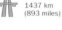 412 ships
2.86m grt

THE TRANSPORTATION NETWORK

163,453 km (101,565 miles)

1437 km (893 miles)

10,799 km (6710 miles)

2052 km (1275 miles)

Maintaining and improving transportation links are of prime concern in what is Europe's fifth-largest country. Swedish governments have traditionally spent large sums on infrastructure, as a way of boosting the economy as a whole.

A 16-km (10-mile) Øresund road and rail link by bridge and tunnel connecting Malmö with Copenhagen opened in 2000, providing Sweden with a new road and rail link with Denmark and the rest of Europe. A new rail link between Arlanda airport and Stockholm is also planned. By law, cars must travel with their headlights on at all times, and there are very strict regulations against drink-driving.

TOURISM
▷ Visitors : Population 1:3.2

2.75m visitors

Up 6% in 2000

MAIN TOURIST ARRIVALS

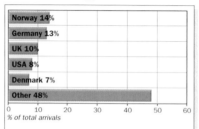

Norway 14%
Germany 13%
UK 10%
USA 8%
Denmark 7%
Other 48%

% of total arrivals

Sweden expanded rapidly as a tourist destination in the 1970s and 1980s. Stockholm is renowned for the beauty of its setting, its Old Town, and the *Vasa*, a magnificent 17th-century warship raised from the harbor bed in the 1960s.

Sweden has fewer lakes than Finland, and lacks Norway's dramatic scenery, but it has many natural attractions. The mountains of the "Midnight Sun" lie north of the Arctic Circle, while the southern coast has many white sandy beaches. The vast tracts of deserted landscape and the simple country communal living also attract visitors, but the cost of travel to Sweden means that Norwegians and Germans top the list of visitors.

A crofter's holding in Dalarna, central Sweden, an area which is still mainly forested. The timber and paper industries play a major role in Sweden's economy.

PEOPLE
▷ Pop. density low

Swedish, Finnish, Sami

21/km² (55/mi²)

THE URBAN/RURAL POPULATION SPLIT

83%   17%

RELIGIOUS PERSUASION

Russian Orthodox 1%   Other Protestant 1%
Muslim 1%   Roman Catholic 2%
Other 6%
Evangelical Lutheran 89%

ETHNIC MAKEUP

Finnish and Sami 3%   Other European 6%
Swedish 91%

As in all of Scandinavia, the nuclear family forms the basis of society. The birthrate is low with, on average, fewer than two children per family. Marriage is declining, and cohabitation outside marriage is common.

Swedish society has an egalitarian tradition. The role of the state is seen as providing conditions allowing each person to gain economic independence through employment. The welfare system is one of the most extensive in the world. However, in the early 1990s, recession reduced benefits; mothers in particular face problems with the closure of child-care facilities. Women make up nearly half the workforce. Over 45% of MPs are women, the highest percentage in the world. In 1999 the Swedish cabinet became the first in the world to have a majority of women ministers.

Sweden has generous asylum laws, and immigrants comprise about 10% of the population. A 17,000-strong minority of Sami live in northern Sweden. Their traditional way of life is protected.

The Evangelical Lutheran Church was disestablished in 2000.

POPULATION AGE BREAKDOWN

Female	Age	Male
3.1%	80+	1.7%
9.3%	60–79	7.9%
13%	40–59	13.3%
13.3%	20–39	13.9%
11.9%	0–19	12.6%

% of population by age group

S

POLITICS   ▷ Multiparty elections

2002/2006   King Carl XVI Gustaf

AT THE LAST ELECTION

Parliament (Riksdag) 349 seats

41% SAP	16% M	14% FP	9% Kd	9% VP	6% CP	5% MpG

SAP = Social Democratic Labor Party   **M** = Moderate Party
FP = Liberal Party   **Kd** = Christian Democratic Party
VP = Left Party   **CP** = Center Party   **MpG** = Green Party

Sweden is a constitutional monarchy with an elected parliament under the leadership of the prime minister.

PROFILE

Politics has traditionally been split between the SAP and trade unions on the left, and a host of moderate center and right-wing parties. Since the 1930s, the SAP has governed every term, except in 1976–1982 and 1991–1994. A shift to the right in 1991 was reversed in the 1994 elections. Ingvar Carlsson, SAP leader, formed a minority government but resigned in 1996 and was replaced by Göran Persson. The SAP lost ground in elections in 1998, increasing its independence on Left Party and Green support. A further shift in the 2002 elections left the Greens holding the balance of power.

MAIN POLITICAL ISSUES
EU membership

Sweden joined the EU in 1995 with Austria and Finland, but, like the UK and Denmark, has opted out of joining the eurozone. It supports EU membership for central European states.

SWEDEN

Total Land Area : 449 964 sq. km
(173 731 sq. miles)

POPULATION

- ▣ over 1 000 000
- ◉ over 100 000
- ○ over 50 000
- ● over 10 000

LAND HEIGHT

- 1000m/3281ft
- 500m/1640ft
- 200m/656ft
- Sea Level

The high cost of the welfare state

The cost of the welfare system contributed to enormous budget deficits in the late 1980s and early 1990s. While this has been brought under control, social security pressures remain from the relatively high level of unemployment and the growing number of pensioners.

Carl XVI Gustaf,
*ascended the throne
in 1973. His role is
purely ceremonial.*

Göran Persson
*of the SAP became
prime minister
in 1996.*

WORLD AFFAIRS   ▷ Joined UN in 1946

 EU   CE   NC   OECD   OSCE

Sweden's main recent foreign policy concern has been adjustment to EU membership, which it achieved in 1995. In 1998 parliament voted to join the Schengen passport-free zone linking nine EU states. In the 1980s, Sweden was a vociferous critic of the antagonistic policy used by the US toward the USSR. Since the collapse of the Soviet Union, and more recently the September 11, 2001 attack on the US, Sweden has altered its traditionally neutral stance. Sweden has WEU observer status, and participates in NATO exercises and several UN peacekeeping operations.

AID   ▷ Donor

 $1.8bn (donations)   Up 10% in 2000

Sweden is one of the few countries to exceed the UN target of 0.7% of GNP in development aid and has declared its intention to increase its allocation.

CHRONOLOGY

Sweden's history has been closely linked to the control of the Baltic Sea and its highly profitable trade routes. Under the house of Vasa, Sweden became a major power, controlling much of the Baltic region. By the 18th century, however, Sweden's position had been eroded by its regional rivals, particularly Russia.

- ❑ **1814–1815** Congress of Vienna. Sweden cedes territory to Russia and Denmark. Period of prolonged peace begins.
- ❑ **1865–1866** Riksdag (parliament) reformed into a bicameral structure.
- ❑ **1905** Norway gains independence from Sweden.
- ❑ **1911** First Liberal government comes to power.
- ❑ **1914** Government resigns over defense policy. ⇨

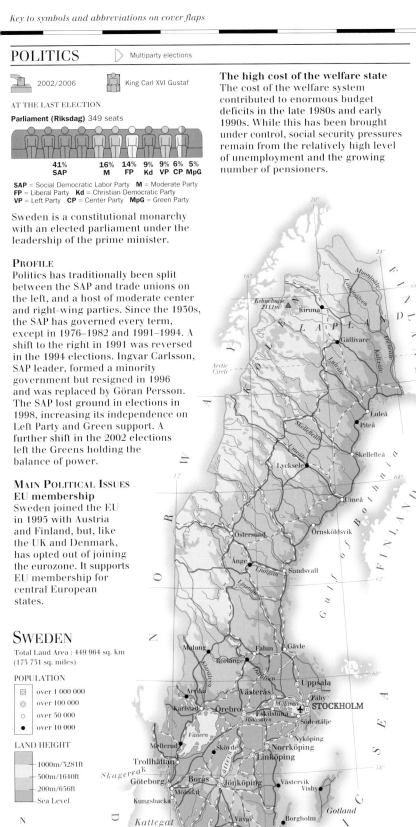

S

CHRONOLOGY *continued*

- ❏ **1914–1917** Sweden remains neutral during WWI but supplies Germany. Allied blockade.
- ❏ **1917** Food shortages. Conservative government falls. Nils Edén forms a Liberal government: limits exports contributing to German war effort.
- ❏ **1919** Universal adult suffrage.
- ❏ **1921** Finland gains Åland Islands as retribution for Sweden's war role.
- ❏ **1932** Severe recession. Social Democrat government under Per Albin Hansson elected.
- ❏ **1939–1945** Sweden neutral. Grants transit rights to German forces.
- ❏ **1945–1976** Continuing Social Democratic rule under Tage Erlander establishes Sweden as world's most advanced welfare state, and one of the most affluent.
- ❏ **1950** Gustav VI Adolf becomes king.
- ❏ **1953** Joins Nordic Council.
- ❏ **1959** Founder member of EFTA.
- ❏ **1969** Erlander succeeded by Olof Palme as prime minister.
- ❏ **1973** Carl XVI Gustaf on throne.
- ❏ **1975** Major constitutional reform. Riksdag (parliament) becomes unicameral with a three-year term. Role of monarchy reduced to ceremonial functions.
- ❏ **1976** SAP loses power. Nonsocialist coalition led by Thorbjörn Fälldin in government.
- ❏ **1978** Fälldin resigns over issue of nuclear power. Ola Ullsten prime minister.
- ❏ **1979** Fälldin prime minister again.
- ❏ **1982** Elections. SAP forms minority government. Palme returns as prime minister.
- ❏ **1986** Palme shot dead. His deputy, Ingvar Carlsson, succeeds him. Police fail to find killer.
- ❏ **1990** Carlsson introduces moderate austerity package, cuts government spending, raises indirect taxes.
- ❏ **1991** Sweden applies to join EU. SAP wins election but is unable to form government; Carlsson resigns. Carl Bildt, leader of MS, forms coalition of nonsocialist parties in middle of serious recession.
- ❏ **1992** Austerity measures succeed in reducing inflation but SAP refuses to support further spending cuts.
- ❏ **1994** EU membership terms agreed. Elections return SAP to power. Referendum favors joining EU.
- ❏ **1995** Joins EU.
- ❏ **1996** Carlsson resigns; replaced by Göran Persson.
- ❏ **1998** Persson remains in office, despite SAP losses in elections; dependent on Left and Greens for parliamentary majority.
- ❏ **2001** Defense reform program.
- ❏ **2002** Elections: SAP still largest party but lacking majority.

DEFENSE

 ▷ Compulsory military service

 $5.19bn   ⬇ Down 1% in 2000

SWEDISH ARMED FORCES

🛡	368 main battle tanks (70 *Centurion*, 40 Strv-103B, 160 Strv-121, 98 Strv-122)	19,100 personnel
⚓	7 submarines, 45 patrol boats	7100 personnel
✈	206 combat aircraft (20 SAAB AJSH-37, 95 SAAB JAS-9, 91 SAAB JA-37)	7700 personnel
🚀	None	

Sweden's sophisticated and powerful military force is supplied with weaponry manufactured by its advanced home defense industry, including Saab fighter jets and Bofors antiaircraft guns. However, with the ending of the Cold War, strategic priorities have changed. Sweden feels less bound to maintain its neutral stance; it has participated in NATO's Partnerships for Peace program since 1994 and has WEU observer status. In 1999 spending cuts were announced, foreshadowing halving the size of the armed forces, because of the reduced military threat in the Scandinavian and Baltic region. A ten-year reform program began in 2001, offering personnel regular office hours and canceling large-scale exercises.

ECONOMICS

▷ Inflation 2.1% p.a. (1990–2000)

📊 $241bn   💵 9.436–10.490 Swedish kronor

SCORE CARD

- ❏ WORLD GNP RANKING 21st
- ❏ GNP PER CAPITA $27,140
- ❏ BALANCE OF PAYMENTS....................$6.62bn
- ❏ INFLATION 1%
- ❏ UNEMPLOYMENT 5%

EXPORTS

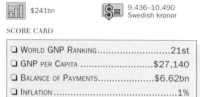

Denmark 6% — Norway 8% — UK 9% — USA 10% — Germany 11% — Other 56%

IMPORTS

Denmark 7% — Netherlands 7% — Norway 9% — UK 9% — Germany 17% — Other 51%

STRENGTHS

Companies of global importance, including Saab, Volvo, Electrolux and SKF, the world's biggest roller bearing manufacturer. Highly developed and constantly updated infrastructure. Sophisticated technology. Skilled labor force virtually bilingual in English.

WEAKNESSES

Labor costs remain uncompetitive. Highest taxation of OECD states, accounting for over 60% of GDP. Peripheral location, raising costs for producers and exporters. Major losses by Ericsson, giant telecoms company.

PROFILE

The state plays a significant role in the economy, particularly the services sector and infrastructure. Sweden's industrial giants have mostly been private-sector companies. The early 1990s saw a shift in economic policy to favor business, but greater growth did not follow, and unemployment and welfare costs drove up the budget deficit in 1994 to one of the OECD's highest. Growth has now resumed, the deficit has been cut back, and unemployment halved, but inflation has started to rise. The world downturn in 2001 brought the krona under pressure. Sweden opted not to introduce the euro from 1999, but the ruling SAP adopted a pro-euro stance in March 2000; the issue is to be decided by national referendum.

ECONOMIC PERFORMANCE INDICATOR

— Consumer Price Index   ▨ GDP

Consumer price index 1995=100 / *GDP 1996=100*

(1996, 1997, 1998, 1999, 2000)

SWEDEN : MAJOR BUSINESSES

- 🚗 Vehicle manufacture
- ☎ Telecommunications
- ⚡ Electrometallurgy
- ⛏ Iron ore mining
- 🔌 Electronics
- 📄 Pulp & paper
- ⚙ Engineering
- 🧪 Chemicals
- 🧵 Textiles

0   200 km
0   200 miles

Kiruna
Gällivare
Umeå
Gävle
Västerås
Stockholm
Göteborg
Norrköping
Linköping
Malmö

RESOURCES

 Electric power 33.9m kw

357,317 tonnes

Low levels of oil production

1.89m pigs,
1.65m cattle,
7.41m chickens

Iron, uranium, copper, lead, zinc, silver

ELECTRICITY GENERATION

Hydro 47% (74bn kwh)

Combustion 6% (10bn kwh)

Nuclear 47% (74bn kwh)

Other 0%

| 0 | 20 | 40 | 60 | 80 | 100 |

% of total generation by type

Sweden is rich in minerals, including iron, copper, and silver. While mining and quarrying account for only 0.3% of GDP, they underpin other industrial sectors. In a referendum in 1980 Sweden decided, on environmental grounds, to abandon nuclear power – which in the 1990s accounted for about half of electricity generation – by the year 2010. However, problems in securing sufficient new energy supplies and in reducing consumption meant that, by the end of 2000, only one of the country's four nuclear reactors had been closed down.

SWEDEN : LAND USE

High mountain regions
Forest
Pasture
Cropland
Pigs
Barley

LAPLAND

| 0 | 200 km |
| 0 | 200 miles |

ENVIRONMENT

 Sustainability rank: 3rd

9% (5% partially protected)

5.5 tonnes per capita

ENVIRONMENTAL TREATIES

Yes   Yes   Yes

Yes   Yes   Yes

Since Sweden's pioneering Environment Protection Act in 1969, it has invested heavily in environmental protection measures. It blames acid rain damage to forests and lakes on airborne sulfur dioxide from factories in western Europe. Swedish nuclear reactors are said to be very safe, with filtered venting systems designed to retain 90% of all radioactivity released if there were a core meltdown. Nonetheless, all reactors are due to be shut down by 2010.

MEDIA

 TV ownership high

Daily newspaper circulation 420 per 1000 people

PUBLISHING AND BROADCAST MEDIA

There are over 100 dailies; the most influential are *Expressen*, *Dagens Nyheter*, *Aftonbladet*, and *Svenska Dagbladet*

6 services: 2 state-owned, 4 independent

3 services: 2 state-controlled, 1 private

Press freedom is strongly entrenched, although radical views are rarely expressed. The influence of the major daily newspapers is largely confined to Stockholm: the provinces have their own strong press.

Two-thirds of the population have cable or satellite TV.

CRIME

 No death penalty

5678 prisoners

Little change in 1999

CRIME RATES

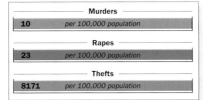

	Murders	
10	per 100,000 population	
	Rapes	
23	per 100,000 population	
	Thefts	
8171	per 100,000 population	

Crime rates are below the European average, although they are the highest among Scandinavian countries. Assault, rape, and theft are growing problems, especially in the cities.

EDUCATION

 School leaving age: 16

99%   275,217 students

THE EDUCATION SYSTEM

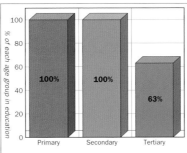

% of each age group in education

| Primary | Secondary | Tertiary |
| 100% | 100% | 63% |

Education spending (public and private) is among the OECD's highest as a percentage of GDP.

Coeducational comprehensive schools are the norm. The higher education system is freely available to most of the population, and many adults return to college to do further courses.

HEALTH

 Welfare state health benefits

1 per 323 people

 Heart and cerebro-vascular diseases, cancers, accidents

Sweden's health care system is comprehensive and of a universally high standard. Spending fell by an average of 2% in real terms in the 1990s, but the trend is now being reversed. Savings have been made by increasing outpatient care, reducing the number of hospital beds, and cutting jobs. Since 1994 individuals have had the right to choose their own doctor, while doctors and specialists can now set up private practices. In 1999 the government agreed to compensate more than 60,000 people subjected to enforced sterilization in 1935–1975.

SPENDING

 GDP/cap. increase

CONSUMPTION AND SPENDING

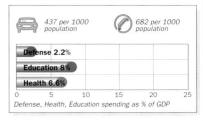

437 per 1000 population

682 per 1000 population

Defense 2.2%

Education 8%

Health 6.6%

| 0 | 5 | 10 | 15 | 20 | 25 |

Defense, Health, Education spending as % of GDP

Sweden has small income differentials, and Swedish executives are paid less than some of their European counterparts. Compared with other European states or the US, social competition and a sense of hierarchy are limited. Despite some cuts in services, the welfare system still rates highly in Europe.

Swedes are keen overseas property buyers, particularly of villas in Italy and France. Net overseas per capita investment remains among the highest in the world and about two-thirds of households own shares. PC ownership is high, and over 50% of Swedes had Internet access by 2001.

S

WORLD RANKING

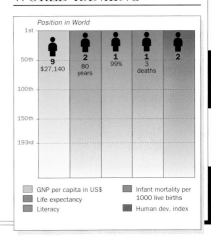

Position in World

9	2	1	3	2
$27,140	80 years	99%	deaths	

GNP per capita in US$
Life expectancy
Literacy

Infant mortality per 1000 live births
Human dev. index

SWITZERLAND

OFFICIAL NAME: Swiss Confederation **CAPITAL:** Bern
POPULATION: 7.2 million **CURRENCY:** Swiss franc **OFFICIAL LANGUAGES:** French, German, and Italian

SWITZERLAND LIES at the center of western Europe geographically, but outside it politically. Sometimes called Europe's water tower, it is the source of all four of the region's major river systems: the Po, the Rhône, the Rhine, and the Inn–Danube. Switzerland has built one of the world's most prosperous economies, aided by the fact that it has retained its neutral status through every major European conflict since 1815. The process of European integration has been the latest and strongest challenge to Swiss neutralism, but it remains outside the EU.

***The Eiger in the Berner Oberland.** In 1994, a referendum voted to ban all truck transit traffic from the Swiss Alps from 2004.*

CLIMATE
Mountain/continental

Temperature and weather fluctuate enormously, not only seasonally but because of the huge variations in altitude and the country's location in central Europe. On the plateau north of the Alps, where most of the population lives, summers are warm and winters dry, cool, and often foggy. South of the Alps, it is warmer and sunnier. Strong southerly winds, or *föhn*, can bring summerlike weather even in winter. Avalanches have been a problem in recent years.

TRANSPORTATION
Drive on right

Kloten, Zürich 22.6m passengers | 16 ships 368,000 grt

THE TRANSPORTATION NETWORK

| 71,059 km (44,154 miles) | 1270 km (789 miles) |
| 3143 km (1953 miles) | 1208 km (751 miles) |

Switzerland is a major freight transit route. Pollution and safety are major concerns, especially after 11 died in a fire in 2001 in the Gotthard road tunnel. The NEAT project, begun in 1996, will provide a high-speed rail link between Basel and Milan, whose trains will carry trucks. Swissair, bankrupt in 2001, has been replaced by a new airline, "swiss."

TOURISM
Visitors : Population 1.5:1

11m visitors | Up 3% in 2000

MAIN TOURIST ARRIVALS

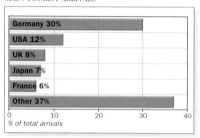

| Germany 30% |
| USA 12% |
| UK 8% |
| Japan 7% |
| France 6% |
| Other 37% |

% of total arrivals

Tourism is Switzerland's third-largest industry. About 350,000 Swiss earn their living from it, and tourism accounts for around 3% of GDP. The Alps are the main attraction, drawing winter and summer tourists from around the world, and Chillon Castle on Lake Geneva continues to be the country's most popular tourist site. In recent years, warmer winters have resulted in a shorter skiing season. The strong Swiss franc makes Switzerland an expensive destination, and Austria offers tough competition.

SWITZERLAND

Total Land Area : 41 290 sq. km. (15 942 sq. miles)

POPULATION
over 100 000
over 50 000
over 10 000

LAND HEIGHT
3000m/9843ft
2000m/6562ft
1000m/3281ft
500m/1640ft
200m/656ft

PEOPLE ▷ Pop. density medium

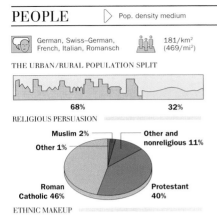

German, Swiss–German, French, Italian, Romansch

181/km² (469/mi²)

THE URBAN/RURAL POPULATION SPLIT

68%   32%

RELIGIOUS PERSUASION

Muslim 2%
Other 1%
Other and nonreligious 11%
Roman Catholic 46%
Protestant 40%

ETHNIC MAKEUP

Romansch 1%
Other 6%
Italian 10%
French 18%
German 65%

Switzerland is composed of distinct German-Swiss, French-Swiss, and Italian-Swiss linguistic groups. About 40,000 in the eastern canton of Grisons speak Romansch. The German-Swiss, in the majority, are a tightly knit community, with a dialect that is impenetrable to most outsiders. In recent years, the three groups have grown further apart. The French-Swiss, in favor of joining the EU, are opposed by the German-Swiss. In Italian-speaking Ticino, a political party has emerged to champion Italian-Swiss interests. Tension between Swiss and immigrant workers gave rise to vocal support for strict limits on the number of foreigners, but in 2000 this was heavily defeated as a referendum proposal.

Society retains strong conservative elements. Two half-cantons granted women the vote at regional level only in 1989 and 1990. Marriage rates are high and divorce is less common than in most other European states.

POPULATION AGE BREAKDOWN

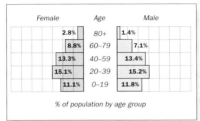

Female	Age	Male
2.8%	80+	1.4%
8.8%	60–79	7.1%
13.3%	40–59	13.4%
15.1%	20–39	15.2%
11.1%	0–19	11.8%

% of population by age group

POLITICS ▷ Multiparty elections

L. House 1999/2003
U. House 1999/2003

President Kaspar Villiger

AT THE LAST ELECTION

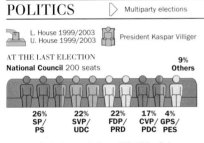

National Council 200 seats

9% Others

| 26% SP/PS | 22% SVP/UDC | 22% FDP/PRD | 17% CVP/PDC | 4% GPS/PES |

SP/PS = Social Democratic Party   SVP/UDC = Swiss People's Party   FDP/PRD = Radical Democratic Party   CVP/PDC = Christian Democratic People's Party   GPS/PES = Green Party of Switzerland

Council of States 46 seats

| 39% FDP/PRD | 33% CVP/PDC | 15% SVP/UDC | 13% SP/PS |

Switzerland is a federal democratic republic with 26 autonomous cantons. The presidency rotates every year.

PROFILE

The same four-party coalition has been in power in Switzerland since 1959. Domestic and foreign policies have changed little. Politics has recently become more contentious, however, with voting patterns becoming more polarized. Divisive issues are those of narcotics and of membership of the EU. Both right-wing and green minority parties have recently gained more seats in parliament. The right-wing SVP/UDC in particular capitalized on growing hostility to immigration to perform strongly in the 1999 elections.

Switzerland's political system is unique in Europe, in that taking important decisions depends on the results of referendums. A petition of more than 100,000 signatures can force a referendum on any issue.

MAIN POLITICAL ISSUES
European integration

Almost all politicians and business leaders favor joining the EU, or at least the European Economic Area (EEA), but voters remain sharply divided. The Swiss fear that their decentralized style of government would be lost within the EU. There are also worries that in a barrier-free Europe, Switzerland's high standards of living would fall because of a large influx of immigrants.

Switzerland and World War II

In 2002 the Bergier report criticized Swiss refusal to admit many Jewish refugees during World War II, but found that the banking system was not built on the assets of Holocaust victims. In 1998 the two largest Swiss banks had agreed to pay $1.25 billion to 31,500 victims and their families in return for agreement that there would be no future claims against Swiss banks or the government. A government Fund for Needy Holocaust Victims was wound up in 2002 after paying $179 million to 309,000 survivors.

WORLD AFFAIRS ▷ Joined UN in 2002

 CE   G10   OECD   OSCE   PfP

The basis of Switzerland's foreign policy remains its neutrality. Geneva has retained its position as a center for many international organizations, including the ICRC and the European headquarters of the UN (although Switzerland only voted to join the UN in 2002). The city has often hosted diplomatic negotiations: those for the START nuclear reduction treaties and peace talks for the former Yugoslavia took place there.

Switzerland has so far not joined the process of closer European integration. In 1992 voters rejected EEA membership, widely seen as the first step toward EU membership. Many advocates of joining the EU believe that the economy will suffer without closer integration. Opponents argue that Switzerland's seeming isolation will enhance its role as an international tax haven. In 2001 a proposal to apply for EU membership was overwhelmingly rejected, although a series of bilateral cooperation agreements with the EU entered into force in 2002.

Ruth Dreifuss, *of the CVP, first woman president, held office in 1999.*

Kaspar Villiger, *of the FDP, who held the one-year presidency in 2002.*

CHRONOLOGY

The autonomy of the Swiss cantons was curtailed by the Habsburgs in the 11th century. In 1291, the three cantons of Unterwalden, Schwyz, and Uri set up the Perpetual League to pursue Swiss liberty. Joined by other cantons, they succeeded in 1499 in gaining virtual independence. The Habsburgs retained a titular role.

❏ **1648** Peace of Westphalia ending Thirty Years' War, in which Switzerland played no active part, recognizes full Swiss independence.
❏ **1798** Invaded by French.
❏ **1815** Congress of Vienna after Napoléon's defeat confirms Swiss independence and establishes its neutrality. Geneva and Valais join Swiss Confederation. ⇨

S

AID ▷ Donor

 $890m (donations) ⬇ Down 10% in 2000

With total disbursements amounting to 0.34% of GNP in 2000, Switzerland ranks above the (OECD) average of 0.22% as an aid donor, although the level of assistance remains below its target of 0.4% of GNP, set in 1994. Good governance and promoting investment are current priorities.

DEFENSE   Compulsory military service

$2.9bn   ⬇ Down 7% in 2000

SWISS ARMED FORCES

	556 main battle tanks (370 Pz-87 *Leopard* 2, 186 Pz-68/88)	3600 personnel (army and air force combined)
	None	
	138 combat aircraft (70 *Tiger* II/F-5E, 26 F/A-18)	
	None	

The army is, in one sense, among the largest in Europe. It is organized so that almost 400,000 conscripts can be called up and armed in a few hours; it still uses skis, bicycles, and horses to protect the Alps. Bridges and tunnels are mined in accordance with a defense strategy drafted in the early 1900s. Military service and further training at intervals are compulsory for males up to the age of 50.

As in the rest of Europe, force numbers are being cut in response to the end of the Cold War. In 1995, legislation allowing civilian service in place of military service was passed. Voters approved in 2001 a referendum proposal allowing Swiss soldiers to bear arms when on international peacekeeping operations.

ECONOMICS ▷ Inflation 1.3% p.a. (1990–2000)

$274bn   1.6205–1.6603 Swiss francs

SCORE CARD

- ❑ WORLD GNP RANKING..........................18th
- ❑ GNP PER CAPITA $38,140
- ❑ BALANCE OF PAYMENTS...................$32.5bn
- ❑ INFLATION 1.6%
- ❑ UNEMPLOYMENT 2%

EXPORTS

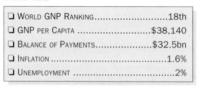

UK 6%
Italy 8%
France 9%
USA 13%
Germany 21%
Other 43%

IMPORTS

UK 6%
USA 8%
Italy 9%
France 10%
Germany 29%
Other 38%

STRENGTHS
Highly skilled workforce. Reliable service provider. Major machine tool and precision engineering industries. Powerful chemical, pharmaceutical, and banking multinationals. Banking secrecy laws attract foreign capital; banking sector contributes 9% of GNP. Ability to innovate to capture mass markets, typified by Swatch watch and Swatch-designed Smart car.

WEAKNESSES
Protected cartels result in many overpriced goods. Highly subsidized agricultural sector.

PROFILE
The economy is widely diversified, with 62% of GDP coming from services and 34% from industry. There are several large multinational enterprises and,

ECONOMIC PERFORMANCE INDICATOR

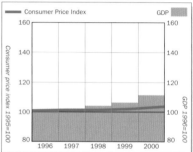

— Consumer Price Index   GDP

notably, a highly successful banking sector managing around one-third of the world's offshore private wealth. During most of the 1990s the economy grew only very slowly, if at all, but performance picked up in 2000 and 2001. In the long term, government finances face pressure to cater for the needs of an aging population.

SWITZERLAND : MAJOR BUSINESSES

Textiles
Chemicals
Electronics
Watch making
Pharmaceuticals
Consumer goods
Light engineering
Research & development
Banking & finance

0   50 km
0   50 miles

* significant multinational ownership

RESOURCES

▷ Electric power 16.7m kw

2975 tonnes

Not an oil producer; refines 132,000 b/d

1.61m cattle, 1.56m pigs, 6.66m chickens

Rock salt, marble, gypsum

ELECTRICITY GENERATION

Hydro 55% (35bn kwh)

Combustion 4% (2.4bn kwh)

Nuclear 41% (26bn kwh)

Other 0%

% of total generation by type

Switzerland is poor in natural resources, having no valuable minerals in commercially exploitable quantities. Over half of its electricity comes from hydropower, while five nuclear plants supply most of the rest, so that spending on imported oil and coal is kept to a minimum – they account for less than 4% of the total import bill. Large-scale antinuclear power demonstrations in the 1980s led to the cancellation of plans for a new nuclear power plant, but a referendum approved continued use of existing plants.

SWITZERLAND : LAND USE

Cropland
Forest
Pasture
High mountain regions
Cattle
Vineyards

ENVIRONMENT

▷ Sustainability rank: 5th

27%

5.9 tonnes per capita

ENVIRONMENTAL TREATIES

Yes   Yes   Yes

Yes   Yes   No

The Swiss are among the most environmentally conscious people in the world and are willing to back their convictions with money. The Basel–Milan link, designed to carry trucks by rail through a tunnel instead of having them cross the Alps by road, was approved by referendum, despite the estimated $13.3bn cost. The planners' aim is a total ban on carrying freight by road through Switzerland by 2004; this may not be necessary if big time savings make the tunnel sufficiently attractive. The Swiss are keen recyclers and taxation is used to encourage this.

MEDIA

▷ TV ownership high

Daily newspaper circulation 376 per 1000 people

PUBLISHING AND BROADCAST MEDIA

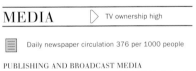

There are 84 daily newspapers. The largest circulations are held by *Tages Anzeiger* and the Zürich-based tabloid *Blick*

3 independent services broadcasting in German, Romansch, French, and Italian

3 independent services broadcasting in German, Romansch, French, and Italian

The Swiss media are organized broadly along regional lines, and reflect the country's linguistic divisions. The German-, Romansch-, French-, and Italian-language TV and radio stations tend to focus on the interests of their specific communities. German, Italian, and French satellite TV is widely available. Few newspapers are distributed throughout the country. *Tribune de Genève* and *Neue Zürcher Zeitung* are exceptions.

CRIME

▷ No death penalty

6390 prisoners

Up 30% in 1999

CRIME RATES

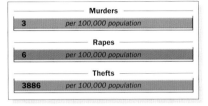

Murders
3   *per 100,000 population*

Rapes
6   *per 100,000 population*

Thefts
3886   *per 100,000 population*

Crime rates are low by international standards. Muggings and burglaries are on the increase and are often related to narcotics. More cases of banking secrecy laws attracting laundered funds are coming to light.

EDUCATION

▷ School leaving age: 16

99%

154,838 students

THE EDUCATION SYSTEM

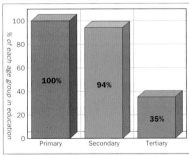

% of each age group in education

Primary **100%**
Secondary **94%**
Tertiary **35%**

Of total government expenditure, around 16% is spent on education. Primary and secondary education are controlled by the cantons, so that there are 26 different systems in operation. Most students after the age of 16 are encouraged to take up vocational studies. Training is thorough and is usually combined with three or four years' apprenticeship in the student's chosen field. The Federal Technological Institute in Zürich has gained an international reputation for its computer programming research.

HEALTH

▷ Welfare state health benefits

1 per 294 people

Heart and cerebro-vascular diseases, cancers, accidents

The health system is among the most efficient and pioneering in the world, and is ranked by WHO second, after Japan, for attainment. High health costs are covered by compulsory insurance schemes.

SPENDING

▷ GDP/cap. increase

CONSUMPTION AND SPENDING

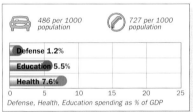

486 per 1000 population

727 per 1000 population

Defense 1.2%
Education 5.5%
Health 7.6%

Defense, Health, Education spending as % of GDP

Immigrant workers do most low-paid and menial jobs. Wages in office jobs are relatively high, although the cost of living is also well above the European average. Many workers choose to live in France and commute across the border. The land market is highly regulated.

WORLD RANKING

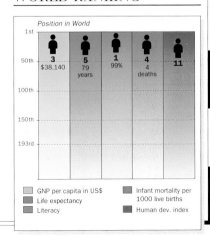

Position in World

3 — $38,140
5 — 79 years
1 — 99%
4 — 4 deaths
11

GNP per capita in US$
Life expectancy
Literacy
Infant mortality per 1000 live births
Human dev. index

S

SYRIA

OFFICIAL NAME: Syrian Arab Republic **CAPITAL:** Damascus
POPULATION: 16.6 million **CURRENCY:** Syrian pound **OFFICIAL LANGUAGE:** Arabic

SYRIA IS REGARDED by many of its people as an artificial creation of French-mandated rule, which lasted from 1920 to independence. They identify instead with a Greater Syria, a successor to the medieval Ummayad caliphate, encompassing Lebanon, Jordan, and Palestine. Since independence, Syria's foreign relations have been turbulent, but the authoritarian Ba'athist regime of Hafez al-Assad (1970–2000) brought a measure of internal stability.

CLIMATE
▷ Steppe/hot desert/ Mediterranean

WEATHER CHART FOR DAMASCUS

The coastal climate is Mediterranean, with mild, wet winters and dry, hot summers. Most of the country gets fewer than 25 cm (10 in) of rainfall a year. Away from the coast, rainfall is very unpredictable and the country becomes increasingly arid, with some desert areas. In the mountains, snow is common in winter.

TRANSPORTATION
▷ Drive on right

 Damascus International 1.5m passengers

 220 ships 427,500 grt

THE TRANSPORTATION NETWORK

26,299 km (16,341 miles)	877 km (545 miles)
1525 km (948 miles)	870 km (541 miles)

The road network is unreliable in rural areas, especially during the winter after rain. Bus services operate to most towns from Damascus and Aleppo. Roads are integrated with the railroads, which carry over four million passengers a year and are vital to freight transportation. The rail link from Aleppo to Mosul, in Iraq, reopened in mid-2000. Damascus is the main international airport and Latakia the main port.

PEOPLE
▷ Pop. density medium

Arabic, French, Kurdish, Armenian, Circassian, Turkic languages, Assyrian, Aramaic

90/km² (234/mi²)

THE URBAN/RURAL POPULATION SPLIT

55% 45%

RELIGIOUS PERSUASION

Christian 10%
Other Muslim 16%
Sunni Muslim 74%

ETHNIC MAKEUP

Armenian, Turkmen, Circassian 2%
Other 3%
Kurdish 6%
Arab 89%

Most Syrians live in the west, where the largest cities are sited. About 90% are Muslim. They include the politically dominant Alawis, a heterodox offshoot of Shi'a, comprising 12% of the population, based in Latakia and Tartous provinces. There is also a sizable Christian minority. In the west and north a mosaic of groups includes Kurds, Turkic-speakers, and Armenians, the latter based in cities. Damascus, Al Qamishli, and Aleppo have small Jewish communities, and there are three villages where Aramaic is spoken. In addition, some 300,000 Palestinian refugees have settled in Syria. Minorities were initially attracted to the ruling Ba'ath Party because of its emphasis on the state over sectarian interests. However, disputes between factions led to the Alawis taking control, creating resentment among the Sunni Muslim majority.

The emancipation of women, promoted initially in the late 1960s, was carried forward under President Hafez al-Assad, whose first woman cabinet minister was appointed in 1976.

TOURISM
▷ Visitors : Population 1:18

 916,000 visitors Up 24% in 1999

MAIN TOURIST ARRIVALS

Lebanon 32%	
Jordan 19%	
Saudi Arabia 10%	
Turkey 7%	
Iran 7%	
Other 25%	

% of total arrivals

Years of political turbulence, allegations of human rights abuses committed under Hafez al-Assad's regime, and strict, complex travel regulations retarded the development of tourism. However, just before the 1990–1991 Gulf War, Syria began to compete with other Middle Eastern states as a tourist destination. Modern hotels were built in most cities and facilities improved to cater for growing numbers of Western visitors. Following the war, tourist numbers dropped sharply, but are now gradually recovering. Syria's main attractions are its antiquities and historic cities, with their covered markets (*soukhs*), baths, and mosques – Damascus, said to be the oldest inhabited city in the world; the ruined desert city of Palmyra; and Aleppo, with its citadel. Syria has a wealth of castles dating back to the Crusades and sites associated with the advent of Islam. In addition, there are as many as 3500 as yet unexcavated archaeological sites. Syria's coastline on the Mediterranean has fine beaches, and there are mountain resorts in Latakia.

The ancient city of Palmyra, in Syria's central region, was once the capital of the kingdom of Queen Zenobia.

POPULATION AGE BREAKDOWN

Female	Age	Male
0%	80+	0%
2.4%	60–79	2.6%
5.4%	40–59	5.7%
13.8%	20–39	13.8%
27.7%	0–19	28.6%

% of population by age group

SYRIA

Total Land Area : 184 180 sq. km
(71 498 sq. miles)

LAND HEIGHT		POPULATION	
2000m/6562ft		over 500 000	⊙
1000m/3281ft		over 100 000	◎
500m/1640ft		over 50 000	○
200m/656ft		over 10 000	●
Sea Level		under 10 000	·

WORLD AFFAIRS

Joined UN in 1945

AL · Damasc · G24 · NAM · OIC

Since Egypt's 1979 accord with Israel, Syria sees itself as the major barrier to Israel's regional dominance. Syria extended its control over Lebanon and hosts radical Palestinian factions, as well as seeking alliances with north African states. The biggest issue with Israel remains the strategically vital Golan Heights, seized by Israel during the Six-Day War in 1967. Peace negotiations foundered when Ariel Sharon became prime minister of Israel in 2001.

There are enduring tensions with Turkey over attitudes to Israel and to Turkish Kurdish guerrillas, access to water, and Syria's desire for the return of its former Alexandretta Province.

Syria, alone among Arab states, backed Iran in the 1980s Iran–Iraq War. Facing international isolation over its alleged backing of terrorists, Syria regained a measure of respectability by securing the release of Western hostages of Shi'a militants in Lebanon. Assad backed the US-led allies in the 1990–1991 Gulf War, and by contributing troops legitimized the action in the eyes of the Arab world. In 2000, however, Syria sent humanitarian aid to Iraq, in defiance of the UN blockade, and in 2002 was accused by the US of continuing to support international "terrorists."

AID

Recipient

$158m (receipts) · Down 31% in 2000

Syria has historically received little aid owing to its human rights record and self-sufficiency in oil. It received one-off payments totaling $2 billion in 1992 and $1.2 billion in 1993 after the Gulf War, mainly from Saudi Arabia and the Gulf states, but with contributions from the West and Japan.

POLITICS

No multiparty elections

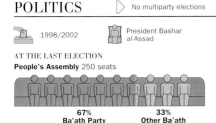

1998/2002

President Bashar al-Assad

AT THE LAST ELECTION
People's Assembly 250 seats

67% Ba'ath Party · 33% Other Ba'ath

Ba'ath Party = Members of the National Progressive Front (allies of the Ba'ath Party) **Other Ba'ath** = Parties allied to the Ba'ath Party

Syria is in effect a single-party state. Its military-backed leader from 1970 to 2000 was Hafez al-Assad, a lifelong Ba'ath Party militant. His personal dominance ensured the succession of his son Bashar after his death in June 2000.

PROFILE

The Ba'athist military swept to power in 1963 with a vision of uniting all Arab nations under a single Syrian-dominated socialist system. The coup ended the power of city elites and promoted citizens from rural areas.

Assad consolidated the Ba'ath Party as the major political force. Unrest among Islamic militants was crushed, notably in the army-led massacre at Hamah in 1982. Assad focused on foreign affairs, bidding to make Syria a major power, but plans to unite with fellow Ba'athist Iraq ended in 1981 amid mutual recriminations.

MAIN POLITICAL ISSUES
Human rights

Martial law has not been rescinded since 1963, but the regime has improved its human rights record in recent years. Political prisoners are released under frequent amnesties, and in 1994 all members of the Jewish minority were granted exit visas to travel abroad.

Political pluralism

Assad dominated Syrian politics for 30 years. His military-backed regime, drawn mainly from his own Alawi minority grouping, kept a tight hold on power, although in his last decade Sunnis gained high political posts. Despite promises made under international pressure, Assad never permitted genuine multipartyism. Shortly before his death, the modernizing Mohammed Miro replaced the long-serving prime minister Mahmoud az-Zoubi, who was forced from office. Assad's death was followed immediately by his son's election to the party leadership, which was approved overwhelmingly by referendum. Initial tentative reforms – such as the spread of "discussion clubs" and a more varied cabinet – were later curtailed.

Hafez al-Assad, ruled for three decades until his death in 2000.

Bashar al-Assad, succeeded his father as president.

S

CHRONOLOGY

Under French mandate from 1920, Syria declared independence in 1941, and achieved full autonomy in 1946. From 1958 to 1961 Syria merged with Egypt to form the United Arab Republic.

❑ **1963** Ba'athist military junta seizes power. Maj. Gen. Amin al-Hafez president.

❑ **1966** Hafez ousted by military coup supported by radical Ba'ath Party members.

❑ **1967** Israel overruns Syrian positions above Lake Tiberias, seizes Golan Heights, and occupies Quneitra. Syria boycotts Arab summit and rejects compromise with Israel.

❑ **1970** Hafez al-Assad seizes power in "corrective coup."

❑ **1971–1999** Assad elected president; reelected four times.

❑ **1973** New constitution confirms dominance of Ba'ath Party. War launched with Egypt against Israel to regain territory lost in 1967. More territory temporarily lost to Israel.

❑ **1976** With peacekeeping mandate from Arab League, Syria intervenes to quell fighting in Lebanon.

❑ **1977** Relations broken off with Egypt after Egyptian president Sadat's visit to Jerusalem.

❑ **1978** National charter signed with Iraq for union.

❑ **1980** Membership of Muslim Brotherhood made capital offense. Treaty of Friendship with USSR.

❑ **1981** Israel formally annexes Golan Heights. Charter with Iraq collapses.

❑ **1982** Islamic extremist uprising in Hamah crushed. Israel invades Lebanon; Syrian missiles in Bekaa Valley destroyed.

❑ **1985** US claims Syrian links to Rome and Vienna airport bombings.

❑ **1986** Alleged Syrian complicity in planting of bomb on Israeli airliner in London. EU states, except Greece, impose sanctions and arms embargo.

❑ **1989** Diplomatic relations reestablished with Egypt.

❑ **1990–1991** Syrian forces crush renegade Gen. Aoun in Beirut. Steers Taif Accords over Lebanon.

❑ **1991** Troops take part in Operation Desert Storm. Damascus Declaration aid and defense pact signed with Egypt, Saudi Arabia, Kuwait, UAE, Qatar, Bahrain, and Oman.

❑ **1995** Inconclusive talks with Israel.

❑ **2000** Forced resignation after 13 years and subsequent suicide of prime minister, Mahmoud az-Zoubi. Death of Hafez al-Assad; succession of son Bashar.

❑ **2001** Israel bombs Syrian radar base in Lebanon.

DEFENSE

 Compulsory military service

 $760m   ⬇ Down 23% in 2000

Syria sees its extensive military capability as a significant deterrent to Israel's territorial expansion. It has fought four wars against Israel since 1948, and is the Arab world's strongest military power after Egypt. Most of its military equipment was obtained from the former Soviet Union.

Throughout 2000, increasing numbers of Lebanese protested against the continuing presence of Syrian troops on their soil – in apparent contravention of the 1989 Taif Accords – and in 2001, Syrian troops were

SYRIAN ARMED FORCES

4700 main battle tanks (T-55/MV, T-62M/K, T-72/72M)		215,000 personnel
2 frigates and 18 patrol boats		6000 personnel
589 combat aircraft (90 Su-22, 20 Su-24, 44 MiG-23BN)		40,000 personnel
None		

withdrawn from Beirut. In May 2002 a US State Department official accused Syria of harboring weapons of mass destruction.

ECONOMICS

 Inflation 6.7% p.a. (1990–2000)

📊 $15.1bn   💲 53.5–45.9 Syrian pounds

SCORE CARD

❑ World GNP Ranking...........................75th
❑ GNP per Capita $940
❑ Balance of Payments....................$1.06bn
❑ Inflation ....–0.4%
❑ Unemployment..................................20%

ECONOMIC PERFORMANCE INDICATOR

EXPORTS

Saudi Arabia 7%
Turkey 10%
Other 34%
France 10%
Italy 12%
Germany 27%

IMPORTS

South Korea 4%
Turkey 4%
France 5%
Germany 7%
Italy 9%
Other 71%

STRENGTHS

Exporter of crude oil – production increasing as a result of new oil strikes. Manufacturing base has grown. Thriving agricultural sector. Low inflation.

WEAKNESSES

High defense spending a major drain. Large black market. Domination of inefficient state-run companies. Lack of foreign investment; foreign currency accounts banned. High population growth and unemployment. Vulnerable water supply. Slow implementation of reforms.

PROFILE

Billions of dollars flowed into the economy from the West and Gulf states after the 1990–1991 Gulf War. This cash injection, along with increased oil revenue, led to rapid growth. Diversion of water from the Euphrates toward fertile plains, rather than poorer land, led to a rise in agricultural output. However, long-term economic prospects remain uncertain. The public sector employs 20% of the workforce, and state controls inhibit private enterprise and investment, and have created a booming black market. Businessmen often channel funds through the freer Lebanese economy. An economic reform package in 2000 created a stock exchange and permitted private banks.

SYRIA : MAJOR BUSINESSES

Ḥamāh
Ḥalab
Al Lādhiqīyah
Bāniyās
Ṭarṭūs
Ḥimṣ
Dayr az Zawr
Damascus
Adra

Cement
Textiles
Chemicals
Metallurgy
Oil refining
Food processing

0   100 km
0   100 miles

S

RESOURCES

▷ Electric power 4.5m kw

14,024 tonnes

12.4m sheep, 979,325 goats, 21m chickens

551,000 b/d (reserves 2.5bn barrels)

Phosphates, oil, natural gas, iron

ELECTRICITY GENERATION

- Hydro 14% (2.7bn kwh)
- Combustion 86% (17bn kwh)
- Nuclear 0%
- Other 0%

% of total generation by type

Oil, mostly good-quality light crude, was discovered along the Euphrates in the 1980s. Gas was found in substantial quantities near Palmyra. Other important minerals are phosphates and iron ore. Hydroelectric power provides a small proportion of electricity. In 2001, a joint project was launched to link the electricity grids of Syria, Jordan, and Egypt. Manufacturing is largely in oil-derived industries, textiles, and foodstuffs. Cotton is the main cash crop; fruit and vegetables are also grown. Livestock, especially sheep and goats, support the rural economy.

SYRIA : LAND USE

Cropland
Pasture
Desert
Sheep
Cotton
Fruits - cash crop

0   100 km
0   100 miles

ENVIRONMENT

▷ Sustainability rank: 107th

None

3.3 tonnes per capita

ENVIRONMENTAL TREATIES

Yes | No | Yes
Yes | Yes | No

The Assad regime's most expensive and controversial environmental project has been the Euphrates Dam, power plant, and irrigation network at Tabaqah. The dam's vast man-made reservoir, Buhayratal al-Assad, engulfed some 300 villages and destroyed 25,000 hectares (62,000 acres) of fertile farmland. A giant cement factory, built by former East Germany at Tartus in the mid-1970s, has increased pollution along a stretch of Syria's Mediterranean coastline. However, its interior remains relatively unspoilt.

MEDIA

▷ TV ownership medium

Daily newspaper circulation 20 per 1000 people

Information became freer after Jordanian papers were again allowed into Syria in 1999, and with the advent of satellite TV, widely watched. Bashar al-Assad encourages Internet use, and in 2001 paved the way for independent publications, including *Ad-domari*, the first satirical journal in 38 years.

CRIME

▷ Death penalty in use

14,000 prisoners

Up 164% 1995–1999

CRIME RATES

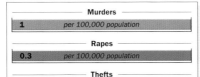

Murders	
1	per 100,000 population

Rapes	
0.3	per 100,000 population

Thefts	
25	per 100,000 population

There is no truly independent judiciary. The powerful security services exercise arbitrary powers of arrest and detention. There are widespread reports of torture in custody. Most politicians removed by President Hafez al-Assad in the 1970s have been released from prison in Damascus.

EDUCATION

▷ School leaving age: 12

74%

227,926 students

THE EDUCATION SYSTEM

Primary 100%
Secondary 42%
Tertiary 6%

% of each age group in education

A modern and universally accessible system of education remains an important objective. Free and compulsory primary education for all was a priority of the Ba'ath Party when it came to power. Coeducation began in the cities and spread to rural areas under the Assad regime. There are seven state universities, notably at Damascus, Aleppo, Tishrin, and Homs. Private universities were allowed from 2001. Education ranks second, though is far behind defense, in government expenditure. UK and US politicians protest at incitement to racism in Syrian school textbooks.

PUBLISHING AND BROADCAST MEDIA

There are 10 daily newspapers, including *Al-Ba'ath*, *Ath-Thawra*, and *Tishrin*.

1 state-controlled service

1 state-controlled service. Independent music stations were permitted from 2002

HEALTH

▷ Welfare state health benefits

1 per 769 people

Heart, respiratory, digestive, infectious, and parasitic diseases

An adequate system of primary health care has been set up since the Ba'ath Party came to power. Treatment is free for those unable to pay. However, hospitals often lack modern equipment and medical services are in need of further investment. Rural areas in particular need assistance to combat the spread of heart, respiratory, and infectious diseases.

SPENDING

▷ GDP/cap. increase

CONSUMPTION AND SPENDING

9 per 1000 population

103 per 1000 population

Defense 5.6%
Education 3.1%
Health 0.9%

Defense, Health, Education spending as % of GDP

Syria is far from the equitable society that early Ba'ath Party thinkers envisioned. The gulf between Syria's rich and poor is widening. The political elite, many of whom live in the West Malki suburb of Damascus, is more numerous and richer than ever before. Palestinian refugees and the urban unemployed make up the poorest groups.

WORLD RANKING

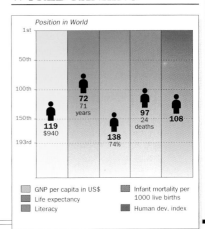

Position in World

1st
50th
100th
150th
193rd

119 $940
72 71 years
138 74%
97 24 deaths
108

- GNP per capita in US$
- Life expectancy
- Literacy
- Infant mortality per 1000 live births
- Human dev. index

S

TAIWAN

OFFICIAL NAME: Republic of China (ROC) **CAPITAL:** Taipei
POPULATION: 22.2 million **CURRENCY:** Taiwan dollar **OFFICIAL LANGUAGE:** Mandarin Chinese

THE ISLAND OF TAIWAN, formerly known as Formosa, lies off the southeast coast of mainland China. Mountains running north to south cover two-thirds of the island. The lowlands are highly fertile, cultivated mostly with rice, and densely populated. In 1949, when the Chinese Communists ousted Chiang Kai-shek's nationalist Kuomintang (KMT) from power on the mainland, he established the Republic of China government on the island. De facto military rule has been democratized progressively since 1986. Mainland China still considers Taiwan a renegade province, and only a few countries now give official recognition to the regime there.

Wen Wu Temple, *on the shores of Sun Moon Lake in the mountains of central Taiwan – a region famous for its many temples. Nearly the whole population is Buddhist.*

CLIMATE

▷ Tropical monsoon

WEATHER CHART FOR TAIPEI

Taiwan has a tropical monsoon climate similar to that of the southern Chinese mainland. Typhoons from the South China Sea between June and October bring the heaviest rains.

TRANSPORTATION

▷ Drive on right

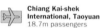 **Chiang Kai-shek International, Taoyuan** 18.7m passengers

 680 ships 5.09m grt

THE TRANSPORTATION NETWORK

31,271 km (19,431 miles)

Sun Yat-sen highway; 538 km (334 miles)

1104 km (686 miles)

None

A railroad encircles the island. A US$16 billion high-speed rail link along the west coast, linking Taipei to Kaohsiung, is under construction with the help of the Japanese Shinkansen Corporation. It is due for completion in 2005. Car ownership is 148 times higher than the 1971 level. The bicycle is not as popular in Taiwan as in mainland China. However, Taiwan is the world's biggest bicycle producer, exporting mostly to Europe and the US.

Access to mainland China has increased as relations slowly improve. By boat, passengers have to travel from the tiny offshore islands of Matsu and Quemoy. The number of visitors is restricted both ways by yearly quotas.

TOURISM

▷ Visitors : Population 1:8.5

2.62m visitors

Up 9% in 2000

MAIN TOURIST ARRIVALS

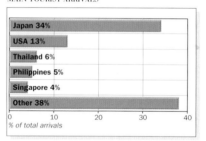

Japan 34%
USA 13%
Thailand 6%
Philippines 5%
Singapore 4%
Other 38%

% of total arrivals

Taiwan is not a major tourist destination, and it has only recently begun promoting itself in the US and Japan. Restrictions on Chinese tourists are being eased. Successive Six-Year Plans focus on upgrading hotels and improving tourist faciities at international airports. The major attraction is the Palace Museum in Taipei, which includes the massive treasure looted by the nationalists from Beijing. Only 5% can be shown at any one time. Sex tourism is an important business in Taipei. Sex establishments often masquerade as barbershops.

PEOPLE

▷ Pop. density high

Amoy Chinese, Mandarin Chinese, Hakka Chinese

688/km² (1782/mi²)

THE URBAN/RURAL POPULATION SPLIT

69%   31%

RELIGIOUS PERSUASION

Other 2%   Christian 5%
Buddhist, Confucian, Taoist 93%

ETHNIC MAKEUP

Aborigine 2%   Mainland Chinese 14%
Indigenous Chinese 84%

Most Taiwanese are Han Chinese, descendants of the 1644 migration of the Ming dynasty from mainland China. The 100,000 nationalists who arrived in 1949 established themselves as a ruling class and monopolized the most prestigious jobs in the civil service.

This caused resentment among the local inhabitants, but as the generation elected on the mainland in 1947 has aged, so local Taiwanese have entered the political process.

There is little ethnic tension in Taiwan, although the indigenous minorities who live in the eastern hills do suffer considerable discrimination.

As in the rest of southeast Asia, the extended family is still important and provides a social security net for the elderly. However, the trend is toward European-style nuclear families. Housing shortages are a major issue.

Women are not well represented in the political process, but are prominent in business and the civil service.

POPULATION AGE BREAKDOWN

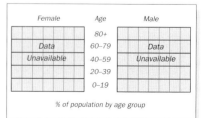

Female	Age	Male
	80+	
Data	60–79	Data
Unavailable	40–59	Unavailable
	20–39	
	0–19	

% of population by age group

T

POLITICS ▷ Multiparty elections

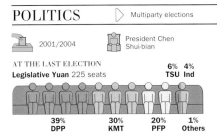

2001/2004

President Chen Shui-bian

AT THE LAST ELECTION
Legislative Yuan 225 seats

| 39% DPP | 30% KMT | 20% PFP | 6% TSU | 4% Ind | 1% Others |

DPP = Democratic Progressive Party **KMT** = National Party of China (Kuomintang) **PFP** = People First Party **TSU** = Taiwan Solidarity Union **Ind** = Independents

From 1986 Taiwan has been a multiparty democracy, though electoral domination by the KMT only ended in 2000. The government is headed by the president, who is answerable to the National Assembly, an additional body which is convened only when constitutional affairs arise.

PROFILE
For nearly four decades from the establishment of the Republic of China government on Taiwan in 1949, Chiang Kai-shek's KMT monopolized political power and ruled by strict martial law. In 1986, Gen. Chiang Ching-kuo, Chiang Kai-shek's son and successor, decided to open the way for more democracy. Later that year the first free multiparty elections were held. In March 1996 Lee Teng-hui became the country's first directly elected president.

The KMT retained power in the 1998 legislative elections, winning an absolute majority of seats. Its 50-year political monopoly came to an end with the presidential elections in March 2000 which saw the KMT candidate coming a poor third. Chen Shui-bian of the pro-independence DPP took office in May 2000. The KMT also lost support in the 2001 legislative elections. The DPP became the largest party in the Yuan but did not obtain a majority.

Chiang Kai-shek, who established the ROC in Taiwan in 1949.

Chen Shui-bian, the first non-KMT president since 1949.

MAIN POLITICAL ISSUES
Relations with China
The DPP has advocated independence from China, despite Chinese threats of military action. Adopting a more prudent stance since his poll victory, Chen Shui-bian promised to make no independence declaration during his term of office. The KMT, officially committed to eventual reunification with China, now favors a more flexible arrangement that presupposes the recognition of a separate Taiwanese national identity.

Political stability
The KMT, in power ever since the Chinese civil war brought Taiwan into existence, has recently been displaced by the DPP. This has meant an unprecedented degree of change within Taiwanese political life. The preservation of political and economic stability, however, remains essential as China intensifies its sovereignty claims on Taiwan after the return of Hong Kong in 1997.

WORLD AFFAIRS ▷ Not a UN member

| APEC | ADB | WTO |

China rejects Taiwan's sovereignty claims, regarding it as a renegade province, and Taiwan conducts overseas relations via trade delegations. It lost its seat on the UN Security Council in 1971, and has since been unable to rejoin the UN. The handover to China of Hong Kong in 1997 deepened fears that China would intensify pressure on countries to sever links with Taiwan. China mounted military exercises in 1999 after President Lee spoke of Taiwan as a separate state. China sought unsuccessfully in 2000 to frighten Taiwanese voters away from backing pro-independence candidate Chen Shui-bian. However, each depends on the other economically; Taiwan is a major foreign investor in China and has recently permitted direct trade. After US rapprochement with China in 1972, Taiwan ceased effectively to be a US client state. US security guarantees are vague, but in practice there are strong bilateral ties with the US and Japan.

TAIWAN

Total Land Area : 35 980 sq. km (13 892 sq. miles)

POPULATION

▣	over 1 000 000
◉	over 500 000
◎	over 100 000
○	over 50 000
●	over 10 000
·	under 10 000

LAND HEIGHT

| | |
| 3000m/9843ft |
| 2000m/6562ft |
| 1000m/3281ft |
| 500m/1640ft |
| 200m/656ft |
| Sea Level |

T

AID Donor

 US$500m (donations)   Up in 2001

Taiwan has a large aid fund devoted to states which have granted it diplomatic recognition. These include the Pacific states of Kiribati and Tonga, which have looked out for Taiwan's interests in the UN since 1971, when it lost its seat to the People's Republic of China. In 1998 Taiwan donated more than US$2 million to seven Central American supporters, including Panama, to promote literacy.

CHRONOLOGY

Following the 1949 communist revolution in China, Gen. Chiang Kai-shek's nationalist KMT party sought refuge in the island province of Taiwan. The KMT saw the revolution as illegal and itself as the sole rightful Chinese government.

❑ **1971** People's Republic of China replaces Taiwan at UN, including on UN Security Council.
❑ **1973** Taipei's KMT regime rejects Beijing's offer of secret talks on reunification of China.
❑ **1975** President Chiang Kai-shek dies. His son Gen. Chiang Ching-kuo becomes KMT leader. Yen Chia-kan succeeds as president.
❑ **1978** Chiang Ching-kuo elected president.
❑ **1979** US severs relations with Taiwan and formally recognizes People's Republic of China.
❑ **1984** President Chiang reelected.
❑ **1986** Political reforms: KMT allows multiparty democracy, ends martial law, and permits visits to Chinese mainland for "humanitarian" purposes for first time in 38 years. From 1988, mainland Chinese are allowed to visit Taiwan on same basis.
❑ **1988** Lee Teng-hui president.
❑ **1990** KMT formally ends state of war with People's Republic of China.
❑ **1991** DPP draft constitution for Taiwan independence opposed by ruling KMT and Beijing. KMT reelected with large majority.
❑ **1995–1996** Legislative elections. KMT majority reduced.
❑ **1996** Lee Teng-hui wins first direct presidential elections.
❑ **1998** KMT secures absolute majority in elections to Legislative Yuan.
❑ **1999** Chinese threats over reference to "separate states" status. Thousands die in earthquake.
❑ **2000** Chen Shui-bian of DPP wins presidency; ends KMT dominance.
❑ **2001** Elections: DPP largest single party. Taiwan admitted to WTO.

DEFENSE Compulsory military service

 US$17.2bn   Up 15% in 2000

China remains the main defense threat, given the recurring tensions over the issue of independence or reunification. Taiwan has the fifth-largest navy in the world and a sizable army, in order to face a possible Chinese invasion: there are over 1.5 million reservists, and military service lasts for two years. Worries about US loyalty have resulted in the purchase of French *Mirage* fighters in addition to US F-16s.

TAIWANESE ARMED FORCES

926 main battle tanks (100 M-48A5, 450 M-48H, 376 M-60A3)	240,000 personnel
4 submarines, 11 destroyers, 21 frigates, 59 patrol boats	62,000 personnel
482 combat aircraft (F-5, *Ching-Kuo*, *Mirage* 2000-5, F-16A/B)	68,000 personnel
None	

ECONOMICS Inflation 5% p.a. (1985–1996)

 US$283bn   33.08–34.99 Taiwan dollars

SCORE CARD

❑ WORLD GNP RANKING.........................16th
❑ GNP PER CAPITA ....US$13,450
❑ BALANCE OF PAYMENTS ....US$9.32bn
❑ INFLATION 1.3%
❑ UNEMPLOYMENT 3%

ECONOMIC PERFORMANCE INDICATOR

EXPORTS

Germany 3% · Netherlands 3% · Japan 10% · Other 38% · Hong Kong 21% · USA 25%

IMPORTS

China 4% · Germany 5% · South Korea 6% · Other 39% · USA 18% · Japan 28%

Taiwan emerged relatively unscathed, and global downturn caused only a brief recession in 2001. Competition from underdeveloped countries with low production costs is dictating a difficult transition toward service industries. This will entail moving from labor-intensive to capital- and technology-intensive industries. Comprehensive Six-Year Plans reflect a strong element of state direction. Heavy investment abroad includes over 60% of inward investment into China since 1990. Taiwan was admitted, along with China, to the WTO in 2001.

STRENGTHS

Highly educated and ambitious workforce, many US-trained and educated, with an inside knowledge of the US market. Manufacturing economy based on small companies which have proved extremely adaptable to changing market conditions. Track record of capturing major markets. Successively the world's biggest TV, watch, PC, and track shoe manufacturer. Economy in strong surplus, allowing it to invest in other southeast Asian economies.

WEAKNESSES

Small economic units lack the muscle of Western multinationals, and are unable to follow predatory pricing policies. Weak research and development: economy has no tradition of generating new products or creating new markets. Unresponsive banking system.

PROFILE

Taiwan's economy has proved resilient. Double-digit growth ended with the Asian financial crisis of 1997–1998, but

TAIWAN : MAJOR BUSINESSES

RESOURCES

▷ Not available

1.29m tonnes

Not an oil producer; refines 542,500 b/d

10.6m ducks, 7.5m pigs, 2.82m geese 118m chickens

Coal, copper, marble, dolomite, gold, silver

ELECTRICITY GENERATION

Hydro 6% (8.8bn kwh)

Combustion 67% (94bn kwh)

Nuclear 27% (36.9bn kwh)

Other 0%

% of total generation by type

Taiwan has few strategic resources and its minerals industry is not a major foreign exchange earner. Oil is imported. Taiwan is a major buyer of South African uranium, but proposals to increase reliance on nuclear power have met strong opposition on safety and waste disposal grounds. However, hydroelectric power has been largely exploited already and combustion is also a controversial option.

Fishing is highly successful, and Taiwan is a major supplier to the huge Japanese market. The fleet is often accused of plundering Atlantic stocks.

TAIWAN : LAND USE

Cropland
Forest
Pasture
Wetlands
Pigs
Rice

0   50 km
0   50 miles

ENVIRONMENT

▷ Not available

8% (3% partially protected)

8.1 tonnes per capita

ENVIRONMENTAL TREATIES

No   No   No

No   No   No

The dash for growth meant the absence of city planning or pollution laws. There is growing opposition to a fourth nuclear power plant, set for completion in 2006, and concern over coal-fired thermal power. Taiwan's fishing industry has been criticized for using longline techniques which trap dolphins, and for plundering other countries' fishing grounds without regard to stock levels.

MEDIA

▷ TV ownership medium

Daily newspaper circulation figures not available

PUBLISHING AND BROADCAST MEDIA

There are 35 daily newspapers. The independent *Lienho Pao* has the largest circulation

1 public-service station, 64 cable companies, and 129 satellite broadcasting channels

110 independent corporations

The rigid state control which used to exist over the media has been relaxed. Opposition parties now have access to the state media. Before the 1990s, press with simplified Chinese characters was banned, thus excluding all publications from the mainland. Taiwan has a large domestic TV and film industry.

CRIME

▷ Death penalty in use

31,000 prisoners

Little change from year to year

CRIME RATES

Most Taiwanese are highly conscious of crime. However, rates are low by US or European standards.

Since the end of martial law in 1986, most political prisoners have been released. Taiwan does not suffer from organized crime to the extent found in Hong Kong or Japan. Multimedia pirating is a major problem.

EDUCATION

▷ School leaving age: 15

94%

1.09m students

THE EDUCATION SYSTEM

100% Primary
99% Secondary
68% Tertiary

% of each age group in education

Since 2000, government funding has given less priority to higher education and more to reforming and improving the antiquated school system, which is rigid and heavily exam-oriented. Free schooling is available from the age of six to 15 and there are also a number of private schools.

Enrollment levels at tertiary and vocational institutions are among the highest in the world.

HEALTH

▷ No welfare state health benefits

1 per 894 people

Cerebrovascular and heart diseases, hypertension

Most health provision in Taiwan is in the private sector. Taiwanese take out elaborate health insurance schemes and it is essential to prove cover before treatment is provided. Health facilities are on a par with the best in the world, and the Taiwanese enjoy a high life expectancy, similar to that in the US. The incidence of AIDS is low, but the infection rate is increasing. An enteroviral epidemic swept Taiwan in 1998, killing scores of babies and affecting thousands of young children.

SPENDING

▷ GDP/cap. increase

CONSUMPTION AND SPENDING

213 per 1000 population

570 per 1000 population

Defense 5.6%

Education 3.6%

Health   No data

Defense, Health, Education spending as % of GDP

Export-led growth over a long period enabled Taiwan to build up large cash reserves. The Taiwanese people share much of the benefit of this success, and average living standards are among the highest in Asia. Inequalities of income distribution are comparatively small, and a high degree of social cohesion has been achieved. In part, this is the result of the land reforms of the 1950s, which gave agricultural workers control of the land while compensating landowners and encouraging them to set up business in the cities. Today, the great majority of Taiwanese would describe themselves as middle class. Consumer goods are widely available, and conspicuous consumption is celebrated.

WORLD RANKING

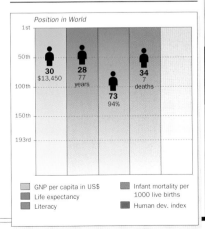

Position in World

1st
50th
100th
150th
193rd

30 $13,450
28 77 years
73 94%
34 7 deaths

GNP per capita in US$
Life expectancy
Literacy
Infant mortality per 1000 live births
Human dev. index

T

TAJIKISTAN

OFFICIAL NAME: Republic of Tajikistan   **CAPITAL:** Dushanbe
POPULATION: 6.1 million   **CURRENCY:** Somoni   **OFFICIAL LANGUAGE:** Tajik

TAJIKISTAN LIES ON the western slopes of the Pamirs in central Asia. Language and traditions are similar to those of Iran rather than those of its northern Turkic neighbors. Tajikistan decided on independence only when neighboring Soviet republics declared theirs in late 1991. Fighting between communist government forces and Islamist rebels, which erupted shortly afterward, has been contained since 1997 by a fragile peace agreement.

A herd of goats in the Varzob Gorge, north of Dushanbe. Livestock drives the rural economy.

CLIMATE
▷ Mountain

WEATHER CHART FOR DUSHANBE

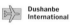

Rainfall is low in the valleys. Winter temperatures can fall below –45°C (–49°F) in mountainous areas.

TRANSPORTATION
▷ Drive on right

Dushanbe International		Has no fleet	

THE TRANSPORTATION NETWORK

11,330 km (7040 miles)		None	
547 km (340 miles)		200 km (124 miles)	

Tajikistan has good cross-border roads and well-maintained airfields, the result of its use by Soviet and US forces during conflicts in neighboring Afghanistan. The best way to reach the mountainous interior is by air.

TOURISM
▷ Visitors : Population 1:12

511,000 visitors		Increasing slowly	

MAIN TOURIST ARRIVALS

Tajikistan does not publish tourism figures by country of origin

0	10	20	30	40

% of total arrivals

Tourism is virtually nonexistent, and insecurity and poverty make it unlikely that it will be developed soon.

PEOPLE
▷ Pop. density low

Tajik, Russian   43/km² (110/mi²)

THE URBAN/RURAL POPULATION SPLIT

28%   72%

RELIGIOUS PERSUASION

Shi'a Muslim 5%
Other 15%
Sunni Muslim 80%

Unlike the other former Soviet -stans, Tajikistan is dominated by a people of Persian (Iranian), rather than Turkic, origin. Ethnic Tajiks make up around 65% of the population. The main ethnic conflict is with the Turkic Uzbeks. As in neighboring Uzbekistan, Russians are discriminated against, and their numbers have thinned from 400,000 in 1989 to fewer than 200,000. By 1990, the 35,000-strong German minority had left. The struggle between Dushanbe-based communists and the Islamist militants in the central and eastern regions displaced more than 50,000 refugees into Afghanistan.

POLITICS
▷ Multiparty elections

L. House 2000/2005   President Imomali
U. House 2000/2005   Rakhmanov

AT THE LAST ELECTION

Assembly of Representatives 63 seats

3% 3%
IRP Others

71%   21%   2%
PDPT   CPT   Vac

PDPT = People's Democratic Party of Tajikistan and allies
CPT = Communist Party of Tajikistan   **IRP** = Islamic Revival Party   **Vac** = Vacant

National Assembly 33 seats

Five deputies are elected by each regional assembly, and a further eight deputies are appointed by the president

The lull in fighting between government forces and Islamist rebels, aided by a 1997 peace accord, has consolidated the regime of former communists led by President Rakhmanov. In 1998, the Islamist United Tajik Opposition (UTO) joined the government in conformity with the accord, which provided for a National Reconciliation Commission along with parliamentary elections. In the 2000 elections the pro-Rakhmanov PDPT, which headed the poll, claimed some support from former UTO members.

TAJIKISTAN

Total Land Area : 143 100 sq. km (55 251 sq. miles)

POPULATION
⊙ over 500 000
◎ over 100 000
○ over 50 000
● over 10 000
• under 10 000

LAND HEIGHT
4000m/13 124ft
3000m/9843ft
2000m/6562ft
1000m/3281ft
500m/1640ft
200m/656ft

WORLD AFFAIRS
▷ Joined UN in 1992

    OSCE

Tajikistan remains heavily dependent on Russia, particularly for economic and military assistance. The introduction of the somoni in 1995 enabled the government to wrest economic control back from Russia. A joint operation with Uzbekistan and Kyrgyzstan was launched in 2000 to combat the Islamic Movement of Uzbekistan, based in northern Tajikistan. Tajik airfields were crucial during the US-led action in Afghanistan in 2001, and in 2002 Tajikistan became the last of the former Soviet Union countries to join NATO's Partnerships for Peace program.

AID
▷ Recipient

 $142m (receipts) — Up 15% in 2000

Russia and Uzbekistan provide military aid. Food aid was required in 2000 and 2001 after severe droughts.

DEFENSE
▷ Compulsory military service

 $80m Down 13% in 2000

Security along the winding southern border with Afghanistan is bolstered by the presence of some 20,000 international (mostly Russian) peacekeepers. The Tajik army numbers only around 6000. There is no navy or air force.

ECONOMICS
▷ Inflation 235% p.a. (1990–2000)

$1.11bn — 2.40–2.55 somoni

SCORE CARD

- ❏ Wᴏʀʟᴅ GNP Rᴀɴᴋɪɴɢ.........................151st
- ❏ GNP ᴘᴇʀ Cᴀᴘɪᴛᴀ $180
- ❏ Bᴀʟᴀɴᴄᴇ ᴏꜰ Pᴀʏᴍᴇɴᴛꜱ......................–$61m
- ❏ Iɴꜰʟᴀᴛɪᴏɴ 33%
- ❏ Uɴᴇᴍᴘʟᴏʏᴍᴇɴᴛ 6%

Sᴛʀᴇɴɢᴛʜꜱ
Few, though Tajikistan has 14% of known world uranium reserves. Hydroelectric power has considerable potential. Carpet-making.

Wᴇᴀᴋɴᴇꜱꜱᴇꜱ
Formal economy precarious. Dependence on barter economy. No central planning. Little diversification in agriculture; only 6% of land is arable. Exodus of skilled Russians. Production in all sectors in decline.

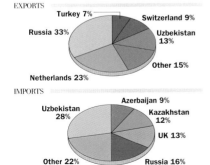

EXPORTS
Turkey 7%
Switzerland 9%
Russia 33%
Uzbekistan 13%
Other 15%
Netherlands 23%

IMPORTS
Azerbaijan 9%
Uzbekistan 28%
Kazakhstan 12%
UK 13%
Other 22%
Russia 16%

RESOURCES
▷ Electric power 4.4m kw

 80 tonnes
381 b/d (reserves 15m barrels)
1.36m sheep, 1.05m cattle, 900,000 chickens
Uranium, gold, iron, coal, lead, mercury, tin

Tajikistan has one key resource – uranium – which accounted for 30% of the USSR's total production before 1990. The end of the nuclear arms race has reduced its value, however. Most of Tajikistan is bare mountain, and just 6% of the land can be used for agriculture. Industry is concentrated in the Fergana Valley, close to the Uzbek border.

ENVIRONMENT
▷ Sustainability rank: 110th

 4%
0.8 tonnes per capita

Landslides, particularly on the lower slopes of the Pamirs, are a serious problem. They are caused as much by the natural geography and by earthquakes as by human activity.

MEDIA
▷ TV ownership high

 Daily newspaper circulation 21 per 1000 people

PUBLISHING AND BROADCAST MEDIA

There are 3 daily newspapers, but the weeklies *Djavononi Todjikiston* and *Tochikiston ovozi* are more influential

3 state-controlled services
1 state-controlled service

Communist control over the media was tightened in early 1994 with the takeover by President Rakhmanov of the press and broadcast media.

CRIME
▷ Death penalty in use

 11,000 prisoners
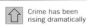 Crime has been rising dramatically

Only remote areas escape the violence perpetrated by armed gangs. Narcotics smuggling along the border with Afghanistan continues to increase.

EDUCATION
▷ School leaving age: 17

 99%
 108,203 students

Literacy rates are high. The university at Dushanbe has been weakened by the departure of its Russian academics.

HEALTH
▷ Welfare state health benefits

 1 per 500 people
Heart, cerebrovascular, respiratory, infectious, and parasitic diseases

Theoretically health provision is free of charge; however, massive underfunding means that patients invariably have to pay to get treatment.

SPENDING
▷ GDP/cap. decrease

CONSUMPTION AND SPENDING

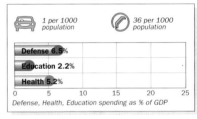

1 per 1000 population
36 per 1000 population

Defense 6.5%
Education 2.2%
Health 5.2%

Defense, Health, Education spending as % of GDP

More than 80% of Tajik people live below the poverty line; the war against the Islamist rebels worsened conditions. The former communist bureaucrats continue to be the wealthiest group.

WORLD RANKING

T

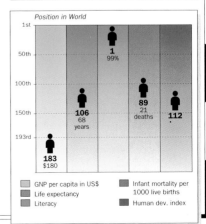

Position in World

1st
50th
100th
150th
193rd

1 — 99%
106 — 68 years
89 — 21 deaths
112
183 — $180

- GNP per capita in US$
- Life expectancy
- Literacy
- Infant mortality per 1000 live births
- Human dev. index

TANZANIA

OFFICIAL NAME: United Republic of Tanzania   **CAPITAL:** Dodoma
POPULATION: 36 million   **CURRENCY:** Tanzanian shilling   **OFFICIAL LANGUAGES:** English and Kiswahili

TANZANIA LIES BETWEEN Kenya and Mozambique on the east African coast. Formed by the union of Tanganyika and the Zanzibar islands, Tanzania comprises a coastal lowland, volcanic highlands, and the Great Rift Valley. It includes Mount Kilimanjaro, Africa's highest peak. Tanzania was led by the socialist Julius Nyerere from 1962 until 1985. His Revolutionary Party of Tanzania (CCM) has won multiparty elections held in 1995 and 2000.

Arusha National Park. *Lying within the Ngurdoto volcanic crater, the park has herds of buffalo, rhino, elephant, and giraffe.*

CLIMATE
▷ Tropical/mountain

WEATHER CHART FOR DODOMA

The coast and Zanzibar are tropical. The central plateau is semiarid and the highlands are semitemperate.

TRANSPORTATION
▷ Drive on left

Dar es Salaam International
621,513 passengers

53 ships
35,500 grt

THE TRANSPORTATION NETWORK

3704 km
(2302 miles)

None

2721 km
(1691 miles)

Lakes Tanganyika, Victoria, and Nyasa are navigable

The roads, railroads, and ports are being upgraded, notably by an $870 million program to improve 70% of Tanzania's trunk roads.

TOURISM
▷ Visitors : Population
1:78

459,000 visitors

Down 27% in 2000

MAIN TOURIST ARRIVALS

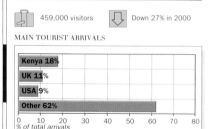

| Kenya 18% |
| UK 11% |
| USA 9% |
| Other 62% |

% of total arrivals

One-third of Tanzania is national park or game reserve. The Ngorongoro Crater and the Serengeti Plain are top attractions. Tourist numbers increased dramatically in the 1990s.

PEOPLE
▷ Pop. density low

Kiswahili, Sukuma, Chagga, Nyamwezi, Hehe, Makonde, Yao, Sandawe, English

41/km²
(105/mi²)

THE URBAN/RURAL POPULATION SPLIT

33%   67%

RELIGIOUS PERSUASION

Other 4%
Muslim 33%
Traditional beliefs 30%
Christian 33%

For many Tanzanians the family is the focus of traditional rural life. About 99% belong to one of 120 small ethnic Bantu groups. The remainder comprise Arab, Asian, and European minorities. The use of Kiswahili as a *lingua franca* has helped make ethnic rivalries almost nonexistent.

POLITICS
▷ Multiparty elections

2000/2005

President Benjamin Mkapa

AT THE LAST ELECTION

National Assembly 296 seats   CUF 7%   Others 7%

82% CCM   TLP 2%   Chadema 2%

CCM = Revolutionary Party of Tanzania   **CUF** = Civic United Front   **TLP** = Tanzania Labor Party   **Chadema** = Party for Democracy and Progress
Others include 5 members chosen by the Zanzibar House of Representatives, 10 appointed by the president, and the attorney general who has a seat ex-officio

Julius Nyerere was the dominant force in Tanzanian politics for over two decades. He founded the ruling party, the CCM, and his philosophy of African socialism guided Tanzania's development. Ali Hassan Mwinyi succeeded Nyerere as president in 1985, introducing a transition to multiparty democracy. Mwinyi stood down in 1995, and Benjamin Mkapa was elected president. Separatism in Zanzibar is a key issue, flaring into violent protests in 2001.

TANZANIA

Total Land Area :
945 087 sq. km
(364 898 sq. miles)

POPULATION

over 1 000 000
over 100 000
over 50 000
over 10 000
under 10 000

LAND HEIGHT

3000m/9843ft
2000m/6562ft
1000m/3281ft
500m/1640ft
200m/656ft
Sea Level

WORLD AFFAIRS Joined UN in 1961

The instability of Tanzania's central African neighbors is a concern. It accepted over half a million Rwandan and Burundian refugees in the 1990s, but instigated mass repatriations in 1996. Improved relations with Uganda and Kenya since 1985 led to the rebirth in 2001 of the East African Community.

AID Recipient

 $1.05bn (receipts) — Up 6% in 2000

Tanzania is heavily dependent on aid to help offset a severe balance-of-payments deficit. The World Bank, Japan, and the UK are the major donors. Most aid is now linked to an IMF-backed economic reform program. In 2001 $3 billion of debt was canceled.

DEFENSE Compulsory military service

 $141m — No change in 2000

Defense accounts for 3.5% of budget spending. The armed forces are closely linked with the ruling CCM. There is an 80,000-strong citizens' reserve force.

ECONOMICS Inflation 22% p.a. (1990–2000)

$9.01bn — 805–917 Tanzanian shillings

SCORE CARD

❏ WORLD GNP RANKING	88th
❏ GNP PER CAPITA	$270
❏ BALANCE OF PAYMENTS	–$517m
❏ INFLATION	5.9%
❏ UNEMPLOYMENT	Not available

STRENGTHS
Coffee, cotton, sisal, tea, cashew nuts. Zanzibar a major producer of cloves. Diamonds, gold. State commitment to effective reforms. Expansion in nontraditional exports. Rise in inward investment. Return to positive growth.

WEAKNESSES
Growth still too low to increase per capita income. Shortage of foreign exchange. Poor credit and equipment limit agricultural development.

EXPORTS

Kenya 5%
Netherlands 6%
Germany 9%
India 13%
Other 47%
UK 20%

IMPORTS

Australia 6%
Kenya 6%
UK 7%
Japan 9%
Other 61%
South Africa 11%

RESOURCES Electric power 543,000 kw

 310,270 tonnes — Limited oil production

14.4m cattle, 10m goats, 4.25m sheep, 30m chickens — Natural gas, oil, iron, diamonds, gold, salt, phosphates, coal

Agriculture accounts for half of GDP and 80% of employment and exports. Forests cover 50% of land. More than 90% of energy demand is met from wood and charcoal. Hydropower provides 70% of electricity and is being expanded. Oil imports take 40% of export earnings, but offshore gas at Songo Songo is being exploited. Oil has been discovered off Pemba Island. The opening of a gold mine near Mwanza makes Tanzania Africa's fourth-largest gold producer.

ENVIRONMENT Sustainability rank: 80th

 16% — 0.1 tonnes per capita

The demand for fuelwood is a threat to forests. Tourism's demands have to be carefully balanced with those of fragile wildlife environments such as the Ngorongoro Crater and the Serengeti.

MEDIA TV ownership low

Daily newspaper circulation 4 per 1000 people

PUBLISHING AND BROADCAST MEDIA

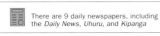

There are 9 daily newspapers, including the *Daily News*, *Uhuru*, and *Kipanga*

3 independent services

5 services: 2 state-owned, 3 independent

The press has grown rapidly since the mid-1990s, with several independent papers. There is no national TV station.

CRIME Death penalty in use

 45,611 prisoners — Down 1% in 1999

Crime levels are low, although theft in Dar es Salaam has risen. Tanzania's human rights record is good.

EDUCATION School leaving age: 14

 76% — 18,867 students

Primary education, which begins at seven and lasts for seven years, is free; secondary students pay fees. In 1999, enrollment was 75% at primary level, but only 6% for secondary education.

HEALTH Welfare state health benefits

 1 per 20,000 people — Diarrheal and respiratory diseases, malaria

There is a national campaign against HIV/AIDS, which affects 7.8% of the adult population. An immunization program for under-twos was introduced in 2001.

SPENDING GDP/cap. increase

CONSUMPTION AND SPENDING

1 per 1000 population — 5 per 1000 population

Defense 1.8%
Education 2.1%
Health 1.3%

Defense, Health, Education spending as % of GDP

The majority of Tanzanians are subsistence farmers. The small wealthy elite is composed mainly of Asian and Arab business families.

WORLD RANKING

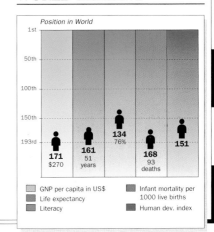

Position in World

171 $270
161 51 years
134 76%
168 93 deaths
151

GNP per capita in US$
Life expectancy
Literacy
Infant mortality per 1000 live births
Human dev. index

T

THAILAND

OFFICIAL NAME: Kingdom of Thailand **CAPITAL:** Bangkok
POPULATION: 63.6 million **CURRENCY:** Baht **OFFICIAL LANGUAGE:** Thai

THAILAND LIES IN THE HEART of southeast Asia. The north, the border with Burma, and the long Isthmus of Kra between the Andaman Sea and the Gulf of Thailand are mountainous. The central plain is the most fertile and densely populated area, while the low northeastern plateau is the poorest region. Thailand has been an independent kingdom for most of its history, and since 1932 a constitutional monarchy, though with frequent periods of military government. Continuing rapid industrialization results in massive congestion in Bangkok and a serious depletion of natural resources.

CLIMATE

▷ Tropical equatorial/monsoon

WEATHER CHART FOR BANGKOK

Thailand's tropical monsoon climate has three seasons – a hot sultry period, rains from May to October, and a dry, cooler season from November to March.

TRANSPORTATION

▷ Drive on left

Bangkok International
29.6m passengers

552 ships
2m grt

THE TRANSPORTATION NETWORK

62,985 km (39,137 miles)	None
4623 km (2873 miles)	3701 km (2300 miles)

Bangkok suffers from huge traffic jams. Its first mass transit system became operational in 1999: as well as the elevated railroad, subway lines are planned. A road corridor through Laos into Vietnam was agreed in 2001.

Island in the Andaman Sea. *Overdevelopment at Thailand's best-known resorts is pushing tourism into new, remoter locations.*

TOURISM

▷ Visitors : Population 1:6.6

9.58m visitors

Up 11% in 2000

MAIN TOURIST ARRIVALS

Japan 12%
Malaysia 12%
China 9%
Taiwan 6%
Singapore 6%
Other 55%

% of total arrivals

Tourism is an important contributor to the Thai economy. Tourist numbers fell in the early 1990s as a result of both the worldwide recession and local overdevelopment during the 1980s boom. Although the number of arrivals has since recovered, visitors are tending to seek the less-developed resorts. Bangkok's hotel occupancy rates continue to fall as yet more hotels are built. Pattaya beach resort, opposite Phetchaburi has seen such uncontrolled development that sea pollution is now a serious problem, while opposition to the intrusion of large numbers of tourists is growing among northern hill tribes.

Although prostitution is illegal, Bangkok and Pattaya are centers for sex tourism, which thrives despite the state's embarrassment at its effect on Thailand's image. Japanese and German men are among the main clients, while Burmese girls are increasingly recruited as prostitutes. Child prostitution is also a major problem.

There has been a boom in golf tourism, especially among the Japanese. The large number of new golf courses have made Thailand one of the largest golf destinations in Asia. The vast amounts of water needed to maintain the courses is aggravating Thailand's serious water shortage.

PEOPLE

▷ Pop. density medium

Thai, Chinese, Malay, Khmer, Mon, Karen, Miao

124/km² (322/mi²)

THE URBAN/RURAL POPULATION SPLIT

22% 78%

RELIGIOUS PERSUASION

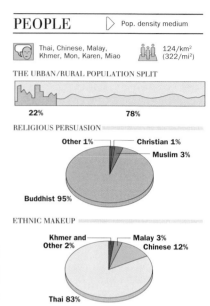

Other 1%
Christian 1%
Muslim 3%
Buddhist 95%

ETHNIC MAKEUP

Khmer and Other 2%
Malay 3%
Chinese 12%
Thai 83%

There is little ethnic tension in Thailand, and Buddhism is a great binding force. The majority of Thais follow Theravada Buddhism, although the reformist Asoke Santi Buddhist sect, which advocates a new moral austerity, is gaining influence. Its principles have been espoused in particular by the Palang Dharma (PD), which seeks to clean up politics.

The far north and northeast hills are home to about 600,000 tribespeople with their own languages, and to permanently settled refugees from Laos, mostly of the Hmong tribal group.

The large Chinese community is the most assimilated in southeast Asia. Sino-Thais are particularly dominant in agricultural marketing. Most of Thailand's one million Muslim Malays live in southern Thailand, bordering Malaysia. They feel stronger affinity with Muslims in Malaysia than with Thai culture, and this has given rise to a secessionist movement.

Women are important in business, but their involvement in national politics is limited.

POPULATION AGE BREAKDOWN

Female	Age	Male
0%	80+	0%
4.5%	60–79	3.9%
10.3%	40–59	9.9%
17.3%	20–39	17.6%
18%	0–19	18.5%

% of population by age group

T

POLITICS ▷ Multiparty elections

L. House 2001/2005
U. House 2000/2006

H.M. King Bhumibol Adulyadej (Rama IX)

AT THE LAST ELECTION
House of Representatives 500 seats

	8% CT	6% CP

49% TRT	26% DP	7% NAP	4% Others

TRT = Thais Love Thais **DP** = Democrat Party
CT = Thai Nation **NAP** = New Aspiration Party
CP = National Development

Senate 200 seats

Under the 1997 constitution,
the members of the Senate
are directly elected
on a nonparty basis.

Thailand is a multiparty parliamentary democracy. The king is the head of state. Despite his position as a constitutional monarch, he has immense personal prestige. Criticism of the king is not tolerated.

PROFILE
The Thai political process, dominated by the military for decades until the 1990s, is highly personalized. Parties seldom have strong ideologies. Coalitions are often unstable, while the lack of coordination between coalition partners is a recurring problem. The NAP was the largest party in the parliament elected in 1996, but its leader Chaovalit Yongchaiyuth was prime minister only for a year. His government, blamed for mismanaging an economic crisis, was then ousted in favor of another coalition under Chuan Leekpai of the DP, who succeeded in retaining office for four years despite the volatility of coalition politics. However, in the 2001 elections the new populist TRT triumphed, winning just short of a majority of seats. TRT leader and former deputy prime minister Thaksin Shinawatra formed a three-party government. The NAP and TRT merged in 2002.

MAIN POLITICAL ISSUES
The military–democratic cycle
Thailand has been ruled by alternating military and civilian governments since 1932. In 1992 promilitary parties chose an unelected army general as prime minister. Violent demonstrations erupted which forced the resignation of the prime minister and precipitated a constitutional amendment obliging future prime ministers to be elected members of the legislature. The military

King Bhumibol.
On the throne since 1946, he is the world's longest-serving ruler.

Thaksin Shinawatra,
a controversial billionaire, elected premier in 2001.

has since maintained a low profile during general elections, and has refrained from interfering in government changes.

Congestion in Bangkok
A major issue is the concentration of industry and commerce in the Bangkok area. Uncontrolled development has left it with traffic congestion which not only is among the world's worst but is also a serious hindrance to economic activity. In December 1999 the first stage of a mass transit system – an elevated railroad – was formally opened.
Since 1993, the government has offered incentives for relocating industry to the provinces. This is also intended to help distribute wealth more evenly – up to 60% of GDP is generated in the Bangkok area.

Water
The national water shortage, caused by rapid industrialization, is so acute that it is affecting farm and industrial output.

WORLD AFFAIRS ▷ Joined UN in 1946

| APEC | ASEAN | Mekong River | NAM | WTO |

Thailand has friendly relations with China. However, relations with neighboring Burma have been strained over Burma's alleged support for Thai ethnic guerrillas operating along the Thai–Burmese border. Many Thai logging companies, often run by the military, have been active in Burma since Thailand's 1988 logging ban at home. Relations with Vietnam are cordial, despite Thailand's opposition to the Vietnamese regime in Cambodia in the 1980s.
Thailand, Indonesia, and Malaysia have liberalized trade to promote development in each country in regions which are distant from their respective capitals.
Thailand maintains close relations with the US, despite some tension over intellectual property rights and minor trade issues, but no longer has any US military bases on its territory.

THAILAND
Total Land Area : 514 000 sq. km (198 455 sq. miles)

LAND HEIGHT	POPULATION
2000m/6562ft	over 5 000 000
1000m/3281ft	over 1 000 000
500m/1640ft	over 100 000
200m/656ft	over 50 000
Sea Level	over 10 000

200 km
200 miles

T

AID

 ▷ Recipient

 $641m (receipts) ⬇ Down 37% in 2000

Japan and the World Bank are the largest aid donors. Thailand has imposed a ceiling on foreign borrowing to keep its debt stable.

CHRONOLOGY

Thailand emerged as a kingdom in the 13th century, and by the late 17th century its then capital, Ayutthya, was the largest city in southeast Asia. In 1782, the present Chakri dynasty and a new capital, Bangkok, were founded.

❑ **1855** King Mongut signs Bowring trade treaty with British – Thailand never colonized by Europeans.

❑ **1868–1910** King Chulalongkorn westernizes Thailand.

❑ **1907** Thailand cedes western Khmer (Cambodia) to France.

❑ **1925** King Prajadhipok begins absolute rule.

❑ **1932** Bloodless military–civilian coup. Constitutional monarchy.

❑ **1933** Military takes control.

❑ **1941** Japanese invade. Government collaborates.

❑ **1944** Pro-Japanese prime minister and prewar military dictator Phibun voted out of office.

❑ **1945** Exiled King Ananda returns.

❑ **1946** Ananda assassinated. King Bhumibol accedes.

❑ **1947** Military coup. Phibun back.

❑ **1957** Military coup. Constitution abolished.

❑ **1965** Thailand allows US to use Thai bases in Vietnam War.

❑ **1969** New constitution endorses elected parliament.

❑ **1971** Army suspends constitution.

❑ **1973–1976** Student riots lead to interlude of democracy.

❑ **1976** Military takeover.

❑ **1980–1988** Gen. Prem Tinsulanond prime minister. Partial democracy.

❑ **1988** Elections. Gen. Chatichai Choonhaven, right-wing CT leader, named prime minister.

❑ **1991** Military coup. Civilian Anand Panyarachun caretaker premier.

❑ **1992** Elections. Gen. Suchinda named premier. Demonstrations. King forces Suchinda to step down and reinstalls Anand. Moderates win new elections.

❑ **1995** CT wins general election.

❑ **1996** Early elections; Chaovalit Yongchaiyuth of NAP becomes prime minister.

❑ **1997** Financial and economic crisis; Chaovalit government falls; DP's Chuan Leekpai prime minister.

❑ **2001** TRT, led by Thaksin Shinawatra, wins elections.

T

DEFENSE

 ▷ No compulsory military service

 $2.46bn ⬇ Down 7% in 2000

THAI ARMED FORCES

333 main battle tanks (50 PRC Type-69, 105 M-48A5, 178 M-60)	190,000 personnel
1 carrier, 12 frigates, and 88 patrol boats	68,000 personnel
153 combat aircraft (14 F-5A/B, 34 F-16A/B, 36 F-5E/F)	48,000 personnel
None	

The military either ruled Thailand, or played a prominent role in politics, for over half a century from 1932. In 1996, its role in the appointed Senate – hitherto a military stronghold – was reduced. Retired military figures are, however, prominent in the major political parties.

Since 1986, defense spending has tended to focus on the navy and air force. Naval vessels have been bought from China, Germany, and Spain, and aircraft from the UK, the US, and Russia.

Thailand's main defense concerns are border disputes with Cambodia, Burma, and Laos; the Muslim secessionist movement in the south; and piracy and fishing disputes in the South China Sea.

ECONOMICS

▷ Inflation 4.2% p.a. (1990–2000)

ECONOMIC PERFORMANCE INDICATOR

$122bn 43.38–44.23 baht

SCORE CARD

❑ WORLD GNP RANKING	31st
❑ GNP PER CAPITA	$2000
❑ BALANCE OF PAYMENTS	$9.2bn
❑ INFLATION	1.5%
❑ UNEMPLOYMENT	4%

EXPORTS

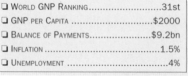

Malaysia 4%
Hong Kong 5%
Singapore 9%
Japan 16%
Other 43%
USA 23%

IMPORTS

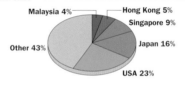

Hong Kong 5%
Malaysia 6%
Singapore 11%
Other 40%
USA 12%
Japan 26%

STRENGTHS

Success of export-based and import-substituting manufacturing. Rapid economic growth. Natural gas. Tourism. Major world exporter of rice and rubber.

WEAKNESSES

Concentration of economic activity in congested Bangkok area. Inadequate water storage facilities. Rapid growth of foreign debt. 60% of population in low-profit farming.

PROFILE

Until the late 1990s, the economy grew at over 9% a year for a decade, driven by a rise in manufacturing and huge overseas investments, especially from Japan. However, as domestic wages rose, Thailand faced stiff competition from China and Vietnam. Thailand also lacked a skilled labor force to develop high-tech production, though it is a big producer of electronics goods.

In 1997 mounting foreign debt and the sharp depreciation of the baht made necessary an IMF-led rescue package. Massive retrenchment and stringent austerity measures followed. By mid-2000 the IMF had ended its direct involvement and was optimistic for future expansion. GDP grew by 4.4% in the following year.

THAILAND : MAJOR BUSINESSES

Khon Kaen
Nakhon Ratchasima
Northanburi
Samut Prakan
Thon Buri
Chon Buri
Bangkok
Si Racha
Pran Buri
Gulf of Thailand

Gas
Textiles
Computers
Oil refining
Petrochemicals
Food processing
Consumer goods
Vehicle manufacture

0   200 km
0   200 miles

RESOURCES ▷ Electric power 21.5m kw

3.61m tonnes

178,000 b/d (reserves 500m barrels)

23m ducks, 8.3m pigs, 6.3m cattle, 190m chickens

Tin, lignite, gas, gems, oil, tungsten, lead, zinc, antimony, gold, copper

ELECTRICITY GENERATION

Hydro 5% (5.2bn kwh)	
Combustion 95% (90bn kwh)	
Nuclear 0%	
Other 0%	

0 20 40 60 80 100

% of total generation by type

Thailand has minimal crude oil and has rejected the nuclear option in favor of speeding up development of its large natural gas fields. It also has significant lignite deposits for power generation. World demand for Thailand's tin has declined, but recent gold and copper finds offer new potential. Thailand has valuable gemstone deposits. It is the world's fourth-largest producer of shrimp.

THAILAND : LAND USE

Cropland
Forest
Pasture
Cattle
Rubber – cash crop
Rice

0   200 km
0   200 miles

ENVIRONMENT ▷ Sustainability rank: 54th

14%

3.2 tonnes per capita

ENVIRONMENTAL TREATIES

Yes		Yes		No	
Yes		Yes		No	

Deforestation, especially of the watersheds in the north, has led to the increasing severity of both floods and droughts. Particularly serious flooding in the south resulted in a total logging ban in 1988. Illegal logging continues, however. Reafforestation projects, some criticized for using single quick-growing species, will not solve the national water shortage. Intensive inland prawn farming leaves a legacy of salination. Mass tourism brings pollution problems to resorts, while backpackers lead the way in exposing remoter locations.

MEDIA ▷ TV ownership high

Daily newspaper circulation 64 per 1000 people

PUBLISHING AND BROADCAST MEDIA

There are 30 daily newspapers, including *Matichon, Siam Rath, Sing Sian Yit Pao, Bangkok Post,* and *Nation*

9 services

4 state-controlled services

Newspapers enjoy a high level of freedom in political reporting. Two of the TV stations are run by the military. There are 60 radio stations around Bangkok, mainly state-run.

CRIME ▷ Death penalty in use

206,011 prisoners

Up 95% 1992–1996

CRIME RATES

Murders	
7	*per 100,000 population*

Rapes	
6	*per 100,000 population*

Thefts	
72	*per 100,000 population*

Political imprisonment is now extremely rare. There is some police involvement in crime, including extrajudicial killings and ill-treatment of prisoners in detention. The king has inspired an opium-substitution crop program. In the south, drug addiction is a major problem and crime is rising. The government has cracked down on music, software, and video piracy.

EDUCATION ▷ School leaving age: 16

96%

1.52m students

THE EDUCATION SYSTEM

100

80

60

40

20

0

% of each age group in education

94%   88%   30%

Primary   Secondary   Tertiary

A poorly developed education system has led to a shortage of skills needed for the expansion of high-tech industries. The Education for All program aims to rectify this situation, and also to improve adult literacy levels.

HEALTH ▷ Welfare state health benefits

1 per 2500 people

Heart diseases, gastroenteritis

High-quality health care is heavily concentrated in Bangkok. Most of the rural population has access to primary health care, and a new scheme to provide care for the poor for just $1 per person was launched in 2001. Trained personnel are aided by village health volunteers, monks, teachers, and traditional healers.

However, estimates suggest that only 30% of users can afford to pay. The poor can apply annually for a certificate entitling them to free health care.

High-profile family planning programs are slowing population growth. An effective AIDS prevention campaign has helped reduce the number of new infections, although the government has been in conflict with international drug companies over its right to produce cheaper generic drugs for AIDS sufferers. Prostitutes have benefited from an extensive sex education program.

SPENDING ▷ GDP/cap. increase

CONSUMPTION AND SPENDING

28 per 1000 population

92 per 1000 population

Defense 2%	
Education 4.7%	
Health 1.9%	

0   5   10   15   20   25

Defense, Health, Education spending as % of GDP

The government is trying to diffuse to the provinces the people and wealth currently concentrated to a very great extent in Bangkok. The northeast in particular is very poor. The gap between rich and poor is greater in Thailand than in other industrializing southeast Asian states.

WORLD RANKING

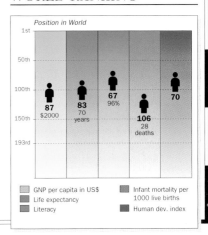

Position in World

1st

50th

100th

150th

193rd

87 $2000

83 70 years

67 96%

106 28 deaths

70

GNP per capita in US$
Life expectancy
Literacy

Infant mortality per 1000 live births
Human dev. index

T

TOGO

OFFICIAL NAME: Republic of Togo **CAPITAL:** Lomé **POPULATION:** 4.7 million
CURRENCY: CFA franc **OFFICIAL LANGUAGE:** French

1960 · 1960 · April 27 · TG · 0 · +228 · .tg

TOGO IS SANDWICHED between Ghana and Benin in west Africa. A central forested region is bounded by savanna lands to the north and south. The port of Lomé is an important entrepôt for west African trade. Togo's president, Gen. Gnassingbé Eyadéma, has been in power since 1967.

CLIMATE

▷ Tropical equatorial/ wet and dry

WEATHER CHART FOR LOMÉ

Togo has a typical Gulf of Guinea climate – very hot and humid on the coast, and drier inland.

TRANSPORTATION

▷ Drive on right

Tokoin, Lomé
232,578 passengers

9 ships
1608 grt

THE TRANSPORTATION NETWORK

2376 km (1476 miles)		None
517 km (321 miles)		50 km (31 miles)

Improving the already good road network and Lomé's port facilities are priorities, given Togo's role as an entrepôt. Air and rail links to the interior, however, are limited.

TOURISM

▷ Visitors : Population 1:78

60,000 visitors

Down 14% in 2000

MAIN TOURIST ARRIVALS

France	14%
Burkina	14%
Benin	11%
Other	61%

0 10 20 30 40 50 60 70 80
% of total arrivals

There is some package tourism, mainly French and German, to coastal tourist villages and hotels built during the expansion program of the 1980s. Tourists have been deterred by the political uncertainty since 1990.

PEOPLE

▷ Pop. density medium

Ewe, Kabye, Gurma, French

86/km² (224/mi²)

THE URBAN/RURAL POPULATION SPLIT

33%   67%

RELIGIOUS PERSUASION

Muslim 15%
Traditional beliefs 50%
Christian 35%

A bitter divide has existed between north and south since before independence. Most southern resentment is directed toward a northern minority, the Kabye people from the Kabye plateau, because of their domination of the military. The Kabye and other northerners in turn resent their own underdevelopment in contrast to the high development, especially educationally, of all southerners. The dominant southern group is the Ewe, who make up more than 40% of the population.

As elsewhere in Africa, the extended family is important and tribalism and nepotism are key factors in everyday life. Some Togolese ethnic groups, such as the Mina, have matriarchal societies. The "Nana Benz," the market-women of Lomé, control the retail trade and have considerable private money. Politics, however, remains a male preserve.

Kabye cultivations near Kara, in northern Togo. The main food crops grown are cassava, yams, and maize.

POLITICS

▷ Multiparty elections

1999/2001 (postponed repeatedly)

President Gnassingbé Eyadéma

AT THE LAST ELECTION

National Assembly 81 seats

95% RPT   3% Ind   2% Vac

RPT = Rally of the Togolese People
Ind = Independents
Vac = Vacant: results of two seats were annulled

Politics has been dominated for over three decades by Gen. Gnassingbé Eyadéma, in power since 1967 and Africa's longest-serving leader.

A democracy movement has been gathering momentum since 1990. Multiparty presidential elections held in 1993 were won by Eyadéma, although some opposition candidates boycotted the poll over the exclusion of Gilchrist Olympio, son of a former president. Eyadéma claimed victory over Olympio in presidential elections in 1998, amid accusations of malpractice and of the killing of hundreds of opposition supporters immediately afterward in the runup to the 1999 Assembly election. (Serious human rights violations were later confirmed by a UN/OAU report.) During subsequent negotiations, the opposition accepted the election results, Eyadéma stated that he would not stand for reelection in 2003, and an accord provided for a new independent electoral body and a political code of conduct. New Assembly elections, set for late 2001, were repeatedly postponed, however, as the deadlock between Eyadéma and the opposition continued.

WORLD AFFAIRS

▷ Joined UN in 1960

OIC · ECOWAS · FZ · AU · UEMOA

The priority is maintaining traditional links, especially with France. In 1998 President Eyadéma became chair of ECOWAS, acting as mediator in the Guinea-Bissau conflict and hosting talks in Sierra Leone. He was also president of the OAU for 2000/2001.

AID

▷ Recipient

 $70m (receipts)   Down 1% in 2000

Development projects and the health of the economy overall have suffered from aid suspensions in the 1990s by donors including the US and the EU.

T

TOGO

Total Land Area : 56 785 sq. km
(21 924 sq. miles)

POPULATION

over 100 000	◉
over 10 000	●
under 10 000	•

LAND HEIGHT

500m/1640ft
200m/656ft
Sea Level

DEFENSE

 Compulsory military service

$30m   Down 12% in 2000

The military has an important role in Togo, and spending on defense is quite high. The army's senior ranks are dominated by loyalists from President Eyadéma's northern Kabré tribe. France guarantees Togo's security through a defense accord, and supplies most military equipment and training.

ECONOMICS

Inflation 7.1% p.a. (1990–2000)

$1.32bn   698.7–736.7 CFA francs

SCORE CARD

❏ World GNP Ranking	149th
❏ GNP per Capita	$290
❏ Balance of Payments	–$106m
❏ Inflation	1.9%
❏ Unemployment	Not available

STRENGTHS

Efficient civil service. Ideal location for role as entrepôt, based on Lomé port. Resourceful entrepreneurs, notably market-women. Proceeds of widespread smuggling. Phosphate deposits have the world's highest mineral content. Self-sufficient in basic foodstuffs.

RESOURCES

Electric power 34,000 kw

 23,074 tonnes   Not an oil producer

1.42m goats, 1m sheep, 289,200 pigs, 8.5m chickens   Phosphates, iron, chromite, bauxite, marble, dolomite

Phosphates are Togo's most important resource. Offshore oil and gas deposits were found in 1999. The Nangbeto Dam, constructed jointly with Benin and opened in 1988, has reduced dependence on Ghana for energy.

ENVIRONMENT

Sustainability rank: 105th

8%   0.2 tonnes per capita

Ecologists have been critical of the transformation of nature reserves into hunting grounds for the military elite. Other problems include coastal erosion around Aneho and desertification.

MEDIA

TV ownership low

 Daily newspaper circulation 4 per 1000 people

PUBLISHING AND BROADCAST MEDIA

There are 2 daily newspapers, *Togo-Presse*, published by the government, and *Les Echos du Demain*	
1 state-owned service	3 services: 1 state-owned, 2 independent

Opposition papers now challenge the government daily *Togo-Presse*, despite some official harassment.

CRIME

Death penalty not used in practice

2043 prisoners   Crime is rising

Togo is normally relatively peaceable, but urban crime generally increased during the 1990s, particularly during periods of political unrest in the capital.

EXPORTS

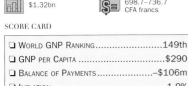

South Africa 4%
Ghana 4%
Other 65%
Colombia 4%
Nigeria 10%
Benin 13%

IMPORTS

Hong Kong 5%
Other 44%
China 7%
Ivory Coast 8%
France 11%
Ghana 25%

WEAKNESSES

Political pariah status led to aid reductions in the 1990s. Low world prices for phosphates. Hydropower generation is vulnerable to drought.

CHRONOLOGY

After colonization by Germany in 1894, Togoland was divided between France and the UK in 1922.

❏ **1960** French sector independent as Togo (UK part joined to Ghana).
❏ **1967** Eyadéma takes power.
❏ **1991–1992** General strike; repression.
❏ **1993** Eyadéma elected president.
❏ **1998** Eyadéma claims victory in disputed election.

EDUCATION

School leaving age: 15

57%   15,028 students

Schooling is based on the French model. The University of Bénin in Lomé has more than 4000 students.

HEALTH

No welfare state health benefits

1 per 10,000 people   Malaria, diarrheal, infectious, and parasitic diseases

Health care suffers from a lack of resources. Around 6% of adults were HIV positive by 2000.

SPENDING

GDP/cap. increase

CONSUMPTION AND SPENDING

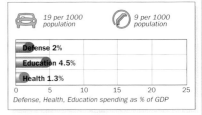

19 per 1000 population   9 per 1000 population

Defense 2%
Education 4.5%
Health 1.3%

Defense, Health, Education spending as % of GDP

Considerable wealth disparities exist between those who work the land and the country's political and business classes. The urban class has been hit by an economic downturn in the late 1990s.

WORLD RANKING

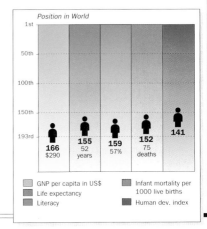

Position in World

166	155	159	152	141
$290	52 years	57%	75 deaths	

☐ GNP per capita in US$
☐ Life expectancy
☐ Literacy
☐ Infant mortality per 1000 live births
☐ Human dev. index

T

TONGA

PACIFIC OCEAN

OFFICIAL NAME: Kingdom of Tonga **CAPITAL:** Nuku'alofa **POPULATION:** 102,200
CURRENCY: Pa'anga (Tongan dollar) **OFFICIAL LANGUAGES:** English and Tongan

1970 1970 June 4 TO +13 +676 .to

LOCATED IN THE SOUTH PACIFIC northeast of New Zealand, Tonga is an archipelago of 170 islands. These are divided into three main groups, Vava'u, Ha'apai, and Tongatapu. Tonga's easterly islands are generally low and fertile. Those in the west are higher and volcanic in origin. Tonga's economy is based on agriculture, especially coconut, cassava, and passion fruit production. Politics is effectively controlled by the king.

CLIMATE

▷ Tropical oceanic

WEATHER CHART FOR NUKU'ALOFA

Tonga has a tropical oceanic climate, with year-round temperatures ranging between 17°C (63°F) and 30°C (86°F).

TRANSPORTATION

▷ Drive on left

Fua'amotu International, Tongatabu
67,000 passengers

18 ships
22,200 grt

THE TRANSPORTATION NETWORK

| 184 km (114 miles) | None |
| None | None |

Improvements at Fua'amotu Airport have led to an increase in flights to and from Tonga.

TOURISM

▷ Visitors : Population 1:2.9

35,000 visitors   Up 13% in 2000

MAIN TOURIST ARRIVALS

| New Zealand 32% |
| USA 19% |
| Australia 19% |
| Other 30% |

% of total arrivals

Tonga's main attractions are its tropical beaches. Flagging tourism has been boosted by political insecurity in the Solomon Islands and Fiji. Fears have been expressed that too many visitors may erode traditional Tongan culture.

Mountainous scenery typical of the westerly islands. Tonga's 170 islands are scattered over a wide expanse of the South Pacific. Only 45 are inhabited.

TONGA

Total Land Area : 748 sq. km (289 sq. miles)

POPULATION
● over 10 000
• under 10 000

LAND HEIGHT
200m/656ft
Sea Level

PEOPLE

▷ Pop. density medium

English, Tongan

142/km² (368/mi²)

THE URBAN/RURAL POPULATION SPLIT

43%   57%

RELIGIOUS PERSUASION

Roman Catholic 15%
Other 21%
Free Wesleyan 64%

Tonga has strong ethnic ties with eastern Fiji, and there has traditionally been considerable population movement between the two states. Tongans tend to see themselves as unique among Pacific islanders, retaining their monarchy and never having been fully colonized.

Respect for traditional values and institutions remains high. Tongans are strong churchgoers; the Wesleyan, Roman Catholic, and Mormon churches are influential and often fund education. A new generation of Western-educated Tongans is querying some traditional attitudes.

POLITICS

▷ No multiparty elections

2002/2005

H.M. King Taufa'ahau Tupou IV

AT THE LAST ELECTION

Legislative Assembly 30 seats

The Legislative Assembly comprises the king, the 11 members of the Privy Council, 9 members indirectly elected by nobles, and 9 directly elected members. There are no political parties: of those elected in 2002, 7 were part of the Human Rights and Democracy Movement.

The main power brokers in Tongan politics are the king, the noble establishment, and the landowners. King Taufa'ahau, on the throne since 1965, effectively heads his government, frequently exercising kingly powers. The Legislative Assembly defers to his judgment and the king has instigated several development projects which have been undertaken without reference to the government.

The king's resistance to growing calls for greater democracy was highlighted in 2000, when he passed over his reformist eldest son and appointed his conservative third son, Prince Ulukalala Lavaka Ata, as premier for life.

WORLD AFFAIRS
 Joined UN in 1999

In 1996, after accepting the Pacific nuclear interests of the US and France in the preceding decade, Tonga finally acceded to the South Pacific Nuclear-free Zone Treaty. It broke its ties with Taiwan in 1998 in return for closer relations with China.

AID
 Recipient

 $19m (receipts) 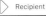 Down 10% in 2000

Aid finances major infrastructure projects; Japan, Australia, and New Zealand are primary donors. After losing aid in 1997, Tonga fought to reclaim its Least Developed Country status, which guarantees funds.

DEFENSE
No compulsory military service

 $2m (estimate)   No significant change

Tonga has a small defense force, which includes both regulars and reserves. Tongan police assisted in security efforts in the Solomon Islands in 2000.

ECONOMICS
Inflation 1.7% p.a. (1990–2000)

 $166m   1.7997–1.9535 pa'anga

SCORE CARD

- ❏ World GNP Ranking..........................186th
- ❏ GNP per Capita $1660
- ❏ Balance of Payments.......................Zero
- ❏ Inflation 5.9%
- ❏ Unemployment...............................13%

Strengths
Agriculture contributes largest percentage of GDP. Tourism main source of hard currency earnings.

Weaknesses
Off main shipping routes. Aid-dependent. Importer of food. High youth unemployment. Corruption scandal in royal court in 2001.

EXPORTS

New Zealand 6% — UK 6% — Other 10% — Japan 50% — USA 28%

IMPORTS

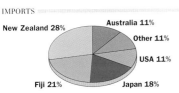

New Zealand 28% — Australia 11% — Other 11% — USA 11% — Fiji 21% — Japan 18%

RESOURCES
Electric power 7000 kw

3663 tonnes Not an oil producer

80,853 pigs, 12,500 goats, 300,000 chickens None

Tonga has no strategic or mineral resources. Electricity is entirely generated from imported fuel. Recent exploration has failed to identify any oil reserves. Tongan waters contain large numbers of tuna.

ENVIRONMENT
 Not available

 None 1.2 tonnes per capita

Tonga does not suffer from serious environmental problems, although it is occasionally afflicted by natural disasters, such as the recurring El Niño effect. Commercial activity has made little impact on the environment.

MEDIA
TV ownership low

 Daily newspaper circulation 72 per 1000 people

PUBLISHING AND BROADCAST MEDIA

There is 1 daily newspaper. Weeklies include the *Conch Shell* and the *Tonga Chronicle*

1 service relaying US programs 4 independent services

In 2000 Cable and Wireless relinquished control of Tonga's telecommunications services, which it had run since 1978, to a local company.

CRIME
Death penalty not used in practice

 68 prisoners 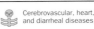 Crime is falling

Crime rates are generally low, partly due to the strong influence of the family. However, offenses such as breaking and entering have increased, along with rising unemployment levels, among young Tongans.

EDUCATION
School leaving age: 14

98% 705 students

Church participation in schools is high. Plans for a new national university were approved in 2000 following violence in Fiji, where Tongan students attend the University of the South Pacific.

HEALTH
No welfare state health benefits

1 per 2176 people Cerebrovascular, heart, and diarrheal diseases

Tonga has some modern health care facilities. However, patients have to be flown out to Australia or New Zealand for sophisticated surgery.

CHRONOLOGY

Originally discovered by the Polynesians, Tonga was visited by the Dutch in the 17th century and Capt. Cook in the 18th century. In the latter half of the 19th century, during the reign of King George Tupou I, the islands became a unified state after a period of civil war.

- ❏ **1875** First constitution established.
- ❏ **1900** Concern over German ambitions in region; Treaty of Friendship and Protection with UK.
- ❏ **1918–1965** Reign of Queen Salote Tupou III.
- ❏ **1958** Greater autonomy from UK enshrined in Friendship Treaty.
- ❏ **1965** King Taufa'ahau Tupou IV accedes on his mother's death.
- ❏ **1970** Full independence within British Commonwealth.
- ❏ **1988** Treaty allows US nuclear warships right of transit.
- ❏ **2000** King appoints third son as prime minister.
- ❏ **2001** Court jester steals US$20 million of state funds.
- ❏ **2002** Election sees strong showing by prodemocracy candidates, as in 1996 and 1999.

SPENDING
GDP/cap. increase

CONSUMPTION AND SPENDING

65 per 1000 population 93 per 1000 population

Defense	No data
Education	4.3%
Health	7.8%

0   5   10   15   20   25
Defense, Health, Education spending as % of GDP

Tongans indulge in few ostentatious displays of wealth. The well-off provide financial support for relatives.

WORLD RANKING

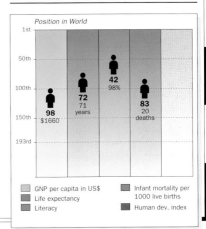

Position in World

1st — 50th — 100th — 150th — 193rd

98 $1660 72 71 years 42 98% 83 20 deaths

- GNP per capita in US$
- Life expectancy
- Literacy
- Infant mortality per 1000 live births
- Human dev. index

TRINIDAD & TOBAGO

OFFICIAL NAME: Republic of Trinidad and Tobago   **CAPITAL:** Port-of-Spain
POPULATION: 1.3 million   **CURRENCY:** Trinidad and Tobago dollar   **OFFICIAL LANGUAGE:** English

CARIBBEAN

THE TWO ISLANDS OF Trinidad and Tobago are the most southerly of the Caribbean Windward Islands and lie just 15 km (9 miles) off the Venezuelan coast. They gained joint independence from Britain in 1962, and Tobago was given internal autonomy in 1987. The spectacular mountain ranges and large swamps are rich in tropical flora and fauna. Pitch Lake in Trinidad is the world's largest natural reservoir of asphalt.

CLIMATE
▷ Tropical oceanic

WEATHER CHART FOR PORT-OF-SPAIN

The islands are a little warmer than others in the Caribbean and escape the hurricanes, which pass by to the north.

TRANSPORTATION
▷ Drive on left

Piarco International, Port-of-Spain
1.93m passengers

50 ships
18,600 grt

THE TRANSPORTATION NETWORK

4252 km (2642 miles)		None	
None		None	

The road network is well developed; there are taxis or minibuses for set routes. National carrier BWIA and Air Caribbean operate Trinidad–Tobago routes. BWIA flies to the US.

TOURISM
▷ Visitors : Population 1:3.9

336,000 visitors

Down 3% in 1999

MAIN TOURIST ARRIVALS

USA 33%	
UK 14%	
Canada 13%	
Other 40%	

% of total arrivals

Oil revenue meant that Trinidad was one of the last Caribbean states to develop tourism. Most is centered on Tobago (said to be the model for the island in *Robinson Crusoe*), famous for its huge variety of South American wildlife, including 210 species of tropical bird.

TRINIDAD & TOBAGO

Total Land Area : 5128 sq. km (1980 sq. miles)

LAND HEIGHT

500m/1640ft
200m/656ft
Sea Level

POPULATION

over 50 000
over 10 000
under 10 000

PEOPLE
▷ Pop. density high

English Creole, English, Hindi, French, Spanish

253/km² (656/mi²)

THE URBAN/RURAL POPULATION SPLIT

74%   26%

ETHNIC MAKEUP

White and Chinese 1%
Mixed 19%
Asian 40%
Black 40%

Trinidad's south Asian community is the largest in the Caribbean and holds on to its Muslim and Hindu inheritance. The open discussion of racial issues goes some way to dissipating latent tensions with the predominantly Christian black Trinidadians.

POLITICS
▷ Multiparty elections

L. House 2001/2002
U. House 2002/2002

President Arthur N. Robinson

AT THE LAST ELECTION

House of Representatives 36 seats

50% PNM   50% UNC

PNM = People's National Movement
UNC = United National Congress

Senate 31 seats

Senators are appointed by the president, including 16 nominated by the prime minister and 6 by the leader of the opposition

The increasingly right-wing PNM dominated politics from independence in 1962 to the 1990s, leading to political fragmentation and an attempted coup by Muslim extremists in 1990. The UNC's Basdeo Panday became the first ethnic Asian prime minister in 1995. He pledged his coalition government to reducing unemployment, crime, and racial discrimination. The UNC's majority was reduced to a tie with the PNM in the 2001 elections. President Arthur Robinson favored PNM leader Patrick Manning as prime minister, but the political stalemate led to new elections being called for late 2002.

Tobago's white sand beaches, verdant landscape, and natural anchorages have enabled it to develop a thriving tourist industry.

WORLD AFFAIRS

Joined UN in 1962

Trinidad withdrew in 1998 and 1999 from the Inter-American and UN Commissions on Human Rights respectively, over appeals against death sentences. New trade agreements include those with Costa Rica, Panama, the Dominican Republic, and Mexico. Sea border disputes with Venezuela relate to fishing and marine oil rights.

AID

Recipient

US$26m (receipts)

Loan repayments exceeded aid received in 2000

Aid is modest: China provided an interest-free loan of US$20 million in 2000 to help small businesses.

DEFENSE

No compulsory military service

US$35m

Down 12% in 2000

Defense forces comprise a land army and a coast guard (with air wing), used to patrol fishing grounds.

ECONOMICS

Inflation 5.5% p.a. (1990–2000)

US$6.42bn

6.240–6.115 Trinidad and Tobago dollars

SCORE CARD

❏ WORLD GNP RANKING	103rd
❏ GNP PER CAPITA	US$4930
❏ BALANCE OF PAYMENTS	–US$644m
❏ INFLATION	3.6%
❏ UNEMPLOYMENT	13%

STRENGTHS

Oil, which accounts for 70% of export earnings. Gas increasingly exploited to support new industries. Methanol, ammonia, iron, and steel exports. Tourism, especially on Tobago.

WEAKNESSES

Insufficiently diversified economy. Reliance on oil – highly sensitive to world price movements. High unemployment.

EXPORTS

France 4%
Spain 3%
USA 47%
Barbados 5%
Jamaica 9%
Other 32%

IMPORTS

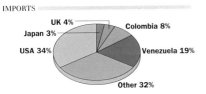

UK 4%
Japan 3%
USA 34%
Colombia 8%
Venezuela 19%
Other 32%

RESOURCES

Electric power 1.2m kw

15,027 tonnes

59,000 goats, 41,000 pigs, 10m chickens

135,000 b/d (reserves 700m barrels)

Oil, natural gas, asphalt, iron

Oil and gas are major resources. Big offshore gas and oil finds in 1998 included the largest discovery of crude oil in 25 years.

ENVIRONMENT

Sustainability rank: 121st

6%

17.4 tonnes per capita

Spillages from oil tankers threaten coastal conservation areas such as the Caroni Swamp, with its many species of butterflies. Forest fires due to periodic drought, and traffic-related pollution and congestion are serious concerns.

MEDIA

TV ownership high

Daily newspaper circulation 123 per 1000 people

PUBLISHING AND BROADCAST MEDIA

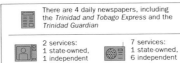

There are 4 daily newspapers, including the *Trinidad and Tobago Express* and the *Trinidad Guardian*

2 services: 1 state-owned, 1 independent

7 services: 1 state-owned, 6 independent

Television schedules are dominated by US programs, while around carnival time radio stations play little but calypso.

CRIME

Death penalty in use

4794 prisoners

Down 63% 1992–1998

Narcotics-related crime increases the murder rate. The country was party to the decision taken in 2001 to replace the authority of the UK's Privy Council with a Caribbean Court of Justice.

EDUCATION

School leaving age: 12

94%

6860 students

Education is based on the former British system. Most students go on to the University of the West Indies; Trinidad hosts the St. Augustine campus. Wealthy Trinidadians, however, go to universities in the US.

HEALTH

Welfare state health benefits

1 per 1250 people

Heart disease, cancers, diabetes, accidents, violence

Oil wealth has given Trinidad a better public health service than most Caribbean states and more private clinics, mainly serving the expatriate community. However, treatment delays are a problem. The spread of HIV/AIDS is of particular concern on Tobago.

SPENDING

GDP/cap. increase

CONSUMPTION AND SPENDING

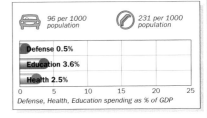

96 per 1000 population

231 per 1000 population

Defense 0.5%
Education 3.6%
Health 2.5%

Defense, Health, Education spending as % of GDP

In Trinidad wealth disparities between the affluent oil-rich business elite, many of whom are expatriate, and farm laborers are particularly marked. Service workers in Tobago's high-value tourism sector are poorly paid. Rural poverty in the interior, particularly among south Asian Trinidadian farmers, is a serious problem.

WORLD RANKING

T

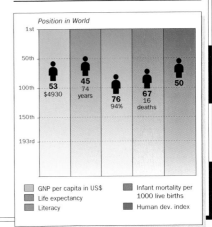

Position in World

53 $4930
45 74 years
76 94%
67 16 deaths
50

- GNP per capita in US$
- Life expectancy
- Literacy
- Infant mortality per 1000 live births
- Human dev. index

TUNISIA

NORTH AFRICA

OFFICIAL NAME: Republic of Tunisia **CAPITAL:** Tunis
POPULATION: 9.6 million **CURRENCY:** Tunisian dinar **OFFICIAL LANGUAGE:** Arabic

NORTH AFRICA'S SMALLEST country, Tunisia lies sandwiched between Libya and Algeria. The populous north is mountainous, fertile in places and has a long Mediterranean coastline. The south is largely desert. Habib Bourguiba ruled the country from independence in 1956 until a bloodless coup in 1987. Under President Ben Ali, the government has moved toward multiparty democracy, but is challenged by Islamic fundamentalists. Closer ties with the EU, whose members include Tunisia's main trading partners, were strengthened through the first Euro-Mediterranean conference held in 1995. Manufacturing and tourism are expanding.

TOURISM

Visitors : Population 1:1.9

5.1m visitors   Up 5% in 2000

MAIN TOURIST ARRIVALS

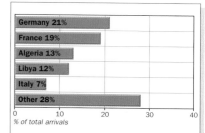

Germany 21%
France 19%
Algeria 13%
Libya 12%
Italy 7%
Other 28%

% of total arrivals

CLIMATE

Mediterranean/ hot desert

WEATHER CHART FOR TUNIS

Tunisia is hot in summer. The north is often wet and windy in winter. The far south is arid. The spring brings the dry, dusty *chili* wind from the Sahara.

TRANSPORTATION

Drive on right

 Skanes Airport, Monastir
3.92m passengers

 78 ships
193,500 grt

THE TRANSPORTATION NETWORK

18,226 km
(11,325 miles)

Highway from Tunis to Carthage airport

1860 km
(1156 miles)

None

Tunisia has six international airports. A highway from Tunis to Carthage airport opened in 1993. A light metro in Tunis and a rail link from Gafsa to Gabès are being built. The southern third of the country has few roads.

Tourists have flocked to Tunisia since the 1960s, attracted by its winter sunshine, beaches, desert, and archaeological remains. One of the Mediterranean's cheapest package destinations, Tunisia attracts almost three million European visitors a year, although some tourists are deterred by the fear of attacks by Islamists. In 2002, 14 German tourists were killed by a suicide bombing on the island of Jerba.

Tourism employs more than 200,000 people and is a focus of investment. However, concern about its environmental impact is growing.

PEOPLE

Pop. density medium

Arabic, French   62/km²
(160/mi²)

THE URBAN/RURAL POPULATION SPLIT

66%   34%

RELIGIOUS PERSUASION

Christian 1%   Jewish 1%

Muslim (mainly Sunni) 98%

ETHNIC MAKEUP

European 1%   Other 1%

Arab and Berber 98%

The population is almost entirely Muslim, of Arab and Berber descent, although there are Jewish and Christian minorities. Many Tunisians still live in extended family groups, in which three or four generations are represented.

Tunisia has traditionally been one of the most liberal Arab states. The 1956 Personal Statutes Code of President Bourguiba gave women fuller rights than in any other Arab country. Further legislation has since given women the right to custody of children in divorce cases, made family violence against women punishable by law, and helped divorced women to get alimony. Family planning and contraception have been freely available since the early 1960s. Tunisia's population growth rate has halved since the 1980s. Women make up 31% of the total workforce and 35% of the industrial workforce. Company ownership by women is steadily increasing; politics, however, remains an exclusively male preserve.

These freedoms are threatened by the growth in recent years of Islamic fundamentalism, which also worries the mainly French-speaking political and business elite who wish to strengthen links with Europe.

The Ben Ali regime has been criticized for its actions against Islamist activists. Despite continuing condemnation of its human rights record, the government's efforts to foster democracy were

praised in 2002 when it received the "Mediterranean Award," given by European human rights leagues.

POPULATION AGE BREAKDOWN

Female		Age	Male	
	0.5%	80+	0.5%	
	3.5%	60–79	3.8%	
	7.4%	40–59	7.2%	
	15.9%	20–39	15.8%	
	22.2%	0–19	23.2%	

% of population by age group

Roman remains *in the western Tozeur region. Diverse archeological remains can be found throughout Tunisia.*

T

POLITICS ▷ Multiparty elections

🏛 1999/2004   🧍 President Zine al-Abidine Ben Ali

AT THE LAST ELECTION
Chamber of Deputies 182 seats

4% 3%
PUP MR

81%
RCD

7% 4% 1%
MDS UDU SLP

RCD = Constitutional Democratic Rally **MDS** = Movement of Social Democrats **PUP** = Popular Unity Party
UDU = Unionist Democratic Union **MR** = Movement for Renewal **SLP** = Social Liberal Party

Formally a multiparty democracy since 1988, Tunisia is still dominated by the RCD and President Ben Ali.

President Ben Ali *became head of state in 1987.*

Mohammed Ghannouchi, *prime minister since 1999.*

PROFILE
President Ben Ali has made some effort to liberalize the political system. The life presidency has been abolished, and political parties and press freedom are encouraged. While allowing for a degree of political plurality, a complex proportional representation system ensured that there was an overwhelming victory for the RCD in the 1994 and 1999 elections. Since 1994 there has been evidence of a renewed crackdown against the left-wing opposition. Measures to promote human rights and further democratization were overshadowed in 2002 when a referendum on constitutional change gave Ben Ali a mandate to stand for a further two terms.

MAIN POLITICAL ISSUES
Fundamentalism
The RCD has clamped down on Islamic fundamentalists, in particular the outlawed Al-Nahda, or Renewal Party. In 1991, 500 Al-Nahda members were arrested following a failed coup, thought to be inspired by fundamentalists. Its leader, Rachid Ghannouchi, is in exile.

Human rights
The RCD has been under increasing attack over its human rights record. The Tunisia League of Human Rights' activities were suspended in January 2001. The RCD is committed to promoting women's rights.

TUNISIA

Total Land Area :
163 610 sq. km
(63 169 sq. miles)

POPULATION

over 500 000 ◉
over 100 000 ◎
over 50 000   ○
over 10 000   ●
under 10 000 ·

LAND HEIGHT

1000m/3281ft
500m/1640ft
200m/656ft
Sea Level

0   100 km
0   100 miles

WORLD AFFAIRS ▷ Joined UN in 1956

 AL   AMU   OIF   NAM   OIC

A foreign policy priority is to strengthen contacts with the West, which have generally been good because of Tunisia's liberal economic and social policies. Attention is focused especially on the EU, Tunisia's main export market. Tunisia played an important role in the run-up to the first Euro-Mediterranean conference, which was held in 1995.

Tunis was host to the Palestine Liberation Organization after it was expelled from Lebanon. Relations with other Arab states, particularly Kuwait and Saudi Arabia, were soured by Tunisia's support for Iraq in the Gulf War. The government regards the political impact of Islamic fundamentalism in neighboring Algeria with concern. Relations with Libya are improving, helped by the fact that Tunisia turned a blind eye to sanctions busters operating through its territory.

CHRONOLOGY
Tunisia has been home to the Zenata Berbers since earliest times and its history is linked to the rise and fall of the Mediterranean-centered empires. Carthage (near present-day Tunis), founded by the Phoenicians in the 9th century BCE, became the hub of a 1000-year trading empire which linked European and African trading networks. Tunisia was then ruled by the Romans, Byzantines, Egyptians, Ottomans, and, finally, the French.

❑ **1883** La Marsa Treaty makes Tunisia a French protectorate, ending its semi-independence. Bey of Tunis remains monarch.
❑ **1900** Influx of French and Italians.
❑ **1920** Destour (Constitution) Party formed; calls for self-government.
❑ **1935** Habib Bourguiba forms Neo-Destour (New Constitution) Party.
❑ **1943** Defeat of Axis powers by British troops restores French rule.
❑ **1955** Internal autonomy. Bourguiba returns from exile.
❑ **1956** Independence. Bourguiba elected prime minister. Personal Statutes Code gives rights to women. Family planning introduced.
❑ **1957** Bey is deposed. Tunisia becomes republic with Bourguiba as first president.
❑ **1964** Neo-Destour made sole legal party; renamed Destour Socialist Party (PSD). Moderate socialist economic program is introduced.
❑ **1969** Agricultural collectivization program, begun 1964, abandoned. ⇨

T

CHRONOLOGY *continued*

- ❑ **1974** Bourguiba elected president-for-life by National Assembly.
- ❑ **1974–1976** Hundreds imprisoned for belonging to "illegal organizations."
- ❑ **1978** Trade union movement, UGTT, holds 24-hour general strike; more than 50 killed in clashes. UGTT leadership replaced with PSD loyalists.
- ❑ **1980** New prime minister Muhammed Mazli ushers in greater political tolerance.
- ❑ **1981** Elections. Opposition groups allege electoral malpractice.
- ❑ **1984** Widespread riots after food price increases.
- ❑ **1986** Gen. Zine al-Abidine Ben Ali becomes interior minister. Four Muslim fundamentalists sentenced to death.
- ❑ **1987** Fundamentalist leader Rachid Ghannouchi arrested. Ben Ali becomes prime minister; takes over presidency after doctors certify Bourguiba senile. PSD renamed RCD.
- ❑ **1988** Most political prisoners released. Constitutional reforms introduce multiparty system and abolish life presidency. Two opposition parties legalized.
- ❑ **1989** Elections: RCD wins all seats, Ben Ali president. Fundamentalists take 13% of vote.
- ❑ **1990** Tunisia backs Iraq in Gulf War. Clampdown on fundamentalists intensifies.
- ❑ **1991** Abortive coup blamed on Al-Nahda; over 500 arrests.
- ❑ **1993** Multiparty agreement on electoral reform.
- ❑ **1994** Presidential and legislative elections. Ben Ali, sole candidate, is reelected.
- ❑ **1996** MDS leader Mohammed Moada imprisoned.
- ❑ **1999** Ben Ali and RCD win elections.
- ❑ **2002** Referendum allows Ben Ali to stand for office after 2004. Suicide bomb attack in Jerba kills 14 German tourists.

AID

 $223m (receipts)   Down 12% in 2000

The EU and France are the largest donors, providing over half of all bilateral aid. Japan, Germany, Italy, Belgium, and the Netherlands are other important sources of assistance. Oil-rich Arab states, including Saudi Arabia and Kuwait, suspended their aid programs to Tunisia after 1990 because of its pro-Iraq stance in the Gulf War. Tunisia's total external debt is estimated at over half of GNP.

DEFENSE

 Compulsory military service

 $350m   ⬆ Up 1% in 2000

Despite its small size – 35,000 troops, of which around two-thirds are conscripts – the military is an important political force, armed mainly with US weapons. Officer training is carried out in the US and France, as well as in Tunisia. Border security with Algeria was tightened in 1995 after Algerian Islamists attacked Tunisian border guards in protest against Tunisian support for Algerian security forces.

TUNISIAN ARMED FORCES

🛡	84 main battle tanks (54 M-60A3, 30 M-60A1)	27,000 personnel
🚢	19 patrol boats	4500 personnel
✈	51 combat aircraft (15 F-5E/F)	3500 personnel
	None	

ECONOMICS

 Inflation 4.5% p.a. (1990–2000)

📊 $20.1bn   💲 1.384–1.461 Tunisian dinars

SCORE CARD

- ❑ WORLD GNP RANKING 62nd
- ❑ GNP PER CAPITA $2100
- ❑ BALANCE OF PAYMENTS .... –$821m
- ❑ INFLATION 2.9%
- ❑ UNEMPLOYMENT 16%

EXPORTS

Spain 5%   Belgium 5%   Germany 12%   Other 30%   Italy 22%   France 26%

IMPORTS

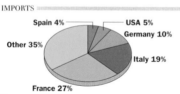

Spain 4%   USA 5%   Germany 10%   Other 35%   Italy 19%   France 27%

STRENGTHS
Well-diversified economy, despite limited resources. Tourism. Oil and gas exports, also agricultural exports: olive oil, olives, citrus fruit, dates. Expanding manufacturing sector, average annual increase was 5.4% in 1990–1998; important sectors are textiles, construction materials, machinery, chemicals. European investment. Ranked as most competitive economy in Africa in World Economic Forum's 2000–2001 report.

WEAKNESSES
Dependence on growth of drought-prone agricultural sector. Growing domestic energy demand on oil and gas resources.

PROFILE
Since it began a process of structural adjustment in 1988, supported by the IMF and the World Bank, Tunisia has become an increasingly open, market-oriented economy. Real GDP growth has averaged 5% since 1987, rising to 6% in

ECONOMIC PERFORMANCE INDICATOR

2001. Annual inflation remained stable, at just below 3% in 2000, despite higher food and energy prices. Prices have been freed, most state companies privatized, and import barriers reduced.

The balance of payments relies on fluctuating tourism receipts to offset a trade deficit. The government must also balance growth with better social provisions. The member states of the EU are Tunisia's main trading partners, accounting for well over 70% of its imports and nearly 80% of its exports; trade has increased significantly since 1999.

TUNISIA : MAJOR BUSINESSES

Oil, Wine, Chemicals, Phosphates, Textiles, Consumer goods, Leather tanning, Vehicle assembly

* significant multinational ownership

RESOURCES

▷ Electric power 2.3m kw

93,170 tonnes

73,000 b/d (reserves 300m barrels)

6.6m sheep, 2.9m turkeys, 43m chickens

Phosphates, iron, zinc, lead, salt, oil, gas

ELECTRICITY GENERATION

Hydro 1% (0.07bn kwh)

Combustion 99% (8.9bn kwh)

Nuclear 0%

Other 0%

% of total generation by type

Tunisia is a leading producer of phosphates for fertilizers, mainly from mines near Gafsa. Oil and gas

TUNISIA : LAND USE

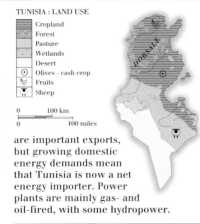

Cropland
Forest
Pasture
Wetlands
Desert
Olives - cash crop
Fruits
Sheep

0   100 km
0   100 miles

are important exports, but growing domestic energy demands mean that Tunisia is now a net energy importer. Power plants are mainly gas- and oil-fired, with some hydropower.

ENVIRONMENT

▷ Sustainability rank: 61st

0.3%

2.4 tonnes per capita

ENVIRONMENTAL TREATIES

Yes   Yes   Yes

Yes   Yes   No

Desertification is a serious problem in the largely arid central and southern regions. However, the dominant environmental issue is the rapid expansion of tourism since the 1980s. Large, insensitively designed hotel and resort developments, which do not fit in with the local architecture, are spoiling coastal areas such as the island of Jerba and Hammamet (although building height restrictions are applied here). Tourism is also making an impact on the fragile desert ecology of the south.

MEDIA

▷ TV ownership medium

☒ Daily newspaper circulation 31 per 1000 people

PUBLISHING AND BROADCAST MEDIA

There are 8 daily newspapers, including *al-Amal*, *La Presse de Tunisie*, and *As-Sabah*

2 state-owned services

1 state-owned service

Reforms since the late 1980s have in theory increased press freedom in Tunisia, traditionally considered a source of liberal ideas in the Arab world. In practice, government restrictions remain. The foreign press is also occasionally banned, but the arrival of satellite TV from Europe has enabled people to receive a wide range of programs. The Internet is heavily censored.

CRIME

▷ Death penalty in use

23,165 prisoners

⬇ Down 4% in 1999

CRIME RATES

Murders
1 per 100,000 population

Rapes
5 per 100,000 population

Thefts
280 per 100,000 population

Street crime is unusual. However, Tunisia's poor human rights record has prompted criticism of its maltreatment of political and other detainees. Arbitrary arrests and torture while in police custody, especially of suspected Islamist activists, are routine.

EDUCATION

▷ School leaving age: 16

71%

180,044 students

THE EDUCATION SYSTEM

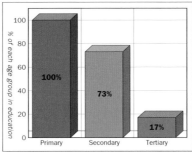

% of each age group in education

Primary 100%
Secondary 73%
Tertiary 17%

Education is compulsory for nine years between the ages of six and 16, with secondary education beginning at the age of 12. Arabic is the first language in schools, but French is also taught, and is used almost exclusively in higher education. There are seven universities; student enrollment has doubled since 1995.

HEALTH

▷ Welfare state health benefits

1 per 1429 people

Heart and cerebrovascular diseases

Well-developed family-planning facilities have almost halved Tunisia's birthrate over the past 30 years. The population growth rate has dropped from 3.2% to 1.9% – the lowest in the region. The mortality rate has been halved, to 5.7 per 1000 population, reflecting the extension of free medical services to over 70% of the population. While services lack sophistication, an umbrella of primary care facilities covers all but the most isolated rural communities. Regional committees organize care for the old, needy, and orphaned.

SPENDING

▷ GDP/cap. increase

CONSUMPTION AND SPENDING

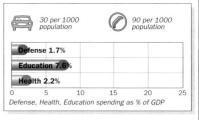

30 per 1000 population

90 per 1000 population

Defense 1.7%

Education 7.6%

Health 2.2%

Defense, Health, Education spending as % of GDP

Today 6% of Tunisians are estimated to live in absolute poverty. In 1970, the figure was 30%. The poorest in the community tend to live in the urban shanty towns, or *bidonvilles*. The Western-oriented elite has links with government or business. Social security covers sickness, old age, and maternity, but not unemployment. The government is concerned that the lack of jobs is encouraging the spread of Islamic fundamentalism; economic growth is its medium-term solution to the problem. Special projects are being set up in the most deprived urban areas to offset the worst effects of poverty.

WORLD RANKING

T

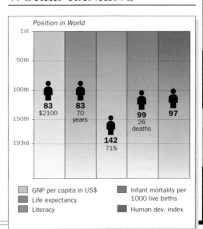

Position in World

1st
50th
100th
150th
193rd

83 $2100
83 70 years
142 71%
99 26 deaths
97

GNP per capita in US$
Life expectancy
Literacy

Infant mortality per 1000 live births
Human dev. index

TURKEY

OFFICIAL NAME: Republic of Turkey **CAPITAL:** Ankara
POPULATION: 67.6 million **CURRENCY:** Turkish lira **OFFICIAL LANGUAGE:** Turkish

TURKEY, MAINLY IN WESTERN ASIA, also includes the region of Eastern Thrace in Europe. It thus controls the entrance to the Black Sea, which is straddled by Turkey's largest city, Istanbul. Most Turks live in the western half of the country. The eastern and southeastern reaches of the Anatolia Plateau are Kurdish regions. Turkey's location gives it great strategic influence in the Black Sea, the Mediterranean, and the Middle East. Lying on a major earthquake fault line, many Turkish towns are vulnerable to earthquakes such as the one which devastated Izmit in 1999.

The Church of the Holy Cross, on Akdamar Island in Lake Van, was built in the 10th century when Christianity was dominant in the region.

CLIMATE

▷ Mountain/ Mediterranean

WEATHER CHART FOR ANKARA

Coastal regions have a Mediterranean climate. The interior has cold, snowy winters and hot, dry summers.

TRANSPORTATION

▷ Drive on right

 Atatürk International, Istanbul
16m passengers

 1135 ships
6.25m grt

THE TRANSPORTATION NETWORK

95,599 km (59,402 miles)	1774 km (1102 miles)
8682 km (5395 miles)	1200 km (746 miles)

Recent rail projects include a high-speed link between Istanbul and Ankara, a line linking Turkey and Georgia, and a light rail system in Istanbul. An extensive network of ports and harbors includes Istanbul and Izmir.

TOURISM

▷ Visitors : Population 1:7.1

 9.59m visitors

 Up 39% in 2000

Visitors are attracted by fine beaches, classical sites such as Ephesus and Troy, and antiquities from the prehistoric to the Ottoman periods. Attacks on foreigners by Kurdish militants in 1994 hit the tourist trade, but business then recovered, and in 2000 tourists spent $7.6 billion. Visitor numbers dropped after the September 2001 attack on the US.

MAIN TOURIST ARRIVALS

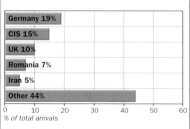

Germany 19%	
CIS 15%	
UK 10%	
Romania 7%	
Iran 5%	
Other 44%	

0 10 20 30 40 50 60
% of total arrivals

PEOPLE

▷ Pop. density medium

Turkish, Kurdish, Arabic, Circassian, Armenian, Greek, Georgian, Ladino

88/km² (227/mi²)

THE URBAN/RURAL POPULATION SPLIT

75%   25%

RELIGIOUS PERSUASION

Other 1%
Muslim (mainly Sunni) 99%

ETHNIC MAKEUP

Arab 2%   Other 8%
Kurdish 20%
Turkish 70%

The Turks are racially diverse. Many are the descendants of refugees, often from the Balkans, but a strong sense of national identity is rooted in a shared language and religion. Most are Sunni Muslim, although a Shi'a community, including the heterodox Alawite sect, is growing fast. The largest minority are the Kurds, and there are some 500,000 Arabic speakers. While women have equal rights in law, men dominate political and even family life. In 2002 women gained the right to an equal portion in the case of divorce.

With a birthrate of 2%, Turkey is projected to have a larger population than any EU country by 2020.

POPULATION AGE BREAKDOWN

Female	Age	Male
0%	80+	0%
4.2%	60–79	3.6%
8.5%	40–59	8.6%
16.3%	20–39	17%
20.5%	0–19	21.3%

% of population by age group

POLITICS ▷ Multiparty elections

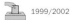

1999/2002

President Ahmet
Necdet Sezer

AT THE LAST ELECTION
Turkish Grand National Assembly 550 seats

| 25% DSP | 23% MHP | 20% Virtue | 16% ANAP | 15% DYP | 1% Ind |

DSP = Democratic Left Party **MHP** = Nationalist Action Party
Virtue = Virtue Party **ANAP** = Motherland Party **DYP** = True
Path Party **Ind** = Independents

Turkey is a multiparty republic. The president, who serves a seven-year term, appoints the prime minister.

PROFILE

The main ideological division in politics is Islamic–secular. In 2002 the powers of the National Security Council were restricted and its membership altered to give civilian dominance.

Bulent Ecevit became prime minister in 1999 after the DSP performed well in the election. He formed a new coalition with the far-right MHP and with ANAP (backed by Istanbul's metropolitan interests), which survived Ecevit's failed attempt to enable President Süleyman Demirel to stand for reelection in 2000. Demirel was replaced by nonparty candidate Ahmet Necdet Sezer. Ecevit's refusal to resign because of ill health prompted a crisis in 2002, when key ministers quit their posts in protest.

MAIN POLITICAL ISSUES
Islamic fundamentalism
Modern Turkey's identity as a secular state was profoundly challenged by the Islamist agenda of the Welfare Party (RP) – the largest parliamentary party after the 1995 election. A secular coalition, formed to keep the RP out of power, disintegrated in mid-1996. RP leader Necmettin Erbakan then forged a coalition with the True Path Party. Ousted in mid-1997, the RP was banned in 1998. Many RP Assembly members allied themselves with the Virtue Party, which in turn was banned in 2001.

Kurdish separatism
Thousands of Kurds have been killed since 1984 in a bitter civil war in southeast Turkey. The secessionist Kurdistan Workers' Party (PKK) has proclaimed three cease-fires since 1992. In 1999 its leader Abdullah Ocalan was sentenced to death. However, he remained in prison as, in April 2002, the PKK disbanded and regrouped as a political organization committed to seeking equal rights for Kurds within Turkey. EU pressure led in 2002 to constitutional amendments which lifted the ban on broadcasting and education in Kurdish which had previously been rigorously enforced.

TURKEY

Total Land Area : 780 580 sq. km (501 382 sq. miles)

LAND HEIGHT	POPULATION
3000m/9843ft	over 5 000 000
2000m/6562ft	over 1 000 000
1000m/3281ft	over 500 000
500m/1640ft	over 100 000
200m/656ft	over 50 000
Sea Level	over 10 000
	under 10 000

Human rights
Turkey's human rights record has been subject to intense international criticism. Concerns have focused on the curtailment of civil liberties, illegal executions, and the treatment of Kurds. Recent reforms have sought to bring Turkey in line with EU standards. Civil liberties were improved in 1995, and in 2002 executions were banned except in time of war.

Prime Minister Bulent Ecevit, whose ill health provoked a political crisis in 2002.

President Ahmet Necdet Sezer, has frequently clashed with Ecevit.

WORLD AFFAIRS ▷ Joined UN in 1945

 CE NATO OECD OIC OSCE

Turkey had great strategic significance during the Cold War as part of NATO's first line of defense. It now has closer ties with former communist neighbors, particularly Bulgaria and Georgia, and with Turkic-speaking central Asian states. It has joined the BSEC, and has tried to mediate between Armenia and Azerbaijan. In 2000 Turkey agreed to send arms to Uzbekistan – its first military involvement in central Asia since the collapse of the Soviet Union. Before 1991 Turkey's chief trading partner was Iraq, and relations were improving until September 2001, when Turkey became the first Muslim country to join the "war on terrorism." In 2002 it took over leadership of the peacekeeping force in Afghanistan.

Negotiations on joining the EU are aided by recently improving relations with Greece, although the Turkish-backed partition of Cyprus remains an obstacle, as are lingering EU concerns over human rights.

AID ▷ Recipient

$325m (receipts)

Up sharply in 1999–2000

Turkey is a net recipient of aid (especially from the Gulf War allies), despite US suspension of aid in 1994 in protest at the treatment of Kurds. The acute economic crisis in 2000–2001 prompted a $10 billion loan from the IMF. Greece, a traditional adversary, offered humanitarian aid after the 1999 Izmit earthquake.

T

CHRONOLOGY

Following the collapse of the Ottoman Empire and Turkey's defeat in World War I, nationalist Mustafa Kemal Atatürk deposed the ruling sultan in 1922, declaring Turkey a republic in 1923.

- ❏ **1924** Religious courts abolished.
- ❏ **1928** Islam no longer state religion.
- ❏ **1934** Women given the vote.
- ❏ **1938** President Atatürk dies. Succeeded by Ismet Inonu.
- ❏ **1945** Turkey declares war on Germany. Joins UN.
- ❏ **1952** Joins CE and NATO.
- ❏ **1960** Military coup; National Assembly suspended.
- ❏ **1961** New constitution.
- ❏ **1963** Association agreement with European Economic Community.
- ❏ **1974** Invades northern Cyprus.
- ❏ **1980** Military coup; martial law.
- ❏ **1982** New constitution.
- ❏ **1983** General election won by Turgut Özal's ANAP.
- ❏ **1984** Turkey recognizes "Turkish Republic of Northern Cyprus." Kurdish separatist PKK launches guerrilla war in southeast.
- ❏ **1987** Applies to join European Communities.
- ❏ **1990** US-led coalition launches air strikes on Iraq from Turkish bases.
- ❏ **1991** Elections won by DYP. Süleyman Demirel premier.
- ❏ **1992** Joins Black Sea alliance.
- ❏ **1993** Demirel elected president. Tansu Çiller becomes DYP leader and heads coalition.
- ❏ **1995** Major anti-Kurdish offensive. Voting age lowered to 18. Çiller coalition collapses. Pro-Islamic RP wins election, but center-right DYP–ANAP coalition takes office. Customs union with EU.
- ❏ **1996–1997** RP leader Necmettin Erbakan heads first pro-Islamic government since 1923.
- ❏ **1997** Mesut Yilmaz reappointed to head minority ANAP government.
- ❏ **1998** RP banned. Yilmaz resigns amid corruption allegations; replaced by Bulent Ecevit of DSP.
- ❏ **1999** Ecevit heads right-wing coalition after elections. Kurdish leader Abdullah Ocalan sentenced to death. Izmit earthquake kills 14,000.
- ❏ **2000** National Assembly refuses to endorse Demirel's reelection. He is replaced by Ahmet Necdet Sezer.
- ❏ **2001** Acute financial crisis. Hunger strikes in high-security prisons. Virtue Party banned.
- ❏ **2002** Democracy and human rights strengthened in constitution. April, PKK replaced by Kurdistan Freedom and Democracy Congress. July, key ministers resign. Early elections promised.

DEFENSE

 ▷ Compulsory military service

 $10.6bn   ⬆ Up 9% in 2000

TURKISH ARMED FORCES

🚗	4205 main battle tanks (2876 M-48, 932 M-60, 397 *Leopard*)	402,000 personnel
🚢	13 submarines, 23 frigates, and 49 patrol boats	53,000 personnel
✈	505 combat aircraft (240 F-16C/D, 87 F-5, 178 F-4E)	60,100 personnel
	None	

Turkey's armed forces are the second-largest in NATO, which it joined in 1952. Turkey is a sizable military power, and it spends a higher percentage of GDP on defense than any other NATO country. NATO membership gives Turkey easy access to Western arms suppliers, although campaigners oppose sales to Turkey on human rights grounds. Israel is an important source for arms. Offensives against Kurdish separatists in northern Iraq and in Turkey's own southeastern provinces have involved over 50,000 troops and repeated incursions into Iraqi territory.

The great majority of Turkey's armed forces personnel are conscripts; 18 months' service is compulsory for all males at the age of 20.

ECONOMICS

▷ Inflation 76% p.a. (1990–2000)

📊 $202bn   💲 670,300–1,455,000 Turkish lira

SCORE CARD

- ❏ WORLD GNP RANKING..........................23rd
- ❏ GNP PER CAPITA $3100
- ❏ BALANCE OF PAYMENTS.................–$9.77bn
- ❏ INFLATION 54.9%
- ❏ UNEMPLOYMENT 6%

ECONOMIC PERFORMANCE INDICATOR

Consumer Price Index ▬▬ GDP ▭

EXPORTS

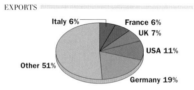

Italy 6% France 6% UK 7% USA 11% Germany 19% Other 51%

IMPORTS

France 6% Russia 7% USA 7% Italy 8% Germany 13% Other 59%

persistently high inflation. Despite structural reforms the economy reached crisis in February 2001, with a collapse of the banking system and the loss of over 500,000 jobs. This was compounded by the economic impact of the September 11 attack on the US. IMF rescue packages were agreed in February, May, and November in return for banking reform and privatizing debt-laden state companies.

STRENGTHS

Liberalization resulted in strong growth in the 1990s. Textiles, manufacturing, and construction sectors competitive on world markets. Tourism industry. Self-sufficient in agriculture. Dynamic private sector. Skilled labor force. Customs union with EU since 1995.

WEAKNESSES

Persistently high inflation. Unsound public finances. Large government bureaucracy. Uneven privatization program. Ailing banking sector. Influence of organized crime. High cost of military action against Kurds.

PROFILE

Turkey has one of the oldest and most advanced of the emerging market economies. In the 1990s it grew strongly, but continued to suffer from

TURKEY : MAJOR BUSINESSES

Istanbul   Ankara   Bursa   İzmit   Kırıkkale   Erzurum   Sivas   İsparta   Adana   Diyarbakır   İzmir   Mersin

- ⚙ Cement
- ❊ Textiles
- ⚗ Chemicals
- 🔌 Electronics
- 🛢 Oil refining
- Iron & steel
- Food processing
- 🚗 Vehicle manufacture

0   200 km
0   200 miles

* significant multinational ownership

T

RESOURCES

 Electric power 23.4m kw

638,097 tonnes

64,621 b/d (reserves 260m barrels)

29.4m sheep, 10.8m cattle, 8.06m goats, 220m chickens

Chromium, oil, copper, borax, coal, gas, bauxite, iron

ELECTRICITY GENERATION

Hydro 38% (42bn kwh)

Combustion 62% (69bn kwh)

Nuclear 0%

Other 0%

% of total generation by type

Under the controversial Southeastern Anatolian Project (GAP) launched in the mid-1980s, Turkey is building 22 dams on the Tigris and Euphrates rivers. In 1999 controversy focused on the Ilisu Dam on the Tigris, which will flood 15 towns and 52 villages, and in 2001 on the Birecik Dam on the Euphrates, which has engulfed the ancient Roman town of Zeugma. Turkey produces oil around Raman, on the Tigris. Eastern provinces are rich in minerals, such as chromium, of which Turkey is a leading producer.

TURKEY : LAND USE

Cropland
Pasture
Forest
High mountain regions
Wheat
Tobacco - cash crop
Sheep

0   200 km
0   200 miles

ENVIRONMENT

 Sustainability rank: 62nd

 1%   3.2 tonnes per capita

ENVIRONMENTAL TREATIES

Yes   Yes   Yes

Yes   Yes   No

Turkey's program of dam building on the Tigris and Euphrates has met with condemnation, particularly from Syria and Iraq, whose rivers will suffer reduced flow rates. Plans for the Ilisu Dam were shelved in 2001. Concern has also been expressed at proposals to build a nuclear power plant. There has been uncontrolled tourist development along the western coast.

MEDIA

 TV ownership high

Daily newspaper circulation 64 per 1000 people

PUBLISHING AND BROADCAST MEDIA

There are 57 daily newspapers. The leaders are the serious *Cumhuriyet* and the sensationalist *Hürriyet*

1 state-controlled service with 5 national channels

1 state-controlled national service and over 50 local stations

The Turkish press is diverse, vigorous, and largely privately owned. Almost all Istanbul newspapers are printed in Ankara and Izmir on the same day. Foreign satellite or cable broadcasts are available, as well as the five national channels of the state Turkish Radio and Television Corporation. The National Assembly has amended censorship laws to ease restrictions on non-Turkish-language broadcasting, but still bans broadcasts in Kurdish and censors Kurdish-related programming. A particularly high number of journalists are imprisoned, and state control over Internet content was increased in 2002.

CRIME

 Death penalty not used in practice

71,860 prisoners   Up 28% in 1999

CRIME RATES

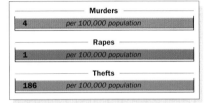

Murders

4   per 100,000 population

Rapes

1   per 100,000 population

Thefts

186   per 100,000 population

The routine use of torture and rape by the police, and the deaths of prisoners in custody, cause concern among human rights groups worldwide. In 2002 the imposition of the death penalty was limited to "times of war" and terrorism.

EDUCATION

 School leaving age: 14

85%   1.43m students

THE EDUCATION SYSTEM

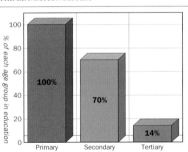

% of each age group in education

100% Primary
70% Secondary
14% Tertiary

After 1923, educational establishments were nationalized. In 1928, a Turkish alphabet with Latin characters was introduced.

In 1997, compulsory education was extended from five to eight years, raising the age for entry into Islamic schools from 11 to 14, in a move seen as designed to reduce attendance at such schools. State schools are coeducational and free. Engineering is usually the strongest faculty in Turkey's many universities.

HEALTH

 Welfare state health benefits

1 per 833 people   Cerebrovascular, heart, respiratory, and digestive diseases

Turkey possesses an adequate national system of primary health care. By Western standards, however, hospitals are underequipped. There are fewer doctors per head than in any European country.

SPENDING

GDP/cap. increase

CONSUMPTION AND SPENDING

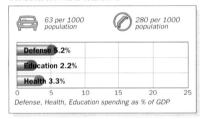

63 per 1000 population   280 per 1000 population

Defense 5.2%
Education 2.2%
Health 3.3%

0   5   10   15   20   25

Defense, Health, Education spending as % of GDP

The economic expansion of the 1980s has created a new class of wealthy entrepreneurs. Urban/rural differences remain pronounced. High inflation in the 1990s eroded earnings of those on fixed incomes, and income inequality has grown. Many Turks take jobs as *Gastarbeiter* (guest workers) in Germany and the Netherlands.

WORLD RANKING

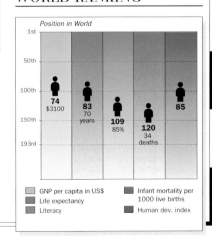

Position in World

1st
50th
100th
150th
193rd

74 $3100
83 70 years
109 85%
120 34 deaths
85

GNP per capita in US$
Life expectancy
Literacy
Infant mortality per 1000 live births
Human dev. index

T

TURKMENISTAN

OFFICIAL NAME: Turkmenistan **CAPITAL:** Ashgabat
POPULATION: 4.8 million **CURRENCY:** Manat **OFFICIAL LANGUAGE:** Turkmen

ONCE THE POOREST of the former Soviet republics, Turkmenistan has adjusted better than most to independence, exploiting the market value of its abundant natural gas supplies. A largely Sunni Muslim area, Turkmenistan is part of the former Turkestan, the last expanse of central Asia incorporated into czarist Russia. Much of life is still based on tribal relationships. Turkmenistan is isolated – telephones are rare and other communications limited.

CLIMATE ▷ Desert/steppe

WEATHER CHART FOR ASHGABAT

Most of Turkmenistan is arid desert, so that only 2% of the total land area is suitable for agriculture.

TRANSPORTATION ▷ Drive on right

Turkmenistan International, Ashgabat

38 ships
38,400 grt

THE TRANSPORTATION NETWORK

19,488 km (12,109 miles)	None
2365 km (1470 miles)	Amu-Darya River

Trains are cramped and chaotic, and buses unreliable. The border with Iran is officially closed to "foreigners."

TOURISM ▷ Visitors : Population 1:16

300,000 visitors   Up 26% in 1998

MAIN TOURIST ARRIVALS

Most visitors are businessmen attracted by Turkmenistan's stability under President Niyazov. Turkmenistan may become a popular tourist destination in future; traditional Turkmen Muslim monuments are slowly being restored.

Karakum Canal zone: *salt flats and the Kopetdag Mountains on the Iranian border. The Karakumy is Turkmenistan's largest desert.*

PEOPLE ▷ Pop. density low

Turkmen, Uzbek, Russian   10/km² (25/mi²)

THE URBAN/RURAL POPULATION SPLIT

45%   55%

RELIGIOUS PERSUASION

Other 2%   Eastern Orthodox 11%
Sunni Muslim 87%

Before czarist Russia annexed Turkmenistan in 1884, the Turkmen were a largely nomadic tribal people. The tribal unit remains strong – the largest tribes are the Tekke in the center, the Ersary on the eastern Afghan border, and the Yomud in the west. Tribal conflicts, rather than tensions with the two main minorities – Russian and Uzbek – are a source of strife. Paradoxically, this has meant that since 1991 nationalism has been less virulent than in other former Soviet republics. Since 1989, Turkmenistan has been rehabilitating its traditional language and culture, as well as reassessing its history. Islam is again central, although few perform the *haj* (pilgrimage) to Mecca and many continue to maintain a cult of ancestors.

Russian remains the language of bureaucracy.

POLITICS ▷ No multiparty elections

L. House 1999/2004
U. House 1998/2003

 President Saparmurad Niyazov

AT THE LAST ELECTION

Parliament 50 seats

In elections to the Parliament in 1999 all of the seats were won by supporters of the ruling Democratic Party of Turkmenistan (**DPT**) – the only registered party.

People's Council 110 seats

The People's Council has 50 directly elected members, the 50 members of the Parliament, 10 appointed regional members and a varying number of ex-officio members.

Officially, Turkmenistan became a multiparty democracy at independence, but President Saparmurad Niyazov has banned the formation of new parties. As in some other ex-Soviet states, former communists, regrouped as the DPT, still dominate the political process, harboring the traditional communist suspicion of Islamic fundamentalism. A main political concern is to prevent the social and nationalistic conflicts that have blighted other CIS republics.

President Niyazov has encouraged an elaborate personality cult, adopting the title of Turkmenbashi (father of all Turkmen), and there are golden statues of him throughout Ashgabat. His spiritual guide to living, *Ruhnama*, has been adopted as a national code, and he has announced his intention officially to rename all 12 months of the year in commemoration of the country's heroes and most potent national symbols.

WORLD AFFAIRS ▷ Joined UN in 1992

 CIS   ECO   EAPC   OIC OSCE

Turkmenistan is concentrating on establishing good relations with Iran and Turkey. It needs investment from both countries, but is wary of Islamic fundamentalism. President Niyazov opposes economic union with the CIS, and has also expressed caution about closer political union with other Turkic-speaking central Asian states.

AID ▷ Recipient

 $32m (receipts)   Up 33% in 2000

Aid is mostly concentrated in the oil and gas industries. Turkmenistan became a member of the ADB in 2000.

T

TURKMENISTAN

Total Land Area : 488 100 sq. km (188 455 sq. miles)

POPULATION

over 100 000	◎
over 50 000	○
over 10 000	●
under 10 000	•

LAND HEIGHT

1000m/1640ft
500m/1640ft
200m/656ft
Sea Level
-200m/-656ft

CHRONOLOGY

The nomadic peoples of western Turkestan came under Russian imperial control from the 1850s.

❏ **1924** Creation of Turkmenistan.
❏ **1991** Independence from USSR. Niyazov retains power, becoming president.
❏ **1994** Former communists win first elections.
❏ **1999** Niyazov's term extended indefinitely by parliament.

EDUCATION
 School leaving age: 17

98%   76,000 students

University students must know the life and works of President Niyazov. They now have to work for two years after their first two years of study.

HEALTH
▷ Welfare state health benefits

1 per 333 people   Cerebrovascular, heart, and respiratory diseases

Spending is now high for the region, but Turkmenistan's health indicators remain among the worst.

SPENDING
▷ GDP/cap. decrease

CONSUMPTION AND SPENDING

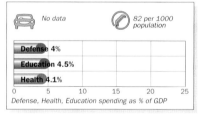

No data   82 per 1000 population

Defense 4%
Education 4.5%
Health 4.1%

Defense, Health, Education spending as % of GDP

The wealthiest group is connected to the government. Transition to a liberal economy has attracted high spending.

DEFENSE
 Compulsory military sevice

$173m   Up 20% in 2000

Turkmenistan relies on Russia for defense. Pilots are trained in Pakistan.

ECONOMICS
 Inflation 408% p.a (1990–2000)

$3.89bn   5200 manats

SCORE CARD

❏ World GNP Ranking.......................119th
❏ GNP per Capita $750
❏ Balance of Payments $412m
❏ Inflation 14%
❏ Unemployment 2%

STRENGTHS
Cotton and gas. Decision to abolish collective farms gradually encouraging private initiative and enterprise.

WEAKNESSES
Cotton monoculture has forced rising food imports. Thriving black market threatens value of manat.

EXPORTS

Turkey 7%   Ukraine 7%
Russia 41%   Iran 10%
Germany 16%
Other 19%

IMPORTS

Japan 8%   UAE 8%
Other 44%   Ukraine 12%
Russia 14%
Turkey 14%

RESOURCES
▷ Electric power 3.9m kw

 9292 tonnes   162,000 b/d (reserves 500m barrels)

 6m sheep, 860,000 cattle, 4.8m chickens   Oil, natural gas, potassium, sulfur, sodium sulfate

During the Soviet years most Turkmen agriculture was turned over to cotton – seen by Moscow as a strategic crop.

ENVIRONMENT
▷ Sustainability rank: 131st

 4%   5.7 tonnes per capita

The building of the Karakum Canal has reduced the flow of water to the Aral Sea by 35%. Plans were announced in 2000 for the construction of a large artificial lake in the Karakumy Desert.

MEDIA
▷ TV ownership medium

✕ Daily newspaper circulation figures are not available

PUBLISHING AND BROADCAST MEDIA

There are 2 daily newspapers, *Neitralnyi Turkmenistan* and *Turkmenistan*, both published in Turkmen

1 state-controlled service   1 state-controlled service

Iranian and Afghan radio stations, beaming in Islamic programs, are popular. TV is only available in cities.

CRIME
▷ No death penalty

 22,000 prisoners   Increasing levels of theft

Levels of crime are generally low. The death penalty was finally abolished in December 1999.

WORLD RANKING

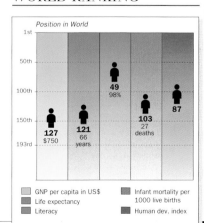

Position in World

127 $750	121 66 years	49 98%	103 27 deaths	87

GNP per capita in US$ Infant mortality per 1000 live births
Life expectancy   Human dev. index
Literacy

T

TUVALU

PACIFIC OCEAN

OFFICIAL NAME: Tuvalu   **CAPITAL:** Fongafale, on Funafuti Atoll   **POPULATION:** 10,800
CURRENCIES: Australian dollar and Tuvaluan dollar   **OFFICIAL LANGUAGE:** English

ONE OF THE WORLD'S smallest, most isolated states, Tuvalu lies 1050 km (650 miles) north of Fiji in the Pacific. A chain of nine coral atolls, it has a land area of just 26 sq. km (10 sq. miles). As the Ellice Islands, it was linked to the Gilbert Islands (now Kiribati) as a British colony until independence in 1978. Politically and socially conservative, Tuvaluans live by subsistence farming and fishing.

CLIMATE
 Tropical oceanic

WEATHER CHART FOR FONGAFALE

Although average humidity exceeds 90%, the climate is pleasantly warm. The mean annual temperature is 29°C (84°F). The October–March hurricane season brings many violent storms.

TRANSPORTATION
 Drive on right

 There is an airstrip on Funafuti atoll

49,000 grt

THE TRANSPORTATION NETWORK

8 km (5 miles)	None
None	None

A ferry links the atolls. There are air links with Kiribati and Fiji. Funafuti and Nukufetau have deepwater berths.

TOURISM
Visitors : Population 1:11

1000 visitors   No change in 2000

MAIN TOURIST ARRIVALS

| Australia 15% |
| New Zealand 10% |
| USA 9% |
| Other 66% |

% of total arrivals

Unspoiled and lapped by some of the world's warmest waters, the remote islands of Tuvalu have surprisingly few visitors. The islands' only paved airstrip and sole hotel are to be found on Funafuti.

PEOPLE
 Pop. density high

Tuvaluan, Kiribati, English   417/km² (1080/mi²)

THE URBAN/RURAL POPULATION SPLIT

45%   55%

RELIGIOUS PERSUASION

Baha'i 1%   Seventh-day Adventist 1%
Other 1%
Church of Tuvalu 97%

Around 95% of Tuvaluans are ethnically Polynesian. Their ancestors came from Tonga and Samoa 2000 years ago. Nui atoll has Micronesian influences. There is an I-Kiribati community on Funafuti; many Tuvaluans who worked in Kiribati took local wives. Over 40% of the population now live on Funafuti, pushing its population density to almost 1600 per sq. km (4000 per sq. mile). Life is still communal, traditional, and hard. Droughts are common and fresh water is precious. Around two-thirds of people depend on subsistence farming, digging special pits out of the coral to grow most of the islands' limited range of crops. Fishing is also important, and Tuvaluans have a reputation as excellent sailors. About 2000 Tuvaluans work overseas, some in Nauru's phosphate mines, others as merchant seamen.

***Tuvalu's soil** is porous, but sufficiently fertile to support coconut palms, pandanus, and salt-tolerant plants. Fresh water supply is limited.*

POLITICS
 Nonparty elections

2002/2006   H.M. Queen Elizabeth II

AT THE LAST ELECTION
Parliament of Tuvalu 15 seats

There are no political parties. All members are independent candidates

The 15 MPs, elected every four years, are independents who work in loose political associations. The prime minister, an MP elected by parliament, works with a cabinet of up to four other MPs. Day-to-day administration lies in the hands of an elected council on each of Tuvalu's islands. From independence until 1998 politics was dominated by Tomasi Puapua (now governor-general) and Bikenibeu Paeniu. Parliamentary defections are unchecked, and brought down the short-lived government of Prime Minister Faimalaga Luka in 2001. Saufatu Sopo'aga was appointed prime minister after elections in 2002.

WORLD AFFAIRS
Joined UN in 2000

 ACP   Comm   PC   PIF   ADB

Tuvalu was admitted as a full member of both the UN and the Commonwealth in 2000, indicating international recognition of its growing involvement in world affairs. Agreements exist with Taiwan, Korea, and the US, whose vessels may exploit Tuvalu's fish-rich territorial waters, which extend for 8.3 million sq. km (3.2 million sq. miles).

AID
Recipient

US$4m (receipts)   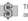 Down 43% in 2000

With a visible trade deficit, aid is crucial to Tuvalu. Most importantly, in 1987 a trust fund was set up, with US$29 million in grants from Australia, New Zealand, and the UK. While support from the UK has reduced, aid from Taiwan and Japan grows. Tuvalu plans to reduce its reliance on aid through public-sector reform and privatization.

DEFENSE
No compulsory military service

There are no armed forces   Not applicable

Tuvalu has no military. Internal security is the responsibility of the small police force.

T

ECONOMICS Not available

US$3m | 1.7997–1.9535 Australian dollars

SCORE CARD

- ❏ WORLD GNP RANKING 192nd
- ❏ GNP PER CAPITA ....US$330
- ❏ BALANCE OF PAYMENTS ....Not available
- ❏ INFLATION 7%
- ❏ UNEMPLOYMENT ....Low

STRENGTHS

Sustainable subsistence economy. EEZ: source of jobs and income. Income from trust fund. US$175 million a year from Internet deal for use of .tv suffix.

WEAKNESSES

World's smallest economy. Physical isolation. Few exports: copra, stamps,

RESOURCES Not available

400 tonnes | Not an oil producer

13,200 pigs, 10,000 ducks, 40,000 chickens | None

Tuvalu's resource potential lies in the waters of its 8.3 million-sq. km (3.2 million-sq. mile) EEZ. Its rich fish stocks are being exploited mainly by foreign boats in return for licensing fees. The salination of soil is increasing, leaving less and less cultivable land. Solar energy is being developed to cut the use of gasoline for power generation. Fuel accounts for about 14% of import costs. The rights to Tuvalu's Internet suffix have been sold to a Canadian media company for up to US$175 million a year.

TUVALU

Total Land Area : 26 sq. km (10 sq. miles)

POPULATION

- ▪ under 10 000

LAND HEIGHT

- 100m/328ft
- Sea Level

0 — 100 km
0 — 100 miles

EXPORTS

- Western Europe 20%
- Other 40%
- Africa 40%

IMPORTS

- Germany 1%
- Japan 3%
- New Zealand 6%
- Other 14%
- Fiji 59%
- Australia 17%

garments. Dependence on imports and aid. Poor-quality soil. Remittances from overseas workers set to fall as Nauru phosphate mines run out.

ENVIRONMENT Not available

 None | Not available

Efforts to protect the environmentally fragile atolls include reafforestation and solar energy projects. On Funafuti, population pressure is leading to overfishing in the atoll lagoon. The "greenhouse effect" is a major concern, since climate changes attributed to it are blamed for a steep rise in cyclone frequency. Any rise in sea levels induced by global warming would quickly submerge the atolls.

MEDIA TV ownership low

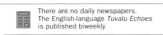 There are no daily newspapers

PUBLISHING AND BROADCAST MEDIA

There are no daily newspapers. The English-language *Tuvalu Echoes* is published biweekly

No TV service | 1 state-owned service

Tuvalu Echoes and its Tuvaluan version, *Sikuleo o Tuvalu*, are published by the government.

CRIME No death penalty

Tuvalu does not publish prison figures | Little change from year to year

Crime is minimal and the result mainly of alcohol-related violence, particularly at the weekends.

EDUCATION School leaving age: 14

95% | Not available

Each island has a primary school. A secondary school and a marine training school are based on Funafuti. Students who attend the University of the South Pacific in Fiji are state-funded.

HEALTH No welfare state health benefits

1 per 1125 people | Malaria, diarrheal, infectious and parasitic diseases

Concerted efforts since independence to improve health care facilities and programs have cut the incidence of communicable diseases. Serious cases of illness or injury are referred to better-equipped hospitals in Australia or New Zealand.

SPENDING ▷ GDP/cap. increase

CONSUMPTION AND SPENDING

No data | 90 per 1000 population

Defense None
Education No data
Health 5.9%

0 5 10 15 20 25
Defense, Health, Education spending as % of GDP

Although living standards are very low, traditional social support systems mean that extreme poverty is rare. Most people rely on subsistence agriculture and fishing, supplemented by remittances from expatriate Tuvaluans.

WORLD RANKING

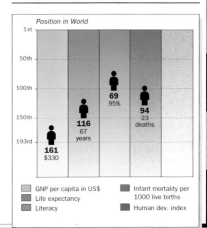

Position in World

- 161 $330
- 116 67 years
- 69 95%
- 94 23 deaths

GNP per capita in US$ | Infant mortality per 1000 live births
Life expectancy | Human dev. index
Literacy

T

UGANDA

EAST AFRICA

OFFICIAL NAME: Republic of Uganda CAPITAL: Kampala POPULATION: 24 million
CURRENCY: New Uganda shilling OFFICIAL LANGUAGE: English

AN EAST AFRICAN COUNTRY of fertile upland plateaus and mountains, Uganda has outlets to the sea through Kenya and Tanzania. Its history from independence in 1962 until 1986 was one of ethnic strife. Since 1986, under President Museveni, peace has been restored and steps have been taken to rebuild the economy and democracy.

Kampala, Uganda's capital. Only a tiny proportion of the city's households are supplied with running water

CLIMATE
▷ Tropical wet & dry

WEATHER CHART FOR KAMPALA

Altitude and the influence of Lake Victoria moderate Uganda's climate. March–May is the wettest period.

TRANSPORTATION
▷ Drive on left

Entebbe International
436,506 passengers

2 ships
5900 dwt

THE TRANSPORTATION NETWORK

| 1800 km (1118 miles) | None |
| 261 km (162 miles) | Lake Victoria and other lakes are navigable |

The government is rebuilding the transportation infrastructure with the help of international aid.

TOURISM
▷ Visitors : Population 1:159

151,000 visitors

Down 37% 1999–2000

MAIN TOURIST ARRIVALS

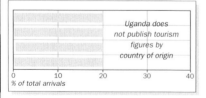

Uganda does not publish tourism figures by country of origin

0   10   20   30   40
% of total arrivals

Major attractions are Uganda's lakes and mountains, notably the rugged Ruwenzori range – the Mountains of the Moon. The brutal murder of eight foreign tourists by Rwandan fighters at the Bwindi national park in March 1999 was a severe setback for Uganda's recovery as a tourist destination.

PEOPLE
▷ Pop. density medium

Luganda, Nkole, Chiga, Lango, Acholi, Teso, Lugbara, English

120/km² (312/mi²)

THE URBAN/RURAL POPULATION SPLIT

14%   86%

RELIGIOUS PERSUASION

Muslim (mainly Sunni) 5%
Other (including Hindu) 11%
Roman Catholic 38%
Traditional beliefs 13%
Protestant 33%

The predominantly rural population consists of 13 main ethnic groups. Traditional animosities were manipulated by ex-rulers Amin and Obote. Since 1986 President Museveni has worked hard for reconciliation, in 1993 allowing the restoration of the four historical monarchies. Uganda now has one of the best human rights records in Africa.

UGANDA

Total Land Area : 236 040 sq. km (91 135 sq. miles)

POPULATION

over 100 000	◎
over 50 000	○
over 10 000	●
under 10 000	·

LAND HEIGHT

3000m/9843ft
2000m/6562ft
1000m/3281ft
500m/1640ft

POLITICS
▷ Nonparty elections

2001/2006

President Yoweri Kaguta Museveni

AT THE LAST ELECTION

Parliament 276 members

Elections to the Parliament took place on a "no-party" basis in June 2001

Since 1986, President Museveni has run a "no-party democracy," with political parties represented in a broadly based government, but banned from campaigning. Continuing to overcome ethnic tension is the main issue after its catastrophic effect in the 1970s and 1980s. Rebel insurgencies in the north and west led to the deaths, kidnapping, and displacement of tens of thousands of people, and destroyed Uganda's economy. A referendum in 2000 on a return to full multipartyism resulted in an overwhelming vote of support for the "no-party" system. In 2001, Museveni won another term in office, taking 69% of the vote, and his supporters maintained a clear majority in the legislative elections.

U

WORLD AFFAIRS
▷ Joined UN in 1962

    OIC

Relations with Sudan and the DRC are strained. Conflicts in these countries and in Rwanda have caused a large influx of refugees into Uganda. Uganda's support for antigovernment rebels in Sudan appeared to end during 1999. Ugandan forces embroiled in the fighting in the DRC were withdrawn from the front line in 2001.

AID
▷ Recipient

 $819m (receipts)   Up 39% in 2000

Aid, mainly from the World Bank and the UK, has risen, its donors encouraged by Uganda's adoption of economic liberalization and private-sector investment policies. Aid has focused on balance-of-payments support, the rehabilitation of the key transportation sector, and the fight against AIDS.

DEFENSE
▷ No compulsory military service

 $247m   Up 24% in 2000

The pre-1986 army was responsible for many atrocities under Amin's rule. Since the 1990s, Uganda was preoccupied with conflicts in neighboring countries, in particular the DRC, where it supported antigovernment rebels. The army has also been deployed to suppress internal rebellions.

ECONOMICS
▷ Inflation 12% p.a. (1990–2000)

 $6.7bn   1767.5–1727.5 new Uganda shillings

SCORE CARD

❏ World GNP Ranking	100th
❏ GNP per Capita	$300
❏ Balance of Payments	–$860m
❏ Inflation	2.8%
❏ Unemployment	Widespread

STRENGTHS
Agriculture. Coffee brings in 93% of export earnings. Potential for more export crops. Road system is being repaired. Proinvestment policies.

WEAKNESSES
Lack of skilled workforce. Instability in the subregion affects confidence. World coffee price fluctuations. High transportation costs.

EXPORTS

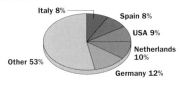

Italy 8%
Spain 8%
USA 9%
Netherlands 10%
Other 53%
Germany 12%

IMPORTS

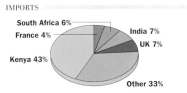

South Africa 6%
France 4%
India 7%
UK 7%
Kenya 43%
Other 33%

RESOURCES
▷ Electric power 189,000 kw

 226,457 tonnes   Not an oil producer

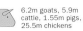 6.2m goats, 5.9m cattle, 1.55m pigs, 25.5m chickens   Copper, cobalt, tin, apatite, magnetite, tungsten, gold

Mineral resources are varied but barely exploited. Uganda has sizable copper deposits. The mines, closed under Obote, are now being reopened. Gold and cobalt mining are also due to resume and oil exploration is under way. Hydroelectric output is being expanded, notably at Owen Falls, with the aim of replacing 50% of oil imports.

ENVIRONMENT
▷ Sustainability rank: 76th

 10% (4% partially protected)   0.1 tonnes per capita

Uganda's priority is economic reconstruction, but ecological issues are not ignored. The construction of a huge hydroelectric power plant at the Kabalega (Murchison) Falls, above Lake Albert, was canceled, following widespread environmental objections.

MEDIA
▷ TV ownership low

 Daily newspaper circulation 2 per 1000 people

PUBLISHING AND BROADCAST MEDIA

There are 5 daily newspapers, *New Vision, The Star, The Monitor, Ngabo,* and *Taifa Uganda Empya*

2 services:
1 state-controlled,
1 independent

4 services:
1 state-controlled,
3 independent

The 13 daily and weekly papers cover the political and religious spectrum; eight are published in English. Only *New Vision* is government-controlled.

CRIME
▷ Death penalty in use

 21,971 prisoners   Up 83% in 1999

Crime levels are low, though theft in Kampala is a growing problem. In 2000, the remains were discovered of 780 followers of the cult of the Restoration of the Ten Commandments of God.

CHRONOLOGY
Uganda's ancient kingdoms were ruled under a British protectorate from 1893 until independence in 1962.

- ❏ **1962–1971** Milton Obote in power.
- ❏ **1971–1986** Ethnic strife, economic collapse under Idi Amin and Obote.
- ❏ **1986** President Museveni in power.
- ❏ **1996** Museveni wins first presidential elections.
- ❏ **2000** Referendum endorses "no-party" system.

EDUCATION
▷ Schooling is not compulsory

 67%   40,591 students

All schools charge fees. Enrollment in primary schools is 94%, but only 11% of pupils go on to secondary school.

HEALTH
▷ Welfare state health benefits

1 per 20,000 people   Malaria, respiratory, and diarrheal diseases, measles

A successful education and prevention campaign has reduced the prevalence of HIV/AIDS to about 5% of adults, from a peak of 14% in the early 1990s.

SPENDING
▷ GDP/cap. increase

CONSUMPTION AND SPENDING

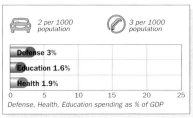

2 per 1000 population   3 per 1000 population

Defense 3%
Education 1.6%
Health 1.9%

0   5   10   15   20   25
Defense, Health, Education spending as % of GDP

Uganda has a small but growing middle class. Those close to the government form the wealthiest group. 44% of the population live below the poverty line.

WORLD RANKING

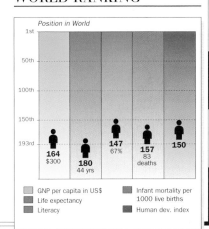

Position in World

1st
50th
100th
150th
193rd

164 $300
180 44 yrs
147 67%
157 83 deaths
150

GNP per capita in US$
Life expectancy
Literacy
Infant mortality per 1000 live births
Human dev. index

U

UKRAINE

EUROPE
Europe
Poland · Russian Fed.
UKRAINE
Romania
Moldova

OFFICIAL NAME: Ukraine   **CAPITAL:** Kiev
POPULATION: 48.4 million   **CURRENCY:** Hryvna   **OFFICIAL LANGUAGE:** Ukrainian

 1991   1991   Aug 24   UA   +2   +380   .ua

UKRAINE IS BORDERED by seven states; to the south it lies on the Black Sea and the Sea of Azov. An independent Ukrainian state was established in 1918, but was overrun in the same year by Soviet forces from the east and Polish forces from the west. In 1991, Ukraine again became an independent state. The country has historically been divided between the nationally conscious and Ukrainian-speaking west (including areas which were part of Poland until World War II) and the east, which has a large ethnic Russian population.

View toward the Cathedral *of the Assumption in Kharkiv. Many Ukrainian cities are equipped with elaborate trolley networks.*

CLIMATE
▷ Continental/steppe/ Mediterranean

WEATHER CHART FOR KIEV

■ *Average daily temperature*   Rainfall ■
°C/°F J F M A M J J A S O N D cm/in
40/104   40/16
30/86   30/12
20/68   20/8
10/50   10/4
0/32   0
-10/14
-20/-4

Ukraine has a continental climate, with the exception of the southern coast of Crimea, which has a Mediterranean climate. There are four distinct seasons.

TRANSPORTATION
▷ Drive on right

✈ **Boryspiel International, Kiev**
1.4m passengers

🚢 966 ships
2.03m grt

THE TRANSPORTATION NETWORK

170,139 km (105,719 miles)	1770 km (1100 miles)
22,473 km (13,965 miles)	4400 km (2734 miles)

There are Soviet-style subways and trolley networks in the major cities. The rail system and the main highway linking Kiev and Lviv are being upgraded. Part of a former submarine port at Sevastopol has been opened to commercial shipping.

TOURISM
▷ Visitors : Population 1:11

4.23m visitors   ⬇ Down 32% in 1999

MAIN TOURIST ARRIVALS

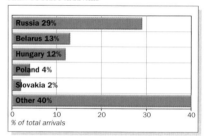

Russia 29%
Belarus 13%
Hungary 12%
Poland 4%
Slovakia 2%
Other 40%

0   10   20   30   40
% of total arrivals

Among potential tourist attractions are warm resort areas in Crimea and the south, and the Carpathian Mountains. The government has maintained a highly regulated system of managing tourism, and bureaucratic hurdles have held up the development of Western-style hotels.

UKRAINE

Total Land Area :
603 700 sq. km (223 089 sq. miles)

POPULATION
⊡ over 1 000 000
⊙ over 500 000
◎ over 100 000
○ over 50 000
● over 10 000

LAND HEIGHT
2000m/6562ft
1000m/3281ft
500m/1640ft
200m/656ft
Sea Level

N

0   100 km
0   100 miles

(Map of Ukraine showing BELARUS, POLAND, SLOVAKIA, HUNGARY, ROMANIA, MOLDOVA, RUSSIAN FEDERATION, Black Sea, Sea of Azov, Crimea, and cities including Kiev, Kharkiv, L'viv, Odesa, Dnipropetrovs'k, Donets'k, Sevastopol', Yalta, Simferopol', Mykolayiv, Zaporizhzhya, Mariupol', Luhans'k, Poltava, Chernihiv, Sumy)

PEOPLE ▷ Population density medium

Ukrainian, Russian, Tatar   80/km² (208/mi²)

THE URBAN/RURAL POPULATION SPLIT

68%   32%

RELIGIOUS PERSUASION

- Jewish 1%
- Other 4%
- Christian (mainly Ukrainian Orthodox) 95%

ETHNIC MAKEUP

- Jewish 1%
- Other 4%
- Russian 22%
- Ukrainian 73%

In the cities and countryside of western Ukraine, Ukrainians make up the vast majority of the population. Russian is spoken by 60% of the people, and in several of the large cities of the east and south, ethnic Russians form a majority, a legacy of 19th-century industrialization and of more recent migration in the Soviet era. At independence, most Russians accepted Ukrainian sovereignty, though tensions remain.

The central government is wary of separatist tendencies in Crimea, however, where Russians make up two-thirds of the population. The Crimea's other main minority, besides ethnic Ukrainians, is the Turkic-speaking Tatar people. Deported en masse to the eastern USSR in 1944 under Stalin, the Tatars have been returning to the Crimea since 1990 and now make up 12% of its population. There is a Romanian-speaking minority in the Odessa region.

POPULATION AGE BREAKDOWN

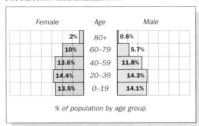

Female	Age	Male
2%	80+	0.6%
10%	60–79	5.7%
13.6%	40–59	11.8%
14.4%	20–39	14.3%
13.5%	0–19	14.1%

% of population by age group

Leonid Kuchma, *who became president in 1994.*

Viktor Yushchenko, *prime minister in 1999–2001, now leads the opposition.*

WORLD AFFAIRS ▷ Joined UN in 1945

 BSEC   CE   CIS   IAEA   OSCE

In the immediate postindependence period Western countries viewed Ukraine as a convenient buffer state between Russia and the rest of Europe. However, their support for Ukraine did not make it strong enough to offset its economic dependence on Russia, especially for fuel. Over time this unqualified backing has been eroded by the West's own relations with Russia, and by distrust of President Kuchma's antidemocratic tendencies. Conversely, Ukraine's ties with Russia, initially seen as a potential aggressor, have been strengthened under Kuchma: a friendship treaty was signed in 1997 and a ten-year economic cooperation accord in 1998. Ukraine signed a trade agreement with the EU in 1995 and was admitted to the Council of Europe in the same year. Despite an ongoing territorial dispute with Romania, a friendship and cooperation treaty was signed in 1997.

POLITICS ▷ Multiparty elections

2002/2006   President Leonid Kuchma

AT THE LAST ELECTION
Parliament 450 seats

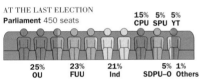

15%	5%	5%
CPU	SPU	YT

25%	23%	21%	5%	1%
OU	FUU	Ind	SDPU–O	Others

OU = Our Ukraine (Viktor Yushchenko bloc)
FUU = For United Ukraine   **Ind** = Independents
CPU = Communist Party   **SPU** = Socialist Party of Ukraine
SDPU–O = United Social-Democratic Party of Ukraine
YT = Yulia Timoshenko Election Bloc

Ukraine introduced a multiparty system in 1991.

PROFILE

Ex-premier Leonid Kuchma, who had defeated Ukrainian nationalist Leonid Kravchuk in the 1994 presidential elections, gained increased powers under constitutional changes in 1996. Multiparty legislative elections were first held in 1994, and the strong position of the communists and allied pro-Kuchma groups was confirmed four years later. Kuchma's own reelection in 1999 was fiercely contested, however, with opponents claiming fraud and violent confrontations taking place in parliament and on the streets. The boost which Kuchma gained in early 2000 from a referendum backing electoral changes, and Western enthusiasm for his new pro-reform government, was dissipated later that year by a scandal linking him with the murder of a journalist and the uncovering of massive financial frauds. His opponents included growing numbers of discarded former ministers, and the pro-Kuchma For United Ukraine was pushed into second place by the opposition Our Ukraine bloc in legislative elections in 2002.

MAIN POLITICAL ISSUES
Economic reform and corruption

There is generally more support for economic reform in western and central Ukraine than in eastern regions. There have been spasmodic efforts to promote the transition to a market economy, notably during 2000 under the then prime minister Viktor Yushchenko. Such radicalism has foundered on lack of sufficient backing from Kuchma, opposition in parliament, and high-level corruption as powerful business and political factions vie to gain advantage.

Separatism among ethnic Russians

The Crimean parliament's declaration of independence in 1994 has been the strongest threat so far to Ukraine's integrity. It was resisted and later rescinded, although Kiev reaffirmed Crimea's status as an autonomous republic. A new Crimean constitution was approved in 1999. Donbass has also pressed for more autonomy, and the Donetsk region has voted to make Russian a joint official language.

CHRONOLOGY

In 1240, Kiev was conquered by the Mongols. The Ukrainian Cossacks later came under the domination of Lithuania, Poland, and Russia.

- ❏ **1918** Independent Ukrainian state after collapse of Russian and Austro-Hungarian Empires. Brest-Litovsk Treaty signed with Germany.
- ❏ **1919** Red Army invades. Ukrainian Soviet Socialist Republic proclaimed.
- ❏ **1920** Poland invades; western Ukraine under Polish occupation.
- ❏ **1922** USSR, and Ukrainian SSR, formed.
- ❏ **1922–1930** Cultural revival under Lenin's "Ukrainianization" policy to pacify national sentiment.
- ❏ **1932–1933** "Ukrainianization" policy reversed. Stalin uses famine to eliminate Ukraine as source of opposition; seven million die. ⇨

U

CHRONOLOGY *continued*

- ❑ **1939** Soviet Union invades Poland and incorporates its ethnic Ukrainian territories into the Ukrainian SSR.
- ❑ **1941** Germany invades USSR. 7.5 million Ukrainians die by 1945.
- ❑ **1942** Nationalists form Ukrainian Insurgent Army, which wages war against both Germans and Soviets.
- ❑ **1954** Crimea ceded to Ukrainian SSR.
- ❑ **1972** Widespread arrests of intellectuals and dissidents by Soviet state. Vladimir Shcherbitsky, a Brezhnevite, replaces moderate reformer Petr Shelest as head of Communist Party of Ukraine (CPU).
- ❑ **1986** World's worst nuclear disaster at Chernobyl power plant.
- ❑ **1989** First major coalminers' strike in Donbass. Pro-Gorbachev Volodymyr Ivashko heads CPU.
- ❑ **1990** Ukrainian parliament declares Ukrainian SSR a sovereign state. Leonid Kravchuk replaces Ivashko.
- ❑ **1991** Full independence declared, conditional on approval by referendum, supported by 90% of voters. CPU banned. Crimea becomes an autonomous republic within Ukrainian SSR.
- ❑ **1993** Major strike in Donbass results in costly settlement, which exacerbates budget deficit and stimulates hyperinflation. CPU reestablished at Donetsk congress.
- ❑ **1994** Crimea elects Yuri Meshkov as its first president. Leonid Kuchma defeats Kravchuk to become first democratically elected president of Ukraine.
- ❑ **1996** Hryvna replaces karbovanets as national currency. New constitution comes into force.
- ❑ **1997** Friendship treaty signed with Russia. Accord on Black Sea fleet.
- ❑ **1998** Ten-year cooperation agreement with Russia. CPU wins largest number of seats in election.
- ❑ **1999** Reelection of Kuchma. Opposition claims of fraud. Kuchma appoints pro-reform government.
- ❑ **2000** Chernobyl site closed.
- ❑ **2001** Kuchma linked with murder of journalist. Reformist premier Viktor Yushchenko replaced after parliamentary defeat.
- ❑ **2002** Opposition parties make large gains in legislative elections.

AID

 Recipient

$541m (receipts)   Down 5% in 2000

US assistance in the 1990s was the world's fourth-largest aid program, while EU assistance totaled $3.5 billion in 1991–1999. In 2000 lenders pressed for better controls on fund allocations.

DEFENSE

 Phasing out conscription

$1.08bn   Down 25% in 2000

UKRAINIAN ARMED FORCES

3937 main battle tanks (149 T-55, 2277 T-64, 1238 T-72, 273 T-80)	151,200 personnel
Black Sea fleet: 1 submarine, 1 cruiser, 2 frigates, 8 patrol boats	13,000 personnel
543 combat aircraft (MiG-23, MiG-29, Su-27, Su-25, Su-24)	96,000 personnel
None	

Ukraine was a center for arms manufacture under the old Soviet system, and now has a major weapons export trade. A member of the CIS, Ukraine finally resolved in 1997 its long-smoldering dispute with Russia over control of the Black Sea fleet, with agreement on the division of the fleet and a 20-year Russian lease on port facilities in Sevastopol. Meanwhile, the Ukrainian parliament had ratified the START-I nuclear disarmament treaty, and Ukraine's nuclear warheads were transferred to Russia under a trilateral weapons dismantling accord also involving substantial US aid.

In 2000 it was decided that compulsory military service, which lasts 18 months, should be ended by 2015. The armed forces have been slimmed down, with further cuts planned to result in a total strength of 285,000 (including central and administrative staff) by 2005.

Ukraine joined NATO's Partnerships for Peace in 1995, signed a security pact with the alliance in 1997, and applied in 2002 for full NATO membership.

ECONOMICS

 Inflation 271% p.a. (1990–2000)

$34.6bn   5.435–5.314 hryvnas

SCORE CARD

- ❑ WORLD GNP RANKING............................55th
- ❑ GNP PER CAPITA $700
- ❑ BALANCE OF PAYMENTS....................$1.48bn
- ❑ INFLATION 22.7%
- ❑ UNEMPLOYMENT...................................12%

EXPORTS

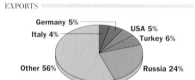

Germany 5%
Italy 4%
USA 5%
Turkey 6%
Other 56%
Russia 24%

IMPORTS

Belarus 4%
Kazakhstan 3%
Turkmenistan 7%
Germany 8%
Russia 42%
Other 36%

ECONOMIC PERFORMANCE INDICATOR

Consumer Price Index   GDP
Consumer price index 1995=100 / GDP 1995=100
1995 1996 1997 1998 1999

STRENGTHS
Well-educated workforce. Good urban transportation infrastructure. Potential for grain and food export. Mineral reserves. Technological potential, especially in aerospace and computers.

WEAKNESSES
Failure to reform centrally planned economy. High inflation. Weak currency. Huge debt. Antireform political elites. Inefficient, subsidized manufacturing industries. Corruption.

PROFILE
Real growth was recorded at last in 2000, the economy having contracted by over half over ten years. Privatization of large enterprises has barely begun, and bureaucracy stifles private enterprise and investment. Lack of land reform holds back agriculture in the "bread basket" of Europe. Controversial legislation passed in 2001 paved the way for the sale of farmland after 2004.

UKRAINE : MAJOR BUSINESSES

Kiev
Kharkiv
L'viv
Donets Basin
Odesa
Dnipropetrovs'k
Kherson
Kryvyy Rih

Iron & steel
Coal mining
Engineering
Iron ore mining
Consumer goods
Food processing
Vehicle manufacture
Chemicals
Electronics
Textiles
Aerospace industry

0   200 km
0   200 miles

U

RESOURCES

 Electric power 53.9m kw

 441,672 tonnes

 78,195 b/d (reserves 1.6bn barrels)

20m ducks, 9.91m cattle, 9.08m pigs, 108m chickens

Coal, iron, oil, natural gas, manganese, lignite, peat, mercury

ELECTRICITY GENERATION

Hydro 9% (16bn kwh)

Combustion 47% (82bn kwh)

Nuclear 44% (75bn kwh)

Other 0%

% of total generation by type

Ukraine imports 80% of its oil and gas (its main source of energy), mostly from Russia. Some gas is in lieu of transit fees for pipelines carrying Russian gas, but it repeatedly fails to keep up payments for the rest. Yet Ukraine has oil and gas reserves of its own. Coal is mined, in appalling safety conditions, mainly in the Donbass–Donetsk region. Just under half of electricity is nuclear-generated.

Ukraine has 5% of global mineral reserves, including the largest titanium reserves, the third-largest deposits of iron ore, and 30% of global manganese ore. There is also mercury, uranium, nickel, and some gold. The metal industry accounted for nearly 20% of GDP in 1997 and 28% of exports. The steel industry has now begun to grow again after seven years of decline.

ENVIRONMENT

 Sustainability rank: 136th

 2%

7 tonnes per capita

ENVIRONMENTAL TREATIES

Yes   Yes   Yes

Yes   Yes   No

As a result of the Chernobyl nuclear disaster in 1986 – the world's worst nuclear accident – over three million Ukrainians live in dangerously radioactive areas and 12% of arable land is contaminated. The last working reactor at Chernobyl closed at the end of 2000, under agreements in which Western countries provided large-scale financial assistance. However, nuclear production continues elsewhere because of the cost of Russian oil and gas imports. Coal-fired power plants are old-fashioned, highly polluting, and inefficient. Industrial pollution is widespread, especially from steel and chemical works in the Donbass region, contributing to the acute problem of low air quality in eight major cities.

MEDIA

 TV ownership high

Daily newspaper circulation 84 per 1000 people

PUBLISHING AND BROADCAST MEDIA

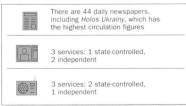

There are 44 daily newspapers, including *Holos Ukrainy*, which has the highest circulation figures

3 services: 1 state-controlled, 2 independent

3 services: 2 state-controlled, 1 independent

Independent, mass-circulation newspapers are published, mainly in Russian. Local TV stations reflect regional political differences.

CRIME

 No death penalty

 219,555 prisoners

Down 2% in 1999

CRIME RATES

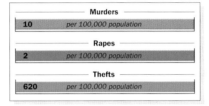

Murders	
10	per 100,000 population

Rapes	
2	per 100,000 population

Thefts	
620	per 100,000 population

Street crime, robberies, violence, and carjackings have increased sharply. The death penalty was abolished in 2000. The country is a major source for people trafficking to the West. Corruption is rampant across the economy. Political killings make headlines, as did the murder in 2000 of journalist Gyorgy Gongadze.

EDUCATION

 School leaving age: 15

 99%

 1.54m students

THE EDUCATION SYSTEM

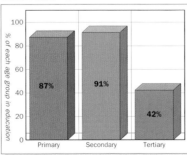

% of each age group in education

Primary 87%   Secondary 91%   Tertiary 42%

Using Ukrainian in schools is the main element in the drive to promote the once-banned language. Some schools in the west no longer teach Russian. Most university teaching is in Russian in eastern regions, and in Ukrainian in those in the west.

UKRAINE : LAND USE

Cropland

Forest

Pasture

Cattle

Wheat - cash crop

Sugar beet

0   200 km

0   200 miles

HEALTH

 Welfare state health benefits

 1 per 333 people

Heart disease, cancers, accidents, violence, tuberculosis

Health care, supposedly free to all, has declined significantly in the post-Soviet period. A $2 million UN program provides treatment and preventive care for the 350,000 people who dealt with the Chernobyl disaster. By 2001, 1% of the population was HIV positive.

SPENDING

GDP/cap. decrease

CONSUMPTION AND SPENDING

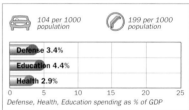

104 per 1000 population   199 per 1000 population

Defense 3.4%

Education 4.4%

Health 2.9%

Defense, Health, Education spending as % of GDP

The gap between rich and poor has widened significantly. Wage arrears – and massive hidden unemployment – are major problems.

WORLD RANKING

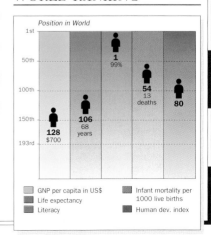

Position in World

128 $700

106 68 years

1 99%

54 13 deaths

80

GNP per capita in US$

Life expectancy

Literacy

Infant mortality per 1000 live births

Human dev. index

U

UNITED ARAB EMIRATES

OFFICIAL NAME: United Arab Emirates **CAPITAL:** Abu Dhabi
POPULATION: 2.7 million **CURRENCY:** UAE dirham **OFFICIAL LANGUAGE:** Arabic

  1971 1972 Dec 2 UAE +4 +971 .ae

THE UNITED ARAB EMIRATES (UAE), created in 1971, is the Arab world's only working federation. Six of its seven emirates cluster around the northeastern corner of the country, while Abu Dhabi has a larger hinterland of semiarid desert relieved by occasional oases. The cities, watered by extensive irrigation systems, have lavish greenery. Prosperity once relied on pearls, but the UAE is now a sizable gas and oil exporter, and has a growing services sector.

CLIMATE ▷ Hot desert

WEATHER CHART FOR ABU DHABI

Although rainfall is minimal, summers are humid. Sand-laden *shamal* winds often blow in winter and spring.

TRANSPORTATION ▷ Drive on right

Dubai International 12.3m passengers 332 ships 933,000 grt

THE TRANSPORTATION NETWORK

| 1088 km (676 miles) | None |
| None | None |

The roads are good. Five of the seven emirates have international airports, of which the busiest is Dubai International.

TOURISM ▷ Visitors : Population 1.3:1

3.42m visitors Up 15% in 2000

MAIN TOURIST ARRIVALS

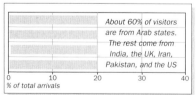

About 60% of visitors are from Arab states. The rest come from India, the UK, Iran, Pakistan, and the US

% of total arrivals

Until the mid-1980s, tourism was minimal. Led by Dubai, the UAE has now launched initiatives to attract Western visitors during the winter for sunshine, heritage, water sports, desert safaris, and duty-free shopping.

PEOPLE ▷ Pop. density low

Arabic, Farsi, Indian and Pakistani languages, English 32/km² (84/mi²)

THE URBAN/RURAL POPULATION SPLIT

86%   14%

ETHNIC MAKEUP

- European 3%
- Other Arab 12%
- Asian 60%
- Emirian 25%

UAE nationals are largely city dwellers, with Abu Dhabi and Dubai the main centers. They are greatly outnumbered by expatriates who arrived in the 1970s during the oil boom, and the Western expatriate community is permitted a virtually unrestricted lifestyle. Most UAE nationals are conservative Sunni Muslims of Bedouin descent, although there is a Shi'a community in Dubai with links to Iran. Islamic fundamentalism is a growing force among the young.

Poverty is rare in the UAE, where the government remains the biggest employer. Women in theory enjoy equal rights with men. A Presidential Marriage Fund discourages UAE men from taking foreign wives.

POLITICS ▷ No legislative elections

Not applicable President Shaikh Zayed bin Sultan al-Nahyan

LEGISLATIVE OR ADVISORY BODIES
Federal National Council 40 seats

There are no political parties. The method of appointment of members of the Federal National Council is determined individually by each of the seven members of the Federation.

The UAE's seven emirates – Abu Dhabi, Dubai, Sharjah, Ras al Khaimah, Ajman, Umm al Qaiwain, and Fujairah – are dominated by their ruling families. The main personalities are the ruler of Abu Dhabi, Shaikh Zayed, who holds the UAE presidency, and the four al-Maktoum brothers who control Dubai. The eldest, Shaikh Maktoum al-Maktoum, is ruler of Dubai as well as vice president and prime minister of the UAE.

President Zayed has relaunched the advisory Federal National Council in response to criticism of the lack of democracy. The growth of Islamic fundamentalism is also a concern. The freedoms granted to Westerners have aroused some anger but, for economic reasons, they are unlikely to be withdrawn.

UNITED ARAB EMIRATES

Total Land Area : 82 880 sq. km (32 000 sq. miles)

POPULATION
- ◎ over 100 000
- • under 10 000

LAND HEIGHT
- 1000m/3281ft
- 500m/1640ft
- Sea Level

U

WORLD AFFAIRS
 Joined UN in 1971

The UAE is well known as an advocate of moderation within the Arab world. It maintains close links with most OECD states, especially the UK and the US. In 1992, conflict flared when Iran seized control of three islands in the Strait of Hormuz. Attempts are still being made to settle the dispute through diplomacy.

AID
 Donor

 $150m (donations)   Up 63% in 2000

Once a generous donor to developing countries, the UAE's contributions have fluctuated with energy prices.

DEFENSE
No compulsory military sevice

$3.34bn   Up 5% in 2000

Training of UAE forces is limited, and personnel are mainly drawn from other Arab states and the Indian subcontinent. US air bases in the UAE supply refueling craft and were used in the 1991 Gulf War, subsequent air patrols over southern Iraq, and during the 2001 air strikes in Afghanistan. The countries of the GCC signed their first defense pact in 2000.

ECONOMICS
Inflation 2.3% p.a. (1990–1999)

 $48.7bn   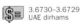 3.6730–3.6729 UAE dirhams

SCORE CARD

- ❑ WORLD GNP RANKING..........................49th
- ❑ GNP PER CAPITA $17,870
- ❑ BALANCE OF PAYMENTS....................$24.6bn
- ❑ INFLATION 4.5%
- ❑ UNEMPLOYMENT ....None

STRENGTHS
Oil and gas reserves are third-biggest in OPEC. Soaring oil prices in 2000. Development of service industries and manufacturing sector. Since 2000, regional tax-free base for e-commerce.

WEAKNESSES
Lack of skilled manpower. Most raw materials and foodstuffs have to be imported. Water resources scarce.

EXPORTS

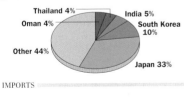

Thailand 4%
Oman 4%
Other 44%
India 5%
South Korea 10%
Japan 33%

IMPORTS

France 6%
Other 67%
USA 6%
Germany 7%
Japan 7%
UK 7%

RESOURCES
Electric power 5.7m kw

 117,607 tonnes   2.42m b/d (reserves 97.8bn barrels)

1.2m goats, 467,281 sheep, 14.7m chickens   Oil, natural gas

The UAE is a major exporter of crude oil and natural gas. Oil production accounts for a great part of export revenue. Mina Jabal Ali in Dubai is the world's largest man-made port and has attracted companies from more than 50 countries. Saadiyat Island off Abu Dhabi is being developed as a financial resort.

ENVIRONMENT
Sustainability rank: 141st

 None   32.4 tonnes per capita

Despite the harsh desert climate, there is a rich variety of plants and animals. Shaikh Zayed champions conservation parks to avert the threat from hunting.

MEDIA
TV ownership high

 Daily newspaper circulation 170 per 1000 people

PUBLISHING AND BROADCAST MEDIA

 There are 7 daily newspapers. The leading Arabic newspaper is *Al-Ittihad. Emirates News* is its English-language counterpart

4 state-owned services   7 state-owned services

Satellite TV is unrestricted. Dubai Media City, opened in 2001, promotes greater press freedom.

CRIME
Death penalty in use

 6000 prisoners   Up sharply in 1999

Street crime and muggings are rare. Dubai is reputed to be a transit point for the smuggling of narcotics and caviar.

An oasis village, inland from Fujairah, now accessible through a well-developed network of new roads.

CHRONOLOGY
The UAE was influenced by the Portuguese and the Ottomans, but British control became dominant in the 19th century.

- ❑ **1971** The UK withdraws as protecting power and the UAE federation is formed.
- ❑ **1991** UAE offers bases to Western forces after Kuwait is invaded.
- ❑ **2000** GCC defense pact signed.

EDUCATION
School leaving age: 12

 76%   16,213 students

UAE citizens enjoy completely free education. Zayed University was set up in three emirates in 1998.

HEALTH
Welfare state health benefits

1 per 556 people   Circulatory and respiratory diseases, cancers

A high-quality system of primary health care is in place for all UAE citizens, with hospitals able to carry out most operations.

SPENDING
GDP/cap. increase

CONSUMPTION AND SPENDING

82 per 1000 population   391 per 1000 population

Defense 5.9%
Education 1.9%
Health 0.8%

0   5   10   15   20   25
Defense, Health, Education spending as % of GDP

UAE nationals enjoy one of the highest per capita incomes in the Arab world. There is no income tax, and oil revenues subsidize public services. Entrepreneurship is encouraged.

WORLD RANKING

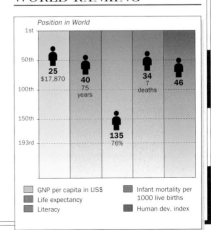

Position in World

1st
50th
100th
150th
193rd

25 $17,870
40 75 years
34 7 deaths
46
135 76%

- ☐ GNP per capita in US$
- ☐ Life expectancy
- ☐ Literacy
- ☐ Infant mortality per 1000 live births
- ☐ Human dev. index

U

UNITED KINGDOM

OFFICIAL NAME: United Kingdom of Great Britain and Northern Ireland **CAPITAL:** London
POPULATION: 59.5 million **CURRENCY:** Pound sterling **OFFICIAL LANGUAGES:** English, Welsh (in Wales)

LYING IN NORTHWESTERN Europe, the United Kingdom (UK) occupies the major portion of the British Isles. It includes the countries of England, Scotland, and Wales, the constitutionally distinct region of Northern Ireland, and several outlying islands. Its only land border is with the Irish republic. The UK is separated from the European mainland by the English Channel and the North Sea. To the west lies the Atlantic Ocean. Most of the population live in towns and cities and, in England, is fairly well scattered. The most densely populated region is the southeast. Scotland is the wildest region, with the Highlands less populated today than in the 18th century. The UK joined the European Communities (EC – later the EU) in 1973, and most of its trade is now with its European partners. Membership of the UN Security Council also gives the UK a prominent role in international politics.

CLIMATE ▷ Maritime

WEATHER CHART FOR LONDON

The UK has a generally mild, temperate, and highly changeable climate. Rain, regarded as synonymous with Britain's weather, is fairly well distributed throughout the year, but recently unusually long dry or wet spells have caused water shortages in some areas, and flooding in others. The west is generally wetter than the east, and the south warmer than the north.

TRANSPORTATION ▷ Drive on left

 Heathrow, London
64.6m passengers

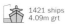 1421 ships
4.09m grt

THE TRANSPORTATION NETWORK

371,603 km
(230,903 miles)

3453 km
(2146 miles)

17,064 km
(10,604 miles)

5700 km
(3542 miles)

The government has not fulfilled its 1997 campaign promise of a more integrated policy, faced with congestion, pollution, and motorists' resentment of high fuel taxes. The rail system, after the rushed privatization of the 1990s, suffers from underinvestment, maintenance problems, and fragmented services. There is still no high-speed rail link to the Channel Tunnel, which opened in 1994.

TOURISM ▷ Visitors : Population 1:2.5

 23.4m visitors

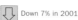 Down 7% in 2001

MAIN TOURIST ARRIVALS

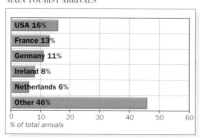

USA 16%
France 13%
Germany 11%
Ireland 8%
Netherlands 6%
Other 46%

% of total arrivals

The UK ranks sixth in the world as a tourist destination. Tourism is among the country's most important industries and a growing source of employment. Heritage is the principal selling point, North Americans, French, and Germans are the main visitors, and London, with its art galleries, theaters, and historic buildings, remains the major destination. Visitors also head for the Roman splendors of Bath, Shakespeare's Stratford-upon-Avon, the medieval buildings of Oxford, Cambridge, and York, and Scotland, where the Highlands are a particular attraction.

Oxford, home to the oldest university in the UK. Teaching began in 1096; the first college was founded in 1249. One of the city's finest buildings is the 17th-century semi-oval Sheldonian Theatre.

PEOPLE   Pop. density high

English, Welsh, Scottish Gaelic, Irish Gaelic

246/km²
(638/mi²)

THE URBAN/RURAL POPULATION SPLIT

90%   10%

RELIGIOUS PERSUASION

Muslim 3%
Methodist 1%
Anglican 47%
Presbyterian 4%
Roman Catholic 9%
Other 36%

ETHNIC MAKEUP

Northern Irish 3%
Welsh 3%
West Indian, Asian, and Other 5%
Scottish 9%
English 80%

The Scottish and Welsh nations remain recognizably distinct, despite forming part of a unified state. Scotland retains its own legal and educational systems, and recent steps in devolution have given both countries greater autonomy.

Britain's ethnic minorities account for 5% of the total population; over 50% of their members were born in Britain. Ethnic minority communities are generally concentrated in the inner cities, where they face problems of deprivation and social stress, and may also suffer from isolation, particularly among women. Although there is little support for overt racist politics, multiethnic recruitment has made little progress in key areas such as policing, and prejudices persist. The level of institutionalized racism was criticized by the UN in 2000.

Two-fifths of all births occur outside marriage, compared with 12% in 1980, but most of these are to cohabiting couples. Single-parent households account for one-fifth of all families with children under 18.

POPULATION AGE BREAKDOWN

Female	Age	Male
2.8%	80+	1.2%
8.9%	60–79	7.5%
12.3%	40–59	12.2%
14.6%	20–39	15.2%
12.3%	0–19	13%

% of population by age group

EUROPE

Black Mount, Rannoch Moor, *in the Scottish Highlands. The Highlands are one of the UK's wildest regions.*

CHRONOLOGY

Great Britain began the 20th century as one of the world's most advanced economies, backed by a massive trading empire.

- ❏ **1906** Reformist Liberal government.
- ❏ **1914** World War I begins.
- ❏ **1918** Armistice signals end of war. Cost to Britain: 750,000 dead.
- ❏ **1922** Irish Free State given dominion status.
- ❏ **1926** General Strike.
- ❏ **1929** World stock market crash. Widespread unemployment.
- ❏ **1931** UK leaves gold standard and devalues pound.
- ❏ **1936** Edward VIII abdicates over marriage to Wallis Simpson.
- ❏ **1938** Prime Minister Neville Chamberlain meets Hitler in Munich over Czech crisis, says threat of war with Germany averted.
- ❏ **1939** Germany invades Poland. UK declares war on Germany. Start of World War II.
- ❏ **1940** Winston Churchill prime minister. Battle of Britain.
- ❏ **1944** 6th June, D-Day invasion of German-occupied France.
- ❏ **1945** End of World War II, costing 330,000 British lives. Labour government elected on social welfare platform.
- ❏ **1946** Nationalization of Bank of England, railroads, coal, utilities. ➪

UNITED KINGDOM

Total Land Area : 244 820 sq. km
(94 525 sq. miles)

POPULATION

over 5 000 000	▣
over 500 000	◉
over 100 000	◎
over 50 000	○
over 10 000	●
under 10 000	•

LAND HEIGHT

1000m/3280ft	
500m/1640ft	
200m/656ft	
Sea Level	

U

N

CHRONOLOGY *continued*

- ❏ **1947** Indian independence.
- ❏ **1948** NHS established.
- ❏ **1949** Founder member of NATO.
- ❏ **1956** Suez crisis: UK intervenes in Canal Zone, withdraws under US pressure.
- ❏ **1957** US nuclear missiles accepted on UK soil.
- ❏ **1961** UK application to EC rejected by French president de Gaulle.
- ❏ **1968** Abortion and homosexuality are legalized.
- ❏ **1969** British troops sent into Northern Ireland.
- ❏ **1973** Joins EC under Conservative Edward Heath. Oil crisis. Industry on three-day week following strikes by power workers and miners.
- ❏ **1974** Labour government, under Harold Wilson, concedes miners' demands. High inflation.
- ❏ **1975** Referendum ratifies EC membership. First North Sea oil pipeline in operation.
- ❏ **1979–1997** Conservative rule, until 1990 under Margaret Thatcher.
- ❏ **1980** Anti-US Cruise missiles protests. Rising unemployment. Inner-city riots.
- ❏ **1981** Privatization program begun.
- ❏ **1982** Argentina invades Falklands. Islands retaken by UK task force.
- ❏ **1983** Tax-cutting policies.
- ❏ **1986** Financial services market deregularized ("Big Bang").
- ❏ **1990** John Major replaces Thatcher.
- ❏ **1991** UK in Gulf War.
- ❏ **1992** Conservatives win fourth consecutive election.
- ❏ **1996** Dunblane primary school massacre; tightening of gun control laws. Health crisis linking "mad cow" disease (BSE) with fatal variant Creutzfeldt-Jakob disease (vCJD).
- ❏ **1997** May, landslide election victory for Labour under Tony Blair. August, Diana, Princess of Wales, killed in car crash in Paris. September, Scottish and Welsh referendums approve creation of own assemblies.
- ❏ **1998–1999** "Good Friday" agreement on political settlement in Northern Ireland, endorsed by referendum but held up by disputes over decommissioning weapons.
- ❏ **1999** March–June, involvement in NATO air war with Yugoslavia over Kosovo crisis. May, Scottish Parliament and Welsh Assembly elected, inaugurated. December, devolution to power-sharing executive in Northern Ireland.
- ❏ **2001** Foot-and-mouth epidemic and mass livestock culling. June, Labour reelected by huge majority. October, military participation in US-led "war on terrorism" in Afghanistan.
- ❏ **2002** 50th anniversary of accession to throne of Queen Elizabeth II.

POLITICS

 Multiparty elections

 2001/2006   H.M. Queen Elizabeth II

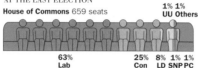

AT THE LAST ELECTION
House of Commons 659 seats

| 63% Lab | 25% Con | 8% LD | 1% SNP | 1% PC | 1% UU | 1% Others |

Lab = Labour Party   **Con** = Conservative and Unionist Party
LD = Liberal Democrats   **UU** = Ulster Unionist parties – (official) Ulster Unionist Party, Democratic Unionist Party, UK Unionist   **SNP** = Scottish National Party   **PC** = Plaid Cymru

House of Lords 679 seats

The House of Lords is an unelected body of just under 100 hereditary peers, 26 spiritual peers (bishops), and over 500 life peers (including lords of appeal – judges), appointed by the monarch

Queen Elizabeth II, *head of state since 1952 and head of the Commonwealth.*

Tony Blair, *prime minister since 1997, and leader of the Labour Party.*

Chancellor of the Exchequer Gordon Brown, *known as the "Iron Chancellor."*

Baroness Thatcher, *the country's only female prime minister (1979–1990).*

The UK is a multiparty democracy. The monarch's power is largely ceremonial.

PROFILE

Margaret Thatcher's 1979 election victory ushered in almost 18 years of Conservative rule, and monetarist and privatization policies. The Labour Party won back power in 1997 as the Conservatives lost impetus and popularity. Tony Blair's "New Labour" government, occupying the political center, retained a massive majority in the June 2001 election. The low turnout, however, reflected a growing perception that it had become "out of touch." The Conservative Party has so far failed to build a credible challenge around opposition to the euro and pledges to reduce taxes.

MAIN POLITICAL ISSUES
Europe

Labour plans a referendum on joining the euro single currency. The Conservatives, in opposition, have become increasingly "eurosceptic," considering that membership of the EU erodes national sovereignty.

Constitutional change

Major changes were made to the UK's system of government in the late 1990s. A separate Scottish Parliament, with substantial devolved powers, was approved by referendum and elected in 1999, as was a new Welsh Assembly. The House of Lords was substantially changed by the abolition of voting rights for the great majority of hereditary peers, pending its complete overhaul; 15 "people's peers" were chosen in 2001. London gained greater autonomy in 2000 with the election both of its own assembly and of a mayor.

In Northern Ireland, the 1998 Good Friday agreement brought unionists and Irish republicans into a power-sharing government. Progress has stalled over the decommissioning of paramilitaries' weapons, and tensions recur when Protestant parades attempt to march through Catholic areas in the summer "Marching Season." The joint executive formed in 1999 has on occasion been suspended, and direct rule reimposed from London. However, in 2001 the IRA agreed to put its weapons definitively "beyond use", and in 2002 apologized for the death of "non-combatants" during its violent campaign.

The economy

Fundamental alternatives on running the economy are no longer argued within mainstream politics. Labour no longer believes in renationalizing privatized industries, and is wary of increasing taxes. Its belief in using private finance and management within publicly owned services, such as health and education, has alienated some traditional Labour supporters, and the financial benefits of this policy are questioned by others.

Canary Wharf, *the centerpiece of the London Docklands development.*

U

WORLD AFFAIRS ▷ Joined UN in 1945

In 2002 prime minister Blair claimed that the UK, if no longer a "great power," could still play a "pivotal role" in world affairs. It holds a permanent seat on the UN Security Council and was a founder member of NATO. The UK joined the EC only in 1973, and remains wary of full integration in Europe. Generally following a pro-US line during and since the Cold War, the UK has supported US-led action against Iraq, and the 1999 NATO bombing to compel Yugoslav forces to pull out of Kosovo. The question of cooperating in a new US missile defense system promised to be a key test of the UK–US relationship, but in September 2001 both gave top priority to waging a "war on terrorism" worldwide.

AID ▷ Donor

 $4.5bn (donations) | Up 31% in 2000

UK foreign aid fell between 1980 and 1997 to below the European average, and well below the nominal target of 0.7% of GNP for industrialized countries. After 1997 the government moved to end the decline, although the figure in 2000 was only 0.31% of GNP. More significant was its concentration on the poorest countries and on partnership with NGOs, building on a change of emphasis already introduced in 1996, when 85% of bilateral aid was directed at 20 states in sub-Saharan Africa and south Asia. The "trade for aid" provision, by which much of the aid budget was tied to contracts for British firms, has been abolished. The program's aims include encouraging good government, widening opportunities for women, and protecting the environment.

DEFENSE ▷ No compulsory military service

 $33.9bn | Down 7% in 2000

BRITISH ARMED FORCES

🚙	636 main battle tanks (294 *Challenger* 2, 338 *Challenger*, 4 *Chieftain*)	113,950 personnel
🚢	16 submarines, 3 carriers, 11 destroyers, 20 frigates, and 23 patrol boats	43,530 personnel
✈	548 combat aircraft (217 *Tornado*, 79 *Jaguar*, 60 *Harrier*, 94 F-3)	53,950 personnel
	58 SLBM in 4 SSBN	

The post-Cold War program Options for Change, implemented in 1993, made significant cuts in army and navy personnel and equipment orders. The UK's independent nuclear deterrent was scaled down. Despite such cuts, defense spending is high, and rose in 2002. Troops remain stationed in Northern Ireland, but the strategic emphasis has increasingly been on developing military capabilities relevant for rapid reaction. UK forces have been prominent in peacekeeping in Afghanistan, the Balkans, Sierra Leone, and elsewhere. Reservists were mobilized for the "war on terrorism."

The UK is a leading arms exporter. Major buyers include Middle Eastern and southeast Asian countries.

ECONOMICS ▷ Inflation 2.9% p.a. (1990–2000)

$1460bn | 0.6694–0.6871 pounds sterling

SCORE CARD

❏ World GNP Ranking	4th
❏ GNP per Capita	$24,430
❏ Balance of Payments	–$24.5bn
❏ Inflation	2.9%
❏ Unemployment	4%

EXPORTS

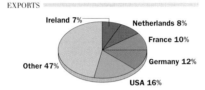

Ireland 7% | Netherlands 8% | France 10% | Germany 12% | USA 16% | Other 47%

IMPORTS

Netherlands 7% | Japan 5% | France 8% | USA 13% | Germany 13% | Other 54%

STRENGTHS
World leader in financial services, pharmaceuticals, and defense industries. Strong multinationals. Precision engineering and high-tech industries, including telecommunications and biotechnology. Energy sector based on North Sea oil and gas production. Innovative in computer software development. Flexible working practices. Success in controlling inflationary tendencies.

WEAKNESSES
Decline of manufacturing sector since 1970s, particularly heavy industries, car manufacturing. Nonparticipation in euro threatens former status as EU's largest recipient of inward investment, has prompted some major investors to close UK factories. Quick-return mentality of many investment decisions.

PROFILE
Manufacturing has been in long-term decline, particularly since the 1980s, when sectors such as financial services expanded rapidly. After sharp recession in 1991, revival was sluggish, but by the late 1990s growth in the UK was faster than that of its European competitors. The government received a £22 billion windfall from the sale in 2000 of "third generation" mobile phone licenses. The rural economy was hit hard by a foot-and-mouth epidemic in 2001 while the wider economy feared recession in 2001–2002 after the US economic downturn and the collapse of the Internet "dot.com" boom. Interest rates were cut to try to boost domestic spending, while the government in 2002 massively increased its own spending on education, health, and defense.

ECONOMIC PERFORMANCE INDICATOR

UNITED KINGDOM : MAJOR BUSINESSES

Oil & gas
Electronics
Computers
Engineering
Vehicle manufacture
Aerospace industry
Banking & finance
Iron & steel
Chemicals
Textiles
Whisky
Media

RESISTANCES

RESOURCES

Electric power
73.4m kw

999,017 tonnes

2.5m b/d (reserves
4.9bn barrels)

36.7m sheep, 10.6m
cattle, 9.5m turkeys,
168m chickens

Coal, oil, limestone,
natural gas

ELECTRICITY GENERATION

Hydro 2% (6.7bn kwh)

Combustion 70% (251bn kwh)

Nuclear 28% (100bn kwh)

Other 0%

0   20   40   60   80   100

% of total generation by type

The UK has the largest energy resources of any EU state, with substantial oil and gas reserves offshore on the continental shelf in the North Sea, and fresh fields in the north Atlantic. Drilled under difficult conditions, North Sea oil is of a high grade. Revenues from taxes on oil companies have been a major contributor to government finances, averaging around $12 billion a year.

Coal reserves are also sizable, but all but a handful of pits have closed, faced with cheap imports and falling demand. Privatization of the electricity industry, and pressure to cut pollution, encouraged the switch from coal- to gas-fired power plants, prompting emergency government measures to save the coal industry in the late 1990s and efforts to boost the role of "cleaner coal" technology. The UK produces few other minerals in significant quantities. Cornwall's last tin mines teeter between closure and rescue. Some very small-scale gold mining survives in Wales and Scotland.

UNITED KINGDOM : LAND USE

Cropland
Pasture
Forest
High mountain regions
Sheep
Cattle
Fruit
Wheat

0   200 km
0   200 miles

ENVIRONMENT

Sustainability
rank: 91st

21%

9.2 tonnes
per capita

ENVIRONMENTAL TREATIES

Yes   Yes   Yes

Yes   Yes   Yes

Apart from destruction of rural environments by road building and sprawling development, the most important issues are health-related. Urban air pollution from traffic is a major focus, as are nuclear safety issues. Food scares have gripped the public since BSE ("mad cow" disease) was linked in the mid-1990s with human deaths. Opposition to genetically modified (GM) foods is widespread, and GM crop trials have been disrupted for fear that modified genes contaminate other species.

MEDIA

TV ownership high

Daily newspaper circulation 331 per 1000 people

PUBLISHING AND BROADCAST MEDIA

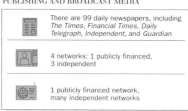

There are 99 daily newspapers, including *The Times, Financial Times, Daily Telegraph, Independent,* and *Guardian*

4 networks: 1 publicly financed,
3 independent

1 publicly financed network,
many independent networks

Newspapers are owned mostly by large media corporations. Many publish Internet editions. Criticized for invasions of privacy, the press presents self-regulation as preferable to legislation. Publication deemed contrary to "national interests" may be banned. Satellite TV and digital terrestrial broadcasting have increased competition with the BBC. The BBC's World Service, despite cutbacks, remains an influential news source internationally.

The Welsh coal industry has virtually disappeared. Wales now has the highest percentage of small business start-ups, relative to the population, of any part of the UK.

CRIME

No death penalty

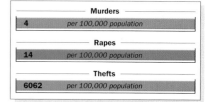

74,091 prisoners

Up 7%
2001–2002

CRIME RATES

Murders	
4	*per 100,000 population*

Rapes	
14	*per 100,000 population*

Thefts	
6062	*per 100,000 population*

Violent crime and domestic abuse are growing problems. Inner-city violence is partly fueled by narcotics dependency and trafficking. Since 1997 the Labour government has maintained a "tough on crime" stance, but sentencing policies place the penal system under serious strain. It has moved toward toleration of marijuana in order to focus on "harder" narcotics. An "antiterrorism" bill introduced in October 2001 gave police extra powers.

EDUCATION

School leaving
age: 16

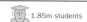

99%

1.85m students

THE EDUCATION SYSTEM

100%   100%   58%

Primary   Secondary   Tertiary

% of each age group in education

The state system is used by 94% of children. Fee-paying private schools include the traditional elite institutions confusingly known as public schools. There is now a new enthusiasm in government for "faith-based" schools.

From the 1960s onward, a two-tier state school system based on academic selection at age 11 was to a great extent replaced by mixed ability comprehensive schools. The 1988 education reforms introduced a national curriculum and weakened the role of local education authorities. The Labour government has focused on testing, assessing teaching standards, and tackling "failing" schools. Public spending targets in 2002 promised significantly increased education expenditure within four years, to 5.6% of GDP.

More colleges were given university status in the 1990s but established centers, particularly Oxford and Cambridge, continue to be the most prestigious and best resourced.

U

REPRESENTATION AND DEVOLUTION

THE 1801 ACT OF UNION unified Great Britain – England, Wales, and Scotland – and Ireland. The principality of Wales had been joined with England in 1536. The English and Scottish crowns had been united in 1603 (James VI of Scotland becoming James I of England), and the 1707 Act of Union united their parliaments. The present full title of the UK – the United Kingdom of Great Britain and Northern Ireland – dates from 1922, when the Irish Free State was created.

The rise of Welsh and particularly Scottish nationalism prompted an abortive initiative to devolve powers to regional governments in the 1970s. The post-1997 Labour government, giving devolution fresh impetus, created new assemblies in Northern Ireland, Scotland, and Wales. It also raised, but has done little to breathe life into, the idea of English regional assemblies.

The Scottish Parliament building, Edinburgh.

SCOTLAND AND WALES
The Scottish National Party's electoral breakthrough in 1974 was fueled by the perception that independence could be viable, if Scotland controlled the oil wealth of the North Sea. The UK government offered devolution to both Scotland and Wales, but in referendums in 1979 the majority in Scotland was insufficient to carry the proposals, while Wales actually voted no. The devolution proposals of 1997, on the other hand, won endorsement both in Scotland and (narrowly) in Wales.

Scotland (pop. 5.13 million, area 78,742 sq. km/30,394 sq. miles, capital Edinburgh) now has a Scottish Parliament of 129 MSPs, elected in 1999. The Scottish Executive, consisting of a First Minister and an 11-member cabinet, is responsible to the parliament, whose powers notably cover education and changing tax rates to generate revenue for Scottish expenditure. One of its first distinctive initiatives was to reject the imposition of university tuition fees, so controversial elsewhere in the UK.

In Wales (pop. 2.9 million, area 20,761 sq. km/8041 sq. miles, capital Cardiff), a 60-member Welsh Assembly was also elected in 1999. It has fewer powers, not including tax-raising, exercised by an eight-member administration also headed by a First Minister.

NORTHERN IRELAND
In Northern Ireland (pop. 1.65 million, area 14,120 sq. km/5450 sq. miles, capital Belfast) the majority Protestant community dominated a "home rule" parliament at Stormont throughout its 50-year existence, until 1972. After that the troubled province was mainly under direct rule from London, interspersed with attempts to create power-sharing institutions, until the 1998 Good Friday agreement. This created a 108-member Northern Ireland Assembly, elected that June by proportional representation, and a 12-member power-sharing executive. However, implementation of the agreement has repeatedly been held up by disputes, mainly about disarming rival Irish republican and unionist "loyalist" paramilitaries.

The extensive upland areas of both Scotland and Wales are used extensively for sheep farming and forestry.

HEALTH

 Welfare state health benefits

1 per 556 people Heart, cerebrovascular, and respiratory diseases, cancers

The National Health Service (NHS) offers universal free health care, but financial pressures have led to shortages, hospital closures, and charges in some areas. In response the government has raised the health budget since 2000, promising $58 billion to the NHS in 2002. Plans involve greater involvement of the private sector.

Recent crises have focused on food safety, from *E. coli* outbreaks to fatal brain disease attributed to eating beef from cattle with "mad cow" disease.

SPENDING

GDP/cap. increase

CONSUMPTION AND SPENDING

373 per 1000 population   589 per 1000 population

Defense 2.4%		
Education 4.7%		
Health 5.8%		

0   5   10   15   20   25
Defense, Health, Education spending as % of GDP

Income inequality in the UK is now greater than in 1884, when records first began. The UN in 2002 recorded 13.4% living below its "poverty" line, half of average income. Average wages for manufacturing workers in 2001 were $29,000 a year, only 4% of the average received by chief executives in large companies. Under Conservative governments in the 1980s and early 1990s, taxation for higher earners was cut, whereas the value of state benefits and pensions fell. Since the mid-1990s, economic growth has helped to bring unemployment down. Labour's 1997 election promises precluded raising income tax. This limited any scope for redistributive action, leaving antipoverty strategies dependent on better targeting of welfare benefits.

WORLD RANKING

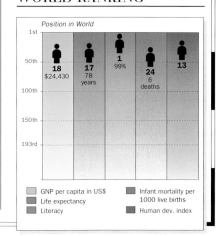

Position in World

18 $24,430	17 78 years	1 99%	24 6 deaths	13

1st
50th
100th
150th
193rd

GNP per capita in US$
Life expectancy
Literacy

Infant mortality per 1000 live births
Human dev. index

U

See also OVERSEAS TERRITORIES *p.640*

UNITED STATES

NORTH AMERICA

OFFICIAL NAME: United States of America   **CAPITAL:** Washington D.C.
POPULATION: 281.4 million   **CURRENCY:** US dollar   **OFFICIAL LANGUAGE:** English

  1776   1959   July 4   USA   -5 to -11 +1   n/a

THE WORLD'S FOURTH-LARGEST country, the United States is neither overpopulated (like China) nor in the main subject to extremes of climate (like much of Russia and Canada). Its main landmass, bounded by Canada and Mexico, contains 48 of its 50 states. The two others, Alaska at the northwest tip of the Americas and Hawaii in the Pacific, became states in 1959. The US was not built on ethnic identity but on a concept of nationhood intimately bound up with the 18th-century founding fathers' ideas of democracy and liberty – still powerful touchstones in both a political and an economic sense. Since the breakup of the Soviet Union, the US holds a unique position – but arouses extreme hatreds – as the sole global superpower.

ALASKA

CLIMATE

Continental/subtropical/mountain/desert/maritime

WEATHER CHART FOR WASHINGTON D.C.

to April. The weather is frequently dramatic, with tornadoes, cyclones, thunderstorms, hurricanes, floods, and droughts. Since 1990, weather-related damage has risen, a trend linked with global climate change.

The Chippendale Block, New York, a notable example of postmodern architecture by the influential US architect Philip Johnson.

Spanning a continent and extending far into the Pacific Ocean, the US displays a wide range of climatic conditions. Mean annual temperatures range from 29°C (84°F) in Florida to –13°C (9°F) in Alaska. Except for New England, Alaska, and the Pacific northwest, summer temperatures are higher than in Europe. Southern summers are humid; in the southwest they are dry. Winters are particularly severe in the western mountains and plains and in the Midwest – where the Great Lakes can freeze. The northeast can have heavy snow from November

HAWAII

U

TRANSPORTATION

▷ Drive on right

Atlanta, Georgia
80.2m passengers

330 ships
10.3m dwt

THE TRANSPORTATION NETWORK

3.73m km
(2.32m miles)

88,727 km
(55,132 miles)

234,131 km
(145,489 miles)

41,009 km
(25,482 miles)

The Mississippi–Missouri river system provided the first transportation network in the US. Today, the US has the world's cheapest, most extensive internal air network and a good system of interstate highways. Railroads, comparatively neglected for years, mainly carry freight, although modern high-speed trains are starting to attract passengers back. Americans have been wedded to the car since Henry Ford began mass production in 1908. By 1919 there were nine million cars in the US. Today the total tops 210 million, including pickups and the ubiquitous "sports utes" (SUVs). Americans make more than half of the world's car journeys. Cheap gasoline underpinned the rise of the car, but problems of congestion and pollution, and the environmental costs of ever more oil production, mean that its role in society needs reviewing.

The Mittens, Monument Valley, Arizona.
These striking natural rock formations are created by erosion of red sandstone. The valley is in the Navajo National Monument.

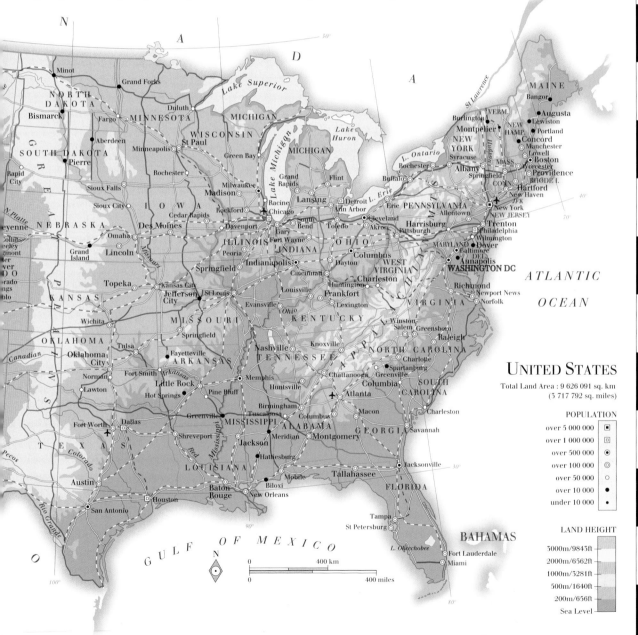

UNITED STATES

Total Land Area : 9 626 091 sq. km
(3 717 792 sq. miles)

POPULATION

over 5 000 000

over 1 000 000

over 500 000

over 100 000

over 50 000

over 10 000

under 10 000

LAND HEIGHT

3000m/9843ft

2000m/6562ft

1000m/3281ft

500m/1640ft

200m/656ft

Sea Level

U

TOURISM ▷ Visitors : Population 1:6.2

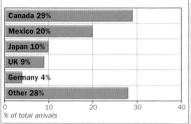

🧳 45.5m visitors   ⬇ Down 11% in 2001

MAIN TOURIST ARRIVALS

Canada 29%	
Mexico 20%	
Japan 10%	
UK 9%	
Germany 4%	
Other 28%	

0   10   20   30   40
% of total arrivals

The US as a destination for international tourism benefited greatly from the deregulation of air fares. Domestic tourism expanded just as rapidly, along with the rise in real incomes. The impact of the September 11, 2001 attacks was complex. While confidence in air travel took time to be rebuilt, the fact that US tourists put safety first meant that over 80% vacationed within the US in 2002. All the states have their attractions, and most court tourists. Top tourist destinations include Florida's Disney World and Disneyland in California, Niagara Falls, Las Vegas, New York, San Francisco, Los Angeles and Hollywood, the Grand Canyon, Death Valley, New Orleans, Atlantic City, and Washington D.C.

Tourism's rapid expansion has also brought some problems. The parks and sites run by the National Parks Service (NPS) have been particular casualties; visitor numbers rocketed in the three decades after 1970. To try and reduce pressure on the most popular areas, there has been a significant expansion in the area of protected land under NPS management since the mid-1970s. Even so, Yellowstone Park has a continuing traffic management crisis, bumper-to-bumper cars plague other high-profile attractions, and those wanting to take a raft ride down the Grand Canyon are likely to spend many months on a waiting list.

PEOPLE ▷ Pop. density low

English, Spanish, German, Arabic, Polish, Korean, Chinese, Armenian, French, Italian, Amerindian languages   31/km² (80/mi²)

THE URBAN/RURAL POPULATION SPLIT

77%   23%

RELIGIOUS PERSUASION

Jewish 2%
Other and nonreligious 12%
Roman Catholic 25%
Protestant 61%

ETHNIC MAKEUP

Native American 2%
Asian 4%
Hispanic 12%
Black American/African 13%
White 69%

Native Americans, the sole inhabitants before white settlers arrived, were dispossessed in the 19th century, and now make up around 2% of the total population. Some of the worst poverty and deprivation in the US can be found in their reservations.

The nation's population increased by some 25 million residents during the 1990s. An immigration boom peaked in the late 1980s and early 1990s, but since then numbers have fallen below a million a year. The new immigrants are disproportionately drawn from Asia and Latin America, with more than two million from Mexico alone in the 1980s.

Census Bureau projections for the year 2050 suggest that 53% of the population will be white non-Hispanics, and just under 14% black; as many as 25% will be Hispanics, while Asians (and Pacific islanders) will total nearly 9%. Blacks increasingly find that they have to compete politically and economically with the newer immigrants. In some communities, such as Los Angeles, this has been a source of tension leading to inner-city riots.

Despite the growth of a black business leadership class, only two black people – television personality Oprah Winfrey and Robert L. Johnson, Black Entertainment Television founder – make the list of the 400 richest Americans. In the mid-1990s, the O. J. Simpson trial, when a black sports hero was acquitted of murdering his white estranged wife, exposed how far the country was divided by race in its perceptions of the justice system. The Nation of Islam, led by the controversial and often provocative Louis Farrakhan, is prominent in emphasizing self-discipline, self-improvement, and community in the quest for empowerment among African-Americans.

POPULATION AGE BREAKDOWN

Female	Age	Male
2.1%	80+	1.1%
7.4%	60–79	6%
12.7%	40–59	12.2%
14.8%	20–39	15%
14%	0–19	14.8%

% of population by age group

POLITICS ▷ Multiparty elections

 L. House 2000/2002 U. House 2000/2002   President George W. Bush

AT THE LAST ELECTION

House of Representatives 435 seats

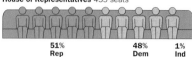

51% Rep   48% Dem   1% Ind

Rep = Republican Party **Dem** = Democratic Party
Ind = Independents

Senate 100 seats

50% Rep   50% Dem

Presidential elections take place every 4 years, House elections every 2 years. One-third of the senators are elected every 2 years for six-year terms.

The US is a democracy with a federal system of government. Many issues are dealt with by the 50 individual states.

PROFILE

Despite a "big business" affiliation, controversial policies (notably on energy), and deep divisions when he was first elected in 2000, Republican president George W. Bush gained overwhelming popular support in September 2001, when a wave of patriotic emotion rallied the nation behind him and his declaration of a "war on terrorism," after the attack on the World Trade Center and the Pentagon. Republicans narrowly control the House but Bush still lacks, by the narrowest of margins in the Senate, a complete congressional majority. His predecessors, whether Democrat (Clinton, elected in 1992) or Republican, had struggled to get major initiatives enacted by a hostile Congress.

MAIN POLITICAL ISSUES
The limits of government

The US has a strong tradition of resisting the extension of government powers. The vigorous defense of constitutional liberties and the rights of citizens, such as freedom of speech or the right to bear arms, is sometimes taken to lengths which appear extreme to other societies. States similarly resist the arrogation of powers by the federal authorities. In areas such as health care

George W. Bush, took office in 2001 after a controversial presidential election.

Gen. Colin Powell, who became the first African–American Secretary of State.

U

POLITICS　*continued*

and education, conservatives oppose as interference what others see as the proper concern of government with social welfare. "Big government" is also denounced in the economic sphere. Opponents of environmental controls, for example, portray them as obstructing free enterprise and wealth creation.

The prestige of the presidency

When George W. Bush took office in 2001, the presidency faced a threefold challenge. The tendentious manner of his election, with fewer votes nationwide than his Democrat rival Al Gore, and controversy over the conduct of polling in Florida, left questions as to the legitimacy of his victory. The narrow Republican majority in the House of Representatives, and a "hung" Senate, with an effective Democrat majority after one elected Republican switched to independent, promised problems. Thirdly, the moral authority of the office of president had suffered serious damage under the outgoing administration of Bill Clinton, beset by personal, financial, and political scandals, and hamstrung toward the end by impeachment proceedings. Bush, initially derided by opponents as an intellectual lightweight, rose in public esteem to appear unassailable within his first year in office. His simple messages of belief in "the American way" rallied a nation shocked by the implications of the terrorist attacks of September 11, 2001.

Energy and the environment

Controversial energy policies introduced by Bush in 2001, at the height of a crisis over Californian electricity shortages, offered much freer rein for the US oil industry to exploit reserves in Alaska. Wilderness conservation activists were appalled. The expansion of nuclear power was also revived as an issue under the Bush energy plan, while the US, alone, chose to repudiate the international Kyoto agreement on cutting carbon dioxide emissions.

Crime, race and poverty

Efforts to regenerate depressed urban areas have relied on new economic opportunities and programs that empower the poor (such as self-management of public housing projects). Both rates of criminality and crime victims are higher in the black community than in any other. Tough anticrime policies in cities such as New York have had a real impact in reducing the level of violence, but have also been accused of unfairly targeting minority ethnic groups; there remains the prospect of a permanently disaffected urban underclass.

WORLD AFFAIRS　▷ Joined UN in 1945

G8	NATO	NAFTA	OAS	OECD

Isolated by two great oceans, the US has for much of its history been able to choose the extent of its involvement in world affairs. Only reluctantly involved in the two world wars, after 1945 it swapped isolationism for involvement. The US took its seat on the Security Council of the UN, based in New York. It helped to set up NATO, although for the US the Cold War was most immediate – and costly – in the Korean and Vietnam wars. The death toll and shock of defeat in Vietnam in the 1970s kept the US out of military involvement overseas for over a decade. Instead, it focused on diplomacy – particularly with China and in the Middle East – and on supporting the opponents of left-wing regimes in developing countries including Nicaragua, Cuba, and Angola.

The collapse of the Eastern bloc after 1989 meant that the US had to redetermine the scope of its foreign responsibilities as the only remaining superpower. It led the intervention in the 1991 Gulf War, but a fiasco in Somalia and a lack of clear policy on Bosnia and Herzegovina and Haiti showed its uncertainty about a role as world policeman. Increasingly isolated in actions against Iraq, the US has shown a preference for air power rather than committing ground forces. Even before the September 11, 2001 attacks, the US had used air strikes to hit back at anti-US terrorism and its alleged sponsors.

The subsequent "war on terrorism" has focused on Islamic extremists, with rumors of imminent attacks on "sponsors" of terrorism, principally Iraq. The US also redefined relations with Russia, prompting further dramatic cuts in both countries' nuclear arsenals and leading Russia to drop its previous objections to the highly controversial national missile defense system (NMD). China, however, remains wary of the NMD, and under Bush the US stance toward China has been notably more confrontational.

Hillary Clinton, *senator for New York and wife of former president Bill Clinton.*

Alan Greenspan, *chairman of the Federal Reserve since 1987.*

9/11

AT 08:48 local time on September 11, 2001, the hijacked American Airlines (AA) flight 11 flew into the North Tower of the World Trade Center (WTC) in New York. The world's most devastating terrorist attack had begun. By 10:29 that morning over 3000 people were dead as three more hijacked planes had been crashed; one into the WTC's South Tower, another into the Pentagon military headquarters, and one in rural Pennsylvania after passengers overpowered the hijackers. The finger of blame was immediately pointed at Islamic terrorists of the extremist al-Qaida network, led by Osama bin Laden.

The attacks, caught on camera and broadcast live around the world, are now known as 9/11. Their impact was profound. The US economy shuddered and the country was effectively placed on a permanent state of alert. Warnings are frequently issued on the basis of security intelligence. The government has responded by restructuring the heavily criticized civil and national defense systems. A new Department of Homeland Security was created and President Bush urged reform of the Central Intelligence Agency (CIA) to refocus on potential terrorist threats.

Perhaps the most significant reaction to 9/11 was the launch of Bush's "war on terrorism," beginning with a massive assault on the *taliban* regime in Afghanistan which was sheltering bin Laden. The "war" has been strongly condemned by those in the Islamic world, who feel that it has been unduly targeted, and point out that little has been done to address the motivation of the attackers: US cultural hegemony and in particular its historic support for the Israeli government.

Designs for a memorial *on the WTC site – Ground Zero – have proved controversial. The site's owners have insisted that 1.1 million sq. m (11.6 million sq. ft) of any designs be reserved for commercial purposes.*

U

THE IMPACT OF INFORMATION TECHNOLOGY

THE UNITED STATES has been, and remains, at the forefront of the so-called information revolution. US society is undergoing profound changes at the beginning of the 21st century as a result of the spread of the personal computer and connectivity through access to electronic mail (e-mail) and the World Wide Web. The Web itself first emerged as a phenomenon of the 1990s, though the Internet on which it is based originated in the 1960s. The Net, developed to allow vital defense communications to be maintained in the event of an attack on a central control site, was first popularized as a medium for exchange within and between universities.

ACCESS TO VIRTUAL SPACE

Mass ownership of the motor car and cheap fuel had conditioned the attitudes of previous generations of Americans to distance, and helped determine the pattern of development of cities and towns. Information technology, the concept of "cyberspace," and the availability of instant "virtual" interaction, are now sometimes described as having abolished distance.

High-speed "broadband" connections to the Internet are becoming increasingly popular, even for individual users. For most people (80% of home users in 2001), however, the link is still by dial-up on an ordinary telephone line to the local number of their Internet service provider. One consequence has been an explosion in phone line use, particularly because the recently deregulated telephone industry offers most Americans the chance to pay for their local calls on a single flat rate rather than by the minute. Computer modems can thus be left connected for long periods.

As with the automobile, part of society is excluded. In 1998, PC ownership stood at some 43% of the population.

Access to the Internet remains out of reach to those who cannot afford a personal computer, such as residents of some of the poverty-stricken inner-city areas like the Bronx in New York.

The Microsoft Campus in Redmond, Washington State. Microsoft is the world's biggest computer software manufacturer; its founder, Bill Gates, is the world's richest man.

By September 2001, 54% of people had Internet access at home. Special initiatives to spread access among disadvantaged groups have had only limited impact.

SECURITY

Activists of different kinds have made increasing use of the Internet to spread their views and maintain contact with supporters. The use of this tool not only by mainstream political organizations and lobbyists but by fringe groups, terrorists, and the pornography and sex industries give rise to serious security concerns. Such concerns underlay the battle in the late 1990s over the use of encryption software to ensure the privacy of electronic communications. The security authorities fought to ensure that all such software should have a key allowing them to decipher messages. Unfortunately, it is notoriously difficult to prevent the rapid spread of software which is made available on the Internet, and impossible to pin down the location of such programs in the usual geographic sense. The policing of a national jurisdiction is compromised by a communications network which recognizes no national boundaries. The same is true with computer viruses, which can infect the programs on computers that come into contact with them and affect the way they function, in some cases very destructively.

COMMERCE

The shopping mall, which has had a major influence in reshaping consumer shopping habits since the 1960s, and the mail-order catalog on which many US households rely so heavily for clothes and similar purchases, are now under challenge from the online "virtual mall." Many companies had to take a trial-and-error approach to developing e-commerce. Initial predictions about its rapid growth have proved overstated. However, it is thought that the number of people in the US who regularly buy necessities such as groceries and related goods online will reach 20 million by 2007. Completing business transactions via the Internet became more feasible with legislation making "e-signatures" legally binding, with effect from 2000.

TRAFFIC AND MOBILITY

The take-up of e-commerce on this scale would replace at least half of Americans' average 17 shopping trips a month made to grocery stores and related outlets. It has the added attraction of shielding shoppers from the perceived dangers of being assaulted or robbed.

Information technology is affecting patterns of personal transportation and mobility in other ways too. At one level, online information services and selling techniques make using the Net a more convenient and efficient way of booking airline tickets. Improvements in communications make it increasingly possible and productive for people to work at home or in decentralized facilities away from company headquarters. So-called "knowledge workers" in particular have had to learn new ways of networking electronically to replace the face-to-face contact of the office environment. The use of special-interest discussion facilities and so-called newsgroups on the Internet has mushroomed accordingly. "Telecommuters" continue to form only a small fraction of the total US working population, but even this is significant in terms of managing peak-time commuter traffic. However, videoconferencing is struggling to make headway as an alternative to traveling long distances to hold business meetings, despite the active encouragement of some gurus of the environmental lobby.

Malls, a typical feature of the suburban landscape, are under threat from Web use.

AID

▷ Donor

💲 $9.96bn (donations)

⬆ Up 9% in 2000

The US gives only 0.1% of GNP in foreign aid, and aid allocations are often held hostage to special pleading in Congress. Egypt and Israel are the major recipients. The strategic aims of the "war on terrorism" prompted an unusual pledge in 2002 of $5 billion for the world's poorest countries.

DEFENSE

▷ No compulsory military service

💲 $295bn

⬆ Up 1% in 2000

AMERICAN ARMED FORCES

	7620 main battle tanks (*Abrams* M-1)	477,800 personnel
	73 submarines, 12 carriers, 27 cruisers, 54 destroyers, 35 frigates, & 21 patrol boats	366,100 personnel
	4147 combat aircraft (B52H, B-1B, F-4, F-15, F-16, A-10A, F-117, OA-10A)	352,500 personnel
	432 SLBM in 18 SSBN, 550 ICBM	

The 9/11 attacks have been popularly cast as a second Pearl Harbor: catching US defenses unprepared and dragging the country into an already simmering global conflict. This time the enemy has been identified as international terrorism. Even before the attacks, emphasis had been shifting away from strategic nuclear deterrence and large warships to "smart" missile systems and "long-range power projection," with rapid intervention capabilities built around air power. Despite setbacks in early tests, a project to create a national missile defense "shield" system was enthusiastically taken up by the incoming Bush administration in 2001.

The enormous US military–industrial complex dates only from the close of the Second World War. In the 1990s, the end of the Cold War and the need to cut the budget deficit combined to slash the defense funds to their lowest level in real terms since 1945. Nuclear weapons tests were superceded by computerized "virtual" tests after the creation of the powerful ASCI White computer.

However, the Republican administration inaugurated in 2001 has steadily increased defense spending. The 2002/2003 budget was up by 11% on the previous year to $350 billion, amounting to more than the combined defense budgets of the world's next nine largest military spenders.

Overseas peacekeeping missions have been put at risk by the US government's objection to the new International Criminal Court.

ECONOMICS

▷ Inflation 2.1% p.a. (1990–2000)

📊 $9602bn

💲 Currency is US dollar

SCORE CARD

❏ WORLD GNP RANKING	1st
❏ GNP PER CAPITA	$34,100
❏ BALANCE OF PAYMENTS	–$445bn
❏ INFLATION	3.4%
❏ UNEMPLOYMENT	4%

EXPORTS

Germany 4%
UK 5%
Japan 8%
Mexico 14%
Other 46%
Canada 23%

IMPORTS

Germany 5%
China 9%
Mexico 11%
Other 44%
Japan 12%
Canada 19%

ECONOMIC PERFORMANCE INDICATOR

— Consumer Price Index   GDP

Consumer price index 1995=100

GDP 1996=100

1996 1997 1998 1999 2000

STRENGTHS

World's largest economy. Wealth of natural resources: energy, raw materials, and food. Strong high-tech base; world-leading research and development. Global leader in computer software. Advanced and competitive manufacturing industry; sophisticated service sector. Entrepreneurial business ethic. World-class multinationals. High quality of postgraduate education, especially in high-tech business. Global dominance of US culture a major boost to US manufacturers. Subsidized crops and favorable tariffs for domestic industries.

WEAKNESSES

Dramatic fall in manufacturing employment over recent decades. Globalization, problem of job losses to lower-wage economies. Tough competition from Asia and EU in leading-edge technologies. Lower savings rate than many competitors. Volatile market values driven by speculation. Weak business regulation and short-termism. Corporate collapses provoking crisis over accounting standards and scope for fraud.

PROFILE

In 1945, the US accounted for about 50% of world output; by the 1990s its share was down to about 25%. That is not, as Americans often think, a sign of failure, but a clear indication that the 1940s and 1950s were unusual periods. A share of 25% is about the same as was claimed by the US in 1914, when it was already the world's biggest economy.

The US has become a great exporter, and continues to have both a stable political system and a uniquely strong combination of skilled labor and natural resources. In 2001, although still ranked the world's most competitive economy, the US entered a downturn after a nine-year boom, its longest ever. The collapse of the massive Enron corporation in 2001, followed by that of WorldCom in 2002, threatened confidence in business values.

U

UNITED STATES : MAJOR BUSINESSES

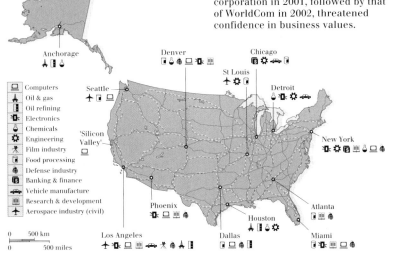

🖥	Computers
⛽	Oil & gas
	Oil refining
	Electronics
	Chemicals
⚙	Engineering
	Film industry
	Food processing
	Defense industry
	Banking & finance
🚗	Vehicle manufacture
	Research & development
✈	Aerospace industry (civil)

Anchorage
Denver
Chicago
St Louis
Detroit
Seattle
'Silicon Valley'
New York
Phoenix
Houston
Atlanta
Los Angeles
Dallas
Miami

0   500 km
0   500 miles

CHRONOLOGY

The original 13 colonies, first established by British settlers on the eastern seaboard in the 17th century, joined to wage a war for independence, 1775–1781, which Britain recognized in 1783. The 1776 Declaration of Independence was followed by the writing of the world's first constitution. A century of westward expansion began. Following the victory of the northern states in the 1861–1865 Civil War, slavery was ended throughout the US, but Native Americans were dispossessed of their land in a series of conflicts.

❑ **1917** US enters World War I.

❑ **1929** New York stock market collapse; economic depression.

❑ **1941** Japanese attack on Pearl Harbor; US enters World War II.

❑ **1950–1953** Korean War.

❑ **1954** Supreme Court rules racial segregation in schools is unconstitutional. Blacks, seeking constitutional rights, start campaign of civil disobedience.

❑ **1959** Alaska, Hawaii become states.

❑ **1961** John F. Kennedy president. Promises aid to South Vietnam. US-backed invasion of Cuba defeated at Bay of Pigs.

❑ **1962** Soviet missile bases found on Cuba; resulting threat of nuclear war narrowly averted.

❑ **1963** Kennedy assassinated. Lyndon Baines Johnson president.

❑ **1964** US involvement in Vietnam stepped up. Civil Rights Act gives blacks constitutional equality.

❑ **1968** Martin Luther King is assassinated.

❑ **1969** Republican Richard Nixon takes office as president. Growing public opposition to Vietnam War.

❑ **1972** Nixon reelected. Makes historic visit to China.

❑ **1973** Withdrawal of US troops from Vietnam; 58,000 US troops dead by end of war.

❑ **1974** August, Nixon resigns following Watergate scandal over break-in to Democrat headquarters. Gerald Ford president.

❑ **1976** Democrat Jimmy Carter elected president.

❑ **1978** US-sponsored Camp David accord between Egypt and Israel.

❑ **1979** Seizure of US hostages in Tehran, Iran.

❑ **1980** Ronald Reagan wins elections for Republicans. Adopts tough anticommunist foreign policy.

❑ **1983** Military invasion of Grenada.

❑ **1985** Air strikes against Libyan cities. Relations with USSR improve; first of three summits held.

❑ **1986** Iran-Contra affair revealed. ▷

RESURCES

 Electric power 793m kw

 5.23m tonnes

7.72m b/d (reserves 30.4bn barrels)

97.3m cattle, 88m turkeys, 59.1m pigs, 1830m chickens

Phosphates, gypsum, oil, coal, sulfur, lead, zinc, copper, gold

ELECTRICITY GENERATION

Hydro 8% (323bn kwh)	
Combustion 72% (2777bn kwh)	
Nuclear 19% (714bn kwh)	
Other 1% (19bn kwh)	

0   20   40   60   80   100

% of total generation by type

The US has an abundance of natural resources, including oil. The 2001 energy plan aimed to step up oil exploration and output, reducing the need for imports. There are massive deposits of coal in the western states – where almost all mining is open-cast – and substantial mineral deposits in the mountains and intramontane basins.

Environmental concerns halted the development of nuclear power after the 1979 accident at Three Mile Island, but expansion is now being considered.

ENVIRONMENT

 Sustainability rank: 45th

 13% (6% partially protected)

19.8 tonnes per capita

ENVIRONMENTAL TREATIES

Yes	Yes	No
No	Yes	No

The US lags far behind other Western countries on environmental issues. The international commitment made at the 1997 Kyoto conference on cutting carbon dioxide emissions was scrapped by President Bush in 2001. The Rockies are a battleground between those who want to maintain their beauty, and those who advocate "wise use" – in practice this often means giving ranchers and miners free rein. In 2002, Congress approved plans to dump nuclear waste in Mt. Yucca, Nevada. Similar issues surround the arguments over extending oil drilling in the Alaskan wilderness. The US is leading the field in genetically modified (GM) food. Huge acreages have been planted with GM cereals, and by 2001 over 60% of soybean production was GM. A consumer backlash, especially in Europe, has worried many farmers.

The timber industry, forced to retreat by conservationists in the Pacific northwest, especially Washington State, has moved to the south, where great stands of pine are harvested as if they were fields of wheat. The US has harnessed hydroelectric power in the past; today, imports of hydropower from Canada are commonplace.

In comparison with western Europe, the US is not intensively farmed. The huge size of farms in the Midwest and west has allowed both arable and livestock farming to be based on a low-input for low-output model.

UNITED STATES : LAND USE

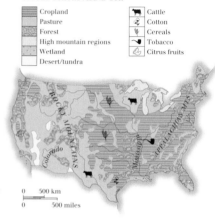

Cropland	Cattle
Pasture	Cotton
Forest	Cereals
High mountain regions	Tobacco
Wetland	Citrus fruits
Desert/tundra	

0   500 km
0   500 miles

MEDIA

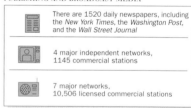 TV ownership high

Daily newspaper circulation 201 per 1000 people

PUBLISHING AND BROADCAST MEDIA

There are 1520 daily newspapers, including the *New York Times*, the *Washington Post*, and the *Wall Street Journal*

4 major independent networks, 1145 commercial stations

7 major networks, 10,506 licensed commercial stations

Mass media as a phenomenon was born in the US. No other society has ever had anything quite like US network TV, or moved so easily into the world of multichannel TV; homes with 50 or more channels are commonplace. The Internet, the most recent in the series of nationwide communication revolutions, is now used regularly by the majority of the population. Newspapers, mostly local rather than national, tend to have very low cover prices, and gain most of their revenue from advertising. They are under increasing threat from cable TV and other outlets. Companies exploring multimedia and ways of providing online news and services are still trying to recover from the crash in their stock values in 2000.

U

CHRONOLOGY *continued*

- ❑ **1987** Intermediate Nuclear Forces Treaty signed by US and USSR.
- ❑ **1988** Republican George Bush Sr. wins presidency.
- ❑ **1989** US overthrows Gen. Noriega of Panama, arrested on drug charges.
- ❑ **1991** January–February, Gulf War against Iraq. US and USSR sign START arms reduction treaty.
- ❑ **1992** Black youths riot in Los Angeles and other cities. Bush–Yeltsin summit agrees further arms reductions. Democrat Bill Clinton defeats Bush in presidential election.
- ❑ **1994** Health care reform legislation defeated in Congress. Midterm elections, Republican majorities in both houses of Congress.
- ❑ **1995** Oklahoma bombing by Timothy McVeigh: over 160 die.
- ❑ **1996** Clinton reelected.
- ❑ **1998** Scandal over Clinton's affair with White House intern leads to impeachment proceedings. August, bombing of US embassies in Kenya and Tanzania; revenge air strikes on Sudan and Afghanistan. December, air strikes against Iraq.
- ❑ **1999** February, Clinton acquitted in Senate impeachment trial. April, Columbine High School shootings by two students. March–June, NATO involvement to end Kosovo conflict, bombardment of Yugoslavia.
- ❑ **2000** Democrat Al Gore concedes tightest presidential election ever to Republican George W. Bush.
- ❑ **2001** January, President Bush takes office. September, world's worst terrorist attack kills thousands as hijacked planes destroy World Trade Center, damage Pentagon. October, US-led military action in "war on terrorism" begins with intensive aerial bombing campaign in Afghanistan. December, accounting scandal at Enron.
- ❑ **2002** July, WorldCom bankruptcy is biggest ever corporate collapse.

Bison in Yellowstone National Park.
The park's ecosystem is under severe strain due to the number of visitors it attracts.

CRIME

▷ Death penalty in use

🏛 2m prisoners　　⬇ Crime is decreasing

CRIME RATES

Murders	
6	*per 100,000 population*

Rapes	
36	*per 100,000 population*

Thefts	
1145	*per 100,000 population*

Violent crime – especially murder – is much more common than in other developed countries, even in relatively well-off areas. However, the murder rate has fallen; six deaths per 100,000 people in 2000 was the lowest for over 30 years. Mass shootings have made gun control a major issue, but a powerful lobby opposes restrictions, basing its arguments on the constitution and the defense of individual liberties.

Imprisonment for narcotics crimes in the US is much more widespread than in most Western countries. Capital punishment has increased since the 1980s, especially in the south. Texas carries out most executions. There are two million people in prison in the US, a quarter of the world total.

EDUCATION

▷ School leaving age: 16

📖 99%　　🎓 14.6m students

THE EDUCATION SYSTEM

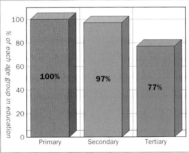

Education in the US is primarily the responsibility of the state governments.

Recent reports critical of standards in US high schools cite problems of discipline, poor structural maintenance, and lack of resources in many areas as driving people away from the public education sector. Private education at secondary level continues to develop rapidly. While the number of Catholic private schools has shrunk, more nondenominational fee-paying schools have been founded.

Four out of every five high school students now go on to some form of tertiary college. The leading US universities are internationally recognized as being of world class.

HEALTH

▷ Limited welfare state health benefits

🧑‍⚕ 1 per 370 people　　Heart and cerebro-vascular diseases, cancers, accidents

There are enormous disparities in US health provision, but a proposed reform of the health care system, high on the Democrats' political agenda in the early 1990s, was blocked by Congress. Sophisticated techniques are available to those with insurance (which they typically receive from their employer); the Texas Medical Center, in Houston, for example, has a budget equivalent to that of some small countries. On the other hand, costs have skyrocketed, and facilities for those dependent on state medical care and aid are woefully underfunded. Preventive care fails to reach all sections of society, and infant mortality statistics in some areas are at near-African levels.

Nearly one in five of the population is clinically obese, and one in every two adults is overweight.

SPENDING

▷ GDP/cap. increase

CONSUMPTION AND SPENDING

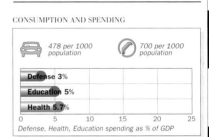

🚗 478 per 1000 population　　💿 700 per 1000 population

Defense 3%	
Education 5%	
Health 5.7%	

0　5　10　15　20　25
Defense, Health, Education spending as % of GDP

Between 1945 and 1973, most Americans got richer. Since then, however, living standards have gone on rising only among those who finish high school. This "education effect" has led to noticeable class divisions, despite the long economic boom of the 1990s. The top 20% had average household incomes of $137,500 by 2000, whereas the incomes of the poorest 20% averaged only $15,000 – and were lower in real terms than in 1980.

WORLD RANKING

URUGUAY

OFFICIAL NAME: Eastern Republic of Uruguay   **CAPITAL:** Montevideo
POPULATION: 3.4 million   **CURRENCY:** Uruguayan peso   **OFFICIAL LANGUAGE:** Spanish

URUGUAY IS SITUATED IN the southeast of South America, sandwiched between its larger neighbors Brazil and Argentina. Its capital, Montevideo, is an Atlantic port on the River Plate, lying on the opposite bank to the Argentine capital Buenos Aires. Uruguay became independent in 1828, after nearly 150 years of Spanish and Portuguese control. Decades of liberal government ended in 1973 with a military coup that was to result in 12 years of dictatorship, during which 400,000 people emigrated. Most have since returned. Almost the entire low-lying landscape is devoted to the rearing of livestock, especially cattle and sheep; Uruguay is a leading wool exporter. Tourism and offshore banking now bring in substantial foreign earnings.

Uruguayan grasslands. *Rich pasture covers three-quarters of the country, ideal for cattle and sheep. Animals and animal products account for over one-third of export earnings.*

CLIMATE ▷ Subtropical

WEATHER CHART FOR MONTEVIDEO

Uruguay has one of the most benign climates in the world. It is uniformly temperate over the whole country. Winters are mild, frost is rare, and it never snows. Summers are generally cool for these latitudes and rarely tropically hot. The moderate rainfall tends to fall in heavy showers, leaving most days sunny.

TRANSPORTATION ▷ Drive on right

Carrasco, Montevideo 1.17m passengers

91 ships 106,900 grt

THE TRANSPORTATION NETWORK

8085 km (5024 miles)	8683 km (5395 miles)
3002 km (1865 miles)	1600 km (994 miles)

The government has sold off its share in the national bus industry – there are extensive internal and international coach and bus services – and has closed down all passenger railroad services. In 1998 the Senate gave the go-ahead for a $1 billion, 45-km (30-mile) road bridge across the River Plate from Colonia to Buenos Aires. Raising international finance for the project will be difficult.

TOURISM ▷ Visitors : Population 1:1.7

2m visitors

Down 5% in 2000

MAIN TOURIST ARRIVALS

Argentina 70%
Brazil 7%
Chile 1%
Other 22%

% of total arrivals

Sandy beaches near the River Plate estuary are a major attraction. The old Spanish fortifications of Montevideo have been destroyed, but the city retains a colonial atmosphere. Punta del Este, 138 km (86 miles) east of the capital, is the main beach resort. Argentinians account for the majority of visitors.

PEOPLE ▷ Population density low

Spanish

19/km² (50/mi²)

THE URBAN/RURAL POPULATION SPLIT

91%   9%

RELIGIOUS PERSUASION

Protestant 2%   Jewish 2%
Roman Catholic 66%   Nonreligious 30%

ETHNIC MAKEUP

Black 4%   Mestizo 6%
White 90%

Most Uruguayans are second- or third-generation European, mostly of Spanish or Italian descent. There are also some *mestizos* (of mixed blood) and a small minority of people descended from Africans or immigrants from Brazil, who live near the Brazilian border, or in or around Montevideo. All indigenous Amerindian groups became integrated in the *mestizo* population by the mid-19th century. More recent immigrants include Jews, Armenians, and Lebanese. Historically, ethnic tensions have been few. The birthrate is low for Latin America.

The considerable prosperity derived from cattle ranching allowed Uruguay to become a welfare state long before any other Latin American country. In spite of Uruguay's serious economic decline since the end of the 1950s, there is still a sizable, if less prosperous, middle class. A clear sign of the country's economic and social deterioration during the years of military dictatorship was the unprecedented growth of shanty towns around Montevideo.

Although Uruguay is a Roman Catholic country, it is liberal in its attitude to religion, and all forms are tolerated. Divorce is legal. Women, who gained the vote in 1932, are regarded as equal to men.

POPULATION AGE BREAKDOWN

Female		Age		Male
1.8%		80+	0.9%	
8.1%		60–79		6.4%
10.9%		40–59	10.2%	
14.3%		20–39		13.9%
16.5%		0–19		17%

% of population by age group

U

POLITICS ▷ Multiparty elections

L. House 1999/2004
U. House 1999/2004

President Jorge Batlle Ibáñez

AT THE LAST ELECTION

Chamber of Representatives 99 seats

41% EPFA	33% PC	22% PN	4% NE

EPFA = Progressive Broad Front **PC** = Colorado Party (Colorados) **PN** = National Party (Blancos)
NE = New Space **Res** = Reserved for the vice president

Senate 31 seats

39% EPFA	32% PC	23% PN	3% NE	3% Res

Uruguay is a presidential multiparty democracy.

PROFILE

Since the return to democracy in 1984, the main Colorado (PC) and Blanco (PN) parties have monopolized power, either alone or in coalitions, despite being traditional opponents. The left-wing Broad Front has been the effective opposition, frequently in alliance with trade unions fighting austerity measures and reform of the social security system. Despite the crowded electoral calendar in 1999, and infighting among Blanco factions, there was broad political consensus on the need for continuing economic reform. In the 1999 elections, Colorado candidate Jorge Batlle won the presidency in the face of an unusually strong left-wing challenge.

MAIN POLITICAL ISSUES
Economic crisis

Recession has left the Batlle government little room for maneuver. The unions oppose tough fiscal targets agreed with the IMF. The currency came under pressure in 2002 due to the turmoil in Uruguay's main single export market, Argentina, coming on top of a foot-and-mouth outbreak which hit beef exports.

The disappeared

An official inquiry into the fate of 160 people who "disappeared" under military rule (1973–1985) aims to give Uruguay a "moral and ethical answer."

Luis Alberto Lacalle Herrera, president from 1990–1995.

Jorge Batlle Ibáñez, who took office as president in 2000.

URUGUAY

Total Land Area : 176 220 sq. km (68 039 sq. miles)

LAND HEIGHT	POPULATION
200m/656ft	over 1 000 000 ◎
Sea Level	over 50 000 ○
	over 10 000 ●
	under 10 000 •

```
0        100 km
0              100 miles
```

WORLD AFFAIRS ▷ Joined UN in 1945

| Geplac | Mercsr | IBRD | OAS | RG |

Regional integration is a major focus, but President Batlle's determination to clinch a bilateral trade deal with the US risked a row with the Brazilian government, which said that it went against the official policy of Mercosur to negotiate trade agreements as a bloc. Argentina also took exception to plans to impose protective tariffs to cushion Uruguayan industry from the effects of Argentine devaluation. Such difficulties threatened diplomatic moves to strengthen Mercosur as a negotiating bloc in upcoming talks to establish a Free Trade Area of the Americas (FTAA).

Uruguay and the US have agreed a legal assistance treaty to allow easier access to bank accounts of those suspected of laundering the proceeds from narcotics trafficking.

AID ▷ Recipient

$17m (receipts)   Down 23% in 2000

Uruguay received an IMF standby loan of $1.5 billion for 2002–2003, but aid remains otherwise modest.

CHRONOLOGY

The Spaniards were the first to colonize the area north of the River Plate. In 1680, the Portuguese also founded a colony there, at Colonia del Sacramento, so starting 150 years of rivalry between the colonial powers for control of the territory.

❑ **1726** Spaniards found Montevideo. By end of the century, whole country is divided into large cattle ranches.
❑ **1808** Montevideo declares independence from Buenos Aires.
❑ **1811** Patriotic rancher and local caudillo, José Gervasio Artigas, fends off Brazilian attack.
❑ **1812–1820** Uruguayans, known as Orientales ("Easterners," from the eastern side of the River Plate), fight wars against Argentinian and Brazilian invaders. Brazil finally takes Montevideo.
❑ **1827** Gen. Lavalleja defeats Brazilians with Argentine help.
❑ **1828** Seeing trade benefits that an independent Uruguay would bring as a buffer state between Argentina and Brazil, Britain mediates and secures Uruguayan independence.
❑ **1836** Start of large-scale European immigration. ⇨

U

CHRONOLOGY *continued*

- ❏ **1838–1865** La Guerra Grande civil war between Blancos (Whites, future conservative party) and Colorados (Reds, future liberals).
- ❏ **1865–1870** President Venancio Flores of Colorados takes Uruguay into War of Triple Alliance against Paraguay.
- ❏ **1872** Peace under military rule. Blancos strong in country, Colorados in cities.
- ❏ **1890s** Violent strikes by immigrant trade unionists against landed elite enriched by massive European investment in ranching.
- ❏ **1903–1907** Reformist Colorado, José Batlle y Ordóñez, president.
- ❏ **1911–1915** Batlle serves second term in office. Batllismo creates the only welfare state in Latin America with pensions, social security, and free education and health service; also nationalizations, disestablishment of Church, abolition of death penalty.
- ❏ **1933** Military coup. Opposition groups excluded from politics.
- ❏ **1942** President Alfredo Baldomir dismisses government and tries to bring back proper representation.
- ❏ **1939–1945** Neutral in World War II.
- ❏ **1951** New constitution replaces president with nine-member council. Decade of great prosperity follows until world agricultural prices plummet. Sharp drop in foreign investment.
- ❏ **1958** Blancos win elections for first time in 93 years.
- ❏ **1962** Tupamaros urban guerrillas founded. Its guerrilla campaign lasts until 1973.
- ❏ **1966** Presidency reinstated. Colorados back in power.
- ❏ **1967** Jorge Pacheco president. Tries to stifle opposition to tough anti-inflation policies.
- ❏ **1973** Military coup. Promises to encourage foreign investment counteracted by denial of political freedom and brutal repression of the left; 400,000 emigrate.
- ❏ **1984–1985** Military step down. Elections. Julio Sanguinetti (Colorado) president.
- ❏ **1986** Those guilty of human rights abuse granted amnesty.
- ❏ **1989** Referendum endorses amnesty in interests of stability. Elections won by Lacalle Herrera and Blancos.
- ❏ **1994–1995** Sanguinetti reelected, forms coalition government. Mercosur membership.
- ❏ **1999** October, presidential election won by Colorado Jorge Batlle.
- ❏ **2002** Uruguay loses investment grade status due to impact of Argentine crisis.

DEFENSE

 No compulsory military service

 $356m

⬆ Up 12% in 2000

The military withdrew from power in 1985 and has since respected civilian rule. "Lodges" operate within the army to promote officers' interests and have displayed opposition to the government's replacements and promotions within the military hierarchy. A 1986 law virtually blocked investigations into killings, torture, and "disappearances" during the dictatorship, but there is still pressure to bring guilty officers to justice. A presidential decree in 1997 granted amnesty to officers punished for political offenses under military rule.

URUGUAYAN ARMED FORCES

🛡	15 main battle tanks (T-55)	15,200 personnel
🚢	3 frigates and 8 patrol boats	5700 personnel
✈	28 combat aircraft (10 A37B, 5 IA-58B)	3000 personnel
	None	

Uruguay's defense budget is modest; the majority of equipment is bought from the US.

ECONOMICS

▷ Inflation 31% p.a. (1990–2000)

📊 $20bn

💲 12.51–14.77 Uruguayan pesos

SCORE CARD

❏ World GNP Ranking	63rd
❏ GNP per Capita	$6000
❏ Balance of Payments	–$593m
❏ Inflation	4.8%
❏ Unemployment	14%

ECONOMIC PERFORMANCE INDICATOR

EXPORTS

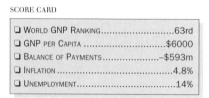

Mexico 4% — Germany 4% — USA 8% — Argentina 18% — Brazil 23% — Other 43%

IMPORTS

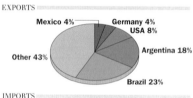

Venezuela 4% — France 4% — USA 10% — Brazil 19% — Argentina 24% — Other 39%

STRENGTHS

Substantial earnings from offshore banking. Tourism. Fertile grasslands. World's second-largest wool exporter. Beef-meat products.

WEAKNESSES

Few natural resources. Dependence on Brazil and Argentine markets. Modest industry. Large public sector deficit. Sluggish economic growth. Loss of investment grade status.

PROFILE

Traditionally an agricultural economy, three-quarters of the country is rich pasture, supporting livestock. Much of the rest is given over to crops. Farming, which formerly brought great wealth, still employs about 15% of the labor force, accounting for some 19% of GDP. Livestock and animal products, especially meat and wool, account for over one-third of export earnings. Manufacturing, accounting for some 18% of GDP, is farm-based. Tourism is increasingly important. Most economic activity – and half the population – is concentrated in Montevideo. Recently growth has been sluggish, making it difficult to achieve GDP and fiscal targets agreed with the IMF. Unions resist spending cuts, and, with public opinion hostile to major privatizations, there has been little progress in necessary structural reforms.

URUGUAY : MAJOR BUSINESSES

Salto
Paysandú
Río Branco
Fray Bentos
Colonia del Sacramento
Durazno
Montevideo

✿ Heavy engineering
❚ Food processing
❘ Oil refining
🧶 Wool spinning
🐂 Meat packing
⬥ Leather
✺ Textiles

0   100 km
0   100 miles

RESOURCES

 Electric power 2.2m kw

 103,043 tonnes   Not an oil producer

 13m sheep, 10.8m cattle, 13m chickens   Gold, iron, gemstones, copper, zinc, lead, manganese

Most of Uruguay is farmland, much of it given over to cattle and sheep. Rice is the country's only other significant export. There are no known oil or natural gas resources. Considerable potential is believed to exist for the mining sector, but only small quantities of building materials and jewelry-quality agate and amethysts are so far extracted. The mining of gold deposits is currently being developed. Hydropower generates most of the electricity.

ELECTRICITY GENERATION

Hydro 90% (8.7bn kwh)
Combustion 10% (0.9bn kwh)
Nuclear 0%
Other 0%

% of total generation by type

URUGUAY : LAND USE
Cropland
Pasture
Forest
Sheep
Cattle
Wheat

0   100 km
0   100 miles

ENVIRONMENT

 Sustainability rank: 6th

 0.3% (0.1% partially protected)   1.8 tonnes per capita

ENVIRONMENTAL TREATIES

Yes   Yes   Yes
Yes   Yes   Yes

Pollution of the main Uruguay and Plate Rivers is a concern, as is traffic density in Montevideo.

MEDIA

 TV ownership high

Daily newspaper circulation 293 per 1000 people

PUBLISHING AND BROADCAST MEDIA

There are 36 daily newspapers, including *El Pais, El Diario,* and *La Mañana*

4 services: 1 state-owned, 3 independent

6 services: 1 state-owned, 5 independent

The press is relatively free. *El País* supports the Blancos (PN), while *La Mañana* backs the Colorados (PC).

CRIME

 No death penalty

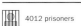 4012 prisoners   Down 11% in 1999

CRIME RATES

Murders	
8	per 100,000 population

Rapes	
10	per 100,000 population

Thefts	
2281	per 100,000 population

Crime levels in Uruguay are low, particularly compared with its neighbors Brazil and Argentina. Domestic theft is the main problem. Bribery is uncommon.

EDUCATION

 School leaving age: 14

98%   77,454 students

THE EDUCATION SYSTEM

100% Primary
88% Secondary
35% Tertiary

% of each age group in education

Education, inspired by the French *lycée* system, is state-funded for 12 years up to secondary level and is compulsory for all children between the ages of six and 14. Both state and private schools follow the same curriculum; private schools are monitored by the government. Facilities are rudimentary in rural areas. Uruguay has two state-funded universities. The children of wealthy Uruguayans tend to complete their studies in the US. Resistance to tax increases and pressure to reduce the fiscal deficit have both placed serious constraints on education spending. Secondary school students continue to stage protests against the resulting effects on the system.

HEALTH

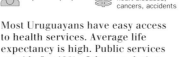 Welfare state health benefits

1 per 270 people   Cerebrovascular and heart diseases, cancers, accidents

Most Uruguayans have easy access to health services. Average life expectancy is high. Public services provide for 40% of the population, while the private sector caters for the remaining 60%. Despite opposition, the government has privatized some of the state medical establishments.

Health spending has in recent years been a victim of the budget cuts and social welfare reforms aimed at controlling the fiscal deficit.

SPENDING

 GDP/cap. increase

CONSUMPTION AND SPENDING

158 per 1000 population   278 per 1000 population

Defense 2.6%
Education 2.5%
Health 1.9%

Defense, Health, Education spending as % of GDP

Uruguay possesses the social mobility which is typical of countries created through decades of large-scale immigration. Many professionals come from modest backgrounds. A 1999 report by the IDB exempted Uruguay (along with Costa Rica and Jamaica) from the regional trend of serious income inequality.

The wealthy tend either to be landowners or are employed in the financial sector. They have traditionally looked toward Europe, rather than the US, for luxury goods.

The most deprived sections of Uruguayan society are the urban poor of Montevideo, a large proportion of whom are of mixed African and European descent, and the rural poor, who own little or no land.

WORLD RANKING

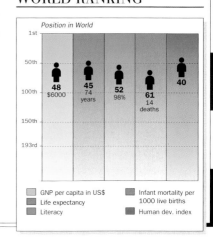

Position in World

1st
50th
100th
150th
193rd

48 $6000
45 74 years
52 98%
61 14 deaths
40

GNP per capita in US$
Life expectancy
Literacy
Infant mortality per 1000 live births
Human dev. index

U

UZBEKISTAN

OFFICIAL NAME: Republic of Uzbekistan **CAPITAL:** Tashkent
POPULATION: 25.3 million **CURRENCY:** Som **OFFICIAL LANGUAGE:** Uzbek

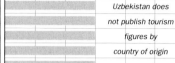

CENTRAL ASIA

S HARING THE ARAL SEA coastline with its northern neighbor, Kazakhstan, Uzbekistan contains the ancient cities of Samarqand, Bukhara (Bukhoro), Khiva, and Tashkent. It is the most populous central Asian republic and has considerable natural resources. The dictatorship of President Karimov has prevented the spread of Islamic fundamentalism.

CLIMATE
▷ Desert/mountain

WEATHER CHART FOR TASHKENT

■ *Average daily temperature*   *Rainfall* ■
| °C/°F | J F M A M J J A S O N D | cm/in |

Uzbekistan has a harsh continental climate. Summers can be extremely hot and dry. Large areas of the country are desert.

TRANSPORTATION
▷ Drive on right

 Tashkent International 1.93m passengers

Has no fleet

THE TRANSPORTATION NETWORK

71,237 km (44,265 miles)

None

3645 km (2265 miles)

1100 km (684 miles)

Uzbekistan has a well-developed transportation system. An extensive network of buses serves country areas, while good Soviet-style systems of trolley buses and trams operate in the major cities. Tashkent's subway system was the first in central Asia. Road and rail networks have deteriorated since 1991, however, and are concentrated in the south and east.

TOURISM
▷ Visitors : Population 1:93

272,000 visitors

Up 196% 1995–1998

MAIN TOURIST ARRIVALS

Uzbekistan does not publish tourism figures by country of origin

| 0 | 10 | 20 | 30 | 40 | 50 | 60 |
% of total arrivals

Uzbekistan has considerable tourist potential. Bukhara, once a trading center on the silk route, is famous worldwide for its architecture and carpet-making. It has great religious significance for Muslims, who are encouraged to make at least one pilgrimage to its holy shrines. Bukhara's Kalyan Mosque is famous for its minaret built of unbaked bricks. The city of Samarqand was expanded in the 14th century by Timur, and contains the monumental gateway of the Shir Dar Madrasa, one of the most beautiful buildings in the Islamic world.

UZBEKISTAN

Total Land Area : 447 400 sq. km
(172 741 sq. miles)

LAND HEIGHT

3000m/9843ft
2000m/6562ft
1000m/3281ft
500m/1640ft
200m/656ft
Sea Level

POPULATION

▣ over 1 000 000
◎ over 100 000
○ over 50 000
● over 10 000

Mosque in Samarqand.
The city remained an Islamic stronghold, despite communist attempts at suppression, when Uzbekistan formed part of the Soviet Union.

U

PEOPLE ▷ Pop. density medium

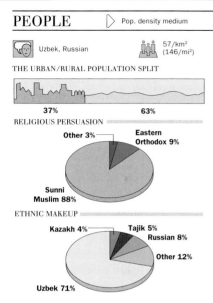

Uzbek, Russian   57/km²
(146/mi²)

THE URBAN/RURAL POPULATION SPLIT

37%   63%

RELIGIOUS PERSUASION

Other 3%
Eastern Orthodox 9%
Sunni Muslim 88%

ETHNIC MAKEUP

Kazakh 4%
Tajik 5%
Russian 8%
Other 12%
Uzbek 71%

Among the former Soviet republics, Uzbekistan has a relatively complex makeup. In addition to the Uzbeks, Russians, Tajiks, and Kazakhs, there are small minorities of Tatars and Karakalpaks. The proportion of Russians has been declining since the

1970s, when net emigration of Russians began. Tensions among ethnic groups have the potential to create regional and racial conflict. The authoritarian nature of the Karimov leadership has so far prevented these antagonisms from becoming violent. Incidents such as the 1989 and 1990 clashes between Meskhetian Turks and Uzbeks are rare. The removal of the dominance of the Communist Party has meant that Uzbek society has reverted to traditional social patterns based on family, religion, clan, and region, rather than on membership of the party. Independence has done little to alter the minor role of women in politics. Arranged marriages are still the custom in the countryside.

POPULATION AGE BREAKDOWN

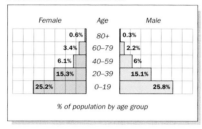

Female		Age	Male	
	0.6%	80+	0.3%	
	3.4%	60–79	2.2%	
	6.1%	40–59	6%	
	15.3%	20–39	15.1%	
25.2%		0–19		25.8%

% of population by age group

POLITICS ▷ Multiparty elections

1999/2004   President Islam Karimov

AT THE LAST ELECTION
National Parliament 250 seats

4% NT   1% Vac

44% Ind   19% PDP   14% F   8% FP   6% LN   4% A

Ind = Independents   **PDP** = People's Democratic Party
F = Fidorkorlar   **FP** = Fatherland Progress Party
LN = Local nominees   **A** = Adolat
NT = National Renaissance   **Vac** = Vacant

Uzbekistan is effectively run by a presidential dictatorship. In 2002 a referendum approved the creation of a second legislative chamber and lengthened the presidential term.

PROFILE

President Islam Karimov's PDP has not been willing to devolve or share power. The 1992 constitution appeared to endorse multipartyism, but Karimov took advantage of greater powers granted to his office by banning several opposition parties, including the nationalist Birlik (Unity) movement and the Islamic Renaissance Party. Erk (Will), the only legal opposition party, was proscribed in 1993, and in 1995, a group of its activists, found guilty of political subversion, received stiff sentences. Opposition is now entirely underground. The intimidation and arbitrary imprison-ment of dissidents are common, and repression has increased since bomb

attacks in Tashkent in 1999. Karimov has kept the support of the Russian minority by avoiding nationalist rhetoric.

MAIN POLITICAL ISSUES
Islamic fundamentalism
Civil war in Tajikistan and the rise of the *taliban* regime in Afghanistan raised fears about Islamic fundamentalism. The constitution stipulates the separation of Islam and the state. A joint operation with Kyrgyzstan and Tajikistan against the pan-regional Islamic Movement of Uzbekistan (IMU) took place in 2000, as cross-border attacks rose. IMU leader Juma Namangani was reported killed alongside *taliban* fighters in Afghanistan in late 2001. A clampdown against the Hizb-ut Tahrir group continues.

Regionalism
The high birthrate puts pressure on limited agricultural resources. Migration from poorer areas have led to calls for secession from some regions. In the densely populated eastern Fergana Valley there has been a number of violent incidents.

Islam Karimov, first elected president in 1990 and Uzbekistan's sole leader since independence.

WORLD AFFAIRS ▷ Joined UN in 1992

 CIS   SCO   OIC   NAM   OSCE

Unlike neighboring Turkmenistan, Kyrgyzstan, and Tajikistan, Uzbekistan has the resources to allow it to follow a relatively independent foreign policy. The Karimov leadership has used this to promote Uzbekistan as the leading central Asian state, a role for which it vies with Kazakhstan. It is a member of the Georgia–Ukraine–Uzbekistan–Azerbaijan–Moldova (GUUAM) group of ex-Soviet states, signifying a wish to maintain a certain distance in relations with Russia, although it suspended its membership in 2002. It has border disputes with its immediate neighbors, in particular over the laying of land mines on its borders with Tajikistan and Kyrgyzstan to prevent terrorist incursions.

Ties with the US were strengthened when Uzbekistan played host to US forces during the war in Afghanistan in 2001. Relations with Turkey are also developing. While Western companies have difficulty in sealing contracts in Uzbekistan, Turkish companies have been commissioned to build vital installations such as telecommunications.

CHRONOLOGY

Part of the great Mongol Empire, present-day Uzbekistan was incorporated into the Russian Empire between 1865 and 1876. Russification of the area was superficial, and it was not until Soviet rule that significant Slav immigration occurred. A further influx of Slavs occurred during Stalin's program of forced collectivization.

❑ **1917** Soviet power established in Tashkent.
❑ **1918** Turkestan Autonomous Soviet Socialist Republic (ASSR), incorporating present-day Uzbekistan, proclaimed.
❑ **1923–1941** Alphabet changes, from Arabic to Latin to Iranized Tashkent, finally to Cyrillic.
❑ **1924** Basmachi rebels who resisted Soviet rule crushed. Uzbek SSR founded (which, until 1929, included the Tajik ASSR).
❑ **1925** Anti-Islamic campaign bans schools and closes mosques.
❑ **1936** Karakalpak ASSR (formerly part of the Russian Soviet Federative Socialist Republic) incorporated into the Uzbek SSR.
❑ **1937** Uzbek communist leadership is purged by Stalin. ➪

U

CHRONOLOGY *continued*

- ❏ **1941–1945** Industrial boom.
- ❏ **1959** Sharaf Rashidov becomes first secretary of Communist Party of Uzbekistan (CPUz). Retains position until 1983.
- ❏ **1966** Tashkent razed by earthquake. Rebuilding brings in large number of Russian and other non-Uzbek migrants.
- ❏ **1982–1983** Yuri Andropov becomes leader in Moscow. His anticorruption purge results in emergence of a new generation of central Asian officials.
- ❏ **1989** First noncommunist political movement, Unity (Birlik), formed but not officially registered. June, clashes erupt between Meskhetian Turks and indigenous Uzbek population of Fergana Valley resulting in more than 100 deaths. October, Birlik campaign leads to Uzbek being declared the official language.
- ❏ **1990** Islam Karimov becomes executive president of the new Uzbek Supreme Soviet. Further interethnic fighting in Fergana Valley; 320 killed.
- ❏ **1991** August, independence is proclaimed and Republic of Uzbekistan is adopted as official name. October, Uzbekistan signs treaty establishing economic community with seven other former Soviet republics. November, CPUz restructured as the People's Democratic Party of Uzbekistan (PDP); Karimov remains its leader. December, Karimov confirmed in post of president. Uzbekistan joins the CIS.
- ❏ **1992** Price liberalization provokes student riots in Tashkent. New post-Soviet constitution adopted along Western democratic lines. All religious parties banned. September, Uzbekistan sends troops to Tajikistan to suppress violence and strengthen border controls.
- ❏ **1993** Growing harassment of opposition political parties, Erk and Birlik.
- ❏ **1994** Introduction of som.
- ❏ **1995** January, Karimov's PDP wins legislative elections. March, referendum extends Karimov's presidential term until 2000. December, Utkur Sultanov replaces Abdulashim Mutalov as prime minister.
- ❏ **1999** Bomb attacks by Islamic terrorists lead to crackdown and arrests of hundreds of opposition activists. Legislative elections.
- ❏ **2000** Karimov reelected.
- ❏ **2002** Referendum extends president's term to seven years.

AID

 ▷ Recipient

 $186m (receipts)　　 Up 20% in 2000

Ukraine's human rights violations and lack of commitment to economic stabilization have generally deterred bilateral aid donors. The World Bank is the largest donor, allocating millions of dollars to support economic reform and a range of agricultural, financial, and social projects.

DEFENSE

 ▷ Compulsory military service

$1.48bn　　Up 20% in 2000

Uzbekistan has a large standing army, mainly conscripts, as well as over 17,000 internal security troops and a 1000-strong National Guard. A military restructuring is under way, with a view to full professionalization of the armed forces.

A policy of mining the land borders with Kyrgyzstan and Uzbekistan to thwart incursions by Islamic militants has drawn international criticism.

UZBEK ARMED FORCES

350 main battle tanks (T-62, T-64, T-72)	40,000 personnel	
None		
135 combat aircraft (Su-17, Su-24, Su-25, Su-27, MiG-29)	15,000 personnel	
None		

ECONOMICS

 ▷ Inflation 247% p.a. (1990–2000)

 $8.84bn　　 322.8–686.9 som

SCORE CARD

- ❏ WORLD GNP RANKING..........................89th
- ❏ GNP PER CAPITA $360
- ❏ BALANCE OF PAYMENTS $184m
- ❏ INFLATION 40%
- ❏ UNEMPLOYMENT...................................10%

EXPORTS

South Korea 4%　Tajikistan 8%　Ukraine 8%　Italy 8%　Other 44%　Russia 28%

IMPORTS

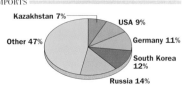

Kazakhstan 7%　USA 9%　Germany 11%　South Korea 12%　Russia 14%　Other 47%

ECONOMIC PERFORMANCE INDICATOR

Consumer Price Index　　GDP

STRENGTHS

Gold. Well-developed cotton market. Considerable unexploited deposits of oil and natural gas. Current production of natural gas makes significant contribution to electricity generation. Manufacturing tradition includes agricultural machinery and central Asia's only aviation factory.

WEAKNESSES

Dependent on grain imports, as domestic production meets only 25% of needs. Very limited economic reform. High inflation. Environmentally damaging irrigation scheme for cotton production.

PROFILE

With the exception of Tashkent, which became an industrial area during World War II, Uzbekistan's economy is predominantly agricultural. Pro-market reforms have been slow, despite fresh assistance from the World Bank to increase the efficiency of privatized companies. The gold sector has attracted investment from US companies. Energy resources are still to be fully exploited. The som was devalued by 50% in 2001.

UZBEKISTAN : MAJOR BUSINESSES

- ❋ Textiles
- ⚱ Fertilizers
- ⟁ Silk weaving
- ⤴ Leather tanning
- ▣ Food processing
- ❁ Engineering

0　200 km
0　200 miles

RESOURCES

 Electric power 11.7m kw

 8536 tonnes

 172,000 b/d (reserves 600m barrels)

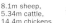 8.1m sheep, 5.34m cattle, 14.4m chickens

Natural gas, coal, oil, gold, uranium, copper, tungsten, aluminum

Uzbekistan has the world's largest gold mine, at Murantau in the Kyzyl Kum Desert, and also large deposits of natural gas, petroleum, coal, and uranium. An important oil field was discovered in 1992 in the Namangan region and production will rise with further investment. Most gas produced is currently used domestically, but it could also become a strong export.

Cotton is the main focus of agriculture: Uzbekistan is the

ELECTRICITY GENERATION

Hydro 12% (5.8bn kwh)
Combustion 88% (40bn kwh)
Nuclear 0%
Other 0%

% of total generation by type

world's fourth-largest producer. A decision after independence to diversify was reversed when the value of cotton as a commodity on the world market became clear. Fruit, silk cocoons, and vegetables for Russian markets are also of rising importance.

UZBEKISTAN : LAND USE

- Cropland
- Pasture
- Forest
- High mountain regions
- Desert
- Wetlands
- Sheep
- Cotton - cash crop

0   200 km
0   200 miles

ENVIRONMENT

 Sustainability rank: 118th

 2%

4.5 tonnes per capita

ENVIRONMENTAL TREATIES

Yes   Yes   Yes
Yes   Yes   No

The irrigation schemes required to sustain the cotton industry have wreaked considerable environmental damage. Soil salination is now a major problem. The Aral Sea has also been seriously depleted. From 61,836 sq. km (23,875 sq. miles) in 1974, it had shrunk to less than half that area by 1997. In 1998 the World Bank approved more than $11 million to save the Aral Sea region. The indiscriminate use of fertilizers and pesticides to raise production has polluted many rivers.

MEDIA

 TV ownership high

 Daily newspaper circulation 3 per 1000 people

PUBLISHING AND BROADCAST MEDIA

There are 3 daily newspapers, including the Uzbek *Khalk suzi* and the Russian *Pravda Vostoka*

2 state-controlled services

1 state-controlled service broadcasting in many languages

Uzbekistan's restrictions on independent publications are designed to encourage the promotion of the personality cult and policies of Karimov, and manifest themselves both in overt censorship and self-censorship by media outlets. Independent journalists face harassment. The expression of Islamic and nationalist opinion is forbidden.

CRIME

 Death penalty in use

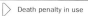 63,900 prisoners

Up 9% in 1999

CRIME RATES

Murders
3 | *per 100,000 population*

Rapes
2 | *per 100,000 population*

Thefts
91 | *per 100,000 population*

Crime has risen as living standards have declined. In 2001 the number of crimes carrying the death penalty was reduced to four. Unofficial Islamic courts in the Fergana Valley are a sign of opposition to the government.

EDUCATION

 School leaving age: 18

 89%

638,200 students

THE EDUCATION SYSTEM

Primary 78% — Secondary 94% — Tertiary 32%

% of each age group in education

The state system still follows the Soviet model, though some instruction is in Uzbek. In the late 1980s, there were a few ethnic Tajik schools and a university in Samarqand. These were closed down in 1992 as relations deteriorated between Uzbekistan and Tajikistan. The rise in Islamic consciousness has led to a growing number of *madaris* – schools attached to mosques. In 1999 the establishment of Tashkent Islamic University was agreed.

HEALTH

 Welfare state health benefits

 1 per 323 people

Circulatory and respiratory diseases, accidents, cancers

The health service has been in decline since the dissolution of the USSR. Some rural areas are not served at all. In 1998 a $69.7 million project to improve health services was announced, with the World Bank providing a loan of some $30 million. Serious respiratory diseases among cotton growers are increasing.

SPENDING

 GDP/cap. decrease

CONSUMPTION AND SPENDING

No data

67 per 1000 population

Defense 8%
Education 7.7%
Health 3.4%

Defense, Health, Education spending as % of GDP

Former communists are still the wealthiest group, since they retain control of the economy. Many rural poor live below the poverty line.

WORLD RANKING

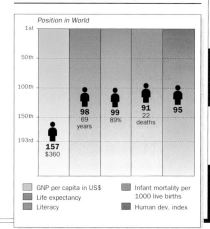

Position in World

98 — 69 years
99 — 89%
91 — 22 deaths
95
157 — $360

- GNP per capita in US$
- Life expectancy
- Literacy
- Infant mortality per 1000 live births
- Human dev. index

U

VANUATU

OFFICIAL NAME: Republic of Vanuatu **CAPITAL:** Port Vila
POPULATION: 200,000 **CURRENCY:** Vatu **OFFICIAL LANGUAGES:** Bislama, English, and French

THE ARCHIPELAGO OF Vanuatu stretches over 1300 km (800 miles) in the South Pacific. Mountainous and volcanic in origin, only 12 of the 82 islands are a significant size – Espiritu Santo and Malekula are the largest. Formerly known as the New Hebrides – ruled jointly by France and Britain from 1906 – Vanuatu became independent in 1980. Politics since then has been democratic but volatile.

CLIMATE
▷ Tropical oceanic

WEATHER CHART FOR PORT VILA

Rainfall in July and September is 96.5 cm/38 inches

The climate is tropical and hot. Rainfall and temperatures decrease north to south. Cyclones occur November–April.

TRANSPORTATION
▷ Drive on right

 Bauerfield, Port Vila   293 ships 1.6m grt

THE TRANSPORTATION NETWORK

256 km (159 miles)		None
None		None

Road quality is generally poor, with routes on some remote islands impassable in the wet season.

TOURISM
▷ Visitors : Population 1:3.5

57,000 visitors   Up 14% in 2000

MAIN TOURIST ARRIVALS

- Australia 62%
- New Zealand 14%
- New Caledonia 8%
- Other 16%

0 10 20 30 40 50 60 70 80
% of total arrivals

Tourism in Vanuatu is facing stiff competition from cheaper regional rivals. Organized tours include sea fishing, sailing, kayaking, and diving.

PEOPLE
▷ Pop. density low

Bislama (Melanesian pidgin), English, French   16/km² (42/mi²)

THE URBAN/RURAL POPULATION SPLIT

20%   80%

RELIGIOUS PERSUASION

- Seventh-day Adventist 6%
- Indigenous beliefs 8%
- Presbyterian 37%
- Roman Catholic 15%
- Anglican 15%
- Other 19%

Indigenous Melanesians – ni-Vanuatu – comprise 94% of the population. Of Vanuatu's 82 islands, 67 are inhabited, but 80% of people live on 12 main islands. The population is becoming more urbanized as one in eight ni-Vanuatu now lives in Port Vila. However, 75% of the population still live by subsistence agriculture.

Vanuatu is home to some of the Pacific's most traditional peoples, and local social and religious customs are strong. With 105 indigenous languages, Vanuatu boasts the world's highest per capita density of languages. Bislama pidgin is the lingua franca.

Women have lower social status than men, and bride price is still commonly paid. Many educated women refuse to marry because of loss of property rights. To boost equality, primary schools are encouraged to take girls.

Vanuatu's unspoilt beaches *are one of the reasons for the upsurge in the tourist industry.*

POLITICS
▷ Multiparty elections

2002/2006   President Fr. John Bani

AT THE LAST ELECTION
Parliament 52 seats

29% UMP	27% VP	11% NUP	10% Ind	4% MPP	19% Others

UMP = Union of Moderate Parties **VP** = Vanua'aku Pati
NUP = National United Party   **Ind** = Independents
MPP = Melanesian Progressive Party

Ni-Vanuatu politics is best described as anarchic. Political allegiances are swiftly changed and governments often toppled. Instability was behind the reluctance of France – not shared by Britain – to grant independence, achieved in 1980, and led to alleged French support, reinforced by the anti-French stance of the VP, for an abortive bid for independent status by Espiritu Santo that year.

The VP ruled under the charismatic Fr. Walter Lini from 1980 to 1991, when he was dismissed as party leader and premier. He defected and formed the NUP, and continued to be a dominant figure until his death in 1999. The UMP headed a succession of coalitions from 1991 until 1998 when the VP returned to power. Its government was based on patched-up relations with Lini, but within months it was back in opposition, unable to command a stable majority. In 2001 a broad-based anti-VP coalition was ousted, bringing back the VP, under Edward Natapei, in coalition with the UMP. Natapei continued to head the coalition after the 2002 polls, despite the UMP's marginal victory.

WORLD AFFAIRS
▷ Joined UN in 1981

OIF   Comm   NAM   PC   PIF

Vanuatu was the first South Pacific nation to gain full membership of the Non-Aligned Movement. It was censured by the OECD in 2002 for failing to improve financial transparency.

AID
▷ Recipient

 $46m (receipts)   Up 24% in 2000

Vanuatu is heavily dependent on aid. Leading donors are Australia, the ADB, the World Bank, France, Japan, New Zealand, and the EU. Vanuatu is classed by the UN as a least developed country (LDC), and as such receives guaranteed aid support.

V

DEFENSE

▷ No compulsory military service

$ There is no army

⇕ Not applicable

There is no army. A small paramilitary force receives training from the US. Papua New Guinean troops helped to end the 1980 secessionist movement on Espiritu Santo.

ECONOMICS

▷ Inflation 2.1% p.a. (1990–2000)

$226m

143.01–146.80 vatu

SCORE CARD

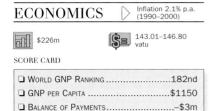

❏ WORLD GNP RANKING	182nd
❏ GNP PER CAPITA	$1150
❏ BALANCE OF PAYMENTS	–$3m
❏ INFLATION	2%
❏ UNEMPLOYMENT	Low

STRENGTHS

Expanding services sector, including tourism and offshore finance. Major economic reforms instituted, including the introduction of a value-added tax and resizing of the public service in return for assistance from the ADB. Large fishing potential.

WEAKNESSES

Large trade and budget deficits. Growth stagnant in recent years and negative in 1999. Dependence on agricultural sector, vulnerable to adverse weather and fluctuating market prices. Shortage of skilled indigenous labor.

EXPORTS

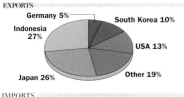

Germany 5%
South Korea 10%
Indonesia 27%
USA 13%
Japan 26%
Other 19%

IMPORTS

Fiji 6%
New Zealand 8%
Australia 26%
Singapore 14%
Other 25%
Japan 21%

RESOURCES

▷ Electric power 11,000 kw

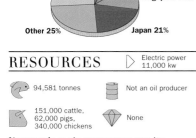

94,581 tonnes

Not an oil producer

151,000 cattle, 62,000 pigs, 340,000 chickens

None

Vanuatu's main resources are its arable land – only partly utilized – and its forests and waters. These could be exploited by the tourist, timber, and fishing industries. New export crops are being explored to offset declining copra and cocoa exports. Beef is of growing importance. Nuclear power development was banned under 1983 legislation.

VANUATU

Total Land Area : 12 200 sq. km
(4710 sq. miles)

POPULATION
over 10 000 ●
under 10 000 ·

Hiu TORRES
Tégua ISLANDS
Loh
Toga
Uréparapara
Mota Lava
Vanua Lava Mota
BANKS ISLANDS
Santa Maria I.
Méré Lava
Big Bay Espiritu Santo
Maéwo
Tabwemasana 1888m
Luganville
Aoba
Passage Lolvavana
Malo
Pentecost
Norsup
Ambrym
Malekula
Paama
Lopévi
Epi
Tongoa
Émaé SHEPHERD
Émaé ISLANDS
Nguna
Émao
Efate
PORT VILA
Bauer Field
Erromango
P A C I F I C   O C E A N
Aniwa
Tanna
Isangel
Futuna
Aneityum
C O R A L   S E A

LAND HEIGHT

1000m/3281ft
500m/1640ft
200m/656ft
Sea Level

N

0   100 km
0   100 miles

ENVIRONMENT

▷ Not available

None

0.3 tonnes per capita

Logging is increasing, but most of the rainforest remains intact, and round wood exports are banned.

Population growth is high, at nearly 3% a year, but is falling. A majority of the population does not have access to a potable and reliable water supply.

MEDIA

▷ TV ownership low

Daily newspaper circulation is very low

PUBLISHING AND BROADCAST MEDIA

There is one daily newspaper, the *Port Vila Presse*, published in French and English

1 state-owned limited service

1 state-owned service

The *Vanuatu Weekly* appears in each official language. Television Blong Vanuatu broadcasts four hours a day.

CRIME

▷ No death penalty

46 prisoners

Little change from year to year

Domestic violence apart, Vanuatu is almost crime-free, with little alcohol-related violence.

CHRONOLOGY

In 1906, Britain and France set up the New Hebrides under joint rule.

❏ **1980** Independence; Walter Lini of VP prime minister. Secession bid by Espiritu Santo.
❏ **1991** UMP coalition with NUP set up by Lini after his expulsion from VP.
❏ **1999** Tidal wave causes extensive damage. Lini dies.
❏ **2001** VP–UMP coalition; Edward Natapei prime minister.

EDUCATION

▷ Schooling is not compulsory

64%

Not available

The abolition of fees has helped to boost primary enrollment. Illiteracy is a major concern.

HEALTH

▷ Welfare state health benefits

1 per 8333 people

Heart diseases, cancers, malaria

A network of rural clinics and village health workers has helped to improve health levels. Nominal fees are charged.

SPENDING

▷ GDP/cap. decrease

CONSUMPTION AND SPENDING

44 per 1000 population

34 per 1000 population

Defense None
Education 4.9%
Health 3.3%

0   5   10   15   20   25
Defense, Health, Education spending as % of GDP

The dominance of subsistence farming and small-scale cash cropping has helped to prevent extreme poverty. Most of the rich are not ni-Vanuatu.

WORLD RANKING

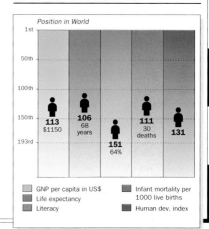

Position in World

1st
50th
100th
150th
193rd

113 $1150
106 68 years
151 64%
111 30 deaths
131

■ GNP per capita in US$
■ Life expectancy
■ Literacy
■ Infant mortality per 1000 live births
■ Human dev. index

V

VATICAN CITY

OFFICIAL NAME: State of the Vatican City **CAPITAL:** Vatican City **POPULATION:** 524
CURRENCY: Euro (lira until 2002) **OFFICIAL LANGUAGES:** Italian and Latin

1929 1929 Oct 22 V +1 +39 .va

THE VATICAN CITY lies close to the Tiber in central Rome and is the world's smallest independent state. It also includes ten other buildings in Rome and the pope's residence at Castel Gandolfo. As the Holy See it is the seat of the Roman Catholic Church, deriving its income from investments and voluntary contributions known as Peter's Pence.

The buildings and gardens of the Vatican City. St. Peter's Basilica was built from 1506–1626 on the traditional site of St. Peter's tomb.

CLIMATE
▷ Mediterranean

WEATHER CHART FOR THE VATICAN CITY

Summers are hot; winters are mild. November is particularly rainy.

TRANSPORTATION
▷ Drive on right

Heliport for official visitors

Has no fleet

THE TRANSPORTATION NETWORK

None		None	
1 km (0.6 miles)		None	

The railroad is only used for carrying freight. Official visitors are transferred from Rome airport by helicopter.

TOURISM
▷ Not available

The Vatican museums can accommodate 20,000 visitors daily

Little change from year to year

MAIN TOURIST ARRIVALS

			Most visitors come from Italy, Germany, Spain, and Central and South America. Numbers from eastern Europe are rising	

0   10   20   30   40
% of total arrivals

Almost all tourists who visit Rome go to the Vatican, while others come as pilgrims. Up to 100,000 hear the pope's annual Easter Message in St. Peter's Square. The Vatican's art collections are among the greatest in the world. Years of restoration work on the Sistine Chapel frescoes were completed in 1999.

PEOPLE
▷ Pop. density high

Italian, Latin

1191/km²
(3082/mi²)

THE URBAN/RURAL POPULATION SPLIT

100%

RELIGIOUS PERSUASION

Roman Catholic 100%

The Vatican has about 500 citizens, including over 100 lay persons. Several hundred more lay staff are employed in the city-state. Citizenship can be acquired through stable residence and holding an office or job within the City. A citizen's family can gain residence only by authorization.

The pope is the spiritual leader of around 17% of the world's population. The countries with the largest number of Roman Catholics are Brazil, Mexico, Italy, the US, and the Philippines.

POLITICS
▷ No legislative elections

On death of reigning pope

His Holiness Pope John Paul II

LEGISLATIVE OR ADVISORY BODIES
College of Cardinals 120 seats

Cardinals under the age of 80 are eligible to elect a new pope. There are no political parties.

The Vatican City operates in the manner of an elected monarchy, as the reigning pope has supreme executive, legislative, and judicial powers, and holds office for life. He is elected by the College of Cardinals, who vote until one candidate for the position of Supreme Pontiff achieves a two-thirds majority.

The administration of the Vatican City State, of which the pope is temporal head, is conducted by the Pontifical Commission. The Holy See, which is the governing body of the Roman Catholic Church worldwide and of which the pope is spiritual head, is governed by the Roman Curia, the Church's administrative network. It is the Holy See that maintains diplomatic relations abroad. Pope John Paul II, elected in 1978, was the first non-Italian pope since 1523. Now in his 80s, he continues to fulfill his duties despite suffering from Parkinson's disease and arthritis.

VATICAN CITY

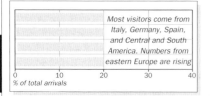

Total Land Area :
0.44 sq. km (0.17 sq. miles)

Main Entrance

Pigna Courtyard

Vatican Museums

Vatican

Radio Vatican

Belvedere Courtyard

Gardens

Sistine Chapel

Raphael Stanza

Papal apartments

Saint Peter's Basilica

St. Peter's Square

Papal Heliport

Vatican Railway Station

ROME

ROME

Monte Vaticano

0   200 metres
0   800 feet

N

WORLD AFFAIRS

▷ Not a member of the UN

The Vatican maintains a neutral stance in world affairs and has observer status in many international organizations. It has mediated in many conflicts, notably achieving the 1993 peace agreement in Mozambique. Pope John Paul II has traveled more extensively than any other pope. His main aims have been to promote political dialogue and bridge religious divides, both within and outside the Catholic Church. In 2000, while on a major tour of the Holy Land, he made an unprecedented apology for the Church's 2000 years of anti-Judaism. In May 2001 he became the first pope to enter (and pray in) a mosque – in Syria – and the first to visit Orthodox Greece.

AID

▷ Donor

 Undisclosed   Undisclosed

Aid is donated through the pope's charities (such as the Holy Childhood Association, which distributes around $15 million a year to children's causes), through funds donated for use at the pope's discretion, and through religious orders acting under papal charter.

DEFENSE

▷ No compulsory military service

 Ceremonial Swiss Guard only   Not applicable

The Vatican is strictly neutral territory. Under the 1954 Hague Convention, it is recognized as "a moral, artistic, and cultural patrimony worthy of being respected as a treasure for all mankind."

ECONOMICS

▷ Not applicable

Not applicable   1.0651–1.1231 euros

SCORE CARD

- ❏ WORLD GNP RANKING ....*The Vatican*
- ❏ GNP PER CAPITA.....................*does not have*
- ❏ BALANCE OF PAYMENTS....................*a national*
- ❏ INFLATION ....*economy in the*
- ❏ UNEMPLOYMENT..........................*usual sense*

STRENGTHS
Istituto per le Opere di Religione has assets of $3–4 billion. Voluntary contributions from Catholics worldwide (Peter's Pence). Interest on investments. Gold reserves in Fort Knox, US. Stamp and coin issues. Budgetary deficit of 23 years reversed to modest surplus in 1993. Receipts from tourists.

WEAKNESSES
Losses incurred by Radio Vaticana and *L'Osservatore Romano*. Cost of foreign papal visits, buildings maintenance, and diplomatic missions.

EXPORTS/IMPORTS

The Vatican produces no goods for export. All commodities are imported, mainly from Italy.

RESOURCES

▷ Electric power: None

 None   None

 None   None

The Vatican imports all its energy. It has no farmland; its area is restricted to buildings and their formal gardens.

ENVIRONMENT

▷ Not available

 None   Not available

The Vatican is increasingly concerned about the need to balance development and conservation. In 1993, the pope urged a gathering of scientists to press colleagues worldwide to inform people on the need to protect the environment.

MEDIA

▷ TV ownership medium

 Daily newspaper circulation figures not available

PUBLISHING AND BROADCAST MEDIA

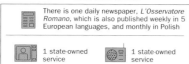
There is one daily newspaper, *L'Osservatore Romano*, which is also published weekly in 5 European languages, and monthly in Polish

1 state-owned service   1 state-owned service

Radio Vaticana's longwave broadcasts are continuing after a brief suspension caused by concern over radiation.

CRIME

▷ No death penalty

 There are no prisons in the Vatican City   Minimal crime levels

The reputation of the 105-strong Swiss Guard was shaken in 1998, when a young guard shot dead his commandant and the latter's wife and then committed suicide. Three Vatican Bank officials were earlier alleged to have been involved in the Banco Ambrosiano affair.

EDUCATION

▷ Not applicable

 99%   14,403 students

The university, founded by Gregory XIII, is renowned for its theological and philosophical learning. There are more than 110,000 primary and secondary Catholic schools around the world.

CHRONOLOGY

The Vatican is located in Rome because tradition held that St. Peter was buried on the site of the Church of Constantine, which was pulled down in the Renaissance to make way for the building of St. Peter's Basilica. The Vatican has been the pope's usual residence since 1417, when the pontiffs returned from Avignon in France at the end of the 39 years of Great Schism.

- ❏ **1870** Italy annexes Papal States.
- ❏ **1929** Lateran Treaty – Fascist Italy accepts Vatican City as independent state.
- ❏ **1978** Cardinal Karol Wojtyla pope.
- ❏ **1981–1982** Attempts on pope's life.
- ❏ **1984** Catholicism disestablished as Italian state religion.
- ❏ **1985** Catholic Catechism revised for first time since 1566.
- ❏ **1994–1995** Opposition to abortion and contraception reiterated at UN conferences in Cairo and Beijing.
- ❏ **1998** Statement repenting Catholic passivity during Nazi Holocaust.
- ❏ **2000** Jubilee Year. Papal apology for Catholic violence and oppression over two millennia.
- ❏ **2001** John Paul II becomes first pope to enter a mosque.

HEALTH

▷ Welfare state health benefits

Pope's own doctor is in permanent residence at Vatican   Heart and cardiovascular diseases, cancers

Pope John Paul II's strong opposition to abortion and contraception has prompted criticism from around the world, and from within the Church.

SPENDING

▷ Not applicable

CONSUMPTION AND SPENDING

No data   No data

No data available

0   5   10   15   20   25
Defense, Health, Education spending as % of GDP

The wealth of the Vatican is primarily that of the Catholic Church. Its art treasures may not be sold. It is not known how much personal wealth its citizens have.

WORLD RANKING

The Pope and his Vatican staff enjoy one of the highest standards of living in the world

V

VENEZUELA

OFFICIAL NAME: Bolivarian Republic of Venezuela **CAPITAL:** Caracas
POPULATION: 24.6 million **CURRENCY:** Bolívar **OFFICIAL LANGUAGE:** Spanish

L OCATED IN THE NORTH of South America, with a long Caribbean coastline, Venezuela has a vast central plain drained by the Orinoco, while the Guiana Highlands dominate the southwest of the country. A Spanish colony until 1811, Venezuela was lauded as Latin America's most stable democracy until its recent political upheavals. Although the country has some of the largest known oil deposits outside the Middle East, much of Venezuela's population still lives in shanty-town squalor.

President Hugo Chávez changed the name of the country as part of his Bolivarian Revolution.

CLIMATE
▷ Tropical wet & dry/ equatorial

WEATHER CHART FOR CARACAS

■ *Average daily temperature*   Rainfall ■
°C/°F J F M A M J J A S O N D cm/in
40/104 — 40/16
30/86 — 30/12
20/68 — 20/8
10/50 — 10/4
0/32 — 0
-10/14
-20/-4

The hot Maracaibo coast is surprisingly dry; the Orinoco Llanos are alternately parched or flooded. Uplands are cold.

TRANSPORTATION
▷ Drive on right

Simón Bolívar International, Caracas
6.66m passengers

246 ships
665,300 grt

THE TRANSPORTATION NETWORK

32,308 km (20,075 miles)	2690 km (1671 miles)
336 km (209 miles)	7100 km (4412 miles)

Massive road building from the 1960s onward has benefited the oil and aluminum industries.The French-designed Caracas subway was completed in 1995. Work on the Centro–Occidental highway and other major roads continues, as does rail development.

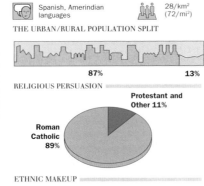

The Orinoco. *Its huge Llanos (plains) are grazed by five million cattle, which are herded close to the river in the dry season.*

TOURISM
▷ Visitors : Population 1:52

469,000 visitors

Down 20% in 2000

MAIN TOURIST ARRIVALS

USA 22%	
Germany 9%	
Netherlands 8%	
Italy 8%	
Argentina 5%	
Other 48%	

0   10   20   30   40   50   60
% of total arrivals

Tourism is still a relatively minor industry in Venezuela, but one with enormous potential. Venezuela has many beaches that are the equal of any Caribbean island's, and a fascinating jungle interior which is a target for more adventurous tourists. For many years, the high value of the bolívar made Venezuela an expensive destination, but, after recent devaluations, it has become one of the cheapest in the Caribbean. Privatizing state-run hotels was part of a drive to attract foreign investment.

PEOPLE
▷ Pop. density low

Spanish, Amerindian languages

28/km² (72/mi²)

THE URBAN/RURAL POPULATION SPLIT

87%   **13%**

RELIGIOUS PERSUASION

Protestant and Other 11%
Roman Catholic 89%

ETHNIC MAKEUP

Amerindian 2%   Black 9%
White 20%
Mestizo 69%

Venezuela is one of the most highly urbanized societies in Latin America, with most of its population living in cities, mainly in the north. A historic "melting pot," it has experienced large-scale immigration from Italy, Portugal, Spain, and all over Latin America. There remains little of the white Hispanic aristocracy that survives in Colombia and Ecuador. The small number of native Amerindians, such as the Yanomami and Pemón, live in remote regions now threatened by illegal settlers. Most of the black population, who are descended from Africans brought over to work in the cacao industry in the 19th century, live along the Caribbean coast.

Oil wealth has brought comparative prosperity, but life in the *barrios* (shanty towns) which sprawl over the hillsides around Caracas is one of extreme poverty. Discontent peaked in the food riots of 1989 and 1991, which left scores dead along with the country's reputation for being a model democracy. The oil boom accelerated change for women, who today are to be found in all the professions. Politics, however, remains a largely masculine preserve. Oil wealth has also brought a measure of Americanization – boxing and baseball are among the most popular sports.

POPULATION AGE BREAKDOWN

Female	Age	Male
0.3%	80+	0.2%
3%	60–79	2.7%
8.1%	40–59	8%
15.6%	20–39	15.9%
22.6%	0–19	23.6%

% of population by age group

V

Key to symbols and abbreviations on cover flaps

VENEZUELA

Total Land Area : 912 050 sq. km
(352 143 sq. miles)

POPULATION

- ⊡ over 1 000 000
- ⊙ over 500 000
- ◎ over 100 000
- ○ over 50 000
- • over 10 000

LAND HEIGHT

- 3000m/9843ft
- 2000m/6562ft
- 1000m/3281ft
- 500m/1640ft
- Sea Level
- - - - Projected Railway

POLITICS

▷ Multiparty elections

2000/2005

President Hugo
Chávez

AT THE LAST ELECTION

National Assembly 165 seats

4% MAS   3% COPEI

56% MVR   19% AD   4% PRVZL   14% Others

MVR = Fifth Republic Movement   **AD** = Democratic Action
PRVZL = Project Venezuela   **MAS** = Movement toward
Socialism   **COPEI** = Social Christian Party
The Patriotic Front (**PP**) comprises the MVR, the MAS,
and some smaller parties

Venezuela is a multiparty democracy.

PROFILE

Corruption, austerity, rising poverty, and
anti-price-rise riots in Caracas were
the backdrop for two coup attempts in
1992, one led by Hugo Chávez, whose
election as president in 1998 broke the
stranglehold on power of the traditional
parties. His increasingly autocratic
style has attracted growing opposition.

MAIN POLITICAL ISSUES
Political stability
President Chávez's fiery rhetoric and
populist stance won him the support of
the poor as he embarked on root and
branch reform, calling elections to a
new Constituent Assembly which
temporarily replaced the existing
Congress. A new "Bolivarian"
constitution, approved by referendum,
allowed fresh elections in 2000 which
gave Chávez a renewed mandate. His
command-style running of the country
increased political opposition, however,
and deterred foreign investment. A
military coup in April 2002 installed
business leader Pedro Carmona as
president for a day, but middle-ranking
officers changed tack following mass
demonstrations and the generally

unfavorable international response.
Restored to office, Chávez offered
a conciliatory response to his
opponents, recognizing the
danger of more upheavals.

Oil policy
Support of oil prices and quotas
within OPEC remain central, despite
Chávez's wish to use oil revenue
to build a broader economy.

WORLD AFFAIRS

▷ Joined UN
in 1945

ACS   AP   OAS   OPEC   RG

A traditionally pro-US orientation
was challenged by Chávez, with his
personal friendship with Cuba's Fidel
Castro, visits to Libya and Iraq, and
opposition to "Plan Colombia." The
Bush administration acknowledged
the new interim government during
the April 2002 military coup, but
denied helping to install it, although
the main coup participants had made
very public visits to the US State
Department in preceding months. The
OAS announced opposition to the coup,
as did all of Venezuela's neighbors.
Better regional economic integration
with the Caribbean and Central and
South America are important.

CHRONOLOGY
Venezuela was the first of the
Spanish imperial colonies to
repudiate Madrid's authority under
the guidance of the revolutionary,
Simón Bolívar, in 1811.

❏ **1821** Battle of Carabobo finally
overthrows Spanish rule and leads
to consolidation of independence
within Gran Colombia (Venezuela,
Colombia, and Ecuador).
❏ **1830** Gran Colombia collapses.
José Antonio Páez rules Venezuela;
coffee planters effectively in control.
❏ **1870** Guzmán Blanco in power.
Rail system constructed.
❏ **1908** Gen. Juan Vicente Gómez
dictator; oil industry developed.
❏ **1935** Gómez falls from power.
Increasing mass participation
in political process.
❏ **1945** Military coup. Rómulo
Betancourt of AD takes power as
leader of a civilian–military junta.
❏ **1948** AD wins elections, with
novelist Rómulo Gallegos as
presidential candidate. Military
coup. Marcos Pérez Jiménez forms
government, with US and
military backing. ⇨

V

CHRONOLOGY *continued*

❑ **1958** General strike. Adm. Larrázabal leads military coup. Free elections. Betancourt, newly returned from exile, wins presidential election for AD. Anticommunist campaign mounted. A few state welfare programs introduced.

❑ **1960** Movement of the Revolutionary Left (MIR) splits from AD, begins antigovernment activities.

❑ **1961** Founder member of OPEC.

❑ **1962** Communist-backed guerrilla warfare attempts repetition of Cuban revolution in Venezuela; fails to gain popular support.

❑ **1963** Raúl Leoni (AD) elected president – first democratic transference of power. Antiguerrilla campaign continues.

❑ **1966** Unsuccessful coup attempt by supporters of former president, Pérez Jiménez.

❑ **1969** Rafael Caldera Rodríguez of COPEI becomes president. Continues Leoni policies.

❑ **1973** Oil and steel industries nationalized. World oil crisis. Venezuelan currency peaks in value against the US dollar.

❑ **1978** Elections won by COPEI's Luis Herrera Campíns. Disastrous economic programs.

❑ **1983** AD election victory under Jaime Lusinchi. Fall in world oil prices leads to unrest and cuts in state welfare.

❑ **1988** Carlos Andrés Pérez of AD wins presidency.

❑ **1989** Caracas food riots; 1500 dead.

❑ **1993–1995** Andrés Pérez ousted on charges of corruption; Caldera Rodríguez reelected. More social unrest.

❑ **1998–2000** Hugo Chávez embarks on radical politcal reform after his Patriotic Front coalition defeats COPEI-led coalition in elections. New controversial Constituent Assembly approves new constitution, later endorsed by referendum. Chávez mandate confirmed by presidential elections. New unicameral National Assembly.

❑ **2002** April, Chávez ousted in military coup. Business leader Pedro Carmona president for a day. Chávez reinstalled a day later after foreign and domestic protests.

V

AID

 Recipient

 $77m (receipts)   Up 75% in 2000

The IDB is supporting the long-term reform of Venezuela's social security system.

DEFENSE

 Compulsory military service

 $1.38bn   ⬆ Up 4% in 2000

Chávez led officers opposed to austerity and corruption in a 1992 coup attempt. Military leaders claimed in April 2002 that civil unrest forced them to oust him, and they backed interim president Pedro Carmona, who dissolved the National Assembly and Supreme Court by decree. However, middle-ranking officers got cold feet and reinstalled Chávez after protests.

VENEZUELAN ARMED FORCES

🛡	81 main battle tanks (AMX-30)	34,000 personnel
🚢	2 submarines, 6 frigates, and 6 patrol boats	18,300 personnel
✈	125 combat aircraft (CF-5A/B, NF-5A/B, *Mirage* 50EV/DV, F-16A/B, EMB-312)	7000 personnel
	None	

ECONOMICS

 Inflation 46% p.a. (1990–2000)

 $104bn   💲 699.8–757.5 bolívares

SCORE CARD

❑ World GNP Ranking	36th
❑ GNP per Capita	$4310
❑ Balance of Payments	$13.2bn
❑ Inflation	16.2%
❑ Unemployment	14%

ECONOMIC PERFORMANCE INDICATOR

EXPORTS

Colombia 3% Brazil 4% Netherlands Antilles 12% Peru 2% USA 51% Other 28%

IMPORTS

Mexico 4% Brazil 5% Italy 4% Colombia 7% Other 44% USA 36%

STRENGTHS

Largest proven oil deposits outside the Middle East and the CIS. Massive reserves of coal, bauxite, iron, and gold, and successful development of new bitumen fuel which has attracted considerable foreign investment. New foreign investment, including telecommunications, banking, iron, and steel. Producer of high-grade aluminum. Labor market becoming more flexible.

WEAKNESSES

Huge, cumbersome state sector; despite some privatization, large areas of state sector still overmanned, inefficient, and subject to widespread corruption. Poor public services which, despite Venezuela's wealth during the oil-boom years, have been badly maintained. Drop in world oil prices in 2001–2002. Major infrastructure renewal now long overdue. Widespread tax evasion. Weak currency.

PROFILE

Venezuela is an economic paradox. While it has one of the strongest economies in Latin America, its government finances have habitually been in crisis due to a culture of nonaccountability and politically motivated patronage in state-owned industries and government bureaucracies. To date, privatizations and government cuts have failed to solve the problem. Promises by President Chávez to deal with excesses and diversify the economy, by lowering dependence on crude oil exports and promoting domestic processing industries, have received a mixed response from investors, who favor more market-oriented reforms.

VENEZUELA : MAJOR BUSINESSES

🛢	Oil
🍾	Rum
🍺	Brewing
⬛	Oil refining
⚙	Ceramics
🐂	Agribusiness
🍃	Tobacco
✿	Engineering
🚃	Metals
🚗	Vehicle assembly

0   200 km
0   200 miles

* significant multinational ownership

RESOURCES

▷ Electric power 21.3m kw

422,607 tonnes

3.42m b/d (reserves 77.7bn barrels)

16m cattle, 5.4m pigs, 4m goats, 115m chickens

Oil, bauxite, iron, natural gas, coal, gold, diamonds, aluminum

ELECTRICITY GENERATION

Hydro 72% (58bn kwh)	
Combustion 28% (23bn kwh)	
Nuclear 0%	
Other 0%	

0   20   40   60   80   100
% of total generation by type

ENVIRONMENT

▷ Sustainability rank: 48th

37%

6.7 tonnes per capita

ENVIRONMENTAL TREATIES

Yes	Yes	Yes
Yes	Yes	No

Flooding and mudflows, exacerbated by overdevelopment of the coastal strip, caused thousands of deaths in late 1999.

MEDIA

▷ TV ownership medium

Daily newspaper circulation 206 per 1000 people

PUBLISHING AND BROADCAST MEDIA

There are 86 daily newspapers. *El Universal* and *El Nacional* are the most prominent

8 services: 2 state-owned, 6 independent

1 state-owned service, 500 independent stations

Private TV stations and newspapers, mainly anti-Chávez, played a major role in the April 2002 coup by working with the military.

CRIME

▷ No death penalty

23,147 prisoners

Little change 1992–1996

CRIME RATES

Murders	
23	per 100,000 population

Rapes	
16	per 100,000 population

Thefts	
173	per 100,000 population

Urban robberies and violence involving young delinquents are major problems, as is narcotics-related crime. Cattle smuggling to Colombia is rife.

Venezuela has a remarkable diversity of resources. It has the world's sixth-largest proven oil reserves, vast quantities of coal, iron ore, bauxite, and gold, and cheap hydroelectric power. Huge investment programs are currently under way to raise production in all these sectors as well as in oil-refining capacity. However, the Chávez government wants to cut the investment budget of the state oil company, PDVSA, reduce its output, and increase its contributions to the exchequer. Such uncertainty has deterred private investors. Venezuela has begun exploitation of Orimulsion, a new bitumen-based fuel from the Orinoco; commercially exploitable reserves are estimated at 270 billion barrels. Venezuela's aim to be the world's largest aluminum producer is threatened after difficulties associated with privatizing the sector.

EDUCATION

▷ School leaving age: 14

93%

550,783 students

THE EDUCATION SYSTEM

% of each age group in education

Primary 100%
Secondary 59%
Tertiary 29%

An extra $1 billion in social spending approved in 2000 includes raising entitlement in the state sector. Education is characterized by teacher shortages and a high drop-out rate; the quality of education at state universities is low. The private sector is growing.

HEALTH

▷ Welfare state health benefits

1 per 417 people

Heart disease, accidents, violence, cancers

The health service suffered along with other public services from poor management in the 1970s and severe cuts in the 1980s and 1990s. Most health care is concentrated in the towns, and people from indigenous communities often have to travel long distances to receive treatment. Medicines, which have to be paid for, are expensive, and preventable diseases are recurring. Hospitals need modernization.

An additional $1 billion in social spending approved in 2000 includes spending on health.

VENEZUELA : LAND USE

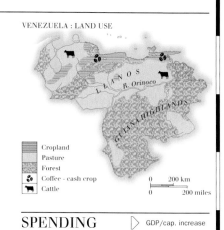

LLANOS
R. Orinoco
GUIANA HIGHLANDS

Cropland
Pasture
Forest
Coffee – cash crop
Cattle

0   200 km
0   200 miles

SPENDING

▷ GDP/cap. increase

CONSUMPTION AND SPENDING

68 per 1000 population

108 per 1000 population

Defense 1.5%	
Education 5.2%	
Health 2.6%	

0   5   10   15   20   25
Defense, Health, Education spending as % of GDP

The oil boom years of the 1970s largely benefited those already rich, with middle-income consumers doing well out of government-sponsored improvements in health and education and subsidized goods, largely at the expense of the poor.

The collapse of world oil prices, economic austerity measures, high inflation, and the devaluation of the bolívar in the 1980s and 1990s has squeezed the middle class and in addition seriously eroded the living standards of working-class households. In 2001, 20.7% of people were living in extreme poverty, according to official figures. Two million were unemployed and 59% of those working were in the informal sector.

WORLD RANKING

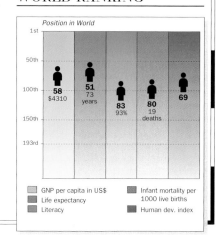

Position in World

1st
50th
100th
150th
193rd

58 $4310
51 73 years
83 93%
80 19 deaths
69

GNP per capita in US$
Life expectancy
Literacy

Infant mortality per 1000 live births
Human dev. index

V

VIETNAM

SOUTHEAST ASIA

Asia

OFFICIAL NAME: Socialist Republic of Vietnam **CAPITAL:** Hanoi
POPULATION: 79.2 million **CURRENCY:** Dông **OFFICIAL LANGUAGE:** Vietnamese

VIETNAM LIES ON the eastern coast of the Indochinese peninsula. Over half the country is dominated by the heavily forested mountain range, the Chaîne Annamitique. The most populated areas, which are also the most intensively cultivated, are along the Red and Mekong Rivers. Partitioned after World War II, Vietnam was not reunited until 1976, a few months after the communist north finally defeated the southern regime and its US allies in the Vietnam War. Vietnam is now a single-party state ruled by the Communist Party. Since 1986, the regime has pursued a liberal economic policy known as *doi moi* ("renovation").

CLIMATE

▷ Tropical monsoon

WEATHER CHART FOR HANOI

Vietnam's geography means that there are sharp local contrasts in the climate. The north has cool winters, while the south is tropical, with even temperatures all year round. The central provinces are affected by typhoons. The northern Red River delta is subject to drought, while the Mekong delta in the south suffers heavy flooding.

TRANSPORTATION

▷ Drive on right

Tan Son Naht International, Ho Chi Minh City

629 ships
784,000 grt

THE TRANSPORTATION NETWORK

23,418 km (14,551 miles)	430 km (267 miles)
2632 km (1636 miles)	17,702 km (10,999 miles)

Rebuilding infrastructure is still a priority. The flagship project, the four-lane Ho Chi Minh Highway linking Hanoi and the south, is unlikely to meet the target of completion by 2003. A major port development plan is under way. Trains travel slowly, with an average speed of around 15 km/h (9 mph), and Hanoi to Ho Chi Minh City takes three days. The bus network is extensive but journeys are also time-consuming. Taxis and cycles provide cheap local transportation. Hanoi has plans for an elevated metro line.

TOURISM

▷ Visitors : Population 1:57

1.38m visitors

Up 14% in 2000

MAIN TOURIST ARRIVALS

% of total arrivals

Until the government opened the way to large-scale tourism in the 1990s, Russians, eastern Europeans, and backpackers from the West made up the bulk of visitors. Other travelers were either on business, or overseas Vietnamese, *Viet Kie*, who were visiting relatives.

Under a "master plan" adopted in 1995, massive investment was channeled into hotels, with an official target of three million tourists a year by 2000. However, arrivals failed to reach even half that figure. Poor transportation infrastructure remains a problem. Vietnam's appeal lies in its unspoiled Asian way of life and in the areas of spectacular natural beauty such as Ha Long Bay on the Red River delta.

Boats moored near Nha Trang. *With 3444 km (2140 miles) of coastline, use of the sea, for transportation and fishing, is vital to Vietnam.*

PEOPLE

▷ Pop. density high

Vietnamese, Chinese, Thai, Khmer, Muong, Nung, Miao, Yao, Jarai

243/km² (630/mi²)

THE URBAN/RURAL POPULATION SPLIT

20%   80%

RELIGIOUS PERSUASION

Christian (mainly Roman Catholic) 7%
Buddhist 55%
Other and nonreligious 38%

ETHNIC MAKEUP

Thai 2%   Chinese 4%
Other 6%
Vietnamese 88%

Family life is strong and is based on kinship groups within village clans. A pronounced north–south cultural split remains evident in the cities. Chinese are the largest minority group. When the victorious communists reunited north and south Vietnam in 1976, they viewed the Saigon Chinese (in what was renamed Ho Chi Minh City), with their Taiwanese links, as a corrupt bourgeoisie. The northern Mountain Chinese were also suspect as a possible fifth column for China's ambitions in Vietnam. Various other mountain minorities (*montagnards*), with a history of collaboration with the French and Americans, were also sidelined by the regime in Hanoi. *Montagnard* resentment over the resettling of lowlanders in mountain regions sparked violent protests in early 2001.

War deaths cause older generations of women to outnumber men. Women form a high proportion of the work-force, and are starting to gain greater political prominence, most notably Vice President Nguyen Thi Binh. Female conscription was reinstated in 2001.

POPULATION AGE BREAKDOWN

Female		Age		Male
	0.5%	80+	0.3%	
	3.9%	60–79	3%	
	6.7%	40–59	5.5%	
	16.2%	20–39	13.5%	
24.6%		0–19		25.8%

% of population by age group

Tran Duc Luong,
*elected president
in 1997.*

Nong Duc Manh,
*powerful general
secretary of the CPV.*

POLITICS

 No multiparty elections

2002/2007

President
Tran Duc Luong

AT THE LAST ELECTION

National Assembly 500 seats

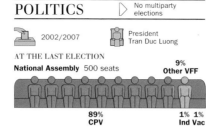

89% CPV	1% Ind	1% Vac	9% Other VFF

The sole permitted political grouping is the Vietnamese Fatherland Front (**VFF**), which is dominated by the Communist Party of Vietnam (**CPV**)
Ind = Independents **Vac** = Vacant

Vietnam is effectively a single-party communist state.

PROFILE

A traditional communist system is still in place, with a powerful politburo elected by the party central committee. Heading the secretariat and politburo, the party general secretary wields much power, alongside the prime minister and president. Changes in senior posts in 1997 left reformers such as new prime minister Phan Van Khai still outnumbered by conservatives. The 2001 party congress balanced the theme of greater democracy with a renewed commitment to socialism.

MAIN POLITICAL ISSUE
Economic reform

Vietnam is trying to move to a market economy without political liberalization. The veteran CPV leaders of the 1950s–1970s resisted democratizing the political process and were slow to transfer power. The hand of the economic reformers was strengthened by Nong Duc Manh's elevation to the party leadership at the 2001 congress, but there remains real concern that opening collective farming and state enterprises to competition will undermine stability, encourage "individualism," and weaken the party's monopoly of power.

VIETNAM

Total Land Area : 329 560 sq. km
(127 243 sq. miles)

POPULATION

▣	over 1 000 000
◉	over 500 000
◎	over 100 000
○	over 50 000
●	over 10 000
•	under 10 000

LAND HEIGHT

2000m/6562ft
1000m/3281ft
500m/1640ft
200m/656ft
Sea Level

0 — 100 km
0 — 100 miles

WORLD AFFAIRS

 Joined UN in 1977

 ASEAN IAEA OIF Mekong River NAM

Economic liberalization has improved relations with the US, with lifting of the aid and trade embargo in 1993, full diplomatic relations in 1995, and a landmark bilateral trade agreement in 2000, ahead of a visit by US president Bill Clinton that November.

Vietnam joined ASEAN in 1995, in the wake of the settlement of the Cambodia issue. Trade and economic cooperation links with Japan have been strengthened. Tension with China was reduced by an agreement in 1999 over their mutual land border, although competing claims to the Spratly Islands remain a source of friction.

AID

 Recipient

 $1.7bn (receipts)   Up 19% in 2000

Vietnam's invasion of Cambodia in 1978 halted all aid from China, Japan, and the West (except for Scandinavian countries), leaving it mostly dependent on the USSR. Western donors resumed assistance in the early 1990s. Their aid rapidly became the main source of capital for improving infrastructure, although foreign investment fell away significantly in the late 1990s.

CHRONOLOGY

From 1825, the brutal persecution of the Catholic community, originally converted by French priests in the 17th century, gave France the excuse to colonize Cochin-China, Annam, and Tonkin, and then merge them with Laos and Cambodia.

❑ **1920** *Quoc ngu* (Roman script) replaces Chinese script.
❑ **1930** Ho Chi Minh founds Indo-China Communist Party.
❑ **1940** Japanese invasion.
❑ **1941** Viet Minh resistance founded in exile in China.
❑ **1945** Viet Minh take Saigon and Hanoi. Emperor abdicates. Republic proclaimed with Ho Chi Minh as president.
❑ **1946** French reenter. First Indochina war.
❑ **1954** French defeated at Dien Bien Phu. Vietnam divided at 17ºN. USSR supports North; US arms South.
❑ **1960** Groups opposed to southern regime unite as Viet Cong.
❑ **1961** US pours in military advisers.
❑ **1964** US Congress approves war.
❑ **1965** Gen. Nguyen Van Thieu takes over military government of South. First US combat troops arrive.
❑ **1965–1968** Operation Rolling

V

CHRONOLOGY *continued*

Thunder – intense bombing of North by South and US.

- **1967** Antiwar protests start in US and elsewhere.
- **1968** Tet (New Year) Offensive – 105 towns attacked simultaneously in South with infiltrated arms. Viet Cong suffer serious losses. Peace talks begin. US eases bombing and starts withdrawing troops.
- **1969** Ho Chi Minh dies. Succeeded by Le Duan. War intensifies in spite of talks.
- **1970** US begins secret attacks in Laos and Cambodia and new mass bombing of North to try to stop arms reaching Viet Cong.
- **1972** 11-day Christmas Campaign is heaviest US bombing of war.
- **1973** Paris Peace Agreements signed, but fighting continues.
- **1975** Fall of Saigon to combined forces of North and Provisional Revolutionary (Viet Cong) Government of South. One million flee after end of war.
- **1976** Vietnam united as Socialist Republic of Vietnam. Saigon renamed Ho Chi Minh City.
- **1978** Invasion of Cambodia to oust Pol Pot regime (by January 1979).
- **1979** Nine-Day War with China. Chinese troops pushed back after destroying everything for 40 km (25 miles) inside Vietnam. "Boat people" crisis. At UN conference, Vietnam agrees to allow legal emigration, but exodus continues.
- **1986** Death of Le Duan. Nguyen Van Linh, new Communist Party general secretary, initiates liberal economic policy of *doi moi* (renovation).
- **1987** Fighting in Thailand as Vietnam pursues Kampuchean resistance fighters across border.
- **1989** Troops leave Cambodia.
- **1991** Open anticommunist dissent made a criminal offense.
- **1992** Revised constitution allows foreign investment, but essential role of Communist Party is unchanged.
- **1995** US–Vietnamese relations normalized. Vietnam joins ASEAN.
- **1997** Legislative elections. Tran Duc Luong president, Phan Van Khai prime minister.
- **1998** Asian financial crisis dampens economic boom.
- **1999** Signing of border treaty with China.
- **2000** Worst flooding along Mekong for 40 years. November, visit by US president Bill Clinton.
- **2001** March, visit by Russian president Vladimir Putin. April, ninth party congress. Nong Duc Manh becomes general secretary.

DEFENSE

 Compulsory military service

 $931m | Up 5% in 2000

Vietnam has large and well-equipped armed forces, notably the world's seventh-largest army. Military service is compulsory, with conscripts serving a two-year term. The army's role in preserving both stability and socialism was reaffirmed in 2001. Increased defense spending on the navy reflects the rising tensions in the South China Sea, where there are disputed claims to the Spratly and Paracel Islands.

VIETNAMESE ARMED FORCES

1315 main battle tanks (T-34/54/55/62, PRC Type-59)	412,000 personnel
2 submarines, 6 frigates, 42 patrol boats	42,000 personnel
189 combat aircraft (53 Su-22, 12 Su-27, 124 MiG-21bis/PF)	30,000 personnel
None	

ECONOMICS

 Inflation 15% p.a. (1990–2000)

$30.4bn | 14,514–15,083 dong

SCORE CARD

- WORLD GNP RANKING.........................58th
- GNP PER CAPITA $390
- BALANCE OF PAYMENTS $507m
- INFLATION ....–0.6%
- UNEMPLOYMENT.................................25%

ECONOMIC PERFORMANCE INDICATOR

EXPORTS
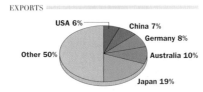
USA 6%, China 7%, Germany 8%, Australia 10%, Japan 19%, Other 50%

IMPORTS

China 11%, South Korea 12%, Taiwan 12%, Japan 14%, Singapore 15%, Other 36%

STRENGTHS

Diverse resource base; unexploited gas reserves. Young, literate, low-cost labor force. Strong light industrial and handicraft export industries.

WEAKNESSES

Weak economic institutions. Weight of bureaucracy. Heavy dependence on aid for reconstruction. Enduring suspicion of entrepreneurial southern attitudes and "individualism." Corruption.

PROFILE

The encouragement of private enterprise began in 1988. Touted by some commentators in the mid-1990s as the next Asian "tiger," Vietnam has aimed at more moderate growth since the Asian crisis of 1997–1998. Annual GDP increases, over 5% again by 2000 and predicted to reach 7% by 2003, are impressive. Inflation, once a huge problem, was held down in the 1990s, and is now under firm control. Increased rice production has boosted incomes, and domestic demand helped the post-1998 upswing. The government promised massive investment in agriculture from 2002.

Attracting foreign investment through reform is essential; government policies regarding state-owned enterprises promise more scope for joint ventures, and new laws on trade-licensing and investment were passed in 2000. Also that year the country's first stock exchange opened. The government has set a target of doubling GDP in the next decade. The potential certainly exists, based on an educated and highly motivated young labor force and on mineral resources, located mostly in the north.

VIETNAM : MAJOR BUSINESSES

Steel, Cement, Textiles, Brewing, Garments, Shipbuilding, Consumer goods, Vehicle manufacture, Oil & gas

* significant multinational ownership

RESOURCES

▷ Electric power
5m kw

1.79m tonnes

20.2m pigs, 57m ducks, 4.2m cattle, 150m chickens

350,000 b/d (reserves 600m barrels)

Coal, oil, tin, zinc, iron, antimony, apatite, salt, bauxite

ELECTRICITY GENERATION

ELECTRICITY GENERATION	
Hydro 51% (11bn kwh)	
Combustion 46% (10bn kwh)	
Nuclear 0%	
Other 3% (0.6bn kwh)	

% of total generation by type

Vietnam is among the world's top ten coffee producers and is the third-largest exporter of rice – to the

ENVIRONMENT

▷ Sustainability rank: 94th

3%

0.6 tonnes per capita

ENVIRONMENTAL TREATIES

Yes	Yes	Yes
Yes	Yes	Yes

In the Vietnam War, seven million tonnes of bombs were dropped, and the defoliant chemical Agent Orange was sprayed over vast areas; a "census" of the continuing health impact was announced in 1999. Half of Vietnam's forests were seriously damaged and some 5% destroyed. Deforestation continued into the 1990s due to logging and expansion of coffee-growing, causing soil erosion and flooding. Floods along the Mekong in 2000 were the worst for 40 years.

MEDIA

▷ TV ownership medium

☒ Daily newspaper circulation 4 per 1000 people

PUBLISHING AND BROADCAST MEDIA

	There are 10 daily newspapers, including *Nhan Dan, Quan Doi Nhan Dan, Hanoi Moi*, and *Sai Gon Giai Phong*
	1 state-owned service with 53 provincial stations
	1 state-owned service with more than 5000 local stations

The media are tightly regulated. TV is the dominant medium. All editors have to be Party members, but criticism of the authorities is still possible. Even *Nhan Dan*, the CPV newspaper, has been known to expose laxity in the system, especially in the judiciary. However, in 2002 the authorities sought out and destroyed books by proscribed authors.

detriment of domestic stocks. Oil production, small by world standards, is sufficient to make it Vietnam's biggest export earner. The Oil and Gas Corporation of Vietnam (PetroVietnam) is involved in joint ventures with international oil firms. Vietnam has unexploited gas reserves in the South China Sea; gas from the only producing field has to be flared off. Timber exports have been banned since 1997 to preserve forests. Northern Vietnam has a surplus of electricity, mainly from hydroelectric schemes.

CRIME

▷ Death penalty in use

▣ 55,000 prisoners

⬇ Down 15% 1997–1999

CRIME RATES

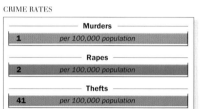

Murders	
1	per 100,000 population

Rapes	
2	per 100,000 population

Thefts	
41	per 100,000 population

The judicial system is based on the Soviet model. The education camps established after liberation have now closed, but religious and political dissidents are still held without trial.

Corruption has risen sharply since economic liberalization, as has the illegal drift of young people to urban areas, where they are blamed for increasing petty crime and "social evils" such as begging, prostitution, and drug-taking. Theft from foreigners is a problem in major cities.

EDUCATION

▷ School leaving age: 15

93%

509,300 students

THE EDUCATION SYSTEM

% of each age group in education

- Primary 100%
- Secondary 61%
- Tertiary 11%

Private sponsorship helps fund education. Vietnamese universities have a strong liberal arts tradition. Social pressure to obtain a degree leads to high levels of cheating among university applicants.

VIETNAM : LAND USE

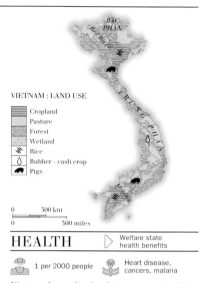

	Cropland
	Pasture
	Forest
	Wetland
	Rice
	Rubber – cash crop
	Pigs

0   300 km
0   300 miles

HEALTH

▷ Welfare state health benefits

1 per 2000 people

Heart disease, cancers, malaria

Vietnam's medical achievements include developing a vaccine for hepatitis B, and extracting an antimalarial drug, artemisinin, from the indigenous thanh hao tree. An extensive campaign is under way to combat the spread of AIDS.

SPENDING

▷ GDP/cap. increase

CONSUMPTION AND SPENDING

No data

32 per 1000 population

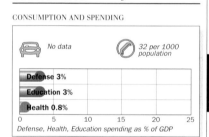

Defense 3%	
Education 3%	
Health 0.8%	

Defense, Health, Education spending as % of GDP

Ostentatious consumerism is rising despite official disapproval, but is beyond most people's reach. Wealth disparities are growing, with rural areas falling deep into poverty.

WORLD RANKING

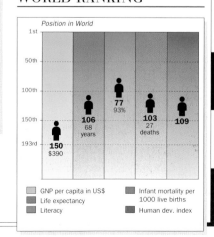

Position in World

- 150 — $390
- 106 — 68 years
- 77 — 93%
- 103 — 27 deaths
- 109

GNP per capita in US$	Infant mortality per 1000 live births
Life expectancy	
Literacy	Human dev. index

V

YEMEN

OFFICIAL NAME: Republic of Yemen **CAPITAL:** Sana
POPULATION: 19.1 million **CURRENCY:** Yemeni rial **OFFICIAL LANGUAGE:** Arabic

YEMEN IS LOCATED in southern Arabia. The west is mountainous, with a fertile strip along the Red Sea. The center and south are largely arid mountains and desert. Until 1990 Yemen was two countries, the Yemen Arab Republic (YAR) in the west and the People's Democratic Republic of Yemen (PDRY) in the south. The YAR was run by successive military regimes; the poorer PDRY was the Arab world's only Marxist state. Post-unification conflict between the two ruling hierarchies, nominally in coalition, led to a two-month civil war in 1994, the ousting of the former Marxists, and a new constitution.

CLIMATE ▷ Hot desert/mountain

WEATHER CHART FOR SANA

The desert climate is modified by altitude, which affects temperatures by as much as 12°C (54°F). Rainfall increases in northwest and central Yemen.

TRANSPORTATION ▷ Drive on right

Sana International
858,843 passengers

45 ships
25,300 grt

THE TRANSPORTATION NETWORK

| 7700 km (4785 miles) | None |
| None | None |

Aden's history as a port stretches back 3000 years. Adequate roads link the main cities, but many rural areas are inaccessible. International airlines, including the modern fleet of Yemenia, serve Sana and Aden.

Hilltop village in northern Yemen, showing traditionally decorated, multistorey houses built from unbaked mud bricks.

TOURISM ▷ Visitors : Population 1:262

73,000 visitors | Down 13% 1999–2000

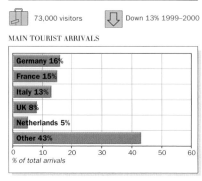

MAIN TOURIST ARRIVALS

Germany	16%
France	15%
Italy	13%
UK	8%
Netherlands	5%
Other	43%

% of total arrivals

Believed to be the home of the legendary Queen of Sheba, Yemen attracts tourists interested in Arab society, architecture, archeology, and historical remains. The Romans called Yemen *Arabia Felix* because of its fertile farmlands and dominance in the frankincense trade. Yemen was the second country, after Saudi Arabia, to convert to Islam.

Southern Yemen has been open to Western visitors only since 1990. Its run-down infrastructure and lack of hotels, especially on the coast, have hindered tourism. Sana, a walled medieval city, is the more interesting center for tourists. It has impressive architecture, particularly tall stone and mud-brick Arab houses, and the palaces of the former imamate. Despite being over 100 km (62 miles) from the capital, the Marib Dam, built in ancient times, is another major attraction.

German and French tourists were the first to travel to northwest Yemen during the 1980s. Hopes of a major rise in tourism following the end of the 1994 civil war were dashed in 1998 after tribesmen kidnapped and killed four tourists.

Tourists are subject to a ban on the consumption of alcohol, except in five-star hotels. Whisky and beer are available on the black market, which operates out of Djibouti.

PEOPLE ▷ Pop. density low

Arabic | 34/km² (88/mi²)

THE URBAN/RURAL POPULATION SPLIT

25% | 75%

RELIGIOUS PERSUASION

- Sunni Muslim 55%
- Shi'a Muslim 42%
- Christian, Hindu, and Jewish 3%

ETHNIC MAKEUP

- Arab 95%
- Indian, Somali, European 2%
- Afro-Arab 3%

Yemenis are almost entirely of Arab and Bedouin descent, though there is a small, dwindling, Jewish minority and people of mixed African and Arab descent along the south coast. The majority are Sunni Muslims, of the Shafi sect. However, Zaydi Shi'a are strong in the north, where many people have close family in Saudi Arabia. Many Yemenis consider Saudi Arabia's Asir province to be part of Yemen.

Agriculture supports more than half the population. Many Yemenis went to work in Saudi Arabia and the Gulf states during the 1970s oil boom. More than a million worked in Saudi Arabia, and their expulsion, due to Yemen's support for Iraq's invasion of Kuwait in 1990, has increased domestic unemployment.

In rural areas and in the western highlands, Islamic orthodoxy is strong and most women wear the veil. In the south, however, women still claim the freedoms they had under the Marxist regime, especially in urban areas.

Tension continues to exist between the south, led by cosmopolitan Aden, and the more conservative west, leading in 1994 to civil war.

POPULATION AGE BREAKDOWN

Female	Age	Male
0.7%	80+	0.7%
1.8%	60–79	1.7%
5.7%	40–59	5.3%
12.9%	20–39	11.9%
28.9%	0–19	30.4%

% of population by age group

Y

YEMEN

Total Land Area : 527 970 sq. km
(203 849 sq. miles)

POPULATION

over 500 000	◉
over 100 000	◎
over 10 000	●
under 10 000	·

LAND HEIGHT

3000m/9843ft
2000m/6562ft
1000m/3281ft
500m/1640ft
200m/656ft
Sea Level

POLITICS

▷ Multiparty elections

1997/2003

President Ali
Abdullah Saleh

AT THE LAST ELECTION

House of Representatives 301 seats

1%
Ba'ath

62% GPC 18% Ind 18% YAR 1% NUPO

GPC = General People's Congress **Ind** = Independents
YAR = Yemeni Alliance for Reform (al-Islah)
NUPO = Nasserite Unionist Popular Organization
Ba'ath = Arab Socialist Ba'ath Party

Yemen is a multiparty democracy. The
president retains executive power.

PROFILE

The merger of the YAR and the PDRY in
1990 united Yemenis under one ruler for
the first time since 1735; free elections
were held in 1993. President Ali Saleh
initially maintained unity. Then, in 1994,
a bloody civil war erupted, fueling a
secessionist movement in the south. By
mid-1994, the southerners, led by ex-
vice president Ali Salem al-Baidh, were
crushed. In elections held in 1997,
Saleh's GPC won an absolute majority,
in part due to a boycott by the former
southern-ruling Yemen Socialist Party,
and in 1999 he won the region's first
direct election of a head of state. Saleh's
regime still faces threats from tribesmen
angry at levels of poverty in their oil-rich
country. Since 1992, tribesmen have
kidnapped more than 100 foreigners,
including diplomats and tourists.

MAIN POLITICAL ISSUES
Relations with Saudi Arabia
Relations have long been strained –
over oil exploration rights, Yemeni
claims on Asir, and accusations that
Riyadh funds insurgent tribesmen.
The two sides clashed violently in 1998
over 1600 km (1000 miles) of disputed
border, despite a 1995 memorandum
of understanding. In early 2001, both
sides withdrew border troops under
a pact reached the previous June.

Instability
Since the end of the civil war, stability
has been threatened by the border
dispute with Saudi Arabia, growing
tribal insurgency, and mounting
popular discontent with President
Saleh's government. The Islamist
al-Islah party, formerly a junior
partner of the GPC, now forms
the chief opposition.

Ali Abdullah Saleh,
former YAR president,
now leader of the
unified Yemen.

Ali Salem al-Baidh,
former vice president,
ousted when south's
resecession failed.

CHRONOLOGY

From the 9th century, the Zaydi
dynasty ruled Yemen until their
defeat by the Ottoman Turks in
1517. The Turks were expelled
by the Zaydi imams in 1636.

❑ **1839** Britain occupies Aden.
❑ **1918** Western Yemen independent.
❑ **1957** Aden made a crown colony,
hinterland a protectorate.
❑ **1962** Army coup in west. Imam
deposed, Yemen Arab Republic
(YAR) declared. Civil war.
❑ **1963** Aden and protectorate united
to form Federation of South Arabia.
❑ **1967** South Arabia independent as
People's Republic of South Yemen.
British troops leave Aden.
❑ **1970** South Yemen renamed
People's Democratic Republic of
Yemen (PDRY). Republican victory
in YAR civil war.
❑ **1972** War between YAR and PDRY
ends in peace settlement.
❑ **1974** Army coup in YAR.
❑ **1978** Lt. Col. Ali Saleh YAR
president. Coup in PDRY. Radical
Abdalfattah Ismail in power.
❑ **1979** PDRY signs 20-year treaty
with USSR.
❑ **1980** Ismail replaced by moderate
Ali Muhammed.
❑ **1982** PDRY peace treaty with Oman.
Major earthquake kills 3000.
❑ **1984** YAR signs 20-year
cooperation treaty with USSR. ⇨

Y

CHRONOLOGY *continued*

- ❏ **1986** Coup attempt in PDRY leads to civil war. Rebels take control of Aden. New PDRY president meets YAR counterpart.
- ❏ **1987** Oil production starts in YAR.
- ❏ **1988** YAR holds elections for consultative council; Muslim Brotherhood gains influence.
- ❏ **1989** Speeding-up of unification process. PDRY publishes a program of free-market reforms. YAR and PDRY sign unification agreement. Constitution of unified Yemen published.
- ❏ **1990** Restrictions on travel between YAR and PDRY lifted. Ali Saleh becomes president of Republic of Yemen. May, formal unification. Pro-Islamic groups oppose secular constitution.
- ❏ **1991** Yemeni guest workers expelled by Saudi Arabia in retaliation for Yemen's position over Iraqi invasion of Kuwait. Arab states boycott independence celebrations.
- ❏ **1992** Assassinations, food riots, and political unrest delay elections.
- ❏ **1994** Southern secessionists defeated in civil war. Amended constitution adopted.
- ❏ **1997** Saleh's GPC wins absolute majority in general election.
- ❏ **1998–1999** Violent border dispute with Saudi Arabia. Kidnapping of tourists, four killed; three members of Islamic Army of Aden (IAA) sentenced to death.
- ❏ **1999** Saleh reelected.
- ❏ **2000** Yemen agrees border with Saudi Arabia after 66-year dispute. October, terror attacks on US naval vessel and UK embassy.
- ❏ **2001** Referendum approves extension of presidential term to seven years.
- ❏ **2002** Government targets suspected al-Qaida allies in tribal areas, expels 100 foreign "scholars."

WORLD AFFAIRS

▷ Joined UN in 1947/1967

| AL | AMF | IBRD | NAM | OIC |

Yemen is internationally regarded as a base for Islamic terrorists and was shunned for its support of Iraq in the Gulf War. The government has tried to break this isolation with active support for the US-led "war on terrorism."

AID

▷ Recipient

 $265m (receipts)   Down 42% in 2000

Since the receipt of $700 million for economic reforms in 1996, aid to Yemen has decreased. The World Bank is the largest donor.

DEFENSE

▷ Compulsory military service

 $489m   Up 14% in 2000

Following unification in 1990, mutual suspicion hampered the integration of the two separate defense forces. Sporadic, bitter clashes have taken place.

The main domestic security concern is insurgent tribesmen and, internationally, anti-Western terrorist activity such as the sinking of USS *Cole* off Aden in late 2000. US military aid has increased accordingly.

YEMENI ARMED FORCES

910 main battle tanks (50 T-34, 500 T-54/55, 250 T-62, 50 M-60A1)	49,000 personnel
9 patrol boats	1500 personnel
111 combat aircraft (10 F-5E, 30 Su-20/22, 20 MiG-21, 5 MiG-29)	3500 personnel
None	

ECONOMICS

▷ Inflation 22% p.a. (1990–2000)

 $6.55bn   164.39–170.70 Yemeni rials

SCORE CARD

❏ WORLD GNP RANKING	102nd
❏ GNP PER CAPITA	$370
❏ BALANCE OF PAYMENTS	$2.06bn
❏ INFLATION	7.9%
❏ UNEMPLOYMENT	30%

ECONOMIC PERFORMANCE INDICATOR

EXPORTS

USA 6% Other 26% India 15% Thailand 17% China 19% South Korea 17%

IMPORTS

Switzerland 5% Kuwait 5% France 5% UAE 10% Other 61% Saudi Arabia 14%

STRENGTHS

Rising oil production. Salt mining. Deposits of copper, gold, lead, zinc, and molybdenum. Industries include oil refining, chemicals, foodstuffs, cement, leather. Improving private sector.

WEAKNESSES

Political instability deters foreign investment. Heavy economic damage caused by civil war. Well-organized black market undermines tax base. Large balance-of-payments deficit. Dependence on subsistence agriculture. Burden of high population growth, leading to unemployment.

PROFILE

Unification in 1990 aimed to transform the economy, particularly through the exploitation of large oil and gas reserves, discovered in 1984; exports of oil began in 1987. Industrial investment around Aden was planned. These policies were severely affected by the 1990–1991 Gulf War. In addition, the expulsion of over one million Yemeni guest workers from

Saudi Arabia imposed a huge burden on the economy and ended the flow of workers' remittances.

The 1994 civil war inflicted further serious damage on the economy – oil refineries, communications centers, and water systems were destroyed. Economic crisis forced the government to reduce expenditure and subsidies on certain staple foods. This provoked widespread civil unrest (there were particularly violent demonstrations in 1998) and encouraged many farmers to switch from food crops, such as wheat, to growing the more profitable narcotic plant *qat*. As a result, Yemen has increasingly had to import food supplies. Lack of central control, poor integration, and the persistence of patronage politics are hampering economic revival.

YEMEN : MAJOR BUSINESSES

Textiles · Cement · Salt mining · Oil refining · Food processing · Light engineering

* significant multinational ownership

RESOURCES ▷ Electric power 810,000 kw

123,252 tonnes

458,000 b/d (reserves 4bn barrels)

4.8m sheep, 4.25m goats, 1.34m cattle, 29.6m chickens

Oil, natural gas, salt, copper, gold, lead, zinc, molybdenum

ELECTRICITY GENERATION

Hydro 0%
Combustion 100% (2.5bn kwh)
Nuclear 0%
Other 0%

0   20   40   60   80   100

% of total generation by type

Oil reserves are considerable, though initial estimates were exaggerated. The 2000 border agreement with Saudi Arabia promises better Yemeni access to oil fields. Salt is the only other mineral to be commercially exploited, and its production continues to grow steadily.

The agricultural sector employs well over half the working population. Cotton is a cash crop. Livestock and livestock products, including dairy products and hides, are mainstays of the north. Yemen's rich fishing grounds in the Arabian Sea now provide a major source of earnings, despite poor equipment. Ambitious plans for an Aden free port, first revealed in 1994, have yet to be realized.

Yemen's population growth, averaging around 3.5 per cent a year and among the highest in the world, is putting severe strain on the country's natural resources, especially water.

YEMEN : LAND USE

RUB 'AL KHALI

0   100 km
0   100 miles

Cropland
Pasture
Desert
Cotton
Grapes
Sheep

ENVIRONMENT ▷ Not available

None

0.9 tonnes per capita

ENVIRONMENTAL TREATIES

No	Yes	Yes
Yes	Yes	No

Large areas remain untouched by development, preserving habitats for rare birds. Problems include water scarcity, overgrazing, and soil erosion.

MEDIA ▷ TV ownership low

Daily newspaper circulation 15 per 1000 people

PUBLISHING AND BROADCAST MEDIA

There are 4 daily newspapers, including *Ath-Thawra, Ar-Rabi' 'Ashar Min Uktubar,* and *Al-Jumhuriyah*

1 state-controlled service

1 state-controlled service

CRIME ▷ Death penalty in use

14,000 prisoners

Up 6% 1997–1999

CRIME RATES

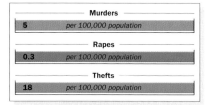

Murders
5   per 100,000 population

Rapes
0.3   per 100,000 population

Thefts
18   per 100,000 population

Political assassinations continue to threaten political stability. There is little formal law enforcement outside the main cities; foreign companies risk kidnappings and theft by Bedouin raiders. There is a proliferation of illicit weapons; the number of firearms has been estimated at 50 million – nearly three times the population size. Some blame lawlessness on the narcotic, *qat.*

EDUCATION ▷ School leaving age: 15

46%

65,675 students

THE EDUCATION SYSTEM

% of each age group in education

100
80
60
40
20
0

78%   Primary
45%   Secondary
10%   Tertiary

Some 80% of the population have had no formal classroom education. Schooling barely extends into the rural areas. Illiteracy is especially high among women: 75% cannot read or write. Only 13% of students at Yemen's two universities – Sana and Aden – are female. Yemen also has some technical colleges. The government's unpopular economic policies have encouraged student activism.

Yemen has a long, distinguished tradition of intellectual debate, and legislation embodies freedom of the press, which, however, is poorly developed. The government keeps tight control of the media and vets the entry of foreign journalists. TV and radio are state-controlled and have a limited range around the principal cities. Satellite TV is not generally available. Ownership of radio and TV receivers is low, with only a tiny minority owning a television set.

HEALTH ▷ No welfare state health benefits

1 per 5000 people

Diarrheal diseases, tuberculosis, malaria, bilharzia

The major cities have an adequate primary health care system. A new 300-bed hospital in Sana is due to be completed in 2004. Rural areas are less well served. Health services are under threat from tribal gangs. In 2002 the World Bank approved a $27 million credit for a Health Reform Support Project to increase health provisions in poor communities.

SPENDING ▷ GDP/cap. increase

CONSUMPTION AND SPENDING

15 per 1000 population

19 per 1000 population

Defense 7.8%
Education 6.7%
Health 2.4%

0   5   10   15   20   25

Defense, Health, Education spending as % of GDP

Most Yemenis suffered a fall in living standards after Saudi Arabia expelled its Yemeni workers. A lack of jobs in other Gulf states has fueled unemployment, estimated at around 30%. Except for a small elite, the ownership of consumer goods is low.

WORLD RANKING

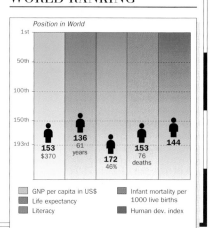

Position in World

1st
50th
100th
150th
193rd

153   136   172   153   144
$370   61 years   46%   76 deaths

GNP per capita in US$
Life expectancy
Literacy

Infant mortality per 1000 live births
Human dev. index

Y

ZAMBIA

SOUTHERN AFRICA

OFFICIAL NAME: Republic of Zambia   CAPITAL: Lusaka
POPULATION: 10.6 million   CURRENCY: Zambian kwacha   OFFICIAL LANGUAGE: English

  1964 1964 Oct 24 Z +2 +260 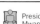 .zm

LYING IN THE HEART of southern Africa, Zambia is a country of upland plateaus, bordered to the south by the Zambezi River. Its economic fortunes are tied to the copper industry. Falling copper prices in the late 1970s, and then the growing inaccessibility of remaining reserves, have led to a severe decline in the economy. In 1991, Zambia achieved a peaceful transition from single-party rule to multiparty democracy.

CLIMATE ▷ Tropical wet & dry

WEATHER CHART FOR LUSAKA

Zambia has a tropical climate, with rains from November to April. The southwest is prone to drought.

TRANSPORTATION ▷ Drive on left

✈ Lusaka International
466,533 passengers   Has no fleet

THE TRANSPORTATION NETWORK

39,700 km (24,668 miles)		60 km (37 miles)
1273 km (791 miles)		2250 km (1398 miles)

The poor rail and road networks, in need of urgent rehabilitation, could sabotage economic recovery. Zambian Airways was liquidated in 1994, and private airlines are now in operation.

TOURISM ▷ Visitors : Population 1:18

 574,000 visitors   Up 26% in 2000

MAIN TOURIST ARRIVALS

Zimbabwe 30%	
South Africa 19%	
UK 9%	
Other 42%	

0   10   20   30   40   50   60
% of total arrivals

Wildlife, the Victoria Falls, and white-water rafting are major attractions. Recent increases in tourism have been at the expense of neighboring Zimbabwe.

PEOPLE ▷ Pop. density low

Bemba, Nyanja, Tonga, Kaonde, Lunda, Luvale, Lozi, English   14/km² (37/mi²)

THE URBAN/RURAL POPULATION SPLIT

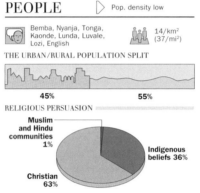

45%   55%

RELIGIOUS PERSUASION

Muslim and Hindu communities 1%
Christian 63%
Indigenous beliefs 36%

Although ethnically heterogeneous, with more than 70 different groups, Zambia has been less affected by ethnic tension than many African states. The largest group, about 34% of the population, is the Bemba, who live in the northeast and also predominate in the central Copperbelt. Other major groups are the southern Tonga, the eastern Nyanja, and the Lozi in the west. There are also some 225,000 refugees, mainly from Angola.

Zambia's main urban area is the Copperbelt, where many third- and fourth-generation town dwellers live. Some half a million children are employed there in hazardous conditions. The rural population lives mainly by subsistence farming.

A National Gender Policy was issued in October 2000 to redress inequalities between the sexes.

Musi-o-Tunya (*The Smoke That Thunders*), *known in English as Victoria Falls. Spray from the falls can be seen 30 km (20 miles) away.*

POLITICS ▷ Multiparty elections

2001/2006   President Levy Mwanawasa

AT THE LAST ELECTION
National Assembly 158 seats

8% FDD   1% Vacant

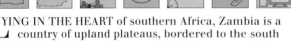

43% MMD   30% UPND   8% UNIP   5% App   5% Others

MMD = Movement for Multiparty Democracy   UPND = United Party for National Development   UNIP = United National Independence Party   FDD = Forum for Democracy and Development   App = Appointed

Up to eight members are appointed by the president, and the speaker is also a member

The 1991 defeat of Kenneth Kaunda, in the first multiparty elections for 19 years, expressed popular discontent with the ailing economy and official corruption. President Frederick Chiluba and the MMD government made little headway in revitalizing the economy, despite socially painful reforms. Chiluba himself, once widely respected as a rallying point for democratic opposition to Kaunda, became much criticized for arbitrary and authoritarian rule, and for failing to address poverty and HIV/AIDS. In 2001 Chiluba purged the MMD of key opponents, prompting the formation of new opposition parties, and made way for his chosen successor, Levy Mwanawasa, to win disputed elections.

WORLD AFFAIRS ▷ Joined UN in 1964

 Comm   ACP   NAM   AU   SADC

Zambia led Africa's opposition to apartheid South Africa and now enjoys close links with Pretoria. It also has a significant role as a mediator in neighboring conflicts.

AID ▷ Recipient

$795m (receipts)   Up 27% in 2000

Regional drought increased the need for aid in 2002. International donors pledged $1.3 billion, returning aid to the yearly levels seen before a freeze in 1997 prompted by state corruption. The IMF agreed a $3.8 billion debt service relief package in 2000.

DEFENSE ▷ No compulsory military service

$65m   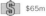 Down 26% in 2000

Despite the relatively small budget, the 21,600-strong armed forces are well equipped. Security along the Angolan border is a main concern.

ECONOMICS
 Inflation 51% p.a. (1990–2000)

 $3.03bn

 4500–3810 Zambian kwacha

SCORE CARD

- ❏ WORLD GNP RANKING.......................131st
- ❏ GNP PER CAPITA $300
- ❏ BALANCE OF PAYMENTS...................–$269m
- ❏ INFLATION 27.3%
- ❏ UNEMPLOYMENT................................50%

STRENGTHS
Potential for self-sufficiency in food. Boom in new export crops, such as cotton and flowers. Minerals, notably copper, cobalt, and coal. Market-oriented reforms and privatization drive attracting foreign private investors.

WEAKNESSES
Falling value of copper, on which Zambia depends for 90% of export earnings. Domestic reserves declining. Shortage of finance for restructuring, exacerbated by aid donors' boycott. High inflation, negative growth, and serious drought in 1998. Arable land underutilized. Delays in privatizing state-owned copper consortium.

EXPORTS

- St. Pierre-Miquelon 7%
- Other 25%
- Malaysia 8%
- Switzerland 10%
- South Africa 25%
- UK 25%

IMPORTS

- USA 5%
- Japan 3%
- Zimbabwe 6%
- UK 8%
- South Africa 57%
- Other 21%

ZAMBIA

Total Land Area : 752 614 sq. km (290 584 sq. miles)

POPULATION

- ◉ over 500 000
- ◎ over 100 000
- ○ over 50 000
- ● over 10 000
- • under 10 000

LAND HEIGHT

- 1000m/3281ft
- 500m/1640ft
- 200m/656ft

0 — 200 km
0 — 200 miles

RESOURCES
 Electric power 2.3m kw

 71,507 tonnes

Not an oil producer

2.4m cattle, 1.27m goats, 30m chickens

 Copper, cobalt, coal, zinc, lead, gold, emeralds, amethysts

Despite declining reserves, copper is still the key resource; Zambia remains one of the world's leading producers. It also has rich hydropower potential.

ENVIRONMENT
Sustainability rank: 69th

 9%

0.2 tonnes per capita

Drought is a recurrent hazard. Rhinos are almost extinct as a result of poaching. Revenues from legal hunting are being channeled into villages to encourage support for conservation.

MEDIA
TV ownership medium

Daily newspaper circulation 14 per 1000 people

PUBLISHING AND BROADCAST MEDIA

There are 3 daily newspapers, including the state-owned *Times of Zambia* and the *Zambia Daily Mail*

2 services: 1 state-controlled, 1 educational

4 services: 1 state-controlled, 3 independent

Broadcasting is dominated by the government. Opposition journalists have been accused of treason.

CRIME
Death penalty in use

12,500 prisoners

Crime is rising

Cases of violent crime, burglary, and rape are rising rapidly. In 1998 Zambia promised to overhaul its prison and police services.

CHRONOLOGY
Northern Rhodesia was developed by Britain solely for its copper. The UNIP, led by Kenneth Kaunda, took power at Zambian independence in 1964.

- ❏ **1972** UNIP one-party government.
- ❏ **1982–1991** Austerity measures and corruption: pressure for democracy.
- ❏ **1991** MMD government elected; Frederick Chiluba defeats Kaunda.
- ❏ **1996** Controversial elections.
- ❏ **2002** Levy Mwanawasa president.

EDUCATION
 School leaving age: 14

78%

22,701 students

Primary education is compulsory. Fees for secondary students have affected the already very low attendance rate.

HEALTH
Welfare state health benefits

1 per 10,000 people

Respiratory infections, diarrheal diseases, AIDS, malaria

An HIV prevention program is reducing rates of infection in Lusaka. More than 25% of town dwellers are HIV-positive.

SPENDING
GDP/cap. increase

CONSUMPTION AND SPENDING

17 per 1000 population

8 per 1000 population

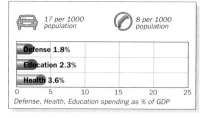

- Defense 1.8%
- Education 2.3%
- Health 3.6%

Defense, Health, Education spending as % of GDP

Standards of living for most Zambians are now lower in real terms than at independence in 1964. Many people lack basic nutrition.

WORLD RANKING

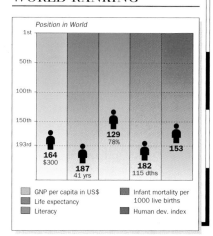

Position in World

- 164 $300
- 187 41 yrs
- 129 78%
- 182 115 dths
- 153

- ☐ GNP per capita in US$
- ☐ Life expectancy
- ☐ Literacy
- ☐ Infant mortality per 1000 live births
- ☐ Human dev. index

Z

ZIMBABWE

OFFICIAL NAME: Republic of Zimbabwe CAPITAL: Harare
POPULATION: 12.9 million CURRENCY: Zimbabwe dollar OFFICIAL LANGUAGE: English

ZIMBABWE IS SITUATED in southern Africa. Its upland center is crisscrossed by rivers flowing into Lake Kariba and the Zambezi River, on which lies the region's most spectacular natural feature, the Victoria Falls (Musi-o-Tunya). Attempts to preserve white rule in the former British colony led to a long guerrilla war before independence in 1980. Robert Mugabe, the country's leader since then and its president since 1987, has become increasingly authoritarian and divisive. Violent seizure of white-owned farmland and severe drought have contributed to virtual economic collapse.

The Kariba Dam, which has created the vast Lake Kariba on the Zambezi River, lies on Zimbabwe's northwest border with Zambia.

CLIMATE

▷ Tropical wet & dry/ steppe

WEATHER CHART FOR HARARE

Because of its altitude, Zimbabwe is comparatively temperate for a country in the tropics; humidity is also low. The rainy season occurs between November and March but, with the exception of the eastern highlands, rainfall is erratic and drought is common. Annual rainfall ranges from 140 cm (55 in) in the eastern highlands to 40 cm (16 in) in the Limpopo valley.

TRANSPORTATION

▷ Drive on left

Harare International
1.02m passengers

Has no fleet

THE TRANSPORTATION NETWORK

8692 km (5401 miles)		None
2592 km (1611 miles)		Lake Kariba

The number of international air links is being increased. Zimbabwe's rail network, among the densest in sub-Saharan Africa, is being updated.

TOURISM

▷ Visitors : Population 1:6.9

1.87m visitors

Down 11% in 2000

MAIN TOURIST ARRIVALS

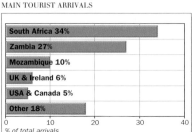

South Africa 34%	
Zambia 27%	
Mozambique 10%	
UK & Ireland 6%	
USA & Canada 5%	
Other 18%	

% of total arrivals

Zimbabwe's principal attractions are the Victoria Falls, the Kariba Dam, numerous national parks, the Great Zimbabwe ruins near Masvingo, and World's View in the Matopo Hills.

Invasions led by "war veterans" of large commercial farms, and the violence in the run-up to the parliamentary elections in 2000, put Zimbabwe on the list of unsafe destinations for many visitors. Fuel and foreign currency shortages have further undermined the tourism sector.

In addition to these factors, Zimbabwe is wary that mass-market tourism might seriously damage the environment. However, the lure of foreign exchange has encouraged the development in Harare of conference facilities and vacation complexes, such as Elephant Hills, around Victoria Falls. State law requires that there be 30% local ownership of tourist ventures.

ZIMBABWE

Total Land Area :
390 580 sq. km
(150 803 sq. miles)

POPULATION

- ⊙ over 500 000
- ◎ over 100 000
- ○ over 50 000
- • over 10 000
- · under 10 000

LAND HEIGHT

- 2000m/6562ft
- 1000m/5281ft
- 500m/1640ft
- 200m/656ft
- 180m/590ft

PEOPLE   ▷ Pop. density low

 Shona, Ndebele, English   33/km² (86/mi²)

THE URBAN/RURAL POPULATION SPLIT

35%   65%

RELIGIOUS PERSUASION

- Other (including Muslim) 1%
- Traditional beliefs 24%
- Syncretic (Christian/Traditional beliefs) 50%
- Christian 25%

ETHNIC MAKEUP

- White 1%
- Asian 1%
- Other African 11%
- Ndebele 16%
- Shona 71%

There are two main ethnic groups, the majority Shona in the north and the Ndebele in the south. Europeans and Asians comprise 2% of the population.

Ethnic tensions plagued the 1980s. 1500 Ndebele were massacred by the army in 1983 alone as the ruling, Shona-dominated, ZANU–PF attempted to suppress the predominantly Ndebele Zimbabwe African People's Union (PF–ZAPU). A Unity Accord in 1987 eased the conflict, and ZAPU leader Joshua Nkomo was appointed vice president in 1990.

As a legacy of colonial rule, whites remain generally far more affluent than blacks. This imbalance has been redressed in part by policies to improve black education and increase office employment. Land redistribution, which was previously slow and dogged by accusations of corruption, was stepped up in 2000, when the

POPULATION AGE BREAKDOWN

Female	Age	Male
0.6%	80+	0.5%
1.6%	60–79	1.6%
5.8%	40–59	5.3%
14.9%	20–39	13.6%
28.3%	0–19	27.8%

% of population by age group

government backed a movement to seize white-owned farms, raising tensions between blacks and whites.

Families are large, and almost half the population is under 15. Zimbabwean society is traditionally patriarchal. In 1999 a Supreme Court ruling provoked protest by according only "junior male" status to black women, especially those marrying under traditional law.

POLITICS   ▷ Multiparty elections

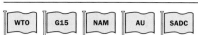 2000/2005   President Robert Gabriel Mugabe

AT THE LAST ELECTION

Parliament 150 seats   1% ZANU–Ndonga

41% ZANU–PF   38% MDC   20% App

ZANU–PF = Zimbabwe African National Union
MDC = Movement for Democratic Change   **App** = Appointed
ZANU–Ndonga = Zimbabwe African National Union – Ndonga

30 seats are set aside for presidential appointments and traditional chiefs

80% of MPs are elected. The president is directly elected every six years.

PROFILE

ZANU–PF emerged victorious in 1980 from the long struggle with Ian Smith's white-supremacist government. In 1987, rivalries with ZAPU were resolved, and Robert Mugabe led the country in his socialist experiment. In reality repression flourished, and the civil service is closed to non ZANU–PF members.

The influence of ZANU–PF was rocked in 2000 when the MDC won a convincing share of votes despite widespread pre-electoral intimidation. The MDC and its supporters have since been the target of government thugs.

MAIN POLITICAL ISSUES
The rule of President Mugabe
Robert Mugabe was appointed president in 1987. His attempts to create a one-party socialist state were called off in 1991 as opposition to white South Africa and support for the Soviet bloc became meaningless. His ZANU–PF has since lost support in the face of economic collapse and a reinvigorated opposition led by the Movement for Democratic Change (MDC). Mugabe's plans to strengthen his position through a referendum were defeated and ZANU–PF only just held on to power in elections in 2000, making his position increasingly precarious.

Land redistribution
Though it has been generally agreed that the distribution of farmland unfairly favors the white minority, the speed and method of what has been largely belated land redistribution has provoked protest. White-owned farms have been confiscated without compensation, and Mugabe gave his full support to the violent occupations led by self-styled "war veterans." The seizure of white-owned farms has been blamed for the huge drop in grain production in 2001–2002.

Robert Mugabe, elected prime minister in 1980 and president in 1987.   *Morgan Tsvangirai, leader of the opposition MDC.*

AID   ▷ Recipient

 US$178m (receipts)   Down 27% in 2000

Bilateral aid fell by a third between 1994 and 1996, after donors including the UK, Denmark, the US, France, and Germany learned that aid intended for small farmers and indigenous enterprises was siphoned off to large industrial projects. In 1998 the IMF approved US$175 million in standby credit. However, political violence since 2000 has prompted most international aid donors to suspend financial support. The UNDP and USAID agreed additional food aid in 2002 to combat massive food shortages.

WORLD AFFAIRS   ▷ Joined UN in 1980

WTO   G15   NAM   AU   SADC

Zimbabwe is an active member of the SADC and the Preferential Trade Area for East and South Africa. Relations with postapartheid South Africa are particularly strong. Zimbabwean troops have been active in the DRC since 1998.

This strong involvement in African affairs and Mugabe's own role as an anticolonial champion have led to regional support for the confrontational policy against white commercial farmers. However, his increasingly antidemocratic stance has brought near-total isolation from the wider international community. The EU and the UK have been particularly

vociferous in their condemnation of Mugabe's regime, and sanctions and aid suspensions have contributed to the perilous state of the economy. In 2002 South Africa and Nigeria were participants in the decision to suspend Zimbabwe from the Commonwealth, in a rare display of African displeasure at Mugabe's "unfree and unfair" reelection.

Z

CHRONOLOGY

In 1953, the British colony of Southern Rhodesia became part of the Federation of Rhodesia and Nyasaland with Northern Rhodesia (now Zambia) and Nyasaland (now Malawi).

❑ **1961** Joshua Nkomo forms ZAPU.
❑ **1962** ZAPU banned. Segregationist Rhodesian Front (RF) wins polls.
❑ **1963** African nationalists in Northern Rhodesia and Nyasaland demand dissolution of Federation. ZANU, offshoot of ZAPU, formed by Rev. Sithole and Robert Mugabe.
❑ **1964** New RF prime minister Ian Smith rejects British demands for majority rule. ZANU banned.
❑ **1965** May, RF reelected. November, state of emergency declared (renewed until 1990). Smith's unilateral declaration of independence. UK imposes economic sanctions. ANC, ZANU, and ZAPU begin guerrilla war.
❑ **1974** RF regime agrees cease-fire terms with African nationalists.
❑ **1976** ZANU and ZAPU unite as Patriotic Front (PF).
❑ **1977** PF backed by "frontline" African states: Mozambique, Tanzania, Botswana, and Zambia.
❑ **1979** Lancaster House talks produce agreement on constitution.
❑ **1980** Independence as Zimbabwe. Following violent election campaign, Mugabe becomes prime minister of ZANU–PF/ ZAPU–PF coalition. Relations severed with South Africa.
❑ **1983–1984** Unrest in Matabeleland, ZAPU–PF's power base.
❑ **1985** Elections return ZANU–PF, with manifesto to create one-party state. Many ZAPU–PF members arrested.
❑ **1987** September, provision for white seats in parliament abolished. December, ZANU–PF and ZAPU–PF sign unity agreement (merge in 1989). Mugabe elected president.
❑ **1990** Elections won by ZANU–PF. Mugabe reelected president.
❑ **1991** Mugabe abandons plan for one-party state.
❑ **1999** Death of Vice President Nkomo. Opposition forms MDC.
❑ **2000** Government loses referendum on new constitution. Expropriations of white-owned farmland by squatters. Strong MDC performance in polls. ZANU–PF accused of using intimidation to retain majority.
❑ **2002** Mugabe reelected in flawed poll. Commonwealth membership suspended. Government deadline for white farmers to leave land. Threat of mass starvation and economic collapse.

DEFENSE

 No compulsory military service

 US$394m   ⬇ Down 6% in 2000

Nationalist guerrillas were the heroes of independence in 1980. By the late 1990s, however, resentment grew when ex-combatants demanded enormous pensions.

Although formally nonaligned, Zimbabwe supported the Mozambican regime against RENAMO guerrillas and backed the US-led operation in Somalia in 1992–1995. The withdrawal of troops from the Democratic Republic of the Congo (DRC), dispatched there in 1998 to help President Laurent Kabila fight rebels, began in April 2001, following the Lusaka peace accord. Although the

ZIMBABWEAN ARMED FORCES

🛡	30 main battle tanks (PRC Type-59, PRC Type-69)	35,000 personnel
	None	
✈	52 combat aircraft (11 *Hunters*, 5 *Hawk*, 9 MiG-21)	4000 personnel
	None	

withdrawal was officially completed in 2002, troops remain in the DRC.

Zimbabwe has in the past received military aid and training from the UK and South Korea.

ECONOMICS

▷ Inflation 26% p.a. (1990–2000)

📊 US$5.85bn   💲 55.10–55.45 Zimbabwe dollars

SCORE CARD

❑ WORLD GNP RANKING.......................105th
❑ GNP PER CAPITA ....US$460
❑ BALANCE OF PAYMENTS................–US$425m
❑ INFLATION 58.5%
❑ UNEMPLOYMENT..................................50%

ECONOMIC PERFORMANCE INDICATOR

— Consumer Price Index   ▬ GDP

EXPORTS

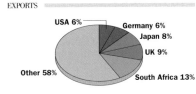

USA 6% · Germany 6% · Japan 8% · UK 9% · South Africa 13% · Other 58%

IMPORTS

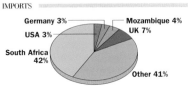

Germany 3% · Mozambique 4% · UK 7% · USA 3% · South Africa 42% · Other 41%

STRENGTHS

Most broadly based African economy after South Africa. Sound infrastructure. Virtual self-sufficiency in energy. Founder member of regional free trade area. Gold, coal, tobacco, horticulture.

WEAKNESSES

Agricultural and hydroelectric output affected by drought. Large budget deficits. High unemployment and inflation. Labor unrest, bank collapses, food price riots. Currency value halved in 1998; devalued again in 2000. Cheap imports damage local industries. Political violence and economic collapse in 2000 scared off investors. Massive food shortages.

PROFILE

The socialist policies of the 1980s were superseded by a more market-oriented economy, introduced in 1991, which increased unemployment and inflation. Prospects for the mining industry appear particularly bleak: privatization of state copper interests was repeatedly delayed, while collapsing mineral prices have forced the closure of diamond, gold, platinum, and chromium mines. The cost of living and inflation soared during the economic chaos of 2000. Prices for basic provisions were fixed in 2001.

ZIMBABWE : MAJOR BUSINESSES

Harare · Kadoma · Gweru · Mutare · Hwange · Kwekwe · Bulawayo · Rutenga

Tobacco	Agribusiness	
Steel	Engineering	
Textiles	Coal mining	
Chemicals	Vehicle assembly	
Footwear		

0   200 km
0   200 miles

RESOURCES   ▷ Electric power 2m kw

12,591 tonnes

Not an oil producer

5.55m cattle,
2.8m goats,
16.5m chickens

Gold, coal, asbestos,
nickel, copper, silver,
iron, emeralds,
lithium, diamonds

ELECTRICITY GENERATION

Hydro 29% (1.9bn kwh)				
Combustion 71% (4.8bn kwh)				
Nuclear 0%				
Other 0%				
0	20	40	60	80	100

% of total generation by type

ENVIRONMENT   ▷ Sustainability rank: 46th

8% (1% partially protected)

1.2 tonnes per capita

ENVIRONMENTAL TREATIES

No	Yes	Yes
No	Yes	No

The 1991–1992 drought left half the population in need of drought relief, and used up 20% of public spending.

In communal areas, the land is suffering from overpopulation and overstocking. Deforestation, soil erosion, and deterioration of wildlife and water resources are widespread.

Measures have been taken to protect the black rhinoceros, including moving animals to safer areas and combating poaching – patrols have killed 150 poachers since 1986. The government also supports a scheme for dehorning rhinos – the horn is the poachers' main target. In 1997 Zimbabwe led the move at the Convention on International Trade in Endangered Species to allow a limited resumption of international trade in ivory. An increase in ivory poaching since 1999 has led to calls for more protection for elephants.

MEDIA   ▷ TV ownership low

⊠ Daily newspaper circulation 19 per 1000 people

PUBLISHING AND BROADCAST MEDIA

There are 2 daily newspapers, the state-controlled *Herald* and the privately run *Daily News*

1 state-controlled service

1 state-controlled service

The state has a controlling interest in the *Herald* newspaper. Persecution of journalists has increased since a new press law was passed in March 2002.

Almost 30% of Zimbabwe's electricity needs are met by hydropower, notably from the Kariba Dam, jointly owned with Zambia. The state power company is seeking to maximize capacity. In 1991, the government agreed to build an extension facility at Kariba South, and with Zambia a joint HEP plant at Bartoka Gorge. An oil pipeline from Beira, Mozambique, to Mutare is being extended to Harare. Coal mining is expanding at Hwange, where Malaysian investments are helping to exploit deposits of 400 million tonnes.

CRIME   ▷ Death penalty in use

🏛 19,376 prisoners   ⬆ Up 8% in 1999

CRIME RATES

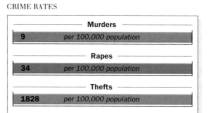

Murders	
9	per 100,000 population
Rapes	
34	per 100,000 population
Thefts	
1828	per 100,000 population

Murder and narcotics-related offenses are rife in urban areas. The illegal occupation of white-owned farms, supported by the government, and electoral violence have led to many deaths. The secret service and the army have been criticized for human rights abuses.

EDUCATION   ▷ School leaving age: 12

📖 89%   🎓 40,850 students

THE EDUCATION SYSTEM

% of each age group in education

- Primary **100%**
- Secondary **49%**
- Tertiary **4%**

Improving education has been one of ZANU–PF's great successes. In barely ten years, primary school attendance rose from 820,000 to some 2.3 million. Education is compulsory and instruction is in English. Fees were introduced after 1992. The government has built two new universities, at Bulawayo and Mutare; it encourages vocational training to create a workforce with skills in agriculture, medicine, and engineering.

ZIMBABWE : LAND USE

- Cropland
- Pasture
- Forest
- ♣ Tobacco - cash crop
- Maize
- 🐂 Cattle

Zambezi
Lake Kariba
Sanyati
Sebi
MAFUNGABUSI PLATEAU
Limpopo

0 — 200 km
0 — 200 miles

HEALTH   ▷ Welfare state health benefits

1 per 10,000 people

AIDS, tuberculosis, accidents, malaria, heart disease, cancers

The largest threat to health is AIDS. It has reduced average life expectancy to just 43 years, created some 800,000 orphans, and kills 700 people a week. A belated AIDS program, offering generic drugs, is now in place. Malaria and tuberculosis account for many other deaths. The beleaguered health system is free for the poor.

SPENDING   ▷ GDP/cap. increase

CONSUMPTION AND SPENDING

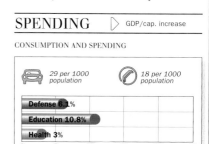

🚗 29 per 1000 population   💿 18 per 1000 population

Defense 6.1%				
Education 10.8%				
Health 3%				
0	5	10	15	20	25

Defense, Health, Education spending as % of GDP

Socialist policies in the 1980s lessened the gap between blacks and whites. But currency depreciation and inflation have since greatly reduced real wages.

WORLD RANKING

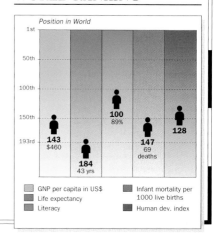

Position in World

- 1st
- 50th
- 100th
- 150th
- 193rd

143 $460
184 43 yrs
100 89%
147 69 deaths
128

- GNP per capita in US$
- Life expectancy
- Literacy
- Infant mortality per 1000 live births
- Human dev. index

Z

OVERSEAS TERRITORIES & DEPENDENCIES

DESPITE THE RAPID process of decolonization since 1945 (pages 52–55), roughly seven million people around the world still live in nonsovereign territories under the protection of the UK, the US, France, Netherlands, Denmark, Norway, Australia, or New Zealand. These remnants of former colonial empires may have persisted for economic, strategic, or political reasons.

Hong Kong and Macao reverted to Chinese control in the late 1990s. Others await political developments, such as referendums, which will determine their future status. Finally, there is a large group of territories that are considered too small, remote, or weak to be able to survive as independent states.

Svalbard
(to Norway)

*BARENTS
SEA*

Jan Mayen
(to Norway)

Faeroe Islands
(to Denmark)

NORTH
SEA

NORWAY

BALTIC SEA

Isle of Man
(to UK)
UNITED
KINGDOM

DENMARK

NETHERLANDS

Channel Islands:
Guernsey and Jersey
(to UK)
FRANCE

E U R O P E

Gibraltar
(to UK)

MEDITERRANEAN SEA

A S I A

*SEA OF
JAPAN*

*YELLOW
SEA*

*EAST
CHINA
SEA*

A F R I C A

*ARABIAN
SEA*

Paracel
Islands
(Disputed)

SOUTH
CHINA SEA

Northern Mariana
Islands (to US)

Guam (to US)

Spratly Islands
(Disputed)

JAVA SEA

*ARAFURA
SEA*

British Indian
Ocean Territory
(to UK)

Cocos (Keeling) Islands
(to Australia)

Ascension
(Administered by
St Helena)

Mayotte (to France)

Christmas Island
(to Australia)

Ashmore &
Cartier Islands
(to Australia)

St Helena
(to UK)

*ATLANTIC
OCEAN*

Réunion (to France)

*INDIAN
OCEAN*

A U S T R A L I A

Europa
(Administered by Réunion)

Bassas da India
(Administered by Réunion)

Tristan da Cunha
(Administered by
St Helena)

Gough Island
(Administered by St Helena)

Amsterdam Island

St. Paul Island

French Southern &
Antarctic Territories
(France)

Crozet Islands

Kerguelen

Bouvet Island
(to Norway)

Heard & McDonald Islands
(to Australia)

UNITED KINGDOM

THE UK STILL HAS THE LARGEST number of overseas territories in the world. What were previously known as Crown colonies and dependent territories are now British overseas territories. Residents are full British citizens. Most territories sustain a large degree of local autonomy, and if they express a consitutional desire for independence then they may have it, as long as they can form a viable independent country. The Isle of Man and the Channel Islands retain their special connection as Crown dependencies, neither a part of the UK nor colonies.

NEW ZEALAND

NEW ZEALAND'S GOVERNMENT has no desire to retain any overseas territories. However, the economic weakness of its dependent territory Tokelau and its freely associated states, Niue and the Cook Islands, has forced New Zealand to remain responsible for their foreign policy and defense.

French Southern and Antarctic territories are not included in the following section. Any territories which involve an Antarctic claim are not shown.

UNITED STATES OF AMERICA

THE OVERSEAS TERRITORIES OF THE US have been seen as strategically useful, if expensive, parts of its backyard. The US has, in most cases, given the local population a say in deciding its own status. Thus, full sovereignty has been granted to three successor states of the former US-administered UN Trust Territory of the Pacific Islands. A US commonwealth territory, such as Puerto Rico, has a greater level of independence than that of a US unincorporated or external territory.

OVERSEAS TERRITORIES AND DEPENDENCIES

- Australia
- New Zealand
- United Kingdom
- United States
- France
- Denmark
- Netherlands
- Norway
- Disputed

ARCTIC OCEAN

BEAUFORT SEA

ATLANTIC OCEAN

Greenland (to Denmark)

NORTH AMERICA

BERING SEA

St Pierre & Miquelon (to France)

UNITED STATES OF AMERICA

Bermuda (to UK)

Gulf of Mexico

Midway Islands (to US)

Johnston Atoll (to US)

PACIFIC OCEAN

Cayman Islands (to UK)

Navassa Island (to US)

CARIBBEAN SEA

Kingman Reef (to US)

Palmyra Atoll (to US)

Clipperton Island (Administered by French Polynesia)

& Howland nds (to US)

Jarvis Island (to US)

Tokelau (to NZ)

Futuna rance)

Cook Islands (to NZ)

American Samoa (to US)

Niue (to NZ)

French Polynesia (to France)

Pitcairn Islands (to UK)

Caledonia rance)

Norfolk Island (to Australia)

NEW ZEALAND

SOUTH AMERICA

French Guiana (to France)

Falkland Islands (to UK)

South Georgia and South Sandwich Islands (to UK)

Caribbean inset

Turks & Caicos Islands (to UK)

Puerto Rico (to US)

British Virgin Islands (to UK)

Anguilla (to UK)

CARIBBEAN SEA

Virgin Islands (to US)

Netherlands Antilles (to Neth.)

Guadeloupe (to France)

Montserrat (to UK)

Aruba (to Neth.)

Martinique (to France)

FRANCE

FRENCH *TERRITOIRES D'OUTRE-MER* are considered an indivisible part of the French Republic. As a result, France has developed economic ties, and stressed the advantage of interdependence over independence, with its overseas territories. A distinct hierarchy has been developed. Overseas *départements*, officially part of France, have their own governments. Territorial *collectivités* are administered by a French-appointed commissioner and a locally elected council, while overseas *territoires* have varying degrees of autonomy.

AMERICAN SAMOA

STATUS: Unincorporated territory of the USA **CLAIMED:** 1900
CAPITAL: Pago Pago **POP.:** 57,291 **DENSITY:** 294/km² (764/mi²)

COMPRISING THE EASTERN half of the Samoan islands, American Samoa sits on the edge of Polynesia in the South Pacific Ocean. Although Christianity, introduced in the 19th century, has taken a very firm hold – Samoa is known as the Bible belt of the Pacific – the traditional and conservative *fa'a Samoa* (Samoan way of life) continues to dominate the islands' culture. At its base is the extended family, the *aiga*, while traditional chiefs, or *matai*, retain their central role in government. Samoa came under the control of the US in 1900, and life there remained largely unchanged until a US-led drive for modernization in the 1960s. Along with better health care and industrial development, *fa'a Amerika* also meant unemployment, pollution, and rising petty crime fueled by alcohol. Tuna processed by Pago Pago's canneries represent 95% of American Samoa's exports. Efforts to diversify include the development of other light industries and tourism.

ANGUILLA

STATUS: British overseas territory **CLAIMED:** 1650
CAPITAL: The Valley **POP.:** 11,561 **DENSITY:** 120/km² (312/mi²)

LYING IN THE CENTER of the Leeward Islands, in the Caribbean, Anguilla has a subtropical climate, the heat and humidity being tempered by trade winds. Lumped by the UK into a joint colonial administration with St. Kitts and Nevis, Anguillans took up arms to protect their dependent status in 1967 when St. Kitts and Nevis were awarded internal self-government. Lacking any major industry, Anguilla relied on the economic stability that came with being an overseas territory. In the 1980s the island's government resolved to introduce a tourist industry and targeted the luxury end of the market. Tourism is now Anguilla's main source of income. Almost 50,000 tourists, around 60% from the US, take their vacations in Anguilla every year.

ARUBA

STATUS: Autonomous part of the Netherlands **CLAIMED:** 1643
CAPITAL: Oranjestad **POP.:** 101,000 **DENSITY:** 523/km² (1347/mi²)

THE MOST DEVELOPED island among the Dutch Caribbean territories, Aruba lies 25 km (15 miles) off the coast of Venezuela. Its tropical climate is moderated by constant trade winds sweeping in from the Atlantic. Formerly the richest island in the Netherlands Antilles, Aruba became a separate dependency of the Netherlands in 1986. Transition to full independence, expected in 1996, was halted in 1994 after an agreement was reached between the governments of the Netherlands, Aruba, and the Netherlands Antilles. The Netherlands voiced concern over the island's security and the danger of its becoming a base for narcotics trafficking, and the Aruban government questioned the desirability of full independence, citing high unemployment and economic instability.

Since 1986, the economy of Aruba, formerly dependent on oil refining, has diversified. Tourism and offshore finance have

Palm Beach, Aruba, also known as the Turquoise Coast, lies on the western side of the island. The beach stretches for 10 km (6 miles) and is the site of a low-rise beach resort.

become the most important sectors of the economy, and there are now more than 700,000 visitors annually; nearly 60% of them come from the US. However, the rapid expansion of tourism has put considerable strain on Aruba's infrastructure, and some attempt has been made to restrict the number of visitors. At the same time facilities have been improved to encourage the growth of a data-processing industry.

Oranjestad, Aruba's capital, contains many Dutch colonial-style buildings. Although first claimed by the Spanish in 1499, Aruba was colonized by the Dutch in the 17th century.

Aruba's cooperation with the US in the region includes support for its actions against narcotics trafficking from South America, and since the closure of the US base in Panama in 1999, US aircraft have used bases on the island to launch reconnaissance flights. Those who oppose this cooperation fear that it could drag Aruba unnecessarily into the civil conflict in Colombia.

LAND HEIGHT ☐ above Sea Level ☐ 200m/656ft ☐ 500m/1640ft ☐ 1000m/3281ft ☐ 1500m/4572ft ☐ above 2000m/6562ft

BERMUDA

STATUS: British overseas territory **CLAIMED:** 1612
CAPITAL: Hamilton **POP.**: 61,688 **DENSITY:** 1164/km² (3084/mi²)

SITUATED MORE THAN 1000 km (650 miles) off the coast of the US, Bermuda consists of a chain of more than 150 coral islands. The Gulf Stream, flowing between Bermuda and

the eastern seaboard of the US, keeps the climate mild and humid. Bermuda is racially mixed; some 60% of the population are of mostly European extraction. Racial tension has lessened since the 1960s and 1970s. A more representative electoral system was established after a Royal Commission visited Bermuda in 1978.

For 30 years after the first general election, held in 1968, Bermuda was ruled by the conservative United Bermuda Party (UBP). Its veteran leader, Sir John Swan, resigned as prime minister and party leader in 1995, when a referendum decisively rejected his campaign for independence from the UK. In a general election in November 1998 the UBP, now under the leadership of Pamela Gordon, was heavily defeated

by the Progressive Labour Party, led by Jennifer Smith, who said that she had no plans to pursue her party's own pro-independence aspirations. Major issues are the social and economic challenges posed by the withdrawal in 1995 of both the US naval base and the British military base, environmental issues, and narcotics trafficking. Bermuda is overwhelmingly a service economy. Lilies are grown for export, but few other agricultural products are grown in sufficient quantity, and the islands are heavily dependent on food imports.

Tourist figures have been falling steadily, but tourism is still a significant industry, most visitors coming from the US. However, financial services have become the most important sector of the economy, helping to maintain one of the highest per capita incomes in the world. The government has attempted to head off international criticism of its financial environment through a series of reforms. Bermuda also operates one of the world's largest flag-of-convenience shipping fleets.

Bermuda has one of the highest densities of golf courses in the world. Eight courses have now been developed.

BRITISH VIRGIN ISLANDS

STATUS: British overseas territory **CLAIMED:** 1672
CAPITAL: Road Town **POP.**: 19,864 **DENSITY:** 130/km² (337/mi²)

AN ARCHIPELAGO of 60 Caribbean islands, 15 of them inhabited, the British Virgin Islands lie at the northwestern end of the Leeward Islands chain. Tourism, now a major economic activity, is suited to the tropical climate, but there is concern about its effect on the environment. There are also fears that traditional place-names are being altered to be more tourist-friendly. The offshore finance sector is important, and has been more tightly regulated since 1990, following scandals involving foreign companies registered in the territory.

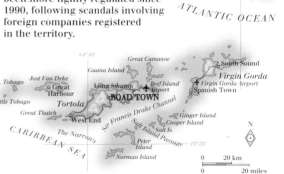

BRITISH INDIAN OCEAN TERRITORY

STATUS: British overseas territory **CLAIMED:** 1814
CAPITAL: Diego Garcia **POP.**: 3200 **DENSITY:** 53/km² (139/mi²)

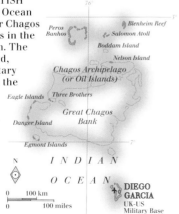

THE BRITISH Indian Ocean Territory, or Chagos Islands, lies in the middle of the Indian Ocean. The coral atolls are uninhabited, except for the US–UK military base on Diego Garcia, and the UK has undertaken to cede the islands to Mauritius when they are no longer required. In 2000 the Ilois people, evicted by the UK from the islands in 1968, won the right in the UK High Court to return, but face strong resistance from the US.

CAYMAN ISLANDS

STATUS: British overseas territory   **CLAIMED:** 1670
CAPITAL: George Town   **POP.:** 39,410   **DENSITY:** 152/km² (394/mi²)

THE LARGEST OF BRITAIN'S territories in the Caribbean, the Cayman Islands lie 225 km (140 miles) west of Jamaica and south of Cuba. The abundance of exotic wildlife, especially marine life, is a powerful draw for tourists. Grand Cayman is credited as the home of modern scuba diving, the first ever specialist shop opening there in 1957. The islanders have rejected greater autonomy, persuaded that their economic stability is linked to their status as an overseas territory. Thanks to the absence of tax and foreign-exchange controls, the islands form one of the world's largest offshore financial centers, but tourism continues to underpin the economy.

CHRISTMAS ISLAND

STATUS: Australian external territory   **CLAIMED:** 1958
CAPITAL: Flying Fish Cove   **POP.:** 2771   **DENSITY:** 21/km² (53/mi²)

SO NAMED because it was sighted on Christmas Day in 1643, Christmas Island lies in the Indian Ocean, 380 km (240 miles) south of Java. Its population is mostly Malay and Chinese, descended from laborers imported to mine rich phosphate deposits. A national park covers some 70% of the island. In 2001 Australia agreed with Russia to begin construction of a major rocket-launching site on the island.

COCOS (KEELING) ISLANDS

STATUS: Australian external territory   **CLAIMED:** 1955
CAPITAL: West Island   **POP.:** 633   **DENSITY:** 45/km² (127/mi²)

IN ALL, 27 coral atolls make up the Cocos (Keeling) Islands. Situated in the Indian Ocean, roughly halfway between Australia and Sri Lanka, they have been part of the Northern Territory electoral district since 1992. The inhabited islands are the European-dominated West Island and Home Island, which has a mainly Cocos Malay community. Coconuts are the sole cash crop.

COOK ISLANDS

STATUS: Territory in free association with New Zealand   **CLAIMED:**
CAPITAL: Avarua   **POP.:** 12,900   **DENSITY:** 54/km² (140/mi²)

LYING IN THE MIDDLE of the South Pacific 3000 km (1900 miles) from New Zealand, the Cook Islands are a combination of 24 coral atolls and volcanic islands. Achieving self-government in 1965, they have adopted a diversified economy focusing primarily on tourism and banking, but with significant trade in giant clams and pearls.

Depopulation is of serious concern as over 40,000 of the indigenous Maori population have migrated, seeing greater opportunities beyond the islands, and now live in New Zealand and Australia. This outflow of labor poses a serious problem for the islands' future development, although remittances to relatives form an important source of income.

The government of the Cook Islands, headed since 2002 by centrist Prime Minister Robert Woonton, is advised by a traditional council known as the House of Ariki. It is seeking a program of political reform including devolving administration to the outer islands and cutting back an expensive bureaucracy.

FAEROE ISLANDS

STATUS: Self-governing territory of Denmark   **CLAIMED:** 1380
CAPITAL: Tórshavn   **POP.:** 46,196   **DENSITY:** 33/km² (86/mi²)

MIDWAY BETWEEN Scotland and Iceland in the North Atlantic, the Faeroe Islands have a moderate climate for their latitude – a result of the warm Gulf Stream current. Home rule since 1948 has given the Faeroese a strong sense of national identity – they voted against joining the European Communities with Denmark in 1973, but now have favorable terms of trade with most EU members. Fishing is the dominant industry, providing over 90% of exports. In the face of international criticism, the Faeroese have continued their traditional cull of pilot whales and bottle-nosed dolphins. Sheep farming is important, and there is a small textile industry which exports traditional woolens, and puffin and eider-duck feathers.

Denmark's moves toward ever closer European integration have strengthened calls in the Faeroes for full independence. Negotiations to establish a "sovereign nation" under the Danish monarchy began in 1998. However, the Danish government's threat in 2001 to suspend subsidies to the islands quashed calls for a referendum.

LAND HEIGHT   ■ above Sea Level   ■ 200m/656ft   ■ 500m/1640ft   ■ 1000m/3281ft   ■ 1500m/4572ft   ■ above 2000m/6562ft

FALKLAND ISLANDS

STATUS: British overseas territory CLAIMED: 1832
CAPITAL: Port Stanley POP.: 2913 DENSITY: 0.24/km² (0.6/mi²)

SITUATED IN THE South Atlantic Ocean, over 12,000 km (7440 miles) from the UK, the Falkland Islands are influenced by the cold Antarctic current. The main islands of East and West Falkland and the hundreds of outlying islands have a cool, temperate climate with frequent strong winds.

The islands gained international attention with the Argentine invasion, and subsequent British recapture, in 1982. Since then, the UK government has invested heavily in a "Fortress Falklands" policy. A new runway and a military base to house an enlarged garrison were built at Mount Pleasant. The islanders, for their part, are determined to maintain the political status quo, but in 1999 improving relations led to the restoration of scheduled air connections with Argentina. Since the Falklands War, the economy of the islands has prospered. Falklanders invested heavily in schools, roads, and tourism in a fresh drive for a strong identity. By 1987, the Falklands had become financially solvent through the sale of fishing licenses. Although sales by Argentina of cheaper, less restrictive licenses caused a fall in fishing revenues, fishing (mostly of squid) is still the major source of income and employment. Depressed wool prices caused a slump in the fortunes of the sheep-farming industry. The UK and Argentina reached agreement in 1995 on oil exploration, and the discovery of oil reserves in the Falklands' territorial waters is revolutionizing prospects for the economy. Tourism, attracting birdwatchers, photographers, and military historians, is steadily growing.

FRENCH GUIANA

STATUS: French overseas department CLAIMED: 1817
CAPITAL: Cayenne POP.: 172,605 DENSITY: 2/km² (5/mi²)

SANDWICHED BETWEEN Brazil and Suriname, French Guiana is the only remaining colony in South America. A belt of coastal marsh, and an interior of equatorial jungle, combine in a location which was, for years, notorious for the offshore penal colony, Devil's Island. The rainforest, which covers 90% of the territory, is particularly rich in flora and fauna. It harbors over 400,000 species, including more different species of bird than in the whole of Europe.

Concentrated near the coast, the population is ethnically mixed. While 40% are creoles, there are some 20,000 Amerindians and a village of 1000 Hmong who fled civil war in Laos in the 1980s.

Kourou was selected for the launch of the Ariane rocket because of its equatorial site. The town has grown from 800 to 15,000 people.

A campaign for greater autonomy in the late 1970s and early 1980s led to limited decentralization of power to a regional council. The grip on local power by the Guianese Socialist Party (PSG) has been undermined since 1993 by a more unified opposition, but it is still the largest party in the regional council.

During the 1990s the people have become increasingly vocal in their condemnation of the French government's perceived indifference to their country's problems, and there were riots in 1996 and 1997 over the education system. The PSG has accordingly campaigned for greater autonomy. As an overseas *département* of metropolitan France, French Guiana is also a region of the EU, but it is heavily dependent on France itself for aid, food, and manufactured goods. It has a number of valuable natural resources, including gold, fishing, and forestry, and also has potential for increased tourism, but these are yet to be fully exploited because of a lack of skilled labor and investment and an underdeveloped infrastructure. The Guiana Space Center, which is situated on the coast at Kourou, has been operational since 1964. From there the Ariane rockets of the European Space Agency (ESA) are launched.

FRENCH POLYNESIA

STATUS: French overseas territory **CLAIMED:** 1843
CAPITAL: Papeete **POP.:** 235,400 **DENSITY:** 67/km² (173/mi²)

A SCATTER OF 130 South Pacific islands and coral atolls over an area the size of Europe combine to form French Polynesia. The average annual temperature varies between 20°C (68°F) and 29°C (84°F), with rainfall of over 150 cm (58 in). Nearly 75% of the population live on the main island of Tahiti. The French administration has developed the islands with little regard for local wishes, and the 70% West Polynesian (Mahoi) majority have seen their simple, self-sufficient economy transformed into one dependent on the French military and tourism. Nuclear testing on Mururoa atoll created many jobs, but there was growing opposition, and a final series of tests, held in 1995-1996 despite widespread international protests, provoked local demonstrations and riots in Papeete.

The Polynesian majority has called increasingly for more autonomy, reduced tourism, and the rebuilding of indigenous trade. Future hopes rest largely on new tuna fishing ventures.

GIBRALTAR

STATUS: British overseas territory **CLAIMED:** 1713
CAPITAL: Gibraltar **POP.:** 27,025 **DENSITY:** 3861/km² (9008/mi²)

G UARDING THE western entrance to the Mediterranean, Gibraltar has survived on military and marine revenues. However, as Britain has cut defense spending, so its military presence on the Rock has declined. In response Gibraltarians have developed a vibrant offshore banking industry. Strict antismuggling legislation, in force since 1995, has curbed extensive smuggling from north Africa into Spain. Gibraltar's relationship with Britain and Spain remains contentious. However, the two governments have undertaken to resolve the territory's status by December 2002, prompting mass protests from the colony's inhabitants, who have felt sidelined in the discussions.

The Rock of Gibraltar. The British built 143 caves, 50 km (30 miles) of roads, and as many km of tunnels, for defensive purposes.

GREENLAND

STATUS: Self-governing territory of Denmark **CLAIMED:** 1380
CAPITAL: Nuuk **POP.:** 56,569 **DENSITY:** 0.03/km² (0.07/mi²)

T HE WORLD'S LARGEST island after Australia, Greenland is situated in the North Atlantic and surrounded by seas that are either frozen or cooled by cold Arctic currents. With an Arctic climate, much of its land is permanently ice-covered. Granted home rule in 1979, Greenlanders are of mixed Inuit and European origin. Younger islanders increasingly reject the traditional, fishing-based, subsistence lifestyle by moving to towns, placing a heavy burden on the advanced welfare system. Proposals to develop the mining of ice for drinking water could create vast revenues, even raising the possibility of economic independence.

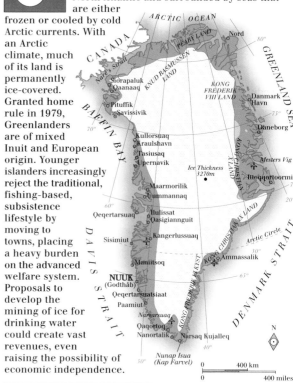

GUADELOUPE

STATUS: French overseas department **CLAIMED:** 1635
CAPITAL: Basse-Terre **POP.:** 431,000 **DENSITY:** 242/km² (627/mi²)

G UADELOUPE lies at the southern end of the Leeward Islands in the Caribbean and includes the north of St. Martin island and St. Barthélémy island near Anguilla. The movement for independence from France has been pronounced and intermittently violent since the 1960s. The economy is largely based on tourism and agriculture, with sugar, rum, and bananas the main exports, but the islands depend on French and EU regional aid. The vulnerability of the banana industry to hurricanes and global trade pacts has prompted efforts to expand sugar production and develop tourism.

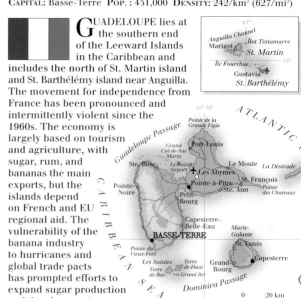

GUERNSEY

STATUS: British Crown dependency **CLAIMED:** 1066
CAPITAL: St. Peter Port **POP.** : 59,807 **DENSITY:** 920/km² (2392/mi²)

LYING 47 KM (29 miles) off the coast of France, Guernsey and its dependencies form the northwestern part of the Channel Islands, historically part of the Duchy of Normandy. English is the language most commonly used, but the Norman patois is spoken in some villages, and some formal business of the legislature is conducted in French. Travel to France is easier than to the UK; Alderney is only 13 km (8 miles) from the French mainland. Residents on the smaller islands have no need for cars, and life continues in an unhurried manner that has changed little through the centuries. The islanders guard this lifestyle with strict residence laws. Guernsey's mild climate has encouraged the development of tourism and market gardening as major industries. Tomatoes and

flowers are produced mainly for the UK market. The low tax system, independent of the UK, has led to a substantial and profitable financial services industry. Many international banks have Guernsey subsidiaries.

Tomato packaging.
Tomatoes and flowers, ripening early in Guernsey's mild climate, are despatched to mainland Britain.

GUAM

STATUS: Unincorporated territory of the USA **CLAIMED:** 1898
CAPITAL: Hågatña **POP.** : 158,000 **DENSITY:** 288/km² (745/mi²)

THE VOLCANIC island of Guam lies at the southern end of the Mariana archipelago in the Pacific. Its tropical climate has encouraged tourism, although it lies in a region where typhoons are common. Guam's indigenous Chamorro people, who comprise just under half the population, dominate the island's political and social life. They are famous for a set of facial expressions, called "eyebrow," which virtually constitutes a language of its own. Although English is the official language, Chamorro is commonly spoken, and in 1998 the spelling of the capital was changed from Agaña to the Chamorran Hagåtña. The US military base, covering one-third of the island, has made Guam strategically important to the

US. Military spending and tourism revenues have failed to benefit all islanders and 23% live below the poverty line. The influx of US culture has also threatened to upset Guam's social stability. Greater independence has been an issue since the early 1980s, with a series of referendums since 1982. A draft Commonwealth Act was rejected by the US Congress after nearly 15 years of deliberation.

ISLE OF MAN

STATUS: British Crown dependency **CLAIMED:** 1765
CAPITAL: Douglas **POP.** : 76,315 **DENSITY:** 133/km² (345/mi²)

LYING HALFWAY BETWEEN England and Northern Ireland in the Irish Sea, the Isle of Man has been inhabited for centuries by the Celtic Manx people. Established by the Vikings in the 9th century, the Manx parliament, the Tynwold, has autonomy from the UK in a number of matters, including taxation, and the death penalty was only officially abolished in 1993. The islanders have used this independence to establish a thriving financial and business sector, which has aided employment as the traditional industries of agriculture and fishing decline. There is still a shellfish industry, specializing in scallops. Tourism is also important: there are more than 200,000 visitors each year. Numbers have been boosted by the growing number of films being made on the island. Manx

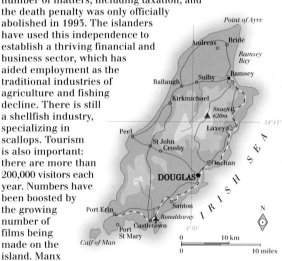

culture received a boost in 1993, when the local language, in danger of dying out, began to be taught in the island's schools once more. The Calf of Man, a small uninhabited island, is administered as a nature reserve.

The annual TT motorbike race *on the Isle of Man. Thousands of people come each year to watch the island's famous Touring Trophy race.*

JERSEY

STATUS: British Crown dependency **CLAIMED:** 1066
CAPITAL: St. Helier **POP.:** 87,186 **DENSITY:** 752/km² (1957/mi²)

THE BAILIWICK OF JERSEY, the largest of the Channel Islands, lies some 22 km (14 miles) from the coast of Normandy in France. The official language (since 1960) is English, but French is still used in the courts. The island has a mild climate owing to the Gulf Stream, fine beaches, and more sunshine than anywhere in the British Isles. Jersey has its own legislative and taxation systems which are a blend of the French and British versions. The Jersey States Assembly is one of the oldest legislative bodies in the world. Members stand as independents, rather than for political parties. It is considered a "Peculiar" of the UK monarchy, and has the right to reject "unacceptable" UK laws.

Historically, agriculture has been Jersey's most important industry, with dairy cows its most famous export, closely followed by early-harvested potatoes, tomatoes, and flowers. By the end of the 20th century, however, farming had been eclipsed by the rise of offshore finance and tourism. The growth of these sectors, and rigid controls on the rights of residence, have ensured high living standards for most of the inhabitants. Jersey is also host to a large Portuguese community which works in the island's tourist industry.

JOHNSTON ATOLL

STATUS: Unincorporated territory of the USA **CLAIMED:** 1858
CAPITAL: *Not applicable* **POP.:** *Not applicable*

JOHNSTON ATOLL LIES 1150 km (714 miles) southwest of Hawaii. The atoll consists of a coral reef, two highly modified natural islands, Johnston and Sand, and two completely man-made islands, Akau (North) and Hikina (East). The islands, which were used by the US for nuclear weapons tests, were seriously contaminated with plutonium in 1962, when a nuclear missile exploded during testing. Regular tests began in 1971, and until 2000 the islands were also used for the storage of nuclear material and the destruction of chemical and biological weapons, including sarin nerve gas and the defoliant Agent Orange. Cleanup operations began in 2000. The only inhabitants left are US government personnel and civilian contractors who maintain the plant. The islands have also been designated by the US as a wildlife refuge, a breeding place for seabirds and green turtles. The US army intends to evacuate the base entirely by 2003.

MARTINIQUE

STATUS: French overseas department **CLAIMED:** 1635
CAPITAL: Fort-de-France **POP.:** 381,427 **DENSITY:** 338/km² (875/mi²)

CHRISTOPHER COLUMBUS described Martinique as "the most beautiful country in the world." It lies in the middle of the Caribbean Windward Islands but has retained remarkably close links both culturally and economically to mainland France. The island is dominated by the Montagne Pelée volcano, which violently erupted and engulfed the old capital, St. Pierre, in 1902. Situated in the Caribbean's hurricane belt, Martinique suffers an average of one natural disaster every five years.

Its long association with France and its status as an overseas *département* have left Martinique with a distinctly French feel; nonetheless nearly 90% of the population are of African or mixed ethnicity, and this influence has created a vibrant Caribbean tradition, particularly in music. Some of Martinique's more famous children include Joséphine Bonaparte (Napoléon's first wife) and Frantz Fanon, the black revolutionary who influenced anticolonial movements in the 20th century.

Economic power remains in the hands of the *Bekes* (descendants of white colonial settlers), who own most of the agricultural land. This situation has led in the past to outbreaks of violence and calls for greater autonomy. However, high living standards depend on French subsidies and a French-style social welfare system. The traditionally agricultural economy, based on the cultivation of sugarcane and bananas, has been forced to diversify as EU subsidy cuts come into effect, and high-class tourism is now the biggest source of income and the largest provider of employment. Almost 80% of the half-million annual visitors come from France. Since the late 1980s unemployment and emigration have been high, with the result that over 30% of Martiniquais nationals are resident in metropolitan France.

Martinique. Tourists are attracted to the island's beaches, its mountainous interior, and the historic towns of Fort-de-France and Saint Pierre.

LAND HEIGHT ⬜ above Sea Level ⬜ 200m/656ft ⬜ 500m/1640ft ⬜ 1000m/3281ft ⬜ 1500m/4572ft ⬜ above 2000m/6562ft

MAYOTTE

STATUS: French territorial collectivity **CLAIMED:** 1843
CAPITAL: Mamoudzou **POP.:** 142,000 **DENSITY:** 380/km² (986/mi²)

PART OF THE COMOROS archipelago, Mayotte lies about 8000 km (5000 miles) from France, between Madagascar and the east African coast. It was the only island in the archipelago to vote against independence from France in a 1974 referendum. The other islands declared unilateral independence in 1975 and laid claim to Mayotte. Despite widespread poverty, endemic unemployment, and a cost of living twice that of France, the Mahorais voted again in 1976 to maintain the link. The main political movement has since unsuccessfully demanded that Mayotte be given the status of a French *département*, hoping that this would bring more aid to develop their largely agricultural economy. France opposes this idea because of the expense involved, but did grant the island *département*-style autonomy in 2000.

The economy is still largely agricultural, producing crops both for internal consumption and for export. However, large quantities of foodstuffs are also imported. France has invested in an airport and port, but tourism has been slow to develop. Nevertheless, the relative prosperity of Mayotte has encouraged separatist movements on the two other small Comoros islands to seek closer relations with France.

MIDWAY ISLANDS

STATUS: Unincorporated territory of the USA **CLAIMED:** 1867
CAPITAL: *Not applicable* **POP.:** 453 **DENSITY:** *Not applicable*

NAMED BECAUSE of its position between California and Japan, Midway is a coral atoll at the western end of the Hawaiian islands; there have been moves to make it part of Hawaii. The site of a major World War II battle, the atoll comprises two large islands, totaling over 4 sq. km (1.5 sq. miles), and several smaller ones. It functions as a naval air base and wildlife refuge. The population is limited to military personnel and civilian contractors, but some tourism is permitted, mainly connected with the wildlife.

MONTSERRAT

STATUS: British overseas territory **CLAIMED:** 1632
CAPITAL: Plymouth **POP.:** 5000 **DENSITY:** 49/km² (128/mi²)

MONTSERRAT IS ONE of the Leeward Islands chain in the eastern Caribbean. It has been devastated by volcanic eruptions which began in 1995 and culminated in massive explosions of the Soufrière Hills volcano in 1997 and 1998. As a result, the southern two-thirds of the island, where Plymouth and Blackburne airport are located, have become uninhabitable and it is illegal to enter the volcano "exclusion zone." Over half of the 10,000 population left permanently for neighboring islands or the UK. Calls for independence, based on a tourist boom in the 1980s, have been largely dropped, as the island is now dependent on UK aid. The disaster soured relations, setting off a bitter dispute over the cost of resettlement and reconstruction. A new capital, tentatively named Port Diana, is planned for the "safe" northern coast. The tourism industry is struggling to rebuild itself, but is hindered by the closure of both the airport and seaport by the eruption. Montserrat can now only be reached via neighboring Antigua.

Montserrat. Known as the Caribbean's "emerald isle" because of its luxuriant flora and Irish heritage.

NETHERLANDS ANTILLES

STATUS: Autonomous part of the Netherlands **CLAIMED:** 1816
CAPITAL: Willemstad **POP.:** 217,000 **DENSITY:** 271/km² (702/mi²)

THE NETHERLANDS Antilles are composed of two Caribbean island groups. Curaçao – the richest and wealthiest island – and Bonaire lie just off the Venezuelan coast, while Saba, St. Eustatius, and Sint Maarten – whose northern half is part of Guadeloupe – lie 800 km (500 miles) to the north. Financial scandals, political instability, and the issue of the federation's future, among other things, have strained relations with the Dutch government, the major aid provider. Refining petroleum using oil from Venezuela is the islands' principal industrial activity.

NEW CALEDONIA

STATUS: French overseas territory CLAIMED: 1853
CAPITAL: Nouméa POP.: 200,000 DENSITY: 10/km² (27/mi²)

NEW CALEDONIA, or, as it is known to the indigenous Kanaks, Kanaky, is an island group 400 km (250 miles) west of Vanuatu and 1350 km (840 miles) off the coast of eastern Australia. Tension over socioeconomic inequalities and independence between the Melanesian Kanaks, who form over half of the population, and the influential expatriate *Caldoches*, resulted in a long history of political violence. Under the 1988 Matignon Accord, France imposed a year of direct rule as the prelude to a new constitutional structure which attempted to address Kanak grievances by providing greater provincial autonomy. Although some racial violence continued after 1988, it has not again reached the same level. The Nouméa accord, signed in 1998, set out a 15-year program for gradual autonomy which would end in a vote on self-determination.

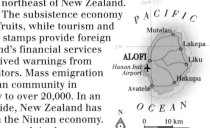

Nickel mining is the territory's most valuable export industry, at over 90% of export income. New Caledonia has over 20% of world reserves, and is the fourth-largest producer in the world, but the industry employs relatively few people, and is vulnerable to fluctuations in the world price. It was seriously affected by the Asian financial crisis of 1997–1998, but recovered on the back of high world prices in 2000. Tourism and agriculture are bigger employers, although less than 1% of total land area is cultivated. Corn, yams, sweet potatoes, and coconuts have traditionally been the main crops, and since the 1990s large numbers of melons have been exported to Japan. Fishing is important, the main products being tuna and shrimps, most of which are also exported to Japan. A project for farming giant clams started in 1996. Unemployment nevertheless remains high among young Kanaks.

A nickel mine, New Caledonia. The importance of the nickel industry to the territory's economy has made the control of reserves a dominant issue in politics, and in negotiations over the island's independence from France.

NIUE

STATUS: Territory in free association with New Zealand CLAIMED: 19
CAPITAL: Alofi POP.: 1857 DENSITY: 7/km² (18/mi²)

THE WORLD'S LARGEST coral island, Niue lies 2400 km (1500 miles) northeast of New Zealand. The subsistence economy produces tropical fruits, while tourism and the sale of postage stamps provide foreign currency. The island's financial services have recently received warnings from international monitors. Mass emigration has seen the Niuean community in New Zealand grow to over 20,000. In an effort to stem the tide, New Zealand has invested heavily in the Niuean economy. Despite this boost, growth is slow.

NORFOLK ISLAND

STATUS: Australian external territory CLAIMED: 1774
CAPITAL: Kingston POP.: 2601 DENSITY: 74/km² (200/mi²)

INHABITED by Australian migrants and descendants of the mutineers of HMS *Bounty*, Norfolk Island lies 1400 km (869 miles) east of Australia. Islanders speak a hybrid language, mixing Westcountry English, Gaelic, and ancient Tahitian. They enjoy substantial autonomy, and in 1991 rejected a plan to become part of the Australian federal state. Tourists, attracted by the climate and unique flora, have brought islanders a relatively high standard of living.

NORTHERN MARIANA IS.

STATUS: Commonwealth territory of the USA CLAIMED: 1947
CAPITAL: Saipan POP.: 69,221 DENSITY: 151/km² (393/mi²)

A FORMER UN trust territory, the Northern Marianas preferred in 1987 to retain links with the US rather than opt for independence. However, local politicians have questioned their current status. US aid fueled a boom during the 1980s, but it depended on immigrant workers who by the early 1990s outnumbered the local Chamorro population. In addition, tourism has speeded the decline of the traditional subsistence economy.

Rota, Northern Marianas. The limestone outcrop of Wedding Cake Mountain overlooks the small village of Songsong.

LAND HEIGHT above Sea Level 200m/656ft 500m/1640ft 1000m/3281ft 1500m/4572ft above 2000m/6562ft

PARACEL ISLANDS

STATUS: *Disputed* **CLAIMED:** *Not applicable*
CAPITAL: *Not applicable* **POPULATION:** *Unknown*

OCCUPIED BY CHINESE FORCES (who call them the Xisha islands), but also claimed by Taiwan and Vietnam, the Paracel Islands are a small collection of coral atolls situated some 325 km (200 miles) east of Vietnam, in the South China Sea. Subject to frequent typhoons and with a tropical climate, the Paracels are at the center of a regional dispute over the vast reserves of oil and natural gas which are believed to lie beneath their territorial waters. China has built port facilities and an airport on Woody Island to support its claim.

PITCAIRN ISLANDS

STATUS: British overseas territory **CLAIMED:** 1887
CAPITAL: Adamstown **POP.:** 43 **DENSITY:** 1.2/km² (3/mi²)

A GROUP OF volcanic South Pacific islands, Pitcairn is Britain's most isolated dependency. Pitcairn Island was the last refuge for the mutineers from HMS *Bounty*. The economy operates by barter, fishing, and subsistence farming, and is reliant on regular airdrops from New Zealand and periodic visits by supply vessels. Postage stamp sales provide foreign currency earnings. In 2002, 20 Pitcairners were charged with sexual assault.

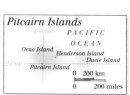

PUERTO RICO

STATUS: Commonwealth territory of the USA **CLAIMED:** 1898
CAPITAL: San Juan **POP.:** 4 million **DENSITY:** 446/km² (1156/mi²)

PUERTO RICO, a US territory since its invasion in 1898, is by far the most populous nonindependent territory. It is the easternmost of the Greater Antilles chain in the Caribbean. The population density, highest around San Juan, is comparable with the Netherlands and is higher than in any US state. The tropical climate attracts growing numbers of tourists, 80% from the US, and there have been major efforts to expand hotel and resort facilities.

Puerto Rico was granted its current commonwealth status in 1952, four years after an abortive pro-independence uprising. The inhabitants have US citizenship but only limited self-government. In three plebiscites, in 1967, 1993, and 1998, the islanders endorsed continued commonwealth status rather than opting for either US statehood or independence. The most recent of these votes was extremely close, but the pro-statehood governor who called the 1993 and 1998 votes, Pedro Rossello, was replaced by the anti-statehood Sila Calderón – the first female governor of Puerto Rico – in 2001.

Although thousands of the mostly Spanish-speaking Puerto Ricans have migrated to the US mainland in search of higher wages, the islanders have one of the highest living

At night, the bright lights of Puerto Rico's well-developed roads, settlements, and busy ports are in sharp contrast to the rest of the Caribbean – notably the dark outline of Haiti, just to the west.

standards in the region. Tax relief, cheap labor, and the island's role as an export-processing zone, mainly for the US market, attracted many businesses. Clothing, electronics, petrochemical, and pharmaceutical industries traditionally dominated, but the decision to phase out tax exemptions for companies reinvesting in the island caused a slump in 1996, and more emphasis is now being placed on the service sector. New industries include health care and clinical testing, biotechnology, and other knowledge-based areas.

Governor Calderón spearheaded the campaign to stop the US navy from using the populated eastern island of Vieques for bombing practice. In 2000 an invasion of the bombing range by protestors led to some high-profile arrests, including that of Robert Kennedy Jr. A year later newly elected US president George W. Bush announced that the navy would not use the island after 2003.

RÉUNION

STATUS: French overseas department **CLAIMED:** 1638
CAPITAL: Saint-Denis **POP.:** 728,400 **DENSITY:** 290/km² (751/mi²)

THE LARGE VOLCANIC ISLAND of Réunion, 800 km (500 miles) east of Madagascar, provides France with an important strategic presence – and a large military base – in the Indian Ocean. Its mountainous interior has forced the majority of the population to live along the coast. Tensions still exist between the very poor black community and the wealthy Indian and European groups, although the violence of 1991 has not been repeated. Despite the introduction of measures applicable to all French overseas *départements*, intended to improve social and economic standards, unemployment remains high and the cost of living very expensive. Réunion's main crop is sugarcane, although Cyclone Dina devastated production in 2002.

ST. PIERRE & MIQUELON

STATUS: French territorial collectivity **CLAIMED:** 1604
CAPITAL: St. Pierre **POP.:** 6516 **DENSITY:** 26/km² (68/mi²)

ST. PIERRE & Miquelon is a group of barren islands lying just off the south coast of Newfoundland, Canada. The islands are surrounded by some of the world's richest fishing grounds. Their inhabitants have traditionally earned a living from fishing, and from servicing foreign trawler fleets off the coast. A long-running and sometimes bitter dispute between Canada and France over fishing and mineral rights was settled in 1992. The ruling, which was generally deemed to be in Canada's favor, has led the French authorities to diversify the economy by developing port facilities and encouraging tourism.

ST. HELENA & DEPENDENCIES

STATUS: British overseas territory **CLAIMED:** 1673
CAPITAL: Jamestown **POP.:** 4647 **DENSITY:** 38/km² (99/mi²)

TOGETHER, the islands of St. Helena, Tristan da Cunha, and Ascension form Britain's main dependency in the south Atlantic. St. Helena is famed for being the final place of exile for Napoléon. Its main economic activities – fishing, livestock farming, and the sale of handicrafts – are unable to support the population; as a result, underemployment on the island is a major problem. Many St. Helenians have been forced to seek work on Ascension, which has no resident population and is operated as a military base and communications center, although civilian flights have been permitted since 1998. Tristan da Cunha, a volcanic island 2000 km (1240 miles) south of St. Helena, is inhabited by a small, closely knit farming community. It was badly hit by severe winter storms in 2001. The 2002 British Overseas Territories Act granted all overseas citizens full British citizenship, assuaging a source of local resentment.

SPRATLY ISLANDS

STATUS: *Disputed* **CLAIMED:** *Not applicable*
CAPITAL: *Not applicable* **POPULATION:** *Unknown*

SCATTERED ACROSS a large area of the South China Sea, the reefs, islands, and atolls that make up the Spratly Islands have become one of South Asia's most serious security issues. Claimed, all or in part, by China, Taiwan, Vietnam, Brunei, Malaysia, and the Philippines, more than 40 of the larger islands now have garrisons from some of the claimant states. The reasons for this interest, and the occasional skirmish, are twofold. Strategically, the islands control some of the world's most important shipping lanes. In addition, surveys suggest that some of the largest oil and gas reserves yet found lie in the Spratlys' territorial waters.

The isolated Chinese occupying force on one of the Spratly Islands.

LAND HEIGHT above Sea Level 200m/656ft 500m/1640ft 1000m/3281ft 1500m/4572ft above 2000m/6562ft

SVALBARD

STATUS: Norwegian dependency CLAIMED: 1920

CAPITAL: Longyearbyen POP.: 2515 DENSITY: 0.04/km² (0.1/mi²)

MORE THAN 150 ice-covered Arctic islands 650 km (400 miles) north of Norway make up Svalbard. In accordance with the Spitsbergen Treaty of 1920, nationals of the treaty powers have equal rights to exploit the coal deposits, subject to regulation by Norway. The only companies still mining are Russian and Norwegian. There has been conflict with Iceland over fishing rights. Over half of the area of the islands is designated as environmentally protected.

TOKELAU

STATUS: New Zealand dependent territory CLAIMED: 1926

CAPITAL: *Not applicable* POP.: 1487 DENSITY: 149/km² (372/mi²)

A 1989 UN REPORT states that in the 21st century these islands in the South Pacific will disappear under the sea, unless action is taken to stop global warming. The economy depends on a tuna cannery and the sale of fishing licenses, postage stamps, and coins; a catamaran link between the atolls has increased tourist potential. Tokelau's small size and economic fragility make independence unlikely, but in 1996 it gained the right to enact its own internal legislation, and since 2001 the local authorities have had full control of the island's public services. Nearly 4000 Tokelauans live in New Zealand.

TURKS & CAICOS ISLANDS

STATUS: British overseas territory CLAIMED: 1766

CAPITAL: Cockburn Town POP.: 16,863 DENSITY: 39/km² (102/mi²)

SITUATED 40 km (25 miles) south of the Bahamas, the Turks and Caicos Islands is a group of 30 low-lying islands, eight of which are inhabited. A traditional salt-based economy was exhausted in 1964, leading to two decades of stagnation. Since the 1980s, however, tourism and offshore banking have led to a dramatic turnaround in the islands' fortunes.

VIRGIN ISLANDS (US)

STATUS: Unincorporated territory of the USA CLAIMED: 1917

CAPITAL: Charlotte Amalie POP.: 108,612 DENS.: 313/km² (811/mi²)

THE US VIRGIN ISLANDS are a collection of 53 volcanic islands, just to the east of Puerto Rico. Most of the population – a mix of African and European ethnic groups – live on the main islands of St. John, St. Thomas, and St. Croix. Tourism is the principal activity, although St. Croix has also used federal aid to develop industry. It has one of the world's largest oil refineries.

St. Thomas, US Virgin Islands, is a major stop-off for Caribbean cruise ships. Tourists are attracted by the island's duty-free shopping.

WAKE ISLAND

STATUS: Unincorporated territory of the USA CLAIMED: 1898

CAPITAL: *Not applicable* POP.: *Not applicable*

WAKE ISLAND, in fact three islands that form the rim of an extinct volcano, has a US air base, whose airstrip can be used in emergencies by trans-Pacific flights. After widespread condemnation a 1998 proposal to store nuclear waste there was dropped. It is claimed by the Marshall Islands.

WALLIS & FUTUNA

STATUS: French overseas territory CLAIMED: 1842

CAPITAL: Matá'Utu POP.: 14,600 DENSITY: 53/km² (138/mi²)

UNLIKE FRANCE'S other South Pacific overseas territories, the inhabitants of Wallis and Futuna have little desire for greater autonomy. The islands' subsistence economy produces a variety of tropical crops, while expatriate remittances and the sale of licenses to Japanese and South Korean fishing fleets provide foreign exchange. Deforestation, leading to soil erosion, is of great concern.

GLOSSARY OF GEOGRAPHICAL TERMS

THE GLOSSARY FOLLOWING lists all geographical terms occurring on the maps and in main-entry names in the Index~Gazetteer. These terms may precede, follow, or be run together with the proper element of the name; where they precede it the term is reversed for indexing purposes – thus Poluostrov Yamal is indexed as Yamal, Poluostrov.

KEY
Geographical term *Language,* Term

A

Å *Danish, Norwegian,* River
Alpen *German,* Alps
Altiplanicie *Spanish,* Plateau
Älv(en) *Swedish,* River
Anse *French,* Bay
Archipiélago *Spanish,* Archipelago
Arcipelago *Italian,* Archipelago
Arquipélago *Portuguese,* Archipelago
Aukštuma *Lithuanian,* Upland

B

Bahía *Spanish,* Bay
Baía *Portuguese,* Bay
Baḥr *Arabic,* River
Baie *French,* Bay
Bandao *Chinese,* Peninsula
Banjaran *Malay,* Mountain range
Batang *Malay,* Stream
-berg *Afrikaans, Norwegian,* Mountain
Birket *Arabic ,* Lake
Boğazı *Turkish,* Lake
Bucht *German,* Bay
Bugten *Danish,* Bay
Buḥayrat *Arabic,* Lake, reservoir
Buḥeiret *Arabic,* Lake
Bukit *Malay,* Mountain
-bukta *Norwegian,* Bay
bukten *Swedish,* Bay
Burnu *Turkish,* Cape, point
Buuraha *Somali,* Mountains

C

Cabo *Portuguese,* Cape
Cap *French,* Cape
Cascada *Portuguese,* Waterfall
Cerro *Spanish,* Mountain
Chaîne *French,* Mountain range
Chau *Cantonese,* Island
Chāy *Turkish,* River
Chhâk *Cambodian,* Bay
Chhu *Tibetan,* River
-chŏsuji *Korean,* Reservoir
Chott *Arabic,* Salt lake, depression
Ch'ün-tao *Chinese,* Island group
Chuôr Phnum *Cambodian,* Mountains
Cordillera *Spanish,* Mountain range
Costa *Spanish,* Coast
Côte *French,* Coast
Cuchilla *Spanish,* Mountains

D

Dağı *Azerbaijani, Turkish,* Mountain
Dağları *Azerbaijani, Turkish,* Mountains
-dake *Japanese,* Peak
Danau *Indonesian,* Lake
Đao *Vietnamese,* Island
Daryā *Persian,* River
Daryācheh *Persian,* Lake
Dasht *Persian,* Plain, desert
Dawḥat *Arabic,* Bay
Dere *Turkish,* Stream
Dili *Azerbaijani,* Spit
-do *Korean,* Island
Dooxo *Somali,* Valley
Düzü *Azerbaijani,* Steppe
-dwīp *Bengali,* Island

E

Embalse *Spanish,* Reservoir
Erg *Arabic,* Dunes
Estany *Catalan,* Lake
Estrecho *Spanish,* Strait
-ey *Icelandic,* Island
Ezero *Bulgarian, Macedonian,* Lake

F

Fjord *Danish,* Fjord
-fjorden *Norwegian,* Fjord
-fjørdhur *Faeroese,* Fjord
Fleuve *French,* River
Fliegu *Maltese,* Channel
-fljór *Icelandic,* River

G

-gang *Korean,* River
Ganga *Nepali, Sinhala,* River
Gaoyuan *Chinese,* Plateau
-gawa *Japanese,* River
Gebel *Arabic,* Mountain
-gebirge *German,* Mountains
Ghubbat *Arabic,* Bay
Gjiri *Albanian,* Bay
Gol *Mongolian,* River
Golfe *French,* Gulf
Golfo *Italian, Spanish,* Gulf
Gora *Russian, Serbian,* Mountain
Gory *Russian,* Mountains
Guba *Russian,* Bay
Gunung *Malay,* Mountain

H

Ḥadd *Arabic,* Spit
-haehyŏp *Korean,* Strait
Haff *German,* Lagoon
Hai *Chinese,* Sea, bay
Ḥammādat *Arabic,* Plateau
Hāmūn *Persian,* Lake

Hawr *Arabic,* Lake
Hāyk' *Amharic,* Lake
He *Chinese,* River
Helodrano *Malagasy,* Bay
-hegység *Hungarian,* Mountain range
Hka *Burmese,* River
-ho *Korean,* Lake
Hô *Korean,* Reservoir
Holot *Hebrew,* Dunes
Hora *Belarussian,* Mountain
Hrada *Belarussian,* Mountains, ridge
Hsi *Chinese,* River
Hu *Chinese,* Lake

I

Île(s) *French,* Island(s)
Ilha(s) *Portuguese,* Island(s)
Ilhéu(s) *Portuguese,* Islet(s)
Irmak *Turkish,* River
Isla(s) *Spanish,* Island(s)
Isola (Isole) *Italian,* Island(s)

J

Jabal *Arabic,* Mountain
Jāl *Arabic,* Ridge
-järvi *Finnish,* Lake
Jazīrat *Arabic,* Island
Jazīreh *Persian,* Island
Jebel *Arabic,* Mountain
Jezero *Serbo-Croat,* Lake
Jiang *Chinese,* River
-joki *Finnish,* River
-jökull *Icelandic,* Glacier
Juzur *Arabic,* Islands

K

Kaikyō *Japanese,* Strait
-kaise *Lappish,* Mountain
Kali *Nepali,* River
Kalnas *Lithuanian,* Mountain
Kalns *Latvian,* Mountain
Kang *Chinese,* Harbor
Kangri *Tibetan,* Mountain(s)
Kaôh *Cambodian,* Island
Kapp *Norwegian,* Cape
Kavīr *Persian,* Desert
K'edi *Georgian,* Mountain range
Kediet *Arabic,* Mountain
Kepulauan *Indonesian, Malay,* Island group
Khalîg, Khalīj *Arabic,* Gulf
Khawr *Arabic,* Inlet
Khola *Nepali,* River
Khrebet *Russian,* Mountain range
Ko *Thai,* Island
Kolpos *Greek,* Bay
-kopf *German,* Peak
Körfäzi *Azerbaijani,* Bay
Körfezi *Turkish,* Bay

Kõrgustik *Estonian*, Upland
Koshi *Nepali*, River
Kowtal *Persian*, Pass
Kūh(hā) *Persian*, Mountain(s)
-kundo *Korean*, Island group
-kysten *Norwegian*, Coast
Kyun *Burmese*, Island

L

Laaq *Somali*, Watercourse
Lac *French*, Lake
Lacul *Romanian*, Lake
Lago *Italian*, *Portuguese*, *Spanish*, Lake
Laguna *Spanish*, Lagoon, Lake
Laht *Estonian*, Bay
Laut *Indonesian*, Sea
Lembalemba *Malagasy*, Plateau
Lerr *Armenian*, Mountain
Lerrnashght'a *Armenian*, Mountain range
Les *Czech*, Forest
Lich *Armenian*, Lake
Liqeni *Albanian*, Lake
Lumi *Albanian*, River
Lyman *Ukrainian*, Estuary

M

Mae Nam *Thai*, River
-mägi *Estonian*, Hill
Maja *Albanian*, Mountain
-man *Korean*, Bay
Marios *Lithuanian*, Lake
-meer *Dutch*, Lake
Melkosopochnik *Russian*, Plain
-meri *Estonian*, Sea
Mifraz *Hebrew*, Bay
Monkhafad *Arabic*, Depression
Mont(s) *French*, Mountain(s)
Monte *Italian*, *Portuguese*, Mountain
More *Russian*, Sea
Mörön *Mongolian*, River

N

Nagor'ye *Russian*, Upland
Nahal *Hebrew*, River
Nahr *Arabic*, River
Nam *Laotian*, River
Nehri *Turkish*, River
Nevado *Spanish*, Mountain (snow-capped)
Nisoi *Greek*, Islands
Nizmennost' *Russian*, Lowland, plain
Nosy *Malagasy*, Island
Nur *Mongolian*, Lake
Nuruu *Mongolian*, Mountains
Nuur *Mongolian*, Lake
Nyzovyna *Ukrainian*, Lowland, plain

O

Ostrov(a) *Russian*, Island(s)
Oued *Arabic*, Watercourse
-oy *Faeroese*, Island
-øy(a) *Norwegian*, Island
Oya *Sinhala*, River
Ozero *Russian*, *Ukrainian*, Lake

P

Passo *Italian*, Pass
Pegunungan *Indonesian*, *Malay*, Mountain range
Pelagos *Greek*, Sea
Penisola *Italian*, Peninsula
Peski *Russian*, Sands
Phanom *Thai*, Mountain
Phou *Laotian*, Mountain
Pi *Chinese*, Point
Pic *Catalan*, Peak
Pico *Portuguese*, *Spanish*, Peak
Pik *Russian*, Peak
Planalto *Portuguese*, Plateau
Planina, Planini *Bulgarian*, *Macedonian*, *Serbo-Croat*, Mountain range
Ploskogor'ye *Russian*, Upland
Poluostrov *Russian*, Peninsula
Potamos *Greek*, River
Proliv *Russian*, Strait
Pulau *Indonesian*, *Malay*, Island
Pulu *Malay*, Island
Punta *Portuguese*, *Spanish*, Point

Q

Qā' *Arabic*, Depression
Qolleh *Persian*, Mountain

R

Raas *Somali*, Cape
-rags *Latvian*, Cape
Ramlat *Arabic*, Sands
Ra's *Arabic*, Cape, point, headland
Ravnina *Bulgarian*, *Russian*, Plain
Récif *French*, Reef
Represa (Rep.) *Spanish*, *Portuguese*, Reservoir
-rettō *Japanese*, Island chain
Riacho *Spanish*, Stream
Riban' *Malagasy*, Mountains
Rio *Portuguese*, River
Río *Spanish*, River
Riu *Catalan*, River
Rivier *Dutch*, River
Rivière *French*, River
Rowd *Pashtu*, River
Rūd *Persian*, River
Rudohorie *Slovak*, Mountains
Ruisseau *French*, Stream

S

Sabkhat *Arabic*, Salt marsh
Şaḥrā' *Arabic*, Desert
Samudra *Sinhala*, Reservoir
-san *Japanese*, *Korean*, Mountain
-sanchi *Japanese*, Mountains
-sanmaek *Korean*,
Sarīr *Arabic*, Desert
Sebkha, Sebkhet *Arabic*, Salt marsh, depression
See *German*, Lake
Selat *Indonesian*, Strait
-selkä *Finnish*, Ridge
Selseleh *Persian*, Mountain range
Serra *Portuguese*, Mountain
Serranía *Spanish*, Mountain
Sha'ib *Arabic*, Watercourse
Shamo *Chinese*, Desert
Shan *Chinese*, Mountain(s)

Shan-mo *Chinese*, Mountain range
Shaṭṭ *Arabic*, Distributary
-shima *Japanese*, Island
Shiqqat *Arabic*, Depression
Shui-tao *Chinese*, Channel
Sierra *Spanish*, Mountains
Sơn *Vietnamese*, Mountain
Sông *Vietnamese*, River
-spitze *German*, Peak
Štít *Slovak*, Peak
Stoeng *Cambodian*, River
Stretto *Italian*, Strait
Su Anbarı *Azerbaijani*, Reservoir
Sungai *Indonesian*, *Malay*, River
Suu *Turkish*, River

T

Tal *Mongolian*, Plain
Tandavan' *Malagasy*, Mountain range
Tangorombohitr' *Malagasy*, Mountain massif
Tao *Chinese*, Island
Tassili *Berber*, Plateau, mountain
Tau *Russian*, Mountain(s)
Taungdan *Burmese*, Mountain range
Teluk *Indonesian*, *Malay*, Bay
Terara *Amharic*, Mountain
Tog *Somali*, Valley
Tônlé *Cambodian*, Lake
Top *Dutch*, Peak
-tunturi *Finnish*, Mountain
Tur'at *Arabic*, Channel

V

Väin *Estonian*, Strait
-vatn *Icelandic*, Lake
-vesi *Finnish*, Lake
Vinh *Vietnamese*, Bay
Vodokhranilishche (Vdkhr.) *Russian*, Reservoir
Vodoskhovyshche (Vdskh.) *Ukrainian*, Reservoir
Volcán *Spanish*, Volcano
Vozvyshennost' *Russian*, Upland, plateau
Vrh *Macedonian*, Peak
Vysochyna *Ukrainian*, Upland
Vysočina *Czech*, Upland

W

Waadi *Somali*, Watercourse
Wādī *Arabic*, Watercourse
Wāḥat, Wâhat *Arabic*, Oasis
Wald *German*, Forest
Wan *Chinese*, Bay
Wyżyna *Polish*, Upland

X

Xé *Laotian*, River

Y

Yarımadası *Azerbaijani*, Peninsula
Yazovir *Bulgarian*, Reservoir
Yoma *Burmese*, Mountains
Yü *Chinese*, Island

Z

Zaliv *Bulgarian*, *Russian*, Bay
Zatoka *Ukrainian*, Bay
Zemlya *Russian*, Bay

GLOSSARY OF ABBREVIATIONS

THIS GLOSSARY provides a comprehensive guide to the abbreviations used.

A

abbrev. abbreviated
ABM antiballistic missile(s)
Afr. Afrikaans
AIDS acquired immunodeficiency syndrome
Alb. Albanian
ALCM air-launched Cruise missile(s)
Amh. Amharic
ANC African National Congress
anc. ancient
APC armored personnel carrier(s)
approx. approximately
Ar. Arabic
Arm. Armenian
ASSR Autonomous Soviet Socialist Republic
Aust. Australian
Az. Azerbaijani
Azerb. Azerbaijan

B

Basq. Basque
BBC British Broadcasting Corporation
BCE Before Common Era
b/d barrels per day
Bel. Belarussian
Ben. Bengali
Ber. Berber
B-H Bosnia and Herzegovina
bn billion (1000 million)
BP British Petroleum
Bret. Breton
Brig. Brigadier
Brit. British
BSE bovine spongiform encephalopathy
Bul. Bulgarian
Bur. Burmese

C

C central
C. Cape
°C degrees (Centigrade)
Cam. Cambodian
Cant. Cantonese
Capt. Captain
CAR Central African Republic
Cast. Castilian
Cat. Catalan
CE Common Era
Chin. Chinese
CIA Central Intelligence Agency
cm centimeter(s)
Cmdr. Commander
CNN Cable News Network
Col. Colonel
Cro. Croat
Cz. Czech
Czech Rep. Czech Republic

D E

Dan. Danish
dept. department
dev. development
Dom. Rep. Dominican Republic
Dr. Doctor
DRC Democratic Republic of Congo
Dut. Dutch
dwt dead weight tonnage
E east
EEC/EC European Community
EEZ Exclusive Economic Zone
ECU European Currency Unit
EMS European Monetary System
Eng. English
est. estimated
Est. Estonian

F G

°F degrees (Fahrenheit)
Faer. Faeroese
Fij. Fijian
Fin. Finnish
Fr. Father
Fr. French
Franc Francophone
Fris. Frisian
ft foot/feet
FYRM Former Yugoslav Republic of Macedonia
FZ Franc Zone
Gael. Gaelic
Gal. Galician
GATT General Agreement on Tariffs and Trade
GDP Gross Domestic Product (the total value of goods and services produced by a country excluding income from foreign countries)
Gen. General
Geor. Georgian
Ger. German
Gk Greek
GNP Gross National Product (the total value of goods and services produced by a country)

H I

Heb. Hebrew
HEP hydroelectric power
HH His/Her Highness
Hind. Hindi
HIPC heavily indebted poor country(ies)
hist. historical
HIV human immunodeficiency virus
H.M. His/Her Majesty
HMS His/Her Majesty's ship
H.R.H His/Her Royal Highness
H.S.H His/Her Serene Highness
Hung. Hungarian
I. Island
ICBM intercontinental ballistic missile(s)
Icel. Icelandic
in inch(es)
In. Inuit (Eskimo)
Ind. Indonesian
Intl International

Ir. Irish
IRBM intermediate-range ballistic missile(s)
Is Islands
It. Italian

J K L

Jap. Japanese
Kaz. Kazakh
kg kilogram(s)
Kir. Kirghiz
km kilometer(s)
km² square kilometer (singular)
Kor. Korean
Kurd. Kurdish
kw kilowatt(s)
kwh kilowatt hour(s)
L. Lake
Lao. Laotian
Lat. Latin
Latv. Latvian
Liech. Liechtenstein
Lith. Lithuanian
LNG liquefied natural gas
Lt. Lieutenant
Lusoph Lusophone
Lux. Luxembourg

M N

m million/meter(s)
Mac. Macedonian
Maced. Macedonia
Maj. Major
Mal. Malay
Malg. Malagasy
Malt. Maltese
MBA Master of Business Administration
mi. mile(s)
mi² square mile(s)
Mong. Mongolian
MP Member of Parliament
MSP Member of Scottish Parliament
Mt. Mountain/Mount
Mts Mountains
MW megawatt(s)
N north
NASA National Aeronautics and Space Administration
Nep. Nepali
Nepad New Partnership for Africa's Development
Neth. Netherlands
NGO Nongovernmental Organization
NIC Newly Industrialized Country
Nic. Nicaraguan
Nor. Norwegian
NPT Non-Proliferation Treaty
NZ New Zealand

P Q R

Pash. Pashtu
Per. Persian
PLO Palestine Liberation Organization
PNG Papua New Guinea
Pol. Polish
Poly. Polynesian

Port. Portuguese
POW prisoner of war
prev. previously
Rep. Represa (Spanish, Portuguese for reservoir)
Rep. Republic
Res. Reservoir
Rev. Reverend
Rmsch. Romansch
Rom. Romanian
Rus. Russian
Russ. Fed. Russian Federation

S

S south
S. & Mon. Serbia & Montenegro
SALT Strategic Arms Limitation Treaty
SCr. Serbo-Croatian
Serb. Serbian
Sinh. Sinhala
SLBM submarine-launched ballistic missile(s)
Slvk. Slovak
Slvn. Slovene
Som. Somali
Sp. Spanish
sq. square
SSBN nuclear-fuelled ballistic-missile submarine(s)
SSM surface-to-surface missile(s)
St. Saint
START Strategic Arms Reduction Treaty
Strs. Straits
Swa. Swahili
Swe. Swedish
Switz. Switzerland

T U

Taj. Tajik
TGV *train à grande vitesse*
Th. Thai
Thai. Thailand
Tib. Tibetan
Turk. Turkish
Turkm. Turkmenistan
UAE United Arab Emirates
Uigh. Uighur
UK United Kingdom
Ukr. Ukrainian
UN United Nations
Urd. Urdu
US/USA United States of America
USS United States ship
USSR Union of Soviet Socialist Republics
Uzb. Uzbek

V W X Y

var. variant
VCR video cassette recorder
Vdkhr. Vodokhranilishche (Russian for reservoir)
Vdskh. Vodoskhovyshche (Ukrainian for reservoir)
Vtn. Vietnamese
W west
Wel. Welsh

GEOGRAPHICAL PLACE-NAMES

THE CHOICES confronting a map-maker when deciding which place-name style to use on a map are surprisingly varied. The criteria adopted may be affected by a range of factors: the existence of foreign and native language forms of a place name (London, Londres, Londra), variant spellings used within the country itself (Gent, Gand), and the existence of completely different language forms for international features (the English Channel, La Manche).

In addition to these, political expedience, simple clarity, and the use to which the published map may be put are all factors that need consideration.

The revision of place-name forms and spellings, which is a continuing administrative activity worldwide, adds a further dimension of complexity to the subject. Since the collapse of Soviet communism, for instance, place-names in Russia have been altered to expunge traces of communist ideology (the most famous being the 1991 reversion of Leningrad to its pre-1914 name, St. Petersburg). In many former Soviet republics, Russian names have been replaced with native language forms (notably in Ukraine, Belarus, Georgia, and Armenia).

Standardized Arabic names have been hindered by the persistent use of French forms in practice.

THE MAPS

The maps in the Nations of the World section of this book have used the most up-to-date reference sources available to provide local name forms and spellings, that is to say those used within the country. In an age when international travel, on holiday or on business, is commonplace, this criterion seems the most appropriate.

English conventional forms have been used for all international features (such as sea areas between countries, and cross-border mountain ranges); for all country names (the Index-Gazetteer provides local forms and spellings, while commonly used alternative names, such as Burma/Myanmar, are also made clear in the national A–Z entry); and for all capital cities. The Index-Gazetteer provides a fully cross-referenced system that will guide the reader the short distance from the English conventional "Florence" to the local "Firenze," as used on the maps.

English conventional forms also appear on all the maps in the World Factfile. These maps have not been indexed, as all contemporary places featured are most usefully and accurately identified on the national maps.

THE INDEX-GAZETTEER

The Index-Gazetteer lists all names that appear on the maps in the Nations of the World section of the book. Physical features are defined as such, as are countries and those administrative or regional names included on the maps; all other names are those of population centers. Location is given by page number, then country, and is narrowed down by positional reference as N(orth), S(outh), E(ast), W(est), or C(entral), or combinations of these as appropriate.

Following each main entry name are given: variant spellings of the name most commonly found; its previous name or names; and such foreign-language forms of the name as are pertinent to modern history since 1940. This is the cut-off date generally adopted, permitting the inclusion of all place-name changes made during or after World War II. Exceptionally, name changes made in Russia and other countries of the former Soviet Union before 1940 are given, since many old names in these countries are now being restored.

INDEX

PAGE NUMBERS SHOWN IN ITALICS

Geographical Terms and Abbreviations on pages 654-657 / ❖ = *capital*

A

Aa *see* Gauja
Aabenraa *see* Åbenrå
Aachen *266 Fr.* Aix-la-Chapelle,
  Dut. Aken. W Germany
Aalborg *see* Ålborg
Aalesund *see* Ålesund
Aaley *364 var.* Âlayh, Aley.
  C Lebanon
Aalsmeer *429* W Netherlands
Aalst *127 Fr.* Alost. C Belgium
Aanaarjävri *see* Inarijärvi
Aanjar *364* C Lebanon
Aarau *550* N Switzerland
Aare *550 var.* Aar. River
  of W Switzerland
Aarhus *see* Århus
Aarlen *see* Arlon
Aarschot *127* C Belgium
Aassi, Nahr el *see* Orantes
Aba *440* S Nigeria
Abaco Island *112* island of N Bahamas
Åbādān *309* W Iran
Abai *see* Blue Nile
Abaiang *348* island of the Gilbert Is,
  W Kiribati
Abakan *485 prev.* Khakassk,
  Ust'-Abakanskoye. C Russian
  Federation
Abancay *463* SE Peru
Abariringa *see* Kanton
Åbaya Hāyk' *245 It.* Abbaia, *Eng.* Lake
  Margherita. Lake of SW Ethiopia
Abay Wenz *see* Blue Nile
Abbeville *255* N France
'Abd al 'Azīz, Jabal *555* mountains
  of NE Syria
'Abdalī *356* S Kuwait
Abd-Al-Kuri *631* island of SE Yemen,
  off the Horn of Africa
Abéché *180 var.* Abécher. E Chad
Abemama *348* island of the
  Gilbert Is, W Kiribati
Abengourou *328* E Ivory Coast
Åbenrå *218 var.* Aabenraa,
  Ger. Apenrade. Jylland,
  SW Denmark
Abeokuta *440* SW Nigeria
Abercorn *see* Mbala
Aberdeen *597* NE Scotland, UK
Aberdeen *603* South Dakota, NC USA
Abergwaun *see* Fishguard
Abersee *see* Wolfgangsee
Abertawe *see* Swansea
Aberystwyth *597* W Wales, UK
Abhā *506* S Saudi Arabia
Abhe, Lake *222, 245 Amh.* Åbhē Bid
  Hāyk'. Lake of Djibouti and Ethiopia
Abidjan *328* S Ivory Coast
Åbo *see* Turku
Aboisso *328* SE Ivory Coast
Abo, Massif d' *180* mountain range
  of N Chad
Abomey *132* S Benin
Abong Mbang *168* SE Cameroon
Abou-Déïa *180* S Chad
Aboudouhour *see* Abū aḍ Ḍuhūr
Abou Kémal *see* Abū Kamāl
Aboumi *260* E Gabon
Abovyan *98 var.* Abovjan. C Armenia
Abra *466* river of Luzon, N Philippines
Abrād, Wādī *631* seasonal river of
  NW Yemen
Abraham Bay *see* Carlton, The
Abruzzese, Appennino *323*
  mountain range of C Italy
'Abs *631 var.* Sūq 'Abs. W Yemen
Abşeron Yarımadası *110*
  Rus. Apsheronskiy Poluostrov.
  Oil-rich peninsula of E Azerbaijan
Abū aḍ Ḍuhūr *555 Fr.* Aboudouhour.
  NW Syria
Abū al Abyaḍ *594* island of N United
  Arab Emirates

Abū al Jirfān, Sha'īb *356*
  var. Sh'ib Abu Jarfan. Dry
  watercourse of N Kuwait
Abū al Khaṣīb *312 var.* Abul Khasib.
  SE Iraq
Abu al Mawj, Ra's *115* cape of
  W Bahrain
Abu Dhabi *594 Ar.* Abū Ẓaby,
  var. Abū Ẓabī. ❖ of United Arab
  Emirates
Abuja *440* ❖ of Nigeria, C Nigeria
Abū Jarjūr, Ra's *115* cape
  of E Bahrain
Abū Kamāl *555 Fr.* Abou Kémal.
  E Syria
Abul Khasib *see* Abū al Khaṣīb
Abuná *136* river of Bolivia
  and Brazil
Abū Thaylah *479* NE Qatar
Abū Ẓabī *see* Abu Dhabi
Abū Ẓaby *see* Abu Dhabi
Abyaḍ, Baḥr al *see* White Nile
Åbybro *218* N Denmark
Abyssinia *see* Ethiopia
Acaill *see* Achill Island
Acajutla *237* W El Salvador
Acapulco *403 var.* Acapulco de
  Juárez. S Mexico
Acaraí, Serra *144 Eng.* Acarai
  Mountains. Mountain range of Brazil
  and Guyana
Acarigua *623* NW Venezuela
Accra *272* ❖ of Ghana, SE Ghana
Achacachi *136* W Bolivia
Acharnés *275 prev.* Akharnaí.
  SE Greece
Acheloos *275 var.* Aspropotamos,
  prev. Akhelóös. River of W Greece
Achénouma *439* NE Niger
Achill Island *316 Ir.* Acaill.
  Island of W Ireland
Achna *see* Athna
Achorstock Point *652* headland of
  W Tristan da Cunha
Achwa *588 var.* Aswa.
  River of N Uganda
Acireale *323* Sicilia, S Italy
Acklins Island *112* island of
  S Bahamas
Aconcagua, Cerro *95* mountain of
  W Argentina
Açores *see* Azores
Açores, Arquipélago dos *see* Azores
A Coruña *534 Cast.* La Coruña.
  NW Spain
Acoua *649* NW Mayotte
Acquaviva *502* NW San Marino
Acre *see* 'Akko
Acurnam *238 var.* Acurenan,
  Akurenan. S Río Muni,
  Equatorial Guinea
Adalia *see* Antalya
Adalia, Gulf of *see* Antalya Körfezi
Adam *448* N Oman
Adama *see* Nazrēt
Adamaoua, Massif d' *168*
  Eng. Adamawa. Plateau of
  West Africa
Adam-jo-Tando *see* Tando Ādam
Adam's Bridge *538* chain of shoals to
  the NW of Sri Lanka
Adamstown *651* ❖ of Pitcairn Islands,
  NE Pitcairn Island, Pitcairn Islands
'Adan *631 Eng.* Aden. SW Yemen
Adana *581 var.* Seyhan. S Turkey
Adapazarı *580 var.* Sakarya.
  NW Turkey
Aḍ Ḍab 'īyah *594* C United Arab
  Emirates
Aḍ Ḍafrah *594* desert region of
  W United Arab Emirates
Ad Dahnā *506* N Saudi Arabia.
Ad Dakhla *414* W Western Sahara
Ad Dalanj *see* Dilling
Ad Dammām *506* desert region of
  NE Saudi Arabia
Ad Dawḥah *see* Doha

Ad Dibdibah *356* mountain
  of W Kuwait
Addīgrat *245* N Ethiopia
Ad Dirāz *115* NW Bahrain
Addis Ababa *245 Amh.* Ādīs Ābeba.
  ❖ of Ethiopia, C Ethiopia
Ad Dīwānīyah *312 var.* Diwaniya.
  C Iraq
Addu Atoll *390* atoll of S Maldives
Adelaide *101* S Australia
Adelphi *498* E St Vincent, St Vincent
  & the Grenadines
Adelsberg *see* Postojna
Aden *see* 'Adan
Aden, Gulf of *222, 528, 631*
  var. Badyarada 'Adméd. Gulf
  connecting the Indian Ocean
  and Red Sea
Adh Dhayd *594 var.* Deira. NE United
  Arab Emirates
Adh Dhirā' *338* W Jordan
Ādī Ārk'ay *245 var.* Addi Arkay.
  N Ethiopia
Adige *323 Ger.* Etsch. River of N Italy
Adi Keyih *240 var.* Adi Keyah.
  SE Eritrea
Adi Kwala *240* S Eritrea
Ādīs Ābeba *see* Addis Ababa
Ādīs Zemen *245* NW Ethiopia
Adi Tekelezan *240* C Eritrea
Adıyaman *581* SE Turkey
Admiralty Islands *458* island group
  of N Papua New Guinea
Ado-Ekiti *440* SW Nigeria
Adola *see* Kibre Mengist
Ado-Odo *440 var.* Ado. SW Nigeria
Adour *255* river of SW France
Adrar des Ifôghas *392* mountainous
  region of C Sahara, NE Mali
Adriatic Sea *323, 512 It.* Mare
  Adriatico, *Slvn.* Jadransko Morje,
  SCr. Jadransko More, *Alb.* Deti
  Adriatik. Area of the Mediterranean
  Sea, between Italy and SE Europe
Adriatik, Deti *see* Adriatic Sea
Adventure Sound *645* bay of the
  South Atlantic Ocean,
  E Falkland Islands
Ādwa *245 var.* Adowa, *It.* Adua.
  N Ethiopia
Adygeya, Respublika *484*
  autonomous republic of
  SW Russian Federation
Adzopé *328* SE Ivory Coast
Aeankan District *396* district of
  Majuro, SE Marshall Islands
Aegean Sea *275, 580*
  Gk Aigaío Pélagos, *Turk.* Ege Denizi.
  Area of the Mediterranean Sea
Aeolian Islands *see* Eolie, Isole
Ærø *218 Ger.* Arrö. Island of
  S Denmark
Afadjado *272 var.* Afadjato, Afadjoto.
  Mountain of SE Ghana
'Afak *312* C Iraq
Afar Depression *see* Danakil Desert
Afghanistan *76-79* officially
  Islamic State of Afghanistan, *prev.*
  Republic of Afghanistan.
  Country of C Asia divided into 30
  admin. units (velayats). ❖ Kābul
Afgooye *528 It.* Afgoi. S Somalia
Afikpo *440* S Nigeria
Afobaka *542* NE Suriname
'Afula *319* N Israel
Afyon *580 prev.* Afyonkarahisar.
  W Turkey
Agadez *439 prev.* Agadès. C Niger
Agadir *414* SW Morocco
Agaña *see* Hagåtña
Aga Point *647* headland on the
  S coast of Guam
Āgaro *245* W Ethiopia
Agat *647* W Guam
Agboville *328* SE Ivory Coast
Ağcäbädi *110 Rus.* Agdzhabedi,
  var. Agdžabedi. C Azerbaijan

Agdam *110* SW Azerbaijan
Agedabia *see* Ajdābiyā
Agen *255* SW France
Agere Hiywet *see* Hägere Hiywet
Agia Napa *see* Ayia Napa
Agigialousa *see* Yenierenköy
Agios Ioannis *see* Ayios Ioannis
Agona Swedru *272 var.* Swedru.
  SE Ghana
Agordat *see* Akordat
Agou Gadzepe *571* SW Togo
Agou, Mont *571 prev.* Pic Baumann.
  Mountain of SW Togo
Āgra *298* N India
Agram *see* Zagreb
Agrigento *323 prev.* Girgenti. Sicilia,
  S Italy
Agrihan *650* island of N Northern
  Mariana Islands
Agrínio *275 prev.* Agrínion. W Greece
Aguachica *195* N Colombia
Aguadilla *651* NW Puerto Rico
Aguadulce *456* S Panama
Aguán *290* river of N Honduras
Aguarico *230* river of Ecuador
  and Peru
Aguascalientes *403* C Mexico
Aguijan *650* island of S Northern
  Mariana Islands
Agusan *466* river of Mindanao,
  S Philippines
Ahaggar *83 var.* Hoggar. Mountain
  range of SE Algeria
Ahja *242 var.* Ahja Jõgi. River of
  SE Estonia
Ahmadābād *298 var.* Ahmedabad.
  W India
Ahmadpur East *451* E Pakistan
Ahuachapán *237* W El Salvador
Ahvāz *309 var.* Ahwāz. W Iran
Ahvenanmaa *see* Åland
Aḥwar *631* SW Yemen
Aibak *see* Āybak
Aigaío Pélagos *see* Aegean Sea
Aiguá *611* S Uruguay
Ai-hun *see* Heihe
Ailigandí *456* E Panama
Ailinginae *396* island of
  NW Marshall Islands
Ailinglaplap *396 prev.* Ailinglapalap.
  Island of S Marshall Islands
Ailuk *396* island of
  NE Marshall Islands
'Aïn Ben Tili *398* N Mauritania
Aïn Oussera *83 var.* Aïn Wessara.
  N Algeria
Aintab *see* Gaziantep
Aïoun el Atroûss *see*
  'Ayoûn el 'Atroûs
Aiquile *136* C Bolivia
Airaī *455* C Palau
Airdrie *170* SW Canada
Airdrie *597* C Scotland, UK
Airlalang *see* Rokan
Aitape *458 var.* Eitape. NW Papua
  New Guinea
Aitos *see* Aytos
Aitutaki *644* island of Southern Cook
  Islands, S Cook Islands
Aix-en-Provence *255* SE France
Aix-la-Chapelle *see* Aachen
Aizu-Wakamatsu *332* Honshū,
  N Japan
Ajaccio *255* Corse, SE France
Ajdābiyā *414 var.* Ajdābiyah,
  Agedabia. NE Libya
Ajeltake District *396* district of
  Majuro, SE Marshall Islands
Ajjinena *see* Geneina
Ajka *292* W Hungary
Ajman *594 Ar.* 'Ajmān, *var.* 'Ujmān.
  NE United Arab Emirates
Ajtos *see* Aytos
Akaba *see* Al 'Aqabah
Akagera *162, 492, 588 var.* Kagera.
  River of E Africa

Al Liwā' *594* oasis region of SW United Arab Emirates

All Saints *92* C Antigua, Antigua & Barbuda

Al Lubnān *see* Lebanon

Al Luḥayyah *631* W Yemen

Al Ma'āmīr *115* NE Bahrain

Almada *474* W Portugal

Al Madīnah *506* *Eng.* Medina. W Saudi Arabia

Al Mafraq *338* *var.* Mafraq. N Jordan

Al Mahdīyah *see* Mahdia

Al Maḥmūdīyah *312* *var.* Mahmudiya. C Iraq

Al Mahrah *631* mountains of E Yemen

Al Majma'ah *506* C Saudi Arabia

Al Mālikīyah *115* W Bahrain

Al Mālikīyah *555* *var.* Dayrik. NE Syria

Almalyk *see* Olmaliq

Al Mamlakah *see* Morocco

Al Manādir *594* *var.* Al Manadir. Desert region of Oman and United Arab Emirates

Al Manāmah *see* Manama

Al Manāṣif *555* mountains of E Syria

Al Manṣūrah *see* El Manṣûra

Al Maqta' *594* C United Arab Emirates

Al Marj *371* *var.* Barka, *It.* Barce. NE Libya

Al Marsá *see* La Marsa

Almaty *340* *var.* Alma-Ata. SE Kazakhstan

Al Mawṣil *312* *Eng.* Mosul. N Iraq

Al Mayādīn *555* *Fr.* Meyadine. E Syria

Al Mazra'ah *338* W Jordan

Almelo *429* E Netherlands

Almendra, Embalse de *535* reservoir of NW Spain

Almere *429* C Netherlands

Almería *535* S Spain

Al Mīnā' *see* El Mina

Al Minyā *see* El Minya

Al Miqdādīyah *see* Al Muqdādīyah

Almirante *456* W Panama

Al Mubarraz *506* NE Saudi Arabia

Al Mudawwarah *338* SW Jordan

Al Muḥammadīyah *115* *var.* Umm aṣ Ṣabbān. Island of NW Bahrain

Al Muḥarraq *115* *var.* Moharek, Muharraq, Jazirat al Muharraq. Bahrain

Al Mukallā *631* *var.* Mukalla. SE Yemen

Al Mukhā *631* *Eng.* Mocha. SW Yemen

Al Muknīn *see* Moknine

Al Munastīr *see* Monastir

Al Muqdādīyah *312* *var.* Al Miqdādīyah. C Iraq

Al Mussayyib *312* C Iraq

Al Obayyid *see* El Obeid

Alofi *650* ❖ of Niue, W Niue

Alofi, Île *653* island of Île Futuna, N Wallis & Futuna

Alohungari *263* E Gambia

Alor, Kepulauan *304* island group of E Indonesia

Alor, Pulau *304* island of E Indonesia

Alor Setar *386* *var.* Alur Setar, Alor Star. NW Peninsular Malaysia

Alost *see* Aalst

Alotau *458* SE Papua New Guinea

Aloupos *215* *var.* Çiftlik Dere. River of NW Cyprus

Alpen *see* Alps

Alpes *see* Alps

Alphen aan de Rijn *429* W Netherlands

Alphonse Group *516* island group of C Seychelles

Alpi *see* Alps

Alps *106, 255, 322, 550* *It.* Alpi, *Fr.* Alpes, *Ger.* Alpen. Mountain range of C Europe

Al Qābil *448* *var.* Qabil. NW Oman

Al Qaḍārif *see* Gedaref

Al Qāhirah *see* Cairo

Al Qal'ah al Kubrá *see* Kalaa Kebira

Al Qāmishlī *555* *var.* Kamishli. NE Syria

Al Qaryah *115* NE Bahrain

Al Qaryāt *371* NW Libya

Al Qaṣr *338* W Jordan

Al-Qaṣrayn *see* Kasserine

Al Qayrawān *see* Kairouan

Al Qubayyāt *see* Qoubaïyât

Al Qubbah *371* NE Libya

Al Quds *see* Jerusalem

Al Qunayṭirah *555* *var.* El Quneitra, Kuneitra. SW Syria

Al Quṣayr *555* *var.* El Quseir, *Fr.* Kousseir. W Syria

Al Quṭayfah *555* *var.* Quṭayfah, Quteife, *Fr.* Kouteifé. SW Syria

Al Quwayrah *338* *var.* Makhfar al Quwayrah, El Quweira. SW Jordan

Als *218* *Ger.* Alsen. Island of S Denmark

Alsace *255* cultural region of NE France

Alsen *see* Als

Al Shahaniyah *see* Ash Shaḥānīyah

Alt *see* Olt

Alta *444* NE Norway

Altai Mountains *412* mountain range of C Asia

Altay *186* *Chin.* A-le-t'ai, *prev.* Ch'eng-hua, *var.* Chenghwa, *Mong.* Sharasume. Xinjiang Uygur Zizhiqu, NW China

Altay *412* W Mongolia

Altay, Respublika *485* autonomous republic of C Russian Federation

Alt de la Coma Pedrosa, Pic *86* mountain of NW Andorra

Altkanischa *see* Kanjiža

Alto Molócuè *419* C Mozambique

Alto Paraná *see* Paraná

Alt-Schwanenburg *see* Gulbene

Altsohl *see* Zvolen

Altun Shan *186* *var.* Altyn Tagh. Mountain range of Xinjiang Uygur Zizhiqu, NW China

Altyn Tagh *see* Altun Shan

Alu *see* Shortland Island

Al Ubayyiḍ *see* El Obeid

Alūksne *362* *Ger.* Marienburg. NE Latvia

Al 'Ulā *506* NW Saudi Arabia

Al 'Umarī *338* C Jordan

Al Uqṣur *see* Luxor

Al Wafrā' *356* SE Kuwait

Al Wāḥāt al Khārijah *see* El Wâhât el Khârga

Al Wajh *506* NW Saudi Arabia

Al Wakrah *479* *var.* Wakra. E Qatar

Al Wukayr *479* *var.* Al Wukair. E Qatar

Alyat *see* Älät

Alyaty-Pristan' *see* Älät

Alytus *376* *Pol.* Olita. S Lithuania

Al Zubayr *see* Az Zubayr

Amadora *474* W Portugal

Amakusa-shotō *332* island group to the W of Kyūshū, SW Japan

Amala *344* river of SW Kenya

Amami-Ō-shima *332* island of Amami-shotō, SW Japan

Amami-shotō *332* island group of Nansei-shotō, SW Japan

Amara *see* Al 'Amārah

Amarapura *159* C Burma

Amasia *see* Amasya

Amasia *581* N Turkey, *var.* Amasija. NW Armenia

Amasya *581* N Turkey

Amatique, Bahía de *280* bay of the Gulf of Honduras

Amaury *400* NE Mauritius

Amazon *145, 195, 463* *Sp.* Amazonas. River of South America

Amazon Basin *144* physical region of C South America

Ambalangoda *538* SW Sri Lanka

Ambalavao *382* S Madagascar

Ambam *168* S Cameroon

Ambanja *382* N Madagascar

Ambato *230* C Ecuador

Ambatondrazaka *382* E Madagascar

Ambergris Cay *130* island of NE Belize

Ambergris Cays *653* island group of S Turks and Caicos Islands

Amblève *127* river of E Belgium

Ambo *see* Hāgere Hiywet

Amboasary *382* S Madagascar

Ambohidratrimo *382* C Madagascar

Ambon *304* *prev.* Amboina. Ambon, C Indonesia

Ambositra *382* C Madagascar

Ambre, Île d' *400* Island of NE Mauritius

Ambriz *88* NW Angola

Ambrym *619* *var.* Ambrim. Island of E Vanuatu

'Amd *631* C Yemen

Ameland *429* island of Waddeneilanden, N Netherlands

American Samoa *642* unincorporated territory of the USA, Pacific Ocean. ❖ Pago Pago

Amersfoort *429* C Netherlands

Amherst *see* Kyaikkami

Amiens *255* N France

Amilḥayt, Wādī *448* *var.* Wādī Umml Ḥayt. Seasonal watercourse of SW Oman

Amīndivi Islands *298* island group of Lakshadweep, SW India

Amioun *364* *var.* Amyūn. N Lebanon

Amirante Islands *516* *var.* Amirantes Group. Island of C Seychelles

Amlamé *571* C Togo

Amman *338* *Ar.* 'Ammān. ❖ of Jordan, NW Jordan

Ammassalik *646* *var.* Angmagssalik. SE Greenland

Ammochostos *see* Gazimağusa

Ammochostos Bay *see* Famagusta Bay

Amnok *see* Yalu

Āmol *309* *var.* Amul. N Iran

Amorgós *275* island of SE Greece

Amouli *642* *var.* Tau. Tau, E American Samoa

Amourj *398* SE Mauritania

Ampara *538* E Sri Lanka

Amphitrite Group *651* island group of N Paracel Islands

'Amrān *631* W Yemen

Amrāvati *298* *prev.* Amraoti. C India

Amritsar *298* N India

Amstelveen *429* W Netherlands

Amsterdam *429* ❖ of the Netherlands, C Netherlands

Amstetten *106* N Austria

Am Timan *180* SE Chad

Amu Darya *77, 562, 581, 614* *Turkm.* Amyderya, *Uzb.* Amudaryo. River of C Asia

Amudat *588* E Uganda

Amul *see* Āmol

Amund Ringnes Island *170* island of Sverdrup Islands, N Canada

Amundsen Gulf *170* gulf of the Beaufort Sea, on the NW coast of Canada

Amundsen-Scott *90* US research station at the South Pole, Greater Antarctica, Antarctica

Amundsen Sea *90* sea of the Pacific Ocean, off Antarctica

Amur *187, 485* *Chin.* Heilong Jiang. River of China and Russian Federation

Amyderya *see* Amu Darya

Amyūn *see* Amioun

An Abhainn Mhór *see* Blackwater

Anaco *623* NE Venezuela

Anadolu Dağları *see* Doğu Karadeniz Dağlari

Anadyr' *485* NE Russian Federation

Anadyr, Gulf of *see* Anadyrskiy Zaliv

Anadyrskiy Zaliv *485* *Eng.* Gulf of Anadyr. Gulf of Bering Sea, bordering NE Russian Federation

Anáfi *275* island of SE Greece

Anaiza *see* 'Unayzah

Analalava *382* N Madagascar

Analamaitsa Plateau *382* plateau of NE Madagascar

Anambas, Kepulauan *304* island group to the NW of Borneo, W Indonesia

Anan *332* Shikoku, SW Japan

Anantapur *298* S India

Anápolis *145* S Brazil

Anarjokka *see* Inarijoki

Anatahan *650* island of C Northern Mariana Islands

Anatolia Plateau *580* plateau of C Turkey

Anatom *see* Aneityum

An Bhearú *see* Barrow

An Bhóinn *see* Boyne

Anchorage *602* Alaska, USA

Ancona *323* C Italy

Andalucia *534-535* autonomous community of S Spain

Andaman Islands *298* island group of SE India

Andaman Sea *159, 304, 567* sea of Indian Ocean, to the SW of Burma and Thailand

'Andām, Wadi *448* seasonal desert watercourse of E Oman

Andapa *382* NE Madagascar

Andaung Pech *see* Bǎ Kêv

Andenne *127* SE Belgium

Andersen Air Force Base *647* NE Guam

Andes *95, 136, 183, 195, 463* mountain range of South America, running the entire length of the west coast

Andfjorden *444* fjord of NE Norway

Andijon *614* *Rus.* Andizhan. E Uzbekistan

Andizhan *see* Andijon

Andkhvoy *101* N Afghanistan

Andoany *382* *prev.* Hell-Ville. N Madagascar

Andong *352* *Jap.* Antō. E South Korea

Andong-ho *352* reservoir of E South Korea

Andorra *86-87* officially Principality of Andorra. Country of SW Europe divided into 7 admin. units (parishes). ❖ Andorra la Vella

Andorra la Vella *86* ❖ of Andorra, W Andorra

Andreas *647* N Isle of Man

Andreas, Cape *see* Apostolos Andreas, Cape

Andria *323* S Italy

Androna, Lembalemba Ambanin' *382* *var.* Plateau de l'Androna. Plateau of N Madagascar

Ándros *275* island of SE Greece

Andros Island *112* island of W Bahamas

Andros Town *112* Andros Island, Bahamas

Andújar *535* SW Spain

Anegada *643* island of NE British Virgin Islands

Aného *571* *var.* Anécho, *prev.* Petit-Popo. S Togo

Aneityum *619* *var.* Anatom, *prev.* Kéamu. Island of S Vanuatu

Änew *see* Annau

An Fheoir *see* Nore

Anfile Bay *240* bay of the Red Sea to the E of Eritrea

Angara *485* river of C Russian Federation

Angarsk *485* C Russian Federation

Angaur *455* island of S Palau

Änge *547* C Sweden

Angel *see* Úhlava

Ángel de la Guarda, Isla *403* island of NW Mexico

Angeles *466* Luzon, N Philippines

Angers *255* NW France

Ängk Tasaôm *165* *prev.* Angtassom. S Cambodia

Anglesey *597* *Wel.* Môn. Island of NW Wales, UK

Angmagssalik *see* Ammassalik

Angoche *419* E Mozambique

Angola *88-89* officially Republic of Angola, *prev.* People's Republic of Angola. Country of Central Africa divided into 18 admin. units (provinces). ❖ Luanda

Angora *see* Ankara

Angoram *458* N Papua New Guinea

Angoulême *255* W France

Angra Pequena *see* Lüderitz

Angren *614* E Uzbekistan

Angtassom *see* Ängk Tasaôm

Anguilla *642* British dependent territory of the Caribbean Sea. ❖ The Valley.

Arraiján *456* C Panama
Arrak District *396* district of Majuro, SE Marshall Islands
Ar Ramādī *312* C Iraq
Ar Rams *594* NE United Arab Emirates
Ar Ramthā *338* N Jordan
Arran, Isle of *597* island of W Scotland, UK
Ar Raqqah *555* *Fr.* Rakka. N Syria
Arras *255* N France
Ar Rawḍah *631* S Yemen
Ar Rawḍatayn *356* N Kuwait
Ar Rayyān *479* *var.* Al Rayyan. E Qatar
Ar Rifā' al Gharbī *115* N Bahrain
Ar Rifā' ash Sharqī *115* NE Bahrain
Ar Riyāḍ *see* Riyadh
Ar Riyān *631* S Yemen
Arrö *see* Ærø
Ar Ru'ays *479* *var.* Ar Ruways, Al Ruweis, Ruwais. N Qatar
Ar Ru'ays *594* *var.* Ruwaisv, Ar Ruways. W United Arab Emirates
Ar Rufayq *479* *var.* Al Rufaig. NW Qatar
Ar Rustāq *448* *var.* Rostak. N Oman
Ar Ruwayshid *338* NE Jordan
Ar Safad *see* Ẓefat
'Arta *222* SE Djibouti
Artashat *98* *var.* Artašat. S Armenia
Artemisa *210* NW Cuba
Artëm-Ostrov *see* Artyom
Artibonite *288* river of Haiti
Artigas *611* *prev.* San Eugenio, *var.* San Eugenio del Cuareim. N Uruguay
Art'ik *98* W Armenia
Art, Île *650* island of Îles Belep, W New Caledonia
Artois *255* cultural region of N France
Artsvashen *98* *var.* Bashkend. NE Armenia
Artvin *581* NE Turkey
Artyom *110* *Rus.* Artëm-Ostrov. E Azerbaijan
Arua *588* NW Uganda
Aruângua *see* Luangwa
Aruba *642* autonomous part of the Netherlands, Caribbean Sea. ❖ Oranjestad
Aru, Kepulauan *304-305* *prev.* Aroe Islands. Island group to the SW of Irian Jaya, W Indonesia
Aruliho *526* NW Guadalcanal, Solomon Is
Årup *218* Fyn, SW Denmark
Arusha *564* N Tanzania
Aruwimi *203* river of NE Dem. Rep. Congo
Arvand Rūd *see* 'Arab, Shaṭṭ al
Arvayheer *412* C Mongolia
Arvika *547* SW Sweden
Aryānah *see* Ariana
Asadābād *77* E Afghanistan
Asad, Buḥayrat al *555* reservoir of Syria
Asahi-dake *333* mountain of Hokkaidō, N Japan
Asahikawa *333* Hokkaidō, N Japan
Asaka *614* *Rus.* Assake, *prev.* Leninsk. E Uzbekistan
Asamankese *272* SE Ghana
Āsau *500* Savai'i, Samoa
Āsbe Teferī *245* *var.* Asba Tafari. C Ethiopia
Ascension *652* dependent territory of St Helena, South Atlantic Ocean
Ascension *see* Pohnpei
Ascoli Piceno *323* C Italy
Aseb *240* *var.* Assab, *Amh.* Āseb. SE Eritrea
'As 'Ēla *222* SW Djibouti
Āsela *245* *var.* Aselle, Asella, Asselle. C Ethiopia
Asenovgrad *152* S Bulgaria
Aseri *242* *Ger.* Asserin. NE Estonia
Ashburton *433* C South Island, New Zealand
Ashdod *319* C Israel
Ashford *597* SE England, UK

Ashgabat *585* *prev.* Ashkhabad, Poltoratsk. ❖ of Turkmenistan, S Turkmenistan
Ashikaya *332* Honshū, SE Japan
Ashkhabad *see* Ashgabat
Ashmyany *122* NW Belarus
Ashots'k *98* *prev.* Ghukasyan, *Rus.* Gukasyan. NW Armenia
Ashqelon *319* C Israel
Ash Shadādah *555* *var.* Ash Shaddādah, Shaddādī, Shedadi, Tell Shedadi. NE ria
Ash Shaḥanīyah *479* *var.* Al Shahaniyah. C Qatar
Ash Shāmīyah *312* C Iraq
Ash Sharāh *338* *var.* Esh Sharā. Mountains of W Jordan
Ash Shaṭrah *312* SE Iraq
Ash Shawbak *338* *var.* Ash Shawbak an Nijil. W Jordan
Ash Shiḥr *631* SE Yemen
Ash Shiṣar *448* SW Oman
Ash Shuqayq *356* *var.* As Shageeg. Desert region of S Kuwait
Ashtarak *98* *var.* Aštarak. W Armenia
Ashton *498* Union I, St Vincent & the Grenadines
Asidonhopo *542* C Suriname
'Asi, Nahr al *see* Orantes
'Asi Oronte, Nahr al *see* Orantes
Asipovichy *122* *Rus.* Osipovichi. C Belarus
'Askar *115* E Bahrain
Asmara *240* *Amh.* Asmera. ❖ of Eritrea, C Eritrea
Aspinwall *see* Colón
Aspropotamos *see* Acheloos
Assab *see* Aseb
Aş Şabīrīyah *356* NE Kuwait
Aş Şabīyah *356* *var.* Sabyah. NE Kuwait
As Sabkhah *555* *var.* Sabkha. NE Syria
Aş Şafāwī *338* N Jordan
Aş Şafī *338* *var.* Safi. W Jordan
'Assa Gaïla *222* N Djibouti
As Sahlat al Ḥadrīyah *115* N Bahrain
Aş Şaḥrā' *see* Sahara
Aş Şaḥrā' al Gharbīyah *see* Sahara el Gharqīya
Aş Şaḥrā' al Lībīyah *see* Libyan Desrt
Aş Şaḥrā' ash Sharqīyah *see* Sahara el Sharqīya
Assake *see* Asaka
As Salimi *356* *var.* Salemy. W Kuwait
As Salīmīyah *356* E Kuwait
'Assal, Lac *222* lake of C Djibouti
As Salṭ *338* *var.* Salt. NW Jordan
As Salwá *479* *var.* Salwah, Salwa. SW Qatar
Assamaka *439* *var.* Assamakka. NW Niger
As Samāwah *312* *var.* Samawa. S Iraq
'Assâmo *222* S Djibouti
Aş Şarīḥ *338* NW Jordan
Assen *429* NE Netherlands
Asserin *see* Aseri
As Shageeg *see* Ash Shuqayq
As Sīb *448* N Oman
Assiout *see* Asyûṭ
Assisi *323* C Italy
Assling *see* Jesenice
Assomption *516* island of the Aldabra Group, SW Seychelles
Aş Şubayḥīyah *356* S Kuwait
As Sufāl *631* S Yemen
As Sukhnah *555* *var.* Sukhne, *Fr.* Soukhné. C Syria
As Sulaymānīyah *312* *var.* Sulaimaniya, *Kurd.* Slêmānī. NE Iraq
As Sulayyil *506* C Saudi Arabia
As Suwaydā' *555* *var.* El Suweida, *Fr.* Soueida. SW Syria
As Suwayq *448* *var.* Suwaik. N Oman
As Suways *see* Suez
Astana *340* *prev* Akmola ❖ of Kazakhstan, N Kazakhstan
Astara *110* S Azerbaijan
Aštarak *see* Ashtarak
Asti *323* N Italy
Astipálaia *275* island of SE Greece
Astove *516* island of the Aldabra Group, SW Seychelles
Astrakhan' *484* SW Russian Federation

Astrakhan-Bazar *see* Cälilabad
Astrida *see* Butare
Asturias *534* autonomous community of NW Spain
Asuka *90* Japanese research station of Greater Antarctica, Antarctica
Āsuisui, Cape *500* cape on the coast of Savai'i, Samoa
Asuncion *650* island of N Northern Mariana Islands
Asunción *460* ❖ of Paraguay, S Paraguay
Asunción Mita *280* SE Guatemala
Aswa *see* Achwa
Aswân *232* *var.* Aswān. SE Egypt
Aswân High Dam *232* dam of SE Egypt
Asyûṭ *232* *var.* Assiout. C Egypt
Ata *572* island of SW Tonga
Atacama Desert *183* desert of N Chile
Atafu Atoll *653* island of NW Tokelau
Atakora, Chaîne de l' *132* *var.* Atakora Mountains. Mountain range of N Benin
Atakpamé *571* C Togo
Aṭâr *398* C Mauritania
Atas, Pulu *526* South Island
Atbara *540* *var.* Nahr 'Aṭbarah. River of NE Sudan
Atbara *540* *var.* 'Aṭbārah. NE Sudan
'Aṭbarah, Nahr *see* Atbara
Atbasar *340* N Kazakhstan
At-Bashy *359* *var.* At-Bashi, At-Baši. C Kyrgyzstan
Athabasca *170* *var.* Athabaska. River of W Canada
Athabasca, Lake *170* lake of W Canada
Athabaska *see* Athabasca
Athens *275* *Gk* Athína, *prev.* Athínai. ❖ of Greece, SE mainland Greece
Athi *344* river of S Kenya
Athienou *215* SE Cyprus
Athína *see* Athens
Athínai *see* Athens
Athlone *316* *Ir.* Áth Luain. C Ireland
Áth Luain *see* Athlone
Athna *215* *var.* Achna, Akhna, *Turk.* Düzce. E Cyprus
Ati *180* C Chad
Atiak *588* NW Uganda
Atitlán, Lago de *280* lake of W Guatemala
Atiu *644* island of Southern Cook Islands, S Cook Islands
Atlanta *603* Georgia, SE USA
Atlantic Ocean *90, 296, 597, 643* *var.* Atlantshaf. Ocean bounded to the W by the Americas, to the E by Europe and Africa and to the S by Antarctica
Atlantshaf *see* Atlantic Ocean
Atlas Mountains *414* mountain range of C Morocco
Atlas Saharien *83* mountain range of N Algeria
Atrak *309, 585* *Rus.* Atrek, *Turkm.* Etrek. River of Iran and Turkmenistan
Atrato *195* river of NW Colombia
Atrek *see* Atrak
Aṭ Ṭaff *594* desert region of C United Arab Emirates
Aṭ Ṭafīlah *338* W Jordan
Aṭ Ṭā'if *506* W Saudi Arabia
At Tall al Abyaḍ *555* *Fr.* Tell Abaid. N Syria
Attapu *360* *var.* Attopeu. SE Laos
Attersee *106* *var.* Kammersee. Lake of N Austria
At Tibnī *555* NE Syria
Attopeu *see* Attapu
At Turbah *631* SW Yemen
Atyrau *340* *prev.* Gur'yev. W Kazakhstan
Auch *255* SW France
Auckland *433* N North Island, New Zealand
Augier *497* S St Lucia
Augila *see* Awjilah
Augsburg *267* S Germany
AUGUSTA *323* SICILIA, S ITALY
Augusta *603* Maine, NE USA
Aujá et Taḥtā *320* E West Bank

Auki *526* NW Malaita, Solomon Is
Auliye-Ata *see* Taraz
Aur *396* island of E Marshall Islands
Aurich *266* NW Germany
Aurillac *255* S France
Aurine, Alpi *see* Zillertaler Alpen
Aurora *286* NW Guyana
Aurora *see* Maéwo
Aus *423* SW Namibia
Aussig *see* Ústí nad Labem
Austin *603* Texas, SC USA
Australes, Îles *646* island group of SW French Polynesia
Australia *100-105* officially Commonwealth of Australia. Country situated between the Indian and Pacific Oceans, divided into 8 admin. units (6 states and 2 territories). ❖ Canberra
Australian Alps *101* mountain range of SE Australia
Australian Capital Territory *101* territory of SE Australia
Austria *106-107* officially Republic of Austria, *Ger.* Österreich. Country of C Europe divided into 9 admin. units (states). ❖ Vienna
Ausuitoq *see* Grise Fiord
Auvergne *255* cultural region of SE France
Auxerre *255* N France
Avarua *644* ❖ of Cook Islands, Rarotonga, S Cook Islands
Avatele *650* S Niue
Aveiro *474* W Portugal
Avellino *323* S Italy
Aves, Islas las *623* island group of N Venezuela
Aveyron *255* river of S France
Avezzano *323* C Italy
Avignon *255* SE France
Avila *535* C Spain
Avilés *534* NW Spain
Avranches *254* NW France
Avuavu *526* SE Guadalcanal, Solomon Is
Awaaso *272* *var.* Awaso. SW Ghana
'Awālī *115* C Bahrain
Awara Plain *344* plain of NE Kenya
Awash *245* *var.* Hawash. River of C Ethiopia
Awbārī *371* W Libya
Awbārī, Şaḥrā' *371* desert of W Libya
Awjilah *371* *It.* Augila. NE Libya
Awled Djellal *see* Ouled Djellal
Axarfjördhur *see* Öxarfjördhur
Axel Heiberg Island *170* var. Axel Heiburg. Island of Sverdrup Islands, N Canada
Axim *272* S Ghana
Axios *see* Vadar
Ayabe *332* Honshū, C Japan
Ayacucho *463* S Peru
Ayaguz *340* *Kaz.* Ayaköz. E Kazakhstan
Ayaköz *see* Ayaguz
Ayamé Reservoir *328* reservoir of E Ivory Coast
Ayamiken *238* NW Río Muni, Equatorial Guinea
Āybak *77* *var.* Aibak, Haibak. NE Afghanistan
Aydarkŭl *614* *Rus.* Ozero Aydarkul. River of E Uzbekistan
Aydın *580* SW Turkey
Ayer Chawan, Pulau *521* island of SW Singapore
Ayer Hitam *386* S Peninsular Malaysia
Ayer Merbau, Pulau *521* island of SW Singapore
Ayers Rock *see* Uluru
Ayeyarwady *see* Irrawaddy
Ayia Napa *215* *var.* Agia Napa. E Cyprus
Áyios Amvrosios *see* Esentepe
Áyios Eustratios *275* island of E Greece
Ayios Ioannis *215* *var.* Agios Ioannis. SW Cyprus
Ayios Seryios *see* Yeniboğazçi
Ayios Theodhoros *see* Çayirova
'Ayn al Ghazāl *371* SE Libya

Banes *211* SE Cuba
Baney *238* N Bioko, Equatorial Guinea
Banfora *157* SW Burkina
Bangalore *298, 303* S India
Bangangté *168* W Cameroon
Bangar *150* NE Brunei
Bangassou *179* SE Central African Republic
Bangfai *360* river of S Laos
Banggai, Kepulauan *304* island group to the E of Celebes, C Indonesia
Banggi, Pulau *387* var. Banggi. Island of NE Borneo, Malaysia
Banghāzī *371* Eng. Benghazi, It. Bengasi. N Libya
Banghiang *360* var. Bang Hieng. River of S Laos
Bangka, Pulau *304* island to the SE of Sumatra, W Indonesia
Bangkok *567* Th. Krung Thep. ❖ of Thailand, C Thailand
Bangladesh *116-119* officially People's Republic of Bangladesh, prev. East Pakistan. Country of S Asia divided into 4 admin. units (divisions). ❖ Dhaka
Bangor *597* Northern Ireland, UK
Bangor *597* N Wales, UK
Bangor *603* Maine, NE USA
Bangui *179* ❖ of Central African Republic, SW Central African Republic
Bangweulu, Lake *635* var. Lake Bengweulu. Lake of N Zambia
Banhā see Benha
Ban Hat Yai see Hat Yai
Ban Houayxay *360* var. Ban Houei Sai. NW Sai Laos
Ban Hua Hin *567* C Thailand
Baní *226* S Dominican Republic
Bani *392* river of S Mali
Banias see Bāniyās
Banī Jamrah *115* NW Bahrain
Banikoara *132* N Benin
Banī Suwayf see Beni Suef
Banī Walid *371* NW Libya
Bāniyās *555* var. Banias. N Syria
Banja Luka *140* NW Bosnia & Herzegovina
Banjarmasin *304* prev. Bandjarmasin. Borneo, C Indonesia
Banjës, Liqeni i *81* lake of C Albania
Banjul *263* prev. Bathurst. ❖ of Gambia, W Gambia
Ban Kengkabao *360* S Laos
Banket *636* N Zimbabwe
Banks, Îles see Banks Islands
Banks Island *170* island of NW Canada
Banks Islands *619* Fr. Îles Banks. Island group of N Vanuatu
Banks Peninsula *433* peninsula of E South Island, New Zealand
Ban Lakxao *360* var. Lak Sao. E Laos
Banmo see Bhamo
Ban Mun-Houamuang *360* SE Laos
Ban Nadou *360* S Laos
Ban Na Môn *360* NE Laos
Ban Napha *360* C Laos
Ban Nasi *360* C Laos
Banningville see Bandundu
Ban Nongsim *360* S Laos
Baños *230* C Ecuador
Bánovce nad Bebravou *523* var. Bánovce. W Slovakia
Ban Phon *360* var. Ban Phone. SE Laos
Ban Phone see Ban Phon
Ban Phou A Douk *360* E Laos
Bansang *263* E Gambia
Banská Bystrica *523* Ger. Neusohl, Hung. Besztercebánya. C Slovakia
Ban Talak *360* E Laos
Bantè *132* W Benin
Ban Thabôk *360* C Laos
Bantry *316* SW Ireland
Bantry Bay *316* Ir. Bá Bheanntraí. Area of the Celtic Sea, SW Ireland
Ban Xot *360* S Laos
Banyak, Kepulauan *304* island group to the NW of Sumatra, W Indonesia
Banyo *168* N Cameroon
Banzart see Bizerte

Baoding *187* var. Pao-ting, prev. Tsingyuan. Hebei, NE China
Baoji *187* var. Pao-chi, Paoki. Shaanxi, C China
Baoro *179* W Central African Republic
Baoshan *186* var. Pao-shan. Yunnan, SW China
Baotou *187* var. Pao-t'ou, Paotow. Nei Mongol Zizhiqu, N China
Ba'qūbah *312* C Iraq
Bar *512* It. Antivari. S Montenegro, Serbia & Montenegro (Yugo.)
Baraawe *528* It. Brava. S Somalia
Baracaldo *535* N Spain
Baracoa *211* SE Cuba
Barahona *226* SW Dominican Republic
Baraka *240* seasonal river of Eritrea and Sudan
Barakī Barak *77* E Afghanistan
Baram *387* var. Barram. River of C Borneo, Malaysia
Barama *286* river of N Guyana
Baramanni *286* N Guyana
Baramita *286* NW Guyana
Baranavichy *122* Rus. Baranovichi, Pol. Baranowicze. W Belarus
Barbados *120-121* island state of the Lesser Antilles, Caribbean Sea divided into 11 admin. units (districts or parishes). ❖ Bridgetown
Bārbār *115* NW Bahrain
Barbuda *92* island of Lesser Antilles which, with Antigua, forms Antigua & Barbuda
Barce see Al Marj
Barcellona *323* Sicilia, S Italy
Barcelona *535* E Spain
Barcelona *623* NE Venezuela
Barclayville *368* SE Liberia
Barcoo see Cooper Creek
Bārda *110* C Azerbaijan
Bardaï *180* N Chad
Bardejov *523* Hung. Bártfa, Ger. Bartfeld. NE Slovakia
Bardera see Baardheere
Bardo *577* var. Bārdaw, Le Bardo. N Tunisia
Bareilly *298* Hind. Bareli. N India
Bareli see Bareilly
Barentsburg *653* Spitsbergen, W Svalbard
Barents Sea *444, 484* Nor. Barentshavet, Rus. Barentsevo More. Sea of Arctic Ocean bordering N Europe
Barentu *240* W Eritrea
Bargny *510* W Senegal
Bari *323* var. Bari delle Puglie. S Italy
Bari delle Puglie see Bari
Barika *83* N Algeria
Barīkowt *77* var. Barikot. NE Afghanistan
Barillas *280* NW Guatemala
Barīm *631* Eng. Perim. Island of SW Yemen
Barinas *623* W Venezuela
Baringo, Lake *344* lake of W Kenya
Barisāl *117* S Bangladesh
Barisan, Pegunungan *304* mountain range of Sumatra, W Indonesia
Barito *304* river of Borneo, C Indonesia
Barkā' *448* var. Birka. N Oman
Barka see Al Marj
Barkly Tableland *101* plateau of N Australia
Bârlad *480* E Romania
Barlavento, Ilhas de *176* northernmost of the two main island groups comprising Cape Verde
Bar-le-Duc *255* NE France
Barletta *323* S Italy
Barnaul *485* C Russian Federation
Barnes Hill *92* N Antigua, Antigua & Barbuda
Barneveld *429* C Netherlands
Barnstaple *597* SW England, UK
Baroda see Vadodara
Barouk, Nahr el *364* river of C Lebanon
Baroul see Salisbury

Barqah *371* Eng. Cyrenaica. Cultural region of NE Libya
Barquisimeto *623* NW Venezuela
Barrage de Nzilo see Nzilo, Lac
Barrancabermeja *195* NW Colombia
Barranquilla *195* N Colombia
Barreiro *474* W Portugal
Barrier Reef *130* coral reef E of Belize
Barrigada *647* C Guam
Barril *176* São Nicolau, N Cape Verde
Barrouallie *498* W St Vincent, St Vincent & the Grenadines
Barrow *316* Ir. An Bhearú. River of SE Ireland
Barrow-in-Furness *597* NW England, UK
Barr, Ra's al *115* cape of S Bahrain
Barsalogho *157* var. Barsalogho. N Burkina
Bartang *562* river of SE Tajikistan
Bártfa see Bardejov
Bartfeld see Bardejov
Bartica *286* NE Guyana
Bartolomé Masó *210* SE Cuba
Baruta *623* N Venezuela
Baruun-Urt *412* E Mongolia
Barú, Volcán *456* var. Volcán de Chiriquí. Volcanic peak of W Panama
Barxudarli *110* W Azerbaijan
Barysaw *122* Rus. Borisov. NE Belarus
Basăk *165* river of Cambodia and Vietnam
Basarabeasca *408* Rus. Bessarabka. SE Moldova
Basel *550* Eng. Basle, Fr. Bâle. NW Switzerland
Bashi Channel *559* Chin. Pa-shih Hai-hsia. Channel connecting the South China Sea and Pacific Ocean
Bashkend see Artsvashen
Bashkortostan, Respublika *484* autonomous republic of W Russian Federation
Basilan Island *466* island of SW Philippines
Basilé, Pico de *238* var. Pico de Santa Isabel. Mountain of C Bioko, Equatorial Guinea
Basle see Basel
Basra see Al Başrah
Bassam see Grand-Bassam
Bassar *571* var. Bassari. NW Togo
Bassein *159* var. Pathein. SW Burma
Basse Santa Su *263* E Gambia
Basse-Terre *646* ❖ of Guadeloupe, SW Guadeloupe
Basseterre *494* ❖ of St Kitts & Nevis, S St Kitts
Bassikounou *398* SE Mauritania
Bassila *132* W Benin
Bass Strait *101* SE strait connecting South Pacific Ocean and Southern Ocean, between Australia and Tasmania
Bastia *255* Corse, SE France
Bastogne *127* SE Belgium
Basutoland see Lesotho
Bata *238* NW Río Muni, Equatorial Guinea
Batabanó, Golfo de *210* gulf of the Caribbean Sea, on the S coast of Cuba
Bataka *224* NE Dominica
Batak, Yazovir *152* reservoir of SW Bulgaria
Batangafo *179* NW Central African Republic
Batangas *466* Luzon, N Philippines
Batan Islands *466* island group of N Philippines
Batavia see Jakarta
Bătdâmbâng *165* prev. Battambang. W Cambodia
Batéké, Plateaux *200* plateau of Congo
Bates, Mount *650* mountain of Norfolk Island
Bath *597* SW England, UK
Bath *331* E Jamaica
Bath *494* SW Nevis, St Kitts & Nevis
Batha *180* seasonal river of S Chad
Bathsheba *121* E Barbados

Bathurst *101* SE Australia
Bathurst see Banjul
Bathurst Island *101* island of N Australia
Bathurst Island *170* island of Parry Islands, N Canada
Bathyz see Badkhyz
Batī *245* NE Ethiopia
Batiki *248* prev. Mbatiki. Island to the E of Viti Levu, C Fiji
Bāţin, Wādī al *356, 506* dry watercourse of SW Asia
Batjan see Bacan, Pulau
Batman *581* var. Īluh. SE Turkey
Batna *83* NE Algeria
Batoe see Batu, Kepulauan
Baton Rouge *603* Louisiana, SC USA
Batouri *168* E Cameroon
Batroûn *364* var. Al Batrūn. N Lebanon
Battambang see Bătdâmbâng
Batticaloa *538* E Sri Lanka
Battowia *498* island of C St Vincent & the Grenadines
Batu Gajah *386* W Peninsular Malaysia
Batu, Kepulauan *304* prev. Batoe. Island group to the W of Sumatra, W Indonesia
Bat'umi *264* W Georgia
Batu Pahat *386* prev. Bandar Penggaram. S Peninsular Malaysia
Bat Yam *319* C Israel
Bauchi *440* NE Nigeria
Baumann, Pic see Agou, Mont
Baures *136* river of NE Bolivia
Bauru *145* S Brazil
Bauska *362* Ger. Bauske. S Latvia
Bavarian Alps *106, 267* Ger. Bayerische Alpen. Mountain range of Austria and Germany
Bawku *272* N Ghana
Baxoi *186* Xizang Zizhiqu, W China
Bayamo *210* SE Cuba
Bayamón *651* NE Puerto Rico
Bayanhongor *412* C Mongolia
Bayano, Lago *456* lake of E Panama
Baydhabo *528* var. Isha Baydhabo, It. Baidoa. SW Somalia
Bayerische Alpen see Bavarian Alps
Bāyir *338* var. Bā'ir. C Jordan
Bāyir, Wādī *338* var. Wādī B'áir. Dry watercourse of C Jordan
Baykal, Ozero *485* Eng. Lake Baikal. Lake of S Russian Federation
Bay, Laguna de *466* lake of Luzon, N Philippines
Baynūnah *594* desert region of W United Arab Emirates
Bayonne *254* SW France
Bayram-Ali see Bayramaly
Bayramaly *585* prev. Bayram-Ali, var. Bajram-Ali. SE Turkmenistan
Bayreuth *267* SE Germany
Bayrūt see Beirut
Baysun see Boysun
Bayt al Faqīh *631* W Yemen
Bayt Lahm see Bethlehem
Bayudha Desert *540* var. Baiyuda, Şaḩrā' Bayyūḑah. Desert NE Sudan
Bayy al Kabīr, Wādī *371* dry watercourse of NW Libya
Bayyūḑah, Şaḩrā' see Bayudha Desert
Bazardüzü Dağ *110* Rus. Gora Bazardyuzyu. Mountain of N Azerbaijan
Bazargic see Dobrich
Bazgrad *152* NE Bulgaria
Bazin see Pezinok
Bcharré *364* var. Bsharrī. NE Lebanon
Beagle Channel *183* channel connecting Pacific Ocean and Atlantic Ocean
Beata, Isla *226* island of SW Dominican Republic
Beatrice *636* NE Zimbabwe
Beau Bassin *400* W Mauritius
Beaufort Sea *170* sea of the Arctic Ocean to the N of North America
Beaufort West *531* Afr. Beaufort-Wes. Western Cape, SW South Africa
Beauvais *255* N France

Béchar *83 prev.* Colomb-Béchar. W Algeria

Bécs *see* Vienna

Bedanda *284* S Guinea-Bissau

Bedok Reservoir *521* E Singapore

Be'ér Sheva' *319* S Israel

Begamganj *117* SE Bangladesh

Begna *444* river of S Norway

Begovat *see* Bekobod

Behagle *see* Laï

Beibu Wan *see* Tongking, Gulf of

Beida *see* Al Baydā'

Beijing *187, 190 Eng.* Peking, *var.* Pei-ching, *prev.* Pei-ping. ❖ and municipaltiy of China

Beira *419* C Mozambique

Beirut *364 var.* Beyrouth, Bayrūt. ❖ of Lebanon, W Lebanon

Beitbridge *636* S Zimbabwe

Beit Lahiya *319, 320* NE Gaza Strip

Beja *474* S Portugal

Béja *577 var.* Bājah. N Tunisia

Bejaïa *83 prev.* Bougie. N Algeria

Bejhi *451 var.* Beji. River of W Pakistan

Bekaa Valley *see* El Beqaa

Bek-Budi *see* Qarshi

Bekdash *585 var.* Bekdaš. NW Turkmenistan

Békéscsaba *292* SE Hungary

Bekobod *614 prev.* Begovat, *Rus.* Bekabad. E Uzbekistan

Bekwai *272* C Ghana

Bélabo *168* C Cameroon

Belait, Sungai *150* river of C Brunei

Belarus *see* Belarus

Bela Vista *419* S Mozambique

Belaya Tserkov' *see* Bila Tserkva

Belbeis *see* Bilbeis

Belcher Is *171* island group in Hudson Bay, C Canada

Beledweyne *528 var.* Belet Huen, *It.* Beket Uen. C Somalia

Belém *145 prev.* Pará. N Brazil

Belep, Îles *650* island group of W New Caledonia

Belfast *597* Northern Ireland, UK

Belfort *255* NE France

Belgium *126-129* officially Kingdom of Belgium. Country of W Europe divided into 9 admin. units (provinces). ❖ Brussels

Belgian Congo *see* Dem. Rep. Congo

Belgorod *484* W Russian Federation

Belgrade *512 SCr.* Beograd. ❖ of Serbia & Montenegro (Yugo.), N Serbia, Serbia & Montenegro (Yugo.)

Belgrano II *90* Argentinian research station of Greater Antarctica, Antarctica

Belice *see* Belize City

Beligrad *see* Berat

Beli Manastir *209 Hung.* Pélmonostor. NE Croatia

Belitoeng *see* Belitung, Pulau

Belitung, Pulau *304 prev.* Belitoeng, Billiton. Island to the SE of Sumatra, W Indonesia

Belize *130* river of Belize and Guatemala

Belize *130-131* Country of Central America divided into 6 admin units (districts). ❖ Belmopan

Belize City *130 Sp.* Belice. E Belize

Beljak *see* Villach

Bellary *298 Hind.* Ballari. S India

Bella Unión *611* N Uruguay

Belle-Anse *288* S Haiti

Belle Île *254* island of NW France

Bellenz *see* Bellinzona

Belleplaine *121* N Barbados

Belle Vue *497* S St Lucia

Bellevue Chopin *224* S Dominica

Belle Vue Maurel *400* NE Mauritius

Bellingshausen *90* CIS research station of South Shetland Islands, Antarctica

Bellinzona *550 Ger.* Bellenz. S Switzerland

Bello *195* NW Colombia

Bello Horizonte *see* Belo Horizonte

Belluno *323* N Italy

Bellville *531* Western Cape, SW South Africa

Belmanier *224* N Dominica

Belmopan *130* ❖ of Belize, C Belize

Belmullet *316* NW Ireland

Belo Horizonte *145 prev.* Bello Horizonte. SE Brazil

Bel Ombre *400* SW Mauritius

Bel Ombre *516* island, Mahé, Seychelles

Belarus *122-125* officially Republic of Belarus, *var.* Belarus, *prev.* Belorussian SSR, *Rus.* Belorusskaya SSR. Country of Europe divided into 6 admin. units (oblasts). ❖ Minsk

Belorusskaya Gryada *see* Byelruskaya Hrada

Beloshchel'ye *see* Nar'yan-Mar

Belostok *see* Białystok

Belo Tsiribihina *382 var.* Belo-Tsiribihina, Belo-sur-Tsiribihina. W Madagascar

Belovár *see* Bjelovar

Beloye More *251, 484 Eng.* White Sea. Sea of Arctic Ocean, bordering NW Russian Federation

Bel'tsy *see* Bălți

Belukha, Gora *340* mountain of E Kazakhstan and Russian Federation

Bembèrèkè *132 var.* Bembéréké. N Benin

Benaco *see* Garda, Lago di

Benares *see* Vārānasi

Ben Arous *577 var.* Bin Arūs. N Tunisia

Bender Beyla *528 var.* Bandarbeyla, Bender Beila. NE Somalia

Bender Cassim *see* Boosaaso

Bendern *374* NW Liechtenstein

Bendery *see* Tighina

Bendigo *101* SE Australia

Bendugu *518* N Sierra Leone

Benduma *368* NW Liberia

Benemérita de San Cristóbal *see* San Cristóbal

Beneški Zaliv *see* Gulf of Venice

Benešov *216* W Czech Republic

Benevento *323* S Italy

Bengal, Bay of *117, 159, 299, 538* bay of the Indian Ocean

Bengasi *see* Banghāzī

Bengbu *187 var.* Peng-pu. Anhui, E China

Benghazi *see* Banghāzī

Benguela *88 var.* Benguella. W Angola

Bengweulu *see* Bangweulu, Lake

Benha *232 var.* Banhā. N Egypt

Beni *136* river of N Bolivia

Beni *203* NE Dem. Rep. Congo

Benidorm *535* SE Spain

Beni Mellal *414* C Morocco

Benin *132-133* officially Republic of Benin, *prev.* Dahomey. Country of W Africa divided into 6 admin. units (departments). ❖ Porto-Novo

Benin, Bight of *132, 440, 571* area of the Gulf of Guinea

Benin City *440* SW Nigeria

Beni Suef *232 var.* Banī Suwayf. N Egypt

Bénitiers, Ile aux *400* island of SW Mauritius

Ben Nevis *597* mountain of C Scotland, UK

Benoni *531* Pretoria-Witwatersrand-Vereeniging, NE South Africa

Be, Nosy *382 var.* Nossi-Bé. Island of N Madagascar

Bénoué *see* Benue

Bénoy *180* S Chad

Benque Viejo del Carmen *130 var.* Benque Viejo. W Belize

Bense *224* N Dominica

Bensonville *368 prev.* Bentol. W Liberia

Bent Jbaïl *364 var.* Bint Jubayl. S Lebanon

Bentol *see* Bensonville

Bentong *386 var.* Bentung. C Peninsular Malaysia

Bên Tre *627 var.* Truc Giang. S Vietnam

Benue *168, 440 Fr.* Bénoué. River of Cameroon and Nigeria

Benxi *187 var.* Pen-ch'i, Penki. Liaoning, NE China

Beograd *see* Belgrade

Beqa *248 prev.* Mbengga. Island to the S of Viti Levu, W Fiji

Bequia *498* island of C St Vincent & the Grenadines

Beragala *538* SE Sri Lanka

Berat *81 var.* Berati, *SCr.* Beligrad. C Albania

Beraun *see* Berounka

Berbera *528* NW Somalia

Berbérati *179* SW Central African Republic

Berbice *286* river of NE Guyana

Berd *98* NE Armenia

Berdyans'k *590 Rus.* Berdyansk, *prev.* Osipenko. SE Ukraine

Bereeda *528 var.* Bareeda, *It.* Bereda. NE Somalia

Berekua *224* S Dominica

Berekum *272* W Ghana

Berettyó *292* river of Hungary and Romania

Berezina *see* Byerazino

Berezino *see* Byerazino

Berezniki *484* W Russian Federation

Bergamo *322* N Italy

Bergen *444* S Norway

Bergen *see* Mons

Berg en Dal *542* NE Suriname

Bergen op Zoom *429* SW Netherlands

Beringen *127* NE Belgium

Bering Sea *485 Rus.* Beringovo More. Sea of Pacific Ocean between NE Asia and NW North America

Bering Strait *485, 602 Rus.* Beringov Proliv. Strait connecting Bering Sea and Chukchi Sea, between NE Russian Federation and Alaska, USA

Berkane *414* NE Morocco

Berkeley Sound *645* area of the South Atlantic Ocean, NE Falkland Islands

Berlin *267, 271* ❖ of Germany, NE Germany

Bermuda *643* British Crown colony of the North Atlantic Ocean. ❖ Hamilton

Bermudian Landing *130* C Belize

Bern *550 Fr.* Berne. ❖ of Switzerland, W Switzerland

Bernal *463* NW Peru

Berner Oberland *550 Eng.* Bernese Oberland. Mountain range of SW Switzerland

Bernina, Passo del *550* mountain pass of SE Switzerland

Béroubouay *see* Gbéroubouè

Beroun *216* W Czech Republic

Berounka *216 Ger.* Beraun. River of W Czech Republic

Berovo *381* E FYR Macedonia

Berry Islands *112* island group of N Bahamas

Bertoua *168* E Cameroon

Beru *348* island of the Gilbert Is, W Kiribati

Beruniy *614 var.* Biruni, *Rus.* Beruni. W Uzbekistan

Berwick-upon-Tweed *597* NE England, UK

Besalampy *382* NW Madagascar

Besançon *255* E France

Beskiden *see* Beskid Mountains

Beskid Mountains *523 var.* Beskids, *Slvk.* Beskydy, *Ger.* Beskiden, *Pol.* Beskidy. Mountain range of C Europe

Beskra *see* Biskra

Bessarabia *see* Basarabeasca

Beszterce *see* Bistrița

Besztercebánya *see* Banská Bystrica

Betanzos *136* S Bolivia

Bétérou *132 var.* Betérou. C Benin

Bethanien *423 var.* Bethanie, Bethany. S Namibia

Bethel *649* E Montserrat

Bethesda *92* SE Antigua, Antigua & Barbuda

Bethlehem *531* Orange Free State, C South Africa

Bethlehem *319, 320 Heb.* Bet Lehem, *Ar.* Bayt Lahm. C West Bank

Betio *348* Tarawa, W Kiribati

Bétou *200* N Congo

Bet Shemesh *319* C Israel

Betsiboka *382* river of C Madagascar

Bette, Pic *371 It.* Picco Bette. Mountain peak of Chad and Libya

Bettembourg *378* S Luxembourg

Beuthen *see* Bytom

Beveren *127* N Belgium

Beverwijk *429* W Netherlands

Bexon *497* C St Lucia

Beyla *283* SE Guinea

Beylul *240* SE Eritrea

Beyneu *340* W Kazakhstan

Beyrouth *see* Beirut

Béziers *255* S France

Bezmein *see* Byuzmeyin

Bezwada *see* Vijayawāda

Bhāgalpur *299* NE India

Bhairab Bāzār *117 var.* Bhairab. C Bangladesh

Bhairahawa *427* C Nepal

Bhaktapur *427* C Nepal

Bhamdoun *364 var.* Bhamdūn. C Lebanon

Bhamo *159 var.* Banmo. N Burma

Bhangtar *134* SE Bhutan

Bharat *see* India

Bhawalpur *see* Bahāwalpur

Bheanntrai, Bá *see* Bantry Bay

Bheri *427* river of W Nepal

Bhimphedi *427* C Nepal

Bhojpur *427* E Nepal

Bhōpal *298* C India

Bhunya *544 var.* Bunya. W Swaziland

Bhutan *134-135* officially Kingdom of Bhutan. Country of S Asia divided into 18 admin. units (districts). ❖ Thimpu

Biabou *498* E St Vincent, St Vincent & the Grenadines

Biafra, Bight of *238, 440 var.* Bight of Bonny. Bay of the Gulf of Guinea, on the coast of W Africa

Biak, Pulau *305* island to the NW of Irian Jaya, E Indonesia

Białystok *471 Rus.* Belostok. E Poland

Biankouma *328* W Ivory Coast

Biarritz *254* SW France

Bibane, Bahiret el *577* lagoon of the Mediterranean Sea on the E coast of Tunisia

Bida *440* C Nigeria

Biel *550 Fr.* Bienne. W Switzerland

Bielefeld *266* NW Germany

Bieler See *550 Fr.* Lac de Bienne. Lake of W Switzerland

Bielitz-Biala *see* Bielsko-Biała

Biella *322* N Italy

Bielsko-Biała *471 Ger.* Bielitz-Biala. S Poland

Biên Đông *see* South China Sea

Biên Hoa *627* S Vietnam

Bienne, Lac de *see* Bieler See

Bienvenue *645* SE French Guiana

Bíe, Planalto do *88 var.* Bié Plateau. Plateau of C Angola

Bié Plateau *see* Planalto do Bíe

Big Bay *619* bay of the Pacific Ocean, off Espiritu Santo, C Vanuatu

Big Bend *544* SE Swaziland

Bigene *284* NW Guinea-Bissau

Bight, The *112* Cat I, Bahamas

Bignona *510* SW Senegal

Bihać *140* NW Bosnia & Herzegovina

Bijagós, Arquipélago dos *284 var.* Bijagos Archipelago. Island group of W Guinea-Bissau

Bijeljina *140* NE Bosnia & Herzegovina

Bijelo Polje *512* E Montenegro, Serbia & Montenegro (Yugo.)

Bikaner *298* NW India

Bikar *396* island of N Marshall Islands

Bikini *396* island of NW Marshall Islands

Bilād Manaḥ *see* Manaḥ

Bilāspur *298* NE India

Bilāsuvar *110 Rus.* Bilyasuvar, *prev.* Pushkino, *var.* Puškino. S Azerbaijan

Bila Tserkva *590 Rus.* Belaya Tserkov'. NW Ukraine

Bonthe *518* Sherbro I, Sierra Leone
Boosaaso *528* var. Bosaso, Bender Qaasim, *It.* Bender Cassim. N Somalia
Boothia Felix *see* Boothia Peninsula
Boothia, Gulf of *170* gulf of the Arctic Ocean between the Boothia Peninsula and Melville Peninsula, N Canada
Boothia Peninsula *170* prev. Boothia Felix. Peninsula of N Canada
Booué *260* NE Gabon
Bophuthatswana Bantustan 'self-governing homeland'; abolished in 1994
Bopolu *368* NW Liberia
Boquete *456* var. Bajo Boquete. W Panama
Boraha, Nosy *see* Sainte Marie, Nosy
Borås *547* SW Sweden
Borborema, Planalto da *145* plateau of NE Brazil
Bordeaux *255* SW France
Borden Peninsula *171* peninsula of Baffin Island, N Canada
Bordergate *544* NE Swaziland
Bordhoy *644* var. Bordo. Island of NE Faeroe Islands
Bordj Bou Arreridj *83* N Algeria
Borgarnes *296* W Iceland
Borgholm *547* Öland, S Sweden
Borgo Maggiore *502* NW San Marino
Borisov *see* Barysaw
Bor, Lagh *344* var. Lak Bor. Dry watercourse of NE Kenya
Borlänge *547* C Sweden
Borneo *304, 386-387* island of SE Asia divided between Brunei, two Malaysian states and part of Indonesia
Bornholm *218* island of E Denmark
Boromo *157* SW Burkina
Borongo *see* Black Volta
Borovo *209* NE Croatia
Borujerd *309* var. Burujird. W Iran
Börzsöny *292* mountain range of N Hungary
Bosanska Gradiška *140* N Bosnia & Herzegovina
Bosanski Brod *140* N Bosnia & Herzegovina
Bosanski Šamac *140* N Bosnia & Herzegovina
Boscobelle *121* NE Barbados
Bösing *see* Pezinok
Boskamp *542* N Suriname
Boskovice *216* SE Czech Republic
Bosna *140* river of N Bosnia & Herzegovina
Bosna I Hercegovina, Federacija *140* republic Bosnia & Herzegovina
Bosnia & Herzegovina *140-141* officially The Republic of Bosnia and Herzegovina. Country of SE Europe. ❖ Sarajevo
Bosporus *see* İstanbul Boğazı
Bossangoa *179* C Central African Republic
Bossembélé *179* C Central African Republic
Boston *603* Massachusetts, NE USA
Boteti *142* var. Botletle. River of C Botswana
Botevgrad *152* prev. Orkhanie, Orkhaniye. W Bulgaria
Bothnia, Gulf of *251, 547* *Fin.* Pohjanlahti, *Swe.* Bottniska Viken. Gulf of the Baltic Sea, between Finland and Sweden
Boti-Pasi *542* C Suriname
Botna *408* river of E Moldova
Botoşani *480* NE Romania
Botrange *127* mountain of E Belgium
Botswana *142-143* officially Republic of Botswana. Country of southern Africa divided into 10 admin. units (districts). ❖ Gaborone
Bottle Creek *653* North Caicos, N Turks and Caicos Islands
Bottniska Viken *see* Bothnia, Gulf of
Bottom, The *650* Saba, N Netherlands Antilles
Bouaflé *328* C Ivory Coast
Bouaké *328* var. Bwake. C Ivory Coast
Bouar *179* W Central African Republic

Bouca *179* C Central African Republic
Bouéni *649* SW Mayotte
Bouéni, Baie de *649* var. Bonéni Bay. Bay of the Mozambique Channel on the SW coast of Mayotte
Bouenza *200* river of S Congo
Boufarik *83* N Algeria
Bougainville Island *458* island of E Papua New Guinea
Bougie *see* Béjaïa
Bougouni *392* SW Mali
Bougouriba *157* river of SW Burkina
Boujdour *414* W Western Sahara
Boukoumbé *132* var. Boukombé. NW Benin
Boukra *414* C Western Sahara
Boulder *603* Colorado, SW USA
Boulogne-sur-Mer *255* var. Boulogne. N France
Bouloupari *650* S New Caledonia
Bouma *248* Taveuni, N Fiji
Boumango *260* SE Gabon
Boumba *168* river of SE Cameroon
Boumbé II *179* river of Cameroon and Central African Republic
Boûmdeïd *398* S Mauritania
Bouna *328* NE Ivory Coast
Boundiali *328* N Ivory Coast
Boungou *179* river of C Central African Republic
Bourail *650* C New Caledonia
Bourgas *see* Burgas
Bourg-en-Bresse *255* E France
Bourges *255* C France
Bourgogne *255* Eng. Burgundy. Cultural region of E France
Bourke *101* E Australia
Bournemouth *597* S England, UK
Boussé *157* C Burkina
Boussouma *157* C Burkina
Boutilimit *398* SW Mauritania
Bouton *497* W St Lucia
Bovec *524* Ger. Flitsch, *It.* Plezzo. NW Slovenia
Boven Kapuas, Pegunungan *386* mountain range of SW Borneo, Malaysia
Boyd's *494* St Kitts, St Kitts & Nevis
Boyne *316* *Ir.* An Bhóinn. River of E Ireland
Boysun *614* Rus. Baysun. SE Uzbekistan
Bozen *see* Bolzano
Bozoum *179* W Central African Republic
Brač *209* var. Brach, *It.* Brazza. Island of S Croatia
Bradford *597* N England, UK
Braga *474* NW Portugal
Bragado *95* E Argentina
Bragança *474* Eng. Braganza. NE Portugal
Braganza *see* Bragança
Brahestad *see* Raahe
Brāhmanbāria *117* C Bangladesh
Brahmaputra *117, 186, 299* var. Tsangpo, *Ben.* Jamuna. River of S Asia
Brăila *480* E Romania
Braine-l'Alleud *127* C Belgium
Branco *144* river of NW Brazil
Branco, Ilhéu *176* island of N Cape Verde
Brandberg *423* mountain of NW Namibia
Brande *218* Jylland, W Denmark
Brandenburg *267* C Germany
Brandon *170* S Canada
Brani, Pulau *521* island of S Singapore
Brasília *145* ❖ of Brazil, C Brazil
Braslaw *122* N Belarus
Braşov *480* prev. Stalin, *Ger.* Kronstadt, *Hung.* Brassó. C Romania
Brasschaat *127* N Belgium
Brassey, Banjaran *387* var. Brassey Range. Mountain range of E Borneo, Malaysia
Brassey Range *see* Brassey, Banjaran
Bratislava *523* Ger. Pressburg, *Hung.* Pozsony, ❖ of Slovakia, SW Slovakia
Bratsk *485* C Russian Federation

Bratskoye Vodokhranilishche *485* Eng. Bratsk Reservoir. C Russian Federation
Braunau am Inn *106* var. Braunau. N Austria
Braunschweig *267* Eng. Brunswick. N Germany
Brava *see* Baraawe
Brava *176* island of SW Cape Verde
Brava, Costa *535* coastal region of E Spain
Bravo Del Norte *see* Grande, Río
Bray *316* Ir. Bri Chuallan. E Ireland
Brazil *146-149* officially Federative Republic of Brazil. Country of South America divided into 28 admin. units (26 states, 1 territory, 1 federal district). ❖ Brasília
Brazos *603* river of SW USA
Brazza *see* Brač
Brazzaville *200* ❖ of Congo, S Congo
Brčko *140* NE Bosnia & Herzegovina
Breda *429* SW Netherlands
Bregalnica *381* river of E FYR Macedonia
Bregenz *106* W Austria
Breidhafjördhur *296* bay of the Denmark Strait, on the coast of W Iceland
Bremen *266* NW Germany
Bremerhaven *266* NW Germany
Bremersdorp *see* Manzini
Brennero, Passo del *see* Brenner Pass
Brenner Pass *106, 323* *It.* Passo del Brennero, *Ger.* Brennerpass, var. Brenner Sattel. Mountain pass of Austria and Italy
Brerton *121* C Barbados
Brescia *322* N Italy
Breslau *see* Wrocław
Brest *122* Pol. Brześć nad Bugiem, prev. Brześć Litewski, Rus. Brest-Litovsk. SW Belarus
Brest *254* NW France
Bretagne *254* Eng. Brittany. Cultural region of NW France
Brewerville *368* W Liberia
Brezhnev *see* Naberezhnyye
Brežice *524* Ger. Rann. E Slovenia
Brezno *523* prev. Brezno nad Hronom, *Ger.* Bries, var. Briesen, *Hung.* Breznóbánya. C Slovakia
Bria *179* C Central African Republic
Bri Chuallan *see* Bray
Bride *647* N Isle of Man
Bridgetown *100* SW Australia
Bridgetown *121* ❖ of Barbados, SW Barbados
Bridgwater *597* SW England, UK
Brienzer See *550* lake of SW Switzerland
Brig *550* SW Switzerland
Brighton *597* SE England, UK
Brikama *263* W Gambia
Brindisi *323* S Italy
Brinstone Hill *494* NW St Kitts, St Kitts & Nevis
Brisbane *101* E Australia
Bristol *597* SW England, UK
Bristol Channel *597* inlet of the Atlantic Ocean, SW England, UK
British Columbia *170* province of SW Canada
British Guiana *see* Guyana
British Indian Ocean Territory *643* British dependent territory of the Indian Ocean. ❖ Diego Garcia
British North America *see* Canada
British Virgin Islands *643* British dependent territory of the Caribbean Sea. ❖ Road Town.
Brittany *see* Bretagne
Brizan *278* W Grenada island, Grenada
Brno *216* Ger. Brünn. SE Czech Republic
Bród *see* Slavonski Brod
Brodeur Peninsula *170* peninsula of Baffin Island, N Canada
Broken Hill *101* S Australia
Brokopondo *542* NE Suriname
Bromberg *see* Bydgoszcz
Brønderslev *218* Jylland, N Denmark

Brooks Range *602* mountain range of Alaska, USA
Broome *100* NW Australia
Brorup *218* Jylland, W Denmark
Broughton Bay *see* Tongjosŏn-man
Brown Hill *494* S Nevis, St Kitts & Nevis
Browns Town *331* N Jamaica
Brownsweg *542* C Suriname
Bruck an der Mur *106* var. Bruck. C Austria
Brufut *263* W Gambia
Brugge *127* Fr. Bruges. NW Belgium
Bruit, Pulau *386* island of W Borneo, Malaysia
Brunei *150-151, 386* officially Sultanate of Brunei, *Mal.* Negara Brunei Darussalam. Country on island of Borneo, SE Asia, divided into 4 admin. units (districts). ❖ Bandar Seri Begawan
Brunei Bay *386* var. Teluk Brunei. Bay of the South China Sea
Brunei Town *see* Bandar Seri Begawan
Brünn *see* Brno
Brunswick *see* Braunschweig
Brusa *see* Bursa
Brussels *127* Fr. Bruxelles, *Dut.* Brussel, *Ger.* Brüssel. ❖ of Belgium, C Belgium
Brüx *see* Most
Bryansk *484* W Russian Federation
Brześć Litewski *see* Brest
Brześć nad Bugiem *see* Brest
Bsharri *see* Bcharré
Bua *385* river of C Malawi
Bu'aale *528* SW Somalia
Buada Lagoon *424* lagoon on the coast of Nauru
Buala *526* SE Santa Isabel, Solomon Islands
Bū al Ḥīdān, Wādī *371* dry watercourse of C Libya
Buba *284* S Guinea-Bissau
Bubanza *162* NW Burundi
Bubaque *284* Ilha de Bubaque, Guinea-Bissau
Bubaque, Ilha de *284* island of SW Guinea-Bissau
Bubi *636* var. Bubye. River of S Zimbabwe
Būbīyan, Jazīrat *356* island of NE Kuwait
Buca *248* prev. Mbutha. Vanua Levu, N Fiji
Bucaramanga *195* N Colombia
Buchanan *368* prev. Grand Bassa. SW Liberia
Bucharest *480* Rom. Bucureşti. ❖ of Romania, S Romania
Budapest *292* ❖ of Hungary, N Hungary
Búdhardalur *296* W Iceland
Budweis *see* České Budějovice
Buea *168* SW Cameroon
Buenaventura *195* W Colombia
Buena Vista *130* N Belize
Buena Vista *136* C Bolivia
Buena Vista *646* S Gibraltar
Buenos Aires *95* ❖ of Argentina, E Argentina
Buenos Aires *206* SE Costa Rica
Buenos Aires, Lago *95, 183* *Sp.* Lago General Carrera. Lake of Argentina and Chile
Buffalo *603* New York, NE USA
Buff Bay *331* E Jamaica
Bug *471, 590* Ukr. Zakhidnyy Buh, *Rus.* Zapadnyy Bug. River of E Europe
Buga *195* W Colombia
Bugala Island *588* island of Sese Islands, S Uganda
Buganda *162* NW Burundi
Bugarama *492* SW Rwanda
Bughotu *see* Santa Isabel
Buġibba *395* N Malta
Bugojno *140* C Bosnia & Herzegovina
Bugumya *492* SW Rwanda
Buin *458* Bougainville I, Papua New Guinea
Buitenzorg *see* Bogor

Bujumbura *162 prev.* Usumbura.
❖ of Burundi, W Burundi
Buka Island *458* island of E Papua
New Guinea
Bukakata *588* S Uganda
Bukasa Island *588* island of Sese
Islands, S Uganda
Bukavu *203 prev.* Costermansville.
E Dem. Rep. Congo
Bukeye *162 var.* Bukaye. C Burundi
Bukhoro *614 var.* Bokhara,
Rus. Bukhara. S Uzbekistan
Bukit Mertajam *386* NW
Peninsular Malaysia
Bukit Panjang *521* area of
C Singapore
Bukit Timah *521* area of C Singapore
Bükk *292* mountain range of
NE Hungary
Bukoba *564* NW Tanzania
Bukum Kechil, Pulau *521* island
of SW Singapore
Bukum, Pulau *521* island of
S Singapore
Bula *284* W Guinea-Bissau
Bulawayo *636 var.* Buluwayo.
SW Zimbabwe
Bulembu *544* NW Swaziland
Bulgan *412* N Mongolia
Bulgaria *152-155* officially Republic
of Bulgaria, *prev.* People's Republic
of Bulgaria. Country of E Europe
divided into 8 admin. units (regions).
❖ Sofia
Buller *433* river of N South Island,
New Zealand
Bulolo *458* SE Papua New Guinea
Bulunghur *614 prev.*
Krasnogvardeysk, *Rus.* Bulungur.
SE Uzbekistan
Bumba *203* N Dem. Rep. Congo
Bumbah, Khalīj *371 var.* Kalīj
Bumbah. Gulf of the Mediterranean
Sea, on the N coast of Libya
Bumbuna *518* C Sierra Leone
Bunbury *100* SW Australia
Bundaberg *101* E Australia
Bundoran *316* N Ireland
Bungoma *344* W Kenya
Bunia *203* NE Dem. Rep. Congo
Bunya *see* Bhunya
Buôn Ma Thuôt *627 var.* Lac Giao.
S Vietnam
Bur Acaba *see* Buurhakaba
Buraimi *see* Al Buraymī
Buraydah *506 var.* Buraida.
C Saudi Arabia
Burco *528 var.* Burao, Bur'o.
NW Somalia
Burdur *580* SW Turkey
Burē *245* NW Ethiopia
Burera, Lac *492* lake of N Rwanda
Būrewāla *see* Mandi Būrewāla
Burgas *152 var.* Bourgas. E Bulgaria
Burgaski Zaliv *152* bay of the Black
Sea on the E coast of Bulgaria
Burgdorf *550* NW Switzerland
Burgos *535* N Spain
Burgundy *see* Bourgogne
Burhou *647* island of N Guernsey
Burias Island *466* island of
C Philippines
Buriram *567* E Thailand
Burkina *156-157* officially Burkina
Faso, *prev.* Upper Volta. Country
of West Africa divided into 30
admin. units (provinces).
❖ Ouagadougou
Burlington *603* Vermont, NE USA
Burma *158-161* officially Union of
Myanmar, *var.* Myanmar. Country of
SE Asia divided into 14 admin. units
(7 states, 7 divisions). ❖ Rangoon
Burnie *101* Tasmania, SE Australia
Burnt Pine *650* C Norfolk Island
Burrel *81 var.* Burreli. C Albania
Burrell Boom *130* E Belize
Bursa *580 prev.* Brusa, *var.* Brusa.
NW Turkey
Būr Sa'īd *see* Port Said
Būr Sūdān *see* Port Sudan
Burtnieku Ezers *362* lake of N Latvia
Buru *304 prev.* Boeroe. Island
of Maluku, E Indonesia

Burujird *see* Borujerd
Burūm *631* SE Yemen
Burundi *162-163* officially Republic of
Burundi, *prev.* Kingdom of Burundi.
Country of Central Africa divided
into 15 admin. units (provinces).
❖ Bujumbura
Bururi *162* SW Burundi
Buryatiya, Respublika *485*
autonomous republic of
C Russian Federation
Busan *see* Pusan
Busengo *492* NW Rwanda
Buševa Planina *see* Baba
Bushire *see* Bandar-e Bushehr
Busia *344* W Kenya
Busing, Pulau *521* island of
SW Singapore
Busira *203* river of
NW Dem. Rep. Congo
Busoro *492* S Rwanda
Busuanga Island *466* island of
W Philippines
Buta *203* N Dem. Rep. Congo
Butare *492 prev.* Astrida. S Rwanda
Butaritari *348* island of the Gilbert Is,
W Kiribati
Butawal *see* Butwal
Butembo *203* NE Dem. Rep. Congo
Butha Buthe *366* N Lesotho
Butiaba *588* NW Uganda
Butterworth *386* NW Peninsular
Malaysia
Butuan *466* Mindanao, S Philippines
Butung, Pulau *304 var.* Buton,
prev. Boetoeng. Island to the
SE of Celebes, C Indonesia
Butwal *427 var.* Butawal. C Nepal
Buurgplaatz *378* mountain of
N Luxembourg
Buurhakaba *528 var.* Bur Hakkaba,
It. Bur Acaba. SW Somalia
Buvuma Island *588* island of S
Uganda
Büyükağrı Dağı *581 Eng.* Mount
Ararat, *var.* Great Ararat. Mountain
of E Turkey
Büyükmenderes *580* river of
SW Turkey
Büyükzap Suyu *see* Great Zab
Buzău *480 Hung.* Bodza.
SE Romania
Buzi *419* river of C Mozambique
Bwake *see* Bouaké
Bwiam *263* SW Gambia
Byahoml' *122* N Belarus
Byala Slatina *152 var.* Bjala Slatina.
NW Bulgaria
Byaroza *122* SW Belarus
Bydgoszcz *471 Ger.* Bromberg.
N Poland
Byelruskaya Hrada *122*
Rus. Belorusskaya Gryada. Hilly
region of N Belarus
Byerazino *122 Rus.* Berezina. River
of C Belarus
Byerazino *122 Rus.* Berezino.
C Belarus
Bykhaw *122 Rus.* Bykhov. E Belarus
Byron Sound *645* area of the South
Atlantic Ocean, NW Falkland Islands
Byrranga, Gory *485* mountain range
of N Russian Federation
Bystrovka *see* Kemin
Bytom *471 Ger.* Beuthen. S Poland
Byumba *492 var.* Biumba. N Rwanda
Byuzmeyin *585 prev.* Bezmein.
S Turkmenistan

C

Caacupé *460* S Paraguay
Caaguazú *460* SE Paraguay
Caála *88 var.* Kaala, *Port.* Vila Robert
Williams. C Angola
Caatinga, Serra da *145* mountain
range of E Brazil
Caazapá *460* S Paraguay
Cabaiguán *210* C Cuba
Cabanatuan City *466* Luzon,
Philippines
Cabháin *see* Cavan
Cabimas *623* NW Venezuela

Cabinda *88 var.* Kabinda.
Non-contiguous region, NW Angola
Cabinda *88 var.* Kabinda. NW Angola
Cabora Bassa, Lake *see* Cahora Bassa,
Lago de
Cabot Strait *171* strait connecting
the Atlantic Ocean and Gulf of
St. Lawrence, between
Newfoundland & Labrador
and Nova Scotia
Cape Verde, Ilhas de *see* Cape Verde
Cabras, Ilha das *505* island to the
N of São Tomé, Sao Tome & Principe
Cabras Island *647* island, W Guam
Čačak *512* C Serbia, Serbia &
Montenegro (Yugo.)
Cacao *645* NE French Guiana
Cáceres *535* W Spain
Cachacrou *see* Scotts Head Village
Cacheu *284 var.* Cacheo.
W Guinea-Bissau
Cacheu *284* river of NW
Guinea-Bissau
Cachimbo, Serra do *145* mountain
range of C Brazil
Cacine *284* S Guinea-Bissau
Cacine *284* river of S Guinea-Bissau
Caconda *88* C Angola
Cadca *523 Hung.* Csaca. N Slovakia
Cadiz *466* Negros, Philippines
Cádiz *534* SW Spain
Cádiz, Golfo de *474, 535* area of the
Atlantic Ocean, on the SW coast
of Spain
Caen *255* NW France
Caerdydd *see* Cardiff
Caerfyrddin *see* Carmarthen
Caergybi *see* Holyhead
Caernarfon *597 var.* Caernarvon.
N Wales, UK
Cagayan *466* river of Luzon,
Philippines
Cagayan de Oro *466* Mindanao,
Philippines
Cagayan Islands *466* island group
of C Philippines
Cagliari *322* Sardegna, W Italy
Caguas *651* E Puerto Rico
Cahabón *280* river of C Guatemala
Cahora Bassa Dam *419* dam
of NW Mozambique
Cahora Bassa, Lago de *419 var.* Lake
Cabora Bassa. Reservoir of
NW Mozambique
Cahors *255* S France
Cahul *408 Rus.* Kagul. S Moldova
Caia *419* C Mozambique
Caibarién *210* C Cuba
Caicos Bank *653* undersea feature
of the Atlantic Ocean
Caicos Passage *112, 653* strait of the
Atlantic Ocean, between the
Bahamas and Caicos Islands
Cai Lan *627* N Vietnam
Caille Island *278* island to the
N of Grenada island, Grenada
Cailungo *502* N San Marino
Caió *284* W Guinea-Bissau
Cairns *101* NE Australia
Cairo *255 Ar.* Al Qāhirah,
var. El Qâhira. N Egypt
Cajamarca *463 prev.* Caxamarca.
NW Peru
Cajón, Represa el *290* reservoir of
W Honduras
Çakilli Dere *see* Yialias
Čakovec *209 Hung.* Csáktornya,
Ger. Csakathurn, *prev.*
Tschakathurn. N Croatia
Calabar *440* S Nigeria
Calabozo *623* C Venezuela
Calabrai *S* S Italy
Calagua Islands *466* island group
of N Philippines
Calais *255* N France
Calama *183* N Chile
Calamian Group *466* island group of
W Philippines
Calandula *88 var.* Kalandula.
NW Angola
Calanscio Sand Sea *see*
Kalanshiyŭ, Sarīr
Călăraşi *480* SE Romania

Călăraşi *408 var.* Cǎlǎras,
Kalarash. C Moldova
Calbayog *466* Samar, W Philippines
Calceta *230* W Ecuador
Calcutta *130* N Belize
Calcutta *299 var.* Kolkata. E India
Calcutta *542* N Suriname
Caldas da Rainha *474* W Portugal
Caledon *366, 531* river of Lesotho and
South Africa
Calf of Man *647* island of S Isle of Man
Calgary *170* SW Canada
Calhau *176* São Vincente,
N Cape Verde
Cali *195* W Colombia
Calibishie *224* N Dominica
Calicut *298 var.* Kozhikode. S India
Calida, Costa *535* coastal region
of SE Spain
California *602* state of W USA
California, Golfo de *403* gulf of the
Pacific Ocean
Cǎlilabad *110 Rus.* Dzhalilabad,
var. Džalilabad,
prev. Astrakhan-Bazar. S Azerbaijan
Callao *463* W Peru
Calliaqua *498* S St Vincent, St Vincent
& the Grenadines
Caltagirone *323* Sicilia, S Italy
Caltanissetta *323* Sicilia, S Italy
Camá *226* river of C Dominican
Republic
Camabatela *88* NW Angola
Camacupa *88 Port.* General Machado.
C Angola
Camagüey *210 prev.* Puerto Príncipe.
C Cuba
Camagüey, Archipiélago de *210*
island group of N Cuba
Ca Mau *627 var.* Quan Long.
S Vietnam
Cambay, Gulf of *see* Khambhat,
Gulf of
Cambodia *164-167* officially State of
Cambodia, *Cam.* Kampuchea,
prev. People's Democratic Republic
of Kampuchea. Country of SE Asia
divided into 20 admin. units
(provinces). ❖ Phnom Penh
Cambrian Mountains *597* mountain
range of Wales, UK
Cambridge *331* NW Jamaica
Cambridge *433* C North Island,
New Zealand
Cambridge *597* E England, UK
Cambridge Bay *170* Victoria Island,
NW Canada
Cambrouze *645* N French Guiana
Cambulo *88* NE Angola
Camden Park *498* SW St Vincent,
St Vincent & the Grenadines
Cameron Highlands *386* highlands of
C Peninsular Malaysia
Cameroon *168-169* officially Republic
of Cameroon, *Fr.* Cameroun.
Country of W Africa divided into 10
admin. units (provinces). ❖ Yaoundé
Camiri *136* S Bolivia
Camopi *645* river of
SE French Guiana
Camopi *645* E French Guiana
Camotes Islands *466* island group of
E Philippines
Campana *95* E Argentina
Campbell River *170* Vancouver
Island, SW Canada
Campeche *403* SE Mexico
Campeche, Bahía de *403 var.*
Gulf of Campeche. Bay of the
Gulf of Mexico
Campeche, Gulf of *see*
Campeche, Bahía de
Câm Pha *627* N Vietnam
Campina Grande *145* E Brazil
Campinas *145* S Brazil
Campine *see* Kempen
Campobasso *323* S Italy
Campo Grande *145* SW Brazil
Campos *145 var.* Campo dos
Goitacazes. SE Brazil
Campossa *284* river of
E Guinea-Bissau
Cam Ranh *627* SE Vietnam

Cerro Chato *611* C Uruguay
Cerro de Pasco *463* C Peru
Cerrón Grande, Embalse *237* reservoir of N El Salvador
Ceru Colorado *642* S Aruba
Červen brjag *see* Cherven-Bryag
Cesena *323* N Italy
Cēsis *362* *Ger.* Wenden. NE Latvia
České Budějovice *216* *Ger.* Budweis. SW Czech Republic
Českomoravská Vrchovina *216* *var.* Bohemian-Moravian Highlands, Ceskomoravská Vrchovina, *Ger.* Böhmisch-Mährische Höhe. Mountain range of S Czech Republic
Český Krumlov *216* SW Czech Republic
Český Les *see* Bohemian Forest
Céspedes *210* SE Cuba
Cess *368* *var.* Rivercess. S Liberia
Cestos *368* *var.* Cess. River of Liberia
Ceuta *414, 535* exclave of Spain, N Morocco
Cévennes *255* mountain range of SE France
Ceylon *see* Sri Lanka
Ceyhan *581* S Turkey
Chacachacare *574* island to the NW of Trinidad, Trinidad & Tobago
Chachapoyas *463* NW Peru
Chachoengsao *567* C Thailand
Chaclacayo *463* W Peru
Chaco *see* Gran Chaco
Chad *180-181* officially Republic of Chad, *Fr.* Tchad. Country of Equatorial Africa divided into 14 admin. units (prefectures). ❖ N'Djamena
Chã da Igreja *176* Santo Antão, N Cape Verde
Chad, Lake *168, 180, 439, 441* *Fr.* Lac Tchad. Lake of C Africa
Chadyr-Lunga *see* Ciadîr-Lunga
Chaeryŏng *351* SW North Korea
Chāgai Hills *450* mountain range of Afghanistan and Pakistan
Chaghcharān *77* C Afghanistan
Chagos Archipelago *643* *var.* Oil Islands. Island group of C British Indian Ocean Territory
Chaguanas *574* W Trinidad, Trinidad & Tobago
Chaguaramas *574* NW Trinidad, Trinidad & Tobago
Chagyl *585* NW Turkmenistan
Chahārborjak *77* SW Afghanistan
Chaillu, Massif du *260* C Gabon
Mongos, Chaîne des *see* Bongo, Massif des
Chainpur *427* W Nepal
Chaiyaphum *567* N Thailand
Chakari *636* S Zimbabwe
Chake Chake *564* Pemba, E Tanzania
Chakwāl *451* NE Pakistan
Chalándri *275* *prev.* Khalándrion. SE Greece
Chalap Dalam *77* mountain range of W Afghanistan
Chalatenango *237* N El Salvador
Chalbi Desert *344* desert region of N Kenya
Chalchuapa *237* W El Salvador
Chalkída *275* *Eng.* Chalcis, *prev.* Khalkís. Evvoia, E Greece
Chalkidikí *275* *prev.* Khalkidhikí, *Eng.* Chalcidice. Peninsula of NE Greece
Challapata *136* SW Bolivia
Challengers *494* S St Kitts, St Kitts & Nevis
Châlons-en-Champagne *255* *prev.* Châlons-sur-Marne. NE France
Chaman *451* NW Pakistan
Chambas *210* C Cuba
Chambéry *255* E France
Chambeshi *635* river of NE Zambia
Chambi, Jebel *577* mountain of W Tunisia
Chamelecón, Río *290* river of NW Honduras
Chamouny *400* S Mauritius
Champagne *255* cultural region of NE France
Champasak *360* S Laos

Chañaral *183* N Chile
Chances Peak *649* mountain peak of S Montserrat
Chan-Chen *130* N Belize
Chan-chiang *see* Zhanjiang
Chandīgarh *298* N India
Chandlers Falls *344* var. Chanlers Falls. Waterfall of C Kenya
Chāndpur *117* E Bangladesh
Changane *419* river of S Mozambique
Changara *419* W Mozambique
Changchun *187* *prev.* Hsinking. Jilin, NE China
Changhua *559* *Jap.* Shōka. W Taiwan
Changi *521* area of E Singapore
Chang Jiang *187* *var.* Yangtze Kiang. River of C China
Changkiakow *see* Zhangjiakou
Chang, Ko *567* island of C Thailand
Changsha *187* Hunan, S China
Chang-tien *see* Zibo
Changuinola *456* W Panama
Ch'angwŏn *352* S South Korea
Changyŏn *351* SW North Korea
Chaniá *275* *Eng.* Canea, *prev.* Khaniá. Crete, S Greece
Channel Islands *254* island group to the NW of France *see* Jersey, Guernsey, Alderney, Sark
Channel, The *see* English Channel
Channel Tunnel *597* tunnel between France and SE England, UK
Chanthaburi *567* C Thailand
Ch'aochou *559* *var.* Chaochow, *Jap.* Chōshū. SW Taiwan
Chao Phraya *567* river of C Thailand
Chaouèn *414* *var.* Chechaouèn, Chefchaouèn, *Sp.* Xauen. N Morocco
Chapada Diamantina *145* plateau of E Brazil
Chapala, Lago de *403* lake of SW Mexico
Chapelton *331* C Jamaica
Chaplin Bay *643* bay of the North Atlantic Ocean, W Bermuda
Chardonnières *288* SW Haiti
Chardzhev *585* *prev.* Chardzhou, *var.* Čardžou, *prev.* Leninsk, *Turkm.* Chärjew. E Turkmenistan
Chardzhou *see* Chardzhev
Ch'arents'avan *98* *var.* Čarencaven. C Armenia
Chari *180* *var.* Shari. River of C Africa
Chārīkār *77* NE Afghanistan
Charity *286* NE Guyana
Chärjew *see* Chardzhev
Charkhliq *see* Ruoqiang
Charleroi *127* S Belgium
Charleston *603* South Carolina, SE USA
Charleston *603* West Virginia, E USA
Charlestown *494* SW Nevis, St Kitts & Nevis
Charlestown *498* Canouan, St Vincent & the Grenadines
Charleville *101* E Australia
Charleville-Mézieres *255* NE France
Charlotte *603* North Carolina, SE USA
Charlotte Amalie *653* ❖ of Virgin Islands (US), S Saint Thomas Island
Charlotte Town *see* Gouyave
Charlottetown *171* Prince Edward Island, SE Canada
Charlotte Ville *224* SW Dominica
Charlotteville *574* N Tobago, Trinidad & Tobago
Charshanga *585* *prev.* Charshangy, *Turkm.* Charshangngy. SE Turkmenistan
Charsk *see* Shar
Chartres *255* N France
Chartres Settlement *645* West Falkland, W Falkland Islands
Chateaubelair *498* NW St Vincent, St Vincent & the Grenadines
Châteauroux *255* C France
Châtelet *127* S Belgium
Chatham Island *433* island of Chatham Islands, E New Zealand
Chatham Islands *433* islands of E New Zealand

Chatkal Range *359, 614* *Rus.* Chatkal'skiy Khrebet. Mountain range of Kyrgyzstan and Uzbekistan
Chāttagām *see* Chittagong
Chattanooga *603* Tennessee, SE USA
Chatyr-Tash *359* E Kyrgyzstan
Châu Đôc *627* *var.* Chau Phu. SW Vietnam
Chauk *159* W Burma
Chaumont *254* NE France
Chau Phu *see* Châu Đôc
Chautara *427* C Nepal
Chaves *474* N Portugal
Chbar *165* E Cambodia
Cheb *267* *Ger.* Eger. W Czech Republic
Chechaouèn *see* Chaouèn
Ch'ech'eng *559* S Taiwan
Che-chiang *see* Zhejiang
Chechenskaya, Respublika *484, 488* autonomous republic of SW Russian Federation
Chech'ŏn *352* *Jap.* Teisen. N South Korea
Checker Hall *121* N Barbados
Cheduba Island *159* island of W Burma
Chefoo *see* Yantai
Chegutu *636* *prev.* Hartley. N Zimbabwe
Cheju *352* *Jap.* Saishū. Cheju-do, South Korea
Cheju-do *352* *prev.* Quelpart, *Jap.* Saishu. Island of S South Korea
Cheju Strait *352* *var.* Chejuhaehyop, Cheju-Haehyop. Strait connecting the Korea Strait and Yellow Sea
Chekiang *see* Zhejiang
Cheleken *585* *var.* Čeleken. W Turkmenistan
Chelkar *340* W Kazakhstan
Chelyabinsk *484* C Russian Federation
Chemin Grenier *400* S Mauritius
Chemnitz *267* *prev.* Karl-Marx-Stadt. E Germany
Chemulpo *see* Inch'ŏn
Ch'eng-chou *see* Zhengzhou
Chengchow *see* Zhengzhou
Chengdu *187* *var.* Chengtu. Sichuan, SW China
Chenghsien *see* Zhengzhou
Ch'eng-hua *see* Altay
Chenghwa *see* Altay
Chenkaladi *538* E Sri Lanka
Chennai *298* *var.* Madras. S India
Cheom Ksan *see* Chŏâm Khsant
Chepo *456* N Panama
Cher *255* river of C France
Cherbourg *255* NW France
Cheren *see* Keren
Cherepovets *484* W Russian Federation
Chergui, Chott ech *83* salt lake of NW Algeria
Chergui, Île *577* *Ar.* Jazirat ash Sharqi. Island of E Tunisia
Cherikaw *122* E Belarus
Cherkasy *590* *Rus.* Cherkassy. C Ukraine
Cherne More *see* Black Sea
Chernivtsi *590* *Rus.* Chernovtsy, *Rom.* Cernăuţi, *Ger.* Czernowitz. W Ukraine
Chernobyl' *see* Chornobyl'
Cherno More *see* Black Sea
Cherry Island *see* Anuta
Cherskiy *485* NE Russian Federation
Cherskogo, Khrebet *485* mountain range of NE Russian Federation
Cherso *see* Cres
Cherson *see* Kherson
Cherven-Bryag *152* *var.* Červen brjag. NW Bulgaria
Chester *597* NW England, UK
Chetumal, Bahía *403* *var.* Chetumal Bay. Bay of the Caribbean Sea
Chetumal Bay *see* Chetumal, Bahía
Cheviot Hills *597* hills of England and Scotland, UK
Ch'ew Bahir *245* *var.* Lake Stefanie. Lake of SW Ethiopia
Cheyenne *603* Wyoming, NW USA
Chhlong *see* Phumĭ Chhlong
Chhuk *see* Phumĭ Chhuk
Chhukha *134* SW Bhutan

Chiai *559* *var.* Chiayi, Kiayi, *Jap.* Kagi. W Taiwan
Chiali *559* *var.* Kiali, *Jap.* Kari. W Taiwan
Chia-mu-ssu *see* Jiamusi
Chiang-hsi *see* Jiangxi
Chiang Mai *567* *var.* Chiengmai. NW Thailand
Chiang-su *see* Jiangsu
Chiat'ura *264* C Georgia
Chiba *332* Honshū, SE Japan
Chibuto *419* S Mozambique
Chicago *603* Illinois, C USA
Chichāwatni *451* E Pakistan
Chichicastenango *280* W Guatemala
Chichigalpa *436* W Nicaragua
Ch'i-ch'i-ha-erh *see* Qiqihar
Chiclayo *463* NW Peru
Chicoutimi *171* SE Canada
Chicualacuala *419* SW Mozambique
Chiemsee *267* lake of SE Germany
Chiesanuova *502* SW San Marino
Chih-fu *see* Yantai
Chihli *see* Hebei
Chihli, Gulf of *see* Bo Hai
Chi-hsi *see* Jixi
Chihuahua *403* NW Mexico
Ch'ikhareshi *264* N Georgia
Chikwawa *385* SW Malawi
Chilanga *635* S Zambia
Chilaw *538* W Sri Lanka
Chile *182-185* officially Republic of Chile. Country of South America divided into 13 admin. units (12 regions and 1 metropolitan area). ❖ Santiago
Chile Chico *183* W Chile
Chilika Lake *298* lake of E India
Chililabombwe *635* C Zambia
Chi-lin *see* Jilin
Chillán *183* C Chile
Chilliwack *170* SW Canada
Chiloé, Isla de *183* *var.* Isla Grande de Chiloé. Island of W Chile
Chilumba *385* *prev.* Deep Bay. N Malawi
Chilung *559* *var.* Keelung, *Jap.* Kirun. N Taiwan
Chilwa, Lake *385* *var.* Lake Shirwa, *Port.* Lago Chirua. Lake of SE Malawi
Chimaltenango *280* W Guatemala
Chimanimani *636* *prev.* Mandidzudzure, *prev.* Melsetter. E Zimbabwe
Chimbay *see* Chimboy
Chimborazo *230* mountain of C Ecuador
Chimbote *463* W Peru
Chimboy *614* *Rus.* Chimbay. NW Uzbekistan
Chimishliya *see* Cimişlia
Chimkent *see* Shymkent
Chimoio *419* C Mozambique
China *186-193* officially People's Republic of China, *Chin.* Zhonghua Renmin Gonghe Guo, *var.* Chunghua Jen-min Kung-ho-kuo, *prev.* Chinese Empire (until January 1912). Country of E Asia divided into 30 admin. units (22 ovinces, 5 autonomous regions, 3 province-level municipalities). ❖ Beijing
Chinan *see* Jinan
Chinandega *436* W Nicaragua
Chincha Alta *463* SW Peru
Chin-chiang *see* Quanzhou
Chin-chou *see* Jinzhou
Chinchow *see* Jinzhou
Chin-do *352* *Jap.* Chin-tō. Island of SW South Korea
Chindwin *159* river of NW Burma
Ch'ing Hai *see* Qinghai Hu
Chinghai *see* Qinghai
Chingola *635* C Zambia
Ch'ingshui *559* *var.* Tsingshui, *Jap.* Kiyomizu. W Taiwan
Ching-Tao *see* Qingdao
Chinguetti *398* C Mauritania
Chinhae *352* *Jap.* Chinkai. S South Korea
Chinhoyi *636* *var.* Sinoia. N Zimbabwe
Chinhsien *see* Jinzhou

Chiniot *451* NE Pakistan
Chinit *165 var.* Chinit. River of
  C Cambodia
Chinju *352 Jap.* Shinshū.
  S South Korea
Chinkai *see* Chinhae
Chink Kaplankyr *585* ridge of
  NW Turkmenistan
Chinko *179* river of E Central
  African Republic
Chintheche *385 var.* Chinteche.
  N Malawi
Chin-tō *see* Chin-do
Chios *275 prev.* Khíos. Island of
  E Greece
Chíos *275 prev.* Khíos, *It.* Scio,
  Turk. Sakis-Adasi. Chios, E Greece
Chipata *635 prev.* Fort Jameson.
  E Zambia
Chipinge *636 prev.* Chipinga.
  E Zimbabwe
Chiquimula *280* SE Guatemala
Chiquimulilla *280* S Guatemala
Chiradzulu *385* S Malawi
Chirang *134* S Bhutan
Chirchik *614 Rus.* Chirchik.
  E Uzbekistan
Chire *see* Shire
Chiredzi *636* SE Zimbabwe
Chirilagua *237* SE El Salvador
Chiriquí, Golfo de *456* gulf of the
  Pacific Ocean to the SW of Panama
Chiriquí Grande *456* W Panama
Chiriquí, Laguna de *456* lagoon of
  W Panama
Chiromo *385* S Malawi
Chirongui *649* S Mayotte
Chirripó Grande, Cerro *206*
  mountain of E Costa Rica
Chirua, Lago *see* Chilwa, Lake
Chirundu *636* N Zimbabwe
Chisenga *385* NW Malawi
Ch'ishan *559 var.* Kishan, *Jap.* Kizan.
  SW Taiwan
Chishtiān Mandi *451* E Pakistan
Chisimaio *see* Kismaayo
Chişinău *408 var.* Kishinev.
  ❖ of Moldova, C Moldova
Chissioua Mtsamboro *649* island of
  NW Mayotte
Chita *485* C Russian Federation
Chitipa *385 prev.* Fort Hill.
  NW Malawi
Chitose *333* Hokkaidō, N Japan
Chitré *456* S Panama
Chittagong *117 Ben.* Chāttagām.
  SE Bangladesh
Chittagong Hills *117* hilly region of
  S Asia
Chitungwiza *636 prev.* Chitangwiza.
  NE Zimbabwe
Chiuta, Lake *385* lake of SE Malawi
Chivhu *636 prev.* Enkeldoorn.
  C Zimbabwe
Chixoy *280 var.* Río Negro,
  Salinas. River of Guatemala
  and Mexico
Chizarira Hills *636* hilly region of
  NW Zimbabwe
Chizumulu Island *385 var.*
  Chisumulu Island. Island of Lake
  Nyasa, E Malawi
Chkalov *see* Orenburg
Chlef *83 prev.* El Asnam, Orléansville,
  var. Ech Cheliff, Ech Chleff.
  NW Algeria
Choa Chu Kang *521* area of
  W Singapore
Chôâm Khsant *165*
  prev. Cheom Ksan. N Cambodia
Choiseul *526 var.* Lauru. Island of the
  W Solomon Islands
Choiseul *526* SW St Lucia
Choiseul Sound *645* area of the South
  Atlantic Ocean, E Falkland Islands
Cholo *see* Thyolo
Choluteca *290* S Honduras
Choluteca *290* river of S Honduras
Choma *635* S Zambia
Chomo Lhari *134* mountain of
  NW Bhutan
Chomutov *216 Ger.* Komotau.
  NW Czech Republic

Ch'ŏnan *352 Jap.* Tenan.
  W South Korea
Chon Buri *567* C Thailand
Chone *230* W Ecuador
Ch'ŏngch'ŏn *351* river of
  W North Korea
Ch'ŏngjin *351* NE North Korea
Chŏngju *351* W North Korea
Chŏngju *352 prev.* Chŏngup,
  Jap. Seiyu. SW South Korea
Ch'ŏngju *352 var.* Chŏngju.
  C South Korea
Chongqing *187 var.* Chungking,
  Ch'ung-ching, Yuzhou. Sichuan,
  SW China
Chongwe *635* E Zambia
Chŏnju *352 Jap.* Zenshū.
  SW South Korea
Chorne More *see* Black Sea
Chornobyl' *590 Rus.* Chernobyl'.
  N Ukraine
Chorzów *471 Ger.* Königshütte.
  S Poland
Chōsen-kaikyō *see* Korea Strait
Chōshū *see* Ch'aochou
Choshui Hsi *559* river of W and
  NE Taiwan
Chota Nagpur Plateau *298* plateau of
  NE India
Choybalsan *412* E Mongolia
Christchurch *433* E South Island,
  New Zealand
Christiana *331* C Jamaica
Christiania *see* Oslo
Christian, Point *651* headland of
  Pitcairn Island, S Pitcairn Islands
Christiansand *see* Kristiansand
Christianshåb *see* Qasigiannguit
Christiansted *653* Saint Croix,
  S Virgin Islands (US)
Christiansund *see* Kristiansund
Christmas Island *644* Australian
  external territory of the Indian
  Ocean. ❖ Flying Fish Cove.
Christmas Island *644* Kiritimati
Chrysochou Bay *see* Khrysokhou Bay
Chu *see* Shū
Chu *359 var.* Ču, *Kir.* Chüy. River of
  Kazakhstan and Kyrgyzstan
Chu *627* river of Laos and Vietnam
Chuādānga *117* W Bangladesh
Ch'uan-chou *see* Quanzhou
Chubek *see* Moskva
Ch'u-chiang *see* Shaoguan
Chucunaque *456* river of E Panama
Chudskoye Ozero *see* Peipus, Lake
Chūgoku-sanchi *332* mountain range
  of Honshū, W Japan
Chuí *see* Chuy
Chukai *386 var.* Cukai. E Peninsular
  Malaysia
Chukchi Sea *485*
  Rus. Chukotskoye More. Sea of Arctic
  Ocean between NE Asia and
  NW N America
Chukotskiy Poluostrov *485*
  Eng. Chukchi Peninsula. Peninsula
  of NE Russian Federation
Chulucanas *463* NW Peru
Chumphon *567* S Thailand
Chunan *559* NW Taiwan
Ch'unch'ŏn *352 Jap.* Shunsen.
  N South Korea
Chungyang Shanmo *559* mountain
  range of C Taiwan
Ch'ung-ching *see* Chongqing
Chungho *559* N Taiwan
Chunghsinghsints'un *559* W Taiwan
Chung-hua Jen-min Kung-ho-kuo
  see China
Ch'ungju *352 Jap.* Chūshū.
  C South Korea
Ch'ungju-ho *352* reservoir of
  C South Korea
Chungking *see* Chongqing
Chungli *559 Jap.* Chūreki. N Taiwan
Ch'ungmu *352* S South Korea
Chunya *564* SW Tanzania
Chuquicamata *183* N Chile
Chur *550 It.* Coira, *Rmsch.* Cuera,
  Fr. Coire. E Switzerland
Church Cay *498* cay of E St Vincent
  & the Grenadines
Churchill *170* C Canada

Church Village *121* SE Barbados
Chūreki *see* Chungli
Chūshū *see* Ch'ungju
Chust *614 var.* Čust. E Uzbekistan
Chuuk Islands *406 var.* Hogoley
  Islands. Island group of C Micronesia
Chuvashskaya *484* autonomous
  republic of W Russian Federation
Chuy *611 var.* Chuí. SE Uruguay
Chykotskoye Nagor'ye *485* mountain
  range of NE Russian Federation
Ciadîr-Lunga *408 var.* Ceadâr-Lunga,
  Rus. Chadyr-Lunga. S Moldova
Cibitoke *162* NW Burundi
Ciceron *497* NW St Lucia
Cicia *248 prev.* Thithia. Island of the
  Lau Group, E Fiji
Cidade Velha *176* Santiago,
  S Cape Verde
Ciego de Ávila *210* C Cuba
Ciénaga *195* N Colombia
Cienfuegos *210* C Cuba
Cieza *535* SE Spain
Çiftlik Dere *see* Aloupos
Cifuentes *210* C Cuba
Cikobia *248 prev.* Thikombia. Island
  to the N of Vanua Levu, N Fiji
Cilacap *304 prev.* Tjilatjap. Java,
  C Indonesia
Cill Airne *see* Killarney
Cill Choinnigh *see* Kilkenny
Cilli *see* Celje
Cill Mhantáin *see* Wicklow
Cimişlia *408 Rus.* Chimishliya.
  S Moldova
Cina Selatan, Laut *see* South
  China Sea
Cincinnati *603* Ohio, NE USA
Ciney *127* SE Belgium
Ciotat *255* SE France
Cirebon *304 prev.* Tjirebon. Java,
  C Indonesia
Cirque de Cilaos *652* mountain range
  of W Réunion
Cirquenizza *see* Crikvenica
Ciskei Bantustan 'self-governing
  homeland' comprising 2 non-
  contiguous territories of E Cape
  Province, South Africa; created in
  1981, abolished in 1994
Citlaltépetl *see* Orizaba,
  Volcán Pico de
Citron *645* NW French Guiana
Citron, River *400* river of NW
  Mauritius
Ciudad Arce *237* W El Salvador
Ciudad Bolívar *623* E Venezuela
Ciudad de Guatemala *see*
  Guatemala City
Ciudad del Este *460 prev.* Puerto
  Presidente Stroessner. SE Paraguay
Ciudad de México *403* Mexico City
Ciudad de Panama *see* Panama City
Ciudad Guayana *623* E Venezuela
Ciudad Juárez *403* NW Mexico
Ciudad Obregón *403* NW Mexico
Ciudad Ojeda *623* NW Venezuela
Ciudad Real *535* C Spain
Ciudad Trujillo *see* Santo Domingo
Ciudad Victoria *403* C Mexico
Civitavecchia *323* C Italy
Clarence *433* river of NE South Island,
  New Zealand
Clarence Island *90* island of South
  Shetland Islands, Antarctica
Clarence Town *112* Long Island,
  Bahamas
Clermont-Ferrand *255* C France
Clervaux *378* N Luxembourg
Cleveland *603* Ohio, NE USA
Clifden *316* W Ireland
Clifton *224* NW Dominica
Clifton *498* Union I, St Vincent
  & the Grenadines
Cloncurry *101* NE Australia
Clonmel *316 Ir.* Cluain Meala.
  S Ireland
Cluj-Napoca *480 prev.* Cluj,
  Hung. Kolozsvár, *Ger.* Klausenburg.
  NW Romania
Clutha *433* river of SW South Island,
  New Zealand
Clyde, Firth of *597* estuary of the
  river Clyde, SW Scotland, UK

Coamo *651* S Puerto Rico
Coast Mountains *170 Fr.* Chaîne
  Côtière. Mountain range of Canada
  and USA
Coast Ranges *602* mountain range of
  W USA
Coatepeque *280* W Guatemala
Coatepeque, Lago de *237* lake of
  W El Salvador
Coatzacoalcos *403 prev.* Puerto
  México. SE Mexico
Cobán *280* C Guatemala
Cobija *136* NW Bolivia
Cochabamba *136* C Bolivia
Cochin *298 var.* Kochi. S India
Cocibolca *see* Nicaragua, Lago de
Cockburn Harbour *653* South Caicos,
  E Turks and Caicos Islands
Cockburn Town *112* Great Exuma I,
  Bahamas
Cockburn Town *653* ❖ Turks and
  Caicos Islands, Grand Turk Island,
  SE Turks and Caicos Islands
Cockpit Country, The *331* physical
  region of NW Jamaica
Coco *290, 436 var.* Wangkí, Segovia.
  River of Honduras and Nicaragua
Cocobeach *260* NW Gabon
Coco, Isla del *206* island of
  SW Costa Rica
Cocoli *see* Corubal
Cocos (Keeling) Islands *644*
  Australian external territory of the
  Indian Ocean. ❖ West Island
Cocos Island *647* island group of
  S Guam
Codrington *92* C Barbuda,
  Antigua & Barbuda
Codrington Lagoon *92* W Barbuda,
  Antigua & Barbuda
Coeroeni *see* Corantijn
Coëtivy *516* island of E Seychelles
Coffs Harbour *101* E Australia
Cogîlnic *408 var.* Cogâlnic,
  Rus. Kogil'nik. River of SE Moldova
Cognac *255* W France
Cogo *238 var.* Kogo, *prev.* Puerto
  Iradier. SW Equatorial Guinea
Cohoha *see* Cyohoha-Sud Lac
Coiba, Isla de *456* island of
  SW Panama
Coihaique *183 var.* Coyhaique. S Chile
Coimbatore *298* S India
Coimbra *474* W Portugal
Coin de Mire *see* Gunners Quoin
Coira *see* Chur
Coire *see* Chur
Coi, Sông *see* Red River
Cojutepeque *237* C El Salvador
Colchester *597* E England, UK
Coleraine *597* Northern Ireland, UK
Colesberg *531* Northern Cape,
  C South Africa
Colihaut *224 var.* Kulihao.
  W Dominica
Collie *100* SW Australia
Collingwood Bay *458* bay of the
  Solomon Sea to the E of Papua
  New Guinea
Colmar *255 Ger.* Kolmar. NE France
Cöln *see* Köln
Cologne *see* Köln
Colomb-Béchar *see* Béchar
Colombia *194-197* officially Republic
  of Colombia. Country of South
  America divided into 32 admin. units
  (departments). ❖ Bogotá
Colombo *538* ❖ of Sri Lanka,
  W Sri Lanka
Colón *210* NW Cuba
Colón *456 prev.* Aspinwall. N Panama
Colón, Archipiélago de *see*
  Galapagos Islands
Colonia *see* Kolonia
Colonia del Sacramento *611*
  SW Uruguay
Colorado *206* river of NE Costa Rica
Colorado *603* river of Texas, SC USA
Colorado *603* river of SW USA
Colorado *603* state of SW USA
Colorados, Archipiélago de los *210*
  island group of NW Cuba
Colorado Springs *603* Colorado,
  SW USA

Douma *see* Dūmā
Douro *see* Duero
Dover *597 Fr.* Douvres.SE England,UK
Dover *603* Delaware, E USA
Dover, Strait of *255 var.* Straits of Dover, *Fr.* Pas de Calais. Strait connecting the English Channel and North Sea between England and France
Dovrefjell *444* mountain of SW Norway
Dowa *385* C Malawi
Dōzen *332* island to the N of Honshū, W Japan
Drâa *414* seasonal river of S Morocco
Drac *see* Durrës
Drač *see* Durrës
Dragon's Mouths, The *574 Sp.* Bocas del Dragón. Strait connecting the Caribbean Sea and Gulf of Paria
Dra, Hamada du *83 var.* Haut Plateau du Dra, Hammada du Drâa. Desert region of W Algeria
Drakensberg *366, 531* mountain range of Lesotho and South Africa
Drake Passage *183* passage connecting Pacific Ocean and Atlantic Ocean between South America and Antarctica
Dráma *275 var.* Dhráma. NE Greece
Drammen *444* S Norway
Drangajökull *296* glacier of NW Iceland
Drava *106, 209, 292, 524 Eng.* Drave, *Hung.* Dráva, *Ger.* Drau, *SCr.* Drava. River of C Europe
Dresden *267* E Germany
Drina *140, 512* river of Bosnia & Herzegovina and Serbia & Montenegro (Yugo.)
Drin Gulf *see* Drinit, Gjiri i
Drinit *81 var.* Drin. River of NW Albania
Drinit, Gjiri i *81 var.* Pellg i Drinit, Drin Gulf. Gulf of the Adriatic Sea, NW Albania
Drinit të Zi *81, 381 var.* Drin i Zi, *Eng.* Black Drin, *SCr.* Crni Drim. River of Albania and FYR Macedonia
Drin i Zi *see* Drinit të Zi
Drinos *81* river of S Albania
Drissa *122* river of Belarus and Russian Federation
Drobeta-Turnu Severin *480 prev.* Turnu Severin. SW Romania
Drochia *408 Rus.* Drokiya. N Moldova
Drogheda *316 Ir.* Droichead Átha. E Ireland
Drontheim *see* Trondheim
Druskininkai *376 Pol.* Druskienniki. S Lithuania
Drysa *122 Rus.* Disna. River of Belarus and Lithuania
Dschang *168* W Cameroon
Duala *see* Douala
Duarte, Pico *226* mountain of C Dominican Republic
Dubai *594 Ar.* Dubayy. NE United Arab Emirates
Dubăsari *408 Rus.* Dubossary. NE Moldova
Dubăsari Reservoir *408* reservoir of NE Moldova
Dubawnt *170* river of C Canada
Dubbo *101* E Australia
Dublane *224* NW Dominica
Dublin *316 Ir.* Baile Átha Cliath. ❖ of Ireland, E Ireland
Dubnica nad Váhom *523 Hung.* Máriatölgyes, *prev.* Dubnicz. NW Slovakia
Dubnicz *see* Dubnica nad Váhom
Dubossary *see* Dubăsari
Dubréka *283* SW Guinea
Dubrovnik *209 It.* Ragusa. SE Croatia
Duc de Gloucester, Îles du *646* island group of C French Polynesia
Ducie Island *651* island of E Pitcairn Islands
Ducos *649* C Martinique
Dudelange *378* S Luxembourg
Dudo, Uadi *see* Dhuudo
Dudwiokahn *see* Dodwekon
Duékoué *328* W Ivory Coast

Duero *474, 534-535 Port.* Douro. River of Portugal and Spain
Duesseldorf *see* Düsseldorf
Duff Islands *526* small island group within Santa Cruz Is, Solomon Is
Dufourspitze *550* mountain of S Switzerland
Dugi Otok *209 It.* Isola Lunga. Island of W Croatia
Duinkerden *see* Dunkerque
Duisburg *266* W Germany
Duitama *195* C Colombia
Duitse Bocht *see* German Bight
Dukhān *479* W Qatar
Dukhan Heights *see* Dukhān, Jabal
Dukhān, Jabal *479 var.* Dukhan Heights. Hilly region of SW Qatar
Dukhān, Jabal ad *see* Dukhān, Jabal ad
Dukhān, Jabal ad *115 var.* Dukhan Heights, Jabal Dukhan. Mountain of C Bahrain
Dukou *see* Panzhihua
Dulce, Golfo *see* Izabal, Lago de
Dulce Nombre de Culmí *290* Honduras
Dulit, Banjaran *see* Dulit Range. Mountain range of W Borneo, Malaysia
Duluth *603* Minnesota, NC USA
Dūmā *555 Fr.* Douma. SW Syria
Dumfries *597* SW Scotland, UK
Dumistān *115* NW Bahrain
Dumont d'Urville *90* French research station of Greater Antarctica, Antarctica
Dumyât *232 Eng.* Damietta. N Egypt
Düna *see* Western Dvina
Dünaburg *see* Daugavpils
Dunai *427* W Nepal
Dunaj *see* Danube
Dunaj *see* Vienna
Dunajská Streda *523 Hung.* Dunaszerdahely. SW Slovakia
Dunapentele *see* Dunaújváros
Dunărea *see* Danube
Dunării, Delta *see* Danube, Mouths of the
Dunaszerdahely *see* Dunajská Streda
Dunaújváros *292 prev.* Sztálinváros, *prev.* Dunapentele. C Hungary
Dunav *see* Danube
Dunavska Ravnina *152 Eng.* Danubian Plain. Lowland region of N Bulgaria
Dundalk *316 Ir.* Dún Dealgan. NE Ireland
Dundalk Bay *316 Ir.* Cuan Dhun Dealgan. Bay of the Irish Sea, to the NE of Ireland
Dundas *see* Pituffik
Dundee *531* Kwazulu Natal, E South Africa
Dundee *597* E Scotland, UK
Dunedin *433* S South Island, New Zealand
Dunfermline *597* E Scotland, UK
Dungarvan *316 Ir.* Dun Garbhain. S Ireland
Dunkerque *255 Eng.* Dunkirk, *Dut.* Duinkerken. N France
Dunkirk *see* Dunkerque
Dunkwa *272* SW Ghana
Dún Laoghaire *316 prev.* Kingstown. E Ireland
Dunqulah *see* Dongola
Dupnitsa *152 prev.* Stanke Dimitrov, *prev.* Marek. W Bulgaria
Duqm *448 var.* Daqm. E Oman
Duque de Caxias *145* SE Brazil
Durán *see* Eloy Alfaro
Durance *255* river of SE France
Durango *403* W Mexico
Durazno *611 var.* San Pedro del Durazno. C Uruguay
Durazzo *see* Durrës
Durazzo, Gulf of *see* Durrësit, Gjiri i
Durban *531* Kwazulu Natal, E South Africa
Durbe *362 Ger.* Durben. W Latvia
Durben *see* Durbe
Durdur *528* seasonal river of NW Somalia
Durham *597* NE England, UK
Durmitor *see* Đeravica

Durrës *81 var.* Durrësi, Dursi, *It.* Durazzo, *SCr.* Drač, *Turk.* Draç. W Albania
Durrësit, Gjiri i *81 var.* Gulf of Durazzo. Gulf of the Adriatic Sea, W Albania
Durūz, Jabal ad *555* mountain range of SW Syria
D'Urville Island *433* island to the NE of South Island, New Zealand
Dusa Mareb *see* Dhuusa Marreeb
Dushanbe *562 var.* Dušanbe, Dyushambe, *prev.* Stalinabad. ❖ of Tajikistan, W Tajikistan
Düsseldorf *266 var.* Duesseldorf. W Germany
Dûstí *562* SW Tajikistan
Dutch New Guinea *see* Irian Jaya
Dutch East Indies *see* Indonesia
Dutch West Indies *see* Netherlands Antilles
Düzce *see* Athna
Dvinsk *see* Daugavpils
Dyanev *see* Deynau
Dyero *see* Dioro
Dyfrdwy *see* Dee
Dyushambe *see* Dushanbe
Džalilabad *see* Cälilabad
Dzaoudzi *649* Petite-Terre, E Mayotte
Džarkurgan *see* Dzharkurgan
Dzaudzhikau *see* Vladikavkaz
Dzavhan *412* river of W Mongolia
Džebel *see* Dzhebel
Dzerzhinskiy *see* Nar'yan-Mar
Dzhalal-Abad *359 var.* Džalal-Abad, *Kir.* Jalal-Abad. SW Kyrgyzstan
Dzhalilabad *see* Cälilabad
Dzhambul *see* Taraz
Džhanak *585* region of W Turkmenistan
Dzharkurgan *see* Jarqŭrghon
Dzhebel *585 var.* Džebel, *Turkm.* Jebel. W Turkmenistan
Dzhelandy *562* SE Tajikistan
Dzhergalan *359 var.* Džergalan, *Kir.* Jyrgalan. NE Kyrgyzstan
Dzhermuk *see* Jermuk
Dzhezkazgan *see* Zhezkazgan
Dzhetygara *see* Zhitikara
Dzhirgatal' *see* Jirgatol
Dzhizak *see* Jizzakh
Dzhugdzhur, Khrebet *485* mountain range of E Russian Federation
Dzhusaly *340* SW Kazakhstan
Dzongsa *134* SW Bhutan
Dzuunmod *412* C Mongolia
Dzüünharaa *412* N Mongolia
Dzvina *see* Western Dvina
Dzyarzhynskaya, Hora *122* mountain of C Belarus

E

Eagle Islands *643* island group of W British Indian Ocean Territory
Eagle Passage *645* passage connecting Falkland Sound and Atlantic Ocean, S Falkland Islands
East Caicos *653* island of N Turks and Caicos Islands
East China Sea *559 Chin.* Nan Hai. Sea of Pacific Ocean, off E Asia
East End *644* Grand Cayman, W Cayman Islands
East End Village *642* E Anguilla
Easter Island *183* Pacific island of Chile
Eastern Cape *531* province of SE South Africa
Eastern Desert *see* Sharqīya, Sahara el
Eastern Ghats *298* mountains of SE India
Eastern Sayans *see* Vostochnyy Sayan
Eastern Scheldt *see* Oosterschelde
Eastern Sierra Madre *see* Sierra Madre Oriental
Eastern Transvaal *see* Mpumalanga
East Falkland *645* island of E Falkland Islands
East Fayu Island *406* island of C Micronesia
East Frisian Islands *see* Ostfriesische Inseln

East London *531 Afr.* Oos-Londen. Eastern Cape, S South Africa
Eastern Cape, S South Africa
East Pakistan *see* Bangladesh
East Malaysia *386-387* eastern part of Malaysia situated on N Borneo
East Siberian Sea *see* Vostochno-Sibirskoye More
East Timor *228-229* Country of SE Asia, ❖ Dili
Eauripik *406* atoll of C Micronesia
Ebebiyin *238* NE Río Muni, Equatorial Guinea
Ebeltoft *218* Jylland, C Denmark
Ebetsu *333* Hokkaidō, N Japan
Ebinayon *see* Evinayong
Eblana *see* Dublin
Ebolowa *168* S Cameroon
Ebon *396* island of S Marshall Islands
Ébrié, Lagune *328* lake of SW Ivory Coast
Ebro *535* river of NE Spain
Ech Cheliff *see* Chlef
Echmiadzin *see* Ejmiadzin
Echternach *378* E Luxembourg
Écija *535* SW Spain
Ečmiadzin *see* Ejmiadzin
Ecuador *230-231* officially Republic of Ecuador. Country of NW South America divided into 20 admin. units (provinces). ❖ Quito
Ed *240* SE Eritrea
Ed Damazin *540 var.* Ad Damazin. E Sudan
Ed Damer *540 var.* Ad Dāmir, Ad Damar. NE Sudan
Ed Dueim *540 var.* Ad Duwaym, Ad Duwêm. C Sudan
Ede *429* C Netherlands
Ede *440* N Nigeria
Edéa *168* SW Cameroon
Eden *597* river of NW England, UK
Edfu *see* Idfu
Edgeøya *653* island of S Svalbard
Edina *368* SW Liberia
Edinburgh *652* N Tristan da Cunha
Edinburgh *597* E Scotland, UK
Edineţ *408 var.* Edineţi, *Rus.* Yedintsy. NW Moldova
Edirne *580* NW Turkey
Edmonton *170* SW Canada
Edward, Lake *588, 203 var.* Lake Rutanzige, Edward Nyanza, Albert Edward Nyanza, Lac Idi Amin. Lake of Uganda and Dem. Rep. Congo
Eems *see* Ems
Eesti Vabariik *see* Estonia
Efate *619 Fr.* Vaté, *prev.* Sandwich Islands. Island group of C Vanuatu
Egadi, Isole *323* island group to the W of Sicilia, S Italy
Ege Denizi *see* Aegean Sea
Eger *292 Ger.* Erlau. NE Hungary
Eger *see* Cheb
Eger *see* Ohře
Egersund *444* SW Norway
Egilsstadhir *296* E Iceland
Egmont, Cape *433* cape of SW North Island, New Zealand
Egmont Islands *643* island group of W British Indian Ocean Territory
Egmont, Mount *see* Taranaki, Mount
Egypt *232-235* officially Arab Republic of Egypt, *prev.* United Arab Republic. Country of NE Africa divided into 26 admin. units (governorates). ❖ Cairo
Eidsvoll *444* S Norway
Eifel *266* plateau of W Germany
Eiffel Flats *636* C Zimbabwe
Eight Degree Channel *390* channel of the Indian Ocean between N Maldives and Lakshadweep, SW India
Eil *see* Eyl
Eilat *see* Elat
Eil Malk *455* island of S Palau
Eindhoven *429* S Netherlands
Einsiedeln *550* NE Switzerland
Eipel *see* Ipeľ, Ipoly
Eire *see* Ireland, Republic of
Eisen *see* Yŏngch'ŏn
Eisenstadt *106* E Austria
Eishū *see* Yŏngju
Eitape *see* Aitape

Eivissa *535 Cast.* Ibiza, *var.* Iviza. Island of the Islas Baleares, E Spain
Ejmiadzin *98 Rus.* Echmiadzin, Ečmiadzin, Etchmiadzin. W Armenia
Ekaterinoslav *see* Dnipropetrovs'k
Ekerem *see* Okarem
Ekeren *127* N Belgium
Ekibastuz *340* NE Kazakhstan
El Alto *136* W Bolivia
El Araïche *see* Larache
El 'Arîsh *232 var.* Al Arîsh. NE Egypt
El Asnam *see* Chlef
Elat *319 var.* Elath, Eilat. S Israel
Elat, Gulf of *see* Aqaba, Gulf of,
Elato *406* atoll of C Micronesia
Elâziğ *581 var.* Elâzir, Elâziğ. E Turkey
Elba, Isola d' *322* island of C Italy
Elbasan *81 var.* Elbasani. C Albania
Elbe *216, 267 Cz.* Labe. River of Czech Republic and Germany
El Beqaa *364 var.* Al Biqâ', Bekaa Valley. Valley of E Lebanon
Elbing *see* Elbląg
Elbistan *581* S Turkey
Elbląg *471 Ger.* Elbing. N Poland
El Boulaida *see* Blida
Elbow Bay *643* bay of the North Atlantic Ocean, C Bermuda
El Bur *see* Ceelbuur
Elburgon *344* W Kenya
Elburz Mountains *see* Alborz, Reshteh-ye Kuhhâ-ye
El Carmen de Bolívar *195* NW Colombia
El Cayo *see* San Ignacio
El Cerro del Aripo *see* Aripo, Mount
Elche *535 Cat.* Elx. SE Spain
El Congo *237* W El Salvador
El Copé *456* C Panama
Elda *535* SE Spain
El Dere *see* Ceeldheere
El Djazaïr *see* Algiers
El Djelfa *see* Djelfa
Eldoret *344* W Kenya
Eleja *362* C Latvia
Elemi Triangle *245, 344* disputed region of Kenya and Sudan
Éléphant, Chaîne de l' *see* Dâmrei, Chuŏr Phnum
Elephant Island *90* island of South Shetland Islands, Antarctica
Eleuthera Island *112* island of C Bahamas
El Faiyûm *232 var.* Al Fayyûm. N Egypt
El Fasher *540 var.* Al Fâshir. W Sudan
El Ferrol del Caudillo *see* Ferrol
El Gal *see* Ceel Gaal
El Gedaref *see* Gedaref
Elgin *597* NE Scotland, UK
El Gîza *232 var.* Gîza, Al Jizah. N Egypt
El Goléa *83* C Algeria
Elgon, Mount *588* mountain of E Uganda
El Hasaheisa *540 var.* Al Ḥuṣayḥiṣah, Al Hasahisa, Hasaheisa. C Sudan
El Haseke *see* Al Ḥasakah
Elías Piña *see* Comendador
Elimina *272* S Ghana
Élisabethville *see* Lubumbashi
El Iskandarîyah *see* Alexandria
Elizabeth *101* S Australia
El Jadida *414 prev.* Mazagan. W Morocco
El Jafr *see* Qā' al Jafr
El Jem *577 var.* Al Djem, Al Jamm. NE Tunisia
El Kef *see* Le Kef
El Kelâa des Srarhna *414* C Morocco
El Khârga *232 var.* Al Khârijah. C Egypt
El Khiyam *364 var.* Khiam. S Lebanon
Elkhovo *152 var.* Elhovo. SE Bulgaria
Ellás *see* Greece
Ellef Ringnes Island *170* island of Sverdrup Islands, N Canada
Ellerton *121* C Barbados
Ellesmere Island *170* island of Queen Elizabeth Islands, N Canada
Ellice Islands *see* Tuvalu

Ellsworth Land *90* physical region of Lesser Antarctica, Antarctica
El Mahalla el Kubra *232 var.* Maḥallah al Kubrá. N Egypt
El Manṣûra *232 var.* Al Manṣûrah. N Egypt
El Mediyya *see* Médéa
El Mina *364 var.* Al Mînâ'. N Lebanon
El Minya *232 var.* Al Minyâ. C Egypt
El Mohammaidia *see* Mohammadia
Elmshorn *266* N Germany
El Nebk *see* An Nabk
El Obeid *540 var.* Al Ubayyiḍ, Al Obayyid. C Sudan
El Oued *83 var.* El Ouâdi, El Wad. NE Algeria
Eloy Alfaro *230 var.* Durán. SW Ecuador
El Palmar *136* E Bolivia
El Paraíso *290* S Honduras
el Pas de las Casa *86* E Andorra
El Paso *602* Texas, SC USA
Elpitiya *538* S Sri Lanka
El Porvenir *456* NE Panama
El Progreso *280 var.* Guastatoya. C Guatemala
El Progreso *290* NW Honduras
El Qâhira *see* Cairo
Elqui *183* river of N Chile
El Quneitra *see* Al Qunayṭirah
El Quseir *see* Al Quṣayr
El Quweira *see* Al Quwayrah
El Rama *436* SE Nicaragua
El Salvador *211* SE Cuba
El Salvador *236–237* officially Republic of El Salvador. Country of Central America divided into 14 admin. units (15 departments, 1 metropolitan area). ❖ San Salvador
El Seibo *226 var.* Santa Cruz de El Seibo. E Dominican Republic
el Serrat *86* N Andorra
Elsinore *see* Helsingør
El Suweida *see* As Suwaydā'
el Tarter *86* NE Andorra
El Tigre *623* NE Venezuela
El Tigre de San Lorenzo *456* SW Panama
El Triunfo *237* SE El Salvador
Elva *242 Ger.* Elwa. SE Estonia
Elvas *474* C Portugal
Elverum *444* S Norway
El Vigía *623* NW Venezuela
El Wad *see* El Oued
El Wâhât el Khârga *232 var.* Al Wâḥât al Khârijah.S Egypt
Elx *see* Elche
Émaé *619* island of C Vanuatu
Emajõgi *242 var.* Emajygi, Ema Jõgi. River of SE Estonia
Emao *619* island of C Vanuatu
Emba *340* W Kazakhstan
Emba *340* river of W Kazakhstan
Embangweni *385 var.* Ephangweni. NW Malawi
Emborcação, Represa da *145* reservoir of S Brazil
Embu *344* C Kenya
Emden *266* NW Germany
Emerald *101* E Australia
Emi Koussi *180* mountain of N Chad
Emlembe *544* mountain of South Africa and Swaziland
Emmen *429* NE Netherlands
Empada *284* SW Guinea-Bissau
Empty Quarter *see* Rub'al Khali
Ems *266, 429 Dut.* Eems. River of Germany and Netherlands
Enareträsk *see* Inarijärvi
Encamp *86* C Andorra
Encarnación *460* S Paraguay
Enchi *272* SW Ghana
Endeavour *121* NW Barbados
Endelave *218* island of C Denmark
Enderbury Island *348* island of the Phoenix Is, C Kiribati
Enderby Land *90* physical region of Greater Antarctica, Antarctica
Enewetak *396 var.* Eniwetok. Atoll of W Marshall Islands
Enghershatu *240* mountain of N Eritrea
England *597* national region of UK

English Channel *255, 597, 647 var.* The Channel, *Fr.* La Manche. Channel connecting the North Sea and Atlantic Ocean between England and France
English Harbour Town *92* S Antigua, Antigua & Barbuda
Engordany *86* C Andorra
Engures Ezers *362* lake of NW Latvia
Enguri *264 Rus.* Inguri. River of NW Georgia
Eniwetok *see* Enewetak
Enkeldoorn *see* Chivhu
Enkhuizen *429* NW Netherlands
Enna *323* Sicilia, S Italy
En Nâqaoûra *364 var.* An Nâqûrah. SW Lebanon
Ennedi *180* plateau of E Chad
Ennis *316 Ir.* Inis. W Ireland
Enniscorthy *316* E Ireland
Enniskillen *597* Northern Ireland, UK
Enns *106* river of C Austria
Enriquillo, Lago *226* lake of SW Dominican Republic
Enschede *429* E Netherlands
Ensenada *403* NW Mexico
Entebbe *588* S Uganda
Enugu *440* S Nigeria
Enu, Ilha de *284* island of SW Guinea-Bissau
Eolie, Isole *323 var.* Isole Lipari, *Eng.* Lipari Islands, *var.* Aeolian Islands. Island group of S Italy
Epéna *200* NE Congo
Eperies *see* Prešov
Eperjes *see* Prešov
Ephangweni *see* Embangweni
Epi *619* island of C Vanuatu
Épinal *255* NE France
Epira *286* E Guyana
Episkopi *215* SW Cyprus
Episkopi Bay *215* bay of the Mediterranean Sea, on the SE coast of Cyprus
Epping Forest *278* S Grenada island, Grenada
Equatorial Channel *390* channel of the Indian Ocean between the atolls of Fuammulah and South Huvadhu Atoll, S Maldives
Equatorial Guinea *238–239* officially Republic of Equatorial Guinea. Country of W Africa divided into 7 admin. units (provinces). ❖ Malabo
Erbil *see* Arbil
Erdenet *412* N Mongolia
Erdi Ma *180* desert region of NE Chad
Ereğli *581* S Turkey
Erenköy *see* Kokkina
Erenköy *215* W Cyprus
Erevan *see* Yerevan
Erfurt *267* C Germany
Erg Chech *83* desert region of SW Algeria
Erg du Djourab *180* desert region of C Chad
Erguig *180* river of SW Chad
Erhlin *559 Jap.* Nirin. W Taiwan
Erie *603* Pennsylvania, NE USA
Erie, Lake *171, 603 Fr.* Lac Érié. Lake of Canada and USA
Erigabo *see* Ceerigaabo
Erik Erikenstretet *653* strait of the Arctic Ocean, E Svalbard
Erikub *396* island of C Marshall Islands
Eritrea *240–241* officially State of Eritrea. Country of E Africa divided into 10 admin. units (regions). ❖ Asmara
Erivan *see* Yerevan
Erlangen *267* S Germany
Erlau *see* Eger
Ermak *see* Yermak
Ermelo *531* Eastern Transvaal, NE South Africa
Er Rachidia *414* E Morocco
Er Rahad *540 var.* Ar Rahad. C Sudan
Er Rif *see* Rif
Erromango *619* island of S Vanuatu
Ersekë *81 var.* Erseka, Kolonjë. SE Albania

Érsekújvár *see* Nové Zámky
Ertis *see* Irtysh
Erzenit *81 var.* Erzen. River of C Albania
Erzgebirge *216, 267 Cz.* Krušné Hory, *Eng.* Ore Mountains. Mountain range of Czech Republic and Germany
Erzincan *581* E Turkey
Erzurum *581 prev.* Erzerum. NE Turkey
Esa'ala *458* Normanby I, Papua New Guinea
Esbjerg *218* Jylland, W Denmark
Esbo *see* Espoo
Escaut *see* Scheldt
Eschen *374* NW Liechtenstein
Esch-sur-Alzette *378* S Luxembourg
Escuintla *280* S Guatemala
Eséka *168* SW Cameroon
Esenguly *see* Gasan-Kuli
Esentepe *215 var.* Ayios Amvrosios. N Cyprus
Eṣfahân *309 var.* Ispahan. W Iran
Esh Sham *see* Damascus
Esh Sharā *see* Ash Sharâh
Esigodini *636 prev.* Essexvale. SW Zimbabwe
Esil *see* Ishim
Eski Dzhumaya *see* Turgovishte
Eskifjördhur *296* E Iceland
Eskije *see* Xánthi
Eskilstuna *547* S Sweden
Eskişehir *580* W Turkey
Esla, Embalse de *535 var.* Embalse de Elsa. Reservoir of NW Spain
Esmeralda *210* E Cuba
Esmeraldas *230* N Ecuador
Esna *see* Isna
Esneux *127* E Belgium
España *see* Spain
Española, Isla *230* island of SE Galapagos Is, Ecuador
Esparta *290* NW Honduras
Esperance *100* SW Australia
Esperanza *90* Argentinian research station of Antarctic Peninsula, Antarctica
Esperanza *226* N Dominican Republic
Espinal *195* C Colombia
Espinhaço, Serra do *145* mountain range of SE Brazil
Espíritu Santo *619 var.* Santo. Island of C Vanuatu
Espoo *251 Swe.* Esbo. SW Finland
Espungabera *419* SW Mozambique
Esquel *95* SW Argentina
Esquipulas *280* SE Guatemala
Éssalou *222* river of N Djibouti
Essaouira *414 prev.* Mogador. W Morocco
Essau *263* W Gambia
Esseg *see* Osijek
Essen *266* W Germany
Essequibo *286* river of Guyana
Essequibo Islands *286* islands of N Guyana
Essexvale *see* Esigodini
Estados, Isla de los *95 Eng.* Staten Island. Island to the E of Tierra del Fuego, Argentina
Estanyó, Pic de l' *86* mountain of N Andorra
Estcourt *531* Kwazulu Natal, E South Africa
Estelí *436* W Nicaragua
Estevan *170* S Canada
Est, Ile de l' *400* E Mauritius
Estonia *242–243* officially Republic of Estonia, *Est.* Eesti Vabariik, *prev.* Estonian SSR, *Rus.* Estonskaya SSR. Country of E Europe divided into 15 admin. units (counties). ❖ Tallinn
Estoril *474* W Portugal
Estrela *474* mountain of C Portugal
Estrela, Serra de *474* mountain range of C Portugal
Eszék *see* Osijek
Esztergom *292 Ger.* Gran. N Hungary
Etchmiadzin *see* Ejmiadzin
Etembue *238* SW Río Muni, Equatorial Guinea

Fon *283* mountainous region of
E Guinea
Fond St. Jean *224* S Dominica
Fongafale *587 var.* Funafuti.
❖ of Tuvalu, Funafuti, Tuvalu
Fonseca, Gulf of *237, 290, 436* gulf
of the Pacific Ocean, on the W coast
of Central America
Fontvieille *410* SW Monaco
Fonuafoʻou *572* island of Nomuka
Group, W Tonga
Fonualei *572* island of N Tonga
Foochow *see* Fuzhou
Forécariah *283* SW Guinea
Forestière *497* N St Lucia
Forlì *323* N Italy
Formentera *535* island of the Islas
Baleares, E Spain
Formosa *95* NE Argentina
Formosa Bay *see* Ungama Bay
Formosa, Ilha *284* island of
Arquipélago dos Bijagós,
SW Guinea-Bissau
Formosa, Serra *145* mountains of
C Brazil
Formosa Strait *see* Taiwan Strait
Forssa *251* SW Finland
Fortaleza *136* N Bolivia
Fortaleza *145 prev.* Ceará. NE Brazil
Fort-Archambault *see* Sarh
Fort-Bayard *see* Zhanjiang
Fort-Cappolani *see* Tidjikja
Fort Charlet *see* Djanet
Fort Collins *603* Colorado, SW USA
Fort-Crampel *see* Kaga Bandoro
Fort-Dauphin *see* Tôlañaro
Fort-de-France *649* ❖ of Martinique,
W Martinique
Fort-Foureau *see* Kousséri
Fort George *see* Grande Rivière, La
Fort-Gouraud *see* Fdérik
Forth *597* river of C Scotland, UK
Fort Hall *see* Murangʼa
Forth, Firth of *597* estuary of the river
Forth, E Scotland, UK
Fort Hill *see* Chitipa
Fortín General Diaz *460*
W Paraguay
Fortín General Eugenio Garay
see General Eugenio A. Garay
Fort Jameson *see* Chipata
Fort Johnson *see* Mangochi
Fort-Lamy *see* NʼDjamena
Fort Lauderdale *603* Florida, SE USA
Fort-Liberté *288* N Haiti
Fort Manning *see* Mchinji
Fort McMurray *170* W Canada
Fort-Repoux *see* Akjoujt
Fort Rosebery *see* Mansa
Fort Saint John *170* W Canada
Fort Saskatchewan *170* SW Canada
Fort-Shevchenko *340* W Kazakhstan
Fort-Sibut *see* Sibut
Fort Smith *170* W Canada
Fort Smith *603* Arkansas, SC USA
Fort-Trinquet *see* Bîr Mogreïn
Fort Victoria *see* Masvingo
Fort Wayne *603* Indiana, C USA
Fort William *597* W Scotland, UK
Fort Worth *603* Texas, SC USA
Fougamou *260* C Gabon
Foulenzem *260* NW Gabon
Foumban *168* NW Cameroon
Foumbouni *198* S Grande Comore,
Comoros
Fourchue, Île *646* island of
N Guadeloupe
Fournaise, Piton de la *652* mountain
of SE Réunion
Four Roads *574* NW Trinidad,
Trinidad & Tobago
Fouta Djallon *283 var.* Futa Jallon.
Mountainous region of W Guinea
Foveaux Strait *433* strait between
South Island and Stewart Island,
New Zealand
Fox Bay East *645* West Falkland,
W Falkland Islands
Fox Bay West *645* West Falkland,
W Falkland Islands
Foxe Peninsula *171* peninsula of
Baffin Island, NE Canada
Foyle *597* river of Ireland and UK

Foyle, Lough *316, 597*
Ir. Loch Feabhai. Inlet of the Atlantic
Ocean, Ireland and UK
Fraile Muerto *611* E Uruguay
Frakštát *see* Hlohovec
France *254-259* officially The
French Republic. Country of Europe
divided into 22 admin. units
(regions, comprising 96 departments).
❖ Paris
Franceville *see* Massoukou
Francistown *142* NE Botswana
Franconian Jura
see Fränkische Alb
Frankfort *603* Kentucky, C USA
Frankfurt am Main *266*
Eng. Frankfort on the Main,
Frankfurt. SW Germany
Frankfurt an der Oder *267*
E Germany
Fränkische Alb *267 Eng.* Franconian
Jura. Mountain range of S Germany
Frantsa-Iosifa, Zemlya *454-485*
Eng. Franz Josef Land. Island group
of N Russian Federation
Fraser *170* river of SW Canada
Fraser Island *101 var.* Great Sandy
Island. Island of E Australia
Frauenburg *see* Saldus
Frauenfeld *550* NE Switzerland
Fray Bentos *611* W Uruguay
Fredericia *218* Jylland,
SW Denmark
Fredericton *171* SE Canada
Frederiksdal *see* Narsaq Kujalleq
Frederikshåb *see* Paamiut
Frederikshavn *218* Jylland,
N Denmark
Frederiksted *653* Saint Croix,
S Virgin Islands
Fredrikshald *see* Halden
Fredrikstad *444* S Norway
Freemans *92* C Antigua,
Antigua & Barbuda
Freeport *112 var.* Freeport-Lucaya.
Bahamas
Freetown *92* SE Antigua,
Antigua & Barbuda
Free State *531 prev.* Orange Free State.
Province of C South Africa
Freetown *518* ❖ of Sierra Leone,
W Sierra Leone
Frégate *516* island of the Inner Islands,
NE Seychelles
Freiburg *see* Fribourg
Freiburg im Breisgau *266*
var. Freiburg. SW Germany
Freistadtl *see* Hlohovec
Fremantle *100* SW Australia
French Guiana *645 var* Guyane.
French overseas department of
N South America. ❖ Cayenne.
French Polynesia *646* French
overseas possession of the Pacific
Ocean.❖ Papeete.
French Somaliland *see* Djibouti
French Sudan *see* Mali
French Territory of the Afars and Issas
see Djibouti
French Togoland *see* Togo
Fria *283* W Guinea
Fribourg *550 Ger.* Freiburg.
W Switzerland
Friedek-Mistek *see* Frýdek-Místek
Friedrichshafen *266* S Germany
Friendly Islands *see* Tonga
Frigate Island *278* island to the
S of Carriacou, Grenada
Frigate Island *498* island of SW
St Vincent & the Grenadines
Frisches Haff *see* Vistula Lagoon
Frobisher Bay *see* Iqaluit
Frome, Lake *101* salt lake of
S Australia
Front Range *see* Maluti
Frunze *see* Bishkek
Frýdek - Místek *216 Ger.* Friedek -
Mistek. SE Czech Republic
Fuammulah *390 var.* Gnaviyani Atoll.
Atoll of S Maldives
Fu-chien *see* Fujian
Fu-chou *see* Fuzhou
Fucht *414* W Western Sahara
Fuenlabrada *535* C Spain

Fuerte Olimpo *460* NE Paraguay
Fuerteventura *535* island of Islas
Canarias, SW Spain
Fuglafjørdhur *644 var.* Fuglefjord.
Eysturoy, N Faeroe Islands
Fugloy *644 var.* Fuglo. Island of
NE Faeroe Islands
Fu-hsin *see* Fuxin
Fujairah *594 Ar.* Al Fujayrah.
NE United Arab Emirates
Fuji *332* Honshū, SE Japan
Fujian *187 var.* Fukien, Fu-chien.
Province of SE China
Fuji-san *332* mountain of Honshū,
SE Japan
Fujisawa *332* Honshū, SE Japan
Fukuchiyama *332* Honshū, C Japan
Fukue *332* island of Gotō-rettō,
SW Japan
Fukue *332* Gotō-rettō, SW Japan
Fukui *332* Honshū, C Japan
Fukuoka *332* Kyūshū, SW Japan
Fukushima *332* Honshū, N Japan
Fukuyama *332* Honshū, W Japan
Fulacunda *284* C Guinea-Bissau
Fulaga *248* island of the Lau Group,
E Fiji
Fulda *266* C Germany
Fullarton *574* SW Trinidad,
Trinidad & Tobago
Funabashi *332* Honshū, SE Japan
Funafuti *see* Fongafale
Funafuti *587* coral atoll of C Tuvalu
Funaota *587* islet of Nukufetau, Tuvalu
Funchal *474* Madeira, Madeira
Islands, Portugal
Fünen *see* Fyn
Fünfkirchen *see* Pécs
Funhalouro *419* SE Mozambique
Furna *176* Brava, S Cape Verde
Furnas, Represa de *145* reservoir
of SE Brazil
Furneaux Group *101* island group
of SE Australia
Fusan *see* Pusan
Fushun *187* Liaoning, NE China
Futa Jallon *see* Fouta Djallon
Futuna *619* island of S Vanuatu
Futuna, Île *653* island of
N Wallis & Futuna
Fuwairet *see* Al Fuwayriṭ
Fuxin *187 var.* Fu-hsin, Fusin.
Liaoning, NE China
Fuzhou *187 var.* Foochow, Fu-chou.
Fujian, SE China
Füzuli *110 Rus.* Fizuli SW Azerbaijan
Fyn *218 Ger.* Fünen. Island of
C Denmark

G

Gaafu Alifu Atoll *see* North Huvadhu
Atoll
Gaafu Dhaalu Atoll *see* South
Huvadhu Atoll
Gaalkacyo *528 var.* Galkaʼyo,
It. Galcaio. C Somalia
Gabela *88* W Angola
Gabès *577 var.* Qābis. C Tunisia
Gabès, Gulf of *577* gulf of the
Mediterranean Sea to the
E of Tunisia
Gablonz an der Neisse *see* Jablonec
nad Nisou
Gabon *260-261* officially The
Gabonese Republic. Country of West
Africa divided into 9 admin. units
(provinces). ❖ Libreville
Gaborone *142 prev.* Gaberones.
❖ of Botswana, SE Botswana
Gabriel, Ilot *400 Eng.* Gabriel Island.
Island of N Mauritius
Gabrovo *152* C Bulgaria
Gabú *284 prev.* Nova Lamego.
E Guinea-Bissau
Gaeta, Golfo di *323 var.* Gulf of
Gaeta. Gulf of the Tyrrhenian Sea, on
the W coast of Italy
Gaferut *406* island of C Micronesia
Gafsa *577 var.* Qafṣah. W Tunisia
Gagnoa *328* C Ivory Coast
Gagra *264* NW Georgia

Gaherré *222* NE Djibouti
Gahnpa *see* Ganta
Gaibānda *117* NW Bangladesh
Gaillimh *see* Galway
Gailtaler Alpen *106* mountain range
of S Austria
Gairdner, Lake *101* salt lake of
S Australia
Gaizin *see* Gaizina Kalns
Gaizina Kalns *362 var.* Gaiziņ.
Mountain of E Latvia
Gâlâfi *222* W Djibouti
Galana *344* river of SE Kenya
Galapagos Islands *230 var.* Tortoise
Islands, *Sp.* Archipiélago de Colón.
Island group of W Ecuador in the
Pacific Ocean
Galaţi *480 Ger.* Galatz. E Romania
Galaymor *see* Kalai-Mor
Galcaio *see* Gaalkacyo
Gales Point *130* E Belize
Galets *652* river of NW Réunion
Galgóc *see* Hlohovec
Galibi *542* NE Suriname
Galicia *535* NW Spain
Galilee, Sea of *see* Tiberias, Lake
Galle *538 prev.* Point de Galle.
SW Sri Lanka
Gällivare *547* N Sweden
Gâlma *see* Guelma
Galomaro *284* C Guinea-Bissau
Galway *316 Ir.* Gaillimh. W Ireland
Galway Bay *316 Ir.* Cuan na
Gaillimhe. Bay of the Atlantic Ocean,
to the W of Ireland
Gamamudo *284* NE Guinea-Bissau
Gamba *260* SW Gabon
Gambia *262-263* officially Republic of
the Gambia. Country of W Africa
divided into 6 admin. units
(divisions). ❖ Banjul
Gambia *263, 283, 510 Fr.* Gambie.
River of W Africa
Gambier, Îles *646* island group of
E French Polynesia
Gambissara *263* E Gambia
Gamboma *200* E Congo
Gamboula *179* SW Central African
Republic
Gamgadhi *427 var.* Gum. W Nepal
Gamlakarleby *see* Kokkola
Gammouda *see* Sidi Bouzid
Gampaha *538* W Sri Lanka
Gamprin *374* NW Liechtenstein
Gâm *627* river of N Vietnam
Gan *390* C Maldives
Ganaane *see* Juba
Gäncä *110 Rus.* Gyandzha,
prev. Kirovabad, Yelisavetpol.
W Azerbaijan
Gand *see* Gent
Gandajika *203* S Dem. Rep. Congo
Gandía *535* E Spain
Ganges *117, 298-299 Ben.* Padma,
Hind. Ganga. River of S Asia
Ganges, Mouths of the *117, 299* large
delta area of Bangladesh and India
Gansu *187 var.* Kansu. Province of
NW China
Ganta *368 var.* Gahnpa.
NE Liberia
Gao *392* E Mali
Gaoua *157* SW Burkina
Gaoual *283* N Guinea
Gap *255* E France
Garabogazköl Bogazy
see Kara-Bogaz-Gol, Proliv
Garagum *see* Karakumy
Garagum Kanaly *see* Karakumskiy
Kanal
Garam *see* Hron
Garamszentkereszt
see Žiar nad Hronom
Garango *157* C Burkina
Garbahaarrey *528 It.* Garba Harre.
SW Somalia
Garda, Lago di *322 var.* Benaco,
Eng. Lake Garda. Lake of N Italy
Gardēz *77 var.* Gardeyz.
E Afghanistan
Gardner Island *see* Nikumaroro
Gardo *see* Qardho
Garissa *344* E Kenya
Garm *see* Gharm

Goettingen see Göttingen
Gogounou 132 N Benin
Goiânia 145 prev. Goyania. S Brazil
Gökdepe see Gëkdepe
Gokwe 636 NW Zimbabwe
Gol 444 S Norway
Golan Heights 319 disputed territory of SW Syria
Gold Coast 101 coastal region of E Australia
Gold Coast 272 coastal region of W Africa
Golden Bay 433 bay on the coast of N South Island, New Zealand
Golden Valley 636 N Zimbabwe
Goldingen see Kuldiga
Golfito 206 SE Costa Rica
Gollel see Lavumisa
Golmud 186 var. Golmo, Chin. Ko-erh-mu. Qinghai, W China
Golungo Alto 88 NW Angola
Goma 203 NE Dem. Rep. Congo
Gombe 440 E Nigeria
Gomel' see Homyel'
Gomera 535 island of Islas Canarias, SW Spain
Gómez Palacio 403 NW Mexico
Gonaïves 288 N Haiti
Gonâve, Canal de la 288 var. Canal de Sud. Channel of the Caribbean Sea between Île de la Gonâve and Haiti
Gonâve, Golfe de la 288 gulf of the Caribbean Sea to the W of Haiti
Gonâve, Île de la 288 island of W Haiti
Gonder 245 var. Gondar. NW Ethiopia
Gondomar 474 NW Portugal
Goodenough Island 458 var. Morata. Island of SE Papua New Guinea
Good Hope 224 E Dominica
Good Hope, Cape of 531 Afr. Kaap die Gooie Hoop. Coastal feature of SW South Africa
Goodlands 400 NE Mauritius
Goose Bay see Happy Valley-Goose Bay
Goose Green 645 East Falkland, C Falkland Islands
Gopālpur 117 N Bangladesh
Gorakhpur 298 NE India
Gorce Island 510 island of W Senegal
Gore 433 S South Island, New Zealand
Goré 180 S Chad
Gorē 245 W Ethiopia
Gorey 648 E Jersey
Gorgān 309 var. Gurgan. N Iran
Gori 264 C Georgia
Goris 98 SE Armenia
Gorki see Horki
Gor'kiy see Nizhniy Novgorod
Görlitz 267 E Germany
Gorlovka see Horlivka
Gorna Dzhumaya see Blagoevgrad
Gorna Oryakhovitsa 152 var. Gorna Orjahovica. N Bulgaria
Gornji Milanovac 512 C Serbia, Serbia & Montenegro (Yugo.)
Gorno Altaysk 485 S Russian Federation
Goroka 458 C Papua New Guinea
Gorontalo, Teluk see Tomini, Teluk
Gorzów Wielkopolski 471 Ger. Landsberg. W Poland
Gosford 101 E Australia
Goshogawara 332 Honshū, N Japan
Gospić 209 C Croatia
Gostivar 381 W FYR Macedonia
Göteborg 547 Eng. Gothenburg. SW Sweden
Gotland 547 island of SE Sweden
Gotō-rettō 332 island group to the W of Kyūshū, SW Japan
Gotse-Delchev 152 var. Goce Delčev. SW Bulgaria
Gottardo, San Passo del 550 Eng. St. Gotthard Pass. Mountain pass of S Switzerland
Göttingen 267 var. Goettingen. C Germany
Gottschee see Kočevje
Gottwaldov see Zlín
Goubétto 222 SE Djibouti
Gouda 59 W Netherlands
Goulburn 101 SE Australia
Goundam 392 NW Mali

Gourcy 157 NW Burkina
Gouré 439 SE Niger
Gourma 157 cultural region of E Burkina
Gouyave 278 var. Charlotte Town. W Grenada island, Grenada
Goverla, Gora see Hoverla
Governador Valadares 145 SE Brazil
Governor's Harbour 112 Eleuthera Island, Bahamas
Govurdak 585 prev. Guardak, Turkm. Gowurdak. SE Turkmenistan
Goya 95 NE Argentina
Goyania see Goiânia
Göyçay 110 Rus. Geokchay. C Azerbaijan
Goz-Beïda 180 SE Chad
Gozo 395 var. Ghawdex. Island of NW Malta
Graaff-Reinet 531 Eastern Cape, S South Africa
Gračanica 140 NE Bosnia & Herzegovina
Gracias 290 W Honduras
Graciosa 474 var. Ilha Graciosa. Island of the Azores, Portugal
Gradačac 140 N Bosnia & Herzegovina
Gradaús, Serra dos 145 mountain range of C Brazil
Gradsko 381 C FYR Macedonia
Grafton 101 E Australia
Graham Bell Island see Greem Bell, Ostrov
Grain Coast 368 coastal region of Liberia
Grampian Mountains 597 mountain range of C Scotland, UK
Gramsh 81 var. Gramshi. C Albania
Gran see Hron
Gran see Esztergom
Gran 542 river of C Suriname
Granada 436 S Nicaragua
Granada 535 S Spain
Gran Canaria 535 var. Grand Canary. Island of Islas Canarias, SW Spain
Gran Chaco 95, 136, 460 var. Chaco. Lowland plain of C South America
Grand Anse 278 SW Grenada island, Grenada
Grand'Anse 516 Praslin, Seychelles
Grand Bahama Island 112 island of N Bahamas
Grand Baie 400 var. Grande Baie. NW Mauritius
Grand Barachois 652 inlet of the Atlantic Ocean on the coast of Miquelon, N Saint Pierre and Miquelon
Grand Bassa see Buchanan
Grand-Bassam 328 var. Bassam. SE Ivory Coast
Grand Bay 278 E Carriacou, Grenada
Grand Bourg 646 Marie-Galante, S Guadeloupe
Grand Caicos 653 var. Middle Caicos. Island of N Turks and Caicos Islands
Grand Canary see Gran Canaria
Grand Canyon 602 canyon of SW USA
Grand Cayman 644 island of W Cayman Islands
Grand Cess 368 SE Liberia
Grand Colombier 652 island of SE Saint Pierre and Miquelon
Grand Cul-de-Sac Marin 646 bay of the Caribbean Sea, N Guadeloupe
Grande-Anse see Portsmouth
Grande, Bahía 95 bay of the Atlantic Ocean, to the SE of Argentina
Grande Comore 198 var. Njazidja. Island of Comoros
Grande, Cuchilla 611 mountain range of E Uruguay
Grande de Buba, Rio 284 river of S Guinea-Bissau
Grande de Chiloé, Isla see Chiloé, Isla de
Grande de Gurupá, Ilha 145 island of N Brazil
Grande de Matagalpa, Río 436 river of C Nicaragua
Grande de Santiago, Río 403 var. Santiago. River of SW Mexico

Grande de Tárcoles, Río 206 river of C Costa Rica
Grande de Térraba, Río 206 river of SE Costa Rica
Grande Prairie 170 W Canada
Grand Erg Occidental 83 desert region of W Algeria
Grand Erg Oriental 83, 577 desert region of Algeria and Tunisia
Grande, Río 403, 602-603 Sp. Bravo Del Norte. River of Mexico and USA
Grande Rivière 497 N St Lucia
Grande Rivière 497 C St Lucia
Grande-Rivière-du-Nord 288 N Haiti
Grande Rivière, La 171 var. Fort George River. River of SE Canada
Grande-Saline 288 W Haiti
Grande-Santi 645 W French Guiana
Grande Terre 516 island of the Aldabra Group, SW Seychelles
Grande Vigie, Pointe de la 646 headland of N Guadeloupe
Grand Falls 344 waterfall of C Kenya
Grand Forks 603 North Dakota, NC USA
Grand Goâve 288 S Haiti
Grand Harbour 395 port of Valletta, E Malta
Grand Ilet 646 bay of the Caribbean Sea, S Guadeloupe
Grand Island 603 Nebraska, C USA
Grand Junction 602 Colorado, SW USA
Grand Lac de l'Ours see Great Bear Lake
Grand Lac des Esclaves see Great Slave Lake
Grand Montagne 400 Rodrigues, Mauritius
Grand Paradis see Gran Paradiso
Grand-Popo 132 S Benin
Grand Rapids 603 Michigan, NC USA
Grand Récif Sud 650 reef of the Pacific Ocean, S New Caledonia
Grand River South East 400 river of E Mauritius
Grand'Rivière 649 N Martinique
Grand Roy 278 W Grenada island, Grenada
Grand Turk Island 653 island of SE Turks and Caicos Islands
Granges see Grenchen
Gran Lago see Nicaragua, Lago de
Gran Paradiso 322 Fr. Grand Paradis. Mountain of NW Italy
Gran San Bernardo, Passo di see Great St Bernard Pass
Grape Bay 643 bay of the North Atlantic Ocean, E Bermuda
Grassy Bay 643 bay of the North Atlantic Ocean, W Bermuda
Gråsten 218 Jylland, SW Denmark
Graudenz see Grudziądz
Grau Roig 86 E Andorra
Graz 106 SE Austria
Great Abaco 112 island of N Bahamas
Great Admiralty Island see Manus Island
Great Ararat see Büyükağrı Dağı
Great Artesian Basin 101 lowlands of C Australia
Great Australian Bight 101 large bay of the Indian Ocean, S Australia
Great Barrier Island 433 island to the NE of North Island, New Zealand
Great Barrier Reef 101, 458 coral reef to the NE of Australia, and the largest in the world
Great Basin 602 physical region of W USA
Great Bear Lake 170 Fr. Grand Lac de l'Ours. Lake of NW Canada
Great Belt see Storebælt
Great Camanoe 643 island of N British Virgin Islands
Great Chagos Bank 643 undersea feature of the Indian Ocean, C British Indian Ocean Territory
Great Coco Island 159 island of SW Burma

Great Dividing Range 101 mountain range of E Australia
Greater Antarctica 90 physical region of Antarctica
Greater Caucasus 110 Rus. Bol'shoy Kavkaz. Mountain range of SW Asia and SE Europe
Great Exhibition Bay 433 inlet of the Pacific Ocean, on the NE coast of North Island, New Zealand
Great Exuma Island 112 island of C Bahamas
Great Falls 602 Montana, NW USA
Great Fish see Groot-Vis
Great Harbour 643 Jost Van Dyke, W British Virgin Islands
Great Hungarian Plain see Alföld
Great Inagua 112 island of S Bahamas
Great Indian Desert see Thar Desert
Great Karoo 531 var. Great Karroo, Afr. Groot Karoo. Plateau region of S South Africa
Great Khingan Range see Da Hinggan Ling
Great Lake see Tônlé Sap
Great Lakes, the 606 see also Erie, Huron, Michigan, Ontario, Superior
Great Mercury Island 433 island to the NE of North Island, New Zealand
Great Nicobar 298 island of Nicobar Islands, to the SE of India
Great Ouse 597 var. Ouse. River of E England, UK
Great Plains 603 plains of Canada and USA
Great Rift Valley 245, 344, 385, 564 var. Rift Valley. Depression of E Africa and SW Asia
Great Ruaha 564 river of C Tanzania
Great Saint Bernard Pass 550 Fr. Col du Grand-Saint-Bernard, It. Passo di Gran San Bernardo. Mountain pass of SW Switzerland
Great Salt Lake 602 salt lake of SW USA
Great Salt Pond 494 SE St Kitts, St Kitts & Nevis
Great Sand Sea. 232, 371 desert of Egypt and Libya
Great Sandy Desert 100 desert region of W Australia
Great Sandy Desert see Rub 'al Khali
Great Sandy Island see Fraser Island
Great Slave Lake 170 Fr. Grand Lac des Esclaves. Lake of W Canada
Great Sound 643 bay of W Bermuda
Great Thatch 643 island of W British Virgin Islands
Great Tobago 643 island of W British Virgin Islands
Great Usutu see Lusutfu
Great Victoria Desert 100 -101 desert region of S Australia
Great Wall 90 Chinese research station of South Shetland Islands, Antarctica
Great Zab 312 Ar. Az Zāb Al Kabir, Turk. Büyükzap Suyu, Kurd. Zê-i Bādīnān. River of Iraq and Turkey
Greco, Cape 215 var. Cape Gkreko. Cape of E Cyprus
Greece 274-277 officially Hellenic Republic, Gk Ellás. Country of SE Europe divided into 51 admin. units (nomos). ❖ Athens
Greeley 603 Colorado, SW USA
Greem Bell, Ostrov 485 Eng. Graham Bell Island. Island of Zemlya Frantsa-Iosifa, N Russian Federation
Green Bay 603 Wisconsin, NC USA
Green Island 92 island to the E of Antigua, Antigua & Barbuda
Green Island 278 island to the N of Grenada island, Grenada
Green Islands 458 var. Nissan Is. Island group of E Papua New Guinea
Greenland 646 var. Grønland. Self governing territory of Denmark, North Atlantic Ocean. ❖ Nuuk.
Greenland Sea 646 sea of the Arctic Ocean, NE Greenland

Green Mountain *652* mountain of C Ascension Island
Greenock *597* W Scotland, UK
Greensboro *603* North Carolina, SE USA
Greenville *368* var. Sino, Sinoe. SE Liberia
Greenville *603* Mississippi, SE USA
Greenville *603* South Carolina, SE USA
Greifswald *267* NE Germany
Grenå *218* NJylland, Denmark
Grenada *278* island which, with the southern Grenadines, comprises the independent state of Grenada
Grenada *278-279* island state of the West Indies, comprising of Grenada island and the southern Grenadine islands to the north. Divided into 6 admin. units (parishes). ❖ St George's
Grenadines *278, 498* group of over 500 small islands between Grenada and St Vincent in the Caribbean Sea. The southern islands are administered by Grenada, while the northern islands form, with St Vincent, the independent state of St Vincent the Grenadines
Grenchen *550* Fr. Granges. NW Switzerland
Grenoble *255* E France
Grenville *278* E Grenada island, Grenada
Grevelingen *429* inlet of the North Sea, on the coast of SW Netherlands
Grevenmacher *378* E Luxembourg
Greymouth *433* W South Island, New Zealand
Grey Range *101* mountain range of E Australia
Gribingui *see* Nana-Grébizi
Griffith *101* SE Australia
Grijalva *403* var. Tabasco. River of SE Mexico
Grimari *179* C Central African Republic
Grimbergen *127* C Belgium
Grimsby *597* NE England, UK
Grímsey Island *296* var. Grimsey. Island of N Iceland
Grímsstadhir *296* E Iceland
Grindavík *296* W Iceland
Grindsted *218* Jylland, W Denmark
Grise Fiord *170* var. Ausuitoq. Ellesmere Island, N Canada
Grmeč *140* mountain range of NW Bosnia & Herzegovina
Grobiņa *362* W Latvia
Grodno *see* Hrodna
Groningen *429* NE Netherlands
Groningen *542* N Suriname
Groote Eylandt *101* island of N Australia
Grootfontein *423* N Namibia
Groot Karoo *see* Great Karoo
Groot-Vis *531* Eng. Great Fish. River of S South Africa
Grootvloer *531* salt flat of Northern Cape, W South Africa
Gros Islet *497* N St Lucia
Gros-Morne *288* N Haiti
Gros Piton *497* mountain of SW St Lucia
Grossbetschkerek *see* Zrenjanin
Grosseto *322* NW Italy
Grosskanizsa *see* Nagykanizsa
Grossmichel *see* Michalovce
Gross-Sankt-Johannis *see* Suure-Jaani
Gross-Steffelsdorf *see* Rimavská Sobota
Grosswardein *see* Oradea
Groznyy *484* SW Russian Federation
Großglockner *106* mountain of W Austria
Grudziądz *471* Ger. Graudenz. N Poland
Grünau *423* S Namibia
Grünberg in Schlesien *see* Zielona Góra
Gruziya *see* Georgia
Guabalá *456* SW Panama
Guacanayabo, Golfo de *210* gulf of the Caribbean Sea, on the S coast of Cuba

Guácimo *206* NE Costa Rica
Guadalajara *403* SW Mexico
Guadalcanal *526* island of C Solomon Is
Guadalquivir *534-535* river of W Spain
Guadalupe *456* C Panama
Guadalupe *see* Canelones
Guadeloupe *646* French overseas department of the Caribbean Sea. ❖ Basse-Terre
Guadeloupe Passage *92, 224* passage connecting the Atlantic Ocean and Caribbean Sea between Guadeloupe and Dominica
Guadiana *474, 534-535* river of Portugal and Spain
Guaico *574* NE Trinidad, Trinidad & Tobago
Guáimero *210* S Cuba
Gualaco *290* C Honduras
Gualán *280* C Guatemala
Gualdicciolo *502* NW San Marino
Gualeguaychú *95* E Argentina
Guam *647* unincorporated territory of the USA, Pacific Ocean. ❖ Hagåtña
Guanabacoa *210* NW Cuba
Guanabo *210* NW Cuba
Guana Island *643* island of N British Virgin Islands
Guanaja *290* Islas de la Bahía, Honduras
Guanajay *210* NW Cuba
Guanajuato *403* C Mexico
Guanare *623* NW Venezuela
Guanare *623* river of W Venezuela
Guangdong *187* var. Kwangtung, Kuang-tung. Province of SE China
Guangju *see* Kwangju
Guangxi *187* Chin. Guangxi Zhuangzu Zizhiqu, Eng. Kwangsi Chuang Autonomous Region, var. Kuang-hsi. Autonomous region of S China
Guangyuan *187* var. Kuang-yuan, Kwangyuan. Sichuan, SW China
Guangzhou *187* var. Kwangchow, Kuang-chou, Eng. Canton. Guangdong, SE China
Guantánamo *211* SE Cuba
Guantanamo Bay *211* US military base, E Cuba
Guaporé *136, 144* var. Iténez. River of Bolivia and Brazil
Guaranda *230* C Ecuador
Guardafui, Cape *see* Caseyr, Raas
Guardak *see* Govurdak
Guárico *623* river of C Venezuela
Guasave *403* W Mexico
Guastatoya *see* El Progreso
Guatemala *280-281* officially Republic of Guatemala. Country of Central America divided into 22 admin. units (departments). ❖ Guatemala City
Guatemala City *280* Sp. Ciudad de Guatemala. ❖ of Guatemala, C Guatemala
Guaviare *195* river of E Colombia
Guayama *651* SE Puerto Rico
Guayaquil *230* var. Santiago de Guayaquil. SW Ecuador
Guayaquil, Gulf of *230, 463* gulf of the Pacific Ocean to the W of Ecuador and Peru
Guayaramerín *136* N Bolivia
Guayuabo *651* N Puerto Rico
Gudara *see* Ghūdara
Gudenå *218* river of Jylland, NW Denmark
Guékédou *283* var. Guéckédou. S Guinea
Guélilé *222* S Djibouti
Guelma *83* var. Gâlma. NE Algeria
Guelta Zemmur *414* E Western Sahara
Guéné *132* NE Benin
Güera *see* Guérou
Guéret *255* C France
Guerguerat *414* SW Western Sahara
Guérin-Kouka *571* NW Togo
Guernsey *254, 647* var. Bailiwick of Guernsey. British Crown dependency, Channel Islands, English Channel. ❖ St Peter Port.
Guérou *398* var. Güera. S Mauritania
Guesneau *497* N St Lucia

Guguan *650* island of C Northern Mariana Islands
Guiana Highlands *144-145, 286, 145, 286, 623* Port. Planalto das Guianas. Mountain range of N South America
Guiana Island *92* island to the NE of Antigua, Antigua & Barbuda
Guiba *see* Juba
Guichón *611* NW Uruguay
Guider *168* var. Guidder. N Cameroon
Guidimouni *439* S Niger
Guier, Lac de *510* var. Lac de Guiers. Lake of N Senegal
Guietsou *260* SW Gabon
Guiglo *328* W Ivory Coast
Güigüe *623* N Venezuela
Güija, Lago de *237* lake of El Salvador and Guatemala
Guillet *224* NW Dominica
Guimarães *474* N Portugal
Guinea *282-283* officially People's Revolutionary Republic of Guinea, prev. People's Republic of Guinea. Country of West Africa divided into 4 admin. units (regions). ❖ Conakry
Guinea-Bissau *284-285* officially Republic of Guinea-Bissau, Port. Guiné-Bissau. Independent state of West Africa divided into 8 admin. units (regions). ❖ Bissau
Guinea, Gulf of *168, 238, 272, 328* Fr. Golfe de Guinée. Gulf of the Atlantic Ocean, on the W coast of Africa
Guinguinéo *510* W Senegal
Güiria *623* NE Venezuela
Guiyang *187* var. Kuei-Yang, Kweiyang. Guizhou, S China
Guizhou *187* var. Kweichow, Kuei-chou. Province of S China
Gujrānwāla *451* NE Pakistan
Gujrāt *451* NE Pakistan
Gukasyan *see* Ashots'k
Gulariya *427* var. Gularia. W Nepal
Gulbene *362* Ger. Alt-Schwanenburg. NE Latvia
Gul'cha *359* var. Gul'ča, Kir. Gülchö. SW Kyrgyzstan
Gülchö *see* Gul'cha
Gulf, The *115, 312, 356, 479, 506* var. Persian Gulf, Ar. Khalij al 'Arabī, Per. Khalīj-e Fars. Gulf of the Arabian Sea between the Arabian Peninsula and Iran
Guliston *614* Rus. Gulistan. E Uzbekistan
Gulja *see* Yining
Gulu *588* N Uganda
Gum *see* Gamgadhi
Gumal *451* river of Afghanistan and Pakistan
Gumdag *585* prev. Kum-Dag. W Turkmenistan
Gümülcine *see* Komotiní
Gümüljina *see* Komotiní
Gümüşhane *581* NE Turkey
Gund *562* Rus. Gunt. River of S Tajikistan
Güneydoğu Toroslar *581* mountain range of SE Turkey
Gunjur *263* W Gambia
Gunner's Quoin *400* var. Coin de Mire. Island of N Mauritius
Gunsan *see* Kunsan
Güntür *298* SE India
Gunzan *see* Kunsan
Gurgan *see* Gorgān
Guri, Embalse de *623* reservoir of E Venezuela
Gurjaani *264* Rus. Gurdzhaani. E Georgia
Gurk *106* river of S Austria
Gurkfeld *see* Krško
Gurktaler Alpen *106* mountain range of S Austria
Gur'yev *see* Atyrau
Gusau *440* N Nigeria
Gushgy *585* prev. Kushka. S Turkmenistan
Gustavia *646* St. Barthélémy, N Guadeloupe
Gutenstein *see* Ravne na Koroškem
Gutland *378* physical region of S Luxembourg

Guwāhāti *299* prev. Gauhāti. NE India
Guyana *286-287* officially Co-operative Republic of Guyana, prev. British Guiana. Country of Central America divided into 10 admin. units (regions). ❖ Georgetown
Guyane *see* French Guiana
Güzelyurt *215* var. Morfou, Morphou. W Cyprus
Güzelyurt Körfezi *see* Morphou Bay
Gwādar *450* SW Pakistan
Gwalior *298* N India
Gwanda *636* SW Zimbabwe
Gwayi *636* river of W Zimbabwe
Gweru *636* prev. Gwelo. C Zimbabwe
Gwy *see* Wye
Gyandzha *see* Gäncä
Gydanskiy Poluostrov *484* Eng. Gyda Peninsula. Peninsula of N Russian Federation
Gympie *101* E Australia
Gyöngyös *292* NE Hungary
Győr *292* Ger. Raab. NW Hungary
Gyula *292* SE Hungary
Gyulafehérvár *see* Alba Iulia
Gyumri *98* Rus. Kumayri, prev. Leninakan, Aleksandropol'. W Armenia
Gyzylarbat *585* prev. Kizyl-Arvat. W Turkmenistan
Gyzylgaya *see* Kizyl-Kaya
Gżira *395* N Malta

H

Ha *134* W Bhutan
Ha'ano *572* island of the Ha'apai Group, Tonga
Ha'apai Group *572* var. Haabai. Island group of C Tonga
Haapsalu *242* Ger. Hapsal. W Estonia
Haarlem *429* W Netherlands
Hab *451* river of S Pakistan
Habarana *538* C Sri Lanka
Ḥabbānīyah, Buḥayrat al *312* var. Hawr al Ḥabbānīyah. Lake of C Iraq
Habiganj *117* NE Bangladesh
Habomai Islands *333* disputed island group of SE Russian Federation
Ḥabshān *594* SW United Arab Emirates
Hachijō-jima *332* island to the SE of Honshū, SE Japan
Hachinohe *332* Honshū, N Japan
Hačmas *see* Xaçmaz
Hadama *see* Nazrēt
Ḥadd al Jamal *115* var. Ra's al Yaman. Cape of E Bahrain
Haddummati Atoll *see* Hadhdhunmathi Atoll
Hadejia *440* N Nigeria
Hadejia *440* river of N Nigeria
Hadera *319* C Israel
Haderslev *218* Ger. Hadersleben. Jylland, SW Denmark
Hadhdhunmathi Atoll *390* var. Haddummati Atoll, Laamu Atoll. Atoll of S Maldives
Hadhramaum *see* Ḥaḍramawt
Hadiboh *631* NE Suqutra, Yemen
Ha Đông *627* N Vietnam
Ḥaḍramawt *631* Eng. Hadhramaut. Mountain range of S Yemen
Haedo, Cuchilla de *611* mountain range of NW Uruguay
Haeju *351* S North Korea
Ha-erh-pin *see* Harbin
Ḥafar al Bāṭin *506* N Saudi Arabia
Hafnarfjördhur *296* W Iceland
Hafnarhreppur *296* S Iceland
Hafren *see* Severn
Hagar Nish Plateau *240* plateau of N Eritrea
Hagåtña *647* var. Agaña. ❖ of Guam, NW Guam
Hagen *266* W Germany
Hāgere Hiywet *245* var. Agere Hiywet, Ambo. C Ethiopia
Hagi *332* Honshū, SW Japan
Ha Giang *627* NW Vietnam
Hago, Lac *492* lake of NE Rwanda

Hague, The *see* 's-Gravenhage
Haibak *see* Âybak
Haidarabad *see* Hyderābād
Hai Dương *627* N Vietnam
Haifa, Bay of *see* Mifraz Ḥefa
Haikou *187 var.* Hoihow. Hainan, S China
Ḥā'il *506* N Saudi Arabia
Hailuoto *251 Swe.* Karlö. Island of W Finland
Hainan *187* province of S China
Hainan Dao *187* island of S China
Hai Phong *627 var.* Haiphong. N Vietnam
Haiti *288-289* officially Republic of Haiti. Country of the West Indies divided into 9 admin. units (departments). ❖ Port-au-Prince
Hajdarken *see* Khaydarkan
Hajdúböszörmény *292* NE Hungary
Ḥājī Ebrāhīm, Kūh-e *312* mountain of Iran and Iraq
Ḥajjah *631* W Yemen
Hakodate *332* Hokkaidō, N Japan
Hakupu *650* E Niue
Ḥalab *555 Eng.* Aleppo, *Fr.* Alep. NW Syria
Ḥalabja *312* NE Iraq
Ḥalānīyāt, Juzur al *448 var.* Jazā'iBin Ghalfān, Jazā'ir Khurīyā Murīyā, *Eng.* Kuria Maria Islands. Island group of S Oman
Ḥalānīyāt, Khalīj al *448 Eng.* Kuria Mur Bay. Bay of the Arabian Sea, S Oman
Halas *see* Kiskunhalas
Haldefjäll *see* Haltiatunturi
Halden *444 prev.* Fredrikshald. S Norway
Halditjåkko *see* Haltiatunturi
Halfa el Gadida *540 var.* New Halfa, Halfa Al Jadida. E Sudan
Halfmoon Bay *433* Stewart Island, SW New Zealand
Half Tree Hollow *652* N St Helena
Halifax *171* SE Canada
Halīl Rūd *309* river of SE Iran
Halla-san *352 Jap.* Kanra-san. Mountain of Cheju-do, S South Korea
Halle *267* C Germany
Hallein *106* N Austria
Halley *90* UK research station of Greater Antarctica, Antarctica
Hall Islands *406* island group of C Micronesia
Hall Peninsula *171* peninsula of Baffin Island, NE Canada
Halls Creek *101* NW Australia
Halmahera *304 prev.* Djailolo, Jailolo, Gilolo. Island of Maluku, E Indonesia
Halmahera, Laut *304* sea of the Pacific Ocean, E Indonesia
Halmstad *547* SW Sweden
Ḥalq al Wādī *see* La Goulette
Hälsingborg *see* Helsingborg
Haltiatunturi *251 Swe.* Haldefjäll, *prev.* Halditjåkko, *Nor.* Reisduoddarhalde. Mountain of Finland and Norway
Ḥamad *see* Madīnat Ḥamad
Hamada *332* Honshū, W Japan
Hamadān *309* NW Iran
Hamada Town *see* Madīnat Ḥamad
Ḥamāh *555* C Syria
Hamamatsu *332* Honshū, C Japan
Hamar *444* S Norway
Hambantota *538* SE Sri Lanka
Hamburg *267* N Germany
Ḥamḍ, Wādī al *506* dry watercourse of W Saudi Arabia
Hämeenlinna *251 Swe.* Tavastehus. SW Finland
Hamersley Range *100* mountain range to the W of Australia
Hamgyŏng-sanmaek *351* mountain range of N North Korea
Hamhŭng *351* C North Korea
Hami *186 Uigh.* Kumul, *var.* Qomul. Xinjiang Uygur Zizhiqu. NW China
Hamilton *643* ❖ Bermuda, C Bermuda
Hamilton *171* SE Canada
Hamilton *433* C North Island, New Zealand

Hamilton *597* C Scotland, UK
Ḥamīm, Wādī al *371* dry watercourse of NE Libya
Hamm *266* W Germany
Hammamet *577 var.* Ḥammāmāt. N Tunisia
Hammamet, Golfe de *577* gulf of the Mediterranean Sea to the E of Tunisia
Hammam Lif *577 var.* Ḥammām al Anf. N Tunisia
Ḥammār, Hawr al *312* lake of SE Iraq
Hammerfest *444* NE Norway
Ḥamrīn, Jabal *312* mountain range of N Iraq
Hamriya *see* Al Ḥamrīyah
Hamrun *395* C Malta
Hāmūn, Daryācheh-ye *see* Sīstān, Daryācheh-ye
Han *352 Jap.* Kan-kō. River of N South Korea
Hanábana *210* river of C Cuba
Hânceşti *see* Hînceşti
Handan *187 var.* Han-tan. Hebei, NE China
Handeni *564* E Tanzania
Handréma, Baie de *649 var.* Mandréma Bay. Bay of the Indian Ocean on the N coast of Mayotte
HaNegev *319 Eng.* Negev. Desert of S Israel
Hanga Roa *183* Easter I, W Chile
Hangayn Nuruu *412* mountain range of W Mongolia
Hangzhou *187 var.* Hangchow, Hang-chou. Zhejiang, E China
Hanka, Lake *see* Khanka, Lake
Hanko *251 Swe.* Hangö. SW Finland
Hankow *see* Wuhan
Hannover *266 Eng.* Hanover. NW Germany
Hanöbukten *547* bay of the Baltic Sea to the S of Sweden
Hanoi *627 Vtn.* Ha Nôi. ❖ of Vietnam, N Vietnam
Hanover *see* Hannover
Hanstholm *218* Jylland, NW Denmark
Han-tan *see* Handan
Hantu, Pulau *521* island of SW Singapore
Hāora *299 prev.* Howrah. E India
Haouach, Ouadi *180* dry watercourse of E Chad
Happy Valley-Goose Bay *171 prev.* Goose Bay. E Canada
Hapsal *see* Haapsalu
Ḥaraḍ *506 var.* Haradh. E Saudi Arabia
Ḥaraḍ *631* N Yemen
Hara Laht *242* bay of the Gulf of Finland, on the coast of N Estonia
Harare *636 prev.* Salisbury. ❖ of Zimbabwe, NE Zimbabwe
Haraze-Mangueigne *180* SE Chad
Harbel *368* W Liberia
Harbin *187 var.* Ha-erh-pin, *prev.* Pinkiang. Heilongjiang, NE China
Harbours, Bay of *645* bay of the South Atlantic Ocean, SE Falkland Islands
Harbour View *331* E Jamaica
Hardangerfjorden *444* fjord of SW Norway
Hardap Dam *423* dam of C Namibia
Haré Meron *319* Mountain of N Israel
Härer *245* E Ethiopia
Hargeysa *528* NW Somalia
Hari *304 var.* Batang Hari, *prev.* Djambi. River of Sumatra, W Indonesia
Ḥarīb *631* W Yemen
Hari Kurk *242* channel of Baltic Sea, between the island of Hiiumaa and Estonia mainland
Ḥārim *555* NW Syria
Ḥarīmā *338* N Jordan
Haringmat *117* river of SW Bangladesh
Harīrūd *77* river of C Asia
Harīrūd *see* Tedzhen
Harlingen *429 Fris.* Harns. N Netherlands

Harmanli *see* Kharmanli
Harns *see* Harlingen
Harper *368 var.* Cape Palmas. S Liberia
Ḥarrah *631* SE Yemen
Harrington Sound *643* bay of the North Atlantic Ocean, N Bermuda
Harris *649* E Montserrat
Harrisburg *603* Pennsylvania, NE USA
Harrismith *531* Orange Free State, E South Africa
Harstad *444* NE Norway
Hartford *368* SW Liberia
Hartford *603* Connecticut, NE USA
Hartley *see* Chegutu
Harz *267 var.* Harz Mountains. Mountain range of C Germany
HaSharon *319 Eng.* Plain of Sharon. Plain of C Israel
Haskovo *see* Khaskovo
Haspengouw *see* Hesbaye
Hasselt *127* NE Belgium
Hassetché *see* Al Ḥasakah
Hastings *121* SW Barbados
Hastings *433* SE North Island, New Zealand
Hastings *518* W Sierra Leone
Hastings *597* SE England, UK
Hatay *see* Antakya
Hātia *117* river and one of the main mouths of the Ganges, S Bangladesh
Hato Mayor *226* E Dominican Republic
Ḥattā *594* E United Arab Emirates
Hattiesburg *603* Mississippi, SE USA
Hattieville *130* E Belize
Hat Yai *567 var.* Ban Hat Yai. S Thailand
Haud *245 var.* Hawd. Plateau of Somalia and Ethiopia.
Haugesund *444* SW Norway
Haukeligrand *444* SW Norway
Haukivesi *251* lake of SE Finland
Hauraki Gulf *433* gulf on the N coast of North Island, New Zealand
Hau *627* river of SW Vietnam
Haut Atlas *414 Eng.* High Atlas. Mountain range of C Morocco
Haute-Sangha *see* Mambéré-Kadéi
Hautes Fagnes *127 Ger.* Hohes Venn. Mountain range of E Belgium
Haute Sûre, Lac de la *378* reservoir of NW Luxembourg
Haut Plateau du Dra *see* Dra, Hamada du
Hauts Plateaux *83* plateau of NW Algeria
Havana *210 var.* La Habana. ❖ of Cuba, NW Cuba
Havířov *216* E Czech Republic
Havlíčkův Brod *216 prev.* Německý Brod, *Ger.* Deutsch-Brod. S Czech Republic
Hawaii *602* island of Hawaiian group, Hawaii, USA, C Pacific
Hawaii *602* non-contiguous state of USA, C Pacific
Ḥawallī *356* E Kuwait
Hawar Islands *115* islands of SE Bahrain
Hawash *see* Awash
Hawea, Lake *433* W South Island, New Zealand
Hawera *433* SW North Island, New Zealand
Hawick *597* S Scotland, UK
Hawke Bay *433* bay of the South Pacific Ocean, on the SE coast of North Island, New Zealand
Hawlēr *see* Arbīl
Ḥawmat as Sūq *see* Houmt Souk
Ḥawrā' *631* C Yemen
Ḥawrān, Wādī *312* dry watercourse of W Iraq
Hawwārah *338 var.* Huwwāra. N Jordan
HaYarden *see* Jordan
Hay River *170* W Canada
Ḥayyān, Ra's *var.* Ra's Ḥayyān. Cape of E Bahrain
Hebei *187 var.* Hopei. Province of NE China
Hebrides, Sea of the *597* sea of the Atlantic Ocean to the NW of UK

Hebron *319, 320 Ar.* Al Khalīl. S West Bank
Heerenveen *429* NE Netherlands
Heerlen *429* S Netherlands
Ḥefa *319* N Israel
Ḥefa, Mifraz *319 Eng.* Bay of Haifa. Bay of the Mediterranean Sea
Hefei *187 var.* Hofei, *hist.* Luchow. Anhui, E China
Heichin *see* P'ingchen
Heidelberg *266* SW Germany
Heihe *187 prev.* Ai-hun. Heilongjiang, NE China
Hei-ho *see* Nagqu
Heilbronn *266* SW Germany
Heiligenkreuz *see* Žiar nad Hronom
Heilong Jiang *see* Amur
Heilongjiang *187 var.* Heilungkiang, Hei-lung-chiang. Province of NE China
Heimaey Island *296 var.* Heimaey, Heimaey. Island of S Iceland
Heitō *see* P'ingtung
Helen *455* island of S Palau
Helena *602* Montana, NW USA
Helgoland *266 Eng.* Heligoland. Island of NW Germany
Helgoländer Bucht *266 var.* Helgoland Bay, Heligoland Bight. Bay of the North Sea
Hell-Ville *see* Andoany
Helmand, Daryā-ye *77* river of Afghanistan and Iran
Helmond *429* S Netherlands
Helsingborg *547 prev.* Hälsingborg. S Sweden
Helsingør *218 Eng.* Elsinore. Sjælland, E Denmark
Helsinki *251 Swe.* Helsingfors. ❖ of Finland, S Finland
Ḥelwân *232 var.* Ḥulwān, Ḥilwān. N Egypt
Henan *187 var.* Honan. Province of C China
Henderson Island *651* island of N Pitcairn Islands
Hendū Kosh *see* Hindu Kush
Hengduan Shan *186* mountain range of SW China
Hengelo *429* E Netherlands
Hengyang *187* Hunan, S China
Hentiesbaai *423* W Namibia
Henzada *159* SW Burma
Heradhsvötn *296* river of C Iceland
Herāt *77* W Afghanistan
Heredia *206* C Costa Rica
Hereford *597* C England, UK
Herisau *550 Fr.* Hérisau. NE Switzerland
Héristal *see* Herstal
Herm *647* island of S Guernsey
Hermannstadt *see* Sibiu
Hermansverk *444* SW Norway
Hermel *364 var.* Hirmil. NE Lebanon
Hermitage *278* C Grenada island, Grenada
Hermon, Mount *555 Ar.* Jabal ash Shaykh. Mountain of SW Syria
Hermosillo *403* NW Mexico
Hernád *see* Hornád
Hernandarias *460 prev.* Tacurupucú. SE Paraguay
Herne *266* W Germany
Herning *218* Jylland, W Denmark
Herstal *127 Fr.* Héristal. E Belgium
Herzliyya *319* C Israel
Herzogenbusch *see* 's-Hertogenbosch
Hesbaye *127 Dut.* Haspengouw. Physical region of C Belgium
Hesperange *378* SE Luxembourg
Hestur *644* island of C Faeroe Islands
Hetauda *427* C Nepal
Hida-sammyaku *332* mountain range of Honshū, C Japan
Hienghène *650* W New Caledonia
Hierro *535 var.* Ferro. Island of Islas Canarias, SW Spain
High Atlas *see* Haut Atlas
Highgate *331* NE Jamaica
Highlands, The *92* highlands of Barbuda, Antigua & Barbuda
High Point *643* headland of W Bermuda

High Veld *see* Northern Karoo
Higüey *226 var.* Salvaleon de Higüey. E Dominican Republic
Hiiumaa *242 var.* Hiiuma, *Ger.* Dagden, *Swed.* Dagö. Island of W Estonia
Hikina *648* island, NE Johnston Atoll
Hildesheim *266* NW Germany
Hilla *see* Al Ḥillah
Hillaby, Mount *121* mountain of Barbados
Hillerød *218* Sjælland, E Denmark
Hillsborough *278* W Carriacou, Grenada
Hilo *602* Hawaii, USA
Hilversum *429* C Netherlands
Hilwan *see* Ḥelwân
Himachal Pradesh *298* state of N India
Himalayas *134, 186, 298, 427* mountain range of S Asia
Himeji *332* Honshū, C Japan
Himora *245 var.* Humera. NW Ethiopia
Ḥimş *555 var.* Homs. W Syria
Hînceşti *408 var.* Hânceşti, *prev.* Kotovsk. C Moldova
Hinche *288* C Haiti
Hindenburg *see* Zabrze
Hindiya *see* Al Hindīyah
Hindu Kush *77, 451 Per.* Hendü Kosh. Mountain range of C Asia
Hingol *450* river of SW Pakistan
Hinson Island *643* island of W Bermuda
Hirmil *see* Hermel
Hirosaki *332* Honshū, N Japan
Hiroshima *332* Honshū, W Japan
Hirschberg in Riesengebirge *see* Jelenia Góra
Hirtshals *218* Jylland, N Denmark
Ḥişbān *338* NW Jordan
Ḥisb, Shaʿib *312 var.* Shaʿib Hasb. Dry watercourse of S Iraq
Hisor *562 Rus.* Gissar. W Tajikistan
Hitachi *332* Honshū, SE Japan
Hitra *444 prev.* Hitteren. Island of W Norway
Hitteren *see* Hitra
Hiu *619* Torres Islands, N Vanuatu
Hjälmaren *547 Eng.* Lake Hjalmar. Lake of S Sweden
Hjalmar, Lake *see* Hjälmaren
Hjørring *218* Jylland, N Denmark
Hkakabo Razi *159* mountain of Burma and China
Hlathikulu *544 var.* Hlatikulu. S Swaziland
Hlohovec *523 prev.* Frakštát, *Ger.* Freistadtl, *Hung.* Galgóc. W Slovakia
Hlotse *366 var.* Leribe. NW Lesotho
Hlybokaye *122 Rus.* Glubokoye. N Belarus
Ho *272* SE Ghana
Hoa Binh *627* N Vietnam
Hoang Liên Sơn *627* mountain range of China and Vietnam
Hoani *198* NW Mohéli, Comoros
Hobart *101* Tasmania, Australia
Hobro *218* Jylland, NW Denmark
Hobyo *528 It.* Obbia. E Somalia
Hồ Chí Minh *627 var.* Ho Chi Minh City, *prev.* Saigon. S Vietnam
Hodeida *see* Al Ḥudaydah
Hódmezővásárhely *292* SE Hungary
Hodonín *216 Ger.* Göding. SE Czech Republic
Hoë Karoo *see* Northern Karoo
Hoeryŏng *351* NE North Korea
Hof *267* SE Germany
Höfðhakaupstadhur *see* Skagaströnd
Hofei *see* Hefei
Hofsá *296* river of E Iceland
Hofsjökull *296* glacier of C Iceland
Hofuf *see* Al Hufūf
Hōgen *see* Fengyüan
Hoggar *see* Ahaggar
Hogoley Islands *see* Chuuk Islands
Hohenems *106* W Austria
Hohes Venn *see* Hautes Fagnes
Hohe Tauern *106* mountain range of W Austria

Hohhot *187 var.* Huhehot, *prev.* Kweisui. Nei Mongol Zizhiqu, N China
Hôi An *627 prev.* Faifo. C Vietnam
Hoihow *see* Haikou
Hoima *588* W Uganda
Hojancha *206* W Costa Rica
Hokitika *433* W South Island, New Zealand
Hokkaidō *333, 337* island of N Japan
Hokkō *see* Peikang
Hokō *see* P'ohang
Hoktemberyan *98 Rus.* Oktemberyan. SW Armenia
Holbæk *218* Sjælland, E Denmark
Holetown *121 prev.* Jamestown. W Barbados
Holguín *210* SE Cuba
Holhol *222* SE Djibouti
Holland *see* Netherlands
Hollandia *see* Jayapura
Hólmavík *296* NW Iceland
Holmsland Klit *218* fjord of Jylland, W Denmark
Holon *319* C Israel
Holonga *572* Uta Vavaʻu, Tonga
Holot Ḥaluza *319* historic site of S Israel
Holstebro *218* Jylland, W Denmark
Holsteinsborg *see* Sisimiut
Holyhead *597* N Wales, UK
Homa Bay *344* W Kenya
Homâyūnshahr *see* Khomeynīshahr
Hombori Tondo *392* mountain of E Mali
Home Island *644* island of C Cocos Islands
Homenau *see* Humenné
Homonna *see* Humenné
Homs *see* Ḥimş
Homs *see* Al Khums
Homyel' *122 Rus.* Gomel'. SE Belarus
Honan *see* Henan
Hondo *130, 403* river of Central America
Honduras *290-291* officially Republic of Honduras. Country of C America divided into 18 admin. units (departments). ❖ Tegucigalpa
Honduras, Gulf of *130, 280* gulf of the Caribbean Sea to the E of Central America
Hønefoss *444* S Norway
Hông Gai *627 var.* Hongay. N Vietnam
Hong Kong *187 var.* Xianggang. S China
Hongwŏn *351* E North Korea
Hongze Hu *187 var.* Hung-tse Hu. Lake of E China
Honiara *526* ❖ of the Solomon Islands, N Guadalcanal, Solomon Islands
Honolulu *602* Oahu, Hawaii, USA
Honshū *332* island of C Japan
Honte *see* Westerschelde
Hoogeveen *429* NE Netherlands
Hoogezand *429* NE Netherlands
Hooker, Cape *531* cape of Marion Island, S South Africa
Hoorn *429* NW Netherlands
Hopei *see* Hebei
Hope Town *112* Great Abaco, Bahamas
Horgen *550* N Switzerland
Hōrin *see* Fenglin
Horki *122 Rus.* Gorki. NE Belarus
Horlivka *590 Rus.* Gorlovka. E Ukraine
Hormuz, Strait of *309, 448, 594 var.* Strait of Ormuz, *Per.* Tangeh-ye Hormoz. Strait connecting the The Gulf and Arabian Sea
Hornád *523 Ger.* Hernad, *Hung.* Hernád. River of Hungary and Slovakia
Horn, Cape *183* cape of S Chile
Horog *see* Khorugh
Horoshiri-dake *333* mountain of Hokkaidō, N Japan
Horowupotana *538* NE Sri Lanka
Horqueta *460* C Paraguay
Horsburgh Atoll *390* atoll of N Maldives
Horsburgh Island *644 var.* Pulu Luar. Island of C Cocos Islands

Horsens *218* Jylland, C Denmark
Horseshoe Bay *643* bay of the North Atlantic Ocean, W Bermuda
Horsham *101* SE Australia
Hørsholm *218* Sjælland, E Denmark
Hortabágny-Berettyó *292* river of E Hungary
Horten *444* S Norway
Horug *see* Khorugh
Hosaʻina *245 var.* Hosseina, *It.* Hosanna. SW Ethiopia
Hose, Penunungan *386 var.* Hose Mountains. Mountain range of Borneo, E Malaysia
Hotan *186 var.* Khotan, *Chin.* Ho-t'ien. Xinjiang Uygur Zizhiqu. NW China
Hot Springs *603* Arkansas, SC USA
Hotte, Massif de la *288* highlands of SW Haiti
Houaïlou *650* C New Caledonia
Houmt Souk *577 var.* Djerba, Ḥawmat as Süd, Jerba, Île de Jerba, Tunisia
Houndé *157* SW Burkina
Houston *603* Texas, SC USA
Hovd *412* W Mongolia
Hoverla *590 Rus.* Gora Goverla. Mountain of W Ukraine
Hövsgöl Nuur *412* lake of N Mongolia
Howe, Cape *101* cape on the SE coast of Australia
Howakil Bay *240* bay of the Red Sea to the E of Eritrea
Howrah *see* Hāora
Hōzan *see* Fengshan
Hradec Králové *216 Ger.* Königgrätz. E Czech Republic
Hrazdan *98 Rus.* Razdan. C Armenia
Hrazdan *98 Rus.* Razdan, Zanga. River of C Armenia
Hrodna *122 Rus.* Grodno. W Belarus
Hron *523 Ger.* Gran, *Hung.* Garam. River of C Slovakia
Hsüeh Shan *559* mountain of N Taiwan
Hsi-an *see* Xi'an
Hsiang-t'an *see* Xiangtan
Hsi Chiang *see* Xi Jiang
Hsinchu *559* NW Taiwan
Hsinchuang *559 var.* Sinchwang, *Jap.* Shinshō. N Taiwan
Hsing-K'ai Hu *see* Khanka, Lake
Hsi-ning *see* Xining
Hsinking *see* Changchun
Hsintien *559 var.* Sintien, *Jap.* Shinten. N Taiwan
Hsin-yang *see* Xinyang
Hsinying *559 var.* Sinying, *Jap.* Shinei. W Taiwan
Hsu-chou *see* Xuzhou
Hsüehshan Shanmo *559* mountain range of N Taiwan
Huacho *463* W Peru
Huainan *187 var.* Hwainan. Anhui, E China
Hualien *559 var.* Hwalien, *Jap.* Karen. E Taiwan
Huallaga *463* river of N Peru
Huambo *88 Port.* Nova Lisboa. C Angola
Huancavelica *463* SW Peru
Huancayo *463* C Peru
Huang Hai *see* Yellow Sea
Huang He *187 Eng.* Yellow River. River of C China
Huánuco *463* C Peru
Huanuni *136* W Bolivia
Huaral *463* W Peru
Huascarán, Nevado *463* mountain of W Peru
Huaraz *463 var.* Huaráz. W Peru
Hubei *187 var.* Hupei. Province of C China
Hubli *298* SW India
Huddersfield *597* N England, UK
Hudson *603* river of NE USA
Hudson Bay *171* bay of the Atlantic Ocean, NE Canada
Hudson Strait *171* strait connecting the Atlantic Ocean and Hudson Bay
Huê *627* C Vietnam
Huehuetenango *280* W Guatemala
Huelva *534* SW Spain
Huesca *535* NE Spain

Hughenden *101* NE Australia
Huhehot *see* Hohhot
Hŭich'ŏn *351* C North Korea
Huizen *429* C Netherlands
Huksan-kundo *352 var.* Huksan-chedo. Island group of SW South Korea
Hull *see* Kingston-upon-Hull
Hull *171* SE Canada
Hullo *242* Vormsi, Estonia
Hulun Nur *187 Chin.* Hu-lun Ch'ih, *prev.* Dalai Nor. Lake of Nei Mongol Zizhiqu, NE China
Ḥulwan *see* Ḥelwân
Humacao *651* E Puerto Rico
Humaitá *460* S Paraguay
Humber *597* river of NE England, UK
Humené *523 Ger.* Homenau, *Hung.* Homonna. E Slovakia
Humera *see* Himora
Húnaflói *296* bay of the Norwegian Sea on the N coast of Iceland
Hunan *187* province of S China
Hundested *218* Sjælland, E Denmark
Hunga Haʻapai *572* island of the Nomuka Group, W Tonga
Hungary *292-295* officially Republic of Hungary, *Hung.* Magyarország, *prev.* Hungarian People's Republic. Country of C Europe divided into 19 admin. units (counties). ❖ Budapest
Hunga Tonga *572* island of the Nomuka Group, W Tonga
Hŭngnam *351* E North Korea
Hung-tse Hu *see* Hongze Hu
Huntington *603* West Virginia, E USA
Huntsville *603* Alabama, SE USA
Hunyani *see* Manyame
Huon Gulf *458* gulf of the Solomon Sea, to the E of Papua New Guinea
Huo-shao Tao *see* Lan Yü
Hupei *see* Hubei
Hurghada *232* E Egypt
Huron, Lake *171, 603* lake of Canada and USA
Hurunui *433* river of NE South Island, New Zealand
Húsavík *644* Sandoy, C Faeroe Islands
Húsavík *296* NE Iceland
Ḥuşayn, Dawḥat al *479 var.* Dauhat al Husein. Inlet of the Gulf of Bahrain on the NW coast of Qatar
Hūth *631* NW Yemen
Huwwāra *see* Hawwārah
Hvammstangi *296* N Iceland
Hvannadalshnúkur *296* mountain of S Iceland
Hvar *209 It.* Lesina. Island of S Croatia
Hvítá *296* river of W Iceland
Hvolsvöllur *296* SW Iceland
Hwach'ŏn-chŏsuji *see* P'aro-ho
Hwainan *see* Huainan
Hwang-Hae *see* Yellow Sea
Hyargas Nuur *412* lake of W Mongolia
Hyderābād *298 Hind.* Hyderabad. C India
Hyderābād *451 var.* Haidarabad. S Pakistan
Hyères *255* SE France
Hyères, Îles d' *255* island group of SE France
Hyesan *351* NE North Korea
Hyvinge *see* Hyvinkää
Hyvinkää *251 Swe.* Hyvinge. S Finland

I

Ialomiţa *480* river of SE Romania
Ialpug *408 Rus.* Yalpug. River of S Moldova
Iaşi *480 Ger.* Jassy. NE Romania
Ibadan *440* SW Nigeria
Ibagué *195* C Colombia

Ibar *512* river of SW Serbia, Serbia & Montenegro (Yugo.)
Ibarra *230 var.* San Miguel de Ibarra. N Ecuador
Ibb *631* W Yemen
Ibenga *200* river of N Congo
Ibérico, Sistema *535 var.* Cordillera Ibérica, *Eng.* Iberian Mountains. Mountains of NE Spain
Ibiza *see* Eivissa
Ibo *see* Sassandra
Iboundji *260* C Gabon
Ibrā' *448* N Oman
Ibrī *448* NW Oman
Irbīl *see* Arbīl
Ibusuki *332* Kyūshū, SW Japan
Içá *144* river of NW Brazil
Ica *463* SW Peru
Iceflavík *see* Keflavík
İçel *see* Mersin
Iceland *296-297* officially Republic of Iceland, *Icel.* Ísland. Country of the North Atlantic Ocean divided into 8 admin. units (regions).
❖ Reykjavík
Ichinomiya *332* Honshū, C Japan
Ichinoseki *332* Honshū, N Japan
Idah *440* S Nigeria
Idaho *602* state of NW USA
Idaho Falls *602* Idaho, NW USA
Idensalmi *see* Iisalmi
Idfu *232 var.* Idfū, Edfu. SE Egypt
Idi Amin, Lac *see* Edward, Lake
Idlib *555* NW Syria
Idrija *524 It.* Idria. W Slovenia
Idzhevan *see* Ijevan
Iecava *362* C Latvia
Ieper *127 Fr.* Ypres. W Belgium
Ifalik *406* atoll of C Micronesia
Ife *440* SW Nigeria
Iferouâne *439* N Niger
Iferten *see* Yverdon
Iganga *588* SE Uganda
Igarka *485* N Russian Federation
Igatimí *see* Ygatimí
Iglau *see* Jihlava
Iglesias *322* Sardegna, W Italy
Igló *see* Spišská Nová Ves
Ignalina *376* E Lithuania
Iguaçu, Salto do *145* *Sp.* Cataratas del Iguazú, *prev.* Victoria Falls. Waterfall of Argentina and Brazil
Iguetti, Sebkhet *398* salt lake of N Mauritania
Ihavandippolhu Atoll *390 var.* Ihavandiffulu Atoll. Atoll of N Maldives
Ihema, Lac *492* lake of Burundi and Rwanda
Ihosy *382* S Madagascar
Iida *332* Honshū, C Japan
Iijoki *251* river of C Finland
Irbil *see* Arbīl
Iisalmi *251 Swe.* Idensalmi. C Finland
Ijebu-Ode *440* SW Nigeria
Ijevan *98 Rus.* Idzhevan, *var.* Idževan. N Armenia
IJssel *429 var.* Yssel. River of C Netherlands
IJsselmeer *429 prev.* Zuider Zee. Lake of N Netherlands
Ikare *440* SW Nigeria
Ikaría *275* island of SE Greece
Ikast *218* Jylland, W Denmark
Ikeja *440* SW Nigeria
Ikerre *440 var.* Ikerre-Ekiti. SW Nigeria
Iki *332* island to the NW of Kyūshū, SW Japan
Ikom *440* S Nigeria
Ikopa *382* river of N Madagascar
Ila *440* SW Nigeria
Ilam *427* W Nepal
Ilan *559 Jap.* Giran. NE Taiwan
Ile *see* Ili
Ilebo *203 prev.* Port Francqui. W Dem. Rep. Congo
Ilesha *440* SW Nigeria
Ilha Solteira, Represa de *145* reservoir of S Brazil
Ili *340 Kaz.* Ile. River of China and Kazakhstan

Iligan *466* Mindanao, S Philippines
Ilirska Bistrica *524* SW Slovenia
Il'jaly *see* Yylanly
Illapel *183* C Chile
Illiassa *263* NW Gambia
Illinois *603* state of C USA
Ilobasco *237* C El Salvador
Ilobu *440* W Nigeria
Iloilo *466* Panay, C Philippines
Ilopango, Lago de *237* volcanic lake of C El Salvador
Ilorin *440* W Nigeria
Iluh *see* Batman
Ilulissat *646 Dan.* Jakobshavn. W Greenland
Il'yaly *see* Yylanly
Imatong Mountains *540* mountains of S Sudan
Imatra *251* SE Finland
Imeni 26 Bakinskikh Komissarov *see* Bakı Komissarı
Imilili *414* W Western Sahara
İmişli *110 Rus.* Imishli, Imišli. C Azerbaijan
Imja-do *352* island of SW South Korea
Imola *323* N Italy
Imperatriz *145* NE Brazil
Imperia *322* N Italy
Impfondo *200* NE Congo
Imphāl *299* E India
Ina *332* Honshū, C Japan
Inakona *526* S Guadalcanal, Solomon Is
In Aménas *83 var.* I-n-Amenas, In Amnas. E Algeria
Inárajan *647* SE Guam
Inarijärvi *251 Swe.* Enareträsk, *Lapp.* Aanaarjávri. Lake of N Finland
Inarijoki *251 Nor.* Anarjokka. River of Finland and Norway
Inawashiro-ko *332* lake of Honshū, N Japan
Inbhear Mór *see* Arklow
Inch'ŏn *352 prev.* Chemulpo, *Jap.* Jinsen. NW South Korea
Inchope *419* C Mozambique
Incles *86* river of NE Andorra
Independence *130* SE Belize
Inderagiri *see* Indragiri
India *298-303* officially Republic of India, *Hind.* Bharat. Country divided into 32 admin. units (25 states and 7 union territories). ❖ New Delhi
Indiana *603* state of C USA
Indianapolis *603* Indiana, C USA
Indian Desert *see* Thar Desert
Indian Ocean *90, 644* ocean bounded to the W by Africa, to the E by Australia and to the S by Antarctica
Indigirka *485* River of NE Russian Federation
Indonesia *304-307* officially Republic of Indonesia, *Ind.* Republik Indonesia, *prev.* United States of Indonesia, Dutch East Indies, Netherlands East Indies. Country of SE Asia divided into 25 admin. units (24 provinces and 1 autonous district). ❖ Jakarta
Indonesian Borneo *see* Kalimantan
Indore *299* NW India
Indragiri *304 var.* Inderagiri. River of Sumatra, W Indonesia
Indre *255* river of C France
Indus *298, 451* river of S Asia
Indus, Mouths of the *451* river delta of S Pakistan
Infante Dom Henrique *505* SE Príncipe, Sao Tomé & Principe
Ingolstadt *267* S Germany
Inguri *see* Enguri
Ingushetiya, Respublika *484* autonomous republic of SW Russian Federation
Ingushskaya, Respublika *484* autonomous republic of SW Russian Federation
Ingwavuma *see* Nggwavuma
Inhambane *419* S Mozambique
I-ning *see* Yining
Inírida *195* river of E Colombia
Inis *see* Ennis
Inland Sea *332 var.* Seto Naikai. Sea of the Pacific Ocean between Honshū and Shikoku,W Japan

Inn *106, 267* river of C Europe
Inner Channel *130 var.* Main Channel. Inlet of W Caribbean Sea
Inner Hebrides *597* island group of NW Scotland, UK
Inner Islands *516 var.* Central Group. Island group of NE Seychelles
Inner Mongolian Autonomous Region *see* Nei Mongol Zizhiqu
Innsbruck *106* W Austria
Inrin *see* Yüanlin
In Salah *83 var.* I-n-Salah. C Algeria
Insein *159* S Burma
Intelewa *542* S Suriname
Interlaken *550* SW Switzerland
Inthanon, Doi *567* mountain of NW Thailand
Intipucá *237* SE El Salvador
Inuvik *170* NW Canada
Invercargill *433* SW South Island, New Zealand
Inverness *597* N Scotland, UK
Inyanga *see* Nyanga
Inyangani *636* mountain of E Zimbabwe
Inyazura *see* Nyazura
Ioánnina *275 var.* Janina, Yannina. W Greece
Iolotan' *see* Yëloten
Ionian Islands *see* Iónioi Nísoi
Ionian Sea *81, 275, 323* *Gk* Iónio Pélagos, *It.* Mar Ionio. Area of the Mediterranean Sea, between Italy and SE Europe
Ionio, Mar *see* Ionian Sea
Iónioi Nísoi *275 Eng.* Ionian Islands. Island group of W Greece
Iori *264* river of Azerbaijan and Georgia
Íos *275* island of SE Greece
Iowa *603* state of C USA
Ipeľ *see* Ipoly
Ipiales *195* SW Colombia
Ipoh *386* W Peninsular Malaysia
Ipoly *292, 523 Slvk.* Ipeľ, *Ger.* Eipel. River of Hungary and Slovakia
Ippy *179* C Central African Republic
Ipswich *101* E Australia
Ipswich *597* E England, UK
Iqaluit *171 prev.* Frobisher Bay. Baffin Island, NE Canada
Iquique *183* N Chile
Iquitos *463* N Peru
Irákleio *275 Eng.* Candia, *prev.* Iráklion. Crete, S Greece
Iran *308-311* officially Islamic Republic of Iran, *prev.* Persia. Country of SW Asia divided into 24 admin. units (provinces). ❖ Tehrān
Iran, Pegunungan *387 var.* Iran Mountains. Mountain range of Borneo, Indonesia and Malaysia
Iran, Plateau of *309* plateau of C Iran
Irapuato *403* C Mexico
Iraq *312-315* officially Republic of Iraq, *Ar.* 'Irāq. Country of SW Asia divided into 18 admin. units (governorates). ❖ Baghdad
Irbe Strait *242 Est.* Kura Kurk, *prev.* Irbe Väin, *Latv.* Irbes Šaurums. Strait connecting the Baltic Sea and Gulf of Riga
Irbid *338* N Jordan
Ireland Island North *643* island of W Bermuda
Ireland Island South *643* island of W Bermuda
Ireland, Northern *see* Northern Ireland
Ireland, Republic of *316-317* officially Republic of Ireland, Éire. Country of W Europe divided into 26 admin. units (counties). ❖ Dublin
Ireng *286 var.* Maú. River of Brazil and Guyana
Irgalem *see* Yirga 'Alem
Iri *352 Jap.* Riri. W South Korea
Irian *see* New Guinea
Irian Jaya *304-305 Eng.* West Irian, *prev.* Dutch New Guinea. Province of W Indonesia
Iringa *564* C Tanzania

Iriomote-jima *332* island of Sakishima-shotō, SW Japan
Iriri *145* river of N Brazil
Irish Sea *316, 597, 647* *Ir.* Muir Eireann. Sea of the Atlantic Ocean between Ireland and UK
Irkeshtam *359 var.* Irkeštam. SW Kyrgyzstan
Irkutsk *485* C Russian Federation
Irmak *581* river of N Turkey
Iroise *254* area of the Atlantic Ocean to the NW of France
'Irqah *631* SW Yemen
Irrawaddy *159 var.* Ayeyarwady. River of C Burma
Irrawaddy, Mouths of the *159* delta area of SW Burma
Irrsee *106* lake of N Austria
Irtysh *340, 484 Kaz.* Ertis. River of Kazakhstan and Russian Federation
Irun *535* N Spain
Iruña *see* Pamplona
Isabela *651* NW Puerto Rico
Isabela, Isla *230* island of SW Galapagos Is, Ecuador
Isachsen *170* Ellef Ringnes Island, N Canada
Ísafdhardjúp *296* inlet of the Atlantic Ocean, NW Iceland
Isangel *619* Tanna, Vanuatu
Isa Town *see* Madīnat 'Īsá
Isalo, Tangorombohitr' *382* mountains of SW Madagascar
Ischia, Isola d' *323* island of S Italy
Ise *332* Honshū, C Japan
Isefjord *218* fjord of Sjælland, E Denmark
Iséré *255* river of SE France
Iseyin *440* W Nigeria
Isfara *562* N Tajikistan
Ísfjördhur *296* NW Iceland
Isha Baydhabo *see* Baydhabo
Isherton *286* S Guyana
Ishigaki-jima *332* island of Sakishima-shotō, SW Japan
Ishikari *333* river of Hokkaidō, N Japan
Ishim *340 Kaz.* Esil. River of Kazakhstan and Russian Federation
Ishinomaki *332* Honshū, N Japan
Ishkoshim *562 Rus.* Ishkashim. S Tajikistan
Ishurdi *117* W Bangladesh
Isidoro Noblia *611* NE Uruguay
Isiolo *344* C Kenya
Isiro *203* NE Dem. Rep. Congo
Iskeçe *see* Xánthi
İskele *215 var.* Trikomo. E Cyprus
İskenderun *581 Eng.* Alexandretta. S Turkey
İskenderun Körfezi *581 Eng.* Gulf of Alexandretta. Gulf of the Mediterranean Sea
Iskŭr *152* river of NW Bulgaria
Iskŭr, Yazovir *152* reservoir of W Bulgaria
Islāmābād *451* ❖ of Pakistan, NE Pakistan
Island Harbour *642* bay of the Caribbean Sea on the N coast of Anguilla
Islay *597* island of Inner Hebrides, W Scotland, UK
Isle *255* river of SW France
Ismâ'ilîya *232 var.* Al Ismā'īlīyah, *Eng.* Ismaila. N Egypt
Isna *232 var.* Isnā, Esna. SE Egypt
Isoka *635* NE Zambia
Isonzo *see* Soča.
Ispahan *see* Eşfahān
İsparta *581* SW Turkey
Israel *318-321* officially State of Israel, *Heb.* Yisra'el. Country of SW Asia divided into 6 admin. units (districts). ❖ Jerusalem
Issano *286* C Guyana
Issia *328* SW Ivory Coast
Issyk-Kul' *see* Balykchy
Issyk-Kul', Ozero *359 var.* Issiq Köl. Lake of NE Kyrgyzstan
İstanbul *581 prev.* Constantinople, *Bul.* Tsarigrad. NW Turkey

Kayts *538* island of N Sri Lanka
Kazakh *see* Qazax
Kazakhskiy Melkosopochnik *340 Eng.* Kazakh Uplands. Uplands of C Kazakhstan
Kazakhstan *340-343* officially Republic of Kazakhstan, *Kaz.* Qazaqstan, *prev.* Kazakh SSR. *Rus.* Kazakhskay SSR. Country of C Asia divided into 19 admin. units (provinces). ❖ Astana
Kazakh Uplands *see* Kazakhskiy Melkosopochnik
Kazan' *484* W Russian Federation
Kazandzhik *see* Gazandzhyk
Kazanlŭk *152 var.* Kazanlăk, Kazanlik. C Bulgaria
Kazan-rettō *332 Eng.* Volcano Islands. Island group to the SE of Honshū, SE Japan
Kazarman *359* C Kyrgyzstan
Kazbek *264* mountain of N Georgia
Kazi Magomed *see* Qazimämmäd
Kazincbarcika *292* NE Hungary
Kazvin *see* Qazvin
Kéa *275* island of SE Greece
Kéamu *see* Aneityum
Kebili *577 var.* Qibilī. C Tunisia
Kebnekaise *547* mountain of N Sweden
Kecskemét *292* C Hungary
Kėdainiai *376* C Lithuania
Kediet ej Jill *398 var.* Kediet Ijill, Kédia d'Idjil. Mountain of NW Mauritania
Kediri *304* Java, C Indonesia
Kédougou *510* SE Senegal
Keeling Islands *see* Cocos Islands
Keelung *see* Chilung
Keetmanshoop *423* S Namibia
Kefallinía *275 Eng.* Cephalonia. Island of W Greece
Kefar Sava *319* C Israel
Keflavík *296 var.* Iceflavik. W Iceland
Kegalla *538 var.* Kegalle. C Sri Lanka
Kegel *see* Keila
Kei Islands *see* Kai, Kepulauan
Keijō *see* Seoul
Keila *242 Ger.* Kegel. NW Estonia
Keila *242 var.* Keila Jõgi. River of NW Estonia
Keishū *see* Kyŏngju
Kéita *180 var.* Doka. River of S Chad
Keïta *439* SW Niger
Keitele *251* lake of C Finland
Këk-Art *359 prev.* Alaykel'. SW Kyrgyzstan
Kékes *292* mountain of N Hungary
Kelang *386 var.* Klang, *prev.* Port Swettenham. W Peninsular Malaysia
Kelantan *386* river of N Peninsular Malaysia
Kelbia, Sebkhet *577 var.* Sabkhat Kalbīyah. Salt flat of NE Tunisia
Këlcyrë *81 var.* Këlcyra. S Albania
Kelifskiy Uzboy *585* region of SE Turkmenistan
Kéllé *200* W Congo
Kelmė *376* NW Lithuania
Kélo *180* SW Chad
Kelowna *170* SW Canada
Keluang *386 var.* Kluang. SE Peninsular Malaysia
Kembolcha *245 var.* Kombolcha. N Ethiopia
Kemerovo *485 prev.* Shcheglovsk. C Russian Federation
Kemi *251* NW Finland
Kemijärvi *251* N Finland
Kemijoki *251* river of NW Finland
Kemin *359 prev.* Bystrovka. N Kyrgyzstan
Kemiö *see* Kimito
Kemmuna *395* island of NW Malta
Kemmunett *395* island of NW Malta
Kempen *127 Fr.* Campine, *Ger.* Kempenland. Heathland of NE Belgium
Kempten *267* S Germany
Kenema *518* SE Sierra Leone
Këneurgench *585 prev.* Kunya-Urgench, Kunja-Urgenč, *Turkm.* Köneür gench. N Turkmenistan

Kénitra *414 prev.* Port Lyautey. NW Morocco
Kenmare *316* SW Ireland
Kentau *340* S Kazakhstan
Kentucky *603* state of C USA
Kenya *344-347* officially Republic of Kenya. Country of E Africa divided into 7 admin. units (provinces). ❖ Nairobi
Kenya, Mount *see* Kirinyaga
Keppel Harbour *521* harbour, S Singapore
Keppel Island *see* Niuatoputapu
Kerava *251 Swe.* Kervo. S Finland
Kerch *590 Rus.* Kerch'. NE Ukraine
Kerema *458* S Papua New Guinea
Keren *240 var.* Cheren. C Eritrea
Kerewan *263* W Gambia
Kericho *344* W Kenya
Kerio *344* river of W Kenya
Kerkenah, Îles *577 var.* Kerkenna Islands, *Ar.* Juzur Qarqannah. Island group of E Tunisia
Kerki *585* SE Turkmenistan
Kerkrade *429* S Netherlands
Kérkyra *275 prev.* Kérkira, *Eng.* Corfu. Island of W Greece
Kérkyra *275 Eng.* Corfu, *prev.* Kérkira. W Greece
Kermān *309 var.* Kirman. SE Iran
Kermānshāh *see* Bākhtarān
Kerora *240* N Eritrea
Kérouané *283* SE Guinea
Kertel *see* Kärdla
Kerulen *412 var.* Herlen Gol. River of China and Mongolia
Kervo *see* Kerava
Keryneia *see* Girne
Kesen'-numa *333* Honshū, N Japan
Késmárk *see* Kežmarok
Kesra *577 var.* Kisrah. NW Tunisia
Keta *272* SE Ghana
Ketchikan *602* Alaska, USA
Kete-Krachi *272 var.* Kete Krakye. E Ghana
Kétou *132* SE Benin
Kettharin Island *see* Kanmaw Island
Keur Massène *398* SW Mauritania
Kévé *571* SW Togo
Kew *653* North Caicos, NW Turks and Caicos Islands
Kežmarok *523 Ger.* Käsmark, *Hung.* Késmárk. NE Slovakia
Khabarovsk *485* SE Russian Federation
Khabura *see* Al Khaburah
Khachmas *see* Xaçmaz
Khairpur *451* S Pakistan
Khakasiya, Respublika *485* autonomous republic of C Russian Federation
Khakassk *see* Abakan
Khakhea *142 var.* Kakia. S Botswana
Khalándrion *see* Chalándri
Khalkidhikí *see* Chalkidikí
Khalkís *see* Chalkída
Khalūf *448 var.* Al Khaluf. S Oman
Khambhat, Gulf of *298 Eng.* Gulf of Cambay. Gulf of Arabian Sea to the W of India
Khamir *631 var.* Khamr. W Yemen
Khamīs Mushayṭ *506* S Saudi Arabia
Khānābād *77* NE Afghanistan
Khānaqin *312* E Iraq
Khānewāl *451* NE Pakistan
Khanh Hung *see* Soc Trăng
Khanka, Lake *187, 485 var.* Lake Hanka, *Rus.* Ozero Khanka, *Chin.* Xingkai Hu, Hsing-K'ai Hu. Lake of China and Russian Federation
Khanka, Ozero *see* Khanka, Lake
Khankendy *see* Xankändi
Khānpur *451* SE Pakistan
Khanty-Mansiysk *484 prev.* Ostyako-Voguls'k. C Russian Federation
Khān Yūnis *319, 320 Ar.* Khan Yunus. Gaza Strip
Kharāb, Ghoubbet el *222* bay at the head of Golfe de Tadjoura, E of Djibouti
Kharanah *see* Al Kir'ānah
Khāriān *451* NE Pakistan

Kharīṭ, Wâdi el *232 var.* Wâdī al Kharīṭ. Dry watercourse SE Egypt
Kharkiv *590 Rus.* Khar'kov. NE Ukraine
Kharmanli *152 var.* Harmanli. S Bulgaria
Khartoum *540 var.* Al Khurṭūm. ❖ of Sudan, C Sudan
Khartoum North *540 var.* Al Khurṭūm al Baḥrī. E Sudan
Khasab *see* Al Khaṣāb
Khāsh Rūd *77* river of W Afghanistan
Khashuri *264* C Georgia
Khaskovo *152 var.* Haskovo. S Bulgaria
Khatt *see* Al Khaṭṭ
Khawr al Bazm *594 var.* Khor al Bizm. Inlet of The Gulf, on the coast of United Arab Emirates
Khawr al 'Udayd *479 var.* Khor al Udeid. Inlet of the The Gulf on the coast of SE Qatar
Duwayhin, Khawr *594* inlet of the The Gulf, on the coast of United Arab Emirates
Khawr Fakkān *594 var.* Khor Fakkan. NE United Arab Emirates
Khaydarkan *359 var.* Khaydarken, Hajdarken. SW Kyrgyzstan
Khazar, Baḥr-e *see* Caspian Sea
Khazar, Daryā-ye *see* Caspian Sea
Khenchela *83 var.* Khenchla. NE Algeria
Khénifra *414* C Morocco
Kherson *590 var.* Cherson. S Ukraine
Khezqazghan *see* Zhezkazgan
Khíos *see* Chíos
Khiwa *614 Uzb.* Khiwa. W Uzbekistan
Khmel 'nyts'kyy *590 Rus.* Khmel'nitskiy, *prev.* Proskurov. W Ukraine
Khodzhent *see* Khujand
Khodzheyli *see* Khujayli
Khoi *see* Khvoy
Khojend *see* Khujand
Kholm *77* N Afghanistan
Khomeynīshahr *309 prev.* Homäyünshahr. W Iran
Khoms *see* Al Khums
Khong Sedone *see* Muang Khôngxédôn
Khon Kaen *567 var.* Muang Khon Kaen. N Thailand
Khor al Udeid *see* Khawr al 'Udayd
Khôr 'Angar *222* NE Djibouti
Khorixas *423* NW Namibia
Khorramābād *309* W Iran
Khorramshahr *309 prev.* Khūnīnshahr. W Iran
Khorugh *562 var.* Horug, *Rus.* Khorog. S Tajikistan
Khotan *see* Hotan
Khouribga *414* C Morocco
Khowst *77* E Afghanistan
Khoyniki *122* SE Belarus
Khrysokhou Bay *215 var.* Chrysochou Bay. Bay of the Mediterranean Sea, on the NW coast of Cyprus
Khujand *562 prev.* Leninabad, Khodzhent, Khojend. NW Tajikistan
Khujayli *614 Rus.* Khodzheyli. W Uzbekistan
Khulna *117* SW Bangladesh
Khūnīnshahr *see* Khorramshahr
Khurīyā Murīyā, Jazā'ir *see* Ḥalānīyāt, Juzur al
Khurramshahr *see* Khorramshahr
Khushāb *451* NE Pakistan
Khvoy *309 var.* Khoi. NW Iran
Khyber Pass *77, 451* mountain pass connecting Afghanistan with Pakistan
Kia *526* SW Santa Isabel, Solomon Is
Kiamusze *see* Jiamusi
Kiangsi *see* Jiangxi
Kiangsu *see* Jiangsu
Kiayi *see* Chiai
Kibondo *564* NW Tanzania
Kibre Mengist *245 var.* Adola. S Ethiopia
Kibungo *492 var.* Kibungu. SE Rwanda

Kibuye *492* W Rwanda
Kičevo *381* W FYR Macedonia
Kidaho *492* NW Rwanda
Kiel *267* N Germany
Kiel Bay *218, 267 Ger.* Kieler Bucht. Bay of the Baltic Sea
Kielce *471* S Poland
Kieler Bucht *see* Kiel Bay
Kieta *458* Bougainville I, Papua New Guinea
Kiev *590 Ukr.* Kyyiv, *Rus.* Kiyev. ❖ of Ukraine, N Ukraine
Kiffa *398* S Mauritania
Kigali *492* ❖ of Rwanda, C Rwanda
Kigembe *492* S Rwanda
Kigoma *564* W Tanzania
Kigwena *162* SW Burundi
Kikila, Lac *653* lake of Île Uvea, S Wallis & Futuna
Kihnu *242* island of SW Estonia
Kikládhes *see* Kyklades
Kikori *458* river of C Papua New Guinea
Kikwit *203* W Dem. Rep. Congo
Kilchu *351* NE North Korea
Kili *396* island of S Marshall Islands
Kilien Mountains *see* Qilian Shan
Kilifi *344* SE Kenya
Kilimanjaro *564* mountain of NE Tanzania
Kilingi-Nõmme *242 Ger.* Kurkund. S Estonia
Kilinochchi *538* N Sri Lanka
Kilis *581* S Turkey
Kilkee *316* W Ireland
Kilkenny *316 Ir.* Cill Choinnigh. SE Ireland
Kilkís *275* N Greece
Kilkoch *316* E Ireland
Killarney *316 Ir.* Cill Airne. SW Ireland
Kilmarnock *597* W Scotland, UK
Kilosa *564* C Tanzania
Kilwa Masoko *564* SE Tanzania
Kimbe *458* New Britain, Papua New Guinea
Kimberley *531* Northern Cape, C South Africa
Kimberley Plateau *100* plateau of NW Australia
Kimch'aek *351 prev.* Sŏngjin. E North Korea
Kimch'ŏn *352* C South Korea
Kimhae *352* SE South Korea
Kimito *251 Swe.* Kemiö. Island of SW Finland
Kimje *352* SW South Korea
Kinabatangan *386* river of NE Borneo, Malaysia
Kinabalu, Gunung *386* mountain of N Borneo, Malaysia
Kindamba *200* S Congo
Kindia *283* SW Guinea
Kindu *203* C Dem. Rep. Congo
King George Bay *645* bay of the South Atlantic Ocean, W Falkland Islands
King George Land *90* island of South Shetland Islands, Antarctica
King Island *101* island of SE Australia
King Island *see* Kadan I.
Kingissepp *see* Kuressaare
King Sejong *90* South Korea research station of Antarctic Peninsula, Antarctica
King's Lynn *597* E England, UK
King's Mills *647* SW Guernsey
Kingston *171* SE Canada
Kingston *331* ❖ of Jamaica, E Jamaica
Kingston *650* ❖ of Norfolk Island, S Norfolk Island
Kingston upon Hull *597 var.* Hull. NE England, UK
Kingstown *498* ❖ of St Vincent & the Grenadines, SW St Vincent
King William Island *170 var.* King William. Island of N Canada
Kinihira *492* N Rwanda
Kinkala *200* S Congo
Kinneret-Negev Conduit *319* canal of S Israel
Kinsale *649* SW Montserrat
Kinshasa *203 prev.* Léopoldville. ❖ of Dem. Rep. Congo, W Dem. Rep. Congo
Kintampo *272* C Ghana

Kuril'skiye Ostrova *333, 485*
  Eng. Kurile Islands. Partially disputed
  island group of E Russian Federation
Kurkund *see* Kilingi-Nõmme
Kurmuk *540* SE Sudan
Kurnool *298* S India
Kurram *451* river of Afghanistan and
  Pakistan
Kuršėnai *376* *var.* Kuršenaj, Kuršenai.
  NW Lithuania
Kursk *484* W Russian Federation
Kuru *134* river of E Bhutan
Kurubonla *518* NE Sierra Leone
Kurume *332* Kyūshū, SW Japan
Kurunegala *538* C Sri Lanka
Kurupukari *286* C Guyana
Kurzeme *362* *Eng.* Courland. Region of
  W Latvia
Kusaie *see* Kosrae
Kushiro *333* Hokkaidō, N Japan
Kushiro *333* river of Hokkaidō,
  N Japan
Kushka *see* Gushgy
Kushmurun *340* N Kazakhstan
Kusho *see* Kwangju
Kushtia *117* W Bangladesh
Kusŏng *351* W North Korea
Kussharo-ko *332* lake of Hokkaidō,
  N Japan
Kustanay *see* Kostanay
Küstendje *see* Constanţa
Kusu Island *see* Tembakul, Pulau
Kütahya *581* *prev.* Kutaiah. W Turkey
Kutai *see* Mahakam
K'ut'aisi *264* W Georgia
Küt al 'Amārah *see* Al Kūt
Kut al Imara *see* Al Kūt
Kutch, Gulf of *see* Kachch, Gulf of
Kuthuleni *544* NW Swaziland
Kutina *209* NE Croatia
Kuūm *351* SE North Korea
Kuusamo *251* E Finland
Kuusankoski *251* S Finland
Kuwait *356-357* officially State of
  Kuwait. Country of SW Asia divided
  into 5 admin units (governorates).
  ❖ Kuwait City
Kuwait Bay *356* bay of the The Gulf,
  on the coast of E Kuwait
Kuwait City *356* *var.* Al Kuwayt.
  ❖ of Kuwait, E Kuwait
Kuybyshev *see* Samara
Kuybyshevskoye Vodokhranilishche
  484 *Eng.* Kuybyshev Reservoir.
  W Russian Federation
Kūysanjaq *see* Koi Sanjaq
Kuyuwini *286* river of S Guyana
Kvændrup *218* Fyn, S Denmark
Kvaløya *444* island of NE Norway
Kvareli *see* Qvareli
Kvarner *209* *It.* Quarnero. Gulf of the
  Adriatic Sea, to the W of Croatia
K'vemo K'edi *264* SE Georgia
Kwa *203* river of W Dem. Rep. Congo
Kwahu Plateau *272* plateau of Ghana
Kwajalein *396* island of
  C Marshall Islands
Kwakoegron *542* N Suriname
Kwakwani *286* E Guyana
Kwale *344* S Kenya
Kwando *see* Cuando
Kwangchow *see* Guangzhou
Kwangju *352* *var.* Kwangchu, Guangju
  Jap. Kōshū. SW South Korea
Kwango *88, 203* *Port.* Cuango. River of
  Angola and Dem. Rep. Congo
Kwangsi Chuang Autonomous Region
  see Guangxi
Kwangtung *see* Guangdong
Kwangyuan *see* Guangyuan
Kwania, Lake *588* lake of C Uganda
Kwanza *see* Cuanza
Kwazulu Natal *531* province of
  E South Africa
Kweichow *see* Guizhou
Kweisui *see* Hohhot
Kweiyang *see* Guiyang
Kwekwe *636* *prev.* Que Que.
  C Zimbabwe
Kwenge *203* river of Angola and Dem.
  Rep. Congo
Kwitaro *286* River of C Guyana
Kwito *see* Cuito
Kyabé *180* S Chad

Kyaikkami *159* *var.* Amherst.
  SE Burma
Kyaiklat *159* S Burma
Kyaikto *159* S Burma
Kyaukpyu *159* W Burma
Kyaukse *159* C Burma
Kyklades *275* *var.* Kikládhes,
  Eng. Cyclades. Island group of
  SE Greece
Kyle, Lake *636* reservoir of
  SE Zimbabwe
Kymijoki *251* river of S Finland
Kyoga, Lake *588* *var.* Lake Kioga.
  Lake of C Uganda
Kyŏnggi-man *351, 352* bay of the
  Yellow Sea off NW South Korea
Kyŏnghŭng *351* NE North Korea
Kyŏngju *351* *Jap.* Keishū.
  SE South Korea
Kyŏngsŏng *see* Seoul
Kyŏsai-tō *see* Kŏje-do
Kyōto *332* Honshū, C Japan
Kyperounda *215* *var.* Kyperounta.
  C Cyprus
Kypros *see* Cyprus
Kyrenia *see* Girne
Kyrgyzstan *358-359* officially Kyrgyz
  Republic, *var.* Kirghizia,
  prev. Republic of Kyrgyzstan,
  prev. Kirghiz SSR, Kirgizskaya SSR.
  Country of C Asia divided into
  6 admin. units (oblasts). ❖ Bishkek
Kýthira *275* island of S Greece
Kýthnos *275* island of SE Greece
Kythrea *see* Degirmenlik
Kyungëy Ala-Too, Khrebet
  see Kungei Ala-Tau
Kyurdamir *see* Kürdämir
Kyūshū *332* island of SW Japan
Kyustendil *152* W Bulgaria
Kyyiv *see* Kiev
Kyyivs'le Vodoskhovyshche *590*
  Rus. Kiyevskoy Vodokhranilische.
  Reservoir of NW Ukraine
Kyzyl *485* C Russian Federation
Kyzyl-Kiya *359* *var.* Kyzyl-Kija,
  Kir. Kyzyl-Kyya. SW Kyrgyzstan
Kyzyl Kum *614* *var.* Kizil Kum,
  Uzb. Qizilqum. Desert region of
  Kazakhstan and Uzbekistan
Kyzylorda *340* *Kaz.* Qyzylorda.
  SW Kazakhstan
Kyzylrabot *see* Qizilrabot
Kyzyl-Suu *359* *prev.* Pokrovka.
  NE Kyrgyzstan
Kyzyl-Suu *359* *var.* Kyzylsu. River of
  Kyrgyzstan and Tajikistan

L

Laagen *see* Lågen
Laaland *see* Lolland
Laamu Atoll *see*
  Hadhdhunmathi Atoll
Laas Dawaco *528* *var.* Las Dawa'o,
  Laz Daua. N Somalia
La Asunción *623* Isla de Margarita,
  Venezuela
Laâyoune *414*
  NW Western Sahara
La Baie *652* channel of the Atlantic
  Ocean between Saint Pierre and
  Miquelon, S Saint Pierre and
  Miquelon
Labasa *248* *prev.* Lambasa. Vanua
  Levu, N Fiji
Labe *see* Elbe
Labé *283* NW Guinea
Laborec *523* *Hung.* Laborca. River of
  E Slovakia
Laborie *497* S St Lucia
Labrador *171* cultural region of
  E Canada
Labrador City *171* E Canada
Labrador Sea *171* area of the Atlantic
  Ocean, off E Canada
La Brea *574* SW Trinidad, Trinidad
  & Tobago
Labuan *see* Victoria
Labuan, Pulau *387* island of
  N Borneo, Malaysia

Labuk *387* river of NE Borneo,
  Malaysia
Labuk, Telukan *387* *var.* Labuk Bay.
  Bay of the Sulu Sea, on the NE coast
  of Borneo, Malaysia
Labutta *159* SW Burma
La Caye *497* E St Lucia
Laccadive Islands *see* Lakshadweep
La Ceiba *290* N Honduras
Lacepede Bay *101* bay on the coast of
  SE Australia
Lac Giao *see* Buôn Ma Thuột
La Chaux-de-Fonds *550*
  W Switzerland
La Chaux, River *400* river of
  SE Mauritius
Lachlan *101* river of SE Australia
La Chorrera *456* C Panama
La Concepción *456* W Panama
La Condamine *410* W Monaco
la Cortinada *86* NW Andorra
La Coruña *see* A Coruña
Lacre Punt *650* headland of Bonaire,
  S Netherlands Antilles
La Croix Maingot *497* NW St Lucia
La Cruz *206* NW Costa Rica
La Désirade *646* island of
  E Guadeloupe
La Digue *516* island of the Inner
  Islands, NE Seychelles
Ladoewani *542* C Suriname
Ladoga, Lake *see* Ladozhskoye Ozero
Ladozhskoye Ozero *251, 484* *Eng.*
  Lake Ladoga. Lake of
  NW Russian Federation
Lae *396* island of W Marshall Islands
Lae *458* E Papua New Guinea
Læsø *218* island of N Denmark
La Esperanza *290* *var.* La Esperanza
  Intibucá. SW Honduras
Lafanga *587* islet of
  Nukufetau, Tuvalu
La Fe *see* Santa Fé
La Fé *210* W Cuba
Lafia *440* C Nigeria
Lafiagi *440* W Nigeria
La Foa *650* S New Caledonia
Lagarfljót *see* Lögurinn
Lagdo, Lac de *168* lake of
  N Cameroon
Lågen *444* *var.* Laagen. River of
  S Norway
Laghouat *83* N Algeria
Lagone *see* Logone
Lagos *440* SW Nigeria
Lagos *474* S Portugal
La Goulette *577* *var.* Ḥalq al Wādī.
  N Tunisia
Lagunillas *136* SE Bolivia
La Habana *see* Havana
Lahad Datu, Telukan *387* *var.* Teluk
  Darvel, Darvel Bay. Bay of the
  Celebes Sea, on the NE coast of
  Borneo, Malaysia
La Haye *see* 's-Gravenhage
Laḥij *631* *var.* Laḥj, *Eng.* Lahej.
  SW Yemen
Lahore *451* NE Pakistan
Lahti *251* *Swe.* Lahtis. S Finland
Laï *180* *var.* Behagle. S Chad
Laibach *see* Ljubljana
Lai Châu *627* NW Vietnam
Lainioälven *547* river of N Sweden
Laisholm *see* Jõgeva
Laitokitok *see* Oloitokitok
Lakamti *see* Nek'emtē
Lakeba *248* *prev.* Lakemba. Island of
  the Lau Group, E Fiji
Lakeba Passage *248* channel of the
  Pacific Ocean E Fiji
Lake District *see* Cumbrian
  Mountains
Lakepa *650* NE Niue
Lakhnau *see* Lucknow
Lakota *328* S Ivory Coast
Lak Sao *see* Ban Lakxao
Lakshadweep *298* *Eng.* Laccadive
  Islands. Island group to the
  SW of India
Lakshām *117* E Bangladesh
Lakshmipur *117* S Bangladesh
La Laguna *535* Islas Canarias,
  SW Spain

Lāla Mūsa *451* NE Pakistan
Lalëzit, Gjiri i *81* *var.* Gji i Lalzës.
  Gulf of the Adriatic Sea,
  SW Albania
La Libertad *280* N Guatemala
La Libertad *237* SW El Salvador
La Libertad *230* W Ecuador
Lalitpur *427* C Nepal
Lālmanir Hāt *117* N Bangladesh
La Louvière *127* SW Belgium
la Maçana *86* W Andorra
Lama-Kara *see* Kara
La Manche *see* English Channel
La Marsa *577* *var.* Al Marsá.
  N Tunisia
La Matepec *see* Santa Ana, Volcan de
Lambaré *460* S Paraguay
Lambaréné *260* W Gabon
Lambi *526* W Guadalcanal,
  Solomon Islands
Lami *248* Viti Levu, W Fiji
Lamía *275* C Greece
Lamin *263* W Gambia
Lamiti *248* Gau, C Fiji
Lamotrek *406* atoll of C Micronesia
Lampang *567* *var.* Muang Lampang.
  NW Thailand
Lamu *344* SE Kenya
Lanao, Lake *466* *var.* Lake Sultan
  Alonto. Lake of Mindanao,
  S Philippines
Lanbi Island *159* *prev.* Sullivan I.
  Island of S Burma
Lancang Jiang *see* Mekong
Lancaster *597* NW England, UK
Lancaster Sound *171* stretch of water
  between Baffin Island and Devon
  Island, N Canada
Lance aux Epines *278* S Grenada
  island, Grenada
Lanchow *see* Lanzhou
Landes *254* cultural region of
  SW France
Landing Bay *649* bay of the
  Caribbean Sea on the SE coast of
  Montserrat
Landsberg *see* Gorzów Wielkopolski
Landshut *267* SE Germany
Langabou *571* C Togo
Langar *614* *Rus.* Lyangar.
  SE Uzbekistan
Langatabbetje *542* E Suriname
Langeland *218* island of S Denmark
Langenthal *550* NW Switzerland
Langjökull *296* glacier of C Iceland
Langkawi, Pulau *386* island to the
  NW of Peninsular Malaysia
Langlade *652* physical region of
  Miquelon, S Saint Pierre and
  Miquelon
Langlade, Isthme de *652* isthmus of
  Miquelon, C Saint Pierre and
  Miquelon
Lang Sơn *627* N Vietnam
Languedoc *255, 258* cultural region of
  SE France
Länkäran *110* *Rus.* Lenkoran'.
  S Azerbaijan
Lansing *603* Michigan, NC USA
Lan Yü *559* *var.* Huo-shao Tao,
  Eng. Orchid Island. Island of
  SE Taiwan
Lanzarote *535* island of Islas
  Canarias, SW Spain
Lanzhou *187* *var.* Lanchow,
  prev. Kaolan. Gansu province,
  C China
Lao Cai *627* NW Vietnam
Laojunmiao *see* Yumen
Laon *255* N France
La Oroya *463* C Peru
Laos *360-361* officially Lao People's
  Democratic Republic. Country of
  SE Asia divided into 16 admin.
  units (provinces, khouèng).
  ❖ Vientiane
La Palma *535* island of Islas Canarias,
  SW Spain
La Palma *456* E Panama
La Palma *237* S El Salvador
La Paz *290* SW Honduras
La Paz *136* *var.* La Paz de Ayacucho.
  ❖ of Bolivia, W Bolivia
La Paz *611* S Uruguay

Liban, Jebel *364 Eng.* Lebanon, Mount Lebanon, *Ar.* Jabal Lubnān. Mountain range of C Lebanon
Libau *see* Liepāja
Liberec *216 Ger.* Reichenberg. N Czech Republic
Liberia *206* NW Costa Rica
Liberia *368-369* officially The Republic of Liberia. Country of West Africa, divided into 9 admin. units (counties). ❖ Monrovia
Liberta *92* S Antigua, Antigua & Barbuda
Libertad *611* S Uruguay
Libertad *130 prev.* Pembroke Hall. N Belize
Librazhd *81 var.* Librazhdi. E Albania
Libreville *260* ❖ of Gabon, NW Gabon
Libya *370-373* officially The Great Socialist People's Libyan Arab Jamahiriya, *prev.* Libyan Arab Republic. Country of N Africa, the current administrative structure is not clear. ❖ Tripoli
Libyan Desert *232, 371, 540 Ar.* Aş Şahrā' al Lībīyah. Desert of N Africa
Libyan Plateau *232, 371 Ar.* Aḍ Ḏiffah. Plateau of Egypt and Libya
Licata *323* Sicilia, S Italy
Lichinga *419* N Mozambique
Lichtenburg *531* North West, N South Africa
Lida *122* W Belarus
Lido di Ostia *323* C Italy
Liechtenstein *374-375* officially Principality of Liechtenstein. Country of C Europe divided into 11 admin. units (communes). ❖ Vaduz
Liège *127 Dut.* Luik, *Ger.* Lüttich. E Belgium
Liegnitz *see* Legnica
Lieksa *251* E Finland
Lien-yun *see* Liangyungang
Lienz *106* W Austria
Liepāja *362 Ger.* Libau. W Latvia
Lier *127 Fr.* Lierre. N Belgium
Liestal *550* N Switzerland
Lievenhof *see* Līvāni
Liezen *106* C Austria
Lifford *316* N Ireland
Lifou *650 var.* Îles Loyauté, E New Caledonia
Lifuka *572* island of Ha'apai Group, Tonga
Līgatne *362* NE Latvia
Ligure, Appennino *322* mountain range of N Italy
Ligure, Mar *see* Ligurian Sea
Ligurian Sea *255, 322 It.* Mar Ligure, *Fr.* Mer Ligurienne. Area of the Mediterranean Sea, between France and Italy
Lihue *602* Kauai, Hawaii, USA
Lihula *242 Ger.* Leal. W Estonia
Liivi Laht *see* Riga, Gulf of
Likasi *203 prev.* Jadotville. SE Dem. Rep. Congo
Likiep *396* island of C Marshall Islands
Likouala *200* river of NW Congo
Likouala aux Herbes *200* river of E Congo
Liku *650* S Niue
Lille *255 Dut.* Rijssel. N France
Lillebælt *218 Eng.* Little Belt, *var.* Lille Bælt. Straits between Fyn and Jylland, SW Denmark
Lillehammer *444* S Norway
Lillestrøm *444* S Norway
Lilongwe *385* ❖ of Malawi, W Malawi
Lilongwe *385* river of W Malawi
Lima *463* ❖ of Peru, W Peru
Limassol *215 var.* Lemesos. SW Cyprus
Limbe *385* S Malawi
Limbe *168 prev.* Victoria. SW Cameroon
Limbé *288* N Haiti
Lim Chu Kang *521* area of NW Singapore
Limerick *316 Ir.* Luimneach. SW Ireland
Limfjorden *218* fjord of Jylland, NW Denmark

Límni Megáli Préspa *see* Prespa, Lake
Limni Prespa *see* Prespa, Lake
Límnos *275 var.* Lemnos. Island of E Greece
Limoges *255* C France
Limón *206* E Costa Rica
Limón *290* NE Honduras
Limon, Mont *400* mountain of Rodrigues, Mauritius
Limousin *255* cultural region of C France
Limpopo *142, 419, 423, 636 var.* Crocodile. River of southern Africa
Limulunga *635* W Zambia
Linakeng *366* E Lesotho
Linares *183* C Chile
Linares *535* S Spain
Lincoln *597* E England, UK
Lincoln *603* Nebraska, C USA
Lincoln Island *651* island of E Paracel Islands
Linden *286* E Guyana
Lindi *203* river of NE Dem. Rep. Congo
Lindi *564* SE Tanzania
Line Islands *348* island group of E Kiribati
Lingayen Gulf *466* gulf of the South China Sea, N Philippines
Lingga, Kepulauan *304* island group to the E of Sumatra, W Indonesia
Linguère *510* N Senegal
Linköping *547* S Sweden
Linyanti *142* river of Botswana and Namibia
Linz *106* N Austria
Lion, Golfe du *255 Eng.* Gulf of Lions. Gulf of the Mediterranean Sea to the S of France
Lions Den *636* N Zimbabwe
Lipa *466* Luzon, N Philippines
Lipari Islands *see* Eolie, Isole
Lipari, Isola *323* island of S Italy
Lipari, Isole *see* Eolie, Isole
Lipetsk *484* W Russian Federation
Lipsk *see* Leipzig
Liptovský Mikuláš *523 Ger.* Liptau-Sankt-Nikolaus, *Hung.* Liptószentmiklós. C Slovakia
Lira *588* N Uganda
Liranga *200* E Congo
Liri *323* river of C Italy
Lisala *203* N Dem. Rep. Congo
Lisbon *474 Port.* Lisboa. ❖ of Portugal, W Portugal
Lisburn *597* Northern Ireland, UK
Lisieux *255* NW France
Lismore *101* E Australia
Lissa *see* Vis
Litani *542, 645 var.* Itany. River of French Guiana and Suriname
Litani *364* river of C Lebanon
Litauen *see* Lithuania
Litavra *see* Lithuania
Lithgow *101* SE Australia
Lithuania *376-377* officially Republic of Lithuania, *Lith.* Lietuva, *Ger.* Litauen, *Pol.* Litwa, *Rus.* Litva, *prev.* Lithuanian SSR, *Rus.* Litovskaya SSR. Country of E Europe divided into 44 admin. units (disicts). ❖ Vilnius
Litla Dimun *644* island of S Faeroe Islands
Little Abaco *112* island of N Bahamas
Little Alföld *292* plain of Hungary and Slovakia
Little Andaman *298* island of Andaman Islands to the SE of India
Little Barrier Island *433* island to the N of North Island, New Zealand
Little Belt *see* Lillebælt
Little Cayman *644* island of C Cayman Islands
Little Coco Island *159* island of SW Burma
Little Inagua *112* island of S Bahamas
Little Minch *597* strait of the Atlantic Ocean, NW Scotland, UK
Little Rock *603* Arkansas, SC USA
Little Scarcies *see* Kaba
Little Sound *643* bay of the North Atlantic Ocean, W Bermuda
Little Tobago *643* island of W British Virgin Islands

Little Tobago *574 var.* Bird of Paradise Island. Island to the E of Tobago, Trinidad & Tobago
Little Zab *312 Ar.* Zāb aş Şaghīr, *Kurd.* Zē-i Kya. River of Iran and Iraq
Litva *see* Lithuania
Litwa *see* Lithuania
Liuch'iu Yü *559* island of SW Taiwan
Liukuei *559* S Taiwan
Liuzhou *187 var.* Liu-chou, Liuchow. Guangxi, S China
Livadhi *215 var.* Leivadi. River of W Cyprus
Līvāni *362 Ger.* Lievenhof. SE Latvia
Lively Island *645* island of E Falkland Islands
Lively Sound *645* area of the South Atlantic Ocean, E Falkland Islands
Liverpool *597* NW England, UK
Lívingston *280* E Guatemala
Livingstone *635 var.* Maramba. S Zambia
Livingstonia *385* N Malawi
Livno *140* SW Bosnia & Herzegovina
Livojoki *251* river of C Finland
Livonia *see* Vidzeme
Livorno *322 Eng.* Leghorn. C Italy
Liwonde *385* S Malawi
Liyāh, Jāl al *356* ridge of NW Kuwait
Ljouwert *see* Leeuwarden
Ljubelj *see* Loibl Pass
Ljubljana *524 var.* Lyublyana, *Ger.* Laibach, *It.* Lubiana. ❖ of Slovenia, C Slovenia
Ljubrlj *see* Loibl Pass
Ljungan *547* river of C Sweden
Ljusnan *547* river of C Sweden
Llallagua *136* SW Bolivia
Lleida *535 Cast.* Lérida. NE Spain
Llolleo *183* C Chile
Lloydminster *170* SW Canada
Lô *627* river of China and Vietnam
Loaita Island *652* island of W Spratly Islands
Loangwa *see* Luangwa
Lobatse *142 var.* Lobatsi. S Botswana
Lobaye *179* river of SW Central African Republic
Lobito *88* W Angola
Lob Nor *see* Lop Nur
Locarno *550 Ger.* Luggarus. S Switzerland
Lôc Ninh *627* SW Vietnam
Locri *323* S Italy
Lőcse *see* Levoča
Lod *319 var.* Lydda. C Israel
Lodge *494* NE St Kitts, St Kitts & Nevis
Lodja *203* C Dem. Rep. Congo
Lodwar *344* NW Kenya
Łódź *471 Rus.* Lodz. C Poland
Loei *567 var.* Muang Loei. N Thailand
Lofa *368 var.* Loffa. River of Guinea and Liberia
Lofoten *444 var.* Lofoten Islands. Islands of NE Norway
Loga *439* W Niger
Logan, Mount *170* mountain of NW Canada
Logone *168, 180 var.* Lagone. River of Cameroon and Chad
Logroño *535* N Spain
Lögurinn *296 var.* Lagarfljót. Lake of E Iceland
Loh *619* Torres Islands, N Vanuatu
Loibl Pass *106 var.* Ljubelj, *Ger.* Loiblpass, *Slvn.* Ljubrlj. Mountain pass of Austria and Slovenia
Loikaw *159* E Burma
Loir *255* river of NW France
Loire *255* river of C France
Loita Hills *344* hilly region of SW Kenya
Loja *230* S Ecuador
Lökbatan *110 Rus.* Lokbatan. E Azerbaijan
Lokeren *127* NW Belgium
Lokitaung *344* NW Kenya
Løkken *218* Jylland, NW Denmark
Lokoja *440* C Nigeria
Lokossa *132* S Benin

Loksa *242 Ger.* Loxa. N Estonia
Lol *540* river of S Sudan
Lola *283* SE Guinea
Lolland *218 prev.* Laaland. Island of S Denmark
Lolotique *237* SE El Salvador
Lolvavana, Passage *619* strait between Maewo and Pentecost, C Seychelles
Lom *152 prev.* Lom-Palanka. NW Bulgaria
Lom *168* river of Cameroon and Central African Republic
Lomahasha *544* NE Swaziland
Lomami *203* river of C Dem. Rep. Congo
Lomas de Zamora *95* E Argentina
Lombok *304* island of Nusa Tenggara, C Indonesia
Lomé *571* ❖ of Togo, S Togo
Lom Sak *567 var.* Muang Lom Sak. N Thailand
Lomond, Loch *597* lake of C Scotland, UK
Londiani *344* W Kenya
London *597* ❖ of United Kingdom
London *171* S Canada
London *348* Kiritimati, E Kiribati
London Bridge *278* island to the N of Grenada island, Grenada
Londonderry *597 var.* Derry. Northern Ireland, UK
Londrina *145* S Brazil
Longa, Proliv *485 Eng.* Long Strait. Strait connecting Chukchi Sea and East Siberian Sea, between NE Asia and NW North America
Long Bay *643* bay of the North Atlantic Ocean, E Bermuda
Long Bay *643* bay of the North Atlantic Ocean, W Bermuda
Longford *316 Ir.* Longphort. C Ireland
Long Island *643* island of W Bermuda
Long Island *see* Arop Island
Long Island *112* island of C Bahamas
Long Island *92* island to the N of Antigua, Antigua & Barbuda
Longmont *603* Colorado, SW USA
Longoni, Baie de *649 var.* Longoni Bay. Bay of the Indian Ocean on the N coast of Mayotte
Longphort *see* Longford
Longreach *101* E Australia
Long Strait *see* Longa, Proliv
Long Swamp *643* Tortola, C British Virgin Islands
Longwood *652* E St Helena
Long Xuyên *627* SW Vietnam
Longyearbyen *653* ❖ of Svalbard, Spitsbergen, W Svalbard
Lonhlupheko *544* NE Swaziland
Lons-le-Saunier *255* E France
Loop Head *316* promontory on the W coast of Ireland
Lop Buri *567* C Thailand
Lopévi *619* island of C Vanuatu
Lopez, Cap *260* W Gabon
Lop Nur *186 var.* Lop Nor, Lob Nor, *Chin.* Lo-pu Po. Lake of Xinjiang Uygur Zizhiqu, NW China
Lo-pu Po *see* Lop Nur
Lora, Hāmūn-i- *450* salt marsh of W Pakistan
Lord Howe Island island of E Australia
Lord Howe Island *see* Ontong Java Atoll
Lorca *535* S Spain
Lorengau *458 var.* Lorungau, Manus I, Papua New Guinea
Lorentz *305* river of Irian Jaya, E Indonesia
Loreto *460* C Paraguay
Lorian Swamp *344* swamp E Kenya
Lorient *254* W France
Lorn, Firth of *597* inlet of Atlantic Ocean, W Scotland, UK
Lorraine *255* cultural region of NE France
Los Amates *280* E Guatemala
Los Andes *183* C Chile
Los Angeles *602* California, W USA
Los Ángeles *183* C Chile
Losap *406* atoll of C Micronesia
Los Chiles *206* NW Costa Rica

Los, Îles de *283* Island group to the SW of Guinea
Lošinj *209* *It.* Lussino. Island of W Croatia
Loslau *see* Wodzisław Śląski
Los Mochis *403* W Mexico
Losonc *see* Lučenec
Losontz *see* Lučenec
Lospalos *228* E East Timor
Los Roques, Islas *623* island group of N Venezuela
Los Teques *623* N Venezuela
Lot *255* river of S France
Lotofaga *500* Upolu, Samoa
Lo-tung *559* *Jap.* Ratō. NE Taiwan
Louang Namtha *360* *var.* Luong Nam Tha. N Laos
Louangphrabang *360* *var.* Luang Prabang. C Laos
Loubiere *224* SW Dominica
Loubomo *200* *prev.* Dolisie. S Congo
Loudima *200* S Congo
Louéssé *200* river of SW Congo
Louga *510* NW Senegal
Loughrea *316* W Ireland
Louis Gentil *see* Youssoufia
Louisiade Archipelago *458* island group of SE Papua New Guinea
Louisiana *603* state of SC USA
Louis Trichardt *531* Northern Transvaal, NE South Africa
Louisville *603* Kentucky, C USA
Louisville *130* N Belize
Loukoléla *200* E Congo
Loum *168* W Cameroon
Louna *200* river of SE Congo
Louny *216* NW Czech Republic
Lourenço Marques *see* Maputo
Lourenço Marques, Baía de *see* Maputo, Baía de
Louvain *see* Leuven
Lovech *152* *var.* Loveč. NW Bulgaria
Lovell Village *498* Mustique, St Vincent & the Grenadines
Lóvua *88* N Angola
Lowell *603* Massachusetts, NE USA
Löwen *see* Leuven
Lower Bann *597* river of Northern Ireland, UK
Lower California *see* Baja California
Lower Carlton *121* NW Barbados
Lower Hutt *433* S North Island, New Zealand
Lower Lough Erne *597* lake of Northern Ireland, UK
Lower Mortlocks *406* island group of C Micronesia
Lower Rhine *see* Neder-Rijn
Lower Tunguska *see* Nizhnyaya Tunguska
Lower West End Point *642* headland on the SW coast of Anguilla
Low Point *644* headland on the E coast of Christmas Island
Loyada *222* W Djibouti
l'Oyapok *see* Oiapoque
Loyauté, Îles *650* island group of E New Caledonia
Loyoro *588* NE Uganda
Loznica *512* W Serbia, Serbia & Montenegro (Yugo.)
Lualaba *203* *var.* Zaire, *Fr.* Loualaba. River of Dem. Rep. Congo
Luampa *635* river of W Zambia
Luanda *88* ❖ of Angola, NW Angola
Luang Prabang *see* Louangphrabang
Luang Prabang Range *360* mountain range of W Laos
Luangwa *635* *Port.* Aruângua. River of Mozambique and Zambia
Luanshya *635* C Zambia
Luapula *203*, *635* river of Dem. Rep. Congo and Zambia
Luar, Pulu *see* Horsburgh Island
Luba *238* *prev.* San Carlos. W Bioko, Equatorial Guinea
Lubānas Ezers *362* lake of E Latvia
Lubang Island *466* island of N Philippines
Lubango *88* *Port.* Sá da Bandeira. SW Angola
Lubao *203* SE Dem. Rep. Congo

Lübeck *267* N Germany
Lübecker Bucht *267* bay of the Baltic Sea, on the N coast of Germany
Lubelska, Wyżyna *471* plateau of SE Poland
Lubiana *see* Ljubljana
Lublin *471* *Rus.* Lyublin. E Poland
Lubnān, Jabal *see* Liban, Jebel
Lubny *590* C Ukraine
Lubumbashi *203* *prev.* Élisabethville. SE Dem. Rep. Congo
Luca *136* SW Bolivia
Lucan *316* *Ir.* Leamhcán. E Ireland
Lucano, Appennino *323* mountain range of S Italy
Lucapa *88* *var.* Lukapa. NE Angola
Lucea *331* NW Jamaica
Lucena *466* Luzon, N Philippines
Lučenec *523* *Hung.* Losonc, *Ger.* Losontz. C Slovakia
Lucerne *see* Luzern
Lucerne, Lake of *see* Vierwaldstätter See
Luchow *see* Hefei
Lucie *542* Suriname
Łuck *see* Luts'k
Lucknow *298* *Hind.* Lakhnau. N India
Lüderitz *423* *prev.* Angra Pequena. SW Namibia
Ludhiāna *298* N India
Ludwigshafen *266* *var.* Ludwigshafen am Rhein. SW Germany
Luebo *203* SW Dem. Rep. Congo
Luena *88* *Port.* Luso. E Angola
Lufira, Lac de Retenue de la *203* *var.* Lac Tshangalele. Lake of SE Dem. Rep. Congo
Lugano *550* *Ger.* Lauis. S Switzerland
Luganville *619* Espiritu Santo, Vanuatu
Lugards Falls *344* waterfall of SE Kenya
Lugenda *419* river of N Mozambique
Luggarus *see* Locarno
Lugh Ganana *see* Luuq
Lugo *535* NW Spain
Lugoj *480* W Romania
Lugusi *see* Rugusye
Luhans'k *590* *Rus.* Lugansk, *prev.* Voroshilovgrad. E Ukraine
Luiana *88* river of SE Angola
Luichow Peninsula *see* Leizhou Bandao
Luik *see* Liège
Luimneach *see* Limerick
Luján *95* C Argentina
Lukang *559* *var.* Lu-chiang, *Jap.* Rokkō. W Taiwan
Lukapa *see* Lucapa
Lukenie *203* river of C Dem. Rep. Congo
Lukhalweni *544* S Swaziland
Lukusashi *635* river of C Zambia
Luleå *547* NE Sweden
Luleälv *547* river of NE Sweden
Lulonga *203* river of NW Dem. Rep. Congo
Lulua *203* river of S Dem. Rep. Congo
Luluabourg *see* Kananga
Lumbo *419* NE Mozambique
Lumi *458* NW Papua New Guinea
Lumphăt *165* *prev.* Lomphat. NE Cambodia
Lumpungu *see* Rumpungwe
Lund *547* S Sweden
Lunga *635* river of Zambia
Lunga, Isola *see* Dugi Otok
Lungi *518* W Sierra Leone
Lungkiang *see* Qiqihar
Lungwebungu *635* river of Angola and Zambia
Luninets *see* Luninyets
Luninyets *122* *Rus.* Luninets. SW Belarus
Lunsar *518* W Sierra Leone
Lunsemfwa *635* river C Zambia
Luong Nam Tha *see* Louang Namtha
Luoyang *187* Henan, C China
Luque *460* S Paraguay
Lúrio *419* NE Mozambique
Lúrio *419* river of NE Mozambique
Lusaka *635* ❖ of Zambia, SE Zambia
Lushnjë *81* *var.* Lushnja. C Albania

Luso *see* Luena
Lussino *see* Lošinj
Lusutfu *544* *var.* Usutu, Great Usutu. River of southern Africa
Lü Tao *559* island of SE Taiwan
Lūt, Baḥrat *see* Dead Sea
Lūt, Dasht-e *309* *var.* Kavīr-e Lūt. Desert region of E Iran
Luton *597* C England, UK
Luts'k *590* *Rus.* Lutsk, *Pol.* Łuck. NW Ukraine
Lüttich *see* Liège
Luud, Waadi *528* *It.* Uadi Lhut. Seasonal river of N Somalia
Luuq *528* *It.* Lugh Ganana. SW Somalia
Luvironza *see* Ruvyironza
Luxembourg *378* ❖ of Luxembourg
Luxembourg *378-379* officially Grand Duchy of Luxembourg, *var.* Lëtzebuerg. Country of W Europe divided into 3 admin. units (districts). ❖ Luxembourg
Luxor *232* *Ar.* Al Uqsur. E Egypt
Luz, Costa de la *535* coastal region of SW Spain
Luzern *550* *Fr.* Lucerne. C Switzerland
Luzon *466* island of N Philippines
L'viv *590* *Rus.* L'vov, *Pol.* Lwów, *Ger.* Lemberg. W Ukraine
Lyallpur *see* Faisalābād
Lyangar *see* Langar
Lycksele *547* N Sweden
Lydda *see* Lod
Lyepyel' *122* *Rus.* Lepel'. N Belarus
Lyme Bay *597* bay of S England, UK
Lyon *255* *Eng.* Lyons E France
Lys *see* Leie
Lysi *215* *var.* Akdoğan. C Cyprus
Lyublin *see* Lublin
Lyublyana *see* Ljubljana

M

Ma'ān *338* SW Jordan
Maanselkä *251* mountain range of NE Finland
Maardu *242* *Ger.* Maart. N Estonia
Maarianhamina *see* Mariehamn
Maarmorilik *646* W Greenland
Ma'arrat an Nu'mān *555* NW Syria
Maarssen *429* C Netherlands
Maart *see* Maardu
Maas *see* Meuse
Maasmechelen *127* NE Belgium
Maastricht *429* S Netherlands
Maaza Plateau *232* plateau of NE Egypt
Mabalane *419* S Mozambique
Mabanda *260* S Gabon
Mabayi *162* NW Burundi
Mabouya Island *278* island to the W of Carriacou, Grenada
Macao *187* *Chin.* Aomen. SE China
Macapá *145* N Brazil
Macará *230* S Ecuador
Macarsca *see* Makarska
MacArthur *see* Ormoc
Macas *230* SE Ecuador
Macassar *see* Ujungpandang
Macdonnell Ranges *101* mountain range of C Australia
Macedonia *380-381* officially the Former Yugoslav Republic of Macedonia, *abbrev.* FYR Macedonia, FYROM, *Mac.* Makedonija. Country of SE Europe divided into 34 admin. units (opcine). ❖ Skopje
Maceió *145* E Brazil
Macenta *283* SE Guinea
Macgillycuddy's Reeks *316* *var.* Macgillicuddy's Reeks Mountains, *Ir.* Na Cruacha Dubha. Mountain range of SW Ireland
Machakos *344* S Kenya
Machala *230* SW Ecuador
Machaneng *142* SE Botswana
Machanga *419* SW Mozambique
Machaze *419* SW Mozambique
Machile *635* *var.* Machili. River of SW Zambia

Luso *see* Luena
Machinga *385* *var.* Kasupe, Kasupi. S Malawi
Machiques *623* NW Venezuela
Machu Picchu *463* S Peru
Macia *419* *var.* Vila de Macia. S Mozambique
Macías Nguema Biyogo *see* Bioko
Mackay *101* NE Australia
Mackay, Lake *101* salt lake of C Australia
Mackenzie *170* river of NW Canada
Mackenzie Mountains *170* mountain range of NW Canada
Macleod, Lake *100* lake of W Australia
Macomer *322* Sardegna, S Italy
Macon *603* Georgia, SE USA
Mâcon *255* E France
Macouba *649* N Martinique
MacRitchie Reservoir *521* reservoir of C Singapore
Macroom *316* SW Ireland
Macuelizo *290* W Honduras
Ma'dabā *338* *var.* Mādabā. NW Jordan
Madagascar *382-383* officially Democratic Republic of Madagascar, *Malg.* Madagasikara, *prev.* Malagasy Republic. Country of SE Africa divided into 6 admin. units (provinces). ❖ Antananarivo
Madan *152* S Bulgaria
Madang *458* E Papua New Guinea
Madaniyin *see* Médenine
Mādārīpur *117* S Bangladesh
Madeira *474* *var.* Ilha de Madeira. Madeira Is, Portugal
Madeira *144* river of Bolivia and Brazil
Madeira, Arquipélago da *see* Madeira Islands
Madeira Islands *474* *Port.* Arquipélago da Madeira. Island group to the SW of Portugal
Madina do Boé *see* Boé
Madīnat al Abyār *371* NE Libya
Madīnat al Ka'bān *479* *var.* Al Ka'aban. N Qatar
Madīnat ash Sha'b *631* *prev.* Al Ittiḩād. SW Yemen
Madīnat ash Shamāl *479* *var.* Madinat el Shamal. N Qatar
Madīnat ath Thawrah *555* N Syria
Madinat el Shamal *see* Madīnat ash Shamāl
Madīnat Ḩamad *115* *var.* Hamada Town, Hamad. W Bahrain
Madīnat 'Īsá *115* *var.* Isa Town. N Bahrain
Madīnat Khalifa *479* E Qatar
Madingo-Kayes *200* S Congo
Madingou *200* S Congo
Madison *603* Wisconsin, NC USA
Madiun *304* *prev.* Madioen. Java, C Indonesia
Madlangampisi *544* N Swaziland
Madona *362* *Ger.* Modohn. E Latvia
Madras *see* Chennai
Madre de Dios *136*, *463* river of Bolivia and Peru
Madrid *535* ❖ of Spain, C Spain
Madriu, Riu *86* river of S Andorra
Madura *304* *prev.* Madoera. Island to NE of Java, C Indonesia
Madurai *298* *prev.* Madura. S India
Maebashi *332* Honshū, SE Japan
Mae Name Khong *see* Mekong
Mae Nam Khong *see* Mekong
Mae Nam Moi *see* Thaungyin
Mae Nam Ping *567* river of NW Thailand
Maévo *619* *prev.* Aurora. Island of C Vanuatu
Mafa'a, Pointe *653* headland of Île Alofi, N Wallis & Futuna
Mafeteng *366* W Lesotho
Mafia *564* island of E Tanzania
Mafou *283* C Guinea
Mafraq *see* Al Mafraq
Mafraq *594* C United Arab Emirates
Mafungabusi Plateau *636* plateau of C Zimbabwe
Mafut *594* NE United Arab Emirates
Magadan *485* NE Russian Federation
Magadi *344* SW Kenya
Magadi, Lake *344* lake of SW Kenya

Magallanes *see* Punta Arenas
Magallanes, Estrecho *see* Magellan, Strait of
Magangué *195* N Colombia
Magaria *439* S Niger
Magat *466* river of Luzon, N Philippines
Magburaka *518* C Sierra Leone
Magdalena *195* river of C Colombia
Magdalena *136* N Bolivia
Magdeburg *267* C Germany
Magelang *304* Java, C Indonesia
Magellan, Strait of *95, 183*
    Sp. Estrecho de Magallanes. Strait connecting the S Atlantic and S Pacific Oceans between Tierra del Fuego and mainland South America
Mageroya *444* *var.* Magerøy. Island of NE Norway
Maggiore, Lake *322, 550* *It.* Lago Maggiore. Lake of Italy and Switzerland
Magh Ealla *see* Mallow
Maglaj *140* N Bosnia & Herzegovina
Magnitogorsk *484* C Russian Federation
Mago *248* *prev.* Mango. Island of the Lau Group, E Fiji
Magt'a' Laḥjar *398* SW Mauritania
Māgura *117* W Bangladesh
Magwe *159* *var.* Magway. W Burma
Magyarkanizsa *see* Kanjiža
Magyarország *see* Hungary
Magyaróvár *see* Mosonmagyaróvár
Mahafaly, Lembalemban' *382* *var.*Plateau Mahafaly. Plateau of SW Madagascar
Mahaicony Village *286* E Guyana
Mahajamba *382* seasonal river of N Madagascar
Mahajanga *382* *prev.* Majunga. N Madagascar
Mahajilo *382* seasonal river of C Madagascar
Mahakali *427* river of India and Nepal
Mahakam *304* *var.* Kutai, Koetai. River of Borneo, C Indonesia
Mahalapye *142* *var.* Mahalatswe. SE Botswana
Maḥallah al Kubrá *see* El Mahalla el Kubra
Mahamba *544* SW Swaziland
Mahanādi *298* river of E India
Mahanoro *382* E Madagascar
Maha Sarakham *567* NE Thailand
Mahaut *224* W Dominica
Mahavavy *382* seasonal river of N Madagascar
Mahaweli Ganga *538* river of C Sri Lanka
Mahdia *577* *var.* Al Mahdīyah, Mehdia. NE Tunisia
Mahdia *286* C Guyana
Mahé *516* island of NE Seychelles
Mahebourg *400* SE Mauritius
Mahenge *564* SE Tanzania
Mahia Peninsula *433* peninsula of E North Island, New Zealand
Mahilyow *122* *Rus.* Mogilëv. E Belarus
Mahina *392* W Mali
Mahiyangana *538* E Sri Lanka
Maḥmūd-e Rāqī *77* NE Afghanistan
Mahmudiya *see* Al Maḥmūdīyah
Mahou *392* S Mali
Mährisch-Ostrau *see* Ostrava
Mährisch-Schönberg *see* Šumperk
Maiana *348* island of the Gilbert Is, W Kiribati
Maicao *195* N Colombia
Mai Ceu *see* Maych'ew
Mai Chio *see* Maych'ew
Maiduguri *441* NE Nigeria
Mailand *see* Milano
Maimāna *see* Meymaneh
Main *266* river of C Germany
Main Camp *see* Banana
Main Channel *648* channel of the Pacific Ocean, S Johnston Atoll
Main Channel *see* Inner Channel
Mai-Ndombe, Lac *203* *prev.* Lac Léopold II. Lake of W Dem. Rep. Congo
Maine *603* state of NE USA
Maine *255* cultural region of NW France

Maïné-Soroa *439* SE Niger
Mainland *597* Shetland, NE UK
Mainland *597* Orkney, NE UK
Maintirano *382* W Madagascar
Mainz *266* *Fr.* Mayence. SW Germany
Maio *176* *var.* Vila de Maio. Maio, S Cape Verde
Maio *176* island of SE Cape Verde
Maiquetía *623* N Venezuela
Maisur *see* Mysore
Maitland *101* E Australia
Maitri *90* Indian research station of Greater Antarctica, Antarctica
Maíz, Islas *436* *var.* Corn Islands. Island group of E Nicaragua
Maizuru *332* Honshū, C Japan
Majardah, Wādī *see* Mejerda, Oued
Majimbini, Réserve Forestière de *649* forest reserve of C Mayotte
Majorca *see* Mallorca
Majunga *see* Mahajanga
Majuro *396* atoll of SE Marshall Islands
Majuro District *396* district of Majuro, SE Marshall Islands
Makamba *162* S Burundi
Makarska *209* *It.* Macarsca. SE Croatia
Makasar *see* Ujungpandang
Makasar, Selat *304* *Eng.* Makassar Strait. Strait connecting the Celebes Sea and Laut Flores, C Indonesia
Makassar Strait *see* Makasar, Selat
Makay, Tangorombohitr' i *382* *var.* Massif du Makay. Mountains of SW Madagascar
Makebuko *162* C Burundi
Makeni *518* C Sierra Leone
Makgadikgadi *142* *var.* Makarikari Pans. Saltpans of NE Botswana
Makhachkala *484* *prev.* Petrovsk-Port. SW Russian Federation
Makharadze *see* Ozurget'i
Makhfar al Quwayrah *see* Al Quwayrah
Makin *348* island of the Gilbert Is, W Kiribati
Makira *see* San Cristobal
Makiyivka *590* *Rus.* Makeyevka, *prev.* Dmitriyevsk. E Ukraine
Makkah *506* *Eng.* Mecca. W Saudi Arabia
Makō *see* MaKung
Makó *292* SE Hungary
Makogai *248* island to the NE of Viti Levu, C Fiji
Makokou *260* NE Gabon
Makona *283* river of S Guinea
Makoua *200* C Congo
Makran Coast *309* coastal region of SE Iran
MaKung *559* *Jap.* Makō. P'eng-hu Tao, W Taiwan
Makunudhoo Atoll *390* *var.* Makunudu Atoll. Atoll of N Maldives
Makurazaki *332* Kyūshū, SW Japan
Makurdi *440* C Nigeria
Makuti *636* N Zimbabwe
Makwate *142* SE Botswana
Mala *see* Malaita
Malabo *238* *prev.* Santa Isabel. ❖ of Equatorial Guinea, N Bioko
Malacca *see* Melaka
Malacca, Strait of *304* strait connecting the Andaman Sea and South China Sea between Malay Peninsula and Sumatra, SE Asia
Malacka *see* Malacky
Malacky *523* *Hung.* Malacka. W Slovakia
Maladzyechna *122* *Rus.* Molodechno, *Pol.* Molodeczno. NW Belarus
Málaga *535* S Spain
Malagasy Republic *see* Madagascar
Malaita *526* *var.* Mala, Island of C Solomon Islands
Malakal *540* *var.* Malakāl. S Sudan
Malambo *195* N Colombia
Malang *304* SE Java, Indonesia
Malange *see* Malanje
Malanje *88* *var.* Malange. NW Angola
Malanville *132* NE Benin

Mälaren *547* lake of SE Sweden
Malatya *580* SE Turkey
Malawi *384-385* officially Republic of Malawi, *prev.* Nyasaland, Nyasaland Protectorate. Country of S Africa divided into 3 admin. units (regions). ❖ Lilongwe
Malawi, Lake *see* Nyasa, Lake
Malaya *see* Peninsular Malaysia
Malāyer *309* NW Iran
Malay Peninsula *567* peninsula of Malaysia and Thailand
Malaysia *386-389* officially Republic of Malaysia, *prev.* the separate territories of Federation of Malaya, Singapore (left 1965), Sarawak and Sabah (North Borneo). Country of SE Asia divided into 15 admin. units (13 states, 2 federal territories). ❖ Kuala Lumpur
Maldegem *127* NW Belgium
Malden Island *348* island of the Line Is, E Kiribati
Maldives *390-391* Officially Republic of Maldives, Maldivian Divehi. Country of the Indian Ocean divided into 19 admin. units (districts). ❖ Male'
Maldonado *611* S Uruguay
Male' *390* *var.* Male. ❖ of Maldives, Male' Atoll, C Maldives
Male *see* Male'
Male' Atoll *390* *var.* Kaafu Atoll. Atoll of C Maldives
Malebo Pool *see* Stanley Pool
Malékoula *see* Malekula
Malekula *619* *var.* Malakula, *prev.* Mallicolo. Island of W Vanuatu
Mali *392-393* officially Republic of Mali, *prev.* Sudanese Republic, French Sudan. Country divided into 8 admin. units (7 regions and 1 capital district). ❖ Bamako
Maliana *228* W East Timor
Malibamatso *366* river of C Lesotho
Mali Hka *159* river of N Burma forming a headstream of the Irrawaddy river
Malindi *344* SE Kenya
Malines *see* Mechelen
Malinga *260* SE Gabon
Malin Head *316* headland on the N coast of Ireland
Mallāq, Wādī *see* Mellègue, Oued
Mallawi *232* *var.* Mallawī. C Egypt
Mallicolo *see* Malekula
Mallorca *535* *Eng.* Majorca. Island of the Islas Baleares, E Spain
Mallow *316* *Ir.* Magh Ealla. SW Ireland
Malmédy *127* E Belgium
Malmö *547* S Sweden
Malmok *650* headland of Bonaire, S Netherlands Antilles
Malo *619* island of W Vanuatu
Maloelap *396* island of E Marshall Islands
Malolo *248* island of the Mamanuca-i-ra Group, W Fiji
Malolos *466* Luzon, N Philippines
Maloma *544* S Swaziland
Malombe, Lake *385* lake of SE Malawi
Małopolska *471* plateau of S Poland
Maloti Mountains *see* Maluti
Malpasso *226* SW Dominican Republic
Mäls *374* S Liechtenstein
Malta *395* island of the Mediterranean Sea, with Gozo and Kemmuna forms the state of Malta
Malta *394-395* officially Republic of Malta. Country of the Mediterranean Sea. ❖ Valletta
Malta Channel *395* *It.* Canale di Malta. Strait of Mediterranean Sea between Malta and Sicily
Maltahöhe *423* S Namibia
Malta, Il-Fliegu ta' *395* *Eng.* South Comino Channel. Strait of Mediterranean Sea between Kemmuna and Malta islands, NW Malta

Maluku *304* *prev.* Spice Islands, *Eng.* Moluccas. Island group of E Indonesia
Maluku, Laut *304* *Eng.* Molucca Sea. Sea of the Pacific Ocean, E Indonesia
Malung *547* C Sweden
Maluti *366* *var.* Maluti Mountains, Maloti Mountains, Front Range. Mountain range of C Lesotho
Malvinas, Islas *see* Falkland Islands
Malyy Kavkaz *see* Lesser Caucasus
Mamanuca-i-ra Group *248* islands of W Fiji
Mamates *366* NW Lesotho
Mambéré *179* river of SW Central African Republic
Mambili *200* river of W Congo
Mamer *378* SW Luxembourg
Mamfe *168* W Cameroon
Mamiku *497* E St Lucia
Mamoré *136* river of Bolivia and Brazil
Mamou *283* W Guinea
Mamoudzou *649* ❖ of Mayotte, N Mayotte
Mampong *272* C Ghana
Mamṭalah, Ra's al *see* Mummaṭalah, Ra's al
Mamuno *142* W Botswana
Man *328* W Ivory Coast
Manado *304* *prev.* Menado. Celebes, C Indonesia
Managua *436* ❖ of Nicaragua, W Nicaragua
Managua, Lago de *436* *var.* Xolotlán. Lake of W Nicaragua
Manaḥ *448* *var.* Bilād Manaḥ. N Oman
Manakara *382* SE Madagascar
Mana *645* river of C French Guiana
Manama *115* *Ar.* Al Manāmah. ❖ of Bahrain, NE Bahrain
Manambaho *382* seasonal river of NW Madagascar
Manambolo *382* river of W Madagascar
Mananjary *382* SE Madagascar
Manantali, Lac de *392* reservoir of W Mali
Manāqish *356* *var.* Manageesh. S Kuwait
Manas, Gora *614* mountain of NE Uzbekistan
Manatí *651* N Puerto Rico
Manatuto *228* N East Timor
Manaus *144* *prev.* Manáos. NW Brazil
Manbij *555* *Fr.* Membidj. N Syria
Manchester *597* N England, UK
Manchester *603* New Hampshire, NE USA
Man-chou-li *see* Manzhouli
Manda Island *344* island of SE Kenya
Mandal *444* SW Norway
Mandalay *159* N Burma
Mandalgovi *412* S Mongolia
Mandali *312* E Iraq
Mandaue *466* Cebu, C Philippines
Mandera *344* NE Kenya
Mandeville *331* SW Jamaica
Mandiana *283* E Guinea
Mandi Būrewāla *451* *var.* Būrewāla. E Pakistan
Mandidzudzure *see* Chimanimani
Mandié *419* NW Mozambique
Mandimba *419* N Mozambique
Mandji *260* C Gabon
Mandouri *571* N Togo
Manfredonia *323* S Italy
Manga *157* C Burkina
Mangai *203* W Dem. Rep. Congo
Mangaia *644* island of Southern Cook Islands, S Cook Islands
Mangalia *480* SE Romania
Mangalmé *180* SE Chad
Mangalore *298* SW India
Mangde *134* river of S Bhutan
Mange *518* NW Sierra Leone
Mango *see* Sansanné-Mango
Mango *see* Mago
Mangoche *see* Mangochi
Mangochi *385* *var.* Mangoche, *prev.* Fort Johnson. SE Malawi
Mangoky *382* river of SW Madagascar
Mangula *see* Mhangura**

Mania *382* river of C Madagascar
Manica *419 var.* Vila de Manica.
 W Mozambique
Manihiki *644* island of Northern Cook
 Islands, N Cook Islands
Maniitsoq *646 Dan.* Sukkertoppen.
 SW Greenland
Manikaraku *526* E Guadalcanal,
 Solomon Is
Manikganj *117* C Bangladesh
Manila *466 var.* Manila City.
 ❖ of the Philippines, Luzon,
 N Philippines
Manisa *580 prev.* Saruhan. W Turkey
Man, Isle of *597, 647* British Crown
 dependency of the Irish Sea.
 ❖ Douglas
Manitoba *170* province of S Canada
Manizales *195* W Colombia
Manjimup *100* SW Australia
Mankayane *544 var.* Mankaiana.
 W Swaziland
Mankono *328* C Ivory Coast
Mankulam *538* N Sri Lanka
Mannar *538 var.* Manar.
 NW Sri Lanka
Mannar, Gulf of *298, 538* gulf of Indian
 Ocean, to the S of India
Mannar Island *538* island to the
 N of Sri Lanka
Mannheim *266* SW Germany
Mano *518* SW Sierra Leone
Mano *518* river of Liberia and
 Sierra Leone
Manombo Atsimo *382 var.* Manombo.
 SW Madagascar
Manono *500* Upolu, Samoa
Manono *203* SE Dem. Rep. Congo
Manorhamilton *316* N Ireland
Manp'o *351 var.* Manp'ojin.
 NW North Korea
Manra *348 var.* Sydney I. Island of
 the Phoenix Is, C Kiribati
Mansa *635 prev.* Fort Rosebery.
 N Zambia
Mansabá *284* NW Guinea-Bissau
Mansajang Kunda *263* E Gambia
Mansa Konko *263* C Gambia
Mansion *494* NE St Kitts,
 St Kitts & Nevis
Mansôa *284* W Guinea-Bissau
Mansôa *284* river of W Guinea-Bissau
Manta *230* W Ecuador
Mantes-la-Jolie *255 prev.* Mantes-sur-
 Seine, Mantes-Gassicourt. N France
Mantova *322 Eng.* Mantua,
 Fr. Mantoue. N Italy
Mantsonyane *366* C Lesotho
Manuae *644* island of Southern Cook
 Islands, S Cook Islands
Manua Islands *642* island group of
 E American Samoa
Manukau Harbour *433* harbour of
 W North Island, New Zealand
Manurewa *433* N North Island,
 New Zealand
Manus Island *458 var.* Great Admiralty
 I. Island of NE Papua New Guinea
Manyame *636 var.* Panhame,
 prev. Hunyani. River of Mozambique
 and Zimbabwe
Manyame, Lake *636 prev.* Robertson,
 Lake. Reservoir of N Zimbabwe
Manyara, Lake *564* lake of
 NE Tanzania
Manyoni *564* C Tanzania
Manzanillo *210* SE Cuba
Manzhouli *187 var.* Man-chou-li.
 Nei Mongol Zizhiqu, NE China
Manzil Bū Ruqaybah *see* Menzel
 Bourguiba
Manzil Tamīm *see* Menzel Temime
Manzini *544 prev.* Bremersdorp.
 C Swaziland
Mao *180* W Chad
Mao *226* NW Dominican Republic
Maoke, Pegunungan *305*
 Dut. Sneeuw-gebergte, *Eng.* Snow
 Mountains. Mountain range of Irian
 Jaya, E Indonesia
Mapoteng *366* NW Lesotho
Mapou *400* N Mauritius
Maputo *419 prev.* Lourenço Marques.
 ❖ of Mozambique, S Mozambique

Maputo, Baía de *419 var.* Baía de
 Lourenço Marques, *Eng.* Delagoa
 Bay. Bay on the coast of Mozambique
Mara *286* E Guyana
Marabá *145* NE Brazil
Maracaibo *623* NW Venezuela
Maracaibo, Lago de *623* inlet of
 Caribbean Sea, NW Venezuela
Maracay *623* N Venezuela
Marada *371* N Libya
Maradi *439* S Niger
Maragarazi *162, 564 var.* Muragarazi.
 River of Burundi and Tanzania
Marāgheh *309 var.* Maragha. NW Iran
Marahoué *see* Bandama Rouge
Marajó, Baía de *145* N Brazil
Marajó, Ilha de *145* island of N Brazil
Marakabei *366 var.* Marakabeis.
 C Lesotho
Marakei *348* island of the Gilbert Is,
 W Kiribati
Maralal *344* C Kenya
Maralik *98* W Armenia
Maramasike *526* island of
 E Solomon Is
Maramba *see* Livingstone
Marambio *90* Argentinian research
 station near Antarctic Peninsula,
 Antarctica
Maramvya *162* SW Burundi
Marandellas *see* Marondera
Marañón *463* river of N Peru
Marash *see* Kahramanmaraş
Maravovo *526* W Guadalcanal,
 Solomon Is
Marāwiḩ *594 var.* Merawwah. Island
 of W United Arab Emirates
Marbella *535* S Spain
Marburg *see* Maribor
Marburg an der Lahn *266*
 W Germany
Marcal *292* river of W Hungary
Marche *255* cultural region of
 C France
Marche-en-Famenne *127* SE Belgium
Marchena, Isla *230* island of
 N Galapagos Is, Ecuador
Marchfield *121* SE Barbados
Mar Chiquita, Lago *95* lake of
 C Argentina
Marcounda *see* Markounda
Marcovia *290* S Honduras
Mardān *451* N Pakistan
Mar del Plata *95* E Argentina
Mardin *581* SE Turkey
Maré *650* island, Îles Loyauté,
 E New Caledonia
Mareeq *528 var.* Mereeg, *It.* Meregh.
 E Somalia
Marek *see* Dupnitsa
Marfa Ridge *395* ridge of NW Malta
Margarita, Isla de *623* island of
 N Venezuela
Margate *531* Kwazulu Natal,
 SE South Africa
Margherita, Lake *see* Ābaya Hāyk'
Margherita Peak *588, 203* mountain of
 Uganda and Dem. Rep. Congo
Marghilon *614 var.* Margelan,
 Rus. Margilan. E Uzbekistan
Mārgow, Dasht-e- *77* desert of
 SW Afghanistan
Mari *215* S Cyprus
Marianao *210* NW Cuba
Marías, Islas *403* island of W Mexico
Maria-Theresiopel *see* Subotica
Máriatölgyes *see* Dubnica nad Váhom
Mar'ib *631* W Yemen
Maribo *218* Lolland, S Denmark
Maribor *524 Ger.* Marburg.
 NE Slovenia
Marid *594* NE United Arab Emirates
Marie Byrd Land *90* physical region of
 Greater Antarctica, Antarctica
Marie-Galante *646* island of
 SE Guadeloupe
Mariehamn *251 var.* Maarianhamina.
 Aland, Finland
Mariel *210* NW Cuba
Marienburg *see* Alūksne
Mariental *423* S Namibia
Marigot *646* St. Martin,
 N Guadeloupe
Marigot *224* NE Dominica

Marigot de Baïla *510* river of
 SW Senegal
Mariguana *see* Mayaguana
Marijampolė *376 prev.* Kapsukas.
 S Lithuania
Marília *145* S Brazil
Marinduque Island *466* island of
 C Philippines
Maringá *145* S Brazil
Marins, Île aux *652* island of
 SE Saint Pierre and Miquelon
Marion Island *531* island of Prince
 Edward Islands, S South Africa
Ionio, Mar *see* Ionian Sea
Maripasoula *645* W French Guiana
Mariscal Estigarribia *460*
 NW Paraguay
Marisule Estate *497* N St Lucia
Maritsa *152, 275 var.* Marica,
 Gk Évros, *Turk.* Meriç. River of
 SE Europe
Mariupol' *590 prev.* Zhdanov.
 SE Ukraine
Mariy El, Respublika *484* autonomous
 republic of W Russian Federation
Märjamaa *242 Ger.* Merjama.
 W Estonia
Marjayoun *364 var.* Marj 'Uyūn.
 S Lebanon
Marka *528 var.* Merca. S Somalia
Marka *385* S Malawi
Market Shop *494* SE Nevis,
 St Kitts & Nevis
Markounda *179 var.* Marcounda.
 NW Central African Republic
Marlánské Lázně *216*
 W Czech Republic
Marmara Denizi *580 Eng.* Sea of
 Marmara. Sea to the NW of Turkey
Marmaris *580* SW Turkey
Marne *255* river of NE France
Marneuli *264* S Georgia
Maro *180* S Chad
Maroantsetra *382* NE Madagascar
Maromokotro *382* mountain of
 N Madagascar
Marondera *636 var.* Marandellas.
 NE Zimbabwe
Maroni *542, 645 Dut.* Marowijne.
 River of French Guiana
 and Suriname
Maros *see* Mureş
Marosvásárhely *see* Târgu Mureş
Marotiri *646* island group of
 S French Polynesia
Maroua *168* N Cameroon
Marovoay *382* NW Madagascar
Marowijne *see* Maroni
Marqūbān *115* NE Bahrain
Marquises, Îles *646* island group of
 N French Polynesia
Marrakech *414 var.* Marakesh,
 Eng. Marrakesh, *prev.* Morocco.
 W Morocco
Marrupa *419* N Mozambique
Marsa *395* C Malta
Marsá al Burayqah
 see Al Burayqah
Marsabit *344* N Kenya
Marsala *323* Sicilia, S Italy
Marsaxlokk *395* SE Malta
Marsaxlokk Bay *395* inlet on the
 SW coast of Malta
Marseille *255 prev. Eng.* Marseilles.
 SE France
Marshall *368* W Liberia
Marshall Islands *396-397* officially
 Republic of the Marshall Islands.
 Country of the Pacific Ocean divided
 into 33 admin. units (districts).
 ❖ Majuro
Marsh Harbour *112* Great Abaco,
 Bahamas
Martaban *159* SE Burma
Martadi *427 var.* Bajura. W Nepal
Martigny *550* SW Switzerland
Martigues *255* SE France
Martin *523 prev.* Turčiansky Svätý
 Martin, *Ger.* Sankt Martin,
 Hung. Turócszentmárton.
 NW Slovakia
Martinique *649* French overseas
 department of the Caribbean Sea.
 ❖ Fort-de-France.

Martinique Passage *224*
 var. Dominica Channel, Martinique
 Channel. Passage connecting the
 Atlantic Ocean and Caribbean Sea
 between Dominica and Martinique
Martuni *98* E Armenia
Marungu *203* mountain range of
 SE Dem. Rep. Congo
Mary *585 prev.* Merv.
 SE Turkmenistan
Maryborough *101* E Australia
Mary Island *see* Kanton
Maryland *603* state of E USA
Marzūq *see* Murzuq
Masai Steppe *564* grassland of
 NW Tanzania
Masaka *588* SW Uganda
Masākin *see* M'saken
Masally *see* Massilı
Masampo *see* Masan
Masan *352 prev.* Masampo.
 S South Korea
Masasi *564* SE Tanzania
Masatepe *436* SW Nicaragua
Masaya *436* S Nicaragua
Masbate *466* island of C Philippines
Mascara *83 var.* Mouaskar.
 NW Algeria
Maseru *366* ❖ of Lesotho,
 W Lesotho
Mas-ha *320* W West Bank
Mashava *636 prev.* Mashaba.
 SE Zimbabwe
Mashhad *309 var.* Meshed. NE Iran
Māshkel *309, 450 var.* Rūd-i Māshkel,
 Māshkid. River of Iran and Pakistan
Māshkel, Hāmūn-i *450* salt marsh of
 Iran and Pakistan
Māshkīd *see* Māshkel
Mashtagi *see* Mastaği
Masīlah, Wādī al *631* dry watercourse
 of E Yemen
Masindi *588* W Uganda
Masinga Reservoir *344* reservoir of
 C Kenya
Masirah, Gulf of *see* Maşīrah, Khalīj
Maşīrah, Jazīrat *448 var.* Masirah,
 Masira. Island of E Oman
Maşīrah, Khalīj *448 var.* Gulf of
 Masirah. Bay of the Arabian Sea,
 E Oman
Masis *98* SW Armenia
Masjed Soleymān *309 var.* Masjed-e
 Soleymān, Masjid-i Sulaiman. W Iran
Maskall *130* NE Belize
Maskanah *555 var.* Meskene. N Syria
Maskin *448 var.* Miskin. N Oman
Mask, Lough *316 Ir.* Loch Measca.
 Lake of W Ireland
Ma *627* river of Laos and Vietnam
Massa *322* N Italy
Massachusetts *603* state of NE USA
Massacre *224* W Dominica
Massawa *240 Amh.* Mits'iwa. E Eritrea
Massawa Channel *240* channel of the
 Red Sea between Dahlak Archipelago
 and mainland Eritrea
Massenya *180* SW Chad
Massif Central *255* plateau region of
 C France
Massilı *110 Rus.* Masally. S Azerbaijan
Massoukou *260 var.* Masuku,
 prev. Franceville. E Gabon
Mastaği *110 Rus.* Mastaga,
 var. Maştaga, Mashtagi.
 E Azerbaijan
Masterton *433* S North Island,
 New Zealand
Masuda *332* Honshū, W Japan
Masunga *142* NE Botswana
Masvingo *636 prev.* Nyanda,
 prev. Fort Victoria. SE Zimbabwe
Mât *652* river of NE Réunion
Matacawa Levu *248* island of the
 Yasawa Group, NW Fiji
Matadi *203* W Dem. Rep. Congo
Matagalpa *436* C Nicaragua
Matale *538* C Sri Lanka
Matam *510* N Senegal
Matamoros *403* E Mexico
Matana *162* C Burundi
Matanzas *210* NW Cuba
Matara *538* S Sri Lanka
Mataró *535* E Spain

Mataura *433* river of SW South Island, New Zealand

Matautu *500* Upolu, Samoa

Matá 'Utu *653* var. Mata Uta. ❖ of Wallis & Futuna, Île Uvea, S Wallis & Futuna

Matela's *366* W Lesotho

Matelot *574* NE Trinidad, Trinidad & Tobago

Matiguás *436* C Nicaragua

Matina *206* E Costa Rica

Matit *81* var. Mat. River of C Albania

Mato Grosso, Planalto de *145* plateau of C Brazil

Matopos *636* SW Zimbabwe

Matosinhos *474* prev. Matozinhos. NW Portugal

Mátra *292* mountain range of N Hungary

Maṭraḥ *448* var. Mutrah. NE Oman

Maṭrûḥ *232* var. Maṭrūḥ. NW Egypt

Matsapha *544* var. Matsapal, Mtsapa. C Swaziland

Matsieng *366* W Lesotho

Matsue *332* Honshū, W Japan

Matsumato *332* Honshū, C Japan

Matsusaka *332* Honshū, C Japan

Matsuyama *332* Shikoku, SW Japan

Matthews Ridge *286* N Guyana

Matthew Town *112* Great Inagua, Bahamas

Mattsee *106* lake of N Austria

Matuku *248* island to the SE of Viti Levu, S Fiji

Maturín *623* NE Venezuela

Mauga Silisili see Silisili, Mount

Maug Islands *650* island group of N Northern Mariana Islands

Maui *602* island of Hawaii, USA, C Pacific

Maun *142* C Botswana

Mauren *374* NE Liechtenstein

Maurice see Mauritius

Maú see Ireng

Mauritania *398-399* officially Islamic Republic of Mauritania, Ar. Mūrītānīyah. Country of W Africa divided into 12 admin. units (regions). ❖ Nouakchott

Mauritius *400-401* Fr. Maurice. Country of Indian Ocean divided into 9 admin. units (districts). ❖ Port Louis

Mavrovsko Ezero *381* lake of W FYR Macedonia

Mawlaik *159* NW Burma

Mawlamyine see Moulmein

Mawr, Wādī *631* dry watercourse of NW Yemen

Mawson *90* Australian research station of Greater Antarctica, Antarctica

Mayaguana *112* Island of S Bahamas

Mayaguana Passage *112* passage between Crooked I and Mayaguana, Bahamas

Mayagüez *651* W Puerto Rico

Mayagüez, Bahia *651* bay of the Caribbean Sea on the W coast of Puerto Rico

Mayaluka *544* SE Swaziland

Maya Mts *130* mountain range of Belize and Guatemala

Mayarí *211* SE Cuba

Maych'ew *245* var. Mai Chio, It. Mai Ceu. N Ethiopia

Maydī see Midi

Mayence see Mainz

Mayenne *255* river of NW France

Maykop *484* SW Russian Federation

Mayli-Say *359* Kir. Mayly-Say. W Kyrgyzstan

Mayly-Say see Mayli-Say

Maymyo *159* N Burma

Mayoko *200* SW Congo

Mayor Pablo Lagerenza see Capitán Pablo Lagerenza

Mayotte *649* French territorial collectivity of the Indian Ocean. ❖ Mamoudzou.

May Pen *331* S Jamaica

Mayreau *498* island of SW St Vincent & the Grenadines

Mayumba *260* S Gabon

Mazabuka *635* S Zambia

Mazagan see El Jadida

Mazār-e Sharīf *77* N Afghanistan

Mazaruni *286* river of N Guyana

Mazatenango *280* SW Guatemala

Mazatlán *403* W Mexico

Mažeikiai *376* NW Lithuania

Mazirbe *362* NW Latvia

Mazowe *636* prev. Mazoe. NE Zimbabwe

Mazowe *636* var. Mazoe. River of Mozambique and Zimbabwe

Mazra'at Turaynā *479* var. Traina Garden. S Qatar

Mazury *471* region of NE Poland

Mazyr *122* Rus. Mozyr'. SE Belarus

Mba see Ba

Mbabane *544* ❖ of Swaziland, NW Swaziland

Mbacké see Mbaké

Mbagne *398* SW Mauritania

Mbaïki *179* var. M'Baiki. SW Central African Republic

Mbakaou, Lac de *168* lake of C Cameroon

Mbaké *510* var. Mbacké. W Senegal

Mbala *635* prev. Abercorn. NE Zambia

Mbalabala *636* prev. Balla Balla. SW Zimbabwe

Mbale *588* E Uganda

Mbalmayo *168* var. M'Balmayo. S Cameroon

Mbalo *526* SE Guadalcanal, Solomon Is

Mbam *168* river of NW Cameroon

Mbanda *162* S Burundi

Mbandaka *203* prev. Coquilhatville. NW Dem. Rep. Congo

Mbanga *168* W Cameroon

M'Banza Congo *88* Port. São Salvador do Congo. NW Angola

Mbanza-Ngungu *203* W Dem. Rep. Congo

Mbarara *588* SW Uganda

Mbatiki see Batiki

Mbé *168* N Cameroon

Mbengga see Beqa

Mbéni *198* NE Grande Comore, Comoros

Mbeya *564* SW Tanzania

Mbigou *260* C Gabon

Mbilua see Vella Lavella

Mbinga *564* S Tanzania

Mbini *238* W Río Muni, Equatorial Guinea

Mbini see Uolo, Río

Mbomo *200* NW Congo

Mbomou see Bomu

Mbour *510* W Senegal

M'Bout see Mbout

Mbout *398* var. M'Bout. S Mauritania

Mbrès *179* C Central African Republic

Mbuji-Mayi *203* prev. Bakwanga. S Dem. Rep. Congo

Mbulu *564* N Tanzania

Mbulungwane *544* S Swaziland

Mbuluzi *544* var. Black Umbeluzi. River of Mozambique and Swaziland

Mbutha see Buca

Mbuye *162* C Burundi

Mchinji *385* prev. Fort Manning. W Malawi

McKean Island *348* island of the Phoenix Is, C Kiribati

M'Clintock Channel *170* var. McClintock Channel. Channel between Prince of Wales Island and Victoria Island, N Canada

McMurdo *90* US research station near Ross Shelf, Antarctica

Mdantsane *531* Eastern Cape, SE South Africa

Mdina *395* W Malta

Mead, Lake *602* reservoir of SW USA

Measca, Loch see Mask, Lough

Mecca see Makkah

Mechelen *127* Fr. Malines. C Belgium

Mecheria *83* var. Mechriyya. NW Algeria

Mecsek *292* mountain range of SW Hungary

Medan *304* Sumatra, E Indonesia

Medawachchiya *538* N Sri Lanka

Médéa *83* var. Lemdiyya, El Mediyya. N Algeria

Medellín *195* NW Colombia

Médenine *577* var. Madanīyīn. SE Tunisia

Medford *602* Oregon, NW USA

Medicine Hat *170* SW Canada

Medina see Al Madīnah

Médina Gonassé see Médina Gounas

Médina Gounas *510* var. Médina Gonassé. S Senegal

Medina Seringe Mass *263* W Gambia

Mediterranean Sea *255, 319, 580* Fr. Mer Méditerranée. Sea of the Atlantic Ocean, enclosed by N Africa, SW Asia and S Europe

Medjerda see Mejerda, Oued

Medoc *254* cultural region of SW France

Médouneu *260* N Gabon

Meekatharra *100* W Australia

Meemu Atoll see Mulaku Atoll

Meenen see Menen

Meerut *298* N India

Meghna *117* river of S Bangladesh

Meghri *98* var. Megri. SE Armenia

Mehdia see Mahdia

Meherpur *117* W Bangladesh

Meheso see Mi'ēso

Me Hka see Nmai Hka

Mehtarlām *77* var. Methariam, Meterlam. E Afghanistan

Meiganga *168* NE Cameroon

Meiktila *159* C Burma

Meissen *267* E Germany

Mejerda, Monts de la *577* var. Monts de la Medjerda, Monts de la Majardah. Mountain range of Algeria and Tunisia

Mejerda, Oued *577* var. Medjerda, Wādī Majardah. River of Algeria and Tunisia

Méjico see Mexico

Mejit *396* island of NE Marshall Islands

Mékambo *260* NE Gabon

Mek'elē *245* var. Makale. N Ethiopia

Mekerrhane, Sebkha *83* var. Sebkra Mekerrhane, Sebkha Meqerghane. Salt flat of C Algeria

Mékhé *510* NW Senegal

Meknès *414* N Morocco

Mekong *159, 165, 186, 361, 567, 627* Chin. Lancang Jiang, var. Lan-ts'ang Chiang, Cam. Mékôngk, Lao. Mènam Khong, Th. Mae Name Khong, Vtn. Sông Tiên Giang, Tib. Za Qu, var. Dza Chu. River of SE Asia

Mekong Delta *627* delta of S Vietnam

Mékrou *132, 157* river of W Africa

Melah, Oued el *577* var. Wādī al Milḥ. Dry watercourse of W Tunisia

Melah, Sebkhet el *577* var. Sabkhat al Milḥ. Salt flat of SE Tunisia

Melaka *386* var. Malacca. SW Peninsular Malaysia

Melbourne *101* SE Australia

Meleda see Mljet

Melekeiok *455* C Palau

Melfi *180* S Chad

Melilla *414, 535* enclave of Spain, NE Morocco

Melitopol' *590* SE Ukraine

Mellègue, Oued *577* var. Wādī Mallāq. River of Algeria and Tunisia

Mellerud *547* SW Sweden

Mellieha *395* NW Malta

Mellieha Ridge *395* ridge of Malta island, Malta

Mělník *216* NW Czech Republic

Melo *611* E Uruguay

Melo, Ilha de *284* var. Melho Island. Island of S Guinea-Bissau

Melsetter see Chimanimani

Melun *255* N France

Melville Hall *224* river of N Dominica

Melville Island *101* island of N Australia

Melville Island *170* island of Parry Islands, N Canada

Melville Islands see St. Giles Islands

Melville Peninsula *171* peninsula of N Canada

Melville Sound see Viscount Melville Sound

Memel see Neman

Memel see Klaipėda

Memphis *603* Tennessee, SE USA

Menabe *382* physical region of W Madagascar

Menado see Manado

Ménaka *392* E Mali

Mènam Khong see Mekong

Menbij see Manbij

Mendawai *304* river of Borneo, C Indonesia

Mende *255* S France

Mendefera *240* S Eritrea

Mendi *458* C Papua New Guinea

Mendip Hills *597* hills of W England, UK

Mendoza *95* W Argentina

Menen *127* prev. Meenen, Fr. Menin. W Belgium

Menongue *88* Port. Serpa Pinto. C Angola

Menorca *535* Eng. Minorca. Island of the Islas Baleares, E Spain

Mentakap *386* var. Mentakab. C Peninsular Malaysia

Mentawai, Kepulauan *304* island group to the W of Sumatra, Indonesia

Mentawai, Selat *304* strait of the Indian Ocean between Pulau Siberut and Sumatra, W Indonesia

Menzel Bourguiba *577* prev. Ferryville, var. Manzil Bū Ruqaybah. N Tunisia

Menzel Temime *577* var. Manzil Tamīm. N Tunisia

Meppel *429* NE Netherlands

Merawwah see Marāwiḥ

Merca see Marka

Mercedes see Villa Mercedes

Mercedes *95* NE Argentina

Mercedes *95* E Argentina

Mercedes *611* W Uruguay

Mercedes Umaña *237* SE El Salvador

Meregh see Mareeq

Méré Lava *619* Banks Islands, N Vanuatu

Mergui *159* SE Burma

Mergui Archipelago *159* island group of S Burma

Meriç see Maritsa

Mérida *403* E Mexico

Mérida *535* W Spain

Mérida *623* W Venezuela

Mérida, Cordillera de *623* var. Sierra Nevada de Mérida. Mountain range of W Venezuela

Meridian *603* Mississippi, SE USA

Merir *455* island of S Palau

Merizo *647* SW Guam

Merjama see Märjamaa

Merlimau, Pulau *521* island of SW Singapore

Merredin *100* SW Australia

Mersa Fatma *240* E Eritrea

Mersa Teklay *240* N Eritrea

Mersch *378* C Luxembourg

Mersey *597* river of NW England, UK

Mersin *581* var. İçel. S Turkey

Mersing *386* SE Peninsular Malaysia

Merthyr Tydfil *597* S Wales, UK

Meru *344* C Kenya

Merv see Mary

Meshed see Mashhad

Meskene see Maskanah

Mesopotamia *312* historical region of SW Asia

Mesopotamia *498* SE St Vincent, St Vincent & the Grenadines

Messalo *419* var. Mualo. River of NE Mozambique

Messina *323* var. Messana. Sicilia, S Italy

Messina *531* Northern Transvaal, NE South Africa

Messina, Stretto di *323* Eng. Strait of Messina. Strait connecting the Ionian Sea and Tyrrhenian Sea, between mainland Italy and Sicilia

Mesta see Néstos

Mestghanem see Mostaganem

Mestia *264* var. Mestiya. N Georgia

Meta *195, 623* river of Colombia and Venezuela

Meta Incognita Peninsula *171* peninsula of Baffin Island, NE Canada
Metangula *419* N Mozambique
Metapán *237* NW El Salvador
Metema *245* NW Ethiopia
Meterlam *see* Mehtarlām
Methariam *see* Mehtarlām
Metković *209* SE Croatia
Metu *245* *var.* Mattu, Mettu. W Ethiopia
Metz *255* NE France
Meuse *127, 255, 429* *var.* Maas. River of W Europe
Mexcala *see* Balsas
Mexiana, Ilha *145* island of N Brazil
Mexicali *403* NW Mexico
Mexicana, Altiplanicie *403* *Eng.* Plateau of Mexico, Mexican Plateau. Plateau of N Mexico
Mexico *402–405* officially United States of Mexico, *Sp.* Estados Unidos Mexicanos, Méjico. Country of North or Central America divided into 31 admin. units (states). ❖ Mexico City
Mexico City *403* *Sp.* Ciudad de México. of Mexico, C Mexico
Mexico, Gulf of *210, 403* *Sp.* Golfo de México. Gulf of the Atlantic Ocean, on the SE coast of North America
Mexico, Plateau of *see* Mexicana, Altiplanicie
Meyadine *see* Al Mayādīn
Meymaneh *77* *var.* Maimana. NW Afghanistan
Mezdra *152* NW Bulgaria
Mfanganu Island *344* *var.* Mfangano Island. Island of Lake Victoria, SW Kenya
Mfouati *200* S Congo
Mhangura *636* *var.* Mangula. N Zimbabwe
Mhlambanyatsi *544* W Swaziland
Mhlosheni *544* S Swaziland
Mhlume *544* NE Swaziland
Mhlumeni *544* NE Swaziland
Miami *603* Florida, SE USA
Miānwāli *451* NE Pakistan
Michalovce *523* *Ger.* Grossmichel, *Hung.* Nagymihály. E Slovakia
Michigan *603* state of NC USA
Michigan, Lake *171, 603* Lake of NC USA
Micomeseng *see* Mikomeseng
Micoud *497* SE St Lucia
Micronesia *406–407* officially Federated States of Micronesia, *prev.* Caroline Islands. Country of the Pacific Ocean divided into 4 admin. units (states). ❖ Palikir
Middelburg *429* SW Netherlands
Middelburg *531* Eastern Cape, S South Africa
Middelburg *531* Eastern Transvaal, NE South Africa
Middelfart *218* Fyn, SW Denmark
Middle Andaman *298* island of Andaman Islands to the SE of India
Middle Atlas *see* Moyen Atlas
Middlegate *650* C Norfolk Island
Middle Island *494* W St Kitts, St Kitts & Nevis
Middlesbrough *597* NE England, UK
Middlesex *130* C Belize
Mīdī *631* *var.* Maydī. NW Yemen
Miercurea-Ciuc *480* *Hung.* Csíkszereda. C Romania
Mieres del Camino *535* NW Spain
Mi'eso *245* *var.* Miesso, Meheso. C Ethiopia
Mikhaylovgrad *see* Montana
Mikhaylovka *484* W Russian Federation
Mikkeli *251* *Swe.* Sankt Michel. S Finland
Mikomeseng *238* *var.* Micomeseng. NE Río Muni, Equatorial Guinea
Mikuni-sammyaku *332* mountain range of Honshū, N Japan
Milagro *230* SW Ecuador
Milange *419* N Mozambique
Milano *322, 327.* *Eng.* Milan, *Ger.* Mailand. N Italy
Milas *580* SW Turkey
Mildura *101* SE Australia

Mil Düzü *110* *Rus.* Mil'skaya Step'. Physical region of C Azerbaijan
Milgis *344* *var.* Malgis. River of C Kenya
Mili *396* island of SE Marshall Islands
Milḩ, Baḩr al *see* Razāzah, Buḩayrat ar
Milḩ, Wādī al *see* Melah, Oued el
Millennium Island *348* *prev.* Caroline Island. Island of the Line Is, E Kiribati
Millet *497* C St Lucia
Millstätter See *106* lake of S Austria
Milo *283* river of E Guinea
Milondo, Mont *260* mountain of C Gabon
Mílos *275* island of SE Greece
Mil'skaya Step' *see* Mil Düzü
Milton Keynes *597* C England, UK
Milwaukee *603* Wisconsin, NC USA
Milyang *see* Miryang
Mimongo *260* C Gabon
Mīnā 'Abd Allāh *356* *var.* Mina Abdulla. E Kuwait
Mīnā' al Aḩmadī *356* *var.* Mina Ahmadi. E Kuwait
Mīnā' Jabal 'Alī *594* NE United Arab Emirates
Minas *611* S Uruguay
Mīnā' Sa'ūd *356* *var.* Mīnā' Su'ud. SE Kuwait
Minas de Corrales *611* N Uruguay
Minas de Matahambre *210* W Cuba
Minatitlán *403* SE Mexico
Minbu *159* W Burma
Minch, The *597* strait of the Atlantic Ocean, between Outer Hebrides and Scotland
Mincivan *110* *Rus.* Mindzhivan. SW Azerbaijan
Mindanao *466* island of S Philippines
Mindanao Sea *see* Bohol Sea
Mindelo *176* *var.* Porto Grande. São Vincente, N Cape Verde
Mindoro *466* island of C Philippines
Mindoro Strait *466* strait connecting South China Sea and Sulu Sea
Mindouli *200* S Congo
Mindzhivan *see* Mincivan
Mingäçevir *110* *Rus.* Mingechaur *var.* Mingeçaur. C Azerbaijan
Mingäçevir Su Anbarı *110* *Rus.* Mingechaurskoye Vodokhranilishche. Reservoir of NW Azerbaijan
Mingala *179* SE Central African Republic
Mingāora *451* *var.* Mingora, Mongora. N Pakistan
Mingechaurskoye Vodokhranilishche *see* Mingäçevir Su Anbarı
Ming-Kush *see* Min-Kush
Minho *see* Miño
Minicoy Island *298* island of Lakshadweep, SW India
Min-Kush *359* *Kir.* Ming-Kush. C Kyrgyzstan
Minna *440* C Nigeria
Minneapolis *603* Minnesota, NC USA
Minnesota *603* state of NC USA
Miño *474, 534* *Port.* Minho. River of Portugal and Spain
Minorca *see* Menorca
Minot *603* North Dakota, NC USA
Minsk *122* ❖ of Belarus, C Belarus
Minto Reef *406* atoll of C Micronesia
Minvoul *260* N Gabon
Minwakh *631* N Yemen
Miquelon *652* N Saint Pierre and Miquelon
Miquelon *652* island of N Saint Pierre and Miquelon
Miquelon, Cap *652* cape of the Atlantic Ocean on the coast of Miquelon, N Saint Pierre and Miquelon
Miragoâne *288* SW Haiti
Miranda de Ebro *535* N Spain
Mirbāṭ *448* *var.* Marbat. SW Oman
Mirebalais *288* C Haiti
Miri *386* NW Borneo, Malaysia
Mirim Lagoon *145, 611* *var.* Lake Mirim. Lagoon of Brazil and Uruguay
Mirim, Lake *see* Mirim Lagoon

Mirny *90* CIS research station of Greater Antarctica, Antarctica
Mirnyy *485* C Russian Federation
Mirpur *see* New Mīrpur
Mirtóo Pelagos *275* *Eng.* Mirtoan Sea. Area of the Mediterranean Sea, S Greece
Miryang *352* *var.* Milyang *Jap.* Mitsuō. SE South Korea
Misery, Mount *see* Liamuiga, Mount
Miskito Coast *see* Mosquito Coast
Miskitos, Cayos *436* island group of NE Nicaragua
Miskolc *292* NE Hungary
Misool, Pulau *304* island of Maluku, E Indonesia
Miṣrātah *371* *var.* Misurata. N Libya
Mississippi *603* river of C USA
Mississippi *603* state of SE USA
Missoula *602* Montana, NW USA
Missouri *603* river of NC USA
Missouri *603* state of C USA
Misurata *see* Miṣrātah
Mitau *see* Jelgava
Mitchell *101* river of NE Australia
Mitèmboni *see* Mitemele, Río
Mitemele, Río *238* *var.* Mitemele, Temboni, Utamboni. River of Equatorial Guinea and Gabon
Mitiaro *644* island of Southern Cook Islands, S Cook Islands
Mito *332* Honshū, SE Japan
Mitre Island *see* Fatutaka
Mitrovica *see* Kosovska Mitrovica
Mitrovicë *see* Kosovska Mitrovica
Mitsamiouli *198* N Grande Comore, Comoros
Mits'iwa *see* Massawa
Mitsoudjé *198* SW Grande Comore, Comoros
Mitsuyō *see* Miryang
Mitú *195* SE Colombia
Mitumba, Monts *203* *var.* Chaîne des Mitumba, Mitumba Range. Mountain range of E Dem. Rep. Congo
Mitzic *260* N Gabon
Miyako *333* Honshū, N Japan
Miyako-jima *332* island of Sakishima-shotō, SW Japan
Miyakonojō *332* Kyūshū, SW Japan
Miyazaki *332* Kyūshū, SW Japan
Miyoshi *332* Honshū, W Japan
Mizdah *371* *var.* Mizda. NW Libya
Mjøsa *444* *var.* Mjøsen. Lake of SE Norway
Mkhondvo *544* *var.* Mkondo. River of South Africa and Swaziland
Mladá Boleslav *216* *Ger.* Jungbunzlau. N Czech Republic
Mlanje *see* Mulanje
Mljet *209* *It.* Meleda. Island of S Croatia
Mmabatho *531* North West, N South Africa
Mmathethe *142* S Botswana
Mnjoli Dam *544* reservoir of NE Swaziland
Mo *444* NW Norway
Moa *518* river of W Africa
Moa *211* SE Cuba
Moabi *260* SW Gabon
Moala *248* island to the SE of Viti Levu, S Fiji
Moamba *419* SW Mozambique
Moanda *260* SE Gabon
Moba *203* E Dem. Rep. Congo
Mobaye *179* S Central African Republic
Mobile *603* Alabama, SE USA
Moca *226* N Dominican Republic
Moçambique *419* island and settlement of NE Mozambique
Moçâmedes *see* Namibe
Moce *248* island of the Lau Group, E Fiji
Mocha *see* Al Mukhā
Mochudi *142* S Botswana
Mocímboa da Praia *419* *var.* Vila de Mocímboa da Praia. N Mozambique
Môco *88* *var.* Serra Môco, Morro de Môco. Mountain of W Angola
Mocoa *195* SW Colombia
Mocuba *419* E Mozambique

Modena *322* NW Italy
Mödling *106* NE Austria
Modohn *see* Madona
Modriča *140* N Bosnia & Herzegovina
Moe *101* SE Australia
Moen *see* Weno
Möen *see* Møn
Moena *see* Muna, Pulau
Moengo *542* NE Suriname
Moesi *see* Musi
Moeskroen *see* Mouscron
Mogadishu *528* *Som.* Muqdisho, *It.* Mogadiscio. ❖ of Somalia, S Somalia
Mogador *see* Essaouira
Mogilëv *see* Mahilyow
Mogotón, Pico *436* mountain of NW Nicaragua
Mohales Hoek *366* SW Lesotho
Mohammadia *83* *var.* El Mohammaidia. NW Algeria
Mohammedia *414* *prev.* Fédala. NW Morocco
Moharek *see* Al Muḩarraq
Mohéli *198* *var.* Mwali. Island of Comoros
Mohn *see* Muhu
Moindou *650* C New Caledonia
Mõisaküla *242* *Ger.* Moiseküll. S Estonia
Moïssala *180* S Chad
Mokhotlong *366* NE Lesotho
Mokil *406* atoll of E Micronesia
Moknine *577* *var.* Al Muknīn. NE Tunisia
Mokp'o *352* *Jap.* Moppo. SW South Korea
Mokra Gora *512* mountain range of SW Serbia, Serbia & Montenegro (Yugo.)
Mokwa *440* W Nigeria
Moldau *see* Vltava
Moldavia *408–409* officially Republic of Moldova, *var.* Moldova, *prev.* Moldavian SSR, *Rus.* Moldavskaya SSR. Country of E Europe divided into 40 admin. units (districts). ❖ Chişinău
Molde *444* SW Norway
Moldo-Too, Khrebet *359* mountain range of C Kyrgyzstan
Moldova *see* Moldova
Molepolole *142* S Botswana
Môle-St-Nicolas *288* NW Haiti
Molineux *494* NE St Kitts, St Kitts & Nevis
Möll *106* river of S Austria
Mölndal *547* SW Sweden
Molodechno *see* Maladzyechna
Molodeczno *see* Maladzyechna
Molodezhnaya *90* CIS research station of Greater Antarctica, Antarctica
Molokai *602* island of Hawaii, USA, C Pacific
Molopo *142, 531* seasonal river of southern Africa
Molotov *see* Severodvinsk
Molotov *see* Perm'
Moloundou *168* SE Cameroon
Moluccas *see* Maluku
Molucca Sea *see* Maluku, Laut
Mombasa *344* SE Kenya
Môn *see* Anglesey
Møn *218* *prev.* Möen. Island of SE Denmark
Mona, Canal de la *226, 651* channel connecting the Atlantic Ocean and Caribbean Sea, between Dominican Republica and Puerto Rico
Monaco *410–411* officially Principality of Monaco. Country of W Europe divided into 4 admin. units (quarters). ❖ Monaco
Monaco *see* München
Monaghan *316* *Ir.* Muineachán. NE Ireland
Monagrillo *456* S Panama
Mona, Isla *651* island of SW Puerto Rico
Monapo *419* NE Mozambique
Monaragala *538* SE Sri Lanka
Monastir *577* *var.* Al Munastīr. NE Tunisia
Monastir *see* Bitola

New Zealand *432-435* officially The Dominion of New Zealand. Country of the Pacific Ocean, divided into 14 admin. units (regions). ❖ Wellington
Nezhyn *590* N Ukraine
Ngabé *200* SE Congo
Ngadda *441* river of NE Nigeria
Ngala *441* NE Nigeria
Ngangerabeli Plain *344* plain of SE Kenya
Ngaoundéré *168* var. N'Gaoundéré, N'Gaundere. N Cameroon
Ngara *564* NW Tanzania
Ngarama *492* N Rwanda
Ngardmau *455* C Palau
Ngaruangl *455* island of N Palau
Ngatik *406* atoll of E Micronesia
Ngau *see* Gau
N'Gaundere *see* Ngaoundéré
Nggamea *see* Qamea
Nggatokae *526* island of the New Georgia Islands, W Solomon Islands
Nggwavuma *544* var. Ingwavuma. River of South Africa and Swaziland
N'Giva *88* var. Ondjiva Port. Vila Pereira de Eça. S Angola
Ngo *200* SE Congo
Ngogolo *544* C Swaziland
Ngoko *168, 200* river of Cameroon and Congo
Ngorongoro Crater *564* crater and conservation area of N Tanzania
Ngororero *492* W Rwanda
Ngounié *260* river of Congo and Gabon
Ngouoni *260* E Gabon
Ngourti *439* E Niger
Ngozi *162* N Burundi
Nguigmi *439* SE Niger
Ngulu *406* atoll of W Micronesia
Ngum *360* river of C Laos
Nguna *619* island of C Vanuatu
Ngundu *636* S Zimbabwe
N'Gunza *see* Sumbe
Nguru *440* NE Nigeria
Ngwempisi *544* river of South Africa and Swaziland
Ngweze *see* Katima Mulilo
Nhacra *284* W Guinea-Bissau
Nhamundá *144* var. Yamundá, Jamundá. River of NW Brazil
Nha Trang *627* SE Vietnam
Nhlangano *544* prev. Goedgegun. SW Swaziland
Niagara Falls *171* SE Canada
Niagassola *283* var. Nyagassola. NE Guinea
Niamey *439* ❖ of Niger, SW Niger
Niamtougou *571* N Togo
Niandan *283* river of E Guinea
Niangay, Lac *392* lake of E Mali
Nianija Bolon *263* river of Gambia and Senegal
Niantanina *283* E Guinea
Niari *200* river of S Congo
Nias, Pulau *304* island to the W of Sumatra, W Indonesia
Niassa, Lago *see* Nyasa, Lake
Nicaragua *436-437* officially Republic of Nicaragua. Country of Central America divided into 16 admin. units (departments). ❖ Managua
Nicaragua, Lago de *436* var. Cocibolca, Gran Lago. Lake of S Nicaragua
Nicastro *323* S Italy
Nice *255* It. Nizza. SE France
Nicholls Town *112* Andros I, Bahamas
Nickerie *542* river of NW Suriname
Nicobar Islands *298* island group to the SE of India
Nicosia *215* var. Lefkosia, Turk. Lefkoşa. ❖ of Cyprus, C Cyprus
Nicoya *206* W Costa Rica
Nicoya, Península de *206* peninsula of W Costa Rica
Nictheroy *see* Niterói
Nidaros *see* Trondheim
Niedere Tauern *106* mountain range of C Austria
Niefang *238* var. Sevilla de Niefang. NW Río Muni, Equatorial Guinea
Niemen *see* Neman
Niéri Ko *510* river of SE Senegal
Nieuw Amsterdam *542* N Suriname

Nieuwegein *429* C Netherlands
Nieuwkoop *429* W Netherlands
Nieuw Nickerie *542* NW Suriname
Niğde *581* C Turkey
Niger *132, 283, 392, 439* river of W Africa
Niger *438-439* officially Republic of Niger. Country of West Africa divided into 7 admin. units (departments). ❖ Niamey
Nigeria *440-441* officially Federal Republic of Nigeria. Country of West Africa divided into 20 admin. units (19 states and 1 federal capital Territory). ❖ Abuja
Niger, Mouths of the *440* delta of the river Niger, on the S coast of Nigeria
Niigata *332* Honshū, N Japan
Niihama *332* Shikoku, SW Japan
Niihau *602* island of Hawaii, USA, C Pacific
Niimi *332* Honshū, W Japan
Nijmegen *429* Ger. Nimwegen. SE Netherlands
Nikki *132* E Benin
Nikolainkaupunki *see* Vaasa
Nikolayev *see* Mykolayiv
Nikol'sk-Ussuriyskiy *see* Ussuriysk
Nikol'skiy *see* Satpayev
Nikopol' *590* SE Ukraine
Nikšić *512* W Montenegro, Serbia & Montenegro (Yugo.)
Nikumaroro *348* var. Gardner I. Island of the Phoenix Is, C Kiribati
Nikunau *348* island of the Gilbert Is, W Kiribati
Nile *232, 540* Ar. Nahr an Nīl. River of N Africa
Nile Delta *232* delta of N Egypt
Nil, Nahr an *see* Nile
Nilphāmāri *117* NW Bangladesh
Nimba, Monts *283* var. Nimba Mountains. Mountain range of W Africa
Nimba, Mount *328, 368* mountain of W Africa
Nimba Mountains *see* Nimba, Monts
Nîmes *255* SE France
Nimwegen *see* Nijmegen
Ningbo *187* var. Ning-po, prev. Ninghsien. Zhejiang, E China
Ning-hsia *see* Ningxia
Ninghsien *see* Ningbo
Ning-po *see* Ningbo
Ningxia *187* Chin. Ningxia Huizu Zizhiqu, Eng. Ningsia Hui Autonomous Region, var. Ning-hsia. Autonomous region of N China
Ninotsminda *264* prev. Bogdanovka. S Georgia
Ninove *127* C Belgium
Niokolo Koba *510* river of SE Senegal
Niono *392* C Mali
Nioro *392* var. Nioro du Sahel. W Mali
Nioro du Rip *510* SW Senegal
Niort *255* W France
Nippon-kai *see* Japan, Sea of
Niquero *210* S Cuba
Nirin *see* Erhlin
Niš *512* Eng. Nish. E Serbia, Serbia & Montenegro (Yugo.)
Nişāb *631* var. Anşāb. SW Yemen
Nisporeni *408* Rus. Nisporeny. W Moldova
Nissan Islands *see* Green Islands
Nissum Bredning *218* inlet of North Sea on the NW coast of Denmark
Nistru *see* Dniester
Niterói *145* prev. Nictheroy. SE Brazil
Nitil *see* Natl
Nitra *523* Ger. Neutra, Hung. Nyitra. River of SW Slovakia
Nitra *523* Ger. Neutra, Hung. Nyitra. SW Slovakia
Niuafo'ou *572* var. Niuafoo. Island of NW Tonga
Niuatoputapu *572* var. Niuatobutabu, prev. Keppel Island. Island of N Tonga
Niuatui *587* islet of Nukufetau, Tuvalu
Niue *650* territory in free association with New Zealand, Pacific Ocean. ❖ Alofi
Niulakita *587* var. Nurakita. Coral atoll of S Tuvalu

Niutao *587* coral atoll of NW Tuvalu
Nizāmābād *298* C India
Nizhnevartovsk *484* C Russian Federation
Nizhniy Novgorod *484* prev. Gor'kiy. W Russian Federation
Nizhniy Pyandzh *see* Panji Poyon
Nizhnyaya Tunguska *485* Eng. Lower Tunguska. River of C Russian Federation
Nizza *see* Nice
Njaba *see* Nja Kunda
Njaiama *518* E Sierra Leone
Nja Kunda *263* var. Njaba. NW Gambia
Njamena *see* N'Djamena
Njardhvík *296* SW Iceland
Njazidja *see* Grande Comore
Njoeng Jacobkondre *542* C Suriname
Njombe *564* S Tanzania
Njoro *344* W Kenya
Nkanini *544* W Swaziland
Nkata Bay *see* Nkhata Bay
Nkayi *200* var. N'Kayi, prev. Jacob. S Congo
Nkhata Bay *385* var. Nkata Bay. N Malawi
Nkhotakota *385* var. Kota Kota, Nkota Kota. C Malawi
Nkonfap *440* S Nigeria
Nkongsamba *168* var. N'Kongsamba. W Cameroon
Nkumekie *238* C Río Muni, Equatorial Guinea
Nkundla *544* W Swaziland
Nkusi *588* river of W Uganda
Nmai Hka *159* var. Me Hka. River of N Burma forming a headstream of the Irrawaddy river
Noākhāli *117* prev. Sudharam. S Bangladesh
Nobeoka *332* Kyūshū, W Japan
Noboribetsu *332* Hokkaidō, N Japan
Nogal, Uadi *see* Nugaal
Noire, Rivière *400* river of SW Mauritius
Noire, Rivière *see* Black River
Noirmoutier, Île de *254* island of W France
Nokia *251* SW Finland
Nokou *180* W Chad
Nokoué, Lac *132* lake of S Benin
Nola *179* SW Central African Republic
Nólsoy *644* island of E Faeroe Islands
Noma, Ra's *see* Nawmah, Ra's
Nomuka *572* island of the Nomuka Group, Tonga
Nomuka Group *572* island group of W Tonga
Nomwin *406* atoll of C Micronesia
Nông Hêt *360* E Laos
Nong Khai *567* NE Thailand
Nonouti *348* island of the Gilbert Is, W Kiribati
Nonsan *352* Jap. Ronzan. W South Korea
Nonsuch Island *643* island of E Bermuda
Noord *642* N Aruba
Noord-Beveland *429* island of SW Netherlands
Noordoewer *423* S Namibia
Noordpunt *650* headland of Curaçao, W Netherlands Antilles
Noordzee *see* North Sea
Nor Achin *see* Nor Hachn
Nor Ačin *see* Nor Hachn
Norak *562* Rus. Nurek. W Tajikistan
Nord *646* N Greenland
Nordaustlandet *653* island of NE Svalbard
Norddeutsches Tiefland *266-267* Eng. North German Plain. Plain of N Germany
Nordfriesische Inseln *266* Eng. North Frisian Islands. Island group of NW Germany
Nordhausen *267* C Germany
Nordishavet *see* Arctic Ocean
Nord, Massif du *288* mountainous region of Haiti
Nord, Mer du *see* North Sea
Nord-Pas de Calais *258* administrative region of N France

Nordsee *see* North Sea
Nordsjøen *see* North Sea
Nordsøen *see* North Sea
Nordtiroler Kalkalpen *106* mountain range of W Austria
Nore *316* Ir. An Fheoir. River of SE Ireland
Norfolk *603* Virginia, E USA
Norfolk Island *650* Australian external territory of the South Pacific Ocean. ❖ Kingston
Norge *see* Norway
Nor Hachn *98* var. Nor Hachyn, Rus. Nor Achin, var. Nor Ačin. C Armenia
Nor Hachyn *see* Nor Hachn
Noril'sk *485* N Russian Federation
Norman *603* Oklahoma, SC USA
Normanby Island *458* island of SE Papua New Guinea
Normandie *255* Eng. Normandy. Cultural region of N France
Normandie, Collines de *255* hilly region of NW France
Norman Island *643* island of S British Virgin Islands
Norrköping *547* S Sweden
Norseman *100* SW Australia
Norskehavet *see* Norwegian Sea
Norsup *619* Malekula, Vanuatu
North Albanian Alps *81, 512* SCr. Prokletije, Alb. Bjeshkët e Nemuna. Mountain range of Albania and Serbia & Montenegro (Yugo.)
Northam *100* SW Australia
Northampton *597* C England, UK
North Andaman *159, 298* island of the Andaman Is, E India
North Battleford *170* SW Canada
North Bay *171* SE Canada
North Caicos *653* island of NW Turks and Caicos Islands
North Carolina *603* state of SE USA
North Channel *597* strait of Atlantic Ocean, between Northern Ireland and Scotland, UK
North Comino Channel *see* Ghawdex, Il-Fliegu ta'
North Dakota *603* state of NC USA
North Devon Island *see* Devon Island
Northeast China *191*
North East Point *644* headland on the NE coast of Christmas Island
Northeast Providence Channel *112* channel between Eleuthera I and Great Abaco I, Bahamas
Northern *531* prev. Northern Transvaal. Province of NE South Africa
Northern Cape *531* province of W South Africa
Northern Cook Islands *644* island group of N Cook Islands
Northern Cyprus, Republic of *see* Cyprus
Northern Dvina *see* Severnaya Dvina
Northern Forest Reserve *224* nature reserve of N Dominica
Northern Ireland *597* var. the Six Counties. Political division of UK
Northern Karoo *531* var. High Veld, Afr. Hoë Karoo. Plateau region of W South Africa
Northern Mariana Islands *650* Commonwealth territory of the USA, Pacific Ocean. ❖ Saipan
Northern Rhodesia *see* Zambia
Northern Sporades *see* Vor eioi Sporades
Northern Transvaal *see* Northern
North Frisian Islands *see* Nordfriesische Inseln
North German Plain *see* Norddeutsches Tiefland
North Huvadhu Atoll *390* var. Gaafu Alifu Atoll. Atoll of S Maldives
North Island *433* northernmost of the two main islands that comprise New Zealand
North Keeling Island *644* island of NW Cocos Islands

Pardubitz see Pardubice
Parecis, Chapada dos 144 var. Serra dos Parecis. Mountain range of W Brazil
Pares 92 E Antigua, Antigua & Barbuda
Parham 92 NE Antigua, Antigua & Barbuda
Paria, Gulf of 574, 623 gulf of the Atlantic Ocean, between Trinidad and Venezuela
Parika 286 NE Guyana
Parima, Serra 144 mountain range of Brazil and Venezuela
Paris 255, 258 ❖ of France, N France
Paris 348 Kiritimati, E Kiribati
Parita, Bahía de 456 bay of the Gulf of Panama
Parkan see Štúrovo
Párkány see Štúrovo
Parkent 614 E Uzbekistan
Parkhar see Farkhor
Parma 322 N Italy
Parnaíba 145 river of NE Brazil
Pärnu 242 Rus. Pyarnu, prev. Pernov, Ger. Pernau. SW Estonia
Pärnu 242 var. Pärnu Jõgi, Ger. Pernau. River of W Estonia
Pärnu Laht 242 bay of the Gulf of Riga, on the SW coast of Estonia
Paro 134 W Bhutan
P'aro-ho 352 var. Hwach'ŏn-chŏsuji. Reservoir of N South Korea
Páros 275 island of SE Greece
Parral 183 C Chile
Parrita 206 S Costa Rica
Parry Islands 170 island group of N Canada
Parry's 494 SE St Kitts, St Kitts & Nevis
Parson's Ground 494 N St Kitts, St Kitts & Nevis
Partizánske 523 prev. Šimonovany, Hung. Simony. W Slovakia
Pasaje 230 SW Ecuador
Pasaquina 237 E El Salvador
Pas de Calais see Dover, Strait of
Pa-shih Hai-hsia see Bashi Channel
Pashmakli see Smolyan
Pasión 280 river of N Guatemala
Pasir Mas 386 N Peninsular Malaysia
Pasir Panjang 521 reservoir of SW Singapore
Paso de los Toros 611 C Uruguay
Passau 267 SE Germany
Passo Fundo 145 S Brazil
Passu Keah 651 island of S Paracel Islands
Pastavy 122 Rus. Postavy, Pol. Postawy. NW Belarus
Pastaza 230, 463 river of Ecuador and Peru
Pasto 195 SW Colombia
Patagonia 95 semi-arid region of S South America
Patchchacan 130 N Belize
Pate 559 N Taiwan
Pate Island 344 var. Patta Island. Island of SE Kenya
Paterna 535 E Spain
Pathein see Bassein
Patía 195 river of SW Colombia
Patience 497 E St Lucia
Pati Point 647 headland on the NE coast of Guam
Patlong 366 S Lesotho
Patna 298 var. Azimabad. NE India
Patos 81 var. Patosi. SW Albania
Patos, Lagoa dos 145 lagoon of S Brazil
Pátra 275 var. Patras, prev. Pátrai. S Greece
Patta Island see Pate Island
Pattani 567 S Thailand
Pattle Island 651 island of W Paracel Islands
Patuākhāli 117 S Bangladesh
Patuca 290 river of E Honduras
Pau 255 S France
Paungde 159 SW Burma
Pāvilosta 362 W Latvia
Pavlodar 340 NE Kazakhstan
Pavlohrad 590 Rus. Pavlograd. E Ukraine

Pavuvu 526 island of C Solomon Islands
Pawai, Pulau 521 island of SW Singapore
Paysandú 611 NW Uruguay
Paz 237 river of Guatemala and El Salvador
Pazardzhik 152 var. Pazardžik, prev. Tatar Pazardzhik. SW Bulgaria
Pazin 209 NW Croatia
Pčinja 381 river of N FYR Macedonia
Pea 572 Tongatabu, Tongatapu Group, Tonga
Peace 170 river of W Canada
Peak, The 652 mountain of C Ascension Island
Pearl Islands see Perlas, Archipiélago de las
Pearl Lagoon see Perlas, Laguna de
Peary Land 646 physical region of N Greenland
Pebble Island 645 island of N Falkland Islands
Peć 512 S Serbia, Serbia & Montenegro (Yugo.)
Pechora 484 River of NW Russian Federation
Pecixe, Ilha de 284 island of W Guinea-Bissau
Pecos 603 river of SW USA
Pécs 292 Ger. Fünfkirchen. SW Hungary
Pedernales 226 SW Dominican Republic
Pedhieos 215 var. Kanli Dere. River of NE Cyprus
Pedhoulas 215 W Cyprus
Pedja 242 var. Pedja Jõgi. River of C Estonia
Pedoulas see Pedhoulas
Pedra Lume 176 Sal, NE Cape Verde
Pedregal 456 W Panama
Pedro Juan Caballero 456 E Paraguay
Pedro Santana 226 W Dominican Republic
Peel 647 W Isle of Man
Pegasus Bay 433 bay of the South Pacific Ocean, on the E coast of South Island, New Zealand
Pegeia see Peyia
Pegu 159 var. Bago. S Burma
Péhonko 132 NW Benin
Pei-ching see Beijing
Peikang 559 var. Pei-chiang, Jap. Hokkō. W Taiwan
Peinan Hsi 559 river of C Taiwan
Peineville 224 N Dominica
Peipsi Järv see Peipus, Lake
Peipus, Lake 242 Est. Peipsi Järv, Rus. Chudskoye Ozero. Lake of Estonia and Russian Federation
Peiraiás 275 prev. Piraiévs, Eng. Piraeus. SE Greece
Peka 366 N Lesotho
Pekalongan 304 Java, C Indonesia
Pekan Muara 150 N Brunei
Pekan Seria 150 W Brunei
Peking see Beijing
Pelée, Montagne 649 mountain of N Martinique
Peleliu 455 island of S Palau
Péligre, Lac de 288 C Haiti
Pelly Bay 170 N Canada
Pelmadulla 538 S Sri Lanka
Pélmonostor see Beli Manastir
Pelopónnisos 275 Eng. Peloponnese. Peninsula of S Greece
Pelotas 145 S Brazil
Pemagatsel 134 SE Bhutan
Pematangsiantar 304 Sumatra, W Indonesia
Pemba 564 island of E Tanzania
Pemba 419 prev. Porto Amélia. NE Mozambique
Pembroke 498 SW St Vincent, St Vincent & the Grenadines
Pembroke Hall see Libertad
Penambo, Banjaran 387 var. Penambo Range, Banjaran Tama Abu. Mountain range of Borneo, Malaysia and Indonesia
Penambo Range see Penambo, Banjaran
Penang see George Town

Peñas Blancas 436 S Nicaragua
Pen-ch'i see Benxi
Pendé 179 river of Central African Republic and Chad
Pendembu 518 E Sierra Leone
Pendjari 132, 157 river of Benin and Burkina
Pendzhikent see Panjakent
P'enghu Liehtao 559 Eng.Pescadores Islands. Island group of W Taiwan
P'eng-hu Shui-tao 559 Eng. Pescadores Channel. Channel connecting South China Sea and Taiwan Strait
P'enghu Tao 559 island of W Taiwan
Peng-pu see Bengbu
Penibético, Sistema 535 Eng. Baetic Cordillera, Baetic Mountains. Mountain range of S Spain
Peniche 474 W Portugal
Peninsular Malaysia 386 var. Malaya, prev. West Malaysia. Western part of Malaysia situated on S Malay Peninsula
Penki see Benxi
Pennine Alps 550 var. Alpes Penninae, Fr. Alpes Pennines, It. Alpi Pennine. Mountain range of SW Switzerland
Pennines 597 var. Pennine Chain. Mountain range of N England, UK
Pennsylvania 603 state of NE USA
Penong 101 S Australia
Penonomé 456 C Panama
Penrhyn 644 island of Northern Cook Islands, N Cook Islands
Penrith 597 NW England, UK
Pentascoinos see Yermasoyia
Pentecost 619 Fr. Pentecôte. Island of C Vanuatu
Penticton 170 SW Canada
Penza 484 W Russian Federation
Penzance 597 SW England, UK
Peoria 603 Illinois, C USA
Pepel 518 W Sierra Leone
Pereira 195 W Colombia
Pergamino 95 E Argentina
Perico 210 NW Cuba
Périgueux 255 SW France
Perim see Barim
Peringat 386 N Peninsular Malaysia
Perkhemahan Berakas 150 N Brunei
Perlas, Archipiélago de las 456 var. Pearl Islands. Island group of SE Panama
Perlas, Laguna de 436 var. Pearl Lagoon. Lagoon of the Caribbean Sea on the E coast of Nicaragua
Perlepe see Prilep
Perm' 484 prev. Molotov. W Russian Federation
Përmet 81 var. Permeti, Premet. S Albania
Pernambuco see Recife
Pernau see Pärnu
Pernik 152 prev. Dimitrovo. W Bulgaria
Pernov see Pärnu
Pérouse see Perugia
Perpignan 255 S France
Perquín 237 E El Salvador
Përrenjas 81 var. Prenjasi, Prenjas. E Albania
Persian Gulf see The Gulf
Persia see Iran
Perth 100 SW Australia
Perth 597 N Scotland, UK
Peru 462-465 officially Republic of Peru. Country of South America divided into 25 admin. units (24 departments and 1 constitutional province). ❖ Lima
Perugia 323 Fr. Pérouse. C Italy
Perugia, Lake of see Trasimeno, Lago
Pesaro 323 N Italy
Pescadores Channel see P'enghu Liehtao
Pescadores Islands see P'eng-hu Ch'ü-tao
Pescara 323 C Italy
Pesek Kechil, Pulau 521 island of SW Singapore
Pesek, Pulau 521 island of SW Singapore

Peshāwar 451 N Pakistan
Peshkopi 81 var. Peshkopia, Peshkopija. NE Albania
Pessons, Pic dels 86 mountain of SE Andorra
Petaḥ Tiqwa 319 C Israel
Petaling Jaya 386 W Peninsular Malaysia
Pétange 378 SW Luxembourg
Petani 386 var. Patani. NW Peninsular Malaysia
Petare 623 N Venezuela
Petauke 635 E Zambia
Petén Itzá, Lago 280 var. Lago de Flores. Lake of N Guatemala
Peterborough 597 E England, UK
Peterborough 171 SE Canada
Peterhead 597 NE Scotland, UK
Peter Island 643 island of S British Virgin Islands
Peters Mine 286 NW Guyana
Pétionville 288 S Haiti
Petit-Bourg 646 C Guadeloupe
Petit Canouan 498 island of S St Vincent & the Grenadines
Petite Butte 400 Rodrigues, Mauritius
Petite Côte 510 coastal region of W Senegal
Petite Dominique 278 island to the NE of Carriacou, Grenada
Petite Martinique 278 island to the NE of Carriacou, Grenada
Petite-Rivière-de-l'Artibonite 288 C Haiti
Petite-Rivière Noire, Piton de la 400 mountain range of SW Mauritius
Petite Savane 224 S Dominica
Petite Soufrière 224 E Dominica
Petite-Terre 649 island, E Mayotte
Petit-Goâve 288 S Haiti
Petitjean see Sidi Kacem
Petit Mustique 498 island of C St Vincent & the Grenadines
Petit Piton 497 mountain of SW St Lucia
Petit-Popo see Aného
Petit St. Vincent Island 278 island to the NE of Carriacou, Grenada
Petra 338 archaeological site of W Jordan
Petre Bay 433 bay of the South Pacific Ocean, on the coast of Chatham Island, New Zealand
Petrich 152 var. Petrič. SW Bulgaria
Petrinja 209 N Croatia
Petroaleksandrovsk see Türtkül
Petropavlovsk 340 N Kazakhstan
Petropavlovsk-Kamchatskiy 485 NE Russian Federation
Petrópolis 145 SE Brazil
Petrosani 480 W Romania
Petrovgrad see Zrenjanin
Petrovsk-Port see Makhachkala
Petrozavodsk 484 Fin. Petroskoi. NW Russian Federation
Pettau see Ptuj
Pevek 485 NE Russian Federation
Peyia 215 var. Pegeia. SW Cyprus
Pezinok 523 Ger. Bösing, Hung. Bazin. SW Slovakia
Phalaborwa 531 Northern Transvaal, NE South Africa
Phangan, Ko 567 island of S Thailand
Phan Rang-Thap Cham 627 SE Vietnam
Phan Thiêt 627 S Vietnam
Phet Buri see Phetchaburi
Phetchaburi 567 var. Phet Buri. C Thailand
Philadelphia 603 Pennsylvania, NE USA
Philip Island 650 island of S Norfolk Island
Philippeville see Skikda
Philippines 466-469 officially Republic of the Philippines. Country of SE Asia divided into 14 admin. units (regions). ❖ Manila
Philippine Sea 406, 466 sea of the Pacific Ocean to the E of the Philippines
Philipsburg 650 St Martin, N Netherlands Antilles
Phillips 494 NE St Kitts, St Kitts & Nevis

Phitsanulok *567 var.* Muang Phitsanulok. N Thailand
Phlórina *see* Flórina
Phnom Penh *165 Cam.* Phnum Pénh. ❖ of Cambodia, S Cambodia
Phnum Aôral *165 prev.* Phnom Aural. Mountain of W Cambodia
Phoenix *602* Arizona, SW USA
Phoenix *400* C Mauritius
Phoenix Island *see* Rawaki
Phoenix Islands *348* island group of C Kiribati
Phôngsali *360 var.* Phong Saly. N Laos
Phong Saly *see* Phôngsali
Phou Bia *360 var.* Pou Bia. Mountain of C Laos
Phrae *567 var.* Muang Phrae. N Thailand
Phra Nakhon Si Ayutthaya *see* Ayutthaya
Phu Cuong *see* Thu Dâu Môt
Phuket *567 Mal.* Ujung Salang. S Thailand
Phuket, Ko *567* island of S Thailand
Phumĭ Chhlong *165* S Cambodia
Phumĭ Chhuk *165* S Cambodia
Phumĭ Chŏăm *165* SW Cambodia
Phumĭ Kâmpóng Trâbêk *165 prev.* Phum Kompong Trabek. C Cambodia
Phumĭ Koŭk Kdoŭch *165* NW Cambodia
Phumĭ Krêk *165* SE Cambodia
Phumĭ Labăng Siĕk *165* NE Cambodia
Phumĭ Mlu Prey *165* N Cambodia
Phumĭ Sâmraông *165 var.* Phumĭ Sâmroŭng, *prev.* Phum Samrong. NW Cambodia
Phumĭ Spoe Tbong *165* C Cambodia
Phumĭ Thmâ Pôk *165* NW Cambodia
Phumĭ Véal Rénh *165* SW Cambodia
Phuntsholing *134* SW Bhutan
Phu Quôc, Đao *627* island of SW Vietnam
Piacenza *322 Fr.* Paisance. N Italy
Piatra-Neamţ *480* NE Romania
Piave *323* river of N Italy
Piaye *497* S St Lucia
Pibor *245* river of Ethiopia and Sudan
Picardie *255 Eng.* Picardy. Cultural region of N France
Pichelin *224* S Dominica
Pico *474 var.* Ilha do Pico. Island of the Azores, Portugal
Picton, Isla *183* island of S Chile
Pidjani *198* SE Grande Comore, Comoros
Pidurutalagala *538* mountain of S Sri Lanka
Piedras *463* river of E Peru
Pielinen *251 var.* Pielisjärvi. Lake of E Finland
Pierre *603* South Dakota, NC USA
Piešťany *523 Ger.* Pistyan, *Hung.* Pöstyén. W Slovakia
Pietermaritzburg *531* Kwazulu Natal, E South Africa
Pietersaari *see* Jakobstad
Pietersburg *531* Northern Transvaal, NE South Africa
Piet Retief *531* Eastern Transvaal, E South Africa
Piggs Peak *544* NW Swaziland
Pigs, Bay of *210* bay of the Caribbean Sea, on southern coast of C Cuba
Pihkva Järv *see* Pskov, Lake
Pikelot *406* island of C Micronesia
Pikine *510* W Senegal
Pikounda *200* C Congo
Piła *471 Ger.* Schneidemühl. NW Poland
Pilar *460 var.* Villa del Pilar. S Paraguay
Pilas Group *466* island group of Sulu Archipelago, SW Philippines
Pilcomayo *95, 136, 460* river of C South America
Pilgrimkondre *542* NE Suriname
Pilis *292 var.* Philis. Mountain range of N Hungary
Pillories, The *498* islands of C St Vincent & the Grenadines
Pillsbury Sound *653* strait of the Caribbean Sea, C Virgin Islands

Pilsen *see* Plzeň
Pimpri *298* W India
Pinang *see* George Town
Pinang, Pulau *386 prev.* Prince of Wales Island. Island of NW Peninsular Malaysia
Pinar del Río *210* W Cuba
Píndos *275 prev.* Píndhos, *Eng.* Pindus Mountains, *var.* Píndhos Óros. Mountain range of C Greece
Pine Bluff *603* Arkansas, SC USA
Pineiós *275 prev.* Piniós. River of C Greece
Pines, Isle of *see* Juventud, Isla de la
Pinetown *531* Kwazulu Natal, E South Africa
P'ingchen *559 Jap.* Heichin. N Taiwan
Pingelap *406* atoll of E Micronesia
P'ingtung *559 Jap.* Heitō. SW Taiwan
Pinkiang *see* Harbin
Pinos, Isla de *see* Juventud, Isla de la
Pins, Île des *650 var.* Kunyé. Island of S New Caledonia
Pinsk *122 Pol.* Pińsk. SW Belarus
Pinta, Isla *230* island of N Galapagos Is, Ecuador
Piracicaba *145* S Brazil
Pirada *284* NE Guinea-Bissau
Piran *524 It.* Pirano. SW Slovenia
Piriápolis *611* S Uruguay
Pirita *242* river of N Estonia
Pirojpur *117* SW Bangladesh
Pirot *512* SE Serbia, Serbia & Montenegro (Yugo.)
Pisa *322* N Italy
Pisco *463* SW Peru
Písek *216* SW Czech Republic
Pishpek *see* Bishkek
Pissila *157* C Burkina
Pistoia *322* N Italy
Pistyan *see* Piešťany
Pita *283* NW Guinea
Pitalito *195* SW Colombia
Pitcairn Island *651* island of S Pitcairn Islands
Pitcairn Islands *651* British dependent territory of the Pacific Ocean. ❖ Adamstown
Pitche *284* E Guinea-Bissau
Piteå *547* NE Sweden
Piteşti *480* S Romania
Pitseng *366* N Lesotho
Pitt Island *433* island of Chatham Islands, New Zealand
Pittsburgh *603* Pennsylvania, NE USA
Pitt Strait *433* strait of Pacific Ocean, between Chatham Island and Pitt Island, New Zealand
Pituffik *646 prev.* Dundas. NW Greenland
Piura *463* NW Peru
Piva *512 var.* Diva. River of C Montenegro, Serbia & Montenegro (Yugo.)
Pivdennyy Bug *590 Rus.* Yuzhnyy Bug. River of W and S Ukraine
Pivsko Jezero *512* lake of NW Montenegro, Serbia & Montenegro (Yugo.)
Pjandž *see* Pyandzh
Placetas *210* C Cuba
Plačkovica *381* mountain range of E FYR Macedonia
Plaisance *288* N Haiti
Plakenska Planina *381* mountain range of SW FYR Macedonia
Plana Cays *112* islets of S Bahamas
Planken *374* C Liechtenstein
Plasencia *535* W Spain
Plate, Île *see* Flat Island
Platte *602-603* river of C USA
Platte, Île *516* island of E Seychelles
Plattensee *see* Balaton
Plauer See *267* lake of NE Germany
Plây Cu *see* Pleiku
Pleasant Island *see* Nauru
Pleebo *see* Plibo
Pleiku *see* Plây Cu
Plenty, Bay of *433* inlet of the Pacific Ocean, on the coast of NE North Island, New Zealand
Pleskau *see* Pskov
Pleven *152 prev.* Plevna. N Bulgaria
Plezzo *see* Bovec

Plibo *368 var.* Pleebo. SE Liberia
Pljevlja *512 prev.* Plevlje. W Serbia, Serbia & Montenegro (Yugo.)
Płock *471* C Poland
Plöcken *106 It.* Passo di Monte Croce Carnico, *var.* Plöcken Pass, *Ger.* Plöckenpass. Mountain pass of SW Austria
Pločno *140* mountain of SW Bosnia & Herzegovina
Ploieşti *480 prev.* Ploeşti. SE Romania
Plovdiv *152 Gk* Philippopolisanc, *prev.* Eumolpias. SW Bulgaria
Plumtree *636* SW Zimbabwe
Plungė *376* NW Lithuania
Plyeshchanitay *122* N Belarus
Plymouth *649* ❖ of Montserrat, SW Montserrat
Plymouth *597* SW England, UK
Plymouth *574* SW Tobago, Trinidad & Tobago
Plzeň *216 Ger.* Pilsen. W Czech Republic
Po *323* river of N Italy
Pô *157* S Burkina
Poabial *368* E Liberia
Pobè *132 var.* Pobé. S Benin
Pobedy, Pik *359 var.* Pobeda Peak, *Chin.* Tomur Feng. Mountain of China and Kyrgzstan
Pocatello *602* Idaho, NW USA
Pocrí *456* S Panama
Podgorica *512 prev.* Titograd. S Montenegro, Serbia & Montenegro (Yugo.)
Podil'ska Vysochyna *590* mountain range of SW Ukraine
Podkamennaya Tunguska *485 Eng.* Stony Tunguska. River of C Russian Federation
Podravska Slatina *see* Slatina
Poeketi *542* E Suriname
Pogradec *81 var.* Pogradeci. SE Albania
P'ohang *352 Jap.* Hokō. E South Korea
Pohjanlahti *see* Bothnia, Gulf of
Pohjanmaa *251 Swe.* Österbotten, *Eng.* Ostrobothnia. Physical region of W Finland
Pohnpei *406 prev.* Ascension, Ponape. Island of E Micronesia
Pohnpei Islands *406* island group E Micronesia
Poindimié *650* C New Caledonia
Point de Galle *see* Galle
Pointe-à-Pitre *646* C Guadeloupe
Pointe-à-Raquette *288* Île de la Gonâve, Haiti
Pointe Michel *224 var.* La Pointe. SW Dominica
Pointe-Noire *646* W Guadeloupe
Pointe-Noire *200* S Congo
Point Fortin *574* SW Trinidad, Trinidad & Tobago
Poitiers *255* C France
Poitou *254-255* cultural region of W France
Poivre Atoll *516* atoll of the Amirante Islands, C Seychelles
Pokhara *427* C Nepal
Pokigron *542* C Suriname
Pokrovka *see* Kyzyl-Suu
Pola *see* Pula
Poland *470-473* officially Republic of Poland, *Pol.* Polska. Country of E Europe divided into 49 admin. units (województwo). ❖ Warsaw
Polatlı *580* C Turkey
Polatsk *122 Rus.* Polotsk. N Belarus
Pol-e Khomrī *77 var.* Pul-i-Khumri. NE Afghanistan
Poliçan *81 var.* Poliçani. S Albania
Polillo Islands *466* island group of N Philippines
Polis *215 var.* Poli. W Cyprus
Polochic *280* river of C Guatemala
Polonnaruwa *538* C Sri Lanka
Poltava *590* NE Ukraine
Poltoratsk *see* Ashgabat
Põltsamaa *242 Ger.* Oberpahlen. C Estonia
Põltsamaa *242 var.* Pyltsamaa. River of C Estonia
Põlva *242 Ger.* Pölwe. SE Estonia

Pomeranian Bay *267, 471 Pol.* Zatoka Pomorska, *Ger.* Pommersche Bucht. Bay of the Baltic Sea, on the coasts of Germany and Poland
Pomio *458* New Britain, Papua New Guinea
Pommersche Bucht *see* Pomeranian Bay
Pomona *130* E Belize
Pomorie *152 var.* Pomoriye. E Bulgaria
Pomorska, Zatoka *see* Pomeranian Bay
Ponape *see* Pohnpei
Ponce *651* S Puerto Rico
Pondicherry *298* S India
Ponérihouen *650* C New Caledonia
Ponferrada *535* NW Spain
Pongo *540* river of S Sudan
Ponta Delgada *474* São Miguel, Azores, Portugal
Ponta Grossa *145* S Brazil
Pontevedra *535* NW Spain
Pontianak *304* Borneo, C Indonesia
Pontian Kechil *386 var.* Puntian Kecil, Pontian Kecil. S Peninsular Malaysia
Pontoise *255* N France
Pontypridd *597* S Wales, UK
Ponziane, Isole *323* island of C Italy
Pooh San *644* NE Christmas Island
Poole *597* S England, UK
Poona *see* Pune
Pooneryn *538* N Sri Lanka
Poopó, Lago *136 var.* Lago Pampa Aullagas. Lake of W Bolivia
Popayán *195* SW Colombia
Popomanaseu, Mount *526* mountain of S Guadalcanal, Solomon Islands
Popondetta *458* SE Papua New Guinea
Popovo *152* N Bulgaria
Poprad *523 Ger.* Deutschendorf, *Hung.* Poprád. NE Slovakia
Poprad *523 Ger.* Popper, *Hung.* Poprád. River of Poland and Slovakia
Pori *251 Swe.* Björneborg. SW Finland
Porirua *433* S North Island, New Zealand
Porlamar *623* Isla de Margarita, Venezuela
Porsangen *444* fjord of N Norway
Porsgrunn *444* S Norway
Portachuelo *136* C Bolivia
Portage la Prairie *170* S Canada
Portalegre *474* E Portugal
Port Alfred *531* Eastern Cape, S South Africa
Port Antonio *331* E Jamaica
Port Augusta *101* S Australia
Port-au-Prince *288* ❖ of Haiti, S Haiti
Port Blair *298* S Andaman, SE India
Port-Bouët *328* SE Ivory Coast
Port d'Envalira *86* zigzag pass of E Andorra
Port-de-Paix *288* N Haiti
Port Dickson *386* SW Peninsular Malaysia
Port Elizabeth *498* Bequia, St Vincent & the Grenadines
Port Elizabeth *531* Eastern Cape, S South Africa
Port Erin *647* SW Isle of Man
Port Étienne *see* Nouâdhibou
Port Florence *see* Kisumu
Port Francqui *see* Ilebo
Port-Gentil *260* W Gabon
Port Harcourt *440* S Nigeria
Port Hedland *100* NW Australia
Port Howard *645* West Falkland, C Falkland Islands
Portimão *474 var.* Vila Nova de Portimão. S Portugal
Port Láirge *see* Waterford
Portland *603* Maine, NE USA
Portland *602* Oregon, NW USA
Portland Bight *331* bay of Caribbean Sea
Portland Point *652* headland on the SW coast of Ascension Island
Port Laoise *316 Ir.* Portlaoighise, Portlaoise, *prev.* Maryborough. C Ireland
Port Lincoln *101* S Australia

Port Loko *518* W Sierra Leone
Port-Louis *646* N Guadeloupe
Port Louis *400 var.* Port-Louis.
❖ of Mauritius, NW Mauritius
Port Lyautey *see* Kénitra
Port Macquarie *101* E Australia
Port Maria *331* N Jamaica
Port Mathurin *400* Rodrigues,
Mauritius
Port Morant *331* E Jamaica
Portmore *331* SE Jamaica
Port Moresby *458* ❖ of Papua New
Guinea, SE Papua New Guinea
Porto *474 Eng.* Oporto. NW Portugal
Porto Alegre *145 prev.* Pôrto Alegre.
S Brazil
Porto Alegre *505* S São Tomé,
Sao Tome & Principe
Porto Alexandre *see* Tombua
Porto Amélia *see* Pemba
Portobelo *456 var.* Porto Bello, Puerto
Bello. N Panama
Porto Edda *see* Sarandë
Port-of-Spain *574* ❖ of Trinidad
& Tobago, NW Trinidad,
Trinidad & Tobago
Porto Gole *284* C Guinea-Bissau
Porto Grande *see* Mindelo
Porto-Novo *132* ❖ of Benin, S Benin
Porto Santo *474 var.* Ilha do Porto
Santo. Island of the Madeira Is,
Portugal
Porto Torres *323* Sardegna, W Italy
Porto Velho *144 prev.* Pôrto Velho.
W Brazil
Portoviejo *230 var.* Puertoviejo.
W Ecuador
Port Pirie *101* S Australia
Port Refuge *644* strait of the Indian
Ocean between Horsburgh Island and
Direction Island, C Cocos Islands
Port Royal *331* SE Jamaica
Port Said *232 Ar.* Bur Sa'îd. N Egypt
Port St Mary *647* S Isle of Man
Portsmouth *597* S England, UK
Portsmouth *224 var.* Grande-Anse.
NW Dominica
Port Stanley *see* Stanley
Port Stephens *645* West Falkland,
W Falkland Islands
Port Sudan *540 var.* Bûr Sûdân.
NE Sudan
Port Swettenham *see* Kelang
Port Talbot *597* S Wales, UK
Portugal *474-477* officially Republic of
Portugal. Country of SW Europe
divided into 18 admin. units
(districts). ❖ Lisbon
Portuguese East Africa
see Mozambique
Port-Vila *619 var.* Vila. ❖ of Vanuatu,
Éfate, Vanuatu
Porvenir *183* Tierra del Fuego, Chile
Porvenir *136* NW Bolivia
Posadas *95* NE Argentina
Posen *see* Poznań
Posŏng *352* river of S South Korea
Postojna *524 Ger.* Adelsberg,
It. Postumia. SW Slovenia
Pöstyén *see* Piešťany
Potaro *286* river of C Guyana
Potchefstroom *531* North West,
N South Africa
Potenza *323* S Italy
Potgietersrus *531* Northern Transvaal,
NE South Africa
Pot House *121* E Barbados
Po'ti *264* W Georgia
Potiskum *440* NE Nigeria
Potoru *518* S Sierra Leone
Potosí *136* S Bolivia
Potsdam *267* NE Germany
Potters Village *92* C Antigua,
Antigua & Barbuda
Pott, Île *650* island of Îles Belep,
W New Caledonia
Pottuvil *538* SE Sri Lanka
Potwar Plateau *450* plateau of
NE Pakistan
Poudre d'Or *400* NE Mauritius
Pouembout *653* W New Caledonia
Poum *650* N New Caledonia
Pout *510* W Senegal
Poutasi *500* Upolu, Samoa

Poûthĭsăt *165 var.* Pursat. River of
W Cambodia
Poûthĭsăt *165 prev.* Pursat.
W Cambodia
Po Valley *322* valley of N Italy
Považská Bystrica *523*
Ger. Waagbistritz, *Hung.*Vágbeszterce.
NW Slovakia
Póvoa de Varzim *474* NW Portugal
Powell, Lake *602* reservoir of
SW USA
Poya *650* C New Caledonia
Poyang Hu *187* lake of E China
Poyan Reservoir *521* reservoir of
W Singapore
Poza Rica *403 var.* Poza Rica de
Hidalgo. C Mexico
Poznań *471 Ger.* Posen. W Poland
Pozo Colorado *460* C Paraguay
Pozsega *see* Slavonska Požega
Pozsony *see* Bratislava
Prábis *284* W Guinea-Bissau
Præstø *218* Sjælland, SE Denmark
Prague *216 Cz.* Praha, *Ger.* Prag.
❖ of Czech Republic,
NW Czech Republic
Praia *176* ❖ of Cape Verde, Santiago,
S Cape Verde
Praslin *516* island of the Inner Islands,
NE Seychelles
Praslin *497* E St Lucia
Prato *322* N Italy
Preguiça *176* São Nicolau,
N Cape Verde
Prenjas *see* Përrenjas
Preparis Island *159* island of
SW Burma
Přerov *216 Ger.* Prerau.
SE Czech Republic
Presidente Prudente *145* S Brazil
Prešov *523 Ger.* Eperies, *var.*
Preschau, *Hung.* Eperjes. NE Slovakia
Prespa, Lake *81, 275, 381 Alb.* Liqen i
Prespës, *Mac.* Prespansko Ezero,
Gk Límni Megáli Préspa, *var.* Limni
Prespa. Lake of SE Europe
Prespës, Liqen i *see* Prespa, Lake
Pressburg *see* Bratislava
Prestea *272* SW Ghana
Preston *597* NW England, UK
Pretoria *531* ❖ of South Africa.
Pretoria-Witwatersrand-Vereeniging,
NE South Africa
Pretoria-Witwatersrand-Vereeniging
531 province of NE South Africa
Préveza *275* W Greece
Prey Vêng *165* S Cambodia
Priboj *512* W Serbia, Serbia &
Montenegro (Yugo.)
Příbram *216* W Czech Republic
Prickly Pear Cays *642* island group
of NW Anguilla
Prieska *531* Northern Cape,
C South Africa
Prievidza *523* C Slovakia
Prijedor *140* NW Bosnia
& Herzegovina
Prilep *381 Turk.* Perlepe.
S FYR Macedonia
Prince Albert *170* SW Canada
Prince Edward Island *171* province
and island of SE Canada
Prince Edward Island *531* island of
the Prince Edward Islands,
S South Africa
Prince Edward Islands *531* island
group of S South Africa
Prince George *170* W Canada
Prince Island *see* Príncipe
Prince of Wales Island *see* Pinang,
Pulau
Prince of Wales Island *170* island of
N Canada
Prince Patrick Island *170* island of
Parry Islands, N Canada
Prince Rupert *170* W Canada
Prince Rupert Bay *224* bay of the
Caribbean Sea, to the
NW of Dominica
Princes Town *574* SW Trinidad,
Trinidad & Tobago
Príncipe *505 var.* Príncipe Island,
Eng. Prince Island. Island to the N of
São Tomé, Sao Tome & Principe

Pripet *122* river of S Belarus
Pripet Marshes *122, 590* forested and
swampy region of Belarus and
Ukraine
Priština *512* S Serbia, Serbia
& Montenegro (Yugo.)
Privas *255* SE France
Privigye *see* Prievidza
Priwitz *see* Prievidza
Prizren *512 Alb.* Prizreni. S Serbia,
Serbia & Montenegro (Yugo.)
Probištip *381* N FYR Macedonia
Probolinggo *304* Java, C Indonesia
Progreso *611* S Uruguay
Prome *159 var.* Pyè. SW Burma
Promissão, Represa de *145* reservoir
of S Brazil
Proskurov *see* Khmel 'nyts'kyy
Prostějov *216 Ger.* Prossnitz.
SE Czech Republic
Provadiya *152 var.* Provadija.
E Bulgaria
Provence *255* cultural region of
SE France
Providence *603* Rhode Island, NE USA
Providence *121* S Barbados
Providence Atoll *516 var.* Providence.
Atoll of the Farquhar Group,
S Seychelles
Providenciales *653* island of
NW Turks and Caicos Islands
Provo *602* Utah, SW USA
Prudhoe Bay *602* Alaska, USA
Prune Island *498* island of
SW St Vincent & the Grenadines
Prut *408, 480, 590 Ger.* Pruth. River
of E Europe
Pruth *see* Prut
Pruzhany *122* SW Belarus
Pryazova'ks Vysochyna *590* mountain
range of SE Ukraine
Prychornomors'ka Nyzovyna *590*
mountain range of S Ukraine
Prydniprovs'ka Nyzovyna *590*
mountain range of NE Ukraine
Prydniprovs'ka Vysochnya *590*
mountain range of NW Ukraine
Przemyśl *471* SE Poland
Przheval'sk *see* Karakol
Pskov *484 Ger.* Pleskau.
W Russian Federation
Pskov, Lake *242 Est.* Pihkva Järv,
Rus. Pskovskoye Ozero. Lake of
Estonia and Russian Federation
Ptsich *122 Rus.* Ptich'. River of
C Belarus
Ptuj *524 Ger.* Pettau. NE Slovenia
Pua'a, Cape *500* cape on the coast of
Savai'i, NW Samoa
Pu'apu'a *500* Savali'i, Samoa
Pucallpa *463* C Peru
Puch'ŏn *352 prev.* Punwŏn.
NW South Korea
Pudasjärvi *251* C Finland
Puebla *403 var.* Puebla de Zaragoza.
S Mexico
Pueblo *603* Colorado, SW USA
Pueblo Nuevo Tiquisate *280*
var. Tiquisate. SW Guatemala
Puente Alto *183* C Chile
Puerto Acosta *136* W Bolivia
Puerto Aisén *183* S Chile
Puerto Armuelles *456* W Panama
Puerto Ayacucho *623* SW Venezuela
Puerto Bahía Negra *460* N Paraguay
Puerto Baquerizo Moreno *230* San
Cristobal I, Galapagos Is.
Puerto Barrios *280* E Guatemala
Puerto Bello *see* Portobelo
Puerto Berrío *195* N Colombia
Puerto Busch *136 var.* Puerto General
Busch. SE Bolivia
Puerto Cabello *623* N Venezuela
Puerto Cabezas *436 var.* Bilwi.
NE Nicaragua
Puerto Carreño *195* E Colombia
Puerto Casado *460* C Paraguay
Puerto Cooper *460* C Paraguay
Puerto Cortés *290* NW Honduras
Puerto El Carmen de Putumayo
230 var. Putumayo. NW Ecuador
Puerto el Triunfo *237* S El Salvador
Puerto Inírida *195 var.* Obando.
E Colombia

Puerto Iradier *see* Cogo
Puerto La Cruz *623* NE Venezuela
Puerto Lempira *290* E Honduras
Puertolland *535* SW Spain
Puerto Maldonado *463* E Peru
Puerto México *see* Coatzacoalcos
Puerto Montt *183* C Chile
Puerto Natales *183* S Chile
Puerto Padre *210* E Cuba
Puerto Pinasco *460* C Paraguay
Puerto Plata *226 var.* San Felipe de
Puerto Plata. N Dominican Republic
Puerto Presidente Stroessner
see Ciudad del Este
Puerto Princesa *466* Palawan,
W Philippines
Puerto Príncipe *see* Camagüey
Puerto Rico *651* Commonwealth
territory of the USA, Caribbean Sea.
❖ San Juan
Puerto Rico Trench *651* undersea
feature of the Caribbean Sea,
N Puerto Rico
Puerto San José *280 var.* San José.
S Guatemala
Puerto Suárez *136* E Bolivia
Puerto Vallarta *403* W Mexico
Puerto Varas *183* C Chile
Puerto Viejo *206* NE Costa Rica
Puertoviejo *see* Portoviejo
Pujehun *518* S Sierra Leone
Pukaki, Lake *433* lake of C South
Island, New Zealand
Pukapuka *644* island of Northern Cook
Islands, N Cook Islands
Pukch'ŏng *351* E North Korea
Pukë *81 var.* Puka. N Albania
Pukekohe *433* NW North Island,
New Zealand
Pukhan *352* river of North Korea and
South Korea
Pula *209 prev.* Pulj, *It.* Pola. W Croatia
Pulangi *466* river of Mindanao,
S Philippines
Pulap *406* atoll of C Micronesia
Pulau *305* river of Irian Jaya,
E Indonesia
Pulau Tekong Reservoir *521*
reservoir of E Singapore
Pul-i-Khumri *see* Pol-e Khomrī
Pully *550* SW Switzerland
Pulusuk *406* island of C Micronesia
Puluwat *406* atoll of C Micronesia
Puná, Isla *230* island to the SW of
Ecuador, in the Gulf of Guayaquil
Punakha *134* C Bhutan
Punata *136* C Bolivia
Pune *298 prev.* Poona. W India
Punggol *521* area of NE Singapore
Púngoè *419 var.* Pungue, Pungwe.
River of C Mozambique
Punkudutivu *538* island of
N Sri Lanka
Puno *463* SE Peru
Punta Arenas *183 prev.* Magallanes.
S Chile
Punta Chame *456* C Panama
Punta del Este *611* S Uruguay
Punta Gorda *130* S Belize
Puntarenas *206* W Costa Rica
Punta Santiago *238* S Bioko,
Equatorial Guinea
Punto Fijo *623* NW Venezuela
Punwŏn *see* Puch'ŏn
Purari *458* river of C Papua New
Guinea
Puri *298* E India
Purmerend *429* NW Netherlands
Pursat *see* Poûthĭsăt
Purus *144* river of Brazil and Peru
Pusan *352 var.* Busan, *Jap.* Fusan.
SE South Korea
Pusat Gayo, Pegunungan *304*
mountain range of Sumatra,
W Indonesia
Pushkino *see* Biläsuvar
Putai *559* W Taiwan
Putorana, Plato *485* mountain range
of N Russian Federation
Puttalam *538* W Sri Lanka
Puttalam Lagoon *538* lagoon of
W Sri Lanka
Putumayo *195, 463* river of
NW South America

Rashū *see* Kŭmsong
Raso, Ilhéu *176* island of NW Cape Verde
Rass Jebel *577 var.* Ra's al Jabal. N Tunisia
Ras Tannūrah *506* E Saudi Arabia
Ratak Chain *396* island group of E Marshall Islands
Ratchaburi *567 var.* Rat Buri. C Thailand
Rathkeale *316* SW Ireland
Rätische Alpen *see* Rhaetian Alps
Ratnapura *538* S Sri Lanka
Ratŏ *see* Lotung
Raub *386* C Peninsular Malaysia
Raufarhöfn *296* NE Iceland
Raukawa *see* Cook Strait
Rauma *251 Swe.* Raumo. SW Finland
Rauna *362* NE Latvia
Răuțel *408 var.* Reuțel. River of N Moldova
Ravenna *323* N Italy
Ravensthorpe *100* SW Australia
Rāvi *451* river of India and Pakistan
Ravne na Koroškem *524 Ger.* Gutenstein. N Slovenia
Rawaki *348 var.* Phoenix Island. Island of Phoenix Islands, C Kiribati
Rāwalpindi *451* NE Pakistan
Rawson *95* SE Argentina
Rayak *364 var.* Riyāq. E Lebanon
Rayong *567* C Thailand
Raysūt *448* SW Oman
Razāzah, Buḥayrat ar *312 var.* Baḥr al Milḥ. Lake of C Iraq
Razdan *see* Hrazdan
Razim, Lacul *480 prev.* Lacul Rezelm. Lagoon of E Romania
Reading *597* SE England, UK
Reăng Kései *165* W Cambodia
Rebun-tō *332* island to the NW of Hokkaidō, N Japan
Rechytsa *122 Rus.* Rechitsa. SE Belarus
Recife *145 prev.* Pernambuco. E Brazil
Recklinghausen *266* W Germany
Reconquista *95* NE Argentina
Redange *378* W Luxembourg
Redcliff *636* C Zimbabwe
Red Deer *170* SW Canada
Redhead *574* NE Trinidad, Trinidad & Tobago
Redon *254* NW France
Red River *170* river of Canada and USA
Red River *603* river of SC USA
Red River *627 var.* Sŏng Coi, *Chin.* Yuan Jiang. River of China and Vietnam
Red Sea *240, 631* sea of Indian Ocean, between the Arabian Peninsula and NE Africa
Red Sea Hills *540* hilly region of NE Sudan
Red Volta *157 Fr.* Volta Rouge. River of Burkina and Ghana
Ree, Lough *316 Ir.* Loch Ri. Lake of C Ireland
Reefton *433* N South Island, New Zealand
Regar *see* Tursunzode
Regensburg *267* SE Germany
Reggane *83* C Algeria
Reggio di Calabria *323 var.* Reggio Calabria. S Italy
Reggio nell' Emilia *322 var.* Reggio Emilia. N Italy
Reghin *480* N Romania
Regina *170* S Canada
Régina *645* E French Guiana
Rehoboth *423* C Namibia
Reḥovot *319* C Israel
Reichenberg *see* Liberec
Reifnitz *see* Ribnica
Ré, Île de *254* island of W France
Reims *255 Eng.* Rheims. NE France
Reine-Charlotte, Îles de la *see* Queen Charlotte Islands
Reine-Élisabeth, Îles de la *see* Queen Elizabeth Islands
Reisduoddarhalde *see* Haltiatunturi
Reisui *see* Yŏsu
Reka *see* Rijeka

Relizane *83 var.* Ghilizane, Ghelîzâne. NW Algeria
Remel el Abiod *577* desert region of S Tunisia
Remich *378* SE Luxembourg
Rendezvous Bay *642* bay of the Caribbean Sea on the S coast of Anguilla
Rendova *526* island of the New Georgia Is, W Solomon Is
Renens *550* SW Switzerland
Rengo *183* C Chile
Rennell *526 var.* Mu Nggava. Island of S Solomon Islands
Rennes *254 Bret.* Roazon. NW France
Reno *602* Nevada, W USA
Réo *157* W Burkina
Republiek *542* N Suriname
Rere *526* E Guadalcanal, Solomon Is
Resen *381* SW FYR Macedonia
Resistencia *95* NE Argentina
Reșița *480 Hung.* Resicabánya, *Ger.* Reschiza. W Romania
Resolute *170* Cornwallis Island, N Canada
Resolution Island *433* island to the SW of South Island, New Zealand
Retalhuleu *280* SW Guatemala
Retan Laut, Pulau *521* island SW Singapore
Retiche, Alpi *see* Rhaetian Alps
Réunion *652* French overseas department of the Indian Ocean. ❖ St Denis
Reus *535* E Spain
Reutlingen *266* S Germany
Reval *see* Tallinn
Rewa *286* river of S Guyana
Rey *309 var.* Shahr Rey. NW Iran
Reyes *136* NW Bolivia
Rey, Isla del *456* island of SE Panama
Reykjahlídh *296* NE Iceland
Reykjavík *296* ❖ of Iceland, W Iceland
Reynosa *403* N Mexico
Reza, Gora *585 var.* Gora Riza. Mountain of SW Turkmenistan
Rezā'īyeh *see* Orūmīyeh
Rezā'īyeh, Daryācheh-ye *see* Orūmīyeh, Daryācheh-ye
Rēzekne *362 Ger.* Rositten, *Rus.* Rezhitsa. E Latvia
Rezina *408* NE Moldova
Rēznas Ezers *362* lake of SE Latvia
Rhadames *see* Ghadamis
Rhaetian Alps *550 Ger.* Rätische Alpen, *Fr.* Alpes Rhétiques, *It.* Alpi Retiche. Mountain range of E Switzerland
Rheden *429* SE Netherlands
Rhein *see* Rhine
Rheinisches Schiefergebirge *266 Eng.* Rhenish Slate Mountains. Mountains of W Germany
Rhenish Slate Mountains *see* Rheinisches Schiefergebirge
Rhétiques, Alpes *see* Rhaetian Alps
Rhine *255, 266, 374, 550 Ger.* Rhein, *Fr.* Rhin, *Dut.* Rijn. River of W Europe
Rhino Camp *588* NW Uganda
Rhode Island *603* state of NE USA
Rhodes *see* Rodos
Rhodesia *see* Zimbabwe
Rhodope Mountains *152, 275 Gk* Orosirá Rodópis, *Bul.* Despoto Planina, *Turk.* Dospad Dagh. Mountain range of Bulgaria and Greece
Rhône *255, 550* river of France and Switzerland
Rhum *597 var.* Rum. Island of Inner Hebrides, W Scotland, UK
Riaba *238 prev.* Concepción. S Bioko, Equatorial Guinea
Riau, Kepulauan *304 var.* Riau Archipelago, *Dut.* Riouw Archipel. Island group to the E of Sumatra, W Indonesia
Riban i Manamby *382* S Madagascar
Ribáuè *419* NE Mozambique
Ribble *597* river of NW England, UK
Ribe *218* Jylland, SW Denmark
Ribeira da Barça *176* Santiago, S Cape Verde
Ribeira Funda *176* São Nicolau, N Cape Verde

Ribeira Grande *176* Santo Antão, N Cape Verde
Ribeirão Preto *145* S Brazil
Riberalta *136* N Bolivia
Ribnica *524 Ger.* Reifnitz. S Slovenia
Rîbnița *408 var.* Râbnița, *Rus.* Rybnitsa. NE Moldova
Richard's Bay *531* Kwazulu Natal, E South Africa
Richard Toll *510* N Senegal
Riche Fond *497* E St Lucia
Richmond *603* Virginia, E USA
Richmond Vale *498* NW St Vincent, St Vincent & the Grenadines
Ridâ *see* Radâa
Ridderkerk *429* SW Netherlands
Rif *414 var.* Riff, Er Rif. Mountain range of N Morocco
Rift Valley *see* Great Rift Valley
Riga *362 Latv.* Rīga. ❖ of Latvia, C Latvia
Riga, Gulf of *242, 362 Est.* Liivi Laht, *prev.* Riia Laht, *Rus.* Rizhskiy Zaliv, *Latv.* Rīgas Jūras Līci Gulf of the Baltic Sea, on the coasts of Estonia and Latvia
Rigestãn *77 var.* Registan. Desert region of S Afghanistan
Riihimäki *251* SW Finland
Rijeka *209 Slvn.* Reka, *Ger.* Sankt Veit am Flaum, *It.* Fiume. NW Croatia
Rijn *see* Rhine
Rijssel *see* Lille
Ri, Loch *see* Ree, Lough
Rimah, Wādī ar *506* dry watercourse of C Saudi Arabia
Rimaszombat *see* Rimavská Sobota
Rimavská Sobota *523 Ger.* Gross-Steffelsdorf, *Hung.* Rimaszombat. SE Slovakia
Rimini *323* N Italy
Rincon *650* Bonaire, S Netherlands Antilles
Ringe *218* Fyn, S Denmark
Ringkøbing *218* Jylland, W Denmark
Ringkøbing Fjord *218* fjord of Jylland, W Denmark
Ringsted *218* Sjælland, SE Denmark
Ringvassøya *444* island of NE Norway
Riobamba *230* C Ecuador
Rio Branco *144* W Brazil
Río Branco *611* E Uruguay
Rio Claro *574* SE Trinidad, Trinidad & Tobago
Río Cuarto *95* C Argentina
Rio de Janeiro *145* SE Brazil
Río Gallegos *95 var.* Puerto Gallegos, Gallegos. S Argentina
Rio Grande *145 var.* São Pedro do Río Grande do Sul. S Brazil
Ríohacha *195* N Colombia
Río Muni *238* mainland region of Equatorial Guinea
Río Negro, Embalse del *611 var.* Lago Artificial de Rincón del Bonete. Reservoir of C Uruguay
Rioni *264* river of W Georgia
Río Sereno *456* W Panama
Riouw Archipel *see* Riau, Kepulauan
Riri *see* Iri
Rîșcani *408 var.* Râșcani. NW Moldova
Rishiri-tō *332* island to the NW of Hokkaidō, N Japan
Rishon Le Ziyyon *319* C Israel
Ritidian Point *647* headland on the N coast of Guam
Rivadavia *95* W Argentina
Rivas *436* S Nicaragua
Rivera *611* N Uruguay
Rivercess *see* Cess
River Sallee *278* NE Grenada island, Grenada
Rivière des Anguilles *400* S Mauritius
Rivière-Pilote *649* SE Martinique
Rivne *590 Pol.* Równe, *Rus.* Rovno. NW Ukraine
Riyadh *506 var.* Ar Riyāḍ. ❖ of Saudi Arabia, C Saudi Arabia
Riyāq *see* Rayak
Rize *581* NE Turkey
Rizhskiy Zaliv *see* Riga, Gulf of
Rizokarpaso *see* Dipkarpaz

Rkîz, Lac *398* lake of SW Mauritania
Road Bay *642* bay of the Caribbean Sea on the W coast of Anguilla
Road Town *642* ❖ of British Virgin Islands, Tortola, C British Virgin Islands
Roanne *255* E France
Roaring Creek *130* C Belize
Roatán *290* Islas de la Bahía, Honduras
Roazon *see* Rennes
Robertson, Lake *see* Manyame, Lake
Robertsport *368* W Liberia
Robinson Crusoe, Isla *183* island of Juan Fernández Islands, W Chile
Rocas, Atol das *145* island of E Brazil
Rocha *611* SE Uruguay
Rochambeau *645* NE French Guiana
Rochester *603* Minnesota, NC USA
Rochester *603* New York, NE USA
Rocheuses, Montagnes *see* Rocky Mountains
Rock, The *646* E Gibraltar
Rockford *603* Illinois, C USA
Rockhampton *101* E Australia
Rockingham *100* SW Australia
Rocklands *544* NW Swaziland
Rock Sound *112* Eleuthera I, Bahamas
Rock Springs *602* Wyoming, NW USA
Rockstone *286* E Guyana
Rocky Mountains *170, 602 var.* Rockies, *Fr.* Montagnes Rocheuses. Mountain range of NW America
Rocky Point *644* headland on the N coast of Christmas Island
Rodez *255* S France
Ródhos *see* Rodos
Rodi *see* Rodos
Rodi Gargancio *323* C Italy
Rodonit, Gjiri i *81* gulf of the Adriatic Sea, NW Albania
Rodópis, Orosirá *see* Rhodope Mountains
Rodos *275 Eng.* Rhodes, *It.* Rodi, *prev.* Ródhos. Island of SE Greece
Ródos *275 Eng.* Rhodes, *It.* Rodi, *prev.* Ródhos. Ródhos, SE Greece
Rodosto *see* Tekirdağ
Rodrigues *400 var.* Rodriquez. Island of E Mauritius
Roermond *429* S Netherlands
Roeselare *127 Fr.* Roulers, *prev.* Rousselaere. W Belgium
Rogachëv *see* Rahachow
Rogaška Slatina *524 prev.* Rogatec-Slatina, *Ger.* Rohitsch-Sauerbrunn. E Slovenia
Rogatec-Slatina *see* Rogaška Slatina
Roger *224* W Dominica
Rogozhina *see* Rrogozhinë
Rohitsch-Sauerbrunn *see* Rogaška Slatina
Roi Et *567 var.* Muang Roi Et. NE Thailand
Roja *362* NW Latvia
Rojo, Cabo *651* cape on the SW coast of Puerto Rico
Rokan *304 var.* Airlalang. River of Sumatra, W Indonesia
Rokel *518 var.* Seli. River of C Sierra Leone
Rokkō *see* Lukang
Rôlas, Ilha das *505* island to the S of São Tomé, Sao Tome & Principe
Roma *see* Rome
Roma *366* W Lesotho
Romang Strait *304* strait connecting the Arafura Sea and Laut Banda, E Indonesia
Romania *480-483 prev.* Socialist Republic of Romania, *var.* Rumania. Country of SE Europe, divided into 40 admin. units (judet). ❖ Bucharest
Romano, Cayo *210* island of NE Cuba
Rombo, Ilhéus do *176 var.* Ilhéus Secos. Island of S Cape Verde
Rome *323 It.* Roma. ❖ of Italy, C Italy
Rømø *218* island of SW Denmark
Ronde *see* Round Island
Ronde Island *278* island to the N of Grenada island, Grenada
Rongelap *396* island of NW Marshall Islands

S

São Tomé *505* ❖ of Sao Tome & Principe, NE São Tomé

Sao Tome & Principe *504-505* officially Democratic Republic of Sao Tome and Principe, *Port.* São Tomé e Príncipe. Country of W Africa divided into 7 admin. units (districts). ❖ São Tomé

São Tomé, Pico de *505* mountain of São Tomé, Sao Tome & Principe

São Vicente *145* S Brazil

São Vicente *176* island of N Cape Verde

Sapele *440* S Nigeria

Sapitwa *385* mountain of S Malawi

Saponé *157* C Burkina

Sappemeer *429* NE Netherlands

Sapporo *332* Hokkaidō, N Japan

Sapta Koshi *427* river of India and Nepal

Sār *115* NW Bahrain

Šara *381* mountain range of FYR Macedonia and Serbia & Montenegro (Yugo.)

Sara Buri *567* C Thailand

Saragossa *see* Zaragoza

Saragt *585* *prev.* Serakhs, *var.* Serahs. S Turkmenistan

Sarajevo *140* ❖ of Bosnia & Herzegovina, SE Bosnia & Herzegovina

Saran' *340* C Kazakhstan

Sarandë *81* *var.* Saranda, *It.* Porto Edda, *prev.* Santi Quaranta. S Albania

Sarandí del Yí *611* C Uruguay

Sarandí Grande *611* S Uruguay

Sarangani Islands *466* island group of SE Philippines

Saratov *484* W Russian Federation

Saravan *360* *var.* Saravane. SE Laos

Sarbhang *134* S Bhutan

Sardegna *322* *Eng.* Sardinia. Island of W Italy

Sardinia *see* Sardegna

Sargodha *451* NE Pakistan

Sarh *180* *prev.* Fort-Archambault. S Chad

Sārī *309* N Iran

Sarigan *650* island of C Northern Mariana Islands

Sarikol Range *562* *Rus.* Sarykol'skiy Khrebet. Mountain range of China and Tajikistan

Sarimbun Reservoir *521* NW Singapore

Sariwŏn *351* SW North Korea

Sark *647* island of SE Guernsey

Sarpsborg *444* S Norway

Sarrebruck *see* Saarbrücken

Sarstoon *130, 280* *var.* Sarstún. River of Belize and Guatemala

Sarstún *see* Sarstoon

Sarthe *255* river of NW France

Saruhan *see* Manisa

Sarykol'skiy Khrebet *see* Sarikol Range

Sary-Tash *359* *var.* Sary-Ta#. SW Kyrgyzstan

Sasebo *332* Kyūshū, SW Japan

Saseno *see* Sazan

Saskatchewan *170* river of C Canada

Saskatchewan *170* province of C Canada

Saskatoon *170* S Canada

Sasolburg *531* Orange Free State, C South Africa

Sassandra *328* S Ivory Coast

Sassandra *328* *var.* Ibo. River of S Ivory Coast

Sassari *322* Sardegna, W Italy

Sassnitz *267* NE Germany

Sasstown *368* SE Liberia

Sataua *500* Savai'i, Samoa

Satawal *406* island of C Micronesia

Satawan *406* atoll of C Micronesia

Sātkhira *117* SW Bangladesh

Satpayev *340* *prev.* Nikol'skiy. C Kazakhstan

Satpura Range *298* mountains of C India

Satu Mare *480* *Hung.* Szatmárnémeti. NW Romania

Satunan-shotō *332* island group of Nansei-shotō, SW Japan

Sau *see* Sava

Sauce *see* Juan L. Lacaze

Saudhárkrókur *296* N Iceland

Saudi Arabia *506-509* officially Kingdom of Saudi Arabia. Country of SW Asia divided into 13 admin. units (provinces). ❖ Riyadh

Saül *645* C French Guiana

Saulkrasti *362* N Latvia

Sault Sainte Marie *171* S Canada

Sauma, Pointe *653* headland of Île Alofi, N Wallis & Futuna

Saûmâtre, Étang *288* lake of SE Haiti

Saurimo *88* *Port.* Vila Henrique de Carvalho. NE Angola

Sauteurs *278* N Grenada island, Grenada

Sava *140, 209, 524, 512* *Eng.* Save, *Hung.* Száva, *Ger.* Sau. River of SE Europe

Savai'i *500* island of NW Samoa

Savalou *132* S Benin

Savan Island *498* island of S St Vincent & the Grenadines

Savannah *603* Georgia, SE USA

Savannakhét *360* S Laos

Savanna-La-Mar *331* W Jamaica

Savave *587* islet of Nukufetau, Tuvalu

Save *419, 636* *var.* Sabi. River of Mozambique and Zimbabwe

Savè *132* SE Benin

Savissivik *646* NW Greenland

Savona *322* N Italy

Savonlinna *251* *Swe.* Nyslott. SE Finland

Savusavu *248* Vanua Levu, N Fiji

Savu Sea *see* Sawu, Laut

Savute *142* river of N Botswana

Sawdā', Jabal *506* mountain of SW Saudi Arabia

Sawdā, Jabal as *371* mountain range of C Libya

Sawdīrī *see* Sodiri

Sawhaj *see* Sohāg

Şawqirah *448* SE Oman

Şawqirah, Ghubbat *see* Suqrah Bay

Sawu, Laut *304* *Eng.* Savu Sea. Sea of the Indian Ocean, C Indonesia

Say *439* SW Niger

Sayaboury *see* Muang Xaignabouri

Sayat *585* E Turkmenistan

Sayhūt *631* E Yemen

Saylac *528* *var.* Zeila. NW Somalia

Saynshand *412* S Mongolia

Say'ūn *631* *var.* Saywūn. C Yemen

Sazan *81* *It.* Saseno. Island of SW Albania

Scaldis *see* Scheldt

Scarborough *597* NE England, UK

Scarborough *574* S Tobago, Trinidad & Tobago

Scarborough *121* S Barbados

Scebeli *see* Shebeli

Schaan *374* W Liechtenstein

Schaanwald *374* NE Liechtenstein

Schaffhausen *550* N Switzerland

Schaulen *see* Šiauliai

Schefferville *171* E Canada

Schelde *see* Scheldt

Scheldt *127* *Dut.* Schelde, *Fr.* Escaut. River of W Europe

Schellenberg *374* N Liechtenstein

Schiedam *429* SW Netherlands

Schiermonnikoog *429* island of Waddeneilanden, N Netherlands

Schifflange *378* S Luxembourg

Schneekoppe *see* Sněžka

Schneidemühl *see* Piła

Schœlcher *649* W Martinique

Schoten *127* N Belgium

Schouwen *429* island of SW Netherlands

Schwäbische Alb *266* *Eng.* Swabian Jura. Mountain range of SW Germany

Schwarzwald *see* Black Forest

Schwaz *106* W Austria

Schweizer Mittelland *see* Swiss Plateau

Schweizer Reneke *531* North West, N South Africa

Schwerin *267* N Germany

Schweriner See *266* lake of N Germany

Schwyz *550* C Switzerland

Schyl *see* Jiu

Sciacca *323* Sicilia, S Italy

Sciasciamana *see* Shashemenee

Scio *see* Chíos

Scoresbysund *see* Ittoqqortoormiit

Scotland *597, 597* national region of UK divided into 12 admin. units (9 regions, 3 island authorities)

Scott Base *90* New Zealand research station near Ross Shelf, Antarctica

Scott Island *90* island to the N of Ross Ice Shelf, Antarctica

Scotts Head Village *224* *var.* Cachacrou. S Dominica

Scrub Island *642* island of NE Anguilla

Scunthorpe *597* NE England, UK

Scutari *see* Shkodër

Scutari, Lake *81, 512* *Alb.* Liqeni i Shkodrës, *SCr.* Skadarsko Jezero. Lake of Albania and Serbia & Montenegro (Yugo.)

Seal Island *642* island of NW Anguilla

Seatons *92* E Antigua, Antigua & Barbuda

Seattle *602, 607* Washington, NW USA

Sébaco *436* C Nicaragua

Sebaiera *414* C Western Sahara

Sebapala *366* S Lesotho

Sebarok, Pulau *521* island S Singapore

Sebastián Vizcaíno, Bahía *403* bay of the Pacific Ocean, on the NW coast of Mexico

Sebastopol *see* Sevastopol'

Sebenico *see* Šibenik

Sébikhoutane *510* W Senegal

Sebou *414* river of N Morocco

Secos, Ilhéus *see* Rombo, Ilhéus de

Sedberat *240* W Eritrea

Sédhiou *510* SW Senegal

Seeheim Noord *423* S Namibia

Seeland *see* Sjælland

Sefadu *518* E Sierra Leone

Sefrou *414* N Morocco

Segamat *386* S Peninsular Malaysia

Ségbana *132* NE Benin

Segewold *see* Sigulda

Segna *see* Senj

Ségou *392* *var.* Segu C Mali

Segovia *535* C Spain

Segovia *see* Coco

Segu *see* Ségou

Séguédine *439* NE Niger

Séguéla *328* W Ivory Coast

Séguénéga *157* NW Burkina

Segura *535* river of S Spain

Sehlabathebe *366* E Lesotho

Seinäjoki *251* *Swe.* Östermyra. SW Finland

Seine *255* river of N France

Seine, Baie de la *254-255* bay of the English Channel to the NW of France

Seiyū *see* Chŏngju

Sejerø *218* island of C Denmark

Şeki *see* Şäki

Sekoma *142* S Botswana

Sekondi-Takoradi *272* S Ghana

Selânik *see* Thessaloníki

Selemia *see* Salamīyah

Selenge *412* river of Mongolia and Russian Federation

Seletar Reservoir *521* reservoir of C Singapore

Selfoss *296* SW Iceland

Seli *see* Rokel

Sélibabi *398* S Mauritania

Selibi Phikwe *142* E Botswana

Selle, Massif de la *288* mountain range of S Haiti

Selle, Pic la *288* *var.* La Selle. Mountain of S Haiti

Sellore Island *see* Saganthit Island

Sellye *see* Skalica

Sellye *see* Šaľa

Sélouma *283* C Guinea

Selukwe *see* Shurugwi

Selvagens, Ilhas *474* island group of the Madeira Is, Portugal

Semakau, Pulau *521* island S Singapore

Semanit *81* *var.* Seman. River of W Albania

Semara *414* N Western Sahara

Semarang *304* Java, C Indonesia

Sembawang *521* area of N Singapore

Sembé *200* NW Congo

Sembehun *518* SW Sierra Leone

Semberong *386* river of SE Peninsular Malaysia

Semendria *see* Smederevo

Semipalatinsk *340* *Kaz.* Semey. E Kazakhstan

Semirara Islands *466* island group of C Philippines

Semliki *588* river of W Uganda

Sên *165* *var.* Sen. River of C Cambodia

Sena *see* Vila de Sena

Senafe *240* SE Eritrea

Senanayake Samudra *538* lake of E Sri Lanka

Senanga *635* SW Zambia

Senang, Pulau *521* island of S Singapore

Sendai *332* Kyūshū, SW Japan

Sendai *332* Honshū, N Japan

Senegal *510-511* officially Republic of Senegal. *Fr.* Sénégal. Country of West Africa divided into 10 admin. units (regions). ❖ Dakar

Senegal *398, 392, 510* *Fr.* Sénégal. River of W Africa

Senica *523* *Ger.* Senitz, *Hung.* Szenice. W Slovakia

Senigallia *323* C Italy

Senj *209* *Ger.* Zengg, *Ital.* Segna. NW Croatia

Senja *247* *prev.* Senjen. Island of NW Norway

Senkaku-shotō *332* island group of Nansei-shotō, SW Japan

Senmonorom *165* E Cambodia

Sennar *540* *var.* Sannār. C Sudan

Senne *127* *Dut.* Zenne. River of C Belgium

Senqunyane *366* river of C Lesotho

Senshin-kō *see* Sŏmjin

Sensuntepeque *237* NE El Salvador

Sentery *203* SE Dem. Rep. Congo

Sentosa *521* island S Singapore

Senye *238* W Río Muni, Equatorial Guinea

Seongnam *see* Sŏngnam

Seoul *352* *Kor.* Sŏul, *prev.* Kyŏngsŏng, *Jap.* Keijō. of South Kore NW South Korea

Sepik *458* river of Indonesia and Papua New Guinea

Sepone *see* Muang Xéphôn

Sept-Iles *170* E Canada

Serahs *see* Saragt

Seraing *127* E Belgium

Serakhis *see* Serrakhis

Serakhs *see* Saragt

Seram *304* *var.* Serang, *Eng.* Ceram. Island of Maluku, E Indonesia

Seram, Laut *304* *Eng.* Ceram Sea. Sea of the Pacific Ocean, E Indonesia

Serang *304* Java, C Indonesia

Serangoon Harbour *521* harbour, E Singapore

Serasan, Selat *304, 386* strait of the South China Sea between Borneo and Kepulauan Natuna, W Indonesia

Seraya, Pulau *521* island of SW Singapore

Serbia *512* *Serb.* Srbija. Republic of Serbia & Montenegro (Yugo.)

Sered' *523* *Hung.* Szered. SW Slovakia

Serekunda *263* W Gambia

Seremban *386* W Peninsular Malaysia

Serengeti Plain *564* plain of N Tanzania

Serenje *635* E Zambia

Sereth *see* Siret

Sérifos *275* island of SE Greece

Serov *484* C Russian Federation

Serowe *142* SE Botswana

Serpa Pinto *see* Menongue

Serpent's Mouth, The *574* *Sp.* Boca de la Serpiente. Strait connecting the Columbus Channel and the Gulf of Paria

Serrakhis *215* *var.* Serrachis, Serakhis. River of NW Cyprus

Serravalle *502* N San Marino

Sérres *275* *prev.* Sérrai. NE Greece

Serule *142* E Botswana

Se San *see* Tônle San

Sesana *see* Sežana

Sese Islands *588* island group of S Uganda

Sesvete *209* N Croatia

Seti *427* river of W Nepal

Sétif *83* *var.* Stif. N Algeria

Settat *414* W Morocco

Setté Cama *260* SW Gabon

Settlement, The *643* Anegada, N British Virgin Islands

Setúbal *474* W Portugal

Setúbal, Baía de *474* bay of the Atlantic Ocean, to the SW of Portugal

Sevan *98* C Armenia

Sevana Lich *98* *Eng.* Lake Sevan. Lake of E Armenia

Sevani Lerrnashght'a *see* Shakh-Dag

Sevan, Lake *see* Sevana Lich

Sévaré *392* C Mali

Sevastopol' *590* *Eng.* Sebastopol. S Ukraine

Severn *597* *Wel.* Hafren. River of England and Wales, UK

Severn *170* river of S Canada

Severnaya Dvina *484* *Eng.* Northern Dvina. River of NW Russian Federation

Severnaya Osetiya-Alaniya *484* autonomous republic of SW Russian Federation

Severnaya Zemlya *485* island group of N Russian Federation

Severnyy Ledovityy Okean *see* Arctic Ocean

Severodvinsk *484* *prev.* Molotov, *prev.* Sudostroy. NW Russian Federation

Severo-Sibirskaya Nizmennost' *485* *Eng.* North Siberian Lowland, *var.* North Siberian Plain. Lowland region of N Russian Federation

Severskiy Donets *see* Donets

Sevilla *534* *Eng.* Seville. SW Spain

Sevilla de Niefang *see* Niefang

Sevlievo *152* C Bulgaria

Sewa *518* river E Sierra Leone

Seychelles *516-517* officially Republic of the Seychelles. Country of the Indian Ocean divided into 23 admin. units (districts). ❖ Victoria

Seydhisfjördhur *296* E Iceland

Seydi *585* *prev.* Neftezavodsk. E Turkmenistan

Seyhan *581* river of S Turkey

Seyhan *see* Adana

Sežana *524* *It.* Sesana. SW Slovenia

Sfântu Gheorghe *480* *prev.* Sfîntu Gheorghe. C Romania

Sfax *577* *var.* Safâqis. E Tunisia

's-Gravenhage *429* *var.* Den Haag, *Eng.* The Hague, *Fr.* La Haye. Seat of government, W Netherlands

Shaanxi *187* *var.* Shensi, Shan-hsi. Province of C China

Shabani *see* Zvishavane

Shabeelle, Webi *see* Shebeli

Shaddādī *see* Ash Shadādah

Shah Alam *386* W Peninsular Malaysia

Shāhbāzpur *117* river of S Bangladesh

Shāhdādkot *451* SW Pakistan

Shaḥḥāt *371* NE Libya

Shahrikhon *614* *Rus.* Shakhrikhan. E Uzbekistan

Shahrisabz *614* *Rus.* Shakhrisabz. SE Uzbekistan

Shahr Rey *see* Rey

Shahrtuz *562* *Rus.* Shaartuz. W Tajikistan

Shahzadpur *117* W Bangladesh

Shakawe *142* NW Botswana

Shakh-Dag *Arm.* Sevani Lerrnashght'a, *Rus.* Shakhdagskiy Khrebet. Mountain range of Armenia and Azerbaijan

Shakhdagskiy Khrebet *see* Shakh-Dag

Shakhrisabz *see* Shahrisabz

Shakhtinsk *340* C Kazakhstan

Shaki *440* W Nigeria

Shām, Bādiyat ash *see* Syrian Desert

Shām, Jabal ash *448* *var.* Jebel Sham. Mountain of N Oman

Shandī *see* Shendi

Shandong *187* *var.* Shantung. Province of E China

Shandong Bandao *187* *var.* Shantung Peninsula. Peninsula of E China

Shangani *636* river of W Zimbabwe

Shanghai *187, 190* city and municipality of E China

Shan-hsi *see* Shaanxi

Shan-hsi *see* Shanxi

Shanhua *559* SW Taiwan

Shannon *316* *Ir.* An tSionainn. River of C Ireland

Shansi *see* Shanxi

Shantar Islands *see* Shantarskiye Ostrova

Shantarskiye Ostrova *485* *Eng.* Shantar Islands. Island group of SE Russian Federation

Shantou *187* *var.* Swatow. Guangdong, SE China

Shantung *see* Shandong

Shantung Peninsula *see* Shandong Bandao

Shanxi *187* *var.* Shansi, Shan-hsi. Province of NE China

Shaoguan *187* *var.* Shao-kuan, *prev.* Ch'u-chiang, *Cant.* Kukong. Guangdong, SE China

Shao-kuan *see* Shaoguan

Shaqrā' *506* C Saudi Arabia

Shaqrā *see* Shuqrah

Shar *340* *var.* Charsk. E Kazakhstan

Sharasume *see* Altay

Shari *see* Chari

Sharīn Gol *412* N Mongolia

Sharjah *594* NE United Arab Emirates

Shark Bay *100* bay to the W of Australia

Sharon, Plain of *see* HaSharon

Sharqī, Jabal ash *see* Anti-Lebanon

Sharqī, Jazīrat ash *see* Chergui, Île

Sharqī, Jebel esh *see* Anti-Lebanon

Shashe *142, 636* *var.* Shashi. River of Botswana and Zimbabwe

Shashemenē *245* *var.* Shashemenne, Shashhamana, *It.* Sciasciamana. S Ethiopia

Shashi *187* *var.* Sha-shih, Shasi. Hubei, C China

Shāṭi', Wādi ash *371* dry watercourse of W Libya

Shaykh, Jabal ash *see* Hermon, Mount

Shaykh 'Uthmān *631* SW Yemen

Shcheglovsk *see* Kemerovo

Shchuchinsk *340* N Kazakhstan

Shea *286* S Guyana

Shebeli *528* *Som.* Webi Shabeelle, *Amh.* Shebele Wenz, *It.* Scebeli. River of Ethiopia and Somalia

Sheberghān *77* *var.* Shibarghan. N Afghanistan

Shedadi *see* Ash Shadādah

Shefar 'am *319* N Israel

Sheffield *597* N England, UK

Shekhem *see* Náblus

Shekhūpura *451* NE Pakistan

Sheki *see* Şäki

Shelikhova, Zaliv *485* *Eng.* Shelekhov Gulf. Gulf of Sea of Okhotsk, bordering NE Russian Federation

Shemakha *see* Şamaxı

Shemgang *134* C Bhutan

Shendi *540* *var.* Shandī. NE Sudan

Shengking *see* Liaoning

Shensi *see* Shaanxi

Shenyang *187* *prev.* Fengtien, *Eng.* Mukden. Liaoning, NE China

Shepherd Islands *619* islands to the C of Vanuatu

Shepparton *101* SE Australia

Sherbro Island *518* island of SW Sierra Leone

Sherbrooke *171* SE Canada

Sheridan *602* Wyoming, NW USA

Sherpur *117* N Bangladesh

's-Hertogenbosch *429* *Ger.* Herzogenbusch, *Fr.* Bois-le-Duc. S Netherlands

Sherwood Ranch *142* SE Botswana

Shetland *597* islands of NE Scotland, UK

Shevchenko *see* Aktau

Shibām *631* C Yemen

Shibarghan *see* Sheberghān

Shibata *332* Honshū, N Japan

Shibh Jazīrat Sīnā' *see* Sinai

Shibīn el Kôm *232* *var.* Shibīn al Kawm. N Egypt

Shihmen *see* Shijiazhuang

Shijak *81* *var.* Shijaku. W Albania

Shijiazhuang *187* *var.* Shihkiachwang, Shih-chia-chuang, *prev.* Shihmen. Hebei, NE China

Shikārpur *451* S Pakistan

Shikoku *332* island of SW Japan

Shiliguri *298* *prev.* Siliguri. NE India

Shimbiris *528* *var.* Shimbir Berris. Mountain of N Somalia

Shimizu *332* Honshū, SE Japan

Shimonoseki *332* Honshū, W Japan

Shimonoseki-kaikyō *332* strait connecting the Sea of Japan and Inland Sea, between Honshū and Kyūsh ū, W Japan

Shinano *332* river of Honshū, N Japan

Shināş *448* NW Oman

Shindand *77* W Afghanistan

Shinei *see* Hsinying

Shinshō *see* Hsinchuang

Shinshū *see* Chinju

Shinten *see* Hsintien

Shinyanga *564* NW Tanzania

Shiogama *332* Honshū, N Japan

Shīrāz *309* SW Iran

Shire *385* *Port.* Chire. River of Malawi and Mozambique

Shire Highlands *385* hilly region of S Malawi

Shirvanskaya Step' *see* Şirvan Düzü

Shirwa, Lake *see* Chilwa, Lake

Shizuoka *332* Honshū, SE Japan

Shkodër *81* *var.* Shkodra, *It.* Scutari, *SCr.* Skadar. NW Albania

Shkodrës, Liqeni i *see* Scutari, Lake

Shkubinit *81* *var.* Shkumbī, Shkumbin. River of C Albania

Shoe Rock *649* headland on the S coast of Montserrat

Shōka *see* Changhua

Sholāpur *see* Solāpur

Shorkot *451* NE Pakistan

Shortland Island *526* *var.* Alu. Island of the Shortland Is, W Solomon Islands

Shortland Islands *526* island group of the W Solomon Islands

Shostka *590* N Ukraine

Shreveport *603* Louisiana, SC USA

Shrewsbury *597* C England, UK

Shū *340* *var.* Chu. SE Kazakhstan

Shu'aybah *356* *var.* Shuaiba. E Kuwait

Shubrâ el Kheima *232* *var.* Shubrā al Khaymah. N Egypt

Shūlgareh *77* N Afghanistan

Shumen *152* *var.* Šumen. E Bulgaria

Shunsen *see* Ch'unch'ŏn

Shuqrah *631* *var.* Shaqrā. SW Yemen

Shurugwi *636* *prev.* Selukwe. C Zimbabwe

Shweli *159* river of Burma and China

Shwe
bo *159* N Burma

Shymkent *340* *prev.* Chimkent. S Kazakhstan

Shyashchytsy *122* C Belarus

Siāhān Range *450* mountain range of W Pakistan

Siāh Kūh *77* mountain range of W Afghanistan

Siālkot *451* NE Pakistan

Siam *see* Thailand

Siam, Gulf of *see* Thailand, Gulf of

Sian *see* Xi'an

Siangtan *see* Xiangtan

Siargao Island *466* island of E Philippines

Šiauliai *362* *Ger.* Schaulen. NW Lithuania

Siazan' *see* Siyäzän

Šibenik *209* *It.* Sebenico. S Croatia

Siberut, Pulau *304* island of Kepulauan Mentawai, W Indonesia

Sibi *451* C Pakistan

Sibiti *200* S Congo

Sibiu *480* *Ger.* Hermannstadt, *Hung.* Nagyszeben. C Romania

Sibu *386* W Borneo, Malaysia

Sibut *179* *prev.* Fort-Sibut. C Central African Republic

Sibutu Passage *386* passage connecting Celebes Sea and Sulu Sea

Sibuyan Island *466* island of C Philippines

Sibuyan Sea *466* sea of the Pacific Ocean

Sichuan *187* *var.* Szechuan, Ssu-ch'uan. Province of SW China

Sicilia *323* *Eng.* Sicily. Island of S Italy

Sicily *see* Sicilia

Sico *290* *var.* Tinto, Río Negro. River of NE Honduras

Sicunusa *544* SW Swaziland

Siders *see* Sierre

Sidi Bel Abbès *83* NW Algeria

Sidi Bouzid *577* *var.* Sīdī bū Zayd, Gammouda. C Tunisia

Sidi el Hani, Sebkhet de *577* *var.* Sabkhat Sīd´ al Hāni´. Salt flat of NE Tunisia

Sidi Kacem *414* *prev.* Petitjean. N Morocco

Sidra *see* Surt

Sidra, Gulf of *see* Surt, Khalīj

Sidvokodvo *544* C Swaziland

Siegen *266* W Germany

Sielo *368* N Liberia

Siĕmréab *165* *prev.* Siem Reap. NW Cambodia

Siena *323* *Fr.* Sienne. C Italy

Sienne *see* Siena

Sierra de Guadarrama *535* mountains of C Spain

Sierra Leone *518-519* officially Republic of Sierra Leone. Country of W Africa divided into 4 admin. units (provinces). ❖ Freetown

Sierra Madre *466* mountain range of Luzon, N Philippines

Sierra Madre *280, 403* mountain range of Guatemala and Mexico

Sierra Madre del Sur *403* mountain range of S Mexico

Sierra Madre Occidental *403* *var.* Western Sierra Madre. Mountain range of NW Mexico

Sierra Madre Oriental *403* *var.* Eastern Sierra Madre. Mountain range of N Mexico

Sierra Maestra *210* mountain range of SE Cuba

Sierra Morena *534-535* mountain range of SW Spain

Sierra Nevada *602* mountain range of W USA

Sierra Nevada de Mérida *see* Mérida, Cordillera de

Sierre *550* *Ger.* Siders. SW Switzerland

Sigatoka *248* *prev.* Singatoka. Viti Levu, W Fiji

Siggiewi *395* S Malta

Sighişoara *480* C Romania

Siglufjördhur *296* N Iceland

Signy *90* UK research station of South Orkney Islands, Antarctica

Sigsig *230* S Ecuador

Siguatepeque *290* W Honduras

Siguiri *283* NE Guinea

Sigulda *362* *Ger.* Segewold. NE Latvia

Sihanoukville *see* Kâmpóng Saôm

Siirt *581* SE Turkey

Sikasso *392* S Mali

Sikwane *142* S Botswana

Silay *466* Negros, C Philippines

Silesia *471* region of SW Poland

Silgadhi *427* *var.* Silgarhi. W Nepal

Silhouette *516* island of the Inner Islands, SE Seychelles

Siliana *577* *var.* Silyānah. NW Tunisia

Silicon Valley *606* business region of SW USA

Siliguri *see* Shiliguri

Silil *528* *var.* Silel. Seasonal river of NW Somalia

Silinhot *see* Xilinhot

Silisili, Mount *500* *var.* Mauga Silisili. Mountain of NW Samoa

Silistra *152 var.* Silistria. NE Bulgaria
Silkeborg *218* Jylland, W Denmark
Sillamäe *242 Ger.* Sillamäggi.
　NE Estonia
Sillein *see* Žilina
Šiluté *376 var.* Šilute. W Lithuania
Silva Porto *see* Kuito
Silver City *644* NE Christmas Island
Silverek *581* SE Turkey
Sima *198* W Anjouan, Comoros
Simanggang *see* Bandar Sri Aman
Simbirsk *see* Ul'yanovsk
Simeto *323* river of Sicilia, S Italy
Simeulue, Pulau *304* island to the
　NW of Sumatra, W Indonesia
Simferopol *590* S Ukraine
Simikot *427* W Nepal
Siminiout *645* S French Guiana
Šimonovany *see* Partizánske
Simony *see* Partizánske
Simplon Pass *550* mountain pass of
　S Switzerland
Simplon Tunnel *550* tunnel of Italy
　and Switzerland
Simpson Desert *101* desert region of
　C Australia
Simunye *544* NE Swaziland
Sinai *232 Ar.* Shibh Jazīrat Sīnā'.
　Desert region of NE Egypt
Sinazongwe *635* S Zambia
Sincelejo *195* NW Colombia
Sinchwang *see* Hsinchuang
Sin Cowe Island *652* island of
　SW Spratly Islands
Sindh *451* administrative region of
　SE Pakistan
Sindhulimadi *427* C Nepal
Sindi *242* SW Estonia
Sine *510* river of W Senegal
Sinendé *132* N Benin
Sines *474* S Portugal
Sinfra *328* C Ivory Coast
Singa *540 var.* Sinjah, Sinja. E Sudan
Singapore *521* river of S Singapore
Singapore *520-521* officially Republic
　of Singapore. Country of SE Asia
　divided into 5 admin. units (districts).
　❖ Singapore City
Singapore Strait *386, 521*
　var. Strait of Singapore. Strait
　connecting Strait of Malacca and
　South China Sea
Singatoka *see* Sigatoka
Singerei *408 var.* Sângerei,
　prev. Lazovsk. N Moldova
Singida *564* C Tanzania
Singora *see* Songkhla
Sining *see* Xining
Sinj *209* SE Croatia
Sinjavina *512 var.* Sinjajevina.
　Mountain range of N Montenegro,
　Serbia & Montenegro (Yugo.)
Sinkiang Uighur Autonomous Region
　see Xinjiang Uygur Zizhiqu
Sinnamary *645* N French Guiana
Sinnûris *232 var.* Sinnūris. N Egypt
Sino *see* Greenville
Sinoe *see* Greenville
Sinoia *see* Chinhoyi
Sinoie, Lacul *480 prev.* Lacul Sinoe.
　Lagoon of E Romania
Sinop *581* N Turkey
Sinp'o *351* N North Korea
Sintien *see* Hsintien
Sint-Niklaas *127 Fr.* St.-Nicolas.
　N Belgium
Sintra *474 prev.* Cintra. W Portugal
Sint-Truiden *127 Fr.* St.-Trond.
　E Belgium
Sinŭiju *351* W North Korea
Sinyang *see* Xinyang
Sió *292* river of W Hungary
Sion *550 Ger.* Sitten. SW Switzerland
Siorapaluk *646* NW Greenland
Sioux City *603* Iowa, C USA
Sioux Falls *603* South Dakota, NC USA
Sipaliwini *542* river of S Suriname
Siparia *574* SW Trinidad, Trinidad
　& Tobago
Siphofaneni *544 var.* Sipofaneni.
　C Swaziland
Siping *187 var.* Ssu-p'ing, Szeping,
　prev. Ssu-p'ing-chieh. Jilin,
　NE China

Siple *90* US research station of South
　Orkney Islands, Antarctica
Siput *386 var.* Sungei Siput.
　NW Peninsular Malaysia
Siquirres *206* E Costa Rica
Siracusa *323 Eng.* Syracuse. Sicilia,
　S Italy
Sirājganj *117* N Bangladesh
Şir Banī Yās *594* island of W United
　Arab Emirates
Sirdaryo *see* Syr Darya
Sir Edward Pellew Group *101* island
　group of N Australia
Siret *480 var.* Siretul, *Ger.* Sereth.
　River of Romania and Ukraine
Sir Francis Drake Channel *643*
　channel connecting the Atlantic
　Ocean and Caribbean Sea, C British
　Virgin Islands
Sirte *see* Surt
Sirte, Gulf of *see* Surt, Khalij
Şirvan Düzü *110 Rus.* Shirvanskaya
　Step'. Mountain range of
　C Azerbaijan
Sirwan *see* Diyālā
Sisak *209 Hung.* Sziszek, *Ger.* Sissek.
　N Croatia
Sisian *98* SE Armenia
Sisimiut *646 var.* Holsteinsborg.
　SW Greenland
Sisôphôn *165* NW Cambodia
Sissek *see* Sisak
Sīstān, Daryācheh-ye *309*
　var. Hāmūşāberī,
　Daryācheh-ye Hāmūn. Lake of E Iran
Sisters, The *278* islands N of Grenada
　island, Grenada
Siteki *544 var.* Stegi. E Swaziland
Sithoniá *275* peninsula of NE Greece
Sitobela *544* S Swaziland
Sitona *240* SW Eritrea
Sitrah *115 var.* Sitra. Island of NE Bahrain
Sittang *159 var.* Sittoung. River of
　C Burma
Sittard *429* S Netherlands
Sitten *see* Sion
Sittwe *159 prev.* Akyab. W Burma
Siuna *436* NE Nicaragua
Sivas *581* C Turkey
Sivers'kyy Donets' *see* Donets
Siyäzän *110 Rus.* Siazan'. NE Azerbaijan
Sjælland *218 Ger.* Seeland,
　Eng. Zealand. Island of E Denmark
Skadar *see* Shkodër
Skadarsko Jezero *see* Scutari, Lake
Skagaströnd *296*
　prev. Höfdhakaupstadhur. N Iceland
Skagen *218* Jylland, N Denmark
Skagerrak *218, 444, 547*
　var. Skagerak. Area of the Baltic Sea
Skalica *523 Hung.* Sellye. W Slovakia
Skeleton Coast *423* coastal region of
　NW Namibia
Skellefteå *547* NE Sweden
Skellefteälv *547* river of N Sweden
Skien *445* S Norway
Skikda *83 prev.* Philippeville.
　NE Algeria
Skive *218* Jylland, NW Denmark
Skjálfandafljót *296* river of C Iceland
Skjern *218* Jylland, W Denmark
Skjern Å *218* river of W Denmark
Skon *165* S Cambodia
Skopje *381 prev.* Skoplje,
　Turk. Üsküb. ❖ of FYR Macedonia,
　N FYR Macedonia
Skoplje *see* Skopje
Skövde *547* S Sweden
Skrunda *362* W Latvia
Skúvoy *644* island of C Faeroe Islands
Skye, Isle of *597* island of
　W Scotland, UK
Skýros *275* island of E Greece
Slagelse *218* Sjælland, SE Denmark
Slaney *316 Ir.* An tSláine. River of
　SE Ireland
Slatina *209 prev.* Podravska Slatina,
　Hung. Szlatina. NE Croatia
Slatina *480* S Romania
Slave Coast *571* coastal region of
　W Africa, Atlantic Ocean
Slavonska Požega *209 prev.* Požega,
　Hung. Pozsega. NE Croatia

Slavonski Brod *209 prev.* Brod,
　Hung. Bród. E Croatia
Slavyansk *see* Slov"yans'k
Sléibhte Chill Mhantáin
　see Wicklow Mountains
Slēmānī *see* As Sulaymānīyah
Sliema *395* N Malta
Sligo *316 Ir.* Sligeach. N Ireland
Sliven *152 var.* Slivno. E Bulgaria
Slobozia *480* SE Romania
Slobozia *408 Rus.* Slobodzeya.
　E Moldova
Slonim *122 Rus.* Slonin. W Belarus
Slovakia *522-523* officially Slovak
　Republika, *prev.* constituent republic
　of Czechoslovakia. Country of
　C Europe divided into 4 admin.
　regions (kraj). ❖ Bratislava
Slovenia *524-525* officially Republic
　of Slovenia, *Slvn.* Slovenija.
　Country divided into 86 admin.
　units (občina). ❖ Ljubljana
Slovenské Rudohorie *523*
　Ger. Slowakisches Erzgebirge,
　var. Ungarisches Erzgebirge.
　Mountain range of C Slovakia
Slov'yans'k *590 Rus.* Slavyansk.
　E Ukraine
Słupsk *471 Ger.* Stolp. N Poland
Slutsk *122* C Belarus
Smallwood Reservoir *171* lake of
　S Canada
Smarhon' *122* NW Belarus
Smederevo *512 Ger.* Semendria.
　N Serbia, Serbia
　& Montenegro (Yugo.)
Smila *590* C Ukraine
Smith's Island *643* island of
　E Bermuda
Smithson Bight *644* bay of the Indian
　Ocean on the S coast of
　Christmas Island
Smolensk *484* W Russian Federation
Smolyan *152 var.* Smoljan,
　prev. Pashmakli. SW Bulgaria
Smyrna *see* İzmir
Snaefell *647* mountain of C Isle of Man
Snake *602* river of NW USA
Sneeuw-gebergte
　see Maoke, Pegunungan
Sněžka *216 Ger.* Schneekoppe.
　Mountain of N Czech Republic
Snow Mountains
　see Maoke, Pegunungan
Snug Corner *278* SW Grenada island,
　Grenada
Snuŏl *165* E Cambodia
Soacha *195* C Colombia
Sobaek-sanmaek *352* mountain range
　of S South Korea
Sobat *540* river of Ethiopia and Sudan
Sobradinho, Represa de *145*
　var. Barragem de Sobradinho.
　Reservoir of E Brazil
Soča *323, 524 It.* Isonzo. River of Italy
　and Slovenia
Socabaya *463* S Peru
Sochi *484* SW Russian Federation
Société, Archipel de la *646* island
　group of W French Polynesia
Socotra *see* Suqutrá
Soc Trăng *627 var.* Khanh,
　Hung. S Vietnam
Sodankylä *251* N Finland
Södertälje *547* SE Sweden
Sodiri *540 var.* Sawdirī, Sodari.
　C Sudan
Sodo *245 var.* Soddo, Soddu.
　SW Ethiopia
Soekaboemi *see* Sukabumi
Soela Väin *242* strait of Baltic Sea,
　between the islands of Hiiumaa and
　Saaremaa, W Estonia
Soembawa *see* Sumbawa
Soerabaja *see* Surabaya
Soerakarta *see* Surakarta
Sofala, Baía de *419* Bay of Indian
　Ocean, off Mozambique
Sofia *382* seasonal river of
　NW Madagascar
Sofia *152 var.* Sofija, *Bul.* Sofiya.
　❖ of Bulgaria, W Bulgaria
Sogamoso *195* C Colombia
Sognefjorden *444* fjord of SW Norway

Sohâg *232 var.* Sawhaj. C Egypt
Sŏjosŏn-man *351* inlet of Korea Bay,
　on W coast of N Korea
Sokch'o *352* N South Korea
Söke *581* SW Turkey
Sokhumi *264 Rus.* Sukhumi.
　NW Georgia
Sokodé *571* C Togo
Sokolov *216* NW Czech Republic
Sokoto *440* NW Nigeria
Sokoto *440* river of NW Nigeria
Sola *444* SW Norway
Solapur *298 var.* Sholapur. SW India
Sol, Costa del *535* coastal region of
　S Spain
Soldeu *86* NE Andorra
Soledad *623* E Venezuela
Soledad *195* N Colombia
Soleure *see* Solothurn
Soligorsk *see* Salihorsk
Solimões *145* local name for a stretch
　of the Amazon river, NW Brazil
Solin *209 It.* Salona. S Croatia
Solingen *266* W Germany
Sollum, Gulf of *232*
　Ar. Khalīj as Sallūm. Gulf of the
　Mediterranean Sea, NW Egypt
Sololá *280* W Guatemala
Solomon Islands *526-527*
　prev. British Solomon Islands
　Protectorate. Country of the South
　Pacific Ocean divided into
　7 admin. units (provinces).
　❖ Honiara
Solomon Sea *458, 526* sea of the
　Pacific Ocean, to the E of Papua
　New Guinea
Solothurn *550 Fr.* Soleure.
　NW Switzerland
Solun *see* Thessaloníki
Solway Firth *597* arm of the
　Irish Sea, W UK
Solwezi *635* NW Zambia
Soma *263* C Gambia
Somalia *528-529* officially Somali
　Democratic Republic, *prev.*
　Somaliland Protectorate, Italian
　Somaliland. Country of E Africa
　divided into 16 admin. units
　(regions). ❖ Mogadishu
Sombor *512 Hung.* Zombor. NW Serbia,
　Serbia & Montenegro (Yugo.)
Somerset *643* Somerset Island,
　W Bermuda
Somerset Island *643* island of
　W Bermuda
Somerset Island *170* island of
　N Canada
Somerset Nile *see* Victoria Nile
Someş *292, 480 Hung.* Szamos,
　Ger. Samosch. River of Hungary and
　Romania
Sŏmjin *352 Jap.* Senshin-kō. River of
　S South Korea
Somme *255* river of N France
Somosomo *248* Taveuni, N Fiji
Somotillo *436* W Nicaragua
Somoto *436* NW Nicaragua
Soná *456* SW Panama
Sonaco *284* NE Guinea-Bissau
Sonda des Vieques *651* bay of the
　Caribbean Sea, E Puerto Rico
Sønderborg *218 Ger.* Sonderburg.
　Als, S Denmark
Søndre Strømfjord *see* Kangerlussuaq
Songea *385* S Tanzania
Songhua Jiang *see* Sungari
Sŏngjin *see* Kimch'aek
Songkhla *567 Mal.* Singora.
　S Thailand
Sŏngnam *352 var.* Seongnam.
　NW South Korea
Songnim *351* SW North Korea
Songo *419* NW Mozambique
Sông Tiên Giang *see* Mekong
Songwe *385* river of Malawi and
　Tanzania
Sonmiāni Bay *451* bay of the Arabian
　Sea, on the S coast of Pakistan
Sonoran Desert *10* desert
　Mexico/USA
Sonsonate *237* W El Salvador
Sonsorol Islands *455* island group of
　Palau

Sue *540* river of S Sudan
Sue Wood Bay *643* bay of the North Atlantic Ocean, C Bermuda
Suez *232* *Ar.* As Suways, *var.* El Suweis. NE Egypt
Suez Canal *232* *Ar.* Qanāt as Suways. Canal of NE Egypt
Suez, Gulf of *232* *Ar.* Khalīj al ʻAqabah. Gulf of the Red Sea, to the NE of Egypt
Sūf *338* NW Jordan
Sugar Loaf *278* *var.* Levera Island. N of Grenada island, Grenada
Şuḩār *448* *var.* Sohar. NW Oman
Sühbaatar *412* N Mongolia
Suigen *see* Suwŏn
Suir *316* *Ir.* An tSiúir. River of S Ireland
Sukabumi *304* *prev.* Soekaboemi. Java, C Indonesia
Sukagawa *332* Honshū, N Japan
Sukarnapura *see* Jayapura
Sukarno, Puntjak *see* Jaya, Puncak
Sukhne *see* As Sukhnah
Sukhumi *see* Sokhumi
Suki *540* E Sudan
Sukkertoppen *see* Maniitsoq
Sukkur *451* S Pakistan
Sukuta *263* W Gambia
Sulaimaniya *see* As Sulaymānīyah
Sulaimān Range *451* mountain range of C Pakistan
Sula, Kepulauan *304* *prev.* Xulla Islands, Soela. Island group to the E of Celebes, E Indonesia
Sulawesi *see* Celebes
Sulawesi, Laut *see* Celebes Sea
Sulby *647* N Isle of Man
Sullana *463* NW Peru
Sullivan Island *see* Lanbi Island
Sultan Alonto, Lake *see* Lanao, Lake
Sulu Archipelago *466* island group of SW Philippines
Sulu Sea *387, 466* sea of the Pacific Ocean, to the NE of Borneo, Malaysia
Sulyukta *359* *Kir.* Sülüktü. SW Kyrgyzstan
Sumatera *see* Sumatra
Sumatra *304* *var.* Sumatera. Island of W Indonesia
Šumava *see* Bohemian Forest
Sumba *644* Sudhuroy, S Faeroe Islands
Sumba *304* *prev.* Soemba, *Eng.* Sandalwood Island. Island of Nusa Tenggara, C Indonesia
Sumba, Selat *304* strait of the Indian Ocean between Sumba and Sumbawa, C Indonesia
Sumbawa *304* *prev.* Soembawa. Island of Nusa Tenggara, C Indonesia
Sumbawanga *564* W Tanzania
Sumbe *88* *Port.* Novo Redondo. W Angola
Sumbuya *518* S Sierra Leone
Šumen *see* Shumen
Sumisu-jima *332* island to the SE of Honshū, SE Japan
Šumperk *216* *Ger.* Mährisch-Schönberg. E Czech Republic
Sumpul *237* river of Honduras and El Salvador
Sumqayıt *110* *Rus.* Sumgait. E Azerbaijan
Sumy *590* NE Ukraine
Sunan *351* SW North Korea
Sunch'ŏn *351* SW North Korea
Sunch'ŏn *352* *Jap.* Junten. S South Korea
Sunda, Selat *304* strait connecting Indian Ocean and Laut Jawa between Java and Sumatra, W Indonesia
Sunderland *597* NE England, UK
Sundsvall *547* N Sweden
Sungai Seletar Reservoir *521* reservoir of N Singapore
Sungari *187* *Chin.* Songhua Jiang. River of NE China
Sun Koshi *427* river of E Nepal
Suntar-Khayata, Khrebet *485* mountain range of NE Russian Federation

Sunyani *272* W Ghana
Sunzu *635* mountain NE Zambia
Suō *see* Suao
Suomenlahti *see* Finland, Gulf of
Suomenselkä *251* physical region of C Finland
Suŏng *165* SE Cambodia
Superior de Tristaina, Estany *86* lake of NW Andorra
Superior, Lake *171, 603* *Fr.* Lac Supérieur. Lake of Canada and USA
Sup'ung-ho *351* reservoir of China and North Korea
Sūq ʻAbs *see* ʻAbs
Sūq ash Shuyūkh *312* SE Iraq
Suqrah Bay *448* Bay of the Arabian Sea, SE Oman
Suquṭrá *631* *Eng.* Socotra. Island of SE Yemen, off the Horn of Africa
Şūr *448* NE Oman
Surabaya *304* *prev.* Surabaja, Soerabaja. Java, C Indonesia
Surakarta *304* *prev.* Soerakarta. Java, C Indonesia
Šurany *523* *Hung.* Nagysurány. SW Slovakia
Sūrat *298* W India
Surat Thani *567* S Thailand
Sûre *378* river of W Europe
Sure, Lagh *344* dry watercourse of NE Kenya
Surfers Paradise *101* Queensland, E Australia
Surin *567* E Thailand
Surinam *400* E Barbados
Suriname *542-543* officially Republic of Suriname, *var.* Surinam. Country of Central America divided into 8 admin. units (provinces). ❖ Paramaribo
Surinam *121* E Barbados
Surkhet *see* Birendranagar
Surkhob *562* river of C Tajikistan
Şūrmān *371* NW Libya
Surt *371* *var.* Sidra, Sirte. N Libya
Surt, Khalīj *371* Gulf of Sirte, Gulf of Sidra. Gulf of the Mediterranean Sea, off N coast of Libya
Sūsah *see* Sousse
Susana *284* W Guinea-Bissau
Susuman *485* Ostrov Sakhalin, E Russian Federation
Sutlej *451* river of India and Pakistan
Suure-Jaani *242* *Ger.* Gross-Sankt-Johannis. C Estonia
Suur Munamägi *242* *var.* Munamägi. Mountain of SE Estonia
Suur Väin *242* strait of the Baltic Sea, between the mainland and the island of Muhu, W Estonia
Suva *248* ❖ of Fiji, Viti Levu, W Fiji
Suwa *240* SE Eritrea
Suwarrow *644* island of Northern Cook Islands, N Cook Islands
Suwayhān *594* E United Arab Emirates
Suways, Qanāt as *see* Suez Canal
Suwŏn *352* *var.* Suweon, *Jap.* Suigen. NW South Korea
Suzhou *187* *var.* Soochow, Su-chou, Suchow, *prev.* Wuhsien. Jiangsu, E China
Suzuka *332* Honshū, C Japan
Svalbard *653* Norwegian dependency of the Greenland Sea
Sväty Kríž nad Hronom *see* Žiar nad Hronom
Svay Chék *165* river of Cambodia and Thailand
Svay Riĕng *165* SE Cambodia
Svendborg *218* Fyn, S Denmark
Sverdlovsk *see* Yekaterinburg
Sverdrup Islands *170* island group of N Canada
Sveti Nikole *381* *prev.* Sveti Nikola. C FYR Macedonia
Svetlogorsk *see* Svyatlahorsk
Svetozarevo *512* *prev.* Jagodina. C Serbia, Serbia & Montenegro (Yugo.)
Svilengrad *152* *prev.* Mustafa-Pasha. SE Bulgaria
Svínoy *644* *var.* Svinø. Island of NE Faeroe Islands

Svishtov *152* *var.* Svištov. N Bulgaria
Svitavy *216* E Czech Republic
Svyetlahorsk *122* *Rus.* Svetlogorsk. SE Belarus
Swabian Jura *see* Schwäbische Alb
Swakopmund *423* W Namibia
Swallow Islands *526* small island group within Santa Cruz Is, E Solomon Is
Swan *100* river of SW Australia
Swansea *597* *Wel.* Abertawe. S Wales, UK
Swatow *see* Shantou
Swaziland *544-545* officially Kingdom of Swaziland. Country of southern Africa divided into 4 admin. units (districts). ❖ Mbabane
Sweden *546-549* officially Kingdom of Sweden, *Swe.* Sverige. Country of Scandinavia divided into 24 admin. units (läns). ❖ Stockholm
Swedru *see* Agona Swedru
Swellendam *531* Western Cape, S South Africa
Swetes *92* S Antigua, Antigua & Barbuda
Swift Current *170* SW Canada
Swindon *597* C England, UK
Swiss Plateau *550* *Ger.* Schweizer Mittelland. Plateau of W Switzerland
Switzerland *550-553* officially Swiss Confederation, *Ger.* Schweiz, *It.* Svizzera. Country of C Europe divided into 26 admin. units (cantons). ❖ Bern
Swords *316* *Ir.* Sórd Choluim Chille. E Ireland
Syaphrubesi *427* *var.* Syabrubensi. C Nepal
Sydney *101, 105* SE Australia
Sydney *171* Cape Breton Island, SE Canada
Sydney Island *see* Manra
Syktyvkar *484* *prev.* Ust'-Sisol'sk. NW Russian Federation
Sylhet *117* NE Bangladesh
Syowa *90* Japanese research station of Greater Antarctica, Antarctica
Syracuse *603* New York, NE USA
Syracuse *see* Siracusa
Syr Darya *340, 562, 614* *Rus.* Syrdar'ya, *Kaz.* Syrdariya, *Uzb.* Sirdaryo. River of C Asia
Syrdar'ya *614* E Uzbekistan
Syria *554-553* officially Syrian Arab Republic, *Ar.* Suriyah. Country divided into 13 admin. units (governorates). ❖ Damascus
Syriam *159* S Burma
Syrian Desert *312, 338, 555* *Ar.* Bādiyat ash Shām. Desert of SW Asia
Syvash, Zatoka *590* inlet of the Sea of Azov
Szabadka *see* Subotica
Szamos *see* Someş
Szatmárnémeti *see* Satu Mare
Szczecin *471* *Ger.* Stettin. NW Poland
Szczeciński, Zalew *see* Oderhaff
Szechuan *see* Sichuan
Szeged *292* *Ger.* Szegedin.SE Hungary
Székesfehérvár *292* *Ger.* Stuhlweissenburg. W Hungary
Szekszárd *292* S Hungary
Szenice *see* Senica
Szentes *292* SE Hungary
Szeping *see* Siping
Szered *see* Sereď
Sziszek *see* Sisak
Szlatina *see* Podravska Slatina
Szolnok *292* C Hungary
Szombathely *292* *Ger.* Steinamanger. W Hungary
Sztálinváros *see* Dunaújváros

T

Tabac, River *400* river of S Mauritius
Ṭabaqah *555* N Syria
Tabaquite *574* C Trinidad, Trinidad & Tobago

Tabarka *577* *var.* Ṭabarqah. NW Tunisia
Tabasará, Serranía de *456* mountain range of W Panama
Tabasco *see* Grijalva
Tabernacle *494* NE St Kitts, St Kitts & Nevis
Tabiteuea *348* island of the Gilbert Is, W Kiribati
Tablas Island *466* island of C Philippines
Table Hill Gordon *92* SE Antigua, Antigua & Barbuda
Tabligbo *571* SE Togo
Tábor *216* SW Czech Republic
Tabora *564* W Tanzania
Tabou *328* *var.* Tabu. S Ivory Coast
Tabrīz *309* NW Iran
Tabuaeran *348* *var.* Fanning Island. Island of the Line Is, E Kiribati
Tabūk *506* NW Saudi Arabia
Tabwémasana *619* mountain of Espiritu Santo, W Vanuatu
Täby *547* SE Sweden
Tachia Hsi *559* river of W Taiwan
Tachoshui *559* E Taiwan
Tacloban *466* Leyte, E Philippines
Tacna *463* SE Peru
Tacoma *602* Washington, NW USA
Tacuarembó *611* N Uruguay
Tacuarembó *611* river of C Uruguay
Tacurupucú *see* Hernandarias
Tademaït, Plateau du *83* plateau of C Algeria
Tadine *650* Maré, Îles Loyauté, E New Caledonia
Tadjoura *222* E Djibouti
Tadjoura, Golfe de *222* inlet of the Gulf of Aden, E of Djibouti
T'aebaek-sanmaek *352* mountain range of South Korea
Taedong *351* river of C North Korea
Taegu *352* *var.* Daegu, *Jap.* Taikyū. SE South Korea
Taehan-haehyŏp *see* Korea Strait
Taejŏn *352* *Jap.* Taiden. C South Korea
Tafahi *572* island of N Tonga
Tafí Viejo *95* NW Argentina
Taftlund *218* Jylland, SW Denmark
Taga *500* Savai'i, Samoa
Taganrog, Gulf of *590* *Ukr.* Tahanroz'ka Zatoka, *Rus.* Taganrogskiy Zaliv. Gulf of the Sea of Azov, SE Ukraine
Tagarzimat *414* W Western Sahara
Tagiura *see* Ṭājūrā'
Tagliamento *323* river of N Italy
Tagtabazar *see* Takhta-Bazar
Taguasco *210* C Cuba
Taguatinga *145* C Brazil
Tagula Island *458* *prev.* Southeast I. Island of SE Papua New Guinea
Tagum *466* river of Mindanao, S Philippines
Tahanroz'ka Zatoka *see* Taganrog, Gulf of
Tahat *83* mountain of SE Algeria
Tahiti *646* island of W French Polynesia
Tahoua *439* W Niger
Taia *518* river of C Sierra Leone
Taiama *518* C Sierra Leone
Taichung *559* *Jap.* Taichū. W Taiwan
Taiden *see* Taejŏn
Taieri *433* river of S South Island, New Zealand
Taihoku *see* Taipei
Taihsi *559* W Taiwan
Tai Hu *187* lake of E China
Taikyū *see* Taegu
Tainan *559* *Jap.* Tainan. SW Taiwan
Tai Pang Wan *see* Mirs Bay
T'aipei *559* *var.* Taipei, *Jap.* Taihoku. ❖ of Taiwan, N Taiwan
Taiping *386* NW Peninsular Malaysia
Taitung *559* *Jap.* Taitō. SE Taiwan
Taiwan *558-561* officially Republic of China (Taiwan). Country of E Asia divided into 16 admin. units (counties). ❖ Taipei

Troppau *see* Opava
Trou-du-Nord *288* N Haiti
Troumaka *498* NW St Vincent,
St Vincent & the Grenadines
Troyan *152 var.* Trojan. NW Bulgaria
Troyes *255* NE France
Tršćanski Zaljev *see* Trieste, Gulf of
Trst *see* Trieste
Truc Giang *see* Bên Tre
Trucial Coast *594* coastal region
of the United Arab Emirates
Trucial States *see* United Arab Emirates
Trujillo *290* N Honduras
Trujillo *463* NW Peru
Trujillo *623* NW Venezuela
Trung Phân *627 prev.* Annam. Cultural
region of Vietnam
Trunk Island *643* island of C Bermuda
Truro *597* SW England, UK
Trutnov *216 Ger.* Trautenau.
NE Czech Republic
Tržaški Zaliv *see* Trieste, Gulf of
Tržič *524 Ger.* Neumarktl. NW Slovenia
Trzynietz *see* Třinec
Tsabong *see* Tshabong
Tsaghkahovit *98* W Armenia
Tsamkong *see* Zhanjiang
Tsangpo *see* Brahmaputra
Tsaratanana, Tangorombohitr' i *382
var.* Massif du Tsaratanana.
Mountains of N Madagascar
Tsarigrad *see* İstanbul
Tsaritsyn *see* Volgograd
Tschakathurn *see* Čakovec
Tschenstochau *see* Częstochowa
Tschernembl *see* Črnomelj
Tschorna *see* Mustvee
Tselinograd *see* Akmola
Tsengwen Hsi *559* river of SW Taiwan
Tsetserleg *412* N Mongolia
Tsévié *571* S Togo
Tshabong *142 var.* Tsabong.
SW Botswana
Tshane *142* C Botswana
Tshaneni *544* NE Swaziland
Tshangalele, Lac *see* Lufira,
Lac de Retenue de la
Tshela *203* W Dem. Rep. Congo
Tshikapa *203* SW Dem. Rep. Congo
Tshohoha Sud, Lac *see* Cyohoha
Sud, Lac
Tshuapa *203* river of C Dem. Rep. Congo
Tsinan *see* Jinan
Tsing Hai *see* Qinghai Hu
Tsinghai *see* Qinghai
Tsingtao *see* Qingdao
Tsingyuan *see* Baoding
Tsinkiang *see* Quanzhou
Tsiroanomandidy *382* C Madagascar
Tsiteli-Tskaro *see* Dedoplistsqaro
Tsitsihar *see* Qiqihar
Tskhinvali *264* C Georgia
Tsna *122* river of S Belarus
Tsodilo Hills *142* mountain range
of NW Botswana
Tsoelike *366* SE Lesotho
Tsu *332* Honshū, C Japan
Tsuchiura *332* Honshū, SE Japan
Tsugaru-kaikyō *332* strait connecting
the Sea of Japan and Pacific Ocean,
between Hokkaidō and Honshū,
N Japan
Tsumeb *423* N Namibia
Tsumkwe *423* NE Namibia
Tsuruga *332* Honshū, C Japan
Tsuruoka *332* Honshū, N Japan
Tsushima *332* island group to the
W of Honshū, W Japan
Tsushima Strait *332 var.* Korea Strait.
Strait connecting the Sea of Japan
and East China Sea, between South
Korea and Japan
Tsuyama *332* Honshū, W Japan
Tuamotu, Îles *646* island group of
N French Polynesia
Tuasivi *500* Savai'i, Samoa
Tuban, Wādī *631* dry watercourse of
S Yemen
Tubmanburg *368* NW Liberia
Ṭubruq *371 Eng.* Tobruk, *It.* Tobruch.
NE Libya
Tucker's Town *643* E Bermuda
Tuckum *see* Tukums
Tucson *602* Arizona, SW USA

Tucumán *see* San Miguel de Tucumán
Tucupita *623* E Venezuela
Tucuruí, Represa de *145* reservoir
of N Brazil
Tudmur *555 var.* Tadmur,
Eng. Palmyra. C Syria
Tugalan *see* Kolkhozobod
Tuira *456* river of SE Panama
Tukangbesi, Kepulauan *304* island
group to the SE of Celebes,
C Indonesia
Tu-k'ou *see* Panzhihua
Ṭūkrah *371* NE Libya
Tukums *362 Ger.* Tuckum. W Latvia
Tukuyu *564* S Tanzania
Tula *642* Tutuila, W American Samoa
Tula *484* W Russian Federation
Tulaghi *526* Florida I. Solomon Islands
Tulcán *230* N Ecuador
Tulcea *480* E Romania
Tuléar *see* Toliara
Tuli *see* Thuli
Ṭūlkarm *319, 320* NW West Bank
Tulle *255* S France
Tulsa *603* Oklahoma, SC USA
Tulsipur *451* W Nepal
Tuluá *195* W Colombia
Tumaco *195* SW Colombia
Tumareng *286* NW Guyana
Tumba, Lac *see* Ntomba, Lac
Tumbes *463* NW Peru
Tumen *351 Rus.* Tumyntszyan. River of
China and North Korea
Tumpat *386* N Peninsular Malaysia
Tumuc-Humac Mountains *145
Port.* Serra Tumuc-Humac. Mountain
range of N South America
Tumuc-Humac, Serra *see*
Tumuc-Humac Mountains
Tunapuna *574* N Trinidad,
Trinidad & Tobago
Tunduru *564* S Tanzania
Tundzha *152 var.* Tundža. River of
Bulgaria and Turkey
Tungabhadra *298* river of S India
Tungaru *348 prev.* Gilbert Islands.
Island group of W Kiribati
Tungkang *559 var.* Tung-chiang,
Jap. Tōkō. SW Taiwan
Tungshih *559 Jap.* Tōsei. W Taiwan
Tung-t'ing Hu *see* Dongting Hu
Tunis *577 var.* Tūnis. of Tunisia,
NE Tunisia
Tunis, Golfe de *577* gulf of the
Mediterranean Sea on the NE coast
of Tunisia
Tunisia *576-579* officially Republic
of Tunisia. Country of North
Africa divided into 25 admin.
units (governorates).❖ Tunis
Tunja *195* C Colombia
Tương Đương *627* NW Vietnam
Tupiza *136* S Bolivia
Turan Lowland *585, 614
var.* Turan Plain, *Rus.* Turanskaya
Nizmennost'. Plain of Turkmenistan
and Uzbekistan
Ṭurayf *506* N Saudi Arabia
Turbat *450* W Pakistan
Turčiansky Svätý Martin *see* Martin
Turda *480* NW Romania
Turgay *340 Kaz.* Torgay.
W Kazakhstan
Turgel *see* Türi
Tŭrgovishte *152 var.* Tărgovište,
prev. Eski Dzhumaya.
NE Bulgaria
Turgutlu *580* W Turkey
Türi *242 Ger.* Turgel. C Estonia
Turin *see* Torino
Turkana, Lake *245, 344 Eng.* Lake
Rudolf. Lake of E Africa
Turkestan *340 Kaz.* Türkistan.
S Kazakhstan
Turkestan Range *562
Rus.* Turkestanskiy Khrebet.
Mountain range of NW Tajikistan
Turkestanskiy Khrebet
see Turkestan Range
Turkey *580-581* officially Republic
of Turkey, *Turk.* Türkiye
Cumhuriyeti. Country of W Asia
divided into 73 admin. units.
❖ Ankara

Turkish Republic of Northern Cyprus
see Cyprus
Turkmenbashi *585 prev.* Krasnovodsk.
W Turkmenistan
Turkmenistan *584-585* officially
Republic of Turkmenistan,
prev. Turkmenskaya Soviet
Socialist Republic. Country divided
into 4 admin. units (oblasts).
❖ Ashgabat
Turkmenskiy Zaliv *585
Turkm.* Türkmen Aylagy. Inlet of
the Caspian Sea, on the W coast
of Turkmenistan
Turks and Caicos Islands *653*
British dependent territory of
the Atlantic Ocean
Turks Islands *653* island group of
SE Turks and Caicos Islands
Turku *251 Swe.* Åbo. SW Finland
Turkwel *344* seasonal river of
NW Kenya
Turmero *623* N Venezuela
Turneffe Islands *130* islands of
E Belize
Turnhout *127* NE Belgium
Turning Basin *648* undersea feature
of the Pacific Ocean,
NW Johnston Atoll
Turnov *216* N Czech Republic
Turnu Severin
see Drobeta-Turnu Severin
Turócszentmárton *see* Martin
Turquino, Pico *210* mountain of
SE Cuba
Turrialba *206* E Costa Rica
Tursunzode *562 prev.* Regar.
W Tajikistan
Tŭrtkŭl' *614 prev.*
Petroaleksandrovsk,
Rus. Turtkul. W Uzbekistan
Turtle Islands *518* island group of
SW Sierra Leone
Tuscaloosa *603* Alabama, SE USA
Tuscan Archipelago
see Toscano, Arcipelago
Tuscany *327.* region of Italy
Tuticorin *298* S India
Tutong *150* NW Brunei
Tutong *150* river of C Brunei
Tutrakan *152* N Bulgaria
Tutuala *228* W East Timor
Tutuila *642* island of
W American Samoa
Tutume *142* E Botswana
Tuvalu *586-587 prev.* The Ellice
Islands. Country of the Pacific Ocean.
❖ Fongafale
Tuvana-i-colo *248* island to the
S of the Lau Group, SE Fiji
Tuvana-i-ra *248* island to the
S of the Lau Group, SE Fiji
Tuvuca *248 prev.* Tuvutha. Island of
the Lau Group, E Fiji
Ṭuwayq, Jabal *506* mountain range
of C Saudi Arabia
Tuxtla Gutiérrez *403 var.* Tuxtla.
SE Mexico
Tuy Hoa *627* SE Vietnam
Tuz Gölü *581 Eng.* Lake Tuz.
Lake of C Turkey
Tuz Khurmātū *312* N Iraq
Tuzla *140* NE Bosnia & Herzegovina
Tver' *484* W Russian Federation
Tvøroyri *644 var.* Tverå. Sudhuroy,
S Faeroe Islands
Twante *159* S Burma
Tweed *597* river of SE Scotland, UK
Twin Falls *602* Idaho, NW USA
Twizel *433* C South Island,
New Zealand
Tychy *471 Ger.* Tichau. S Poland
Tynda *485* E Russian Federation
Tyne *597* river of NE England, UK
Tyneside *573*
Tyrnau *see* Trnava
Tyrol *see* Tirol
Tyrrhenian Sea *323 It.* Mare Tirreno.
Area of the Mediterranean Sea
Tyumen' *484, 489* C Russian Federation
Tyup *359 Kaz.* Tjup. NE Kyrgyzstan
Tyva, Respublika *484* autonomous
republic of C Russian Federation
Tzekung *see* Zigong

U

Uaco Cungo *88 var.* Waku Kungo,
Port. Santa Comba. C Angola
UAE *see* United Arab Emirates
Uamba *see* Wamba
Uanle Uen *see* Wanlaweyn
Ubangi *179, 200, 203 Fr.* Oubangui.
River of C Africa
Ubarts' *122* river of Belarus
and Ukraine
Ubayyiḍ, Wādī al *312* dry
watercourse of SW Iraq
Ube *332* Honshū, W Japan
Uberaba *145* S Brazil
Uberlândia *145* S Brazil
Ubin, Pulau *521* island of
NE Singapore
Ubon Ratchathani *567* E Thailand
Ucar *110 Rus.* Udzhary. C Azerbaijan
Ucayali *463* river of C Peru
Uchquduq *614 Rus.* Uchkuduk.
C Uzbekistan
Udine *323* N Italy
Udmurtskaya Respublika *484*
autonomous republic of
W Russian Federation
Udon Thani *567* N Thailand
Udzhary *see* Ucar
Ueda *332* Honshū, C Japan
Uele *203 var.* Welle. River of
N Dem. Rep. Congo
Ufa *484* W Russian Federation
Ugāle *362* NW Latvia
Uganda *588-589* officially
Republic of Uganda. Country of
E Africa divided into 39 admin.
units (districts). ❖ Kampala
Ugum *647* river, S Guam
Uherské Hradiště *216
Ger.* Ungarisch-Hradisch.
SE Czech Republic
Úhlava *216 Ger.* Angel. River of
W Czech Republic
Uíge *88 Port.* Vila Marechal Carmona,
Carmona. NW Angola
'Uiha *572 var.* Uiha. Island of the
Ha'apai Group, Tonga
Ŭijŏngbu *352 Jap.* Giseifu.
NW South Korea
Uil *340* NW Kazakhstan
Uis *423* NW Namibia
Uitenhage *531* Eastern Cape,
S South Africa
Ujae *396* island of W Marshall Islands
Ujelang *396* island of
W Marshall Islands
Újgradiska *see* Nova Gradiška
'Ujman *see* Ajman
Ujungpandang *304 prev.* Makasar,
Macassar. Celebes, C Indonesia
Ujung Salang *see* Phuket
Újvidék *see* Novi Sad
UK *see* United Kingdom
Ukerewe *564* island of
NW Tanzania, situated at the S of
Lake Victoria
Uki Island *526* island of
E Solomon Islands
Ukmergė *376* E Lithuania
Ukraine *590-593* officially Republic
of Ukraine, *Ukr.* Ukrayina,
Rus. Ukraina, *prev.* Ukrainian Soviet
Socialist Republic, Ukrainskaya S.S.R.
Country of E Europe divided into
25 admin. units (24 regions and 1
republic). ❖ Kiev
Uku-jima *332* island of Gotō-rettō,
SW Japan
Ula *122 Rus.* Ulla. River of N Belarus
Ulaangom *412* NW Mongolia
Ulan Bator *412 var.* Ulaanbaatar.
❖ of Mongolia, C Mongolia
Ulan-Ude *485 prev.* Verkhneudinsk.
C Russian Federation
Ulawa Island *526* island of
E Solomon Is
Uleåborg *see* Oulu
Uleälv *see* Oulujoki
Uleträsk *see* Oulujärvi
Uliastay *412* W Mongolia
Ulithi *406* atoll of W Micronesia

Ulla *see* Ula
Ullapool *597* N Scotland, UK
Ulm *267* S Germany
Ulonguè *419 var.* Ulongwé.
NW Mozambique
Ulsan *352 Jap.* Urusan. SE South Korea
Ulster *597* province of Ireland, mostly
included within Northern Ireland, UK
Ulúa *290* river of NW Honduras
Uluru *101 var.* Ayers Rock.
Rocky outcrop of C Australia
Ulverstone *101* Tasmania, Australia
Ul'yanovsk *484 prev.* Simbirsk.
W Russian Federation
Uman' *590* C Ukraine
'Umān, Khalīj *see* Gulf of Oman
Umatac *647* SW Guam
Umboi *458 var.* Rooke I. Island of
E Papua New Guinea
Umbro-Marchigiano, Appennino *323*
mountains of C Italy
Umeå *547* NE Sweden
Umeälv *547* river of NE Sweden
Umm al Ḥayt, Wādī *see* Amilḥayt, Wādī
Umm al Qaiwain *594 Ar.* Umm al
Qaywayn. NE United Arab Emirates
Umm an Na'sān *115* island of
W Bahrain
Umm aş Şabbān *see* Al Muḥammadīyah
Umm as Samīn *448* seasonal desert
lake of W Oman
Umm Bāb *479* W Qatar
Umm Durmān *see* Omdurman
Umm Ruwaba *540 var.* Umm
Ruwābah, Um Ruwāba. C Sudan
Umm Sa'id *see* Musay'īd
Umm Şalāl 'Alī *479*
var. Umm Silal Ali. NE Qatar
Umm Şalāl Muḥammad *479*
var. Umm Silal Mohammed. E Qatar
Umtali *see* Mutare
Umtata *531* Eastern Cape,
SE South Africa
Umvukwes *see* Mvurwi
Umvuma *see* Mvuma
Umzingwani *636* river of
S Zimbabwe
Una *140, 209* river of Bosnia
& Herzegovina and Croatia
'Unayzah *506 var.* Anaiza.
C Saudi Arabia
Ungama Bay *344 var.* Formosa Bay.
Bay of the Indian Ocean to the
SE of Kenya
Ungarisches Erzgebirge
see Slovenské Rudohorie
Ungarisch-Hradisch
see Uherské Hradiště
Ungava Bay *171* bay of the Labrador
Sea, E Canada
Ungava, Péninsule d' *171* peninsula of
E Canada
Ungheni *408 Rus.* Ungeny. W Moldova
Ungvár *see* Uzhhorod
Union *278* NW Grenada island, Grenada
Union Island *498* island of
SW St Vincent & the Grenadines
United Arab Emirates *594-595*
prev. Trucial States, *abbrev.* U.A.E.
Country of SW Asia comprised of
7 admin units (states). ❖ Abu Dhabi
United Kingdom *596-601* officially
United Kingdom of Great Britain
and Northern Ireland, *abbrev.* UK.
Country of NW Europe comprising
from 1st April 1998, England (149
admin. units -35 counties, 45 unitary
authorities, 36 metropolitan bor-
oughs, 32 London boroughs and the
City of London), Wales (22 admin.
units - unitary authorities), Scotland
(22 admin. units - local authorities)
and Northern Ireland (26 admin.
units - districts). ❖ London
United States of America *602-609*
officially United States of America,
abbrev. USA. Country of North
America divided into 51 admin. units
(50 states, 1 federal district).
❖ Washington DC
Uno *284* Ilha de Uno, Guinea-Bissau
Uno, Ilha de *284* island of
SW Guinea-Bissau
Uoleva *572* island of the Ha'apai
Group, Tonga

Uolo, Río *238 var.* Mbini, Woleu. River
of Equatorial Guinea and Gabon
Uozu *332* Honshū, C Japan
Upaar Lagoon *538* lagoon of
E Sri Lanka
Upala *206* NW Costa Rica
Upata *623* E Venezuela
Upernavik *646* W Greenland
Upington *531* Northern Cape,
W South Africa
'Upolu *500* island of SE Samoa
Upper Bann *597* river of Northern
Ireland, UK
Upper Conaree *494* E St Kitts,
St Kitts & Nevis
Upper Hutt *433* S North Island,
New Zealand
Upper Lough Erne *597* lake of
Northern Ireland, UK
Upper Mortlocks *406* island group
of C Micronesia
Upper Peirce Reservoir *521* reservoir
of C Singapore
Upper Volta *see* Burkina
Uppsala *547* SE Sweden
Ural *340, 484 Kaz.* Zayyq. River of
Kazakhstan and Russian Federation
Ural Mountains *484 var.* Ural'skiy
Khrebet, *var.* Ural'skiy Gory. Mountain
range of W Russian Federation
Ural'sk *340 Kaz.* Oral. NW Kazakhstan
Ura-Tyube *562* NW Tajikistan
Ureca *238 prev.* San Antonio de Ureca.
S Bioko, Equatorial Guinea
Uréparapara *619* Banks Islands,
N Vanuatu
Urfa *see* Şanlıurfa
Urganch *614 Rus.* Urgench, *prev.*
Novo-Urgench. W Uzbekistan
Urgench *see* Urganch
Urgut *614* SE Uzbekistan
Urlings *92* SW Antigua, Antigua
& Barbuda
Urmia *see* Orūmīyeh
Urmia, Lake *see* Orūmīyeh,
Daryācheh-ye
Uroševac *512 Alb.* Ferizaj. S Serbia,
Serbia & Montenegro (Yugo.)
Uruapan *403 var.* Uruapan del
Progreso. SW Mexico
Uruguaiana *144* S Brazil
Urubamba *463* river of C Peru
Uruguay *95, 145, 611 Port.* Uruguai.
River of S South America
Uruguay *610-613* officially Eastern
Republic of Uruguay. Country of
South America divided into 19 admin.
units (departments). ❖ Montevideo
Urukthapel *455* island of C Palau
Ürümqi *286 var.* Urumchi,
Wu-lu-mu-ch'i, *prev.* Ti-hua. Xinjiang
Uygur Zizhiqu, NW China
Urusan *see* Ulsan
USA *see* United States of America
Uşak *580 prev.* Ushak. W Turkey
Usakos *423* C Namibia
Usborne, Mount *645* mountain of East
Falkland, E Falkland Islands
Ushant *see* Ouessant, Île d'
Ushuaia *95* S Tierra del Fuego,
Argentina
Usk *597 Wel.* Wysg. River of S Wales, UK
Üsküb *see* Skopje
Usmas Ezers *362* lake of W Latvia
Usol'ye-Sibirskoye *485* C Russian
Federation
Ussuriysk *485 prev.* Voroshilov,
Nikol'sk-Ussuriyskiy. SE Russian
Federation
Ust'-Abakanskoye *see* Abakan
Uster *550* NE Switzerland
Ustica, Isola de *323* island of S Italy
Ústí nad Labem *216 Ger.* Aussig.
NW Czech Republic
Ustinov *see* Izhevsk
Ust'-Kamchatsk *485* NE Russian
Federation
Ust'-Kamenogorsk *340*
Kaz. Öskemen. E Kazakhstan
Ust'-Sisol'sk *see* Syktyvkar
Ustyurt Plateau *340, 614*
Uzb. Ustyurt Platosi. Plateau of
Kazakhstan and Uzbekistan
Usulután *237* SE El Salvador

Usumacinta *280, 403* river of
Guatemala and Mexico
Usumbura *see* Bujumbura
Usutu *see* Lusutfu
Uta Vava'u *see* 'Uta Vava'u
Utah *602* state of SW USA
Utamboni *see* Mitemele, Río
'Uta Vava'u *572* island of the Vava'u
Group, Tonga
Utena *376* E Lithuania
Utirik *396* island of N Marshall Islands
Utrecht *429* C Netherlands
Utsunomiya *332* Honshū, SE Japan
Uttaradit *567 var.* Utaradit. N Thailand
Utuado *651* C Puerto Rico
Utupua *526* island of the Santa
Cruz Is, E Solomon Is
Uummannaq *646* W Greenland
Uummannarsuaq *see* Nunap Isua
Uvea, Île *653* island of
S Wallis & Futuna
Uvs Nuur *412* lake of NW Mongolia
Uwajima *332* Shikoku, SW Japan
Uyo *440* S Nigeria
Uyuni *136* W Bolivia
Uzbekistan *614-617* officially Republic
of Uzbekistan. Country of C Asia
divided into 12 admin. units
(oblastey). ❖ Tashkent
Uzgen *359 Kir.* Özgön. W Kyrgyzstan
Uzhhorod *590 Rus.* Uzhgorod,
Cz. Užhorod, *Hung.* Ungvár.
W Ukraine
Užice *512 var.* Titovo Užice. W Serbia,
Serbia & Montenegro (Yugo.)

V

Vaal *531* river of C South Africa
Vaasa *251 prev.* Nikolainkaupunki,
Swe. Vasa. W Finland
Vác *292 Ger.* Waitzen. N Hungary
Vaches, Île aux *516 var.* Bird Island.
Island group of the Inner Islands,
N Seychelles
Vacoas *400* W Mauritius
Vadar *275 prev.* Axios. River of Greece
and FYR Macedonia
Vadile *see* Vatili
Vadodara *298 prev.* Baroda. W India
Vaduz *374* ❖ of Liechtenstein,
W Liechtenstein
Vágar *644 var.* Vágo. Island of
W Faeroe Islands
Vágbeszterce *see* Považská Bystrica
Vágújhely *see* Nové Mesto nad Váhom
Vágur *644 var.* Våg. Sudhuroy,
S Faeroe Islands
Váh *523 Ger.* Waag, *Hung.* Vág. River
of W Slovakia
Väike-Emajõgi *242* river of S Estonia
Väinameri *242 prev.* Muhu Väin.
Area of Baltic Sea, off the coast
of W Estonia
Vaitogi *642* Tutuila, W American
Samoa
Vaitupu *587* coral atoll of C Tuvalu
Vakhah *562 var.* Vahš. SW Tajikistan
Vākhān, Kūh-e *77*
Per. Kowtal-e Khaybar. Mountain
range of C Asia
Vakhsh *562 var.* Vahš. River of
SW Tajikistan
Valdecañas, Embalse de *535*
reservoir of W Spain
Valdez *602* Alaska, USA
Valdia *see* Weldiya
Valdivia *183* C Chile
Valence *255* SE France
Valencia *535 Cat.* València.
Autonomous community of NE Spain
Valencia *623* NW Venezuela
Valencia, Golfo de *535* area of the
Mediterranean Sea, E of Spain
Valera *623* W Venezuela
Valga *242 Ger.* Walk. S Estonia
Valira *86* river of Andorra and Spain
Valira del Nord *86* river of
NW Andorra
Valira d'Orient *86* river of C Andorra
Valkeakoski *251* SW Finland
Valladolid *535* NW Spain
Valle de la Pascua *623* C Venezuela
Valledupar *195* N Colombia

Vallée de Mboun *510* river of
C Senegal
Vallée du Ferlo *510 var.* Ferlo.
River of N Senegal
Vallegrande *136* C Bolivia
Vallenar *183* N Chile
Valletta *395 prev.* Valetta. ❖ of Malta,
E Malta
Valley *121* C Barbados
Valley, The *642* ❖ of Anguilla,
E Anguilla
Vallgrund *251* island of W Finland
Valmiera *362 Ger.* Wolmar. NE Latvia
Valona *see* Vlorë
Valona, Bay of *see* Vlorës, Gjiri i
Valparaíso *183* C Chile
Valsayn *574* NW Trinidad,
Trinidad & Tobago
Van *581* E Turkey
Vanadzor *98 prev.* Kirovakan.
N Armenia
Vanard *497* NW St Lucia
Vana-Vändra *see* Vändra
Vancouver *170* SW Canada
Vancouver Island *170* island of
SW Canada
Vanda *see* Vantaa
Vanderbijlpark *531*
Pretoria-Witwatersrand-Vereeniging,
NE South Africa
Vändra *242 prev.* Vana-Vändra,
Ger. Fennern. C Estonia
Vaner, Lake *see* Vänern
Vänern *547 Eng.* Lake Vaner,
prev. Lake Vener. Lake of
SW Sweden
Vangaindrano *382* S Madagascar
Van Gölü *580 Eng.* Lake Van. Lake
of E Turkey
Vangunu *526* island of the New
Georgia Is, Solomon Is
Vanikolo *526 var.* Vanikoro. Island of
the Santa Cruz Is, E Solomon Islands
Vanimo *458* NW Papua New Guinea
Van, Lake *see* Van Gölü
Vannes *254* W France
Vantaa *251 Swe.* Vanda. SW Finland
Vanua Balavu *248*
prev. Vanua Mbalavu. Island of the
Lau Group, E Fiji
Vanua Lava *619* Banks Islands,
N Vanuatu
Vanua Levu *248* island of N Fiji
Vanuatu, Republic of *618-619*
officially Republic of Vanuatu,
prev. New Hebrides. Country of the
Pacific Ocean divided into 11 admin.
units (districts). ❖ Port-Vila
Vanua Vatu *248* island of the Lau
Group, E Fiji
Vao *650* Île des Pins, S New Caledonia
Varadero *210* NW Cuba
Vārānasi *298 prev.* Benares. NE India
Varangerfjorden *444* fjord of
NE Norway
Varannó *see* Vranov nad Topľou
Varaždin *209 Hung.* Varasd,
Ger. Warasdin. N Croatia
Vardar *381* river of Greece and
FYR Macedonia
Varde *218* Jylland, W Denmark
Vardenis *98* E Armenia
Varėna *376 Pol.* Orany. S Lithuania
Varesa *322* N Italy
Vârful Moldoveanu *480*
prev. Vîrful Moldoveanu. Mountain of
C Romania
Varkaus *251* SE Finland
Varna *152 prev.* Stalin. E Bulgaria
Varnenski Zaliv *152*
prev. Stalinski Zaliv. Bay of the Black
Sea to the E of Bulgaria
Varnensko Ezero *152* lake of
E Bulgaria
Vasa *see* Vaasa
Vaslui *480* E Romania
Västerås *547* S Sweden
Västervik *547* S Sweden
Vasto *323* C Italy
Vaté *see* Efate
Vatican City *323, 620-621* officially
Vatican City State. City state at the
C of Rome, C Italy
Vatili *215 var.* Vadili. C Cyprus
Vatnajökull *296* glacier of SE Iceland

ACKNOWLEDGMENTS

DORLING KINDERSLEY would like to express their thanks to the following individuals, companies and institutions for their help in preparing this atlas:

ADDITIONAL CARTOGRAPHY
Advanced Illustration (Congleton, UK)
Andrew Bright
Cosmographics (Watford, UK)
Malcolm Porter
Swanston Publishing (Derby, UK)
Andrew Thompson

DESIGN
Boyd Annison, Icon Solutions (Chesham, UK) *for Macintosh consultancy and chart templates*
Bruno Maag, Dalton Maag (London, UK) *for font consultancy and production*

RESEARCH AND REFERENCE
Dr D Alkhateeb, Organization of Petroleum Exporting Countries (OPEC, Vienna, Austria)
Amnesty International (London, UK)
Caroline Blunden
CNN International (New York, USA)
DATAQUEST EUROPE SA (PARIS, FRANCE)
CSL Davies
Department of Trade and Industry Export Market Information Centre (London, UK)
The Flag Institute (Chester, UK)
Foreign and Commonwealth Office (London, UK)
Alexander Fyges-Walker
Christel Heideloff, Institute of Shipping Economics and Logistics (Bremen, Germany)
International Bank for Reconstruction and Development (World Bank, Washington, DC, USA)
International Committee of the Red Cross (ICRC, Geneva, Switzerland)
International Civil Aviation Organization (ICAO, Montreal, Canada)
International Criminal Police Organization (INTERPOL, Lyon, France)
International Institute for Strategic Studies, for information from *The Military Balance* (London, UK)
International Boundaries Research Unit, University of Durham
Institute of Latin American Studies, University of London (London, UK)
Intermediate Technology Development Group (Rugby, UK)
Chris Joseph, United States Travel and Tourism Administration (USTTA, London, UK)
Latin American Bureau (London, UK)
Patrick Mahaffey, Ohio European Office (Brussels, Belgium)
Peter Mansfield
Robert Minton-Taylor
National Meteorological Library and Archive (Bracknell, UK)
Oil and Gas Journal (Houston, Texas)
Organization for Economic Cooperation and Development (OECD, Paris, France)
Penal Reform International (London, UK)
Matt Ridley
Screen Digest (London, UK)
William Smith, Chicago Sun-Times (Chicago, USA)
Tourism Concern (London, UK)
United Nations Crime Prevention and Criminal Justice Branch (UNCPC, Vienna, Austria)
United Nations Development Programme (UNDP, New York, USA)
United Nations Environment Programme (UNEP, Nairobi, Kenya)
United Nations Food and Agriculture Organization (UNFAO, Rome, Italy)
United Nations International Labour Organization (UNILO, Geneva, Switzerland)
United Nations Population Fund (UNFPA, New York, USA)
Westminster Reference Library (London, UK)
World Conservation Monitoring Centre (Cambridge, UK)
World Health Organization (WHO, Geneva, Switzerland)
World Tourism Organization (Madrid, Spain)

The many embassies, High Commissions, airports, national information and tourist offices in London and around the world.

PICTURE CREDITS

t=top, b=below, a=above, l=left, r=right, c=center

Agence France Presse: 300crb, 332tr, 337tc, 445cr; Gustavo Amarelle 95crb; Dong-A Ilbo 353bcr; Victor Drachev 123bcr; Vassil Donev 153bcr; Gonzalo Espinoza 137cr; Eric Feferberg 188br; Francois Guillot 256crb; Martyn Hayhow 256bcr; Doug Kanter 601tr; Attila Kisbenedek 293br; John MacDougall 77tr; Shah Marai 78br; Tatiana Munoz 195tcr; Keld Navntoft 219tcr; Inacio Rosa 475bcr; Dibyangshu Sarkar 303tc; Bernd Settnik 269br; Sergei Supinsky 587tr; Weda 229tc; Greg Wood 337bl; **Ancient Art & Architecture Collection:** 44bcr; 45bl; 45cr; 47tc; G Tortoli 45tcb; **Arcaid:** P Mauss Esto 598bc; **Art Archive:** 44bl; 49bl; 49cr; 51tcl; **Aspect Picture Library:** 202tr; Brian Seed 418c; D Bayes 370bl; Fiona Nichols 504tr; **Associated Press AP:** 245bcr; AFP 415bcr; Aaron Favila 467crl; Humberto Pradera/Agencia Estado 149cb; Bullit Marquez 191bl; Jan Bauer 271cb Alexander Zemlianichenko 488cb; **Bridgeman Art Library, London/New York:** Hermitage, St Petersburg 46bc; Lauros - Giraudon / Château de Malmaison 48br; National Maritime Museum, London 47br; Private Collection 48bcl; **D Donne Bryant Stock Picture Agency:** 461tc; Byron Augustin 456tr; **Dale Buckton:** 594br; **Camera Press:** A Pucciano 95cr; F Goodman 649bca; H Andrews 627bcr; S Smith 245bcl; T Charlier 607bl; **The Canadian Alliance:** 172crb; **The J Allan Cash Photolibrary:** 55tl; 106c; 116bc; 146br; 149tl; 168tr; 171tr; 241tr; 291tc; 308tr; 320br; 339cr; 342bl; 383tc; 386c; 458bc; 440tr; 451bc; 474tr; 566bc; **CDA:** 429tcr; **Bruce Coleman Ltd:** 498bl; B&C Calhoun 174bc; Dr MP Kahl 94tr; F Prenzel 104tc; Gerald Cubitt 88tr; Gerald Cubitt 244bc; Gordon Langsbury 262bc; J Fry 511tc; J Jurka 542bc; K. Maj 470c; Kim Taylor 515tr; L Lee Rue 91tl; LC Marigo 281tc; M Berge 396tr; MPL Fogden 462ca; O Langrand 370tr; P Davey 214bc; S Prato 274bc; **Colorific:** J Polleross / M Kreiner 54cl; M Rogers 536tr; Sandro Tucci 159bcl; **Comstock:** 194bc, 24tl, 535tc; Tor Eigeland 399tc; **Corbis UK Ltd:** AFP 387bcl; Bettmann 146car, 233bcr, 555bcr; Bob Krist 642br; Francoise de Mulder 141bc; Jack Fields 648bl; Jack Fields 650bl; James L Amos 647tr; Miki Kratsman 486br; Nik Wheeler 175tc; Franz-Marc Frei 261tcb; Peter Wilson 327cb; Joseph Sohm 603bl; **James Davis Travel Photography:** 144bl; 145tl; 218c; 250bc; 330bc; 379tl; 390tr; 401tc; 411tl; 428bc; 432bc; 493tc; 494bl; 496bl; 502tr; 550bc; 558tr; 586tr; 626bl; 643cr; Prisma 86cla; Prisma / Schwarz 226tr; S Begawan 151tl; S Thingeyjar 297tc; World View - Fototheek Amsterdam 177tc; **Mary Evans Picture Library:** 47bl, 47trb; **Chris Fairclough Colour Library:** 51cr; 575tl; 572br; **Robert Harding Picture Library:** 76tr; 180c; 225tl; 273tc; 282bl; 293tr; 512tr; 556bl; 558bc; 449tc; 466tr; 480tr; 484tr; 501tc; 513tc; A Woolfitt 114bc; C Martin 266bc; C Rennie 580ca; D Hughes 259bc; Explorer 377tc; Explorer / Roy 110tr; F Dubes 148bl; Frerck / Odyssey 402tr; G Hellier 217tr; G Hellier 518tr; G Roli 444tr; Gascoine 610bl; P Craven 254tr; Photri 222ca; R Rainford 593tl; Rosehaven Management Ltd 653ccr; Sassoon 134cb; Sassoon 264cl; **Paul Harris Photography:** Paul Harris Photography 491bl; **Hulton Getty:** 50bc; 51br; 53br; **Robert Hunt Library:** 55tc; **Hutchison Library:** 156bl; 369tc; 596bc; 656tr; Andrew Hill 654bc; Bernard Gérard 261bc; Christine Pemberton 210tr; J Henderson 286ca; JG Fuller 650bcr; L Taylor 484bl; M Macintyre 568ca; Robert Francis 560tr; Trevor Page 524ca; **Image Bank:** 594bl; A Rippy 186tr; G Jung 576tr; GA Rossi 646bl; M Beebe 209cl; ME Newman 324tr; P Trummer 252tr; T Madison 187br; **Images Colour Library:** 426bl; **Image Select:** Ann Ronan 42bc; **Impact Photos:** A le Garsmeur 412ca; Alain leGarsmeur 175bl; Ben Edwards 450tr; C Penn 584tr; G-J Norman 556tr; J Arthur 526tcr; Mark Henley 188tr; Robin Lubbock 236tr; Ben Edwards 303bl; B Babarov/Vika 489bl; **David King Collection:** 55tcb **Magnum:** H Cartier-Bresson 53tcb; **Panos Pictures:** B Tobiasson 329bc; Chris Stowers 353bl; D Hulcher 360tr; Jeremy Hartley 156bc; Marc French 289tc; Morris Carpenter 365bcr; Neil Cooper 130tr; R Giling 558tr; S Sprague 160tr; Sean Sprague 136c; Penny Tweedie 105tc; Michael Harvey 149cl; John Miles 190tr; JC Tordai 320cl; 488tr; 597cb; 602bl; **Pa News Photo Library:** AFP 419cbl; **Popperfoto:** 51trb; 52br; 55bl; AFP / Armand 54br; David Mercado / Reuters 137crl; Will Burgess / Reuters 105bl; Reuters 271cl; Jeremy Piper / Reuters 102bcr; John Cobb / Reuters 345cr; Official U.S. Air Force Photo 55cr; **Popperfoto / Reuters:** Peter Andrews 637cr; Simon Baker 453tlb; Denis Balibouse 203cl; Russell Boyce 594cra; Andreas Comas 415bcl; Bogdan Cristel 481cr; Jack Dabaghian 551br; Dimitar Dilkoff 153bcl; Larry Downing 601bl; Marcus Gyger 107bcl; Mohamed Hammi 573tcr; Hyungwon Kang 601bcr; Sergei Karpukhin 487bcl; Kamal Kishore 300cbr; Francois Lenoir 127bcr; Havakuk Levison 318bcr; David Loh 623tcl; Christinne Muschi 172bcbr; Alexander Natruskin 487bl; Patrick de Noirmont 159bcl; Enny Nuraheni 505bcr; Pilar Olivares 463tcl; Hrvoje Polan 293bcr; Romero Ranoco 467bcr; Molly Riley 268bcr; Oswaldo Rivas 555tr; Henry Romero 403bcl; Jayanta Shaw 623tl; Ruben Sprich 547cr; STR 451tr, 577car; Ray Stubblebine 251tcr, 293bcl, 577trb, 607bcl; Eriko Sugita 334cbr; Sukree Sukplang 563tr; Supri 305bcl; Susumu Takahashi 334cbl; Martin Thomas 182tr; Goran Tomasevic 631bcl; Pierre Virot 476bcl; Haydn West 594tcr; Kimberly White 618tcr; Darren Whiteside 165tcr; Rick Wilking 600bcr, 600br; **Reuters:** 123bcl, 146tcr; 507br; **Rex Features:** 116tcbr; 172bcr; 251tr; 256cb; 275tcr; 275tr; 341bcl; 481cbr; 486bcr; 516bl; 526tr; 543tcl; 587tcr; 594trb; 611bc; 627bcl; Andy Hernandez 488cl, J Sutton Hibbert 597cbl; Paul Brown 602cr; David Hartley 219tr; F Stevens 211bcl; Farnood 309tcr; Ken McKay 551tr; Sipa-Press 77tcr; Sipa-Press 95cbr; Sipa-Press 83bcl; Sipa-Press 127bcl; Sipa-Press 188bcr; Sipa-Press 315tc; Sipa-Press 318bcr; Sipa-Press 429tcl; Sipa-Press 445tc; Sipa-Press 565tcr; Sipa-Press 651bcr; Sipa-Press 645bl; Tony Kyriacou 555bcl; Torregano 419cl; **Harry Smith Collection:** 46clb; **South American Pictures:** Jevan Berrange 206br; P Dixon 456tr; T Morrison 606tr; **Sovfoto/Eastfoto:** 652br; **Frank Spooner Pictures:** 55bl; 83bl; 324bc; 545tcr; 543tr; A Denize 371tcr; A Sassaku 146tr; Alain Morvan / Gamma 547c; Alexis Duclos 182tcr; Bob Stern / Liason 243tr; C Hires 551bcr; C Poulet 647cr; Eric Vandeville 195tr; Tim Crosby / Gamma - Liaison 602cl; Gamma 471tr; Gamma / B Iverson 253bcl; Gamma / E de Keerle 471tcr; Gamma / F Apesteguy 507bcr; Gamma / Iliona - Figaro Magazine 113tc; Gamma / K Al Arab 573tcl; Gamma / L Chaperon 268cbr; Gamma / N Jallot 408tr; Gamma / Najer 587bcl; Gamma / Xinhua 188bcl; Georges Merillon 447bcl; KJ Eddy 107bcl; L Anticoli 524cbr; Liason / Peterson 441bcr; Liason / Markel 637c; N Saganssky 203cr; P Perrin 531tcr; Patrick Piel 268bra; Reglain 165tr; Victoria Brynner 451btcr; **Tony Stone Images:** 179bc; 198tr; 354bl; 444br; A Cassidy 302br; Alan Kearney 182bc; Alan Smith 562c; C Waite 592bc; D Armand 172tr; D Hanson 562bl; D Schultz 50bl; Dennis Stone 344tr; G Allison 605bl; H Kurihara 270bc; J Pragen 616tr; Joe Cornish 317bc; Marcus Brooke 158tr; O Benn 93tc; 279tr; P Chesley 406ca; Penny Tweedie 384bc; R Evans 152tr; R. Everts 415tr; R Smith 102tr; R Smith 231tl; S Egan 546tr; Steven Rothfeld 393tc; **Sygma:** aldev 550ca; R Reuter 502bl; **Telegraph Colour Library:** 100tr; Ford Motor Company Ltd 52clb; **Topham Picturepoint:** 153tc; 238tr; 520tr; 527cl; 571tr; 453cal; 478bc; 487tr; 504bc; 642c; Keystone 524cbl; **Trip:** 506bl; G Spenceley 580tr; T Goodman 285tc; V Shuba 122c; V Sidoropolev 340tr; **World Pictures:** 83cla; 322bl; 422ca; 554tr; **Zefa Picture Library:** 80bc; 120bl; 200ca; 526bl; 570bc; 618bl; Everts 630ca; F Lanting 142tr; Streichan 268bl.

NOTES

For the very latest information, visit:
www.dk.com and click on the Maps & Atlases icon

NOTES

For the very latest information, visit:
www.dk.com and click on the Maps & Atlases icon

NOTES

For the very latest information, visit:
www.dk.com and click on the Maps & Atlases icon

NOTES

For the very latest information, visit:
www.dk.com and click on the Maps & Atlases icon

NOTES

For the very latest information, visit:
www.dk.com and click on the Maps & Atlases icon